Instrumentation and Techniques for Radio Astronomy

OTHER IEEE PRESS BOOKS

Instrumentation and Techniques for Radio Astronomy

Edited by

Paul F. Goldsmith

**Professor of Physics and Astronomy
University of Massachusetts
  and
Vice President, Research and Development
Millitech Corporation**

A volume in the IEEE PRESS Selected Reprint Series,
prepared under the sponsorship of the
IEEE Microwave Theory and Techniques Society.

**IEEE
PRESS**

The Institute of Electrical and Electronics Engineers, Inc., New York

IEEE PRESS
1988 Editorial Board
Leonard Shaw, *Editor in Chief*
Peter Dorato, *Editor, Selected Reprint Series*

IEEE Order Number: PC0236-0

Library of Congress Cataloging-in-Publication Data

Instrumentation and techniques for radio astronomy.

  (IEEE Press selected reprint series)
  Includes bibliographies and indexes.
  1. Radio astronomy—Instruments. 2. Radio astronomy—Technique.
I. Goldsmith, Paul F., 1948-   .
QB476.5.I57   1988   522'.682   88-12983
ISBN 0-87942-240-8

Contents

Introduction

RADIO astronomy is no longer among the *newest* of the *new astronomies*. X-ray astronomy, gamma-ray astronomy, and neutrino and gravitational wave astronomy have all come on the scene more recently and in succession taken the title of newest window onto the universe. Thus, it is entirely reasonable that radio astronomy should now be considered as another tool available to astronomers in their study of distant objects, one whose findings are often used in conjunction with results obtained at other wavelengths.

In the more than 50 years that have passed since the first detection of radio waves from space, the study of emission at radio frequencies has contributed much to our knowledge of the universe on the largest scale, galaxies, our Milky Way, the birth and death of stars, and the solar system. In his book *Cosmic Discovery,* Harwit (1981) has identified 43 astronomical discoveries that he considers especially important; these span the interval from prehistory to the present time. Of these, 7 were made initially using radio-astronomy techniques including the discovery of cosmic masers, pulsars, radio galaxies, unidentified radio sources, quasars, superluminal sources, and the microwave background. Other discoveries made with a partial involvement of radio astronomy not too long after observation at other wavelengths include cold interstellar clouds, flare stars, and interstellar magnetic fields. While no such compilation can be a definitive assessment of the contributions of a particular scientific discipline, it is apparent that radio astronomy has made important contributions to a wide range of areas in astronomy.

What is radio astronomy? There are two ways in which the field can be defined: the first is in terms of the wavelength range observed, and the second in terms of the techniques that are utilized. For ground-based observations, the radio window extends from a low frequency cutoff of approximately 10 MHz imposed by the ionospheric plasma to a high frequency limit of between 400 and 900 GHz resulting from absorption by water vapor. Of course, if we consider observations from space, these restrictions are relaxed. Even the yet lower cutoff imposed by the interplanetary medium could be circumvented by observing from a spacecraft launched into the appropriate orbit! But considering for the moment only the window for ground-based radio astronomy, we find that there are at least 16 octaves in frequency available (*i.e.*, the frequency varies by a factor of 2^{16} over the range). Thus, it is hardly surprising that in this wide range there is a vast assortment of emission mechanisms that are operative and phenomena that can be studied. The great extent of the radio-astronomical frequency range is particularly striking when compared to the 1 octave of the optical band or the 7 octaves of the infrared region.

The radio region of the electromagnetic spectrum is commonly divided in terms of wavelength ranges, with the categories (in order of decreasing wavelength or increasing frequency) being:

(a) meter wavelengths covering wavelengths > 1 m,

(b) centimeter wavelengths covering wavelengths between 1 m and 1 cm,

(c) millimeter wavelengths covering wavelengths between 1 cm and 1 mm,

(d) submillimeter wavelengths covering wavelengths < 1 mm.

There is considerable overlap in defining the subregions of the radio range; portions of (a) and (c) together with (b) are generally referred to as microwaves, while (d) is often used interchangeably with the term far-infrared wavelengths.

Given this large extent in terms of frequency coverage, it is perhaps surprising that there is any unity at all in terms of the technology used in radio astronomy. There are at the present time, however, two salient characteristics of radio-astronomical technology that, while not unique to this discipline, serve together to define quite effectively its particular features. The distinguishing characteristics of radio-astronomical instrumentation are the use of (1) *coherent receivers,* and (2) *diffraction-limited optics.* Each of these characteristics is significant and merits a brief discussion.

Coherent receivers are sensitive to the electric field *amplitude,* rather than *intensity.* This is in contrast to most receivers used in the infrared and optical regions (as well as at shorter wavelengths) which respond to the intensity (or power) of the incident radiation. This means that they are not sensitive to the phase of an incident signal. The fact that the coherent receivers are sensitive to the phase of the incident radiation has fundamental implications for the limits to their sensitivity, as well as determining in a general sense their configuration. In many radio-astronomical receiving systems, the phase information is not ultimately used, but it is critical for the operation of interferometric systems. The requirements of high sensitivity and extreme stability imposed by radio-astronomical observations have been major stimuli to the advancement of the art of low-noise coherent receiver design.

The relatively long wavelengths of the radio region have compelled radio astronomers to extract the maximum angular resolution possible from the antennas available; this generally means operating at the limit set by diffraction. The general relationship for diffraction-limited operation is angular resolution = wavelength/aperture size. No antenna or telescope can have significantly higher angular resolution. This has encouraged a general upward trend in observing frequency (*i.e.*, to smaller wavelengths) together with the construction of relatively large telescopes. The requirement for diffraction-limited operation that the surface errors be less than approximately 1/20 wavelength has promoted the development of new techniques of telescope structure design and surface construction. The trade-off between minimum sidelobe level and maximum efficiency is a special characteristic of diffraction-limited optics, and has resulted in the design of special types of feed systems which have found wide application in related areas.

Although these characteristics do help define the special

characteristics of radio-astronomical instruments, it is also clear that coherent receivers and diffraction-limited optics are also employed in instrumentation used for many other purposes. It is impossible to list all areas in which related technology plays an important role, but in a broad sense, microwave communications, radar, and remote sensing are among the areas that have most clearly benefited from technical contributions made by radio astronomy. This has, of course, been a two-way street, and many important discoveries in radio astronomy have depended critically on equipment initially designed for other purposes. Presently, in areas as diverse as geodesy, image processing, and quantum-noise-limited detectors, radio astronomy is continuing its productive interaction with other branches of astronomy and other areas of forefront technology.

This volume has several purposes. First, given the relative maturity of the field of radio astronomy, it appeared appropriate to pull together a representative collection of the technical contributions that it has made. Second, persons entering this field, be they students or professional scientists or engineers, find it extremely difficult to gain an overview of the development of radio astronomy. This difficulty arises because the published articles are distributed over a diverse collection of publications. One testimony to this is the fact that the articles included here have come from a total of 21 scientific journals and conference proceedings. I hope that a mixture of articles selected for their historical value and chosen to represent the present state of the art will clarify the broad trends in radio-astronomical instrumentation and highlight some of the interactions with related technological areas.

In selecting the articles to be included many difficult choices had to be made. I have attempted to maintain a reasonable balance among articles of indisputable historical significance, those with the most complete discussion of general principles, and those which give a good feeling for the present state of the art. Since these qualities are only rarely found together, and since the size of the volume is limited, this posed a vexing problem. The breadth of the different areas of radio astronomy became increasingly apparent as different leads were tracked down, and the size pressure seemed continually to grow. In particular, it seemed important to include a selection of articles on the special techniques for data processing developed by radio astronomers, since these methods of handling information are to an increasing extent becoming inseparable from the hardware used to gather the photons themselves.

I wish to thank my colleagues at the Five College Radio Astronomy Observatory and the Department of Physics and Astronomy, particularly N. Erickson, C. R. Predmore, and R. Snell, for providing the ambience as well as the information that was necessary for the completion of this work. I am indebted to Drs. D. Backer and T. Cornwell for suggesting some very interesting articles that have survived the final pruning. The suggestion to include a historical summary of radio astronomy was made by one of the reviewers of the original proposal. Drs. J. D. Kraus, A. R. Thompson, and M. Harwit made valuable suggestions about the manuscript, but the responsibility for what emerged (albeit not for the content of the articles themselves) is mine. My deepest gratitude is to S. Reiss who not only read through the manuscript and made many improvements, but who also provided constant encouragement and support. I apologize to the very many people whose efforts to advance the instrumentation and techniques for radio astronomy could not be included in this necessarily limited reprint volume. Omissions in this regard are remedied in part by the list of specific publications and review articles given at the end of the introduction to each part.

PAUL F. GOLDSMITH
University of Massachusetts
Amherst, Massachusetts

REFERENCES AND BIBLIOGRAPHY

1981: Harwit, M., *Cosmic Discovery*. New York, NY: Basic Books, 1981.

In the sixties, a number of reviews appeared which summarized the state of the art in radio-astronomical instrumentation and techniques. While the field has now diversified beyond this point, these articles still give interesting overviews of the field, and contain some exceptionally clear discussions as well.

1962: Bracewell, R. N., *Radio Astronomy Techniques,* Handbuch der Physik Vol. LIV. Berlin: Springer-Verlag, 1962, pp. 42–129.

1964: Findlay, J. W., ''Antennas and receivers for radio astronomy,'' in *Advances in Radio Research*, J. A. Saxon, Ed. London: Academic Press, 1964, pp. 37–119.

1966: Wild, J. P., ''Instrumentation for radio astronomy,'' *Phys. Today*, vol. 19, no. 7, pp. 28–40, 1966.

There are also texts which discuss instrumentation and techniques in some detail. The book by Christiansen and Hogbom covers a wide variety of antenna systems. The classic text by Kraus has recently appeared in a second edition in which the discussion of receivers has been updated. The book by Rohlfs is primarily devoted to astrophysical analysis, but contains discussions of antennas, interferometers, and receivers. The work by Thompson *et al.* is a thorough discussion of all aspects of interferometry and aperture synthesis.

1985: Christiansen, W. N. and J. A. Hogbom, *Radio Telescopes,* 2nd ed. Cambridge: Cambridge University Press, 1985.

1986: Kraus, J. D., *Radio Astronomy,* 2nd ed. Powell, OH: Cygnus-Quasar Books, 1986.

Rohlfs, K., *Tools of Radio Astronomy*. Berlin: Springer-Verlag, 1986.

Thompson, A. R., J. M. Moran, and G. W. Swenson, Jr., *Interferometry and Synthesis in Radio Astronomy*. New York, NY: John Wiley and Sons, 1986.

The appearance of four texts devoted, at least in part, to radio-astronomical instrumentation within a period of 2 years is anomalous since there had been no book published on this topic in the preceding decade. The articles contained in the present collection fill in some of the historical details not covered by these works.

Historical Overview of Radio Astronomy

ALTHOUGH having a history of only slightly over 50 years, radio-astronomical investigations have been carried out at a rapid pace by many researchers with varied backgrounds working in countries throughout the world. Since the range of astronomical objects that can be studied at radio wavelengths is extraordinarily large, even the early work seems to have leaped from one topic to another, with little evidence for any orderly progression. The developments in radio astronomy, particularly in the earliest years, were driven by the appearance of new technology. In many cases, it was the astronomers themselves, although often trained as physicists or electrical engineers, who developed or applied "state of the art" instrumentation to a particular area of radio astronomy with exciting new results.

Consequently, the suggestions by reviewers of the proposal for this reprint volume that there should be a historical review of radio astronomy seemed appropriate, despite the editor's lack of qualifications as a historian of science. The general references given at the end of this part provide an enticing introduction to the development of this field, but also, as is characteristic of historical retrospectives, leave many interesting questions unanswered. Articles cited are listed in chronological order. A number of the early papers have been reprinted in the collection *Classics in Radio Astronomy* edited by W. T. Sullivan.

THE INDIVIDUAL PIONEERS

Although there were a number of earlier unsuccessful attempts to detect radio radiation from outside the earth, Karl Jansky deserves credit for making the first unambiguous detection of extraterrestrial radio waves, and he can, with justice, be called the founder of radio astronomy. Jansky was trained as a physicist, and started work at Bell Laboratories in 1928. Commercial transatlantic radio telephony service had begun in 1927 with a link between New York and London operating at 60 kHz, but higher frequencies were desirable to reduce the antenna size required for a given gain, as well as to obtain access to increased bandwidth. The issue of interference was not a new one, but investigation of this problem at shorter wavelengths was considered sufficiently important that Jansky was assigned to work on it full-time shortly after starting his job.

In order to track down the origin of the interference, Jansky needed a directional antenna, and in 1929 a two-wavelength-long wire antenna was constructed. Each element consisted of serpentined quarter-wave sections with 90° relative orientation, the long dimension being parallel to the ground. In addition to the driven element, there was one parasitic element to give the antenna front-to-back discrimination. The antenna rotated in azimuth once per 20 minutes, and the output of the receiver was displayed on a chart recorder. The antenna was designed to operate at 14.5 m wavelength (frequency 20.7 MHz), but the wavelength was changed to 14.6 m after tests revealed that local interference was present at the initial wavelength. Serious observations began in the winter of 1930–31 after the antenna was moved to Holmdel, New Jersey.

Jansky obtained data by recording the output of the radio receiver as the antenna was rotated. The antenna pattern had a beamwidth of 24° FWHM (Full Width to Half Maximum) in azimuth, and 36° in elevation. Due to the action of the earth as a ground plane, the effective peak response of the antenna pointed 26° above the horizon. A stable receiver system enhanced the ability to determine variations in the intensity of the received radiation over long periods of time.

Jansky published his first paper giving details of the antenna and receiver system in 1932 (Jansky, 1932). In it, he also described three different types of static: from local thunderstorms, from distant thunderstorms, and "a steady hiss type static of unknown origin." In these initial measurements, the origin of the hiss component of the static was clearly felt to be extraterrestrial. Over the period of the observations (December 1931 to February 1932), the direction of the maximum coincided with that of the sun, and Jansky concluded that the sun or the subsolar point on the earth was the origin of the hiss-type static. However, the ongoing observations throughout 1932 showed something quite different. The time of day of the maximum signal advanced steadily at 2 hours per month, exactly as would be the case if the source were located outside the solar system. This motion of the source at a sidereal rate of one circuit every 23 hours and 56 minutes rather than once every 24 hours was convincingly demonstrated by Jansky's data, and he was able to fix the celestial coordinates of the source of radiation (assumed to be pointlike) as Right Ascension 18 hours and declination −10° (Jansky, 1933a).

Fixing the source as being located toward a distant region in space was certainly surprising, but even more exciting was the (approximate) coincidence of the direction of the source of the radio waves with the center of the Milky Way. In fact, Jansky had detected the emission produced by energetic electrons trapped in regions of intense magnetic fields near the center of our galaxy. The first association of the source of radiation with the galactic center was given in a 1933 paper (Jansky, 1933b), and was presented in a talk to URSI (International Union of Radio Science) in April of that year.

There are many interesting questions associated with Jansky's work. Among these are the fact that his observations took place near a minimum of the 11-year sunspot cycle, and hence radio emission from the sun was at such a low level that it could not be detected with his system. At a different time, solar activity would have been far more intense, and bursts of radiation from the sun would probably have been much more readily accepted as being of immediate astronomical interest, and also of relevance to communications problems. This would presumably have resulted in greater support from Bell Telephone Laboratories' management for Jansky's research.

Jansky did follow up on the "interstellar interference," and in a subsequent paper more accurately determined the direction of the maximum of the emission, and that the source was, in fact, the entire plane of the Milky Way (Jansky, 1935).

In this paper, Jansky discussed some possible mechanisms which could be responsible for the observed radio emission. Most of these are dependent on stars, and the nondetection of the sun was raised as an objection. The existence of interstellar matter was known, but the only component identified was gas at a temperature of approximately 15,000 K. Jansky indicated that the distribution of radiation is consistent with an interstellar origin, but quantitative comparison of the observations with theory indicated that thermal emission from this material could not be the correct explanation. This is undoubtedly one of the reasons why astronomers, in general, did not pay much attention to Jansky's work; this question has been discussed in some detail by Greenstein (1983).

One person who was interested in the observations that Jansky carried out was Grote Reber, who lived in Wheaton, Illinois. Reber had been trained as an electronics engineer, but was also an enthusiastic amateur radio operator. Reber reasoned that if the radiation detected by Jansky were of thermal origin, its spectrum should rise as the square of the frequency, and consequently would be much easier to detect at a shorter wavelength. He appreciated the importance of being able to locate the source of the emission precisely, and that observing at a high frequency would be advantageous in this regard, since the beamwidth of a radio antenna is inversely proportional to the observing frequency. Reber chose a very different type of antenna than that used by Jansky, one much more closely related to optical telescopes: the parabolic reflector. Reber was working for a major radio receiver manufacturer, and consequently had expertise in radio techniques, but this radio telescope design and construction work was an "extracurricular" effort which involved many novel ideas as well as his own time and money (Reber, 1958, 1983).

Reber's paraboloid was 31.5 ft in diameter and was constructed of metal sheets attached to a wooden frame. The feed system consisted initially of a dipole in a section of circular waveguide. The mounting system was adjustable in elevation only, so that sources would drift through the antenna beam as a result of the earth's rotation.

The telescope was constructed in 1937, and Reber made his initial observations at the extraordinarily (for the time) short wavelength of 10 cm—a frequency almost a factor of 150 higher than that used by Jansky. Consequently, the intensity of a thermal signal from the galactic center should have been much greater than that observed by Jansky, but Reber was unable to detect anything. One major advantage of the parabolic reflector antenna is evident; its essentially frequency-independent operation meant that to change frequency, only modification of the feed system was required. Reber changed to a wavelength of 33 cm in the fall of 1938, but was still unable to detect any signals. At this point it was apparent that much higher sensitivity was required (although the possibility of a nonthermal spectrum was being suggested), so Reber again lowered his observing frequency, this time to the highest frequency at which a radio frequency (RF) amplifier could be constructed. The frequency employed was 162 MHz (wavelength 187 cm), and the beamwidth of the antenna must have been approximately 12.5° (the 3° reported by Reber is almost certainly erroneous as it is much less than the wavelength divided by the aperture diameter; see the discussion by Sullivan (1978) and that in Section B of the book *Classics in Radio Astronomy*).

In April 1939 Reber definitely detected the galactic plane. Preliminary results were published in 1940 in the *Proceedings of the Institute of Radio Engineers* (Reber, 1940a) and, indicative of the beginning of interactions with astronomers, in the *Astrophysical Journal* (Reber, 1940b). In the former paper, the received intensity was given in units of watts per square centimeter (much more comprehensible to astronomers than the units of microvolts per meter employed by Jansky), and the characteristic intensity was converted to a stellar magnitude. In addition, a quantitative (although unfortunately inaccurate) explanation for the intensity observed in terms of free-free emission was attempted, and astronomical conclusions were drawn. These factors may help explain the interest taken by other astronomers in radio work, with important publications starting in 1940 (see, for example, the papers reprinted in Section C of *Classics in Radio Astronomy*).

Reber continued his investigations, and with improved sensitivity was able to move up to a frequency of 480 MHz, thus gaining a factor of 3 in angular resolution. One of his major contributions was making the first radio maps of a large portion of the sky (Reber, 1944, 1948). Although relatively crude by the standards of optical astronomy, these did succeed in making many astronomers aware of the value of radio observations for studies of galactic structure, and suggested the line of much research in the future.

SOLAR SYSTEM RADIO ASTRONOMY

Early Observations of the Sun

The pioneering observations of Jansky and Reber were of emission from high-energy electrons in the interstellar medium of our galaxy, and no discrete sources were identified. This emission is called synchrotron emission as a result of the repetitive circular orbits of electrons in regions with magnetic fields. The process is referred to as nonthermal, since the electrons are far from being in thermal equilibrium. Given that the sun is so clearly the dominant extraterrestrial source of radiation at optical wavelengths, it is perhaps surprising that it remained a relatively elusive object for the first radio astronomers. This is due in part to the quiescent state of the sun at the time of Jansky's work, its relatively small angular size compared to the central region of the Milky Way, and also to the relative weakness of its thermal emission compared to galactic nonthermal emission at the very low frequencies initially used. In fact, Reber did succeed in detecting the sun in 1943 (Reber, 1944), but by that time at least two other groups involved in radar research during the Second World War had made detections of very different aspects of solar radio emission, although their results were not published until later.

The earliest recording of emission from the sun in the radio region came in February 1942, when radar systems in Britain operating at 4 to 6 m wavelength detected a very high intensity of radio waves from the sun on two successive days. The signal was very broadband, and was confirmed by observations from several locations. The peak intensity was one hundred thousand times greater than that expected from a 6000-K blackbody (the apparent temperature of the solar photosphere seen at optical wavelengths). J. S. Hey, who was working on jamming threats to the radar systems, was able to identify the sun as the source of the interference. Furthermore, an active sunspot group happened to be crossing the central meridian of the sun at the time. While it was clear that the sun was the source of the radiation, it was not possible to unambiguously associate the emission with a particular region, due to the limited angular resolution available. The search for methods of achieving the enhanced resolution required for detailed study of solar radio emission was one of the driving forces in the development of interferometers and aperture synthesis in the following years. The 1942 result was published by Hey only after the end of the war (Hey, 1946).

Like the emission from the Milky Way detected by Jansky and Reber, the solar emission analyzed by Hey was of nonthermal origin. However, almost at the same time, a group led by G. Southworth at Bell Telephone Laboratories also working on radar-related questions, but at much shorter wavelengths, had detected thermal emission from the sun. The same argument that had led Reber (incorrectly) to assume that the Milky Way would be a more easily detectable radio source at higher frequencies, in this case made the thermal emission from the quiescent sun readily detectable with a 5-ft-diameter parabolic antenna. The improvement in receiver sensitivity as a result of wartime research efforts was also an important factor. Southworth reported measurements at wavelengths of 1.25, 3.2, and 9.8 cm, with intensities somewhat higher than, but in general agreement with, the values predicted by the Rayleigh-Jeans (long wavelength) tail of a 6000-K blackbody (Southworth, 1945). In addition to the solar data, some very forward-looking statements about the apparatus and prospects for radio-astronomical studies in the future, as well as the practical point that fluorescent lights are a source of centimeter wavelength thermal emission, were included in his report.

The postwar years saw a great deal of activity concerned with determining the origin of the nonthermal solar radiation. A critical requirement was obtaining higher resolution than could be obtained from the single antennas available. Occultation of the sun by the limb of the moon during a solar eclipse is one method of obtaining higher angular resolution. It was first utilized by A. E. Covington in Canada during the eclipse of November 1946. He was able to determine that a large sunspot present had an effective temperature of 1.5 million degrees at a wavelength of 10.7 cm (Covington, 1948). This technique was also employed by a number of scientists in Australia at 50 cm wavelength. Monitoring of the "quiet sun" revealed that the whole disk effective temperature at meter wavelengths was close to 1 million degrees. It was apparent that a wide variety of emission mechanisms were at work, but details remained obscure.

Both in Australia and in Great Britain, astronomers studying solar activity wished to free themselves from the restriction of waiting for a solar eclipse in order to obtain high angular resolution. An antenna of greater directivity was required, and this led to the use of the first interferometric systems, which had quite varied forms. The Australian group consisting of L. L. McCready, J. L. Pawsey, and R. Payne-Scott utilized the interference between the beam of their single antenna located on a cliff above the ocean and that of the image antenna produced by the water (which acted as a nearly perfect reflector). This "sea interferometer" produced an interference pattern which varied with time since the projected distance between the antenna and its image varied as a function of elevation of the sun. The report of the initial observations at 200 MHz (1.5 m wavelength) made in 1946 was published in the following year (McCready et al., 1947), and confirmed several important aspects of solar radio activity—that the intensity is highly time variable, and that it is concentrated in regions of sunspots.

The high angular resolution achieved by the sea interferometer was a powerful stimulus to solar studies in Australia, and encouraged various lines of investigation. Observations revealed that the very irregular emission was circularly polarized. Some of the emission could best be described as pulsed, and the pulses were found to exhibit a frequency-dependent delay in their arrival, with the highest frequencies arriving first. The interpretation of solar bursts being produced in the solar corona above sunspots, and emitting radio radiation as they traveled outward at velocities from a few hundred to many thousands of kilometers per second, gradually evolved during the late 1940s and early 1950s. Considerable specialized instrumentation for solar observations was developed, culminating in the Culgoora radioheliograph developed by J. P. Wild and his co-workers (Wild, 1967). This instrument, consisting of 96 antennas arranged in a circle 3 km in diameter, functioned as an interferometer capable of forming once per second an image of the sun at multiple frequencies. The Culgoora radioheliograph and other systems capable of image formation on a very rapid time scale, such as that in Nancay, France, have contributed much to our understanding of short-term solar activity.

Astronomers at Cambridge University in Great Britain were among those also searching for a method of obtaining high angular resolution. M. Ryle and D. Vonberg, at roughly the same time as the Australian sea interferometer work, also developed a two-element interferometer, one in which *both* elements were tangible antennas (Ryle and Vonberg, 1946). While initially more costly than the sea interferometer, this approach is obviously far more versatile. The first observations with the two-element interferometer were made at a frequency of 175 MHz. The distance between the two elements, which determined the angular resolution, could be varied, and results for spacings between 10 and 140 wavelengths were reported in 1946. The upper limit to the diameter of the source apparently coincident with a large sunspot present during the observations in August 1946 was 10 arcminutes, and the equivalent temperature was 2×10^9 K. This enormous value was taken as evidence that the emission must be of nonthermal origin. The polarization of the radiation was investigated, and the radiation found to be highly circularly polarized, but with handedness changing on a time scale of a few days.

As indicated in the preceding paragraphs, the Australian observations were contemporaneous with, or even had started earlier than, those in Great Britain, but the latter results were published first. The phenomenon of several groups pursuing the same line of research in parallel and obtaining results at almost the same time was to occur repeatedly in the early years of radio astronomy. In the present case, probably the most important contribution of Ryle and Vonberg was introducing the Michelson interferometer with variable element separation into the radio region of the electromagnetic spectrum. This had a major effect on the development of subsequent systems for almost all areas of radio-astronomical research, extending far beyond the solar observations for which it was initially intended.

Observations of the Moon

The moon figured prominently in early postwar research at radio wavelengths. R. H. Dicke and R. Beringer (1946) measured the effective blackbody temperature of the moon to be 292 K at a wavelength of 1.25 cm. Measuring the emission from the moon at different wavelengths developed into a powerful tool for determining the properties of the lunar surface material; the low value inferred for the thermal conductivity led to the picture of a highly porous or fluffy surface before any spacecraft had landed on it. The moon was also the target of the first radar-astronomy experiments. In 1946 the first radar echoes were obtained by J. DeWitt of the U.S. Army signal corps at 2.7 m wavelength (DeWitt and Stodola, 1949). The first measurement of the earth–moon distance had an accuracy of only 1000 miles, but indicated the potential of radar astronomy for distance determinations within the solar system.

Emission from Other Planets

The first detection of radio emission from another planet was a serendipitous discovery which occurred in 1955. B. Burke and K. Franklin, who were observing at a frequency of 22 MHz with the purpose of studying radio sources, detected strong bursts of radiation which appeared to come from a direction near the Crab Nebula in the constellation Taurus. The sun was not in this direction, and the possibility of terrestrial interference being the origin was quite remote since the location of the source of the bursts appeared to follow a sidereal pattern. However, careful studies of the source position over a 3-month period indicated that its celestial coordinates were not exactly fixed. Although initially it seemed improbable that Jupiter could be the source, its position was found to agree with that inferred for the origin of the bursts (Burke and Franklin, 1955).

The association of Jupiter with very intense bursts of radio radiation having durations of minutes to hours was confirmed, and various groups of astronomers used a wide variety of techniques to obtain additional information concerning this surprising result. The bursts are thought to be produced in the Jovian ionosphere or magnetosphere. The mechanism responsible is not well understood, but the bursts are influenced by the position of Jupiter's satellite Io, which is also thought to be responsible to a large extent for the production of charged particles in Jupiter's equatorial belt. These particles are required for the steady but intense synchrotron emission observed at wavelengths greater than 10 cm. The wobbling of the plane of polarization of the steady low-frequency emission was the first indication that the magnetic axis of the planet is tilted by approximately 10° relative to the axis of rotation.

Detection of the thermal emission from other planets had to wait until a somewhat later date, one reason being that the emission from these objects, which have moderate temperatures (a few hundred Kelvin) and subtend angles of 1 arcminute or less, is relatively weak at radio frequencies. The first success occurred in 1956, when C. Mayer, T. McCullough, and R. Sloanaker of the Naval Research Laboratory, using a 3.2-cm-wavelength radiometer system and a 50-ft-diameter parabolic antenna, were able to detect Venus, Mars, and Jupiter (Mayer et al., 1958). The value of radio measurements is particularly evident in the study of planets with thick atmospheres, since the dense gas is still largely transparent to radio waves. Hence the temperature of the planetary surface is directly measured, while in the infrared region, the temperature seen is that characteristic of some altitude up in the atmosphere.

The radio-wavelength temperatures determined at 3 cm for Mars and Jupiter (which does have a thick atmosphere) are in reasonable agreement with the infrared measurements. Venus, on the other hand, was found to emit like a blackbody at 600 K, as compared to a 225-K effective temperature in the infrared. This large difference was the first indication of the extra heating of the surface by the "greenhouse" effect resulting from infrared radiation from the surface being trapped by the thick cloud layer surrounding the planet. The emission from Mercury was first reported 4 years later (Howard et al., 1962). The temperature of 400 K was higher than would be found under the assumption that Mercury rotates synchronously with its revolution about the sun. The value found for the temperature indicated that no part of the planetary surface seen at maximum elongation (including both the sunlit and the dark hemispheres) could be continuously facing away from the sun. The rotation of Mercury was measured from radar-astronomical observations discussed below.

Other Solar System Objects

Thermal emission from all of the planets in the solar system as well as from many of their satellites has been detected, providing us with information about the nature of their surfaces and atmospheres. Spectroscopic observations have also provided valuable information about the composition and circulation of atmospheres of solar system objects. Carbon monoxide has been extensively studied in the atmosphere of Venus (cf. Wilson et al., 1981). It has also been detected in the atmospheres of Mars and of Saturn's moon Titan. The HCN molecule has been detected as well in the atmosphere of this satellite.

Comets are thought to be formed of material characteristic of the primordial solar system. Determination of their composition has proven difficult. The participation of radio astronomy in this work has steadily grown as the sensitivity and angular resolution of the instruments available have increased. The apparitions in recent years of comets Kohoutek (1973), West (1976), and Halley (1986) have been the focus of searches for a wide variety of molecular species. At the present time, OH, HCN, and NH_3 have definitely been detected in comets, while the presence of a number of other species has been indicated but not verified (Snyder, 1982). Continuum emission has been detected from a number of comets, which has helped to determine the diameter of the nucleus as well as limit the number of relatively large, icy grains present in the cometary halo (cf. Altenhoff, 1985). The status of cometary radio astronomy has been summarized in the conference proceedings edited by Irvine *et al.* (1987b).

Radar Astronomy

Determination of the rotation rate of Mercury has long been a challenge for astronomers. Radar astronomy offered the possibility of directly measuring the rotational velocity of the surface, which had proven impossible to do definitively at optical wavelengths. Using the Arecibo antenna at a frequency of 430 MHz, G. Pettengill and R. Dyce (1965) found the rotation period to be 59 days, while its orbital period is 89 days. Discoveries using radar astronomy also contributed significantly to our knowledge of the rotation of Venus whose surface is entirely obscured in the optical and infrared portions of the spectrum. R. Carpenter and R. Goldstein measured the rotation period to be 250 days, but in a retrograde sense (Carpenter, 1964; Goldstein, 1964). Radar-astronomical measurements have been critical for determining the distance to objects in the solar system. After the first results for the moon discussed above, Venus was the next target for radar echoes. This work, begun in 1958, has resulted in increasingly accurate determination of the Astronomical Unit, the separation of the sun and the earth (cf. Pettengill, 1965). Radar-astronomical mapping of planetary surfaces also provided the only detailed source of information about surface features until the advent of planetary space probes.

Radar astronomy has played a major role in the study of meteors. As they pass through the atmosphere, these rapidly moving objects produce trails of ionized material, which were studied relatively early by scientists using military radar systems. The Giacobinid meteor shower of 1946 provided a copious source of radar echoes and demonstrated the value of radar observations (Hey, 1973, pp. 32–34). Simultaneous observation from several sites can accurately determine the orbit of the meteor, and confirm that these objects originate from orbits around the sun. The study of the ionization trails has also improved our understanding of the structure of the earth's atmosphere at heights on the order of 100 km above the surface.

STELLAR RADIO ASTRONOMY

The radio emission from individual stars is extremely difficult to detect. In the early years of radio astronomy, unidentified radio sources were often referred to as "radio stars," but this proved to be essentially a misnomer, and

stellar radio astronomy has been restricted to unusual categories of stars that emit much more intensely than do normal stars such as the sun.

Radio astronomy has contributed significantly to the study of flare stars, which, as the name indicates, undergo large and generally erratic outbursts. Starting with observations by several groups in 1963, it has been found that radio flares often occur simultaneously with optical events, but that the increase in the radio emission can be dramatic (as in the case of the sun) while the change in the optical output is less than an order of magnitude. As the angular diameters of stars are extremely small, this field of research has benefited dramatically from the increased angular resolution available from connected arrays such as the very large array (VLA), and from very long baseline interferometry (VLBI). The flare emission from dwarf stars appears to be associated with stellar features including plages (bright regions) and starspots (dark regions).

Stellar radio emission occurs preferentially in binary and multiple star systems. One of the best-studied of these is β Persei (Algol), which is an eclipsing and spectroscopic triple system. Observations by R. M. Hjellming, E. Webster, and B. Balick (1972) with the NRAO three-element interferometer could only set an upper limit to the angular diameter of the emission region of 0.5″. However, the large variations on a very short time scale seen for an event on July 11, 1972 implied a size of less than a few light-minutes (one light-minute is equal to 0.004″ at the 30-parsec distance of Algol). VLBI observations (Clark *et al.*, 1976) were able to resolve the angular diameter of the source to be 1.7 milliarcseconds or 0.05 Astronomical Unit, comparable to the sizes of the individual stars. The very small size of the emission region implies a brightness temperature of 8×10^9 K, and indicates that the emission is almost certainly of nonthermal origin.

Pulsars

Certainly the most surprising and important contribution of observations at radio wavelengths to stellar astronomy has been the discovery of pulsars. In 1967, observations with an 81.5-MHz array started at Cambridge University. The goal of this work was to distinguish between small diameter (<1 arcminute) and larger diameter radio sources; the latter were mostly "ordinary" radio galaxies, while the former were expected to be quasars (discussed below). The approach adopted was to use interplanetary scintillation as an angular size filter, since for the larger diameter sources this effect is smeared out. To detect the scintillations, a short time constant of 0.1 second was employed on the receiver output, and the particular signature of a small-diameter source drifting through the fixed pattern of the array could be used to identify the sources of interest. In August 1967, Jocelyn Bell, the graduate student involved with this project, noticed that the chart recording of the receiver output contained signals that did not exactly resemble the signature expected from a small-diameter source, but which also did not appear to be interference. The unusual signals, called "scruff" by Bell, suggested variation on a shorter time scale than could be readily seen with the chart recorder speed that was being used. In November, observations were made at higher speed, and the scruff was resolved into a series of pulses with a periodicity of $1\frac{1}{3}$ seconds.

The most obvious explanation was man-made interference, but analysis of the data by Antony Hewish, Bell's thesis advisor, indicated that the Right Ascension, 19 hours and 19 minutes, had remained constant to within 10 seconds over a 3-month period. The apparently fixed celestial coordinates of the source of the pulses, together with the lack of any plausible terrestrial source, gave the Cambridge scientists confidence that they had detected a new type of astronomical radio emitter. By the time a paper was submitted reporting the results for the first "pulsar," three other pulsating radio sources had been detected, with the fastest having a period of one quarter second (Hewish et al., 1968). The existence of a number of sources argued against the possibility (which had been seriously considered) that the pulses were produced by an extraterrestrial civilization as a form of signal. The excitement of this discovery has been wonderfully described by Bell Burnell (1983).

The relatively short period and great regularity of the pulses suggest that the mechanical motion of a condensed object was associated with their production. The initial hypothesis of white dwarf stars was followed shortly by the suggestion of T. Gold (1968) and of F. Pacini (1968) that pulsars are rotating neutron stars. This has been confirmed by the detection of pulsars in the Vela and Crab Nebulae having periods of 88 and 33 milliseconds, respectively; these rates of rotation could not be tolerated by white dwarfs due to their much larger size. The identification of pulsars as extremely condensed objects consisting of essentially pure neutrons confirmed the existence of this state of matter, and has led to great advances in the understanding of how matter behaves under these extreme conditions. The subject of pulsars has been covered in detail in a book by Manchester and Taylor (1977).

Although the nature of pulsars is well established, the details of the radiation mechanism are still not understood. The general picture that has been developed is that the magnetic field of a star is trapped as its core collapses following explosion as a supernova. The strength of the magnetic field is enormously increased as the core shrinks from stellar size to approximately 20 km in diameter. The rapidly rotating magnetic field acts as a dynamo, with the intense electromagnetic fields accelerating charges which radiate by the synchrotron process. The great regularity of the pulses is a result of the moment of inertia of the neutron star. Extremely precise measurements of pulse periods have revealed that pulsars do slow down as a result of the energy radiated. In addition, "glitches" in the period have been observed, which are thought to arise from starquakes which change the moment of inertia of the condensed object. The radiation from pulsars is responsible for energizing the cloud of material remaining from the exploded star. This material forms a supernova remnant such as the Crab Nebula, which can be observed at optical as well as at radio wavelengths.

In addition to providing a wealth of information concerning the products of supernovae, and about electrodynamics under extreme conditions, pulsars have provided the first evidence of gravity waves. R. Hulse and J. Taylor (1975) discovered a pulsar in a binary system. The acceleration of the pulsar and its companion in their orbits produces gravitational radiation; from Einstein's Theory of General Relativity the power radiated is calculated to be 8×10^{24} W, and this energy loss affects the orbital motion. The data collected to date show a change in orbital period in extremely good agreement with that predicted, and thus provide a valuable confirmation of this important theory.

Recently a class of superfast pulsars has been identified. The first of these was discovered by D. Backer and collaborators (1982); this "millisecond pulsar" has a pulse period of 0.0016 second (1.6 ms). In general, a rapidly rotating neutron star is one recently formed, and will be slowing down relatively quickly. Characteristics of the millisecond pulsar—such as the modest rate at which it is slowing down—indicate that it is relatively old. Theoretically, if a relatively old pulsar has a companion star (i.e., is in a binary system), mass can be ejected from the companion and transferred to the pulsar with sufficient angular momentum to increase the rate of rotation of the neutron star. This spin-up process leaves a rapidly spinning pulsar which otherwise has characteristics of an older object, as is observed. Studies of these objects are expected to yield new information about mass transfer between objects bound together in multiple star systems.

SPECTRAL LINES AND THE INTERSTELLAR MEDIUM

The 21-cm Line of Atomic Hydrogen

The possibility of observing spectral lines from atomic hydrogen at radio wavelengths was predicted prior to their detection. In fact, neither the electronic transitions of hydrogen with large principal quantum number, nor the hyperfine transition in the electronic ground state had ever been observed when H. Van de Hulst discussed their detectability and relevance for astronomy in a seminar that took place in Holland in 1944. His presentation to a group of amateur and professional radio engineers and astronomers dealt first with explaining the continuum emission from the galaxy which had been detected by Jansky and Reber. Van de Hulst developed an accurate calculation for the intensity of free-free emission from ionized hydrogen, which agreed with other theoretical analyses as well as with the experimental data. In addition, he discussed the possibility of detection of recombination lines from neutral hydrogen which are produced by changes in the principal atomic quantum number, n. For a change of one unit, the resulting emission occurs in the radio region of the spectrum when n exceeds approximately 25. Unfortunately, Van de Hulst's calculation of the expected linewidths was in error and this resulted in excessively pessimistic estimates for the line-to-continuum ratio and thus the detectability of this radiation. This work was published in the following year (Van de Hulst, 1945); the error in the theoretical analysis was rectified by N. S. Kardashev (1959).

Certainly the most important part of Van de Hulst's 1945 paper concerned the hyperfine transition of the ground state of atomic hydrogen. The absence of any laboratory data necessitated a calculation of the frequency, which was accurate to better than 1%, and also an estimate of the line strength. For the latter, Van de Hulst derived a minimum value of the spontaneous emission rate that would be required for detectability of the spectral line emission. Later work showed this estimate to be very conservative and that emission at 21 cm from atomic hydrogen (HI) in interstellar clouds is almost ubiquitous both in the Milky Way and other galaxies.

Wartime conditions were not conducive to carrying out an observational program, so there was no immediate experimental follow-up to Van de Hulst's theoretical work. However, after the end of the Second World War several groups started efforts to detect the 21-cm hydrogen line. H. Ewen and E. Purcell at Harvard University, encouraged in part by a visit from Van de Hulst, were the first to detect this emission in March 1951. C. Muller and J. Oort in Holland were attacking this problem simultaneously. W. Christiansen and J. Hindman in Australia, after hearing about the American results, began a program to detect the 21-cm line. In an impressive example of deliberateness, Ewen and Purcell waited for the Dutch report of their relatively high angular resolution results and confirmation of the success of the Australian effort, to publish their detection. The American and Dutch reports with a notice of the Australian results appeared consecutively in one volume of the journal *Nature* (Ewen and Purcell, 1951; Muller and Oort, 1951).

Study of the 21-cm transition of atomic hydrogen has the advantage (shared with all spectral lines as compared to continuum emission) that dispersion and systematic shifts in the velocity of the material emitting the radiation can be accurately determined. This has led to the widespread use of the HI hyperfine transition as a tracer of the structure of interstellar clouds, and of their distribution in space. In addition, many studies have employed the 21-cm line to deduce the structure of the Milky Way. This work has now been extended to other galaxies, and observation of the 21-cm HI line has become one of the most important tools of extragalactic research.

Recombination Lines

The emission from hydrogen in the form of recombination lines mentioned above was first detected by Z. Dravskikh, A. Dravskikh, and V. Kolbason (1964) in the USSR; lines have subsequently been detected throughout the centimeter and millimeter wavelength ranges. The transitions between states of high quantum number are natural steps in the return to the ground state of an atom which has recombined following ionization, and thus recombination line radiation is generally produced in regions where hydrogen is ionized (HII regions). Study of recombination lines has yielded information about the electron temperature in these HII regions, and high-angular-resolution observations have revealed a complex structure. The relative intensities of recombination lines from different elements can yield important information about their relative abundance, and extensive observations have been carried out to determine the helium-to-hydrogen abundance ratio which is of crucial importance in determining conditions in the earliest moments in the universe.

Molecular Line Emission

Spectral line emission from molecules in the interstellar medium was also predicted well in advance of observation. I. Shlovsky (1949) discussed the emission from OH and CH in some detail, and C. Townes (1954) presented a number of molecular species as likely candidates. Three diatomic molecules, CH, CH^+, and CN, were known to be present in interstellar space from observations at optical wavelengths, but sufficiently precise measurements of the frequency of radio frequency transitions of OH to justify a search were not available until 1960. S. Weinreb *et al.* (1963) detected OH in absorption (gas temperature below that of the background) at 1667 MHz against the continuum emission from Cassiopeia A. They used the Millstone Hill antenna of MIT Lincoln Laboratory and a newly developed autocorrelation spectrometer. This success was followed by other detections of OH in absorption in many regions of the galaxy.

Another transition of the OH molecule at 1665 MHz was soon found to radiate in emission. The antenna temperature measured with a filled aperture antenna exceeded 1000 K when pointed toward the Orion Nebula, but higher angular resolution interferometric observations revealed that the source was unresolved, and that the true brightness temperature was much greater. Further observations, including those employing VLBI, revealed that the brightness temperature for some sources exceeds 10^9 K. This is so greatly in excess of any plausible physical temperature that the emission must be nonthermal. The idea of microwave amplification in space—an interstellar maser—was proposed shortly thereafter. This idea received support from the detection of a high degree of polarization for the intense, compact sources.

Thermal emission from interstellar ammonia, NH_3, was discovered by A. Cheung and collaborators (1968). This was the first indication that the complexity of interstellar molecules could extend well beyond the simplest diatomics. In 1969, formaldehyde, H_2CO, was detected in absorption and H_2O emission was found, again a masering line. The first bonanza year for interstellar molecules was 1970, with six new species discovered. These included HCO^+, the first ion detected at radio wavelengths, and carbon monoxide (Wilson *et al.*, 1970). While much simpler than some previously detected molecules, CO has come to play a unique role in opening up the molecular interstellar medium. This is a result of its widespread distribution, and the relative ease with which it can be excited to emit radiation from the lower rotational transitions which occur at short millimeter wavelengths. Observations of CO thus are an excellent tracer of H_2, which although the most abundant molecule, cannot readily emit since it has no permanent dipole moment.

Primarily through observations of CO, an entire new class of the most massive objects in the Milky Way—Giant Molecular Clouds—has been discovered. These nearly completely molecular objects dominate the mass of the interstellar medium between the center of the Milky Way and the radius of the sun. Currently, observations of CO in the Milky Way and in other galaxies are yielding information about the formation rate of stars, and the cycling of material between stars and the interstellar medium (see, e.g., Solomon and Edmunds, 1980).

A variety of millimeter wavelength radio telescopes with diameters between 1 and 45 m are in use around the world to study the distribution of CO and of other molecules. The first data produced by millimeter wave interferometers appeared in the early 1980s and currently these instruments can obtain an angular resolution of a few arcseconds. This is limited primarily by the sensitivity of these two- or three-antenna instruments. More powerful instruments under construction and those being designed will certainly allow millimeter wavelength interferometry to play a major role in the studies of the details of the star formation process and in the determination of the distribution of molecular material in other galaxies.

The years following the detection of carbon monoxide have witnessed major improvements in the sensitivity of millimeter wavelength receiver systems for radio astronomy. This enhanced capability has played a significant role in expanding the number of chemical species identified in interstellar molecular clouds. More and more complex molecules have been found, as illustrated by the long-chain linear molecules HC_7N, HC_9N, and $HC_{11}N$. At the present time, over 70 different molecular species exclusive of isotopic variants have been identified. The distinctive and complex interstellar chemistry required to produce them is thought to include a combination of grain surface and gas-phase reactions and thus provides a valuable probe of molecular cloud conditions and evolution (cf. Irvine *et al.*, 1987a).

RADIO GALAXIES AND QUASARS

Radio Galaxies

The radio-wavelength emission observed by Jansky and by Reber was extended in angular size, and appeared to coincide with the plane of the Milky Way with a peaking toward the galactic center. As discussed above, emission from individual stars is not a plausible explanation, since this is in general quite weak. In the early years of radio astronomy, there did not appear to be any clear expectation of nonstellar or non-solar-system pointlike sources. Thus, J. Hey and his collaborators were surprised to find that the signal from the direction of the constellation Cygnus varied irregularly on a time scale of a few seconds, for it appeared a reasonable assumption that such variations could be produced only by a source of very small angular size (Hey *et al.*, 1946). The 12° beamwidth of the antenna they were using allowed location of the source to an accuracy of only 2°. This work was followed by attempts to improve the positional accuracy, and J. Bolton and G. Stanley (1948) were able to set an upper limit of 8 arcminutes using the sea interferometer discussed earlier. Their determination of the position of the source was unfortunately erroneous due to incorrect correction for refraction introduced by the earth's atmosphere.

This early work established that sources of small angular size were present in the radio sky. The list was soon enlarged with the discovery by Ryle and Smith (1948) of radio emission from the constellation Cassiopeia. Until this point the angular resolution and positional accuracies had been sufficiently poor that there was no clear association of the radio sources with optical objects. Bolton, Stanley, and Slee (1949) changed this situation by measuring the positions of three small-diameter sources to better than 10 arcminutes. Using the system of denoting the strongest radio source in a given constellation as

"A," the next strongest, "B," etc., these were Taurus A (the Crab Nebula supernova remnant) and Virgo A and Centaurus A, which were nearly coincident with external galaxies. F. Graham Smith (1951) used the interferometer at Cambridge to measure the positions of Cassiopeia A and Cygnus A to better than 1 arcminute accuracy. The source in Cassiopeia was identified with a supernova remnant produced by a star which exploded about 1700, but which had not been observed optically due to attenuation of starlight by intervening interstellar dust. The source in Cygnus A coincided with a galaxy hundreds of millions of light-years away. The relatively intense radio waves detected at the earth imply that the radio luminosity of this object is on the order of 10^{38} W, millions of times greater than that of the Milky Way, while its optical luminosity is comparable. Thus the term "radio galaxy" is appropriate to describe systems such as this which are exceptionally luminous at radio frequencies.

The efforts to understand the structure of radio galaxies have resulted in important progress, although not without occasional setbacks. For example, the fluctuations in the signal from Cygnus A which were assumed to be intrinsic to the source, and thus led to its being considered to have a very small angular size, were later shown to be produced in the earth's atmosphere. Despite a few identifications of radio point sources, the general uncertainty about their nature in the early 1950s led to the assumption that they were radio stars. Since a star has a very small angular diameter, R. Hanbury Brown, working at Jodrell Bank, decided that a system capable of resolving stellar angular diameters, which are hundredths of an arcsecond or less, was required. At the radio wavelengths then in use, a baseline of thousands of kilometers was necessary, so that bringing the radio signals together for correlation as required for a Michelson-type interferometer was not practical. Hanbury Brown was thus led to the idea of the intensity interferometer, in which the *intensities* of the two signals rather than their *amplitudes* are correlated. The first intensity interferometer was used to study the structures of the radio sources Cygnus A and Cas A, both of which were found to have a size scale of several arcminutes (Hanbury Brown *et al.*, 1952). The high angular resolution potential of the intensity interferometer was thus not required for the study of these objects, and its relatively low sensitivity restricted its subsequent use in radio astronomy. It has played an important role in determining stellar diameters in the optical region of the spectrum, where coherent signal processing techniques are only now beginning to be developed.

The "typical" radio galaxy has emission coming from two lobes which lie on opposite sides of the optical galaxy located between them, and which are separated from the central object by 30,000 to over 1 million light-years. Extensive and impressively detailed maps of their radio emission have been made using aperture synthesis instruments such as the VLA in the United States, the Westerbork Synthesis Radio Telescope in the Netherlands, and the Merlin Array in Great Britain. This work has revealed that these lobes, which range from a few to over 10^5 light-years in size, have a complex structure. The emission is highly polarized, indicating that the synchrotron radiation process is operative. This requires the presence of highly energetic electrons, together with a somewhat tangled magnetic field. The energy radiated must be replenished, and it is the central galaxy which appears to be the source, as indicated by the jets and streamers extending from the central object out toward the radio lobes in a number of radio galaxies.

Many radio galaxies are resolved on an angular scale of 1 arcminute. However, in systematically studying the sizes of radio galaxies, Hanbury Brown found a number that appeared much smaller. B. Mills (1952) and Hanbury Brown, Palmer, and Thompson (1954, 1955) developed the radio-linked interferometer in order to study the structure of these small-angular-diameter radio sources. This is a Michelson-type system, but the signals picked up by two widely separated radio antennas are brought together by being retransmitted and received over a ground radio link rather than just using cable or waveguide. The spacing of the two antennas can thus be readily varied, and a succession of efforts were made during the late 1950s and early 1960s with ever-increasing separation (Morris *et al.*, 1957). A baseline of 100 km was attained in 1962; the corresponding angular resolution was close to 1 arcsecond, but there were still 7 sources (out of the 384 studied) that were smaller than this limit and remained unresolved.

Quasars

Correlating the positions of small-diameter radio sources with objects seen at visible wavelengths was a goal for astronomers trying to understand these sources. In order to accomplish this, accurate positions were needed, since there was no expectation that the radio source would be coincident with a particularly prominent optical object. Lunar occultations (Hazard *et al.*, 1963) and interferometry (Matthews and Sandage, 1963) were used to determine accurate absolute positions. Two radio sources, 3C48 and 3C273 (numbers 48 and 273 in the third Cambridge catalog), were found to be associated with moderately faint, blue, star-like objects. The optical astronomers had a powerful tool at their disposal—spectroscopy—and had measured the wavelengths of a number of spectral lines, but these did not agree with those of the atoms and ions expected to be present. J. Greenstein and T. Matthews (1963) and Greenstein and M. Schmidt (1964) realized that the lines observed did agree with those expected, if extremely large Doppler shifts were included. The observed shifts indicated that both of these objects are receding, with velocities equal to 15 and 31% of the speed of light for 3C48 and 3C273, respectively. Using Hubble's law to relate velocities and distances to far objects, the two sources are found to be 1.9 and 4.5 million light-years away, at least a thousand times further than comparably bright objects in our galaxy. These radio sources certainly cannot be stars, since they are emitting far too much energy. On the basis of their appearance, but in deference to their starlike optical appearance, they were denoted quasi-stellar objects, or quasars.

The small angular size, together with great distance of quasars results in a major problem in understanding the means by which they produce the power they radiate, since the volume available is relatively small. This problem has been rendered more acute by detection of variability in both the radio and optical outputs. Time scales are as short as a few days, with major variations occurring over a few months. This rapid variability gives upper limits on the size of the emission regions significantly smaller than those implied by the arcsecond angular resolution limit discussed above.

To make further progress, further extension of interferometer baselines is necessary, but bringing the signals together for correlation becomes impractical at distances greater than a few hundred kilometers. The intensity interferometer discussed above lacks sufficient sensitivity for this work. Instead, the signals from the source can be separately recorded on magnetic tape, if a suitable reference signal is available to keep track of the phase of the incoming data. The tapes can later be played back together and the correlation between the signals recovered. The hydrogen maser has sufficient phase stability to allow this process to work for frequencies at least as high as 100 GHz. This combination of separate antennas and data recording systems is now referred to as very long baseline interferometry (VLBI), since the separation of the antennas can be made arbitrarily large.

A Canadian group led by N. Broten carried out the first extragalactic VLBI observations in 1967. The observing frequency was 448 MHz and the baseline 3074 km, corresponding to an angular resolution well below 0.1″. Even with this extremely high angular resolution, the sources 3C273 and 3C45 were still unresolved (Broten *et al.*, 1967). Other groups of scientists rapidly entered this field; baselines were extended to intercontinental distances and frequencies raised to obtain even higher resolution. Many quasars still show considerable unresolved structure on a scale of 0.001″ (1 milliarcsecond), which corresponds to an angular resolution of 1 part in 2×10^8. Other quasars break up into a group of blobs, suggesting a very inhomogeneous emission source. The study of the small-scale structure of radio galaxies and quasars will be one of the prime missions of the very long baseline array telescope (VLBA) which is presently under construction. This array of 10 antennas spread across the United States will produce radio images with high dynamic range over the frequency range from 300 MHz to 43 GHz, with synthesized beamsizes from 24 to 0.2 milliarcseconds.

One of the most exciting discoveries so far from VLBI observations is the detection of apparently faster than light (superluminal) motions in a number of quasars. As is the case in most VLBI work, the coordinated efforts of many astronomers were necessary to obtain the results. Starting with the publication in 1971 (Cohen *et al.*, 1971; Whitney *et al.*, 1971) of data covering the previous 5 years, it has become increasingly clear that the apparent separation of the emission peaks within at least three quasars is approximately a linearly increasing function of time, and that the rate of physical motion across the sky implied by multiplying the angular motion by the distance given by the red shift of the sources greatly exceeds the speed of light. Since motion of physical objects at speeds greater than that of light is forbidden by relativity theory, other explanations must be sought. The apparent superluminal motion is now generally interpreted as due to radiating material moving nearly parallel to the line of sight at a velocity *close to* but less than the speed of light. The finite signal travel time results in a distortion of the emission as seen from the earth, and the motions that are at less than light speed in the frame of the quasar appear to us as faster than light transverse motions on the sky. Although there is not thought to be any violation of relativity in the process, the superluminal sources do raise a number of questions. These center around how the jetlike motions of blobs of material ejected from the central source are produced. This is thought to be closely related to the intriguing questions of how the jets in radio galaxies are formed and collimated, and how the immense energy for all these sources is obtained in relatively small regions. In the most extreme cases, the power radiated from a source on the order of the solar system in size is comparable to the output of an entire galaxy! Extreme conditions are undoubtedly present in the central parts of quasars, and the most plausible scenarios involve radiation from material being drawn into a massive black hole.

COSMOLOGY

Radio Source Counts

In the years following the recognition that a fraction of the sources of radio emission are discrete objects, it became clear that they represented a population of objects distributed over vast distances. Providing that certain assumptions are made, they should be capable of yielding information about the cosmological evolution of the universe. This is essentially the same idea as observing galaxies at optical wavelengths and studying their distribution as a function of intensity as a means of learning about the structure of space. At optical wavelengths, the technique is plagued by a number of problems including variable obscuration and sky brightness, contamination by emission lines, and effect of galaxy color. Initially, it was hoped that radio source studies might be free of these problems; certainly the absorption is negligible at radio wavelengths, and the sky background is low and uniform. It has gradually become clear, however, that radio source counts have their own difficulties due to instrumental performance and limitations in our understanding of radio galaxies.

An analysis of the distribution of flux densities from radio galaxies was published by M. Ryle and P. Scheuer in 1955 based on data taken with the Cambridge four-element interferometer operating at a wavelength of 3.7 m (Ryle and Scheuer, 1955). They concluded that there was a great excess of weak relative to strong sources, when compared to what one would expect for a static universe uniformly populated with radio sources. This result was taken to support the "Big Bang" model of the universe in which conditions have been continuously evolving as opposed to the uniform "Steady State" models. The cosmological implications of the radio data were obviously significant and generated considerable excitement.

In 1957, B. Mills and O. Slee published results based on observations with the Mills Cross instrument in Australia operating at 3.5 m wavelength, which had a smaller and more symmetric beam with a cleaner response pattern than the Cambridge instrument (Mills and Slee, 1957). The results of this survey were quite different; in particular, for sources located away from the plane of the galaxy, the number with

flux density greater than F varied as $1/F^{3/2}$, exactly as expected for a uniform, static universe. The differences are now understood as largely due to confusion produced by the antenna response pattern of the Cambridge interferometer.

The ever-increasing sensitivity of radio instrumentation has led to extension of the flux density range to the range of tens of microjanskys (1 jansky $= 10^{-26}$ W/m^2/Hz), extending a factor of a million below the strongest sources. The large number of sources counted has enabled the use of differential rather than integral source counts, i.e., the number of sources per unit solid angle at a given flux density level is analyzed rather than the number of sources weaker than a particular level. Compared to the two order of magnitude range of the early work referred to above, the situation appears far more complex, with the distribution of sources being consistent with a static Euclidean universe within certain flux density intervals separated by regions in which the distribution rises and falls. It is generally accepted that the evolution of radio galaxies plays a significant role in determining the flux density distribution, as is also the case at optical wavelengths. It also appears necessary to use optical identifications and red shifts to analyze the radio source count information. There is still some evidence for an increase in the number of radio sources per unit volume in the early universe. There is, as well, evidence for a surprising discontinuity in the number density at the level of 10 Jy. Both of these points are the subject of ongoing investigations.

The Cosmic Background Radiation

Radio astronomy has had a profound impact on observational cosmology as a result of the discovery of what is generally referred to as the "Cosmic Microwave Background" radiation. This work was a direct result of an effort at Bell Laboratories to accurately calibrate the noise power collected by an antenna originally built for satellite communications experiments. The antenna was a horn reflector, a design which has very low pickup from directions other than that of the main beam, and which consequently should be capable of being accurately calibrated. A. Penzias and R. Wilson addressed this problem at a frequency of 4.08 GHz, continuing the work of E. Ohm who had published measurements indicating that there was an unexplained antenna temperature of 3.3 K at a frequency of 2.3 GHz. This result (Ohm, 1961) was, within the experimental uncertainty, consistent with there being no excess, so that more accurate measurements were required. By using a very carefully calibrated reference load cooled to liquid helium temperature, together with a low loss switch to compare the antenna and the reference source, Penzias and Wilson were able to determine that there was an *excess* antenna temperature of 3.5 ± 1 K, once the contributions of the antenna, feed system, and sky emission were taken into account (Penzias and Wilson, 1965). The work leading up to this discovery has been described from various perspectives by Wilson (1983) and D. Wilkinson (1983).

The observed excess was studied over a 9-month period in 1964–65, and was found to be isotropic, unpolarized, and time independent. Known astronomical sources were incapable of producing these characteristics, and Penzias and Wilson proposed that the radiation was a remnant of the cosmic explosion that resulted in our presently expanding universe. This effect had been predicted by G. Gamow, and R. Alpher and R. Herman about 15 years previously (Gamow, 1948; Alpher and Herman, 1948), and was evaluated in an article by R. Dicke and his collaborators from Princeton which accompanied Penzias and Wilson's results (Dicke *et al.*, 1965). The observations of P. Roll and D. Wilkinson (1966) at a frequency of 9.3 GHz were consistent with the excess emission having a blackbody spectrum. This result has been subsequently confirmed throughout the centimeter and millimeter wavelength ranges, despite formidable experimental difficulties. The present best estimate of the temperature of the cosmic microwave background radiation is close to 2.8 K.

The cosmic background is significant because it represents unique information about the early universe. Studies of this primordial radiation are one of the strongest pieces of evidence for the "Big Bang," a giant explosion which initiated the expansion of the universe. In addition, they place critical constraints on the formation of galaxies, an event thought to have taken place somewhat after the radiation we now detect last interacted with matter. High-precision searches for anisotropies in the cosmic background have been carried out over a wide range of radio frequencies, achieving limits approaching 1 part in 10^4 on small angular scales, on which perturbations due to material in the process of forming galaxies should be reflected. On larger angular scales, motion of the earth with respect to the rest frame of the background radiation has been detected. The accuracy of the measurements of the cosmic background radiation has been continuously improved and the range of measurements continuously widened using the earth's surface, airplanes, and balloons as observing platforms. A review of the subject has been given by Weiss (1980), and other references in Part V can be consulted for more recent progress in the instrumentation.

The study of the spectrum and distribution of the cosmic background radiation is the mission of the Cosmic Background Explorer (COBE) satellite. By getting above the earth's atmosphere, and having a long-lived, dedicated observing facility, we should gain important information about the early history of the universe.

REFERENCES AND BIBLIOGRAPHY

1932: Jansky, K. G., "Directional studies of atmospherics at high frequencies," *Proc. IRE*, vol. 20, p. 1920, 1932.

1933: Jansky, K. G., "Radio waves from outside the solar system," *Nature*, vol. 132, p. 66, 1933. (a)
Jansky, K. G., "Electrical disturbances apparently of extraterrestrial origin," *Proc. IRE*, vol. 21, p. 1387, 1933. (b)

1935: Jansky, K. G., "A note on the source of interstellar interference," *Proc. IRE*, vol. 23, p. 1158, 1935.

1940: Reber, G., "Cosmic static," *Proc. IRE*, vol. 28, p. 68, 1940. (a)
Reber, G., "Cosmic static," *Astrophys. J.*, vol. 91, p. 621, 1940. (b)

1944: Reber, G., "Cosmic static," *Astrophys. J.*, vol. 100, p. 279, 1944.

1945: Southworth, G. C., "Microwave radiation from the sun," *J. Franklin Inst.*, vol. 239, p. 285, 1945.
Van de Hulst, H. C., "Radio waves from space," *Ned. Tij. Natuurkunde*, vol. 11, p. 210, 1945. (A translation of this paper is available as Paper 34 in *Classics in Radio Astronomy*, W. T. Sullivan, Ed. Dordrecht: Reidel, 1982.)

1946: Dicke, R. H. and R. Beringer, "Microwave radiation from the sun and moon," *Astrophys. J.*, vol. 103, p. 375, 1946.

Hey, J. S., "Solar radiations in the 4–6 metre radio wave-length band," *Nature*, vol. 157, p. 47, 1946.
Hey, J. S., S. J. Parsons, and J. W. Phillips, "Fluctuations in cosmic radiation at radio frequencies," *Nature*, vol. 158, p. 234, 1946.
Ryle, M. and D. D. Vonberg, "Solar radiation on 175 Mc./s," *Nature*, vol. 158, p. 339, 1946.

1947: McCready, L. L., J. L. Pawsey, and R. Payne-Scott, "Solar radiation at radio frequencies and its relation to sunspots," *Proc. Roy. Soc., Ser. A*, vol. 190, p. 357, 1947.

1948: Alpher, R. and R. Herman, "Evolution of the universe," *Nature*, vol. 162, p. 774, 1948.
Bolton, J. G. and G. J. Stanley, "Observations on the variable source of cosmic radio frequency radiation in the constellation of Cygnus," *Aust. J. Sci. Res.*, vol. A1, p. 58, 1948.
Covington, A. E., "Solar noise observations on 10.7 cm," *Proc. IRE*, vol. 36, p. 454, 1948.
Gamow, G., "The origin of elements and the separation of galaxies," *Phys. Rev.*, vol. 74, p. 505, 1948.
Reber, G., "Cosmic static," *Proc. IRE*, vol. 36, p. 1215, 1948.
Ryle, M. and F. G. Smith, "A new intense source of radio-frequency radiation in the constellation of Cassiopeia," *Nature*, vol. 162, p. 462, 1948.

1949: Bolton, J. G., G. J. Stanley, and O. B. Slee, "Positions of three discrete sources of galactic radio-frequency radiation," *Nature*, vol. 164, p. 101, 1949.
DeWitt, J. H. and E. K. Stodola, "Detection of radio signals reflected from the moon," *Proc. IRE*, vol. 37, p. 229, 1949.
Shlovsky, I. S., "Monochromatic radio emission from the galaxy and the possibility of its observation," *Astron. Zh.*, vol. 26, p. 10, 1949. (A translation of this paper is available as Paper 35 in *Classics in Radio Astronomy*, W. T. Sullivan, Ed. Dordrecht: Reidel, 1982.)

1951: Ewen, H. I. and E. M. Purcell, "Radiation from galactic hydrogen at 1420 Mc/s," *Nature*, vol. 168, p. 356, 1951.
Graham Smith, F., "An accurate determination of the positions of four radio stars," *Nature*, vol. 168, p. 555, 1951.
Muller, C. A. and J. H. Oort, "The interstellar hydrogen line at 1420 Mc/s and an estimate of galactic rotation," *Nature*, vol. 168, p. 357, 1951.

1952: Hanbury Brown, R., R. C. Jennison, and M. K. Das Gupta, "Apparent angular sizes of discrete radio sources," *Nature*, vol. 170, p. 1061, 1952.
Mills, B. Y., "Apparent angular sizes of discrete radio sources," *Nature*, vol. 170, p. 1063, 1952.

1954: Hanbury-Brown, R., H. P. Palmer, and A. R. Thompson, "Galactic sources of large angular diameter," *Nature*, vol. 173, p. 945, 1954.
Townes, C. H., "Microwave spectra of astrophysical interest," *J. Geophys. Res.*, vol. 59, p. 191, 1954.

1955: Burke, B. F. and K. L. Franklin, "Observations of a variable radio source associated with the planet Jupiter," *J. Geophys. Res.*, vol. 60, p. 213, 1955.
Hanbury Brown, R., H. P. Palmer, and A. R. Thompson, "A rotating lobe interferometer and its application to radio astronomy," *Phil. Mag., Ser. 7*, vol. 46, p. 857, 1955.
Ryle, M. and P. A. G. Scheuer, "The spatial distribution and the nature of radio stars," *Proc. Roy. Soc., Ser. A*, vol. 230, p. 448, 1955.

1957: Mills, B. Y. and O. B. Slee, "A preliminary survey of radio sources in a limited region of the sky at a wavelength of 3.5 m," *Aust. J. Phys.*, vol. 10, p. 162, 1957.
Morris, D., H. P. Palmer, and A. R. Thompson, "Five radio sources of small angular diameter," *The Observatory*, vol. 77, p. 103, 1957.

1958: Mayer, C. H., T. P. McCullough, and R. M. Sloanaker, "Observations of Venus at 3.15 cm wavelength," *Astrophys. J.*, vol. 127, p. 1, 1958. (a)
Mayer, C. H., T. P. McCullough, and R. M. Sloanaker, "Observations of Mars and Jupiter at a wavelength of 3.15 cm," *Astrophys. J.*, vol. 127, p. 11, 1958. (b)
Reber, G., "Early radio astronomy at Wheaton, Illinois," *Proc. IRE*, vol. 46, p. 14, 1958.

1959: Kardashev, N. S., "On the possibility of detection of allowed lines of atomic hydrogen in the radio frequency spectrum," *Sov. Astron. AJ*, vol. 3, p. 813, 1959.

1961: Ohm, E. A., "Project Echo Receiving System," *Bell Syst. Tech. J.*, vol. 40, p. 1065, 1961.

1962: Howard, W. E., A. H. Barrett, and F. T. Haddock, "Measurements of the microwave radiation from the planet Mercury," *Astrophys. J.,* vol. 136, p. 995, 1962.

1963: Greenstein, J. L. and T. A. Matthews, "Red-shift of the unusual radio source 3C48," *Nature,* vol. 197, p. 1037, 1963.

Hazard, C., M. B. Mackey, and A. J. Shimmins, "Investigation of the radio source 3C 273 by the method of lunar occultations," *Nature,* vol. 197, p. 1037, 1963.

Matthews, T. A. and A. R. Sandage, "Optical identification of 3C 48, 3C 196, and 3C 286 with stellar objects," *Astrophys. J.,* vol. 138, p. 30, 1963.

Weinreb, S., A. H. Barrett, M. L. Meeks, and J. C. Henry, "Radio observations of OH in the interstellar medium," *Nature,* vol. 200, p. 829, 1963.

1964: Carpenter, R. L., "Study of Venus by CW radar," *Astron. J.,* vol. 69, p. 2, 1964.

Dravskikh, Z. V., A. F. Dravskikh, and V. A. Kolbason, "An excited hydrogen line profile in the Omega Nebula," *Astr. Tsikr.,* no. 305, p. 2, 1964.

Goldstein, R. N., "Venus characteristics by Earth-based radar," *Astron. J.,* vol. 69, p. 12, 1964.

Greenstein, J. L. and M. Schmidt, "The quasi-stellar radio sources 3C 48 and 3C 273," *Astrophys. J.,* vol. 140, p. 1, 1964.

1965: Dicke, R. H., P. J. E. Peebles, P. G. Roll, and D. T. Wilkinson, "Cosmic blackbody radiation," *Astrophys. J.,* vol. 142, p. 414, 1965.

Penzias, A. A. and R. W. Wilson, "A measurement of excess antenna temperature at 4080 Mc/s," *Astrophys. J.,* vol. 142, p. 419, 1965.

Pettengill, G., "Planetary radar astronomy," Chapter 18 in *Solar System Radio Astronomy,* J. Aarons, Ed. New York, NY: Plenum, 1965.

Pettengill, G. H. and R. B. Dyce, "Radar determination of the rotation of the planet Mercury," *Nature,* vol. 206, p. 1240, 1965.

1966: Roll, P. G. and D. T. Wilkinson, "Cosmic background radiation at 3.2 cm—Support for cosmic blackbody radiation," *Phys. Rev. Lett.,* vol. 16, p. 405, 1966.

1967: Broten, N. W., J. L. Locke, T. H. Legg, C. W. McLeish, R. S. Richards, R. M. Chisholm, H. P. Gush, J. L. Yen, and J. A. Galt, "Observations of quasars using interferometer baselines up to 3074 km," *Nature,* vol. 215, p. 38, 1967.

Wild, J. P. (Ed.), "The Culgoora radioheliograph," *Proc. Inst. Radio Electron. Eng. (Australia),* vol. 28, p. 277, 1967.

1968: Cheung, A. C., D. M. Rank, C. H. Townes, D. D. Thornton, and W. J. Welch, "Detection of NH_3 molecules in the interstellar medium by their microwave emission," *Phys. Rev. Lett.,* vol. 21, p. 1701, 1968.

Gold, T., "Rotating neutron stars as the origin of the pulsating radio sources," *Nature,* vol. 218, p. 731, 1968.

Hewish, A., S. J. Burnell, J. D. Pilkington, P. F. Scott, and R. A. Collins, "Observation of a rapidly pulsating radio source," *Nature,* vol. 217, p. 709, 1968.

Pacini, F., "Rotating neutron stars, pulsars, and supernova remnants," *Nature,* vol. 219, p. 145, 1968.

1970: Wilson, R. W., K. B. Jefferts, and A. A. Penzias, "Carbon monoxide in the Orion Nebula," *Astrophys. J. (Lett.),* vol. 161, p. L43, 1970.

1971: Cohen, M. H., W. Cannon, G. H. Porcell, D. B. Shoffer, J. J. Broderick, K. I. Kellerman, and J. D. Jauncey, "The small-scale structure of radio galaxies and quasi-stellar sources at 3.8 centimeters," *Astrophys. J.,* vol. 170, p. 207, 1971.

Whitney, A. R., I. I. Shapiro, A. E. E. Rogers, D. S. Robertson, C. A. Knight, T. A. Clark, R. M. Goldstein, G. E. Marandino, and N. R. Vandenberg, "Quasars revisited: Rapid time variations observed via very-long-baseline interferometry," *Science,* vol. 173, p. 225, 1971.

1972: Hjellming, R. M., E. Webster, and B. Balick, "Radio behavior of β Persei," *Astrophys. J. (Lett.),* vol. 178, p. L139, 1972.

1973: Hey, J. S., *The Evolution of Radio Astronomy.* New York, NY: Science History Publications, 1973.

1975: Hulse, R. A. and J. H. Taylor, "Discovery of a pulsar in a binary system," *Astrophys. J. (Lett.),* vol. 195, p. L51, 1975.

1976: Clark, T. A., L. K. Hutton, C. Ma, I. I. Shapiro, J. J. Wittels, D. S. Robertso, H. F. Hinteregger, C. A. Knight, A. E. E. Rogers, A. R. Whitney, A. E. Niel, G. M. Resch, and W. J. Webster, Jr., "An unusually strong radio outburst in Algol: VLBI observations," *Astrophys. J. (Lett.),* vol. 206, p. L107, 1976.

1977: Manchester, R. N. and J. H. Taylor, *Pulsars.* San Francisco, CA: Freeman, 1977.

1978: Sullivan, W. T., III, "A new look at Karl Jansky's original data," *Sky and Telescope,* vol. 56, p. 101, Aug. 1978.

1980: Solomon, P. M. and M. G. Edmunds (Eds.), *Giant Molecular Clouds in the Galaxy.* Oxford: Pergamon, 1980.

Weiss, R., "Measurements of the cosmic background radiation," *Annu. Rev. Astron. Astrophys.,* vol. 18, p. 489, 1980.

1981: Wilson, W. J., M. J. Klein, R. K. Kakar, S. Gulkis, and E. T. Olsen, "Venus I. Carbon monoxide distribution and molecular-line searches," *Icarus,* vol. 45, p. 624, 1981.

1982: Backer, D. C., S. R. Kulkarni, C. Heiles, M. M. Davis, and W. M. Goss, "A millisecond pulsar," *Nature,* vol. 300, p. 615, 1982.

Snyder, L. E., "A review of radio observations of comets," *Icarus,* vol. 51, p. 1, 1982.

Sullivan, W. T. (Ed.), *Classics in Radio Astronomy.* Dordrecht: Reidel, 1982.

1983: Bell Burnell, J., "The discovery of pulsars," in *Workshop Proc., Serendipitous Discoveries in Radio Astronomy,* K. Kellerman and B. Sheets, Eds. Green Bank, WV: National Radio Astronomy Observatory, 1983, p. 160.

Greenstein, J. L., "Optical and radio astronomers in the early years," in *Workshop Proc., Serendipitous Discoveries in Radio Astronomy,* K. Kellerman and B. Sheets, Eds. Green Bank, WV: National Radio Astronomy Observatory, 1983, p. 79.

Reber, G., "Radio astronomy between Jansky and Reber," in *Workshop Proc., Serendipitous Discoveries in Radio Astronomy,* K. Kellerman and B. Sheets, Eds. Green Bank, WV: National Radio Astronomy Observatory, 1983, p. 71.

Wilkinson, D. T., "Discovery of the 3 K radiation," in *Workshop Proc., Serendipitous Discoveries in Radio Astronomy,* K. Kellerman and B. Sheets, Eds. Green Bank, WV: National Radio Astronomy Observatory, 1983, p. 175.

Wilson, R. W., "Discovery of the cosmic microwave background," in *Workshop Proc., Serendipitous Discoveries in Radio Astronomy,* K. Kellerman and B. Sheets, Eds. Green Bank, WV: National Radio Astronomy Observatory, 1983, p. 185.

1985: Altenhoff, W. J., "The solar system: (Sub)MM continuum observations," in *Proc. ESO-IRAM-Onsala Workshop on "(Sub)Millimeter Astronomy,"* P. A. Shaver and K. Kjar, Eds. Munich: European Southern Observatory, 1985, p. 591.

1987: Irvine, W. M., P. F. Goldsmith, and A. Hjalmarson, "Chemical abundances in molecular clouds," in *Interstellar Processes,* D. J. Hollenbach and H. A. Thronson, Eds. Dordrecht: Reidel, 1987, p. 561. (a)

Irvine, W. M., F. P. Schloerb, and L. E. Tacconi-Garman (Eds.), *Cometary Radio Astronomy.* Green Bank, WV: National Radio Astronomy Observatory, 1987. (b)

General

Information on the state of radio-astronomical observations in 1958 together with a considerable number of articles on their interpretation can be found in:

Bracewell, R. N. (Ed.), *Proc. Paris Symp. on Radio Astronomy,* IAU Symposium No. 9 and URSI Symposium No. 1. Stanford, CA: Stanford University Press, 1958.

A number of articles in the issue of the *Proceedings of the IRE* (now IEEE) devoted to radio astronomy contain interesting historical information:

Proc. IRE, vol. 46, no. 1, Jan. 1958.

A very clearly written early overview of the development of radio astronomy at a nontechnical level is:

Graham Smith, F., *Radio Astronomy.* Baltimore, MD: Penguin Books, 1960.

Considerable historical information on the development of instrumentation and astrophysical theory is contained in:

Shlovsky, I. S., *Cosmic Radio Waves.* Cambridge, MA: Harvard University Press, 1960.

Both technical aspects and astronomical advances are treated in:

Steinberg, J. L. and J. Lequeux, *Radio Astronomy.* New York, NY: McGraw-Hill, 1963.

Techniques of solar radio astronomy, as well as an historical overview and discussion of the current state of this complex field, are given in:

Kundu, M. R., *Solar Radio Astronomy*. New York, NY: Interscience, 1965.

The development of the Jodrell Bank facility and radio astronomy at the University of Manchester is discussed in detail in several books by Sir Bernard Lovell:
The Story of Jodrell Bank. Oxford: Oxford University Press, 1968.
Out of the Zenith. Oxford: Oxford University Press, 1973.
The Jodrell Bank Telescopes. Oxford: Oxford University Press, 1985.

The historical development of radio astronomy through the early 1970s is treated in this book which also includes a valuable list of references:
Hey, J. S., *The Evolution of Radio Astronomy*. New York, NY: Science History Publications, 1973.

A second special issue of the *Proceedings of the IEEE* devoted to radio and radar astronomy which contains both astronomical and technical reviews is:
Proc. IEEE, vol. 61, no. 9, Sept. 1973.

The development of radio astronomy in Great Britain is covered in considerable depth in the following sociological and scientific study:
Edge, D. O. and M. J. Mulkay, *Astronomy Transformed*. New York, NY: John Wiley, 1976.

The design and construction of a large radio telescope at Ohio State University and some of the discoveries made with it are described in:
Kraus, J. D., *Big Ear*. Powell, OH: Cygnus-Quasar Books, 1976.

The important developments in radio astronomy made at the Dover Heights field station in Australia in the late 1940s and early 1950s are described by one of the important participants in:
Bolton, J. G., "History of Australian radio astronomy," *Proc. Astron. Soc. Australia,* vol. 4, p. 349, 1982.

Important articles on observations, techniques, and theory in the first 20 years of radio astronomy are reprinted with valuable commentary in:

Sullivan, W. T. (Ed.), *Classics in Radio Astronomy*. Dordrecht: Reidel, 1982.

The major role played by accidental discovery in radio astronomy was emphasized in the talks given at a conference to commemorate the 50th anniversary of Jansky's discovery. The talks are reproduced in:
Kellerman, K. and B. Sheets (Eds.), *Workshop Proc., Serendipitous Discoveries in Radio Astronomy*. Green Bank, WV: National Radio Astronomy Observatory, 1983.

The discovery and subsequent study of pulsars are described in nontechnical terms in:
Greenstein, G., *Frozen Star*. New York, NY: Plume, 1983.

Commentaries on the development of radio astronomy at a number of institutions can be found in:
Sullivan, W. T. (Ed.), *The Early Years of Radio Astronomy*. Cambridge: Cambridge University Press, 1984.

An historical overview, as well as a brief discussion, of many topics of current research interest is found in the classic textbook on radio astronomy:
Kraus, J. D., *Radio Astronomy,* 2nd ed. Powell, OH: Cygnus-Quasar Books, 1986.

Chapter 1 of the book listed below by Thompson *et al.* gives historical information related to the use of interferometers. This comprehensive text also contains a detailed discussion of the instrumentation of Part II and the techniques of Parts VIII and IX of the present collection of reprinted articles.
Thompson, A. R., J. M. Moran, and G. W. Swenson, Jr., *Interferometry and Synthesis in Radio Astronomy*. New York, NY: John Wiley, 1986.

The following book is primarily a nontechnical overview of the "radio-astronomical" view of the universe, but Chapter 16 reviews the history of radio astronomy:
Verschuur, G. L., *The Invisible Universe Revealed; The Story of Radio Astronomy*. New York, NY: Springer, 1987.

Part I
Filled Aperture Antennas

FILLED aperture antennas have been the workhorses of radio astronomy. Starting with Reber's use of a parabolic antenna, this design has dominated instruments intended for single-dish work as well as elements of interferometric arrays. This preeminent position is challenged only at the longest wavelengths, at which wire antennas are used due to their lower cost. The antennas of Mills Cross-type arrays have consisted of cylindrical parabolas, since the beam in the long direction is designed to be electrically steered; this same antenna shape was employed in the University of Illinois telescope described by Swenson (1986).

The basic parabolic antenna system has evolved from a prime focus design generally used at meter and centimeter wavelengths (cf. the first and second papers in this part, by Lovell, and Bowen and Minnett, respectively), to secondary focal arrangements which have become more popular at higher frequencies. These include both the Gregorian system using an elliptical secondary reflector (third paper, by Hachenberg *et al*.) and the Cassegrain system which employs a hyperbolic secondary reflector (fourth paper, by Chu *et al*.). In recent designs, the Nasmyth focal arrangement, which has the focal point free from motion in one coordinate, has been widely adopted. This arrangement is generally employed to obtain a location for receivers which does not need to move in elevation, and which rotates with the complete telescope in azimuth. Examples of the use of a Nasmyth focus include the AT&T Bell Laboratories millimeter wavelength antenna described in the fourth paper, and the submillimeter telescope described by Baars *et al*. (1984).

Off-axis imaging capability has not been a major concern in the design of radio telescopes. This is not a consequence of field of view limitations of most secondary focal arrangements, since the number of independent resolution elements (pixels) that can be efficiently imaged is *large* ($\geqslant 10$) for focal ratios greater than unity, as discussed in a paper by Ruze (1965). Rather, it is the cost of receivers and associated signal processing hardware that has precluded simultaneous observations of an appreciable number of pixels. This is one area in which radio astronomy has lagged far behind shorter wavelength observational disciplines, but this is beginning to change with the design of a number of focal plane array systems for radio telescopes operating in the 1 to 3 mm wavelength range. In the absence of such systems, maps are built up by combining observations of many individual positions, with an accompanying slow data rate and associated calibration problems. Some techniques to overcome these limitations are described in Part V and off-line image restoration techniques are discussed in Part IX.

The present selection of papers cannot do justice to all of the specialized forms of antennas that have been developed for

astronomical research. The texts by Kraus (1986) and by Christiansen and Hogbom (1985) contain considerably more information. The articles by Khaikin and Kaidanovskii (1959) and by Fourikis (1978) describe some unusual antenna designs.

The first five papers in this part describe milestones in the development of radio-astronomical antenna systems. The Jodrell Bank 250-ft telescope described in the first paper was the first very large fully steerable antenna, and one in which the use of an altazimuth mounting system set the standard for almost all future large antennas. This antenna was resurfaced and improved in 1971 so that it can be used to wavelengths as short as 6 cm. The Australian 210-ft telescope described in the second paper was originally specified to work efficiently to wavelengths as short as 21 cm, but in fact, due to improvements over the years, has been used at wavelengths as short as 7 mm. This instrument also introduced the master collimator system for antenna pointing which has been used in other large antennas, notably the 45-m millimeter telescope at Nobeyama, Japan.

The 100-m radio telescope at Effelsberg (third paper) which is operated by the Max-Planck-Institut für Radioastronomie in Bonn remains the world's largest fully steerable radio telescope. Like the two antennas designed for centimeter wavelengths described above (and many other radio telescopes), it has far exceeded its original design specifications, and is now used in the 7 mm wavelength range, with tests at shorter wavelengths in progress. The adjustment of the surface panels of the Effelsberg telescope is discussed in the seventh paper of Part III. This instrument was the first to employ the principle of homologous deformation introduced by von Hoerner and described in the sixth paper, which allows the telescope shape to remain a parabola even under varying strain produced by changes in elevation. The only adjustment required is a motion of the feed or secondary mirror to compensate for change in the focal length of the primary reflector.

The AT&T Bell Laboratories 7-m-diameter telescope in Holmdel, New Jersey, is described in the fourth paper included here. It is an unusual design in that it is an offset Cassegrain system. Thus, there is no blockage of the aperture from the secondary mirror or support legs, which results in an exceptionally clean and highly efficient beam. The 7-m telescope allows receivers in the normal Cassegrain focal position, or at a Nasmyth focus; large receivers employing liquid cryogens can be advantageously situated at the latter position since it does not move in elevation.

The state of the art in large millimeter radio telescopes is illustrated by the IRAM 30-m antenna on Pico Veleta, Spain, which is described by Baars *et al*. in the fifth paper. This instrument uses the principle of homologous design together with very careful thermal control to achieve extremely small beamwidths (minimum wavelength/diameter approximately 3×10^{-5}; beamwidth 7 arcseconds) and high gain. The receivers are mounted in a cabin fixed in elevation, and the beam is brought over the elevation axis to the Cassegrain focal point by two flat mirrors.

Cassegrain or Gregorian antennas of the symmetric type have the secondary mirror on axis, and as a result suffer from aperture blockage both of the reflector and of the support legs. This topic is discussed in the seventh paper, by Ruze. The exact calculation of the effects of the blockage on the amplitude distribution and on the polarization of the radiation pattern is an involved subject, and is discussed in a paper by Lamb and Olver (1986), and in several of the articles in a reprint volume edited by Love (1978).

A novel approach to correcting for primary reflector deformations by deliberately changing the shape of the secondary mirror is described by von Hoerner and Wong in the eighth paper. This is the closest that radio astronomy has come to utilizing the "rubber mirror" approach developed for the optical range. Radio telescope design to date has, with the exception of the principle of homologous deformation discussed earlier, concentrated on developing as completely rigid a structure as possible, and not on making any real-time adjustments. However, this approach may be changing; a number of recent radio antennas (including the 45-m Nobeyama telescope and the 15-m UK-Netherlands-Canada James Clerk Maxwell telescope on Mauna Kea, Hawaii) have motorized adjustments of the surface panels. It seems plausible that in the future these will be computer-interfaced, and corrections will be made by the on-line control system.

The only antenna among those described in this selection of reprints which is not a parabola is the 1000-ft-diameter spherical antenna at Arecibo, Puerto Rico. This unique instrument is described in the ninth paper, by Gordon and LaLonde. The scanning capability of the spherical reflector is used to compensate for the antenna being fixed, by means of a movable feed which allows sources to be tracked within 20° of the zenith. The accompanying complication is that a line feed must be used to illuminate the spherical mirror; this is a complex subject which is covered in several of the papers in the reprint volume edited by Love (1978). The Arecibo antenna surface has also been upgraded since the instrument was built with the result that operation at wavelengths below 20 cm is now possible; further improvements to the surface and feed system are presently being considered.

The obvious overlap between antennas used for radio astronomy and those used for communications systems, radiometry, remote sensing, and other applications is in part responsible for the enormous literature on this topic, to the point that a moderately thorough listing of relevant references would result in an unwieldy compendium. Thus, the references given below are ones that are thought to be of particular interest for radio astronomy, or which are rich sources of information leading deeper into the literature on this subject.

REFERENCES AND BIBLIOGRAPHY

1950: Kraus, J. D., *Antennas*. New York, NY: McGraw-Hill, 1950.

1959: Khaikin, S. E. and N. L. Kaidanovskii, "A new radio telescope of high resolving power," Paper 29 in *Proc. IAU/URSI Paris Symp. on Radio Astronomy*, R. N. Bracewell, Ed. Stanford, CA: Stanford University Press, 1959, pp. 166–170.

Kraus, J. D., "The Ohio State University 360-ft. radio telescope," *Nature*, vol. 184, pp. 669–672, 1959.

1963: Blum, E. J., A. Boischot, and J. Lequeux, "Radio astronomy in France," *Proc. IRE Australia*, vol. 24, pp. 208–213, 1963. Information about the large Nancay telescope employing a fixed spherical reflector 300 m × 35 m and a tiltable plane reflector 200 m × 45 m is given in this article, along with a summary of radio-astronomical research taking place in France at the time of publication.

1965: Ruze, J., "Lateral-feed displacement in a paraboloid," *IEEE Trans. Antennas Propagat.*, vol. AP-13, pp. 660–665, 1965.

1966: *Deep Space and Missile Tracking Antennas*, Symposium held in conjunction with the ASME Winter Annual Meeting, 1966. New York, NY: American Society of Mechanical Engineers. This symposium proceedings contains a number of interesting articles, particularly those about the MIT Lincoln Laboratory 120-ft Haystack antenna and the JPL Deep Space Network 210-ft antenna.

1969: Mar, J. W. and H. Liebowitz (Eds.), *Structures Technology for Large Radio and Radar Telescope Systems*. Cambridge, MA: MIT Press, 1969.

1970: Cogdell, J. R., J. J. G. McCue, P. D. Kalachev, A. E. Salomonovich, I. G. Moiseev, J. M. Stacey, E. E. Epstein, E. E. Altshuler, G. Feix, J. W. B. Day, H. Hvatum, W. J. Welch, and F. T. Barath, "High resolution millimeter reflector antennas," *IEEE Trans. Antennas Propagat.*, vol. AP-18, pp. 515–529, 1970.

1971: Findlay, J. W., "Filled-aperture antennas for radio astronomy," *Annu. Rev. Astron. Astrophys.*, vol. 9, pp. 271–292, 1971.

Meeks, M. L. and J. Ruze, "Evaluation of the Haystack antenna and radome," *IEEE Trans. Antennas Propagat.*, vol. AP-19, pp. 723–728, 1971.

1972: Findlay, J. W. and S. von Hoerner, *A 65-Meter Telescope for Millimeter Wavelengths*. Charlottesville, VA: National Radio Astronomy Observatory, 1972.

Vu, T. B., "Antenna tolerance theory—A survey of basic methods and recent developments," *Proc. IREE Australia*, vol. 33, no. 6, pp. 268–274, 1972.

1975: von Hoerner, S., "Radio telescopes for millimeter wavelength," *Astron. Astrophys.*, vol. 41, pp. 301–306, 1975.

1976: Love, A. W. (Ed.), *Electromagnetic Horn Antennas*. New York, NY: IEEE Press, 1976.

Meeks, M. L. (Ed.), *Methods of Experimental Physics: Vol. 12 Astrophysics, Part B: Radio Telescopes*. New York, NY: Academic Press, 1976. Of particular relevance are Chapter 1.2, "Types of astronomical antennas" by W. J. Welch; Chapter 1.3, "Analysis of paraboloidal-reflector systems" by W. V. T. Rusch; and Chapter 1.4, "Feed systems for paraboloidal reflectors" by J. Ruze.

1977: Padman, R., "Reduction of baseline ripple on the spectra recorded with the Parkes radio telescope," *Proc. Astron. Soc. Australia*, vol. 3, pp. 111–113, 1977.

1978: Fourikis, N., "A new class of millimetre-wave telescope," *Astron. Astrophys.*, vol. 65, pp. 385–388, 1978.

Love, A. W. (Ed.), *Reflector Antennas*. New York, NY: IEEE Press, 1978.

Morris, D., "Chromatism in radio telescopes due to blocking and feed scattering," *Astron. Astrophys.*, vol. 67, pp. 221–228, 1978.

1979: Love, A. W., "Quadratic phase error loss in circular apertures," *Electron. Lett.*, vol. 15, pp. 276–277, 1979.

1980: Goldsmith, P. F. and N. Z. Scoville, "Reduction of baseline ripple in millimeter radio spectra by quasi-optical phase modulation," *Astron. Astrophys.*, vol. 82, pp. 337–339, 1980.

1981: Clarricoats, P. J. B., "Feeds for reflector antennas—A review," in *Proc. Second Int. Conf. on Antennas and Propagation*, Heslington, York, United Kingdom, Apr. 13–16, 1981, pp. 309–317.

1983: Akabane, K., M. Morimoto, N. Kaifu, and M. Ishiguro, "Large MM-wave telescopes in Japan," Nobeyama Radio Observatory Report No. 16, 1983.

Baars, J. W. M., "Technology of large radio telescopes for millimeter and submillimeter wavelengths," in *Infrared and Millimeter Waves*, vol. 9, K. J. Button, Ed. New York, NY: Academic Press, 1983, pp. 241–281.

1984: Baars, J. W. M., P. G. Mezger, B. L. Ulich, W. F. Hoffmann, and R. E. Parks, "Design features of a 10 m telescope for submillimeter astronomy,'' in *Advanced Technology Optical Telescopes II* (SPIE vol. 444). Bellingham, WA: SPIE, 1984, pp. 65–71.

1985: Bregman, J. D. and J. L. Casse, ''A simulation of the thermal behaviour of the UK-NL millimeter wave telescope,'' *Int. J. Infrared and Millimeter Waves*, vol. 6, pp. 25–40, 1985.

Christiansen, W. N. and J. A. Högbom, *Radio Telescopes,* 2nd ed. Cambridge: Cambridge University Press, 1985.

Smith, D. H., ''The submillimeter giants,'' *Sky and Telescope*, pp. 119–123, Aug. 1985.

1986: Kraus, J. D., *Radio Astronomy*, 2nd ed. Powell, OH: Cygnus-Quasar Books, 1986. Chapter 6 contains an excellent introduction to the subject together with an extensive bibliography.

Lamb, J. W. and A. D. Olver, ''Blockage due to subreflector supports in large radiotelescope antennas,'' *IEE Proc.*, vol. 133, pt. H, pp. 43–49, 1986.

Swenson, G. W., Jr., ''Reminiscences: The Illinois 400-foot radio telescope,'' *IEEE Antennas and Propagation Society Newsletter*, vol. 28, no. 6, pp. 13–16, 1986.

Winnewisser, G., M. Bester, R. Ewald, W. Hilberath, K. Jacobs, W. Krotz-Vogel, M. Miller, M. Olberg, T. Pauls, E. Pofahl, G. Rau, R. Schieder, H. Stubbusch, B. Vowinkel, C. Wieners, and W. Zensen, ''The University of Cologne 3-m radio telescope,'' *Astron. Astrophys.*, vol. 167, pp. 207–213, 1986.

1987: Hudson, J. A., R. L. Plambeck, and W. J. Welch, ''Aperture efficiency enhancement in a Cassegrain system by means of a dielectric lens,'' *Radio Sci.,* vol. 22, pp. 1091–1101, 1987.

1988: Baars, J. W. M., A. Greve, B. G. Hooghoudt, and J. Penalver, ''Thermal control of the IRAM 30-m radio telescope,'' *Astron. Astrophys.,* vol. 195, pp. 364–371, 1988.

Castets, A., R. Lucas, B. Lazareff, J. Cernicharo, A. Omont, G. Duvert, B. Fouilleux, T. Forveille, L. Pagani, G. Beaudin, A. Deschamps, P. Encrenaz, S. Lebourg, H. Gheudin, H. Pérault, G. Ruffié, B. Clavelier, J. Lacroix, R. Lauque, G. Montignac, A. Baudry, and M. Champion, ''The 2.5-m millimeter telescope on Plateau de Bure,'' *Astron. Astrophys.,* vol. 194, pp. 340–343, 1988.

THE JODRELL BANK RADIO TELESCOPE

By Prof. A. C. B. LOVELL, O.B.E., F.R.S.

THE large radio telescope which has been built at the Jodrell Bank Experimental Station of the University of Manchester is nearly complete. It is frequently stated that the instrument is the largest radio telescope in existence. This is quite erroneous without a good deal of qualification. The telescope is essentially a paraboloidal steel bowl 250 ft. in diameter, with the focus in the aperture plane, built so that it can be directed towards any part of the sky. It is, therefore, relatively small compared with many radio telescopes elsewhere. For example, aerials of the 'Mills Cross' type have been built with an extension of more than a mile, and even one element of Ryle's adjustable cylindrical parabolic interferometer system which is nearing completion near Cambridge has a 50 per cent greater surface area. Nevertheless, the instrument at Jodrell Bank is believed to be by far the largest completely steerable pencil-beam type of radio telescope in existence, although considerable uncertainty exists as to the Russian facilities in this type of work. Its nearest operational competitors in size are the 80-ft. diameter telescopes at Dwingeloo in Holland, and near Bonn in Western Germany, although both these instruments are figured to a much higher degree of accuracy and are designed for work on much shorter wave-lengths than the Jodrell Bank telescope. In the next few years it is expected that 90-ft. aperture radio telescopes will be built at the California Institute of Technology, a 140-ft. instrument for the U.S. National Radio Observatory of Associated Universities Incorporated at Greenbank, Virginia, and one of 200- to 300-ft. aperture near Sydney. Hence, even although the British instrument has taken two years longer to build than originally specified, it is likely to rank as the major instrument of this type for radio astronomical research for a considerable time.

The engineers responsible for the telescope are Messrs. Husband and Co., of Sheffield and London. When the idea for the instrument was first put forward in 1948 and 1949 many engineers refused to consider the possibility of its construction. Mr. H. C. Husband, however, believed that the task could be accomplished, and the results of his work and of his team of designers now stand in the shape of more than two thousand tons of steel on the Cheshire plain. In principle the motion of the telescope is altazimuth. The bowl, which weighs about seven hundred tons, is driven

in elevation by a Ward–Leonard system through two 27-ft. racks from the dismantled battleship *Royal Sovereign*. These are mounted 170 ft. above ground on two towers which themselves rotate on a 350-ft. railway track to give the motion in azimuth. The drive is through two bogies under each tower, again through a Ward–Leonard system. Four additional bogies, which are undriven, serve as wind carriages on each side of the structure. The towers are connected near ground-level through a heavy diametral girder system supported on the central pivot which is the fundamental locating part of the telescope. The power and instrumental cables come through this central pivot into a motor room situated within the diametral girder immediately above the central pivot. This room contains the motor generator sets and controls for the Ward–Leonard system.

The 17-ft. gauge double railway track on which the telescope rotates is mounted on deep-piled foundations

Fig. 1. The Radio Telescope at Jodrell Bank nearing completion. In the foreground is Prof. A. C. B. Lovell

Reprinted with permission from *Nature*, vol. 180, pp. 60–62, July 13, 1957.
Copyright © 1957, MacMillan Journals Ltd.

which extend in some places to 90 ft. underground. The various power, control and instrumental cables are taken into an annular laboratory underneath the central pivot and then through an underground tunnel to the control room. This control room houses the main control racks and console. The computer system consists of magslip resolvers working in servo loops to solve the necessary equations in order that the telescope can be given a sidereal movement. A wide range of movements can be selected at the control desk, for example, automatic sidereal motion at a given right ascension and declination, motion in galactic latitude and longitude, straightforward motion in azimuth and elevation, and various automatic scanning movements with a choice of rasters. Parallax corrections can also be introduced when it is desired to track a body in the solar system. There are no slip rings so that the danger of creating electrical interference is avoided, and the limit of motion is 420 degrees, after which an automatic reversal takes place.

The specification calls for a tracking accuracy of at least 12 minutes of arc at speeds up to 4 degrees per minute. The maximum slewing speed is about 22 degrees per minute in azimuth and elevation. The position of the telescope in azimuth and elevation is repeated back to the control room through magslips driven by accurately machined chain racks independently of the driving system. The specification requires these positions to be repeated to an accuracy of ± 1 minute of arc.

The reflecting membrane is of $\frac{1}{16}$ in. thick steel sheet welded from 7,000 individual sections of about 3 ft. × 3 ft. on to the purlins of the steel framework. It has been essential to ensure good conductivity across these welded sections otherwise the membrane would become very lossy at certain wave-lengths. The primary aerial feed is carried at the focus on a steel tower built up $62\frac{1}{2}$ ft. from the apex of the paraboloid.

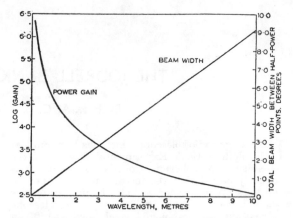

Fig. 3. Calculated power gain and beam width of the 250-ft. aperture radio telescope as a function of wave-length

This tower diminishes in cross-section rapidly with height in order to avoid obscuration and scattering from the primary feed. It was, however, essential to design it with sufficient stiffness so that no displacement occurred as the bowl turns over. An important scientific requirement is that easy access can be attained to the primary feed so that the operational wave-length can be changed readily. Originally it was intended to have an access tower which could move out on tracks on top of the diametral girder to put the operator in reach of the aerial when the bowl was inverted. This was later abandoned in favour of a system whereby the aerial is mounted on a 50-ft. steel tube which slides into the top of the aerial tower. With the bowl inverted this can be winched down to ground-level and replaced by another 50-ft. tube complete with aerial system. The radio-frequency cables from the aerial run inside this tube and can be reached from a small platform near the base of the tower when the bowl is facing towards the zenith.

In much of the work it will be necessary to mount the radio-frequency pre-amplifiers and other parts of the receiving equipment as close as possible to the aerial. These essential units will be contained in a small laboratory which swings underneath the bowl. Further laboratory space is available at the tops of the two towers, but even from those the minimum length of cable run to the primary feed is about 200 ft. Other scientific apparatus will be installed on the base girders ; but the main recording apparatus will be in laboratories adjacent to the control room.

From the astronomical point of view the success of the engineers will be judged by the accuracy with which the telescope moves and the extent to which the reflecting membrane retains its paraboloidal shape in different aspects and under various wind conditions. The theoretical curves showing the beam width and power gain as a function of wave-length are shown in Fig. 3. Calculations have been made of the effect of irregularities in the membrane on the power gain, from which it is clear that, except at very short wave-lengths for which the telescope is not primarily designed, large distortions are permissible without any appreciable effect on the performance of the telescope. During the next few months it is hoped to begin the scientific programme by establishing these curves on an experimental basis by observing certain radio sources at different azimuths and elevations over a range of wave-lengths.

Fig. 2. Underside of the bowl showing the stabilizing wheel, which is 280 ft. in diameter, and the hanging laboratory

Those interested in the immediate scientific programme of the radio telescope may be referred to the forty-seventh Kelvin Lecture[1], where its tasks are set against the current background of knowledge in radio astronomy. It is sufficient to remark at this stage that all the developments which have occurred in radio astronomy during the past years have served to underline the importance of this telescope. With this instrument, together with the large interferometric systems in Cambridge, Great Britain can reasonably look forward to a most fruitful period in radio astronomy.

[1] Lovell A. C. B., *Proc. Inst. Elec. Eng.*, **103**, 711 (1956).

The Australian 210-foot Radio Telescope

E. G. BOWEN* AND H. C. MINNETT**

Summary

The history of the C.S.I.R.O. 210-ft radio telescope project at Parkes is briefly reviewed. The principal performance requirements for the instrument are summarized and the means adopted to overcome the major design problems are discussed. Features of the telescope, including the novel drive and control system are described and some preliminary results of performance measurements are given.

1. Introduction

Since the end of World War II, Australia has been one of the leading countries in the new science of radio astronomy. Whilst the earliest radio telescopes were usually simple directional aerials and sometimes discarded radar arrays, the rapid advance of the subject soon led to the development of new and improved instruments.

In one line of development, the search for higher and higher resolution resulted in the invention of many ingenious telescopes of the interferometer type. The Mills Cross and Christiansen Crossed-Grating Interferometer developed in the Radiophysics Laboratory are outstanding examples of these.

At the same time, the pressing need for increased sensitivity particularly at centimetric wavelengths led to the building of larger and larger parabolic reflector aerials to collect as much of the received energy as possible. In fully steerable form, this is the most versatile of all radio telescopes, since it can be pointed at will over a large range of angles and the wavelength and polarization can easily be altered.

In Australia, this line of development has culminated in the building of the C.S.I.R.O. 210-ft radio telescope near Parkes in N.S.W. The new telescope which was officially commissioned on 31 October, 1961, by the Governor-General of Australia is now the principal instru-

*Chief of the Division of Radiophysics, C.S.I.R.O., Sydney, N.S.W.
**Radiophysics Laboratory, C.S.I.R.O., Sydney, N.S.W.
Manuscript received by The Institution December 5, 1962.
U.D.C. number 523.162 : 621.396.677.1 (94).

ment of the Australian National Radio Astronomy Observatory, operated by the Division of Radiophysics.

2. Basic Requirements

Funds for the project first became available in 1955 with generous donations from the Carnegie Corporation, the Rockefeller Foundation and from a number of private donors in Australia. Together with an equal contribution from the Australian Commonwealth Government, the initial sum totalled about £500,000.

At the beginning of the project the following broad requirements for the giant telescope were drawn up by C.S.I.R.O. :

(a) The diameter D of the paraboloid or " dish " was to be as large as possible consistent with the funds available and the engineering problems involved. Since the received energy is proportional to D^2 and the beamwidth to λ/D, where λ is the operating wavelength, both sensitivity and resolution improve with D.

(b) The surface of the dish was to conform to a paraboloid of revolution to within $\lambda/16$, for proper focussing of the received signals. It was essential for the aerial to operate down to at least 21 cm, the wavelength of the important hydrogen-line radiation, so that a surface error of not more than $\pm\frac{1}{2}$ in. was specified.

(c) The pointing error when tracking celestial objects was to be better than one tenth of the beamwidth, which at $\lambda = 21$ cm is between 10 and 15 minutes of arc. Thus a pointing accuracy of better than one minute of arc was required when tracking objects in celestial coordinates.

To minimize the cost of the telescope, several design compromises were proposed. The first was to restrict the coverage of the telescope to angles greater than 30° above the horizon. This limitation did not significantly reduce the accessibility of the regions of most interest to astronomers, and in any case atmospheric refraction effects are troublesome at lower angles.

Second ; it was recognized that it would be unnecessarily severe to insist that surface distortions should be measured relative to the nominal design shape. It was therefore specified that the surface error given under (b) above could be measured relative to the paraboloid of revolution which best fitted the actual surface at any time.

Third ; there was no need for the telescope to operate in high winds and it was decided that observations would be confined to periods when the average velocity was less than 20 m.p.h. (a gust velocity of about 30 m.p.h.). Safety from damage in the stowed condition was, of course, essential up to about 100 m.p.h.

Finally, advantage was taken of the fact that suitable sites can be found in Australia free from ice and snow. Thus wire mesh, a very good reflector of radio waves, could be used for the dish surface which would therefore remain porous under all expected climatic conditions. This factor significantly reduced the aerodynamic problem.

3. Design Problems

When the above requirements were formulated in 1955, many basic engineering problems were unsolved and there was little precedent on which to base a design. Although construction of the 250-ft telescope at Jodrell Bank, England, had commenced, the C.S.I.R.O. requirements on accuracy were more severe. The next biggest radio telescope in the world, at Dwingeloo in Holland, was 90 ft in diameter.

Freeman Fox and Partners, consulting engineers of London, were therefore asked to undertake a comprehensive feasibility study of the basic problems and to recommend a suitable design. This work commenced early in 1956 under the direction of Mr. Gilbert Roberts. Mr. M. H. Jeffery, later resident engineer during construction work at Parkes, supervised the design team. Owing to the novelty and scope of the problems, very close collaboration with C.S.I.R.O. was essential and one of us (H.C.M.) was attached to the consultants for this purpose throughout the project. This arrangement enabled engineering considerations to be related continuously to the detailed scientific requirements and comcompromise solutions, when necessary, to be reached expeditiously.

From these investigations a feasible design was evolved and submitted to C.S.I.R.O. in December, 1957. After careful assessment in Sydney, the main recommendations were accepted. Some of the major conclusions of the design study follow.

3.1 *Type of dish structure*

It was apparent that a dish structure and mount of the size required would have to be designed for minimum deflection rather than on a stress basis. It followed that steel should be used throughout rather than one of the light alloys in order to minimize deflections from wind loads and differential temperature effects. Even so, it was not clear that a rigid structure would meet the required tolerances and some form of automatic device might have been needed in the structure to maintain the correct shape under all conditions of gravity, wind and temperature.

However, the design study showed that the proposed dish structure would meet the tolerances for diameters up to about 300 ft. Moreover, for the diameter finally selected it was found possible to halve the specified tolerance so that operation down to a wavelength of about 10 cm became feasible.

3.2 *Form of mount*

Initially there was considerable controversy among radio astronomers regarding the type of mounting which should be used for very large telescopes. For dishes up to 100 ft, the polar axis or equatorial mount, as used by optical astronomers, had obvious attractions, because a relatively simple constant-rate drive system was sufficient for tracking distant celestial objects.

For larger dishes, this form of mount poses serious geometrical and counterweighting problems in maintaining the desired coverage and the pointing accuracy becomes increasingly difficult to meet. It was not known whether these difficulties would prove more serious than the drive and control problems inherent in the structurally simpler altazimuth mounting. Obviously, to track in equatorial coordinates a very accurate coordinate converter would have to be developed together with a servo-drive system capable of precise and stable operation even under gusty wind conditions.

As part of the overall design study the servo problems were investigated in consultation with the Electrical Engineering Department at Imperial College, London[1]. It was shown that a highly accurate and stable servo-drive system was feasible provided advantage was taken of the low acceleration requirements of the telescope in normal operation. Thus the bandwidth should be narrow enough to avoid the complex stability problems arising from structural and mechanical resonances which tend to be low in large telescopes in spite of the most rigid possible design. At the same time, the low servo impedance to wind-torque frequency components outside this bandwidth may be compensated by the use of high drive-system inertia augmented, if necessary, by acceleration feedback techniques.

With this approach to the servo-drive design and the use of the novel control system described in the next section, Freeman Fox and Partners concluded that an altazimuth-mounted dish would meet the pointing-accuracy requirements better and more cheaply when the dish diameter exceeded 150 ft.

3.3 *Method of control*

A unique feature of the proposed design was the system suggested by Dr. Barnes Wallis of Vickers-Armstrong (Aircraft) for controlling the telescope in equatorial coordinates[2]. Fig. 1 illustrates the concept as developed in the final design.

1. Minnett, H. C., " Design study of servo drive system for an altazimuth radio telescope ", Appendix C of Design Study Report by Freeman Fox and Partners, September 1957.
2. " Improvements in telescope mountings ", Vickers-Armstrong (Aircraft) Limited, British Patent Application No. 29248/1955.

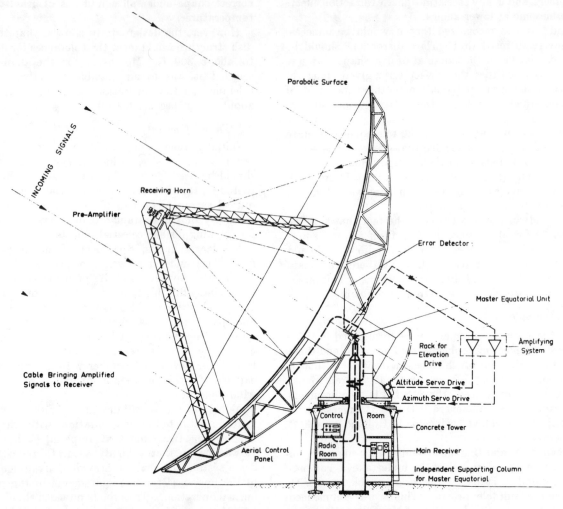

Figure 1.—Method of controlling the telescope for motion in equatorial coordinates.

Concentric with the azimuth axis of the mount is a long column which is structurally independent of the rest of the telescope and has separate foundations. On top of this column at the intersection of the altitude and azimuth axes is a relatively small and precisely made equatorial mount which carries a flat mirror. The equatorial may be driven from the control room of the telescope so that the mirror normal follows any desired equatorial path on the sky.

Attached to the bottom of the dish structure adjacent to the equatorial is a two-axis autocollimator or error detector which projects a beam of light onto the flat mirror and receives the reflected beam on a photocell system. Any misalignment between the dish axis and the mirror normal generates altitude and azimuth error signals which, after amplification and processing are fed back to the appropriate motors of the altazimuth mount. The dish is therefore constrained to follow accurately the motions of the equatorial.

A very important advantage of this system of coordinate conversion is the automatic correction of many of the sources of pointing error in the altazimuth mount. These include errors due to gravitational, temperature and steady-wind distortions, gearing inaccuracies and track-level variations. This feature was an important factor in the comparison of altazimuth and equatorial mounts.

3.4 *Size of dish*

One of the most important problems during the design study was to determine the relationship between the size and cost of the telescope. The need to keep the cost within the limit of the funds available was a controlling factor in the investigations. Considerable effort was therefore needed to devise the simplest possible solutions and to exploit wherever possible the potential economies of ideas such as the equatorial control system.

It was found that the cost of the proposed instrument would increase as the 5/2 power of the diameter. On the basis of this relationship, the C.S.I.R.O. selected 210 ft as the diameter of the dish.

3.5 *Final design*

Following the acceptance of the design study recommendations, the detailed design and specification of the 210-ft telescope was undertaken by Freeman Fox and

Partners and tenders called for its construction.

In July 1959, the prime contract for the construction was awarded to the West German firm of Maschinenfabrik Augsburg-Nürnberg, generally known as M.A.N. Work at the site on the base tower was commenced a few months later by the Sydney firm of Concrete Constructions Ltd., under sub-contract to M.A.N.

Responsibility for the detailed design and supply of the servo-drive system and the auxiliary electrical equipment was given to Associated Electrical Industries of Manchester. This firm had been actively associated with the project during the design study phase. The sub-contract for the whole of the master equatorial control system was undertaken by Askania-Werke of West Berlin.

4. Features of the Telescope

The telescope is located at a geographic latitude of 32° 59′ 55″ south and longitude 148° 15′ 50″ east. This site is twelve miles north of the town of Parkes and 200 miles west of Sydney.

The region meets the important requirement of very low radio-frequency noise level because of its remoteness from Sydney and the shielding provided by hills from local sources of interference. At the same time the land in the immediate vicinity is flat enough to allow for the future installation of a second smaller telescope to form a variable-spacing interferometer system for precise position finding. Power for the installation is available from the N.S.W. grid system and the town of Parkes provides good supporting facilities with air, road and rail connections to Sydney and other centres.

The general arrangement of the telescope is shown in the frontispiece and in Fig. 1. The parabolic reflector has a three-legged feed support and is attached centrally to a very compact and rigid altazimuth mount. The lower section of the mount consists of a concrete tower which also houses personnel and equipment.

4.1 *Parabolic reflector*

The focal length to diameter ratio (f/D) of the reflector was determined mainly by radio considerations. Although beam squint due to sagging of the feed support increases slowly with focal length, none of the structural factors have a decisive influence on the choice of f/D.

The selected value of 0.41 permits the use of reasonably simple feeds having axially symmetrical response patterns. A horn-type feed can be used at centimetric wavelengths, while at longer wavelengths a double dipole with plane reflector can be designed to give equivalent results. When the feed response is tapered to between 10 and 20 db at the rim of the dish, a good compromise between low sidelobe level and aperture efficiency is feasible and the paraboloid beam is very nearly circular.

A higher f/D ratio would have necessitated larger and more elaborate feeds to achieve the same performance, although secondary radio factors such as cross-polarization rejection and off-axis coma distortion would have been slightly improved. For f/D ratios below 0.41, these secondary factors are somewhat worse and, more important, it becomes increasingly difficult to design a feed which makes efficient use of the available aperture and produces a circular main beam. The value of 0.41 was

therefore chosen as a good compromise between these various factors[3].

Supported by three legs above the focal point of the reflector is a large cabin which has proved to be one of the most useful features of the telescope (Fig. 2). The cabin is about 10 ft across and provides shelter for about 1000 lb of front-end radio receiving equipment and working space for maintenance. Access to the cabin is by a small elevator in one of the tripod legs and by ladders in the other two. The tripod legs and cabin block about two per cent of the energy incident on the dish and the legs obscure a further three per cent of the energy reflected to the feed.

Below the cabin is a platform suspended on threaded rods so that it may be moved along the radio axis relative to the focal plane. In the centre of the platofrm is a mounting which locates the feed in an accurately defined position and permits the plane of polarization to be rotated by a motor operated from the control desk.

Moderately directive aerials, one for each of the operating wavelengths, are fastened to the top of the cabin so that they point to a suitable part of the sky. These can be used as comparison aerials in switched receiving systems. One of these aerials, a 75 cm horn, is visible in the photograph.

The reflecting surface of the paraboloid comprises two sections (Fig. 2). For structural reasons, the central 55-ft diameter zone is formed from segments of thick steel

Figure 2.—Inside the 210-foot paraboloid showing the solid central section of the surface and the surrounding panels of galvanized steel mesh.

3. Gardner, F. F., " Influence of f/D ratio on the performance of paraboloid aerials ", Radiophysics Laboratory Report No. RPL 135/1, Dec. 1958.

plate welded together. During erection the individual segments were adjusted carefully on threaded studs before welding, and template tests have shown that the final surface now lies within a few millimetres of the nominal shape. The remainder of the dish surface consists of panels of $\frac{5}{16}$ in. galvanized steel mesh which reflects 98 per cent of the incident radio power at 10 cm wavelength. The loss of energy through the mesh is comparable with the spillover of energy at the rim with the feeds discussed above and therefore little significant reduction of the ground radiation entering the feed would result from the use of a finer mesh.

The mesh panels are attached to radial purlins which are supported at intervals on screw devices to permit the shape of the reflector to be adjusted. These adjusters are attached to a system of left and right-handed spiral members forming a shell membrane which adds considerable stiffness to the dish structure. The membrane is carried in a system of thirty ribs, tied together by ring girders, and cantilevered from a strong cylindrical hub in the centre.

Inside the hub are two rooms, the upper of which houses a four inch optical telescope, observer's chair and a selection of controls and indicators for operating the radio-telescope. This is a useful facility for calibration tests and special tracking procedures. The lower room is available for housing radio equipment.

4.2 *Altazimuth mount*

The mount provides the means for moving the 300-ton dish so that it can follow a celestial object accurately and smoothly anywhere within the coverage of the instrument. It is a compact and extremely rigid structure in order not only to reduce static deflections but also to assist the design of the servo-drive system by raising the frequencies of resonance.

The dish structure is carried on two altitude bearings connected at points near the bottom of the hub, which is extended to form two counterweights (see Fig. 3). The bearings are supported by a " turret " which rotates on four rollers on a 37-ft diameter circular track under the lateral restraint of a central bearing. The track and lateral bearing are attached to the top of a three-storied concrete tower so that the dish can be tilted about the altitude axes to an angle of 60° from the zenith. The driving torque for the motion is applied to racks attached to the bottom of the counterweight structures, each rack engaging with a pinion powered by a dc servo motor through a reduction gearbox.

The counterweights have been made heavier than necessary to balance the dish weight in order to remove backlash which is an important consideration in the servo-drive stability. The extra weight also ensures that the dish can always be stowed manually in the " safe " zenith position if the electrical supply fails. In this position a massive pin locks the counterweight structure to the turret thus unloading the altitude gears.

Two of the four azimuth rollers are driven by gearbox/motor units very similar to those used for the altitude drive. These have been designed with extremely small backlash and together with the friction torque of the rollers, this is sufficient to ensure stability of the servo.

Figure 3.—The altazimuth mount and dish during construction. Each of the ribs is cantilevered from the central hub which is carried by the altitude bearings.

In the stowed position of the dish, the load can be removed from the rollers by raising them a few millimetres above the track surface with a hydraulic jacking system. This reduces the pressure on the track and prevents damage under very high wind loads.

The motion in azimuth is restricted to $\pm 225°$ by the cable " twister " which carries electrical and radio circuits from the tower to the turret. The neutral direction of the azimuth range is placed 70° east of north so that neither of the azimuth limits of rotation is encountered when tracking a celestial object provided initial account is taken of whether the path crosses the meridian north or south of the zenith.

The turret structure houses a number of rooms, the two principal ones being outboard of the counterweights. One of these accommodates servo-amplifying machines and the other is available for radio equipment. The space between the counterweights and the equatorial column is occupied by a three-level structure. The top level forms the floor of an enclosure protecting the equatorial, while the cable twister is housed in the bottom section. The middle level, which is a junction box room for the twister cabling, has access to the base tower via a spiral staircase in the equatorial column.

The concrete base tower is fully airconditioned. The top floor is the control centre of the telescope and accommodates the control desk, servo-drive cubicles and data-handling equipment.

The second floor of the tower houses the main units of the radio receiving equipment. The room is connected by a system of coaxial and multicore cables to the aerial cabin, lower hub room and turret radio room. Eight of the coaxial circuits are of the low-loss, helical-membrane type and are maintained a few pounds above atmospheric pressure by nitrogen cylinders to prevent the entry of moisture. Since these lines are rigid, they are transposed to flexible sections in the azimuth twister and zenith-angle " bender " to carry the circuits around the axes of rotation of the altazimuth mounting.

The ground floor of the tower is a general purpose room which houses airconditioning and power distribution equipment and staff facilities. It is also used as a store and light workshop.

4.3 *Drive and control equipment*

The master equatorial unit which guides the altazimuth mount in equatorial coordinates is shown in Fig. 4. In this view the declination axis is almost horizontal and the tube supporting the plane mirror is almost vertical.

At the left hand end of the declination axis, the eye-piece of an optical telescope is visible. This telescope, when used with an autocollimating attachment allows various internal adjustments to be checked. It may also be used to sight through a transparent part of the plane mirror in order to align the equatorial by tracking visible stars. For this purpose a removable hatch is provided in the wall of the room so that a clear view of the sky is available when the altazimuth mount is properly posi-

tioned. The instrument panel attached to the column may be used to control the equatorial during such tests.

At the top of the photograph the error detector unit is visible projecting from the base of the dish hub. The principle of this device is shown diagrammatically in Fig. 5[4]. A lamp and lens system in the unit projects a beam of parallel light on to the equatorial mirror. After reflection, the beam is collected by the lens (for clarity two separate lenses are shown) and is then divided into two components by a half-silvered mirror.

Consider first the altitude component. The light comes to a focus on a splitting prism and is divided into two parallel beams which are equal in intensity if the error detector is aligned correctly relative to the normal to the equatorial mirror. However if the dish has an altitude error, the reflected light beam is deviated from the edge of the prism and one of the emergent beams is increased in intensity and the other decreased. The beams are interrupted alternately by a shutter rotating at 70 revs/sec before falling on a photocell. The output of the cell therefore contains a 70 c/s modulation component having an amplitude determined by the magnitude of the altitude error.

The azimuth light component is treated in an identical way, but the edge of its splitting prism is perpendicular to the edge of the altitude prism. Thus an error in azimuth produces an output only from the azimuth channel, because the light beam reflected into the altitude channel moves parallel to the edge of the splitting prism and the emergent beams remain equal in intensity.

Figure 4.—The master equatorial unit and local control panel. The error detector unit projects from the bottom of the hub at the top of the photograph.

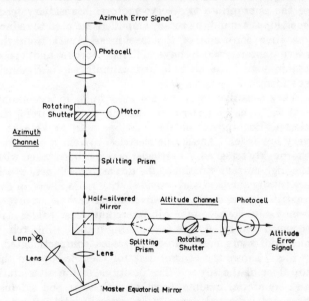

Figure 5.—Principle of the error detector.

The output of each channel after selective preamplification in the band 70 ± 5 c/s to eliminate harmonic responses is fed to the main servo amplifier in the control room[5]. (Fig. 6 is a simplified block diagram of the overall servo system of each of the channels.) The preamplified

4. This is a simplified representation. See Kalweit, C., " The Askania control system for the Australian 210-ft radio-telescope ", *Askania-Warte*, **59**, April 1962, pp. 1-7 (in German).
5. Rothwell, J., " Controlling the big Australian radio telescope ", *Control*, **4**, March 1961, pp. 84-87.

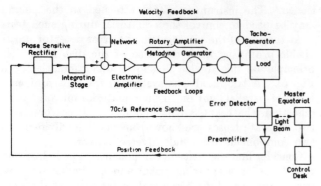

Figure 6.—*Block diagram of the basic servo system for the altitude and azimuth drives.*

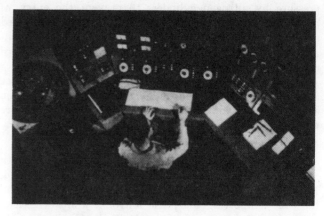

Figure 7.—*Control desk on the top floor of the tower.*

signal is mixed in a phase-sensitive rectifier with a 70 c/s reference signal synchronized to the rotation of the shutter in the error detector. A low-pass filter following the rectifier selects the error-dependent dc output signal and heavily attenuates other frequencies including noise components generated by vibrations of the error detector. After passing through an integrating stage the signal is combined with feedback components proportional to motor velocity and acceleration. This processing is designed to give the servo the required dynamic characteristics.

After further electronic amplification, the signal excites the control winding of the metadyne unit of a three-stage rotary amplifier housed in a turret room (Section 4.2). The metadyne output drives the field of a dc generator whose armature is connected in series with the armatures of the appropriate two servo motors. Subsidiary feedback loops around the rotary amplifiers are used to reduce the time constants of the machines. Apart from the servo motors, which have different torque and speed requirements, the altitude and azimuth servo channels are identical.

Very sensitive receivers are used in radio astronomy and RF noise interference reaching the feed from the electrical equipment on the telescope must be kept to a very low level. The whole electrical system is therefore thoroughly screened. Particular care has been taken with the high-current circuits of the dc servo machines, which are totally enclosed and provided with radio filters on the ventilating air ducts and woven-mesh gaskets on every removable coverplate. All interconnecting cables are screened and fitted with terminating glands, while relays and contactors are enclosed in sealed compartments.

Fig. 7 shows the control desk for the telescope on the top floor of the tower. The positions of the altazimuth and equatorial mounts, together with solar and sidereal time are indicated digitally on optical-projection units. These indicators together with controls for setting the driving rates of both mounts appear on the central panel of the desk.

The plexiglass hemisphere on the left of the desk represents the sky above the telescope and is engraved with equatorial and altazimuth coordinates. It contains two pointers to indicate the positions of the dish and equatorial mount respectively, and is used to bring the two into coarse alignment before activating the servo-feedback loop. Adjacent to the model are controls for stowing and

locking the dish in the zenith position and various warning and interlock indicators.

The right-hand side of the desk is occupied by controls which allow selected areas of the sky to be scanned automatically, and with controls for printing out position and time data on an electric typewriter.

An extensive intercommunication system connects different parts of the telescope. A push-button selector panel on the control desk permits the controller to communicate by microphone and loudspeaker with any one of twelve stations or to call all stations together. The controller can also connect any one station to any other when requested.

5. Performance

5.1 Surface accuracy

The most important aspect of the telescope's performance is the quality of the reflector and feed as a radio aerial. This depends to a very large extent on the accuracy with which the surface conforms to a paraboloid of revolution.

Each of the adjustment points described in Section 4.1 carries a permanent survey target whose distance from the vertex has been accurately measured. A rigid pyramid mounting (visible in Fig. 2) is also provided at the vertex for holding survey theodolites. An angular change of 10 seconds of arc at this point corresponds to a target movement of 1.5 mm normal to the dish surface.

With the dish pointing to the zenith it has been found possible, with proper precautions, to set each of the eight hundred adjustment points to this order of accuracy on a given occasion. Repetition of this process at intervals of a few months to study the stability of the settings show that individual points may vary by a few millimetres. Errors in the shape of the mesh panels between adjustment points have not yet been studied in detail, but peak values probably do not exceed 6 mm. Preliminary measurements as the dish is tilted from the zenith suggest that the rms deviation of the target points from the paraboloid of best fit at 60° tilt is about 3 mm.

The overall magnitude of the surface deviations is within the mechanical tolerance specified and the first radio observations at once demonstrated the promising quality of the reflector. Standard methods of measuring gain and pattern would require a separation between telescope and test aerial of the order of D^2/λ to limit the phase error to $\lambda/8$. This corresponds to 25 miles at 10 cm wavelength and the test aerial would have to be at a height of at least 70,000 ft. Such measurements are therefore difficult and to date radio tests have been confined to the use of celestial radio sources.

Observations on celestial radio sources have shown that the half-power beamwidths are about 14 minutes of arc at 20 cm wavelength and 7 minutes of arc at 10 cm in good agreement with predicted values for the feeds in use. At each wavelength, the beam has a high degree of circular symmetry which is a major advantage in observations of linear polarization. Measurements on the strongest cosmic sources are in progress to investigate the sidelobe pattern. Present indications are that the first lobes are between 20 and 30 db below the main beam response.

Theoretical curves of dish gain relative to an isotropic source are shown in Fig. 8 as a function of wavelength, assuming an aperture efficiency of 0.65 and a mesh surface which reflects 98 per cent of the incident power at 10 cm. At shorter wavelengths the gain decreases due to surface irregularities. Two curves are shown for rms surface errors of 5 mm and 10 mm respectively, on the assumption that the irregularities are random and many wavelengths

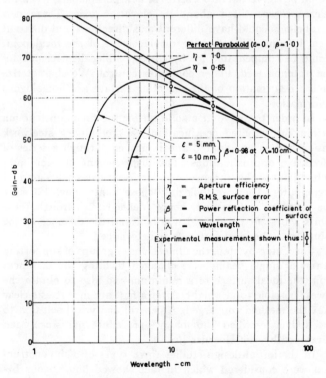

Figure 8.—Predicted variation of the gain of the paraboloid with wavelength for different surface errors.

in extent[6]. Gain values deduced from preliminary measurements made by B. F. Cooper at 10 and 20 cm on sources of known intensity are shown by the circled points and suggest that the overall rms error for the dish is about 5 mm. The accuracy of this determination however is not high, and further measurements are planned at wavelengths below 10 cms where the effect of the surface errors is greater.

5.2 *Pointing accuracy*

The absolute pointing accuracy of the telescope in equatorial coordinates depends in part on the correct initial adjustment of the master equatorial. During installation, optical tests with inbuilt autocollimating devices were made to ensure the accurate relative alignment of altazimuth and equatorial axes, the plane mirror and the optic axis of the internal telescope. The polar axis was then carefully aligned parallel to the earth's axis of rotation by photographing star trails near the south celestial pole with the telescope. Subsequently, a series of star position measurements distributed over the coverage of the equatorial have shown that the rms deviation of the optic axis from the indicated direction is about 25 seconds of arc.

A second component of pointing inaccuracy is contributed by servo-following error between the error detector and equatorial directions. Tests have shown that under maximum conditions of angular velocity and wind loading, this error is less than the specified value of 10 seconds of arc.

Finally, the radio axis of the dish must be correctly related to the error detector axis. Radio sources of known position are being used for this purpose. Owing to gravitational deflections of the dish and tripod, the position of the radio axis varies with altitude angle. Once calibrated, this component of pointing error together with that due to atmospheric refraction will be corrected automatically in the altitude servo drive.

6. Conclusion

The new instrument at Parkes represents a considerable advance in the design of very large steerable radio telescopes and embodies a number of novel solutions to the basic problems. The Radiophysics Laboratory has therefore undertaken a comprehensive investigation of the instrument's performance as a guide to the designers of future large telescopes. The results of tests carried out to date indicate very satisfactory performance and demonstrate that a powerful new tool is now available for the study of radio astronomy in Australia.

6. Ruze, J., " Physical limitations on antennas ", Technical Report No. 248, Electronics Research Laboratory of M.I.T.

The 100-Meter Radio Telescope at Effelsberg

OTTO HACHENBERG, BERND H. GRAHL, AND RICHARD WIELEBINSKI

Abstract—The 100-m fully steerable radio telescope of the Max-Planck-Institut für Radioastronomie (MPIfR) has now been in service for over a year. The telescope was designed to allow observations in the decimeter and centimeter wavelength regions, with the vastly improved resolution given by the 100-m diameter. During construction of the telescope measurements were made at a wavelength of 75 cm. Since the completion of the adjustments, tests and observations were made at 11-cm and 2.8-cm wavelengths. In particular, at the 2.8-cm wavelength, tests have fully confirmed the design goals. In fact, we hope to extend the operational range of the telescope at least to the wavelength of 1.2 cm.

A number of problems, new in this field, had to be solved in the construction of an instrument of this size and precision. Measuring methods are described which allowed adjustment of the surface to ±1 mm. A computer-controlled servo loop allows setting of the telescope to 5″. Experience of operation under various conditions is described.

I. INTRODUCTION

FROM the earliest days of radio astronomy, the development of radio telescopes took a number of distinct paths. No single type of instrument can meet the various observational requirements which are desirable for studies of the radio sky. The parabolic dish, already used in the pioneering observations by Reber [1], is a general-purpose instrument capable of operation over a wide range of frequencies. When fully steerable radio telescopes are required, to follow one position in the sky for many hours, the most suitable antenna is a parabolic dish.

The evolution of this type of antenna took the path of increased size, or increased operating frequency (or both); these factors determine the resolution of the radio telescope. The construction of the Jodrell Bank Mark I radio telescope [2] marked a big step in the increase in the resolution achievable by a single-dish reflecting antenna. The next big step was the commissioning of the Parkes telescope [3], which, although smaller in size, was usable at higher frequencies. The completion of the Max-Planck-Institut für Radioastronomie (MPIfR) 100-m telescope in Effelsberg was a step affording both increased size and increase in the highest operating frequency. As a result, observational capability over a large range of frequencies has been added along with vastly increased resolution.

The construction of the steelwork necessitated development of a new design principle. A stage has been reached in the design of large reflecting antennas where the additional rigidity required to guarantee higher frequency operation increased the weight of the structure beyond economical considerations. The new design principle was to allow elastic deformations, but it required a design of steelwork which would guarantee that a paraboloid shape would be maintained on tipping from zenith to horizon [4], [5]. Only a movement of the focus is required to counteract the continuously

Manuscript received March 1, 1973; revised May 1, 1973.
The authors are with Max-Planck-Institut für Radioastronomie, Bonn, Germany.

flexing structure. This design has now been checked out and indeed appears to be performing according to the computations.

In the present paper, the original conception of the design is described. The design is then followed through the various construction phases. Finally, the telescope's performance is discussed after one year of operation.

II. THE CONSTRUCTION PRINCIPLE

A. The Structure

From the very first discussions in 1964 it was clear that a telescope with 100-m diameter and with a surface accuracy of 1 mm (rms) could not be achieved with types of telescope constructions in use at that time. A new approach was necessary. An alternative to the rigid structure was to allow a certain elastic deformation of the supporting steel structure. However, it would have to be of such a kind that the parabolic surface, once adjusted at zenith position, by deformation, would be converted into a series of new paraboloids, when the telescope was tilted about its elevation axis. These new paraboloids could have different axis directions and different focal lengths. The resulting displacement of the focal point would be compensated for by corresponding displacement of the antenna feed, under computer control. We characterize such a deformation of the reflecting surface as "homologous deformation."

In order to arrive at a solution for the final construction within the required specifications, a preliminary steelwork design had to be adopted and improved through a series of successive approximations. The displacement of each individual joint of the framework under gravity was calculated by a computer. Then we took only those joints which lay near the reflector surface and, hence, defined it. Through these joints a best fit paraboloid was laid and the distance of the joints from the ideal surface was determined.

Then, step by step, the dimensions of groups of supporting beams or single beams were changed to bring the distances from the ideal surface to a minimum and also to obtain the lowest possible total weight. After a first attempt it was clear that the method converged. A general analytical solution to the above described "homology" procedure has since been developed by von Hoerner [6].

In the initial design studies three types of dish construction were considered which in fact showed homologous behavior. From these three, a design [4], [5] which can best be described as a dish with axial support was finally chosen. A cross section of that dish is shown in Fig. 1. The reflecting surface is rotationally symmetric. It has 24 radial bearers which are joined by means of ring members. The radial bearers from point C in Fig. 1 to the ballast box at B are tubes. The whole 100-m diameter shell construction is held only at points B and A.

The advantage of an axially symmetric structure can be explained by comparing the behavior of a flat circular disk

Reprinted from *Proc. IEEE*, vol. 61, no. 9, pp. 1288–1295, Sept. 1973.

32

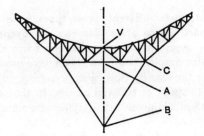

Fig. 1. Cross section of the 100-m telescope.

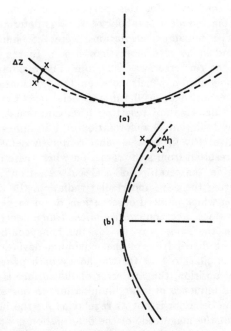

Fig. 2. Geometry of deflections for a telescope. (a) In zenith.
(b) Horizontal.

held at its central point. With the disk in a horizontal position (corresponding to the zenith position of a telescope) under gravity, a paraboloidal deformation is observed. On tilting the disk to horizon (disk to vertical position), a series of paraboloids are observed with a final flat disk (telescope at elevation 0°).

A similar situation occurs in a paraboloidal telescope. With the telescope at the zenith position, a deformation due to gravity moves a point x on the surface to x' as shown in Fig. 2(a). A deflection $\overline{\Delta z}$ occurs in this case. By tilting the telescope to the horizon, as shown in Fig. 2(b), another deflection vector $\overline{\Delta h}$ is observed. If the surface is adjusted in zenith position, then the general formula for the deflection Δ at any zenith angle α is

$$\overline{\Delta} = - \overline{\Delta z}(1 - \sin \alpha) + \overline{\Delta h} \cos \alpha.$$

For a deep dish (the 100-m telescope has focus/diameter ratio of 0.3) the same basic consideration can be applied. However, asymmetric deviations occur, which can be minimized by application of the previously described iterative procedure. Although the method can be applied in principle to any telescope design, the advantage of an axially symmetric design is seen from the fact that after only three iteration computations a satisfactory design could be achieved. The maximal elastic deflections for the 100-m telescope are 76 mm while the deviations from the best fit paraboloid are less than

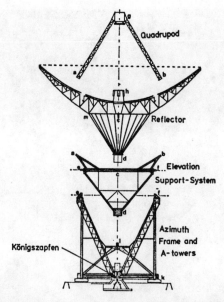

Fig. 3. Exploded view of the telescope showing the various parts.

3 mm. Other designs which were investigated showed a much slower convergence. Thus it was accepted that the design would give a surface with rms deflections ∼1.2 mm.

The adoption of the rotationally symmetric design had many other advantages. The cantilever members and girders were identical, allowing simplified construction procedure. Great precision in manufacturing could be achieved.

To support the reflecting surface, an independent carrier system is needed. It is an octahedron, a figure of eight triangular faces, having its central plane horizontal when the dish points at the zenith. The frame consists of a square and a heavy cross, one beam of which bridges the two elevation bearings and forms the elevation axis. This beam penetrates the reflector construction without contact with the framework and is fastened to the hub of the spoke wheel by a pin. The second arm of the cross joins the ends of the elevation wheel. Two beams of the octahedron run from the elevation bearings down to the apex of the cone of 24 tubes, to support the reflector weight that is concentrated at the counterweight, especially in the zenith position. The upper half of the octahedron forms the quadrupod that supports the focal-point cabin free of the reflector surface. An "exploded" view of the telescope is shown in Fig. 3, which shows the details separately. Fig. 4 shows clearly the telescope's steelwork.

The design and construction of the steelwork were carried out by Arge Star, a consortium of the firms Krupp and MAN.

B. Pointing and Steering

During the design phase of the telescope a careful investigation was made to determine whether it would be necessary to have a constructive device for measuring the angular position of the telescope directly at the central part of the reflector. (This method was chosen at Parkes.) This technique would introduce major difficulties for the steel construction in the case of the 100-m telescope. The decision was made to measure the position angles of the telescope at the axes and to take into account the fact that the measured angles (especially the elevation angle) would not indicate the exact position of the axis of the parabola. It was clearly

Fig. 4. The 100-m telescope, showing the backup structure.

foreseen in the design of the telescope that the structural elastic deformations would not permit a unique point, lying always on the parabola axis, to be identified. In all operating conditions, an automatic correction of the telescope position (by means of a computer) was essential. In particular, the correction of the focus movement due to homology had to be continuously carried out. For the dynamics of the designed system it was important that wind effects be considerably smaller than gravitational effects. This was verified in practice, with a considerable beneficial effect being derived from the sheltered valley chosen for the telescope site.

III. TELESCOPE ANCILLARIES

A. The Reflector Surface

The deviation of the reflector surface from the ideal paraboloid, as a result of the elastic deformation of the steel structure when the telescope is tilted, can be minimized. For the inner part of the 100-m telescope, in particular up to a diameter of 85 m, an error of about 0.65 mm (rms) is to be expected when the telescope points at the horizon. It is of great importance that the construction of the surface does not lead to an unnecessary increase of this error, i.e., by no more than about 20 percent. This means that the surface should have a precision of approximately 0.4 mm (rms) arising both from manufacturing as well as from adjustment errors. To achieve this value, the reflector surface was constructed using single plates or panels of about 120 by 300 cm. These are fixed to the backing structure at their four corners where they are adjustable. The surface consists altogether of 2372 single panels.

Considering the above requirements as well as the high manufacturing costs and, finally, the influence of wind on the

reflector rim, three different panel types were provided. In the inner part, up to 65-m diameter, honeycomb-sandwich panels were used. In the ring between 65-m diameter and 85-m diameter simple aluminum plates were used. These were reinforced in the rear by ribs. Since squares were formed, this induced us to speak of coffered panels. In the outer reflector ring from 85-m diameter up to 100-m diameter, wire mesh panels were used.

When the manufacturing process of the honeycomb-sandwich panels was completed, the surface precision of each panel was checked. For this purpose, the deviation of the surface from the ideal parabolic sector was determined by 31 points. The maximum deviations were 0.5 mm for most panels, and in no case more than 1 mm. In the horizontal position, the rms error was 0.24 mm. The maximum elastic deformation of the panels between the horizontal position and the vertical one was 0.6 mm. Special tests were made to study the deformation resulting from additional loads and from solar radiation. A deformation of 1.06 mm was caused by a load of 100 kg on the panel center. A deformation of $\sim$1 mm resulted from solar radiation when there could be a difference in temperature of 5°C between front and rear. Thus, during the day, under solar radiation, the panels had a precision which allowed observations down to $\lambda = 1$ cm.

The surface precision control of the frame panels was performed in the same way as for the honeycomb-sandwich panels. The distribution of the maximum deviations of the panels is similar to that for the honeycomb panels. In the horizontal position, the rms error of the surface is 0.27 mm. Under the influence of solar illumination on one side of the panels, the behavior is not as regular as for the honeycomb panels, but the magnitude of the deformation is not greater.

The mesh size of the wire netting is 6 mm. The maximum deviation from required shape was measured to be 1 mm (rms). These panels are still usable at 2.8-cm wavelength.

B. Control System

The telescope control uses two different drive systems. The main drive system is for driving the telescope in azimuth and elevation. In addition, there is a control system for focus displacements. All functions can be initiated from the control desk either for simple manual operation, or for normal automatic operation of the telescope, by means of the computer.

For manual operation of the telescope, the operator sets a demanded value for the telescope speed. An analog servo system controls the corresponding telescope motion which is measured at the motors by tachometers. An accurate setting to a certain altazimuthal position can be performed by visual control of a direct display of the angle position at the control desk. The normal astronomical operation of the telescope requires an automatic position control for which the computer serves as a part of the position feedback loop. The parts of the computer program, called "position control," read the position of the encoders for comparison with the demanded position and then calculate the telescope speed for optimal arrival at the required position. This position is normally a function of time corresponding to the astronomical needs of tracking a radio source or scanning the sky. Thus the required position in azimuth and elevation is generated by a coordinate transformation program which needs the input of a clock and the astronomical position as required by the observer. This program also allows display of the telescope position in astronomical coordinates.

For focus displacement (i.e., axial focus motion, lateral motion, and tilting of the whole Gregory reflector) and for polarization rotation of the receiver box in the primary focus, the principles of control are similar to those of the telescope drive system. Simple positioning can be done by hand from the control desk by switching the motors on and off at discrete constant speeds, while a visual control of the position readout of encoders shows the focus position or the polarization angle. When automatic focus correction or automatic polarization is required, the required function can be controlled by the computer. The main difference in these control systems with respect to the normal drive control of the telescope in position is that there is no velocity feedback loop in the system.

The task of driving the telescope in the two coordinate systems presents more problems than the rather simple motions required for the focus adjustment. Although the principles of telescope control in both coordinates are the same, there are differences in detail which result from the structural peculiarities of the telescope. In this respect, the points at which the angle encoding systems are coupled to the telescope axis must be considered. The azimuth encoder is mounted at the "Königszapfen," a bearing at the center of the lowest movable frame, and measured the telescope rotation in relation to the telescope foundations. The elevation encoders read the angle of rotation of the elevation axis at both bearings in relation to the A-towers which support the bearings. For azimuth there is a friction drive which consists of sixteen drives, while for elevation a gear coupling is provided. Two drives are floating on the elevation gear. These have a flexible mounting to a platform of the azimuth rotation structure.

It is obvious that for both coordinates the dynamics of positioning is different. The feedback loop for positioning the telescope in azimuth has to take into account only the very stiff part of the telescope structure between the drives and the encoding system. The feedback loop for positioning the telescope in elevation must allow for the dynamics of big parts of the telescope structure, the towers of the azimuth rotation system, the octahedron, and also the platform on which the drives are mounted. For both coordinates there is no direct dynamic (or static) control of the axis of the parabola or the main beam of the antenna as no monopulse feed operation of the antenna is provided.

Vibrations of the telescope structure are of significant importance in the design of the correct drive servo-loop systems. In our case, the major importance is the rather low resonant frequency of 1.3 Hz for the elevation system, which caused some difficulty in the design of the feedback loop. The corresponding resonant frequency for the azimuth is 2.9 Hz and has given no difficulties. The electrical systems were delivered by AEG-Telefunken company which also adjusted the servo loops to give a final satisfactory performance.

C. Encoders

The encoders which give the azimuth and elevation position are incremental systems using photoelectric marker reading heads. The angular resolution is 2″. To allow the absolute position determination, additional markers are placed every 5°. The system was delivered by Firma Heidenhain, Traunreut.

The azimuth encoder is in a foundation in the center of the movable frame of the telescope. A wheel 2 m in diameter runs on a large bearing fixed in the ground which connects through bellows to the telescope. Inside the bellows, which are about 1 m in diameter, run all the cables to the movable part of the telescope. Four scanning heads at 90° intervals read the markers from a tape. Each scanning head has four separate scanning fields which are shifted in phase in steps of 90°. The sinusoidal signals from the photoelements are converted into square waves and transmitted to the counters through 125 m of cable. A special interference suppression logic is employed to make the chance of sporadic pulse counts negligible.

The elevation encoders are of a compact design connected to each end of the elevation axis. While the azimuth encoders are read every 8″ with reflected light, the elevation encoders make use of light transmitted through a radial grating etched on a glass plate. The pulses are obtained every 40″. Interpolation gives 2″ readout.

For direct and immediate control of the position of the telescope, important when the telescope is driven by hand, the readings are displayed in degrees, minutes, and seconds of arc at the control desk. These digital displays are supplemented by coarse analog displays for use in case of main power supply failure. The initial setting of the counters can be performed automatically by crossing a zero marking on the tapes.

D. The Cabins

The decision to use the Gregorian configuration was made to allow for the possibility of low-frequency observations (in the decimeter wavelength range) as well as centimeter wavelength capability. The Gregorian subreflector is thus ellipsoidal and the prime focus is at a point about 2 m below the apex of the subreflector.

The prime focus cabin is an octagonal structure 5.5 m high. Inside the cabin, two levels have been constructed for the equipment racks. Below the floor of the prime focus cabin a heavy movable frame is placed to which the 6.5-m diameter subreflector is attached. This fame can make movements in three directions: 20-cm axial movement (relative to the fixed prime focus cabin), 20-cm lateral movement in the direction of the elevation motion; it can also perform a 1° tilt. All these motions are required for the focus correction which is under computer control.

In the center of the Gregory subreflector a 1-m diameter opening can support a tube which allows a receiver to be lowered into the prime focus position. This procedure requires an internal crane and a fully remote-controlled receiver box. A second crane supported on the outside of the prime focus cabin allows servicing of the cabin and easy removal of equipment from the telescope. The procedure during servicing is to lift the equipment from the cabin with the telescope in the zenith position, tip the telescope to 7° elevation, and lower the equipment. The prime focus cabin is temperature stabilized.

The secondary focus cabin is a complicated structure some 6 m in diameter and 7 m high. A Teflon sheet cover protects the feed horns from rain. Receivers can be installed in the

Fig. 5. Photograph showing prime focus and secondary focus cabin.

middle of the cabin. The lowest section of the cabin, seen in Fig. 5, is left free and has numerous hatches which can be opened. They are required for surveying the telescope surface. A rigid support for the surveying theodolite, which defines the apex of the paraboloid, is at the center of the cabin at its lowermost level.

E. Cables

The cables are of the low-loss air dielectric type Flexwell Cu2Y. For each cabin, two 5/8-in cables and thirteen 3/8-in cables are laid in the telescope. To allow for elevation and azimuth movements, the low-loss cable runs are interrupted by short lengths of flexible double-screened coaxial cable of the type RG214N. Correction filters and booster amplifiers are necessary to transfer the signals from a satisfactory passband response and signal-to-hum ratio. All line drivers for switching impulses are earth-free. For signaling and monitoring, four cables with six quads each and one run of 12 twisted pairs, individually shielded, were provided for each cabin. More recently, a decision has been made to transfer signals and monitoring levels digitally through a pulse-code modulated system, since the number of functions desired has outstripped the available cable facility.

F. Feeds

The two telescope receiving positions require two separate types of feed and different receiver packages. The prime focus package requires a small feed illuminating up to 160°, which is integrally built into the package. The feed for secondary focus operation must illuminate the 15° angle which the elliptical subreflector subtends at the second focus.

The development of the feed for the secondary focus is [7] based on existing knowledge of field distribution in a plane at the focus and on the properties of hybrid modes to match such a field. A narrow-flare-angle corrugated conical horn was chosen, fed by the dominant HE_{11} hybrid mode. The aperture of the horn was chosen to give −10-dB illumination at the edge of the subreflector. The length of the horn is limited by the physical space in the secondary focus cabin. For example, for an 11-cm wavelength the secondary focus hybrid mode feed is 3.5 m long with an aperture of 1 m.

There are a variety of feeds available for the prime focus [8], [9]. Although work is proceeding on the development of a hybrid-mode feed, no such working feed for the 160° illumination angle has yet been made. The present feed configuration has been empirically developed from an open circular waveguide with λ/4 chokes around the aperture. It was furthermore discovered that moving a set of chokes away from the aperture increased the illumination angle. A high-efficiency feed horn designated the Mark IV feed was first used. A further development, a Mark V horn, gives poorer efficiency but lower sidelobes.

G. Receivers

Two receiving cabins require different types of receiver arrangement. In the prime focus the receiver box is lowered through the floor into a tube 3.5 m long. The receiver front end must be completely remote controlled. A box configuration, shown in Fig. 6, has been developed [10], [11] and cooled parametric amplifiers successfully operated at 11- and 2.8-cm wavelength. The box is temperature stabilized to ±0.1°C by means of heaters and Peltier elements. The box weighs some 300 kg. Fans circulate the air so that a high degree of receiver stability is obtained. The system noise temperatures measured at the zenith are 90 K for the 11-cm receiver and 120 K for the 2.8-cm receiver. Short-term instabilities in the total power record of the receivers are less than ∼0.01 dB. The universal local oscillator (ULO) signal for the receivers is brought up from the receiver room in the frequency range 1.3–1.7 GHz. Multiplication by 2 gives, for example, the LO signal for the 2.695-MHz receiver which can then be used for recombination line measurements. The ULO signal is derived from a rubidium clock which synchronizes a frequency synthesizer and phase-locked multipliers.

The IF signal at 60 MHz from the front end is sent after amplification to a high level down to the 350 m of cable. Correction filters are necessary due to the amplitude–frequency response of the long cables. The signals are led, after detection, to the various items of processing equipment and finally onto a magnetic tape for further processing.

At present, only one receiver configuration is available for testing the secondary focus. More effort will be devoted in the future to the receivers for the secondary focus. Receivers can be controlled from the control desk, shown in Fig. 7, where the operator drives the telescope, while the astronomer has control of the whole observing process.

H. Driving Programs

The control of the telescope has always been planned to be by the use of an on-line computer. A Ferranti Argus 500 computer is used with various input/output peripherals. The purpose of this computer is to perform

a) coordinate transformation
b) telescope-drive control

Fig. 6.   A receiver box as used in the primary focus.

Fig. 7.   Control desk of the 100-m telescope. Telescope drive control is on the left, receiver and computer control on the right.

c) receiver control
d) real-time data processing.

The computer has a 24K memory (1-μs speed) with 24 b per word. Disk storage of 650K words is also available. A description of all aspects of the telescope drive program has been given [12] so that here a summary will suffice.

An observer calls through the Teletype the receiver he desires—a digital back end for continuum observations or an autocorrelation spectrograph for line measurements. The observer can then (through a Teletype) call on an observing procedure. He can choose the number of subscans in a scan, the extent of scanning, drive velocity, markers, etc. A descriptive coordinate system is used so that in addition to the

various astronomical coordinates one can use other more complicated coordinates, for example, a system centered on the nebula M31 and defined by the major axis of that object.

The procedure is to store the necessary driving requirements into a "passive field" in the computer. An operator can call back these instructions and, if they are correct, allow the procedure to be executed.

IV. Mechanical Adjustment

The adjustment of the telescope was made with two aims: 1) the correct setting of the axes to minimize instrumental errors and 2) the best adjustment of the reflector surface to provide optimum antenna gain.

Though it was clearly understood from the beginning that instrumental errors can be corrected by means of computer control, we tried to keep these corrections as small and simple as possible. Of major importance in this respect was the leveling of the azimuth rail. This was done with an optical level after allowing some two years for foundation settlement. For this purpose the telescope was lifted off the rails and the screws were adjusted. The adjustment reached an accuracy of 0.2 mm rms. A precorrection of 0.4 mm was added to allow for final foundation settlement.

The adjustment of the horizontal axis was made by fitting thin metal shims onto the supports of the elevation bearings. An optical level, positioned near the center of the elevation axis, was used and with it observations were made of marks fixed at the centers of both elevation bearings. Both marks were observed as a function of azimuth position of the telescope. The deviations between both axes of rotation were 5″ after these adjustments had been made.

The adjustment of the axis of the parabola included the setting of the focus position and it could not be separated from the adjustment of the dish surface. All these adjustments were made using a platform at the center of the central hub of the reflector structure within the apex cabin. The center of the reflector was defined by the innermost ring of the telescope surface. For radial positioning of the theodolite mounting, the distance to this ring had to be measured and adjusted.

After a coarse adjustment of the panels of the reflector surface, optical targets were fixed at circles of constant distances from this ring. A special steel tape, stretched radially within the reflector surface, was used for careful positioning of the targets. The reflector was adjusted during the winter months of 1971–1972 with the telescope in zenith position. The reflector consists of 2372 panels arranged in 17 rings concentric with the apex. Each panel is fastened with four screws adjustable in height and carries at least one target. In the total there are 5016 target points.

A photographic survey of the targets was made using the DKM 3A theodolite delivered by Kern, Aarau (Switzerland). This was done during the night so that the effects of unknown thermal influences would be reduced. The deviations were measured on the photographs for each target, and then each screw had to be accordingly adjusted. However, this photographic system was rejected since visual control of each adjustment was necessary, and a method was developed to do the measurements and adjustments visually over long time intervals while remaining independent of thermal effects.

After carefully defining the best fit axis of the coarsely adjusted reflector, eight standard points at ring 13 were adjusted according to the required parabolic figure. In a three-week period with a temperature variation of 20°C, these points were measured to investigate thermal effects. It was found that for all theodolite settings and temperatures, the particular deviations could be explained and easily eliminated. The targets on 24 radials (12 meridian sections) were adjusted, which gave a secondary set of 408 standard points. During the adjustment of the whole surface, the theodolite position was continuously referred to the eight standard points. Some points were adjusted by referring their position to adjacent radials. The lower rims of the panels were adjusted with a ruler.

After the reflector was adjusted, a complete visual survey was done over a period of 10 days. The residual surface inaccuracies were determined using a computer program. Roughly 200 points were readjusted later. As a result of these measurements the best fit paraboloid at the zenith was shown to have an rms deviation of <1.0 mm.

V. Observations at 11 cm and 2.8 cm

The first tests of the telescope were started in June 1972 at a wavelength of 11 cm. The radio tests determined the following parameters:

1) pointing measurements and repeatability of the pointing;
2) measurements of the hysteresis effects;
3) focusing measurements;
4) instrumental polarization;
5) measurements of aperture and beam efficiency;
6) measurements of the beam pattern and sidelobe level.

To start a series of measurements in a telescope where numerous parameters were variable required a decision on the steps to be followed. The first measurements were made with a fixed focus, and on a fixed polarization. The values of the focus were set to be approximately optimum for the elevation of 45°. Scans were made under computer control in elevation and azimuth through known radio sources. The averaged pointing effects were fitted with those expected from the pointing theory of an altazimuth instrument [13]. A total of seven constants, in addition to refraction, were found to be necessary to fit the data with an accuracy comparable with the scatter of the measurements. The final result gives confidence that the absolute pointing accuracy of the 100-m telescope is about 5″ rms.

The pointing error of the telescope might depend on the direction in which it is being, or has been, driven. Tests were made by scanning a source in elevation and azimuth with differing scan velocities in both directions. The conclusion reached on the basis of these measurements was that the hysteresis effect in azimuth is negligible while in elevation it is some 15″ and direction dependent. The on-line computer is programmed to correct automatically for this effect.

There are three focus movements necessary to correct for the continuously changing telescope surface. These are the radial focus, the axial focus, and the tilt. Computations of the steelwork deflections provided a good basis for determining the order in which the three focal movements were to be determined. With fixed axial focus and fixed tilt, the radial focus was first investigated. Scans were made in elevation and azimuth of sources with radial focus setting changing in steps of 20 mm from 0 to 140 mm. The largest movement which the mechanism permits is 200 mm. For the defocused condition a coma lobe was clearly observed in elevation scans.

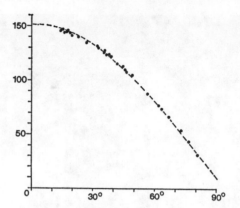

Fig. 8. Radial movement of the prime focus (in millimeters) versus elevation (in degrees).

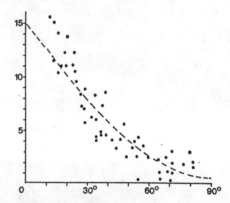

Fig. 9. Axial movement of the prime focus (in millimeters) versus elevation (in degrees).

TABLE I
Telescope Performance

λcm	Feed	El HPBW	Az HPBW	First Sidelobe −dB	n_e %
11	MkIV	4.4′	4.5′	18	56% ± 4
11	MK V	4.8′	4.8′	27	50% ± 3
2.8	Single choke	1.3′	1.3′	19	45% ± 5

The coma lobe could be seen above or below the main beam depending on the focus setting. In fact, the correct focus was experimentally fixed as the setting for which the coma lobe was a minimum. A collection of observational points and the best fit curve which is now incorporated in the on-line program are shown in Fig. 8. It should be noted that the steelwork computations gave exactly the same curve shape with slight variations depending on the assumptions of the feeds illumination characteristic.

For the measurements of the axial focus, the radial focus was adjusted to the optimum value, and again azimuth and elevation scans were made. These measurements were only approximately at 11-cm wavelength since the maximum movement expected is only some 20 mm. A more exact determination of the axial focus variation was possible after a 2.8-cm wavelength receiver was installed in October 1972. These measurements with an expanded axial focus setting scale are shown in Fig. 9.

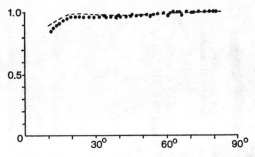

Fig. 10. Relative gain at λ = 2.8 cm for various elevations.

The tilt of the Gregory subreflector which holds the prime focus receiver box in its center was also checked. The setting is not critical for prime focus continuum observations and it was decided to leave the tilt position fixed for the time being.

Preliminary measurements of instrumental polarization were also made. Two calibration sources—3C286 and 3C295—were used. The results thus obtained indicated that instrumental polarization was ≥0.5 percent.

Measurements of aperture efficiency, the half-power beamwidth, and sidelobe level were made for two different feed horns at 11 cm and for a single feed horn at 2.8 cm. The results are collected in Table I. The measurement of the overall aperture efficiency is difficult. However, it was possible to determine the variation in telescope gain relative to the zenith with very good accuracy. This is shown in Fig. 10 which shows that after accounting for extinction at 2.8-cm wavelength the relative gain does not vary by more than ±1 percent from zenith to elevation of 30°.

ACKNOWLEDGMENT

Data used in this paper were provided by numerous people who contributed to make the 100-m telescope such a success. In particular, the authors wish to acknowledge the "Test Commission" headed by Dr. I. I. K. Pauliny-Toth who made the measurements described in Section V.

REFERENCES

[1] G. Reber, "Cosmic static," *Astrophys. J.*, vol. 100, pp. 279–287, Nov. 1944.
[2] A. C. B. Lovell, "The Jodrell Bank radio telescope," *Nature*, vol. 180, pp. 60–62, July 13, 1957.
[3] E. G. Brown and H. C. Minett, "The Australian 210-ft radio telescope," *Proc. IRE* (Austr.), vol. 24, p. 98, Feb. 1963.
[4] O. Hachenberg, "Studien zur Konstruktion des 100-m Teleskops," *Beitr. Radioastron.*, vol. 1, no. 2, pp. 31–61, 1968.
[5] ——, "The new Bonn 100-m radio telescope," *Sky and Telescope*, vol. 40, pp. 338–343, Dec. 1970.
[6] S. von Hoerner, *Structures Technology for Large Radio and Radar Telescope Systems*, J. W. Mar and H. Liebowitz, Eds. Cambridge, Mass.: MIT Press, 1969, pp. 311–333.
[7] W. Spika, "Realization of a hybrid-mode secondary focus feed for the 100-m radio telescope at Effelsberg," in *Proc. European Microwave Conf.*, 1971, Paper B6/1:1.
[8] R. Wohlleben, R. Wielebinski, and H. Mattes, "Feeds for the 100-m Effelsberg telescope," in *Proc. European Microwave Conf.*, 1971, Paper B5/5:1.
[9] R. Wohlleben, H. Mattes, and O. Lochner, "Simple small primary feed for large opening angles and high aperture efficiency," *Electron. Lett.*, vol. 8, no. 19, pp. 474–476, 1972.
[10] N. J. Keen, "First receiver systems for the 100-m Effelsberg radio telescope," *Nachrichtentech. Z.*, no. 3, pp. 168–174, 1971.
[11] N. J. Keen and P. Zimmermann, "The first two receivers for the radio astronomy program on the 100-meter radio telescope," *Nachrichtentech. Z.*, vol. 26, no. 3, pp. 124–128, 1973.
[12] P. Stumpff, H. G. Girnstein, W. Voss, and J. Schraml, "Prozeßsteuerung des 100-m Teleskops," *Mitt. Astron. Ges.*, vol. 31, pp. 101–117, 1972.
[13] P. Stumpff, "Astronomische Pointingtheorie für Radioteleskope," *Kleinheubacher Ber.*, vol. 15, pp. 431–437, 1972.

The Crawford Hill 7-Meter Millimeter Wave Antenna

By T. S. CHU, R. W. WILSON, R. W. ENGLAND, D. A. GRAY, and W. E. LEGG

(Manuscript received January 27, 1978)

A 7-meter offset Cassegrainian antenna with a precise surface has been built and tested. Measurements using a terrestrial source were made and compared with calculations for 19, 28.5, and 99.5 GHz. Low sidelobe level (≤ -40 dB) at one degree off the main beam and low cross polarization (≤ -40 dB) throughout the main beam are achieved using a quasi-optical 19/28.5-GHz feed system that also demonstrates very low multiplexing loss (~ 0.1 dB). The prime-focus gain measurement at 99.5 GHz found the difference between the measured and calculated gains to be (0.79 ± 0.45) dB, which is consistent with the expected rms surface error (~ 0.1 mm). Multiple-beam operation accommodates both COMSTAR beacons at 19 and 28.5 GHz and millimeter wave radio astronomy observations without physical disturbance of equipment.

I. INTRODUCTION

The Crawford Hill 7-meter antenna (Fig. 1) was built for propagation measurements with the COMSTAR beacons at 19 and 28.5 GHz, and for radio astronomy at frequencies from 70 to 300 GHz. It demonstrates that low sidelobes and high cross-polarization rejection can be obtained in an earth station antenna serving several satellites simultaneously. The antenna will also serve as a test bed for future propagation and antenna measurements.

The main part of this paper is contained in the following three sections. Section II describes the 7-meter antenna starting with the initial requirements. Section III gives a detailed description of the 19/28.5-GHz

Fig. 1—The 7-meter offset Cassegrainian millimeter wave antenna. The subreflector is mounted below antenna aperture, and key structural parts are covered with insulation.

feed system for the COMSTAR beacon experiment. Section IV describes measurements of the completed antenna at 19, 28.5, and 99.5 GHz using terrestrial sources on a tower at a distance of 11 km.

II. DESCRIPTION

2.1 Requirements

The major requirements imposed by the satellite beacon experiment on the 7-m antenna are for cross polarization to be less than −35 dB within the main beam, sidelobes to be below −40 dB at 1 degree or more from the main beam, multi-beam operation to be possible, and performance in all weather to be good, especially during summer thunder

Reprinted with permission from *Bell Syst. Tech. J.*, vol. 57, no. 5, pp. 1257–1288, May–June 1978.

7-METER MILLIMETER WAVE ANTENNA

Fig. 2—The subreflector with a machined aluminum surface of 20 μm rms. The oval shape is provided for azimuthal off-axis beams.

showers. The attainment of at least minimal performance at the high end of the radio astronomy band further requires a surface accuracy of 0.1-mm rms and a maximum pointing error of 10″ arc. The expectation of performing beam-feed experiments and of possibly accommodating large cryogenically cooled radio astronomy receivers made it desirable to provide for a Nesmyth focus (feed along the elevation axis) in addition to the Cassegrainian focus.

The requirements for low sidelobes imposes severe limitations on the amount of aperture blocking that can be tolerated.[1] Thus, an offset paraboloid was chosen with almost no aperture blockage. This can be made compatible with the cross-polarization and multibeam requirements by choosing a sufficiently large secondary focal ratio.[2,3,4] An additional benefit of the offset design is a large return loss. The low sidelobe requirements also place restrictions on the allowable surface errors. The magnitude of the allowable errors depends on their correlation lengths and the angles at which sidelobes can be tolerated, but a tolerance of about 0.01λ is needed to avoid degradation of the far sidelobes.[5] Thus a surface tolerance of 0.1-mm rms was chosen to simultaneously satisfy the sidelobe requirements at 30 GHz and allow radio astronomical operation in the 200 to 300 GHz band.

The main reflector has a focal length of 6.5659 m. It is offset such that its bottom is 1.2074 m above the reflector axis. The subreflector is a 1.2-m by 1.8-m oval portion of a hyperboloid offset 8.258 cm above its axis (Fig. 2). It has focal lengths of 0.8697 m and 5.2262 m giving a magnification ratio of 6.01. The central ray from the feedhorn to the subreflector makes an angle of 6.828 degrees with the axis and the illumination cone has a half angle of 5.06 degrees. The main reflector is subtended by a circular cone of 26.75-degree half angle, and the axis of the cone makes an offset angle of 37.26 degrees with the axis of the main reflector. The geometry results in a blocking of 4.4 cm of the main reflector by the subreflector and an expected contribution to the cross-polarization level of −56 dB in the main beam region, well below the required level.

The narrow beam pattern required for the feed is the main price which must be paid for the advantages of a large effective F/D ratio. Small-cone-angle corrugated horns have been rejected for both radio astronomy and propagation feeds because of their excessive length. In the radio astronomy feed, cryogenically cooled receivers are in use, and the best performance is obtained by cooling the feedhorn with the receiver and radiationally coupling out of the Dewar. Thus, a small-cone-angle horn would not only be a large mass to cool, but a large low-loss Dewar window would be difficult to make. Instead, a quasi-optical feed system[6] is used for radio astronomy to couple a small corrugated horn in the receiver Dewar to the 7-meter antenna. This feed system incorporates image rejection, local oscillator injection, and calibration.

In the case of the 19/28.5-GHz propagation feed, the two widely separated frequencies would lead to problems with frequency-dependent phase patterns of a single corrugated feed and with higher order modes generated in a low-loss waveguide polarization diplexer. These problems were circumvented by using quasi-optical methods[7] for both polarization and frequency diplexing as shown in Fig. 3. Four launchers, each consisting of an offset ellipsoid and a corrugated horn, are used, each for a single frequency and polarization. A quasi-optical frequency diplexer[8] first combines two frequencies at each polarization, then a polarization grid[9] combines two orthogonal polarizations. A 45-degree mirror finally reflects the combined beam to the subreflector, leaving adjacent areas in the focal plane available for other spherical roller bearing. The other end has a 1.4-m bore radial roller bearing sized to provide a 1.2-m diameter path to the side cab from the vertex area. The surface of the box girder facing the subreflector has a 1.6-m opening for the RF beam, completing the path from the subreflector to the elevation axis. A truss girder above the box girder supports the nine steel trusses that are radial to the axis of the main reflector and support its surface panels. The center truss is the longest and has been made deeper than the others by connecting its outer member to the elevation wheel support structure. The stiffness of the other trusses was adjusted using computer calculations so that gravity and wind deflections are expected to introduce mainly pointing errors with only small deviations of the antenna surface from its parabolic shape.

The 27 surface panels are arranged in four approximately equal width rings. The panels are A356 aluminum castings containing 17-cm deep ribs on the back side near the edges. These are joined by 9-cm deep ribs running in the circumferential direction with spacings of approximately 30 cm. The castings were tempered to T51, rough machined on a numerically controlled milling machine, stress relieved, and then machined to the final contour on the same machine. The manufacturer tested all the panels on his milling machines and found the surfaces to be accurate to better than 50-μm rms with an average value of about 40 μm. One of the early panels was tested on a Portage measuring machine and was found to have the same rms error (37 μm) as determined by the manufacturer. It was then subjected to 10 rapid temperature cycles from −40°C to 60°C and back. Remeasurement on the Portage machine showed that the surface error had increased to 50-μm rms.

After machining, the panels were cleaned and painted with a 30 to 50-μm coat of an Alkyd base, TiO_2 pigment, flat white paint. After final alignment of the panel corners to the alignment template (see below), five panels near the center were measured against the template on an 18-cm grid. One of these panels was found to have a much larger surface error (100-μm rms) than expected with one bad area and a general warp. A visual inspection showed this bad area to be by far the worst on the antenna. However, it is only a small contribution to the total surface error of 100-μm rms.

The panels are held to the back-up structure by 2.54-cm diameter adjusting bolts. At the panel end, these bolts attach to machined pads in the corners of the peripheral ribs with ball and socket joints to avoid transmitting torques to the panels. These adjusting bolts are perpendicular to the surface and bend to take up the differential expansion between the aluminum panels and the steel backup structure.

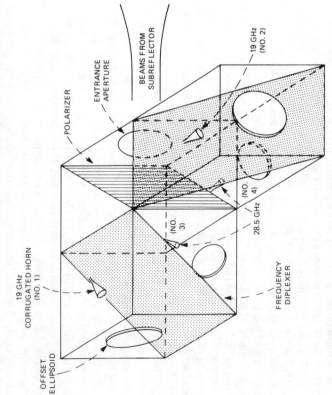

7-METER MILLIMETER WAVE ANTENNA

19 GHz CORRUGATED HORN (NO. 1)

OFFSET ELLIPSOID

POLARIZER

ENTRANCE APERTURE

BEAMS FROM SUBREFLECTOR

19 GHz (NO. 2)

(NO. 3)

(NO. 4)

28.5 GHz

FREQUENCY DIPLEXER

Fig. 3—Sketch of a 19/28.5 GHz dual-polarization feed system. The beam from the subreflector passing through the entrance aperture is first split by the polarizer and then subdivided by two frequency diplexers.

receivers. This feed system was first tested on axis, but operates in a position to produce a beam 0.5 degrees off-axis in azimuth, leaving the on-axis feed position free for the higher frequency radio astronomy feed. This off-axis operation results in very little degradation in the antenna pattern.

Past experience has shown that, after aging, radome surfaces hold thick water films during rain, causing large microwave attenuation.[10] The 7-m antenna was therefore designed without a radome, but with thermal insulation of critical parts of the structure to allow operation in the sun without significant degradation of pointing or surface accuracy. The support structure affords stiffness sufficient for operation at 30 GHz in winds up to 70 mph.

2.2 Mechanical description

The mount for the 7-m antenna is a conventional elevation-over-azimuth design with a single 2-m diameter cross-roller azimuth bearing and a yoke holding the elevation bearings. The elevation moving structure is built around a steel box girder. One end of the girder has a 32-cm bore

Computer calculations of thermal distortions of the structure coupled with temperature measurements on a test fixture showed that the sun shining on the structure from some angles would produce unacceptably large pointing and surface errors if the only thermal control were white paint. A strategy was adopted of insulating much of the structure and circulating ambient air through the critical volumes. The back of the surface panels has been sprayed with foam and the remainder of the main reflector backup structure enclosed by 5-cm-thick foam panels. A large blower floods this insulated volume with ambient air. The main members of the subreflector support structure are made of 10.16-cm-square steel tubing and are insulated. Ambient air is blown along them by blowers in the elevation girder. Most of the remaining steel structure is covered by 5-cm-thick panels of foam to increase thermal time constants, but no air circulation is provided.

In addition to shielding the backup structure from solar heat, the insulation of the back of the main reflector surface panels reduces the heat flux through the panels and hence the steady-state temperature difference between their surface and ribs. The penalty, however, is an increase ($\sim 2\times$) in the temperature rise of the surface panels in the full sun. This uniform temperature rise has the same effect as an absolute temperature change on the differential thermal expansion between the steel backup structure and the aluminum panels. It causes the panels to expand or contract in a direction parallel to the surface of the paraboloid defined by the backup structure. This has an almost negligible effect on the performance of the antenna. Warping of the panels due to thermal gradients perpendicular to their surface or thermal warping of the backup structure is much more important.

Two equipment rooms are on the antenna structure (Fig. 1). The vertex equipment room and the hollow elevation girder behind it provides a mounting space for receivers at the Cassegrainian focus. They move in elevation angle with the antenna. A double window of 50-μm Mylar* sheet allows RF energy to pass into this area with negligible loss (~ 0.1 dB) at 100 GHz. The side cab is provided for the Nesmyth focus. It moves only in azimuth. At present, it contains only receiver support equipment and the main elevation cable wrap. An 8-m by 6-m control building 15 meters from the antenna contains the rest of the receivers, the control computer, and a work area.

The volume under the azimuth bearing contains the azimuth cable wraps. A maypole wrap is used for most of the cables, but an auxiliary clock spring wrap carries the phase-sensitive RF cables.

The drive for each axis uses a single bull gear with two motors connected to separate speed reducers and pinions. The amplifiers for the two motors are biased so that, for low torque output, they torque in op-

* Registered trademark of E. I. Dupont.

Fig. 4—Panel alignment template mounted in place of subreflector support tower above 7-meter reflector prior to installation of tent. The template consists of two sections joined by a link.

support tower and the vertex equipment room. The alignment template was machined in two parts, each of which has a reference hole at each end. The reference holes at the outer end of the inner template and the inner end of the outer template are at the same height and were joined by an accurate link. A counterweighted boom built on an adjustable vertical bearing supported the template. Another link adjusted the inner hole of the inner template to the correct radius from the rotational axis. Before each use, the vertical axis was set using an electronic tiltmeter, and the four template reference holes were set to their required relative height using gravity-oriented links on an optical level. Twelve transducers were attached to the template at radii corresponding to the panel mounting screws. They were referred to the accurate bottom surface of the template and used to measure panel corner positions.

The subreflector support structure has been kept well below the axis of the main reflector, especially near the secondary focus. This, coupled with the oversized subreflector, allow beams up to 2 degrees off axis in azimuth or 1 degree in elevation to be launched from the vertex equipment room. The subreflector itself is a machined aluminum casting of the same material and temper as the panels. It was machined on a vertical lathe and has a measured surface accuracy of 20-μm rms.

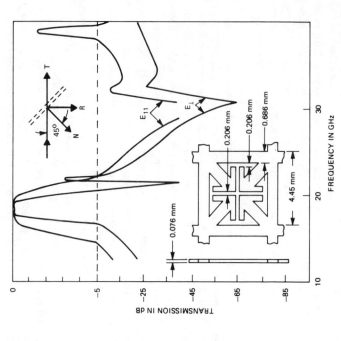

Fig. 5—Transmission of a double-self-supported frequency diplexer (see Fig. 3) with the indicated dimensions of the grid mask at 6.3 mm spacing and 45-degree angle of incidence for two polarizations.

Each of the two frequency diplexers is a pair of self-supported grids made of beryllium copper 75-μm thick. The 6.3-mm spacing between the girds is maintained by mounting on opposite sides of an aluminum frame with an oval aperture of 30.5 cm by 45.7 cm. Figure 2 shows that the polarization for Feeds No. 1 and No. 3 is perpendicular to the plane of incidence at the frequency diplexer, while that for Feeds No. 2 and No. 4 is parallel to the plane of incidence. The transmission characteristics for both polarizations and the dimensions of this double-self-supported grid are given in Fig. 5, which has been taken from Arnaud and Pelow's article.[8] The measured insertion losses for both transmitting 19.04 GHz and reflecting 28.56 GHz are less than 0.1 dB. Figure 5 also shows essentially perfect transmission and rejection frequency bands of well over 2 GHz each. This performance should be compared with a minimum insertion loss of 0.2 dB and a 2-GHz band loss of up to 1 dB for a typical waveguide diplexer at 19 and 30 GHz.

Several alternate quasi-optical frequency diplexers were tested. For a large (45-degree) angle of incidence, none of them achieved the desired transmission (−0.1 dB) and rejection (−20 dB) simultaneously for fre-

---

posite directions to eliminate backlash, but when more than 15 percent of the maximum torque is required the opposing motor reverses. Each of the four dc motors is rated at 3 horsepower with phase control excitation. The gearing is chosen to give a slew speed of 1 degree/s in elevation and 2 degree/s in azimuth with 1 degree/s² acceleration in both axes and the ability to hold position or drive to the stow position in a 70-mph wind. This gearing results in the motor inertia dominating the antenna's inertia. Each motor has a brake with the same torque rating as the motor. When the drive system returns to standby from an active condition, the brakes are set before the motors relax so that the anti-backlash preload of the gear system is maintained.

The analog portion of the drive system is set up as a velocity control loop. Each drive motor has a tachometer generator. The sum of the tachometer signals is used in the main feedback loop, and the difference is used to damp possible oscillations in which the motors move in opposite directions.

Each axis has a direct drive Inductosyn* system which has an angular accuracy of 0.001 degree and a resolution of 21 bits. The antenna's minicomputer reads the position of each axis every 10 ms, subtracts it from the desired position, and applies the scaled difference to the drive system as a velocity command. Drive system overshoots are minimized by compressing the gain of the feedback loop within the computer by a factor of 4 for command velocities greater than $\frac{1}{4}$ of the maximum velocity. This strategy keeps the commanded velocity below the deceleration limit of the drive system when approaching the final position. When a source is tracked that moves at the sidereal rate or slower, the servo error has not been observed to exceed 0.001 degree with winds up to approximately 50 mph.

III. 19/28.5 GHz DUAL POLARIZATION FEED SYSTEM

3.1 Quasi-optical diplexers

The polarization diplexer (see Fig. 2) is simply a polarization grid made by photo-etching a copper-covered Mylar sheet. Copper strips 0.25-mm wide and 0.018-mm thick are spaced 0.25-mm apart on a Mylar sheet 0.013-mm thick. The grid is mounted on an aluminum supporting frame with an oval-shaped aperture of 33.02 cm by 48.26 cm. To achieve a flat grid, the supporting frame is made of jig plate instead of regular aluminum stock, which shows warping after machining. The plane of the grid is oriented at 45 degrees with respect to the incident beam from the subreflector. The conducting copper strips are perpendicular to the plane of incidence to avoid cross-polarized radiation[9] for both transmitting and reflecting orthogonal polarizations being diplexed.

---

* Registered trademark of Farrand Industries, Inc.

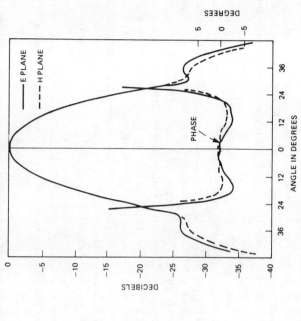

Fig. 7—Measured radiation patterns of 28.5-GHz corrugated horn. Rotation center 1.42 cm behind horn aperture.

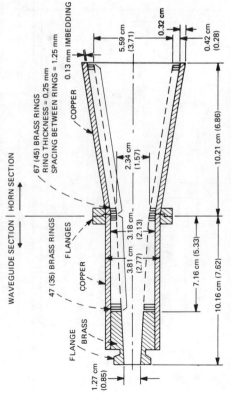

Fig. 6—The 19(28.5)-GHz conical corrugated horn and its hybrid mode launcher.

quency bands in the ratio of 1.5 to 1. A single gridded Jerusalem-cross diplexer[8] showed a transmission loss of 0.2 dB at 19 GHz for the polarization perpendicular to the plane of incidence and a rejection of only −10 dB at 28 GHz for the polarization parallel to the plane of incidence. When Mylar substrates were added to the double self-supported grid for improving its mechanical strength, transmission losses of 0.4 and 0.2 dB were observed at 19 GHz for the perpendicular and parallel polarizations.

The mechanical resonance frequency of the quasi-optical grids has been measured to be around 80 Hz, which is well above the passband of the baseband data in the COMSTAR beacon experiment.

3.2 Corrugated horns

Four corrugated horns (two each at 19 and 28.5 GHz) were constructed to illuminate the offset ellipsoids. The dimensions of the 19-GHz corrugated horn are illustrated in Fig. 6, where the numbers inside the parentheses are essentially scaled designs for 28.5 GHz. They are shortened versions of a very long corrugated horn built by Dragone.[11] The impedance matching between the smooth wall of the circular waveguide and the quarter wavelength slots of the corrugated horn is provided by a linear taper from half-wavelength corrugations that behave like a conducting surface. The 19-GHz transition from 1.27-cm diameter circular waveguide to 1.067 cm by 0.432 cm rectangular waveguide is hardware from the DR-18A terrestrial 18-GHz digital radio system. This transition consists of a tapered section from circular to square shape and a quarter-wavelength transformer. The 28.5-GHz transition from

0.846-cm diameter circular waveguide to 0.711 cm by 0.356 cm rectangular waveguide was also similarly designed. The measured return loss of all horns is better than 35 dB at the beacon frequencies 19 GHz and 28.5 GHz, and remains better than 31 dB over a 2-GHz band.

The measured far-field radiation patterns of the corrugated horns are shown for 28.5 GHz in Fig. 7, which are in excellent agreement with the calculations. The measured patterns at 19 GHz are essentially the same as those in Fig. 7, with the phase center located at 2.13 cm behind the horn aperture. The offset ellipsoid intercepts the horn radiation in the Fresnel region. It is difficult to measure the radiation pattern at a distance corresponding to the ellipsoid location. However, the calculated 20-dB half-beamwidth of the Fresnel zone radiation pattern is 2 degrees broader than those of the far-field patterns, and will be the same as the 28-degree half-cone angle of the ellipsoid subtended at the focus as shown in Fig. 8.

3.3 Offset ellipsoids

To be mirror-imaged at the Cassegrainian focal region (secondary geometrical focus of the subreflector for the on-axis case), the common phase center of the four offset launchers should be located in the middle was started with estimates by Gaussian beam approximation[12-14] and finalized by computer simulations which were essentially numerical integration of diffraction integrals,[2,15] assuming various design parameters.

Dimensions of both 19-GHz and 28.5-GHz offset ellipsoidal launchers are given in Fig. 8. Since the distance between the phase center and the 28.5-GHz ellipsoid is much greater in wavelengths than that of 19 GHz, the 28.5-GHz ellipsoid has larger size and greater curvature than a scaled version of the 19-GHz ellipsoid. The offset ellipsoid is subtended by a circular cone at the focus. The offset angle of the feedhorn axis is 3 degrees greater than that of the cone axis; thus approximately equal illumination taper can be achieved on the top and bottom edges of the reflector. The intersection of an ellipsoid and a circular cone subtended at one focus is a plane ellipse subtended by another circular cone at the other focus. The radiating beam from the ellipsoid will lie along the latter cone axis which deviates from the major axis of the ellipse by 4.16 degrees at 19 GHz and 5.28 degrees at 28.5 GHz. Aluminum jig plates 2.54-cm thick are first cut into ellipses of 32.41 by 30.43 cm and 23.85 by 22.40 cm; then they are positioned and oriented correctly with respect to ellipsoidal axes for computer-controlled numerical machining.

of the feed box as shown in Fig. 9. Therefore, ellipsoids instead of paraboloids were used in the design of the feed system. One notes that a small movement of the phase center can be accomplished by the defocusing of a paraboloid. However, to place the phase center at a considerable distance in front of or behind a reflector with a small F/D ratio, an ellipsoid or a hyperboloid should be used. The design of an offset launcher

3.4 Mechanical mounting

All the components are mounted in an L-shape frame of 106.7 × 106.7 × 61 cm as illustrated schematically in Fig. 2. The frame is made of structural aluminum angles with sufficient diagonal bracing to insure rigidity without blocking the radiating beams. The mounting of corrugated horns and offset ellipsoids was designed to allow adjustment for positioning, orientation, and focusing.

Owing to the uncertainty of the orbital position assignment for the COMSTAR satellite, the polarizations of the beacon signals were not precisely given. Therefore physical rotation of the feed frame around the beam axis is needed to match an arbitrary pair of orthogonal linear polarizations.* This required rotation is accomplished by a thrust ball bearing with 33-cm diameter circular opening in the front and a floating bearing in the back. These bearings are mounted on an aluminum supporting frame as shown in Fig. 10. This bearing mount is directly attached to steel I-beams through holes in the floor of the vertex equipment room of the 7-m antenna. The connection between the bearing mount and the steel I-beam allows differential thermal expansion between the indoor aluminum structure and the outdoor steel structure.

---

* This rotation is also needed for calibrating the amplitude and phases of the cross-polarized signals received from the satellite beacons.

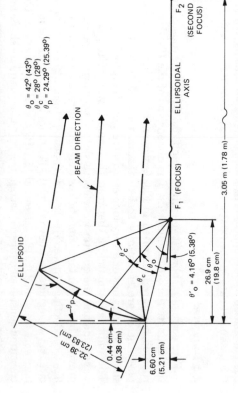

7-METER MILLIMETER WAVE ANTENNA

$\theta_o = 42°$ (43°)
$\theta_c = 28°$ (28°)
$\theta_p = 24.29°$ (25.39°)

BEAM DIRECTION

ELLIPSOID

ELLIPSOIDAL AXIS

θ_c θ_o θ_p

F_1 (FOCUS)

F_2 (SECOND FOCUS)

$\theta'_o = 4.16°$ (5.38°)

26.9 cm (19.8 cm)

3.05 m (1.78 m)

6.60 cm (5.21 cm)

0.44 cm (0.38 cm)

32.39 cm (23.83 cm)

Fig. 8—Schematic diagram of the 19(28.5)-GHz offset ellipsoid.

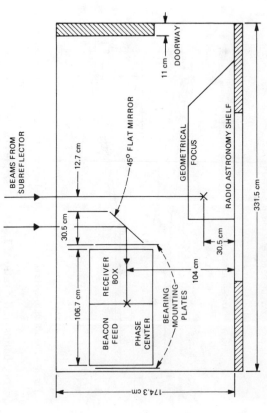

BEAMS FROM SUBREFLECTOR

DOORWAY

11 cm

12.7 cm

45° FLAT MIRROR

30.5 cm

106.7 cm

30.5 cm

BEACON FEED

RECEIVER BOX

PHASE CENTER

BEARING MOUNTING PLATES

GEOMETRICAL FOCUS

RADIO ASTRONOMY SHELF

104 cm

331.5 cm

174.3 cm

Fig. 9—Top view of vertex equipment room which has a height of 2.3 m and moves with the antenna.

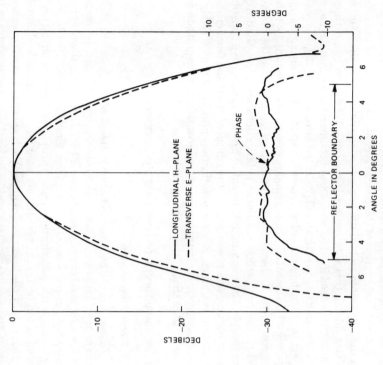

Fig. 11a—Measured radiation patterns of Feed No. 1 at 19 GHz (see Fig. 3).

choic chamber. To simulate the illumination of the 7-m offset Cassegrain antenna, the distance between the transmitting source horn used in the pattern measurements and the phase center of the receiving feed system was the same (526 cm) as that between the subreflector and its geometrical focus. Mechanical alignment was established before electrical measurements.

The measured insertion loss for a quasi-optical polarization or frequency diplexer was found to be 0.1 dB or less in each case. Amplitude and phase patterns were measured with respect to the common rotation center and the common optical boresight. After a few iterative adjustments of orientations and locations for both corrugated horn and ellipsoid, good agreement between measured and calculated patterns was achieved for each offset launcher. In particular, the measured patterns verified the calculated beamwidth, the effective phase center of the offset ellipsoids, and the approximate pattern symmetry predicted in the asymmetrical offset plane.

7-METER MILLIMETER WAVE ANTENNA

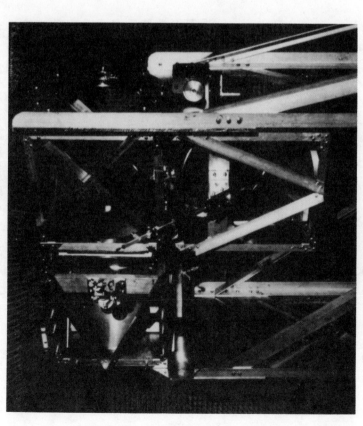

Fig. 10—A 19/28.5 GHz duo-polarization quasi-optical feed system (see Fig. 3) for the beacon measurements.

The 45-degree flat mirror, which consists of a 0.794-cm-thick aluminum jig plate, is mounted in front of the 33-cm bearing opening as shown in Fig. 9. Both azimuth and elevation orientations of the mirror can be adjusted around its center to facilitate experimental search for proper illumination of the subreflector. An oversized mirror of 38.10 by 53.34 cm oval shape was used in the initial measurements of the 7-m antenna. However, a smaller mirror of 30.48 by 43.18 cm shows very little truncation effect and is used in the beacon propagation experiment.

3.5 Measured results of the feed system

The design objective of the feed system is for each of the four offset launchers to illuminate the subreflector with a spherical wave of 15 dB taper over a 10-degree sector from a common phase center, with very low cross-polarized radiation as well as very low insertion loss of the quasioptical diplexers. To achieve and demonstrate the desired performance, we aligned the components through pattern measurements in an ane-

The measured phase patterns of Fig. 11 imply an alignment accuracy of about $1/100$ beamwidth among the beams for the beacon receiver of the 7-m antenna. One notes that any moderate asymmetry or misalignment of the amplitude pattern of the feed system, which has little effect on the beam-pointing accuracy, can be tolerated in a communication system. However, good alignment of the feed amplitude patterns is essential to the stability of the differential phase between the beams of the 7-m antenna.

After the alignment measurements, the RF stages of the receiver[16] for the 19- and 28.5-GHz beacon experiment were mounted on the feed frame. The alignment was then checked using the beacon receiving system. Tests were made on the thermal stability of the differential phase between the two orthogonally polarized 19-GHz feeds. The change of the differential phase with respect to temperature was about 0.1 degree per 1°F and could be partially explained by the difference in waveguide lengths. Local heating of the polarization-diplexing grid shows a 0.2-degree change of the differential phase.

Since the calibration of the beacon receiver makes use of the rotation of the feed frame around the beam axis, the behavior of the differential phase during this rotation was examined. When the two orthogonal linearly polarized incident waves are of comparable magnitude (i.e, when the transmitting polarization is oriented at roughly 45 degrees with respect to the orthogonal polarizations of two 19-GHz receiving feeds), the measured differential phase remains essentially constant (within 0.1 degree) over a 10-degree rotation of the feed frame around the beam axis. When the two orthogonal linear polarizations are of vastly different magnitude, the measured differential phase can involve a substantial error because of a phase quadrature component arising from cross-polarization coupling. Even a polarization grid cannot effectively discriminate against an incident cross polarization of a magnitude much greater than that of the in-line polarization.

IV. MEASUREMENTS OF THE 7-METER ANTENNA

4.1 Transmitting sources

To measure the gain and radiation patterns of the Crawford Hill 7-m antenna, a weatherproof box, which contains transmitting sources at 19, 28.5, and 99.5 GHz was placed on an AT&T Long Lines tower at Sayreville, N.J., approximately 11 km from Crawford Hill. Two horn-lens antennas with polarization grids (providing cross-polarization discrimination better than 50 dB) are used for vertical and horizontal polarizations at the lower frequencies. Each antenna has a frequency diplexer and can transmit 19 and 28.5 GHz. Waveguide switches independently control the polarization or turn off each transmitter. The

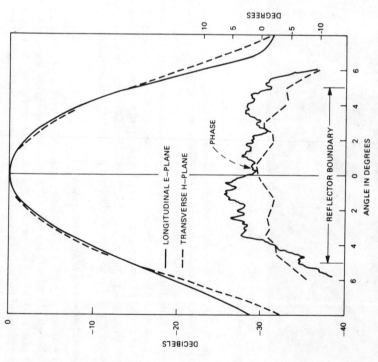

7-METER MILLIMETER WAVE ANTENNA

Fig. 11b—Measured radiation patterns of Feed No. 3 at 28.5 GHz (see Fig. 3).

The final measured patterns of Feeds No. 1 (19 GHz) and No. 3 (28.5 GHz) are shown in Fig. 11 for both E and H planes. Essentially identical measured patterns were obtained for the orthogonally polarized pair of Feeds No. 2 (19 GHz) and No. 4 (28.5 GHz). Following the nomenclatures used for horn reflector antennas, the longitudinal plane is the asymmetrical offset plane which divides the ellipsoid into two symmetrical halves and the transverse plane is the orthogonal principal plane. The measured phase patterns are less than ±5 degrees from a common spherical phase front while the measured amplitude patterns are within ±0.5 dB of perfect coincidence over the 15-dB pattern-width illuminating a 10-degree sector with respect to the common boresight. The cross-polarized radiation remained below −45 dB for all directions. It was found necessary to cover both the transmitting horn and the front bearing mount of the receiving feed system with absorbers to avoid excessive interactions. The ripples in the measured phase patterns of Fig. 11 were identified as the effects of the residual interactions.

7-METER MILLIMETER WAVE ANTENNA

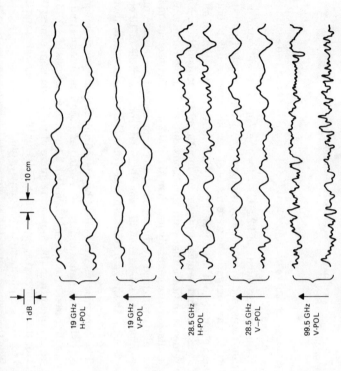

	19 GHz	28.5 GHz	99.5 GHz
FAR FIELD DISTANCE =	6.2 KM	9.3 KM	32.7 KM
3dB HALF BEAMWIDTH =	0.08°	0.055°	0.016°

Fig. 12—Terrain profile of boresight range for 7-meter antenna.

antenna gains are 30 and 33 dB with half-power beamwidths of 6.5 and 5 degrees at 19 and 28.5 GHz, respectively. The sources at 19.04 and 28.56 GHz are 100-mw Gunn oscillators.

An antenna consisting of two cylindrical reflectors[17] with a dual-mode feed and a vertically polarized grid is used to transmit a 99.5-GHz signal from a 10-mw IMPATT source with only about 50-KHz FM noise. The oscillator is connected through an isolator directly to the feed horn. The antenna has a gain of 41.5 dB with a half-power beamwidth of 1 degree in the elevation plane.

Because of the distant location, the source box is remotely controlled from Crawford Hill. The beams from three source antennas were initially co-aligned before installation on the Sayreville tower. The expected power received by the 7-m antenna was about −30 dBm at all three frequencies.

4.2 Probing measurements of the incident field

In evaluating measured results of a large aperture antenna, it is necessary to know the field distribution incident on the aperture from a distant transmitting source. The path profile of the Sayreville to Crawford Hill measuring range is shown in Fig. 12. Although the well-elevated transmitting and receiving sites provide a clear line of sight,

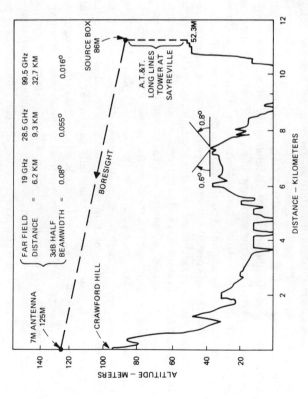

Fig. 13—Sample vertical scans for each of the five transmissions from source at Sayreville. Repeatable scans at 19 and 28.5 GHz indicate small reflections from the terrain. Uncorrelated scans at 99.5 GHz indicate atmospheric scintillations.

there still exist potential reflectors and scatterers of the millimeter-wave energy transmitted from Sayreville. A carriage and radial track mechanism was designed to permit a probe antenna to be moved along a diameter of a circular aperture to obtain field amplitude measurements. The incident field along four diameters of the 7-m aperture with 45-degree angular separation was scanned for each of the five possible states of transmission, i.e., vertical and horizontal polarizations at both 19 and 28.5 GHz and vertically polarized 99.5 GHz. Each scan was made twice to check on repeatability.

Figure 13 shows sample segments of scan pairs for all five transmissions. The good duplications of fluctuations at 19 and 28.5 GHz indicate that the systematic deviations in the field are results of specular reflections and diffractions. At 99.5 GHz, very little correlation exists between scans on a given diameter. Here the fluctuations are mostly atmospheric scintillations rather than terrain reflection. One notes that the 99.5-GHz transmitting antenna at Sayreville has a much narrower beamwidth than those of lower frequencies and the electrically rougher terrain is much less specular.

7-METER MILLIMETER WAVE ANTENNA

The spatial variations of the scan data suggest that most of the scattering comes at large angles with respect to the transmission path. This observation has been indeed confirmed by angular spectra obtained from a Fourier transform of the data. Hence, the terrain reflections should be negligible in the received power of the 7-m antenna pointing directly toward the transmitting source. The effect on the measured patterns will be only minor disturbance on sidelobes in the elevation plane at 19 and 28.5 GHz when the antenna is pointing below the source.

The received power in a gain-standard horn, which has a gain of the same order as that of our probe horn, will exhibit similar fluctuations across the aperture as those of Fig. 13. Since the fluctuations at 99.5 GHz are mostly atmospheric scintillations, the uncertainty arising from this fluctuation can be suppressed by taking an average of a number of comparisons between the gain standard and the 7-m antenna. However, at 19 and 28.5 GHz, the fluctuations are caused by terrain reflections; to reduce the uncertainty here, an average needs to be taken of numerous horn locations over the aperture. Analysis of the data resulting from probing measurements indicates that scanning the horn over a 1.4-m aperture segment gives a standard deviation of 0.4 dB at 19 GHz and 0.34 dB at 28.5 GHz.

4.3 Prime focus measurements

To provide an evaluation of surface tolerance as well as to locate the primary focal point of the 7-m reflector for subsequent installation of the subreflector, we first conducted 99.5-GHz prime focus measurements using a dual-mode feedhorn with 20-dB taper at the reflector boundary. The measured feed patterns are shown in Fig. 14. The focal region was probed with the feed until the best patterns were obtained. The measured patterns in the azimuth and elevation planes are shown in Fig. 15 together with the calculated pattern envelope assuming a perfect reflector surface.[2] The calculated patterns are approximately the same for azimuth and elevation. Expanded patterns not shown here indicated good agreement between measured and calculated half-power beamwidths (0.032 degree); however, the measured sidelobe levels are higher than the calculated values especially in the elevation plane. It is of interest to note that only near sidelobes in the elevation plane are higher than the corresponding lobes in the azimuth plane, whereas the far sidelobes in the two planes are essentially similar. Furthermore, the measured far sidelobe levels are consistent with an rms surface roughness of 0.1 mm. The excessive near sidelobe level in the elevation plane appears to be caused by a surface distortion of large-scale size. Since the reflector panels were set using a two-section template, a relative misalignment of these two sections could cause the elevation patterns we observed. This conjecture has been confirmed by pattern calculations

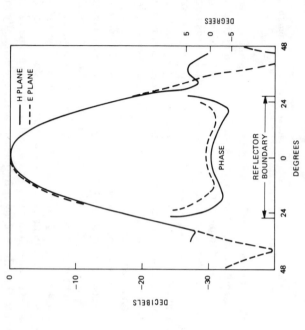

Fig. 14—Measured patterns of 99.5-GHz dual mode horn (0.98-cm diameter aperture) for prime-focus feed.

assuming a relative displacement (~0.15 mm) between two sections of the template.

Gain measurements of 99.5 GHz for the 7-m offset reflector using prime focus feed were made by comparison with a calibrated gain standard.[18] The comparisons were made at various times of day to obtain some estimate of the effects of scintillation. A time constant of about 1 s was used in the measuring set to smooth out the scintillation. Measurements indicate that gain variations due to diurnal variations of scintillations are generally less than about 0.2 dB. This observation has been also confirmed by the repeatability of measured patterns. A measured gain of 74.63 ± 0.45 dB was obtained from a sample of 10 comparisons with the gain standard taken on a clear quiet evening shortly after sunset.

The comparisons were made by measuring the difference in signal received by the gain standard (30.77 dB) and the 7-m antenna padded by a calibrated attenuator (40.05 dB). The error estimates of ±0.45 dB were obtained from the root-sum-square of the 3σ random errors: 0.14 dB for the calibrated attenuator, 0.16 dB for the gain standard, and 0.40 dB for the sample mean of 10 comparisons.

The theoretical gain of the antenna at 99.5 GHz having no roughness

7-METER MILLIMETER WAVE ANTENNA

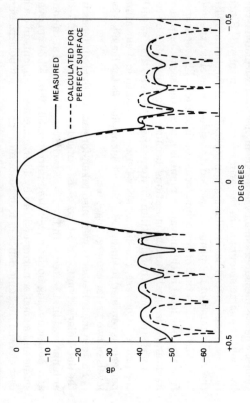

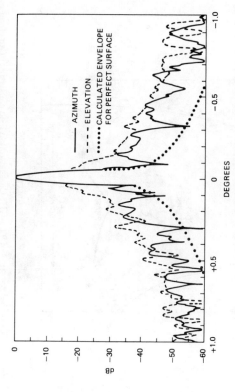

Fig. 16a—Measured 28.5-GHz azimuthal pattern with prime focus feed of 20-dB taper showing excellent agreement with calculated pattern for perfect surface.

follows:

66.42 dB	= Area Gain
−1.56 dB	= Illumination Taper
−0.08 dB	= Spillover
−0.1 dB	= Feed Loss (Estimated)
64.68 dB	= Calculated Gain for Perfect Surface

4.4 Subreflector alignment and measured results

The hyperboloidal subreflector is required to be confocal and coaxial with the paraboloidal main reflector. The primary focal point has been given by pattern measurements using prime focus feeds. Before installation of the subreflector, a laser beam was first fixed along the reflector axis. The subreflector was oriented with the aid of the laser beam reflected from a mirror attached to the bottom of the subreflector, centered on and perpendicular to its axis. The position of the subreflector was adjusted by interpreting the measured 99.5-GHz patterns until they were consistent with the prime-focus-fed patterns.

The Cassegrainian feed, used in the 99.05-GHz pattern measurements of the complete 7-m antenna including subreflector, consists of an offset ellipsoid and a dual-mode horn. The feed is essentially a scaled model of the 19-GHz offset launcher in the 19/28.5-GHz duo-polarization feed. Figure 17 shows that the measured 99.5-GHz feed patterns are almost the same as those of the 19/28.5-GHz feeds. Thus we can make direct comparison of measurements on the 7-m antenna at 99.5 GHz with the 19- and 28.5-GHz performance.

Fig. 15—99.5-GHz scan over a 2 degree range measured with prime-focus dual-mode feed of 20-dB taper. The difference between measured far sidelobe levels and the calculated pattern envelope for perfect surface is consistent[5] with 0.1 mm rms surface tolerance. The higher near-sidelobe level in the elevation plane stems from large-scale surface distortion.

is calculated as follows:

77.26 dB	= Area Gain
−1.56 dB	= Illumination Taper
−0.08 dB	= Spillover
−0.2 dB	= Feed Loss (Estimated)
75.42 dB	= Calculated Gain for Perfect Surface

Using the formula $e^{-(4\pi\epsilon/\lambda)^2}$, where ϵ is the rms surface tolerance, we see the difference between measured and calculated gains, (0.79 ± 0.45) dB, corresponds to an rms roughness of (0.1 ± 0.03) mm.

Using a 20-dB taper corrugated horn (see Figs. 6 and 7) as a prime focus feed, we also measured the patterns of the 7-m offset reflector at 28.5 GHz as shown in Fig. 16. Excellent agreement between measured and calculated patterns were obtained in the azimuth plane. However, as with 99.5 GHz, there was a noticeable discrepancy between measured and calculated sidelobe levels in the elevation plane. The first sidelobe is −25 dB compared with −16 dB at 99.5 GHz.

The measured gain at 28.5 GHz was obtained by padding the 7-m reflector with a calibrated attenuator (39.93 ± 0.05) dB, and comparing with a calibrated gain standard (24.98 ± 0.08) dB.[18] Using 15 different locations for the gain standard, the measured gain was determined to be (64.7 ± 0.6) dB. The large 3σ error limit was the consequence of the data spread and is consistent with the results of the probing measurements. The expected prime focus gain at 28.5 GHz can be calculated as

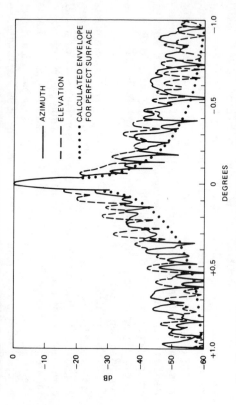

Fig. 18—99.5-GHz scan over a 2-degree range measured with Cassegrainian feed of 14-dB taper and compared with the calculated pattern envelope for perfect surface. The above patterns are consistent with those of prime-focus feed in Fig. 15.

Measured patterns are found to be insensitive to the location of the feed phase center as expected from the very long effective focal length of the antenna. With the feed phase center located on axis, the beam pointing of the Cassegrainian configuration agrees with that of the prime focus configuration to within 0.02 degree. Measured 99.5-GHz patterns are shown in Fig. 18 together with the calculated pattern envelope[15] for comparison. Measured half-power beamwidths agree with calculated values, whereas measured sidelobe levels are higher than calculated levels by about the same amount as in the prime focus measurements. Since the Cassegrainian feed has an illumination taper of 14 dB in contrast with the 20-dB taper of the prime focus feed, the sidelobe level is expected to be higher than that of the prime focus configuration. The measured first sidelobe level in the elevation plane is almost the same for prime focus and Cassegrainian configurations, because it is dominated by reflector distortion rather than illumination taper. It is seen that, as with the prime-focus case, the discrepancy between azimuth and elevation patterns is confined to the near sidelobe region, whereas the far sidelobe levels of two patterns merge together.

Pattern measurements were also taken for each of the four feeds in the duo-polarization 19/28.5-GHz quasi-optical feed assembly. The 7-m antenna was first tested with the 19/28.5-GHz feeds located at the on-axis position using both an oversized mirror of 38.10 by 53.34 cm oval shape and a smaller mirror of 30.48 by 43.18 cm. Measurements in each case showed the coincidence of four beam maxima for each polarization and frequency of the quasi-optical feed network. The smaller mirror

7-METER MILLIMETER WAVE ANTENNA

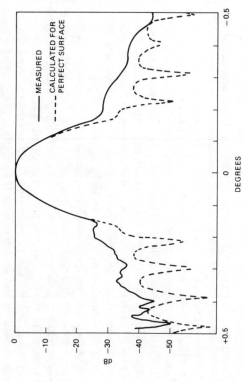

Fig. 16b—28.5-GHz elevation pattern with prime-focus feed of 20-dB taper. The measured first sidelobe level is −25 dB compared with −16 dB at 99.5 GHz in Fig. 15.

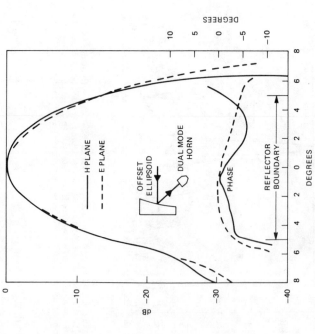

Fig. 17—Measured patterns of 99.5-GHz Cassegrainian feed consisting of an offset ellipsoid and a dual-mode horn.

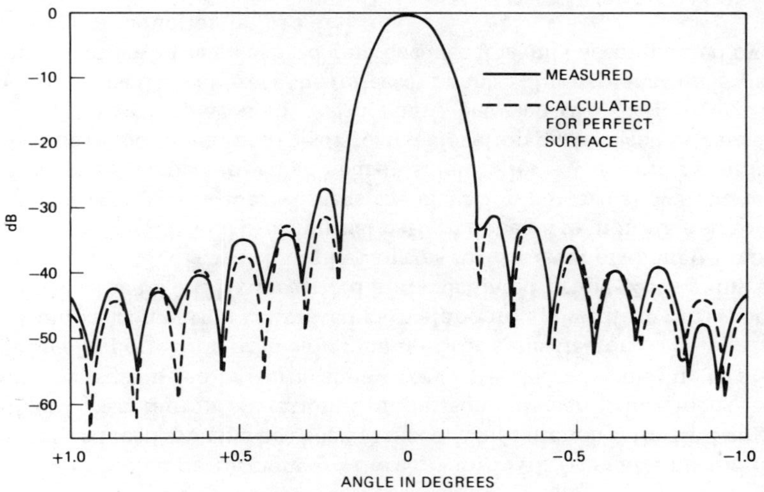

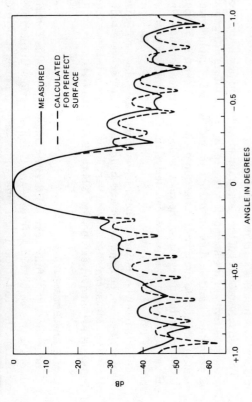

Fig. 19a—Measured 19-GHz azimuthal pattern with Cassegrainian feed of 15-dB taper and vertical polarization showing excellent agreement with the calculated pattern for perfect surface.

Fig. 19b—Measured 19-GHz elevation pattern with Cassegrainian feed of 15-dB taper and vertical polarization showing fair agreement with the calculated pattern for perfect surface.

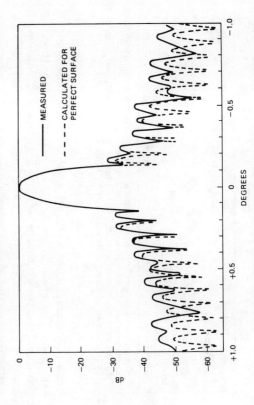

Fig. 20a—28.5-GHz azimuth pattern with Cassegrainian feed of 15 dB taper and vertical polarization. The measured pattern is consistent with that in Fig. 16a for prime focus feed of 20-dB taper.

showed very little truncation effect in measured patterns. Following these on-axis measurements, the feed box for the beacon receiver together with the smaller mirror was moved to a position 0.5 degree off axis from the center of the vertex equipment room to allow clearance for an on-axis beam for millimeter-wave radio astronomy as shown in Fig. 9. As expected,[3,4] measurements showed very little difference between the 0.5-degree off-axis and on-axis beams for both 19 and 28.5 GHz.

Measured cross-polarized radiation for each on-axis beam remained below −40 dB in all directions with respect to the in-line polarization maximum, while that for each 0.5-degree off-axis beam is smaller than −39 dB. One notes that the optimum orientations of the polarization-grid diplexer for nulling the cross polarization are about 0.7 degree apart between two orthogonally polarized feeds. The above cross-polarization data were measured using a compromise orientation of the polarization grid.

Figures 19 and 20 show comparisons between measured and calculated patterns[15] at 19 and 28.5 GHz. The measured patterns were obtained from vertically polarized feeds at on-axis position with the smaller 45-degree mirror, and remained essentially the same for other combinations of polarization and mirror at both on-axis and 0.5-degree off-axis positions. Good match between calculated and measured azimuthal patterns is illustrated in Figs. 19(a) and 20(a), whereas the agreement between calculated and measured elevation patterns is, again, less sat-

isfactory as shown in Figs. 19b and 20b. The measured sidelobes of all elevation patterns have generally shown, especially in Fig. 20b, a period twice that of the calculated value. This observation supports the conjecture about a relative misalignment of two sections of the template

Table I — Derived Cassegrainian gain in decibels with 15-dB feed taper

Frequency (GHz)	99.5	28.5	19
Area Gain	77.26	66.4	62.88
Illumination Taper	−0.93	−0.93	−0.93
Spillover*	−0.35	−0.4	−0.45
Feed and Multiplexing†	−0.2	−0.2	−0.2
Surface Tolerance	−0.8	−0.1	−0.05
Derived Gain	74.98	64.77	61.25
Error Estimates	±0.5	±0.25	±0.25
Gain Efficiency	59.2%	68.7%	68.7%

* The estimates for spillover loss are higher at 19 and 28.5 GHz because of truncations in the quasi-optical 19/28.5 GHz duo-polarization feed.
† No multiplexing is involved at 99.5 GHz, whereas both frequency and polarization diplexing losses are included in the estimates at 19 and 28.5 GHz.

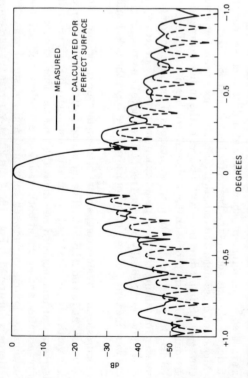

Fig. 20b—28.5-GHz elevation pattern with Cassegrainian feed of 15-dB taper and vertical polarization. The measured pattern is consistent with that in Fig. 16b for prime focus feed of 20-dB taper.

as discussed in Section 4.3. Measured 19-GHz sidelobe levels 1 degree away from the beam maximum are −43 dB in the azimuth plane and −38 dB in the elevation plane. These levels are of interest in avoiding interference from future adjacent satellites in synchronous orbits.

V. CONCLUDING REMARKS

The expected performance of the Crawford Hill 7-m antenna and its associated feed systems has been realized. This antenna is the first large offset Cassegrain in operation. Comparison between prime-focus-fed and Cassegrain-fed patterns shows very little degradation due to any surface imperfection or misalignment of the subreflector.

Comparison with a calibrated gain standard showed the difference between the 99.5-GHz measured and calculated prime-focus gains to be (0.79 ± 0.45) dB, which implies an rms surface error of about 0.1 mm. The measured 99.5-GHz azimuthal pattern appears to indicate a surface error of this magnitude, whereas the elevation pattern shows skewed shape and unexpectedly high near-in sidelobe level. Computer simulations have indicated that the distorted elevation pattern can be caused by a relative displacement (~0.15 mm) between two sections of the template used to calibrate the reflector panels. This explanation is also consistent with the measured gain because it is only accompanied by a small gain reduction (~0.15 dB).

Pattern measurements using the quasi-optical 19/28.5 GHz dual-polarization feed assembly have shown the coincidence of the four beams.

Cross-polarized radiation is −40 dB or less in all directions throughout each beam. The measured results have now confirmed the theoretical prediction[2] that there should be very little cross-polarized radiation from an offset Cassegrainian antenna with a large effective F/D ratio if the feed radiation is free of cross polarization.

Good agreement between calculated and measured patterns is shown in the azimuthal plane, whereas the comparison is less satisfactory in the elevation plane. At 1 degree away from the beam maximum, the 19-GHz sidelobe level is −43 dB in the azimuth plane and −38 dB in the elevation plane. Since the synchronous orbit is seldom close to the elevation plane of a ground station antenna, these measured sidelobe levels have practically achieved the objective of 40-dB discrimination between adjacent synchronous satellites at 1-degree spacing.

The gain measurement of the Cassegrainian configuration is hampered by the sensitivity of the harmonic mixer to the temperature difference between indoor and outdoor environments and by a lot of cable movement in addition to the terrain reflection problem at 19 and 28.5 GHz. However, having determined the main reflector surface tolerance by prime focus measurements, we can derive the Cassegrainian gains as shown in Table I.

Multiple-beam operation has been achieved with a 0.5-degree off-axis beam for beacon feed and an on-axis beam for millimeter-wave radio astronomy. Measurements showed very little difference between 0.5-degree off-axis and on-axis beams at both 19 and 28.5 GHz.

VI. ACKNOWLEDGMENTS

The authors gratefully acknowledge the following valuable contributions to this work. Ford Aerospace and Communications Corporation carried out the mechanical design, built, and installed the main structure. M. J. Grubelich and K. N. Coyne consulted on many mechanical design problems, including the main structure, subreflector, and feed system. M. J. Gans designed the vertex room windows and calculated the surface distortions from mechanical measurements. H. H. Hoffman collaborated on the antenna-receiver interface. D. C. Hogg proposed the antenna configuration. E. A. Ohm recognized the large beam scanning capability of the antenna configuration and modified the vertex equipment room and subreflector support structure design to facilitate launching beams at large off-axis angles while keeping the gravitational pointing error small. He also designed and supervised the fabrication of the subreflector. F. A. Pelow supplied the quasi-optical diplexers. A. Quigley and members of the shop built most of the feed hardware. H. E. Rowe examined detailed surface tolerance effects. J. Ruscio and T. Fitch took care of receiver cables in the cable wraps. R. A. Semplak provided the 99.5-GHz Cassegrainian feed. R. H. Turrin designed the carriage and radial track mechanism for probing measurements. D. Vitello programmed the pattern calculations. G. V. Whyte implemented the subreflector mount.

REFERENCES

1. C. Dragone and D. C. Hogg, "The Radiation Pattern and Impedance of Offset and Symmetrical Near-Field Cassegrainian and Gregorian Antennas," IEEE Transactions, AP-22, May 1974, pp. 472–475.
2. T. S. Chu and R. H. Turrin, "Depolarization Properties of Offset Reflector Antennas," IEEE Transactions, AP-21, May 1973, pp. 339–345.
3. E. A. Ohm, "A Proposed Multiple-Beam Microwave Antenna for Earth Stations and Satellites," B.S.T.J., 53, No. 8 (October 1974), pp. 1657–1666.
4. E. A. Ohm and M. J. Gans, "Numerical Analysis of Multiple-Beam Offset Cassegrainian Antennas," AIAA Paper #76-301. AIAA/CASI 6th Communication Satellite Systems Conference, Montreal, Canada, April 5–8, 1976.
5. C. Dragone and D. C. Hogg, "Wide Angle Radiation Due to Rough Phase Fronts," B.S.T.J., 42, No. 7 (September 1963), pp. 2285–2296.
6. P. F. Goldsmith, "A Quasi-Optical Feed System for Radio Astronomical Observations at Millimeter Wavelengths," B.S.T.J., 56, No. 8 (October 1977), pp. 1483–1501.
7. T. S. Chu and W. E. Legg, "A 19/28.5 GHz Quasi-Optical Feed for an Offset Cassegrainian Antenna," IEEE/APS Symposium, Amherst, Mass., October 1976.
8. J. A. Arnaud and F. A. Pelow, "Resonant-Grid Quasi-Optical Diplexers," B.S.T.J., 54, No. 2 (February 1975), pp. 263–283.
9. T. S. Chu, M. J. Gans, and W. E. Legg, "Quasi-Optical Polarization Diplexing of Microwaves," B.S.T.J., 54, No. 10 (December 1975), pp. 1665–1680.
10. I. Anderson, "Measurements of 20 GHz Transmission Through a Radome in Rain," IEEE Transactions, AP-23, September 1975, pp. 619–622.
11. C. Dragone, "Reflection and Transmission Characteristics of a Broadband Corrugated Feed: A Comparison Between Theory and Experiment," B.S.T.J., 56, No. 6 (July-August 1977), pp. 869–888.
12. T. S. Chu, "Geometrical Representation of Gaussian Beam Propagation," B.S.T.J., 45, No. 2 (February 1966), pp. 287–299.
13. H. Kogelnik and T. Li, "Laser Beams and Resonators," Proc. IEEE, 54, October 1966, pp. 1312–1329.
14. M. J. Gans and R. A. Semplak, "Some Far-Field Studies of an Offset Launcher," B.S.T.J., 53, No. 7 (September 1974) pp. 1319–1340.
15. J. S. Cook, E. M. Elam, and H. Zucker, "The Open Cassegrain Antenna: Part 1—Electromagnetic Design and Analysis," B.S.T.J., 44, No. 7 (September 1965), pp. 1255–1300.
16. H. W. Arnold, D. C. Cox, H. H. Hoffman, R. H. Brandt, R. P. Leck, and M. F. Wazowicz, "The 19 and 28 GHz Receiving Electronics for the Crawford Hill COMSTAR Beacon Propagation Experiment," B.S.T.J., this issue, pp. 1289–1329.
17. C. Dragone, "An Improved Antenna for Microwave Radio Systems Consisting of Two Cylindrical Reflectors and a Corrugated Horn," B.S.T.J., 53, No. 7 (September 1974), pp. 1351–1377.
18. T. S. Chu and W. E. Legg, "Gain of Corrugated Conical Horns," IEEE/APS Symposium, College Park, Md., May 1978.

The IRAM 30-m millimeter radio telescope on Pico Veleta, Spain

J.W.M. Baars [1], B.G. Hooghoudt [1], P.G. Mezger [1], and M.J. de Jonge [2]

[1] Max Planck Institut für Radioastronomie, Auf dem Hügel 69, D-5300 Bonn, Federal Republic of Germany
[2] Institut de Radio Astronomie Millimetrique (IRAM), Grenoble, France

Received September 29, accepted October 29, 1986

Summary. In the Spanish Sierra Nevada near 2900 m altitude, the new 30-m telescope for millimeter astronomy is now operational. Here we describe the original design features, which resulted in the high reflector and pointing accuracy, necessary for operation near 1 mm wavelength. The open air telescope is thermally insulated and the temperature of critical sections is controlled to better than 1 K day and night. A reflector surface error, measured with radio-holographic techniques, of about 80 µm and a pointing and tracking accuracy of about 1″ in wind velocities of $12 \, \text{m s}^{-1}$ and under stable atmospheric conditions have been reached. These can be further improved. Receivers are available for the 3, 2, and 1.2 mm atmospheric windows. First tests at 0.87 mm have confirmed the high quality of this instrument, which is the most powerful filled-aperture telescope in the short millimeter wavelength range.

Key words: radio telescope – millimeter wavelength astronomy – instruments

1. Introduction

Since the discovery in 1970 of carbon monoxyde in the interstellar medium by its rotational transition at a wavelength of 2.6 mm, radio astronomy at millimeter wavelengths has undergone an enormous development. This has been made possible, firstly by improvements of more than an order of magnitude in receiver sensitivity, secondly by the construction of larger and more accurate telescopes and finally by locating the telescopes at high mountain sites to minimize the effects of the troposphere.

In this paper we present a description of the new Millimeter Radio Telescope (MRT) of IRAM (Institute for Radio Astronomy in the Millimeter Range), the French-German institute, established in 1979 by the Centre National de la Recherche Scientifique (CNRS) and the Max Planck Gesellschaft (MPG), to provide these countries with first class facilities for millimeter wavelength astronomy. The complementary IRAM instrument is a three element array of 15-m antennas, which is under construction on Plateau de Bure, 90 km south of Grenoble, where IRAM has its headquarters.

The MRT is located on Pico Veleta in the Spanish Sierra Nevada, chosen for its relatively low geographic latitude, height and dryness of the atmosphere. A low water vapor content has been confirmed by actual measurements. The pertinent data of the site are assembled in Table 1. Spain's contribution to the project is channeled through an agreement between IRAM and the Instituto Geografico Nacional and entails the free availability of the site and offices in Granada. In exchange Spanish scientists obtain preferred access to the telescope for 10 % of the time.

After some preliminary studies in the early seventies, detailed design work for a large millimeter radio telescope was started at MPIfR in 1975 and, after the foundation of IRAM, the MPIfR was contracted to take resposibility for the construction of the 30-m telescope.

A high accuracy of the reflector, which would allow observations near 1 mm wavelength, was given the highest priority, rather than the largest possible diameter. Preliminary design work led to the conclusion that a 30 m telescope with a reflector accuracy of 0.1 mm would be technically and financially feasible. A consortium of the companies Krupp Industrietechnik and MAN, which had already designed the Effelsberg 100-m radio telescope, undertook the detailed design and construction of the MRT.

Table 1. Characteristics of the Pico Veleta site

Location
Pico Veleta, Sierra Nevada,
45 km from Granada, Spain

Telescope coordinates	
Altitude	2870 m (elevation axis)
Geographical longitude	03°23′58″.1 West
Geographical latitude	37°04′05″.6 North

Climate	
Number of days and nights with >75 % clear sky	200 per year
Average wind velocity	$8 \, \text{m s}^{-1}$
Maximum wind velocity	$50 \, \text{m s}^{-1}$
Maximum summer temperature	20 °C
Minimum winter temperature	−15 °C
Average diurnal variation	5 °C
Precipitable water vapor to zenith	
Average summer	4 mm
Average winter	2 mm
Typical good zenith transmission at 1.2 mm wavelength	0.90

Send offprint requests to: J.W.M. Baars

2. Technical features

2.1. The telescope

The telescope has a paraboloidal main reflector, placed on an alt-azimuth mounting (Figs. 1 and 2). The total movable weight is approximately 800 t. The basic specifications of the telescope, together with the values achieved in practice, are assembled in Table 2, while Table 3 shows the geometrical data of the optics system. Some features of the design have been described earlier by Baars and Hooghoudt (1980) and Baars (1983). Apart from minimizing gravitational deformations, two other aspects have ab initio been given equal importance in the design, i.e. the control of thermal deformations and of wind-induced pointing errors. At the required performance level these are determining factors in the operation of the telescope.

Gravitational and wind deformations are controlled by a homologous, computer-optimized structural design (von Hoerner, 1967; Brandt and Gatzlaff, 1981). In addition the influence of the wind on the pointing stability must be minimized. Thus a stiff

Fig. 1. Photograph of the MRT with control- and residence building. Note the thermal insulation enclosing the structure. The outside can be heated to avoid the deposit of ice during occasional icestorms

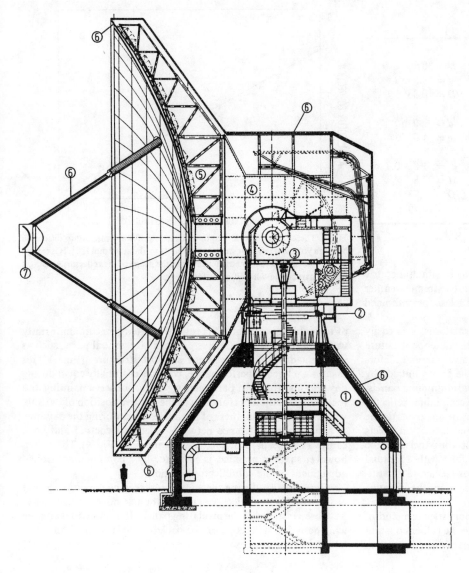

Fig. 2. Cross section of the MRT. A concrete pedestal (1) carries a 5 m diameter azimuth bearing (2). A two-storey cabin (3) contains all drive systems on the lower floor and the radio astronomy receivers in the upper part between the elevation bearings. The yoke and conesection (4) is supported at the elevation bearings and carries the reflector structure (5). Thermal insulation (6) covers the outside of the telescope and is also present between reflector surface and its support structure. The quadrupod and prime focus cabin are also insulated. The reference antenna for radio-holography is mounted in the prime-focus cabin (7)

57

Table 2. Telescope specifications and performance

	Tolerances (rms, in micrometer)	
	Specified	Achieved
Reflector panels	50	26
Measurement/setting of surface	50	35[a]
Residual deformations:		
Gravity	50	40
Wind $(12\,\mathrm{m\,s^{-1}})$	35	30
Temperature	25	20
Secondary reflector	25	15
Total (root sum squared)	100	70[a]
Tracking accuracy (wind $12\,\mathrm{m\,s^{-1}}$)	2″	1″ [b]
Blind pointing accuracy	1″	2–3″ [c]
Temperature gradients	1 K	1 K

[a] Expected in near future
[b] In absence of large-scale atmospheric disturbances
[c] Being improved, goal of 1″ rms seems feasible

Table 3. Geometry of the telescope optics

Paraboloidal main reflector:		
diameter	$D =$	30 m
focal length	$f =$	10.5 m
focal ratio	$f/D =$	0.35
Hyperboloidal subreflector:		
diameter	$d =$	2.0 m
eccentricity	$e =$	1.0746
Nasmyth reflectors (flat):		
size	$d_n =$	1.0×0.7 m
Cassegrain magnification factor:	$M =$	27.8
Effective focal ratio of Nasmyth:	$f_e/D =$	9.73
Distance from prime to secondary focus:	$f_c =$	19.79 m

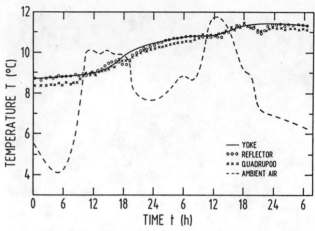

Fig. 3. Temperature variation of different telescope sections measured on two days in October 1985 with relatively large variations in ambient air temperature. Reflector and quadrupod temperatures are controlled to equal that of the yoke. Note the slow and retarded increase of the yoke temperature and the good performance of the temperature control system with temperature deviations of typically less than 0.5 K

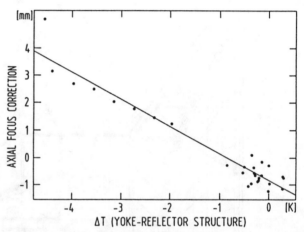

Fig. 4. The change in focal length as function of the temperature difference between yoke and reflector structure amounts to about $0.9\,\mathrm{mm\,K^{-1}}$. It is caused by a systematic and symmetrical change in the shape of the reflector, yielding a best fit paraboloid with different focal length

and relatively heavy structure results, which simultaneously fulfills the requirement of surviving the most extreme weather situation specified: 30 cm of solid ice on the telescope combined with a wind speed of 200 km per hour.

Given the very small gravitational deformations, it is easily possible that temperature gradients in an open telescope structure cause the largest deformations. To minimize thermal deformations, the exposed MRT is covered on the entire outside by thermal insulation, consisting of double-walled aluminium panels, filled with 40 mm of polyurethane. Only the aluminium reflector panels, coated with a heat reflecting paint, are exposed to the ambient atmospheric conditions. The thermal insulation sheets are also installed between the aluminium reflector panels and the steel support structure (Fig. 2). Five large fans with 15 radial venting tubes create a circumferential airflow to avoid stratification of the air and temperature gradients in the support structure.

In addition, the temperature of the support structure and of the quadrupod with prime focus box is tied to the rather constant temperature of the very massive yoke by means of an active system of temperature control. This results in a temperature uniformity well within one degree Celsius between all essential parts, such as yoke, quadrupod and reflector support structure. (Fig. 3) This assures a telescope performance, which is essentially equal during day and nighttime. In particular we do not observe pointing and focussing variations as function of the relative position of the Sun, which limits the usefulness of most telescopes during the daytime. A temperature difference between reflector structure and yoke causes a change in focal length of $0.9\,\mathrm{mm\,K^{-1}}$ (Fig. 4). This effect, however small in practice, is corrected by a computer-controlled adjustment of the subreflector position.

Occasionally, during the winter, severe icing conditions occur at the site. To avoid ice growing on the telescope, heating wires have been installed in the insulating panels. It is found that after an icing period the telescope can be back into operation within a few hours.

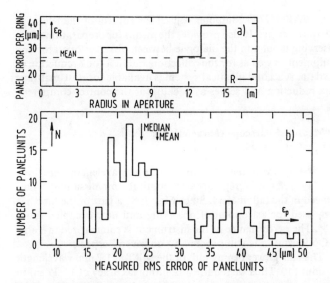

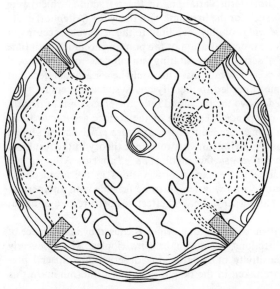

Fig. 5a and b. Histograms of the distribution of panel errors. In **a** the average rms error per panel ring is shown, which gives an impression of the distribution of panel errors over the aperture area. **b** shows the distribution of the rms error of all 210 panel units. The average, unweighted error is 26 μm; weighting with the surface area and the illumination function gives a similar average value

Fig. 6. Map of reflector deviations from the best fit paraboloid, derived from radio-holography measurements in August 1985. The contour interval is 50 μm, the rms deviation is 85 μm. The thick line is zero, dashed contours are negative. The large deflection, denoted C, is a panel, purposely displaced by 200 μm for calibration purposes

The design of the servo-control system takes into account the strong wind velocity (more than 12 m s^{-1} for 20% of the time). The pointing and tracking accuracy, specified at 1″, can only be achieved with an advanced state-controller which uses a multiple set of input parameters, such as angular position and velocity of both telescope- and motor-axes and the inclination angle of the tower, and which are read into the control system with a cycletime of 6 ms. With this new servo-system an accuracy and stability of 1″ in the tracking mode has been achieved with wind velocities to 12 m s^{-1}.

The reflector surface of the MRT consists of a total of 420 panels, arranged in seven concentric rings. The panels are made of stretched aluminium skins, which are bonded to aluminium honeycomb of 40 mm thickness. Their average size is about 1 m × 2 m. Two panels are mounted on one steel subframe, using 15 adjustable bolts for each panel (Eschenauer et al., 1980). This allows the optimization of the shape of the panels to the prescribed paraboloidal form, by adjusting them with the aid of an accurate 3D-measuring machine. The distribution of achieved panel accuracies is shown in Fig. 5; the average value is 26 μm rms.

The subframes are inside the temperature controlled space of the reflector support structure (Fig. 2). They are connected to the backup structure through adjustable bolts at their corners.

We have applied two different methods to measure the position of the subframes with respect to the best fitting paraboloid on the completed telescope. The first is an improved and modified version of the classical theodolite – tape method, used only, while the telescope was in the zenith position. A measuring accuracy of about 70 μm rms was achieved. The measurement mainly served as a check on the results of our second method, which is inherently more flexible, albeit less straightforward.

With this method – radio holography – one derives the shape of the reflector surface from the phase distribution in the telescope aperture, which in turn is obtained by Fourier Transformation of the antenna pattern, measured both in amplitude and phase with the aid of a test transmitter. The method has been pioneered by Bennett et al. (1976), and used in various forms by several groups (Morris, 1984). We employ the strong water-vapor maser-source in Orion as the test transmitter at 13.3 mm wavelength. The phase reference is provided by a small antenna, parallel to the axis of the telescope on top of the prime focus box. These measurements of the reflector deviations were obtained at an elevation angle of about 45 degrees, which is close to that, where most observations are made. In an alternative scheme, called phase-retrieval holography (Morris, 1984), the reflector profile is deduced from measurements of the intensity pattern only, obtained at 3.5 mm wavelength with the aid of a transmitter, located on top of the Pico Veleta (Morris et al., 1987).

The final setting of the panels is achieved in a number of iterative steps with intermediate measurements. One such a result is shown in Fig. 6 in the form of a contour map of deviations from the best fitting paraboloid; the present rms-value is 85 μm, obtained with a measuring accuracy of about 35 μm. We expect to reach a reflector accuracy of about 70 μm after further panel adjustments.

2.2. Receiver systems

The telescope operates in a Nasmyth-focus configuration. After reflection at the hyperbolic secondary reflector the beam is directed into an equipment room, which is located between the elevation bearings in the azimuth section. Two flat mirrors placed under 45 degrees with respect to the elevation axis bring the beam to a focal point in the cabin, which is fixed in relation to the elevation movement of the telescope. A system consisting of flat mirrors, polarisation splitter and dichroic mirror creates several focus locations, allowing the simultaneous operation of two or three receivers. This offers a great flexibility in the operation of the telescope and in the selection of the main observing frequency. Thus it is possible to react quickly to changing sky transmissivity by switching to the most appropriate frequency. An important feature of any observation is the calibration of the antenna temperature scale. This is done by insertion of hot and cold

Table 4. Receiver characteristics

Front-end type	Frequency range	Receiver temp.
Schottky mixer (2x)	75–115 GHz	200 K (DSB)
SIS mixer (2x)	75–115 GHz	125 K (DSB)
SIS mixer	140–170 GHz	175 K (DSB)
Schottky mixer (MPIfR)	216–236 GHz	330 K (DSB)

Back-end type	Bandwith	Channels
Filter bank	512 MHz	512 × 1 MHz
Filter bank	25.6 MHz	256 × 0.1 MHz
Spectrum expander [a]	12.8/n MHz	128 × 0.1/n MHz
Digital autocorrelator	8 × 40 MHz	2048 [b]
Continuum	500 MHz	2

[a] On loan from Bordeaux Observatory, $n = 2, 4, 8$ or 16
[b] Several resolutions from less than 1 to 150 kHz 1024 channels available summer 1987

matched loads in front of the receiver. At the MRT these are provided by rf-absorbing material at room (293 K) and liquid nitrogen (77 K) temperature. They can be inserted in the beampath between the second Nasmyth mirror and the polarisation splitter. Consequently we use this location as the plane to which measured quantities, like antenna temperature and aperture efficiency are referred. The loads can be switched under program control. Moreover the ambient load is configured as a rotating chopper, enabling observations switching between sky and load. These features significantly ease the calibration of the temperature scale and the measurement of the sky transmissivity through sky-dips.

The receivers for the MRT are being developed and built in the IRAM laboratories in Grenoble. At the present time the telescope is equipped with cooled Schottky-mixer frontends for the wavelength-bands 2.6–4 and near 1.3 mm (the last one supplied by the MPIfR). Improved systems with SIS-mixers are coming on-line for the 3 mm and 2 mm bands, while a similar system for 1.3 mm is under construction. Several spectrometer backends are available. Table 4 presents the major present characteristics of the receivers.

A VAX 11/780 computer is used for control of the telescope and the receivers. It also provides the means for preparation of the observing input and the dialogue between observer/operator and equipment as well as for the collection and first-look reduction of the data. A second identical system is available for post-real-time data reduction and serves as backup for the control computer.

3. Measured telescope characteristics and first operational experience

The first tests with the telescope at 3.5 mm wavelength were done in May 1984 and regular astronomical commissioning commenced in the fall of 1984. Since May 1985 a part of the time has been used for astronomical observing and now occupies about 70%. The performance of the instrument is summarized in Table 5. Optimization is still proceeding on several aspects.

The half-power beamwidth of the MRT at 1.2 mm wavelength is about 11″. Thus we require a pointing accuracy of 1″. Position determinations of a set of about 30 sources, distributed over the sky, are used to derive the nine parameters of a pointing model. The measurements are normally done near 86 GHz and employ extragalactic point sources, planets, as well as SiO-maser stars. The rms residuals from the model are 1″–3″, the higher value applying to the elevation coordinate. We believe that these are caused by short term variations in the refraction, which were identified as such for the first time with the MRT (Altenhoff et al., 1987). These will probably set a limit to the achievable accuray of pointing. A slow deterioration of the pointing accuracy with time has been observed, necessitating an update of the model parameters every few months. This could be a seasonal effect, but a longer time is needed to identify any systematic trends.

As stated above the measured reflector precision is at present 85 μm. An independent estimate of the surface accuracy can be made from a measurement of the aperture efficiency at several wavelengths. Applying antenna tolerance theory to the results of Table 5, we derive a surface error of 88 ± 5 μm rms. After further panel adjustments we expect observations in the atmospheric window near 0.9 mm wavelength to be feasible with an aperture efficiency of about 20%, i.e. and effective aperture area of about 150 m².

Chromatism of a telescope, which is caused by interference of multiple reflections, results in baseline-ripples, which severely limit the sensitivity of spectroscopy observations. Special precautions were taken in the design of the MRT to minimize this

Table 5. Measured telescope parameters (status spring 1986)

	Receiver type				
	Schottky-mixer			Bolometer	
Wavelength in mm	3.3	2.6	1.3	1.2	0.87
Aperture efficiency (%) [a]	53	50	25 [b]	24 [b]	9 [b]
Main beam efficiency (%)	67	59	47	38	
Effective collecting area (m²)	380	355	170	170	
Half-power beam-width (arcsec)	25.3	21.0	12.3	11.2	7

[a] Illumination function differs for different wavelengths
[b] Significant improvement expected after further panel adjustment

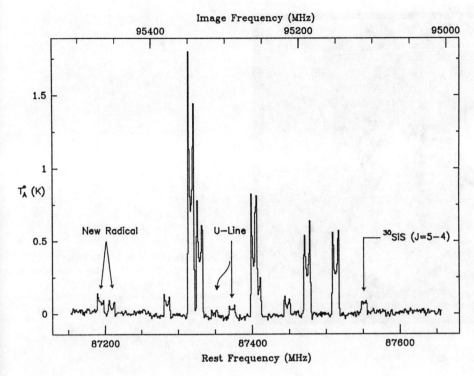

Fig. 7. A 500 MHz wide spectrum near 87 GHz, obtained with the SIS-receiver, of the carbon star IRC + 10216. The integration time was 14 min. This spectrum shows the detection of a new radial and some unidentified lines. Note the excellent baseline, to which no correction has been applied (Guélin et al., 1986)

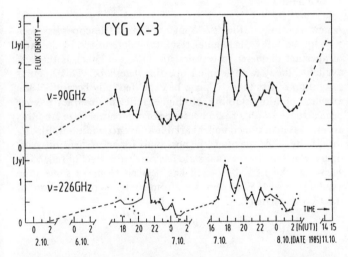

Fig. 8. Simultaneous observation at 90 and 226 GHz of some outbursts of Cyg X-3 during October 1985, obtained with the Schottky-mixer receivers and the focal-plane chopper in an "on-off" observing mode (Baars et al., 1986)

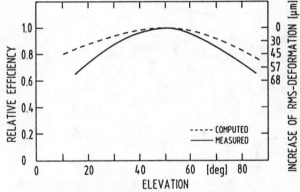

Fig. 9. The relative gain of the telescope as function of elevation angle, measured in March 1986 with the MPIfR Bolometer-receiver at 1.2 mm wavelength (solid) and that predicted from computed structural deformations (dashed). The effective additional surface deviation is also inserted along the righthand scale. The measured deviation (52±5 μm at elevation 80°) is comparable to that predicted (40± 10 μm). (Measurements by Salter and Steppe, priv. comm.)

effect (Morris, 1978). The spectrum, shown in Fig. 7 without any baseline corrections, as it was observed, indicates the quality of the baseline (Guélin et al., 1986).

The observing technique of beam-switching is routinely used at millimeter wavelengths to minimize the influence of atmospheric noise fluctuations, particularly on observations of continuum sources. To enable this the telescope is equipped with a chopping secondary. The lightweight secondary mirror, made of carbon-fibre reinforced plastic, can be chopped over four arc minutes on the sky with a frequency of about 3 Hz. This system is not yet fully operational and a focal plane chopper, i.e. a butterfly-wheel, which rotates in front of the second Nasmyth reflector, is used at present. An example of a beam-switched observation at frequencies of 90 and 226 GHz simultaneously is presented in Fig. 8 (Baars et al., 1986).

Test observations have also been done with the broadband [3]He-cooled bolometer-system of the MPIfR (Kreysa, 1984). In the wavelengthband near 1.2 mm the sensitivity was about an order of magnitude better than with the Schottky-mixer receiver. Some 10 min of integration time were sufficient to detect for the first time the flux density of Pluto at a level of about 15 mJy (Altenhoff et al., 1987). The elevation dependence of the antenna gain, as measured with this receiver, is shown in Fig. 9. One sees clearly that the surface has been set optimally near 45° elevation, the

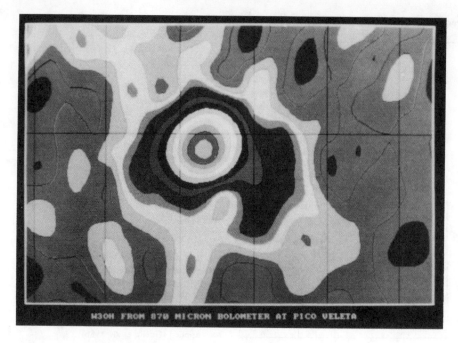

Fig. 10. A map of the compact H II-region W3−OH, obtained with the MPIfR bolometer-receiver at 0.87 mm wavelength. With the beamwidth of 7″ and a source size of about 2″, this map essentially represents the antenna beam. (Kreysa and Salter, priv. comm.)

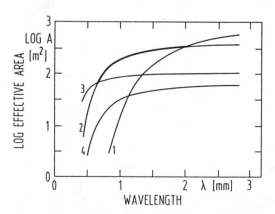

Fig. 11. Plot of the effective area (log-scale) as function of wavelength for the major millimeter telescopes: *1* Nobeyama ($D = 45$ m, $\varepsilon = 160\ \mu$m), *2* MRT (30 m, goal 70 μm), *3* JCMaxwell (15 m, goal 35 μm), *4* NRAO (12 m, 70 μm). The MRT is the most sensitive between 150 and 500 GHz

4. Conclusion

A new milestone in the development of radio telescopes has been set with the MRT. Its angular resolution, determined by the ratio of aperture diameter to wavelength, $D/\lambda_{min} = 35000$, is the best available today with a filled aperture telescope. This angular resolution of less than 10″, can be exploited fully because of the excellent pointing stability. No higher reflector accuracy has yet been achieved with a radio telescope of comparable size, and this accuracy is maintained under variable ambient conditions. Finally the quality of the site offers the possibility for an efficient, all-year, operation. The operational state has reached a level of reliability such that the IRAM-Council has decided to make some time available as of 1987 to observers from countries outside those of the IRAM partners.

elevation at which the holographic measurement was done. The decline in gain at the extreme elevations is somewhat larger than predicted by the structural computations.

Some tests have been made with the bolometer in the atmospheric window at 0.87 mm in May 1986. Figure 10 shows the result of a mapping of the source W3−OH with a beamwidth of only 7″. The telescope parameters measured with the bolometer system are also assembled in Table 5.Clearly the MRT shows already a satisfactory performance at 0.87 mm, well beyond the original design goal of 1.3 mm. A comparison of the effective areas of the major short mm-wavelength telescopes is shown in Fig. 11. The MRT is the most sensitive telescope in the frequency range from about 150 to 500 GHz. At the long mm-wavelengths it is surpassed by the Japanese 45-m telescope, while only at frequencies above 500 GHz will the new J.C. Maxwell telescope on Hawaii be more sensitive.

Acknowledgements. A project of this size is a cooperative effort by many people, in this case from industrial contractors, special consultants, MPIfR and IRAM. It is impossible to mention all who have contributed to the successful completion of the MRT. We thank Dr. Mäder and Mr. Gatzlaff (Krupp) and Dr. Kärcher (MAN.) for their project management. A special acknowledgement is due to our colleagues Greve, Harth and Morris (reflector measurement), Schraml, Beckmann, Brunswig and Bardenheuer (servo control, software), Blundell, Carter, Gundlach, Haas and Hein (receivers), Lucas, von Kap-Herr and Schmidt (software) and Guélin, Thum, Salter, Steppe, Altenhoff, Wink, Cernicharo and Gomez Gonzales (telescope commissioning). We thank the engineers, technicians and operators of the contractors, MPIfR and IRAM for their contributions. This project was made possible through a substantial grant from the "Stiftung Volkswagenwerk", for which we are grateful. We thank our colleagues for providing unpublished results.

References

Altenhoff, W.J., Baars, J.W.M., Downes, D., Wink, J.E.: 1987, *Astron. Astrophys.* (submitted)

Altenhoff et al., 1987 (in preparation)

Baars, J.W.M.: 1983, Infrared and Millimeter Waves, ed. K. Button, 9, Chap. 5, pp. 241–281, Academic Press, New York

Baars, J.W.M., Hooghoudt, B.G.: 1980, Proc. *Optical and Infrared Telescopes for the 1990s*, ed. A. Hewitt, Tucson, p. 458

Baars, J.W.M., Altenhoff, W.J., Hein, H., Steppe, H.: 1986, *Nature* 324, 39

Brandt, P., Gatzlaff, H.: 1981, *Techn. Mitt. Krupp Forschungs-Ber.* 39, 111

Bennett, J.C., Anderson, A.P., McInnes, P.A., Whitaker, A.J.T.: 1976, *IEEE Trans. Antennas Propag.* **AP-24**, 295

Eschenauer, H., Gatzlaff, H., Kiedrowski, H.W.: 1980, *Techn. Mitt. Krupp Forschungs-Ber.* **38**, 43

Guélin, M., Cernicharo, J., Kahane, C., Gomez-Gonzales, J.: 1986, *Astron. Astrophys.* **157**, L17

von Hoerner, S.: 1967, *Astron. J.* **72**, 35

Kreysa, E.: 1984, *Proc. URSI Int. Symp. Millimeter and Submillimeter Wave Radio Astron.*, Granada, p. 153

Morris, D.: 1978, *Astron. Astrophys.* **67**, 221

Morris, D.: 1984, *Proc. URSI Int. Symp. Millimeter and Submillimeter Wave Radio Astron.*, Granada, p. 29

Morris, D., Hein, H., Steppe, H., Baars, J.W.M.: 1987, *Proc. IEE* (submitted)

Design of Large Steerable Antennas

SEBASTIAN VON HOERNER

National Radio Astronomy Observatory, Green Bank, West Virginia

(Received 22 September 1966)

The design of large, steerable, single reflectors is investigated in general, in order to find the basic principles involved and the most economical solutions. (Let λ be shortest wavelength, D the antenna diameter, W the antenna weight; ρ the material density, S the maximum stress, E the modulus of elasticity, C the thermal expansion coefficient; p_s the survival surface pressure, p_0 the observational wind pressure, Δ the temperature differences; and γ dimensionless constants.) There are three natural limits for tiltable antennas: first, the thermal deflection limit, $\lambda = \gamma C \Delta D = 2.4$ cm $(D/100$ m$)$; second, the gravitational deflection limit, $\lambda = \gamma(\rho/E)D^2 = 5.3$ cm $(D/100$ m$)^2$; third, the stress limit, $D = \gamma S/\rho = 620$ m. Let each antenna be a point in a D, λ plane. The part of the plane permitted by the three limits then can be divided into four regions according to what defines the antenna weight. First, the gravitational deflection region (W governed by ρ/E); second, the wind deflection region (governed by p_0/E); third, the survival region (governed by p_s/S); fourth, the minimum structure region (stable self-support). Formulas are derived for the regional boundaries, and for $W(D,\lambda)$ within each region.

Some aspects of economy are considered. There is a most economical λ for any D. Radomes can give advantage within the wind deflection region only, which means for $D \leq 50$ m for conventional antennas. The azimuth drive should be on standard railroad equipment for $\lambda \geq 8$ cm. An economical antenna with $D = 150$ m and $\lambda = 20$ cm should cost about four million dollars.

There are four means of passing the gravitational limit. First, avoiding deflections by not moving in elevation angle (fixed-elevation transit telescope). Second, fighting the deflections with motors (Sugar Grove). Third, canceling the deflections with levers and counterweights (large optical telescopes). Fourth, designing a structure which deforms unhindered, but which deforms one paraboloid of revolution into another one (homologous deformation). It is proved that *exact* homology solutions do exist, and an explicit solution is given for two dimensions.

INTRODUCTION

THERE is a growing need in radio astronomy for building very large antennas for 10 or 20 cm wavelength, and another need for observing as short a wavelength as possible with antennas of moderate size. Since both these demands soon run into structural as well as financial limitations, a general survey of the whole problem seems indicated.

Before the astronomer can ask the engineer to design a telescope of diameter D and for wavelength λ, he ought to know which D, λ combinations are possible at all, which are at the limit of his funds, and which combinations might be considered the most economical ones. On the other side, it might help the engineer to know that antennas with certain D, λ combinations are completely defined by survival conditions and nothing else, others by gravitational deflections and nothing else, and so on. Furthermore, it might give the engineer a helpful challenge if he knows with what weight a near-to-ideal design is supposed to meet the specifications.

The present investigation is held as general as possible in order to make it applicable to telescopes of any diameter. It seeks the natural limits of antennas, the most economical type of design, and useful approximation formulas giving the weight of a telescope as a function of diameter and wavelength. Finally, we ask whether the gravitational limit can be passed by designing a structure which deforms unhindered if tilted, but which still gives some exact paraboloid of revolution at the surface for any angle of tilt.

The paper was originally submitted (June 1965) as an LFST report. The LFST group, led by Dr. Findlay of NRAO, investigates the possibilities of the "Largest Feasible Steerable Telescope." All LFST reports, and two summaries, can be obtained on request from NRAO, Green Bank, West Virginia.

I. NATURAL LIMITS

A. Gravity and Elasticity

Even with gravity as the only force (no load or wind) one could not build indefinitely high structures. A limit is reached when the weight of the structure gives a pressure at its bottom equal to the maximum allowed stress of the material used. S is the maximum allowed stress of material, ρ the density of material, h_0 the maximum height of structure, and γ_1 the geometrical shape factor ($\gamma_1 = 1$ for standing pillar or hanging rope). The maximum height of a structure, no matter what its purpose, is then

$$h_0 = \gamma_1 S/\rho. \qquad (1)$$

A second limit applies to any structure which, while being tilted, shall maintain a given accuracy, defined in our case by the shortest wavelength to be used. Even a standing pillar gets compressed under its own weight, the lower parts more than the upper ones. We call E the modulus of elasticity, h the height of structure, Δh the change of height under its own weight, and γ_2 the geometrical shape factor ($\gamma_2 = \frac{1}{2}$ for standing

Reprinted with permission from *Astron. J.*, vol. 72, no. 1, pp. 35–47, Feb. 1967.

TABLE I. Some material constants.

	Density ρ (g/cm^3)	Maximum stress S (kg/cm^2)	Elasticity E (10^3 kg/cm^2)	Maximum height S/ρ (km)	Gravitat. deflection ρ/E [cm/(100 m)^2]	Thermal expans. C_{th} (10^{-6}/°C)	Price p ($/kg)
Steel	7.8	1400	2100	1.79	0.37	12	1.5
Aluminum	2.7	910	700	3.37	0.38	24	4.5
Wood	0.7	130	120	1.86	0.58	3.5	0.5
Concrete	2.4	200	200	0.83	1.20	8	0.08

pillar or hanging rope). Integrating the compression from bottom to top yields

$$\Delta h = \gamma_2 h^2 \rho / E. \tag{2}$$

The deformations increase with the *square* of the size. For antennas of given wavelength and increasing size, this second limit is reached much earlier than the first one. Both limits are easily understood, since the weight goes with the third power of the size, but the strength only with the second power. Both limits are not ultimate but can be surpassed with certain tricks. As to the first limit, one could start at the bottom with a large cross section and taper it toward the top, but this structure cannot be tilted. Passing the second limit is discussed later.

Both limits depend on the combination of only three material constants: maximum stress, density, and elasticity. Table I gives four examples, together with the coefficient of linear thermal expansion, and with a rough estimate of price including erection. The largest structure can be made from aluminum, about two miles high. All four materials give the same order of magnitude for this maximum height, which could be increased only by tapering, and we understand why even mountains cannot be higher than a few miles. Steel, aluminum, and wood are about equal with respect to deflections under their own weight, while concrete is worse by a factor of 3. Thermal deflections are worse for aluminum and best for wood (but wood has too much deformation with humidity). Since the second limit will be reached first, there is no need to go to the more expensive aluminum, and we arrive at normal *steel* as the best material. The largest block of steel could be a mile high, but a block only 400 ft high is already compressed under its own weight by 3 mm.

B. Octahedron

What should be the over-all shape of a large structure for minimizing the deflections from its own weight and from wind, if the structure is to be held at a few points and to be turned in all directions? If a structure is cantilevered with length a and width b, a lateral force will give a deflection proportional to $(a/b)^2$. Since this is a rapid increase with decreasing width, and since any external force can become lateral for a turning structure, we get the requirement

Equal diameters in any direction. (3)

Small deviations from this rule do not matter much, but for greater deviations the deflections increase with the *square* of the diameter ratio.

The simplest structure we can think of, approaching requirement (3), which can be held at two points and turned from a third one, and which provides a flat surface through its center with a point normal to it for the focus, is the octahedron. Furthermore, its deflections are easily calculated. Thus, we adopt the octahedron as a near-to-ideal model for the basic structure of an antenna. Compared with usual designs, it avoids cantilevering and it includes the feed support as part of the basic structure.

If all members shown in Fig. 1(a) have equal cross section Q, the weight of the whole structure gives rise to the force $F = 2.88dQ\rho$ in one of the outer members, and from Eq. (1) we find the diameter of the largest possible octahedron from steel as

$$d_0 = (1/2.88)S/\rho = 620 \text{ m}. \tag{4}$$

The numerical value of γ_2 depends on where we measure the deflection and with respect to which point. With respect to the focal point at the top, we get the values shown in Fig. 1(b) for a diameter of 100 m. The rms deflection in the horizontal plane, as seen from the top, is 0.34 cm; but since we have neglected any lateral sagging of the members, we multiply by a safety factor of 1.5 in order to be on the safe side, and we obtain for the root mean square deflection

$$\text{rms } (\Delta h) = 0.51 \text{ cm } (d/100 \text{ m})^2. \tag{5}$$

We call D the diameter of the antenna surface, and we call $C = D/d$ the cantilevering factor. The latter should not be much larger than unity because of require-

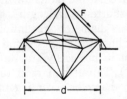

FIG. 1 (a) Octahedron with diagonals. (b) Deflections (in cm) in the horizontal plane of an octahedron of 100-m diameter, as seen from the top point of the octahedron.

ment (3) and should be chosen such that strong torques around the basic structure are avoided. The best value of C will depend on the actual design, and after some estimates we adopt

$$C = D/d = 1.25. \qquad (6)$$

Calling λ_{gr} the shortest possible wavelength with respect to gravitational deflections, we demand rms (Δh) $= \lambda_{gr}/16$. The gravitational limit of a telescope then is

$$\lambda_{gr} = 5.3 \text{ cm } (D/100 \text{ m})^2. \qquad (7)$$

C. Active and Passive Weight

Since the next point is a crucial one for large antennas, and since no suitable terminology seems to exist, I would like to introduce my own, calling:

Active weight $= W_{ac} =$ weight of those parts of the structure which oppose deflections to the same extent that they add weight. In our case, only the main chords of the octahedron members are active. If we have nothing but active weight, the gravitational deflections are completely independent of the cross section of the members, and thus, for a given diameter, the deflections of the structure are given by (5) and do not depend on the weight.

Passive weight $= W_{ps} =$ weight of everything else, such as braces and struts in the octahedron members, the antenna surface and the structure beneath it, and any part of the drive mechanism being fixed to the octahedron. Passive weight adds to the total weight without opposing the deflections it causes.

Total weight $= W_{ps} + W_{ac}$.

Passivity factor $= K = $ (total weight)/(active weight) $= 1 + W_{ps}/W_{ac}$.

With the help of this terminology we obtain a very quick estimate for the deflections of any given structure; because if any passive weight is present and is distributed about evenly, we simply have to multiply both Eqs. (5) and (7) with K. The connection between antenna diameter and shortest wavelength with respect to gravitation then is

$$\lambda = 5.3 \text{ cm } K(D/100 \text{ m})^2. \qquad (8)$$

Since passive weight always is present, at least in the antenna surface and its holding structure, K can approach unity only if the active weight approaches infinity. Practically, K will be between, say, 1.2 and 1.8. Table II gives some examples for the utmost limit, $K = 1$.

D. Thermal Deflections

If an outer member of the octahedron is ΔT degrees (centigrade) warmer than the rest of the structure, its length will increase by $C_{th} \Delta T d / \sqrt{2}$. A rough estimate gives for the rms deflection of the antenna surface, with

TABLE II. Shortest wavelength λ for an antenna of diameter D. λ_{gr} with respect to gravitational deflections if tilted by 90°, see (7); λ_{th} with respect to thermal deflections in sunshine, see (10).

D (m)	λ_{gr} (cm)	λ_{th} (cm)
25	0.33	0.60
50	1.32	1.2
75	2.98	1.8
100	5.3	2.4
150	11.9	3.6
200	21.2	4.8
300	47.7	7.2

C_{th} from Table I for steel,

$$\text{rms } (\Delta h) = 0.03 \text{ cm } \Delta T D / 100 \text{ m}. \qquad (9)$$

A large antenna will most probably stand *in the open*; ΔT then is given by sunlight and shadow but is independent of the antenna diameter. The thermal deflections then increase with D and will dominate in small antennas, while the gravitational deflections, increasing with D^2, dominate in large antennas.

ΔT is negligible during nights and cloudy days, and a good reflecting paint keeps it rather low even in sunshine. Measurements at Green Bank on sunny summer days gave a maximum difference of 8°C between a painted metal surface in full sunshine, and some structure in the shadow. Since this is the most extreme case (and since the surface itself should "float" on the structure), the average difference in the main structure will be considerably less; adopting $\Delta T = 5$°C should be safe enough. Calling $\lambda_{th} = \text{rms } (\Delta h)/16$ the shortest wavelength to be used, with respect to thermal deflections alone, we have from (9)

$$\lambda_{th} = 2.4 \text{ cm } (D/100 \text{ m}). \qquad (10)$$

Comparing (10) with (8), we find:

Gravitational deflections $\geqslant$ thermal deflections,

$$\text{if } DK \geqslant 45 \text{ m}. \qquad (11)$$

Around an antenna *in a radome*, a vertical temperature gradient builds up, and ΔT will increase with D. The thermal deflection then is proportional to D^2, just as the gravitational one, and the question of which one is larger depends only on the gradient, but not on the diameter. An estimate shows that both deflections are equal if the gradient is about 10°C/100 m. A cooling system must keep it below this limit.

In summary, we have three natural limits for the size of steerable antennas if the shortest wavelength is given. First, antennas below 45-m diameter are limited by thermal deflections according to (10). Second, the diameter of larger antennas is limited by gravitational deflections according to (8). Third, the largest tiltable structure has about 600-m diameter according to (4), independent of wavelength.

For antennas between 45- and 600-m diameter, the second limit applies as given in Table II. This limit is

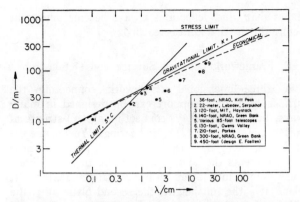

FIG. 2. Three natural limits for conventional, tiltable antennas, with nine actual examples for comparison. No. 1 is inside a dome. (Broken line see Sec. IIIA.)

not final. It can be pushed a little by adjusting the surface at an elevation angle of 45°, for example. But it cannot be surpassed considerably without applying special tricks (discussed in Sec. IV). As shown in Fig. 2 by a comparison with nine conventional telescopes, our theoretical limits (although derived on general grounds only) describe the actual designs quite well.

II. SOME FORMULAS FOR ESTIMATES

After having derived the limit of an antenna, we next want to know its weight. There are four items that can define the weight: first, gravitational deflections; second, wind deflections; third, survival conditions; fourth, the minimum stable structure. We need general formulas in order to learn which of these items is the defining one, and in order to estimate the resulting weight. One could use a model design which can be scaled up and down, but if we do not ask for more accuracy than, say, ±30%, these formulas can be derived on general grounds without a special model.

A. Weight of Members

Each member needs a certain minimum diameter in order to prevent buckling and sagging. There are two opposing criteria for the design of a member: the passivity factor should be close to unity in order to keep the deflections down, but the total weight should be low to keep the costs down. Starting with the first extreme, we could avoid any passive weight for the octahedron if we build each of its members from a single steel pipe of proper diameter and wall thickness. But then an octahedron of 400-ft diameter would weigh over 3000 tons, much more than we want to pay for, and much more than is needed against wind loading. This means we must split up the members into three or four main chords connected by braces; for very long members and small forces even a *multiple splitting* is necessary, where the main chords again are split up into three thinner chords. Going again to the extreme, we arrive at a certain minimum structure just for

stable self-support, no matter what its purpose. A rough estimate shows that if we do not care at all about deflections and wind forces, an octahedron of 400-ft diameter would have a minimum structure of about 130 tons (but would deform under its own weight by about 5 cm).

This calls for a careful compromise between the two opposing criteria. Since the same type of problem must arise in communication towers, we have taken the data quoted for 10 towers with a nonguyed length between 40 and 140 ft, and with longitudinal forces between 7 and 120 tons; in addition, some examples with double splitting were calculated for a length up to 300 ft and forces up to 1500 tons. The result can roughly be approximated by the formula (W is the weight in tons, F the force in tons, l the length in 100 m):

$$W = 0.06Fl + 8l^2 \text{ (for normal steel).} \quad (12)$$

Struts perpendicular to the main chords are passive, diagonals at 45° are half passive and half active. As an approximation we assume that the first term plus $\frac{1}{3}$ of the second term is active, while $\frac{2}{3}$ of the second term is passive weight.

B. Weight of Surface

For wavelengths above 5 cm, we do not need a closed surface and adopt a simple galvanized wire mesh (transmission 15–20 dB down). The weight then shows only a very slow variation with wavelength which we neglect. Some available examples of wire mesh show a weight of 0.3 lb/ft² which we should multiply by 2 or 3 to allow for some light aluminum frames. To be on the safe side, we adopt 1.2 lb/ft² = 5.8 kg/m² for the weight of the surface; this allows for a closed aluminum skin of 2.15 mm thickness if $\lambda < 5$ cm. A circular surface of diameter D then has the weight

$$W_{sf} = 46 \text{ tons } (D/100 \text{ m})^2. \quad (13)$$

C. Survival Condition, and Forces during Observation

We adopt a maximum wind speed of 110 mph, giving a pressure of 50 lb/ft² = 242 kg/m². In stow position (looking at zenith) the antenna projects perpendicular to the wind only a fraction of its surface, for which we adopt $\frac{1}{4}$. Some experiments with wire mesh show that the wind force varies roughly as $1/\lambda$ and, compared to a closed surface, is down by a factor 5.5 for $\lambda = 20$ cm, if the surface is perpendicular to the wind. Since this is not the case in stow position, and since we have neglected wind forces on the structure, we multiply by a safety factor of 2.9 and adopt for the maximal wind force 500 tons $(10 \text{ cm}/\lambda)(D/100 \text{ m})^2$ for $\lambda \geqslant 5$ cm, and 1000 tons $(D/100 \text{ m})^2$ for $\lambda \leqslant 5$ cm. This would allow for a solid layer of ice 6.4 cm thick, or for a snow load of 13 lb/ft², if the antenna is built for $\lambda = 10$ cm. This seems to be enough from our present experience at Green Bank, where large accumulations of snow always can be avoided by carefully tilting the dishes, and where ice can be melted off with a jet engine from the ground. In order to be a little more safe, we increase the maximum snow load to 20 lb/ft² for $\lambda = 10$ cm (or 40 lb/ft² for $\lambda \leqslant 5$ cm) and obtain as the survival force in stow position

$$F_{sv} = \begin{cases} 760 \text{ tons } (10 \text{ cm}/\lambda)(D/100 \text{ m})^2 \\ \qquad\qquad\qquad\qquad \text{for } \lambda \geqslant 5 \text{ cm,} \quad (14) \\ 1400 \text{ tons } (D/100 \text{ m})^2 \quad \text{for } \lambda \leqslant 5 \text{ cm.} \end{cases}$$

If built for this specification defined by snow loads, the antenna will withstand a wind velocity of 136 mph in stow position, and of 85 mph in observing positions.

If we observe only in winds up to 25 mph (going to stow position for higher winds), we derive, with a safety factor of 1.5 to include the forces on the structure, the maximum horizontal force in observing positions F_{oh} as

$$F_{oh} = \begin{cases} 75 \text{ tons } (10 \text{ cm}/\lambda)(D/100 \text{ m})^2 \\ \qquad\qquad\qquad\qquad \text{for } \lambda \geqslant 5 \text{ cm,} \quad (15) \\ 150 \text{ tons } (D/100 \text{ m})^2 \quad \text{for } \lambda \leqslant 5 \text{ cm.} \end{cases}$$

The maximum uplifting force F_{ou} (at elevation angle 45°) is about $2^{-\frac{3}{2}}$ of the maximum horizontal force, and we adopt

$$F_{ou} = \begin{cases} 25 \text{ tons } (10 \text{ cm}/\lambda)(D/100 \text{ m})^2 \text{ for } \lambda \geqslant 5 \text{ cm,} \\ 50 \text{ tons } (D/100 \text{ m})^2 \qquad\quad \text{for } \lambda \geqslant 5 \text{ cm.} \end{cases} \quad (16)$$

D. Wind Deflections

Let a structure of length l be built for survival under force F_{sv}. If it is used under force F_{oh}, its length will change by

$$\Delta l = (S/E)(F_{oh}/F_{sv})l. \quad (17)$$

Here we have another important combination of material constants S/E, which for normal steel is 6.7×10^{-4}. We demand $\Delta l = \lambda/16$ for the shortest wavelength, and we assume $l = D/\sqrt{2}$ as the average distance a force has to travel from the surface to a main bearing. If an antenna is built for survival, the shortest wavelength as defined by wind deflections then is

$$\lambda = 7.5 \text{ cm } (D/100 \text{ m}). \quad (18)$$

This limit can be surpassed by multiplying the active weight of the structure by a factor α_w, and the shortest wavelength is then

$$\lambda = (1/\alpha_w)7.5 \text{ cm } (D/100 \text{ m}). \quad (19)$$

A comparison of Eqs. (18) and (8) shows that wind deflections are more important for small antennas than for large ones; they can be neglected for diameters above about 100 m. A comparison of (18) and (10)

shows that wind deflections always become important before the thermal limit is reached.

E. Framework between Surface and Octahedron

A three-dimensional framework connecting the surface to a main support can be designed in many ways, but it will always connect many (N) structural surface points to few (2) main bearings. We imagine the surface as being the base of a quadratic pyramid whose top represents the holding point or bearing. We divide this pyramid into layers by horizontal planes; the first plane at $\frac{1}{2}$ the full height, the second plane at $\frac{1}{4}$, the third at $\frac{1}{8}$, and so on. At the top we have 1 structural point, in the first plane we assume 4 points, in the second plane 16 points, and so on. If plane j is the surface, we have $N = 4^j$ surface points. In the layer between plane i and $i+1$ we need $n_i = 2 \times 4^i$ main members ($i = 1, 2, \cdots, j-1$), and we assume a maximum force along each member of $F_{sv}\sqrt{2}/n_i$. From (12) we obtain the weight of each member, and we multiply by two to include horizontal members and bracing diagonals. We neglect the four members in the first layer as being part of the octahedron. In this way the weight of the whole framework turns out to be (for $\lambda \geqslant 5$ cm)

$$W_{fr} = 46 \text{ tons } (10 \text{ cm}/\lambda)(D/100 \text{ m})^3$$
$$+ 16.6 \text{ tons } (j-1)(D/100 \text{ m})^2. \quad (20)$$

How many (j) planes do we need? We include the structure of surface panels (if any) in our estimate, and we demand that the distance between neighboring surface points, $l_0 = D/2^j$, can be covered by a *straight* line without deviating at its center more than $\lambda/16$ from the ideal surface, which means $l_0^2 = \frac{1}{2}D\lambda$. Both equations for l_0 give together

$$j = 5.48\{1 + 0.303 \log[(10 \text{ cm}/\lambda)(D/100 \text{ m})]\}. \quad (21)$$

The logarithmic term can be neglected for a wide range of D and λ, and we obtain for the weight of framework plus panel structure, as defined by survival,

$$W_{fr} = \begin{cases} 46 \text{ tons } (10 \text{ cm}/\lambda)(D/100 \text{ m})^3 \\ \quad + 75 \text{ tons } (D/100 \text{ m})^2 \quad \text{for } \lambda \geqslant 5 \text{ cm,} \\ 92 \text{ tons } (D/100 \text{ m})^3 \\ \quad + 75 \text{ tons } (D/100 \text{ m})^2 \quad \text{for } \lambda \leqslant 5 \text{ cm.} \end{cases} \quad (22)$$

F. Octahedron and Total Weight for Survival

The forces acting on the octahedron are the survival force, the weights of framework and surface, and the octahedron's own weight; in order to be safe, we add all contributions directly. We multiply by $\sqrt{2}$ since the octahedron legs are tilted, we divide by four since the forces are distributed over four legs, and we obtain the force along one leg. Formula (12) then gives the weight of one leg, which we multiply by 12 (diagonals already being represented by the framework). We solve the resulting equation for the weight of the octahedron, W_{os}, as defined by survival conditions, and as a good approximation for the range $0 \leqslant D \leqslant 300$ m we obtain

$$W_{os} = \begin{cases} 170 \text{ tons } (10 \text{ cm}/\lambda)(D/100 \text{ m})^3 \\ \quad + 39 \text{ tons } (D/100 \text{ m})^2 \quad \text{for } \lambda \geqslant 5 \text{ cm,} \\ 340 \text{ tons } (D/100 \text{ m})^3 \\ \quad + 39 \text{ tons } (D/100 \text{ m})^2 \quad \text{for } \lambda \leqslant 5 \text{ cm.} \end{cases} \quad (23)$$

Including the surface from Eq. (13), the framework from (22) and the octahedron from (23), we obtain for the total weight of the moving structure, as defined by survival

$$W_{ms} = \begin{cases} 216 \text{ tons } (10 \text{ cm}/\lambda)(D/100 \text{ m})^3 \\ \quad + 160 \text{ tons } (D/100 \text{ m})^2 \quad \text{for } \lambda \geqslant 5 \text{ cm,} \\ 432 \text{ tons } (D/100 \text{ m})^3 \\ \quad + 160 \text{ tons } (D/100 \text{ m})^2 \quad \text{for } \lambda \leqslant 5 \text{ cm.} \end{cases} \quad (24)$$

The passive part includes the surface plus $\frac{2}{3}$ of the D^2 terms of octahedron and framework, the rest being active:

$$W_{ac} = \begin{cases} 216 \text{ tons } (10 \text{ cm}/\lambda)(D/100 \text{ m})^2 \\ \quad + 38 \text{ tons } (D/100 \text{ m})^2 \quad \text{for } \lambda \geqslant 5 \text{ cm,} \\ 432 \text{ tons } (D/100 \text{ m})^3 \\ \quad + 38 \text{ tons } (D/100 \text{ m})^2 \quad \text{for } \lambda \leqslant 5 \text{ cm,} \end{cases} \quad (25)$$

$$W_{ps} = 122 \text{ tons } (D/100 \text{ m})^2, \quad (26)$$

and the passivity factor is derived as

$$K = \begin{cases} 1 + \dfrac{\lambda/10 \text{ cm}}{1.77 D/100 \text{ m} + 0.31 \lambda/10 \text{ cm}} \quad \text{for } \lambda \geqslant 5 \text{ cm,} \\[2mm] 1 + \dfrac{1}{3.54 D/100 \text{ m} + 0.31} \quad \text{for } \lambda \leqslant 5 \text{ cm.} \end{cases} \quad (27)$$

G. Increased Rigidity

In Eq. (24) the weight of the antenna is defined entirely by survival conditions. This is all we need for longer wavelengths, but for shorter ones we must increase the active weight for reducing the deflections caused by gravitation and wind.

First, the *gravitational deflections.* Given D and λ, we check whether K obtained from (27) is smaller than or equal to K needed for (8); if it is, the gravitational deflections of the survival structure are small enough, and the total weight is given by (24). If not, we must multiply the active weight of (25) by a factor α_{gr} so

TABLE III. Characteristic points for antennas of diameter D. λ is the wavelength, W the weight of moving structure.

D (m)	λ_{gr} (cm)	λ_{th} (cm)	λ_{gw} (cm)	W_{gw} (tons)	λ_{ws} (cm)	W_{ws} (tons)	λ_{gs} (cm)	W_{gs} (tons)	W_{mn} (tons)
25	0.33	0.60	0.39	52	1.88	17			10
50	1.32	1.20	1.59	181	3.75	94			40
75	2.95	1.80	3.63	383	5.63	252			90
100	5.30	2.40	7.30	460	7.30	460	7.3	460	160
150	11.9	3.60					18.9	740	360
200	21.2	4.8					38.5	1090	640
300	47.7	7.2					105	2020	1440

limits:
gr = gravitational limit ($W \to \infty$)
th = thermal limit ($\Delta T = 5°C$)
mn = minimum structure ($\lambda \to \infty$)
boundaries:
gw = gravitational deflections and wind defl.
ws = wind deflections and survival
gs = gravitational deflection and survival

that by multiplying the denominator of (27) by α_{gr} we make K from (27) equal to K from (8), which gives

$$\alpha_{gr} = [\lambda_{gr}/(\lambda - \lambda_{gr})](W_{ps}/W_{ac}), \quad (28)$$

where λ_{gr} is the gravitational limit of (7), and W_{ps} and W_{ac} are given by (26) and (25). The total weight, as defined by gravitational deflections, then is

$$W_{mg} = [\lambda/(\lambda - \lambda_{gr})]122 \text{ tons } (D/100 \text{ m})^2. \quad (29)$$

Second, the *wind deflections.* Given D and λ, we check whether α_w obtained from (19) is smaller than or equal to one. If it is, the wind deflections of the survival structure are small enough. If not, we multiply the active weight of (25) by α_w, add the passive weight of (26), and obtain the total weight of the moving structure, as defined by wind deflections (W_{mw} measured in tons, D in 100 m, and λ in 10 cm):

$$W_{mw} = \begin{cases} 162D^4/\lambda^2 + 29D^3/\lambda + 122D^2 \quad \text{for } \lambda \geqslant 5 \text{ cm,} \\ 324D^4/\lambda + 29D^3/\lambda + 122D^2 \quad \text{for } \lambda \leqslant 5 \text{ cm.} \end{cases} \quad (30)$$

H. Regions of Different Weight Definitions

We have now derived the weight of an antenna, as defined by survival in Eq. (24), as defined by gravitational deflections in (29), and by wind deflections in (30). Each of these equations holds within a certain region of a D, λ diagram (see Fig. 3), and we now ask for the boundaries between the different regions.

The boundary between the wind deflection region and the survival region was already given in (18) as $\lambda = 7.5$ cm $(D/100$ m). The boundary between the survival region and the gravitational deflection region

TABLE IV. Comparison of estimated and actual weights.

Antenna	D	λ	Antenna weight From Fig. 4	Actual	Material
300-ft, Green Bank	92 m	15 cm	300 tons	450 tons	steel
210-ft, Parkes	64	6	200	270	steel
120-ft, Haystack	37	1	80	50	aluminum

69

we obtain by letting $\alpha_{gr}=1$ in (28) and solving the resulting quadratic equation for λ. The boundary between the wind deflection region and the gravitational deflection region we obtain by letting α_{gr} from (28) equal α_w from (19) and solving for λ. Finally, we observe that only the D^3 term in (24) is defined by survival, while the D^2 term is the weight of the minimum structure (for $\lambda \to \infty$); as the boundary between survival region and the minimum structure region we define (arbitrarily) the value of λ where the D^3 term is $\frac{2}{3}$ of the D^2 term.

The results are shown in Fig. 3, together with the three natural limits derived in Sec. I, and some values are given in Table III. For example, an antenna of 50-m diameter (164 ft) cannot be built for wavelengths below 1.32 cm (gravitational limit); within the region 1.32 cm $\leqslant \lambda \leqslant$ 1.59 cm, the weight of the antenna is defined by keeping the gravitational deflections down; within the region 1.59 cm $\leqslant \lambda \leqslant$ 3.75 cm, the weight is defined by fighting the wind deflections; from 3.75 cm to 10 cm, the weight is defined by survival conditions, and above 10 cm the minimum structure dominates.

III. ECONOMY

A. Economical Designs

In Fig. 4 we give the weight of an antenna as function of its diameter and of the shortest wavelength to be used. The weight of an actual antenna will depend on its special design, but any type of design will show natural limits and characteristic boundaries qualitatively similar to those of Fig. 4. The quantitative values of Fig. 4 belong to a "near-to-ideal" design and depend, first, on using an *octahedron* as the basic structure (more depth than usual, feed supports are part of main structure); second, on taking simple *wire mesh* with low wind resistance (instead of expanded metal); third, on the validity of Eq. (12) for long members, which actually means *many trials*, even for details, until the best solution is found. Under these conditions, Fig. 4 is supposed to be a realistic estimate.

That our theoretical weight estimates are not too optimistic is shown in Table IV by a comparison with some of the more economical existing antennas. As a counterexample, one could mention the 140-ft telescope at Green Bank, which should have a weight of 90 tons for the dish structure according to Fig. 4, and actually has 370 tons (of aluminum). By the way, no great weight difference is to be expected between antennas from steel and those from aluminum; both materials give the same gravitational deflections, and both give the same weight in the wind deformation region, since the ratio ρ/E is the same; see Table I. Regarding the difference in price, steel then is more economical.

B. Economical Wavelengths

As to the choice of D and λ, an economical antenna should be close to the boundary of the gravitational

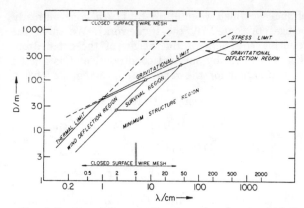

FIG. 3. Regions of diameter D and wavelength λ, in which the weight of the structure is defined by different conditions, and the three limits of Fig. 2.

deflection region, because of the steep increase of the weight in this region. This is especially obvious for $D \geqslant 100$ m, where the boundary is given by λ_{gs} in Table III. Below λ_{gs}, the weight is entirely defined by the rigidity needed for observation (governed by the quantity ρ/E), while the structure is stronger than needed for survival. Above λ_{gs}, the weight is entirely defined by survival (governed by p/S, with p the wind pressure on the wire mesh), while the structure is more rigid than needed for observation. For $\lambda \gg \lambda_{gs}$, the stable self-support of the structure becomes more important than its purpose (governed by Parkinson's law). It follows that λ_{gs} is the most economical wavelength, any other wavelength giving a waste of either strength or rigidity or both.

C. Radome or Not?

Observing in severe weather is not important in radio astronomy. The 300-ft telescope at Green Bank is not used above 25 mph wind nor in heavy snowfall,

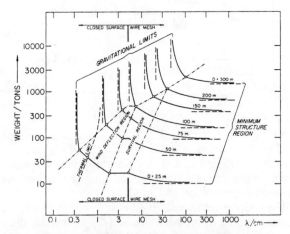

FIG. 4. Weight of dish structure (surface, framework, octahedron) as function of diameter D and wavelength λ. Economical antennas should be close to the uppermost boundary. - - - - natural limits of Fig. 2. --·--·--·-- boundaries between regions of Fig. 3.

which gives a completely negligible loss of 43 h per year. The only task of a radome is to shield the telescope against wind forces. A radome is thus of advantage in the wind deflection region of Fig. 4, since the wind deflection of the radome does not matter. But it is of only little advantage in the survival region, since counteracting the survival forces just needs a given amount of construction material, no matter whether we put it into the telescope or into the radome. A radome gives obviously no help in the gravitational deflection region, nor in the minimum structure region. We thus are left with the wind deflection region (and only with its center part, away from the limits, regarding the additional costs of the radome), which means

No radomes for conventional telescopes

above 50 m diameter.   (31)

But if a large antenna passes the gravitational limit, then a radome might be considered again; see Sec. IVA.

D. Foundations

Most economical for a large antenna seems to be an altazimuth mount, with two towers standing on wheels on a circular track on the ground. The tower legs should be wide astride in order to decrease the uplifting forces at the ground, the best basic shape being a regular tetrahedron (slightly modified for more clearance for the rotating dish). In order to have only one circular track, we put one leg of each tower at the center of the circle on a strong pintle bearing. The track, then, has no lateral force, which is a great advantage.

Foundations are very expensive if made for a special purpose. A single steel track embedded in concrete, taking 300 tons downward and 80 tons laterally and upward, would cost $700 000/mile. Thus, we recommend using standard railroad equipment, with normal roadbed, ties and rails; this costs $80 000/mile; the maximum load is 30 tons per axle or 450 tons per 100 ft. An estimate shows for $500 \leqslant D \leqslant 600$ ft, that two large steel gondolas per tower leg are sufficient for load and counterweight during observation, if filled with rock and gravel. A small piece of fortified foundation is needed for the stow position.

The deviation of a normal railroad track is about 0.5 in. after one year of normal use; comparing the speeds of trains and telescopes, we may safely assume half of this value. With the cantilevering from (6) and tetrahedral towers, one edge of the dish will deviate by ± 6.3 mm (relative to the opposite edge), independent of diameter. If we demand a pointing accuracy of $\pm \frac{1}{16}$ of a beamwidth, the deviations of the rails after one year of use will allow observation down to 8.2 cm:

Normal railroad for wavelengths above 8 cm.   (32)

Limitation (32) holds if positions are read off at the drive system, and/or if deviations of the rails deform the surface. But if we have, for example, the dish held

TABLE V. Price estimate.

Item	Amount	Price (incl. erection)	Million dollars
Dish structure	750 tons	2000 $/ton	1.50
Surface	200 000 ft²	1.5 $/ft²	.30
Towers	600 tons	1500 $/ton	.90
Bearings+pintle b.			.10
Drives			.20
Controls			.10
Miscellaneous			.05
Railroad	0.9 mile	100 000 $/mile	.09
Gondolas	10	35 000 $	.35
Stow foundation			.15
Pintle foundation			.10
		total	3.84

with two bearings on top of two towers which sit on wheels on a circular horizontal track, then the deviations of the rails cannot deform the surface, they only shift the position; and if the pointing direction of the antenna is measured, say, at its apex by some optical device looking at several light beacons on the ground, then the pointing accuracy is completely independent of the shape of the rails:

Normal railroad for *any* wavelength, if pointing is measured independently, and if deviations of the rails do not deform the antenna surface.   (32a)

E. Price Estimate (for $\lambda = 20$ cm and $D = 500$ ft)

The following estimate is certainly very approximate but still tries to be without bias. With respect to hydrogen-line observations, we choose $\lambda_{gs} = 20$ cm and obtain from Table III a diameter of 500 ft. For the azimuth drive, we assume friction wheels on the tracks. For the elevation drive, we definitely want to avoid any additional passive weight at the dish, and we recommend guiding the antifocus point of the octahedron along a curved leg ($\frac{1}{4}$ circle) of a third tower. All three towers are connected close above ground by long horizontal members. After calculating wind, snow, and dead loads for legs and connections, we find a total weight of about 600 tons for the azimuthal structure.

The result of Table V might seem low. But since many safety factors have already been included in our estimates, we regard 4 million dollars as a realistic figure for the total price, provided that the design really uses optimization in every detail, and that economy in fact is the leading principle.

IV. PASSING THE GRAVITATIONAL LIMIT

A. In General

Equation (7) gives the limit of a tiltable telescope, and is derived from Eq. (2), giving the compression of a structure under its own weight. Details of the design do not matter much; any structure must be compressed under its own weight, and by changing amounts if tilted. For a diameter of 150 m, the largest deformation at the rim is about 1.5 cm, no matter whether we hold the structure at two bearings, or support it at many points from the bottom, and even a floating sphere deforms by the same amount. If we want to pass this limit, we have four possibilities: (1) Avoiding the deformations by not moving in elevation angle; (2) Fighting the deformations with strong motors in the structure; (3) Canceling the deformations with levers and counterweights; (4) Allowing deformations which do not hurt the performance.

The first possibility is verified by the Arecibo telescope in Puerto Rico, where a spherical reflector is fixed into a round valley; observation is limited to within 20° from the zenith. Some other ways are discussed in the next section. The second possibility was tried at Sugar Grove; we think it will always be complicated and expensive. The third possibility is used in large optical telescopes, for example in the 200-in. at Mt. Palomar; but a rigorous application to a large radio telescope would either add too much weight, or would make the structure too soft against wind deformations. Our main emphasis will be on the fourth possibility, with special regard to those observations where full sky coverage is needed, which a fixed-elevation telescope cannot give.

Suppose we pass the gravitational limit in some way or other. The next natural limit then is the thermal limit from Eq. (10), assuming $\Delta T = 5°C$ in full sunshine with a good protective paint. But we might go one step further and introduce a second thermal limit, without sunshine, say, $\Delta T = 2°C$. In Fig. 5 we show the weight of a telescope as function of diameter and wavelength, if the gravitational deflections are omitted in some way. Since we have used for Fig. 5 the same formulas as before, the values given in Fig. 5 will hold for structures not too different from the previous one, which means mainly that the *height* of the structure is comparable to its *diameter*. The most economical wavelength now is given by the boundary between the wind deflection region and the survival region; especially for antennas above 100 m diameter, the weight increases very steeply to the left of this boundary. Here, a radome might be reconsidered, or placing the antenna in a valley shielded by mountains against wind, or limiting the observation to lower wind velocities, or designing a structure which sits flatter to the ground.

B. Fixed-Elevation Transit Telescopes

Although most radio astronomers would prefer full steerability about two axes, they would be sufficiently satisfied with a transit instrument if it gives them one or two hours observing time per transit of a source; this means full steerability about one axis, and a very limited steerability about a second one. Since gravitational deflections are our main problem, and since movement in azimuth does not change these deflections,

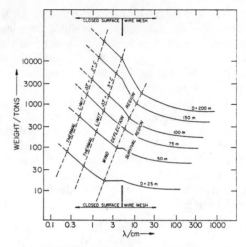

FIG. 5. Weight of dish structure as in Fig. 4, if gravitational deflections are omitted.

we arrive at a telescope turning 360° in azimuth but only about 10° in elevation. The beam then describes a circle around the zenith, and a radio source of proper declination will give two transits per day through this circle. The choice of the best elevation angle is a compromise between two opposing demands: we want a large sky coverage (low elevation), but we should not observe too close to the horizon (high elevation). The best way might be to build two such telescopes with elevation angle 45°, one situated at +45° and one at −45° geographical latitude. The mirror could either be a paraboloid which turns by 10° in elevation, or it could be a fixed-elevation sphere, where a small secondary connecting mirror and feed move 10° in elevation. The azimuth movement in any case would be on circular tracks. The feed could be either fixed to the dish with feed supports, or it could be on a separate, nonmoving tower. Several designs have been worked out by the LFST group.

The least expensive telescope of this kind would be the fixed-elevation sphere with the feed tower fixed to the dish. The most flexible telescope would be the one from the previous paragraphs, with full steerability in elevation; we adjust the surface for a given elevation angle and use it only within ±5° of this elevation; whenever desired later on, the surface can be adjusted to any other elevation angle. The weight of this telescope is given by Fig. 5 as function of diameter and wavelength, while the fixed-elevation sphere might be lower by 30 or 40% (height somewhat lower, no elevation drive).

Large mirror flat on the ground. If we want a large telescope of, say, 200-m diameter, and want to use it for as short a wavelength as possible, we see from Fig. 5 that the weight increases very rapidly with decreasing wavelength for $\lambda < 15$ cm. This increase is due to the D^4/λ^2 term in Eq. (30) where the weight is defined by wind deflections. The increase is so steep that we should

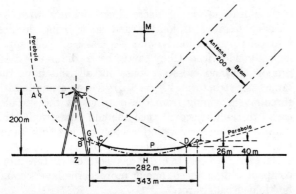

FIG. 6. Fixed-elevation transit telescope with large surface flat on the ground. The parabolic mirror CPD moves on wheels in a flat cylindrical trough GHI by 10° around M. The trough moves on horizontal circular tracks by 360° around Z. In stow position, the rim of the mirror is only 26 m above ground.

look for a better solution, where the antenna sits flat on the ground and does not pick up so much wind force.

A possible solution is sketched in Fig. 6, using a parabola at 45° elevation with its focus at F. In a usual design, we would use part AB, where a large surface is high above ground. Now, we use part CD which is 40% larger but is never more than 40 m above ground. This parabolic mirror P (282 m long and 200 m wide) is mounted on wheels on a flat cylindrical trough GHI (343 m long) with its center line through M. Moving the mirror in this trough around axis M gives the 10° of movement in elevation. The trough sits on wheels on horizontal circular tracks around center point Z, giving the 360° of movement in azimuth. The feed is mounted movable along a track T about 50 m long, which is 10° of a circle around M. The track can be turned by 360° around a vertical axis and is mounted on a tower 200 m high.

Having the feed at the primary focus would give the following disadvantage. In order to illuminate the antenna beam symmetrically, the feed must have an asymmetric pattern. This could be done, but then the feed cannot be rotated for polarization measurements. This problem is resolved by using a small, tilted secondary mirror of Gregorian type. Feed and secondary mirror then are moved together along track T. Some examples of secondary mirrors have been calculated; it is possible to achieve a symmetrically illuminated aperture with a symmetrically illuminating feed.

C. Homologous Deformation, in General

Although radio astronomers would be satisfied with a transit instrument, they still would prefer full steerability if it can be achieved (within reasonable costs). There is one observing technique, lunar occultations, where full steerability is crucial in order to obtain the brightness distribution across a source in many directions and for very accurate position determinations. It seems worthwhile to design a telescope with special re-

gard to lunar occultations; half of the time (moon below horizon) it can be used for other observations. But can we pass the gravitational limit, with 90° tilt in elevation, and without strong servo motors in the structure?

The laws of physics tell us that a structure, under the influence of gravitation, deforms into a state of minimum energy; the center of gravity must move down. The material constants ρ and E tell us the amount by which it must move down. But there is no law of nature telling us that a parabolic surface must deform into something different from a parabolic surface. We thus look for a structure which deforms down whatever it must, but still gives a perfect paraboloid of revolution for any elevation. The focal length might change a bit, but this can be taken care of by servo-adjusting the feed according to elevation. A deformation of this kind, deforming a surface of given type into another surface of same type, we call "homologous deformation."

Equal softness. As a first approach, let us consider, in Fig. 7(a) the cross section through a large telescope of conventional design. The heavy vertical lines above the main bearings represent the main frame which usually is a heavy, flat square box. Figure 7(b) shows the deformation of the surface; the two "hard" points above the bearings stick out considerably, while the "soft" points at the rim hang down. This disadvantage of having soft and hard surface points can be avoided by a structure shown in Fig 7(c), where the way from each surface point to the next bearing is of the same length, giving equal softness to all surface points. A structure of this type is not an exact solution of our problem, but it might be a fairly good first approximation.

Exact solutions One can easily show that the problem of homologous deformations is solvable in a mathematical sense. First, we assume an exact paraboloid of revolution for the surface in the absence of gravity. For small gravitational deformations, then, homology holds in any elevation angle if it holds in two (looking at zenith and at horizon, for example), since small deformations can be superimposed. Second, we let the

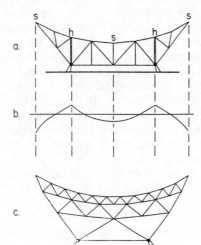

FIG. 7. Equal softness. (a) Conventional design, with hard (*h*) and soft (*s*) surface points. (b) Deformation of this telescope, looking at zenith. (c) Structure, where all surface points have equal softness.

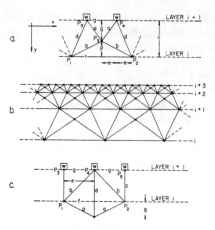

FIG. 8. Two types of cells, where layer $i+1$ is parallel to layer i for any elevation. (a) Most simple type, general solution. (b) Particular solution $(c=2e=3q)$ of same type, three layers. (c) Pressure-stable cell, which keeps height constant if P_1P_2 is compressed.

surface be defined by a finite but arbitrary number N of structural points (joints of members), where N must be chosen so large that any gravitational deformations of surface pannels in between neighboring points can be neglected. Homology then demands that all N surface points must be on an exact paraboloid of revolution (being defined by 6 points) in zenith as well as in horizon position. Homology, thus, is a set of $2(N-6)$ conditions to be fulfilled. Third, a structure of p points needs a minimum of $3(p-2)$ members just for stability. Even if our structure had no additional members at all, it still must have at least $3(N-2)$ members, and each member has one structural degree of freedom, its cross section. Since $3(N-2)-2(N-6)$ $=N+6>0$, we have more degrees of freedom than conditions, which means the problem is solvable in a mathematical sense: For any stable structure, there is a family of homology solutions with at least $N+6$ free parameters.

Of course, a mathematical solution is not always a useful one. In order to make it a physical solution, each cross section must be within $0<Q<\infty$. Also, what we might call a "practical solution" must fulfill some additional conditions, for example: the total weight should not be much greater than needed for a normal telescope; the structure should not be too soft against wind deflections; it should be not critical against small deviations of cross sections from the theoretical value; the bearings should have clearance for rotation, and so on. But, on the other hand, having a large number of free parameters to play with should give us enough freedom for practical solutions, too.

Layers and cells. Even if there are solutions, how do we get one? We certainly cannot play at random with $N+6$ free parameters (not even on a fast computer) until we hit something useful. We need some logical principle to guide us during the design. For this purpose we suggest dividing the space between bearings and surface into layers of decreasing thickness with increasing number of joints, each layer being divided into cells by the joints, and we make all cells topologically identical (they may have different sizes, proportions and cross sections, but all cells have the same basic structure). Figure 8(b) gives an example with three layers. Going from the bearings to the surface, each three-dimensional cell should quadruple the number of joints such that each layer has four times more cells than the previous one, until we arrive at the surface with N cells. In order to keep the structure more compact and balanced, one or more of the layers could be "folded backwards."

The basic idea of this arrangement is to let the single cell fulfill a certain set of conditions such that the structure as a whole fulfills the same conditions. One might formulate, for example: provided that layer i is on a surface of given type for any elevation angle, layer $i+1$ then must be on a surface of the same type for any elevation, too. Solving the problem for one cell then would give a solution for the whole telescope.

D. Homology, Solutions in Two Dimensions

In order to learn whether the suggested procedure works and whether there are reasonable solutions, we try it for two dimensions; and for simplicity we replace parabolas by straight lines. The simplest cell we could find is shown in Fig. 8(a), where load w replaces the weight of the structure in all higher layers. We let cross sections $Q_a=Q_b$ and call $U=Q_a/Q_d$. Given w, ρ, Q_d, and e, the cell has three degrees of freedom: c, q, and U.

For solving our homology problem, we impose two conditions:

(1) If gravity has direction y, point P_3 shall
not move in x direction. (33)

(2) If gravity has direction x, point P_3 shall
not move in y direction. (34)

With three degrees of freedom and two conditions, the cell has one free parameter left. But we might go one step further and use it up for fulfilling a third condition:

(3) For any direction of gravity, point P_3 shall
move the same amount. (35)

Under conditions (33) and (34), P_3P_4 is parallel to P_1P_2 for any elevation angle, and P_3P_4 keeps constant length if P_1P_2 does. (Members f and g are redundant but should be included.) Under the addition of condition (35), the cell can easily be applied to a curved line, too. If layer i keeps its shape under any elevation, layer $i+1$ keeps its shape, too, and the gravitational deformation just gives it a *parallel shift*. Applied to a parabola, even the focal length would stay constant for varying elevation.

We omit the tedious calculations and just quote the results. The cell of Fig. 8(a) has two types of solutions, a general solution and a particular one. First, we give the general solution. Calling

$$\eta = \tfrac{1}{4}q(2e-c)/(ep-cq), \tag{36}$$

conditions (33) and (34) take the form

$$U = (c/qd^3)(a^3+\eta b^3) \tag{37}$$

and

$$\frac{2w}{d\rho Q_d} = U\left[\frac{(c+q)(a+b)}{4cd(\eta-1)}\frac{a}{d}\right]-1, \tag{38}$$

while condition (35) leads to

$$e^2 = qc. \tag{39}$$

Imposing all three conditions leaves no freedom, but the solution depends on $w/\rho Qd$. Calling W the weight of the cell, and $\Omega = 2w/W$ the weight ratio of all upper cells to this one, we find the full range $0 \leqslant \Omega \leqslant \infty$ is already covered if c/e varies over the narrow range $1.62 \leqslant c/e \leqslant 1.66$, and if $q/c \approx 0.371$. For various Ω, the solution for U is as follows:

Ω	0	0.5	1.0	2.0	∞	
U	8.53	5.90	5.48	5.20	4.91	(40)

We see that not only *are* there solutions, they are physical solutions $(0 < U < \infty)$; but we hesitate to call them practical because the values obtained for U are so large that our homology principle would introduce too much extra weight. Even by dropping the third condition, we could not find any geometrically reasonable solution with U below 3.

Second, there is a particular solution for

$$ep = cq \quad \text{and} \quad 2e = c, \tag{41}$$

shown in Fig. 8(b) which does not fulfill the third condition, but where the geometry does not depend on the weight ratio. For the particular solution we obtain

$$\Omega = 0.167\frac{-U^2+1.68U+7.75}{U^2-3.57U-2.60}. \tag{42}$$

The full range $0 \leqslant \Omega \leqslant \infty$ is now covered if U varies within

$$3.75 \leqslant U \leqslant 4.19, \tag{43}$$

but these values, again, are too high. Two things could be considered: by introducing an additional degree of freedom, $Q_a \neq Q_b$, we might bring the cross sections down, also a structure with smaller U might be a good enough approximation. But there is the following additional difficulty.

Even if we had a good solution for the single cell, it would be a solution for the whole structure only if the structure were indefinitely long. In a structure of finite size we get *boundary distortions*. They result mainly in pressure (or tension) along line g, and partly in torques such that the forces at P_3 and P_4 have different

directions. It seems possible to counteract or smooth these distortions by varying all cross sections within a layer from the center to the end of the structure, especially since the layers must keep their shape but can deviate from being parallel to each other. An investigation of this type would go beyond the scope of this paper, however. We therefore look for another type of cell where at least the influence of pressure is removed.

The simplest *pressure-stable cell* we could find is given in Fig. 8(c). Redundant members like f and g can be included in three dimensions (as diagonals) although not in two. Under various tilts, P_3P_4 and P_4P_5 will not keep their lengths constant, but this does not matter since the next layer is pressure stable, too. Conditions (33) and (34) are fulfilled automatically by reasons of symmetry for P_3, P_4, and P_5. We add condition (35) for P_4 in order to guarantee application to a curved structure. Given w, ρ, Q_b, and e, the cell has five degrees of freedom: c, q, Q_a, Q_c, and Q_d. Homology is reached by imposing three conditions:

(1) If P_1P_2 is compressed, P_4 shall
> not move in y direction. (44)

(2) If gravity has direction y, point P_3 shall
> move as much as point P_4. (45)

(3) For any direction of gravity, point P_4 shall
> move the same amount. (46)

This leaves two free parameters, and we use both for demanding:

(4) The additional weight introduced by
> homology shall be a minimum. (47)

Condition (44) is fulfilled if

$$(cdeq/Q_d)+(a^3c/Q_a) = b^3q/Q_b. \tag{48}$$

The following treatment is restricted to $\Omega \gg 1$ for simplicity. Condition (45) then gives

$$2c^3/Q_c = b^3/Q_b. \tag{49}$$

Conditions (44) and (46) give

$$2e^3 = c(e^2+bc) \tag{50}$$

which is an equation of sixth order for c/e and has only one physical solution:

$$c/e = 0.903. \tag{51}$$

We let $U_a = Q_a/Q_b$, $U_c = Q_c/Q_b$, $U_d = Q_d/Q_b$ and $\varphi = q/e$, then conditions (44), (45), and (46) lead to

$$U_c = 0.603 \tag{52}$$

and

$$U_a = (1/\varphi)(1+\varphi^2)^{\frac{3}{2}}[U_d/(2.70U_d-\varphi-0.903)], \tag{53}$$

where the two free parameters are φ and U_d.

In order to fulfill condition (47), we call

$$W = \rho(2Q_a a + 2Q_b b + Q_c c + Q_d d)$$

the total weight of the cell, and $W_0 = 2\rho Q_b(b+c+e)$ the weight if we do not ask for homology, letting $q=0$ and letting all Q be equal to Q_b. Now, we take the ratio $\tau = W/W_0$, and we let $\partial \tau / \partial \varphi = 0$ and $\partial \tau / \partial U_d = 0$. The solutions are $\varphi = 0.441$ and $U_d = 1.44$. With these values, we obtain $\tau = 1.37$, which is a very fortunate result. Compared to a normal structure, our homology condition is fulfilled with only 37% increase of the total weight. Altogether, the cell is now completely defined and we have:

$$c/e = 0.903, \quad Q_a/Q_b = 1.67,$$
$$q/c = 0.489, \quad Q_c/Q_b = 0.603, \tag{54}$$
$$Q_d/Q_b = 1.44.$$

With the values (54), the cell of Fig. 8(c) seems to be a very good, practical solution. The geometrical shape is convenient, and a weight increase of only 37% is a very low price to be paid for homologous deformation. This cell is an exact solution for an indefinite structure; and for a finite structure, the main part of the boundary distortion is removed. The remaining distortion by torques should be removable, too, by varying the cross sections toward the boundary. The application to a three-dimensional cell is easy; for example, let three lines f be the three sides of an equilateral triangle, and let member d go perpendicular through the center of the triangle. This cell will double the number of joints from one layer to the next. If we want to quadruple the number of joints, we let three members d be perpendicular on the center of each side of the triangle. Instead of a triangular structure, we could also take a quadratic structure with the same two possibilities.

In summary, we have proved that the problem of homologous deformations has exact mathematical solutions, and we have shown that also practical solutions do exist. With this finding, the present investigations has reached its goal.

Application to telescopes is possible in two ways. First, one starts with an "equal-softness structure" (Sec. IV C), and then improves it gradually by trial and error until the deviations from homology become tolerable. Good results can be achieved this way see Rohlfs (1966). Second, a mathematical method was developed for obtaining exact homology solutions for a telescope as a whole, within one or two hours on a fast computer (von Hoerner 1965, 1967). This method has been programmed and successfully applied to several structures [Biswas, Jennings, and von Hoerner (in preparation)].

REFERENCES

Rohlfs, K. 1966, *Sterne und Weltraum* **5**, 104.
von Hoerner, S. 1965, LFST-Report No. 4.
——. 1967, *J. Struct. Div.* (to be published).

FEED SUPPORT BLOCKAGE LOSS IN PARABOLIC ANTENNAS[*]

JOHN RUZE, M.I.T. Lincoln Laboratory
Lexington, Massachusetts

The gain reduction of a parabolic antenna due to the feed supports has received little theoretical attention and no systematic experimental investigation insofar as the literature discloses.[1-4] The method generally used is to find the shadow area by graphical projection and then subtract this area, weighted by the aperture illumination, from the unobstructed dish to obtain the reduction in axial field. Alternately this shadow area may be found by photographic means. Here a small shielded light source is placed at the focal point and a shadowgraph taken (Fig. 1).

It should be pointed out that there is considerable question as to the theoretical justification of this procedure when it is applied to the divergent spherical wave emanating from the focal region. The shadow area obtained is the optical shadow. The feed supports are normally of the order of a wavelength so that the scattered energy is only moderately directive. The *radio frequency* shadow is therefore much wider and may at the longer wavelengths encompass, with some transparency, the entire reflector. Nevertheless, the region of the optical shadow is a stationary phase point which detracts directly from the axial field.

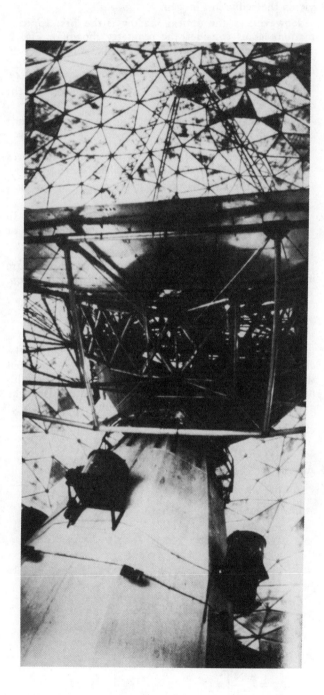

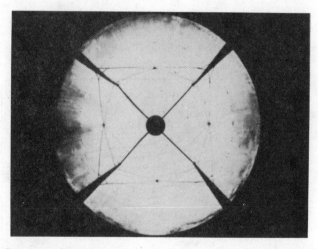

Fig. 1 — Shadowgraph of typical feed support blockage with cassegrain sub-reflector.

[*] *Based on CAMROC Technical Memo No. 19 dated 10 Feb. 1967, Cambridge Radio Observatory Committee.*

The proper way to handle this program would be to find the field scattered by the feed supports in phase and magnitude; find the currents induced on the reflector, and then the axial field by integrating over the surface and also the field scattered in the axial direction by the feed supports themselves. However, since the supports extend from the immediate vicinity of the feed (near field) to the reflector surface this calculation is extremely difficult.

It should be noted that the calculation of loss due to a Cassegrain subreflector or a metal space frame radome is much more direct. Here the incident field is known and there is little reflector interaction. Calculation of the forward scattered field is fairly exact and yields the reduction in gain.

However, as the optical shadow is the first approximation, let us consider the geometry shown in Fig. 2. All quantities are normalized to a unit antenna radius. There are "N" struts of width "w" and of depth "d." The subreflector or feed area has a diameter "2M." The struts extend from "M_o" at an angle "α" to the dish surface.

The optical shadow projected on the aperture plane consists of three parts (Fig. 2b), namely,

1) A_1 — the shadow of the subreflector or feed area caused by the outgoing plane wave.

2) A_2 — the shadow of the struts caused by the outgoing plane wave.

3) A_3 — the shadow of the struts caused by the diverging spherical wave from the feed or effective focal point.

These areas and that of the unobstructed aperture must be weighted by the aperture illumination function. They have the general form

$$A = \iint f(r)\, r\, dr\, d\varphi$$

where the integration is over the aperture to obtain the effective aperture or over the mathematically projected shadow to obtain the blockage. If we assume an illumination taper of the form $(1-ar^2)$ the integral can be evaluated in closed form with the result:

effective aperture

$$A_o = \pi[1-a/2]$$

subreflector blockage

$$A_1 = \pi M^2$$

struts out-going

$$A_2 = wN\left[(r_o-M) - \frac{a}{3}(r_o^3-M^3)\right]$$

struts diverging

$$A_3 = \frac{wN}{M_o} \int_{r_o}^{1} \frac{2f \sin(\theta-\alpha)\, f(r)\, dr}{(1+\cos\theta)\cos\alpha}$$

or

$$A_3 = \frac{wN}{2M_o}\left[\left\{(1-r_o^2) - \frac{a}{2}(1-r_o^4)\right\}\right.$$

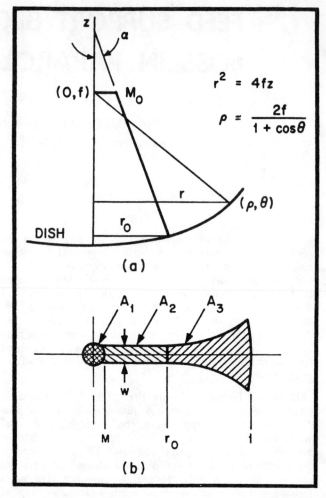

Fig. 2 — a) Geometry of feed supports; b) Projection of aperture plane.

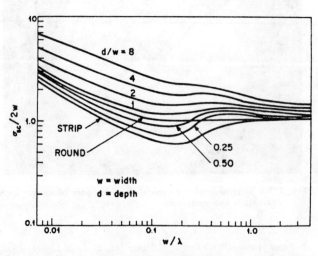

Fig. 3 — Total average scattering cross-section of rectangular cylinders.

78

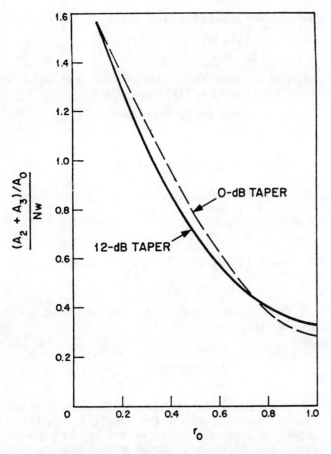

Fig. 4 — Feed support blockage vs. attachment point illumination taper 0 and 12 dB. $f/D = 0.4$; $M = M_0 = 0.1$.

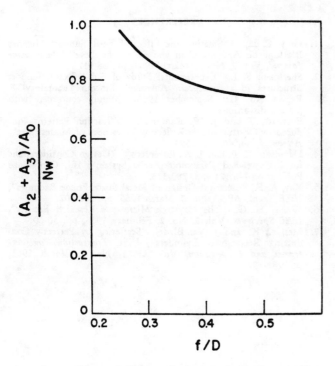

Fig. 5 — Effect of f/D on feed support blockage uniform illumination; $r_0 = 0.5$; $M = M_0 = 1$.

$$- 2f \tan \alpha \left\{ (1-r_0) - \frac{a}{3} (1-r_0{}^3) \right\}$$

$$+ \frac{\tan \alpha}{2f} \left\{ \frac{(1-r_0{}^3)}{3} - \frac{a}{5} (1-r_0{}^5) \right\} \Big]$$

The antenna blockage efficiency factor is then given by

$$\frac{G}{G_0} = \left[1 - \frac{A_1}{A_0} - \overline{ICR}\frac{(A_2 + A_3)}{A_0} \right]^2$$

where $\overline{ICR}$ is the average induced current ratio[5] or the average scattering efficiency of the members. The average is taken of the two polarizations and implies an angularly symmetric feed support geometry. It is unity when the members are large in terms of wavelength and is larger when they are small. It depends on member width and depth and is given in Fig. 3.[7]

NUMERICAL CALCULATIONS

From the above equations a number of useful curves can be obtained.

1) Effect of Angle of Feed Supports — Fig. 4

Here we plot the effective feed support blockage as a function of the point of attachment at the dish surface. We note the marked advantage of placing the feed supports at the reflector rim. Physically this is due to the fact that the large fan shaped area is eliminated.

2) Effect of Aperture Illumination — Fig. 4

We note that this is small as the support shadow forms closely an angular segment of the aperture.

3) Effect of Focal Length (f/D) — Fig. 5

This effect is also small and indicates that long focal length dishes are to be preferred. It is due to the magnification of the fan-shaped region at short focal lengths. Further, when the supports are at the rim there is no effect as there is no fan-shaped region.

4) Effect of Subreflector Size (M) — Fig. 6

For supports close to the focal region the effect is marked and indicates that large subreflector regions are to be preferred. This is because, for small subreflectors, the supports are close to the radiating source and therefore they scatter strongly or create large fan-type shadows. The effect becomes small for supports at the reflector rim. In this connection it must be remembered that although large sub-reflectors are desirable as far as support blockage is concerned, they are undesirable from the standpoint of subreflector blockage.

NUMERICAL EXAMPLE

The 120' diameter HAYSTACK[6] antenna has four feed supports 5" x 21" attached to the reflector at a radius of 30 feet. The focal length is 48 feet and the subreflector is 9.33' in diameter. In addition there are 535 feet of 0.5" steel rope concentrated in the center region, bracing the feed supports.

For the feed supports we use the equations with a 12 dB illumination taper (a = 0.75) and M = 0.078, hence

$$\frac{A_2 + A_3}{A_0} = 0.815 \frac{4 \times 5}{60 \times 12} = 0.0226$$

The induced current ratio (from Fig. 3) is:

	$\overline{ICR}$
@ 7750 MHz	1.3
@ 1650 MHz	1.55

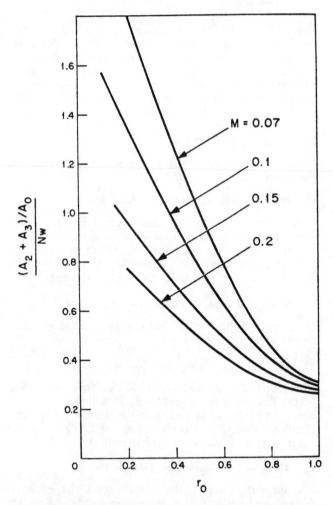

Fig. 6 — Effect of sub-reflector size (M) on feed support blockage. Uniform illumination f/D = 0.4; M = M$_o$.

The subreflector has an effective blockage of

$$\frac{A_1}{A_0} = \frac{1}{1-a/2}\left[\frac{9.33}{120}\right]^2 = 0.0097$$

The wires are inclined to the aperture at about 45 degrees so we multiply by 0.707 to obtain their projected area; they cast a double shadow so there is an additional factor of two; as they are in the center of the

reflector we include the taper factor

$$\left[\frac{535 \times 0.5}{60 \times 60 \times 12}\right] \times \frac{0.707 \times 2}{\pi (1-a/2)} = 0.0045$$

As they are round their induced current ratio is 1.15 at 1650 MHz and 1.0 at 7750 MHz (Fig. 3).

Finally we have for the loss of gain
@ 1650 MHz

$$\frac{G}{G_0} = [1 - 0.0226(1.55) - 0.0097 - 0.0045(1.15)]^2$$

$$= 0.903 (0.44 \text{ dB}).$$

@ 7750 MHz

$$\frac{G}{G_0} = [1 - 0.0226(1.3) - 0.0097 - 0.0045]^2$$

$$= 0.915 (0.39 \text{ dB}).$$

It is of interest to note that the total blockage is divided approximately as:

Struts	69%
Subreflector	21%
Wires	10% .

Similar calculations for uniform illumination of the reflector yields essentially the same gain loss but a slightly different distribution of the various contributions. Admittedly the proposed method lacks rigorous theoretical foundation and no experimental confirmation. However, it should prove useful in evaluating on a relative basis various structural designs.

REFERENCES

1. Gray, C. L., "Estimating the Effect of Feed Support Member Blocking on Antenna Gain and Side Lobe Level", *Microwave Journal*, Vol. 7, No. 3, March 1964, pp. 88-91.
2. Sheftman, F. I., "Experimental Study of Sub-reflector Support Structures in a Cassegrain Antenna", Lincoln Laboratory-TR Report No. 416, September 1966. Mainly concerned with pattern distortions.
3. Everhart, E. and J. W. Kantorski, "Diffraction Patterns Produced by Obstructions in Reflecting Telescopes of Modest Size", *Astron. J.* 64, 455, (1959).
4. Davidson, C. F. and I. A. Ravenscroft, "Design Considerations for a Center-Fed Paraboloid Aerial System", IEE Conference Publication No. 21, pp. 289-313.
5. Kay, A. F., "Electrical Design of Metal Space Frame Radomes", *IEEE Trans.* AP-13, No. 2, March 1965, pp. 188-202.
6. Weiss, H. G., "The Haystack Microwave Research Facility", *IEEE Spectrum*, Vol. 2, No. 2, February 1965, pp. 50-69.
7. Mei, K. K. and J. Van Bladel, "Scattering by Perfectly Conducting Rectangular Cylinders", *IEEE Transactions on Antennas and Propagation*, Vol. AP-11, No. 2, March 1963, pp. 185-192.

Improved Efficiency with a Mechanically Deformable Subreflector

SEBASTIAN VON HOERNER AND WOON-YIN WONG

Abstract—Like most conventional radio telescopes, the 140-ft at the National Radio Observatory (NRAO), Green Bank, WV, is limited in its short-wave performance by gravitational deformations, whose main part is of an astigmatic shape with elevation-dependent amplitude. This surface degradation could be corrected if a subreflector were deformed in a similar shape by the same amount. For testing this possibility an experimental Cassegrain subreflector was built which can be deformed in an astigmatic mode by simple means: two stiff diagonals and four points in-between where a motor pushes or pulls normal to the surface, servo-controlled by the on-line computer. Although the amount of deformation is too limited in the present setup, good results have already been obtained. Seven unresolved radio sources were observed at the water vapor line (22.3 GHz, λ = 1.345 cm), at various telescope pointings, and with different amounts of subreflector deformation, scanning in both directions for complete beam maps. The astigmatic deformation of the subreflector gave considerable improvements, especially far south and east: fairly strong secondary beams disappeared; the beamshape became more symmetrical, narrow, and round; and at 20° elevation the aperture efficiency increased by factors between two and three.

I. INTRODUCTION

In a previous paper [1] we explained that most conventionally built radio telescopes are expected to have a strong gravitational astigmatism as a function of the elevation angle, and we suggested that this surface deterioration could be improved if a subreflector were mechanically deformed in the proper way. Then von Hoerner [2] showed that *any* shape of the main reflector can be exactly corrected by a similarly shaped subreflector if the rays between the primary and Cassegrain reflector do not cross each other (see also [3]). This condition is well fulfilled by the large-scale gravitational deformations of the 140-ft backup structure but not for the small-scale internal bumpiness of its panels.

Furthermore, the following was shown in [2] and turned out to be applicable to the 140-ft gravitational deformations. If the surface errors of the main reflector are far enough below the critical case of crossing rays, then the correcting subreflector must have deviations from a hyperboloid which are practically identical in shape and amount with the deviations of the main reflector from a paraboloid.

The gravitational astigmatism of the 140-ft telescope was directly measured at the λ = 2.8-cm wavelength with an elongated feed horn illuminating a central strip of the aperture [4]. We define astigmatism as

$$\Delta z = A(\beta)r^2 \cos (2\Omega),\qquad(1)$$

where Δz is the deviation of the surface from a paraboloid of revolution and is measured parallel to the telescope axis, and β is the elevation (equal to the angular height above the horizon); the radial distance r is normalized (dimensionless) to $r = 1$ at the rim and $\Omega = 0$ in the south rotating clockwise about the axis (see Fig. 1). The rim aplitude A should depend

Manuscript received February 10, 1979; revised June 7, 1979.

The authors are with the National Radio Astronomy Observatory, Green Bank, WV 24944.

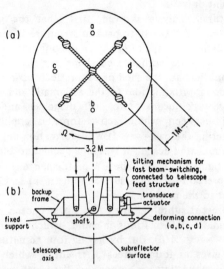

Fig. 1. Deformable subreflector. (a) Back view showing two stiff diagonals, five fixed supports at the filled circles, and four motors pushing or pulling at the open circles (a, b, c, d). (b) Side view with rough sketch of mechanical design. Closed-loop servo contains transducer (measuring linear distance), actuator (motor), and on-line computer (for elevation-dependent deformations).

on elevation only and should do so in the following way:

$$A(\beta) = -A_0 + (1 - \sin \beta)B.\qquad(2)$$

The measurements agreed well with this equation, yielding the two constants as

$$A_0 = 1.72 \text{ mm},$$

$$B = 8.70 \text{ mm}.\qquad(3)$$

The astigmatic rim amplitude then is, for example,

$$A = -1.72 \text{ mm at zenith } (\beta = 90°),$$

$$= +4.00 \text{ mm at } \beta = 20° \text{ elevation,}$$

$$= +6.98 \text{ mm at the horizon } (\beta = 0).\qquad(4)$$

The correcting subreflector then should be a hyperboloid of revolution which is mechanically deformed such that the deformations follow the shape of (1) with the amounts A as given above.

The last adjustment of the surface panels, in 1973, was intended to give the best performance at $\beta_0 = 60°$ elevation, on the meridian. From (1) and the measured values (3) we derive zero astigmatism at an elevation of

$$\beta_0 = \arcsin (1 - A_0/B) = 53.3°\qquad(5)$$

in sufficient agreement with the intended adjustment. However, beam mapping with point sources was done at λ = 1.345 cm (22.3 GHz) by B. Turner and L. Rickard with the old Cassegrain and also at prime focus (unpublished), and in all cases the best beams were *not* obtained on the meridian but

Reprinted from *IEEE Trans. Antennas Propagat.*, vol. AP-27, no. 5, pp. 720–723, Sept. 1979.

about one or two hours east. This is an unfortunate chance result of the last adjustment. Also, very odd multiple beams occurred far west, thus demonstrating strong additional non-astigmatic deformations which result from the hour-angle component of gravity, while astigmatism results from the z component only.

Structural analysis showed, first, that the gravitational deformation of the 140-ft backup structure is mainly of the simple astigmatic shape of (1) when moving along the meridian but is of a more complicated shape for the hour-angle deformation. Second, the shell of a subreflector can be deformed in a good approximation to the astigmatic shape of (1) by relatively easy means: two stiff diagonals and only four points in-between where active force is applied; whereas counteracting the east–west (EW) deformation might require a more complicated design. It was thus decided to build only the simple astigmatic design. The EW performance is hoped to improve appreciably if a new surface adjustment would bring the best point to the meridian.

The present subreflector of Fig. 1 was chosen mainly with the astigmatic deformations in mind, which means that $a = b = -c = -d = A(\beta)$. But by having four actuating motors it actually gives us four degrees of freedom, which adds three additional nonastigmatic modes of deformation. We hoped that maybe some nonastigmatic deformations of the telescope could be corrected as well.

II. THE PRESENT DESIGN

The detailed design of this deformable subreflector is described at length in two NRAO internal reports [5], [6]. A brief description of the engineering design is included in the following. A special feature of the 140-ft subreflector is the tilting mode of fast beam switching, where the Cassegrain mirror is tilted by 5.8 arcmin within 40 ms, rocking back and forth with up to four cycles per second. The deformable subreflector thus must have low weight and low moment of inertia.

The deformable subreflector consists of three main components: 1) the subreflector surface, which is a shell structure of fiberglass aluminum honeycomb sandwich construction; 2) four actuators together with electromechanical parts, control, interface, and computer programming; and 3) an aluminum framework acting as a stiffener to the surface shell, as support for the actuators and as interface to the tilting mechanism for the beam switching technique. Schematic drawings are shown in Fig. 1 for the arrangement of actuators as well as the design concept of this deformable subreflector.

The shell has an aluminum core 2.5-cm thick, sandwiched by two layers of fiberglass cloths impregnated, and hence held tightly to the honeycomb core by epoxy. The required curing temperature of the epoxy is room temperature, so that the problem of requiring a large oven and the thermal distortion of the structure become nonexistent. The subreflector is fabricated on a mold by layers. The manufacturing tolerance was held within 0.25 mm rms. The front face of the subreflector is coated by a thin layer of aluminum film for reflection and then covered with white paint.

As is shown in Fig. 1, the four actuators are arranged so that they are 90° apart, with each one located on the north, east, south, and west directions. The locations are derived by the investigation of the analytical results of the backup structure model and its astigmatic effects, which were later verified by direct measurement. The radial distance of these actuators

from the center depends on the stiffness of the shell structure and the available torque of the actuator's motors.

We encountered a difficulty in securing four lightweight (≈ 1.5 kg) and tight tolerance (backlash $\leqslant 0.2$ mm) actuators from industry within a certain time schedule and a certain cost. Due to the experimetnal nature of the project we decided to use four surplus actuators from the armed forces, with a slight compromise in performance but swift in procurement and low in price. In the present experimental setup the rim of the subreflector cannot be deformed by the full range of (4) but only within the range

$$-0.439 \text{ mm} \leqslant A \leqslant +1.809 \text{ mm} \tag{6}$$

for the astigmatic mode and less for the other three modes (against which the shell is stiffer and which were not considered in the original design). However, the electronic design is done in such a way that only little modification is required for future more powerful actuators. The actuators are controlled, monitored, and driven by the on-line computer (Honeywell-316). The amount of actuation is regularly updated, depending on the position of the telescope, and is checked by a closed-loop control system. There are four identical control systems, one four each actuator. The location of these actuators are so far apart from each other that the coupling effects are negligible.

The subreflector is connected mechanically by two bolts at the two push rods and two removable bearing housings at the center shaft. The subreflector is designed to withstand the dynamical force of periodic tilting with an amplitude of 5.8 arcmin in 40 ms without undue surface degradation. The entire package weighs 198 kg (436 lb).

III. OBSERVATIONS

The purpose of the observations was mainly, by comparing beam shapes as well as peak values of antenna temperature, to find out whether or not this method of mechanically deforming a subreflector can give significant and useful improvements. Observations were made at $\lambda = 1.345$ cm (22.3 GHz) of the water vapor line, during a total of 88 h, in June 1978. We used the following seven radio sources: W3, DR 21, 3C 84, W 51, Orion A, R Crt, W Hya; ranging in declination from $+61°$ to $-28°$. The receiver was a reflection-type ruby maser with a clear-sky system temperature of 90 K. The illumination of the subreflector had about 11-dB edge taper, 13 dB at the edge of the aperture, and the effective focal ratio was about 4.5 with the 3.2-m subreflector.

Beam maps were taken by scanning a rectangular field of sky centered on the radio source. The size of this field was always 8 arcmin in declination δ (declination is the angular distance of the source above the celestial equator) and twice as large, $(16/\cos \delta)$ arcmin, in right ascension (equal to the longitude in the equatorial system). Each map consisted of 16 equally spaced scans, 0.5 arcmin apart in declination, moving through the range of right ascension in about $\frac{1}{2}$ min, and with a receiver integration time of 1 s. The (undistorted) half-power beamwidth is $1.2 \lambda/D = 1.30$ arcmin.

We obtained 196 beam maps. In every case the telescope was first "peaked up" with its main beam on the best position and then on the best focal length. We used all four modes and various degrees of subreflector deformation, always with an undeformed case interspersed for comparison. Several beam maps (with deformed and undeformed subreflectors) were

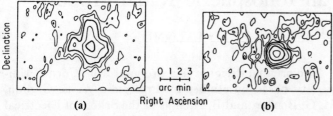

(a)   Right Ascension   (b)

Fig. 2.   Beam shape with unresolved point source (W Hya, declination −28.1°), far south on meridian, and at elevation 23.3°. Contour lines are in geometric progression, at 1, 2, 4, 8, 16, ⋯, K antenna temperature. Using the largest available astigmatic deformation, A = 1.81 mm. (a) Undeformed subreflector, peak = 19.9 K. (b) Deformed subreflector, peak = 40.0 K.

repeated with the subreflector in its tilting mode of fast beam switching, and no problem was encountered.

Whereas the astigmatic mode of deformation gave strong and obvious changes of beam shapes and of peak values, the other three modes unfortunately did not give significant changes. Thus, the question of their improvment value cannot be answered at present. This will need stronger motors and some other changes which are planned for the future.

IV. RESULTS

With the astigmatic mode of deformation, in spite of its limited range (6), we obtained considerable improvements of beam shape and efficiency, especially for pointings east and/or far south, where the efficiency was increased by factors between two and three for the maximum available deformation, A = 1.809 mm.

Fig. 2 shows an example far south on the meridian, which is the pointing where any improvement is most important regarding all observations of sources with southern declinations. Deforming the subreflector makes the beam more symmetric and narrow, and it improves the aperture efficiency by a factor of 40.0/19.9 = 2.0. Another source, close to the equator and in an eastern pointing, is shown in Fig. 3, where the deformation even removes a fairly strong secondary beam and yields an improvement factor of 149/66 = 2.3.

Fig. 4 shows two things. First, for pointings east and south, the efficiency would still increase if we could employ stronger deformations beyond the present limit of (6). This needs new motors with more force and range. Second, the further west we point, the less improvement we obtain. This asymmetry should disappear with an improved and more symmetrical new surface adjustment. But also, stronger motors might help if we could force the subreflector into the other nonastigmatic modes of deformation, and if these would turn out to be useful again. Although an improved version with stronger motors and some other changes are planned for the future, the present setup will meanwhile be used by the observers.

ACKNOWLEDGMENT

It is a pleasure to thank R. Lacasse for the design and manufacture of the electronic servo mechanism and its interface with the 140-ft on-line computer, B. Vance for making the computer program which regulates the deformation automatically according to the telescope pointing, and R. Fisher for his helpful assistance during the observations and data reduction.

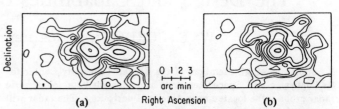

(a)   Right Ascension   (b)

Fig. 3.   Beam shape with unresolved point source (Ori A, declination −5.4°), close to equator and east. Hour angle 3 h 30 min east, elevation 22°. Contour lines and astigmatic deformation as in Fig. 2. (a) Undeformed subreflector, peak = 66 K. (b) Deformed subreflector, peak = 149 K.

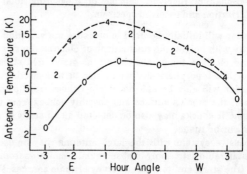

Fig. 4.   Antenna temperature far south, as function of hour angle; tracking radio source W Hya at declination −28.1°.

Symbol	Rim Deformation	
0	0.00 mm	undeformed
2	0.90 mm	half of maximum
		astigmatic deformation
4	1.81 mm	present maximum

REFERENCES

[1] S. von Hoerner and W-Y. Wong, "Gravitational deformation and astigmatism of tiltable radio telescopes," *IEEE Trans. Antennas Propagat.*, vol. AP-23, no. 5, pp. 689–695, Sept. 1975.

[2] S. von Hoerner, "The design of correcting secondary reflectors," *IEEE Trans. Antennas Propagat.*, vol. AP-24, no. 3, pp. 336–340, May 1976.

[3] P. R. Cowles and E. A. Parker, "Reflector surface error compensation in Cassegrain antennas," *IEEE Trans. Antennas Propagat.*, vol. AP-23, no. 3, pp. 323–328, May 1975.

[4] S. von Hoerner, "Measuring the gravitational astigmatism of a radio telescope," *IEEE Trans. Antennas Propagat.*, vol. AP-26, no. 2, pp. 315–318, Mar. 1978.

[5] W-Y. Wong, "The design of the deformable subreflector for the 140-ft radio telepscope," Nat. Radio Astron. Obser., Green Bank, WV, Eng. Div. Internal Rep. No. 110, 1979.

[6] R. J. Lacasse, "Correctable subreflector controller," Nat. Radio Astron. Obser., Green Bank, WV, Electron. Div. Internal Rep. No. 193, 1978.

The Design and Capabilities of an Ionospheric Radar Probe*

W. E. GORDON†, SENIOR MEMBER, IRE, AND L. M. LaLONDE†

Summary—Staff members of the Cornell University Center for Radiophysics and Space Research have designed an ionospheric radar probe to be located near Arecibo, Puerto Rico. The radar will have the following general specifications:

1) Antenna reflector, 1000-foot-diameter spherical bowl, illuminated by a 430-Mc dual-polarized feed.
2) Transmitter of 2.5 Mw peak, 150 kw average power, or 100 kw CW power.
3) Dual-channel receiver, capable of measuring total power, polarization and received spectrum.

The radar will initially be used to measure the variation of electron density with height, the fluctuations of electron density at fixed heights and electron temperatures and magnetic field strengths at various heights. Ionospheric drifts may also be measured.

The radar will also be able to obtain echoes from planets, information of the moon's surface and possibly echoes from the sun. Hydromagnetic shocks may also be detected and a study of cislunar ionization can be made.

The passive system with the large antenna may be used as an instrument in radio astronomy to observe radio emission from planets and from true stars, and to make a survey of radio sources. With additional facilities, many radio astronomy measurements can be made taking advantage of the large antenna aperture and resulting high resolving power.

I. INTRODUCTION

STAFF members of the Cornell University Center for Radiophysics and Space Research have designed a radar to probe the ionosphere. The radar is the basic instrument of a radio observatory being constructed near Arecibo, Puerto Rico. The specifications for the radar were determined by the desire to observe and measure characteristics of the ionosphere using radio wave scattering by free electrons [1]–[7]. The antenna required for this purpose is a 1000-foot-diameter spherical dish using a suitably corrected feed.[1]

A summary of the radar specifications is given in Section II. The expected capabilities of the instrument in probing the ionosphere are described in Section III. A radar that is designed with the specifications given in Section II for the purpose of studying the ionosphere can make many other measurements in the solar system. These are included in Section IV. The uses of the radio telescope formed from the radar's antenna and a sensitive receiver or by the addition of antenna feeds and receivers at other frequencies are indicated in Sections V and VI.

The original design was a proposal from Cornell University based on studies by Professors W. E. Gordon, H. G. Booker, and B. Nichols of the School of Electrical Engineering and Professor W. McGuire of the School of Civil Engineering.

II. SUMMARY OF SPECIFICATIONS

The reflector will be a 1000-foot-diameter aperture of a spherical bowl with radius of curvature 870 feet and a surface tolerance of $+0.1$ foot.

The feed support will be a rotating azimuth truss supported on a triangular platform suspended from three towers by prestressed cables (Fig. 1). A movable carriage on the truss plus the rotation of the truss will provide sky coverage within 20° of the zenith. A reference point on the supported end of the line feed is to be positioned and held within a six-inch cube. The pointing and deflection of the line feed combined must not permit a departure of the unsupported end of the feed from the desired radial position of more than five inches.

Fig. 1—Artist's conception of the ionospheric radar probe being built near Arecibo, Puerto Rico.

The feed will be dual-polarized at a frequency of 430 Mc, capable of fully illuminating the aperture. This will yield a beamwidth of $\frac{1}{6}°$, a gain of 60 db, sidelobes more than 17 db below main lobe, and polarization crosstalk below 30 db. The bandwidth will be at least 1 Mc, with power handling in one channel of 2.5 Mw peak. The weight will be less than five tons, and the feed will be rigid enough to operate in 30-mile winds and to survive 140-mile winds.

The transmitter will have a peak power output at 430-Mc frequency of 2.5 Mw, and an average power of 150 kw, with pulse-to-pulse stability of 0.2 db; or 100 kw CW power. The frequency stability of the primary source is 2.5 parts in 10^9 over periods from one second

* Received by the PGAP, September 15, 1960. The work described in this paper was sponsored by the Advanced Res. Projects Agency under Contract AF19 (604)-6158 with Electronics Res. Directorate of the Air Force Cambridge Research Center.
† Center for Radiophysics and Space Res., Cornell University, Ithaca, N. Y.
[1] The Antenna Laboratory of the Air Force Cambridge Research Center has contributed to the field of spherical antennas for over ten years. The basic fundamentals appear in: R. C. Spencer, C. J. Sletten, and J. E. Walsh, "Correction of spherical aberration by a phased line source," *Proc. NEC*, vol. 5, pp. 320–333; 1950.

Reprinted from *IRE Trans. Antennas Propagat.*, vol. AP-9, pp. 17–22, Jan. 1961.

to one hour. The frequency deviation of the RF output from the source frequency will be less than 3 cps. Pulse lengths of 2 to 10,000 μsec will be available with a repetition rate of 1 to 1000 pps. Provision will be made for coherent pulses, pulse shapes square or "Gaussian" and external modulation.

The duplexer will have a recovery time of 150 μsec, receiver isolation of 65 db, and an insertion loss in the receive direction of 0.3 db.

The receiver will be dual-channel at a frequency of 430 Mc, with a noise figure of less than 2 db. RF-IF conversion will be effected at the feed. The 30-Mc IF will pass through 1100 feet of cable to the operations building, where IF amplifiers, bandwidth adjustment, detection, and data recording/processing equipment will be located. The gain stability of the equipment will be 0.2 db. The oscillator frequencies will be derived from the primary source in the transmitter. Data recording will include a range-gated total power receiver for electron density measurements, spectrum analysis for temperature measurements and polarization indication for magnetic field measurements.

III. Ionospheric Measurements

A. The Upper Ionosphere

The first use of the radar will be to observe the nature of the echoes obtained with a beam directed at the zenith. Echoes will be obtained from the bottom of the D region up to the high F region of the ionosphere. The maximum height from which echoes will be obtainable will depend upon the pulse length, bandwidth, and the integration time used. In and above the F region, the strength of the echoes will give the variation of the electron density with height [1], [2]. Continuous observations over a period of time will give the profile of electron density as a function of time of day, season of year, and degree of solar activity.

Measurement of electron density by this method requires either absolute calibration of the radar, for example by means of a spherical satellite, or a local measurement of the maximum electron density in the F region by means of a conventional ionospheric sounder. A sounder is available at Ramey Air Force Base, 30 miles west of the Arecibo radar.

Observations will also be made of the frequency spectrum of the returns as a function of height. In the main part of the F region, this is expected to take the form of a narrow peak superimposed upon a weaker and broader maximum. There is little doubt that the narrow peak will be observed and that its width will give the thermal velocity of the positive ions. From this, it will be possible to deduce the temperature and the Debye distance as a function of height.

At great heights where the wavelength is of the order of the Debye distance, the spectrum is expected to take a more complicated form. At extremely great heights, the spectrum may take the form of three lines, the outer lines being separated from the central line by the plasma frequency. It is doubtful whether the two outer lines will be observable. If they are, however, their frequencies will give a direct measure of the electron density and their width will be a measure of the electron temperature.

The probable effect of the earth's magnetic field upon the returns from the high ionosphere is unclear. A search will be made for sidebands with a separation equal to the gyromagnetic frequency. An experiment should be carried out to see how the returns depend upon the angle between the axis of the beam and the direction of the earth's magnetic field. If these results provide a means of measuring the strength of the earth's magnetic field, this will then be measured as a function of height and time.

The polarization of the returns as a function of the height from which the return is observed should be examined for Faraday rotation. It is quite likely that, in this way, information will be available concerning the variation of the strength of the earth's magnetic field with height.

B. The Lower Ionosphere

With the beam vertical, measurements can be made of the strength of returns from the D and E regions of the ionosphere. These returns will include echoes from meteoric ionization. By making studies as a function of time of day and with the beam at various angles to the vertical, it should be possible to subtract out the returns from undiffused meteor trails. The remainder of the echo can then be interpreted as due to fluctuations of electron density. The spectrum of the returns can be observed and the Doppler spread can be interpreted in terms of the fluid mechanical velocities involved. Such observations probably refer to the small scale in the spectrum of fluid mechanical irregularities present in the lower ionosphere. It would be desirable to make observations in such a way that an appreciable portion of the spectrum of fluid mechanical irregularities is observed. This involves observing back scattering either as a function of scattering angle at a fixed frequency, or as a function of frequency at a fixed scattering angle. At 430 Mc the fixed frequency method observations would have to be made with receiving antennas at sites other than Arecibo.

C. Ionospheric Drifts

By maintaining the axis of the beam at a fixed angle to the vertical and scanning in azimuth, ionospheric drift should produce a sinusoidal variation of Doppler shift. From this the magnitude and direction of the horizontal drift of the irregularities responsible for the returns can be deduced. With the beam at an angle of 17° off the zenith, a horizontal drift of 10 meters per second would produce a maximum Doppler shift of 10 cps at 430 Mc.

IV. Capabilities in the Solar System

A. Planetary Observations

The power of the Arecibo radar at 430 Mc will be some 43 db above that of the Millstone radar with which echoes from Venus have already been observed [7]. There will be no difficulty therefore in obtaining echoes; 5 msec pulses could be used, transmitted with a repetition time of approximately 80 msec for periods of several minutes, but less than the radar transit time. The received signals would be recorded in digital form on tape and subsequently analyzed with an electronic computer.

In addition to obtaining echoes from Venus, it will be possible to obtain a fine range discrimination, dividing the planet into 10 or more range rings. The scattered power in each of these rings can then be measured separately for one angular position of the planet. Unless the surface of Venus is of great uniformity (if, for example, it were covered entirely with ocean), such a measurement will show the presently unknown rotation speed of Venus and the degree of roughness in the various discriminable range zones. Much information can be gathered concerning the nature of the planet's surface. If, for example, such an experiment were carried out with a planet like the earth, it would be possible to discover that a certain fraction of the area, approximately one half, consisted of a much smoother type of surface than the remainder. It would even be possible to observe the variation in the scattering properties of the sea that result from different degrees of storm disturbances on the surface, and through this variability infer the existence of a sea. The existence of solid ground inclined at different angles to the horizontal could be inferred, and the major mountain ranges could be detected.

Similar roughness measurements will be possible for Mars. In that case, it will be possible to observe the correlation between the visible features and the scattering properties. In particular, it will be valuable to know whether the large average inclinations to the horizontal, which are commonly associated with vegetation, occur on Mars and whether they are perhaps associated with the dark areas suspected of being vegetation.

For Jupiter, an approximate knowledge of the reflection coefficient at 430 Mc would serve to define more closely than is possible at present the main features of the Jovian atmosphere and surface. The Arecibo radar performance will suffice to give these data, unless the albedo of the planet is abnormally low.

Ranges to Mars and Venus will be measurable to very high accuracy. Velocity measurements can be made to high accuracy by means of the Doppler shift. These will contribute not only to precise definition of the scale of the solar system, but also to precise knowledge of the planetary orbits. The advance of the perihelion of the inner planet can be checked to great precision when range as well as angle is available, and this is of great consequence as a test of the general theory of relativity and other theories of gravitation.

Many other detailed radar studies of the planets will become possible, but planning studies beyond the observations described will depend, to a considerable extent, on the findings of the first phase of the work.

B. Lunar Observations

The object of obtaining radar reflections from the moon is to discover the reflecting properties and roughness of the lunar surface as measured at the radar wavelength of 70 cm. It will be possible to investigate the different types of lunar ground and to obtain detailed correlation between radar roughness and optical appearance. This is desirable both for eventual lunar exploration and for theories of the formation of the moon.

At 430 Mc, the Arecibo radar will be so powerful that range discrimination can be used limited only by the shortest pulse length that the transmitter will generate. With a beamwidth of $\frac{1}{6}°$ and a range discrimination of perhaps 1 km, it will be possible to measure in detail the correlation between the roughness of the ground and the optical appearance of the moon. The areas of the lunar surface that contribute at any instant to the echoes are in general in the shape of part of a ring. It should be possible to recognize the contribution of optically prominent features such as the more recent craters, for they are likely, in some cases, to make an overwhelming contribution to the particular patch in which they are included. A great deal, therefore, can be learned about the roughness on approximately the scale of greatest interest in planning transport on the moon.

If the gain of the Arecibo radar can be accurately calibrated, for example by the use of a spherical satellite, then the reflection coefficient of the surface material of the moon at 430 Mc can be measured more accurately perhaps than it is at present. This would help to determine some, though not all, of the electrical properties of the lunar surface.

C. Solar Observations

At 430 Mc, the absorption in the solar atmosphere before reflection occurs is very large, making it difficult to estimate the radar performance to be expected. The reflecting surfaces in general will be smooth and scattering will occur only from regions where the surfaces are normal to the direction of observations; that is, from "glints." In disturbed regions on the sun, such glints may, however, be plentiful, much as the glints in which the sun is reflected over a rough sea. If, in such regions of the sun, there are high gradients of electron density so that the attenuation is not too high, observable reflections may occur. The magnitude, angular extent and, especially, the time variation in the Doppler shifts associated with such reflections would contribute greatly to our present knowledge of the solar atmosphere.

The reflection of a 430-Mc wave from the sun is predicted by Cohen [8] to occur from levels sufficiently high so that severe absorption does not take place. In the presence of a magnetic field of the order of 140 gauss, he predicts an echo some 3 db greater than from Venus at closest approach, assuming similar reflectivities for the two bodies. The Arecibo radar will have sufficient power to see such a signal against the high background noise of the solar cloud.

D. Hydromagnetic Disturbances and Shock Waves

The manner in which the Arecibo radar might be able to detect the passage of a hydromagnetic wave or shock is by recording the increase of electron density averaged over the volume determined by the beamwidth and pulse length. This volume is of the order of one cubic kilometer at a height of 100 km. It is clear therefore, that the radar is most likely to see disturbances on a scale of the order of 1 km or larger.

The radar can be regarded as a pressure-sensitive microphone whose sensitive volume is any of the slabs defined by the beamwidth and the pulse length. The sensitivity diminishes with distance and with decreasing electron density. Maximum sensitivity to a fractional change of electron density will lie between 100 and 300 km and will amount to about one part in 300.

It should be noted that, while waves and shocks in the upper ionosphere are likely to be hydromagnetic in character and to concern only the ionized constituents of the atmosphere, in the lower ionosphere, waves and shocks are more likely to be of the type encountered in conventional fluid dynamics.

The present-day interpretation of magnetic storm phenomena implies rapidly-changing electron densities by amounts of the order of a few per cent for big storms. The Arecibo radar should, therefore, have no difficulty in detecting these phenomena. It will be important to correlate these ionospheric observations with magnetometer observations made at ground level. It is expected, however, that use of the magnetic observatory at San Juan will be adequate for this purpose. The interpretation of a number of known types of fluctuations on ground level magnetometers may be simplified by ionospheric observations made by the Arecibo radar.

The natural background of disturbances due to rapid hydromagnetic waves is not well known. It is clear that magnetic disturbances that have been recorded up to a frequency of $\frac{1}{10}$ cps are associated with such waves in the upper ionosphere, but no work exists at present that indicates the magnitude and localization of the density changes that are involved.

An experimental determination of the background level from natural causes of the density fluctuations at each height will be essential to give estimates of the measurable effects resulting from explosive disturbances at various altitudes and distances. It will also be possible then to give an estimate of the distance over which the shock wave from a high-velocity object, such as a missile, will give a sufficiently marked change in the electron density.

Hydromagnetic shock waves propagating in the space of the solar system may have fronts as narrow as 1000 km or so. If, across such a front, the electron density changes abruptly from a low value to one of the order of 10^3 or 10^4 per cubic centimeter, then this should lead to detectable effects with the Arecibo radar in some geometric circumstances. Such shock waves are of great consequence in the dynamics of the plasma of the solar system, and every effort should be made to detect them. In this connection, one is reminded of the long-delay radio echoes which appear to have been observed on occasion in the past and for which no explanation exists as yet [9]–[11].

E. Cislunar Ionization

Radar echoes from the moon suffer Faraday rotation of the plane of polarization. A good deal of this rotation is associated with the earth's ionosphere. In the light of the observations already suggested, particularly those made in Section III A, the effect of the earth's ionosphere should be known up to a considerable height. By subtracting the effects of the known ionosphere from the observed Faraday rotation on lunar echoes, it should be possible to make deductions about the remaining ionization out to the moon's orbit.

V. RADIO ASTRONOMICAL CAPABILITIES WITH INITIAL FACILITIES

A. Radio Emission from Planets

Emission from the planets at sufficiently high frequencies consists of thermal radiation. Thermal radiation has already been detected from Venus, Mars, and Saturn [12], [13]. Bursts of radiation have been detected from Jupiter [14], and there is clear evidence that there is a nonthermal component in the radiation from the planet.

The temperature derived from thermal radio measurements agree reasonably well with the infrared values except for Venus, where the radio temperatures are substantially higher. These radio measurements are very important for the study of the Venusian atmosphere, and they should be extended as much as possible. The Arecibo dish should be able to observe Venus at 430 Mc and this will greatly extend the known spectrum of Venus.

B. Radio Emission from True Stars

The popular term "radio star" is a misnomer, for the sun is the only true star that has ever been seen in the radio spectrum. The Arecibo dish, however, will provide so much sensitivity that a search for radio emission from stars will be worthwhile. Assuming that the minimum sensitivity for the Arecibo dish is 10^{-27} watts at 100 Mc, and since the flux from the quiet sun at 100 Mc is 5×10^{-21} watts, the quiet sun would be marginally detectable if it were 450 times farther away. The nearest star, however, is 10^6 times farther away. We expect therefore, that any signals received will be nonthermal in origin; probably they will be like Type II solar bursts.

C. Survey of Radio Sources

There is some disagreement between surveys of radio sources in England and in Australia. The latitude of the Arecibo site is such that a survey conducted at Arecibo would overlap both the English and the Australian surveys and could thus play an important role in resolving the discrepancies.

VI. Radio Astronomical Capabilities with Additional Facilities

It would be of great interest to perform most of the experiments and observations already listed at some different frequency, such as 40 Mc. Additional receiving equipment would also permit the performance of the following interesting experiments.

A. Study of Spectrum and Polarization of Solar Bursts in the Frequency Band 10–20 Mc

Study of solar bursts in the frequency band 10–20 Mc would give important information on the outer corona, but the use of these frequencies is now difficult because of interference caused by oblique scattering from the ionosphere. The narrow beam of the Arecibo system could give it a significant advantage over an interferometer in rejecting this type of interference.

B. Measurements of the Spectrum and Polarization of Bursts from Jupiter in the Frequency Band 15–25 Mc

In connection with the study of bursts from Jupiter, a measurement of spectrum and polarization should be made in the frequency band 15–25 Mc. Approximately this frequency band should be used, since it is known that the spectrum decreases rapidly with frequency from 18–22 Mc.

C. Two-Wavelength Comparison of Radio Sources

It is not known, at present, how many different classes of sources are involved in the multitude of radio sources detected thus far. An analysis of the radio spectrum seems a good way to obtain information on this. The argument here is that it is improbable that radio noise generation on enormously different scales and with different physical circumstances would result in similar spectra. It is expected that a two-wavelength comparison would suffice to show up different classes if they exist. This experiment would involve the construction of a smaller, accurately-scaled antenna. It is suggested that the frequencies 430 Mc and 108 Mc be used.

D. Interferometric Observations

Use of an interferometer would permit an investigation of the lunar surface on an extremely small scale. It would be possible to combine the two dishes of the previous experiment to form an interferometer. However, the interferometer does not necessarily have to be located at Arecibo.

E. Observations Associated with Deep Space Probes

A satellite would be in the narrow beam of the Arecibo radar for an extremely short period of time. The rates of change of angular position of a deep space probe would be much smaller, so that the Arecibo radar should be able to perform a function similar to that of the Jodrell Bank dish in connection with Pioneer V.

F. Scattering from Field-Aligned Irregularities

Many observations have been made under both auroral and nonauroral conditions of scattering from the ionosphere which appears to come from irregularities aligned along the earth's magnetic field [15]. Among such observations are those of forward scatter in the southwestern United States at 200 Mc [16]. The physics of the irregularities involved is not yet understood, and additional observations of this phenomenon would be very worthwhile. Since the beam of the Arecibo radar cannot be directed normal to the magnetic field, receiving equipment would have to be set up at locations that satisfy the requirement of specular reflection with respect to the magnetic lines of force.

G. Hydrogen-Line Observations

If at least a part of the big dish turns out to be good enough for use at the hydrogen-line frequency of 1420 Mc, a substantial program of observations at that frequency could be made.

H. Light of the Night Sky

The comparatively inexpensive equipment needed for measuring the light of the night sky should be installed at Arecibo for purposes of correlation with other ionospheric observations.

VII. Acknowledgment

The following organizations have contributed to the design and are participating in the construction of the radar.

Center for Radiophysics and Space Research, Cornell University.
Electronics Research Directorate, Air Force Cambridge Research Laboratory.
Joint Venture of Von Seb Inc.; Developmental Engineering Corporation; Severud-Elstad-Krueger Associates; Praeger-Kavanagh (design and field supervision during construction of antenna reflector and feed support).
Technical Research Group (antenna feed).
Levinthal Electronic Products, Inc. (transmitter).
U. S. Corps of Engineers (buildings, roads, utilities).

Many individuals, both in the organizations listed above and in other organizations, have contributed to the design and are contributing to the construction. The list is too long for inclusion here.

The material in this paper was developed by staff members of the Center for Radiophysics and Space Research (H. G. Booker, M. H. Cohen, J. P. Cox, T. Gold, W. E. Gordon, L. M. LaLonde, B. Nichols, E. L. Resler, Jr., and E. E. Salpeter). The task of the authors has been largely editorial.

BIBLIOGRAPHY

[1] W. E. Gordon, "Incoherent scattering of radio waves by free electrons with applications to space exploration by radar," PROC. IRE, vol. 46, pp. 1824–1829; November, 1958.

[2] K. L. Bowles, "Observations of vertical incidence scatter from the ionosphere at 41 Mc/s," *Phys. Rev. Lett.*, vol. 1, pp. 454–455; December, 1958.

[3] J. P. Dougherty and D. T. Farley, "A theory of incoherent scattering of radio waves by a plasma," to be published *Proc. Roy. Soc. (London)*.

[4] T. Laaspere, "On the effect of a magnetic field on the spectrum of incoherent scattering," to be published in *J. Geophys. Res.*

[5] E. E. Salpeter, "Scattering of radio waves by electrons above the ionosphere," *J. Geophys. Res.*, vol. 65, pp. 1851–1852; June 1960.

[6] J. A. Fejer, "Scattering of radio waves by an ionized gas in thermal equilibrium," *Can. J. Phys.*, vol. 38, pp. 1114–1133; August, 1960.

[7] R. Price, P. E. Green, Jr., T. J. Goblick, Jr., R. H. Kingston, L. G. Kraft, Jr., G. H. Pettengill, R. Silver, and W. B. Smith, "Radar echoes from Venus," *Science*, vol. 129, pp. 751–753; March, 1959.

[8] M. H. Cohen, "High-frequency radar echoes from the sun," (Correspondence), PROC. IRE, vol. 48, p. 1479; August, 1960.

[9] C. Stormer, "Short-wave echoes and the Aurora Borealis," *Nature*, vol. 122, p. 681; November, 1928.

[10] Balth. van der Pol, "Short-wave echoes and the Aurora Borealis," *Nature*, vol. 122, pp. 878–879; December 8, 1928.

[11] E. V. Appleton, "Short-wave echoes and the Aurora Borealis," *Nature*, vol. 122, p. 879; December 8, 1928.

[12] C. H. Mayer, T. P. McCullough, and R. M. Sloanaker, "Observations of Venus at 3.15-cm wave length," *Astrophys. J.*, vol. 127, pp. 1–10; January, 1958.

[13] C. H. Mayer, T. P. McCullough, and R. M. Sloanaker, "Observations of Mars and Jupiter at a wave length of 3.15 cm," *Astrophys. J.*, vol. 127, pp. 11–16; January, 1958.

[14] A. G. Smith and T. D. Carr, "Radio-frequency observations of the planets in 1957–1958," *Astrophys. J.*, vol. 130, pp. 641–647; September, 1959.

[15] B. Nichols, "Evidence of elongated irregularities in the ionosphere," *J. Geophys. Res.*, vol. 64, pp. 2200–2202; December, 1959.

[16] J. L. Heritage, S. Weisbrod, and W. J. Fay, "Evidence for 200-megacycles per second ionospheric forward scatter mode associated with the earth's magnetic field," *J. Geophys. Res.*, vol. 64, pp. 1235–1241; September, 1959.

THE relatively low frequencies at which observations were carried out during the early years of radio astronomy made the task of improving the available angular resolution a difficult one. For example, at a frequency of 30 MHz, which corresponds to a wavelength of 10 m, a telescope 100 m in diameter produces a beamwidth of over 5°. Thus, as discussed in the Historical Overview, interferometric techniques—the use of more than one antenna simultaneously—were introduced relatively quickly. The first interferometric observations were made by McCready *et al.* (1947) in Australia using the sea interferometer. While minimizing the additional equipment required to convert a single antenna to an interferometer, the range of applications of this system is quite limited. The first two-antenna instrument was constructed by Ryle and Vonberg and the results of the first observations together with a short description of the principle of operation are contained in the first paper included in this part, which was published in 1946.

Interferometers of many sorts began to have a major impact on radio-astronomical observations almost immediately. In the quest for higher angular resolution, the interferometer baseline (separation of the two radio telescopes) was extended by relaying the downconverted signal over a radio link between the two antennas. The first use of this technique is described in the following paper, by Mills, together with measurements of three radio sources at baselines up to 10 km.

Much of the early work with interferometers was directed toward determining the location of pointlike sources. However, McCready *et al.* were concerned with solar radio emission, and consequently had to deal with the response of an interferometer to an extended source. The fact that the response of the interferometer is related to the Fourier transform of the brightness distribution of the source along a direction perpendicular to the interferometer axis is discussed in Mills' paper, along with the corollary that to recover the brightness distribution of a source, measurements with a range of baselines are required. The requirements on the projected baseline (which is the vector antenna separation projected onto the sky) are critical for understanding the design, operations, and limitations of interferometric systems.

The concept of *synthesizing* a large aperture by combining the results of many measurements with an interferometer is the central point of the paper by Ryle and Hewish, which is the third included in this part. It is helpful to recall that the output of an antenna is the sum of the contributions reflected by the elements of the surface. The power collected by the feed (being proportional to the square of the electric field) can be thought of as the sum of the products taken pairwise, of the flux intercepted by the elements constituting the antenna surface. A filled aperture antenna has all element spacings, from zero to some upper limit fixed by its maximum dimension, with their relative weighting determined by the geometry of the antenna surface, and by the feed system.

A two-element interferometer has only a single baseline, but this can be changed in time either by deliberately varying the relative positions of the antennas, or as a result of the earth's motion changing the baseline projected onto the sky. One can obtain a particular system response pattern by combining the outputs of many antennas with a given set of baselines, as described by Christiansen and Warburton (1953). This type of antenna system is called an array. The term is now also applied to a collection of antennas whose outputs are correlated pairwise, which is essentially a large number of two-element interferometers operating simultaneously. With this latter type of instrument, independent information from different baselines is obtained at any particular moment. The baseline coverage is augmented both by the earth's rotation, and by changing the relative positions of the antennas themselves if desired. Missing baselines are among the problems which have to be addressed by post-observation processing techniques discussed in Part IX.

The fourth paper included here, by Fomalont, describes the basics of aperture synthesis in a concise fashion. This subject is also treated in articles by Bracewell (1962), Rowson (1963), Swenson and Mathur (1968), and in texts and review articles included under References and Bibliography.

The following two articles (the fifth and sixth papers, by Baars *et al.* and Napier *et al.* respectively) describe major interferometer systems that are in operation at the present time. The great power of the Westerbork Array and the Very Large Array (VLA) is their ability to make "radio pictures." That is, by virtue of the large number and good distribution of baselines obtained during an observation of a few hours duration (or sometimes less), a reasonably good map of the source brightness distribution can be made. The interferometer at Nobeyama, described in the seventh paper by Ishiguro *et al.*, carries out the same task at millimeter wavelengths, albeit with only five antennas. Other millimeter interferometer projects are described in articles by Welch *et al.* (1977) and Masson *et al.* (1984).

The VLA has a maximum baseline of approximately 35 km; the signals are distributed over this distance by an oversized waveguide. For much greater antenna separations, the radio-linked interferometer (described in the second paper) is more effective. The eighth paper, by Davies *et al.*, describes the Multi Telescope Radio Linked Interferometer (MTRLI), also called the Multi Element Radio Linked Interferometer (MERLIN), at Jodrell Bank, which has a maximum baseline of approximately 130 km. At its shortest operating wavelength of 1.35 cm, the angular resolution is 0.02 arcsecond. The six antennas have been situated to provide optimum distribution of baselines for sources at a wide range of declinations.

The ninth paper, by Mills and Little, describes a somewhat different type of instrument, developed in the early 1950s in Australia. It consists of two long but narrow antennas arranged in a cross; each has a highly asymmetric pattern, with the two having a 90° relative orientation. By multiplying the voltage outputs of the two elements of the cross, one obtains an output which is the product of the electric field patterns of the two elements, which consequently has a very narrow width in both dimensions. In order to perform satisfactorily, the sidelobe levels of each element must be quite low (since they enter into the final power pattern without being squared); this can be achieved by tapering the illumination along each element. This variant of the interferometer, generally referred to as the *Mills Cross,* played a particularly important role in the analysis of radio source counts discussed in the Historical Overview.

The final paper in Part II is concerned with the intensity interferometer developed by R. Hanbury Brown and his collaborators. The basic idea is to multiply together the output *power* (proportional to incident intensity) from each of the two elements of an interferometer, rather than the output *voltage* (proportional to incident field amplitude). The advantage is that the bandwidth requirement is negligible compared to that required for predetection correlation, and so the post-detection signal from one antenna can very easily be transmitted in order to be correlated with that from the other element of the intensity interferometer. Details of the theory can be found in the papers by Hanbury Brown and Twiss (1954, 1956).

As discussed in the Historical Overview, the intensity interferometer was developed to make transoceanic baselines feasible, for which the reduced bandwidth appeared to be a crucial advantage. However, the intensity interferometer suffers from a much lower signal-to-noise ratio for the same signal intensity and system parameters as a conventional amplitude (or Michelson) interferometer, and consequently has not been widely used in radio astronomy. It has found application in optical astronomy, where correlating two signal beams while keeping track of the phase is relatively more difficult than at radio wavelengths. A second factor which undoubtedly limited the use of the intensity interferometer for extremely long baseline work was the development of tape recording and phase reference systems which made possible amplitude-correlation very long baseline interferometry (VLBI), which is discussed in Part VIII.

These ten articles only suggest the richness and diversity of radio interferometry. Many very important articles have been omitted due to space constraints. One example of this is a paper of major practical importance by Ryle (1952) on the phase switched interferometer. It has, however, been reprinted in *Classics in Radio Astronomy* edited by Sullivan (see the references following the Historical Overview).

REFERENCES AND BIBLIOGRAPHY

1947: McCready, L. L., J. L. Pawsey, and R. Payne-Scott, "Solar radiation at radio frequencies and its relation to sunspots," *Proc. Roy. Soc., Ser. A,* vol. 190, pp. 357-375, 1947.

1952: Ryle, M., "A new radio interferometer and its application to the observation of weak radio stars," *Proc. Roy. Soc., Ser. A,* vol. 211, pp. 351-375, 1952.

1953: Christiansen, W. N. and J. A. Warburton, "The distribution of radio brightness over the solar disk at a wavelength of 21 centimetres," *Aust. J. Phys.,* vol. 6, pp. 190-202, 1953.

1954: Hanbury Brown, R. and R. Q. Twiss, "A new type of interferometer for use in radio astronomy," *Phil. Mag., Ser. 7,* vol. 45, pp. 663-682, 1954.

O'Brien, P. A., "The distribution of radiation across the solar disk at metre wavelengths," *Mon. Notic. Roy. Astron. Soc.,* vol. 113, pp. 597-612, 1954.

1955: Hanbury Brown, R., H. P. Palmer, and A. R. Thompson, "A rotating lobe interferometer and its application to radio astronomy," *Phil. Mag., Ser. 7,* vol. 46, pp. 857-866, 1955.

1956: Hanbury Brown, R. and R. Q. Twiss, "Correlations between photons in two coherent beams of light," *Nature,* vol. 177, pp. 27-29, 1956.

1958: Christiansen, W. N. and D. S. Mathewson, "Scanning the sun with a highly directional array," *Proc. IEEE,* vol. 46, pp. 127-131, 1958.

1961: Swarup, G. and K. S. Yang, "Phase adjustment of large antennas," *IRE Trans. Antennas Propagat.,* vol. AP-9, pp. 75-81, 1961.

1962: Bracewell, R. N., *Radio Astronomy Techniques,* Handbuch der Physik Vol. LIV. Berlin: Springer-Verlag, 1962, pp. 42-129.

Ryle, M., "The new Cambridge radio telescope," *Nature,* vol. 194, pp. 517-518, 1962.

1963: Rowson, B., "High resolution observations with a tracking radio interferometer," *Mon. Notic. Roy. Astron. Soc.,* vol. 125, pp. 177-188, 1963.

1968: Moffet, A. T., "Minimum-redundancy linear arrays," *IEEE Trans. Antennas Propagat.,* vol. AP-16, pp. 172-175, 1968.

Swenson, G. W. and N. C. Mathur, "The interferometer in radio astronomy," *Proc. IEEE,* vol. 56, pp. 2114-2130, 1968.

1972: Chow, Y. L., "On designing a supersynthesis antenna array," *IEEE Trans. Antennas Propagat.,* vol. AP-20, pp. 30-35, 1972.

1976: Pooley, G., "Connected-element interferometry," Chapter 5.2 in *Methods of Experimental Physics: Vol. 12 Astrophysics, Part C: Radio Observations,* M. L. Meeks, Ed. New York, NY: Academic Press, 1976, pp. 158-173.

Rogers, A. E. E., "Theory of two element interferometers," Chapter 5.1 in *Methods of Experimental Physics: Vol. 12 Astrophysics, Part C: Radio Observations,* M. L. Meeks, Ed. New York, NY: Academic Press, 1976, pp. 139-157.

1977: Welch, W. J., J. R. Forster, J. Dreher, W. Hoffman, D. D. Thornton, and M. C. H. Wright, "An interferometer for millimeter wavelengths," *Astron. Astrophys.,* vol. 59, pp. 379-385, 1977.

1980: Thompson, A. R., B. G. Clark, C. M. Wade, and P. J. Napier, "The very large array," *Astrophys. J. Suppl.,* vol. 44, pp. 151-167, 1980.

1982: Erickson, W. C., M. J. Mahoney, and K. Erb, "The Clark Lake Teepee-Tee telescope," *Astrophys. J. Suppl.,* vol. 50, pp. 403-420, 1982.

1984: Ishiguro, M., K.-I. Morita, T. Kasuga, T. Kansawa, H. Iwashita, Y. Chikada, J. Inatani, H. Suzuki, K. Handa, T. Takahashi, "The Nobeyama millimeter-wave interferometer," in *Proc. URSI Int. Symp. on Millimeter and Submillimeter Wave Radio Astronomy,* Granda, Spain, 1984, pp. 75-84.

Jones, I. G., A. Watkinson, P. C. Egau, T. M. Percival, D. J. Skellern, and G. R. Graves, "The FST-a 20 arc second synthesis telescope," *Proc. Astron. Soc. Australia,* vol. 5, pp. 574-578, 1984.

Masson, C. R., G. L. Berge, M. J. Claussen, G. M. Heiligman, R. B. Leighton, K. Y. Lo, A. T. Moffet, T. G. Phillips, A. I. Sargent, S. L. Scott, D. P. Woody, and A. Young, "The Caltech millimeter wave interferometer," in *Proc. URSI Int. Symp. on Millimeter and Submillimeter Wave Radio Astronomy,* Granda, Spain, 1984, pp. 65-74.

1985: Christiansen, W. N. and J. A. Hogbom, *Radiotelescopes,* 2nd ed. Cambridge: Cambridge University Press, 1985, ch. 5-7.

1986: Thompson, A. R., "The interferometer in practice," Chapter 2 in *Synthesis Imaging* (NRAO Summer School Notes, 1985), R. A. Perley, F. R. Schwab, and A. H. Bridle, Eds. Charlottesville, VA: National Radio Astronomy Observatory, 1986, pp. 9-30.

Thompson, A. R., J. M. Moran, and G. W. Swenson, Jr., *Interferometry and Synthesis in Radio Astronomy.* New York, NY: John Wiley and Sons, 1986.

Solar Radiation on 175 Mc./s.

Appleton[1] and Hey[2] have directed attention to the fact that radio-frequency energy, with some of the characteristics of random 'noise', is emitted with greatly increased intensity from the sun under the conditions of violent disturbance associated with a large sunspot. These observations were confined mainly to the region of frequencies near 60 Mc./s.

Pawsey, Payne-Scott and McCready[3], who have made observations on 200 Mc./s., suggested that radiation of this type is also observable under less disturbed conditions.

In order to investigate other aspects of this phenomenon, we have constructed a device which automatically records and measures the 'noise' received on 175 Mc./s., and which has a sensitivity such that a power of 3×10^{-15} watts (approximately 1 per cent of the receiver noise power) can be detected. This sensitivity corresponds to a thermal energy temperature of 30° K., and it has been possible to record the 'noise' received from the galaxy on a small broadside aerial consisting of eight half-wave dipoles.

For the purpose of investigating solar radiation under conditions of low solar activity, it is necessary to discriminate against the background of galactic radiation. While this could be achieved by building an aerial to give a sufficiently narrow beam, a very large structure would be required, and observation would be restricted to a short time every day unless arrangements were made for moving the polar diagram of the aerial. An alternative method was therefore used, analogous to Michelson's method for determining stellar diameters. Two aerial systems were used with a horizontal separation of several wave-lengths, and their combined output was fed to the receiving equipment. Such an arrangement produces a polar diagram of the form shown in Fig. 1 where the angle between zeros is governed by the spacing of the two aerials and the envelope is determined by the polar diagram of each individual aerial system. If the angle between minima is sufficiently large compared with the solar angular diameter, then, as the aerial polar diagram is swept past the sun by the earth's rotation, any radiation from the sun should be recorded as an oscillatory trace.

Fig. 2 shows a typical record obtained with an aerial separation of 10 λ, and with only slight solar activity (July 17). The oscillatory contribution due to radiation from the sun can be seen superimposed on the slowly varying background of the galactic radiation. Records of this type enable an estimate to be made of the level of solar radiation even when it is only about one quarter the galactic contribution, and at the present time we have found that the sun is usually sufficiently disturbed to give such records. The power is indicated on the diagram in terms of an 'equivalent aerial temperature', and is the power which has to be fed to an aerial in a black-body enclosure of this temperature, to maintain equilibrium. The temperature of a distant source whose radiation obeys a black-body distribution may be estimated from the observed equivalent aerial temperature by correcting for the ratio of solid angles of source and aerial polar diagram.

During the appearance of a large sunspot between July 20 and August 1, the solar radiation was much increased, and the opportunity was taken to use the apparatus to determine the angular diameter of the source, by observing the ratio of maximum to minimum intensity as the polar diagram of the two aerials with a separation of many wave-lengths was swept past the sun. This experiment was carried out with a series of different aerial spacings, the final value being 140 λ, and a sample of the records obtained with this spacing

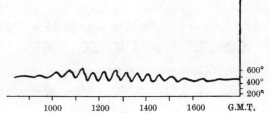

Fig. 2. RECORD OBTAINED WITH 10 λ SEPARATION (JULY 17, 1946)

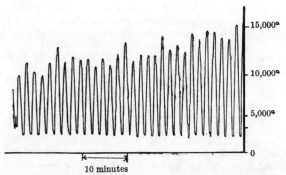

10 minutes

Fig. 3. RECORD OBTAINED WITH 140 λ SEPARATION (JULY 26, 1946)

is shown in Fig. 3. The maximum/minimum ratio obtained under these conditions corresponds to a source diameter of 10 minutes of arc. Any inequalities in the two aerial systems would result in an over-estimate of diameter, and this is therefore a maximum value.

Since the value obtained does not greatly exceed the diameter of the visual spot, it is reasonable to relate the source of this radiation with the visual spot itself, or a region closely associated with it.

During the afternoon of July 25 the observed intensity attained a value which would correspond, in the case of black-body radiation, from a source of this diameter, to a temperature greater than $2 \cdot 10^9$ ° K.

Since the existence of such temperatures in a region from which radiation of this wave-length would escape seems improbable, we considered that the radiation was non-thermal in origin, and the possibility of ordered electron motion was therefore investigated by an examination of the polarization of the radiation. This was carried out by arranging the two aerial systems of the 'Michelson' device to be polarized in planes at right angles to each other. If the radiation were emitted by a completely random 'thermal' source, the two perpendicularly polarized components would not be phase-coherent and no interference effects would be observed. The existence of interference effects would show the presence of phase coherence, and hence prove that the radiation was not of 'thermal' origin. Further, by noting the direction of the sun relative to the aerial systems when an interference maximum was produced, it would be possible to differentiate between plane and right- and left-handed circular polarization.

Using such a system it was found that during periods of intense radiation the polarization was, within the accuracy of measurement, completely circular. (Inequalities in the aerial system limit the accuracy, but at least 90 per cent of the incident energy was circularly polarized.)

Measurements taken over the period July 27–August 3 showed the polarization to be anti-clockwise, viewed along the positive direction of propagation (left-handed). Between August 3 and August 7 the degree of polarization diminished, being virtually completely random on August 7. On August 8, 40 per cent polarization was observed again, but with right-handed polarity—the result, presumably, of increased activity in a subsidiary sunspot.

Any theory of the emission of circularly polarized radiation from sunspots must presumably be given in terms of the magnetic field known to be present in those spots. In considering the mechanism of such a process account must be taken of the magnetic field and electron density not only in the region appropriate to the observed frequency, but also in the overlying layers, where selective absorption of the radiation will occur, in a manner similar to the 'gyro-magnetic' phenomena familiar in the terrestrial ionosphere.

It will be necessary to collect more experimental data before possible theories can profitably be considered in detail.

M. Ryle
D. D. Vonberg

Cavendish Laboratory,
Cambridge.
Aug. 22.

[1] Appleton, *Nature*, **156**, 534 (1945).
[2] Hey, *Nature*, **157**, 47 (1946).
[3] Pawsey, Payne-Scott and McCready, *Nature*, **157**, 158 (1946).

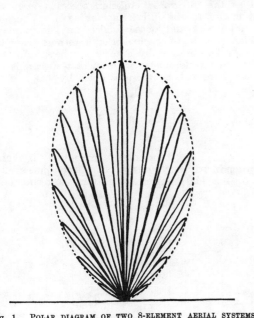

Fig. 1 POLAR DIAGRAM OF TWO 8-ELEMENT AERIAL SYSTEMS WITH SEPARATION OF 10 λ

Reprinted with permission from *Nature*, vol. 158, pp. 339–340, Sept. 7, 1946.

APPARENT ANGULAR SIZES OF DISCRETE RADIO SOURCES

Observations at Sydney

The first observers of the discrete sources of cosmic noise were unable to determine their size, and many considered that they might be of stellar dimensions. However, recent observations have shown that a number of these apparently 'point' sources have angular extensions of the order of 1°[1-3]. Position measurements have also suggested that the stronger sources might be identified with nebulæ of an angular size ranging from half a minute to several minutes of arc[2,4,5].

In order to check the hypothesis that the discrete source radiation originates in nebulæ rather than in individual stars, an interferometer with a lobe spacing as small as 1 minute of arc was considered necessary. At a frequency of 100 Mc./s., which is convenient because small low-gain aerials may be used, a Michelson-type interferometer requires an aerial separation of 10 km. to achieve this resolution. Such an interferometer was constructed using one small, easily portable aerial and one large fixed aerial. The signal from the portable aerial is transmitted back to the main receiver by a radio link. A simplified diagram of the equipment is shown in Fig. 1.

The local oscillator frequency must be transmitted together with the converted signal, as it is necessary to preserve the phase of the reconstituted signal at the receiving end of the link. Also a time-delay must be inserted in the signal-path from the fixed aerial to equalize the propagation times from each aerial to the mixing point. This time-delay is obtained with a mercury-filled acoustic delay line. Automatic gain controls are used to ensure equality of the signal-levels from each aerial and pre-amplifier system before they are combined in the receiver. The equipment has shown a variation in sensitivity of about 15 per cent over the period of six months that it has been in operation.

Preliminary measurements have been made on the four most intense sources observable from Sydney with various aerial spacings in an east–west direction. With the range of aerial spacings used, the four sources show a reduction in the amplitude of their interference patterns as the aerial spacing is increased, indicating that all four sources have been resolved by the equipment. The effects of aerial separation on the amplitudes of the interference patterns are summarized in Table 1.

Table 1

Aerial spacing	0·29 km.	1·02 km.	5·35 km.	10·01 km.
Relative amplitude of interference pattern for Cygnus A	1	1	0·3	0·05
Taurus A	1	0·55	*	*
Virgo A	1	0·4	*	*
Centaurus A	1	0·3	*	*

* In these cases the amplitudes of the interference patterns were less than the noise fluctuations of the equipment and could not be observed. The relative amplitudes are less than 0·1–0·2.

The probable errors in the relative amplitudes are thought to be about 10 per cent. The errors will be discussed in detail elsewhere; but it is worth mentioning one potential source of error. This is phase instability in the equipment and propagation link. It was checked using 101 Mc./s. transmissions from a small transmitter, which were received by both aerials and compared in phase at the mixing point. The resulting error was found to be negligible.

An inspection of the table reveals that the amplitudes of the Taurus A, Virgo A and Centaurus A sources are all significantly reduced with an aerial spacing of 1 km., whereas a spacing of 5 km. is required to produce a comparable reduction in the amplitude of the Cygnus source. The first-mentioned sources are all, therefore, of similar size and approximately five times the size of Cygnus A.

It has been shown that the amplitude and phase of the interference pattern produced by a source when observed with a given aerial spacing is related to the radio brightness distribution across the source by a Fourier transform[6] (the brightness being measured as the integrated emission across a strip at right angles to the interferometer axis). Thus, if the amplitudes and phase angles are known at a sufficient number of spacings, the brightness distribution may be obtained by Fourier synthesis. In the case of an interferometer with an east–west axis as we have here, this synthesis is given by

$$B_t = \int_0^\infty A_\omega \cos(\theta_\omega + \omega t) \, d\omega,$$

where B_t is the brightness distribution in right ascension in arbitrary units, ω denotes the pulsatance of the sinusoidal interference pattern, corresponding

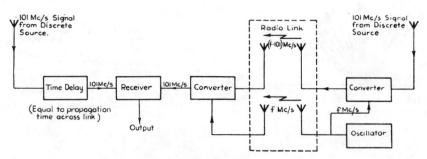

Fig. 1. Simplified diagram of the interferometer

to the aerial spacing, and A_ω and θ_ω the amplitude and phase of the pattern.

The present interferometer cannot be easily adapted to measure phase, so that the true distribution B_t is not easily obtained. If, however, the distribution is assumed to be symmetrical, θ_ω is then zero and an 'equivalent symmetrical distribution' can be obtained. This equivalent distribution is merely the cosine transform of the amplitude–pulsatance spectrum.

For the Cygnus source we have four points from which to determine this spectrum. In Fig. 2(a) these points are shown and a smooth curve is drawn through them. Because of the absence of observations in the region between 1 and 5 km., there is some uncertainty in the exact shape of the curve, but its general features are clear. The equivalent brightness distribution corresponding to this spectrum is shown in Fig. 2(b). Because of the uncertainty in the spectrum, there is a corresponding uncertainty in the shape of this curve, particularly with regard to the skirts.

In the case of the other sources, there are not enough points to determine the spectra; but, if the reasonable assumption is made that their brightness distributions are similar to that of Cygnus A, it is possible to estimate their sizes. Defining an 'effective' size as the angle between half-brightness points on the equivalent distribution curve, we have the results of Table 2.

Table 2

Source	Effective size (E–W) (minutes of arc)
Cygnus A	1·1
Taurus A	4
Virgo A	5
Centaurus A	6

The size of the Centaurus source has been given previously as about $\frac{1}{3}°$ [1,2]. It has been shown, however, that it consists of an extended source of nearly 2° in size with a strong concentration near its centre[3]. The size quoted in Table 2 is that of the central concentration. Measurements with very close spacings have confirmed the interpretation in terms of two apparently quite distinct contributions.

The four sources have been found to have positions very close to nebulæ, and it is now found that their angular sizes are comparable with those of the nebulæ in question. We may therefore say that the suggested identifications of the four sources concerned have been strengthened, almost to the point of certainty. The confirmation is particularly welcome in the case of Cygnus A, as when it was first pointed out that this nebula was near the radio position of the source[4] it appeared to be a typical galaxy, too faint to be responsible for the radiation. Minkowski has now shown, however, that it has very peculiar features (personal communication). Taking these sources in conjunction with the others for which angular sizes have been obtained, there is now strong evidence that the discrete sources are largely, if not entirely, nebulæ. Both galactic and extra-galactic nebulæ have been identified.

This work is continuing and will be described in more detail elsewhere.

B. Y. MILLS

Division of Radiophysics,
Commonwealth Scientific and
Industrial Research Organization,
Australia.

[1] Mills, B. Y., *Aust. J. Sci. Res.*, A5, 266 (1952).
[2] Mills, B. Y., *Aust. J. Sci. Res.*, A5, 456 (1952).
[3] Bolton, J. G., "Extended Sources of Galactic Noise", U.R.S.I. Report (1952).
[4] Bolton, J. G., Stanley, G. J., and Slee, O. B., *Nature*, 164, 101 (1949).
[5] Mills, B. Y., and Thomas, A. B., *Aust. J. Sci. Res.*, A4, 158 (1951).
[6] McCready, L. L., Pawsey, J. L., and Payne-Scott, Ruby, *Proc. Roy. Soc.*, A, 190, 357 (1947).

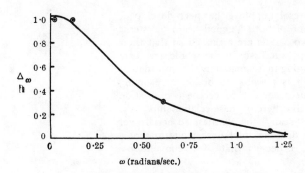

(a) Amplitude–pulsatance spectrum of Cygnus A

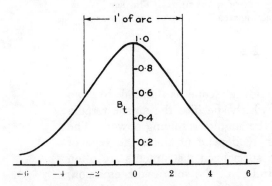

(b) Derived equivalent brightness distribution across Cygnus A

Fig. 2

THE SYNTHESIS OF LARGE RADIO TELESCOPES

M. Ryle and A. Hewish

(Received 1959 August 24)

Summary

Many investigations in radio astronomy are limited by the resolving power which can be achieved by conventional methods of aerial construction.

A new method of obtaining increased resolving power has been developed, which has been applied to the construction of both " pencil-beam " systems and interferometers. In this method two aerials are arranged so that their relative position may be altered to occupy successively the whole area of a much larger equivalent aerial. By combining mathematically the information derived from these different positions, it is possible to obtain a resolving power equal to that of the large equivalent aerial. Since the combination of the individual records may be done with different phase relationships, it is possible, without extra observations, to " scan " the synthesized aerial over an appreciable solid angle; because of this the total observing time of a synthesized instrument is of the same order as that of a conventional instrument.

An interferometric system designed for the study of radio stars has been built which has an equivalent area for resolution of $8 \cdot 10^5$ sq. ft as well as a " pencil-beam " system with an equivalent area of $3 \cdot 10^6$ sq. ft. The sensitivity of both systems corresponds to a " collecting area " of about $2 \cdot 10^5$ sq. ft.

---

1. *Introduction.*—The operation of a radio telescope is limited by two dominating factors. One is the sensitivity, which specifies the smallest radio flux which may be detected, and the other is the angular resolving power. The relative importance of these factors is strongly dependent upon the type of investigation undertaken and for this reason specialized instruments are often developed for particular applications. In studies of the solar radio emission, for example, where high resolving power is needed at metre wave-lengths the primary requirement is that of angular resolution; the use of conventional aerials such as paraboloids is clearly impracticable. Even if structures large enough to give the required resolution were mechanically feasible their energy-collecting area would be far in excess of that needed to detect the relatively intense solar radio emission. Instruments have therefore been developed in which high angular resolution has been obtained by dispersing small aerials over a wide area of ground. Such systems include the diffraction grating type of array (Christiansen 1953; Blum, Boischot and Ginat 1958) and the interferometer of variable spacing (Stanier 1950, Machin 1951, O'Brien 1953); these two methods achieve the same result but in the former case all the aerials are present simultaneously, while in the latter case two aerials only are moved, successively, to the different positions of the grating.

Similar techniques have been developed for the study of the distribution of the galactic emission. Conventional systems have been successfully employed for the short wave-lengths, but in many problems, particularly those involving the distribution of radio "brightness" at high galactic latitudes, it is important to make observations at wave-lengths of several metres; by dispersing the available collecting area it is again possible to obtain much larger resolving powers with adequate sensitivity. Two systems have been used, one of which employs a pair of mutually perpendicular thin apertures (Mills and Little 1953); the other utilizes a single thin aperture in conjunction with a small movable aerial (Blythe 1957). An earlier version of the latter instrument was used by Scheuer and Ryle (1953) to obtain increased resolving power in one coordinate for studying the distribution of radio emission across the galactic plane.

The observation of radio stars presents a more complex problem in which three main approaches have been made. (*a*) The use of conventional systems of moderate size at a wave-length sufficiently short to provide the required angular resolution (Piddington and Trent 1956; Seeger, Westerhout and Conway 1957; Roman and Yaplee 1958). These observations have been characterized by very small signal/noise ratios, and in consequence long integration times are used and extensive surveys have not been possible. (*b*) The use of conventional aerials of larger collecting area at longer wave-lengths (Hanbury Brown and Hazard 1953; Ryle and Hewish 1955); the signal/noise ratio of such systems has usually been large and the observations have been limited by resolution. (*c*) The use of "unfilled" systems such as the Mills Cross (Mills and Little 1953) at a long wave-length. The angular resolution of such a system may be very large but the collecting area is smaller.

If any major increase in the depth of radio star surveys is to be made it is necessary to construct systems with a collecting area exceeding 50 000 ft² capable of operating at shorter wave-lengths or to use, at long wave-lengths, "unfilled" systems having larger collecting areas than the Mills Cross.

Since the flux density of the sources falls with wave-length and since the difficulties of constructing an aerial of given physical size get very much greater if it is to operate at short wave-lengths, it is possible that even with the new receiver techniques now being developed (Adler 1958, Giordmaine *et al.* 1958) it may be difficult to attain a sufficient improvement by using conventional systems at short wave-lengths. The alternative approach, which utilizes a long wave-length and correspondingly large collecting areas, appears to offer greater possibilities.

In this paper an account is given of a new principle in the design of large radio telescopes. The method, which has already been used successfully by Blythe (1957), makes use of two aerials arranged successively in different configurations to provide information equivalent to that obtained from the use of an aerial of much greater physical size. Apart from a considerable economy of structure this method avoids some of the difficulties associated with the physical achievement of a graded excitation of amplitude and phase which is required in the case of large extended arrays such as the Mills Cross. Besides allowing greater collecting areas to be realized, the shape of the reception pattern can be adjusted, by computation alone, to suit different types of observation. The method necessarily involves considerable computation but this does not present a serious problem with the large electronic computers now available.

2. *The principle of aperture synthesis.*—It is well known that a simple interferometer, consisting of a pair of non-directional receiving elements connected to a receiver, gives a response proportional to one Fourier component of the two-dimensional distribution of radio brightness across the sky. By taking measurements in which both the spacing and the orientation of the interferometer are varied it is possible to determine the two-dimensional transform of the brightness distribution within certain limits and hence, by Fourier inversion, to derive the brightness distribution as observed with a specified resolving power (O'Brien 1953). Aperture synthesis may be regarded as a logical extension of this method but it will here be discussed in a somewhat different, but essentially equivalent, way which may give a more direct understanding of the method.

Consider, for example, the operation of a receiving aperture such as a large two-dimensional array, or paraboloidal reflector. The signal which is delivered to a receiver from such an aerial may be regarded as the vector addition of the currents induced in each portion of the aperture. In the case of a paraboloid this vector addition is achieved, automatically, at the focal point, while for an array the same result is achieved by connecting all the elements to the receiver in the same phase. When the direction of an incident wave does not coincide with the normal to the aperture plane the induced current in the elements of the aperture will suffer a progressive phase shift and vector addition then gives rise to the usual directional properties of the aperture. In the case of an array, it it possible to add the currents with their phases suitably adjusted, and in this way to vary the direction of the principal response of the aperture; this feature is of great importance in large aerial systems where physical tilting of the aperture plane may be impossible.

Now if it were possible to measure, separately, the current induced at each portion of the aperture by moving a small aerial successively across it, vector addition of the currents would give, for a constant source, exactly the same result as that obtained by using the complete large aperture. This process is not possible using a single receiving element since measurement of the phase of the induced current raises difficulties. By using two small aerials arranged as an interferometer, however, the relative phases are readily obtained and the successive determination of the current in this way forms the basis of the synthesis method.

(*a*) *The synthesis of a single aperture.*—Consider a single rectangular aperture such as that shown in Fig. 1. The resultant current induced in the aperture by an incident plane wave is $\sum I_n e^{i\phi n}$ where ϕ_n is the phase of the wave at the position of the nth element. The power P delivered to a receiver is then given by

$$P \propto \sum I_n e^{i\phi n} \sum I_n e^{-i\phi n} \propto \sum I_n{}^2 + \sum I_m I_n \cos(\phi_m - \phi_n). \qquad (1)$$

If all the elements are of the same size, the first term in expression (1) is simply N times the power derived from a single element; the terms of type $I_m I_n \cos(\phi_m - \phi_n)$ are just the output obtained when a phase-switching receiver is connected to the elements m and n through equal cables (Ryle 1952). A single measurement of the power induced in one element, suitably combined with the outputs from a phase-switching receiver connected to a pair of elements, arranged successively to cover all possible combinations of positions, then gives a result exactly equivalent to using the whole large aperture. By performing the addition with the value of $(\phi_m - \phi_n)$ adjusted to give a progressive phase shift across the aperture it is possible to vary the direction of the maximum response of the aperture in a manner analogous to tilting the aperture plane. It is thus possible, without the necessity for further observations, to scan the synthesized aperture through an angle which is only limited by the directional properties of the small element. To perform the addition with modified values of $(\phi_m - \phi_n)$ it is also necessary to measure terms of the type $I_m I_n \sin (\phi_m - \phi_n)$, and these are obtained by connecting the pair of elements in phase quadrature to the phase-switching receiver.

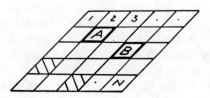

FIG. 1.—*The synthesis of a rectangular aperture.*

The measurements might proceed in the following way: having determined the power induced in a single element, an interferometer is set up with one element A fixed at position 1 and the sine and cosine terms are obtained by placing the other element B successively in all the remaining positions. Element A is then moved to position 2, etc., and the same procedure carried out. Now if this were done it is immediately obvious that many arrangements of A and B, such as those shaded in Fig. 1 are identical. A procedure for obtaining the different terms with no repetition is indicated in Fig. 2 (a) where the element A is fixed and element B is arranged to cover, once only, all the positions contained in a rectangle of approximately twice the area. It is of course possible to move both elements, in which case all arrangements can be obtained without moving the elements outside a rectangle of side D. It may be noted that if a total of N observations was required initially, the number now needed with no repetition is $(2N)^{1/2}$.

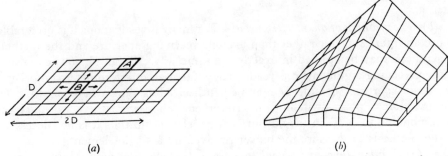

(a)   (b)

FIG. 2.—(a) *One method of synthesizing a rectangular aperture without repeating any element configuration.*

(b) *The weighting function for each position of the movable element.*

To obtain results equivalent to using the complete aperture the different terms must now be added with a weighting factor appropriate to the number of repetitions of each configuration of the elements. The form of the weighting function appropriate to each position of the movable element is sketched in Fig. 2 (*b*). Direct addition of the terms without such a tapered weighting function gives a result equivalent to the use of an "optimum" array as discussed by Arsac (1955) and Barber (1958). While such systems give the greatest ratio of information to noise they are sometimes undesirable in practical measurements owing to the large subsidiary maxima inherent in their reception patterns.

When the elements used in the synthesis have a width much greater than one wave-length, addition of the different terms with a progressive phase-shift is equivalent to the use of an aperture having a discontinuous "stepped" phase distribution across it. As in the case of an echelon diffraction grating this gives rise to more than one principal maximum in the reception pattern. This difficulty can be avoided if it is arranged that the different locations of the movable element overlap slightly.

It is not necessary for the elements *A* and *B* to have the same shape and there are sometimes practical reasons why it is preferable for one of them to extend over the complete width of the required aperture so that synthesis is only performed in one dimension. A system of this type, which is exactly equivalent to that shown in Fig. 2 (*a*) is shown in Fig. 3 (*a*). When this method is adopted it is necessary to arrange that the current induced in each element of the extended array has the value appropriate to the weighting function shown in Fig. 2 (*b*). The aperture must consequently have a tapered distribution of excitation along its length.

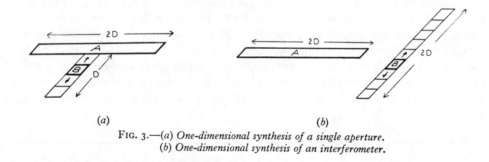

(*a*)   (*b*)

FIG. 3.—(*a*) *One-dimensional synthesis of a single aperture.*
(*b*) *One-dimensional synthesis of an interferometer.*

(*b*) *The synthesis of an interferometer.*—For many investigations it is preferable to use an interferometer rather than a single receiving aperture and the method of synthesis may be applied in exactly the same way. By an obvious extension of the preceding arguments it is readily shown that an interferometer comprising two large apertures of side *D* may be synthesized as shown in Fig. 3 (*b*). In the case of a phase-switching interferometer, however, there is no term corresponding to $\sum I_n{}^2$ in the output of the receiver and thus it is not necessary in the synthesis to measure the power received by a single element.

3. *Observation time and signal-to-noise ratio.*—Before describing practical applications of the synthesis method it is necessary to consider how it compares with conventional methods in regard to the observation time and signal to noise

ratio. Since a synthesis using small elements requires many different measurements it might be expected that the observing time is necessarily longer. The time required for the repeated observations is, however, almost exactly compensated by the increased scanning rate which may be employed. It is useful to define a quantity which gives a direct comparison of the performance of synthesized and conventional instruments having the same angular resolution. In principle, the observation of a given area of sky can, by suitably arranging the scanning rate and the receiver time constant, be made to occupy a certain interval of time. We can therefore define the "efficiency" of a synthesized instrument as the ratio of its absolute sensitivity (i.e. signal/noise ratio) as compared with that of a conventional instrument of the same resolving power when the same observing time is allowed in each case. The efficiencies of various types of synthesized instrument will now be discussed.

(*a*) *Scanning rate and observation time.*—During a survey a radio telescope is generally scanned continuously on one coordinate, either by steering the instrument or by using the Earth's rotation, and a given region of the sky is investigated strip by strip. For our purpose it is more convenient to imagine that the scanning is performed in discrete intervals in both coordinates, the instrument being directed towards a certain point in the sky for the required integration time and them moved instantaneously to the next point. The angular distance between adjacent points in the scanning lattice depends upon the resolving power of the instrument and Bracewell (1956) has shown that the greatest interval allowed when the aerial is a rectangular aperture of sides a and b is given by $\lambda/2a$ and $\lambda/2b$ in the respective coordinates, where λ is the wave-length. No additional information is gained by using a finer scanning lattice but information is lost if a coarser lattice is adopted.

Suppose, now, that a given solid angle Ω of the sky is scanned using a square aperture of side D. The total observing time is given by $4\Omega D^2\tau/\lambda^2$ where τ is the integration time of the receiver. If the sky is scanned using the equivalent synthesized instrument in which the small elements are square apertures of side d, the time required to scan the solid angle for each of the arrangements is $4\Omega d^2\tau/\lambda^2$ where it is assumed that the receiver time constant is the same as before. The number of different arrangements is approximately $2D^2/d^2$ and the total observing time for the same receiver time constant is therefore $8\Omega D^2\tau/\lambda^2$. This is seen to be only twice that required had the complete aperture been used and it is important to notice that it is independent of the size of the small elements.

To calculate the observing time when synthesis is performed in only one dimension, consider the system shown in Fig. 4 (*a*). If one of the elements is a rectangle of sides d_1 and d_2, and the other an extended aperture of width d_1 and length $2D$, then the scanning intervals are $\lambda/2d_1$ and $\lambda/2D$ *. The total number of arrangements is approximately D/d_1 and the observing time is therefore $4\Omega D^2\tau/\lambda^2$. The observing time is thus half that needed when both elements were small and the same as if a complete aperture had been used. It is also of interest to consider the system shown in Fig. 4 (*b*) which is similar to an aerial constructed by Mills except that one arm of the cross is omitted. Although no

---

* The extended aperture of length $2D$, when used with the appropriate tapered excitation is equivalent, when scanning, to a uniform aperture of length D. More exactly, the scanning interval is $\lambda/2D+d_2$ but we assume $D\gg d_2$.

16

synthesis is required such a system yields the same (non-repetitive) information as do the synthesized apertures under discussion and it will be shown that it has approximately the same efficiency. We note here that the scanning interval is $\lambda/2D$, as for the complete aperture, so that the observation time is again $4\Omega D^2\tau/\lambda^2$.

It should be noted that this analysis assumes that the bandwidth of the receiver is not too great and that the area of sky under investigation is large enough to contain more than one coarse scanning interval $\lambda/2d$. The effect of a finite bandwidth is to narrow the reception pattern of the individual elements at large spacings (Ryle and Vonberg 1948), and this may necessitate the use of a finer scanning lattice. In the case of a single aperture it may be shown that the effect is negligible provided that $\Delta f/f$ is appreciably less than d/D; for an interferometer the ratio $\Delta f/f$ must be less than the ratio of d to the maximum element spacing.

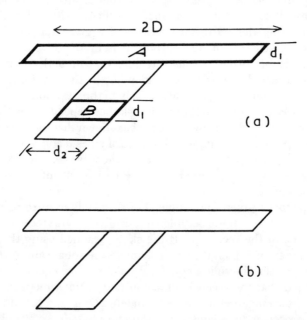

Fig. 4.—(a) *The synthesis of a rectangular aperture.* (b) *A similar system which may be compared to the Mills Cross. It can be regarded as a synthesis in which all configurations are present simultaneously.*

(b) *The efficiency of synthesized systems.*—To derive the efficiency of a synthesized aperture we note that the signal obtained with a complete aperture of side D is directly proportional to its area D^2. If, for an integration time τ, the root mean square receiver noise is n then the signal/noise ratio is proportional to D^2/n.

When the aperture is synthesized using small elements of area d^2 the signal obtained by adding the $2D^2/d^2$ different observations with appropriate weights is proportional to D^2. Since the receiver noise is unrelated for the different observations the resultant noise is $2^{1/2}nD/3d$ where weighted addition has been performed. The signal/noise ratio is thus given by $3Dd/2^{1/2}n$. Since however, the observing time for a fixed receiver time constant was twice that required when using a conventional aperture, the efficiency is given by $3d/2D$.

In the case where a one-dimensional synthesis is used, as in Fig. 4 (a), the signal obtained for each arrangement of A and B is proportional to $2(Dd_1^2 d_2)^{1/2}$ since the equivalent area is given by the geometric mean of the two apertures (Ryle 1952). The number of arrangements is D/d_1 and the signal/noise ratio is $D(3d_1 d_2)^{1/2}/n$. In this case the observing time was the same as with a conventional aperture and so the efficiency is $(3d_1 d_2)^{1/2}/D$. For the system of Fig. 4 (b), in which no synthesis is employed, the signal is $D(2d_1 d_2)^{1/2}$ and so the signal/noise ratio is $D(2d_1 d_2)^{1/2}/n$ and hence the efficiency is $(2d_1 d_2)^{1/2}D$. All the systems shown in Figs. 3 and 4 thus have a comparable efficiency and the maximum possible efficiency in a synthesized aperture is $(d/D)^{1/2}$ which is attained when $d_2 = D$.

While the efficiency of a synthesized aperture is necessarily less than that of a conventional aperture of the same resolving power this should not be regarded as an immediate disadvantage. As mentioned earlier, the efficiency of a conventional aperture is often far higher than is required for a particular investigation so that the same physical aperture, arranged as a synthesized system of adequate efficiency and increased angular resolution, would have provided a much more powerful instrument.

4. *The practical application of the method*

(a) *General considerations*.—Since systems derived by one- or two-dimensional synthesis have approximately the same performance there might seem to be considerable advantages with a two-dimensional system arising from the economy of physical structure. There are, however, a number of practical disadvantages with two-dimensional synthesis and these are discussed below.

(i) Data handling and computation: for a one-dimensional synthesis involving N arrangements of the elements, the corresponding two-dimensional system would require N^2 arrangements. The computational and data-handling problems are thus enhanced.

(ii) Scanning: the angular scanning rate of a two-dimensional system must be considerably higher than for a one-dimensional system of the same resolving power for a survey to occupy a given observing time (cf. Section 3). It is a great practical convenience to employ the Earth's rotation for scanning in right ascension, but this fixed scanning rate may be too slow for the desired observational programme. Facility for pointing the elements in both coordinates is then necessary which raises practical difficulties which may be avoided in a one-dimensional system.

(iii) Receiver linearity: in a two-dimensional synthesis each intense source will be received over a solid angle λ^2/d^2, and it may be difficult to ensure a sufficiently linear overall response to allow observation of weak sources within this area. In the case of one-dimensional synthesis the area of sky affected in this way is reduced by the receptivity of the long aerial to λ^2/Dd.

(iv) Bandwidth: a limitation particularly relevant to the synthesis of an interferometer is set by the condition $d/D > \delta f/F$ (cf. Section 3). Since the overall spacing of the synthesized apertures will be several times larger than their individual aperture D, the maximum bandwidth permitted in a two-dimensional synthesis is necessarily smaller than for the corresponding one-dimensional system in which the long aerial runs East–West.

(b) *The design of systems utilizing the Earth's rotation*.—For the reasons outlined above systems adopting one-dimensional synthesis have a number of practical advantages. In current applications long thin apertures running

16*

East–West, with provision for rotation about the long axis, may be constructed relatively economically and so the remaining discussion will be restricted to the design of such systems; it is possible, however, that other problems might reach a simpler solution if two-dimensional methods were employed.

At first sight the restriction of the design to what is essentially a transit instrument, in which the scanning rate is fixed by the Earth's rotation, might be thought to lead either to inadequate sensitivity due to a too short integration time, or to an unnecessarily long observing time. While it is true that the observing time required to survey a given area of sky is determined entirely by the angular resolution of the system and the rotation speed of the Earth, it is, however, possible to adjust the efficiency so that this scanning rate is always the optimum for a particular application. It is assumed that the area of sky involved includes 24^h in right ascension and an angle of at least λ/d in declination.

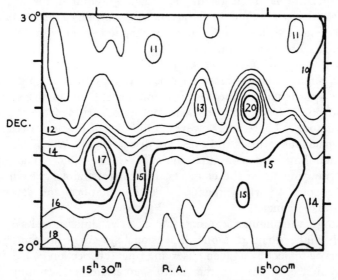

FIG. 5.—*Part of a map obtained with the 7·9 m " pencil-beam " system.*
The contours are in units of 1000 °K.

In Section 3 the efficiency of a one-dimensional system such as that in Fig. 3 was shown to be proportional to the quantity $(d_1 d_2)^{1/2}/D$. Once the efficiency has been determined, the relative values of d_1 and d_2 may be decided by considerations of structural economy provided that their product is constant.

5. *The design of two practical systems*

(a) *A pencil beam aerial for a wave-length of 7·9 m.*—This instrument was designed to provide a high resolution survey of the general galactic background radiation at a long wave-length so that, in conjunction with other observations at shorter wave-lengths, reliable information would be available on the spectral distribution of the emission from many galactic features.

The original instrument built by Blythe (1957) employed one-dimensional synthesis as in Fig. 4 (a) with an East–West array of length $D = 1200$ ft and a movable element of length $d_1 =$ one wave-length. The effective width d_2 was less than the wave-length so that a single series of observations provided a complete map of the accessible sky. However, the presence of ground reflections led to an uncertainty in the receptivity in declination and gave an uncertainty in the overall scaling of brightness at southerly declinations.

The long fixed element of the radio star interferometer.

M. Ryle and A. Hewish, The synthesis of large radio telescopes

In the new instrument, which was designed to give an angular resolution of $0°\cdot8 \times 0°\cdot8$, a fixed aerial of length 3300 ft and width 40 ft was used so that a well-defined receptivity in declination was achieved at declinations above $-10°$. This increase in width and the use of a movable element of length 100 ft gave an efficiency of 7 per cent.

A preliminary survey of limited resolving power $(0°\cdot8 \times 1°\cdot6)$, in which spacings of the movable aerial up to 30λ were used, has been carried out and part of the map is shown in Fig. 5.

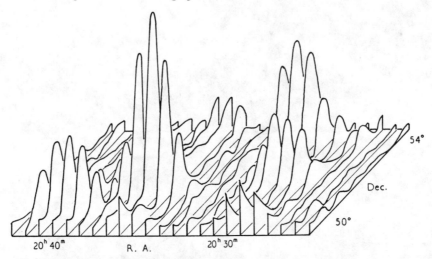

FIG. 6.—*A section of the first map obtained with the radio star interferometer. The most intense source at* $20^h 37^m$ *has a flux density of* 11×10^{-26} *w.m.*$^{-2}$ $(c/s)^{-1}$.

(*b*) *A radio star interferometer on a wave-length of* $1\cdot7$ *m.*—A system capable of carrying out a deeper survey of radio sources than those made hitherto Mills and Slee 1957, Edge *et al.* 1959) was required, and considerable increases of both resolution and sensitivity were necessary. In addition the design was influenced by the importance of obtaining the positions of sources with greater accuracy, particularly in declination, and by the necessity for obtaining good measurements of the angular diameter of the most intense sources.

An interferometric system having an approximately symmetrical primary reception pattern was adopted, in which the resolution in right ascension was provided by a 1450 ft parabolic trough running East–West (as in Plate 7). The required efficiency (about 25 per cent) was decided from an extrapolation of the number versus flux density counts of previous surveys; the ratio of the width d_1 of both aerials, and the length d_2 of the movable aerial were governed (cf. Section 4) by consideration of the relative costs of the parabolic elements and of the railway track supporting the movable aerial (Ryle 1960).

A section of the first map obtained with this instrument is shown in Fig. 6. Some observations to establish accurate positions of 64 of the most intense radio stars have already been described (Elsmore, Ryle and Leslie 1960).

Mullard Radio Astronomy Observatory,
    Cavendish Laboratory,
        Cambridge:
1959 August.

References

Adler, R., 1958, *Proc. I.R.E.*, **46**, 1300.

Arsac, J., 1955, *Opt. Acta.*, **2**, 112.

Barber, N. F., 1958, *New Z. J. Sci.*, **1**, 35.

Blum, E. J., Boischot, A., and Ginat, A., 1957, *Ann. d'Astrophys.*, **20**, 115.

Blythe, J. H., 1957, *M.N.*, **117**, 644.

Bracewell, R. N., 1956, *Aust. J. Phys.*, **9.**, 297.

Brown, R. Hanbury, and Hazard, C., 1953, *M.N.*, **113**, 123.

Christiansen, W. N., 1953, *Nat re*, **171**, 831.

Edge, D.O., McAdam, W. B., Shakeshaft, J. R., Baldwin, J. E., and Archer, S., 1959, *Mem. R.A.S.*, **68**, 37.

Elsmore, B. E., Ryle, M. and Leslie, P. R. R., 1959, *Mem. R.A.S.*, **68**, 61.

Giordmaine, J. A., Alsop, L. E., Mayer, C. H., and Townes, C. H., 1958, *Proc. I.R.E.*, **47** 1062.

Machin, K. E., 1951, *Nature*, **167**, 889.

Mills, B. Y., and Little, A. G., 1953, *Aust. J. Phys.*, **6**, 272.

Mills, B. Y. and Slee, O. B., 1957, *Aust. J. Phys.*, **10**, 162.

O'Brien, P. A., 1953, *M.N.*, **113**, 597.

Piddington, J. H., and Trent, G. H., 1956, *Aust. J. Phys.*, **9**, 481.

Roman, N. G., and Yaplee, B. S., 1958, *Proc. I.R.E.*, **46**, 199.

Ryle, M., 1952, *Proc. Roy. Soc.* A, **211**, 351.

Ryle, M., 1960, *J.I.E.E.*, **6**, 14.

Ryle, M., and Hewish, A., 1955, *Mem. R.A.S.*, **67**, 97.

Ryle, M., and Vonberg, D. D., 1948, *Proc. Roy. Soc.* A, **193**, 98.

Seeger, C. L., Westerhout, G., and Conway, R. G., 1957, *Ap. J.*, **126**, 585.

Scheuer, P. A. G., and Ryle, M., 1953, *M.N.*, **113**, 3.

Stanier, H. M., 1950, *Nature*, **165**, 354.

Earth-Rotation Aperture Synthesis

EDWARD B. FOMALONT

Abstract—Resolution for radio astronomy in the order of 1″ is necessary for the study of distant radio galaxies and quasars, for detecting faint sources, and for the mapping of clouds of hydrogen and other molecules. To obtain these resolutions many new or planned radio instruments use arrays of moderate size radio telescopes to synthesize large physical apertures. These instruments are generally composed of one or several linear arrays and utilize the rotation of the earth to change the relative orientation of the array and the radio source.

The techniques used for earth-rotation aperture synthesis are discussed. The response of a two-element interferometer and the geometry associated with earth-rotation synthesis are reviewed; the current and proposed designs for these instruments and their performance are described; and, finally, the inversion methods for determining the angular power distribution of a radio source from the array response (visibility function) are outlined.

I. Introduction

THE BASIC MEASUREMENT in radio astronomy is the determination of the distribution of power $I_{ij}(\nu, \delta, t)$ from a celestial source as a function of frequency ν, angular position δ, polarization state ($i, j = 1, 2$ represent the four polarization states), and a possible variation with time t. With the use of a simple large telescope with narrow bandwidth capabilities and proper polarimeters all of the characteristics of the radio radiation can be measured.

For astronomical applications, however, the useful sensitivity of a simple large telescope is often more limited by its finite angular resolution than by its sensitivity. Thousands of discrete radio sources, many associated with distant galaxies

Manuscript received April 19, 1973. The National Radio Astronomy Observatory, Green Bank, W. Va., is operated by Associated Universities, Inc., under Contract with the National Science Foundation.

The author is with the National Radio Astronomy Observatory, Green Bank, W. Va. 24944.

and quasars, with incident power levels of 10^{-27} W·m^{-2} Hz^{-1} have been cataloged, but only a few percent can be resolved with the beamwidth of several arc minutes available using even the largest simple telescopes operating at their highest frequency ($\sim$30 GHz). Detection of faint sources also requires high resolution to avoid confusion between the many sources at low power levels in the reception area of a telescope. Radiation, both line and continuum, from the Milky Way Galaxy and nearby galaxies can generally be resolved with large telescopes, but higher resolution would be useful even here in understanding the physical processes concerning regions of ionized hydrogen, neutral hydrogen, cold dense molecular clouds, or concerning the mass and velocity distributions in galaxies.

Many methods have been used by radio astronomers to obtain high resolution since it is not economically feasible to construct a simple steerable telescope larger than about 100 m or a fixed telescope of about 400 m. Because it is unnecessary to completely fill an area with a reflecting surface to provide the resolution of that area, skeleton instruments have been designed to obtain high resolution without using large amounts of collecting area. A skeleton instrument may contain several long thin pieces of a parabolic reflector, several cylindrical reflectors, or an array of many small elements suitably connected. Configurations such as a "tee," cross, or ring using these reflectors produce pencil beams comparable with the spanned aperture. Skeleton instruments are limited to sizes, however, to about a kilometer. A detailed analysis of these instruments is given elsewhere [1].

It is not even necessary that all of the elements of an aperture be present simultaneously. A large aperture may be synthesized by using several moveable elements to occupy in

Reprinted from *Proc. IEEE*, vol. 61, no. 9, pp. 1211–1218, Sept. 1973.

turn the relative positions that occur in a large aperture. The relative positions can also be changed by utilizing the rotation of the earth. If a radio source is followed in its diurnal motion by all of the elements, the relative positions between the elements as viewed by the radio source change both in length and orientation. Thus by utilizing both the rotation of the earth and the mobility of the elements, very large apertures can be synthesized. Observations of transient phenomena such as pulsars and the sun are not suited for aperture synthesis.

The techniques used for earth-rotation aperture synthesis (sometimes called supersynthesis) are discussed in this paper. The application of coherence theory to radio interferometers and arrays has been fully developed elsewhere [2]–[6], but a brief geometric derivation is given in Section II. In Section III the characteristics of various types of arrays and array design criteria with the emphasis on earth-rotation synthesis methods are discussed. Fourier inversions and other techniques used to obtain the angular power distribution are described in Section IV.

II. Basic Interferometry

The radiation produced by virtually all astronomical objects is statistical in nature and can be completely described in terms of the average power distribution $I_{ij}(\nu, \delta, t)$. Using the theory of partial coherence, all of the characteristics of the power distribution are related to cross correlations of the incident radiation field at different points in space (spatial coherence), in time (temporal coherence), and between the field components (polarization coherence) [7]. It can be shown that the spatial coherence function and the angular power distribution are two-dimensional Fourier transforms and the temporal coherence function and the frequency power distribution (spectrum) are one-dimensional Fourier pairs. Thus by measuring the coherence of the incident radiation over a wide range of spatial and temporal separation in various polarizations, the power distribution may be reconstructed with much higher angular or spectral resolution than is possible with a large simple telescope.

Arrays of antenna elements are essentially probes of the spatial coherence of a radiation field. With a large number of elements, suitably connected, a sufficient number of spatial coherence components can be measured in a reasonably short period of time in order to obtain a satisfactory reconstruction of the angular power distribution.

A simple two-element interferometer, the building block of all arrays, is shown in Fig. 1. Consider a monochromatic signal of frequency ν emanating from a point source whose direction is given by δ. Two elements separated by a fixed baseline B intercept the radiation, two samples of which are then transported without loss of characteristics to a common point where they are correlated; i.e., their time averaged sum or product is taken. In most arrays the radio frequency ν is converted to an intermediate frequency (IF) ν_{IF}, by a superheterodyne process using a common local oscillator of frequency ν_0, before transmitting the signals to the central point. Time delay τ_D is usually inserted in the IF line to compensate for the geometric time delay difference τ_g in the arrival of the radiation to each element.

The time-variable part of the response after correlation is

$$R(\nu, \delta) = S_\nu \exp\left\{j2\pi[\nu\tau_g - \nu_{IF}\tau_D]\right\} \tag{1}$$

where S_ν is proportional to the source power, usually denoted as flux density with one flux unit defined as $10^{-26} \, \text{W} \cdot \text{m}^{-2} \, \text{Hz}^{-1}$.

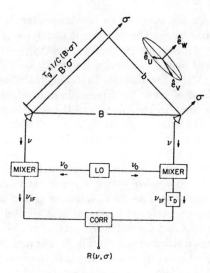

Fig. 1. Schematic diagram of a simple two-element interferometer.

The usual complex (phasor) notation has been used for the response. The phase of the response is equal to the *phase* (not time!) path length difference between the two radiation paths.

The diurnal motion of an extraterrestrial radio source continuously varies the geometric delay causing the response to have a quasi-sinusoidal behavior—commonly called fringes. Alternatively, the fringes can be described as a "beating" of the two signals which are Doppler shifted to slightly different frequencies at the correlator due to the relative motion of the two elements with respect to the source.

For observations over a large frequency bandwidth $\Delta\nu$ the differential time delay $\Delta\tau$ between the signals at the correlation point must be smaller than $(\Delta\nu)^{-1}$ to insure that the entire bandwidth adds in phase. Thus the inserted delay τ_D is usually varied (tracked) to compensate for the change of τ_g. This can be done with digital or analog techniques to an accuracy of better than 10^{-9} s. Assuming accurate delay tracking with respect to an angular position δ_0, the interferometer response becomes

$$R(\nu, \delta) = S_\nu \exp\left\{j2\pi\frac{\nu_0}{c}B\cdot\delta_0\right\}$$
$$\cdot \exp\left\{j2\pi\frac{\nu}{c}B\cdot(\delta - \delta_0)\right\}. \tag{2}$$

The response to an extended source $I(\delta - \delta_0)$ with a wide bandwidth interferometer having a frequency characteristic of $\beta(\nu)$ and a primary power pattern $G(\delta - \delta_0)$[1] is given by integrating (2) over δ and ν

$$R = \exp\left\{j2\pi\frac{\nu_0}{c}B\cdot\delta_0\right\} \int\int \beta(\nu)G(\delta - \delta_0)I(\delta - \delta_0)$$
$$\cdot \exp\left\{j2\pi\frac{\nu}{c}B\cdot(\delta - \delta_0)\right\} d\delta \, d\nu$$
$$\equiv \exp\left\{j2\pi\frac{\nu_0}{c}B\cdot\delta_0\right\} \mathcal{U}(B). \tag{3}$$

For simplicity, the angular dependence of the source and the

---

[1] The primary power pattern of a two-element interferometer equals the product of the voltage pattern of each element.

power pattern are assumed to be independent of frequency and normalized at δ_0. The response is composed of two parts: the fast time dependence (fringes) is equal to the response of a point source at δ_0 at an observing frequency ν_0. A slower varying quantity $\mathcal{V}(B)$, the visibility function, describes the amplitude and phase offset of the fringes. In general, the visibility function is a complicated function of the bandwidth, emission extent, and baseline geometry. The visibility function is closely related to the spatial mutual coherence function used in coherence theory.

The various quantities are generally expressed in terms of an astrometric coordinate system fixed to the radio source. A unit vector e_w is taken as the direction of δ_0 and the two perpendicular unit vectors are e_u and e_v with ground projections towards the east and the north, respectively. This coordinate system is shown schematically in Fig. 1. The angular displacement $(\delta - \delta_0)$ can be expressed in terms of the direction cosines (l, m, n)

$$(\delta - \delta_0) = e_u l + e_v m + e_w (n - 1). \quad (4)$$

Using the astronomical coordinates of δ_0 as (h_0, δ_0) where h_0 is the hour angle and δ_0 is the declination and (h, δ) are the coordinates of δ, the direction cosines become

$$l = -\cos \delta \sin (h - h_0)$$

$$m = \cos \delta_0 \sin \delta - \sin \delta_0 \cos \delta \cos (h - h_0) \approx (\delta - \delta_0)$$

$$n = \sin \delta_0 \sin \delta + \cos \delta_0 \cos \delta \cos (h - h_0) \approx 1 - \tfrac{1}{2}(l^2 + m^2).$$

$$\approx (h_0 - h) \cos \delta_0 = (\alpha - \alpha_0) \cos \delta_0 \quad (5)$$

The right ascension of a source α is related to the hour angle by the sidereal time t where $\alpha = t - h$. The baseline B can be decomposed into a transverse projected baseline $b = e_u u + e_v v$ and a longitudinal baseline $e_w(B \cdot \delta_0)$ which is equal to $c\tau_g$ where c is the speed of light. In the astrometric frame of reference, the components of the baseline change as a source moves in its diurnal track, giving

$$u = \frac{\nu_0}{c} |B| \cos D \sin (h_0 - H)$$

$$v = \frac{\nu_0}{c} |B| \{\cos \delta_0 \sin D - \sin \delta_0 \cos D \cos (h_0 - H)\}$$

$$w = \frac{\nu_0}{c} |B| \{\sin \delta_0 \sin D + \cos \delta_0 \cos D \cos (h_0 - H)\} \quad (6)$$

where the baseline direction is given by (H, D), the hour angle and declination of the interferometer end point. The loci of points in the $(u-v)$ plane for the projected baseline b are ellipses centered on the v axis. The units of the baseline are most naturally expressed in units of wavelengths at the local-oscillator frequency. Examples of tracks are given in Fig. 2. Many synthesis arrays make use of the earth rotation to cover a wide range of projected spacing.

With certain simplifications, the visibility function defined in (3) becomes

$$\mathcal{V}(u, v) = \iint G(l, m) I(l, m) \exp \{j2\pi (ul + vm)\} \, dl \, dm \quad (7)$$

the two-dimensional transform of the angular power distribution. An array of N elements can be decomposed into a maximum number of $N(N-1)/2$ interferometer pairs each sampling the visibility function on the two-dimensional $(u-v)$

plane. With sequential movement of some of the elements and the use of the earth rotation, if necessary, the $(u-v)$ plane can be "adequately" covered and a radio map $I'(l, m)$ can be reconstructed using the inverse relation

$$I'(l, m) = G(l, m) I(l, m)$$

$$= \iint \mathcal{V}(u, v) \exp \{-j2\pi (ul + vm)\} \, du \, dv. \quad (8)$$

The major assumptions and simplifications used in deriving (7) and (8) are [4] as follows.

1) The linearization of $B \cdot (\delta - \delta_0)$ into u and v is only valid for small angular displacements. The neglected term $(n-1)w$ can be incorporated by a redefinition of m only for an east-west oriented baseline [8] where an aperture is defined by the array. Otherwise, the Fourier pair relationship between the visibility function and the angular power distribution is not true. For a baseline of 10^6 wavelengths, the phase term can be as large as $70°$ at a displacement of $10'$.

2) The effect of a large frequency bandwidth has not been included. For a bandwidth $\Delta\nu$, the exponential in the integral of (3) does not add coherently for emission at large angular displacements in the direction of the projected baseline. For a baseline of 10^6 wavelengths and $\Delta\nu = 100$ MHz the coherence angle is about $2'$. For some arrays, large bandwidths are chosen to produce a coherence area which is smaller than the primary beam area [9], [10]. The resultant radio maps using (8) produce a "radial smearing" due to the large bandwidths [11].

3) The radio map includes the effect of the primary power pattern which must remain unchanged for all interferometer pairs at all times. This requires accurate pointing of each element as the source moves in its diurnal track and the use of identical element pairs. It also assumes that the elements be polar mounted so that there is no mutual rotation between $G(l, m)$ and $I(l, m)$.

4) The $1/n$ term in the solid angle $d\delta = (1/n)dl \, dm$ can be incorporated in the definition of $G(l, m)$.

5) There is no angular coherence of the radiation. A treatment for partially coherent radiation [4] shows that little can be learned of the angular power distribution by interferometric techniques.

6) Only identically polarized pairs should be included in (8) for reconstruction of the radio map in the particular polarization state measured by the pairs.

III. ARRAY DESIGN AND PERFORMANCE

The first synthesis array designed to use the rotation of the earth was erected in 1955 [12] and in the last ten years increasing use and understanding of earth-rotation synthesis arrays have been made [13]. These arrays consist of a modest number (2 to $\sim$30) of elements, each a paraboloidal antenna of significant size on a suitable mount for mechanically following a radio source in its diurnal track. In most arrays some of the telescopes are moveable on railroad track or roadbeds in order to change the array resolution or fill-in factor. All of

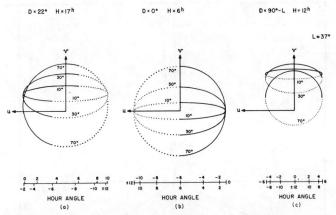

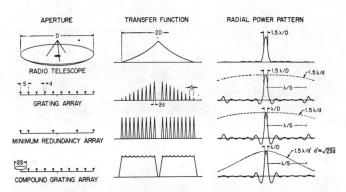

Fig. 3. The radial transfer function and radial power pattern with a) a simple parabolic telescope, b) a grating array, c) a minimum redundancy array, and d) a compound grating array.

Fig. 2. The loci of points in the $(u–v)$ plane produced by a tracking interferometer. (a) A skew baseline. (b) An east–west baseline. (c) A north–south baseline. The loci are drawn for declinations 70°, 30°, and 10°. The solid portion of each curve corresponds to the hour angle range −6h to +6h; the dotted portion for the hour angle range of +6h to +18h. The hour angle scale, which is only a function of u, is given at the bottom of each diagram. For declinations south of the equator, use the curve with positive declination but reflect the curve about $v = v_0$. The baseline orientation is given by H and D. (Taken from [11].)

these arrays require hours or days of observing time of an area of sky in order to obtain a good angular reconstruction of the source.

Nearly all of the recent earth-rotation synthesis arrays consist of a linear array of antennas from which a satisfactory one-dimensional distribution can be synthesized [14, table I]. The elements are usually placed at integral multiples of a basic interval called the stepping interval. As demonstrated in Fig. 2, the direction of maximum resolution rotates as a source is tracked so that relatively complete angular coverage of the $(u–v)$ plane may be obtained in about 12 h. The most suitable coverage for high-declination sources is obtained with an east–west oriented array where the $(u–v)$ tracks are centered at the origin. Only half of the $(u–v)$ plane need be sampled since the angular power distribution in any polarization must be a real quantity. For arrays with only a few moveable elements, many configurations are necessary to cover the $(u–v)$ plane.

The response of an array, or in fact any radiating surface, can be calculated using the transfer function $c(b)$ [1]. The transfer function is a measure of the density of mutual separations b contained in any radiating surface and is equal to the autocorrelation of the aperture current distribution. The aperture may be continuous as in an antenna or piecewise continuous as in an array. The Fourier transform of the transfer function is the synthesized power pattern $A(\delta)$ where

$$A(\delta) = \int c(b) \exp\{j2\pi\delta\cdot b\}\,db. \qquad (9)$$

Exploitation of the circular symmetry of east–west earth-rotation synthesis gives

$$A'(r) = \int \rho c'(\rho) J_0(2\pi\rho r)\,d\rho \qquad (10)$$

where $A'(r)$ is the radial angular power distribution and $c'(\rho)$ is the transfer function of the linear array.

In Fig. 3 the transfer function and radial power distribution are given for a parabolic antenna and a grating array, a minimum redundancy array, and a compound grating array using earth-rotation synthesis. The parabolic antenna con-

tains a large redundancy of small separations which results in a moderately large beamwidth but very low sidelobes. The grating array (assuming all pairs are correlated) has a "lumpy" transfer function because of the gaps between the elements; the envelope of the transfer function and beamwidth is similar to that of a parabolic antenna. However, the sidelobe level is much larger. Sizeable grating lobes occur at radii intervals of λ/S, where S is the array stepping interval and λ is the wavelength of the radiation. The finite aperture size of the array elements rejects radiation at displacement angles larger than $\sim\lambda/d$. A minimum redundancy array optimizes the number of independent consecutive spacings available from a given number of elements [15]. The transfer function has a uniform envelope resulting in a small beamwidth but substantially larger sidelobes than a grating array. In order to suppress grating lobes, compound grating arrays are used. These arrays combine one or several large elements with a physical size greater than the stepping interval of the remaining smaller elements. Only element pairs involving the larger element(s) are used. The grating sidelobes are suppressed by the narrow primary pattern of the element pairs.

Since the visibility function for each element pair is measured separately, the transfer function may be changed by suitably weighting the data associated with each spacing. Thus the response of the grating array may be changed to that of the minimum redundancy array by weighting more heavily the large spacings before inverting the data. Various methods of weighting are discussed in Section IV.

The arrays at Westerbork [16] and Cambridge [17] contain several fixed equi-spaced antennas and several moveable antennas, all on an east–west line. With observations on several subsequent days with different configurations, the stepping interval can be made sufficiently small, if desired, so that the grating lobes are placed outside of the relevant source emission or the primary beam area. Also these arrays could be expanded in resolution with the addition of a few antennas. Usually the redundant spacings are not correlated with a resultant loss in the signal-to-noise ratio. The minimum redundancy array at Stanford [14] and that proposed by Cal Tech [18] give maximum coverage in the $(u–v)$ plane for a given number of antennas. However, changes in resolution and stepping interval cannot be made unless all of the antennas are moveable.

The transfer function obtained with non east–west linear arrays [19] does not generally fill a half-plane. However, useful earth-rotation aperture synthesis can still be done. In this case the grating rings and beamwidth are not elliptical and

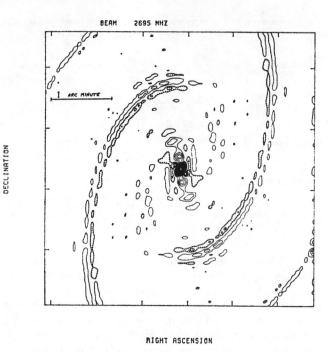

Fig. 4. The synthesized power pattern for a source at declination 40° using the NRAO interferometer. The spacings of 300-m units of 1, 2, 3, 4, 5, 6, 7, 8, and 9, at a frequency of 2695 MHz, were used with -6^h to $+6^h$ tracking. 5-percent contour levels are drawn with the 0 contour omitted. The grating lobes and the half-power beamwidth are asymmetric. 20-percent sidelobes are produced by the missing sector in the $(u-v)$ plane.

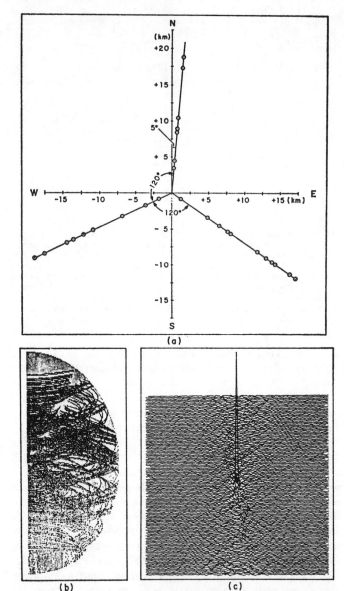

Fig. 5. The VLA. (a) Approximate location of the 27 elements for the 21-km configuration. (b) The transfer function for a source at $\delta = 20°$, 12-h tracking. (c) The synthesized power pattern shown using line profiles. (Taken from [10].)

larger sidelobes occur because of missing wedges in the $(u-v)$ plane. In Fig. 4 the synthesized power pattern for a source at declination 40° using the National Radio Astronomy Observatory, Green Bank, W. Va., three-element interferometer is shown. The array is oriented 28° south of west. Sample tracks in the $(u-v)$ plane for this baseline are shown in Fig. 2(a).

The disadvantages of a linear array are the lack of resolution in some angles for sources near the equator or in the opposite hemisphere and the requirement of 12 h of tracking. The array at Fleurs [20] contains both an east–west and a north–south arm. Although each arm is used independently, good coverage in the $(u-v)$ plane can be obtained for low declination sources. At the Owens Valley Radio Observatory in California a 1600-ft north–south rail track complements a 4000-ft east–west track enabling effective aperture synthesis with three elements as far south as the galactic center.

The VLA [10], an array of 27 elements, each a 25-m paraboloid, is a Y-shaped array having three equiangular linear arms of 21 km. The optimum element positions were obtained empirically in an attempt to minimize the number of gaps in the $(u-v)$ plane for a variety of source declinations with about 10 h of tracking time [21]. For a high-declination source, the maximum sidelobe level is about the same as a simple telescope, and for a source south of declination 5° sidelobes of 5 percent are present. All antennas are moveable so that four similar arrays with a maximum arm length of 0.5 to 21 km can be obtained. The transfer function and power pattern for the VLA are shown in Fig. 5. No grating lobes occur since the element positions on each arm do not have a stepping interval.

General properties of two-dimensional arrays using earth-rotation synthesis have been investigated [22]. In order to cover uniformly the $(u-v)$ plane, three guidelines are suggested: 1) The north–south extent should be longest to com-

pensate for baseline foreshortening in that direction. 2) A higher density of baselines should be oriented in the east–west direction to compensate for the slow rate of $(u-v)$ coverage by an east–west interferometer near 0^h hour angle. 3) The element position should follow a square law for which successive element separations in any radial direction are consecutive odd integers. The element locations for the VLA, although obtained in an empirical manner, do follow the above criteria reasonably well.

The theory of randomly placed elements [23] has not been seriously considered in the design of present arrays. A major advantage is the suppression of grating responses. However, many elements are necessary to avoid statistical clumping.

IV. INVERSION METHODS

The basic output of a synthesis array is a vast collection of numbers. A sample of the visibility function $\mathcal{V}(b_j)$ is associated with the response of each correlated pair of antennas over a certain averaging interval and defined by an average projected spacing b_j in the $(u-v)$ plane. The averaging interval should be sufficiently short so that the change of projected spacing over the observation is less than an element radius. Many calibrations are necessary to convert the "raw" visibility functions into a calibrated set and these have been described elsewhere [11]. In brief, the calibrations are obtained by observing strong small-diameter sources of known flux density, angular position, and polarization. The measurement of the system gain, various path length changes affecting the visibility function phase, precise location of each element, and the system polarization response can all be determined by suitable astronomical measurements.

A. Normal Fourier Methods

The angular power distribution $I(\sigma)$ can be determined from the radio map $I'(\sigma)$ using the discrete form of (8)

$$I'(\sigma) = G(\sigma)I(\sigma) = \sum_{j=1}^{N} w_j \mathcal{U}(b_j) \exp\left\{-j2\pi b_j \cdot \sigma\right\} \quad (11)$$

where w_j is the observation weight. The power pattern $A(\sigma)$ corresponding to a set of observations is found by setting $\mathcal{U}(b_j) = 1$.

A straightforward application of (11) is possible with no more than a few hundred points in the $(u-v)$ plane but requires a computing time proportional to N^4 complex additions and multiplication, where N is the number of sampled points. With the use of linear arrays in earth-rotation synthesis, a polar coordinate representation of the data is natural and (11) can be separated into a radial sum and an angular sum with a decrease in the computing time.

The fast Fourier transform (FFT) [24] algorithm makes it possible to obtain radio maps from an extensive and detailed set of observations using but several minutes of execution time, proportional to $2N^2 \log_2 N$, on a medium-size computer. The use of the algorithm requires the "gridding" of the visibility function; that is, the determination of the value on a rectangular lattice of points in the $(u-v)$ plane. This requires transformation of the data which slightly modify the radio maps.

The effects of the various manipulations in making a radio map $I'(\sigma)$ using the (FFT) and its relationship to $I(\sigma)$ are shown in the following equation:

$$I'(\sigma) = \mathcal{F}\left\{q(b) \cdot \text{III}(\Delta) \cdot [(w(b) \cdot \mathcal{U}(b)) * s(b)]\right\}$$
$$= Q(\sigma) * \text{III}(1/\Delta) * \left\{S(\sigma) \cdot [W(\sigma) * (G(\sigma) \cdot I(\sigma))]\right\} \quad (12)$$

where $\mathcal{F}$ is the two-dimensional Fourier transform; $\cdot$ is multiplication; $*$ is convolution; and (s, S), (q, Q), and (w, W) are Fourier pairs.

First, the visibility[2] function $\mathcal{U}(b)$ is weighted by $w(b)$ and then convolved with a smoothing function $s(b)$. This convolution is necessary to define a continuous visibility function from the discrete sampled points. The function is then sampled on a grid of interval Δ—mathematically indicated by multiplication with a "shah" function $\text{III}(\Delta)$ [25]. The gridded data may then be further weighted by $q(b)$. The fast Fourier transform $\mathcal{F}$ is then applied. In usual practice most of the grid points contain no data so the saving on computing time as compared with more direct transforming methods is not very significant for a small number of observations.

The desired radio map $I(\sigma)$ is transformed in the following manner. First, the Fourier transform of the visibility function is assumed equal to $G(\sigma) \cdot I(\sigma)$ as discussed in connection with (7). The visibility function weighting $w(b)$ modifies the synthesis beamwidth and the sidelobe levels as desired by $W(\sigma)$. Further "tapering" by $S(\sigma)$ results from the convolution of $s(b)$ in the $(u-v)$ plane. The tapering effect is strictly true only for a high density of sampled points. With a small number of points interaction between the $(u-v)$ sampling and the grid sampling can produce a variety of effects. The grid sampling of Δ in the $(u-v)$ plane produces a radio map which repeats at intervals $1/\Delta$; the inner region of $1/\Delta$ by $1/\Delta$ is called the "field of view" of the map and all radiation outside of this area is aliased into the field of view. Thus the grid interval Δ should be as small as possible to prevent overlapping in the map. Finally, a further convolution of the radio map $Q(\sigma)$ is produced by any post-gridded weighting. The synthesized power pattern is obtained by replacing the measured visibility function by unity, i.e., determining the response to a point source at the phase center

$$A(\sigma) = Q(\sigma) * \text{III}(1/\Delta) * \left\{S(\sigma) \cdot W(\sigma)\right\}. \quad (13)$$

Three types of convolution functions are generally considered. The one requiring the least amount of calculations is the "pill-box function" $s(b) = 1$, $|b| < D/Z$; $s(b) = 0$, $|b| > D/Z$. If the pill-box size D is chosen equal to the grid size Δ, the two weighting functions q and w are interchangeable. The tapering function then has a form $\sin x/x$ in each dimension where $x = \pi D l$. A Gaussian smoothing function is most commonly used because the tapering function, also Gaussian, decreases smoothly to zero. However, lengthy calculating time, often longer than that for the FFT, is needed. The interesting convolution function $\sin(\pi l\Delta)/\pi l\Delta$ in each dimension produces no effect in the radio map within the field of view $1/\Delta$. However, calculation time is even more lengthy in this case.

The choice of weighting depends on many criteria. For high resolution, uniform weighting is chosen. Equal weight is given to each sampled $(u-v)$ cell regardless of the amount of integration time in the cell. This weighting produces a nearly uniform transfer function with generally large sidelobe levels. The radio map associated with uniform weighting is generally called the "principal solution." However, there is nothing intrinsically superior to this particular weighting. Conventional weighting adds an additional Gaussian-shaped taper which decreases the transfer function at the outer spacings to about 30 percent. The beamwidth is slightly broadened compared with uniform weighting but near-in sidelobe levels are significantly reduced. For optimum signal to noise the data should be weighted in proportion to the square of the signal-to-noise ratio. For noise-limited data of a point source, the weight is proportional to the integration time. This produces a transfer function which varies approximately inversely with spacing and thus produces poorer resolution, but gives the best signal to noise for a detection of a point source.

B. Other Inversion Methods

The Fourier inversion processes typically result in sidelobe levels of about 5 percent even with good coverage in the $(u-v)$ plane. With incomplete coverage, sidelobe levels of 30 percent are common. Thus faint features which are still more intense than the noise fluctuations can be obscured by sidelobes. Two techniques are commonly used for reducing the effects of sidelobes: "cleaning," and source subtraction.

---

[2] Effects due to averaging the visibility function over a finite interval are not considered here.

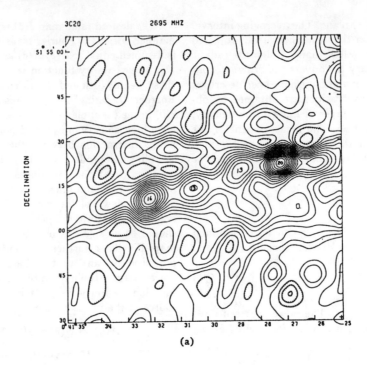

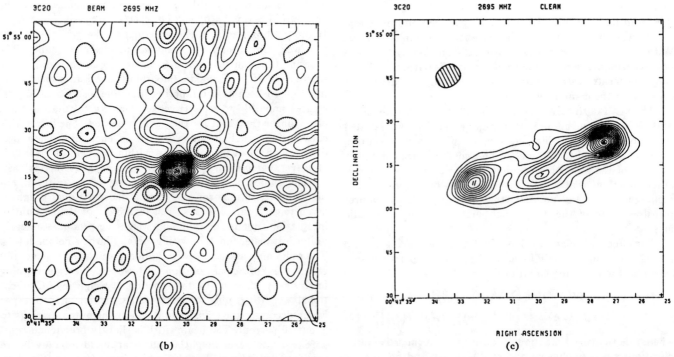

Fig. 6.   The results of using the "cleaning" deconvolution method on the radio source 3C 20. (a) The radio map drawn with 5-percent contour levels. Peak levels are shown. (b) The synthesized power pattern showing the relatively poor coverage in the (u–v) plane; 5-percent contours. (c) The clean map. The clean beam is shown in the upper left. The same contour interval of the radio map is used as in (a). No sidelobes larger than 5 percent remain.

"Cleaning" is a type of band-limited deconvolution in which the radio map is decomposed into a sum of synthesized power patterns [26]. The technique is also related to problems of pattern recognition. If $I'(\delta)$ is the radio map and $A(\delta)$ is the associated synthesized beam pattern, then a set of numbers $C_i(\delta_i)$ are determined such that

$$I'(\delta) = \sum_i C_i A(\delta - \delta_i) + I_r(\delta) \qquad (14)$$

where $I_r(\delta)$ is a residual map. The decomposition cannot be done analytically and an iterative scheme has been developed.

The technique searches for the maximum value of the radio map and then subtracts the synthesized beam pattern, suitably scaled, centered at the position of maximum. The remainder map is then searched for the maximum point and further subtractions of the beam are made. Termination of the process occurs when the residual map is no larger than the expected signal to noise, the number of iterations exceeds a certain value, or the lack of convergence.

It is important in the cleaning process that the radio map be a true convolution of the angular power distribution and a unique synthesized beam pattern. For direct inversion methods such is the case (except for the effects discussed in connection with (7) and (8)). For the FFT inversion method, the aliasing effect limits accurate cleaning to the center of the field of view.

The method works because most radio maps contain nearly empty sky with relatively few regions of radio emission. Thus the number of parameters needed to describe the radio map is often much less than the number of independent samples in the (u–v) plane. For limited observations of very complex emission regions, the cleaning technique breaks down.

In order to display the result, a reconvolved map is obtained using a "clean" beam $A_c(\sigma)$ to convolve the set of point responses

$$I'(\sigma) = G(\sigma)I(\sigma) = \sum_{i=1}^{N} C_i A_c(\sigma - \sigma_i) + I_r(\sigma). \quad (15)$$

The clean beam is somewhat arbitrary but should reflect the actual coverage in the (u–v) plane. Often it is given an elliptical-Gaussian shape which matches that of the central part of $A(\sigma)$ but containing none of the sidelobes. The result of cleaning a source is shown in Fig. 6.

Often a radio map contains obvious sidelobe structure from a few strong sources. A useful method of removing these sidelobes is to subtract the corresponding visibility function of the strong sources from the total visibility function. The Fourier inversion of the remainder will then produce a radio map with the intense features and all of their sidelobes removed. The parameters of the source to be subtracted can be obtained from the original radio map or from model fitting, to be discussed next.

The angular power distribution can be obtained by model fitting methods in which the visibility functions of simple models are compared directly with the observed visibility function. The technique is commonly used for specialized observations of strong, isolated, small-diameter radio sources where only very sparse coverage of the (u–v) plane is needed to define the relevant angular power distribution; such as, accurate flux densities, positions, and approximate sizes. The emission associated with extragalactic objects can be approximately described with simple models and often interferometric data of planets are compared with circular disk models.

Iterative methods are generally used in model fitting whereby residuals between the model visibility function and the measured visibility function are used to obtain a better model, etc. Additional details are given elsewhere [11].

Conventional Fourier mapping techniques are of little value for determining the angular power distribution from data with poor or no phase data. The data obtained from most very-long baseline interferometers, intensity interferometers, and some radio-link interferometers are phase unstable so that only the visibility amplitude can be obtained. For these data model fitting techniques can be used. A method has been developed in which it is possible to generate a finite set of angular power distributions which are compatible with the measured visibility amplitudes [27]. With accurate data, some phase information and a priori knowledge of the source parameters, a unique solution is possible.

A possible method for determining the angular power distribution using a "maximum entropy analysis" has been developed [28], [29]. The analysis uses only the measured visibility data and derives a radio map which is the most random

(contains a minimum of extra information) and consistent with the data. The technique is successful with uniformly sampled one-dimensional data with improved resolution, virtually no sidelobes, and little cost in computing time. However, the generalization of the method to two-dimensional data has not been worked out.

ACKNOWLEDGMENT

The author would like to thank Dr. J. R. Fisher and Dr. E. W. Greisen for their comments on the manuscript.

REFERENCES

[1] W. N. Christiansen and J. A. Högbom, *Radiotelescopes*. London, England: Cambridge Univ. Press, 1969.
[2] R. N. Bracewell, "Radio interferometry of discrete sources," *Proc. IRE*, vol. 46, pp. 97–105, Jan. 1958.
[3] H. C. Ko, "Coherence theory of radio-astronomical measurements," *IEEE Trans. Antennas Propagat.*, vol. AP-15, pp. 10–20, Jan. 1967.
[4] G. W. Swenson, Jr., and N. C. Mathur, "The interferometer in radio astronomy," *Proc. IEEE*, vol. 56, pp. 2114–2130, Dec. 1968.
[5] G. W. Swenson, Jr., "Synthetic-aperture radio telescopes," *Ann. Rev. Astron. Astrophys.*, vol. 7, pp. 353–374, 1969.
[6] T. Hagfors and J. M. Moran, Jr., "Detection and estimation practices in radio and radar astronomy," *Proc. IEEE*, vol. 58, pp. 743–759, May 1970.
[7] L. Mandel and E. Wolf, "Coherence properties of optical fields," *Rev. Mod. Phys.*, vol. 37, pp. 231–287, Apr. 1965.
[8] W. N. Brouw, "Data processing for the Westerbork radio telescope," Ph.D. dissertation, Univ. of Leiden, Leiden, The Netherlands, 1971.
[9] J. N. Douglass, F. N. Bash, F. D. Ghigo, G. F. Moseley, and G. W. Torrence, "First results from the Texas interferometer: Positions of 605 discrete sources," *Astron. J.*, vol. 78, pp. 1–17, Feb. 1973.
[10] National Radio Astronomy Observatory, *A Proposal for a Very Large Array Radio Telescope*, vol. III, Jan. 1969.
[11] E. B. Fomalont and M. C. H. Wright, "Interferometry and aperture synthesis," in *Galactic and Extragalactic Radio Astronomy*. New York: Springer, 1973, ch. 10.
[12] W. N. Christiansen and J. A. Warburton, "The distribution of radio brightness over the solar disk at a wavelength of 21 centimeters. III. The quiet sun—Two-dimensional observations," *Aust. J. Phys.*, vol. 8, pp. 474–486, Dec. 1955.
[13] M. Ryle and A. Hewish, "A synthesis of a large radio telescope," *Mon. Notices Roy. Astron. Soc.*, vol. 120, pp. 220–230, 1960.
[14] R. N. Bracewell *et al.*, "The Stanford five-element radio telescope," this issue, pp. 1249–1257.
[15] A. T. Moffet, "Minimum-redundancy linear arrays," *IEEE Trans. Antennas Propagat.*, vol. AP-16, pp. 172–175, Mar. 1968.
[16] J. W. M. Baars *et al.*, "The synthesis radio telescope at Westerbork," this issue, pp. 1258–1266.
[17] M. Ryle, "The 5-km radio telescope at Cambridge," *Nature*, vol. 239, pp. 435–438, Oct. 1972.
[18] California Institute of Technology, "Construction of a multi-element interferometer at the Owens Valley Radio Observatory," proposal submitted to the National Science Foundation by the Calif. Inst. Tech., Pasadena, Calif., Apr. 1966.
[19] D. E. Hogg, G. H. Macdonald, R. G. Conway, and C. M. Wade, "Synthesis of brightness distribution in radio sources," *Astron. J.*, vol. 74, pp. 1206–1213, Dec. 1969.
[20] W. N. Christiansen, "A new Southern Hemisphere synthesis radio telescope," this issue, pp. 1266–1270.
[21] N. C. Mathur, "A pseudodynamic programming technique for the design of correlator supersynthesis arrays," *Radio Sci.*, vol. 69, pp. 235–244, Mar. 1969.
[22] Y. L. Chow, "On designing a supersynthesis antenna array," *IEEE Trans. Antennas Propagat.*, vol. AP-20, pp. 30–35, Jan. 1972.
[23] Y. T. Lo, "A mathematical theory of antenna arrays with randomly spaced elements," *IEEE Trans. Antennas Propagat.*, vol. AP-12, pp. 257–268, May 1964.
[24] W. T. Cochran *et al.*, "What is the fast Fourier transform?" *Proc. IEEE*, vol. 55, pp. 1664–1674, Oct. 1967.
[25] R. N. Bracewell, *The Fourier Transform and Its Applications*. New York, McGraw-Hill, 1965.
[26] J. A. Högbom, "Aperture synthesis with a non-regular distribution of interferometer baselines," submitted to *Astron. Astrophys. Suppl.*
[27] R. H. T. Bates, "Contributions to the theory of intensity interferometry," *Mon. Notices Roy. Astron. Soc.*, vol. 142, pp. 413–428, 1969.
[28] J. P. Burg, "Maximum entropy spectral analysis," presented at the 39th Meeting of the Soc. of Exploration Geophysicists, Oklahoma City, Okla., 1970.
[29] J. Ables, submitted to *Astron. Astrophys. Suppl.*

The Synthesis Radio Telescope at Westerbork

JACOB W. M. BAARS, J. FREDERIK VAN DER BRUGGE, JEAN L. CASSE, J. P. HAMAKER,
L. H. SONDAAR, J. J. VISSER, AND KELVIN J. WELLINGTON

Abstract—The synthesis radio telescope (SRT) is an array of 20 simultaneous interferometers formed by 10 fixed and 2 movable antennas on an E–W baseline of 1600 m, operating on the principle of rotational aperture synthesis. It is capable of observing continuum radiation in 4 Stokes parameters and line radiation at 50-, 21-, and 6-cm wavelengths. This paper centers on a description of the electronic system with emphasis on those features by which its exceptional stability has been realized. Included are sections on the mechanical construction and adjustments, the data processing, and some examples of astronomical results.

I. INTRODUCTION

THE Westerbork synthesis radio telescope (SRT) constitutes one of world's largest and most sensitive instruments based on the principle of rotational aperture synthesis. It is unique in its capability to produce high-resolution maps in four polarization parameters of very faint objects with only a few days' observation. The phase and amplitude stability achieved in the mechanical and electronic system result in a high-quality synthesized beam, giving the instrument a dynamic range of more than 20 dB. Its high speed and sensitivity make it the only instrument capable of hydrogen line observations on more than the nearest few external galaxies. As a result of these qualities it has produced, since its inauguration in June 1970, a series of exciting astronomical discoveries.

During the first two years of its operation, the SRT has been equipped with 1415-MHz front ends and a back end designed for continuum observations. For some periods of time a modified version has been used to observe hydrogen line radiation [1]. In September 1972, a set of front ends for observations in the 6- and 50-cm bands was installed. The salient characteristics of the instrument from the user's point of view are shown in Table I. Future plans call for lowering of the system noise at all frequencies and the installation of a 5120-channel digital correlation back end.

This paper briefly describes the construction of the instrument with particular emphasis on the features to which it owes its stability and accuracy. Some examples of results are included as a last section. A more extensive description of the instrument, directed at an astronomer's audience, appears in a series of papers in *Astronomy and Astrophysics*.

II. PRINCIPLES OF OPERATION

Aperture synthesis [2], [3] is based on the measurement of the individual spatial Fourier components of a sky field by interferometers of various spacings and orientations. From these components, the original brightness distribution can be reconstructed by Fourier inversion. The individual components may be measured simultaneously or sequentially; in

Manuscript received February 4, 1973; revised April 30, 1973. The Netherlands Foundation for Radio Astronomy is financially supported by The Netherlands Organization for the Advancement of Pure Research (ZWO).

The authors are with The Netherlands Foundation for Radio Astronomy, Radio Observatory, Dwingeloo 7514, The Netherlands.

Reprinted from *Proc. IEEE*, vol. 61, no. 9, pp. 1258–1266, Sept. 1973.

TABLE I
CHARACTERISTICS OF THE SRT WITH THE PRESENTLY AVAILABLE FRONT ENDS AND A CONTINUUM BACK END

Frequency (MHz)	610	1415	4995
Bandwidth (MHz)	4	4	4
Half-power primary antenna beam (arc minutes)	83	36	11
Half-power synthesized beam in right ascension (arc seconds)	56	24	6.8
Grating ring interval in right ascension (arc minutes) (after 1×12 h)	23	10	2.8
System temperature (K)	400	260	200
Aperture efficiency	0.59	0.54	0.44
Rms noise on 12-h map:			
Flux density S (1 mfu = 10^{-29} W·m^{-2}Hz^{-1})	1.3	0.9	0.9
Brightness temperature T_b (K)	1.7	1.2	1.2

Note: The theoretical sensitivities quoted for 1415 MHz are attained in practice. For the other frequencies, empirical figures are not yet available.

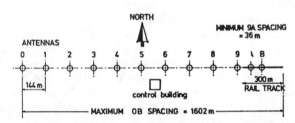

Fig. 1. Geometrical layout of SRT. The ten fixed antennas numbered 0 to 9 and the two movable antennas *A* and *B* are placed on a straight E–W baseline. Antennas *A* and *B* can be moved to any arbitrary position on the 300-m long rail track.

the latter case, the basic assumption is made that the source does not change in appearance during the time required for the measurements. In rotational synthesis only the spacing(s) of the interferometer(s) need to be varied, the diurnal rotation of the earth providing the required 180° range of orientations in a 12-h period.

The SRT consists of 10 fixed 25-m steerable parabolic antennas (numbered 0 through 9) which form a set of 20 simultaneous E–W interferometers with 2 similar antennas (*A* and *B*) movable on a rail track (Fig. 1). A single 12-h observation with *A* and *B* at 72- and 144-m distances from 9 covers a set of concentric rings at 72-m intervals in the aperture plane, corresponding to a synthesized telescope beam of 24″×24″ csc δ at 1415 MHz, with grating rings at multiples of 10′×10′ csc δ away from the center. These rings can be removed by making more observations with telescopes *A* and *B* moved to different postions to fill in some or all of the missing spacings in the aperture plane.

The absence of error sidelobes in the beam depends on the accuracy with which the contributions of the various measured Fourier components cancel outside the main beam. We therefore aimed at a phase and amplitude stability for each interferometer on the order of 1° and 1 percent, respectively. The phase of an interferometer equals the difference between the

TABLE II
CHARACTERISTICS OF THE SRT ANTENNAS

Diameter	25 m
Focal length	8.75 m $\pm$ 0.5 mm $\Big\}f/d = 0.35$
Mounting	equatorial
Sky coverage	$\pm 90°$ in hour angle, $-38°$ to $90°$ in declination
Drive speed	slew 18°/min; track, scan 0.125–0.5°/min
Pointing repeatability	0.01°
Reflector	8-mm mesh of 0.8-mm stainless-steel wire
Surface deviation	1.4 mm rms $\Big\}$ for all antennas
Distance polar-declination axis	5.00 m $\pm$ 0.3 mm
Angle polar-declination axis	90° $\pm$ 3"
Total weight	100 tons

Fig. 3. A view of the SRT during a testing period. Not all antennas are pointing in the same direction, as they do during astronomical observations. The two mobile antennas are in the foreground on the railtrack.

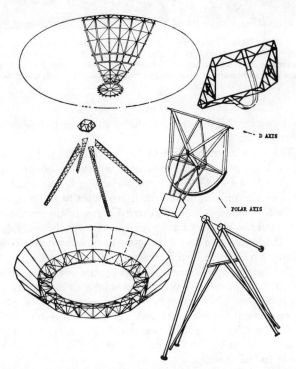

Fig. 2. The SRT antennas "exploded" into the main components. Right-hand part from bottom: pedestal, fixed to concrete foundation; polar axis house with polar gearrack and counterweight; declination construction with gearrack. Left-hand part, bottom: reflector supporting ring girder, attached to the four corners of the declination construction; top: reflector surface consisting of three concentric rings of panels; middle: focus cage and supporting quadrupod, also attached to corners of the declination construction.

phases of its constituting elements. Therefore, absolute phase variations in individual telescopes are not important as long as symmetry is maintained in the interferometers. This has been the leading principle in the construction of the whole instrument.

To monitor and, where possible, correct for the residual errors, about 20 percent of the observing time is spent in calibrations on celestial point sources of precisely known positions.

III. MECHANICAL CONSTRUCTION

The characteristics of the paraboloidal antennas are compiled in Table II. The large number that had to be built made possible the extensive use of templates in the construction and assembly of the parts, resulting in tight dimensional tolerances at a moderate cost. The major components, shown in Fig. 2, were assembled in a temperature-controlled hall (75 by 30 m²)

which was erected for this purpose on the site. To avoid the deformation occurring in welding operations, extensive use was made of epoxy bonding resins. With the production in full swing, one antenna was completed every five weeks. Fig. 3 shows the completed antennas.

In all production phases, a large effort was put into checking dimensional tolerances, using the available measuring methods to their limits of precision. The horizontal and vertical coordinates of the reflector panels were measured to accuracies of 0.3 and 0.1 mm, respectively. The peak deviations of the assembled reflectors from the ideal paraboloid are ± 2 mm. In other parts of the construction, the critical dimensions have a spread of about 0.5 mm between antennas.

For the array, an E–W reference baseline was established, using a Wild T3 theodolite and a leveling instrument, with circumpolar stars as a reference. By repeated measurements under various atmospheric conditions, accuracies of 0".1, corresponding to about 1 mm over the 1600-m baseline, were achieved. Distances were determined with the aid of Invar steel wires, allowing an accuracy of ± 0.1 mm on the basic 144-m interval. The antennas were positioned with respect to this reference line with a 1-mm accuracy using standard geodetic techniques.

The rail track was designed to enable arbitrary positions for the moving antennas to be chosen, and, therefore, it had to be aligned with high accuracy over its full length. This was done by observing, from one end of the track, a zone plate mounted on the rail and illuminated by a laser at the other end. Most of the rail deviates less than 0.3 mm from a straight line, with a few 0.5-mm peaks. More details on construction and measurement are presented by Baars and Hooghoudt [4].

A comparison of geodetic and radio-astronomical baseline determinations is given in Table III.

IV. PRINCIPLES OF THE RECEIVER SYSTEM

The receiver system is designed to provide complete information on the intensity and polarization distribution in the observed field. This requires the observation of the 4 complex correlation coefficients XX, XY, YX, and YY between pairs of orthogonal dipoles X and Y in the fixed and moving antennas. Thus we have to deal with $10 + 2$ antennas, each hav-

TABLE III
BASELINE PARAMETERS AS DETERMINED BY GEODETIC SURVEYING AND BY OBSERVATION OF RADIO POINT SOURCES OF KNOWN POSITION

	Geodetic	Radio Astronomical
Total baseline length (m)	1620	
Length of rail track (m)	300	
Mean distance between fixed antennas (m)	143.9910 ± 0.0003	143.9895 ± 0.0001
Baseline declination (″)	0.0 ± 0.5	-0.65 ± 0.10
Baseline hour angle (epoch 1968)	$270°00'00''.63 \pm 0''.50$	$270°00'01''.71 \pm 0''.06$
Maximum deviation of fixed antennas from nominal position (mm)		
East–West		0.8
North–South		2.1
Vertical		3.0

Note: The accuracies quoted for the radio measurements are internal mean errors. Systematic errors not detectable by the calibration procedures are suspected to be considerably larger and may account in part for the observed position deviations of the individual antennas.

Fig. 5. A 6/50-cm front-end unit. The rings of the dual-frequency feed are visible. The central circular waveguide of the inner 6-cm feed extends into the insulated receiver box. The opened box has been rotated to show the parametric amplifier drawers and the external Peltier heater/cooler units.

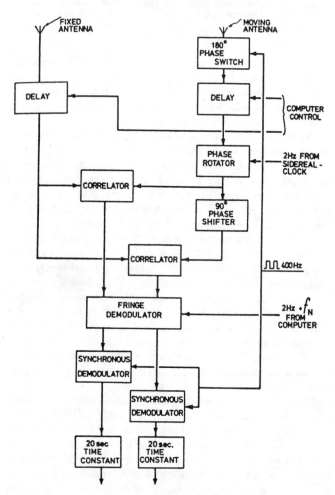

Fig. 4. Functional block diagram of the main receiver for a single interferometer.

ing two dipole channels; these are combined in the central receiver backend to form a total of 80 complex correlator channels, 4 for each interferometer. Each correlator channel consists of a cosine and a sine correlator providing the real and imaginary parts, respectively, of the complex visibility. The full designation of each of the 160 correlators consists of the antenna and dipole labels involved and the letter S or C for sine/cosine; e.g., $5XAYC$. In the sequel we shall use the term

interferometer for a correlator channel with its two associated dipole channels.

The interferometers are of the single-sideband type. We discuss their principles of operation with reference to Fig. 4. The standard technique of phase switching is employed. The signal in the moving antennas is periodically switched 180° in phase. This modulation is carried along to the very end of the receiver, where it is finally removed by synchronous detection. This makes the system immune to correlated interference picked up anywhere in the receiver.

The delays, which are adjusted by the control computer according to the relative position of source and antennas, provide the time-of-arrival equalization necessary to obtain proper correlation. Simultaneous sine and cosine correlations are obtained by the use of two correlator units, the moving antenna's input for one being shifted in phase by 90° with respect to the other.

The continuous motion of the earth with respect to the source results in continuous rotation of the complex correlation vector. The rate of rotation, the natural fringe frequency f_N, is a function of the source–baseline geometry. To obtain a steady output, an opposite rotation in 90° steps is applied by the control computer in the fringe demodulator. The ripple produced by the discrete phase steps is filtered out by the 20-s time constants. To avoid difficulties in those cases where f_N is close to zero, the natural rotation is speeded up by a constant "artificial fringe" frequency in the phase rotator, and the fringe demodulation frequency adjusted accordingly.

Small offsets in the synchronous detector outputs manifest themselves as a small perturbation in the center of the sky map, with its associated grating rings. In those cases where this effect is disturbing, the fringe rotation is not stopped completely, the residual rotation being removed in the data processing. This results in the perturbation being smeared over the whole map and disappearing in the noise.

V. THE 6- AND 50-cm FRONT ENDS

The front-end units serve to amplify and convert the weak signals received by the X and Y dipoles to make them suitable for transmission over about 1 km of cable to the main building. In this process, the short- and long-term gain and phase drifts

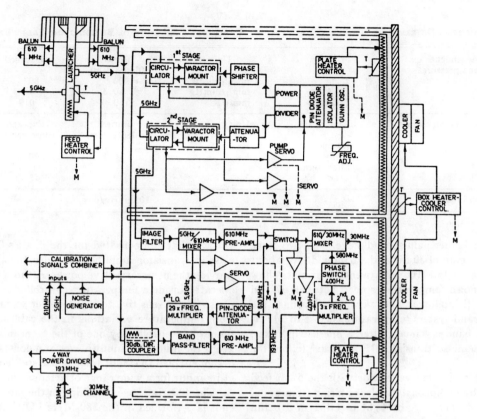

Fig. 6.   Block diagram showing half (one polarization) of a 6- and 50-cm front-end receiver. The long dashed lines indicate the outer thermostat and two of the four "drawers." The top drawer contains the low-noise 6-cm parametric amplifier and its associated hardware; the bottom one contains the remainder of the receiver for both 6 and 50 cm. Components on the left are common to both polarizations. The calibration-signals combiner provides the possibility of CW and noise signal injection into the four receivers for noise, amplitude, and phase calibration of the interferometers. The latter calibrations are not yet implemented for the 6/50-cm system. *T*, temperature sensor; *M*, monitor point.

are to be kept at a minimum. Our state of the art is represented by the recently installed 6/50-cm dual-frequency front ends, which we discuss here. The older 21-cm front ends have been described by Casse and Muller [5].

Each front end is constructed in a standard frame, about 1 m long by 0.5 m square. This forms an interchangeable self-contained unit, capable of being installed at the focus within a half hour (Fig. 5). The block diagram of the electronics is shown in Fig. 6.

A concentric dual-frequency dual-polarization scalar feed was developed in cooperation with a group at the Technological University of Eindhoven. The illumination at the edge of the reflector is −17.5 and −14.7 dB, respectively, for 6 and 50 cm, including space attenuation. This gives feed efficiencies of 69 and 72 percent and spillover losses of 3.7 and 8.3 percent. A major constraint imposed on the design of the 50-cm feed was the space available for accommodating both feed and launcher. This was held to a diameter of 47 cm and a length of 36 cm.

Situated at the focus of the reflector, the receiver must meet the very exacting stability requirements in temperatures ranging from −15 to +45°C. This is accomplished by enclosing all electronic components in a double thermostat. Peltier units mounted on the outer box can either heat or cool it, maintaining any point of the box at (19 ± 3)°C. Mounted inside and thermally insulated from this coarsely regulated

enclosure are four "drawers" containing the receiver proper. Each has its own servo system which maintains the temperature at (23.0 ± 0.1)°C.

The center section of the 6-cm feed, in the form of a circular waveguide, is brought directly into the receiver box. A short section of it is made of polypropane covered with thin layers of copper and silver. This thermal insulating section together with a servo-controlled heating transistor enables the following waveguide-to-coaxial transition to be temperature stabilized.

The transition has two orthogonal coaxial probes separated longitudinally by a few wavelengths and coupled to the first preamplifiers via 4 cm of semirigid coaxial cable. The whole arrangement brings the signal from the focus to the preamplifier with a minimum of loss. An important feature in the construction was the precisely circular waveguide. This enabled a polarization isolation of 45 dB to be obtained in the feed–launcher combination.

The preamplifier is an uncooled two-stage degenerate parametric amplifier made by Airborne Instruments Laboratory. Each two-stage unit is mounted separately, with its associated servo, monitor, and pump hardware, in one of the four drawers. A free-running Gunn oscillator at 10 GHz supplies 60 mW of pump power for both stages via a waveguide power divider, phase shifter, and attenuator. A p-i-n diode attenuator is used to servo the pump power. The input to the servo is the var-

TABLE IV
QUANTITIES DETERMINING THE PHASE STABILITY OF A CABLE BETWEEN CONTROL BUILDING AND FRONT END

Length and its dependence on temperature and pressure	Property	Unit	Buried (depth 1 m)	Exposed	Jumper
	L	m (physical)	900	80	5
	$(\partial L/\partial P)_T$	mm$_{el}$/mmHg	0.3	0.03	0
	$(\partial L/\partial T)_P$	mm$_{el}$/K	10	0.9	−0.7
Typical variations in temperature	Time Scale	Unit	Buried (depth 1 m)	Exposed (unscreened)	Jumper
	Annual	K/day	0.3		
	Daily	K	0.1	20	20
	Short-term (<1 h)	K	0	5	5
Typical variation in pressure			10 mmHg/day		

actor current of the higher gain second stage. The two-stage paramp units have a gain of 30 dB and a bandwidth of about 120 MHz centered at 4995 MHz. The twenty-four units show a double-sideband noise temperature spread between 47 and 75 K. Together with spillover and sky contributions, the average single-sideband system temperature is close to 200 K.

A 50-MHz wide bandpass image filter, a microstrip mixer, and a 610-MHz low-noise transistor IF amplifier follow the paramp.

The 50-cm feed, a circular waveguide with the 6-cm feed in its center, is fed by orthogonal dipoles entering from opposite sides of the feed. They are brought together in a balun by semirigid coaxial cables. In Europe the interference level of TV signals adjacent to the radio-astronomy band is such that a narrow-band filter directly after the feed is essential. Unfortunately, this degrades the system noise temperature to approximately 350 K at 50 cm.

The first intermediate frequency (610 MHz) of the 6-cm system was chosen to be the same as the signal frequency for the 50-cm system. This enables a high isolation p-i-n diode microstrip switch at 610 MHz to be used to choose between observations at 6 or 50 cm. Following this, a printed circuit mixer converts the signal to a 30-MHz intermediate frequency with 14-MHz bandwidth.

As seen from Fig. 5, the feed and receiver box form a unit which, supported in large-diameter bearings, can be rotated inside the frame. This rotation, spanning 140°, is desirable to enable high-accuracy observations of both linear and circular polarization to be made and for the determination of some of the instrumental polarization parameters. A stepping motor and worm drive control the rotation to an accuracy of 0.01°. Stainless-steel semirigid cables wound in a spiral bring the RF signals in and out of the box while preserving phase stability during rotation.

Nearly all components were produced in-house. Printed-circuit and microstrip techniques were extensively used.

VI. THE LOCAL-OSCILLATOR SYSTEM

The local-oscillator (LO) signal is distributed from the control building at 193 MHz through coaxial cables; the required 580- and 5605-MHz signals are generated in each front end by ×3 and ×29 multiplication, respectively. The performance of the latter multiplier is of critical importance and we describe this unit in some more detail. The multiplier proper consists of a cheap glass-packaged step recovery diode inductively coupled to a two-section cavity filter. Good efficiency and stability over a 30-dB power range were found to be attainable by a combination of self-bias and a fixed-bias

voltage that compensated for the diode's 0.6-V threshold. A class B transistor amplifier precedes the multiplier stage to provide the necessary drive power. Although the whole unit is operated inside a temperature-controlled frontend drawer, its phase is still sensitive to input power variations which result from temperature variations in the cables. A p-i-n diode attenuator, controlled by one of the 6-cm mixer diode currents, keeps the multiplier input power at a constant level. A special design enabled that unit to be phase stable as well as matched at its input for a wide range of attenuation values.

The 400-Hz phase switching in the moving antenna frontends is performed in the 580-MHz LO circuit.

VII. THE CABLE SYSTEM

Having taken all conceivable precautions to insure adequate phase stability in the electronics of the system, one is left with the cables connecting the front ends and the control building as the main source of drifts. The electrical length of the "heliax"-type gas-filled cables is a function of temperature and gas density, both of which will tend to change with time. The major source of error resides in the first LO cables, all other cables operating at much lower frequencies.

Each cable consists of a) a long run from the control building to the base of the antenna, b) three shorter runs connecting the antenna base via two axis crossings to the focus, and c) two short flexible jumper cables at the axis crossings. The main factors that determine the phase stability of such a connection are listed in Table IV. The pressure stability quoted does not represent the ultimate attainable, but with a total of about 600 connectors in the system, one is not likely to get rid of all gas leaks.

The effect of the large variations indicated by Table IV can be considerably reduced by making the distribution system symmetrical. Thus 900-m lengths of buried cable were used for all antennas, regardless of their physical distances from the control building. Furthermore, all cables are pneumatically connected together to obtain uniform pressure throughout the system. Since the rapid ambient temperature variations of the exposed cables cannot be symmetrized, these cables are shielded over most of their length by a double pipe providing a thermal time constant of several hours.

An entirely different problem was encountered in the axis crossings, which had to be constructed on the outside of bearings about 60 cm in diameter. RG223 braided coax jumpers are used, supported by a spring blade that controls the bending of the cables and constrains the jumper to a plane perpendicular to the axis. In its extreme positions, the shapes of the jumper in that plane are a C and an Ω, respectively. The phase change versus shaft position varies from cable to cable, being about 1 mm for the best ones and reproducible to within 1 mm for most of them. The jumpers are thermally screened by about 3 cm of foam plastic.

VIII. The Continuum Back End

The back end has to perform the various functions outlined in Section IV and Fig. 4. We shall describe here the successive circuit blocks.

Each of the 24 delay systems consists of a binary series of carefully adjusted lengths of air–dielectric coaxial cable: 10, 20, $\cdots$, 320, 640, and 320 m; the unit step corresponds to one wavelength at the operating frequency. Through the operation of diode switches, each length can be bypassed through a short lossy cable having the same attenuation as the delay cable. The maximum residual delay error in an interferometer is 10 m, corresponding to a loss of 1.5 percent in correlation for our 4-MHz bandwidth. The delay system, with its 160 λ maximum length, forms a major element of asymmetry in our interferometers. It is installed in the basement of the control building where the temperature is fairly constant. Phase deviations are regularly measured and used to apply corrections in the off-line data reduction. They are found to be no more than 10° in the worst cases, and show small seasonal and secular variations. Preceding the delay system are booster amplifiers, the skew bandpass of which compensates for the attenuation-frequency characteristics of the IF return and delay cables.

The 2-Hz phase rotation and the 90° phase shift for the moving antennas are applied via the LO signal in a 30- to 11-MHz frequency conversion. The phase rotator consists of a binary series of 41-MHz lumped-element delays operated in a way analogous to the delay system, with a 9° unit step. Spurious sideband frequencies at harmonics of the 2-Hz drive frequency are below −40 dB. At 11 MHz, each dipole channel is fed to a unity gain dividing amplifier, the parallel outputs of which are connected to the correlators.

The correlator (Fig. 7) has to perform the basic correlation operation which is equivalent to multiplying the real input voltages V_{i1} and V_{i2} and selecting the low-frequency component in the product. The multiplication is realized in the form

$$V_{i1}V_{i2} = \tfrac{1}{4}\{(V_{i1} + V_{i2})^2 - (V_{i1} - V_{i2})^2\}$$

by the circuit shown. This method has the advantage of combining the dipole channels before passing them through the quasirectangular bandpass amplifiers, the steep phase characteristics of which might otherwise impair the system's phase stability.

Thanks to the low signal-to-noise ratios usually applicable in radio astronomy, a simple linear detector can perform the squaring operation. Let P_s be the signal power $(V_{i1} \pm V_{i2})^2$ and P_r the system noise power at the detector input; then the output is

$$V_0 = (P_r + P_s)^{1/2} \simeq P_r^{1/2} \cdot (1 + P_s/2P_r).$$

The approximation is valid for almost all practical cases. It can be improved by the introduction of off-line corrections, if necessary.

The output of the AF differential amplifier, which still carries the 400-Hz phase-switch modulation, is band-filtered by a switching filter [6]. This circuit has been integrated with the fringe demodulator and synchronous detector on a pair of

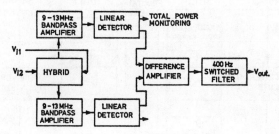

Fig. 7. Block diagram of a correlator module. A broad-band hybrid combines the signals into their sum and difference. The receiver bandwidth is defined by the 9–13-MHz bandpass amplifiers which consist of three cascode stages using double-tuned maximally flat transformers as interstage networks. The switched filter has a time constant of 0.6 ms.

AF printed circuit boards for each complex channel. These boards had to be designed with meticulous care to avoid interference from the high-level 400-Hz drive voltages. The detector employs MOSFET transistors which provide low offsets.

The 160 correlated outputs are filtered by 20-s time constants (equalized within 1 percent for each sine–cosine pair) and sampled by a reed-relay scanner and digitized in a 10-s cycle.

IX. The 21-cm Phase Calibration System

The principal blocks of the system are shown in Fig. 8. The phase of the main receiver is measured during astronomical observations by observing the receiver's response to correlated CW signals injected at the paramp inputs of each interferometer. To separate the astronomical and calibration signals, the latter are modulated by phase-switching in the moving antennas prior to injection, at half the rate at which the main receiver is switched.

The injection signals are carried to the front ends over separate calibration cables, the phase lengths of which also have to be determined. This is done by the well-known technique of modulating the end reflection of each cable by a switched load. By switching the loads in the two arms of an interferometer at different frequencies (of about 10 kHz), the phase difference between the cables can be measured directly through correlation of a pair of modulation sidebands [7]. The special calibration receiver built for this purpose produces analog outputs proportional to the sine and cosine of the phase difference. The 20 interferometers are measured sequentially, interferometer selection being effected by modulating only two cable loads at a time. Thus microwave switches can be avoided and the main receiver calibration circuits can operate without interruption. The extra loss due to distributing the calibration signal over 12 telescopes simultaneously and recombining the return signals into the calibration receiver is acceptable. The overall loss from oscillator to calibration receiver input amounts to about 135 dB.

Even with careful tuning the isolation between oscillator and calibration receiver cannot be expected to be stable to better than 50 dB. A specially developed system minimizes the sine and cosine components of the unmodulated stray signal by injecting compensating signals, the signs and magnitudes of which are automatically adjusted by independent narrow-band feedback loops. A 60-dB reduction of the stray signal is obtained in this way, with the modulated return signals from the antennas passing unaffected.

The various output signals are sampled and recorded along with the astronomical data. They can be used for off-line correction of the receiver phase.

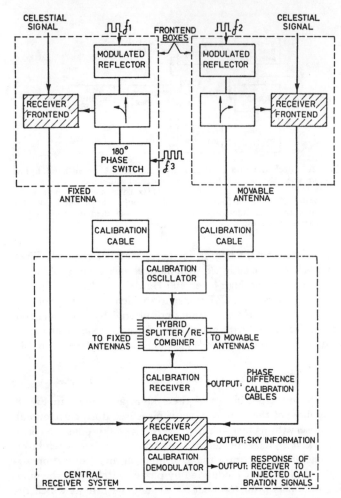

Fig. 8. Block diagram of the 21-cm phase calibration system. The hatched blocks represent the main receiver discussed in previous sections. On its way from the points of injection to the correlator, the calibration signal undergoes the various operations discussed and shown in Section IV and Fig. 4.

Owing to the excellent behavior of the main receiver, the phase calibration system has so far only been used in special receiver tests. Evaluations made to date indicate that, with some improvements still to be made, it is capable of an overall accuracy of 1° at 1415 MHz, or 0.6 mm.

X. Computer Programs

A Philips P9200 on-line computer controls the antenna steering, delay switching, fringe stopping, and data acquisition described in previous sections. It also does a limited amount of receiver status checking. It accepts commands and informs the operator on system status and malfunctions in real time through a dialogue teleprinter. This relatively simple control system is being integrated into a much larger one centered on an HP2100 RTE system. When completed, the system will be capable of unattended telescope control over prolonged periods of time and of handling the data flow from the 5120-channel line back end. In addition, a substantial part of the off-line data processing will be transferred to Westerbork; this will be particularly useful for an early detection of equipment malfunctions [8].

The off-line data reduction [9], [10] involves a great variety of functions and options, a schematic diagram of which is shown in Fig. 9. All of these are implemented on an IBM 360/65 computer at the University of Leiden. A reduction and software group is responsible for the maintenance, updating, and further expansion of the software package. In the

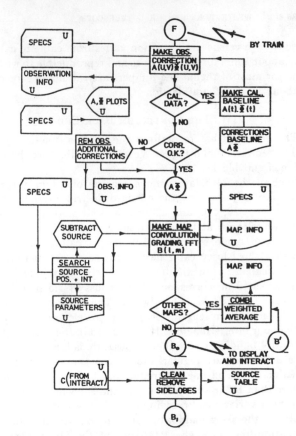

Fig. 9. A flow chart of the reduction procedure. Tapes with observational data (F) travel by train to the computer center. The names of the program modules are underlined. Blocks with a U indicate an activity by the user. Tapes (A, φ) contain the corrected visibility data; B tapes, the sky brightness map. Display and Interact are program modules, described in the text.

standard reduction procedure, they are also responsible for the routine processing of observations. This includes all the instrumental corrections based on information supplied on the observation tapes or by the SRT telescope group, and the baseline corrections which are derived weekly from a series of calibration observations.

The produced map can be displayed in different ways (contour plot, profile plot, etc.). This enables the astronomer to decide on further reduction of certain areas in the field, as, for instance, subtracting strong sources or cleaning an area of sidelobe effects. For this purpose, interactive programs have been developed by astronomers both at Groningen and Leiden, for a PDP9 and an IBM1800 disk-operating system, respectively. These form essentially duplications of the lower half of Fig. 9, differing from it mainly in the data volume that they can handle. They enable the observer to find the most suitable form of representation for his data before invoking the more powerful but much slower procedures using the IBM360.

The system as a whole minimizes the routine work on the part of the user, leaving him most of his time for interpretation. Högbom and Brouw [11] give additional data on telescope performance and data reduction.

XI. Examples of Results

We conclude with some results which illustrate the performance and capabilities of the SRT.

Fig. 10 shows some plots of interferometer amplitudes and phases versus hour angle obtained on a calibration point source. Both short- (several minutes) and long- (several hours) term stabilities are good. The similarity of some of the larger phase fluctuations in different interferometers and the

122

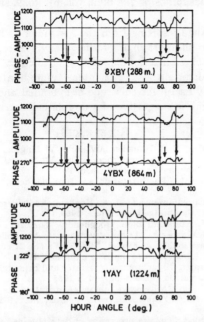

Fig. 10. Some examples of the amplitude and phase stabilities of some of the 6-cm wavelength interferometers for 12 h of continuous observation of the radio source 3C147. For each interferometer the top scan is the amplitude (vertical scale as indicated) and the bottom scan the phase (vertical scale as indicated in bottom diagram). Averaging time is 15 min. Many of the shorter variations in phase (arrowed) are present in all interferometers and, being dependent on baseline length, are interpreted as atmospheric fluctuations. Slow variations apparent in the phase are due to small errors in the off-line corrections for baseline parameters. Slow variations in amplitude are due to pointing errors.

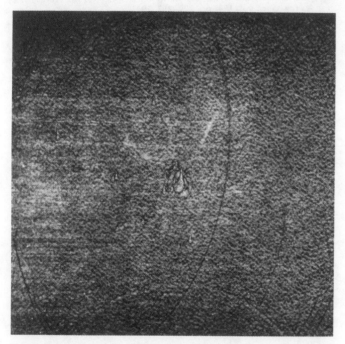

Fig. 11. Profile map of a full synthesis observation of a field around the galaxy M51. The main object shows a central source and the spiral arms of the galaxy. Thus the ridges in M51 are not sidelobe effects. The weak ridges running across the field are grating lobes at 40′ from their source of origin. More than 15 other sources are visible with intensities of a few milliflux units or more. The rms noise level is about 0.5 mfu.

baseline dependence of their sizes indicate that they originate in the atmosphere. Fluctuations of this type form a natural limitation on telescope performance [12].

Fig. 11 is a profile plot of a 1°×1° field centered on the galaxy M51. This observation took 4×12 h. The plot gives a

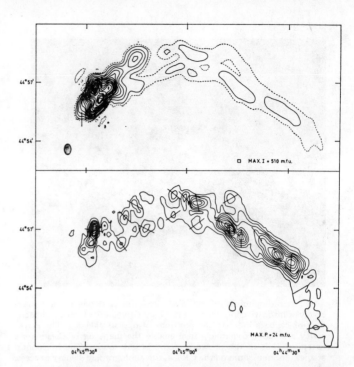

Fig. 12. Contour map of total (I) and polarized (P) intensity of the radio galaxy 3C129. Full drawn contours are in steps of 5 percent (I) and 10 percent (P) of the maximum intensity expressed in milliflux units per synthesized beam area. Straight lines in the P map indicate angle and percentage of polarization. The hatched ellipse represents the half-power beam area.

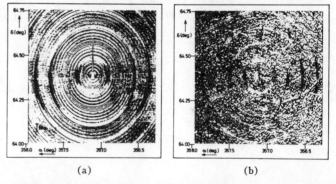

(a)   (b)

Fig. 13. The strong point source in (a) has been subtracted in (b). Note the large difference in intensity scale between the figures (factor 25). The grating rings have almost disappeared. Weak sources, obscured in the original picture, are now seen. Instrumental stability generally allows subtraction to a residue of less than 1 percent.

good impression of the noise level. Point sources of 5 milli-flux units (mfu) are easily discernable by eye. M51 shows a strong central source and spiral arms. This is the first observation of a spiral galaxy to show the arms in the continuum radio radiation.

Fig. 12 shows a contour map of a source, 3C129, in total intensity and linear polarization. Note the intricate structure and large variations in brightness. The map of polarized radiation shows even more components than that of total intensity. The high sensitivity and full polarization capabilities of the SRT make the study of these objects very fruitful.

The stability in amplitude and phase determines to what extent point sources may be subtracted from the observations, since the subtraction procedure is based on a fit of the error-free antenna pattern. Both the source response and all its sidelobe responses are subtracted. An example is given in Fig. 13. The strong point source of this 12-h observation is subtracted in the right-hand part. Several weak sources are now

123

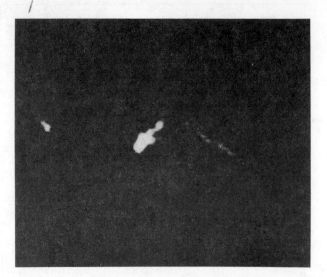

Fig. 14. A radiophotograph of 3C129, showing a strong main source, resolved into two components, and a long curved tail of low brightness. The black bars indicate the positions of optical galaxies. The sensitivity and large dynamic range enable the mapping of these weak extended tails in the presence of the bright source. The horizontal stripes represent hum on the CRT in the photographic display process.

apparent, which were obscured by the sidelobes of the strong source in the original picture. Source subtractions with SRT data generally are possible to a residue of about 1 percent. In "full synthesis" (4×12 h) observations, the beam quality is so good that subtractions are often not necessary.

The final example (Fig. 14) illustrates the sensitivity and dynamic range resulting from the instrument's high stability. It is a radio photograph of a radio galaxy. A strong main component, resolved in a double source, is trailed by a long curved tail of weak brightness. The intensity-modulated photo display is very useful for this type of extended and weak radiation. The tail is suggestive of a double-helix structure.

References

[1] R. J. Allen, J. P. Hamaker, and K. J. Wellington, "The synthesis radio telescope at Westerbork; The 80 channel spectrometer," *Astron. Astrophys.*, to be published, 1973.
[2] E. Fomalont, "Earth-rotation aperture synthesis," this issue, pp. 1211–1218.
[3] W. N. Christiansen and J. A. Högbom, *Radiotelescopes*. London, England: Cambridge Univ. Press, 1969.
[4] J. W. M. Baars and B. G. Hooghoudt, "The SRT; The layout and mechanical aspects," *Astron. Astrophys.*, to be published, 1973.
[5] J. L. Casse and C. A. Muller, "The SRT; The 21 cm continuum receiver," *Astron. Astrophys.*, to be published, 1973.
[6] R. H. Frater, "Synchronous integrator and demodulator," *Rev. Sci. Instrum.*, vol. 36, pp. 634–637, May 1965.
[7] T. H. Legg, "Microwave phase comparators for large antenna arrays," *IEEE Trans. Antennas Propagat.*, vol. AP-13, pp. 428–432, May 1965.
[8] E. Raimond, "On site processing of data obtained with the SRT," *Astron. Astrophys.* (Suppl.), to be published, 1973.
[9] W. N. Brouw, "Data processing for the Westerbork synthesis radio telescope," Ph.D. dissertation, University of Leiden, Leiden, The Netherlands, 1971.
[10] H. W. van Someren Greve, "Data handling of the Westerbork synthesis radio telescope," *Astron. Astrophys.* (Suppl.), to be published, 1973.
[11] J. A. Högbom and W. N. Brouw, "The SRT; Principles of operation, performance, and data reduction," *Astron. Astrophys.*, to be published, 1973.
[12] R. Hinder and M. Ryle, "Atmospheric limitations to angular resolution of aperture synthesis radio telescopes," *Mon. Notices Roy. Astron. Soc.*, vol. 154, pp. 229–253, 1971.

The Very Large Array: Design and Performance of a Modern Synthesis Radio Telescope

PETER J. NAPIER, A. RICHARD THOMPSON, SENIOR MEMBER, IEEE, AND RONALD D. EKERS

Invited Paper

Abstract—Since its development in the 1960's, the technique of obtaining high-resolution radio images of astronomical objects using Fourier synthesis has advanced sufficiently so that today such images often provide better angular resolution than is obtainable with the largest optical telescopes. A synthesis array measures the Fourier transform of the observed brightness distribution by cross-correlating the signals from antennas separated by distances up to tens of kilometers. The antennas must be equipped with low-noise receiving systems and connected together by phase-stable transmission links. Wide-bandwidth digital correlators are used to perform the cross correlation. The data-reduction algorithms and computing system play a critical role in determining the quality of the images produced by the array.

The Very Large Array (VLA) synthesis telescope, recently constructed in New Mexico, consists of twenty-seven 25-m-diameter antennas arranged in a Y-shaped array. Each arm of the Y is approximately 21 km long and the antennas can be moved to various positions on the arms by a rail-mounted transporter. The antennas are equipped with cryogenically cooled receiving systems and are interconnected by low-loss, TE_{01}-mode, large-diameter waveguide. The cross-correlation products for each of the 351 pair combinations of antennas are measured for 4 IF signals by a 50-MHz bandwidth digital correlator.

In this paper we discuss the design of synthesis arrays in general, and describe the design and performance of the VLA in particular, under the seven headings: array geometry design, sensitivity considerations, phase stability requirements, signal transmission system, delay and correlator system, control system, and data-reduction requirements. In each section, we review the underlying instrumental requirements and provide details of how the VLA was designed to meet them. Recently developed data-reduction algorithms provide effective ways of correcting synthesis images for the effects of missing Fourier components and instrumental and atmospheric amplitude and phase errors. The power of these algorithms is demonstrated using actual VLA images.

I. INTRODUCTION

THE INCREASE in the performance of radio telescopes from the late 1940's through the present time has led not only to the discovery of many thousands of radio sources but, more importantly, to a spectacular increase in our appreciation of the nature and variety of objects in the astronomical universe. An essential element in this progress has been the ability to compare the observed features of objects over a wide spectral range; the radio, infrared, visible, ultraviolet, and X-ray domains. Because of the enormous number of detectable objects, particularly in the visible range, accurate position measurements are essential to enable a source detected in one spectral domain

to be identified without ambiguity in another. Furthermore, intercomparison of structural detail calls for comparable angular resolution. Until recent times, the angular resolution of large, ground-based optical telescopes, which is usually limited to about 1″ by atmospheric effects, was unapproached in other parts of the spectrum. At centimeter wavelengths, however, advances in imaging techniques based on interferometry have led to the design of arrays with resolution finer than 1″ and capable of synthesizing images with 10^6 to 10^7 resolution elements. The technique used has been variously described as Fourier synthesis, aperture synthesis, and earth-rotation synthesis, and was demonstrated by Christiansen and Warburton [1], and Ryle [2], and developed by Ryle at Cambridge. The Very Large Array (VLA), construction of which was completed at the end of 1980 at a cost of $78M (1977 dollars) is the latest and most powerful synthesis array to come into operation. It is routinely producing radio images with an angular resolution as fine as a few tenths of a second of arc.

In a Fourier synthesis array, signals from different antennas are combined in pairs in voltage-multiplier circuits, the outputs of which are averaged over time for periods of the order of a few seconds. The combination of a multiplying and time averaging circuit is usually termed a correlator. As is well known, combining signals from spaced antennas results in a reception pattern which is characterized by quasi-sinusoidal fringes, the angular spacing of which depends upon the spacing of the antennas in units of the observing wavelength. The response to a radio source on the sky is given by the product of the sinusoidal fringe function and the radio brightness pattern of the source, integrated in angle over the antenna beams. The result is proportional to the amplitude and phase of a component of the Fourier transform of the brightness pattern of the source, with angular frequency equal to that of the fringes. This function, known as the complex visibility of the source, is obtained from the correlator output by suitable calibration. Thus by observing with many different spacings and angles between the antennas, the Fourier transform of the source pattern can be measured in sufficient detail that an inverse transformation yields an accurate radio picture. Derivations of the Fourier transform relationship in synthesis imaging based on the concept of fringe patterns, as outlined above, are given by Bracewell [3] and Fomalont and Wright [4]. The image obtained by a synthesis array is also commonly called a map or a brightness distribution. In this paper we use these terms interchangeably.

An alternative description of synthesis imaging can be given in terms of the mutual coherence function of the radiation field. The

Manuscript received April 11, 1983; revised August 19, 1983. The National Radio Astronomy Observatory (NRAO) is operated by Associated Universities, Inc., under contract with the National Science Foundation.

The authors are with the National Radio Astronomy Observatory, Socorro, NM 87801.

Reprinted from *Proc. IEEE*, vol. 71, no. 11, pp. 1295–1320, Nov. 1983.

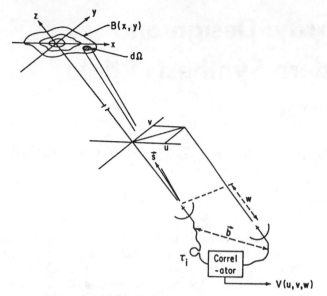

Fig. 1. Relationship between the (x, y, z) and (u, v, w) coordinate systems in which the source brightness B and the antenna baseline $\vec{b}$ are specified. The z axis points towards the center of the field to be mapped. The complex visibility $V(u, v, w)$ is derived from the correlator output by removal of the fringe pattern. (Adapted from [71].)

relative phases of the signals at two spaced antennas vary with the position angle of the point from which they originate on the source, and this variation depends upon the antenna spacing. Thus the measured coherence of the radiation at the two antennas varies with the antenna spacing in a manner that depends upon the angular distribution of the radio brightness of the source [3], [5].

In this paper we shall review the technique of synthesis imaging and its present capability, and show how the design of the VLA was derived from the astronomical performance goals. Heeschen [6] has described the history of the VLA Project and, in addition to other authors [7], [8], has considered the scientific capabilities of the telescope. Here we will be more concerned with the design of the instrument. In the remainder of this paper we will discuss the design of synthesis arrays in general and the VLA in particular with respect to the seven topics: array geometry, sensitivity, phase stability, signal transmission, signal delaying and correlation, the control system, and data reduction.

II. ARRAY GEOMETRY CONSIDERATIONS

A. Coordinate Systems in Synthesis Imaging

The correlator output $r(\vec{b}, \vec{s}_0)$ is related to the brightness of the source on the sky, $B(\vec{s})$, by

$$r(\vec{b}, \vec{s}_0) = \int B(\vec{s}) \exp\left(j2\pi\vec{b}\cdot\vec{s}\right) d\Omega$$

$$= V(\vec{b}, \vec{s}_0) \exp\left(j2\pi\vec{b}\cdot\vec{s}_0\right). \qquad (1)$$

With reference to Fig. 1, $\vec{b}$ is the antenna spacing vector measured in wavelengths, $\vec{s}$ is a unit vector indicating the direction of the element of solid angle $d\Omega$, and s_0 is a unit vector in the direction of the center of the field to be mapped (the z direction). V is the complex visibility of the source. The two antennas track the radio source, and the effects of the beam patterns are assumed to be included within $B(\vec{s})$. To make an image of an area of sky we set up the Cartesian coordinate system (u, v, w) with the w axis pointing towards the central position of the field, and

the u and v axes in a plane normal to this direction with u measured towards the east and v towards the north. The antenna base line vector $\vec{b}$ has components (u, v, w) where u and v represent the antenna spacing as seen from the direction of the field center, and w represents the relative delay between the signals at the two antennas. The unit vector $\vec{s}$ has direction cosines (x, y) relative to the axes (u, v). In terms of the visibility V, which represents the amplitude and phase of the correlator output fringes, we have

$$V(u, v, w) = \int\int B(x, y)$$

$$\cdot \exp\left[j2\pi\left(ux + vy + \sqrt{(1 - x^2 - y^2)}\, w\right)\right] \frac{dx\, dy}{\sqrt{1 - x^2 - y^2}}.$$

$$(2)$$

The (x, y) plane, as defined above, represents the projection of the celestial sphere onto a tangent plane at the field center. For imaging a field of small angular dimensions this is usually the most convenient projection, with least distortion of the brightness profiles. Distances from the (x, y) origin at the image center are proportional to the sines of the corresponding distances on the celestial sphere, so for small images, distances in x and y are closely proportion to true angular distances.

If all of the measurements could be made with the antennas in the (u, v) plane so that $w = 0$, (2) would conveniently reduce to an exact, two-dimensional Fourier transform. In general, however, an observation lasts for several hours, and as the earth's rotation carries the antennas through space, the spacing vector for any pair moves on a conical surface centered on the rotation axis. The measurements must thus be defined in terms of the three-dimensional system (u, v, w). However, visibility values in the (u, v) plane can be derived from measurements outside the plane provided that x and y are small enough to enable us to write

$$\sqrt{1 - x^2 - y^2}\, w \approx \left[1 - \frac{1}{2}(x^2 + y^2)\right] w \sim w. \qquad (3)$$

Then from (2) we obtain

$$V(u, v, w) \exp\left(-j2\pi w\right)$$

$$= \int_{-\infty}^{\infty}\int_{-\infty}^{\infty} \frac{B(x, y)}{\sqrt{1 - x^2 - y^2}} \exp\left[j2\pi(ux + vy)\right] dx\, dy.$$

$$(4)$$

The factor $\exp(-j2\pi w)$ modifies the argument of V, to provide the approximate value that would be observed in the (u, v) plane. Thus the left-hand side of (4) can be written as $V(u, v, 0)$, and the inverse transform of (4) becomes

$$\frac{B(x, y)}{\sqrt{1 - x^2 - y^2}} = \int_{-\infty}^{\infty}\int_{-\infty}^{\infty} V(u, v, 0)$$

$$\cdot \exp\left[-j2\pi(ux + vy)\right] du\, dv. \qquad (5)$$

The modification of $V(u, v, w)$ to the (u, v) plane is only exact for radiation from the direction $x = y = 0$, and for points away from this origin the term $\frac{1}{2}(x^2 + y^2)w$ that was omitted from (3) results in a phase error equal to $\pi(x^2 + y^2)w$. This places a limit on the size of the image that can be obtained. If the area to be imaged is observed down to low elevation angles, w takes values comparable to the spacings of the antennas on the ground. Thus, for example, for an image with a resolution of $1''$, the width of the field must not exceed about $2'$ if the maximum phase errors introduced by the above approximation are not to exceed $5°$.

In the particular case of an east–west linear array, the spacing vectors remain in a plane normal to the direction of the pole as the earth rotates. In imaging with such an array the approximation in (3) leads to a predictable distortion of the coordinates which can be corrected [9], [10]. Unfortunately, an array in which the measurements all lie in a plane parallel to the celestial equator has vanishingly small north–south resolution for direction close to the celestial equator. For such directions, arrays with baselines that are more widely distributed in azimuth are required, as will be discussed in the following section. Then the effect of tracking over large hour angles is that the distribution of the spacing vectors is three-dimensional, and the phase errors resulting from the approximate conversion of the visibility data to the (u, v) plane are not so easy to correct for. Possible approaches to circumvent the restriction of the field size are discussed by Clark [11] and include and building up of large images from mosaics of smaller ones, implementing (2) as a three-dimensional Fourier transform, and the use of an image with no correction for phase errors to provide a basis for an improved estimate of $V(u, v, 0)$ from $V(u, v, w)$. This last idea has been discussed by Bracewell [12] and could be applied iteratively. There appears to be no reason to doubt that such techniques will allow large images to be synthesized at the cost of some additional computation. Up to the present, however, there has been sufficient important research to be performed using small fields that little has been done towards development of large-field techniques.

B. General Considerations in Array Design

We now consider how to obtain a satisfactory distribution of visibility measurements in the (u, v) plane. At any instant a measurement with one pair of antennas provides visibility values at two points since $V(-u, -v) = V^*(u, v)$ because $B(x, y)$ is a purely real function. As the antennas track a radio source, the spacing vector in the (u, v) plane traces out an arc of an ellipse. For a source at the pole the ellipse becomes a circle, and as the declination of the source is decreased the ellipse becomes narrower in the v direction and degenerates to a straight line for a source on the celestial equator. This behavior is easily visualized by considering how the relative positions of two antennas vary with earth's rotation, as seen by an observer in space looking down towards the earth. Expressions for u and v in terms of the antenna spacing and the source position can be found in [4] and [5]. Tracking a source across the sky for 6 to 12 h provides one way of increasing the coverage of the (u, v) plane. The other principal means is, of course, using pairs of antennas with different spacings. With just two antennas it is possible to synthesize an image to any desired complexity by moving one antenna to a new position for each successive observation of the source. Synthesis imaging with the centimeter-wavelength interferometer at the Owens Valley Radio Observatory [13] is performed in this manner. On the other hand, an array of n_a antennas allows up to $n_a(n_a - 1)/2$ pairs to be operated simultaneously, and in the case of the VLA, for which $n_a = 27$, an image can be synthesized in a single transit of a source. In between there are arrays with combinations of fixed and moveable antennas such as the one at Westerbork [14] which takes up to 4 days to perform a synthesis. Highly successful instruments operating in each of the above modes have been constructed.

The behavior of the spacing loci in the (u, v) plane with declination, i.e., the circularity of the loci at declinations near the pole, allows satisfactory high-declination images to be obtained

with one-dimensional, east–west arrays. The design of such linear arrays for minimum redundancy in the spacings has been discussed by Moffett [15] and Ishiguro [16]. At a declination of 30°, the synthesized beam of an east–west array becomes broadened by a factor of two in the north–south direction. Thus for imaging sources within about 30° of the celestial equator, a two-dimensional configuration of antennas with spacings in a range of azimuthal directions is necessary. For non-east–west baselines, the spacing loci also shrink to straight lines as the declination approaches zero, but the lines are not coincident with the u axis as they are for east–west baselines.

The required spacing of loci within the (u, v) plane can easily be appreciated by considering the use of a two-dimensional, discrete Fourier transform to obtain a radio image from visibility data. If the chosen dimensions of the image are x_0 by y_0 radians, the visibility data in the (u, v) plane must be specified at points on a rectangular grid with spacing $\Delta u = x_0^{-1}$ and $\Delta v = y_0^{-1}$ in the u and v directions, respectively. The visibility values at these points are then interpolated from the observed values on the tracking loci. At this point we consider only the simplest interpolation procedure, sometimes referred to as cell averaging [17]. A more detailed discussion of interpolation procedures is given in Section VIII-C. The (u, v) plane is divided into rectangular cells of dimensions Δu by Δv and the mean of all measured values within each cell is assigned to the central point. Clearly, it is desirable that as many cells as possible be intersected by one or more loci, which requires that the increments in antenna spacings should be comparable with Δu and Δv. The choice of antenna configuration in a synthesis array thus depends mainly upon optimizing the coverage in the (u, v) plane as outlined above, taking account of the range of declination required, the number of antennas available, the acceptable number of antenna moves, etc. Imaging of broader sources requires larger fields and smaller values of Δu and Δv. Smaller antenna spacings are then required, and the ability to change the scale of the antenna configuration enhances the versatility of the instrument. Also, as will be explained in Section III-A, the lower resolution provides greater sensitivity to broad, low-brightness sources.

Table I lists the parameters of most currently operational synthesis arrays, or arrays that are currently under construction. The range of size and complexity of these instruments illustrates the development of synthesis arrays over the past two decades. Note that in this paper we are considering only what are sometimes referred to as connected-element arrays, i.e., those in which the antennas are directly connected to a central electronics complex and the cross correlation of the pairs of signals is done in real time. We exclude very-long-baseline interferometry in which antenna spacings can be as large as several thousand kilometers. For such observations, it is usual to derive the local oscillator signals from an independent frequency standard at each antenna and to record the signals on magnetic tape for subsequent playback at a correlator location; see, for example, [42].

C. Antenna Configurations for the VLA

Listed in Table II are the performance requirements of the VLA that principally shaped the design of the instrument, and the main features that resulted from them. The first five listed requirements are those that affected the layout of the antennas. The resolution calls for spacings up to approximately 350 000 wavelengths, or 21 km for 6 cm. The requirement that the resolution be maintained over a wide range of declinations, including the celestial equator, means that earth rotation cannot

TABLE I
PRINCIPAL PARAMETERS OF CURRENTLY OPERATIONAL SYNTHESIS ARRAYS

Common Name	Institution	Frequency (Ghz)	T=Total no. elements (M)=no.movable d=size (m) T (M)	d	Total Geometrical Collecting Area (m²)	Array Size and Geometry (km)	λ/d for Longest Baseline Resolution at Highest Frequency (arc sec)	References*
Owens Valley Centimeter Interferometer	Caltech, USA	0.6-10.7	2 (2) 1 (1)	27 40	2401	0.5 N/S, 1.2 E/W	4.6	13
Owens Valley Millimeter Interferometer	Caltech, USA	88-120	3 (3)	10.4	85	0.43 T	1.2	18
Culgoora Radioheliograph	CSIRO, Australia	0.04, 0.08, 0.16, 0.327	96	13	12742	3 circular	63	19
Australia Telescope	CSIRO, Australia	0.3 - 44	7 (6) 1	22 64	5800	6 E.W. 300 Irregular	.01	20
Penticton Interferometer	Dominion RAO, Canada	0.408, 1.4	4 (2)	9	254	0.6 E/W	74	21
Gauribidanur Decameter Wave Telescope	Raman Res. Inst. Indian Inst. Astphys.	0.035	1000 dipoles		250λ²	T, 1.5 E/W, 0.5 N/S	1260	22
IRAM Array	IRAM, France	70-375	3 (3)	15	530	0.39, Approx. T	0.4	23
One-Mile Telescope	Mullard RAO, U.K.	0.408, 1.42 2.7, 5.0	3 (1)	18.3	790	1.5 E/W	.8	24
Half-Mile Telescope	Mullard RAO, U.K.	1.42	4 (2)	9	254	0.73 E/W	60	25
5 km Telescope	Mullard RAO, U.K.	2.7, 5.0, 15.4, 32	4 (4)	13	1062	4.6 E/W	0.7	26
5 km, 151 MHz Telescope	Mullard RAO, U.K.	0.151	60	4 element Yagi	2000	5.0 E/W	60	27
VLA	NRAO, USA	1.4, 5, 14.4, 23	27 (27)	25	13200	21, 21, 19 Y	0.07	7
Green Bank Interferometer	NRAO, USA	2.7, 8.1	3 (2) 2	26 14	1750	35 Irregular	0.2	28,29
Westerbork Synthesis Radiotelescope(WSRT)	Netherlands Foundation for Radio Astronomy	0.608, 0.327, 1.4, 5.0	14 (4)	25	6872	2.8 E/W	4.4	30
Ooty Radiotelescope	Tata Institute, India	0.327	4 1 5	132 x 30 100 x 9 25 x 9	17865	4.7 Irregular	40	31
Nobeyama Array	Tokyo Astrophys. Obs., Japan	22, 115	5 (5)	10	392	.68 T	.8	32
Bologna Cross	University of Bologna, Italy	0.408	6 8	20 x 95 24 x 80	26800	E/W, 0.4 N/S	150	33
Hat Creek Millimeter Interferometer	University of California, USA	80-115	3 (3)	6	85	.3 E/W, .2 N/S	1.8	34
MERLIN	University of Manchester, U.K.	0.408, 1.66, 5.0, 22	1 2 3	76 15 25	6362	134 Irregular	.02	35
Clark Lake Array	University of Maryland, USA	0.0015 - 0.0125	720 Spirals	Log	250λ²	3.0 T	160	36,37
Fleurs Synthesis Telescope	University of Sydney (E.E.), Australia	1.4	64 6	6 14	2700	1.6EW, 0.8NS, 3.6 Irregular	12	38,39
Molonglo Obs. Synthesis Telescope	University of Sydney (Physics), Australia	0.843	88	11.6x17.7	18068	1.6 E/W	4.6	40
UTRAO Interferometer	University of Texas, USA	0.335, 0.365, 0.380	5	Helix Array	650	3.58 E/W, 3.38 N/S	45	41

(In many cases the listed references describe the original instrumental parameters only, and values given in the table have been supplemented through recent personal communications.)

be relied upon to provide two-dimensional coverage of the (u, v) plane and the array must contain spacings at many azimuths. Thus a linear array will not suffice. At first sight, it may seem unnecessary to cover so much of the available sky since large numbers of stars and galaxies are visible in most directions. However, the history of astronomy has shown that often some particular object offers a vital clue to new understanding because of its proximity or orientation relative to the earth, or the unusual prominence of some important feature. Thus we should not ignore any regions of the sky. Furthermore, the sun, most of the planets, and the central region of our galaxy must be observed at declinations south of $+30°$. Obtaining the full resolution in 8 h of tracking and keeping sidelobes of the synthesized response less than -16 dB relative to the main beam requires a large enough number of antennas to give sufficiently uniform sampling of the (u, v) plane without combining results taken with different configurations of the antennas.

For a two-dimensional array there is no uniquely optimum choice for the antenna configuration. Economic and practical considerations strongly favor some arrangement of straight lines of antennas which simplifies the layout of power and signal cables and requires the acquisition of land in narrow corridors only. Symmetrical-cross arrays consisting of intersecting north–south and east–west linear configurations fulfill the above re-

quirement and are well known in radio astronomy [43], [44]. They followed from the development by Mills [45] in which the outputs of the two antennas producing orthogonal narrow fan beams are combined in a voltage multiplier to produce a narrow pencil-beam response. Symmetrical crossed arrays, however, contain redundant spacings: for example the vector spacing between the antennas at the north and west ends of the arms is the same as that for the antennas at the south and east ends, and similar redundant pairings can be found all down the arms. The redundancy can be removed by omitting one of the four arms leaving a T-shaped array which has almost the same resolution. Optimization of the parameters of a T-array next leads to an equiangular Y with angles of $120°$ between the arms. The Y can be inverted so that one arm points north rather than south without affecting the performance. The optimization procedure is essentially one of trial and error, in which the (u, v) loci are computed for the desired tracking range, which for the VLA is ± 4 h, and for various declinations. The (u, v) plane is divided into cells as described earlier and the aim is to minimize the percentage of holes, i.e., cells that are not intersected by one of the spacing loci. The equiangular Y can hardly be bettered with regard to simplicity and economy, and no configuration with significant improvement of (u, v) coverage was found for the VLA. At the latitude of $34°$ where the VLA is located, the chosen design has arms of

TABLE II
PERFORMANCE REQUIREMENTS AND RESULTING DESIGN FEATURES
OF THE VLA

Performance Requirement	Design Feature
1. Angular resolution ≤0.6" at 6 cm wavelength.	Array arms 21 km long; maximum spacing 36.4 km.
2. Specified resolution to be achieved over declination range +90° to -20°.	Equiangular Y configuration and tracking range -4^h to 4^h of hour angle.
3. Variable resolution to allow observation of sources with a range of angular scales.	Four configurations with total range 36:1 in linear scale, requiring rail track and transporters.
4. Map with full resolution to be obtainable from one 8-hour observation.	Use of 27 antennas and configuration designed for optimum (u,v) coverage.
5. Peak sidelobes of synthesized beam not to exceed -16 dB except at declination -0°.	As in 4 above.
6. Antenna beams to allow circular field 1' diameter at shortest wavelength.	Antennas are 25m-diameter shaped parabolic reflectors.
7. Sensitivity to be sufficient for detection of source of flux density 10^{-4} Jy.	Size of antennas, system noise temperatures, and 8^h observing time.
8. Ability to measure complete polarization characteristics.	Oppositely polarized feeds with separate receiving channels for each output.
9. Several wavelength bands to be available under computer control.	Offset feeds, rotating sub-reflector, and front ends for several bands mounted in one Dewar.
10. Spectroscopic as well as continuum observations to be possible.	Digital spectral correlator using recirculation principle.

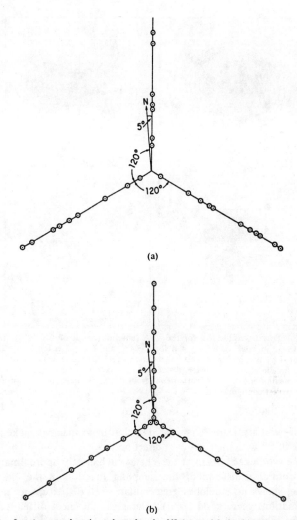

Fig. 2. Antenna-location plans for the VLA. (a) Mathur's [46] empirically optimized design study. (b) Power-law design adopted for the VLA.

equal lengths. An array designed for higher latitude might be elongated in the north–south direction to offset the effect of foreshortening when observing at low declinations.

The positions of the antennas along the arms should not be uniformly spaced but should be chosen so that the pair combinations provide the most effective range of different spacings. This problem was investigated by Mathur [46] who devised a pseudo-dynamic computation technique in which arbitrarily chosen initial positions are adjusted by a computer. In the array model, a small change is made in the position of one antenna to sense the direction that will improve the (u, v) coverage, and it is then moved to obtain maximum improvement. This is done to each antenna in turn and the cycle repeated until no further improvement is forthcoming. The result, shown in Fig. 2(a), is a distribution of rather random appearance which does not have the same number of antennas on each arm. The coverage of the (u, v) plane with this arrangement is very good. A more analytical approach was attempted by Chow [47] who considered the shape of the volume swept out by an ellipse, which represents the boundary of the array as a source is tracked. This did not lead directly to an optimum configuration, but provided some valuable insights. One of these was the suggestion of a power-law distribution in which the distance of the nth antenna on each arm from the center of the Y is the proportional to n^α. Chow's original analysis [47] suggested a value of 2 for α. In a further study he concluded that the optimum value of α depends on the tracking time and latitude of the array, and found by a largely empirical investigation that the performance is broadly optimized with a value of $\alpha = 1.6$. Comparison of the 1.6 power-law array

with Mathur's solution showed the two to be approximately equally good. The great advantage of the power-law design is found in purely practical considerations which involve requirement 3 in Table II. The variability in resolution calls for the ability to move antennas between several series of fixed locations to obtain different configurations of antennas with the same basic pattern but different scales of spacings. With the power-law arrangement, appropriate choice of the parameters allows an appreciable number of antenna stations to be shared between different configurations. The design chosen for the VLA incorporates four scaled configurations in which the most distant antenna on an arm is 21, 6.4, 1.95, or 0.59 km from the junction of the Y. There are nine antennas on each arm and $\alpha = 1.716$. The total number of antenna foundations required is 72 which may be compared with 108 (4×27) that would be required for four configurations with the design of Fig. 2(a). This difference amounted to more than $1M in construction costs. The power-law spacings are shown in Fig. 2(b), and the corresponding (u, v) plane coverage for several declinations in Fig. 3. In the latter diagrams it is easy to see the increasing ellipticity of the (u, v) loci as the declination is decreased, resulting in straight east–west lines at zero declination. For all antenna pairs where the separation is exactly east–west, the loci lie along the u axis for observations at zero declination. To eliminate this redundancy in the pairings of corresponding antennas in the south–east and

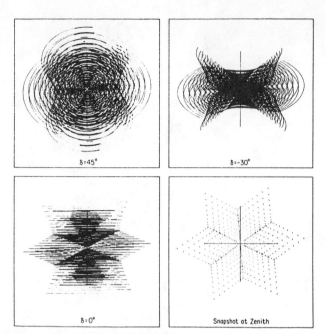

Fig. 3. Spacing-vector loci for the VLA (antenna layout of Fig. 2(b)). The coverage in hour angle is ± 4 h for $\delta = 45°$ and $0°$, ± 3 h for $\delta = -30°$ limited by the minimum tracking elevation of $9°$ for the antennas, and ± 5 min for the snapshot. The lengths of the (u, v) axes from the origin represent the maximum distance of an antenna from the array center, i.e., 21 km for the largest configuration.

Fig. 4. An aerial view of the VLA with the antennas positioned in the most compact configuration. 22 of the 27 antennas are shown. The north, south–east, and south–west arms of the array are, respectively, in the bottom right, left, and top right parts of the photograph. The central Control Building is above the center of the Y and the large building on the south–west arm is used for antenna maintenance. The extension of the north arm to the south of the Y center is needed to prevent blockage of the railway track in the most compact configuration.

south–west arms, the Y is rotated so that these spacings make an angle of $5°$ with the east–west direction.

The synthesized beam at the VLA, which closely approximates the Fourier transform of the sampling function in the (u, v) plane, shows no sidelobes greater than -16 dB except at zero declination where good (u, v) coverage is the most difficult to obtain. This result is in accordance with requirement 5 of Table II, and is one of the most important factors in determining the number of antennas in the array. During early studies of the VLA [48] a figure of -20 dB was used for the peak sidelobes, and 36 antennas were required to satisfy this condition. Since the initial design phase of the VLA was completed, the development of image processing techniques, in particular the CLEAN algorithm which will be described in Section VIII-D, have enabled the effects of sidelobes to be largely removed from an image. The level of the sidelobes is, therefore, regarded as somewhat less critical nowadays, but the uniformity of (u, v) coverage that low sidelobes implies remains a good criterion of the ability of the array to gether information efficiently.

During the first year of operation with the complete VLA some thought and experimentation have been devoted to examining whether the range of spacings in any given configuration is adequate for all imaging purposes. A single configuration can adequately cover a range of 43 in angular scales and for very complex objects significant improvement may be obtained by combining images taken in two adjacent configurations. In cases involving sources south of $-15°$ declination it is better to use stations on the north arm that correspond to the next larger configuration to that of the other arms, to offset the effect of foreshortening.

Much of what can be said about developments in antenna configuration for two-dimensional synthesis arrays has been covered in describing the case of the VLA. The circular ring array should, however, be mentioned. An array in which the antennas are located at uniform intervals around the circumference of a circle 2 mi in diameter has been constructed at Culgoora,

Australia, specifically for observations of the sun [19] (see Table I). This instrument was not originally designed to use the synthesis technique with which we are concerned here, but its performance as a synthesis array has been studied [49] and it is being partially adapted to that mode. The array was designed as a beam-scanning instrument using an ingenious phasing scheme devised by Wild [50] to reduce the sidelobes. Although this array has been highly successful for solar observations, the ring configuration has disadvantages for a general-purpose instrument like the VLA. It does not lend itself to scaled configurations, and it cannot be expanded like an array with open-ended arms.

D. The VLA Rail System and Site

Fig. 4 shows an aerial view of the VLA. Each antenna is supported by three concrete foundation piers, the tops of which are 1.9 m above the ground. There are 24 sets of piers along each arm to provide for the four configurations. A double rail track runs parallel to each arm of the array at a distance of 30.48 m (100 ft) from the center line of the foundations. The rail system consists of two standard-gauge tracks spaced 5.486 m (18 ft) apart. Two special transport vehicles have been designed to run on this track, to carry antennas between foundations. At each antenna location a spur of the double track runs from the main line to the foundation. The wheels of the transporters are mounted on trucks that can turn through $90°$ to allow the vehicle to move between the main track to the spur track which intersect at right angles. The wheels can be raised and lowered hydraulically through a range of 15 cm which enables a transporter to lift an antenna clear of the foundation and is also used in making the right-angle turns. The speed of a transporter is 8 km $\cdot$ h^{-1} when carrying an antenna, and with the two transporters the antennas can be moved between configurations in four to ten days depending upon the configurations involved.

The location of the VLA is in west-central New Mexico, on the Plains of San Augustin, 80 km west of the town of Socorro. The area is a semi-arid plateau about 2100 m above sea level, otherwise used only for ranching. The average annual rainfall is 27 cm. The site was chosen because the high, dry location tends to minimize phase fluctuations associated with atmospheric water vapor, the southern location within the U.S. improves the coverage of the southern sky, and the low level of development in the area minimizes both electromagnetic interference and the impact of the array upon other activities. A general description of the array which includes further details of the site, antennas, etc., is given by Thompson *et al.* [7].

III. SENSITIVITY CONSIDERATIONS

A. Hardware Design for Good Sensitivity

Several authors have examined the sensitivity theory for synthesis arrays [4], [51], [52]. An expression for the rms noise level in a synthesis image is [52]

$$\Delta B = \frac{\sqrt{2}\, kT_{\text{sys}}}{\eta_c \eta_a A \sqrt{n_p \cdot t \cdot \Delta f}} \qquad (6)$$

where ΔB is the noise level in the synthesized image in units of flux density per synthesized beam area for one polarization, k is Boltzmann's constant, T_{sys} (K) is the system temperature of the receivers on the array elements. η_c ($\leqslant 1$) is the efficiency of the correlator compared with a perfect analog correlator (see Section VI-B), A (m^2) is the geometrical collecting area (and η_a ($\leqslant 1$) is the aperture efficiency) of each antenna in the array, n_p is the number of pairs of antennas whose outputs are multiplied together in the correlator, t (s) is the total observing time, and Δf (Hz), for a single-sideband receiving system, is the bandwidth at the input to the correlator. A number of simplifying assumptions have been made in the derivation of (6). All antennas in the array are assumed to be identical and to have receivers with identical noise temperatures. The bandpass responses of all transmission paths from the antennas to the correlator are also assumed to be identical and rectangular: see [53] for a discussion of the importance of this latter assumption. Finally, it is assumed that all data points in the (u, v) plane are given equal weight when the Fourier transform to the image plane is computed: see [4] for a discussion of the effect on sensitivity of weighting in the (u, v) plane.

Expression (6) shows that, for best sensitivity in a synthesis array, the correlator should provide the product of all possible pairs of antennas. For an array of n_a antennas, n_p has a maximum value of $n_a(n_a - 1)/2$. As n_a becomes large, n_p approaches $n_a^2/2$ and (6) becomes

$$\Delta B = \frac{2kT_{\text{sys}}}{\eta_c \eta_a n_a A \sqrt{t \cdot \Delta f}}. \qquad (7)$$

With $\eta_c = 1$, (7) is identical to the sensitivity expression for a total power radiometer, with system temperature T_{sys} and bandwidth Δf attached to a single antenna having the same total effective collecting area, $\eta_a n_a A$, as the synthesis array. The synthesis array has the advantage, however, that all points in the field of view are observed with sensitivity ΔB, while a single antenna must observe each beam area separately for time t [51].

Expression (7) indicates that, as with all radio telescopes, the sensitivity of a synthesis array is increased by reducing the system temperature and increasing the total effective collecting area. In general, one wishes to maximize the product $n_a A$ for the available amount of funding, subject to constraints such as: A should not be too large or the field of view of the array will be small; n_a should be large enough so that there is adequate filling of the (u, v) plane and so that effective use can be made of closure phase data (see Section VIII-B); n_a should not be so large that the total cost of the receivers, correlators, and data reduction computer becomes excessive. Typically, the cost of a reflector

antenna increases approximately as diameter raised to the 2.6th power and rms surface accuracy raised to the 0.7th power [54]. Examination of Table I indicates that antenna sizes in the centimeter- and millimeter-wavelength bands range from 200 to 4000 wavelengths reflector diameter. An important consideration in the antenna design is the question of whether or not to shape the antenna for high efficiency when a Cassegrain geometry is used. The modern technique of making the main reflector non-parabolic and the subreflector nonhyperbolic to obtain a 20- to 30-percent improvement in η_a, the aperture efficiency, is fully reviewed in the collection of papers in [55]. In general, this type of shaping is not appropriate for a single-antenna telescope because of the loss of prime focus operation capability [56] and because of the high first sidelobe that results from uniform aperture illumination. A third problem is the very rapid deterioration in the gain and sidelobe behavior of a shaped-reflector system when the beam is scanned off axis. As an example of this last point, the VLA shaped geometry described in Section III-B will produce a 30-percent loss of gain and 10-dB increase in coma lobe level if the beam is scanned 3.5 beamwidths off axis by laterally displacing the feed away from the secondary focus, whereas for an equivalent unshaped geometry these degradations would be quite negligible. For the elements of a synthesis array the loss of the beam scanning capability and the relatively high first sidelobe are not usually a problem. If the deviation of the main reflector from a paraboloid can be kept small enough to allow very-long-wavelength operation from the prime focus, shaping the profiles of synthesis antennas is probably worthwhile. The profile of the VLA main reflector deviates in a systematic way from a parabola by only 1.0 cm rms.

Apart from this question of reflector shape, the design of antennas, feeds, and low-noise receivers for synthesis arrays is not significantly different from single-antenna designs except that, because of the larger quantities needed, cost is relatively more important. The antennas for the Owens Valley Millimeter Interferometer [18] and the feeds for the Molonglo Observatory Synthesis Telescope [57] are examples of innovative designs driven by the need for low construction cost. In the area of low-noise receivers for centimeter-wavelength arrays, the development of cryogenically cooled GaAs FET amplifiers [58]–[60] is a significant recent development because of their relatively low cost.

An important property of a synthesis array is the difference between its sensitivity to point sources and to extended sources. The image of a point source of flux density S has an apparent brightness of $S(\text{W} \cdot \text{m}^{-2} \cdot \text{Hz}^{-1})$ per synthesized beam area, independent of the size of the synthesized beam. Therefore, the signal-to-noise ratio for a point source, $S/\Delta B$, where ΔB is given by (7), is independent of the synthesized beamwidth. For an extended source that is larger than the synthesized beam, having constant brightness $B(\text{W} \cdot \text{m}^{-2} \cdot \text{Hz}^{-1} \cdot \text{sr}^{-1})$ the flux density per synthesized beam is $B\Omega_s$, where Ω_s is the area of the synthesized beam in steradians. The signal-to-noise ratio for this extended source is $B\Omega_s/\Delta B$. If Ω_s is increased by reducing the extent of the array, and the number of antennas is not reduced so that ΔB is not increased, the signal-to-noise ratio will increase so long as the beam is smaller than the smallest detectable source structure [4]. For this reason, it is desirable that the size of a synthesis array be made variable to optimize the brightness sensitivity to extended objects. For the VLA, the scale factor between adjacent configurations is 3.3. Since Ω_s varies approximately as the inverse square of the array extent, each VLA configuration is an order of magnitude more sensitive to extended emission than the next large one. The four configurations thus provide a range of 10^3 in brightness sensitivity.

Fig. 5. One of the VLA antennas on the south–east arm of the array tipped to a low elevation angle. The cluster of Cassegrain feeds in the center of the reflector, and the three concrete support pillars of the antenna foundation can be seen.

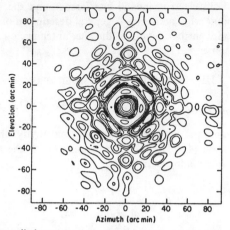

Fig. 7. The radiation pattern of a VLA antenna measured in right circular polarization at 4.86 GHz. This plot was obtained by interpolating between data points measured on a grid having 8.0' spacing. Only the amplitude is shown here. The contours are −3, −6, −10, −15, −20, −25, −30, −35, and −40 dB.

Fig. 6. A view of the four Cassegrain feeds located in the center of the antenna reflector. The annulus of circular waveguides around the groove-matched circular dielectric lens comprises the hybrid lens [61] of the 1.35–1.73-GHz feed. This lens, which is approximately 1.6 m in diameter, is illuminated by a small corrugated horn whose circular aperture can be seen underneath. Above the lens are the two square and one circular apertures of the 15-, 23-, and 5-GHz feeds. Surrounding the higher frequency feeds are heat lamps used to remove precipitation from the feed apertures.

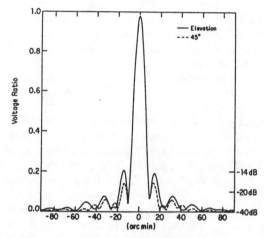

Fig. 8. Profiles in the elevation and 45° directions of the radiation pattern shown in Fig. 7.

B. The VLA Antennas and Low-Noise Receivers

To satisfy the requirements for the sensitivity and field of view in Table II, the antennas chosen for the VLA are shaped parabolic reflectors of 25-m diameter. The half-power beamwidth in minutes of arc is closely equal to 1.4 times the wavelength in centimeters and is circular in cross section. The antennas are instrumented for reception in the wavelength bands 17–22, 6, 2, and 1.3 cm. Thus the size of the antennas is about as large as could be allowed to obtain a satisfactory field of view. When matched with low-noise receivers the antennas provide the required sensitivity. Each antenna is supported by an altazimuth mount that provides complete sky coverage down to 9° elevation angle, as shown in Fig. 5. The reflectors are designed to have an rms surface accuracy of one-sixteenth of a wavelength at a frequency of 22 GHz. The feed system has a Cassegrain geometry with off-axis secondary focal point. The subreflector is asymmetrical so that the off-axis focal point does not result in any geometrical phase error or loss of performance. To satisfy the frequency diversity requirement of Table II, the feeds for all four wavelength bands are permanently mounted on a circle around the main reflector axis, as shown in the photograph in Fig. 6. Frequency changes are made by simply rotating the subreflector, under computer control, about the main-reflector axis to focus the radiation on the appropriate feed. The feeds provide both left

and right circular polarization. The feed for the 18–21-cm wavelength range is a small corrugated horn illuminating a large lens composed of dielectric and waveguide elements [61]. The feed for the 6-cm range is a lens-corrected corrugated horn and for the 2- and 1.3-cm ranges multiflare, multimode horns are used. Table III shows the antenna aperture efficiency achieved at each band, and estimates of the various contributing efficiency losses. The relatively low efficiency of the 18–21-cm feed results from the difficulty of using a Cassegrain geometry at this long wavelength, and was accepted to allow a significant saving in construction cost to be made [61]. The asymmetric geometry results in a pointing difference of 0.06 beamwidths between right and left circularly polarized beams [62] due to cross-polarization effects [63]. More details on the VLA antennas and feed system are available in [7], [61], and [64].

A synthesis array is ideally suited for measuring the surface accuracy of its antennas by using what has become known as the "holographic" technique [65]. In this method, the antenna radiation pattern is measured in amplitude and phase and the complex aperture distribution is then obtained as the Fourier transform of the complex radiation pattern. The profile of the reflector can then be inferred from the phase of the aperture distribution. In the application of this method to a synthesis array [66], one reference antenna tracks a strong astronomical point source while the rest of the antennas in the array scan backwards and forwards

TABLE III
APERTURE EFFICIENCY CONTRIBUTIONS FOR THE VLA ANTENNAS

(The spillover and illumination efficiencies were estimated by integrating measured feed radiation patterns. Blockage includes quadrupod, subreflector, and feed blockage. Diffraction loss was predicted using a physical-optics scattering analysis of the subreflector. Surface efficiency indicates loss due to reflector surface irregularities under conditions of minimum gravity, wind, and thermally induced deformations. Miscellaneous losses include mismatch and resistive losses.)

f_{GHz}	Spillover	Illumination	Blockage	Diffraction	Surface	Misc.	Total Predicted	Measured
1.35-1.73	.82	.98	.85	.86	1.00	.94	.55	.52
4.5-5.0	.92	.98	.85	.96	.97	.94	.67	.65
14.4-15.4	.90	.95	.85	.98	.85	.94	.56	.53
22-24	.99	.95	.85	.99	.68	.94	.46	.46

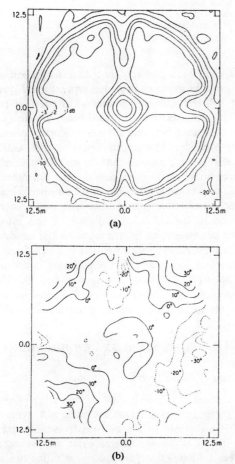

Fig. 9. The complex aperture distribution of a VLA antenna obtained at 4.86 GHz using the holographic measurement method. The resolution of the measurement across the aperture is approximately 1.4 m. (a) Aperture amplitude distribution showing the expected features of uniform illumination produced by the shaped reflector system and subreflector and feed-leg blockage. (b) Aperture phase distribution. This phase distribution is dominated by linear and quadratic phase errors caused by small pointing and subreflector defocussing errors.

across the source. The complex cross correlation between the reference antenna and each scanning antenna then gives directly the complex voltage radiation pattern of the scanning antenna. The complex voltage aperture distributions can be obtained using exactly the same Fourier transform computer programs that are used for making astronomical synthesis images. The VLA antennas have been measured at 6-cm wavelength using this technique and some typical results are shown in Figs. 7-9. These were obtained by sampling the radiation pattern on a 23 by 23 grid

with a spacing between points of 8.0′ which corresponds to 0.94 of the Nyquist sampling interval for a 25-m aperture at 4.86 GHz. The radio source 3C84 was used for the measurements.

The original VLA low-noise receiver front-ends consisted of cryogenically cooled parametric upconverters for the 17–22-cm band, two cooled parametric amplifiers in cascade for the 6-cm band, and cooled Schottky diode mixers for the 2- and 1-3-cm bands. This receiver system has been described previously [64] but is currently undergoing extensive improvements. A block diagram of the most recent receivers is shown in Fig. 10. The 17–22-cm parametric upconverter is being replaced with a cooled GaAs FET amplifier [60] principally to avoid a spurious (second harmonic of the pump) response present in the upconverter and also to obtain better reliability and slightly improved sensitivity. The second-stage 6-cm parametric amplifier is being replaced with a cooled GaAs FET amplifier to improve reliability. A cooled 2-cm GaAs FET amplifier is being installed as a preamplifier ahead of the Schottky-diode mixer to improve sensitivity and a similar upgrade for the 1.3-cm receiver is planned for the future. Table IV shows typical values for the VLA system temperatures and estimates of the contributing noise sources. The values are appropriate to the block diagram shown in Fig. 10. A T_{rx} of 150 K is expected when the cooled 1.3-cm GaAs FET amplifier is installed. The T_{spill}, T_{loss}, and T_{rx} contributions are somewhat higher than is usual for low-noise receivers due to the compromises that were made to have all receivers and feeds available simultaneously. Optimized systems can obtain better performance [67], [68]. T_{spill} includes the computed spillover to the ground due to subreflector diffraction, and approximately 5 K of scattering off the feed legs which was inferred from measurements of the subreflector noise temperature and is consistent with feed leg scattering on other antennas [69]. Using the values of η_a and T_{sys} given in Tables III and IV, Table V shows the sensitivity of the 27-antenna array for an 8-h observation. ΔB is calculated using (7) with $\eta_c = 0.81$ and Δf equal to the widest VLA bandwidth of 46 MHz.

C. Radio Frequency Interference

Modern high-speed digital correlators can provide observing bandwidths of 50 MHz or more. Since such bandwidths are much wider than the frequency bands allocated to radio astronomy [70], man-made radio frequency interference may limit the sensitivity of synthesis arrays especially at frequencies of 1.7 GHz and lower. Synthesis arrays are less sensitive to radio interference than single-antenna radio telescopes because of two interference-reducing effects [71]. The first effect results from the fact that a source of man-made interference does not move in the same way

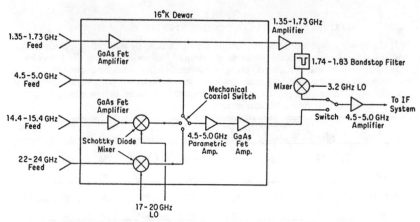

Fig. 10. A schematic diagram of the most recent VLA low-noise receivers. Two systems of the type shown, combined in the same dewar, are used at each antenna to provide for two oppositely polarized outputs from each feed. In the future, a cooled GaAs FET amplifier will be installed in front of the 22–24-GHz cooled mixer.

TABLE IV
NOISE CONTRIBUTIONS IN THE VLA RECEIVERS

(The contributions are: T_{bg}—cosmic microwave background; T_{sky}—atmospheric emission; T_{spill}—spillover to the ground due to subreflector diffraction and feed leg scattering; T_{loss}—resistive attenuation in feeds and input waveguides; T_{cal}—injected calibration signal; T_{rx}—receiver temperature at input to cryogenic dewar; T_{sys}—total system temperature.)

Wavelength Band	T_{bg}	T_{sky}	T_{spill}	T_{loss}	T_{cal}	T_{rx}	T_{sys}
20 cm	3	3	14	8	2	30	60
6 cm	3	3	7	5	2	30	50
2 cm	3	8	6	13	7	75	112
1.3 cm	3	17	6	21	7	296	350

TABLE V
SENSITIVITY OF THE VLA AFTER AN 8-h INTEGRATION USING ONE IF WITH BANDWIDTH 50 MHz ON ALL 27 ANTENNAS

f_{GHz}	ΔB mJy per synthesized beam*
1.3 - 1.7	0.03
4.5 - 5.0	0.02
14.4 - 15.4	0.05
22 - 24	0.08

* 1 mJy = 10^{-29} w/m²/Hz

with respect to the array as does the observed astronomical source. The phase of the astronomical signal arriving at the two elements of an interferometer varies in a characteristic way due to the motion of the astronomical source in the sky. The phase of one of the elements is continuously varied to correct for this phase change. The signal from a source of interference that does not have the same motion will, therefore, have a continuous change of phase superimposed on it. When the multiplier output is averaged in the correlation integrator and subsequent data processing the continuous phase change will cause a significant reduction in the interference amplitude. Since the rate of change of phase due to source motion goes to zero for $u = 0$, interference will tend to be largest along the v axis in the (u, v) plane, resulting in generally east–west oriented structure in the image. Estimates of the amount of interference suppression provided by this averaging effect for the VLA [71] are 22 and 28 dB at 21- and

1.3-cm wavelength, respectively for a 12-h integration in the most compact array. In the most extended array, these figures become 30 and 36 dB, respectively. The second interference-suppressing effect is the decorrelation of broad-band interference due to relative time delays between antennas. Since the astronomical signals are broad-band noise, the delays in each antenna in an interferometer are adjusted so that the signals received from the radio source arrive at the correlator in time synchronism. These delays will, in general, not be correct for an interfering source so that, if the interference is broad-band, it will be decorrelated. Computation of the reduction provided by this effect for the VLA in the presence of an interfering source of bandwidth 25 to 50 MHz in geostationary orbit predicts between 5 and 20 dB in the most compact array and between 15 and 35 dB in the most extended array [71]. In addition to these effects, the VLA makes use of switchable filters at the output of the low-noise receivers, and at the inputs to the correlator, to reduce interfering signals.

IV. PHASE STABILITY CONSIDERATIONS

A. The Effect of Phase Errors

Several authors have investigated the degrading effects of phase errors on the quality of observations made with synthesis arrays [72]–[76]. Phase errors in the measured visibility data cause a reduction in gain of the synthesized beam and an increase in its sidelobe levels. These effects are analogous to the beam degradation resulting from the phase errors in the aperture of reflector antennas, where the phase errors are caused by inaccuracies in the reflector surface. If the phase errors in the (u, v) plane of a synthesis array are random with Gaussian statistics we can use this analogy to show [77] that if the phase errors have rms value $\Delta\phi$ (radians), then the loss of gain of the synthesized beam is given by

$$\frac{\text{gain with phase errors}}{\text{gain without phase errors}} = e^{-(\Delta\phi)^2/2}. \qquad (8)$$

The factor of two in the exponent of (8) results from the fact that, for a single antenna the power beam is given by the square of the Fourier transform of the phase errors, while for a synthesis array, the power beam is given by the Fourier transform of the phase errors directly. Equation (8) is useful for phase errors up to approximately 1 rad. Note that it is in good agreement with a directly measured curve of gain loss versus rms phase error [73]. The assumption that the distribution of phase errors in the (u, v) plane is random is, in general, not accurate for a linear array [78] or for an array with a small number of antennas. However, for two-dimensional arrays with a large number of antennas, such as the Culgoora array [19] or the VLA, it is a reasonable assumption except for a systematic increase in atmospheric phase errors in the longer baselines [75]. The theory of errors in reflectors [77] can also be used to provide an estimate of the sidelobes imposed on the synthesized beam by phase errors in the (u, v) plane, or alternatively, these effects can be determined empirically [73].

Phase errors result primarily from variations in the length of the signal path through the earth's atmosphere and from phase variations in the instrumentation. These effects are discussed in more detail below. Data processing techniques to reduce phase errors are discussed in Section VIII-B.

B. Atmospheric Phase Errors

Atmospheric phase errors in synthesis arrays operating at frequencies of approximately 1.7 GHz and lower are dominated by differential delays introduced by propagation through the charged particles of the ionosphere [72], [73], [79]. The time delays introduced by the ionosphere vary inversely as the square of the frequency. Irregularities in the ionosphere have scale sizes in the range 20–360 km and drift at velocities of approximately 300 km/h [73]. A typical value for the rms differential phase error introduced by the ionosphere on a baseline of 24 km at a frequency of 1400 MHz is 0.16 rad [73], although this can vary significantly as a function of solar activity and time of day. Over the range of baselines 1 to 100 km, phase errors due to the ionosphere increase approximately linearly with baseline length [72].

Interferometer phase errors caused by propagation through the earth's troposphere have been measured at several observatories including Cambridge [80] with baselines up to 5 km, Green Bank [81] with a baseline of 11 km, Westerbork [78] with baselines up to 1.3 km, in the USSR [82] with baselines up to 40 km, and at the VLA [75] with baselines up to 30 km. The troposphere causes differential time delays that are independent of frequency and are due predominantly to the water vapor and, to a lesser extent, to the other components of the atmosphere. The VLA measurements [75] show that typically the rms value of the phase errors caused by propagation through the troposphere are proportional to the 0.7th power of the baseline for baselines up to 30 km. In the series of 23 test observations, reported in [75], carried out over a variety of times of day and seasons of the year, 35 percent of the observations did not show atmospheric phase errors above the level of phase errors expected due to position errors of the calibrator sources observed. This implies atmospheric phase errors in these cases of less than 11° rms at 4.9 GHz on a baseline of 10 km. For the 65 percent of test observations where the atmosphere was clearly seen, the rms phase error at 4.9 GHz on a baseline of 10 km averaged 31° for sources observed at many different places in the sky. Phase fluctuations are less when a single source is tracked continuously.

Although the "self-calibration" techniques described in Section VIII-B can significantly reduce the effect of atmospheric phase errors when the observed radio source is sufficiently strong, for weak sources or for astrometric observations it is desirable to have other ways of correcting for the atmosphere. Several investigators [83]–[85] have examined the possibility of using dual-channel microwave water-vapor radiometers to provide a correction for tropospheric phase errors. Recent tests of this concept at

the VLA [86] using microwave water-vapor radiometers developed at JPL [85] appear encouraging. During periods when the corrections are particularly effective it has been found possible to reduce phase errors of 80° rms at 4.89 GHz down to 20° rms. The fact that the radiometers provide less effective phase corrections at other times is probably due, in part, to inadequate radiometer stability and tropospheric events involving excessive amounts of liquid water.

C. Electronics System Design for Good Phase Stability

Clearly, the goal of the electronics system designer should be to make phase errors that result from the equipment small compared with the phase errors caused by the ionosphere or troposphere. Instrumental phase errors need to be controlled only for time scales shorter than the time intervals between phase calibration observations of point sources of known position. For the VLA, an overall phase stability specification for the electronics of 1° of rms phase error per gigahertz of observing frequency was chosen. This specification would not be good enough for the new millimeter interferometers currently under development [23], [32] and would be unnecessarily tight at longer wavelengths such as 327 or 408 MHz where the ionosphere causes phase errors of more than 10° rms. At the VLA, the time period between calibration observations is typically 10 to 30 min. On much shorter time scales, say 30 s to 1 min, a tighter phase stability specification is required if instrumental phase and amplitude "closure errors" [53] are to be kept sufficiently small. This is necessary if the amplitude and phase error removal procedures described in Section VIII-B are used. A phase stability specification of a few degrees rms, independent of observing frequency, is appropriate for very short time periods. The key to obtaining good instrumental phase stability is to provide a phase-stable local oscillator (LO) distribution system to the antennas and a stable IF transmission path from the antennas back to the central location where the cross correlations are performed. Radio astronomers have settled on three principal techniques for obtaining this phase stability with antennas located at the ends of long transmission systems. These techniques can be used to distribute the LO in a phase-stable manner to the elements of an array. They can also be used to provide a phase-stable calibration signal which is returned along the same transmission path as the received astronomical signals to allow corrections for changes in the length of the path to be made. In the first technique, originally suggested by Swarup and Yang [87], a small amount of the LO is reflected back along the transmission path over which it traveled to the remote antenna. At the location of the master oscillator, the phases of the reflected and master signals are compared to give a measurement of phase shift introduced in twice traversing the long transmission path. The key to the technique is to be able to recognize the reflection from the end of the transmission line in the presence of many other unavoidable reflections. This is achieved by switching the reflector at the end of the line on and off at a low frequency (typically a few kilohertz) so that its effect can be synchronously detected back at the master location. This simple concept has been extended and applied to many instruments [19], [30], [34], [88]–[91].

The second commonly used technique, which we will call the "round-trip-phase-measurement" technique, is similar to the method of Swarup and Yang described above in that a sample of the LO from a distant antenna is returned along the same or a parallel transmission path as the outgoing signal. A direct measurement of twice the phase shift due to the transmission path can then be made. Instead of being generated by a reflection,

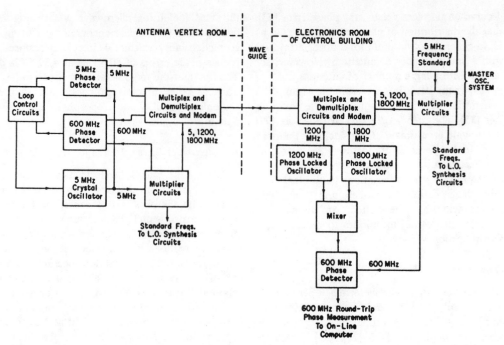

Fig. 11. Simplified block diagram of the oscillator system. Except for the two Master Oscillator blocks, each antenna requires a separate system of the type shown.

however, the returned LO, usually with a small frequency offset added, is generated using phase-locked loops at the antenna and is injected back into the transmission path using a directional coupler. An advantage of this method over the simple reflection technique is that more of the circuitry at the remote antenna can be included in the round-trip-phase-measurement loop. Several successful applications of this technique have been described [29], [32], [35], [73], [92].

The third common technique for distributing phase-stable signals over long distances was considered by J. Granlund as a possible LO distribution system during early VLA design studies [93]. In this method, signals at approximately half the LO frequency travel in opposite directions along a transmission line which extends from one end of the array to the other. If these two signals are mixed at each antenna, it can be shown that the resulting sum signal has the same phase at each antenna. This method has the advantage of simplicity and lower line attenuations because signals at only one half of the LO frequency are transmitted. Applications of the method are described in [31] and [94].

Finally, for completeness, we reference two more proposed techniques [95], [96] which resemble one or more of the principal methods described above, and note that fiber-optics communication systems show promise of high phase stability [97].

D. VLA Phase Stabilized LO System

The VLA uses the "round-trip-phase-measurement" technique mentioned in the previous subsection to meet its phase stability specifications of 1° rms phase error per gigahertz between corresponding signals on any two antennas in the array. At the VLA use is made of a TE_{01}-mode circular waveguide system described in Section V to transmit a reference signal out to each antenna from the master oscillator and a sample of the antenna LO back to the central electronics building. The round-trip-phase-measurement method will provide good phase stability only if the outgoing and return transmission paths introduce identical phase shifts onto the LO signal. Reflections in the long low-loss waveguide

runs could give rise to phase responses which vary rapidly as a function of frequency so it is essential that the outgoing and returned LO signals are very close to the same frequency. For propagation in the TE_{01} waveguide, LO and IF signals are single-sideband modulated onto carrier frequencies in the range 26 to 50 GHz. The difficulty and expense of providing electronic components for these high frequencies that would allow simultaneous propagation in both directions with good isolation between outgoing and returned signals led to the choice of a time-shared communications system. The master oscillator signal is transmitted to each antenna for 1 ms out of every 52 ms. For the remaining 51 ms, the LO and IF data are returned to the central location.

A simplified block diagram of the VLA oscillator system is shown in Fig. 11. The basic LO at each antenna is a high-stability 5-MHz crystal oscillator. During the 1 ms out of every 52 ms that the master oscillator is available at the antenna, the LO is phase-locked to it. The 5-MHz crystal oscillator has a fractional frequency stability of better than 10^{-11} on time scales of 100 ms, so for the 51 ms out of every 52 ms that it must run free, adequate phase stability is maintained. The accuracy with which the phase at the antenna can be controlled is determined, among other things, by the signal-to-noise ratio on the phase reference signal arriving at the antenna. For a given signal-to-noise ratio, the higher the frequency of the reference signal, the better. The primary reference signal sent to each antenna on the waveguide system is 600 MHz. The phase of the 5-MHz crystal oscillator is controlled so that the 120th harmonic of the oscillator is phase-locked to the 600-MHz reference signal. Although this technique of phase referencing to a high-order harmonic results in more accurate phase control than phase locking directly to a 5-MHz reference, it results in phase ambiguities at intervals of 3° in the 5-MHz phase. To resolve this ambiguity, a 5-MHz reference signal is also sent to the antenna. Phase lock is initially acquired using the 5-MHz phase detector. The time constant on the phase-lock loop is 1 s and another locked crystal oscillator is used at 50 MHz to help reduce noise and unwanted sidebands.

Fig. 12 shows the spectrum of the signals that are sent from an

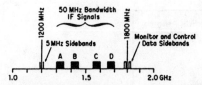

Fig. 12. The spectrum of the modulation in the waveguide transmission system. The 50-MHz bandwidth IF signals A, B, C, D are present in transmission from the antennas only, but the other signals travel in both directions. IF's A and B are taken from the right circularly polarized feed output and C and D from the left circularly polarized output.

antenna, via the waveguide system, to the central electronics building. The spectrum for the signal path from the central building to the antenna is the same except that the four 50-MHz-wide IF bands are absent. The spectrum shown in Fig. 12 is modulated onto a high-frequency carrier signal whose frequency f_c is given approximately by the expression

$$f_c = 24.0 + 2.4m \text{ GHz}, \qquad m = 1 \text{ to } 9. \qquad (9)$$

For any arm of the Y, the carrier frequency allocated to each of the nine antennas on the arm is determined by a different value of m. The 600-MHz reference signal is obtained as the difference between the 1200- and 1800-MHz signals and the 5-MHz reference is amplitude modulated onto the 1200-MHz signal. Also, the 1200- and 1800-MHz signals are used in downconversion of the four IF bands to baseband. With this scheme, the phases of the 600- and 5-MHz signals and of the four IF signals at baseband are all unaffected by variations in the phase of the waveguide carrier signal, although they are affected by changes in waveguide length.

The round-trip-phase measurement for each antenna is performed in the central electronics building by comparing the phases of the master 600-MHz reference and the 600-MHz signal returned from the antenna. This measurement is used to correct for the effect of variations of waveguide length on the phase of the LO at the antenna and on the phase of the IF bands returned in the waveguide. Typically, the 600-MHz round-trip phase for any antenna varies quasi-sinusoidally with a 24-h period and peak-to-peak variation of 10° to 20°. This implies peak-to-peak changes of 0.7 to 1.4 cm in the waveguide path length. The round-trip-phase variations are strongly correlated with ambient air temperature and are due mainly to changes in the length of the run of waveguide up the antenna and to the temperature sensitivity of the regulators that keep the dry-nitrogen pressure constant in the waveguide. Astronomical tests show that the effects of these waveguide path-length changes are removed by the phase correction system with an accuracy of better than 0.3 mm rms.

At the antenna, LO signals are required in the frequency range 2.0 to 4.0 GHz for the 20- and 6-cm wavelength observing bands and in the range 17.0 to 20.0 GHz for the 2- and 1.3-cm wavelength bands. The electronic components that synthesize these frequencies from the primary 5 and 600 MHz are outside the 600-MHz phase correction loop so particular care must be taken to make these components phase-stable. Phase variations in them are due primarily to variations in their operating temperature. To minimize temperature effects, filters and amplifiers with narrow fractional bandwidths, which have high temperature coefficients of phase, are avoided. In several areas, phase-locked loops are used instead of narrow-band filters to improve the purity of LO signals. The temperature of the electronics room on the antenna containing the LO and low-noise receiver equipment is controlled to within ±1°C. Astronomical phase stability tests show that, after the 600-MHz phase correction has been applied,

a residual phase error of 2.8° of phase at an observing frequency of 4.9 GHz results from a 1°C change in antenna equipment-room temperature.

V. SIGNAL TRANSMISSION SYSTEM

The signal transmission system for a synthesis array provides a path to send phase reference signals out to the antenna and to return LO and IF data signals from the antenna. The primary requirements for signal transmission are the same as those of conventional communication systems; sufficient bandwidth for the data rate and sufficiently low attenuation to communicate over the required distance at the desired frequency, preferably without the need for repeaters. The added requirement of high phase stability significantly increases the design problem. For synthesis instruments designed to perform spectral line observations, the transmission system must also be designed to have an amplitude and phase response which is flat to better than, say, 0.1 dB and 0.5° phase on frequency scales of a few tens of kilohertz [53]. The phase stability and spectral flatness requirements can be avoided if the IF data are digitized at the antenna. In the past, transmission systems for arrays have utilized open-wire transmission line [19], rectangular waveguide [87], circular waveguide [91], radio links [31], [35], and most commonly coaxial cables [26], [30]. The Australia Telescope which is currently being designed [20] will probably use a fiber-optic transmission line to link its antennas together.

The task facing the transmission-system designer for the VLA was to provide a path of up to 22-km length from each antenna to the Control Building for IF data having a total bandwidth of 200 MHz, and a path in both directions for several narrow-band, phase-reference signals at frequencies above 1 GHz. Use of a conventional coaxial cable system would have necessitated the use of many repeater amplifiers because of cable attenuation. This would have significantly increased the complexity of the system and made the phase stability specification more difficult to meet. Similarly, conventional waveguide systems would have had too much loss over the 22-km distance. Radio links were rejected because of the very large total bandwidth needed for 27 antennas and because of possible interference and blockage problems. At the time of the design in 1972, optical communication links were insufficiently developed to be considered. The transmission medium eventually chosen was the TE_{01}- mode, large-diameter, circular waveguide which was being developed by U.S. and Japanese telephone companies for high-bandwidth, long-distance communication links [98]. The telephone companies have since abandoned this type of waveguide in favor of fiber-optics systems but its use at the VLA has been very successful. Since details of the waveguide system have been published elsewhere [99]–[102] only an overview will be provided here.

A 60-mm-diameter circular waveguide runs the length of each of the three arms of the array. The inside surface of the waveguide is helically wound insulated copper wire which allows only the TE_{0N} modes to propagate. At all except the last of the 24 antenna stations on each arm, a directional coupler is inserted in the main waveguide run to allow an antenna to communicate with the Control Building. The first 22 couplers are sector couplers [99] and the twenty-third station uses a beam-splitter coupler to obtain increased coupling. A length of 20-mm-diameter helix-lined waveguide connects the directional coupler to the modem located in the vertex room of the antenna. Waveguide rotary joints are located on the azimuth and elevation axis of the antenna. In the Control Building a signal distributor, comprising eleven-channel dropping filters, on each waveguide run, provides

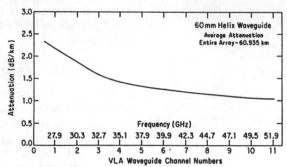

Fig. 13. The measured attenuation in the VLA TE_{01}-mode 60-mm helix waveguide.

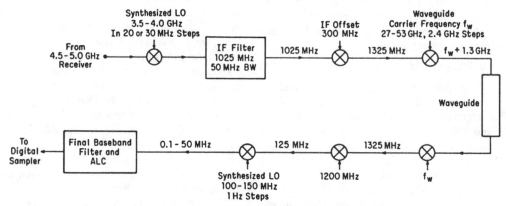

Fig. 14. The frequency conversion scheme for the A IF signal shown in Fig. 12. IF's B, C, and D undergo similar conversions but have different IF offset and final LO frequencies. In addition, IF's C and D use 1800 MHz instead of 1200 MHz for the second frequency conversion in the Control Building.

output ports for each of the eleven channels on each arm. Nine of the channels are used to communicate with the antennas on the arm and two are spare. At the signal distributor output and at the end of the 20-mm-diameter waveguide in the antenna vertex room, TE_{01} circular to TE_{10} rectangular mode converters provide convenient ports to the waveguide system.

The mainline 60-mm-diameter waveguide is directly buried in the ground after being covered with a waterproof coating. The direct burial technique, which was used instead of installation in a conduit for reasons of economy, required significant development before a burial procedure was perfected that ensured that the waveguide straightness did not degrade with time [100]. The 60-mm waveguide loss is a combination of resistive loss and mode-conversion loss due to lack of straightness in the installed waveguide. At lower frequencies, the total loss decreases with increasing frequency as the waveguide becomes more overmoded. At high frequencies, the loss increases with frequency as the mode-conversion loss becomes dominant. The waveguide is pressurized with dry nitrogen to prevent loss due to oxygen absorption at frequencies above 50 GHz and to prevent ingress of moisture. A measured attenuation curve for the complete run of 60-mm waveguide in the array is shown in Fig. 13. The total waveguide system loss budget, which is discussed in [101], requires that the waveguide attenuation be below 1.4 dB/km for communication to the most remoted stations. Clearly, the attenuation specification was successfully met at frequencies above 40 GHz.

The specification on the flatness of the TE_{01}-mode transfer function, between any antenna and the Control Building, is 1-percent rms variation in power gain and 0.3° rms departure from phase linearity over bandwidths of 10 MHz. These specifi-

cations were determined primarily to enable measurements to be made of weak narrow-band spectral features in the presence of strong continuum (broad-band) radiation. Fine-scale ripples in the waveguide frequency response can be caused by mode conversion between the desired TE_{01} mode and higher order TE_{0N} modes which propagate at different velocities, or by reflections from discontinuities in the waveguide and couplers. A theoretical and experimental study of these problems [100] shows that the dominant source of the ripples is TE_{01}-to-TE_{02}-mode conversion in the sector couplers and in the signal distributor. The ripples resulting from these effects were reduced to acceptable levels by the installation of a TE_{02}-mode filter between the mainline waveguide and the signal distributor [102].

The need for versatility in the choice of observing frequency, the rejection of spurious responses, and the use of high carrier frequencies in the TE_{01}-mode waveguide result in a transmission system with numerous frequency conversions. As an example, Fig. 14 shows the frequency conversions in the transmission path of one of the four 50-MHz bandwidth IF signals returned from the antenna during an observation in the 4.5–5.0-GHz band. The first LO, tunable in 20- or 30-MHz steps, converts the received signal down to a first IF centered at 1025 MHz. The signals transmitted along the waveguide lie in the 27–53-GHz range. LO and IF signals are single-sideband amplitude modulated onto millimeter-wave carriers in modem units which incorporate Gunn-diode oscillators and diode mixers working as modulators. The same modems are used in both transmit and receive modes, and in the latter case the Gunn diode acts as an LO. Diode switches and a switched circulator are used to control the transmit and receive functions. More information about the millimeter frequency modems is available in [103].

In the central Control Building the broad-band IF signals are mixed down to baseband in the range 0.1–50 MHz. Computer selectable filters having bandwidths of $50/2^M$ MHz, $M = 0$ to 8, are used to select the part of the IF signal to be passed on to the digital delay and correlator system. The baseband signals are passed through an automatic level control circuit before digitization.

VI. DELAY AND CORRELATOR SYSTEM

A. Analog and Digital Techniques

The signals from the baseband IF amplifiers of the VLA are sampled and digitized, and the introduction of compensating time delays and the computation of cross correlation are performed digitally. The compensating delays are adjusted under computer control, and for a given direction of observation the last antenna that an approaching wavefront encounters is assigned zero delay. The maximum compensating delays are required when the source under observation is close to the horizon, the largest one being approximately equal to the light travel time over the longest antenna baseline. Compensation for the different travel times in the transmission lines from the antennas must also be included, and for the VLA the total range of delay adjustment is 164 μs. To implement such a delay in cable would require large numbers of amplifiers and frequency-compensating networks to overcome the attenuation, and would hardly be practicable. Acoustic delay lines have been used for the longer elements in some arrays, but with any analog system two problems arise when the delays are large. First, maintaining precise calibration requires either temperature control or incorporation of an automatic delay-measuring system. Second, it is practically impossible to avoid variations in the frequency response when large analog units are switched into and out of a signal path. Such variations introduce calibration errors, and for the VLA the tolerable variations in the frequency responses are no more than ± 1 dB [53].

Spectral line correlator systems, in which the correlation is measured at numerous frequencies across the IF band, can be implemented in analog circuitry by using filter banks with separate sets of correlators for each filter channel. Again there are some disadvantages. First, care must be taken that the filters do not introduce temperature-related phase variations. Second, the filter banks are expensive, especially if several sets are required to allow choice of the number and width of the channels to suit different spectral lines. With a digital system, correlation is measured as a function of time offset and Fourier transformation provides the spectral data. Variation of the frequency resolution is largely a matter of changing clock frequencies.

The advantages of digital techniques are less pronounced for millimeter-wavelength arrays. With current techniques signal bandwidths up to 50 MHz can easily be sampled digitally. However, at millimeter wavelengths, bandwidths of several hundred megahertz are often used to obtain sufficient sensitivity, since both the sizes of the antennas and the flux densities of the sources are less than at centimeter wavelengths. Also at millimeter wavelengths the antenna spacings are proportionally smaller and the required delays are less.

In a simple interferometer or array in which the LO signals at each antenna are approximately maintained in constant relative phase, the signals from a radio source produce quasi-sinusoidal oscillations at the correlator output as the source moves through the fringe pattern. The variable delay that compensates for the changing relative path lengths to the antennas is usually inserted

at an intermediate frequency that is small compared with the observing frequency, so the phase changes that result from the adjustment of the delays have only a small effect upon the frequency of these oscillations. In the VLA such fringe-frequency oscillations can exceed 200 Hz for the longest antenna spacing and the highest observing frequency. To measure the amplitude and phase of such oscillations would require sampling the waveform at twice the fringe frequency and fitting a sine wave by computer. The computing task can be greatly lessened by introducing a phase change, usually through an LO, to reduce the fringe frequency oscillations at the correlator output to zero. This requires a continuous phase shift with a rate and initial value that can be set individually for each antenna under computer control. In the VLA this is accomplished by digitally synthesizing at each antenna a frequency of 100 kHz with the required phase changes, and then adding this frequency to the 3.5–4.0-GHz LO shown in Fig. 14. To measure the output amplitude and phase in the continuum mode of observation, two correlator circuits are used with each antenna pair. One of these measures the correlation with a quadrature phase shift applied to one signal. The outputs represent the real and imaginary parts of (1), and reducing the fringe frequency to zero removes the factor $\exp(j2\pi \vec{b} \cdot \vec{s}_0)$, so the complex visibility is measured directly. The combination of two correlators is often referred to as a complex correlator. In the spectral-line observing mode it can easily be shown that the quadrature correlator is unnecessary, so long as the correlation is measured with both positive and negative time offsets of one antenna with respect to the other in each pair. Note that the phase changes required to slow the fringe oscillations to zero, plus the phase changes resulting from adjustment of the compensating delays, are exactly those required by the factor $\exp(-j2\pi w)$ in (4) to take out the effect of the changing path length represented by w.

In any correlating radio interferometer there are likely to be low-level, slowly varying voltages at the correlator outputs that arise by unwanted mechanisms. Such mechanisms include inexact adjustment of analog correlator circuits, errors in threshold settings of digital samplers, crosstalk between IF channels, and low-level signals from the LO system that infiltrate the IF amplifiers. These signals would degrade sensitivity if not removed. An effective way of suppressing them in a two-element interferometer is to reverse periodically the phase of the signal at one antenna, in the early stages of the receiving system. The wanted component of the correlator output then reverses in sign and can be synchronously demodulated. This phase-switching technique is highly effective, and was originally devised to obtain the response of a correlator from a power-law detector [104]. To apply the technique to a multielement array it is necessary to drive the phase switches at the antennas with different two-state waveforms that are mutually orthogonal, i.e., the expectation of the product of any pair of such waveforms is zero. In the VLA this is achieved by using Walsh functions that are generated by the control computer [105]. Each phase reversal takes place during one of the blanking intervals of the received signal described in Section IV-D, so transient effects of the switching are eliminated. Since the digital correlators do not introduce offset errors, the second reversal is performed on the digital data at the sampler outputs rather than on the larger number of correlator outputs.

B. Quantization and Sampling Schemes

Digital processing in radio astronomy has been practiced since the late 1950's [106], [107]. Signals are usually coarsely quantized, and each sample is represented by 1 or 2 bits only. Since the IF

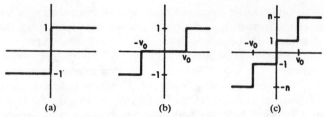

Fig. 15. Characteristic curves for signal quantization, (a) two-level, (b) three-level, and (c) four-level. In each case, the abscissa is the input voltage and ordinate is the quantized output. Values for the threshold level v_0 and the weighting factor n are derived in [110]–[112] for optimum sensitivity of the correlator output.

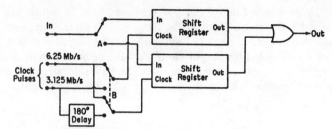

Fig. 16. The variable delay unit. The bit frequency of the input and output data streams is 6.25 Mbits/s. When the shift registers are clocked at 6.25 Mbits/s, switch A is in the position shown, and all input bits go to the top register. When the clock frequency of 3.125 Mbits/s is selected by switch B, switch A is driven at 3.125 MHz so that the input bits are divided alternately between the two registers.

TABLE VI
DEGRADATION IN SENSITIVITY FOR VARIOUS QUANTIZATION SCHEMES

No. of Quantization Levels	Sensitivity Relative to Analog Correlator	
	Sampling rate = Nyquist rate[1]	Sampling rate = 2xNyquist rate[1]
2	0.64	0.74
3	0.81	0.89
4	0.88	0.94

[1]Nyquist rate is that for unquantized waveform.

signal is band-limited it can be sampled at the Nyquist rate, i.e., twice the bandwidth. Sampling at the Nyquist rate without quantization would result in no loss of performance [107], but the quantization causes some loss in sensitivity. The simplest form of quantization involves distinguishing only two levels of the signal, i.e., the sign of the voltage, and the quantization characteristic is shown in Fig. 15(a). Only 1 bit is needed to represent each sample. The correlation coefficient of the quantized signals ρ_q is related to the correlation of the unquantized signals ρ by the well-known formula derived by Van Vleck [108]

$$\rho_q = \left(\frac{2}{\pi}\right) \sin^{-1} \rho. \qquad (10)$$

For $\rho \ll 1$, the signal-to-noise ratio is reduced by a factor $2/\pi$ ($= 0.64$). Since quantization generates new frequency components the waveform that is sampled is no longer band-limited, and sampling at the Nyquist rate appropriate to the unquantized waveform does not conserve all of the information. Sampling at twice the Nyquist rate increases the degradation factor from 0.64 to 0.74 [109]. However, an even greater increase in sensitivity is obtained by using 2 bits to represent each sample value rather than by doubling the sampling rate. Schemes with 2-bit quantization have been investigated by Cooper [110] and the two most important ones can be described as three-level and four-level quantization, the characteristics for which are shown in Fig. 15(b) and (c). The degradation factors are given in Table VI which contains results by several authors [110]–[112]. In four-level quantization, the performance is optimized if the weighting factor n defined in Fig. 15(c) has a value of 3 or 4 [110]. The products in the multiplication take values of ± 1, $\pm n$, and $\pm n^2$, i.e., numbers up to 9 or 16. In three-level quantization, only products of high-level samples produce a nonzero output from the correlator, and n is unity. The counter which averages the multiplier output can thus be significantly simpler in the three-level scheme, and this type of quantization was chosen for the VLA as a compromise between sensitivity and complexity.

Maintaining correct adjustment of the threshold levels is more difficult in three-level sampling than in two-level where the

threshold is 0 V. Corrections for threshold errors can be based on counts of the relative frequencies of occurrence of the different levels in the output of each sampler. The level tolerances and the correction scheme used for the VLA, as well as the determination of the true signal correlation from measurements made after three-level quantization are discussed by D'Addario et al. [113].

C. Implementation in the VLA

For the 50-MHz bandwidths of the VLA, the Nyquist frequency is 100 MHz and the samplers run at that rate. Emitter-coupled logic is used for the 100-MHz circuits, but in the delay system each 100-MHz bit stream is converted to 16 bit streams at 6.25 MHz. The main delay functions can then be implemented in MOS logic for economy and compactness. To obtain variability in the delays, a circuit shown in Fig. 16 is used which was suggested by B. G. Clark and developed by R. P. Escoffier, both of NRAO. The two shift registers each have 512-bit capacity, and the input data rate is 6.25 MHz. Suppose that switch A is held in position 1 while N bits are entered into the top register, both registers being clocked at 6.25 MHz. Next, both the input data stream and the clock pulses are divided alternately between the two registers until a further $(512 - N)$ bits have been entered into each register. The cycle is then repeated continuously until the delay is changed. Each bit suffers a delay of $(1024 - N)$ intervals of the 6.25-MHz clock, and N can be varied from zero to 512 giving a delay range of 82 to 164 μs. A further 82-μs delay using one shift register can be inserted as required, and delays of one clock interval can be obtained at the intermediate bit rates of 100, 50, 25, and 12.5 MHz. The smallest delay increment is 625 ps, and this is obtained by varying the phase of the 100-MHz pulses that trigger the samper in 16 steps relative to the main 100-MHz clock pulses, as shown in Fig. 17. The 625-ps increment corresponds to 5.6° of phase at the 25-MHz center frequency of the baseband signal, which is small enough to avoid introducing serious phase errors in the visibility [53]. The delayed data streams are recombined into the original 100-MHz streams and then passed to the correlators.

The correlators contain multipliers to form the products of data samples, and counters to integrate the product values. The products −1, 0, and 1 are counted as 0, 1, and 2 to avoid the use of bidirectional counters, and the results are converted back to the true values in the control computer. Two specially designed integrated circuits (IC's) [114] are used in the correlator. These are a dual three-level multiplier using emitter-coupled circuitry, and a 12-bit integrating counter using low-power Schottky logic.

Measurements of correlation made with the delays adjusted to equalize the signal propagation times provide images of the average radio brightness across the IF band, and this is referred

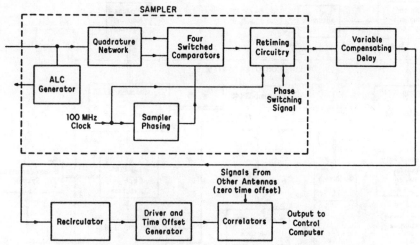

Fig. 17. The VLA sampler, delay, and correlator system for one digital channel. The phase of the 100-MHz sampling pulses can be adjusted in 625-ps steps to provide fine adjustment for the compensating delay. Four comparators are used in the sampler, two for each of the sine and cosine components of the signal. The recirculator and time offset generator are only used in the spectral line mode.

to as the continuum mode of operation. In the continuum mode, the recirculator and time-offset blocks in Fig. 17 are bypassed. The alternate mode is the spectral-line mode in which spectral features within the IF band are measured. The VLA contains provision for measuring the correlation at 16 points across a 50-MHz-wide band. This requires combining signals from each antenna pair using 32 time offsets from -16 to $+15$ times the 10-ns sampling interval. The number of correlators required is $351 \times 32 = 11\,232$. Fourier transformation converts visibility as a function of time offset to visibility as a function of frequency, and the 16 spectral points are commonly referred to as frequency channels. Finer frequency resolution is obtained for narrower signal bandwidths by using a recirculation technique [115]. With the 25-MHz IF bandwidth, the sampling frequency can be halved, but if the correlators are operated at the full 100-MHz frequency the rate at which multiplications occur is sufficient to allow the data to be put through the correlators twice. This is performed with appropriate timing to double the number of frequency offsets. The recirculation requires only additional memory to accumulate data at the reduced sampling frequency and allow readout at 100 MHz. Thus with the 25-MHz bandwidth, 32 frequency channels are available, and as the IF bandwidth is successively reduced by factors of 2, the number of channels increases to a limit of 256 that is set by the capacity for the subsequent data processing. The numbers of channels quoted above are obtained when only one IF band is processed from each antenna. If two bands are processed without polarization information the number of channels is decreased by two. If polarization measurements are required, four products must be formed for each antenna pair; i.e., RR, LL, LR, and RL where R and L refer to the signals obtained at the right and left circular polarization outputs of the antennas. The number of channels is then reduced by a factor of four. The ability to tune the IF bands from the antennas to different frequencies, to use them for polarization measurements, and to choose IF bandwidths and recirculation factors provides a large number of modes of operation. The correlator outputs are transferred at intervals of 92.16 μs to a large storage and integration system using random-access memory circuits. This allows the correlator outputs to be averaged for periods up to 10 s.

In continuum operation, a total of 11 664 12-bit integration products are developed each 92.16 μs in the 27 000 IC's of the driver and correlator subsystems; thus there are 2.3 IC's per product as a figure-of-merit parameter for circuit economy. In spectral line operation, up to 373 248 products can result from 39 500 IC's of the recirculator, driver, and correlator subsystems; this gives 0.11 IC's per product. In the output integrating system, the high-density, 4096-bit memory circuits permit a design using only 0.03 IC's per 24-bit product. Further details about the VLA delay and correlator system are available in [116].

For some types of observations it is useful to be able to combine the signals from the antennas in an additive manner and produce a single output from the whole array. These include very-long-baseline interferometry (VLBI) in which the VLA acts as one element, the signals being recorded on magnetic tape for subsequent correlation with signals recorded at other antennas in North America or other continents [42]. Observations of pulsars in which the signal pulses are gated into a special processor also make use of the phased-array mode [117]. To operate the array in this manner the delayed samples representing the real outputs of the complex correlators are converted to a three-level analog form, and combined in an analog summing network. This is done independently for each of the IF channels from the antennas. To adjust the phasing of the array, an unresolved calibration source is observed using the correlators in the usual manner to determine the instrumental phase values. Corrections for this instrumental component are then applied through the fringe-rotation phase shifters. In VLBI observations the source is usually unresolved by the VLA, and can itself be used as the calibrator, thereby allowing continuous adjustment of the phase in real time.

D. Recent Developments in Digital Correlators

A more recent development in spectral correlator systems is being proposed for the millimeter-wavelength array in Nobeyama, Japan [118]. Here the output from each sampler is fed to circuitry that performs a fast Fourier transform on 1024 consecutive samples before data from pairs of antennas are correlated. For each of 1024 spectral channels that are produced, the data rate at the output of the transform circuits is less than the sampling

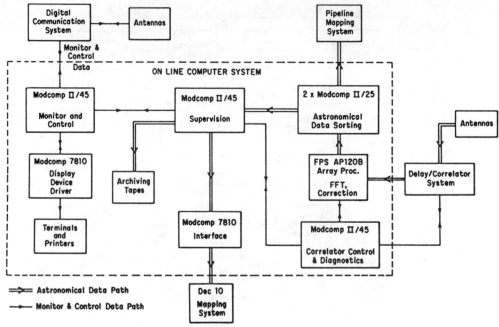

Fig. 18.   Block diagram of the VLA on-line computer system.

frequency by a factor of 1024. Thus the total rate at which correlations are performed does not increase in proportion to the number of channels. Compared with the VLA spectral correlator, the Nobeyama plan reduces the correlator circuits and does not require that the number of channels decrease as the signal bandwidth increases. These advantages are obtained at the expense of the additional circuitry for the fast Fourier transform. It is planned to use 3 bits (8 levels) to represent each sample in the Nobeyama system. The initial sampler bandwidth will be 80 MHz and serial-to-parallel conversion will be used immediately following the samplers to reduce the frequency of the bit streams down to 10 MHz for processing in large-scale CMOS circuitry. In another new millimeter array under construction at IRAM (see Table I) the use of 4-bit representation of the samples (16 levels) is planned. Only 2–3 percent of sensitivity is lost in the quantization, but there is some subsequent degradation from truncation after multiplication. Digital signal processing is an area of synthesis-array technology which is continuing to evolve quite rapidly.

VII.   VLA CONTROL SYSTEM

The VLA control system is designed to allow the instrument to operate completely unattended except for a single array operator who operates the on-line control computers. Typically, operations personnel other than this are present on site for only 8 h each day. The unattended antenna elements have various safety features including automatic circuitry that commands the antenna to its safe stow position if communication with the on-line computer is lost or if the wind velocity in the vicinity of the antenna exceeds 80 km $\cdot$ h^{-1}.

A highly centralized monitor and control system is used in which all digital commands originate in, and monitor data are returned to, the single central on-line computer system. The on-line system consists of 3 Modcomp II/45s, 2 Modcomp II/25s, 2 Modcomp 7810 computers, and an FPS AP-120B array processor linked together in a network as shown schematically in Fig. 18. An observer initiates an observing program with the VLA by preparing a computer file containing the names of the astronomical sources to be observed, their positions, the start and stop

times for the observations, the observing frequency, and the bandwidth desired for the observation. This file is read by the supervisory Modcomp II/45 computer which is programmed to generate all the commands necessary for the instrument to perform the desired observations. At the start of the observation of each radio source commands are sent to all antennas to set receivers to the correct frequency and bandwidth. Every 0.052 s, commands are sent to the digital delay lines to update their delay setting. Every 0.10 s, new altitude and azimuth commands are sent to each antenna. Every 1.25 s, new phase and phase rate commands are sent to the phase rotators in the LO path at each antenna to keep the relative phases of the signals at the correlators constant for radiation from the center of the synthesized field. This phase command includes a correction term which is derived from the round-trip-phase measurement described in Section IV-D to remove the effect of changes in length of any waveguides and cables in the communication path to each antenna. Also included, because of the large extent of the array, is a correction term for phase shifts resulting from the differences in atmospheric path length due to earth curvature. This latter correction makes use of a model atmosphere derived from ground-based measurements of pressure, temperature, and dew point provided by an on-line weather station. This same atmospheric model is used to provide the correction for refraction which is included in the antenna pointing commands. Correction terms for precession, nutation, aberration, and repeatable pointing errors due to antenna structural deformation are also included in the pointing commands. Besides the geometric delay term due to the rotation of the earth, the delay command includes corrections for predetermined delays in cables, waveguides, or electronic components in the signal path to each antenna.

The on-line computer software is designed to support three separate simultaneous observations. Thus the 27 antennas can be divided into two or three subsets if the observer desires and each of these "subarrays" can observe a different list of radio sources.

All commands are distributed from the supervisory computer to the hardware via a Modcomp II/45 computer whose primary tasks are to act as an interface with the hardware for the transmission and reception of commands and monitor data, and to process the monitor data through a variety of threshold-detec-

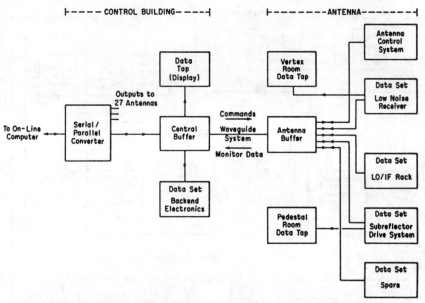

Fig. 19. Block diagram of the VLA Digital Communication System. One each of the units shown is required per antenna except for the serial-to-parallel converter.

tion algorithms to detect the occurrence of failures in the hardware. Communication between the on-line computer and the equipment in the array is provided by the Digital Communication Subsystem (DCS), a block diagram of which is shown in Fig. 19. A serial-to-parallel converter distributes the commands to one of 27 central buffers, one for each antenna, located in the central electronics room. If the command is addressed to a piece of equipment located at the antenna it is sent, via the waveguide communication system, to another buffer located at the antenna. Commands are distributed from the buffers to the appropriate hardware subsystem via modules known as data sets. There are four data sets at each antenna to control and monitor the antenna drive system, the low-noise receiver, the LO system, and the subreflector-drive system. A fifth data set is spare. In the central electronics room, a data set provides monitor and control for the back-end equipment for each antenna. The flow of monitor data is essentially the reverse of the command flow. For maintenance and trouble-shooting purposes, modules known as data taps provide binary or decimal display of any command or monitor word selected by manually entering the desired address. Words representing analog voltages are reconverted to analog to allow monitoring by oscilloscope or chart recorder.

For transmission down the waveguide, data are biphase modulated onto a 500-kHz signal which is modulated onto the 1800-MHz LO signal. Transmission in the two directions in the waveguide follows the 19.2-Hz sequence described in Section IV-D, with bit lengths of 4 μs for transmission out to the antennas and 50 μs for transmission back to the Control Building. During each cycle, up to four command words are transmitted to each antenna and up to eight to each local data set. Two monitor words per cycle are received from each data set. An analog-to-digital converter in the data sets converts analog monitor point inputs to 12-bit digital format. For each antenna, the five data sets provide the capacity for up to 240 different 24-bit digital commands, 320 24-bit digital monitor words, and 640 analog monitor voltages, not including the antenna pointing commands. Only about 30 percent of the capacity is currently used. Further information about the DCS is available in [119].

The analog monitor points include the antenna position, drive motor currents, cryogenic temperatures, signal levels, power supply voltages, phase-lock voltages, etc. The data are read back in cycles no longer than 10 s and are compared with maximum and minimum tolerable values. In addition, all commands sent to the electronics to select frequency bands, oscillator frequencies, etc., are read back for checking. The results of the threshold checking are used to initiate maintenance by technicians who have access to several days worth of monitor data stored in the computer.

Other diagnostic features include a switch at the antenna which allows the IF signals selected from the low-noise front-ends to be interchanged between the *A* and *C* or *B* and *D* sideband channels in the waveguide. This facility helps trace anomalous behavior of any IF signal to the front-end or the transmission and final IF stages. In the digital signal-processing area, pseudo-random test signals are injected at the delay unit inputs during the intervals of the 19.2-Hz cycle in which no data are received from the antennas. The multiplier outputs are examined for the expected results and a malfunction can be traced to a delay unit, recirculator, driver, or multiplier. If the fault is in a delay unit or recirculator, a self-repairing action takes place in which a spare unit is switched in to replace the malfunctioning one.

To facilitate maintenance, almost all of the electronics is constructed in modular units which are of convenient size for handling and slide into slotted bins in the electronics racks. Electrical connections to the modules are through rear-mounted connectors that automatically engage when the units are inserted. The only exception to the modular construction are the cooled front-end components that are mounted in an evacuated dewar, and the digital delay and correlator system in which the circuitry is on plug-in boards.

Apart from monitor and control, the other important function of the on-line computer system is the collection and correction of the data provided by the correlator. Control and diagnostics for the correlator are provided by a dedicated Modcomp II/45 computer. The integrated data from the correlators are read out every 10 s via the FPS AP-120B array processor shown in Fig. 18. The array processor performs various data corrections and, in the case of spectral line observations, carries out the lag to frequency Fourier transform. Two further Modcomp II/25 computers provide buffering and sorting before passing the data on to the data-processing computer systems. The data rate from the correlator into the on-line computer can be high, depending on the

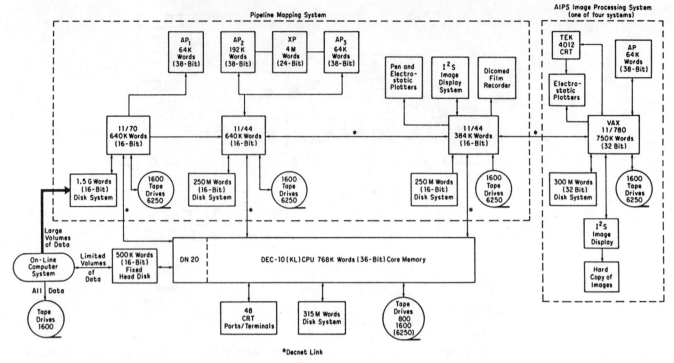

Fig. 20. Block diagram of the calibration, mapping, and image processing
computers at the VLA.

type of astronomical observation being performed. In the extreme
case of a high-resolution spectral line observation, the correlator
produces 256 complex frequency channels for each of 351 baselines every 10 s, giving a continuous data input rate of approximately 18000 16-bit words/s.

VIII. DATA REDUCTION REQUIREMENTS

A. General Considerations

The output from the on-line computer is the stream of data
from the correlator which has been corrected for known instrumental effects and averaged for the maximum time consistent
with the expected rate of change of the visibility function. This
visibility data must then be calibrated using astronomical standards, edited, and merged into databases for each object observed. In order to make an image of an observed region of sky,
the Fourier transform of the visibilities must be computed. Such
an image may be degraded by errors and by the lack of complete
spatial coverage in the (u, v) plane. New and powerful image
restoration techniques are now available to generate better quality images. Finally, images must be displayed and analyzed to
extract the information of astronomical interest. Fig. 20 is an
overall block diagram of the network of computers used at the
VLA to perform these functions. A more detailed description of
each of these steps follows.

B. Calibration and Self-Calibration

The instrumental stability requirements discussed in Section IV
provide stable measurements of the relative amplitude and phase
over short periods of time. In order to determine the absolute
amplitude calibration and the zero of phase, and to provide the
long-term stability, it is necessary to make astronomical observations of reference sources of known flux densities and positions.
For some synthesis arrays (e.g., those at Cambridge and Westerbork) the instrumental stability is such that astronomical
calibration need only be done on a time scale of days [14], [24].

However, for arrays working at high angular resolution the atmosphere alone will introduce serious phase variations on much
shorter time scales. In this case it is necessary to interleave
calibration observation throughout the astronomical observing
program. These calibration sources must be as close in the sky as
possible to maximize the correlation in the atmospheric phase
fluctuations, and to minimize time spent moving the antennas to
the calibration source. To relate the phases of the calibrator and
the source under observation it is necessary to know the relative
positions of the antennas to a small fraction of the wavelength.
This baseline calibration requires observation of a number of
calibration sources widely distributed over the sky, and is performed as a special operation whenever antennas are relocated.
For the VLA, a network of about 400 point-source calibrators
with positions known to better than 0.1″ has been established.
During high-resolution observations, one of these may be observed as frequently as once every 10 min. For measuring polarization of radio sources it is necessary to include a calibrator with
a known component of polarized radiation [120], [121].

The calibration observations can be used to set the complex
gain of each of the antenna pairs. The complex gains relate the
uncalibrated (primed) visibility values to the calibrated (unprimed) values, for antennas j and k, as follows:

$$V_{jk}(t) = G_{jk}(t)V'_{jk}(t). \qquad (11)$$

By choosing calibration sources that are unresolved and have
known flux density and position, V_{jk} can be predicted and $G_{jk}(t)$
determined at the times of the calibrator observations. $G_{jk}(t)$ is
assumed to change smoothly and slowly in accordance with the
instrumental specifications for amplitude and phase stability. The
values of G_{jk} can thus be interpolated for all t and used to correct
the observed V'_{jk} for the source under investigation.

Since the VLA contains 351 antenna pairs there are 351 values
of G_{jk} to be determined. This is a sufficiently large number that it
was computationally expedient to express the calibration in terms
of the individual antenna gains $g_j = |g_j|e^{i\phi_j}$ where

$$G_{jk} = |g_j||g_k|e^{i(\phi_j - \phi_k)} + \epsilon_{jk}. \qquad (12)$$

The term ϵ_{jk} represents effects that cannot be attributed to multiplicative antenna gain terms [53]. If they are small, the $n_a(n_a - 1)/2$ complex correlator gains are reduced to n_a amplitude and $(n_a - 1)$ phase calibration values (one antenna can be defined to have zero phase since only phase differences between antennas are meaningful). Although in the case of the VLA this was done for expediency, it led to the very powerful concept of antenna-based self-calibration to which we will return after a historical digression.

In 1958, Jennison realized that in a triangle of three antennas it is possible to form a phase-closure relation which is free of instrumental phase errors [122]. This technique was used to determine information about radio source structure at a time when it was not possible to build phase-stable radio interferometers. Later when phase-stable interferometers were developed this technique fell out of use and was largely forgotten. In 1968, Rogstad [123] extended the phase-closure relations to an array, and pointed out that since this technique also removes the atmospheric errors there might be applications to optical imaging. The technique is in fact the same as that of adaptive optics [124]–[126], but the application to optics is technically much more difficult than to radio observations where the corrections can be made digitally. In VLBI arrays, which have not been included in this review, large instrumental phase drifts occur because the LO is derived from independent atomic time standards. This inability to measure the phase directly led again to the use of the phase-closure relations [127], [128]. For an n_a-element array there are only $(n_a - 1)(n_a - 2)/2$ closure relations, which is $n_a - 1$ too few to determine the phase on all the $n_a(n_a - 1)/2$ measured correlations. To address this problem for VLBI, more complex algorithms were developed in which additional constraints were placed on the image to remove the extra degrees of freedom [128], [129]. The technique is also known as hybrid mapping [128], [130] because it uses measured amplitudes but only indirectly deduced phases.

Independent of this application of the phase-closure technique to VLBI arrays, a "self-calibration concept" was (concurrently) evolving at the VLA. It was noticed that since the number of measured visibilities (351) greatly exceeds the number of antenna gains (27) it is possible to make an antenna-based calibration of an observation using the actual source being observed. This requires some additional assumption about the source structure; for example, if a large complex source contains a small-diameter component which dominates the visibility at long spacings, the visibility can be assumed constant for these long spacings and used to calibrate *all* the antennas. By using this technique, atmospheric and instrumental errors can be almost completely removed, instead of only partially removed by referencing to a calibration source which is observed at a slightly different time and direction. Eventually it was realized [131], [132], that the two techniques; phase closure and antenna-based self-calibration, are identical, and that the phase-closure relations can be deduced from (11) and (12). The term ϵ_{jk} in (12) is complex and called the closure error.

Further developments, which involve the image-restoration techniques to be discussed later, have now made it possible to apply self-calibration to relatively complex sources with great success [128], [129], [131]–[133]. Fig. 21(a) and (b) shows an example of the use of self-calibration with the VLA. Fig. 21(a) is a normal high-quality image of a radio quasar which has errors at a level of 1 percent of the peak brightness. After self-calibration (Fig. 21(b)) the errors in the map are reduced to less than 0.2 percent, and the low-level jet of radio emission becomes clearly visible. A variation on the self-calibration technique is employed with the Westerbork Synthesis Radio Telescope in which redun-

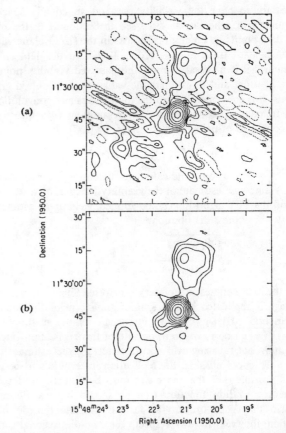

Fig. 21. Map of the radio quasar 1548 + 115. (a) Without self-calibration. (b) With self-calibration. The lowest contour level is 0.6 percent in each case.

dant baselines (identical baselines using different antennas) are used to obtain additional constraints on the calibration which are independent of assumptions about source structure [125], [134].

Self-calibration is now used routinely at the VLA to improve image quality by more than an order of magnitude. Since the errors in a self-calibrated image are determined by errors which are not antenna-dependent [53], for example correlator errors or effects of unequal bandpass responses, emphasis should be placed upon reducing such effects in future designs.

The implementation of the visibility calibration at the VLA is done in the DEC10/KL general-purpose mainframe computer. It has 3 Mbytes of fast memory and 1.2 Gbytes of disk storage. This computer is used either interactively or in batch processing to inspect and edit data, and to determine antenna-based calibration coefficients and apply them to the visibility data. The calibrated visibility data are either written onto magnetic tape or transferred through a computer link to the mapping systems.

C. Imaging

The next step in the data reduction is computation of the sky brightness distribution from the calibrated values of the complex visibility [4], [135]. This requires evaluation of the Fourier transform relation described by (5) in Section I-A. Equation (5) can be applied directly to the individual sample points in the (u, v) plane to determine each pixel in the image but this process is time-consuming since the number of sine and cosine multiplications required is $2MN_xN_y$, where there are M visibility measurements and the image is of size N_x by N_y pixels. In general, $N_x = N_y = N$ and $M \simeq N_xN_y$, so the number of multiplications required goes as N^4. However, if the fast Fourier transform (FFT) algorithm [136] can be used, the number of multiplications goes as $2N^2 \log_2 N$ which gives an enormous computational advantage for large N. Unfortunately, in order to use the FFT it is necessary to have the visibility data on an equally spaced rectangular grid, whereas, as discussed in Section II-B, the observed data points lie on elliptical tracks in the (u, v) plane with the data in each track in time order. To obtain the data on a uniform grid it is necessary to sort the observed visibility points into (u, v) order and to estimate the values at the required grid positions. It is convenient to express this process as a convolution with a function $C(u, v)$ followed by resampling with the two-dimensional bed-of-nails function $^2\mathrm{III}(u, v)$ [137], that is

$$V_g(u, v) = {}^2\mathrm{III}(u, v)\big(C(u, v) ** V(u, v)\big) \tag{13}$$

where the double asterisk indicates two-dimensional convolution. $V_g(u, v)$ is now an estimate of the visibility function on a regular grid, and the FFT algorithm can be applied, yielding an estimate of the sky brightness distribution

$$B(x, y) = \overline{V_g(u, v)}$$
$$= \overline{{}^2\mathrm{III}(u, v)} ** \big(\overline{C(u, v)} \; \overline{V(u, v)}\big) \tag{14}$$

where the bar denotes a Fourier transform. Since the resulting image is replicated by the Fourier transform of the sampling function $^2\mathrm{III}(u, v)$, it is important that the function $\overline{C(u, v)}\,\overline{V(u, v)}$ be nearly zero outside of this replication interval. If it is not, aliasing will occur, distorting the estimate of $B(x, y)$. To avoid aliasing, the replication interval has to be as large as possible (i.e., the image size must be numerically large) and the convolution function $C(u, v)$ should have a Fourier transform that goes to zero as fast as possible outside the replication interval. In this respect, the two-dimensional sinc function would be an ideal convolving function, but the computation of the convolution would have to extend over the entire (u, v) plane and would be just as time-consuming as the direct Fourier transform. Thus $C(u, v)$ must also be a function that can be approximated by a small number of values. Two classes of function now considered best for this application are the spheroidal functions and products of exponential and sinc functions [138]. These convolving functions are sufficiently good that no significant errors are added by the use of the FFT. Examples are given in [139]. An alternative hardware approach in order to avoid the need for using the FFT has been explored for the Fleurs Synthesis Telescope [140].

In addition to computing the radio image it is also important to determine the synthesized beam. This is the Fourier transform of the instrumental transfer function and is obtained by replacing the measured visibilities by unity (or by the weights the visibilities that were assigned in the computation of the image). This synthesized beam is used heavily in the subsequent image processing.

Two independent mapping systems are available for the production of VLA images. The high-capacity mapping system uses a pipeline of tasks to sort and grid the visibility data and compute the FFT. These tasks are executed using a tightly coupled network of two PDP11/44 computers and a PDP11/70 CPU with three FPS AP-120B array processors and a specially designed 4-megaword transpose memory to perform the matrix transpose step in the two-dimensional FFT computation. This system has an estimated capacity sufficient to produce up to 256 1024×1024 images from a database containing 2×10^8 spectral line visibility data values in real time (i.e., ~ 8 h). The other system uses the general-purpose image reduction software system called astronomical image processing system (AIPS) which is designed to run in a super-minicomputer such as a VAX780 with an FPS AP-120B array processor [141]. Although somewhat slower than the pipeline mapping system just described, it has much more flexibility. The AIPS software is designed to be

exported and is currently implemented in many other processing systems in the VLA-user community.

The final outputs of the mapping systems at the VLA are the two-dimensional arrays representing images. These may vary in size from 256×256 to 4096×4096, and from a set of four images describing the full polarization characteristics of a continuum source to the 256 images corresponding to the different frequency channels of a spectral line observation. These images are written onto magnetic tape using the widely accepted flexible image transport system (FITS) format [142].

D. Image Restoration

Although the images generated by the foregoing procedures meet the specifications that we have described for the VLA in Table II, significant further enhancement of the image quality is now routinely achieved through the use of digital image restoration techniques. The quality of the synthesized image is degraded by three factors: incomplete (u, v) coverage, errors in the data, and approximations in the analysis.

First consider the effects of not being able to measure the entire (u, v) plane. There remain uneven gaps between the spacing loci, there is no information beyond the longest spacings, and there is a small hole near the origin since it is not possible to observe with overlapping antennas. The gaps in the (u, v) plane generate sidelobes (both positive and negative) that are distributed over the entire image. In the VLA, gaps between the spacing loci result in irregular sidelobes at the -16-dB level (Section II-C). The missing short spacings cause bowl-shaped depressions in the image. The finite maximum spacing results in a smearing of the image by a beam of finite width and near sidelobes which are caused by the abrupt cutoff at the maximum spacing. A simple form of image improvement which is used with most synthesis arrays is the introduction of weighting associated with the measured visibility data. This can compensate for the nonuniform density of samples in the (u, v) plane and taper the sharp boundary (e.g., [12]). Such a taper reduces the near-sidelobe levels but causes a loss in resolution. To obtain a higher quality image it is necessary to allow data in the unmeasured (u, v) areas to take more realistic values than zero. To obtain this interpolation and extrapolation requires some additional constraints. For example, since the sky intensity cannot be negative the new values in the (u, v) plane can be made to satisfy this constraint. A further powerful boundary condition which can be applied to almost all radio images is that the sky consists of discrete regions of emission surrounded by large areas of nearly zero brightness. This is a strong constraint since the sidelobe patterns due to missing information introduce ripples across large areas of the image which violate this condition. The first successful image restoration algorithm which has come into routine use with synthesis arrays is the method CLEAN due to Hogbom [143]. This algorithm assumes that the radio image can be represented by a finite set of delta functions in an otherwise empty field. It uses a simple iterative procedure to decompose the image into delta functions and finally convolves them with a clean beam (usually a Gaussian) to de-emphasize the higher spatial frequencies which correspond to unjustifiably large extrapolations. Although this algorithm is now being used extensively there are only a few published discussions of its properties [144], [145]. For large images that require many thousands of CLEAN iterations, a faster version of the algorithm based on the use of the FFT has been developed by Clark [146]. Another approach has been taken at the University of Sydney Electrical Engineering Department by implementing CLEAN using special-purpose hardware. Cornwell [147] has improved the algorithm by applying an addi-

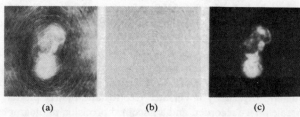

Fig. 22. Radiographs of the radio galaxy 3C310. (a) Fourier transform of calibrated visibility data, with no special processing. (b) Synthesized beam corresponding to image in (a). (c) Image produced by CLEAN algorithm.

tional smoothness constraint to avoid an instability which occurs for very extended radio sources. Fig. 22 shows an example of the use of this algorithm. Fig. 22(a) is a VLA image of the radio galaxy 3C310 which has not been processed beyond Fourier transformation of the visibility. This radio galaxy has a bright nucleus and weak lobes, but these low-brightness regions are distorted by the sidelobe patterns which are present at the 2-percent level. As can be seen in Fig. 22(b), this same pattern is present in the synthesized beam and is due to the incomplete (u, v) coverage. After application of the smoothness-stabilized CLEAN algorithm [147] this sidelobe pattern is suppressed well below the noise and the quality of the low-brightness part of the image is greatly improved. This powerful algorithm has had a further important effect on the use of synthesis arrays, especially the VLA. Since the sidelobes from incomplete (u, v) coverage can be suppressed, it has become possible to reduce observing time for objects with sufficient brightness, and which are sufficiently small for the CLEAN algorithm to give a reliable result. Some VLA observing programs are now executed in periods as short as 5 min instead of the 8 h originally specified in Table II. Other algorithms, notably the maximum-entropy method (MEM) [148] are also being successfully employed for the improvement of images from synthesis arrays. Although MEM is sometimes justified by erudite discussions on the meaning of entropy for spatial brightness distributions, in practice it appears that the positivity and the smoothness constraints inherent in this method are sufficiently strong to provide high-quality image restoration [149].

So far we have discussed the correction of errors due to limited (u, v) coverage. Algorithms using similar *a priori* information on the image properties are also available to help suppress effects of errors in the data. Again the positivity constraint on intensity provides a good example. Since a phase error in the data generally results in asymmetric positive and negative error lobes, an algorithm which suppresses the negative lobes effectively discriminates against the phase error. An application to the case where the phase errors are very large has been demonstrated by Baldwin and Warner [130], who succeeded in obtaining good-quality radio images using almost no phase information. The extremely powerful self-calibration technique already discussed in Section VIII-B uses a combination of constraints both on the nature of the error and the properties of the image.

E. Image Analysis

After producing an image in the computer it is necessary to display it to see the structures present, to monitor the various processing steps, and to produce the final hard copy for storage or publication. Images generated by modern synthesis arrays can contain enormous amounts of information by most radio astronomy standards. Typical VLA images contain from 10^4 to 10^5 resolution elements and in extreme cases as much as two orders of magnitude more. For such large amounts of information

careful attention has to be given to the image display techniques in order to obtain good impressions of what is present in the data and to recognize possible errors. The traditional contour plot, although quantitative, becomes less useful for complex images and is difficult to interpret by visual inspection. Ruled-surface plots are visually cleaner, but can hide information and are not satisfactory in regions of low signal-to-noise ratio. Radiographs or the gray-scale displays in which the radio intensity is translated into brightness or darkness give the best visual impression and are very good for quick video displays of images with many resolution elements. However they are not quantitative and cannot represent a large range of brightness. The gray-scale display is ideal in situations of low signal-to-noise ratio where the ability of human perception to recognize patterns is a great advantage [150]. If the radio intensity is mapped into color instead of brightness we have a pseudo-color display. If the color transitions are made at discrete levels, the quantitative advantage of the contour plot is obtained, but some of the advantages of the continuous gray-scale, such as the ability to follow a ridge line, are lost. Since the optimum display depends on the image properties and the type of information desired, it is an advantage to have many different types of display. Fig. 23 compares three types for the same astronomical image, the radio galaxy 3C310 which has low-brightness radio lobes and a strong source at the nucleus [151]. The contour map can be used to obtain quantitative information such as the strength of the central source or the average brightness of the lobes. The ruled-surface plot shows much more dramatically the large range in brightness between the nucleus and the radio lobes. It also shows a central circular depression in the southern lobe which is a new feature not previously seen in a radio galaxy. The gray-scale representation gives no idea of the strength of the nucleus but it does show the recently discovered filamentary arcs better than the other displays. Further discussion and examples of display types are found in [150], [152], [153].

Modern digital display technology has now made it possible to manipulate displays interactively. For the gray-scale or pseudo-color displays, the transfer characteristic between the radio intensity and the displayed brightness can be changed continuously and interactively. This circumvents the dynamic range limitation of the gray-scale display. Vector graphic display systems can present ruled-surface displays with interactive control of the scaling and perspective. When the image contains more resolution elements than the display, features such as zooming and panning become necessary.

Three-dimensional images of brightness as a function of two spatial coordinates and one frequency coordinate, obtained from spectral-line data, present even more serious display problems. The simplest technique is to form a set of two-dimensional slices through the three-dimensional block and to display them either as a mosaic, or in sequence utilizing time as the third dimension. This cinematographic type of display is more powerful if the frame direction and speed can be controlled interactively. Another possibility is to use a false-color display in which one variable (e.g., velocity) is mapped into color. This works quite well when the brightness is a simple function of the third dimension. Image displays which can operate on intensity, hue, and saturation (instead of traditional red, green, and blue) have considerable advantages for interactive manipulation of such false-color displays [154]. Examples of the use of false-color displays are given in [150], [155]. A third possibility is use of vector graphics displays to show complex three-dimensional surfaces in perspective. As with the two-dimensional displays, the method which works best depends on the nature of the image and the information desired.

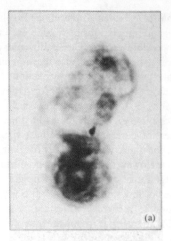

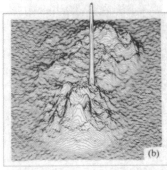

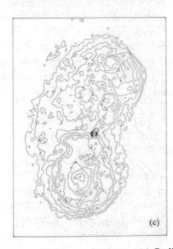

Fig. 23. Three ways to display a radio image. (a) Radiograph. (b) Ruled surface plot. (c) Contour plot.

Image display is a necessary facility in analysis of data but is not normally the desired end result. Further numerical processing is usually required to extract the astronomically relevant information [152], [153], [155]. If accurate comparison is to be made between images generated by different types of instruments, such as optical photographs and synthesis images, it is necessary to correct for some instrumental characteristics. These include the brightness variations over the radio image caused by the primary beam of the array elements, the angular resolution, and the geometric distortion discussed in Section II-A. It may also be necessary to estimate parameters other than the radio brightness of the image. In some cases these are single values such as the position, size, or intensity of a discrete radio source. In other cases a new parameter may be determined at each point in the image, generating an entirely new image in the new parameter space. Examples of such parameters are the radio spectral index for a continuum radio source, or the velocity in a rotating galaxy determined from the Doppler shift of the neutral hydrogen line.

For the VLA all of the types of image processing discussed above are included in the AIPS software system mentioned in Section VIII-B [141]. This system has been implemented at NRAO in four separate hardware systems, three of which use VAX780 computers and one a MODCOMP Classic. Each system is equipped with an FPS array processor and an I^2S image display system with four 512×512 memory planes. Hard copy output is provided by a Versatec electrostatic matrix plotter, a gray-scale image recorder, and a Dicomed precision photographic recorder. A large amount of disk storage, 1 Gbyte, is included in each system to handle large three-dimensional spectral line images. Image storage and interactive cinematographic display is provided by a special-purpose digital disk storage device designed by R. P. Escoffier and capable of storing 512 512×512 images and displaying them at 6 frames/s.

ACKNOWLEDGMENT

A large number of scientists and engineers contributed to the design and construction of the VLA. As an example of this, we note that the series of VLA scientific and technical reports which document the design and performance of the VLA has an authorship list in excess of 120 people. Much of the information we have presented here was obtained from these reports and from discussions with many of our colleagues. We wish to thank T. Cornwell for reading and commenting on parts of the manuscript and D. Retallack for assistance with photography.

REFERENCES

[1] W. N. Christiansen and J. A. Warburton, "The distribution of radio brightness over the solar disk at a wavelength of 21 cm, III. The quiet sun—two dimensional observations," *Australian J. Phys.*, vol. 135, pp. 151–174, 1955.

[2] M. Ryle, "The new Cambridge radio telescope," *Nature*, vol. 194, pp. 517–518, 1962.

[3] R. N. Bracewell, "Radio interferometry of discrete sources," *Proc. IRE*, vol. 46, pp. 97–105, 1958.

[4] E. B. Fomalont, and C. H. Wright, "Interferometry and aperture synthesis," in *Galactic and Extragalactic Radio Astronomy*, G. L. Verschuur and K. I. Kellermann, Eds. New York: Springer, 1974, ch. 10, pp. 256–290.

[5] G. W. Swenson and H. C. Mathur, "The interferometer in radio astronomy," *Proc. IEEE*, vol. 56, pp. 2114–2130, 1968.

[6] D. S. Heeschen, "The Very Large Array," in *Telescopes for the 1980's*, G. Burbidge and A. Hewitt, Eds. (Annual Reviews Inc., Palo Alto, CA), pp. 1–61, 1981.

[7] A. R. Thompson, B. G. Clark, C. M. Wade, and P. J. Napier, "The Very Large Array," *Astrophys. J. Suppl.*, vol. 44, pp. 151–167, 1980.

[8] R. M. Hjellming and R. C. Bignell, "Radio astronomy with the VLA," *Science*, vol. 216, pp. 1279–1285, 1982.

[9] W. N. Brouw, "Data processing for the Westerbork synthesis radio telescope," Ph.D. dissertation, Leiden University, Leiden, The Netherlands, 1971.

[10] A. H. Rots, "Distribution and kinematics of neutral hydrogen in the spiral galaxy M81," Ph.D. dissertation, Leiden University, Leiden, The Netherlands, 1974.

[11] B. G. Clark, "Large field mapping," in *Proc. NRAO Workshop No. 5, Synthesis Mapping* (NRAO, Socorro, NM) pp. 10.1–10.18, June 1982.

[12] R. N. Bracewell, "Computer image processing," *Ann. Rev. Astron. Astrophys.*, vol. 17, pp. 113–134, 1979.

[13] R. B. Read, "Two-element interferometer for accurate position determination at 960 MHz," *IRE. Trans. Antennas Propagat.*, vol. AP-9, pp. 31–35, 1961.

[14] J. A. Högbom and W. N. Brouw, "The synthesis radio telescope at Westerbork. Principles of operation, performance, and data reduction," *Astron. Astrophys.*, vol. 33, pp. 289–301, 1974.

[15] A. T. Moffett, "Minimum-redundancy linear arrays," *IEEE Trans. Antennas Propagat.*, vol. AP-16, pp. 172–175, 1968.

[16] M. Ishiguro, "Minimum redundancy linear arrays for a large number of antennas," *Radio Sci.*, vol. 15, pp. 1163–1170, 1980.

[17] A. R. Thompson and R. N. Bracewell, "Interpolation and Fourier transformation of fringe visibilities," *Astron. J.*, vol. 79, pp. 11–24, 1974.

[18] R. B. Leighton, "A 10 m telescope for millimeter and submillimeter astronomy," California Institute of Technology, Tech. Rep. for NSF

Grant AST 73-04908, May, 1978.

[19] K. V. Sheridan, N. R. Labrum, and W. J. Payten, "Three-frequency operation of the Culgoora radioheliograph," *Proc. IEEE*, vol. 61, pp. 1312–1317, 1973.

[20] D. H. Smith, "Australia's bicentennial bonanza," *Sky Telesc.*, vol. 65, pp. 120–121, 1983.

[21] R. S. Roger, C. H. Costain, J. D. Lacey, T. L. Landecker, and F. K. Bowers, "A supersynthesis radio telescope for neutral hydrogen spectroscopy at the Dominion Radio Astrophysical Observatory," *Proc. IEEE*, vol. 61, pp. 1270–1275, 1973.

[22] K. S. Dwarakanath, R. K. Shevgaonkar, and Ch. V. Sastry, "Observations of the supernova remnants HB9 and IC443 at 34.5 MHz," *J. Astrophys. Astron.*, vol. 3, pp. 207–216, 1982.

[23] Status Report of Array Design Study Group, Institut de Radio Astronomie Millimetrique, Grenoble, France, Oct. 1981.

[24] B. Elsmore, S. Kenderdine, and M. Ryle, "The operation of the Cambridge one-mile diameter radio telescope," *Mon. Not. Roy. Astron. Soc.*, vol. 134, pp. 87–95, 1966.

[25] J. E. Baldwin, J. E. Jennings, J. R. Shakeshaft, P. J. Warner, D. M. A. Wilson, and M. C. H. Wright, "Maps of the distribution of polarization in five radio sources at 21.1 cm wavelength," *Mon. Not. Roy. Astron. Soc.*, vol. 150, pp. 253–270, 1970.

[26] M. Ryle, "The 5 km radio telescope at Cambridge," *Nature*, vol. 239, pp. 435–438, Oct. 1972.

[27] J. E. Baldwin, personal communication.

[28] D. E. Hogg, G. H. MacDonald, R. G. Conway, and C. M. Wade, "Synthesis of brightness distribution in radio sources," *Astron. J.*, vol. 74, pp. 1206–1213, Dec. 1969.

[29] J. C. Coe, "NRAO interferometer electronics," *Proc. IEEE*, vol. 61, pp. 1335–1339, Sept. 1973.

[30] J. W. M. Baars, J. F. van der Brugge, J. L. Casse, J. P. Hamaker, L. H. Sondaar, J. J. Visser, and K. J. Wellington, "The synthesis radio telescope at Westerbork," *Proc. IEEE*, vol. 61, pp. 1258–1265, 1973.

[31] G. Swarup and D. S. Bagri, "An aperture-synthesis interferometer at Ooty for operation at 327 MHz," *Proc. IEEE*, vol. 61, pp. 1285–1287, 1973.

[32] M. Ishiguro, "The Nobeyama mm-wave 5-element synthesis telescope," Nobeyama Radio Observatory Tech. Rep., no. 7, Aug. 1981.

[33] A. Breccesi and M. Ceccarelli, "The Italian cross radiotelescope, I. Design of the antenna," *Nuovo Cimento*, vol. 23, pp. 208–215, 1962.

[34] R. E. Hills, M. A. Janssen, D. D. Thornton, and W. J. Welch, "The Hat Creek millimeter-wave interferometer," *Proc. IEEE*, vol. 61, pp. 1278–1282, 1973.

[35] J. G. Davies, B. Anderson, and I. Morison, "The Jodrell Bank radio-linked interferometer network," *Nature*, vol. 288, pp. 64–66, Nov. 1980.

[36] W. C. Erickson and J. R. Fisher, "A new wideband, fully steerable, decametric array at Clark Lake," *Radio Sci.*, vol. 9, pp. 387–401, 1974.

[37] W. C. Erickson, M. J. Mahoney, and K. Erb, "The Clark Lake teepee-tee telescope," *Astrophys. J. Suppl.*, vol. 50, pp. 403–419, 1982.

[38] *Fleurs Synthesis Telescope* (A Special Issue of *Proc. IREE Aust.*), vol. 34, no. 8, Sept. 1973, contains papers which discuss all aspects of the instrument.

[39] R. H. Frater, R. G. Gough, and A. Watkinson, "The synthesized beamshapes for the Fleurs extension," *Proc. Astron. Soc. Aust.*, vol. 4, pp. 24–25, 1980.

[40] B. Y. Mills, "The Molonglo observatory synthesis telescope," *Proc. Astron. Soc. Aust.*, vol. 4, pp. 156–159, 1981.

[41] J. N. Douglas, F. N. Bash, F. D. Ghigo, G. F. Moseley, and G. W. Torrence, "First results from the Texas interferometer: Positions of 605 discrete sources," *Astron. J.*, vol. 78, pp. 1–17, Feb. 1973.

[42] M. H. Cohen, "Introduction to very-long-baseline interferometry," *Proc. IEEE*, vol. 61, pp. 1192–1197, 1973.

[43] R. N. Christiansen and R. F. Mullaly, "Solar observations at a wavelength of 20 cm with a crossed-grating interferometer," *Proc. IRE Aust.*, vol. 24, pp. 165–173, 1963.

[44] R. N. Bracewell and G. Swarup, "The Stanford microwave spectroheliograph antenna, a microsteradian pencil beam interferometer," *IRE Trans. Antennas Propagat.*, vol. AP-9, pp. 22–30, 1961.

[45] B. Y. Mills, "Cross-type radio telescopes," *Proc. IRE Aust.*, vol. 24, pp. 132–140, 1963.

[46] N. C. Mathur, "A pseudodynamic programming technique for the design of a correlator supersynthesis array," *Radio Sci.*, vol. 4, pp. 235–243, 1969.

[47] Y. L. Chow, "On designing a supersynthesis antenna array," *IEEE Trans. Antennas Propagat.*, vol. AP-20, pp. 30–35, 1972.

[48] *A Proposal for A Very Large Array Radio Telescope, vol. I.* Green Bank, WV: National Radio Astronomy Observatory, 1969, ch. 6.

[49] G. W. Swenson and N. C. Mathur, "The circular array in the correlator mode," *Proc. IRE Aust.*, vol. 28, pp. 370–374, 1967.

[50] J. P. Wild, "A new method of image formation with annular apertures and an application to radio astronomy," *Proc. Roy. Soc., A.*, vol. 286, pp. 499–509, 1965.

[51] W. N. Christiansen and J. A. Hogbom, *Radiotelescopes*. Cambridge, England: Cambridge Univ. Press, 1969, ch. 8, pp. 206–209.

[52] P. J. Napier and P. C. Crane, "Signal-to-noise ratios," in *Proc. NRAO Workshop No. 5, Synthesis Mapping* (NRAO, Socorro, NM), pp. 3.1–3.28, June 1982.

[53] A. R. Thompson and L. R. D'Addario, "Frequency response of a synthesis array: Performance limitations and design tolerances," *Radio Sci.*, vol. 17, pp. 357–369, 1982.

[54] R. Stevens, Jet Propulsion Laboratory, personal communication.

[55] A. W. Love, Ed., *Reflector Antennas.* New York: IEEE Press, 1978.

[56] G. James, "Primary-focus operation of shaped dual-reflector antennas," *IEEE Trans. Antennas Propagat.*, vol. AP-31, pp. 537–538, 1983.

[57] B. Y. Mills and A. G. Little, "Future plans for the Molonglo radio telescope," *Proc. Astron. Soc. Aust.*, vol. 2, pp. 134–135, 1972.

[58] S. Weinreb, "Low-noise cooled GASFET amplifiers," *IEEE Trans. Microwave Theory Tech.*, vol. MTT-28, pp. 1041–1054, Oct. 1980.

[59] G. Tomassetti, S. Weinreb, and K. Wellington, "Low-noise 10.7 GHz cooled GaAsFet amplifier," *Electron. Lett.*, vol. 17, pp. 949–950, 1981.

[60] S. Weinreb, D. Fenstermacher, and R. Harris, "Ultra low-noise, 1.2–1.7 GHz cooled GaAsFET amplifiers," *IEEE Trans. Microwave Theory Tech.*, vol. MTT-30, pp. 849–853, 1982.

[61] J. J. Gustincic and P. J. Napier, "A hybrid lens feed for the VLA," in *Dig. IEEE PGAP Int. Symp.* (Stanford, CA), pp. 361–364, June 1977.

[62] P. J. Napier and J. J. Gustincic, "Polarization properties of a Cassegrain antenna with off-axis feeds and on-axis beam," in *Dig. IEEE PGAP Int. Symp.* (Stanford, CA), pp. 452–454, June 1977.

[63] T. S. Chu and R. H. Turrin, "Depolarization properties of offset reflector antennas," *IEEE Trans. Antennas Propagat.*, vol. AP-21, pp. 339–345, 1973.

[64] S. Weinreb, M. Balister, S. Maas, and P. J. Napier, "Multiband low-noise receivers for a very large array," *IEEE Trans. Microwave Theory Tech.*, vol. MTT-25, pp. 243–247, 1977.

[65] M. P. Godwin, A. P. Anderson, and J. C. Bennett, "Optimization of feed position and improved profile mapping of a reflector antenna from microwave holographic measurements," *Electron. Lett.*, vol. 14, pp. 134–136, 1978.

[66] P. F. Scott and M. Ryle, "A rapid method for measuring the figure of a radio telescope reflector," *Mon. Not. Roy. Astron. Soc.*, vol. 178, pp. 539–545, 1977.

[67] J. L. Casse, E. E. M. Woestenberg, and J. J. Visser, "Multifrequency cryogenically cooled front-end receivers for the Westerbork synthesis radio telescope," *IEEE Trans. Microwave Theory Tech.*, vol. MTT-30, pp. 201–209, 1982.

[68] G. Peter, "Low noise GaAsFet dual channel front end," *Microwave J.*, vol. 25, pp. 153–156, 1982.

[69] A. G. Cha, "Wide-band diffraction improved dual-shaped reflectors," *IEEE Trans. Antennas Propagation*, vol. AP-30, pp. 173–176, 1982.

[70] V. Pankonin and R. M. Price, "Radio astronomy and spectrum management: The impact of WARC-79," *IEEE Trans. Electromag. Compat.*, vol. EMC-23, pp. 308–317, 1981.

[71] A. R. Thompson, "The response of a radio astronomy synthesis array to interfering signals," *IEEE Trans. Antennas Propagat.*, vol. AP-30, pp. 450–456, 1982.

[72] R. Hinder and M. Ryle, "Atmospheric limitations to the angular resolution of aperture synthesis radio telescopes," *Mon. Not. Roy. Astron. Soc.*, vol. 154, pp. 229–253, 1971.

[73] R. S. Warwick, R. J. Davis, and R. E. Spencer, "Phase stability, angular structure and position measurements with a radio-link interferometer," *Mon. Not. Roy. Astron. Soc.*, vol. 177, pp. 335–347, 1976.

[74] A. E. Rogers and J. M. Moran, "Coherence limits for very long baseline interferometry," *IEEE Trans. Instrum. Meas.*, vol. IM-30, pp. 283–286, 1981.

[75] J. W. Armstrong and R. A. Sramek, "Observations of tropospheric phase scintillations at 5 GHz on vertical paths," *Radio Sci.*, vol. 17, pp. 1579–1586, 1982.

[76] P. J. Napier and R. H. T. Bates, "Identification and removal of phase errors in interferometry," *Mon. Not. Roy. Astron. Soc.*, vol. 158, pp. 405–424, 1972.

[77] J. Ruze, "Antenna tolerance theory—A review," *Proc. IEEE*, vol. 54, pp. 633–640, 1966.

[78] J. P. Hamaker, "Atmospheric delay fluctuations with scale sizes greater than one kilometer, observed with a radio interferometer array," *Radio Sci.*, vol. 13, pp. 873–891, 1978.

[79] T. Hagfors, "The ionosphere," in *Methods of Experimental Physics*, vol. 12. New York: Academic Press, 1976 pt. B, ch. 2.1, pp. 119–134.

[80] P. J. Hargrave and J. L. Shaw, "Large scale tropospheric irregularities and their effect on radio astronomical seeing," *Mon. Not. Roy. Astron. Soc.*, vol. 182, pp. 233–239, 1978.

[81] J. P. Basart, G. K. Miley, and B. G. Clark, "Phase measurement with an interferometer baseline of 11.2 km," *IEEE Trans. Antennas Propagat.*, vol. AP-18, pp. 375–379, 1970.

[82] A. F. Dravskikh and A. M. Finkelstein, "Tropospheric limitations in phase and frequency coordinate measurements in astronomy," *Astrophys. Space Sci.*, vol. 60, pp. 251–265, 1979.

[83] D. C. Hogg, F. O. Guiraud, and M. T. Decker, "Measurement of excess radio transmission length on earth-space paths," *Astron. Astrophys.*, vol. 95, pp. 304–307, 1981.

[84] J. M. Moran and B. R. Rosen, "Estimation of the propagation delay through the troposphere from microwave radiometer data," *Radio Sci.*, vol. 16, pp. 235–244, 1981.

[85] G. M. Resch, M. C. Chavez, and N. I. Yamane, "Description and overview of an instrument designed to measure line-of-sight delay due to water vapour," Jet Propulsion Lab., TDA Progress Rep. 42–72, pp. 1–19, Oct. 1982.

[86] G. M. Resch, D. E. Hogg, and P. J. Napier, "Radiometric correction of atmospheric path length fluctuations in interferometric experiments,"

Radio Sci., in publication.

[87] G. Swarup and K. S. Yang, "Phase adjustments of large antennas," *IRE Trans. Antennas Propagat.*, vol. AP-9, pp. 75–81, 1961.

[88] A. G. Little, R. W. Hunstead, and G. G. Calhoun, "A constant phase local oscillator system for a cross type radio telescope," *IEEE Trans. Antennas Propagat.*, vol. AP-14, pp. 645–646, 1966.

[89] R. N. Bracewell, R. S. Colvin, L. R. D'Addario, C. J. Grebenkemper, K. M. Price, and A. R. Thompson, "The Stanford five-element radio telescope," *Proc. IEEE*, vol. 61, pp. 1249–1257, 1973.

[90] T. H. Legg, "Microwave phase comparators for large antenna arrays," *IEEE Trans. Antennas Propagat.*, vol. AP-13, pp. 428–432, 1965.

[91] J. Delanroy, J. Lacroix, and E. J. Blum, "An 8-mm interferometer for solar radio astronomy at Bordeaux, France," *Proc. IEEE*, vol. 61, pp. 1282–1284, 1973.

[92] T. L. Landecker and J. F. Vaneldik, "A phase-stabilized local oscillator system for a synthesis telescope," *IEEE Trans. Instrum. Meas.*, vol. IM-31, pp. 185–192, 1982.

[93] NRAO, "A proposal for the Very Large Array radio telescope," NRAO, Socorro, NM, vol. II, p. 14.3, Jan. 1967.

[94] W. N. Christiansen, "A new southern hemisphere synthesis radio telescope," *Proc. IEEE*, vol. 61, pp. 1266–1270, 1973.

[95] M. C. Thompson, L. E. Wood, D. Smith, and W. B. Grant, "Phase stabilization of widely separated oscillators," *IEEE Trans. Antennas Propagat.*, vol. AP-16, pp. 683–688, 1968.

[96] M. Morimoto, "New calibration method for large aerial arrays," *Electron. Lett.*, vol. 1, pp. 192–193, 1965.

[97] G. Lutes, "Development of optical fiber frequency and time distribution systems," in *Proc. 13th Annu. Precise Time and Time Interval Application and Planning Meet.*, pp. 243–262, Dec. 1981.

[98] T. E. Abele, D. A. Alsberg, and P. T. Hutchinson, "A high-capacity digital communication system using TE_{01} transmission in circular waveguide," *IEEE Trans. Microwave Theory Tech.*, vol. MTT-23, pp. 326–333, 1975.

[99] J. W. Archer, M. Ogai, and E. Caloccia, "The sector coupler—Theory and performance," *IEEE Trans. Microwave Theory Tech.*, vol. MTT-29, pp. 202–208, 1981.

[100] J. W. Archer, E. M. Caloccia, and R. Serna, "An evaluation of the performance of the VLA circular waveguide system," *IEEE Trans. Microwave Theory Tech.*, vol. MTT-28, pp. 786–791, 1980.

[101] S. Weinreb, R. Predmore, M. Ogai, and A. Parrish, "Waveguide system for a very large antenna array," *Microwave J.*, vol. 20, pp. 49–52, 1977.

[102] J. W. Archer, "TE_{ON} mode filters for the VLA circular waveguide system," *Electron. Lett.*, vol. 15, pp. 343–345, 1979.

[103] W. E. Dumke, "Module T1 Modem," NRAO, VLA Tech. Rep. 28, Oct. 1976.

[104] M. Ryle, "A new radio interferometer and its application to the observation of weak radio stars," *Proc. Roy. Soc.*, vol. 211A, pp. 351–375, 1952.

[105] J. Granlund, A. R. Thompson, and B. G. Clark, "An application of Walsh functions in radio astronomy instrumentation," *IEEE Trans. Electromag. Compat.*, vol. EMC-20, pp. 451–453, 1978.

[106] R. M. Goldstein, "A technique for the measurement of the power spectra of very weak signals," *IRE Trans. Space Electron. Telem.*, vol. SET-8, pp. 170–173, 1962.

[107] S. Weinreb, "A digital spectral analysis technique and its application to radio astronomy," Research Lab of Electronics, MIT, Cambridge, MA, Tech. Rep. 412, 1963.

[108] J. H. Van Vleck and D. Middleton, "The spectrum of clipped noise," *Proc. IEEE*, vol. 54, pp. 2–19, 1966.

[109] W. R. Burns and S. S. Yao, "Clipping loss in the one-bit autocorrelation spectral line receiver," *Radio Sci.*, vol. 4, pp. 431–436, 1969.

[110] B. F. C. Cooper, "Correlators with two-bit quantization," *Aust. J. Phys.*, vol. 23, pp. 521–527, 1970.

[111] J. B. Hagen and D. T. Farley, "Digital-correlation techniques in radio science," *Radio Sci.*, vol. 8, pp. 775–784, 1973.

[112] F. K. Bowers and R. J. Klinger, "Quantization noise of correlation spectrometers," *Astron. Astrophys. Suppl.*, vol. 15, pp. 373–380, 1974.

[113] L. R. D'Addario, A. R. Thompson, F. R. Schwab, and J. Granlund, "Complex cross-correlators with three-level quantization: Design tolerances," submitted to *Radio Sci.*

[114] "Giant radio telescope uses 19,500 chips for complex signal processing at 100 MHz," *Electronics*, *vol.* 51, *no.* 21, *pp.* 44–46, *Oct.* 1978.

[115] J. A. Ball, "The Harvard minicorrelator," *IEEE Trans. Instrum. Meas.*, vol. IM-22, pp. 193–196, 1973.

[116] R. P. Escoffier and C. M. Broadwell, "The correlator system for the Very Large Array radio telescope," *Radio Sci.*, in publication, 1983.

[117] J. H. Taylor, "Pulsar receivers and data processing," *Proc. IEEE*, vol. 61, pp. 1295–1298, 1973.

[118] Y. Chicada, "Techniques for spectral measurements," Nobeyama Radio Observatory Tech. Rep. 8, Tokyo Astronomical Observatory, Mitaka, Tokyo, Japan, 1981.

[119] D. W. Weber, "An overview of the monitor and control system," NRAO, Socorro, NM, VLA Tech. Rep 44, Mar. 1980.

[120] K. W. Weiler, "Methods of polarization measurement," *Astron. Astrophys.*, vol. 26, pp. 403–407, 1973.

[121] R. C. Bignell, "Polarimetry," in *Proc. NRAO Workshop No. 5, Synthesis Mapping* (NRAO, Socorro, NM), pp. 6.1–6.28, June 1982.

[122] R. C. Jennison, "A phase sensitive interferometer technique for the measurement of the Fourier transforms of spatial brightness distributions of small angular extent," *Mon. Not. Roy. Astron. Soc.*, vol. 118, pp. 276–284, 1958.

[123] D. H. Rogstad, "A technique for measuring visibility phase with an optical interferometer in the presence of astmospheric seeing," *Appl. Opt.*, vol. 7, pp. 585–588, 1968.

[124] W. T. Rhodes and J. W. Goodman, "Interferometric technique for recording and restoring images degraded by unknown aberrations," *J. Opt. Soc. Amer.*, vol. 63, pp. 647–657, 1973.

[125] J. P. Hamaker, J. D. O'Sullivan, and J. E. Noordam, "Image sharpness, Fourier optics and redundant-spacing interferometry," *J. Opt. Soc. Amer.*, vol. 67, pp. 1122–1123, 1977.

[126] T. J. Cornwell, "Adaptive optics in radio astronomy," in *Proc. OSA/AAS Meet.*, St. Paul, MN, June 1983.

[127] A. E. E. Rogers *et al.*, "The structure of radio sources 3C273B and 3C84 deduced from the 'closure' phases and visibility amplitudes observed with three-element interferometers," *Astrophys. J.*, vol. 193, pp. 293–301, 1974.

[128] A. C. S. Readhead and P. N. Wilkinson, "The mapping of compact radio sources from VLBI data," *Astrophys. J.*, vol. 223, pp. 25–36, 1978.

[129] W. D. Cotton, "A method of mapping compact structure in radio sources using VLBI observations," *Astron. J.*, vol. 84, pp. 1122–1128, 1979.

[130] J. E. Baldwin and P. J. Warner, "Phaseless aperture synthesis," *Mon. Not. Roy. Astron. Soc.*, vol. 182, pp. 411–422, 1978.

[131] T. J. Cornwell and P. N. Wilkinson, "A new method for making maps with unstable radio interferometers," *Mon. Not. Roy. Astron. Soc.*, vol. 196, pp. 1067–1086, 1981.

[132] T. J. Cornwell, "Self-calibration," in *Proc. NRAO Workshop No. 5, Synthesis Mapping* (NRAO, Socorro, NM) pp. 13.1–13.15, June 1982.

[133] F. R. Schwab, "Adaptive calibration of radio interferometer data," *Proc. SPIE*, vol. 231, pp. 18–24, 1980.

[134] J. E. Noordam and A. G. de Bruyn, "High dynamic range mapping of strong radio sources, with application to 3C84," *Nature*, vol. 299, pp. 597–600, 1982.

[135] W. N. Brouw, "Aperture synthesis," *Methods Comput. Phys.*, vol. 14, pp. 131–175, 1975.

[136] J. W. Cooley and J. W. Tukey, "An algorithm for the machine calculation of complex Fourier series," *Math. Comput.*, vol. 19, pp. 297–301, 1965.

[137] R. N. Bracewell, *The Fourier Transform and its Applications*. New York: McGraw-Hill, 1978.

[138] F. R. Schwab, "Optimum gridding," NRAO, Socorro, NM, VLA Sci. Memo. 132, 1980.

[139] E. W. Greisen, "The effects of various convolving functions on aliasing and relative signal-to-noise ratios," NRAO, Socorro, NM, VLA Sci. Memo 131, Dec. 1979.

[140] R. H. Frater and D. J. Skellern, "Direct transform hardware. Processing of rotational synthesis data—I and II," *Astron. Astrophys.*, vol. 68, pp. 391–396, and vol. 68, pp. 397–403, 1978.

[141] E. W. Greisen, "An overview of the NRAO–AIPS Image Processing System," presented to URSI meeting, Boulder, CO, Jan. 1982.

[142] D. C. Wells, E. W. Greisen, and R. H. Harten, "FITS: A flexible image transport system," *Astron. Astrophys. Suppl.*, vol. 44, pp. 363–370, 1981.

[143] J. Högbom, "Aperture synthesis with a non-regular distribution of interferometer baselines," *Astrophys. J. Suppl.*, vol. 15, pp. 417–426, 1974.

[144] U. J. Schwartz, "Mathematical-statistical description of the iterative beam removing technique (Method CLEAN)," *Astron. Astrophys.*, vol. 65, pp. 345–356, 1978.

[145] T. J. Cornwell, "Image restoration and the CLEAN technique," in *Proc. NRAO Workshop No. 5, Synthesis Mapping* (NRAO, Socorro, NM), pp. 9.1–9.14, June 1982.

[146] B. G. Clark, "An efficient implementation of the algorithm CLEAN," *Astron. Astrophys.*, vol. 89, pp. 377–378, 1980.

[147] T. J. Cornwell, "A method of stabilizing the CLEAN algorithm," *Astron. Astrophys.*, vol. 121, pp. 281–285, 1983.

[148] S. F. Gull and E. Daniell, "Image reconstruction from noisy and incomplete data," *Nature*, vol. 272, pp. 686–690, 1978.

[149] R. Nityananda and R. Narayan, "Maximum entropy image reconstruction—A practical non-information-theoretic approach," *J. Astrophys. Astron.*, vol. 3, pp. 419–450, 1982.

[150] R. J. Allen, "Exploring methods of data display in an interactive astronomical data-processing environment," in *Image Formation From Coherence Functions in Astronomy*, C. van Schooneveld, Ed. New York: Reidel, 1979, pp. 143–155.

[151] W. van Breguel and E. B. Fomalont, "Radio observations of 3C310," in preparation, 1983.

[152] B. G. Clark, "Information-processing systems in radio astronomy and astronomy," *Ann. Rev. Astron. Astrophys.*, vol. 8, pp. 115–138, 1970.

[153] R. D. Ekers, R. J. Allen, and J. R. Luyten, "Interactive processing of map data produced by the Westerbork supersynthesis radio telescope," *Astron. Astrophys.*, vol. 27, pp. 77–83, 1973.

[154] M. Buchanan, "Effective utilization of colour in multidimensional data presentations," *Proc. SPIE*, vol. 199, pp. 9–18, 1979.

[155] E. B. Fomalont, "Image display, processing and analysis," in *Proc. NRAO Workshop No. 5, Synthesis Mapping* (NRAO, Socorro, NM), pp. 11.1–11.39, 1982.

THE NOBEYAMA MILLIMETER-WAVE INTERFEROMETER.

M. Ishiguro, K.-I. Morita, T. Kasuga, T. Kansawa, H. Iwashita, Y. Chikada,
J. Inatani, H. Suzuki, K. Handa, T. Takahashi.

Nobeyama Radio Observatory. Nobeyama, Minamisaku, Nagano 384-13, Japan.

H. Tanaka.
Department of Electrical Engineering. Toyo University.
2100, Nakanodai, Kujirai, Kawagoe, Saitama 350, Japan.

H. Kobayashi, R. Kawabe.
Department of Astronomy. University of Tokyo.
7-3-1, Hongo, Bunkyo-ku, Tokyo 113, Japan.

ABSTRACT.

Antennas, receiving system and computer system of the Nobeyama millimeter-wave interferometer are briefly described. A preliminary result of mapping Cygnus A at 22 GHz is also shown. The dynamic range of the cleaned map was 50 : 1 without self-calibration.

INTRODUCTION.

The construction work of the Nobeyama millimeter-wave interferometer has almost completed. The interferometer has five 10-m diameter anntenas which have high surface accuracy(70 μm rms) sufficient for the millimeter wave operation. The antennas are equipped with low-noise receivers at 1.3 cm and 2.6 mm wavelengths which contain important spectral lines such as H_2O, NH_3 and CO. The goal of this interferometer is the high-resolution mapping of molecular sources and continuum sources at millimeter wavelengths.

The interferometer receiver is equipped with an extremely phase-stable local-oscillator system and a large digital correlator from which 1024 maps corresponding to the independent frequency channels can be produced. The maximum length of the baseline is 560 m in east-west and 520 m in north-south(tilted by 33 degrees from the true north). The location of thirty stations was determined to achive good U-V coverage at low declinations(Ishiguro 1978; see figure 1). The system parameters of the interferometer is shown in Table 1.

ANTENNAS.

The cassegrain coude optics was adopted for the 10-m antennas. Wide band radio waves from 22 GHz to 115 GHz are transmited to the ground level through the low-loss beam wave-guide system. The surface accuracies of the sub-reflector and the mirrors for the beam wave-guide are better

Fig. 1. An aerial view of the Nobeyama Radio Observatory showing the 45-m telescope and the five antennas of the interferometer on the east-west baseline.

Table 1. The system parameters of the interferometer

	22GHz	115GHz
Field of view	5.8'	1.1'
Maximum resolution for natural weighting	4.1"	0.8"
System temperature	100 K	700 K
Continuum correlator Maximum bandwidth	250 MHz	250 MHz
Spectral correlator Maximum bandwidth	320 MHz (±2000 Km/s)	320 MHz (±400 Km/s)
Number of channel	1024 ch.	1024 ch.
Number of Correlation	10 complex correlations and 5 auto correlations	
Sensitivity ΔS$_{rms}$	0.16 mJy (250 MHz, 12H X 6 configurations)	1.4 mJy
ΔT$_{brms}$	0.75 K (320 KHz, 12H X 6 configurations)	6.3 K

than 30 μm rms. Honeycomb panels reinforced with carbon fiber are used for these mirrors. Cross polarization generated by the second curved mirror is compensated by the third curved mirror, and is measured less than 0.1%.

The reflector surface of the main dish is composed of 36 aluminium honeycomb panels, the surface accuracies of which are 60 μm on an average(Ishiguro 1981). The surface error of the main dishes were measured with a resolution of 10 μm by a template which is eqipped with fourteen magnetic sensors. The template was calibrated using the laser interferometer before measuring the main dishes, and the repeatability of the measurements was about 30 μm rms. Figure 2 shows the example of the surface error of a 10-m antenna. The surface error measured by the template was 66 μm rms for this antenna. Average surface accuracy for five antennas is 71 μm rms.

We made radiometric measurement of Jupiter to estimate the aperture efficiency and the beam efficiency at 115 GHz. We obtained 50% for the aperture efficiency and 66% for the beam efficiency at the elevation angle of 30 degrees. The measured aperture efficiency agrees well with the surface error measured by the template, when spill over, illumination and the gravitational deformation are included.

The entire antenna structure including supporting legs are covered with heat insulating plates. The inside air is ventilated with electric fans to equalize the temperature gradient of the antenna structure. Anti back-lash precision gear drive and high resolution(20 bits) encoders attached directly to AZ and EL axes guarantee a pointing accuracy of better than 10" rms for the wind speed less than 7 m/s.

We have determined the pointing characteristics of the 10-m antennas using strong H_2O maser sources at 22 GHz. With a high signal to noise ratio of each measurement, the measured pointing residual after best fit is 7" rms and is sufficiently accurate for the observation at 22 GHz. The pointing accuracy at 115 GHz may be a problem ; the daily change of the tilt of AZ axis is about 10 arc sec p-p. To correct for this change, we are considering to use the data from X-Y inclinometer for the real-time pointing correction.

RECEIVING SYSTEM.

Figure 3 shows the overall block diagram of receiving system. Inside the 10-m antenna, dual-frequency frontend(22 GHz / 115 GHz), phase locked local oscillaters and down convertors are located. For 22 GHz band(21.8 - 23.8 GHz), a two-stage cooled parametric amplifier of 100 K is used. For 115 GHz band(105 - 117 GHz), a cooled mixer of 700 K is used . So, the effective use of the interferometer at 115 GHz band will be limited not only by the atmospheric phase stability but also by the receiver sensitivity. SIS receiver which is under development at Nobeyama will take place of the present receiver in the near feature.

Local oscillators at the remote antennas are phase-locked to the common reference signal of 1.7 GHz at the control building through the phase-stabilized local-oscillator system shown in figure 4 . A round-trip phase correction system based on the early design of the VLA is used. The reference signal of about 830 MHz is sent out from the control building to the antenna through the underground cable with a frequency multiplexer. A return signal of slightly different frequency(by 30 MHz)is sent back from the antenna. From comparison of the two signals the phase of the original reference signal is adjusted so that the phase at the antenna is constant. The compensation loop is designed so as to accomodate the tuning range of at least ± 0.2%. Doppler tracking is performed through the 1.7 GHz reference signal. Fringe-stopping and phase-switching are performed through the reference signal of 25 MHz for the third local oscillator. Cable lengths are made equal at any station for the sake of symmetry.

Phase-stabilized coaxial cable was developed for the Nobeyama interferometer. The temperature coefficient of the cable is less than 1 ppm/°C. Thirty cables of equal length are installed inside the underground tunnel, the top of which is 1.5m under the ground level. The temperature variation of this depth was measured less than 0.1°c/day and greatly helps in reducing the phase variation due to the cables. The phase stability of the cable can be improved by controlling the internal gas pressure according to the change in the surrounding temperature. By careful adjustment of the gas-pressure, a temperature coefficient of 0.1 ppm/°C can be achived. The reference signal for the local oscillators(830 MHz, 860 MHz, 10 MHz,25 MHz) and IF return signal(110 - 430 MHz) are multiplexed onto a single cable. The resolution in frequency and the phase of fringe tracking is 0.001 Hz and 0.36°, respectively.

Analog delay lines of 14-bits are used for the delay system. Long cables are housed in the basement of the control building where air-conditioning provides temperature stability of 0.3°C/day. Phase jumps caused by the switching of delay lines are corrected by offsetting the phase of fringe rotator. Amplitude errors in delay lines are less than 0.2 dB and are corrected in the off-line data reduction.

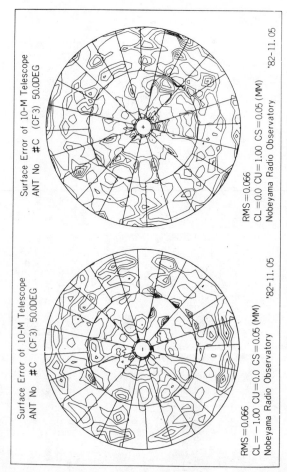

Surface Error of 10-M Telescope
ANT No #C (CF3) 50.0DEG

RMS=0.066
CL =-1.00 CU=0.0 CS=0.05 (MM)
Nobeyama Radio Observatory   '82-11.05

Surface Error of 10-M Telescope
ANT No #C (CF3) 50.0DEG

RMS=0.066
CL =0.0 CU=1.00 CS=0.05 (MM)
Nobeyama Radio Observatory   '82-11.05

Fig. 2. Surface accuracy of a 10-m antenna(EL = 50°). Right and left maps show the deviations from the best-fit paraboloid in positive and negative directions, respectively, with the contour interval of 50 μm. The rms error is 66 μm for this antenna.

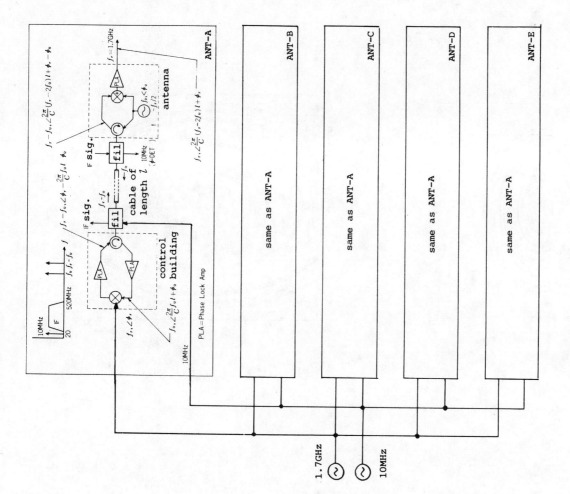

Fig. 4. Phase-stabilized local-oscillator system. Cable-lengths are equal at any stations for the sake of symmetry. Fringe-stopping and phase-switching are performed through the refernce signal of 25 MHz for the third local oscillator.

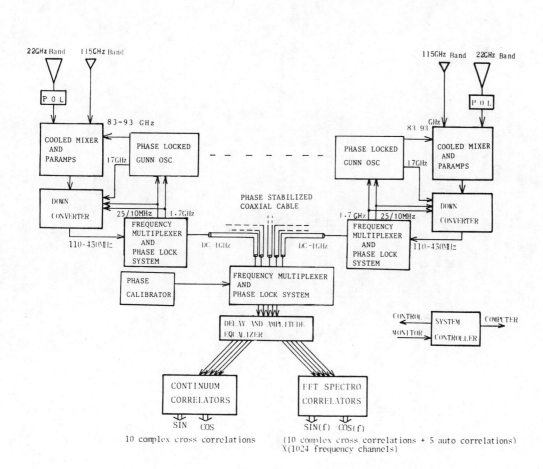

Fig. 3. Block diagram of the receiving system.

153

There are two types of correlator system. One is wide-band analog correlator, and the other is digital FFT spectro-correlator(Chikada et al.). Orthogonality between the sin and cos components from the analog correlator is measured down to the S/N -30 dB, and was 90 °± 3° on an average. Long term stability of the analog correlator is satifactory.

The digital FFT spectro-correlator processes five inputs of 320 MHz instantaneous bandwidth, which is presently limited to 80 MHz by the speed of A/D converters. It has 1024 complex frequency channels for each of the fifteen correlators. The highest resolution in frequency is 4.9 KHz. Construction of the correlator can realized by the use of CMOS LSI's having low power consumption(100 mW per chip or less). The four kinds of CMOS gate arrays are butterfly, multiplier, accumulator and corner-turner. The number of bits in A/D conversion and FFT are 3 and 6 to 8, respectively. For the FFT, a parallel pipeline scheme is implemented in which LSI's operate at 10 MHz clock rate. One serial and thirty-two parallel conversion reduces the clock rate to 10 MHz. This year we have initiated a project of expanding the speed of A/D converters up to 320 MHz to match the speed of correlators. Initial test observation of H_2O maser source W49 with this spectro-correlator was successfully conducted in August 1983.

COMPUTER SYSTEM.

FACOM M-180II AD which has 4 MB of memory and three small computers are used for the on-line computer. The interface between the small computers and the various instrumentation is standardized to IEEE-488 and RS232C. Programs for antenna tracking, fringe rotators, delay lines and synthesizer settings, and programs for data aquisition for the correlator backends have been completed. Data reduction is done by a FACOM M-200(8 MB) in the main building. Adapting of AIPS to the FACOM M-200 computer is under way.

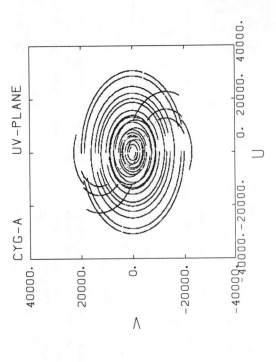

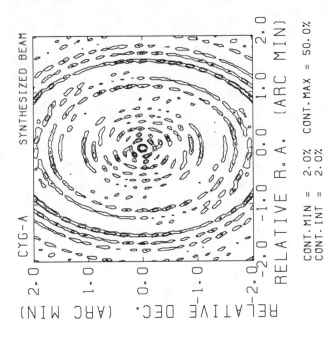

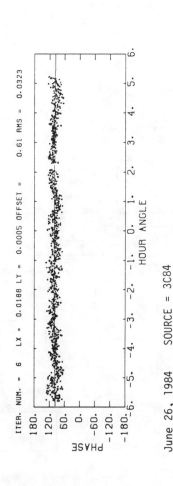

June 26, 1984   SOURCE = 3C84

Fig. 5. A typical phase stability of the interferometer at the frequency of 22 GHz in summer season for the 100 m baseline. Each data point is obtained by averaging the results from an analog correlator over 10 seconds.

Fig. 6. UV-plane and synthesized beam for the observation of Cygnus A.

PRELIMINARY RESULTS.

Overall phase stability of the interferometer system was confirmed by tracking point sources. Figure 5 shows the typical phase stability of the interferometer at 22 GHz in summer season.

Aperture synthesis mapping of Cygnus A has been performed during the period July - August , 1984. We used three configurations to obtain 23 east-west spacings from 27 m to 447 m, with some supplementary north-south spacings. The center frequency was 22.243 GHz and the bandwidth was 50MHz. Visibility data of Cygnus A were calibrated with respect to the nearby point source 2005+403 at intervals of 2 hours. Total observation time was 78 hours including the phase calibration. Figure 6 shows the U-V plane coverage and the synthesized beam for the observation of Cygnus A. Figure 7 shows the dirty map and the clean map of Cygnus A. If we assume the flux density of 2005+403 as 2.98 Jy(Dreher, 1984), peak brightnesses for components NW hotspot, SE hotspot and central source are approximately 7.4 ± 0.2, 10.0 ± 0.2 and 0.9 ± 0.2 Jy/beam, respectively. The central source coincides with the position at 89 GHz(Wright and Birkinshaw,1984) to ± 0.4". In spite of the summer weather, we obtained a beautiful map of dynamic range 50:1 as our first result from the Nobeyama interferometer.

REFERENCES.

Chikada, Y. et al.(1984), A digital FFT spectro-correlator for radio astronomy, INDIRECT IMAGING, ed. J.A. Roberts, 387-404, CAMBRIDGE UNIVERSITY PRESS.
Dreher, J. W.(1981), High-resolution maps of the hotspots of several class II radio galaxies, A. J., 86, 833-847.
Ishiguro, M.(1978), A design study of the array configuration for the 10-m 5-element super-synthesis telescope, Nobeyama Radio Observatory Report No.1.
Ishiguro, M.(1981), The Nobeyama mm-wave 5-element synthesis telescope, Nobeyama Radio Observatory Report No.7.
Wright, M. and Birkinshaw M.(1984), Hot spots in Cygnus A at 89 GHz, Ap. J., 281, 135-140.

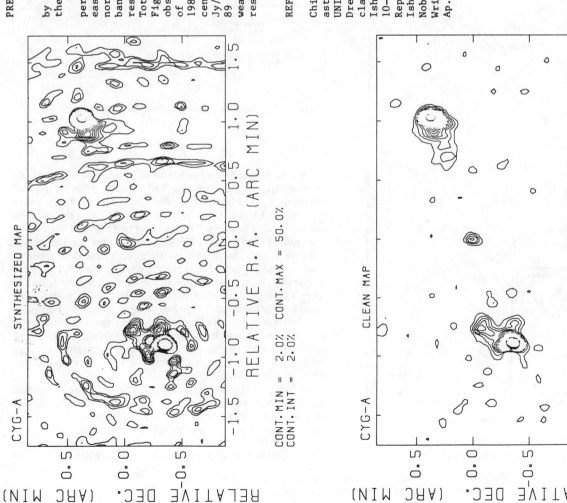

Fig. 7. Synthesized map and clean map of Cygnus A.

The Jodrell Bank radio-linked interferometer network

J. G. Davies, B. Anderson & I. Morison

University of Manchester Nuffield Radio Astronomy Laboratories, Jodrell Bank, Macclesfield, Cheshire SK11 9DL, UK

The Multi Telescope Radio Linked Interferometer (MTRLI) has just been brought into operation at Jodrell Bank and here we introduce the instrument, describe its capabilities, and present some of the first maps to be made with it. MTRLI produces high quality maps of radio sources with resolutions varying from ~1 arc s to ~0.02 arc s depending on the frequency of operation. The obvious astronomical use of the instrument is for the mapping of powerful extragalactic radio sources to investigate the physical processes within them, but it will also have the capability to map galactic line as well as galactic continuum sources. The maps presented here were made at 408 MHz and are all of extragalactic sources. They illustrate the ability of MTRLI to map at low frequencies the steep spectrum emission which tends to be overlooked with existing synthesis instruments which have to work at much higher frequencies to obtain the same resolution.

The MTRLI which has been under construction since 1975 and is an extension of the long baseline interferometers first operated at Jodrell Bank[1] in 1954 over a baseline of 0.91 km (480 wavelengths). Subsequently the baselines were suc-cessively increased by using radio links[2] to 127 km (2 million wavelengths) by 1967 (ref. 3). The present MTRLI concept was proposed in 1974 by Palmer as a further improvement and extension of these two telescope interferometer systems. The MTRLI uses of one of the two large telescopes at Jodrell Bank, (the MK IA 76-m telescope or the MK II 25×37-m telescope) and the existing outstations at Defford and Wardle, together with three new 25-m telescopes sited at Knockin Darnhall and Tabley which are of the E-Systems design as used in the VLA. The first phase of the instrument using four telescopes is now working; the second phase in which the Darnhall and Tabley telescopes are added to the array will be complete by the end of 1980. When all telescopes are in use, data can be obtained from 15 baselines simultaneously ranging from 133 to 6 km in length. The system will be operated at a variety of wavelengths ranging from 73 to 1.35 cm, although at the higher frequencies not all telescopes are usable. Figure 1 indicates the position of the telescopes and the orientation of the baselines. The three new telescopes were sited to provide reasonably uniform coverage of the spatial frequency components of the sky brightness dis-tribution. This is shown in Fig. 2 for declinations 10° and 50° north for the 15 baselines of the array. Because of the large N–S component in many of the baselines the resolution in the N–S direction remains comparable to that in the E–W direction even at low declinations.

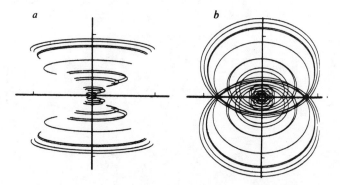

Fig. 2 The projected spacings of the MTRLI baselines for sources of declination: a, 10°; b, 50°. One division represents 100 km.

Each outstation telescope has an on-line computer to provide local control of the telescope and receiver or allow remote operation by land line from Jodrell Bank. Standard microwave links with a bandwidth of 10 MHz carry the radio signals from the outstations to Jodrell Bank, using two repeater stations in the case of Defford and Knockin, and direct for the nearer telescopes. Local oscillators at all sites are phase locked using a single frequency near 1,500 MHz. Pulses at this frequency are transmitted from Jodrell Bank to the outstations over the same paths as are used for the microwave links, locking high quality crystal oscillators at each repeater site and telescope. Pulses are then returned over the same paths and the received phase compared with that transmitted. In this way changes in the electrical path lengths are continuously measured with an accuracy of about 1 mm in 150 km, and corrections made to the phase of the received signals.

At Jodrell Bank the signals from each telescope are passed through phase rotators to remove the effects of the Earth's rotation, an analogue delay continuously variable between 0 and 75 ns and a digital delay of up to 1 ms in 50-ns steps to compensate for the varying geometry of the telescope baselines. The signals are then taken in pairs and correlated in a set of

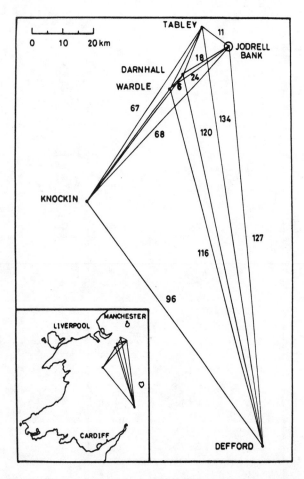

Fig. 1 The relative positions of the MTRLI telescopes with the baseline lengths given in kilometres.

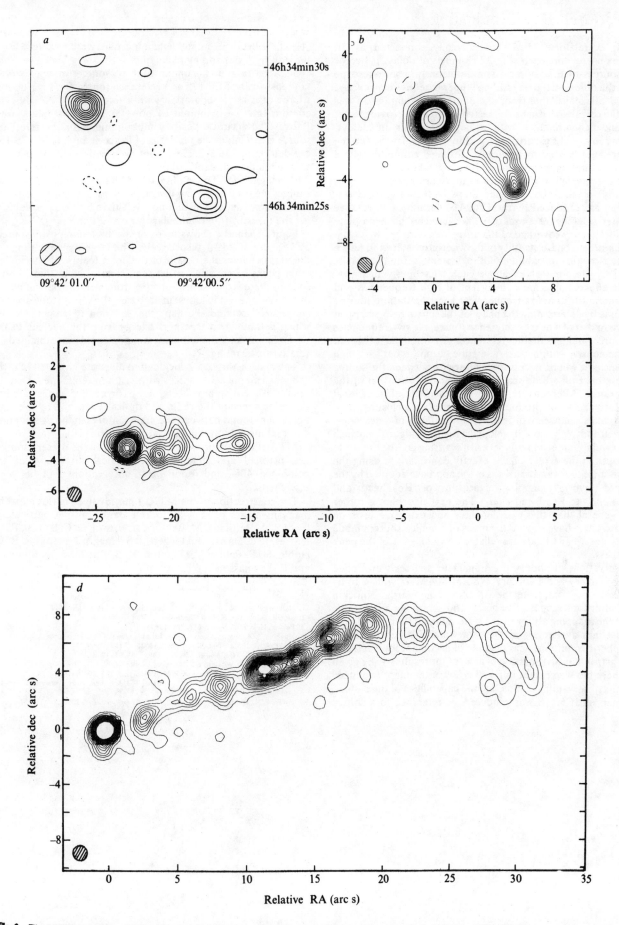

Fig. 3 The MTRLI maps. *a*, 5C5.168 at 408 MHz, the contour interval is 2 mJy per beam area. *b*, 3C153 at 408 MHz; the contour levels are 1, 3, 5 ... 33, 35, 50, 75% of peak brightness; *c*, 3C249.1 at 408 MHz; the contour levels are 2, 4, 6, 8 ... 28, 30, 40, 50, 60, 70, 80% of peak brightness. *d*, Virgo A at 408 MHz; the contour levels are 2, 4, 6, 8 ... 38, 40% of peak brightness. The half-power beamwidth of the circular gaussian restoring beam is shown in each case in the lower left-hand corner.

one-bit correlators. These are normally connected as 16 complex correlators each with 32 channels of delay. If two or more sources are to be observed simultaneously in one telescope beam, the differential delay for the longer baselines may exceed the 32×50 ns available from one of these correlators. In this event the correlators can be connected together to provide fewer independent correlators with a larger number of delays. Integration over 1 s is performed within the correlators, and the data are then passed to a purpose designed computer where further integration, processing and storage takes place.

Sources are typically observed at all hour angles where the elevation remains above 5° and maps of the brightness distribution are produced by the CLEAN[4] technique to remove unwanted sidelobe responses of ~5% caused by the incomplete spatial frequency coverage of the array.

The inherent phase stability of the interferometers and hence the map making capability of the system is fundamentally limited by the atmosphere above each telescope, with tropospheric effects being most important at high frequencies and ionospheric effects at low frequencies. These limitations may be overcome by determining the phase of the source being mapped relative to that of a nearby reference source. At low frequencies the source and a suitable reference can often be observed simultaneously within the telescope beam[4], but at high frequencies it will be necessary to alternate between the source and reference on a time scale short compared with that of the fluctuations. Alternatively, with the sacrifice of positional information, one can produce maps which use the closure phase relationships[6] between different combinations of telescopes. Cornwell and Wilkinson[7] developed a mapping algorithm which corrects amplitude and phase errors arising at the separate telescopes in the array. This is exactly equivalent to using the closure amplitudes and phases as in the approach adopted by the standard 'hybrid' mapping algorithms of Readhead and Wilkinson[8] and Readhead et al.[9]. The new algorithm is rather more general than these earlier ones and extensive tests on simulated data have shown that maps of complex sources made from typical MTRLI data are reliable down to ~1% of the peak brightness.

Observations of about 30 extragalactic sources were made during January and February 1980 at 408 MHz with a four-telescope array comprising the MK IA and the Wardle, Knockin and Defford telescopes. The observations provide an extreme test of the mapping ability of the system in that there were only six baselines available and the data were obtained in severe ionospheric conditions near the time of maximum solar activity. A full astronomical discussion of most of the results will be given elsewhere but we present in Fig. 3 a few of the maps produced from the observations to illustrate the capabilities of the system. The map of 5C5.168 was produced by reference to a source

elsewhere in the beam, the remaining maps were produced using the new hybrid mapping technique.

5C5.168 (Fig. 3a) is one of four 5C sources mapped simultaneously using 5C5.175 as a reference source. This is a double source of 6 arc s separation with a total flux of 70 mJy. The contour interval is limited by noise at 2 mJy per beam area. Because an accurate position measurement for the reference source is available, the positions of the components of 5C5.168 are determined to ~0.2 arc s, but even with such accuracy the source remains unidentified on the Palomar Sky Survey down to 20th magnitude. The ability of MTRLI to map such weak sources means that we can investigate how the morphologies of radio sources depend on luminosity. This in turn is a prerequisite for the investigation of angular size evolution.

Figure 3b, and c shows maps of two high luminosity sources 3C153 and 3C249.1 respectively. The large dynamic range of the maps enables the details of the point features between the main peaks of emission to be seen clearly. Note that in both sources there is emission stretching from near the optical object out to just one of the lobes; in each case the second lobe appears completely isolated. Perhaps the emission is from the beam which is thought to transport energy from the nucleus to the outer lobes, and then the asymmetry could be explained by relativistic beaming.

The map of Virgo A is included to illustrate how well MTRLI can map complicated sources with just four telescopes. That the details of the map are substantially correct is demonstrated by the coincidence of the radio and optical knots and by the good agreement found between the main features in the present map and that made with the VLA at 5 GHz (ref. 10).

MTRLI observations at 1,666 MHz of all the present sources are planned and detailed astrophysical discussion of the combined 408- and 1,666-MHz observations will be given elsewhere.

The construction of the MTRLI has involved major effort by many members of staff at Jodrell Bank, but we particularly thank J. A. Battilana, M. Bentley, A. Brown, D. C. Brown, R. D. Davies, R. J. Davis, M. Doggett, J. S. Haggis, J. Hopkins, R. G. Noble, H. P. Palmer, L. Pointon, R. S. Pritchard, D. Stannard and P. Thomasson.

Received 14 July; accepted 18 September 1980.

1. Hanbury Brown, R., Palmer, H. P. & Thompson, A. R. Phil. Mag. 46, 857 (1955).
2. Morris, D., Palmer, H. P. & Thompson, A. R. The Observatory 77, 103, (1957).
3. Palmer, H. P. et al. Nature 213, 789 (1967).
4. Hogbom, J. A. Astr. Astrophys. Suppl. 15, 417–426 (1974).
5. Peckham, R. J. Mon. Not. R. astr. Soc. 165, 25–38 (1973).
6. Jennison, R. C. Mon. Not. R. astr. Soc. 188, 276–284 (1958).
7. Cornwell, T. J. & Wilkinson, P. N. (in preparation).
8. Readhead, A. C. S. & Wilkinson, P. N. Astrophys. J. 223, 25–36 (1978).
9. Readhead, A. C. S., Walker, R. C., Pearson, T. J. & Cohen, M. H. Nature 285, 137–140 (1980).
10. Owen, F. N., Hardee, P. E. & Bignell, R. C. Astrophys. J. Lett. 239, L11–15 (1980).

A HIGH–RESOLUTION AERIAL SYSTEM OF A NEW TYPE

By B. Y. Mills* and A. G. Little*

[*Manuscript received May 6, 1953*]

Summary

A method of constructing an aerial system of high resolution but small area and low cost is described. Its application to the production of narrow pencil beams at metre wavelengths for investigations in radio astronomy is discussed. A small-scale model has been constructed to test the principle.

I Introduction

Recent studies of cosmic radio-frequency radiation have shown that its brightness distribution over the sky is complex. Sources of an angular size less than about $\frac{1}{4}°$ have been known to exist for some years and, now, extended sources of considerably greater size have been observed (Bolton 1952; Mills 1952), some of which appear to merge with the general background radiation. Interferometric methods, which were so useful in the early days of radio astronomy, have encountered serious difficulties when used for observing such a complex distribution. It therefore appears desirable to rely mainly on the use of pencil beam aerials of high resolving power for future work, and to reserve the use of interferometric methods for special applications.

A study of the available information suggests that a beam width of the order of 1° or less is desirable for such a pencil beam. For an aerial of conventional form at metre wavelengths this beam width would require a prohibitively large and costly structure so that an alternative solution has been sought.

A satisfactory solution is possible because the number of randomly distributed discrete sources which can be individually detected at metre wavelengths with a large aerial is determined by the beam width (or the resolution) rather than the gain of the aerial. This follows from the fact that in these circumstances the number of discrete sources with intensities above the detectable threshold will normally greatly exceed the number which may be separately resolved. Advantage can be taken of this to construct an aerial system of high resolution but relatively low gain, that is, small effective area, which sacrifices very little of the usefulness of a conventional aerial but which can be made at a fraction of the cost.

Such a system can be constructed from two long aerials arranged in the form of a cross. At the wavelengths we are considering, these aerials preferably consist of arrays of dipoles. In Figure 1 this is shown schematically, together with an idealized diagram of the outline of the aerial beam produced when the

* Division of Radiophysics, C.S.I.R.O., University Grounds, Sydney.

arrays comprising the two arms of the cross are connected together in the same phase. An aerial diagram of this form is unsatisfactory for, although there is an enhanced response in the central region where a source is received by both arrays, the total solid angle over which reception occurs is very large. However, advantage may be taken of the presence of signals in both aerials from the central region and their phase coherence for, if the arrays are now connected in antiphase, there will be no response from this central region, while the " spokes " of the diagram will be unaffected. If, therefore, the connections between the arms of the cross are switched rapidly between the two conditions, a source which is in the solid angle common to both beams will deliver a modulated signal at the switching frequency, while the signal from a source which is received by one aerial alone has no modulation imposed. After amplification and detection the modulated signal may be picked out by a phase-sensitive detector and used to

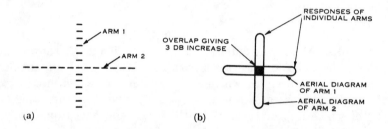

Fig. 1 (a).—Plan view of dipoles in cross arrangement.

Fig. 1 (b).—Idealized response of the cross arrangement, plan view.

deflect a pen recorder. The recorder then gives the integrated signal from within the central region so that, in effect, a pencil beam is produced which has a size determined by the *maximum* dimension of each array.

A similar method of using the common portion of overlapping beams to produce a narrower effective beam has already been used by one of the authors in an interferometer (Mills 1952).

II. Elementary Theory

In order to appreciate the possibilities of an aerial of this type, it is necessary to examine its operation in more detail. Consider the response of the system to a signal of unit intensity originating in any arbitrary direction and, as a first approximation, assume that there is no interaction between the two arrays. We shall also assume that for each array the response in any direction may be fully represented by the voltage polar diagram, that is, the phase of the response is everywhere either equal or opposite to that in the direction normal to the array. This is true for aerials in which the current distribution is symmetrical about the mid point.

When the arrays are connected together in phase addition the received power is given by $P_1 = k(S_1 + S_2)^2$, where S_1 and S_2 are the amplitudes of the voltage polar diagrams of each aerial in the selected direction.

c

When the arrays are connected in phase opposition the received power is given by $P_0 = k(S_1 - S_2)^2$.

The modulated power delivered to the receiver is given by the difference of these quantities, that is,

$$P_m = 4kS_1S_2.$$

The composite power polar diagram is therefore given by the product of the voltage polar diagrams of each array. One consequence is that the recorder deflection will be negative where the voltage diagrams are of opposite sign. A further consequence is that trouble from side lobes will be accentuated, since in the planes in which one of the arrays has a wide angle of reception, the side lobes of the composite beam will be those appropriate to the voltage rather than the power diagram. For example, if the current distribution is uniform along each array, the first side lobes in the planes of the arrays are 20 per cent in amplitude instead of the normal 4 per cent. As any complexity in the composite diagram leads to great difficulty in the interpretation of the records, it is therefore imperative to minimize the side lobes.

A well-known method of eliminating side lobes in an aerial is to employ a current distribution across the aperture in the form of a normal error curve which produces a polar diagram of the same shape. Such a distribution has another advantage in the present case, for, consider two idealized polar diagrams

$$S_1 = e^{-k_1\theta^2}, \qquad S_2 = e^{-k_2\varphi^2},$$

where θ and φ are the two zenith angles in the planes of the arms, then the composite diagram is

$$S_c{}^2 = S_1S_2 = e^{-(k_1\theta^2 + k_2\varphi^2)}.$$

Thus a beam is produced without side lobes and of simple elliptical section everywhere. When $k_1 = k_2$ the beam section will be circular, which is the simplest possible case to deal with.

In practice a beam of exactly this shape cannot be produced because an infinite aperture is required. If, however, the aperture is finite and extended until the current has fallen to about 10 per cent., it is found that a sufficiently close approximation to an error curve is obtained. To obtain the same beam width this requires an aperture some 50 per cent. larger than that for a uniform distribution, but the advantages are obvious.

The above analysis of the operation is approximate only, for in general there will be an interaction between two arrays placed so close together. The effect of such an interaction may be obtained from a thermodynamical argument. Consider such an aerial in a constant temperature enclosure. No modulated output will then be produced, as the power received by the arrays will be a function of the enclosure temperature only and will be independent of any method of connecting them together. If the composite diagram formed from the cross product of the two voltage diagrams were to have an average value of zero this result would be given by the previous analysis. In general, however, the average value will not be zero as can be seen, for example, by integrating the idealized composite diagram above. The neglected interaction between the arrays must then be just sufficient to produce zero output. This point is considered in a practical example later.

From the thermodynamical argument it can be seen that the aerial, as described, measures only *differences* in brightness temperature between the complete reception angle of both arrays and the central solid angle common to both. However, a little consideration will show that the absolute value of the temperature over the central angle may be obtained by adding the temperature calculated from the amplitude of the modulated signal to the average temperature of the two arrays.

When the aerial is used in this way the results are similar to those which would be obtained from a conventional aerial of the same beam width with an attenuator connected between it and the receiver, the attenuation being roughly

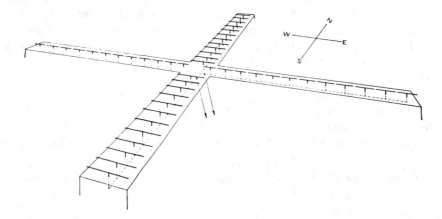

Fig. 2.—The experimental aerial system.

equal to the ratio of the area of the conventional aerial to that of the arrays forming the cross. When using a narrow pencil beam at wavelengths of a few metres for observing discrete sources, this attenuation is not very important because the resolution and not the sensitivity is the limiting factor. It does, however, lead to reduced sensitivity for observing the brightness temperature of extended distributions.

III. An Experimental Model

The construction of an aerial at metre wavelengths to produce a 1° beam, even when using the above method, is a large undertaking. It was decided, therefore, to construct first a small-scale model to test the principle and to allow experiments with possible designs. A sketch of this model is shown in Figure 2. It operates at a frequency of 97 Mc/s and the arms of the cross are 120 ft in length. They are arranged in the north-south and east-west directions. The beam width is 8°.

The east-west arm consists of a line of folded dipoles, end to end, backed by a wire-mesh reflecting screen. The dipoles are fed from a twin-wire transmission line stretching the length of the arm which is itself fed in the centre and terminated at each end by matching resistors. Quarter-wave resonant stubs are used to couple the dipoles to the feed line, the currents in the dipoles being adjusted by changing the point at which they are connected to the stub as shown in Figure 3 Standing waves on the feed line are kept low by adding capacity at appropriate places. The north-south arm consists of an array of full-wave dipoles similarly fed The beam is swung in declination by changing the phases of the currents in the dipoles of this arm. Phase changing is performed by changing the points of connection of the stubs to the feed line.

The currents are adjusted in each arm to coincide with the normal error curve which has a value at the ends of the arrays of about 10 per cent. of that at the centre. Only about one-half of the power collected by an array is fed to the receiver, the remainder being dissipated in the matching resistors at the end of each feed line. Each arm is connected to its own preamplifier, the outputs of which are combined through a phase switching arrangement as described above. Further amplification, detection, and recording are accomplished in a conventional manner (Mills 1952).

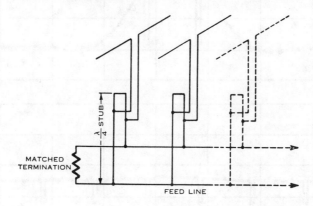

Fig. 3.—Arrangement of dipole feeds in experimental aerial.

The factors involved in observing uniform temperature distributions can be seen very clearly with this aerial. In order to reduce complications due to cross coupling between the arrays the central dipoles of each were omitted. This results in relatively large distances between the closest dipoles so that cross coupling is small. An attenuation of more than 40 db was measured between arrays. The effect of omitting the central dipole, however, is to subtract its radiation field from the total, so that the voltage polar diagrams of each array, which would normally be zero outside the central beam, are now negative in that region. Sources in the spokes of the diagram of Figure 1 therefore produce a negative deflexion. It is easily shown that the average value of the composite diagram is now zero so that, if the aerial is pointed at a uniform temperature distribution, the positive deflexion over the central beam is counterbalanced by the negative deflexion produced over the much larger solid angle of the spokes of the diagram, and the net deflexion is zero. If a different construction were utilized to eliminate this depression of the polar diagram, then, as was shown before, the cross coupling between aerials which is introduced produces a similar effect.

Sample records obtained with this experimental model are shown in Figures 4 and 5. Two quantities are recorded in each case, the output from the phase-sensitive detector in the upper graph and the receiver output level in the lower. The latter is a measure of the average temperature of the two arrays The addition of the two temperatures derived from these graphs gives the actual temperature averaged over the central beam Calibration marks are shown every 20 min when the preamplifiers are connected to cold resistors for a period of 1 min.

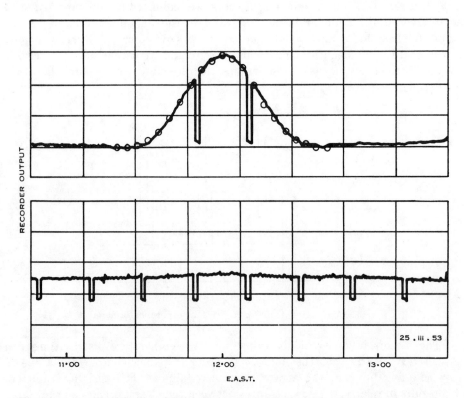

Fig. 4.—A record of the Sun showing the agreement between the observed and computed polar diagrams.

A record obtained on the Sun at a declination of $+1°$ is shown in Figure 4. The sensitivity was much reduced for observing such a strong source. Points derived from the computed polar diagram are shown superimposed on the record for comparison ; the agreement is excellent. The lower graph shows only a slight increase as the Sun passes through the aerial beams, the major part of the deflexion being due to the galactic background. The declination of $+1°$ represents an inclination of the beam of $35°$ from the vertical.

Figure 5 shows portion of a record obtained at a declination of $-34°$, that is, with the beam pointing vertically upwards. The comparatively strong " point " source 03–3 (Mills 1952) is shown first, then the weaker source 04–3. The effect of an " extended " source is illustrated when the Galaxy at about galactic longitude 220° crosses the beam. The record illustrates one advantage of this system, for the sky temperature is divided into a slowly varying background which is recorded at relatively low sensitivity on the bottom graph and a detailed structure which is recorded at high sensitivity on the upper. A conventional aerial operating at the same sensitivity would require some system of backing off the output to keep the deflexion within the range of the recorder.

A survey of the southern sky is now in progress and an analysis of this and similar records will be given in a subsequent paper. It is interesting to note, however, that the resolution of the model aerial has been sufficient to detect radiation from the large Magellan Cloud and to show that the centre of the Galaxy is narrower than previously suspected from earlier low-resolution surveys.

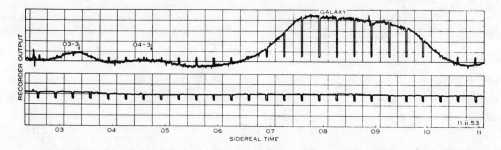

Fig. 5.—A record centred on declination —34°, showing two discrete sources and the Galaxy at longitude 220°.

The experimental work which has been performed with the small-scale model has shown that the principle of operation is sound and has demonstrated the feasibility of a full-size aerial. Work is in progress on the design and construction of an aerial which will operate at a frequency of about 80 Mc/s and have a beam width of less than 1°. It will be capable of surveying approximately one-half of the celestial sphere.

IV. ACKNOWLEDGMENTS

The authors are indebted to Dr. J. L. Pawsey for valuable discussions on the many problems encountered in this work and to Mr. K. Sheridan for his able assistance with the technical details of the aerial.

V. REFERENCES

BOLTON, J. G. (1952).—Extended sources of galactic noise. U.R.S.I. Report 1952.
MILLS, B. Y. (1952).—*Aust. J. Sci. Res.* A **5** : 266-87.

APPARENT ANGULAR SIZES OF DISCRETE RADIO SOURCES

Observations at Jodrell Bank, Manchester

THE existence of discrete sources of extra-terrestrial radio-frequency radiation is now well established[1,2] and the positions of more than one hundred sources have been published[3-5]. Attempts to identify these sources with any particular class of visual object have so far failed, and the origin of the radiation remains unexplained. One of the fundamental requirements in the study of these sources is a knowledge of their apparent angular size, and although attempts to make this measurement have been made by several observers[1-3], it has proved to be beyond the resolving power of their equipment. The present communication gives a preliminary account of a successful attempt to measure the angular size of the two most intense sources the positions[6] and intensities[4] of which are given in Table 1.

Table 1. CO-ORDINATES AND INTENSITY OF THE TWO MOST INTENSE RADIO SOURCES

Source	Right Ascension epoch 1950	Declination epoch 1950	Intensity* at 81 Mc./s. (watts/sq.m./ c.p.s.)
Cygnus	19h. 57m. 45·3s. ±1s.	N 40° 35·0′ ±1′	13·5 ×10⁻²⁶
Cassiopeia	23h. 21m. 12·0s. ±1s.	N 58° 32·1′ ±0·7′	22·0 ×10⁻²⁶

* The intensity given is twice that observed in one plane of polarization.

The co-ordinates are those given by Smith (ref. 6), and the intensities are those given by Ryle, Smith and Elsmore (ref. 4).

Previous measurements have been made with interferometers. Bolton and Stanley[1], using an interferometer mounted on a cliff and operating in a manner analogous to 'Lloyd's mirror', have shown that the apparent size of the source in Cygnus is less than 8 minutes of arc. More recently, Stanley and Slee[2], using a similar instrument, claim to have reduced this limit to 1½ minutes of arc. Ryle and Smith[3], using an instrument analogous to Michelson's stellar interferometer and with a base-line of 500 metres, have shown that the source in Cassiopeia has an angular size of less than 6 minutes of arc.

The resolving power of an interferometer depends primarily on the ratio of the wave-length to the base-line, and the limits quoted above represented the best performance obtained with the instruments. These limits cannot be reduced significantly without a corresponding extension of the base-line.

In 1950 it was decided at Jodrell Bank to attempt to measure the angular size of the two sources shown in Table 1, or at least to reduce the upper limits given for their size. It was assumed that this angular size might lie anywhere between the limit of a few minutes of arc and the diameter of the visible stars, and an instrument of the highest possible resolving power was therefore sought. While it appeared to be possible to extend the base-lines of existing inter-

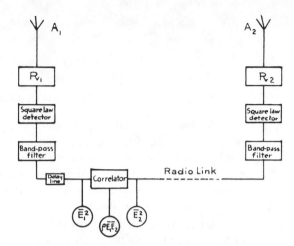

Fig. 1. Schematic diagram of the equipment

ferometers to lengths of the order of 10–50 km., it was considered that for much longer base-lines the problem of maintaining an adequate stability of phase in the transmission of signals along the base-line would prove to be difficult. For this reason an interferometer of completely new design was developed.

Fig. 1 shows a simplified block diagram of the instrument. Two aerial systems A_1 and A_2, each with an aperture of 500 square metres, are connected to two independent superheterodyne receivers R_1 and R_2. These receivers are both tuned to 125 Mc./s. and have a band-width of 200 kc./s. The intermediate frequency output of receiver R_1 is rectified in a square-law detector and is fed to a low-frequency filter the pass-band of which extends from 1 to 2 kc./s. The output from receiver R_2 is treated in an identical manner and the two low-frequency outputs are then multiplied together in a 'correlator' and their 'cross-correlation coefficient' is obtained. When a long base-line is used, one low-frequency output is transmitted over a radio-link by modulating a high-frequency carrier. The time of arrival of the two low-frequency outputs at the 'correlator' is equalized by inserting in one output a delay equal to the time of transmission along the base-line. It is essential that all the components should preserve the relative phase of the two low-frequency signals. In practice, this problem is considerably simpler than the corresponding one of preserving the relative phase of the

Table 2. EXPERIMENTAL RESULTS

	Base-line		Cygnus		Cassiopeia	
	Length* (km.)	Bearing†	Correlation coefficient	Angular width of equivalent strip	Correlation coefficient	Angular width of equivalent strip
A	0·30	349·5°	0·99±0·10	< 5′	0·96±0·09	3′ 46″(< 5′ 50″)
B	2·16	113·0°	0·30±0·03	2′ 10″ ±4″	0·08±0·02	2′ 55″ ±10″
C	2·16	235·5°	0·79±0·08	1′ 00″ ±7″	< 0·01	≮ 3′ 36″
D	3·99	177·0°	0·79±0·07	0′ 34″ ±8″	0·07±0·01**	

* The value given is the actual distance between the two stations. The effective length of the base-line is calculated from the elevation of the source and the orientation of the base-line.

† The bearing is measured east from north, and is the relative bearing of the base-line from the fixed station at Jodrell Bank.

** This value is not taken into account in Fig. 3, since the other results indicate that the point may lie on a secondary maximum of the curve relating correlation coefficient to base-line.

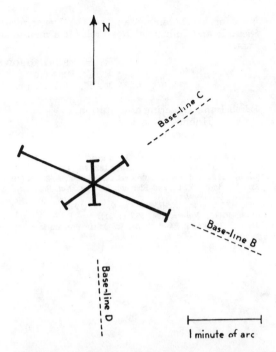

Fig. 2. Equivalent angular width of the source in Cygnus observed from different base-lines. The value shown is the width of an equivalent rectangular strip of constant surface intensity

radio-frequency signals in a conventional interferometer, and hence it should be easier to use the new type of instrument with long base-lines.

In operation, the two aerials (A_1 and A_2) are fixed in azimuth (due north or south) and at an elevation corresponding to the declination of the source. As the source transits the aerial beam, the increase in the mean square value of the two low-frequency outputs ($\overline{\varepsilon_1^2}$ and $\overline{\varepsilon_2^2}$) is recorded together with their product ($\rho\,\overline{\varepsilon_1\varepsilon_2}$). The value of the 'cross-correlation' coefficient' (ρ) is then found from an analysis of the three records.

The theory of the instrument is involved, and it will be given in detail elsewhere[7]. It can be shown that the value of the cross-correlation coefficient (ρ) is given by an expression similar to that for the visibility of the fringes in a Michelson stellar interferometer :

$$\rho = \frac{\sin^2 (\pi\alpha b/\lambda)}{(\pi\alpha b/\lambda)^2} , \qquad (1)$$

where α is the angular width of an equivalent rectangular source of constant surface intensity ; b is the effective length of the base-line ; and λ is wavelength.

Equation (1) shows that by a suitable choice of the base-line it is possible to obtain a value for the angular width of an equivalent rectangular source of constant surface intensity. The shape and angular width of the actual source, together with the distribution of brightness across its disk, can only be found from a large number of observations made with base-lines of different length and orientation.

A model of the instrument, built in 1950, was used to measure the apparent diameter of the sun at 125 Mc./s. As the results of this test were satisfactory, a full-scale instrument was built in 1951, and this

has since been applied to the two sources shown in Table 1. Measurements have so far been made on both sources with four different base-lines and the results are given in Table 2. The length and orientation of the base-lines are given in the table, together with the observed values of the cross-correlation coefficient (ρ). The angular size of the equivalent rectangular strip has been calculated for each base-line and the results are also shown in Table 2.

The interpretation of these results in terms of the precise actual size and shape of the sources must be treated with caution, since the three extended base-lines differed in length ; and until more results are available it is uncertain what effect variations in the length of the base-line may have on the equivalent angular size of the source.

Figs. 2 and 3 show a plot of the measurements given in Table 2. Fig. 2 shows clearly that the source in Cygnus exhibits a pronounced asymmetry. The value of its equivalent angular size varies from 35″ to 2′ 10″ of arc with the orientation of the base-line, and the major axis of the source appears to be inclined at an angle of between 90° and 120° to the celestial meridian. A preliminary analysis indicates that the results are incompatible with a source of simple elliptical shape and constant surface intensity, and that a more complicated model must be used. Further progress in this analysis must await more observations. The results in Fig. 3 suggest that the source in Cassiopeia is not markedly asymmetrical. The equivalent angular size of the source appears to be about 4′ of arc.

These preliminary measurements establish two major points. First, the apparent angular size of the two most intense radio sources is thousands of times greater than that of the visible stars and is of the order of a few minutes of arc. Secondly, the source in Cygnus exhibits a pronounced asymmetry

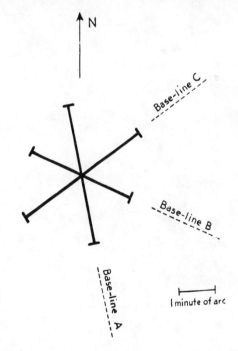

Fig. 3. Equivalent angular width of the source in Cassiopeia observed from different base-lines. The value shown is the width of an equivalent rectangular strip of constant surface intensity

in angular size, whereas the source in Cassiopeia appears to be roughly symmetrical. The measurements are not yet adequate to define satisfactorily the shape of the sources or the distribution of intensity across their disks. Further observations are now being made with the present apparatus using different base-lines, and the results will be published later.

The work was carried out at the Jodrell Bank Experimental Station of the University of Manchester. The construction of the apparatus was made possible by a grant from the Department of Scientific and Industrial Research. We wish to thank Dr. R. Q. Twiss for his assistance with the mathematical theory, and Prof. A. C. B. Lovell for making the necessary facilities available and for his interest in the investigation. One of us (R. H. B.) is indebted to Messrs. Imperial Chemical Industries for a fellowship, and one (M. K. D. G.) to the Government of India for a scholarship and to the Department of Scientific and Industrial Research for a maintenance grant.

R. HANBURY BROWN
R. C. JENNISON
M. K. DAS GUPTA

Jodrell Bank Experimental Station,
    Holmes Chapel,
        Cheshire.

[1] Bolton, J. G., and Stanley, G. J., *Nature*, **161**, 312 (1948).
[2] Ryle, M., and Smith, F. G., *Nature*, **162**, 462 (1948).
[3] Stanley, G. J., and Slee, O. B., *Aust. J. Sci. Res.*, A, **3**, 234 (1950).
[4] Ryle, M., Smith, F. G., and Elsmore, B., *Mon. Not. Roy. Astro. Soc.*, **110**, 508 (1950).
[5] Mills, B. Y., *Aust. J. Sci. Res.*, A, **5**, 266 (1952).
[6] Smith, F. G., *Nature*, **168**, 555 (1951).
[7] Hanbury Brown, R., and Twiss, R. Q. (in preparation).

Part III
Antenna Measurement and Calibration

THE papers in this part are designed to give an overview of the techniques developed for measuring the performance of radio-astronomical antennas. Since satellite communication and radar systems use large antennas similar to (and sometimes identical with) those employed by radio astronomers, it is not surprising that this topic has received considerable attention from a wide spectrum of the antenna engineering community. Radio astronomers have made considerable contributions to this area. Since radio telescopes have had the highest gains of any large antennas built, they have advanced the state of the art in large antenna design and fabrication. Many aspects of antenna construction developed by radio astronomers to solve their particular needs have not been published. However, a number of important articles have appeared on the measurement and adjustment of the surface panels which make up the surface of large radio telescopes, some of which are included under References and Bibliography.

The issues addressed in this part are intertwined with those of Part VII. The intended distinction is that the measurements described here, in general, refer to the antenna only, with an emphasis on the determination of its performance and imperfections. Those discussed in the latter part refer to calibration of the astronomical data and include topics such as the determination of the atmospheric absorption. It is true that the topics of the two parts are often indistinguishable, so the papers as well as the references included with each should be cross-checked.

The first paper included in this part, by Baars, serves as an introduction to the use of cosmic sources to measure the basic characteristics of antenna performance. Additional information can be found in the texts on antennas referenced in Part I, and also in that by Kuz'min and Salomonovich (1966).

The second paper, by Ruze, reviews the main sources of less than perfect antenna performance. The model for surface errors used assumes that the deviations are random and can be characterized by a root-mean-square (rms) dispersion and a correlation length. This leads to a relatively simple form for the antenna gain, which is widely known as the "Ruze formula." If the surface errors are not random, the situation is more complex. Some methods of dealing with the situation are discussed in the references. The third paper included here, by Kaufmann *et al.*, presents some measurements on a 45-ft antenna at a wavelength of 7 mm. Despite the residuals being distributed in two Gaussian distributions with different mean values, the Ruze formula appears to give a value for the efficiency consistent with that measured. The Ruze formula is often applied even without knowledge of the form of the surface errors; non-Gaussian errors are discussed by Davis and Cogdell (1971) and by Kaufmann *et al.* (1987).

The fourth paper included here, by Greve and Hooghoudt,

emphasizes that the illumination of the antenna aperture determines the weighting function for calculating the rms error of the surface. This is particularly significant if the errors are highly variable over the aperture, as is often the case with traditional surveying methods.

The subsequent three papers are concerned with the "holographic" method of measuring the surface of an antenna. This approach is based on the Fourier transform relationship between the far-field angular pattern and the field in the antenna aperture. If the amplitude and phase of the former can be determined, then in principle we can recover the aperture field amplitude and phase as well. If the feed system is assumed to illuminate the antenna with a spherical wave, then any small-scale deviations from a plane wave in the aperture field distribution can be attributed to the phase perturbations produced by errors in the antenna surface compared to its ideal shape. A variety of techniques using both astronomical (fifth and seventh papers, by Scott and Ryle, and Godwin *et al.*, respectively) and terrestrial (sixth paper, by Mayer *et al.*) sources have been employed. The accuracy of the holographic method appears to be superior to that of any other system yet devised; the great advantage of this method is that with an astronomical source, measurements at different elevation angles can readily be made. As a result, most radio observatories now employ, or are planning to use, this method of measuring the shape of their antennas.

The final paper in this part, by Morris, presents a method of surface error measurement which does not require measurement of the *phase* of the far-field radiation pattern of the antenna. Instead, *intensity* measurements at different focus positions are required. There is obviously a considerable advantage in the simplicity of this system—no additional electronics whatever is required—but the required signal-to-noise ratio limits its applicability with astronomical sources. Its use with terrestrial sources in the near field is the subject of investigations presently under way.

REFERENCES AND BIBLIOGRAPHY

1955: Matt, S. and J. D. Kraus, "The effect of the source distribution on antenna patterns," *Proc. IRE,* vol. 43, pp. 821–825, 1955.

1961: Bracewell, R. N., "Tolerance theory of large antennas," *IEEE Trans. Antennas Propagat.,* vol. AP-9, pp. 49–58, 1961.

Ko, H. C., "On the determination of the disk temperature and the flux density of a radio source using high-gain antennas," *IRE Trans. Antennas Propagat.,* vol. AP-9, pp. 500–501, 1961.

1966: Kuz'min, A. D. and A. E. Salomonovich, *Radioastronomical Methods of Antenna Measurements.* New York, NY: Academic Press, 1966.

1967: Zarghamee, M. S., "On antenna tolerance theory," *IEEE Trans. Antennas Propagat.,* vol. AP-15, pp. 777–781, 1967.

1971: Davis, J. H. and J. R. Cogdell, "Reflector efficiency evaluation by frequency scaling," *IEEE Trans. Antennas Propagat.,* vol. AP-19, pp. 58–63, 1971.

1972: Hartsuijker, A. P., J. W. M. Baars, S. Drenth, and L. Gelato-Volders, "Interferometric measurement at 1415 MHz of radiation pattern of paraboloidal antenna at Dwingeloo Radio Observatory," *IEEE Trans. Antennas Propagat.,* vol. AP-20, pp. 166–176, 1972.

Vu, T. B., "Antenna tolerance theory—A survey of basic methods and recent developments," *Proc. IREE Australia,* vol. 33, pp. 268–274, 1972.

1973: Cogdell, J. R. and J. H. Davis, "Astigmatism in reflector antennas," *IEEE Trans. Antennas Propagat.,* vol. AP-21, pp. 565–567, 1973.

1974: Findlay, J. W. and J. M. Payne, "An instrument for measuring deformations in large structures," *IEEE Trans. Instrum. Meas.,* vol. IM-23, pp. 221–226, 1974.

Stutzman, W. L. and H. C. Ko, "On the measurement of antenna beamwidth using extraterrestrial radio sources," *IEEE Trans. Antennas Propagat.,* vol. AP-22, pp. 493–495, 1974.

1975: von Hoerner, S. and W.-Y. Wong, "Gravitational deformation and astigmatism of tiltable radio telescopes," *IEEE Trans. Antennas Propagat.,* vol. AP-23, pp. 317–323, 1975.

1976: Bennett, J. C., A. P. Anderson, P. A. McInnes, and A. J. T. Whittaker, "Microwave holographic metrology of large reflector antennas," *IEEE Trans. Antennas Propagat.,* vol. AP-24, pp. 295–303, 1976.

Payne, J. M., J. M. Hollis, and J. W. Findlay, "New method of measuring the shape of precise antenna reflectors," *Rev. Sci. Instr.,* vol. 47, pp. 50–55, 1976.

1977: Ulich, B. L., "A radiometric antenna gain calibration method," *IEEE Trans. Antennas Propagat.,* vol. AP-25, pp. 218–223, 1977.

1978: Harris, A. B., "Contribution of reflector profile errors to antenna sidelobe radiation," *Electron. Lett.,* vol. 14, pp. 343–345, 1978.

von Hoerner, S., "Measuring the gravitational astigmatism of a radio telescope," *IEEE Trans. Antennas Propagat.,* vol. AP-26, pp. 315–318, 1978. (a)

von Hoerner, S., "Telescope surface measurement with two feeds," *IEEE Trans. Antennas Propagat.,* vol. AP-26, pp. 857–860, 1978. (b)

1982: Rusch, W. V. T. and R. Wohlleben, "Surface tolerance loss for dual-reflector antennas," *IEEE Trans. Antennas Propagat.,* vol. AP-30, pp. 784–785, 1982.

1984: Ellder, J., L. Lundahl, and D. Morris, "Test of phase-retrieval holography on the Onsala 20 m radiotelescope," *Electron. Lett.,* vol. 20, pp. 709–710, 1984.

Rahmat-Samii, Y., "Surface diagnosis of large reflector antennas using microwave holographic metrology: An iterative approach," *Radio Sci.,* vol. 19, pp. 1205–1217, 1984.

1985: Zarghamee, M. S. and J. Antebi, "Surface accuracy of Cassegrain antennas," *IEEE Trans. Antennas Propagat.,* vol. AP-33, pp. 828–837, 1985.

1986: Greve, A., "Reflector surface measurements of the IRAM 30-m radio telescope," *Int. J. Infrared and Millimeter Waves,* vol. 7, pp. 121–135, 1986.

Pelyushenko, S. A., "Measurement of antenna directional patterns using extended radio sources," *Radiophys. and Quantum Electron. (USA),* vol. 29, pp. 249–258, 1986.

1987: Kaufmann, P., Z. Abraham, E. Scalise, Jr., R. E. Schaal, J. L. M. DoVale, and J. W. S. Vilas Boas, "Aperture efficiency of Itapetinga 45-ft antenna at $\lambda = 3.3$ mm," *IEEE Trans. Antennas Propagat.,* vol. AP-35, pp. 996–1000, 1987.

The Measurement of Large Antennas with Cosmic Radio Sources

JACOB W. M. BAARS

Abstract—Strong cosmic radio sources provide a constant broad-band and accurately positioned test transmitter for measurements of large antennas. Some sources have their flux density determined absolutely and can be used to calibrate the antenna gain. This paper presents up-to-date data on the radio sources which are useful for antenna measurements. The measurement of pointing and focusing corrections is discussed. The main part of the paper is concerned with the derivation of major antenna parameters such as aperture and beam efficiency, beam solid angle, sidelobe levels, error pattern characteristics from measurements on radio sources. The effects of a finite angular source size are discussed, and it is shown how measurements on sources of different size increase the information on the derived antenna parameters. The methods to measure very weak sidelobes are treated and the external factors, solar and galactic radiation, influence of the earth and atmosphere, which might limit the accuracy of the measurement are described. The paper takes a practical approach to the subject and contains graphs with numerical data.

I. Introduction

RADIO astronomers have used large reflector antennas from the earliest days of their science [27]. The radar research during the Second World War has led to a strong development of the theory and techniques of microwave antennas (e.g., Silver [33]). In the first two decades after the war the development and construction of ever larger antennas has been mainly for application to radio and radar astronomy [8], [10]. Recently tracking stations for space probes and ground terminals for communication via satellites have increased the number of large reflectors significantly. The calibration of antennas with apertures of some 100–3000 wavelengths poses a special and difficult problem, which is a result of the large size. The technique of using celestial radio sources for this calibration has been developed almost entirely by radio astronomers.

Consider, for instance, the determination of the antenna gain. A reliable theoretical calculation based on Maxwell's equations and the appropriate boundary conditions can only be done for some very simple geometries of the antenna as a dipole and a horn. The theory for horns can be applied to apertures of the order of 100 wavelengths as shown by very accurate measurements of Hogg and

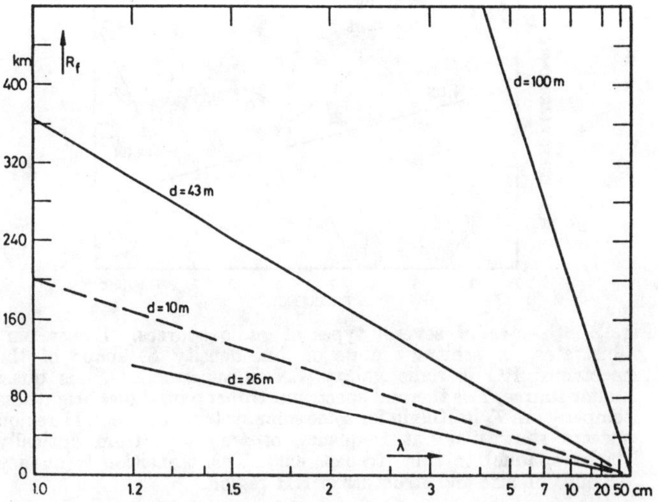

Fig. 1. Far-field distance $R_f = 2d^2/\lambda$ as function of wavelength λ for several antenna diameters d. For dashed line of the millimeter antenna ($d = 10$ m) wavelength scale should be read in mm.

Wilson [17], and Jull and Deloli [20]. An accuracy of 1–2 percent appears to be realistic for a gain calculation. This result is of great importance for the application of radio sources to gain measurements of large antennas.

With an entirely theoretical calculation of the gain of a reflector antenna, one cannot expect an accuracy of better than 5–10 percent due to the large number of small effects, which have to be taken into account [4], [28].

The experimental measurement of the gain on a test range becomes impractical or even impossible for large antennas. The required length of the range R becomes several tens of kilometers to satisfy the far-field criterion

$$R \geq R_f = \frac{2d^2}{\lambda} \qquad (1)$$

where R_f is the far-field distance, d is the aperture linear size, and λ is the wavelength. Fig. 1 shows R_f as a function of wavelength for several values of d.

For some restricted measurements this criterion can be eased, if the necessary corrections are applied [31]. For accurate work, this is not satisfactory, while moreover the use of an earthbound range of low elevation angle will increase the risk of spurious ground reflections causing calibration errors.

A number of strong point like cosmic radio sources radiating over a broad continuous frequency band provides

Manuscript received October 11, 1972; revised December 22, 1972.
J. W. M. Baars is with the Dwingeloo Radio Observatory, Dwingeloo, the Netherlands. (The Dwingeloo Radio Observatory is operated by the Netherlands Foundation for Radio Astronomy with the financial support of the Netherlands Organisation for the Advancement of pure Research (Z.W.O.).)

Reprinted from *IEEE Trans. Antennas Propagat.*, vol. AP-21, no. 4, pp. 461–474, July 1973.

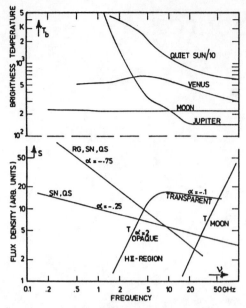

Fig. 2. Spectra of several types of radio sources. Lower part illustrates, in arbitrary units of flux density S, shape of the spectrum; RG is radio galaxy, SN is supernova, QS is quasi stellar source, T is thermal spectrum. Upper part shows brightness temperature T_b in Kelvin for some solar system objects. HII-region spectra show bend at frequency of transition from optically thick (opaque) to thin (transparent). This transition frequency depends on size and structure of HII region.

a set of almost ideal test sources for measurements of the parameters of large antennas. This paper is concerned with the specific advantages and limitations of the use of radio sources.

We discern the following aspects of antenna calibration, where radio sources can be used.

1) Pointing of the electrical axis and determination of the true focal point. For the former we need radio sources with accurately known positions in the sky.

2) Absolute antenna characteristics, as gain, aperture and beam efficiency, and antenna noise temperature. For these we require an accurate knowledge of the absolute flux density of the radio source and a calibration of the received noise power. The absolute flux density of a few very strong sources has been measured with small antennas (mainly horns), whose gain is known from calculation or measurement. The special techniques involved in absolute gain and receiver calibration have been described by Findlay [9].

3) Relative antenna pattern parameter, as the shape of the main beam, the relative sidelobe level, and the half-power beamwidth (HPBW).

It should be noted that the accuracy and detail of antenna measurements is limited by the strength of the source and the sensitivity of the antenna–receiver combination. We shall see that the available radio sources and contemporary electronics allow a detailed measurement of the characteristics of antennas to be made.

Kuz'min and Salomonovich [22] have written on the general subject of this paper. Although many useful details can be found in their book, the emphasis of this paper is on subjects which are not treated there in detail.

II. Radio Sources for Antenna Calibration

It is clear that a celestial radio source always satisfies the far-field criterion. To be an ideal test source for our purpose the source would have the following properties.

1) A precisely known position in the sky. Preferably it should cover a large range in elevation angle on its daily path along the sky as seen from the location of the antenna.

2) Very small angular size, so as to appear as a point-source to the narrowest beam.

3) An absolutely and accurately known power flux over a large frequency range, without showing time variation or a polarized component.

4) A large flux density to allow a large dynamic range in pattern measurements with equipment of moderate sensitivity.

No cosmic source exhibits all properties simultaneously. However, there is a great variety of source types and there are several sources satisfying one or more of our requirements. For a general discussion of radio sources we refer to Kraus [21] or Smith [34]. The power flux S is expressed in terms of *flux units* (1 fu $= 10^{-26}$ W m^{-2} Hz^{-1}) or in equivalent brightness temperature T_b, where

$$S = \frac{2k}{\lambda^2} \int_{\Omega_s} T_b \, d\Omega \qquad (2)$$

with Ω_s the solid angle of the source and k Boltzmann's constant.

The spectrum, i.e., the relation between flux S and frequency ν, can in most cases be expressed by the simple formula $S \propto \nu^\alpha$, where α is called the *spectral index*. A *thermal source* (T) has $\alpha = 2$. The majority of radio sources, as radio galaxies (RG), quasars (QS), and supernova remnants (SN), are *nonthermal* with $\alpha = -0.25$ to -1. Fig. 2 shows the different forms of the spectrum. The choice of source depends on the specific purpose of the measurement and on the operating frequency. In Table I, we have assembled the data on a set of sources, which have one or more of the required properties. The table is self-explanatory to a large extent, but some additional comments are in order.

The first three are the strongest sources of relatively small angular diameter. Their *absolute* spectrum has been determined over a frequency region of about 30 MHz to 16 GHz from observations with calibrated antennas. An up to date summary has been given by Baars and Hartsuijker [3]. The spectra are shown in Fig. 3.

None of the three sources is ideal. They have an angular size which becomes significant for antenna beamwidths of less than 10′. Cassiopeia A has a flux density decrease of 0.9 ± 0.1 percent per year [3], Cygnus A has a curvature in its spectrum, Taurus A shows a significant degree of linear polarization. However, all these aspects have been thoroughly studied. Because of the high accuracy of the absolute spectrum these sources are the best for gain calibration of large antennas, assuring a large signal-to-noise ratio (SNR).

TABLE I

RADIO SOURCES FOR ANTENNA MEASUREMENTS

A. FLUX DENSITY AND POLARIZATION DATA

Source	Type	Spectral Index (α)	Flux Density (fu)			Linear Polarization (percent, angle)			Remarks
			1 GHz	3 GHz	10 GHz	1 GHz	3 GHz	10 GHz	
Cassiopeia A (Cas A, 3C461)	SN	−0.787	3185 ±30	1340 ±15	520 ±5	<1	<1	1.5, 40°	Annual flux decrease of 0.9 percent.
Cygnus A (Cyg A, 3C405)	RG	−1.205*	2270	690 ±20	162 ±4	0	0.3, 170°	8, 146°	*Spectrum flattens below 2 GHz, polarization strongly frequency dependent.
Taurus A (Tau A, 3C144)	SN	−0.263	986 ±12	739 ±10	537 ±8	1.0, 100°	3.6, 135°	7.0, 143°	
Virgo A (Vir A, 3C274)	G	−0.853	285 ±5	112 ±3	40 ±2	0	0	0	*Curved, thermal spectrum.
Orion A (Ori A, 3C145)	T	*	330	420	420	0	0	0	
Hydra A (3C218)	G	−0.91	60	22	7				
Hercules A (3C348)	G	−1.02	62	20	6	2.4, 50°	6.0, 25°		
3C353	G	−0.71	73	33	14	2.5, 170°	6.6, 100°		
DR21	T	1.75, −0.13*	4	20	21	0	0	0	*Compact HII region, spectrum breaks at about 3 GHz.
Moon	T	+2.0		T_b = 225 K		Small at limb [36]			Compute flux density using daily value of angular size.
Venus	T	—	Read T_b from Fig. 2			See [24]			Compute flux density using daily value of angular size.
Jupiter	—	—	Read T_b from Fig. 2			See [24]			Compute flux density using daily value of angular size.
Quiet Sun	—	—	Read T_b from Fig. 2			—			Compute flux density using daily value of angular size.

B. POSITION AND ANGULAR SIZE

Source	Type	Position (deg) (1950.0)		Annual Precession (10^{-2} deg)		Source Shape*	Angular Size ($RA \times DEC$, position angle)	Remarks
		Right Ascen.	Declination	ΔRA	ΔDEC			
Cas A	SN	350.290	58.540	1.127	0.549	Disc (with limb brightening)	4′ × 4′, 0°	Background structure
Cyg A	RG	299.438	40.583	0.865	0.274	Double "point"	1.6′ × 1′, 110°	Background structure
Tau A	SN	82.880	21.982	1.504	0.059	Gaussian	3.3′ × 4.0′, 135°	
Vir A	G	187.080	12.667	1.265	−0.052	Gaussian	1′ × 1.8′, 10°	
Ori A	T	83.208	−05.420	1.228	−0.066	Gaussian	3.5′ × 3.5′, 0°	
Sagittarius A	—	265.613	−28.975	1.588	−0.043	Gaussian	4.0′ × 2.5′, 45°	Complex background
3C123	RG	68.480	29.571	1.574	0.204	"point"	5″	
3C147	QS	84.682	49.830	1.937	0.052	"point"	0.5″	
3C273	QS	186.637	02.328	1.278	−0.553	Double "point"	20″, 43°	
3C295	RG	212.390	52.437	0.893	−0.470	"point"	4.5″	
DR21	T	309.309	42.152	0.891	0.353	Gaussian	20″	
Planets	T	see tables in *Astronomical Ephemeris* for daily values				disc		Parallax correction for accurate position.

* Source shape is approximate, but applicable for a HPBW larger than the source size.

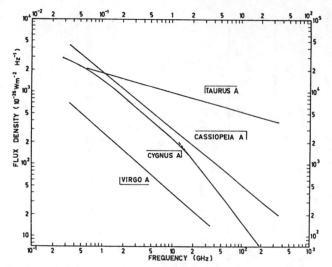

Fig. 3. Absolute spectra of strongest radio sources derived from absolute flux density measurements and some accurate relative measurements against Cas A. Small bars next to source name indicate which of coordinate scales apply to source. Further data in Table I.

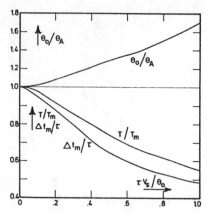

Fig. 4. Effect of time constant τ on measured result of scan of antenna beam of HPBW θ_A over point source with scan-velocity V_s. Curve θ_0/θ_A indicates broadening of half-power width, T/T_M is relative decrease in measured peak signal, and $\Delta t_M/\tau$ delay in units of τ in reaching peak signal. Note that abscissa used measured θ_0 not theoretical θ_A.

The group of weaker sources in Table I has less accurate flux densities but generally a very small angular size. They are useful for pointing calibration and focusing of large antennas with very narrow beams. Their position in the sky is accurately known from interferometer measurements.

We have included the nearer planets, because they become the stronger objects at millimeter wavelengths. Although the position and angular size varies, it can be found accurately at any time from the American Ephemeris and Nautical Almanac, published yearly by the Government Printing Office. For details on the brightness temperature variations as function of frequency, see Mayer [24].

The Moon has a well known brightness temperature [36] but a large angular size. This can be exploited to measure certain pattern parameters, as discussed in Section V. The Sun is by far the strongest object in the sky, but of very limited use for our purpose. The main reasons are its large angular size and the strong variations in strength due to solar activity. Sometimes it is necessary to use it in order to obtain a sufficient SNR.

The flux densities in Table I are "extra-atmospheric", i.e., the actual value at the antenna must be corrected for attenuation by the earth's atmosphere. A summary of the most reliable data on the attenuation has been presented by Howell and Shakeshaft [19].

III. Antenna Pointing and Focusing Calibration

A. Pointing

Rarely will an antenna be built to a precision and with a rigidity that it can be pointed perfectly at any angular position by merely controlling it to the correct encoder readings. Errors in the true direction of the beam originate in deviations from orthogonality between the telescope axes, gravitational deflections in the reflector and feed support structure, and also by the atmospheric refraction. The necessary corrections can be obtained by measuring the apparent position of a radio source which true position is known. Point sources are ideal for this purpose. With a good SNR the pointing errors can be determined to an accuracy of about 5 percent of the HPBW. An excellent account of these errors, together with experimental results, has been given by Meeks et al. [25], to which we refer for further details.

In deriving the errors in antenna axis direction, one must first correct the measurements for *refraction* by the atmosphere. Tropospheric radio refraction is essentially independent of frequency up to about 100 GHz. Bean [6] treats this subject in detail. Working at frequencies below about 500 MHz, refraction by the ionosphere should be taken into account.

All measurements in which the antenna beam moves with respect to the source have to be corrected for the influence of a finite time constant in the receiver. The effect is a broadening of the HPBW, a decrease in the maximum signal and a delay in reaching the peak intensity. Fig. 4 presents these effects (adapted from Howard [18] and Kuz'min and Salomonovich [22]). If $\tau < 0.06\,\theta_A/V_s$, where τ is the time constant of a simple RC network and V_s the angular velocity of the source with respect to the antenna beam of width θ_A, the resulting measurement is essentially undistorted and delayed by an amount τ.

In cases of insufficient sensitivity the Sun can be used for pointing measurements with somewhat impaired accuracy. Special precautions have to be taken, which have been described by Graf et al. [11].

B. Focusing

The effects of gravity will in general influence the position of the focal point, of the feed or both as a function of antenna pointing direction. A defocusing will impair the performance of the antenna.

Axial defocusing along the reflector axis causes a strong decrease in the antenna gain, a slight broadening of the mainbeam and a strong increase in the near sidelobes. The change in gain can be used as a sensitive indicator of the correct focus. It is easy to show that axial defocusing causes a phase error over the aperture proportional to $\cos \Theta$, where Θ is the angle between a ray from the feed to the reflector at a radius r and the reflector axis. With a paraboloidal reflector of focal length f we have

$$\cos \Theta = 1 - 2\left(\frac{r}{2f}\right)^2 + 2\left(\frac{r}{2f}\right)^4 - \cdots.$$

The phase error contains even powers of the aperture radial coordinate.

For not too deep a reflector we need retain only the first term to obtain for the phase error over the aperture

$$\Psi(r) = \beta r^2 \qquad (3)$$

with $\beta = (2\pi/\lambda)\Delta_a(1 - \cos \Theta_0)$, where Δ_a is the axial focus displacement and Θ_0 is the aperture angle of the reflector.

Let us introduce this result into the scalar radiation integral of a circular aperture with an aperture illumination function

$$A(r) = p + (1 - p)(1 - r^2)$$

the parabolic illumination on a pedestal p (e.g., Hansen [12, p. 64]). The gain G, normalized to the gain G_0 in the focused case, is

$$\frac{G}{G_0} = p^2 \left(\frac{\sin(\beta/2)}{\beta/2}\right)^2 + (1 - p)^2$$

$$\cdot \left\{\left(\frac{\sin(\beta/2)}{\beta/2}\right)^4 + \frac{4}{\beta^2}\left(\frac{\sin \beta}{\beta} - 1\right)^2\right\} \qquad (4)$$

Fig. 5 shows the case of constant $(p = 1)$ and parabolic $(p = 0)$ illumination. The rise in sidelobe level is also given.

Experimentally, one measures the antenna temperature T_A of a point source for different values of Δ_a. The symmetry line of the resulting curve $T_A(\Delta_a)$ indicates the correct focus. An extended source will give a less pronounced result because the beam broadening and increased sidelobe level partially offset the loss of on-axis signal. As an example Fig. 6 shows experimental focusing curves on a point source and the Moon.

The calculation of the beam broadening is tedious but straightforward. Cheng [7] has indicated an upper limit. The result is sensitive to the illumination function. We have found that the linear beam broadening can be approximated by

$$\frac{\theta_A'}{\theta_A} \simeq 1 + 0.02\beta^2, \qquad \beta < 1.$$

Radial defocusing causes a phase error proportional to $\sin \Theta$, i.e., in odd powers of the radial aperture co-

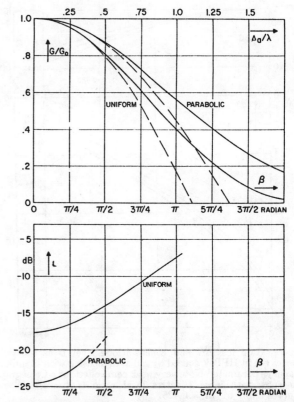

Fig. 5. Relative gain G/G_0 and level of first sidelobe L as function of phase error β at aperture edge caused by axial defocusing Δ_a. Top scale gives Δ_a in wavelengths and is valid for paraboloid with $\Theta_0 = 60°$ (focal ratio $f/d = 0.425$). Dashed lines are quadratic approximation to (4).

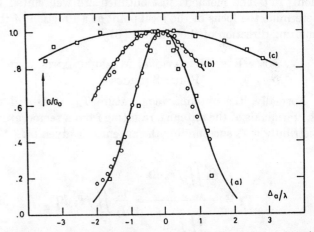

Fig. 6. Experimental focusing curves; relative gain as function of axial defocusing. Circles are measurements at $\lambda = 6$ cm, squares at $\lambda = 2$ cm with HPBW of 10′ and 3.4′, respectively. Curve (a) is obtained with point source, curves (b) and (c) with Moon. Effect with extended object, Moon, is strongly reduced due to increased beamwidth and sidelobe level of defocused reflector. Note that symmetry line of result on Moon is slightly shifted. Focusing on source much larger than HPBW appears hazardous.

ordinate. The general effects on the pattern have been described by Ruze [29]. The gain decrease is very slow, but the rise in the level of the *coma* sidelobe is large, as shown in Fig. 7. If one measures an asymmetry in the sidelobe pattern, radial defocusing will generally be the cause.

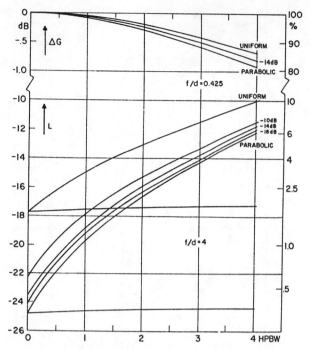

Fig. 7. Gain loss ΔG and coma-sidelobe level L as function of beam tilt in HPBW caused by radial defocusing with illumination taper as parameter. Note very weak coma effect for a Cassegrainian antenna with large effective focal ratio ($f/d = 4$). All other curves apply to $f/d = 0.425$.

Very large antennas are sometimes designed to deflect under the influence of gravity, but to maintain a parabolic reflector surface with varying focal length and axis direction [16]. The methods just outlined are well suited to determine the locus of the focal point as a function of the pointing direction of the antenna [37].

IV. Some Basic Relations; Measurements with Point Sources

Formally, the antenna temperature T_A, measured at the terminals of the antenna, resulting from a temperature distribution T_b surrounding the antenna, is given by

$$T_A = \frac{\eta_R \iint_{4\pi} f T_b \, d\Omega}{\iint_{4\pi} f \, d\Omega} = \frac{\eta_R}{\Omega_A} \iint_{4\pi} f T_b \, d\Omega \tag{5}$$

where η_R is the radiation efficiency, accounting for the ohmic loss in the antenna, f denotes the normalized antenna power pattern, $d\Omega$ is an infinitesimal increment in solid angle, and $\Omega_A = \iint_{4\pi} f \, d\Omega$ defines the antenna pattern solid angle.

When the antenna beam is scanned across a source with a brightness distribution $T_b(x',y')$, the antenna temperature $T_A(x,y)$ is given by the antenna convolution integral

$$T_A(x,y) = \frac{\eta_R}{\Omega_A} \iint f(x - x', y - y') T_b(x',y') \, dx' \, dy'. \tag{6}$$

Thus the measured distribution T_A is the true distribution T_b smeared by the finite width of the beam. A radio astronomer must solve this integral equation to recover T_b. Reversely, if one has a simple functional form for T_b, the antenna pattern f may be found by a closed-form solution of (6). For sources with an angular size $\theta_S \ll \theta_A$, the HPBW of the antenna, T_b can be approximated by a delta-function and the measurement yields the beam shape directly.

We separate the antenna solid angle in two parts as

$$\Omega_A = \Omega_m + \Omega_l \equiv \iint_{\text{main beam}} f \, d\Omega + \iint_{\text{sidelobes}} f \, d\Omega \tag{7}$$

where we call Ω_m the main beam solid angle and Ω_l the sidelobe solid angle. We define the *main beam efficiency* $\eta_B = \Omega_m/\Omega_A$; it is the percentage of all power received which enters the main beam, assuming the antenna is surrounded by a source of uniform temperature. Similarly $\eta_l = \Omega_l/\Omega_A$ and clearly $\eta_B + \eta_l = 1$.

For a true point source one easily derives from (2) and (6)

$$S = \frac{2kT_A}{A} \tag{8}$$

using the well-known relations

$$\frac{\eta_R}{\Omega_A} = \frac{\eta_R D}{4\pi} = \frac{G}{4\pi} = \frac{A}{\lambda^2}. \tag{9}$$

Here A denotes the absorption area of the antenna, D is its directivity, and G its gain. Introducing the *aperture efficiency* $\eta_A = A/A_g$, where A_g represents the geometrical aperture area, the following useful relation can be derived:

$$\eta_R \eta_B = \frac{\eta_A \Omega_m A_g}{\lambda^2}. \tag{10}$$

If η_R is known, the beam efficiency can be found from (10) without the need to measure the entire antenna pattern. Only η_A and Ω_m must be determined from a measurement on a point source and an integration of the mainbeam, respectively.

On the other hand, a complete pattern measurement would yield D, a point source measurement G, and the radiation efficiency could be obtained as $\eta_R = G/D$. An accurate determination along the preceding lines requires an extreme accuracy in the measurement of the total pattern.

Another method to find η_R employs the absolute measurement of the antenna temperature T_A, while the antenna is pointed at the "cold sky." It has been described by Kuz'min and Salomonovich [22, ch. 5]. With extreme care it appears feasible to determine η_R to an accuracy of a few percent at microwave frequencies. Since in a well designed antenna η_R will be close to one, we shall ignore it and consider it to be incorporated into the beam efficiency.

176

V. MEASUREMENTS WITH EXTENDED SOURCES

A. Measurement of Absorption Area and HPBW

We now turn to the effects of the finite angular size of the source on the measurement result. Although the direct measurement of the important parameter A [see (8)] becomes more complicated, we shall show that measurements with sources of different angular size can increase our knowledge about other antenna parameters. This is a direct result of the fact that the source "fills" a larger portion of the antenna pattern. Fig. 8 illustrates *qualitatively* the change in measured antenna temperature for a source with increasing solid angle Ω_s and constant brightness temperature T_b. We write the source brightness distribution as $T_b\psi(x,y)$ with $\psi(0,0) = 1$. Now the source solid angle $\Omega_s = \iint \psi(x,y)\,dx\,dy$. If $\Omega_s \ll \Omega_m$ we can approximate the beam pattern $f(x,y) = 1$ over Ω_s and T_A will increase linearly with Ω_s. For $\Omega_s < \Omega_m$ we define a weighted source solid angle $\Omega_\Sigma = \iint f\psi\,d\Omega$. Now T_A will increase more slowly, as indicated in Fig. 8, until it reaches $T_A = \eta_B T_b$ for $\Omega_\Sigma = \Omega_m$. If the antenna were free of sidelobes, T_A would stay constant with a further increase in Ω_s. The actual presence of the sidelobes causes T_A to increase slowly until it reaches T_b, if the source completely surrounds the antenna. In the region beyond $\Omega_\Sigma = \Omega_m$, we can define a modified beam efficiency η_B' which incorporates the power in the sidelobes up to a convenient solid angle.

Experience has shown that the main beam can be well approximated by a Gaussian function. Limiting ourselves to a circularly symmetric beam, we find $\Omega_m = 1.133\theta_A^2$, where θ_A is the HPBW of the beam. Most radio sources can be represented either by a Gaussian or disc distribution (see Table I) with half-power width θ_s and diameter $2R$, respectively.

The measured antenna temperature for a finite width is

$$T_A = \frac{\eta_B T_b \Omega_\Sigma}{\Omega_m} = \left(\frac{SA}{2k}\right)\left(\frac{\Omega_\Sigma}{\Omega_s}\right) < \frac{SA}{2k}. \quad (11)$$

To derive the absorption area A from an observation of an extended source the measured T_A must be multiplied by a correction factor $K = \Omega_s/\Omega_\Sigma$. The factor becomes

$$K = \begin{cases} 1 + x^2, & \text{for Gaussian source} \\ \dfrac{x^2}{\{1 - \exp(-x^2)\}}, & \text{for disc source} \end{cases} \quad (12)$$

where $x = \theta_s/\theta_A$ in the former and $x = R/0.6\,\theta_A$ in the latter case.

From the measured half-power width θ_0 of a scan through the source, we can derive the HPBW for a Gaussian source distribution as

$$\theta_A = (\theta_0^2 - \theta_s^2)^{1/2}. \quad (13)$$

For a disc distribution the convolution integral does not produce a simple formula for the half-power width. Fig. 9 presents the results of scans by an antenna beam

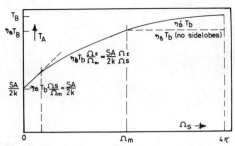

Fig. 8. Qualitative sketch of increase in measured antenna temperature of source of constant brightness temperature T_b with increasing source solid angle Ω_s.

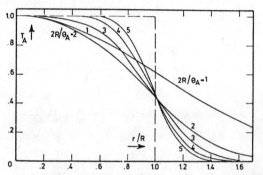

Fig. 9. Normalized antenna temperature versus normalized scan angle r/R of antenna beam with HPBW θ_A over disc source with angular radius R (dashed line) for several values of the ratio $2R/\theta_A$. Note that resulting half-power width can be smaller than source width (unpublished results by P. Stumpff).

TABLE II
CORRECTION FACTORS FOR MEASUREMENT WITH
EXTENDED SOURCES

x	$T_A(\text{true})/T_A(\text{meas})$		θ_0/θ_A	
	Gaussian	disc	Gaussian	disc
0.0	1.00	1.00	1.00	1.00
0.05	1.0025	1.0013	1.0025	1.0013
0.1	1.01	1.005	1.005	1.0025
0.2	1.04	1.020	1.020	1.010
0.3	1.09	1.046	1.044	1.022
0.5	1.25	1.130	1.118	1.061
0.7	1.49	1.265	1.221	1.116
1.0	2.00	1.582	1.414	1.225

over a disc source for several values of the parameter $2R/\theta_A$. Note that for certain values of this parameter the measured θ_0 is somewhat *smaller* than the disc diameter $2R$. If $2R/\theta_A < 1$ the result of the convolution is approximately Gaussian and θ_A can be found from

$$\theta_A = \left(\theta_0^2 - \frac{\ln 2}{2}(2R)^2\right)^{1/2} \quad (14)$$

Table II contains the values for K and the linear beam broadening for Gaussian and disc sources.

In the case that $2R/\theta_A \geq 6$ the convolution can be approximated by that of a Gaussian beam with a straight edge. The result will be in the form of the error integral. The antenna beam can in principle be recovered by

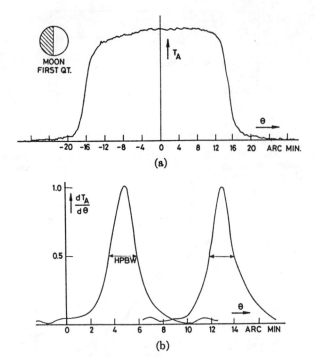

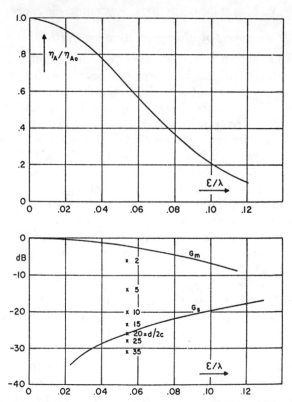

Fig. 10. (a) Scan over Moon made with 2′ HPBW at $\lambda = 2$ cm. Phase of Moon is indicated; note slight asymmetry due to cooler dark area. (b) Differentiating steep slopes of curve (a) delivers a reasonable curve for antenna beam. Derived HPBW is 2.1′ which should be compared with an expected value of 2.0′. Difference might be due to beam-broadening caused by axial defocusing.

Fig. 11. Relative aperture efficiency (or gain G_m) and level of error pattern G_s as function of rms surface error ϵ/λ. Parameter is correlation length in terms of $d/2c$. Curves of G_s for other values of $d/2c$ run through cross parallel to drawn curve.

differentiation of the measured scan. Fig. 10 gives an example of such a procedure. We have found in practice that good results can be obtained with a HPBW of 3–5′ and the Sun or Moon ($2R = 32'$) as a source. This method is important at millimeter wavelengths, where receiver sensitivity is still low, the beamwidth small and the Sun or Moon might be the only sources strong enough to give a good SNR to several tens of dB below the beam peak.

B. Measurement of Sidelobes and Error Pattern

The availability of sources with different angular size can be used to obtain data on the *average* sidelobe level near the main beam without the need to measure the sidelobes individually with a high SNR. For a reflector with significant random errors over the surface the correlation length of the errors and the angular size of the error pattern can be derived.

Let us first consider a perfect reflector. We assume the radiation pattern to be circularly symmetric, consisting of a main beam and several ring shaped sidelobes (the Airy pattern). We represent the antenna pattern, as function of the radial coordinate u, by

$$f(u) = f_m(u) + \sum_i f_{li}(u).$$

A reasonably good approximation for the half-power width θ_{li} and the radius of maximum intensity r_i of the ith sidelobe is $\theta_{li} = \theta_A/2$ and $r_i = \theta_A(\frac{1}{2} + i)$. These relations have been determined from the $\Lambda_1^2(u)$-function,

which represents the power pattern of a uniformly illuminated circular aperture [33, ch. 6].

We can write for the pattern solid angle up to and including i sidelobes

$$\Omega = \Omega_m + \sum_i \Omega_{li}$$

$$= 1.133\theta_A^2 + \sum_i \pi\theta_A^2(\tfrac{1}{2} + i)f_{li}(\text{max})$$

$$= \Omega_m \left\{ 1 + \frac{\pi}{1.133} \left(\frac{3}{2}f_1 + \frac{5}{2}f_2 + \cdots \right) \right\}.$$

We now define an effective beam efficiency η_B' as the percentage of power entering the main beam *and* the sidelobes up to the solid angle equal to that of the observed source. Thus by measuring η_B' on an extended source and η_B on a point source, we can obtain an estimate of the sidelobe level from

$$\frac{\eta_B'}{\eta_B} = 1 + 2.77 \left\{ \frac{3}{2}f_1 + \frac{5}{2}f_2 + \cdots \right\}.$$

As an example, we mention a measurement made with a 92 m diameter radio telescope at $\lambda = 10$ cm. We have $\theta_A = 5'$, while the Moon's size $\theta_s = 32'$. Thus two sidelobes are subtended by the Moon, when measuring its intensity. We found $\eta_B' = 0.173$, and $\eta_B = 0.129$. Taking $f_1 = f_2$, we derived an average sidelobe level $f_{1,2} = -15$ dB, which was in good agreement with the directly measured value.

We now turn to a case of great practical value. Every reflector has imperfections in its surface, which influence the performance depending on the ratio of the surface deviations from a perfect paraboloid to the wavelength. To the extent that these errors are randomly distributed over the reflector their effect on the antenna pattern has been treated by Ruze [30] and independently by Scheffler [32]. The reader, if unfamiliar with this subject, is advised to read Ruze's [30] excellent review first. The effects of random errors can be treated through the mean-square reflector deviation $\langle \epsilon^2 \rangle$ and the correlation radius c. The power pattern of an imperfect reflector is composed of the *diffraction pattern* f_D and the *error pattern* f_E. Thus

$$f(x,y) = f_D(x,y) + f_E(x,y,\langle \epsilon^2 \rangle,c). \qquad (15)$$

The errors $\langle \epsilon^2 \rangle$ cause a root-mean-square phase fluctuation over the aperture of

$$\delta = (\langle \delta^2 \rangle)^{1/2} = \left(\frac{4\pi}{\lambda}\right)(\langle \epsilon^2 \rangle)^{1/2} \qquad (16)$$

where ϵ should be weighted by the aperture illumination function.

The treatment by Scheffler allows the general result for the error pattern, derived by Ruze in the form of an infinite series, to be written as

$$f_E(\theta) = \frac{1}{\eta_{A0}}\left(\frac{2c}{d}\right)^2 \{\exp(\delta^2) - 1\}$$

$$\cdot \begin{cases} \exp\left\{-\left(\frac{\pi c \theta}{\lambda}\right)^2\right\}, & \text{for } \delta^2 \leq 1 \\[2mm] \dfrac{1}{\delta^2}\exp\left\{-\left(\frac{\pi c \theta}{\lambda \delta}\right)^2\right\}, & \text{for } \delta^2 \geq 1 \end{cases} \qquad (17)$$

where η_{A0} is the aperture efficiency of the perfect reflector $(\lambda \to \infty)$, d the diameter of the reflector, c the correlation radius, and θ is the angular coordinate of the rotationally symmetric error pattern.

The boundary between the two expressions lies at $\delta^2 = 1$, or $\epsilon = \lambda/12.5$. In practice, one will not operate an antenna at such a short wavelength that $\delta^2 > 1$. Thus, in the following, we limit ourselves to the case $\delta^2 \leq 1$.

The error pattern has a Gaussian shape with a half-power width

$$\theta_E = 2(\ln 2)^{1/2}\left(\frac{\lambda}{\pi c}\right) = \frac{0.53\lambda}{c}. \qquad (18)$$

The HPBW of a normally tapered aperture (edge taper 16–18dB) has a value $\theta_A \simeq 1.20\lambda/d$. Thus the ratio

$$\frac{\theta_E}{\theta_A} \simeq \frac{0.44d}{c}.$$

Integrating (15) over all space we obtain the antenna solid angle as

$$\Omega_A = \Omega_0 + \Omega_0\{\exp(\delta^2) - 1\} = \Omega_0 \exp(\delta^2)$$

where Ω_0 is the antenna solid angle of the perfect reflector. From this we derive the relative change in aperture efficiency

$$\frac{\eta_A}{\eta_{A0}} = \exp(-\delta^2) + \frac{1}{\eta_{A0}}\left(\frac{2c}{d}\right)^2 \{1 - \exp(-\delta^2)\}. \qquad (19a)$$

For the case that $2c \ll d$, i.e., many independent regions of deviation over the surface, the contribution of the second term is negligible.

The important relation between aperture efficiency and random reflector errors becomes

$$\frac{\eta_A}{\eta_{A0}} = \exp(-\delta^2) = \exp\left(-\frac{16\pi^2\langle \epsilon^2 \rangle}{\lambda^2}\right). \qquad (19b)$$

By measuring η_A at several wavelengths both η_{A0} and ϵ can be found. This measurement is often easier to perform than an actual determination of ϵ on the surface. A change in ϵ with antenna position caused by gravitational influences can also be measured.

The peak level of the error beam with respect to that of the main beam follows from (19a) as

$$\frac{f_E(0)}{f_D(0)} = \frac{1}{\eta_{A0}}\left(\frac{2c}{d}\right)^2 \{\exp(\delta^2) - 1\}. \qquad (20)$$

Fig. 11 shows the results of (19b) and (20). For $\epsilon = \lambda/16$, a widely used criterion for specifying the limiting useful wavelength, the aperture efficiency has dropped to 0.57 times the peak value. With $d/2c = 10$ the error pattern level is then at $-18\ dB$.

We have already seen that, provided $c \ll d$, the error beam is much broader than the main beam. Thus although its intensity is much smaller than that of the main beam, by virtue of its large width it may contain a significant part of the received power, when an extended source is observed. Some of the difficulties due to this effect in properly calibrating radio astronomical observations have been discussed by Heiles and Hoffman [14].

We shall now indicate how a measurement on both a small and an extended source can be used to derive the correlation radius of the surface irregularities. When measuring an extended source of angular size θ_s, where $\theta_A < \theta_s < \theta_E$, we can write for the effective beam efficiency

$$\eta_B' = \frac{\eta_r}{\Omega_0}\exp(-\delta^2)\iint\limits_{\Omega_s}(f_D + f_E)\, d\Omega.$$

The integral over the diffraction beam f_D only yields η_B and can be measured on a small source, the contribution of the error beam being negligible. Integrating the error beam f_E over the source solid angle and substituting

TABLE III
MEASURED PARAMETERS OF A 25-m RADIO TELESCOPE

Wave-length λ (cm)	Aperture efficiency		Main beam solid angle Ω_m (sr)	Beam efficiency	
	η_A	η_A(corr)		η_B(corr)	η_B'(moon)
49.2	0.59	0.59	66.0×10^{-5}	0.79	
21.2	0.54	0.54	12.4×10^{-5}	0.73	
6.0	0.44	0.48	1.06×10^{-5}	0.70	
2.9	0.27	0.38	0.24×10^{-5}	0.53	0.70

(17), we derive the expression

$$\eta_B' - \eta_B = \{1 - \exp(-\delta^2)\} \left\{1 - \left[\exp - \left(\frac{\pi c \theta_s}{2\lambda}\right)^2\right]\right\}.$$

Writing η_{B0} for the beam efficiency of a perfect reflector, which can be found from η_{A0} with the aid of (10), we easily find

$$\eta_{B0} - \eta_B = 1 - \exp(-\delta^2).$$

Combining these two relations, we obtain the useful relation

$$\frac{\eta_B' - \eta_B}{\eta_{B0} - \eta_B} = 1 - \exp\left\{-\left(\frac{\pi c \theta_s}{2\lambda}\right)^2\right\} \qquad (21)$$

where it has been assumed that the diffraction beam has no significant sidelobes compared to the level of the error beam. Solving for the correlation radius c we find

$$c = \frac{2\lambda}{\pi\theta_s}\left\{-\ln\left(\frac{\eta_{B0} - \eta_B'}{\eta_{B0} - \eta_B}\right)\right\}^{1/2} \simeq \frac{2\lambda}{\pi\theta_s}\frac{\eta_B' - \eta_B}{\eta_{B0} - \eta_B}, \quad (22)$$

where the approximation is the first term in the series expansion of the logarithm.

As an example we apply the results of this section to a practical case. We use several measurements of a 25-m antenna at the Westerbork Radio Observatory, Wester-bork, The Netherlands, (Table III). The column η_A (corr) gives the aperture efficiency after correction for the transmission loss of the reflector mesh. The beam efficiency is computed with (10). Both $\eta_{A0} = 0.60$ and $\eta_{B0} = 0.80$ are found from an extrapolation to $\lambda \to \infty$. At $\lambda = 2.9$ cm a measurement on the Moon yields $\eta_B' = 0.70$. From (22) we deduce $c = 2.0$ m, which is in reasonable agreement with the average size of the reflector panels. Clearly $d/2c \gg 1$ and (19b) can be used to find the rms surface error ϵ. We obtain $\epsilon = 1.5$ mm from the radio source measurements, while direct measurements on the reflector surface yields $\epsilon = 1.4$ mm (Baars and Hooghoudt [5]). The level of the error pattern at $\lambda = 2.9$ cm with respect to the peak of the main beam follows from (20) as $f_E(0)/f_D(0) = -16.5$ dB. For the HPBW of the error pattern we find from (18) $\theta_E = 26'$. These are amenable for direct measurement on a small source with a sensitive receiver, but this has not been done yet.

VI. MEASUREMENT OF ANTENNA PATTERN

In Section V we have dealt mainly with the measurement of those pattern parameters for which the power flux of the test source must be known. The knowledge of these parameters is essential for an evaluation of the capabilities of the antenna.

The determination of the detailed *shape*, i.e., the relative sidelobe level of the antenna pattern over a large solid angle will be discussed now. For many applications this is important, e.g., for the radio astronomer to correct his observations for the power received from the surrounding, for the communication engineer to estimate the possible interference from a transmitter radiating in the sidelobes.

Since we want to measure the strength of the pattern with respect to the main beam maximum, the flux density of the source need not be known. Ideally the source should have the following properties:

1) located in the far field (see (1));

2) point like, or at least small enough in angular size in order not to broaden the features in the antenna pattern;

3) sufficiently strong to enable a measurement of all parts of the antenna pattern; for a large antenna this may require measuring levels of 60 dB or more below the main beam response.

An earth-bound transmitter will easily satisfy requirement 2) and can be made strong enough to satisfy 3). However, for a large antenna (in excess of several hundreds of wavelengths in size) it will be difficult to place it in the farfield. If one can manage to do this the transmitter will be at a very low elevation angle. This causes problems with reflection at and radiation from the ground. Perhaps the most extensive set of measurements with a transmitter has been done by Higgs (15) on the pattern of a 25-m antenna at 820 MHz. His results suffer from the need to apply large corrections, which are not accurately known. These corrections are a direct result of the low elevation angle of the source.

A geostationary satellite with a transmitter would satisfy all three criteria. A serious drawback is the likely requirement of a large range of frequencies with the danger of mutual interference between users. Especially radio astronomers would be very reluctant to such a solution. In this context it is of interest to mention the measurement of the pattern of the Goldstone 64-m deep-space antenna with the aid of a transmitter located in the Surveyor I spacecraft on the surface of the Moon. Sidelobes down to -60 dB could be measured [23].

The use of cosmic radio sources offers a solution to the problem, although it has of course its limitations. The sources are certainly in the far field of the largest antenna and their daily movement in azimuth and elevation enables useful measurements to be done with an antenna of limited steerability. The angular size varies from less than $10^{-3}{''}$ for some quasars to about 0.5° for the Sun and the Moon (Table I). Unfortunately the smallest sources are not the strongest, so requirements 2) and 3) will not always be fulfilled simultaneously. Generally speaking the source size can be up to 0.1–0.2 times the antenna HPBW without introducing significant errors (compare Table II). Fig. 12 gives an indication of the

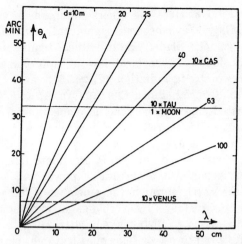

Fig. 12. Antenna HPBW θ_A as function of wavelength λ with reflector diameter as parameter. Angular size of several sources is indicated, in some cases multiplied by factor ten.

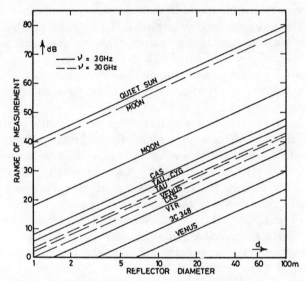

Fig. 13. Range of measurement of antenna patterns (in dB) as function of the diameter d of reflector for several strong radio sources at frequencies of 3 and 30 GHz. Assumed receiver noise fluctuation is $\Delta T = 0.01$ K rms, and 0 dB gives SNR = 3.

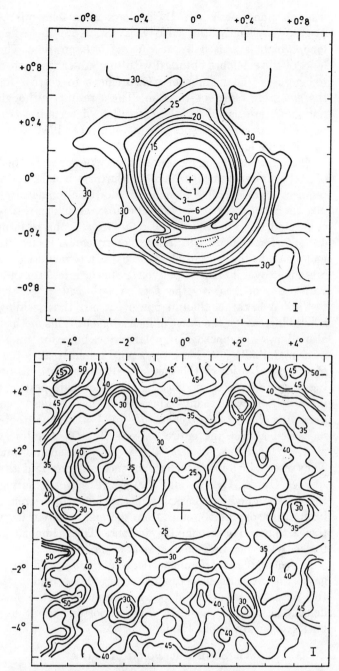

Fig. 14. Example of measurement of main beam (top) with Cas A and larger area (bottom) with Sun. HPBW is 18' so angular structure in large area has been smeared by the Sun's large angular extent. Note coma sidelobe which results from radial offset position of feed horn. Level of -17 dB for offset of 2 HPBW fits curves of Fig. 7 (Unpublished results by Kemper and Schouten).

range over which several sources can be used to satisfy requirement 2).

The range of measurement in intensity depends, for given receiver sensitivity and antenna size, on the flux density of the sources. In Fig. 13 we have indicated this for the strongest sources. The figure applies to a measurement with a single antenna. With a special method, to be discussed presently, a larger range of measurement is possible. For other frequencies the figure can be modified according to the spectrum of the source (Table I). The adjustment for a different receiver sensitivity is made through the formula for the rms noise fluctuation ΔT

$$\Delta T = \frac{T_s}{(\tau \Delta \nu)^{1/2}}$$

where T_s is the system noise temperature, $\Delta \nu$ is the receiver bandwidth, and τ is the integration time.

The Sun is by far the most powerful radio source, but nevertheless of restricted use in accurate pattern measurements because of its relatively large angular size and strongly varying intensity with solar activity. If one uses it to determine very low sidelobe levels, one must make the measurement in a period of solar quietness and preferably monitor the total flux with a small auxiliary antenna at the same frequency. A certain smearing of the angular sidelobe structure will generally be unavoidable. This applies also to the use of the Moon. Fig. 14 shows an example of a sidelobe pattern measured with the Sun and with Cassiopeia A for the main beam. It was obtained with the Dwingeloo's 25-m telescope at 11-cm wavelength; the HPBW is 18'. Note the coma lobe, which is caused by an off-axis feed position. The level of the sidelobes obtained with the Sun can be scaled to those of the mainbeam measurement to correct for the smearing effect of the Sun. These results, although not ideal, are useful for the correction of certain astronomical observations and give a reasonable idea about the sidelobe level.

There remains to be mentioned a very powerful method of pattern measurement with radio sources, which however requires a considerable amount of additional equipment, viz the *interferometric method*. The antenna under test is connected to a relatively small reference antenna to form an interferometer. The reference antenna keeps its mainbeam at the source, while the pattern to be measured is scanned over it. The output voltage of the interferometer varies proportional to the *field-strength pattern* of the test antenna, rather than its power pattern, thus yielding a large increase in the range of measurement. This method also allows the phase and polarization behavior to be measured. Smith [35] has given a good general description of the capabilities of the method, although his data on radio sources need modification. A thorough practical investigation with many experimental details has been presented by Hartsuijker *et al.* [13]. These authors measured almost the entire pattern of a 25-m antenna at λ = 21 cm to a level of 60 dB below the mainbeam. They used a 7.5-m diameter reference antenna and the sources Cas A, Cyg A, and Tau A. It is worth noting that this method is remarkably insensitive to interference from other sources as well as from thermal radiation from the earth entering through the sidelobes. This is important for accurate measurements of weak sidelobe levels.

If one of the interferometer elements has a calibrated gain, the measurement can be arranged to yield both the absolute gain of the other antenna and the flux density of the radio source. This important result has been treated by Wyllie [38].

VII. Some Additional Remarks

The foregoing sections have demonstrated the great potential of the use of cosmic radio sources for the measurement of large antennas at microwave and millimeter waves. The techniques have been mainly developed by radio astronomers and have been scarcely published. The success of the method depends to a large extent on a good knowledge and a proper evaluation of the properties of the radio sources used as test transmitters. The material on the sources collected in Table I and Figs. 2 and 3 now appears well established and unlikely to be subject to significant changes in the future.

In the limited space of this paper, it is impossible to present an exhaustive summary of all the possibilities and limitations of the technique. It is good to remember that the radio sky away from the test source may contaminate the measurements, especially at frequencies below 1 GHz. Here the general radiation of the Galaxy (our Milky Way) becomes significant, particularly close to the galactic plane. The brightness temperature of the Galaxy follows a spectral law of approximately $\nu^{-2.5}$, leading to strong radiation ($T_b \simeq 15$ K at 1 GHz) over *large solid angles*. Since Cas A, Cyg A, and Tau A are all located close to the galactic plane, special care must be taken for "background" effects at the lower frequencies.

A similar danger is presented by the thermal radiation of the earth, which is at an approximate temperature of 290 K. A significant amount of power will be received from the earth through the sidelobes. Changes in this power, for instance by a change in elevation angle of the antenna may mistakenly be attributed to variations in sky brightness or sidelobe features close to the mainbeam.

The radiation of the earth can be used to find the major antenna parameters in an approximate manner. The beam efficiency can be found from a measurement of the value of the antenna temperature with the antenna pointed at the horizon and at the zenith. This method has been developed by Mezger [26]. The accuracy is rather limited, but it is a useful method in those cases where low receiver sensitivity does not allow the use of radio sources.

In very accurate measurements with radio sources it might be necessary to avoid the daytime because of possible influence from solar radiation entering a region of relatively strong sidelobe levels. Especially in times of solar activity the Sun can be more of a disadvantage than an asset in antenna measurements.

In making accurate measurements the influence of the earth's atmosphere must be considered. The attenuation by oxygen and water vapor in the troposphere has already been mentioned in Section I. The attenuation is proportional to the total amount of atmosphere traversed which is proportional to sec z, where z is the zenith angle, and valid to $z = 80°$. Due to the dynamic behavior of the atmosphere there will be a fluctuating component in the attenuation, which leads to a fluctuating atmospheric emission. This can be serious at frequencies of 10 GHz and higher and impair the accuracy of the measurements. These effects have been studied by Baars [1],[2].

Finally, we mention one interesting method of antenna calibration, which does not use cosmic radio sources, but applies the same techniques. As test source a black disc of well known temperature is used located in the farfield of the antenna. From the known temperature and angular size of the disc, the antenna gain and beam efficiency can be derived. The method, also known as the "artificial moon" method has been developed by Troitskii *et al.* at the Gorki Institute of Radiophysics, USSR. Kuz'min and Salomonovich [22, ch. 5] briefly discuss the basics of the technique. Its advantage over a test transmitter is that it enables absolute gain calibration without the usual substitution method.

VIII. Concluding Summary

There exists a set of cosmic radio sources which can be used as a test transmitter for the measurement of antenna parameters. They provide a remarkably constant source of broad-band radiation, traversing the sky daily along a precisely known path. Thus they are ideal for the measurement of antenna pointing and feed focusing corrections. The strongest sources have accurately known *absolute* spectra, which makes the calibration of antenna gain possible in a simple and straightforward manner. By the use of sources of different angular size additional information can be obtained.

The sources are always in the farfield and problems with ground reflections because of low elevation can be avoided.

For the sources of small angular size the intensity is the main limiting factor in their use. For only the most sensitive antenna installations do they provide sufficient signal to measure the weakest features in the pattern. The use of an interferometer improves that situation significantly.

To use the method effectively one needs a good knowledge of the source and to properly evaluate the effects of the atmosphere and surface of the earth and other discrete (Sun) or distributed (Galactic plane) radio sources in the sky. The aim of this paper has been to guide in this evaluation and to show the great capabilities of cosmic radio sources in the measurement of large antennas.

Nomenclature

A — Absorption area of antenna aperture.
c — Correlation radius of reflector surface deviations.
D — Directivity of antenna.
d — Diameter of antenna aperture.
$f(x,y)$ — Normalized antenna power pattern.
f_D — Diffraction pattern.
f_E — Error pattern.
G — Antenna gain.
k — Boltzmann's constant (1.38×10^{-23} J K^{-1}).
S — Flux density of radio source.
T_A — Antenna temperature.
T_b — Brightness temperature.
δ — Phase error due to reflector surface deviation.
ϵ — Reflector surface deviation.
η_A — Aperture efficiency.
η_B — Mainbeam efficiency.
η_l — Sidelobe efficiency.
η_R — Radiation efficiency.
θ_0 — Measured HPBW of antenna.
θ_A — True HPBW of antenna.
θ_E — Half-power width of error pattern.
θ_s — Half-power angular size of radio source.
λ — Wavelength.
Ω_0 — Antenna pattern solid angle of perfect reflector.
Ω_A — Antenna pattern solid angle of actual reflector.
Ω_l — Solid angle of sidelobes.
Ω_m — Mainbeam solid angle.
Ω_s — Solid angle of radio source.
Ω_Σ — Source solid angle weighted by antenna pattern.

Acknowledgment

The author wishes to record his gratitude to Dr. P. G. Mezger, who introduced him to the subject of this paper and with whom several extensions of the technique were developed.

References

[1] J. W. M. Baars, "Reduction of tropospheric noise fluctuations at centimetre wavelengths," *Nature*, vol. 212, p. 494, 1966.
[2] ——, "Dual beam parabolic antennae," Dissertation Delft University of Technology. Groningen, The Netherlands: Wolter-Noordhoff, 1970.
[3] J. W. M. Baars and A. P. Hartsuijker, "The decrease of flux density of Cassiopeia A and the absolute spectra of Cassiopeia, Cygnus A and Taurus A," *Astron. Astrophys.*, vol. 17, p. 172, 1972.
[4] ——, in preparation, 1973.
[5] J. W. M. Baars and B. G. Hooghoudt, "The synthesis radio telescope at Westerbork. II, general lay-out and mechanical aspects," *Astron. Astrophys.*, to be published.
[6] B. R. Bean and E. J. Dutton, *Radio Meteorology*. Washington, D. C.: NBS Monog. 92, 1966.
[7] D. K. Cheng, "Effect of arbitrary phase errors on the gain and beamwidth characteristics of radiation pattern," *IRE Trans. Antennas Propagat.* (Commun.), vol. AP-3, pp. 145–147, July 1955.
[8] J. W. Findlay, "Radio telescopes", *IEEE Trans. Antennas Propagat.*, vol. AP-12, pp. 853–864, Dec. 1964.
[9] ——, "Absolute intensity calibrations in radio astronomy," *Ann. Rev. Astron. Astrophys.*, vol. 4, p. 77, 1966.
[10] ——, "Filled-aperture antennas for radio astronomy," *Ann. Rev. Astron. Astrophys.*, vol. 9, p. 271, 1971.
[11] W. Graf, R. N. Bracewell, J. H. Deuter, and J. S. Rutherford, "The sun as a test source for boresight calibration of microwave antennas," *IEEE Trans. Antennas Propagat.*, vol. AP-19, pp. 606–612, Sept. 1971.
[12] R. C. Hansen, "Aperture theory," in *Microwave Scanning Antennas*. New York: Academic Press, 1964.
[13] A. P. Hartsuijker, J. W. M. Baars, S. Drenth, and L. Gelato-Volders, "Interferometric measurement at 1415 MHz of radiation pattern of paraboloidal antenna at Dwingeloo radio observatory", *IEEE Trans. Antennas Propagat.* vol. AP-20, pp. 166–176, Mar. 1972.
[14] C. Heiles and W. Hoffman, "The beam shape of the NRAO 300-ft telescope and its influence on 21 cm line measurements," *Astron. J.*, vol. 73, p. 412, 1968.
[15] L. A. Higgs, "The antenna characteristics at 820 MHz of the Dwingeloo radio telescope," *Bull. Astron. Inst. Neth.*, (BAN) Suppl. vol. 2, p. 59, 1967.
[16] S. von Hoerner, "Design of large steerable antennas," *Astron. J.*, vol. 72, p. 35, 1967.
[17] D. C. Hogg and R. W. Wilson, "A precise measurement of the gain of a large horn-reflector antenna," *Bell Syst. Tech. J.*, vol. 44, p. 1019, 1965.
[18] W. E. Howard, "Effects of antenna scan rate and radiometer time constant on receiver output," *Astron. J.*, vol. 66, p. 521, 1961.
[19] T. F. Howell and J. R. Shakeshaft, "Attenuation of radio waves by the troposphere over the frequency range 0.4–10 GHz," *J. Atmos. Terr. Phys.*, vol. 29, p. 1559, 1967.
[20] E. V. Jull and E. P. Deloli, "An accurate absolute gain calibration of an antenna for radio astronomy", *IEEE Trans. Antennas Propagat.*, vol. AP-12, pp. 439–447, July 1964.
[21] J. D. Kraus, *Radio Astronomy*. New York: McGraw-Hill, 1966.
[22] A. D. Kuz'min and A. E. Salomonovich, *Radioastronomical Methods of Antenna Measurement*. New York: Academic Press, 1966.
[23] G. S. Levy, D. A. Bathker, A. C. Ludwig, D. E. Neff, and B. L. Seidel, "Lunar range radiation patterns of a 210-foot antenna at S-band," *IEEE Trans. Antennas Propagat.* (Commun.), vol. AP-15, pp. 311–313, Mar. 1967.

[24] C. H. Mayer, "Thermal radio emission of the planets and the moon," in *Surfaces and Interiors of Planets and Satellites*, A. Dollfus, Ed. New York: Academic Press, 1970.

[25] M. L. Meeks, J. A. Ball, and A. B. Hull, "The pointing calibration of the Haystack antenna," *IEEE Trans. Antennas Propagat.*, vol. AP-16, pp. 746–751, Nov. 1968.

[26] P. G. Mezger, "Der absolute Strahlungsfluss einiger Radioquellen bei 1419 MHz," *Z. Astrophys.*, vol. 46, p. 234, 1958.

[27] G. Reber, "Cosmic static," *Proc. IRE*, vol. 28, pp. 68–70, Feb. 1940.

[28] W. V. T. Rusch and P. D. Potter, *Analysis of Reflector Antennas*. New York: Academic Press, 1970.

[29] J. Ruze, "Lateral feed displacement in a paraboloid," *IEEE Trans. Antennas Propagat.*, vol. AP-13, pp. 660–665, Sept. 1965.

[30] ——, "Antenna tolerance theory—a review," *Proc. IEEE*, vol. 54, pp. 633–640, Apr. 1966.

[31] A. Y. Salomonovich, B. V. Braude, and N. A. Esepkina, "Measuring parameters of pencil-beam antennas in the near-zone," *Radio Eng. Electron. Phys.*, vol. 9, p. 876, 1964.

[32] H. Scheffler, "Über die Genauigkeitsforderungen bei der Herstellung optischer Flächen für astronomische Teleskope," *Z. Astrophys.*, vol. 55, 1, 1962.

[33] S. Silver, *Microwave Antenna Theory and Design*. New York: McGraw-Hill, 1949.

[34] F. G. Smith, *Radio Astronomy*. Baltimore, Md.: Penguin, (Pelican A 479), 1962.

[35] P. G. Smith, "Measurement of the complete far-field pattern of large antennas by radio-star sources," *IEEE Trans. Antennas Propagat.*, vol. AP-14, pp. 6–16, Jan. 1966.

[36] V. S. Troitsky, "Investigation of the surfaces of the moon and planets by the thermal radiation," *Radio Sci.*, vol. 69D, p. 1585, 1965.

[37] R. Wielebinski, personal communication, Aug. 1972.

[38] D. V. Wyllie, "An absolute flux density scale at 408 MHz," *Mon. Notic. Roy. Astron. Soc.*, vol. 142, p. 229, 1969.

Antenna Tolerance Theory—A Review

JOHN RUZE, FELLOW, IEEE

Abstract—The theoretical basis of antenna tolerance theory is reviewed. Formulas are presented for the axial loss of gain and the pattern degradation as a function of the reflector surface rms error and the surface spatial correlation.

Methods of determining these quantities by astronomical or ground-based electrical measurements are described. Correlation between the theoretical predictions and the performance of actual large antenna structures is presented.

I. Introduction

THE REQUIREMENT of precise optics for good image quality is well known in optical technology, and methods of testing and contour shaping have been developed to obtain precisions in excess of one part in 10^7. Optical systems of very large D/λ (diameter to wavelength), ratio are therefore common. Large antennas, such as required for radio astronomy or interplanetary probes, are engineering structures subject to gravity, wind, and thermal strains. Contour measurement and adjustment to the accuracy desired is also extremely difficult. Normal civil engineering structures have a precision of about one part in a thousand. Significant progress has been made in recent large parabolic antennas both in precision of construction (one part in 30 000) and in the computer prediction of deformation under various loads. Nevertheless, the tolerance of the structure sets a limit on the highest frequency of operation and thereby on the D/λ ratio. It is desirable to review the theory of aperture errors and their effect on the antenna radiation pattern.

We begin with a simple approach and attempt to develop a tolerance theory in an heuristic manner. The axial gain of a circular aperture with an arbitrary phase error or aberration $\delta(r, \phi)$ may be written as

$$G(0) = \frac{4\pi}{\lambda^2} \frac{\left| \int_0^{2\pi} \int_0^a f(r, \phi) e^{j\delta(r,\phi)} r\,dr\,d\phi \right|^2}{\int_0^{2\pi} \int_0^a f^2(r, \phi) r\,dr\,d\phi}, \quad (1)$$

where $f(r, \phi)$ is the in-phase illumination function in terms of the aperture coordinates r, ϕ.

For small phase errors, the exponential may be expanded in a power series with the result that the ratio

of the gain to the no-error gain G_0 is

$$\frac{G}{G_0} \approx 1 - \overline{\delta^2} + \overline{\delta}^2 \quad (2)$$

where

$$\overline{\delta^2} = \frac{\int_0^{2\pi} \int_0^a f(r, \phi) \delta^2(r, \phi) r\,dr\,d\phi}{\int_0^{2\pi} \int_0^a f(r, \phi) r\,dr\,d\phi}$$

$$\overline{\delta} = \frac{\int_0^{2\pi} \int_0^a f(r, \phi) \delta(r, \phi) r\,dr\,d\phi}{\int_0^{2\pi} \int_0^a f(r, \phi) r\,dr\,d\phi} .$$

In general, the phase reference plane can be chosen so that $\overline{\delta}$, the illumination weighted mean phase error, is zero. The loss of gain is then simply

$$\frac{G}{G_0} = 1 - \overline{\delta_0^2}, \quad (3)$$

where $\overline{\delta_0^2}$ is calculated from the mean phase plane.

This simple relation (3), that the fractional loss of gain is equal to the weighted mean-square phase error was probably first pointed out by Marechal [1] and, in antenna technology, by Spencer [2]. It is valid for any illumination and reflector deformation, provided the latter is small in wavelength measure. It indicates that for a one dB loss of gain the rms phase variation about the mean phase plane must be less than $\lambda/14$ or, for shallow reflectors, the surface error must be less than $\lambda/28$.

We next seek a more exhaustive analysis valid for large phase errors and one that would give information on the radiation pattern. The problem is common to a class of problems, illustrated in Fig. 1, where a plane wave is distorted into an error phase front. Alternately, we can say that to a narrow or "diffracted limited" direction of transmission is added a wider angular spectrum of scattered energy.

If we have detail knowledge of the phase front error, the radiation pattern or angular spectrum can be obtained by machine computation of the standard Kirchhoff integral [3]. Unfortunately, such detail knowledge is not available, and we must fall back on various statistical estimates of the character of the surface distortion and obtain a probable radiation pattern.

Manuscript received October 1, 1965; revised December 1, 1965. This paper was presented at the 1965 Symposium of the IEEE Group on Antennas and Propagation.

The author is with the Lincoln Laboratories, Massachusetts Institute of Technology, Lexington, Mass. (Operated with the support of the U. S. Air Force.)

Reprinted from *Proc. IEEE*, vol. 54, no. 4, pp. 633-640, Apr. 1966.

We can begin the analysis by subdividing the aperture into N subregions, each with a phase error and with no relation or correlation with contiguous regions. This crude model is shown in Fig. 2, where the aperture phase front is represented by a number of hatboxes of random heights. The axial field is the sum of these individual vector contributions. With no phase error the power sum of N unit vectors is N^2 (see Fig. 3). If we now assume that the phase of each vector is randomly in error by an amount taken from a Gaussian population of standard deviation δ, in radians, then the expected or average power sum is

$$\overline{P} = N^2 e^{-\overline{\delta^2}} + N(1 - e^{-\overline{\delta^2}}). \qquad (4)$$

The first term may be considered as the coherent power and the second as the incoherent. For small or large errors, we get the limiting forms for the coherent or incoherent addition of waves. The distribution of the sum is also of interest [4]. For small phase errors, the distribution is Gaussian in voltage with a standard deviation $\sqrt{N}\delta$, so that the distribution becomes relatively more peaked with a larger number of vectors and smaller phase errors. For large phase errors, we have the well-known Rayleigh distribution characterized by the mean power N.

The expected or average radiation function of the crude model shown in Fig. 2 can be derived. The procedure is briefly as follows [5]: the field at a general far-field point is expressed as a Kirchhoff surface integral. The power pattern is obtained by multiplying by the conjugate integral yielding a double surface integral of the two running surface vector variables. A correlation function is defined as a function of the vector difference of these variables. The average or expected value is then obtained. To perform the integration, assumptions must be made on the spatial nature of the correlation and on the frequency distribution of the phase errors. For the model chosen, these assumptions are that the phase values are completely correlated in a diameter "$2c$" and completely uncorrelated for larger distances. In addition, the various phases come from a Gaussian population of rms error "δ." As in all statistical problems, the number of components must be large so that

$$N \approx \left(\frac{D}{2c}\right)^2 \gg 1. \qquad (5)$$

The result of this process is

$$G(\theta, \phi) = G_0(\theta, \phi) e^{-\overline{\delta^2}} + \left(\frac{2\pi c}{\lambda}\right)^2 (1 - e^{-\overline{\delta^2}}) \Lambda_1\left(\frac{2\pi c u}{\lambda}\right), \qquad (6)$$

where

$G_0(\theta, \phi)$ is the no-error radiation diagram whose axial value is $\eta(\pi D/\lambda)^2$

η is the aperture efficiency

u is $\sin \theta$

$\Lambda_1(\ )$ is the Lambda function.

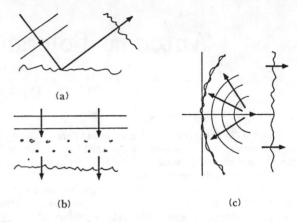

Fig. 1. A class of problems. (a) Reflection from a rough surface. (b) Transmission through a random medium. (c) Diffraction from an imperfect paraboloid.

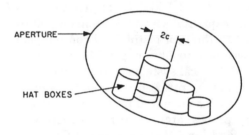

Fig. 2. Aperture subdivided into a number of hatboxes.

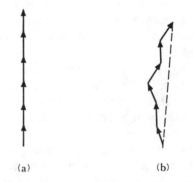

Fig. 3. Addition of vectors. (a) No phase error. (b) rms phase error "δ."

Although the model chosen is a crude one, (6) illustrates the changes in the radiation pattern and its similarity to (4) should be noted. We see that the no-error radiation diagram has been reduced by an exponential tolerance factor. A broad scattered field has been added whose "beamwidth" is inversely proportional to the size of the correlated region in wavelengths, so that smooth reflectors (large c) scatter more directively and rough reflectors (small c) more diffusely. For small phase errors, the relative magnitude of the axial scattered field is

$$\frac{1}{\eta}\left(\frac{2c}{D}\right)^2 \overline{\delta^2}. \qquad (7)$$

The model chosen can be considerably improved by replacing the hatboxes with hats as shown in Fig. 4. If

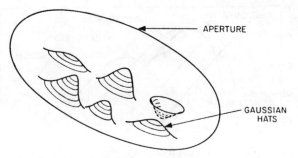

Fig. 4. Aperture subdivided into a number of hats.

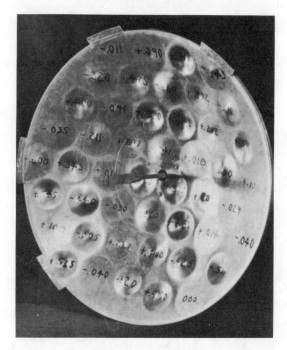

Fig. 5. Special model constructed to test theory.

the phase front distortions are assumed to be of Gaussian shape, the required integrations can again be performed [5] with the following result:

$$G(\theta, \phi) = G_0(\theta, \phi)e^{-\bar{\delta}^2}$$
$$+ \left(\frac{2\pi c}{\lambda}\right)^2 e^{-\bar{\delta}^2} \sum_{n=1}^{\infty} \frac{\bar{\delta}^{2^2}}{n \cdot n!} e^{-(\pi c u/\lambda)^2/n}. \quad (8)$$

Although (8) is more complex, the general effects are similar to those discussed previously.

We have considered a two-dimensional distribution of errors. It is of interest to present the one-dimensional case derived by Bramley [6] in our notation

$$G(\theta) = G_0(\theta)e^{-\bar{\delta}^2} + \frac{\sqrt{\pi}c}{\lambda} e^{-\bar{\delta}^2} \sum_{n=1}^{\infty} \frac{\bar{\delta}^{2n}}{\sqrt{n} \cdot n!} e^{-(\pi c u/\lambda)^2/n}. \quad (9)$$

The gain reduction and pattern degradation predicted by (8) was checked in the original reference [5] by the construction of a special model, Fig. 5, which fulfilled the statistical assumptions necessary for the theoretical development.

II. Discussion

From (8), we can write the reduction of axial gain as

$$\frac{G}{G_0} = e^{-\bar{\delta}^2} + \frac{1}{\eta} \left(\frac{2c}{D}\right)^2 e^{-\bar{\delta}^2} \sum_{n=1}^{\infty} \frac{\bar{\delta}^{2n}}{n \cdot n!}. \quad (10)$$

In the region of interest, i.e., reasonable tolerance losses, and for correlation regions that are small compared to the antenna diameter, the second term may be neglected and we have for the gain

$$G = G_0 e^{-\bar{\delta}^2} = \eta \left(\frac{\pi D}{\lambda}\right)^2 e^{-(4\pi\epsilon/\lambda)^2}, \quad (11)$$

where we define "ϵ" as the effective reflector tolerance in the same units as λ; i.e., that rms surface error on a shallow reflector (large f/D), which will produce the phase front variance $\bar{\delta}^2$. In Fig. 6 we plot the loss of gain (11) as a function of the rms error and the peak surface error. The ratio used, 3:1, is one found experimentally for large structures and results, in part, from the truncation used in the manufacturing process (i.e., large errors are corrected).

It should be noted that for small errors (11) is identical with (3), with the exception that the former is independent of the illumination function and the latter is not. For the statistical analysis, it was necessary to assume a uniform distribution of errors, for which case the illumination dependence factors out in (3) and becomes identical to (11).

For deep (nonshallow) reflectors, the surface tolerance is not exactly equal to the effective tolerance "ϵ." In addition, structural people at times measure the reflector deformations normal to the surface and at times in the axial direction. The relation between these quantities is

$$\epsilon = \frac{\Delta z}{1 + (r/2f)^2} \quad (12a)$$

$$\epsilon = \frac{\Delta n}{\sqrt{1 + (r/2f)^2}}. \quad (12b)$$

The result is that the tolerance gain loss in dB, as computed from the reflector axial or normal mean square error, is too high by a factor A. This factor is given in Fig. 7. For shallow reflectors, this correction factor approaches unity.

Equation (11) indicates that if a given reflector is operated at increasing frequency, the gain, at first, increases as the square of the frequency until the tolerance effect take over and then a rapid gain deterioration occurs. Maximum gain is realized at the wavelength of

$$\lambda_m = 4\pi\epsilon, \quad (13)$$

where a tolerance loss of 4.3 dB is incurred. This maximum gain is

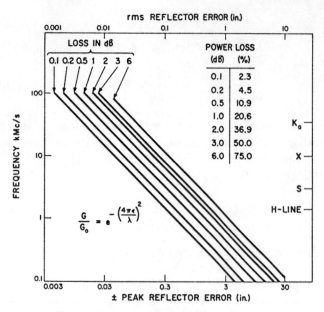

Fig. 6.   Gain loss due to reflector tolerance.

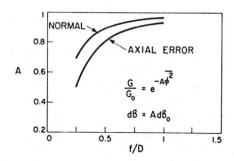

Fig. 7.   Correction factor due to reflector curvature.

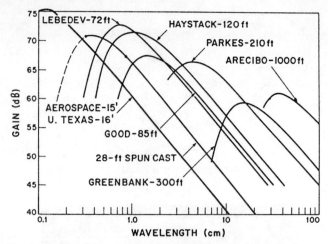

Fig. 8.   Gain of large paraboloids (based on published estimates).

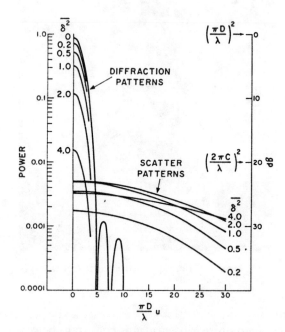

Fig. 9.   Radiation patterns of phase distorted circular aperture, 12 dB illumination taper, $D = 20c$.

$$G_{\max} \approx \frac{\eta}{43}\left(\frac{D}{\epsilon}\right)^2 \qquad (14)$$

and is proportional to the square of the precision of manufacture (D/ϵ).

This behavior is illustrated in Fig. 8, where we show some of the world's large antennas. The frequency region where the smaller and more precise structure is superior to the larger and coarser antenna, and the converse, is evident.

Next we consider the effect of surface errors on the radiation diagram. In Fig. 9, we show the pattern of a 12-dB tapered circular aperture with random phase errors and with $D = 20c$. We plot from (8) the expected power diffraction and scatter patterns for mean-square phase errors of 0.2, 0.5, 1.0, 2.0, and 4.0 in radian squared measure. These correspond to tolerance gain losses of 0.87, 2.2 4.3, 8.6, and 16.6 dB, respectively. The complete radiation diagram is the *power sum* of the diffraction and scatter patterns. It should be noted that the diffraction pattern is reduced by the exponential tolerance factor and that the energy lost appears in the scattered pattern, which broadens as the surface error increases.

With further increase in loss, the diffraction pattern is submerged in the scattered energy and disappears. Scheffler [7],[1] in a similar analysis, has pointed out that for large phase errors the scattered pattern approaches

$$G_s(\theta) = \left(\frac{2\pi c}{\lambda}\right)^2 \frac{[1 - e^{-\overline{\delta}^2}]}{\overline{\delta}^2} e^{-(\pi cu/\lambda)^2/\overline{\delta}^2}, \qquad (15)$$

so that for extremely large phase errors the radiated energy is scattered over an angular region with the intensity equal to

$$G_s(\theta) = \left(\frac{c}{2\epsilon}\right)^2 e^{-(cu/4\epsilon)^2}. \qquad (16)$$

[1] This reference was first brought to my attention by Dr. P. Mezger of the National Radio Astronomy Observatory.

We note that under these extreme conditions the beamwidth is defined by the average surface slopes and is wavelength independent, a result we would have expected from geometric optics.

Before leaving the theoretical discussion, it should be recalled that the distribution in the focal plane has the same shape as the radiated angular spectrum. Therefore, the same relation (8) can be used to determine the spot size due to surface imperfections or small scale atmospheric inhomogeneities.

III. APPLICATION TO ANTENNA STRUCTURES

The experimental check of the theory afforded by the specially constructed model (Fig. 5) merely verifies the mathematical development. We turn now to practical structures and list those factors which deviate from the theoretical assumptions.

1) The surface errors are not random, but to a large part are due to calculable gravity, wind, and thermal strains. However, analysis of actual antenna photogrammetric measurements indicates that the reflector deviations, if not strictly random and Gaussian, are distributed in a bell-shaped curve [8], [9].
2) The actual reflector errors are not uniformly distributed over the aperture. Again, photogrammetric measurements and deformation calculations after structural compensation indicate that this condition is not grossly violated [8], [9].
3) The theory assumes a fixed, circular correlation region. As the contour adjustment points are normally spaced in a uniform grid, there is a tendency for this condition; however, various structural factors such as pie-panel segments would yield elliptical correlated regions of varying size.
4) The theory also requires that the number of uncorrelated regions in the aperture be large, that is $D \gg 2c$. It has been found that for compensated structures the number of regions is related to the panel size or spacing of the target points.
5) It was also assumed that the spatial phase correlation function had a particular shape, namely Gaussian. Another smooth deformation surface would have yielded slightly different functional forms in the shape of the scattered power.
6) Finally, we have developed a statistical theory and obtained the average power pattern of the ensemble of such antennas. We apply the theory to one sample.

Therefore, a check of the performance of actual antenna structures with the above theory is necessary. Correlation has been obtained between frequency-gain measurements and optical photogrammetric measurements [9]. We present here other confirmation.

In Fig. 10, we show a horn reflector antenna [10]. The

Fig. 10.   Horn reflector antenna.

gain of this antenna was precisely measured over 6:1 range of frequencies [11]. Equation (11) can be written as

$$10 \log G\lambda^2 = 10 \log \eta(\pi D)^2 - \left(\frac{4\pi\epsilon}{\lambda}\right)^2 10 \log e, \quad (17)$$

which is the straight line

$$y = a - bx$$

when $G\lambda^2$ in dB is plotted against reciprocal wavelength squared. The vertical intercept is a measure of the aperture efficiency and the reflector tolerance can be obtained from the slope.

The experimental data is shown in Fig. 11, where outside of a gain droop at low frequencies, due to diffraction effects, the data follow a straight line with a mean deviation of 0.166 dB. The indicated aperture efficiency also lies between the calculated efficiencies of 78.34 and 76.13 percent for the two polarizations used. The predicted surface tolerance is an effective value of 33 mils or 48 mils normal to the parabolic surface. The agreement of the measured data with the predicted straight line relationship is a confirmation of antenna tolerance theory. In addition, this procedure, combined with a linear regression analysis of the experimental data, to establish confidence limits, is probably the most convenient and accurate method of determining the surface precision [12].

We next consider the determination of the size of the correlation region by means of electrical measurements. The temperature measured on an extended astronomical source is equal to the product of the fractional enclosed power and the source brightness temperature. With no surface errors, practically all the radiated power is enclosed by the source if it is at least several beamwidths in extent. With reflector errors, some of the scattered energy is outside of the source and the measured temperature is decreased. This reduction depends on both

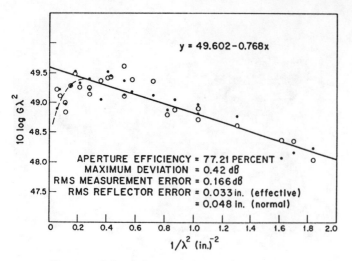

Fig. 11. Gain vs. frequency-horn reflector antenna.

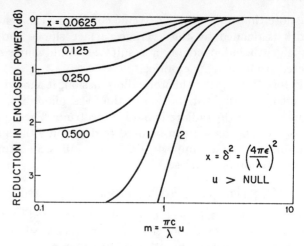

Fig. 12. Reduction in enclosed power.

the rms surface error and the size of the correlation region.

The fractional enclosed power in a cone angle $u_0 = \sin \theta_0$ (several beamwidths) can be obtained by integration of (8) with the result:

$$EP(\theta_0) = [1 - S]\left[1 - e^{-\bar{\delta}^2} \sum_{n=1}^{\infty} \frac{\overline{\delta^{2^n}}}{n!}\, e^{-(\pi c u_0/\lambda)^2/n}\right] \quad (18)$$

where S is the fractional energy very widely scattered by aperture blockage (feed supports, etc.). It should be noted that, either with no surface error or with large cone angle, the enclosed power is a constant.

Fig. 12 shows the reduction of enclosed power as a function of the rms error and the enclosed cone angle. If temperature measurements are now made of the same source at two different frequencies, the enclosed power is different. If we enter the temperature ratio (after correction for atmospheric effects and spectral index) as an ordinate into Fig. 12 and the frequency ratio as an abscissa, then we can obtain a set of values of tolerance error and correlation intervals which satisfy this condition (with a known source cone angle). If the reflector tolerance is known from point source gain measurements, the required correlation radius is determined.

This type of measurement was applied to the HAYSTACK radio telescope (120-foot diameter in a metal space frame radome) at the frequencies of 7750 and 15 500 Mc/s. The moon was used as an extended source and the planet Jupiter as a point source. By means of the procedure outlined, it was concluded that the rms surface error ϵ was 0.053 inch and that the correlation radius c was 4.4 feet.

A check of antenna tolerance theory is obtained by comparison of the predicted antenna pattern based on the astronomically determined values of (ϵ, c) and the experimentally measured pattern with a ground-based transmitter. Figure 13 shows this comparison at the frequency where the tolerance effect is significant. The

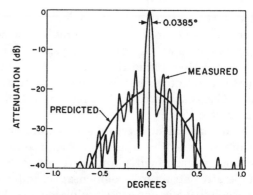

Fig. 13. Comparison of measured and predicted patterns, HAYSTACK (15.745 Gc/s).

agreement of the predicted pattern and the actual measured characteristic is excellent considering the statistical nature of the problem and that the sidelobe peaks should be 3 dB higher than the average intensity. A corresponding pattern, taken with a similar feed, at 7750 Mc/s where the tolerance effects are not significant showed sidelobe levels of about 25 dB down.

From the above measurements and those cited in the references, it may be inferred that the present status of antenna tolerance theory is such that the behavior of large antennas may be determined by the specification of two quantities: the rms surface error and the correlation interval. These quantities may be determined from electrical ground-based or astronomical data. Detail correlation of these electrically determined values and actual mechanical measurements is lacking. However, available photogrammetric or other structural estimates are not in variance with the theoretical predictions.

APPENDIX

A. Derivation of (4)

Consider the power sum of N unit vectors whose phases "δ" come from a normal distribution of zero mean and variance $\overline{\delta^2}$

$$P = \sum_i^N \sum_j^N e^{j(\delta_i - \delta_j)} = \sum_i^N \sum_j^N e^{jv_{ij}}. \qquad (19)$$

y is another statistic, defined as the difference of two samples taken from the original distribution. It can be readily shown that it is normally distributed with zero mean and variance $2\overline{\delta^2}$ or

$$W(y) = \frac{1}{\sqrt{4\pi\delta^2}} e^{-y^2/4\overline{\delta^2}}, \qquad (20)$$

now

$$\overline{P} = \sum_i^N \sum_j^N \overline{\cos y} + i\overline{\sin y},$$

for $i \neq j$

$$\overline{\cos y} = \int_{-\infty}^{\infty} \cos y W(y) dy = e^{-\overline{\delta^2}} \qquad (21)$$

$$\overline{\sin y} = \int_{-\infty}^{\infty} \sin y W(y) dy = 0. \qquad (22)$$

Therefore

$$\overline{P} = N^2 e^{-\overline{\delta^2}} + N(1 - e^{-\overline{\delta^2}}). \qquad (23)$$

B. Derivation of (8)

The gain function of an aperture is written:

$$G(\theta_1 \phi) = \frac{4\pi}{\lambda^2} \frac{\left| \int f(\bar{r}) e^{j\bar{k} \cdot \bar{r}} e^{j\delta(r,\phi)} dS \right|^2}{\int f^2(\bar{r}) dS}, \qquad (24)$$

where

$\bar{r}$ is an aperture vector position variable,
$\bar{k} = (2\pi/\lambda)\overline{p_0}$ is a vector in the direction of observation,
$\delta(r, \phi)$ is the aperture phase perturbation function,
dS is an elemental aperture area,
$f(\bar{r})$ is the aperture illumination function.

The numerator may be written as:

$$\iint f(\bar{r}_1) f(\bar{r}_2) e^{j\bar{k} \cdot (\bar{r}_1 - \bar{r}_2)} e^{j(\delta_1 - \delta_2)} dS_1 dS_2,$$

denoting $\bar{\tau} = \bar{r}_1 - \bar{r}_2$ as the vector difference between the two aperture running variables and $y(\bar{\tau}) = \delta_1 - \delta_2$ as the phase difference of the two points. We have

$$\iint f(\bar{r}_1) f(\bar{r}_1 + \bar{\tau}) e^{j\bar{k} \cdot \bar{\tau}} e^{jy(\overline{\tau})} dS_1 dS_\tau$$

defining $\phi(\bar{\tau})$ as the illumination correlation function

$$\phi(\bar{\tau}) = \frac{\int f(\bar{r}_1) f(\bar{r}_1 + \bar{\tau}) dS_1}{\int f^2(\bar{r}_1) dS_1}. \qquad (25)$$

We now rewrite (24) as:

$$G(\theta_1 \phi) = \frac{4\pi}{\lambda^2} \int \phi(\bar{\tau}) e^{j\bar{k} \cdot \bar{\tau}} e^{jy(\tau)} dS_\tau$$

and the average or expected value

$$\overline{G(\theta_1 \phi)} = \frac{4\pi}{\lambda^2} \int \phi(\bar{\tau}) e^{j\bar{k} \cdot \bar{\tau}} [\overline{\cos y(\tau)} + i\overline{\sin y(\tau)}] dS_\tau \qquad (26)$$

for τ large compared to "c," the phase correlation distance where the two phase samples are uncorrelated, $y(\tau)$ is normally distributed with zero mean and variance $2\overline{\delta^2}$. When τ approaches zero, $y(\tau)$ approaches zero with zero variance. Some convenient form must be assumed for the variance function. Taking

$$\overline{y^2}(\tau) = 2\overline{\delta^2}[1 - e^{-\tau^2/c^2}] \qquad (27)$$

from (21) and (22), we have

$$\overline{\cos y(\tau)} = e^{-\overline{\delta^2}[1 - e^{-\tau^2/c^2}]}$$

$$\overline{\sin y(\tau)} = 0.$$

Equation (26) may be rewritten as

$$\overline{G(\theta, \phi)} = \frac{4\pi}{\lambda^2} e^{-\overline{\delta^2}} \int \phi(\bar{\tau}) e^{j\bar{k} \cdot \bar{\tau}} e^{\overline{\delta^2} e^{-\tau^2/c^2}} dS_\tau$$

$$= \frac{4\pi}{\lambda^2} e^{-\overline{\delta^2}} \sum_{n=0}^{\infty} \int \phi(\bar{\tau}) e^{j\bar{k} \cdot \bar{\tau}} \frac{\overline{\delta^{2n}}}{n!} \cdot e^{-n\tau^2/c^2} dS_\tau.$$

The first term is the unperturbed pattern $G_0(\theta_1 \phi)$

$$\overline{G(\theta_1 \phi)} = G_0(\theta_1 \phi) e^{-\overline{\delta^2}}$$
$$+ \frac{4\pi}{\lambda^2} e^{-\overline{\delta^2}} \sum_{n=1}^{\infty} \int \phi(\bar{\tau}) e^{j\bar{k} \cdot \bar{\tau}} \frac{\overline{\delta^{2n}}}{n!} e^{-n\tau^2/c^2} dS_\tau.$$

Due to the exponential factor, the remaining terms have their principal contribution for $\tau < c$. As we have assumed that c is small compared to the aperture dimensions, the illumination correlation function (25) may be assumed as unity in evaluating these terms. The angular integration can be immediately performed with the result:

$$G(\theta_1 \phi) = G_0(\theta_1 \phi) e^{-\overline{\delta^2}}$$
$$+ \frac{8\pi^2}{\lambda^2} e^{-\overline{\delta^2}} \sum_{n=1}^{\infty} \frac{\overline{\delta^{2n}}}{n!} \int J_0\left(\frac{2\pi}{\lambda} u\tau\right) e^{-n\tau^2/c^2} \tau d\tau.$$

The integral can be evaluated by extending the limits and recalling that

$$\int_0^\infty J_0\left(\frac{2\pi}{\lambda}u\tau\right)e^{-n\tau^2/c^2}\tau d\tau = \frac{c^2}{2n}e^{-(\pi cu/\lambda)^2/n},$$

with the final result

$$G(\theta_1\phi) = G_0(\theta_1\phi)e^{-\overline{\delta^2}}$$

$$+\left(\frac{2\pi c}{\lambda}\right)^2 e^{-\overline{\delta^2}}\sum_{n=1}^\infty \frac{\overline{\delta^{2n}}}{n\cdot n!}e^{-(\pi cu/\lambda)^2/n}. \qquad (28)$$

In this derivation, we have assumed that we are dealing with highly directive antennas. The obliquity factor has, therefore, been suppressed and we have used the small angle formulation of the Kirchhoff integral.

C. The Function

$$S(m,x) = e^{-x}\sum_{n=1}^\infty \frac{x^u}{n\cdot n!}e^{-m^2/n}$$

x \ m	0.0	0.5	1.0	2.0	3.0	4.0
0.2	0.1723	0.1351	0.0655	0.0042	0.0001	—
0.5	0.3458	0.2739	0.1379	0.0120	0.0007	—
1.0	0.4848	0.3907	0.2093	0.0263	0.0026	0.0002
2.0	0.4986	0.4162	0.2505	0.0521	0.0087	0.0013
3.0	0.4111	0.3549	0.2367	0.0690	0.0159	0.0031
4.0	0.3242	0.2877	0.2091	0.0779	0.0227	0.0054

D. Data for Fig. 12

RATIO ENCLOSED POWER TO ANGLE u_0 dB

TOTAL POWER

x \ m	0.25	0.5	1.0	1.414	2.0	2.83
0.125	0.50	0.42	0.20	0.08	0.01	0.001
0.25	1.00	0.83	0.40	0.16	0.03	0.003
0.50	2.02	1.65	0.80	0.35	0.09	0.011
1.00	4.00	3.21	1.59	0.76	0.25	0.046
2.00	7.80	6.03	3.07	1.67	0.69	0.185

ACKNOWLEDGMENT

The author is indebted to Dr. J. L. Meeks and Dr. S. Weinreb for the radiometric measurements of the HAYSTACK antenna. Also, to Dr. J. W. Findlay and Dr. Mezger of the National Radio Astronomy Observatory for many helpful discussions.

REFERENCES

[1] A. Marechal, "The diffraction theory of aberrations," *Rep. Progr. in Phys. (GB)*, vol. XIV, p. 106, 1951; for English summary see E. Wolf.

[2] R. C. Spencer, "A least square analysis of the effect of phase errors on antenna gain," Air Force Cambridge Research Center, Bedford, Mass., AFCRC Rept. E5025, January 1949.

[3] A. R. Dion, "Investigation of effects of surface deviations on HAYSTACK antenna radiation patterns," M.I.T. Lincoln Lab., Lexington, Mass., Rept. 324, July 1963.

[4] P. Beckman, "The probability distribution of the vector sum of N unit vectors with arbitrary phase distributions," *ACTA Tech. (Czechoslovakia)*, vol. 4, no. 4, pp. 323–334, 1959.

[5] J. Ruze, "The effect of aperture errors on the antenna radiation pattern," *Suppl. al Nuovo Cimento*, vol. 9, no. 3, pp. 364–380, 1952. This work used for the theoretical basis of antenna tolerance, was first prepared by the author as a Ph.D. dissertation under the direction of Prof. L. J. Chu at M.I.T. in 1952.

[6] E. N. Bramley, "Some aspects of the rapid directional fluctuations of short radio waves reflected from the ionosphere," *Proc. IEE (London)*, vol. 102B, pp. 533–540, 1955.

[7] H. Scheffler, "Uber die Genauigkeitsforderungen bei der Herstellung optischer Flachen fur astronomische Teleskope," *Z. Astrophys. (Germany)*, vol. 55, pp. 1–20, 1962.

[8] J. W. Findlay, "Operating experience at the National Radio Astronomy Observatory," *Ann. N. Y. Acad. Sci.*, vol. 116, pp. 25–40, June 1964.

[9] P. G. Mezger, "An experimental check of antenna tolerance theory using the NRAO 85-foot and 300-foot telescopes," *1964 Internat'l Symp. on Antennas and Propagation*, pp. 181–185.

[10] R. W. Friis and A. S. May, "A new broadband microwave antenna system," *Trans. AIEE (Communication and Electronics)*, vol. 77, pp. 97–100, March 1958.

[11] A. Sotiropoulos and J. Ruze, "HAYSTACK calibration antenna," M.I.T. Lincoln Lab., Lexington, Mass., Tech. Rept. 367, December 1964.

[12] J. Ruze, "Reflector tolerance determination by gain measurement," *1964 NEREM Conv. Rec.*, pp. 166–167.

[13] R. H. T. Bates, "Random errors in aperture distributions," *IRE Trans. on Antennas and Propagation*, vol. AP-7, pp. 369–372, October 1959.

[14] H. G. Booker, J. A. Ratcliffe, and D. H. Shinn, "Diffraction from an irregular screen with application to ionospheric problems," *Phil. Trans. (GB)*, vol. 242, ser. A, pp. 579–609, 1950.

[15] R. N. Bracewell, "Tolerance theory of large antennas," *IRE Trans. on Antennas and Propagation*, vol. AP-9, pp. 49–58, January 1961.

[16] B. V. Brande, N. A. Esepkina, N. L. Kaidanovskii, and S. E. Khaikin, "The effects of random errors on the electrical characteristics of high-directional antennae with variable-profile reflectors," *Radioteknika i Elektronika (USSR)*, vol. 5, no. 4, pp. 75–92, 1960.

[17] D. K. Cheng, "Effect of arbitrary phase errors on the gain and beam width characteristics of radiation pattern," *IRE Trans. on Antennas and Propagation*, vol. AP-3, pp. 145–147, July 1955.

[18] A. Consortini, L. Rouchi, A. M. Scheggi, and G. Toroldo DiFrancia, "Gain limit and tolerances of big reflector antennas," *Alta Frequenza (Italy)*, vol. 30, pp. 232–276, March 1961.

[19] C. Dragone and D. C. Hogg, "Wide angle radiation due to rough phase fronts," *Bell Sys. Tech. J.*, vol. 42, pp. 2285–2296, September 1963.

[20] J. Robieux, "Influence of the precision of manufacture on the performance of aerials," *Am. Radio Elect.*, vol. 11, pp. 29–56, January 1956.

[21] Ya. S. Shifrin, "The statistics of the field of a linear antenna," *Radio Engrg. Electronic Phys.*, vol. 8, pp. 351–358, March 1963.

[22] R. A. Shore, "Partially coherent diffraction by a circular aperture," in *Electromagnetic Theory and Antennas*, E. C. Jordan, Ed. New York: Pergamon, 1963, pp. 787–795.

The Effect of Slowly Varying Surface Errors in Large Millimeter Wave Antennas: A Practical Verification in the Itapetinga 45-ft Reflector

PIERRE KAUFMANN, MEMBER IEEE, R. E. SCHAAL,
AND J. C. RAFFAELLI

Abstract—The aperture efficiency of the 45-ft Itapetinga reflector was found to be about 50 percent at $\lambda = 7$ mm. The reflector surface panels were mechanically measured in 2664 points displaying the presence of slowly varying errors that negligibly effect the aperture efficiency. We confirm that the reflector's aperture efficiency seems to be predictable from mechanical measurements of the surface when the distribution of errors is non-Gaussian, when the surface is only slightly rough in terms of a wavelength, and when the slowly varying errors across the surface are known [1], [2].

I. INTRODUCTION

The 45-ft radome-enclosed Itapetinga radio telescope was installed at the end of 1971 and since 1972 has been used in radio astronomical research at a relatively larger wavelength (i.e., 13.5 mm) [3]. At that time, very little could be learned about its characteristics at shorter wavelengths. In the past years, however, a number of improvements have been added to the system. The reflector's 72 panels were reset, new counterweights were added, and some minor modifications were introduced in the driving system (both at mechanical and electronic servo sections). Tracking accuracy was of the order of 5" rms in both elevation and azimuth axis, up to elevation angles close to 85° (see Fig. 1).

During the past (local) winter we performed the first systematic 7 mm (43 GHz) measurements with the Itapetinga dish. The radiometer utilized a room temperature mixer with balanced Schottky barrier diodes, providing a system temperature of about 1000 K (including the sky noise). Two rectangular horns of conventional design, producing 12 dB tapering, were constructed for the dual-horn feed which produced two beams in space separated by nearly four half-power beamwidths (the beams were of about 2' in size).

Due to the existence of other mechanical mounts at the antenna Cassegrain focus, the 7-mm feed mount was laterally displaced from the main reflector axis by about five half-power beamwidths. This may have produced a gain reduction factor of about 0.8 [4] for the main beam and slightly smaller for the comparison beam. The measurements were performed using the beam switching on–on technique, and we used the conservative gain reduction factor of 0.8 for correcting the measurements.

A narrow neon lamp, placed in front of the main horn, was calibrated as a reference noise source. The radome attenuation at 7 mm (i.e., radome membrane plus space frames blockage) was verified by placing an absorber in front of the feed horn and comparing the absorber noise contribution to the emission of a neighbouring hill through the radome. We found a gain reduction factor of the order of 0.67, which is comparable to

Manuscript received October 10, 1977; revised May 9, 1978. This work was supported by grants from Brazilian research agencies BNDE-FUNTEC (contract 87), FAPESP (proc. 67/514), and CNPq (proc. 5618/72 and proc. 5796/75).

The authors are with CRAAM/ON/CNPq—Conselho Nacional de Desenvolvimento Científico e Tecnologico, São Paulo. SP, Brazil.

Fig. 1. Rear view from 45-ft mm wave reflector, radome-enclosed, from the Itapetinga Radio-Observatory Atibaia, Brazil.

the manufacturer's specifications. At $\lambda = 7$ mm, the radome behavior is close to the maximum resonant attenuation (predicted for 5 mm).

II. FORMULAS AND SKY TRANSMISSION

For the measurement of aperture efficiency ($\eta_a = A_e/A_p$, where A_e is the effective aperture and A_p is the physical aperture), we used some well-known formulas. For point sources the effective aperture is

$$A_e = 2kT_a/S, \tag{1}$$

where k is the Boltzmann constant, T_a is the corrected antenna temperature, and S is the flux density. Using S in Janskys (1 Jy $= 10^{-26}$ Wm^{-2} Hz^{-1}), $k = 1.38 \times 10^{-23}$ J/K, and T_a in K, we have

$$A_e = 2.76 \times 10^3 T_a/S \qquad \text{m}^2. \tag{2}$$

For planets with a finite disk of angular size $\theta_d = (\theta_e + \theta_p)/2$ (where θ_e and θ_p are the equatorial and polar diameters, respectively), we must take into account their solid angles $\Omega = 1.85 \times 10^{-11} \theta_e \times \theta_p$ sr (with the diameters in arc seconds). For a planet with brightness temperature T_b, the corresponding flux density will be

$$S = \frac{2k}{\lambda^2} \iint\limits_{\text{disc}} T_b \, d\Omega \tag{3}$$

or, for a uniform brightness temperature distribution across

Reprinted from *IEEE Trans. Antennas Propagat.*, vol. AP-26, no. 6, pp. 854–857, Nov. 1978.

the planet disk,

$$S \sim \frac{2kT_b}{\lambda^2} \Omega, \qquad (4)$$

where Ω is the planet's solid angle. Substituting in (1) we obtain

$$A_e = (T_a/T_b)(\lambda^2/\Omega)K \qquad \text{m}^2. \qquad (5)$$

K is a correction taking into account the planet size in relation to the antenna half-power beamwidth θ_a [5], i.e.,

$$K = [1 + 0.18(\theta_d/\theta_a)^2]^2. \qquad (6)$$

The corrected antenna temperature was obtained from the measured antenna temperatures $T_a(0)$ such that

$$T_a = T_a(0) \exp(\tau/\sin H)/\eta, \qquad (7)$$

where τ is the optical depth toward zenith (i.e., $10 \log e^{-\tau}$ is the attenuation in decibels), H is the elevation angle of the observation, and η is the product of the known gain reduction factors. This expression also assumes that the antenna beamwidth is very narrow.

All observations were carried out under clear sky conditions. The optical depth was estimated using the known tipping method, where sky temperature is measured at different elevation angles. The log of different temperatures is plotted against sec (90-H), where the slope is τ [6], [7]. Another tipping method [8] using interative numerical solutions for the radiative transfer equation, in some cases, produced larger estimates of τ. This method uses measurements taken at higher elevation angles, assuming that the sec (90-H) law is not accurate at low angles. On certain occasions, when the values of the calculations are higher (say, $\tau > 0.25$), we found that the values of τ are different, at different elevation angle ranges. In some cases, the measured temperature differences for different elevation angles have not shown the expected changes, and the solutions for the radiative transfer equation could not be found. On the other hand, the first method [6], [7] uses sky temperature measurements from low elevation angles to the zenith and provides a single value of τ. When the calculated τ is smaller (say, $\tau < 0.25$), the two methods seem to agree. A comparison with the direct measurement of extinction of the solar emission was in disagreement with the tipping methods. In this case, however, the measurements are taken only at very low elevation angles ($H < 20°$), and the results are affected by unknown parameters, such as the antenna beam efficiency degradation at very low elevation angles and whether the sec (90-H) law for sky transmission still holds close to the horizon.

The problem is known to be controversial and deserves further research. For the correction of the present measurements we used the smaller estimates derived from the first tipping method. It ranged typically from $\tau \sim 0.2$ (0.87 dB) to $\tau \sim 0.35$ (1.52 dB), at night and during the day, respectively.

The effects of clear air turbulence were noted on certain occasions, especially after sunset hours. The dual beam separation of four half-power beamwidths was not close enough to entirely suppress the effect of the turbulence, which in fact constituted the ultimate limit to the system noise flutuations in those hours [4]. The pointing errors, due to ray bending in the atmosphere, were effectively corrected by the driving computer program, which takes into account the refractivity as a function of ground level temperature and relative humidity [9].

III. MEASUREMENTS OF APERTURE EFFICIENCY AT 7 MM

We performed several measurements on different days using different celestial sources for calibration. Virgo A is well-known as a strong and stable point radio source, with $S = 11.5$ Jy at $\lambda = 7$ mm, as indicated by recently published spectra [10], [11]. The planets, at $\lambda = 7$ mm, were Jupiter, for which we assumed $T_b \sim 140$ K [12]–[14]; Venus, with $T_b \sim 370$ K [7], [15]; Saturn, with $T_b \sim 125$ K [12], [13], [16]; and Uranus, with $T_b \sim 135$ K [13].

The measurements were performed at elevation angles ranging from 23–55 degrees, and the experimental results are shown in Fig. 2. The indicated error bars correspond to the standard deviation from the mean for each observation. The curve plotted in Fig. 2 was derived independently by observing the strong SiO celestial maser WHya, integrating the channels that contained the line emission, and normalizing. Both sets of results appear to be consistent and indicate that the Itapetinga reflector has maximum aperture efficiency towards higher elevation angles, attaining about 50 percent. The gain degradation with elevation angle is small, reducing to about 35 percent efficiency at 20 degrees of elevation.

IV. THE APERTURE EFFICIENCY AND THE REFLECTOR SURFACE MEASUREMENTS

The 72 panels from the Itapetinga reflector were measured in 37 areas each (i.e., 2664 points) utilizing a special measuring device made by the manufacturer. The measurements and the data obtained were described in detail in an internal report [17]. In Fig. 2 we show the distribution of the deviations in terms of the percentage of the total number of measurements. The distribution cannot be fitted to a Gaussian curve, which can be attributed to the systematic appearance of "low areas" in the panels.

It is known [1] that when there are slowly varying errors across a surface that is only slightly rough compared to a wavelength, a redistribution of radiation is produced at small angles with respect to the antenna axis, which influences the main beamshape but not the aperture efficiency. The well-known statistical models developed by Ruze [18] can only be applied in the case of a purely random distribution of deviations. When there are correlated errors over localized regions in the reflector's surface, that model cannot be applied directly, and we cannot predict any precise law for the efficiency versus wavelength dependence [2].

In a simplified approach we may associate the overall data, as displayed in Fig. 3, to the overlap of at least two major distributions, approximately Gaussian, separated by about -0.38 mm (i.e., -0.05λ at $\lambda = 7$ mm), each one presenting an rms deviation of $\sigma \sim 0.2$ mm. The entire reflector equivalent rms deviation should then be of the order of $\sigma \sim 0.2$ mm, for which the application of Ruze's [18] theory may be applicable. The separation of the two distributions is very small compared to a wavelength and would have only a marginal effect on the shape of the main beam of the antenna at 7 mm [1]. From Ruze's [18] theory the predicted efficiency is

$$\eta_a = \eta_0 \exp\left[-\left(\frac{4\pi\sigma}{\lambda}\right)^2\right]. \qquad (8)$$

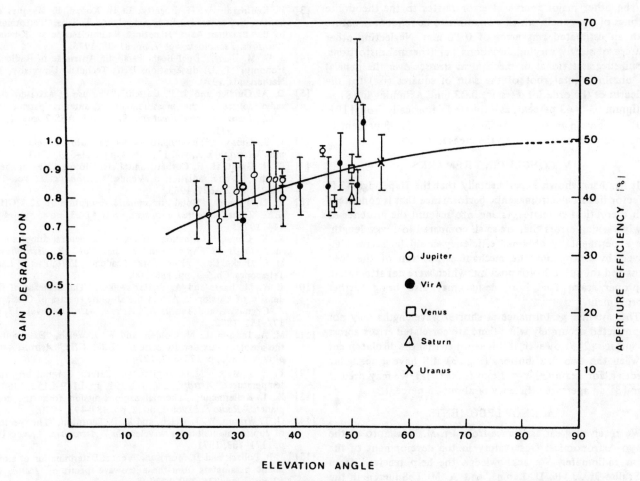

Fig. 2. Aperture efficiency measurements of the Itapetinga 45-ft reflector at 7 mm using various calibrators. The full line was determined independently, with the use of the strong SiO line emitter WHya, and fit well with the gain dependence on elevation effect suggested by the calibrators' measurements.

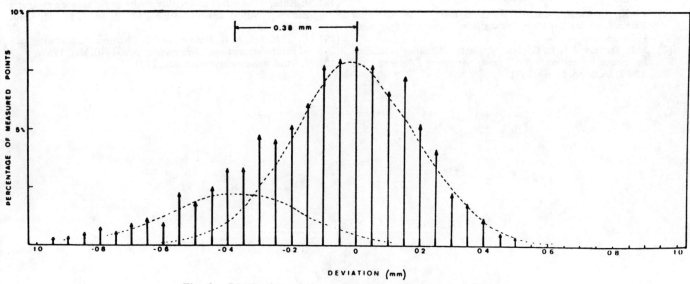

Fig. 3. Distribution of the surface errors measured over 37 areas on 72 panels (i.e., 2664 measurements) from the Itapetinga reflector. Distribution shows the systematic contribution of "low areas", and might be fitted, in a first approximation, to two overlapping random distributions, represented by approximate gaussian curves separated by 0.38 mm (i.e., 0.05 at λ = 7 mm).

The other major source of error relates to the theodolite settings of the panels, which we assume to be random in nature with an estimated rms error of 0.12 mm. Neglecting other sources of slowly varying deviations [1] (thermal distortions, localized gravitational or mechanical stresses over the panels) we obtain a total root of the sum of squares (rss) for the reflector of the order of σ (rss) $\sim$ 0.23 mm. Assuming in (8) an optimum $\eta_0 \sim$ 63 percent, as estimated empirically from [19], [20], we obtain an expected efficiency of nearly 53 percent at 7 mm, which is comparable to the measurements.

V. CONCLUDING REMARKS

It has been shown experimentally that the Itapetinga 45-ft reflector has an electromagentic performance that is consistent with theoretical estimates, taking into account the presence of slowly varying errors that are small compared to a wavelength. The experimentally obtained efficiency could be further improved by optimizing the mechanical position of the feed horns and the subreflector position, which were not attempted. A proper scalar feed horn design may also bring further improvement.

The antenna performance at shorter wavelengths may not be predicted accurately when there are correlated errors across the surface [2]. However, if it is assumed that the displacement between the two distributions (Fig. 3) still have a negligible effect at, for example, $\lambda = 3$ mm (100 GHz), we may predict from [8] an aperture efficiency of about 25 percent.

ACKNOWLEDGMENTS

We received great assistance from Dr. M. Morimoto of the Tokyo Astronomical Observatory in the development of the 7-mm radiometer. We acknowledge the help received from E. Scalise Jr., J. R. D. Lépine, and A. M. LeSqueren in the course of this work. We are grateful to one referee for his helpful suggestions and corrections.

REFERENCES

[1] R. N. Bracewell, "Antenna tolerance theory," in *Statistical Methods of Radio Wave Propagation*. London: Pergamon, 1960, pp. 179–183.

[2] J. H. Davis and J. R. Cogdell, "Reflector efficiency evaluation by frequency scaling," *IEEE Trans. Antennas Propagat.*, vol. AP-19, no. 1, pp. 58–63, 1971.

[3] P. Kaufmann, W. G. Fogarty, E. H. Koppe, P. Marques dos Santos, E. Scalise, Jr., R. E. Schaal, and T. Tiba, "Performance of the Brazilian 45-ft Itapetinga Radiotelescope at K-band," *Rev. Bras. Tecnologia*, vol. 7, pp. 81–88, 1976.

[4] J. W. M. Baars, "Dual Beam Parabolic Antennae in Radio Astronomy," Ph.D. dissertation, Delft Technical University, The Netherlands, 1970.

[5] D. A. Guidice and J. P. Castelli, "The use of extraterrestrial radio source in the measurement of antenna parameters," *IEEE Trans. Aerosp. Electron. Syst.*, vol. AES-7, no. 2, pp. 226–234, 1971.

[6] J. E. Gibson, "The brightness temperature of Venus at 8.6 mm," *Astrophys. J.*, vol. 137, no. 2, pp. 611–619, 1963.

[7] C. R. Grant, H. H. Corbett, and J. E. Gibson, "Measurements of the 4.3-mm radiation of Venus," *Astrophys. J.*, vol. 137, no. 2, pp. 620–627, 1963.

[8] W. G. Fogarty, "Total atmospheric absorption at 22.2 GHz," *IEEE Trans. Antennas Propagat.*, vol. AP-23, no. 5, pp. 441–444, 1975.

[9] R. K. Crane, "Refraction effects in the neutral atmosphere," in *Methods of Experimental Physics Vol. 12, Astrophysics*, M. L. Meeks, Ed., New York: Academic, 1976, Part B: Radio Telescopes, Ch. 2.5, pp. 186–200.

[10] J. W. M. Baars and A. P. Hartswijker, "The decrease of flux density of Cassiopeia A, and the absolute spectra of Cassiopeia A, Cygnus A, and Taurus A," *Astron. Astrophys.* vol. 17, pp. 172–181, 1972.

[11] M. A. Janssen, L. M. Golden, and W. J. Welch, "Extension of the absolute flux density scale to 22.285 GHz," *Astron. Astrophys*, vol. 33, pp. 373–377, 1974.

[12] E. E. Epstein, "Mars, Jupiter, and Saturn: 3.4-mm brightness temperatures," *Astrophys. J.*, vol. 151, pp. L149–L152, 1968.

[13] K. I. Kellermann, "Thermal radio emission from the major planets," *Radio Sci.*, vol. 5, no. 2, pp. 487–493, 1970.

[14] G. T. Wrixon, W. J. Welch, and D. D. Thornton, "The spectrum of Jupiter at millimeter wavelengths," *Astrophys. J.*, vol. 169, pp. 171–183, 1971.

[15] J. B. Pollack and D. Morrison, "Venus: Determination of atmospheric parameters from the microwave spectrum," *Icarus*, vol. 12, no. 3, pp. 376–390, 1970.

[16] G. T. Wrixon and W. J. Welch, "The millimeter wave spectrum of Saturn," *Icarus*, vol. 13, pp. 163–172, 1970.

[17] V. Miselis, "Surface accuracy of 45-ft diameter reflector panels," Rep. no. D71-19, ESSCO, Concord, MA, Oct. 1971.

[18] J. Ruze, "The effect of aperture errors on the antenna radiation pattern," *Nuovo Cimento Suppl.*, vol. 9, no. 3, pp. 364–380. 1953.

[19] K. E. Mckee, A. G. Holtum, and T. Charlton, "Optimizing gain of parabolic antennas," *Microwaves*, pp. 34–40, Mar. 1967.

[20] J. Ruze, private communication, 1971.

Quality Evaluation of Radio Reflector Surfaces

A. Greve and B. G. Hooghoudt

Max-Planck-Institut für Radioastronomie, Auf dem Hügel 69, D-5300 Bonn 1,
Federal Republic of Germany

Received February 11, accepted June 3, 1980

Summary. For the evaluation of radio reflector surfaces we consider three rms-values σ: σ_n represents the geometrical quality of the surface, σ_p, σ_{TP} take into account the corresponding phase deviations and the illumination function of the horn. The effective rms-value σ_{TP} represents the electrical quality of the reflector; this quantity and the corresponding surface weighting function should be considered in design calculations, adjustment procedures, and for the comparison with radio gain measurements. The Effelsberg 100 m antenna is taken as an example of a steep parabolic reflector.

Key words: radio reflectors – antenna tolerances – gain measurements

Introduction

In the design phase of radio telescopes, and during the adjustment of reflectors, much effort is attributed to the minimization of surface deviations in order to achieve accuracies of root mean square (rms) value $\sigma \simeq \lambda_{min}/20$, with λ_{min} the shortest wavelength of observation. For the evaluation of the reflector and beam quality usually the "geometrical" rms-value (σ_n) is derived from the deviations (δ_n) measured *normal* to the surface. As pointed out by Ruze (1966) and Wested (1966), for steep reflectors the phase deviations (δ_p) of the reflected wavefront, weighted by the illumination function (T) of the feed horn, must be used for computing the effective rms-value σ_{TP}. The phase and taper effects may be accounted for (Ruze, 1966) by applying a reduction factor ($R \lesssim 1$) to σ_n, i.e. $\sigma_{TP} = R\sigma_n$. The detailed analysis, however, reveals the pronounced influence on σ_{TP} of the zone between $\simeq 1/3$ and $\simeq 2/3$ the radius of the reflector. Hence, the radial dependence of the surface weighting function S_{TP} [defined below, Eq. (6)] should be considered in the design and the adjustment procedures.

In Sect. I we evaluate theoretically a parabolic reflector of focal ratio $f = 0.30$; in Sect. II we discuss the Effelsberg 100 m reflector as a typical example.

I. Theoretical Evaluation

Assume that a parabolic reflector is surveyed at M target positions; each target represents an area element of the same size in the aperture plane. Assume k rings of targets with

$N_j (j = 1, 2, ..., k)$ targets on ring j, $M = \sum N_j$. The condition of equal area elements requires N_j to be proportional to $2\pi\varrho_j$, with $\varrho_j (0 \leq \varrho_j \leq 1)$ the fractional radius of ring j. Assume that the deviations $\delta_n(i)$, $i = 1, 2, ..., M$, are measured *normal to the surface of a best fit paraboloid*, hence $\langle \delta_n \rangle = \sum\limits^{M} \delta_n(i) = 0$. The geometrical rms-value σ_n then is

$$\sigma_n^2 = M^{-1} \sum^{M} \delta_n(i)^2 . \tag{1}$$

It is convenient to split Eq. (1) into the contributions of the various rings; for ring j we define

$$\sigma_n(j)^2 = N_j^{-1} \sum_{i=1}^{N_j} \delta_n(i,j)^2 . \tag{2}$$

In case N_j is sufficiently large, $\sigma_n(j)$ represents the geometrical rms-value of ring j. Hence, for Eq. (1) we write

$$\sigma_n^2 = \sum_{j=1}^{k} (N_j/M)\sigma_n(j)^2 = \sum_{j=1}^{k} S_n(j)\sigma_n(j)^2 \tag{3}$$

with S_n the surface weighting function.

For reflectors of large focal ratio ($f = F/D \gtrsim 1.5$), as used for optical mirrors which also have uniform illumination, the geometrical rms-value σ_n is a valid parameter for the image quality (Väisälä, 1922; Ruze, 1966; Greve and Hunt, 1974). For this case the deviations δ_n are nearly identical with the corresponding phase deviations δ_p of the reflected wavefront. For reflectors of small focal ratio ($f \lesssim 1$), as used for radio antennas, Fig. 1 shows that the phase deviation δ_p is identical with the projection of the deviation δ_n in the direction of the incident and reflected ray. For this case, $\delta_p \leq \delta_n$ and the rms-value to be used is derived from δ_p so that $\sigma_p < \sigma_n$. In the application to radio reflectors, the deviations δ_p must be weighted by the illumination function T of the feed horn, for this case $\delta_{TP} \leq \delta_p \leq \delta_n$ so that $\sigma_{TP} < \sigma_p < \sigma_n$. The value σ_{TP} is a direct measure of the wavefront deviations detected by the horn.

Hence, we distinguish three types of rms-values:

a) geometrical rms-value σ_n, derived from the deviations $\delta_n(i,j)$ measured *normal* to the surface of a best fit paraboloid, and calculated from Eqs. (1) and (3);

b) phase rms-value σ_p, derived from the phase deviations

$$\delta_p(i,j) = \delta_n(i,j) \cos\gamma(i,j) \tag{4}$$

with $\operatorname{tg}\gamma(i,j) = r(i,j)/2F$ (F = focal length, Fig. 1), and calculated from the equation

$$\sigma_p^2 = \sum_{j=1}^{k} (N_j/M) \cos^2\gamma(j)\sigma_n(j)^2 = \sum_{j=1}^{k} S_p(j)\sigma_n(j)^2 ; \tag{5}$$

Send offprint requests to: A. Greve

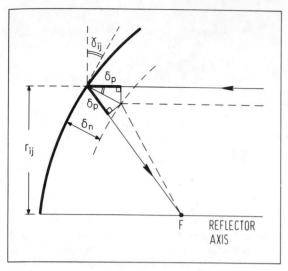

Fig. 1. Geometrical relation between surface deviation δ_n, normal to the surface, and the corresponding phase deviation δ_p, i.e. the projection of δ_n in the direction of the ray path

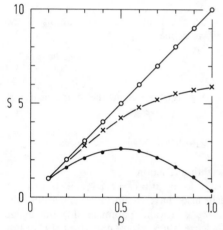

Fig. 2. Plot of normalized surface weighting functions $S_n(\bigcirc)$, $S_p(\times)$, $S_{TP}(\bullet)$ defined by Eqs. (3), (5), and (6), calculated for $f = 0.30$ and 10 dB taper function. Values are normalized to the corresponding values S at $\varrho = 0.1$; ϱ is the normalized radius of the aperture

c) the effective (tapered phase) rms-value σ_{TP}, derived from the deviations δ_p weighted by the power illumination function $T(\varrho_j)$, is calculated from the equation

$$\sigma_{TP}^2 = \sum_{j=1}^{k} N_j T(\varrho_j)^{1/2} \cos^2 \gamma(j) \sigma_n(j)^2 \bigg/ \sum_{j=1}^{k} N_j T(\varrho_j)^{1/2}$$

$$= \sum_{j=1}^{k} S_{TP}(j) \sigma_n(j)^2. \tag{6}$$

The Eq. (6) is in agreement with Eq. (2) of Ruze's (1966) derivations. In Fig. 2 we show the surface weighting functions S_n, S_p, S_{TP} defined by Eqs. (3), (5), and (6); the plotted values are normalized to the corresponding values $S_n(\varrho = 0.1)$, $S_p(\varrho = 0.1)$, $S_{TP}(\varrho = 0.1)$. The values are calculated for $f = 0.30$ and a 10 dB feed horn taper function $T(\varrho_j)^{1/2} = p + (1 - p)(1 - \varrho_j^2)$, $p = 0.315$. The effective surface weighting function S_{TP} has maximum values in the region $\varrho \simeq 0.3$–0.7, hence the accuracy of the reflector

Table 1

Zone	$\sigma_n(\varrho)$	σ_{TP}
$0 \leq \varrho \leq 1.0$	σ_0	$0.88\sigma_0$
$0 \leq \varrho \leq 0.3$	$2\sigma_0$	$1.13\sigma_0$
$0.3 < \varrho \leq 1.0$	σ_0	
$0 \leq \varrho < 0.4$	σ_0	
$0.4 \leq \varrho \leq 0.7$	$2\sigma_0$	$1.41\sigma_0$
$0.7 < \varrho \leq 1.0$	σ_0	
$0 \leq \varrho \leq 0.7$	σ_0	$1.20\sigma_0$
$0.7 < \varrho \leq 1.0$	$2\sigma_0$	

surface in this region contributes significantly to the resulting rms-values σ_{TP}. In order to illustrate quantitatively the influence of the zone between $\sim 1/3$ and $\sim 2/3$ the radius of the reflector, we have assumed various distributions of $\sigma_n(j)$ as given in Table 1. The corresponding values σ_{TP}, calculated from Eq. (6) for $f = 0.30$ and the given taper function, i.e. S_{TP} as shown in Fig. 2, are given in the same table. Consequently, in design calculations and for adjustment procedures, minimum deformations should be obtained for this particular surface region. The location and the width of this region of maximum influence is nearly insensitive of the taper value at the rim.

II. Evaluation of the Effelsberg 100 m Reflector

Taking the Effelsberg 100 m antenna as an example of a steep parabolic reflector ($f = 0.30$), the quantitative analysis indicates the necessity of using the effective rms-value σ_{TP} when interpreting survey measurements and radio measurements.

From the survey measurements (Greve, 1980) we deduced the values $\sigma_n(j)$ which are shown in Fig. 3. Because of the large values $N_j(N_1 = 48, N_{33} = 480)$, the $\sigma_n(j)$ represent the geometrical rms-values of the individual rings. Using the values $\sigma_n(j)$, in Fig. 3 we show the geometrical rms-values $\sigma_{n,k}$ calculated from Eq. (3) for increasing *surface areas confined within ring k*. The corresponding values $\sigma_{p,k}$ and the effective rms-values $\sigma_{TP,k}$, calculated from Eqs. (5) and (6) for $f = 0.30$ and the 10 dB taper function used for the feed horns, are also shown in Fig. 3.

Table 2

λ (cm)	Diam. (d) of illum. aperture (m)	$\eta_A(\%)$	From gain measurements σ_{TP} [Eq. (7)]	From survey measurements (Fig. 3) σ_{TP}	σ_n
11	100	47			
6	100	45	± 3	0.82	1.2
3.3	100	46			
2.8	$d \simeq 100$	45			
1.2	$d \pm 85$	26	$0.75 \gtrsim 0.09$	0.80	1.1
0.9	85	21	± 3 $0.65 \gtrsim 0.07$	0.70	0.87
0.4[a]	60	5.6 ± 1.4	$0.47 \gtrsim 0.04$	0.60	0.75

[a] Harth et al. (1978)

198

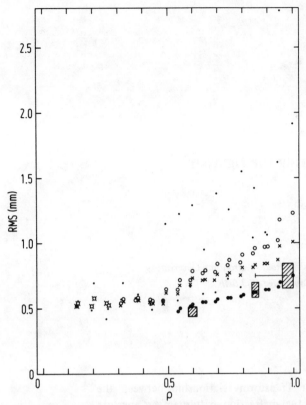

Fig. 3. RMS-values for zenith position of the Effelsberg 100 m reflector, derived from survey measurements. Geometrical rms-value of individual rings $\sigma_n(j)$: ·, the rms-values for the surface area confined within ring k are $\sigma_{n,k}$: $\bigcirc$, $\sigma_{p,k}$: $\times$, $\sigma_{TP,k}$: ●

Empirical values of the aperture efficiency η_A as function of the wavelength λ of observation may be used to derive σ_{TP}. The values η_A, listed in Table 2, are obtained with prime focus feed horns and refer to high elevation angles of the reflector. The listed values η_A are deduced from measurements prior to 1978 and correspond to the initial setting of the reflector surface and the associated evaluation for zenith position (Greve, 1980, see Fig. 3). Note that for the shorter wavelengths there exists an uncertainty of the aperture illuminated by the feed horns, representative values are given in Table 2.

The values η_A are related to σ_{TP} by (Ruze, 1966)

$$\eta_A = \eta_A^{(0)} \exp\{-(4\pi\sigma_{TP}/\lambda)^2\} \tag{7}$$

with $\eta_A^{(0)}$ a constant, i.e. the aperture efficiency of the perfect reflector ($\lambda \to \infty$). From the measurements at $\lambda \gtrsim 2.8$ cm we adopt from Table 2 the value $\eta_A^{(0)} = 45 \pm 3\%$. Using this value $\eta_A^{(0)}$ and the values $\eta_A(\lambda)$, $\lambda \leq 1.2$ cm, given in Table 2, we calculate from Eq. (7) the corresponding values σ_{TP} which are given in Table 2. In this table we also list the corresponding values σ_n and σ_{TP} derived from the survey measurements. Note that the survey measurements, and hence the derived values σ_n, σ_{TP}, are inaccurate by approximately ± 0.05 mm. In the derivation of the values of Table 2 we have taken into account the variation of $f = F/D$ with wavelength. We find good agreement between the radio and survey measurements.

Conclusion

Theory, and the analysis of the Effelsberg 100 m antenna, shows the necessity of using the effective rms-value σ_{TP} in the evaluation of reflector surfaces. The value σ_{TP} takes into account the radial dependence of all weighting functions. The surface weighting function S_{TP} has maximum values in the zone between $\sim 1/3$ and $\sim 2/3$ the radius of the reflector, this fact should be taken into account in design calculations and adjustment procedures.

Acknowledgement. Dr. W. Altenhoff (MPIfR) provided the values of the aperture efficiency, extracted from test observations made by various members of the MPIfR. The comments by Drs. J. E. Wink and J. W. M. Baars (MPIfR) helped to improve the presentation of the results. We thank the referee, Dr. J. Ruze, for giving the correct form of Eq. (6); his detailed comments on the manuscript were very much appreciated.

References

Greve, A., Hunt, G.C.: 1974, *Optik* **40**, 18
Greve, A.: 1980, *Z. f. Vermessungswesen* (in press)
Harth, W., Altenhoff, W., Wohlleben, R., Steffen, P., Vowinkel, B., Gebler, K.H.: 1978, MPIfR Techn. Report No. 48
Ruze, J.: 1966, *Proc. IEEE* **54**, 633
Väisälä, Y.: 1922, Ann. Univ. Fennicae Aboensis, Turku, Ser. A1, No. 2
Wested, H.: 1966, Design and Constr. of Large Steerable Aerials, IEE Conf. Publ. 21, London

A rapid method for measuring the figure of a
radio telescope reflector

P. F. Scott and M. Ryle *Mullard Radio Astronomy Observatory,*
Cavendish Laboratory, Madingley Road, Cambridge CB3 0HE

Received 1976 August 10

Summary. The well-known Fourier Transform relationship between the
complex illumination function and the diffraction pattern of an aperture
has been applied to measurements of the reception patterns of dishes of the
5-km telescope; the results have provided contour maps showing the distribution of reflector surface errors with an accuracy of ~ 0.1 mm. By repeating
the measurements at different zenith angles the structural deflections due to
gravity were investigated.

1 Introduction

Findlay (1971) has reviewed methods for determining the figure of a radio telescope. For
small dishes designed for operation at wavelengths < 1 cm, radial templates have been used,
while for instruments with diameters > 20 m distance measurements, or distance and theodolite angle measurements, from the axis have given accuracies of ~ 1 mm. Greater accuracy
has been obtained by replacing the theodolite with pentaprisms having different angles of
reflection, or a single pentaprism whose distance along the axis may be varied (Kühne
1966; Kelly, Neate & Shinn 1970; Slater 1971). For large reflectors the use of modulated
light beams (Froome & Bradsell 1966; Payne 1973) has provided accuracies of $\sim 0.1-0.2$
mm. Measurements of the curvature over the surface by means of a small trolley, followed
by a double integration to obtain the profile (Payne, Hollis & Findlay 1976), have given
accuracies of better than 0.05 mm. Deflections of a dish have been determined by means of
a microwave transmitter with a transponder attached to points on the surface (Findlay &
Payne 1974) but not absolute measurements of the figure.

Most of these methods are time-consuming, particularly when repeated measurements
are necessary during alignment of the reflector surface. Many of them necessitate the telescope being pointed to the zenith, and so the gravitational deflections when the telescope is
steered to other zenith angles may be difficult to establish. Some methods require the presence of an operator *in situ*, whose weight may introduce a significant deflection, and with
some it may be difficult to test distortion caused by unsymmetrical heating by the Sun,
because of the time needed and the effect of solar heating on the measuring equipment itself.

In order to assess the probable performance of the eight dishes of the 5 km telescope at
wavelengths shorter than those at present in use, we have adopted an alternative method

which has proved convenient and quick, and by which the surface profile can be checked at any zenith angle. Four of the dishes were measured simultaneously and automatically to an accuracy of about 0.1 mm in only 5 hr. The method employs the well-known Fourier Transform relationship between the aperture illumination and the corresponding diffraction pattern, and provides a map showing the distribution of surface errors across the aperture. The presence of focusing or pointing errors is revealed as well as the surface errors.

If the complex reception pattern of a dish were measured over a full half-sphere it would, in principle, be possible to determine the complete distribution of amplitude and phase across the equivalent aperture plane. If the measurements are restricted to the region surrounding the main response and extending to n beamwidths, i.e. an angle $n\lambda/D$ from the axis, the results can provide information on the aperture distribution (and thus the shape of the reflector surface) with a spatial resolution $\sim D/n$ where D is the diameter of the instrument, details of structure on smaller scales being lost. This restriction may be acceptable in practice, since the magnitude of any small-scale irregularity is very easy to determine with a template, and it is the large-scale departures, such as those caused by errors in the setting of individual panels or by gravitational deflections of the backing structure, which are difficult to measure by conventional means.

The method is analogous to that described by Bennett *et al.* (1976), but does not use holographic techniques.

2 The measurement of the complex reception pattern

Consider a pair of dishes which are directed to a distant point source of known position and connected to a phase-switching receiver (Ryle 1952). The output of such a receiver is proportional to the product of the complex voltages produced by the source in the two dishes. If now one of the dishes is directed to a point (θ, ϕ) away from the source, the output of the phase-switching receiver is modified and provides a measure of the complex receptivity of the dish in this off-axis direction. By pointing the dish successively with different values of θ, ϕ, the complex reception pattern may be measured over any desired range of θ and ϕ. Allowance must, of course, be made for the progressive phase change resulting from the diurnal movement of the source in the sky. It should be noted that, since the signal-to-noise ratio only decreases as the voltage response, smaller values of receptivity can be measured than when investigating the pattern of a single dish with a total-power receiver.

It has here been assumed that the offset pointing of the dish does not introduce additional phase shifts. The measurements will define the equivalent aperture distribution over a plane passing through the rotation axis. This distribution correctly reflects errors in the actual dish surface, provided that the separation between the dish surface and this plane is much smaller than the Fresnel distance appropriate to the smallest size of irregularity under investigation. Where the distance between the dish and rotation axis is large compared with the depth of the dish, appropriate phase corrections may be applied before the analysis (Fig. 1). The effects can, in any case, be minimized by making the measurements at the shortest convenient wavelength.

3 The limitations of the method

Suppose that it is required to measure the errors, ϵ, in the positioning of the panels of a dish, each of which occupies a fraction $1/\alpha^2$ of the aperture area, with an uncertainty $\Delta\epsilon$. Since

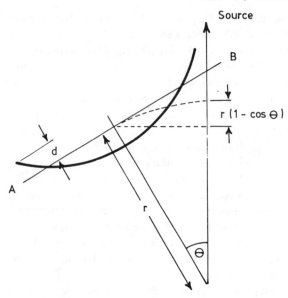

Figure 1. The phase-path errors arising from imperfections in the reflector surface are similar to those in an equivalent aperture plane A–B provided that the depth d of any part of the dish from this plane is small compared with the Fresnel distance appropriate to the observing wavelength and scale of the imperfections. Rotation of the dish about an axis displaced r from this equivalent aperture plane introduces a path error $r(1 - \cos\theta)$ which may be used to correct each sample before the Fourier inversion.

the measurements are likely to be made at a wavelength, λ, larger than that at which the performance of the dish is seriously degraded, ϵ will typically be less than $\lambda/20$.

Two main limitations in the measurement of ϵ occur in practice: (i) that arising from the finite signal-to-noise ratio, and (ii) that due to errors in phase-measurement caused by tropospheric irregularities altering the relative phase path from the source to the test and reference dishes. Both effects introduce errors in the measurement of the complex reception pattern and hence in the derived aperture distribution.

3.1 THE SIGNAL-TO-NOISE RATIO REQUIRED

To find the location of individual panels each of which has linear dimensions $\sim D/\alpha$, the reception pattern must be determined over a range of angles $\alpha\lambda/D$; over this range the contribution of a single panel to the output of the phase-switching receiver is $\sim 1/\alpha^2$ of that produced when both dishes are directed at the source. An error ϵ in the figuring ($\equiv 4\pi\epsilon/\lambda$ in phase) will therefore introduce a relative component $\sim 4\pi\epsilon/\alpha^2\lambda$. However, in determining the value of ϵ for this particular panel, observations in α^2 directions are combined in the Fourier inversion, so that if the signal-to-noise ratio is R when the source is observed in the main beam for the same integration time as that used for each sample, the error ($\Delta\epsilon$) in determining ϵ is given by $\Delta\epsilon = \alpha\lambda/4\pi R$.

3.2 TROPOSPHERIC IRREGULARITIES

Two classes of tropospheric irregularity exist which are relevant here; in the first, scale sizes of ~0.7 km occur, predominantly in day-time and summer, which can introduce rms path errors of up to 2 or 3 mm (Hinder & Ryle 1971); a second class of irregularity with typical scale-sizes of 5–15 km can occur at all times of year and day and may introduce rms errors of up to 1 mm of path per km of telescope aperture (Hargrave & Shaw, in preparation). On some occasions the contribution of both classes is < 0.04 mm per km. The phase errors likely to occur thus depend on the time of day and year, and on the spacing between the dish being measured and the reference dish. For spacings < 1 km and periods excluding summer daytime the instantaneous path errors are unlikely to exceed 1 mm and, by selecting good observing periods, errors below 0.3 mm can readily be obtained. Since the errors are largely uncorrelated from sample to sample their effect will be further reduced by $1/\alpha$ due to the averaging occurring in the analysis. Under good observing conditions and for the case $\alpha = 10$ (100 panels per dish) the error in the measurement of a panel position due to tropospheric irregularities is likely to be < 0.02 mm.

4 Results of some preliminary observations on four of the dishes of the 5-km telescope

In order to investigate the gain and beam shape likely to be obtained if the 5-km telescope were used in the 33.4-GHz band, measurements of the figures of four of the 13-m dishes have been made at a frequency of 15.4 GHz. In normal operation the instrument determines the amplitudes and phases of the correlated signals from pairs of aerials, the phases being continuously adjusted for the motion of the source, atmospheric refraction and instrumental collimation error, so that a source at the centre of the beam will have a constant phase of zero (Ryle & Elsmore 1973). Observations were made of 3C 84, which at 15.4 GHz is effectively a very compact source, and the four dishes under test were scanned automatically over a square grid of 17 × 17 points separated by 4 arcmin and centred on the source, the signals from these dishes being combined with that from a dish which tracked the source normally. The source was observed for 1 min at each offset position; with this integration time the signal-to-noise ratio on-source is ~ 250 so that it should be possible to measure the error in the mean position of an area occupying 1 per cent of the total aperture (approximately one panel) with an accuracy $\Delta\epsilon \sim 0.1$ mm.

Small corrections (of up to 8° of phase) for the variation of phase path with offset angle, due to the separation between the effective centre of a dish and its axes of rotation, were applied to the observations before the Fourier inversion. For observations at 15.4 GHz and a panel size of ~ 1.5 m, the Fresnel distance is ~ 100 m, so that the errors arising from the depth of the dish (~ 3.5 m) are unimportant. Although the tropospheric conditions were not ideal during these observations, the effects were unimportant for the dish closest to the reference dish, whereas for the most distant dish (~ 1.2 km spacing) the errors introduced were comparable with those due to noise.

The observations, which were made with linearly-polarized feeds with the *E*-vector in pa 0°, occupied two periods of ~ 2½ hr on each of two successive days. On each day 3C 84 was observed near the meridian (mean zenith angle ~ 10°) and again at an hour angle of ~ 5 hr (mean zenith angle ~ 50°) in order to investigate any deterioration of the figure due to gravity. (The dishes had been set up originally at the zenith by a pentaprism method.) Although improved accuracy could have been obtained by extending the length of the observations and by selecting periods of better 'seeing', the results obtained have errors $\Delta\epsilon \sim 0.1$ mm and are already of value in establishing the performance of the dishes at 33.4 GHz.

The results of Fourier inversion of the first set of measurements (at zenith angle ~ 10°) are illustrated in Figs 2(a) and (b) which are contour maps of the derived aperture distributions in amplitude and phase for one of the dishes. Fig. 2(a) reveals clearly the shadow of the Cassegrain secondary reflector and also those of the three support legs. For clarity, the phase

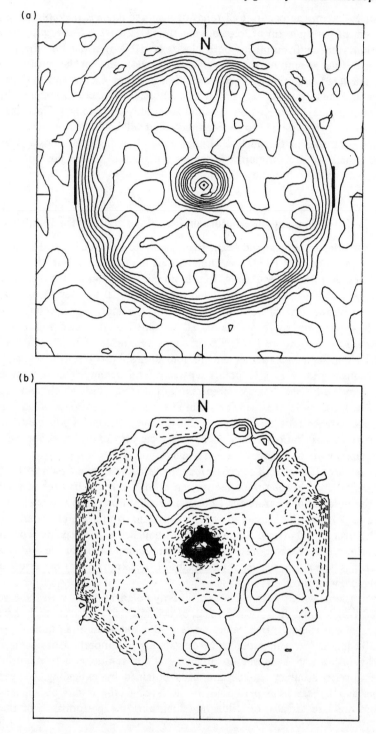

Figure 2. The aperture distribution derived for one of the dishes when directed close to the zenith. (a) Amplitude. The contours indicate equal intervals of voltage. The solid bars indicate the physical diameter of the dish. (b) Phase-path. One contour corresponds to a path error of 0.5 mm, equivalent to a surface error of 0.25 mm near the centre of the dish and 0.35 mm near the edge.

204

diagram of the aperture distribution, Fig. 2(b), has not been plotted beyond the radius at which the amplitude has fallen to 10 per cent of that across most of the aperture; the phase-path errors are expressed in mm and have an rms of 0.8 mm, corresponding to errors in the reflector surface of the order of 0.5 mm.

The corresponding results for observations at a mean zenith angle of 50° are shown in Fig. 3. Whilst there are clear changes in the distribution of path errors with zenith angle, it is evident that many of the features are common to both zenith angles, so that resetting some of the panels should give a significant improvement at all zenith angles.

The expected forward gain and reception pattern at a frequency higher than that of the observations can readily be estimated by converting the surface errors to phase angles at the new frequency and carrying out another Fourier inversion. Computations for the distributions shown in Figs 2 and 3 when operating at 33.4 GHz indicate a reduction in forward power gain relative to that with an ideally-adjusted reflector surface by a factor of 0.65 at the zenith and 0.43 at a zenith angle of 50°.

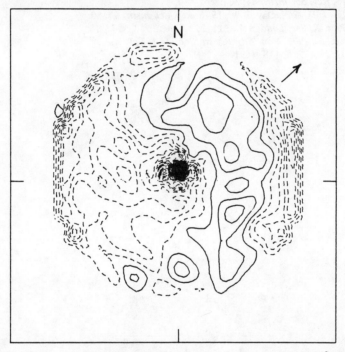

Figure 3. The distribution of phase path as in Fig. 2(b), but for zenith angles ~ 50°. The instrument is equatorially mounted and the arrow indicates the direction of the zenith.

Conclusion

The method described allows the surface errors due to panel setting and deflection of the backing structure to be measured rapidly and at a range of zenith angles. By increasing the area of the reception pattern scanned, with a corresponding increase in measuring time, it would be possible to extend the method to examine errors in the surface on a smaller scale.

The method could be applied to the measurement of an individual large radio telescope if a smaller telescope were mounted nearby, and could provide an economical way of checking the surface accuracy regularly. It may also be appropriate for measuring dishes designed for mm-wave astronomy, by using the intense sources of H_2O maser emission.

Acknowledgments

We are indebted to Dr R. E. Hills for discussions from which the present method evolved, and to Dr S. Kenderdine for writing the program for scanning the telescope over the sampled area.

References

Bennett, J. C., Anderson, A. P., McInnes, P. A. & Whitaker, A. J. T., 1976. *IEEE Trans. Ant. Propag.,* **24**, 295.
Findlay, J. W., 1971. *A. Rev. Astr. Astrophys.,* **9**, 271.
Findlay, J. W. & Payne, J. M., 1974. *IEEE Trans. Instr. Measur.,* **23**, 221.
Froome, K. D. & Bradsell, R. H., 1966. *J. Sci. Instr.,* **43**, 129.
Hinder, R. & Ryle, M., 1971. *Mon. Not. R. astr. Soc.,* **154**, 229.
Kelly, J. B., Neate, P. R. & Shinn, D. H., 1970. IEE Conf., *Earth station technology,* Conf. Publ., **72**, 227.
Kühne, C., 1966. IEE Conf., *Design and construction of large steerable aerials,* Conf. Publ., **21**, 187.
Payne, J. M., 1973. *Rev. Sci. Instr.,* **44**, 304.
Payne, J. M., Hollis, J. M. & Findlay, J. W., 1976. *Rev. Sci. Instr.,* **47**, 50.
Ryle, M., 1952. *Proc. R. Soc. London A.,* **211**, 351.
Ryle, M. & Elsmore, B., 1973. *Mon. Not. R. astr. Soc.,* **164**, 223.
Slater, R. H., 1971. *Proc. IEE,* **118**, 1691.

A Holographic Surface Measurement of the Texas 4.9-m Antenna at 86 GHz

CHARLES E. MAYER, STUDENT MEMBER, IEEE, JOHN H. DAVIS, MEMBER, IEEE,
WILLIAM L. PETERS, III, AND WOLFHARD J. VOGEL, MEMBER, IEEE

Abstract—An instrument has been built which allows the electromagnetic measurement of the surface accuracy of a large millimeter-wavelength antenna. The University of Texas 4.9-m radio telescope has been measured with this technique at 86.1 GHz to an accuracy of 4 μm at the surface.

Our technique is an interferometric one which is fast, accurate, and able to measure the whole antenna surface at once. While the technique is illustrated by its use on a large antenna, it could be used in a near-field measurement of a smaller antenna.

Several antenna surface maps are presented. A comparison of run-to-run repeatability was made. The technique itself was tested by deforming the antenna surface in a known way and subsequently detecting the deformation.

I. INTRODUCTION

ACCURATE MEASUREMENT of large reflector antennas has become increasingly important as higher gain antennas have come into use. Several measurement methods have been used, including laser theodolite ranging, mechanical position measurements, planar sampling of near fields, and spherical sampling of both near and far fields.

Laser methods, Baars [1] and Moromoto [2], require measuring the range and angle to reflecting targets placed on the surface of the antenna. The laser methods appear to show promise although the technique has not been used on an operational antenna. Several mechanical methods have been used for the field testing of large antennas. Findlay [3] has demonstrated a bar-level apparatus which seems to give repeatable

Manuscript received August 19, 1982; revised November 1, 1982.

The authors are with the University of Texas at Austin, College of Engineering, Department of Electrical Engineering, Electrical Engineering Research Laboratory, Austin, TX 78712.

results, but a full antenna map has not been made. Payne [4] built and tested a surface curvature measuring chart which was used to measure the 36-ft millimeter antenna at Kitt Peak.

The electromagnetic techniques can generally be divided into categories by the type of measurement surface need. Planar electromagnetic scanning techniques probe a set of points which lie in a plane perpendicular to the axis of the antenna, Repjar [5]. Raquet [6] has reported a NASA test facility of this type of measure antennas for space application. A significant advantage of planar scanning is that the antenna can remain fixed. However, the fixture which carries the probe must be larger than the antenna under test. Bennett [7] has shown that spherical scanning can be used successfully on large antennas. In this technique, samples are taken on a spherical surface surrounding the antenna. Experimentally, spherical scanning is more attractive for a large antenna because the antenna probably already has a two-axis mount. However, Hess [8], Bennett [9], and Joy [10] show that spherical scanning leads to more theoretical and computational difficulty.

Greve [11] showed that in most cases the mechanical measurements made on large antennas have predicted better performance than the electromagnetic performance achieved. Thus the trend toward electromagnetic measurements seems justified. A number of carefully done electromagnetic measurements have been reported. However, Hess [12] has shown that the usual instrumental techniques restrict the antenna size to less than 150λ.

In addition to the instrumental problems, there are data-handling problems for a large antenna. Baars [1] has defined a large antenna to be an antenna with a gain maximum in ex-

Reprinted from *IEEE Trans. Instrum. Meas.*, vol. IM-32, no. 1, pp. 102–109, Mar. 1983.

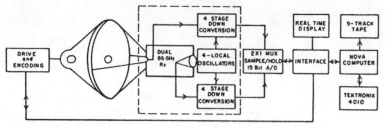
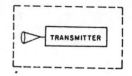

Fig. 1.   Data acquistion system.

cess of 75 dB. This requirement means that the antenna must be larger than 2500λ in size. If an antenna surface of this size is sampled at half-wavelength intervals by any means whatever, the data collection is too massive to deal with in a practical way. Thus a large antenna system by this definition must, of necessity, be undersampled.

If the evaluation of a large antenna is undertaken, one must carefully choose the information to be measured. Practical experience shows that typical large antennas concentrate the radiated power close to the main beam even if there are significant surface errors. In addition, the validity of choosing only the main beam area for sampling can be experimentally verified if the directivity can be measured. The directivity and the antenna solid angle are not independent, so the amount of radiated power outside the sampling solid angle can be evaluated. It should be pointed out that it is easy to imagine an antenna which has a high gain and also a large amount of broadly scattered power. Ruze [13] showed that a normal reflector antenna with a rough surface can scatter almost any fraction of its power over almost any solid angle while still forming a diffraction beam.

If the sampling solid angle is restricted to a solid angle of only a few square degrees, then the geometry of the spherical sampling can be simplified by using the usual small-angle approximations. Under this restriction, the use of the spherical scanning for a large antenna is simpler than the use of the spherical scanning technique for a small antenna.

This paper reports the design of an instrumental technique appropriate for spherically scanning a large antenna. Our goal is to produce an aperture plane image of the fields near the antenna. The aperture plane image is the Fourier transform of the far zone fields. These far zone fields are estimated by recording the amplitude and relative phase for a grid of angular positions of the antenna. Phase information requires two independent receivers, one coupled to the reflector antenna under test, the other coupled to a reference antenna. At each angular position, the downconverted output of each receiver is sampled directly by a fast A/D converter. A time series fast Fourier transform (FFT) is performed on each of the sampled voltages, yielding the voltage spectrum for each receiver. A cross correlation of the two spectra gives the amplitude and phase information desired. The power from each antenna is obtained from the autocorrelations. After repeating this process for each point of the grid, a spatial FFT produces the antenna aperture plane amplitude and phase, from which the reflector surface can be readily evaluated. A sequence of experiments are presented which are designed to test the accuracy and repeatability of the data acquired. In addition, two known pertur-

bations were introduced to the antenna surface so that the electromagnetic measurements could be carefully tested. The data are analyzed by using the small-angle approximations outlined above; however, the instrumental technique is not restricted to this situation. Other data reduction schemes such as Larsen [14], Wood [14], and Paris [16] could be adapted to the instrumentation presented here.

The instrumentation has been used to characterize the University of Texas' Millimeter-Wave Observatory (MWO) 4.9-m parabolic reflector antenna located on Mount Locke, near Fort Davis, TX. The radiation pattern of the antenna is an important part of the interpretation of radio astronomical data since many radio sources are significantly larger than the antenna beam. An additional objective was to improve the performance of the antenna so that it could be used effectively at frequencies up to 300 GHz. The MWO had an existing pattern range with a transmitter located at a distance of about 12.9 km. Since the 4.9-m antenna is regularly used for astronomy, it already had a computer-controlled drive and pointing system. The combination of the drive system and the pattern range allowed convenient use of spherical scanning techniques.

II. INSTRUMENTATION

A block diagram of the experimental arrangement is shown in Fig. 1. A transmitter, a millimeter-wavelength receiver package, an IF section, and a computing section are included. The dotted line in Fig. 1 enclose all equipment unique to the pattern range experiment. Most of the equipment is used for normal radio astronomical observations.

The transmitter produces a CW signal generated by a phase-locked klystron at 86.160 GHz. The signal must be stable to 100 Hz in the scanning time, which is up to 3 h, i.e., a reference oscillator must have stability exceeding 10^9 in 3 h. An oven-controlled 100-MHz crystal oscillator provides this stability. The transmitter has an automatic-gain control loop which utilizes a voltage-controlled variable attenuator driven by a monitor of the transmitter power. An elevation profile of the pattern range is shown in Fig. 2. A ridge between the transmitter site and the antenna provides a natural block to multipath interference. Cogdell [18] was unable to find any multipath interference diffracted from the top of this ridge.

The receiver, which is a modified version of a dual-channel receiver built by Ulich [17] and contains all of the millimeter-wavelength components, was mounted at the prime focus of the 4.9-m antenna. One receiver channel is connected to the $f/0.5$ rectangular waveguide feed horn which collects the

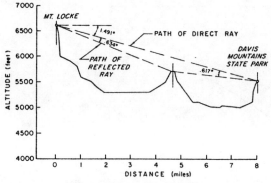

Fig. 2. Pattern range profile.

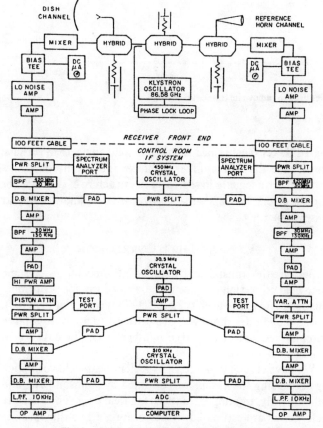

Fig. 3. Dual-channel receiver block diagram.

signal reflected off the antenna; the other receiver channel is connected to a reference horn which observes the transmitter directly. The reference horn is 1.4 m long with a 2° flare angle and is mounted coaxially with the 4.9-m antenna but directed out the opposite end of the receiver package. Ulich [19] measured the characteristics of the reference horn and found that it compared very favorably with theoretical predictions. Thus the two millimeter receiver channels, the reference horn, and the feed for the 4.9-m antenna are housed in a single receiver package, rigidly mounted at the prime focus position. This arrangement can be seen in Fig. 6. A significant advantage of this scheme is that the reference antenna is carried by the test antenna, so that rapid phase shifts caused by differential motion of the two antennas are avoided. Another advantage is that the same phase-locked klystron can be used as a local oscillator for both receiver channels. A problem caused by the common local oscillator is possible leakage between the two channels;

however, isolation of 85 dB was finally achieved by using three-ring hybrids in a reflection canceling circuit (Mayer [20]).

The dual IF system is shown in Fig. 3. There are four heterodyne stages starting with the 86.16-GHz range signal and ending with a baseband signal in the 0–10-kHz band. The IF system, designed by Mayer [20], has a dynamic range which exceeds 80 dB, does not add significant noise to that from the millimeter mixer, and affords isolation between the channels in excess of 85 dB. In addition, a high-level output is provided in the 0–10-kHz band to drive an A/D converter.

Perhaps the most unusual feature of the system is that the 0–10-kHz signal from both channels is sampled by a fast 15-bit A/D converter without any previous detection. Thus one has a digital sampling of the IF voltage which is proportional to the field intensity at the output of the two antennas. The 15-bit voltage sampling corresponds to 30 bits of power or $10 \log_{10} 2^{30}$ = 90 dB of dynamic range. Not all of this dynamic range is available because of roundoff error, but 80 dB is easy to achieve. The large dynamic range is absolutely necessary since the antenna radiation pattern falls off rapidly as one moves away from the peak. Because this lower level pattern may subtend a large solid angle, its integrated contribution may be quite significant.

III. Data Acquisition

The basic data required are estimates of the cross correlation between the test antenna and the reference horn at a large number of antenna positions with respect to the transmitter. In a typical experiment the antenna is pointed at ~7000 positions which takes ~3 h.

At each angular position of the antenna, 1024 samples of IF voltages are taken. The samples require 25 μs each, and the two channels are sampled alternately. Thus the Nyquist frequency for each channel is 10 kHz. The computer then Fourier transforms the test antenna and reference antenna data separately. The delay caused by the alternate sampling is corrected by adding a phase shift to the reference horn channel. The cross-correlation spectrum can now be formed by multiplying the two Fourier transforms together frequency by frequency. Finally, the cross correlation is found by numerically integrating the correlation spectrum over a small range of frequencies which includes the transmitter signal. This windowing procedure ensures a maximum signal-to-noise ratio.

In addition to the cross correlation, which comprise the primary data, the power in each channel is computed also. First, each autocorrelation spectrum or power spectrum is formed by multiplying the Fourier transform by itself. Then the spectrum is integrated over the same range of frequencies which is used for the cross correlation. Thus after observing each antenna position, the power form each channel and the amplitude and phase of the cross correlation is computed. These four quantities are not independent. If there were no receiver noise, each of these quantities could be derived from the other three. The consistency of these quantities is then a measure of the signal-to-noise ratio of the receiver. This test provides excellent feedback about the data quality while the data are being taken.

Each of the power spectra, as well as the cross-correlation

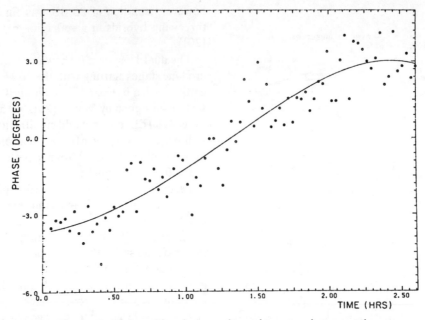

Fig. 4. Boresight position phase samples used to remove instrumental drift.

spectrum can be displayed in "real time." These displays have proved exceedingly useful in diagnosing system problems. In addition, the power in both the horn and test antenna channels can be plotted against antenna position. This, of course, is the normal antenna pattern.

The antenna is scanned over a grid of points approximately centered on the transmitter site. Data grids of 83 by 83 points were taken in a raster fashion. The boresight position was observed first, then a line of 83 points, and then the boresight position was observed again. This process was continued until the spherical section grid was completed. The series of boresight observations enabled instrumental phase drift ($\sim$3°/h) to be removed from the data. The boresight phases are shown in Fig. 4.

IV. DATA PRESENTATION

We have chosen to present our data as aperture field maps rather than far-field gain patterns. The aperture field amplitude and phase both have useful interpretation for large antennas. The deviation from constant aperture phase can be interpreted as antenna surface deviations since geometric optics can be used. The intensity of the aperture fields is also of interest. The intensity of some feeds, such as small horns, can be readily calculated. However, the experimental determination of the illumination function of more complicated feeds may be highly desirable.

Given the size of the antenna and deciding the number of phase and amplitude samples desired across the aperture, the spacing of the angular samples is fixed by their usual Fourier transform relationships. In fact, we mapped a square about 30 percent larger than the antenna to ensure that the sampling theorem had been satisfied. This implies that the angular spacing of the measurement points is fixed as the ratio of the wavelength to the side of the square. We have assumed that the small-angle approximations hold. The width of the angular area studied is then constrained by the number of samples chosen and the angular data point spacing. If most of the ra-

diated power is contained within this solid angle, then we have confidence that the number of phase and amplitude points sampled are adequate to represent the structure of the reflector surface errors. We, of course, have no *a priori* knowledge that this will be the case.

V. NEAR-FIELD CORRECTION

Because the transmitter cannot be infinitely far away, some near-field correction must be made. Procedures for performing this correction have been the subject of much work (Joy [10]). These techniques, however, are difficult to apply to a large antenna because of the large number of coefficients which must be calculated.

We have chosen to make an expedient near-field correction by defocusing the antenna as outlined by Johnson [21], and subsequently, treating the data as far-field data. The aperture field points are estimated by performing a two-dimensional discrete Fourier transform on the array of sample points. Since our range is quite long, 1.9 D^2/λ, we feel that little error results from this procedure. In the near future, we plan to implement a more rigorous correction which is appropriate for large antennas.

VI. EXPERIMENTAL RESULTS

Several measurements have been obtained. Maps of the aperture amplitude and phase have been made for the undisturbed antenna system, as well as maps with the surface disturbed. In one case, a microwave reflecting shim was added to a small area of the surface and microwave absorbing material was added to a different area. In another case, the antenna was rotated so that a different gravitational loading would prevail. The differential effects of the absorber, the shim, and the gravitational loading can be seen using the unperturbed map as a reference.

The data presented here are angular squares 83 points by 83 points. The aperture plane sampling is 128 × 128 points of

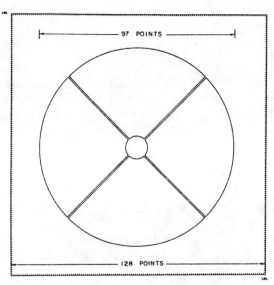

Fig. 5.　Aperture plane sampling arrangement.

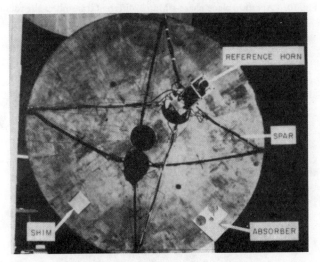

Fig. 6.　4.9-m antenna with prime focus mounted receiver and diagnostic distortions.

which 97 points correspond to the 4.9-m antenna diameter. The geometry of the sampling arrangement is shown in Fig. 5.

Since there are many possible sources of error, experimental verification of the overall method was desired. Therefore, known surface irregularities were introduced as shown in Fig. 6. On the lower left side is a brass shim about 8 mils thick with dimensions 12 in by 13 in. This shim should shift the phase by 40 at 86.1 GHz but leave the amplitude unchanged. On the right side is an odd-shaped piece of absorbing material. The absorber should reduce the field amplitude to near zero in this section of the dish.

On Jan. 1, 1981, the first successful measurements were obtained at the MWO on the 4.9-m antenna. These are shown in Fig. 7. The shim and absorber shown in Fig. 6 were added to the surface and a second measurement was obtained that evening, as shown in Fig. 8.

Figs. 7 and 8 show the two resulting aperture plane phase maps. The contour interval is 20° of phase which corresponds to a surface deviation of about 100 μm. The shim clearly shows up in the second map. The position of the absorber is indicated by the absence of contours due to low-field levels. The solid contours represent negative phase angles which correspond to

areas that are high on the surface. The dashed contours represent positive phase angles which correspond to low areas of the surface where the incoming wave has traveled farther than it would have if the antenna were a true parabola. The similarity of the two maps over the regions which do not have the shim and absorber represents a measure of the repeatability of the experiment. Clearly, on the scale of the errors there is little difference between the two maps.

Subtracting one map from the other map along the cut line in Figs. 7 and 8 reveals several interesting details, shown in Fig. 9 (see also Fig. 10, later, for comparison). The difference over the unaltered reflector area indicates the measurement error. The rms phase excursions correspond to about 4 μm of surface deviation. The phase shift at the shim corresponds to the expected amount given the shim's thickness. The phase change at the edge of the shim indicates the lateral measurement resolution, which is about 3 in. The presence of many contour lines to the upper left of the receiver blockage is due to a local surface irregularity. The electroplating in this area was not performed properly, resulting in poor reflectivity which causes low-field amplitudes with rapidly varying phase. The detection of this problem is possible only by electromagnetic measurements; mechanical and ranging methods would fail to uncover this type of problem.

Fig. 11 shows the amplitude of the antenna aperture fields. The z direction is the relative magnitude of the electric-field strength. The flat portion values were not set to zero, but are the results of the transformation of the data. The slight ripples in the area outside of the antenna show the noise in the data as well as the tails of the two-dimensional sinc resolution element. One can also see the reduced amplitude where the receiver support legs block the aperture. The absorbing material can easily be seen, but the shim area is not visible in the amplitude data. In Fig. 12, a more detailed view of the absorber section is given by expanding the array size of the Fourier transform and adding an offset. The plot is the amplitude map of the first measurement minus the amplitude map of the second measurement normalized to the first measurement map. The edge of the dish is indicated by the curved solid line and all information outside the surface is meaningless. The physical dimensions of the piece of absorber, shown in Fig. 6, are superimposed over the amplitude contours. A perspective view of the absorber area along with the resolution element afforded by the square windowing function are shown in Fig. 13.

Another issue of importance is the gravitational loading of the antenna. Since the 16-ft antenna is polar mounted, the antenna structure rotates with respect to the vertical as the antenna moves along the horizon. A 90° rotation should induce a maximum differential distortion. A second transmitter site was located at a distance of 3 D^2/λ and at a position which produced a 94° rotation from that of the previous site. A third experiment like the previous two was performed to measure the gravitational loading effect. Fig. 10 shows the phase contour map obtained using the alternate transmitter site. It may be compared to Fig. 8 to see differences induced by gravity loading. Even though there are differences, the maps are remarkably similar. Fig. 14 shows the difference of the maps shown in Figs. 8 and 10. The scale has been expanded to 8° contour levels which correspond to about 40-μm surface steps.

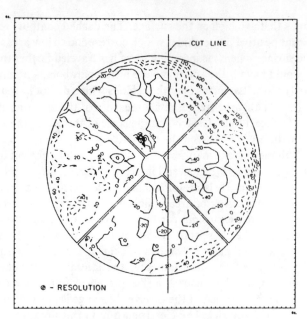

Fig. 7.   Aperture plane phase—Map 1, normal surface.

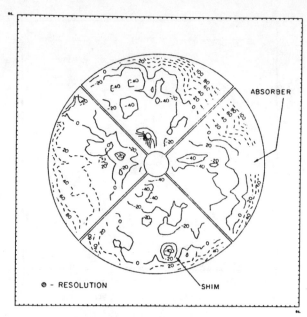

Fig. 10.   Aperture plane phase—Map 3, different transmitter site rotates gravity vector 94°.

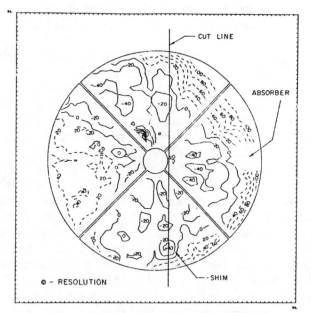

Fig. 8.   Aperture plane phase—Map 2, after diagnostic distortions of Fig. 6 added.

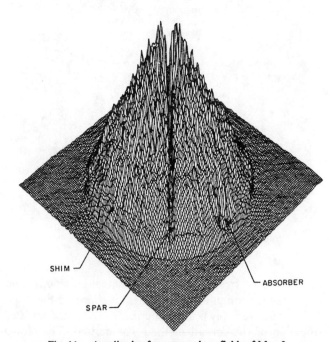

Fig. 11.   Amplitude of aperture plane fields of Map 2.

Although this map is noisy, one can see there are clusters of contours around the spar attachment points, as one would expect. The rms phase deviation of this difference map is 8°. This implies surface deviations of 40 μm due to gravity.

VII. ATMOSPHERIC LIMITATIONS

The waves from the transmitter site propagate through the lower atmosphere which is, in general, a turbulent medium that introduces fluctuations in amplitude and phase across the receiver aperture. In order to assess this effect on our measurements, we use the theory of Tatarski [22] and a review of electromagnetic beam propagation in turbulent media by Fante [23].

The turbulent medium is characterized by the index of refraction structure constant C_n^2, a function of pressure, tem-

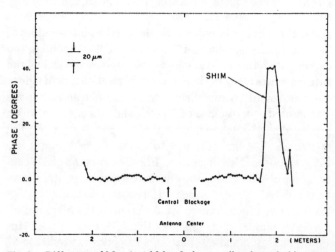

Fig. 9.   Difference of Map 1 and Map 2 along cut line through shim area.

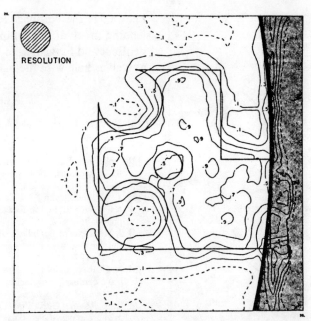

Fig. 12. Amplitude contour map of absorber area (Map 1 − Map 2)/ Map 1.

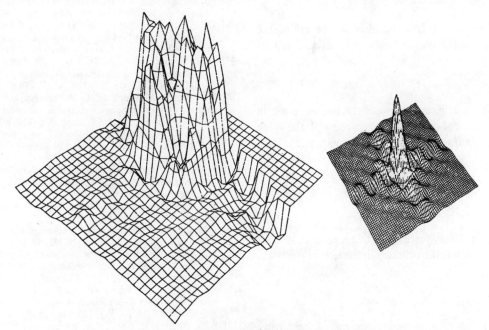

Fig. 13. Amplitude perspective of absorber area (Map 1 − Map 2)/Map 1 and resolution element.

perature, wind speed, and—especially at millimeter wavelengths—of humidity. It can be estimated from scintillation data using the equation

$$C_n^2 = \frac{\sigma_A^2}{(0.31)\, k^{7/6} L^{11/6}}$$

where σ_A^2 is the mean square of the normalized logarithmic amplitude, k the wavenumber, and L the path length (12.9 km). This equation is valid for weak scattering ($\sigma_A^2 \ll 1$) and for geometries in which the outer scale size L_o of the turbulence is greater than the geometric mean of the wavelength λ and path length L ($L_o > (\lambda L)^{1/2}$). The latter condition requires $L_o > 6.5$ m and is well satisfied when the usual assumption is

made that the outer scale size is about one third of the path altitude above ground. The first condition is satisfied because a typical value of σ_A^2 measured on either aperture while scans were made is 2.1×10^{-4}. This yields $C_n^2 = 3.08 \times 10^{-15}$ m$^{-2/3}$ and is only an order of magnitude larger than the microwave measurements reported by Fante [23].

Tatarski's theory also predicts that the phase deviation between any two points on a wavefront is

$$\sigma_d = 1.7\, C_n k L^{1/2} \rho^{5/6}$$

where σ_d is the rms phase deviation in radians between the two points on the wavefront separated by ρ, k is the wavenumber in m^{-1}, and L is the range distance in meters. The phase de-

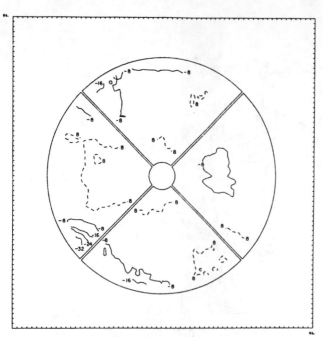

Fig. 14.   Aperture plane phase, Map 2 — Map 3, gravitationally induced distortions.

viation between the two receiver channels can be calculated from this formula and the measurement geometry. The predicted value is

$$\sigma_R = 4.27 \, C_n k L^{1/2}$$

where σ_R is the phase deviation between the two channels in radians. This value can, of course, be compared with the value directly measured. In this case the predicted value is

$$\sigma_R = 0.022 \text{ rad}$$

and the measured value is

$$\sigma_R = 0.026 \text{ rad}.$$

This agreement is probably fortuitous, but it does show that atmospheric turbulence is the primary limitation of the measurement.

VIII. CONCLUSIONS

An instrument has been built for the purpose of measuring the surface of a large millimeter-wavelength antenna by electromagnetic means. The instrument was tested by repeated measurements of the same antenna and by modifying the antenna surface in a known way. We detected the modifications, and their measured effects on amplitude and phase were as expected. Our technique, which is fast, measures the whole antenna at once and has a demonstrated accuracy of 4 μm rms, making it attractive for almost any antenna in use or contemplated today.

ACKNOWLEDGMENT

The authors would like to thank B. Ulich for helping to formulate this program. In addition, they would like to acknowledge the National Radio Astronomy Observatory for lending much of the equipment used for this work. P. VandenBout and F. Bash have lent invaluable administrative help as well as help with the manuscript.

REFERENCES

[1] J. W. M. Baars, "Technology of large radio telescopes for millimeter and submillimeter wavelengths," in *Reviews in Infrared and Millimeter Waves*, vol. 2, K. J. Button, Ed.   New York: Plenum, in publication.
[2] Nobeyama Radio Observatory—Staff, "Surface measurement of NRO 45-M telescope," Nobeyama Radio Observatory Tech. Rep. 5, pp. 1–6, 1981.
[3] J. W. Findlay and J. N. Ralston, "Testing the stepping method on the test track and the 140-foot telescope," *NRAO MM-Telescope Memos*, pp. 94–95, 1977.
[4] J. M. Payne, M. J. Hollis, and J. W. Findlay, "New method of measuring the shape of precise antenna reflectors," *Rev. Sci. Instrum.*, vol. 47, pp. 50–55, 1976.
[5] A. G. Repjar and D. P. Kremer, "Accurate evaluation of a millimeter wave compact range using planar near-field scanning," *IEEE Trans Antennas Propagat.*, vol. AP-30, pp. 419–425, May 1982.
[6] C. A. Raquet, G. R. Sharp, and R. E. Alexovich, "High resolution 6.7 m × 6.7 m near-field planar scanner," presented at the URSI National Science Meeting (Jan. 1982).
[7] J. C. Bennet, A. P. Anderson, P. A. McInnes, and A. J. T. Whitaker, "Microwave holographic metrology of large reflector antennas," *IEEE Trans. Antennas Propagat.*, vol. AP-24, pp. 295–303, May 1976.
[8] D. W. Hess, "An algorithm with enhanced efficiency for computing the far-field from near-field data on a partial spherical surface," presented at the URSI National Science Meeting (Jan. 1982).
[9] J. C. Bennet and E. P. Schoessow, "Antenna near-field/far-field transformation using a plane-wave-synthesis technique," *Proc. Inst. Elec. Eng.*, vol. 125, pp. 179–184, Mar. 1978.
[10] E. B. Joy, W. M. Leach, Jr., G. P. Rodrigue, and D. T. Paris, "Applications of probe-compensated near-field measurements," *IEEE Trans. Antennas Propagat.*, vol. AP-26, pp. 379–389, May 1978.
[11] A. Greve, "Antenna reflector measurement methods," in *Abstracts: U.R.S.I. XXth General Assembly* (Washington, D.C., August 10–19, 1981), p. 62.
[12] D. W. Hess, "Practical considerations for near-field measurements and the use of probe correction," presented at the URSI National Science Meeting (Jan. 1982).
[13] J. R. Ruze, "Antenna tolerance theory—A review," *Proc. IEEE*, vol. 54, no. 4, pp. 633–640, Apr. 1966.
[14] F. H. Larsen, "Probe correction for spherical near-field measurements," *Electron. Lett.*, vol. 13, no. 14, pp. 393–395, 1977.
[15] P. J. Wood, "The prediction of antenna characteristics from spherical near field measurements," *Marconi Rev.*, First Quarter, pp. 42–68, 1977.
[16] D. T. Paris, W. M. Leach, Jr., and E. B. Joy, "Basic theory of probe-compensated near-field measurements," *IEEE Trans. Antennas Propagat.*, vol. AP-26, pp. 373–379, May 1978.
[17] B. L. Ulich, J. H. Davis, P. J. Rhodes, and J. M. Hollis, "Absolute brightness temperature measurements at 3.5-mm wavelength," *IEEE Trans. Antennas Propagat.*, vol. AP-28, pp. 367–377, May 1980.
[18] J. R. Cogdell, "Calibration program for the 16-foot antenna," Elec. Eng. Res. Lab., Univ. of Texas at Austin, Tech. Rep. NGL-006-69-1, Jan. 1969.
[19] B. L. Ulich, "A radiometric antenna gain calibration method," *IEEE Trans. Antennas Propagat.*, vol. AP-25, pp. 218–223, Mar. 1977.
[20] C. E. Mayer, "A methodical approach to a large microwave system design," Master's thesis, Univ. of Texas at Austin, July 1981, and Elec. Eng. Res. Lab., Univ. of Texas at Austin, Tech. Rep. ST-7920966-81-1.
[21] R. C. Johnson, H. A. Ecker, and J. S. Hollis, "Determination of far-field antenna patterns from near-field measurements," *Proc. IEEE*, vol. 61, pp. 1668–1694, Dec. 1973.
[22] V. I. Tatarski, *Wave Propagation in a Turbulent Medium*,   New York: McGraw-Hill, 1961.
[23] R. L. Fante, "Electromagnetic beam propagation in turbulent media: An update," *Proc. IEEE*, vol. 68, pp. 1424–1443, Nov. 1980.

Improvement of the Effelsberg 100 meter telescope based on holographic reflector surface measurement

M.P. Godwin[1], E.P. Schoessow[1], and B.H. Grahl[2]

[1] Minalloy House, 18–22 Regent Street, Sheffield S1 4DA, UK
[2] Max-Planck-Institut für Radioastronomie, Auf dem Hügel 69, D-5300 Bonn 1, Federal Republic of Germany

Received December 13, 1985; accepted May 28, 1986

Summary. Microwave holography was applied to measure the surface of the 100 meter telescope of the MPIfR. An array of 195×195 samples of the complex antenna pattern was obtained using a geostationary satellite beacon as the source, allowing the image of the dish to be formed with a resolution of 0.56 m. As a result of this measurement the adjustment of the inner 60 m portion of the reflector was made. Subsequent gain tests confirmed the rms deviation now to be 0.4 mm.

Key words: radio telescope – surface accuracy – holography

1. Introduction

The accuracy of the surface is an important limiting factor for high frequency observations with a reflector telescope. The 100 meter telescope at Effelsberg was designed for use down to 2 cm wavelength (Hachenberg et al., 1973). This was achieved by the following:
– Computer aided design of the reflector structure to minimize the residual gravitational deformation of the surface (homology design).
– Installation of electromechanical devices for compensation of focus displacement due to gravitational deformation.

The panel size was chosen relatively small (about $3 \text{ m} \times 1.2 \text{ m}$), for various reasons including stiffness and ease of manufacture. Therefore the number of panels is very large: the inner 80 m diameter solid surface alone comprises 1492 panels whilst the overall surface, that is including the outer mesh panels, has 2356 panels. A consequence of this large number of panels is that surface measurement and adjustment are very tedious.

The first measurement of the surface was performed by mechanical optical methods (Greve, 1981). Targets were placed at the panel corners at defined distances from the vertex of the reflector which were measured by steel tape. The angle of the targets measured from the antenna axis at the vertex was then determined using a theodolite. This method could only be applied in the zenith position of the reflector. Thus the reflector had its ideal shape at zenith and, due to residual gravitational defor-

mation, the antenna performance decreased with distance from zenith. In 1978 the inner part of the surface was remeasured by the same method (Grahl, 1978). This led to an improved adjustment including corrections for the residual gravitational deformation effects according to calculations of deformation. Radioastronomical tests at $\lambda = 7$ mm (Altenhoff et al., 1980) showed that some of the expected improvements were obtained but the attempted correction of gravitational deformation effects yielded only a minor increase in performance. The gain maximum was now at about 60° elevation (see Fig. 1, curve number 1).

During 1980 and 1981 a large number of the reflector panels needed to be replaced. Although this was done with extreme care an increase of the r.m.s. deviation of the surface was unavoidable (see Fig. 1, curve number 2). This led to a plan to readjust the surface once more. Holographic techniques (Scott, P.J. and M. Ryle, 1977) (Godwin, 1982) were considered as an alternative to the classical surveying methods. However, it was clear that the special needs for the 100 m dish were difficult to fulfil. The requirements were
– Measurement accuracy 0.2 mm
– Resolution 1 m D/100
– High reliability in view of enormous workload involved in surface readjustment.

Measurements of the Chilbolton dish, (Godwin et al., 1981) which used the beacon signal of the geostastionary satellite OTS-2 seemed very promising. Hence this method was taken into consideration for the measurement of the inner 80 m diameter solid portion of the surface of the 100 m dish.

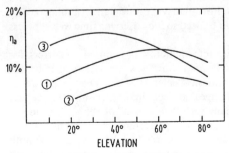

Fig. 1. Antenna efficiency (ref. to 100 m diameter) as a function elevation angle

---
Send offprint requests to: B.H. Grahl

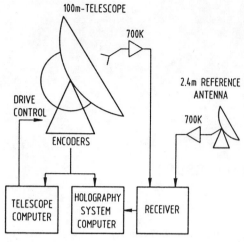

Fig. 2. Holographic antenna test system

2. Holographic measurement system

Holographic reflector surface measurement is based on the principle that deviations from the desired paraboloid shape can be calculated from the phase component of the complex reflector surface current distribution, which can in turn be derived from the measured, complex antenna pattern.

At any point in the map of surface errors the value indicated is essentially an average of the true error over a small surrounding area. The size of this averaging area is the resolution of the measurement and is determined by the size of angular region over which the antenna pattern is measured. In order to obtain accurate readjustment of panels it is necessary that the resolution be small compared to the panel dimensions. A resolution of 0.56 m, equal to about half the panel edge dimension, was chosen. With a small oversampling allowance this required the pattern to be sampled at 195 × 195 points. The instrumentation used to record the data is shown schematically in Fig. 2. The source used was the 11.786 GHz beacon radiated by the geostationary satellite OTS-2 which appeared at an elevation angle of about 32° for the 100 m antenna. A 2.4 m antenna was used to provide a phase reference signal. The reference antenna can be seen alongside the 100 m telescope in Fig. 4. The use of a primary focus feed on the 100 m antenna was advantageous for two reasons: easy installation and elimination of additional phase effects of a subreflector. This feed was chosen to provide an illumination of 15 dB towards the edge of the 80 m diameter region of interest. The telescope was driven in a raster scan mode with the subscans taken moving in the positive elevation direction. That is, 195 separate pattern cuts of 195 samples were acquired, with the antenna moving upward in elevation for each. Between each cut the antenna elevation was reset and the azimuth pointing incremented. Fortunately the automatic scanning facility of the 100 m dish could be used so the observing time was kept less than 12 hours.

Additional steering information was required to correct for movement of the satellite and this was kindly supplied by the European Space Operation Center (ESOC). Over a period of several hours, small changes can occur in the r.f. equipment, cables, etc. Hence to accompany each "main scan" of 195 × 195 points, a small "calibration scan" was performed, each comprising a number of 195 point subscans recorded whilst moving in azimuth across the main scan. Comparison of the main scan and calibration scan data allowed the equipment drift to be determined and

removed. Correction was also required for the phase difference between the antenna outputs that arise due to the satellite motion. With the satellite EIAP of 31 dbW the highest signal-to-noise ratio achieved in the system bandwidth of 30 Hz was 79 db.

3. Observations

The measurements took place in October 1982. The total activity including installation of the test equipment, observations, demounting an preliminary data processing were performed within 10 days. Three high resolution scans were taken overnight between 20th and 23rd October. For the first of these scans a number of panels were displaced from their normal positions. The panels were restored to their normal positions prior to the second and third scans. The purpose of this was to demonstrate the repeatability and accuracy of the measurement technique by comparison of the various results.

Slight difference in signal to noise ratio occurred between the scans, the second of which provided highest on axis signal to noise of 79 dB. Weather conditions during the tests were good with low wind and, especially during the second night, clear skies.

4. Results

The principle plane amplitude radiation patterns from the second scan are shown in Fig. 3. Whilst the inner sidelobe structure of the azimuth pattern exhibits reasonable symmetry, the first sidelobes in the elevation pattern are noticeably unbalanced. Figure 5 shows the surface current distribution over the solid portion of the reflector (79.462 m diameter). Shadows cast on the dish surface by the struts are well defined and clearly visible. Slightly higher illumination can be seen at the edges of the shadows and this is thought to be a diffraction effect.

Diffraction might also be expected to result in some "in-filling" of the shadows. Close inspection shows this to be present but it is noticeable that the current in the left most shadow is more uniformly low than in the other three. Since the struts are of open space frame construction and the left most one carries the steps and the walkway to the focus cabin, the cause of the surface current in the shadowed region is thought to be predominantly transmission through the struts.

Analysis of the systematic phase component of the surface current distribution showed the feed to be displaced from its optimum gain position by the following amounts.

$$Dx = -2.10 \text{ mm},$$
$$Dy = 4.66 \text{ mm},$$
$$Dz = 0.78 \text{ mm},$$

where Dx and Dy are lateral displacements parallel to and normal to the antenna axis respectively, and Dz refers to axial displacement. (The Dy and Dz errors are readily corrected using the electromechanical focus correction system. However, this system makes no provision for adjustment to the Dx component. An alternative approach was, therefore, adopted to correct the error in the x coordinate and this is described later). Since the overall position error is small compared to the measurement wavelength, its correction realizes a small gain increase of only 0.02 dB at the test frequency. Furthermore the error in the y coordinate is too small to account for the asymmetry in the elevation pattern referred to earlier.

216

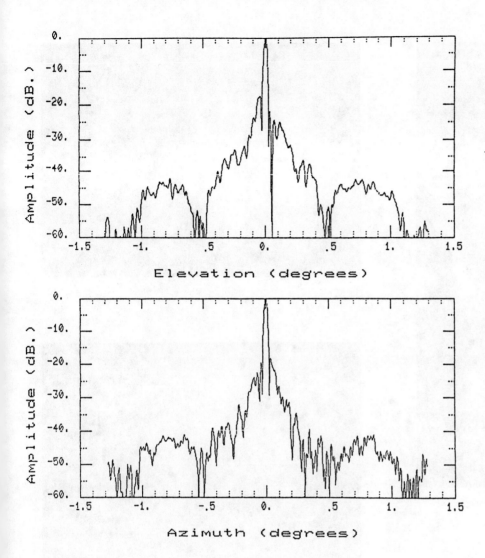

Fig. 3. Principal plane patterns from scan 2.
Measuring interval azimuth: 50″
Measuring interval elevation: 42″
Pointing accuracy: ~ 4″

Figure 6 shows the map of reflector surface errors derived from scan 2. The errors shown are measured normal to the surface of the best fit paraboloid. Data in blocked and shadowed areas is meaningless and has, therefore, been suppressed. The tick marks on the horizontal and vertical diameters indicate the panel ring boundaries.

The r.m.s. error of 0.73 mm derived from scan 2 is only slightly above the design goal of 0.6 mm. The major features revealed by the map are thought to relate to gravitational deformation. Some qualitative relation to the calculated deformations (Hachenberg et al., 1973) can be observed but it is difficult to make a quantitative comparison especially because of the readjustment of the inner part of the surface in 1978 (Grahl, 1978) which took calculated deformation into account. It seems that the 1978 adjustment has led to some overcorrection for the deformation of the inner rings. The asymmetry between the upper and lower part is clearly visible and is very probably responsible for the side lobe asymmetry in the elevation pattern discussed earlier.

The accuracy of the reflector surface error map is limited by noise in the receiver system. Due to feed illumination taper the effects of receiver system noise become more severe close to the edge of the dish. The standard deviation in determining the mean position of the surface over each resolution cell has been estimated as shown in Table 1.

Table 1

	SNR (dB)	Standard deviation (mm)	
		Reflector center	Reflector edge
Scan 1	76	0.01	0.08
Scan 2	79	0.005	0.04
Scan 3	76	0.01	0.08

Propagation effects were observed to cause only one or two degrees fluctuation of the phase data on the beam under good conditions. Computer simulation suggests that the error due to scintillation is considerably less than that due to receiver system noise.

A proof of the repeatability of the measurements is possible by examining the differences between the surface error maps. Figure 7 shows the difference pattern of map 1 and map 2. The four positive panel displacements of 2 mm which were introduced for scan 1 are observable. Minor deviations are also evident in the outer areas. These may be the results of day to day variations in the antenna shape but it should be noted also that each surface error map used

Fig. 4. The reference antenna alongside with the 100 m telescope

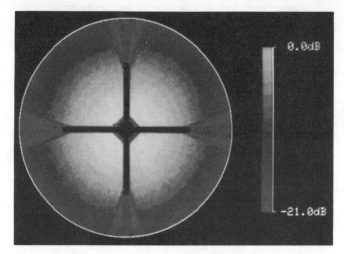

Fig. 5. Map of reflector current distribution

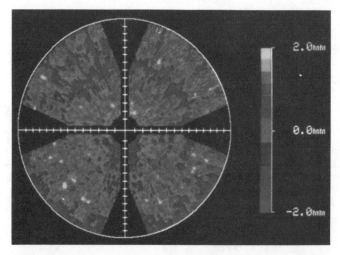

Fig. 6. Map of reflector surface errors from scan 2

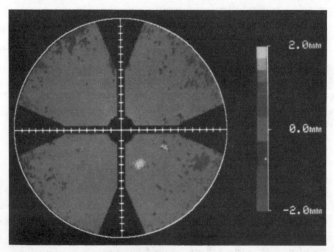

Fig. 7. Map of profile differences: Scan 2 – Scan 1

in the subtraction is referenced to its own best fit paraboloid and these are slightly different in view of the deliberately displaced panels. The r.m.s. value of the difference map is 0.21 mm.

The differences map for scan 2 and scan 3, in between which no panels were adjusted has an r.m.s. value of 0.17 mm. Color displays of the differences maps give a clear indication that these residual differences are connected with the structure of the reflector, or in some areas of the surface, spring loading effects of panels.

5. Panel adjustment

The panels of the 100 m dish were designed to be a stiff construction and deflection due to gravitation should be below 0.2 mm. They are fixed to the telescope structure with four screws at each corner. These screws are especially designed for the purpose of panel adjustment. For the calculation of numerical values for readjustment of the surface stiff panels were assumed. Samples falling within masked areas were rejected for the purposes of calculating the corrections. In addition a resolution "safety margin" was allowed around the edge of each panel. This was because, due to the finite resolution, the correction recommended for a given panel could be affected by the neighbouring panels. The margin minimizes this undesirable effect. If on any panel, after rejection of samples falling within the masked areas or the margin, three or more profile samples remain, a set of adjustments was

218

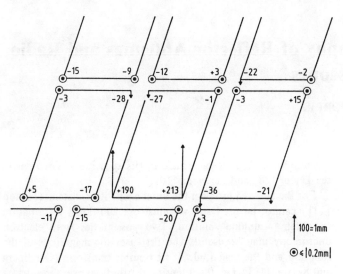

Fig. 8. Difference between panel adjustments from scan 1 and scan 2 showing a tilted panel (Units 0.01 mm)

6. Improvement of the telescope

The readjustment of the panels was started in 1983. Within this year the staff of the Effelsberg Station was able to adjust panels of the inner 60 m diameter part of the reflector. This job required great care. For the adjustment of a panel each of the four corners were checked by dial gauges which were mounted to the fixture of neighbouring panels. The working process was quite accurate but is probably limited to an r.m.s. error of 0.1 to 0.2 mm due to various contributions including spring and load effects. Also the surface suffers from temperature variations over the year due to differences in thermal expansion of the steel structure and aluminum panels.

The status of the antenna was tested in December 1983 at 43.1 GHz (Altenhoff, Wink, 1984). The measured variation of efficiency is shown in Fig. 1, curve number 3. Due to the illumination pattern (15.8 dB attenuation at 60 m diameter) these results provide a clear indication of the improvement of the readjusted part of the reflector. The maximum efficiency is now at about 32° elevation as expected due to the elevation of the satellite which served as signal source for the holographic measurements.

The improvement of the efficiency at this elevation from curve number 2 to curve number 3 is about a factor of 2.8 which leads to an estimated improvement of the r.m.s. error of the inner part of the dish. This error of the inner portion is now thought to be ∼ 0.40 mm. The elevation of 32° for the maximum gain is extremely valuable, for most astronomical observations are performed in the range of 10° to 50° elevation. Further improvement of the antenna will be achieved by ajustment of the outer rings of the solid part of the surface.

calculated. If only one or two acceptable samples remain on the panel, a mean displacement correction was calculated in which the panel is raised or lowered but not allowed to tilt. Where no acceptable samples remain, there was of course no adjustment recommended. Because the feed position adjustment mechanism at the focus of the Effelsberg antenna incorporates no movement in the X-direction the data used for the calculation of panel corrections were derived with the x-coordinate feed position optimization suppressed. Thus the feed position error removal for this coordinate is embodied in the panel corrections.

Correction numbers were calculated for all three scans although those of scan 2 were regarded as optimum for adjustment. Comparison of the panel corrections derived from scan 1 and 2 in the areas surrounding those panels which had been especially lifted for test purposes provides an interesting insight into the overall performance of the holographic measurement and data processing. The schematic diagram of panels (Fig. 8) shows for a testfield the difference between scans 1 and 2 indicated by vectors. In the center is a panel where the lower two corners were lifted by 2 mm during scan 1. Better than any error analysis this picture indicates the reliability of the results with respect to the readjustment of the surface and confirms that the outstandingly high resolution required for readjustment of the 100 m telescope has been achieved.

References

Altenhoff, W.J., Baars, J.W.M., Downes, D., Pankonin, V., Wink, J.E., Winnberg, A., Matthew, H.E., Genzel, R., Olnan, F.M.: 1980, *Astron. J.* **85**, 9

Altenhoff, W.J., Wink, J.E.: 1984, Technischer Bericht Nr. 18, Max-Planck-Institut für Radioastronomie

Godwin, M.P., 1982 "Microwave Diagnostics for Large Reflector Antennas", Ph.D Thesis, University of Sheffield

Godwin, M.P., Whitaker, A.J.T., Bennett, J.C., Anderson, A.P.: 1981, IEEE Conf. Publ. **195**, 232

Grahl, B.-H.: 1978, Technischer Bericht Nr. 51, Max-Planck-Institut für Radioastronomie

Greve, A.: 1981, Zeitschrift für Vermessungswesen, 106. Jahrgang S. 308, Juni

Hachenberg, O., Grahl, B.-H., Wielebinski, R.: 1973, *Proc. IEEE* **61**, Nr. 9, September, 1288

Scott, P.J., Ryle, M.: 1977, *Monthly Notice Roy. Astron. Soc.* **178**, 539

Phase Retrieval in the Radio Holography of Reflector Antennas and Radio Telescopes

D. MORRIS

Abstract—Methods of phase retrieval from simulated intensity information have been tested for use in the radio holography of reflector antennas. In numerical simulations the Misell algorithm has been used successfully to retrieve the aperture phase distribution from two numerically simulated power polar diagrams, one in focus and the second defocused. The technique uses no auxilliary reference antenna. However, it does need a high signal to noise ratio, typically 50 dB if a 60 × 60 array is to be measured to a precision such that the gain is within 1 percent of ideal. It should be most useful where no direct phase measurements are possible and ground-based or satellite transmitters can be used as sources. The use of astronomical maser sources (22 GHz) can give information on large scale deformations.

I. INTRODUCTION

THE ACCURACY needed in the setting and measurement of the surface of large radio-telescopes designed for operation at millimeter and shorter wavelengths is now at the limit of conventional surveying methods. This is particularly true if the surface is to be checked at several angles of tilt. "Holographic" methods have been used with some success but until now have required additional reference antennas and phase stable receiving systems.

This paper explores the possibility of using measurements of intensity only and relying on "phase retrieval" methods to obtain the necessary phase information. As a first step we have made a numerical simulation to test the feasibility of the method and its sensitivity to noise and measurement errors.

The idea of using measurements of the far-field pattern of an antenna to deduce the phase errors present in its aperture is not new (see for example Blum *et al.* [6]). However, its practical use dates from the work of Bennet *et al.* [4] and Scott and Ryle [26]. A "holographic" recording technique was used by Bennet *et al.* [4], and this name has remained even though recent measurements have used direct phase recording (Scott and Ryle [26]), Godwin *et al.* [16], Anderson *et al.* [1], Bennett and Godwin [5], Mayer *et al.* [21]). A similar method has been analyzed by Von Hoerner [31]. It was first proposed by Shenton and Hills [27] and uses the focal plane fields.

The need for accurate phase measurements makes these methods relatively difficult and expensive if additional equipment such as auxillary reference antennas and phase sensitive receivers must be found. On the contrary, most reflector antennas have receivers capable of measuring the power polar diagram. What can be deduced about the reflector shape from such power measurements alone? So posed, the problem becomes the well-known "phase retrieval problem."

II. THE PHASE RETRIEVAL PROBLEM

There is extensive literature on calculating the phase of a complex function from knowledge of only its magnitude, or the magnitude of its Fourier transform (its spectrum). For reviews see Taylor [28] and Ferwerda [10].

For the antenna problem the usual positivity constaint (Fienup [11]) cannot be used and we must use additional information to obtain a unique solution. Two possibilities from electron microscopy may be useful. The first uses the magnitude of the function and the magnitude of its Fourier transform (Gerchberg and Saxton [13]), i.e., the intensity distributions over the antenna aperture and in the far-field diffraction pattern. The second method, due to Misell [22], uses the intensity distributions in two defocused images, i.e., two far-field patterns recorded with different axial focus settings. This second method seems *a priori* more suitable for antenna diagnosis. Both measurements (in-focus and defocused radiation patterns) use the same experimental technique. Furthermore the solution for the aperture phase distribution has been shown to be unique for the one-dimensional case (Huiser and Ferwerda [18], Hoenders [17]). This result can probably be extended to two dimensions (Drenth *et al.* [9]). In contrast, the measurement of aperture and far-field intensities involves different techniques and the calculated solution for the phase is not always unique (Huiser *et al.* [19], Huiser *et al.* [20]).

III. POSSIBLE ALGORITHMS

Three classes of algorithms have been proposed for solving the phase retrieval problem. The direct methods (Van Toorn and Ferwerda [30], Gerchberg and Saxton [13]) are recursive and are so sensitive to errors and noise in the input data that most authors have dismissed them for practical use. The remaining two methods can be regarded as error reduction techniques. They include gradient search methods such as the method of steepest descent, and the iterative Fourier transform method (Gerchberg and Saxton [14], Misell [22]). The relation between the two has been discussed by Fienup [12]. Boucher [7], [8], in one-dimensional tests, has concluded that the Misell algorithm is slightly superior to the gradient search technique in terms of sensitivity to errors and noise. In two-dimensional tests Saxton [25] found that both methods (gradient search and iterative transform) were reasonably insensitive to noise.

Both methods have the disadvantage that they may find "local" minima instead of the desired "global" minimum. In the interative transform method this has been termed "locking" (Gerchberg and Saxton [14]). It is readily detected in computer simulations but may be a practical limitation since measurement errors produce similar effects on the convergence of the algorithm.

In a survey of tests of the iterative transform algorithm, Taylor [28] has concluded that success depends on the particular function involved, the number of samples, and the initial trial function. It thus seemed useful to us to make computer trials simulating, as closely as possible, the measurement of a reflector antenna. We report here on tests of the iterative transform method

Manuscript received November 30, 1982; revised September 2, 1983.
The author is with IRAM, Domaine Universitaire de Grenoble, 38406 St. Martin d'Heres, France.

Reprinted from *IEEE Trans. Antennas Propagat.*, vol. AP-33, no. 7, pp. 749–755, July 1985.

applied to defocused pairs of far-field power patterns, that is, the Misell algorithm.

IV. MEASUREMENT ERRORS

Even if the phase retrieval algorithm itself introduces little error, it is clear that a penalty in precision must be paid if intensity (power) measurements alone are used to deduce the phase distribution across the antenna aperture.

The influence of multiplicative errors (e.g., photographic grain noise or photon statistical noise) on the Misell algorithm has been studied numerically by Boucher [8] and similar studies were also made by Misell himself [23]). In both cases only one-dimensional tests were made. For radio measurements, the dominant error is likely to be additive receiver noise. We now analyze its contribution to the phase errors in the aperture plane.

Consider for simplicity a square antenna of diameter D wavelengths with an aperture electric field distribution $E(x, y)$; the harmonic time variation is suppressed. This can be represented by a square array of $N \times N$ equally spaced complex numbers. Its far-field pattern is then given over a finite data window, whose diameter is N/D rad, by the discrete Fourier transform (DFT)

$$A(a, b) = \sum_{i=0}^{N-1} \sum_{j=0}^{N-1} \{E(x_i, y_j) e^{i(x_i a + y_j b) 2\pi/N}\}$$

$$= \text{DFT} \{E(x, y)\}$$

$$E(x, y) = \sum_{i=0}^{N-1} \sum_{j=0}^{N-1} \{A(a_i, b_j) e^{-i(x_i a + y_j b) 2\pi/N}\}/N^2$$

$$= \text{DFT}^{-1} \{A(a, b)\}.$$

Thus $N \times N$ complex samples of the far-field pattern spaced at the critical interval $1/D$ rad will define the aperture distribution with a surface resolution of D/N wavelengths.

The final trial function at the end of the iterations of the Misell algorithm is, in the in-focus far-field plane, $T(a, b)$. Inverse Fourier transformation yields the solution for the aperture distribution

$$S(x, y) = \text{DFT}^{-1} \{T(a, b)\}.$$

In the ideal case when the algorithm introduces negligible error, $T(a, b)$ differs in amplitude from the correct value $A(a, b)$ by the measurement error $dA(a, b)$ of the in-focus measurements. Similarly, the phase of $T(a, b)$ differs from that of $A(a, b)$ by an angle $dP(a, b)$ introduced during the previous iteration and depending on both the measurement error in the defocused plane and the history of previous iterations. This is equivalent to the addition of an error vector $dT(a, b)$.

We assume, from the symmetry of the algorithm, that its operation is such that, overall, the error contributions to $dT(a, b)$ from $dA(a, b)$ and $dP(a, b)$ are equal. So the total error power when summed over the $N \times N$ measured points in the far field is just twice the contribution due to $dA(a, b)$ alone:

$$\sum_{i=0}^{N-1} \sum_{j=0}^{N-1} \{A(a_i, b_j) - T(a_i, b_j)\}^2$$

$$\cong \sum_{i=0}^{N-1} \sum_{j=0}^{N-1} dA^2(a_i, b_j). \tag{1}$$

After transformation to the aperture plane the total error power is

$$\sum_{i=0}^{N-1} \sum_{j=0}^{N-1} \{dE^2(x_i, y_j)^2\} \cong \sum_{i=0}^{N-1} \sum_{j=0}^{N-1} \{S(x_i, y_j) - E(x_i, y_j)\}^2$$

and is given by Parseval's theorem as

$$\sum_{i=0}^{N-1} \sum_{j=0}^{N-1} dE^2(x_i, y_j) = N^{-2} \sum_{i=0}^{N-1} \sum_{j=0}^{N-1} A(a_i, b_j) - T(a_i, b_j)^2$$

$$\cong 2N^{-2} \sum_{i=0}^{N-1} \sum_{j=0}^{N-1} dA^2(a_i, b_j). \tag{2}$$

The root mean square (rms) phase errors dQ in the aperture plane are given by

$$dQ^2 = 0.5 N^{-2} \sum_{i=0}^{N-1} \sum_{j=0}^{N-1} \{dE(x_i, y_j)/E(x_i, y_j)\}^2$$

and substituting for $dE(x, y)$ from (2)

$$dQ^2 \cong K^2 \sum_{i=0}^{N-1} \sum_{j=0}^{N-1} \{dA^2(a_i, b_j)\}/[A^2(0, 0)] \tag{3}$$

where $A(0, 0)$ is the maximum value (on boresight) of the far-field amplitude, assumed to be approximately $N^2 E(0, 0)$ if the phase errors are small. K is a factor to allow for any taper in the aperture distribution and any associated weighting of the phase errors. For a Gaussian amplitude taper t then $K^2 = (1/t - 1)/(1 - t)$ if amplitude weighting is used (for -14 dB taper $K = \sqrt{5}$).

In the case of the interferometer analyzed by Scott and Ryle [26], the electric field amplitude in the far field is proportional to the output V of the detector whose rms noise fluctuations dV are constant during measurements. Then substituting in (3)

$$dQ^2 \cong K^2 N^2 dV^2 / V^2(0, 0)$$

$$dQ \cong KN/R \tag{4}$$

which differs from the value given by Scott and Ryle [26] by the normalization factor K. (Here $R = V/dV$ is the signal to noise ratio on boresight.)

However, in the present case when only power can be measured, the amplitude is taken as the square root of the detector output, or in the presence of noise, the square root of the absolute value of the detector output. For dA/A small we have approximately $dA = dV/2A(a, b)$ which varies within the polar diagram. The aperture plane rms phase errors are, from (3),

$$dQ \cong KdV \left[\sum_{i=0}^{N-1} \sum_{j=0}^{N-1} \{1/A(a_i, b_j)\}^2 \right]^{0.5} /2A(0, 0)$$

or

$$dQ \cong KS(N)/2R. \tag{5}$$

Since the sum $S(N)$ can become very large in the far sidelobes where $A(a, b)$ is small, the method is increasingly susceptible to the errors introduced by noise at large scan angles. The exact value of $S(N)$ and its variation with N depend on the sidelobe structure which in turn will be a function of taper, blocking, and surface errors. In practice a variation faster then N^2 can be expected. Thus in the regime where the amplitude errors are small the Misell algorithm will result in larger phase errors than when phase can be directly measured (4). This is particularly true for large N (high surface resolution).

Another limiting case corresponds to measurements made near nulls where the signal is negligible with respect to receiver noise. In this case

$$dA(a, b) = \{|dV(a, b)|\}^{0.5}$$

and

$$\sum_{i=0}^{N-1} \sum_{j=0}^{N-1} dA^2(a_i, b_j) \cong N^2 dV(2/\pi)^{0.5}$$

where we have used the result for the mean output of a full wave linear detector when fed with Gaussian noise (Bennett [3]).

Then (3) becomes

$$dQ \cong K[2/\pi]^{0.25} N/R^{0.5}. \tag{6}$$

Hence, in the worst case the Misell algorithm demands a signal-to-noise ratio which is approximately the square of that needed when phase can be directly measured. In fact the numerical trials described below show that this regime is more nearly applicable in practice. For small phase errors the fractional loss in antenna gain is

$$dG = dQ^2.$$

Then the signal-to-noise ratio needed is

$$R \cong (2/\pi)^{0.5} K^2 N^2/dG.$$

V. THE MISELL ALGORITHM

The algorithm is supplied with a first guess at the aperture distribution—the initial trial function—and an in-focus and a defocused far-field-amplitude distribution. For the present tests, the latter were derived from a model antenna.

The trial function is in turn repeatedly compared with the two far-field amplitude patterns. At each iteration the amplitude of the trial function is replaced by the simulated far-field amplitude but its phase is retained, and thus is progressively modified as the iterations proceed. The connection between the two far-field patterns is made via the aperture plane (by fast Fourier transform) where focussing and defocusing is achieved by subtracting or adding a phase correction which varies as radius squared. One further constraint, not strictly necessary for the algorithm, has been included. During each change of focus in the aperture plane, the trial function is put to zero beyond the edge of the telescope, i.e., a mask is applied. At each comparison, in either focused or defocused far-field plane, the mean square difference, expressed as a fraction, is taken between the trial amplitude and that simulated by the model. This error can at worst remain constant during the iterations (a "locked" solution at a local minimum) or decrease toward zero at the correct solution (Gerchberg and Saxton [14], Boucher [8]). In the presence of noise or measurement error the algorithm converges to or "locks onto" a solution with nonzero error since the two far-field patterns, when corrupted by noise, are no longer consistent with a single aperture distribution.

VI. COMPUTER SIMULATIONS

The model antenna consisted of a two-dimensional complex array of $N \times N$ elements simulating the aperture distribution.

The amplitudes were given a Gaussian taper (usually −14 dB) and this distribution could be offset to produce a slight asymmetry. It could be masked to zero within regions "blocked" by a central obstruction supported by four "legs," and also beyond the edge of the circular aperture (see Fig. 1). The radius of this aperture was usually about 0.45 of the array diameter. Thus the simulated far-field values obtained by DFT correspond to slightly oversampled measurements, to avoid aliasing. The phase distribution could simulate defocus errors, astigmatism, a displaced panel or random surface errors.

Far-field patterns were calculated by the discrete Fourier transform (Arembepola [2]) for the in-focus and defocused cases. For the latter, the phase error in the aperture plane was assumed to vary as radius squared. In practice, when low F/D systems are measured, higher order terms would be necessary. The amplitude changes produced by defocusing are small for high gain antennas and have been neglected. In the tests described below, a maximum defocus amounting to 9.4 rad at the antenna edge was used. However, other tests indicate that this is not critical as long as it can reduce the antenna gain significantly bearing in mind the measurement errors. Thus with a 40 dB signal to noise ratio a defocus of about 1.6 rad was needed but with 50 dB signal to noise ratio even 0.393 rad gave satisfactory convergence. For a typical prime focus telescope these values correspond to motions of the order of a wavelength in the receiver (or the secondary mirror in Cassegrain systems). The magnitude of the far-field patterns were corrupted by simulated measurement errors before being input to the Misell algorithm. The effects of additive noise (simulating receiver noise), multiplicative noise (simulating random gain variations), and nonlinearity were studied. To avoid negative amplitudes, the absolute values of the corrupted amplitudes were taken before being input to the algorithm.

The errors of the solution were estimated by comparing the aperture distribution at the output of the Misell algorithm with that supplied by the model. Phase errors have been expressed as amplitude weighted rms values.

VII. RESULTS

An initial trial function with a Gaussian taper (either −7 dB or −20 dB) and random phase (rms = 0.5 rad) was used but these values were not critical. Usually, after several tens of iterations the algorithm converged to a solution close to the desired function originally supplied by the model. Thereafter convergence was slow (Boucher [8]) and calculations were usually stopped after some hundreds of iterations. The deviations between the model and solution (the errors) were apparently random and near to those expected from the simulated measurement errors (calculated using (3)). As a measure of the performance of the algorithm, the ratio of the rms phase error in the aperture plane to that expected from the simulated measurement errors has been calculated. It was of the order unity (range 0.5 to 2.0) depending on the noise level. For this purpose the actual amplitude errors $dA(a, b)$ were used in (3), and the assumptions of (1) can be tested.

To illustrate the algorithm's performance we give the results for a simulated antenna having astigmatism (0.5 rad) and a displaced panel (0.5 rad). The illumination pattern was offset from centre by about 0.15 radii. A central blockage of about one tenth of the antenna diameter has been assumed. Random surface errors made little change in performance.

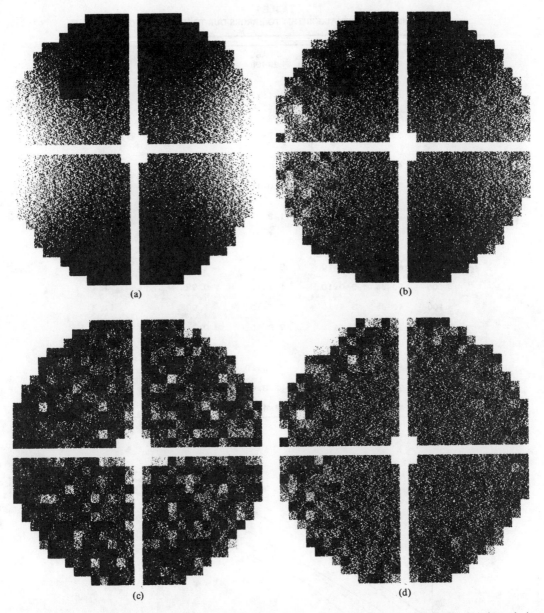

(a)

(b)

(c)

(d)

Fig. 1.   Grey scale plots showing typical results from the Misell algorithm for a signal to noise ratio of 45 dB. Fig. 1(a) shows the input phase distribution supplied by the model antenna. It has an astigmatism of $+-0.5$ rad and a panel displaced in phase by 0.5 rad (1/25 wavelengths surface displacement). Fig. 1(b) gives the output solution phase. The errors in amplitude and phase (difference solution minus model) are displayed in Figs. 1(c) and 1(d), respectively. The extremes of phase error are $+0.82$ and -0.51 rad and the extreme amplitude errors are $+9.0$ and -1.0 percent. The amplitude weighted rms phase error is $+-0.094$ rad corresponding to a gain loss of 0.87 percent.

For most tests a 32 × 32 array was used and the antenna diameter was taken as 30 resolution elements. The results are summarized in Table I for this case and for a limited series of tests on 64 × 64 and 128 × 128 arrays (60 and 120 resolution elements per diameter, respectively). For each signal-to-noise ratio (column 3), the sum-square error (percent) reached after the number of iterations listed in column 2 is given in column 4. Three performance estimators are used. They are the weighted rms phase error in the aperture plane (column 5), the ratio of this is that expected from the noise level (column 6), and the antenna gain loss (percent) due to the phase errors (column 7).

The results for a 45 dB signal to noise ratio are displayed in the "grey scale" plots of Fig. 1. The phase distribution across the aperture of the model antenna is shown in Fig. 1(a). The

solution after 200 iterations is given in Fig. 1(b), and the phase and amplitude error distributions in Figs. 1(c) and 1(d), respectively. The random nature of the residual phase errors (solution minus model input) is demonstrated in these latter two plots. Note that the phase errors are not distributed with circular symmetry since the illumination pattern has been offset.

The dependence of the amplitude weighted phase errors on the signal to noise ratio and the number of samples used (which determines the surface resolution achieved) is illustrated graphically in Fig. 2. In these tests no "blocking" was present in the aperture. Fig. 2 shows that the form of the dependence is in agreement with (6) over most of the range considered. (A steepening and separation of the curves at high signal-to-noise ratios indicates the transition to the regime of (5)). However, the magnitude of

TABLE I
RESPONSE OF MISELL ALGORITHM TO ERRORS DUE TO RECEIVER NOISE

ARRAY SIZE(N)	ITER. NO.	S/N RATIO	SUM SQUARE ERROR (%)	PHASE ERROR	PHASE ERROR/ EXPECTED	GAIN LOSS (%)
32	50	inf.	9.2×10^{-5}	4.16×10^{-3}	-------	<0.01
	200	60 db	4.6×10^{-4}	5.4×10^{-3}	1.32	<0.1
	200	50 db	1.5×10^{-2}	3.7×10^{-2}	1.42	0.14
	200	40 db	1.5×10^{-1}	1.7×10^{-1}	1.26	2.7
	200	30 db	8.6×10^{-1}	6.0×10^{-1}	1.04	27.0
	(200	20 db	2.4	1.3	0.58	79.0)
64	200	60 db	6.3×10^{-3}	3.3×10^{-2}	1.96	0.1
	100	50 db	5.0×10^{-2}	1.1×10^{-1}	0.99	1.1
	100	40 db	3.9×10^{-1}	3.7×10^{-1}	1.003	11.7
	(30	30 db	1.46	9.8×10^{-1}	0.71	57.1)
128	200	60 db	2.1×10^{-2}	7.0×10^{-2}	1.28	0.48
	200	50 db	1.5×10^{-1}	2.3×10^{-1}	0.93	5.0
	(200	40 db	9.0×10^{-1}	7.4×10^{-1}	0.84	37.0)

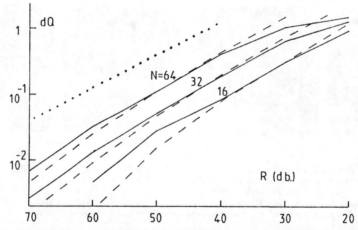

Fig. 2. The dependence of the amplitude weighted rms phase errors dQ (rad) on the signal to noise ratio R (dB) for several sizes of the data window (diameter N pixels). The expected errors calculated using (3) are shown dashed. The dotted line represents the predictions of (6) in the "small signal approximation" ($N = 64$).

the errors is overestimated by about four times. Thus the dependence found in these tests can be approximated by (6) with $k = 0.6$.

On the other hand, the errors predicted from (3) (shown dashed in Fig. 2) agree with those observed over most of the range. This gives some confirmation to the assumptions of (1). The discrepancies at low signal to noise ratios (e.g., 20 dB) can be attributed to failure of the small angle approximations. At high signal-to-noise ratio the deviations from the predictions of (3) may be due to "locking" onto local minima near the "global" minimum. It has been noticed that the solutions depend in their details on the initial random phase distribution assumed for the initial trial function. Furthermore, smaller phase errors can be

obtained by averaging several solutions made with different initial trial functions. Table II summarizes a limited number of tests. It shows that a reduction in phase error by a factor of about two can be obtained in this way. This suggests that the "locked" local minima are randomly spaced about the "global" minimum. Averaging in this way may be useful in reducing the "computational noise" and arriving at a solution independent of the initial trial function.

The influence of random receiver gain fluctuations and of non-linearities in the measuring devices has also been investigated. As expected from the tests by Misell [23], [24] and Boucher [7], [8] the algorithm is not sensitive to random gain errors. For example, an rms fluctuation of 10 percent in amplitude

TABLE II
AVERAGES OF SOLUTIONS MADE WITH DIFFERENT INITIAL TRIAL FUNCTIONS

S/N RATIO	−57 db	−47 db	−37 db
NO. AVERAGED	PHASE ERROR	PHASE ERROR	PHASE ERROR
	10^{-2}	10^{-2}	10^{-1}
1	2.657	8.988	3.193
4	1.748	5.991	1.896
16	1.408	5.044	1.586
64	1.340	4.812	
expected	1.058	6.087	2.753

(0.83 dB) gave an rms phase error of 0.072 rad, corresponding to an antenna gain loss of 0.5 percent. Similarly, a quadratic term in the system response amounting to a maximum of 10 percent in amplitude (0.83 dB) on boresight led to an rms phase error of 0.038 rad with a gain loss of 0.34 percent. This insensitivity is presumably because a high dynamic range is only needed in a small region in the main lobe of the antenna pattern. Furthermore, to a first approximation, both in-focus and out-of-focus beam patterns will be distorted in similar ways. This suggests that the optimum observing method would use measurements taken with equal but opposite defocus values.

VIII. CONCLUSION

From the present tests the Misell algorithm appears to be satisfactory for use in antenna diagnostics. No spurious solutions or solutions corresponding to local minima ("locked solutions") which differed greatly from the true solution were found. However, in tests with simulated receiver noise, the errors in the solutions were greater than expected on the basis of noise alone. This effect was most pronounced at low noise levels (high signal-to-noise ratios). It may indicate "locked solutions" very close to the correct solution. Their influence can be reduced be averaging several solutions obtained with different initial trial functions.

The main limitation to the method for antenna diagnosis seems to be the high signal-to-noise ratio which is required (about the square of that needed when phase can be directly measured). However, where satellite or ground-based transmitters can be used as sources, the method may prove useful when direct phase measurements are impossible. A peak signal-to-noise ratio of 50 dB allows about 60 × 60 points to be measured (or set in position) such that the antenna gain is within one percent of ideal at the measurement wavelength (Table I). Values of 40 dB should be attainable with large telescopes observing the astronomical sources of 22 GHz maser emission. This is sufficient for studies of large scale (e.g., gravitational) deformations. Some practical tests on real antennas are desirable.

Preliminary tests of the Gerchberg-Saxton algorithm to analyze intensity data in the aperture plane and in the far field have given similar performance. In principle this method may fail since the solutions are not always unique (Huiser et al. [19], [20]).

ACKNOWLEDGMENT

I thank B. Arambepola for advice on the use of his FFT routine, D. Emerson and U. Schwarz for computing help, and L. Wieliachev for directing me to the literature on speckle interferometry. Thanks are due to D. Downes for improving the text and to S. Halleguen for photography.

REFERENCES

[1] A. P. Anderson, J. C. Bennett, and A. J. T. Whitaker, "Measurement and optimisation of a large reflector antenna by microwave holography," Inst. Elec. Eng. Conf. Proc., 169, pp. 128–130, 1978.
[2] B. Arambepola, "Fast computation of multidimensional discrete Fourier transforms," Proc. Inst. Elec. Eng., vol. 127F, pp. 49–52, 1980.
[3] W. R. Bennett, "Methods of solving noise problems," Proc. IRE, vol. 44, pp. 609–638, 1956.
[4] J. C. Bennett, A. P. Anderson, P. A. McInnes, and A. J. T. Whitaker, "Microwave holographic metrology of large reflector antennas," IEEE Trans. Antennas Propagat., vol. AP-24, pp. 295–303, 1976.
[5] J. C. Bennett and M. P. Godwin, "Necessary criteria for the diagnosis of panel misalignments in large reflector antennas by microwave holography," Electron. Lett., vol. 13, pp. 463–465, 1977.
[6] E. J. Blum, J. Delannoy, and M. Joshi, "Method pour mettre en phase les elements d'un reseau d'antennes," Comptes Rendus, vol. 252, pp. 2517–2519, 1961.
[7] R. H. Boucher, "Convergence of algorithms for phase retrieval from two intensity distributions," SPIE, vol. 231, pp. 130–141, 1980.
[8] ——, "Phase retrieval techniques for image and wavefront reconstruction," Ph.D. dissertation, Univ. Rochester, Rochester, NY, 1980.
[9] A. J. J. Drenth, A. M. J. Huiser, and H. A. Ferwerda, "The problem of phase retrieval in light and electron microscopy of strong objects," Opt. Acta, vol. 22, pp. 615–618, 1975.
[10] H. A. Ferwerda, Inverse Source Problems in Optics, H. P. Baltes, Ed. New York: Springer-Verlag, 1978.
[11] J. R. Fienup, "Reconstruction of an object from the modulus of its Fourier transform," Opt. Lett., vol. 3, pp. 27–29, 1978.
[12] ——, "Phase retrieval algorithms: A comparison," Appl. Opt., vol. 21, pp. 2758–2769, 1982.
[13] R. W. Gerchberg and W. O. Saxton, "Phase determination from image and diffraction plane pictures in the electron microscope," Optik, vol. 34, pp. 275–284, 1971.
[14] ——, "A practical algorithm for the determination of phase from image and diffraction plane pictures," Optik, vol. 35, pp. 237–246, 1972.
[15] M. P. Godwin, A. P. Anderson, and J. C. Bennett, "Optimisation of feed position and improved profile mapping of a reflector antenna from microwave holographic measurements," Electron. Lett., vol. 14, pp. 134–136, 1978.
[16] M. P. Godwin, A. J. T. Whitaker, J. C. Bennett, and A. P. Anderson, "Microwave diagnostics of the Chilbolton 25 m antenna using the OTS satellite," Inst. Elec. Eng. Conf. Pub., 195, pp. 232–236, 1981.
[17] B. J. Hoenders, "On the solution of the phase retrieval problem," J. Math Phys., vol. 16, pp. 1719–1725, 1975.

[18] A. M. J. Huiser and H. A. Ferwerda, "The problem of phase retrieval in light and electron microscopy of strong objectes. II: On the uniqueness and stability of object reconstruction processes using two defocussed images," *Opt. Acta,* vol. 23, pp. 445–456, 1976.

[19] A. M. J. Huiser, A. J. J. Drenth, and H. A. Ferwerda, "On phase retrieval in electron microscopy from image and diffraction patterns," *Optik,* vol. 45, pp. 303–316, 1976.

[20] A. M. J. Huiser, P. Van Toorn, and H. A. Ferwerda, "On the problem of phase retrieval in electron microscopy from image and diffraction patterns. III: The development of an algorithm," *Optik,* vol. 47, pp. 1–8, 1977.

[21] C. E. Mayer, J. H. Davis, W. L. Peters, and W. J. Vogel, "Electromagnetic measurements of large reflector antennas," *IEEE Trans. Inst. and Measurement,* vol. IM-32, p. 102, 1983.

[22] D. L. Misell, "A method for the solution of the phase problem in electron microscopy," *J. Phys. D.* (Appl. Phys.), vol. 6, pp. L6–L9, 1973.

[23] ——, "An examination of an iterative method for the solution of the phase problem in optics and electron optics. I: Test calculations," *J. Phys. D.* (Appl. Phys.), vol. 6, pp. 2200–2216, 1973.

[24] ——, "An examination of an iterative method for the solution of the phase problem in optics and electron optics. II: Sources of error," *J. Phys. D.* (Appl. Phys.) vol. 6, pp. 2217–2225, 1973.

[25] W. O. Saxton, *Computer Techniques for Image Processing in Electron Microscopy.* New York: Academic, 1978.

[26] P. F. Scott and M. Ryle, "A rapid method for measuring the figure of a radio telescope reflector," *Monthly Nat. Roy. Astron. Soc.,* vol. 178, pp. 539–545, 1977.

[27] D. B. Shenton and R. E. Hills, "A proposal for a U.K. mm wavelength astronomy facility," Sci. Res. Council, Appleton Lab., Nov. 1976.

[28] L. S. Taylor, "The phase retrieval problem," *IEEE Trans. Antennas Propagat.,* vol. AP-29, pp. 386–391, 1981.

[29] P. Van Schiske, "Ein und mehrdeutigkeit der Phasenbestimmung aus bild und Beugungfigur," *Optik,* vol. 40, pp. 261–275, 1974.

[30] P. Van Toorn and H. A. Ferwerda, "The problem of phase retrieval in light and electron microscopy of strong objects. IV: Checking of algorithms by means of simulated objects," *Opt. Acta,* vol. 23, pp. 469–481, 1976.

[31] S. Von Hoerner, "Telescope surface measurement with two feeds," *IEEE Trans. Antennas Propagat.,* vol. AP-26, pp. 857–861, 1978.

Part IV
Receivers and Radiometers

THE two terms in the title of this part are used interchange-ably to describe radio-wavelength equipment used to measure the intensity of incident radiation. As such, they include all of the equipment associated with the radio telescope or antenna. The equipment for spectroscopic observations is considered separately in Part VI. Here, we are concerned primarily with the "front ends," or the first stages of the complete radiometric system. Even in a broadband system, the "back end," or detector and video amplifier, does not in general pose the same challenge as nor have the impact of the radio frequency (RF) stages of the receiver. However, obtaining full advantage from the nominal receiver sensitivity often requires using special input switching and signal processing techniques. These topics are included with the discussion of more specialized techniques in Part V.

Due to the weakness of the signals that are typically encountered in radio astronomy, the receiver noise perfor-mance is generally a critical parameter. Since the overall system noise performance is established by the first stages having appreciable power gain, it is the front end noise which must be minimized. Much of the effort devoted to radio-astronomical receiver development has thus concentrated on reducing the noise added by the input stages of the receiver. There have been many important successes, and while the development work has been carried out by a mixture of astronomers, physicists, and engineers, there is no doubt that radio astronomy has been a major force behind the development of parametric amplifiers, maser amplifiers, cryogenic Schottky mixers, and most recently, superconducting tunnel junction mixers.

The papers in this part have been chosen to represent the sweep of radio-astronomical efforts. A few papers have been selected for their historical perspective, and several for their overview of radio-astronomical receiver systems (as distin-guished from just low-noise front ends). Due to the progress of technology, there are some types of receivers, which, while having been at one time of considerable significance, appear at the present time not likely to be important for developments in the near future. Since this review volume is not intended to be a historical compendium, topics such as tunnel diode and parametric amplifiers are omitted, but the interested reader can find information in the references given below. It is impossible to do justice to the wide range of contributions to low-noise receiver development, but the references cited in the articles included, as well as those given below, should help to remedy any omissions.

The first paper, by Drake and Ewen, describes a particular system, but is included here primarily for its treatment of the issues of gain variations and radiometer stability. The use of comparison radiometry to circumvent these problems is widespread in radio astronomy. Other references on this topic are the pioneering article by Dicke (1946, reprinted in *Classics in Radio Astronomy* edited by Sullivan; see the references following the Historical Overview) and papers by Orhaug and Waltman (1962) and Selling (1964).

The second and third papers, by Weinreb *et al.* and Cong *et al.* respectively, describe systems for the VLA and for the Columbia-GISS 4-ft millimeter radio telescope, respectively. It is an indication of rapid progress that the first stages of both these receivers have been considerably improved since these articles were written; the VLA receivers have largely been changed to use cooled field effect transistor (FET) amplifiers, and the 4-ft telescope has been equipped with a superconduct-ing mixer.

FET amplifiers have had a dramatic impact on radio astronomy, as a result of their broad bandwidth, stability, and ease of operation, compared to existing types of front ends for the few hundred megahertz to few gigahertz frequency range. In addition, they have been almost universally adopted as the intermediate-frequency (IF) amplifier of choice for higher frequency heterodyne systems. The fourth and fifth papers included here, by Weinreb and Weinreb *et al.* respectively, describe a number of very successful designs. Of course, low-noise transistor amplifier design has an enormous range of applicability. Present-day commercially available units in-tended for communications and electronic warfare offer noise figures at ambient temperature that just a few years ago required cryogenic operation, and bandwidths that were only optimistic dreams. However, the insatiable demand of radio astronomers for lower noise has played a significant role in the advancement of the state of the art of FET amplifiers. At the present time, cryogenically cooled high electron mobility transistors (HEMTs) are being evaluated by radio astrono-mers, and appear to offer even lower noise than FETs. This may ultimately be most important for IF amplifiers used with extremely low noise mixers, since when used directly as first-stage RF amplifiers, other factors such as galactic emission, the earth's atmosphere, and ground pickup can make the improvement over FETs relatively insignificant.

The sixth paper, by Moore and Clauss, describes a maser (*m*icrowave *a*mplification by *s*timulated *e*mission of *r*adiation) amplifier with exceedingly low noise and good tuning charac-teristics. While maser amplifiers have played an important role in radio astronomy, their complexity and cost of operation have resulted in relatively limited use, and they are being successfully challenged by the lowest noise transistor amplifi-ers.

The seventh, eighth, and ninth papers, by Kerr, Predmore *et al.*, and Erickson respectively, describe millimeter wave-length receivers using Schottky diode mixers. This has been the dominant type of receiver in this wavelength range since its beginning in the mid-1970s. Again, since this spectral region has great potential importance for communications, radar systems, and industrial control applications, it is not surprising that there is a vast literature on mixer systems. The particular contributions of radio astronomy have been the optimization of mixer and diode performance for cryogenic operation, and the extension of this technology to higher frequencies. The comprehensive reprint volume edited by Kollberg (1984) contains many papers on mixer theory and practice.

Submillimeter astronomy is as yet only in its infancy, but high-spectral-resolution systems employed to date have essentially been extensions of radio-astronomical techniques. At these very short wavelengths, waveguide transmission generally used throughout the millimeter and centimeter range is excessively lossy, and a different method of propagation, such as quasi-optical transmission, is employed. Discussions of general techniques and some particular types of antennas and receivers are given in the references below.

The tenth, eleventh, and twelfth papers, by Feldman, Sutton, and Pan *et al.* respectively, are on the topic of superconducting mixers. It had been recognized for some time that different phenomena present at very low temperatures, such as the Josephson effect, have potential for high-performance mixers and amplifiers. Yet despite years of effort, there were no practical devices forthcoming. This is in contrast to the case of the maser amplifier, where only 3 years separated its discovery in 1955 and the first observational results obtained by Giordmaine *et al.* (1959) using a system at 3.2 cm wavelength.

Successful use of superconductor–insulator–superconductor (SIS) tunnel junctions began in the late 1970s. These devices basically behave in a manner similar to that of Schottky diodes, although quantum mechanical effects can result in mixers with gain. A number of groups have developed the ability to fabricate SIS devices, and many more have incorporated them into very low noise mixer receivers at frequencies of 40 GHz and higher. The paper by Feldman gives a very useful introduction to quantum mixers. A more detailed treatment can be found in a paper by Tucker and Feldman (1985). The two articles on actual systems included here give an introduction to the microwave circuits employed to take advantage of the performance of SIS devices, and to the results obtained. There is considerable activity in this area at the present time, and neither the noise nor the frequency limitations of SIS mixers have been thoroughly probed.

REFERENCES AND BIBLIOGRAPHY

Reviews and Discussion of Radiometer Theory

1946: Dicke, R. H., "The measurement of thermal radiation at microwave frequencies," *Rev. Sci. Instr.*, vol. 17, pp. 268–275, 1946.
1962: Orhaug, T. and W. Waltman, "A switched load radiometer," *Publ. NRAO*, vol. 1, no. 12, pp. 179–204, 1962.
1964: Selling, T. V., "The application of automatic gain control to microwave radiometers," *IEEE Trans. Antennas Propagat.*, vol. AP-12, pp. 636–639, 1964.
    Tiuri, M. E., "Radio astronomy receivers," *IEEE Trans. Antennas Propagat.*, vol. AP-12, pp. 930–938, 1964.
1965: Oliver, B. M., "Thermal and quantum noise," *Proc. IEEE*, vol. 53, pp. 436–454, 1965.
1973: Penzias, A. A. and C. A. Burrus, "Millimeter-wavelength radio-astronomy techniques," *Annu. Rev. Astron. Astrophys.*, vol. 11, pp. 51–71, 1973.

1976: Price, R. M., "Radiometer fundamentals," Chapter 3.1 in *Methods of Experimental Physics: Vol. 12 Astrophysics, Part B: Radio Telescopes*, M. L. Meeks, Ed. New York, NY: Academic Press, 1976.
1977: Evans, G. and C. W. McLeish, *RF Radiometer Handbook.* Dedham, MA: Artech House, 1977.
1984: Okwit, S., "An historical view of the evolution of low-noise concepts and techniques," *IEEE Trans. Microwave Theory Tech.*, vol. MTT-32, pp. 1068–1082, 1984.
1986: Kraus, J. D., *Radio Astronomy*, 2nd ed. Powell, OH: Cygnus-Quasar Books, 1986, ch. 7.

Receiver Systems

1973: Phillips, T. G. and K. B. Jefferts, "A low temperature bolometer heterodyne receiver for millimeter wave astronomy," *Rev. Sci. Instr.*, vol. 44, pp. 1009–1014, 1973.
1977: Wilson, W. J., "The aerospace low-noise millimeter-wave spectral line receiver," *IEEE Trans. Microwave Theory Tech.*, vol. MTT-25, pp. 332–335, 1977.
1982: Archer, J. W., "All solid-state low-noise receivers for 210-240 GHz," *IEEE Trans. Microwave Theory Tech.*, vol. MTT-30, pp. 1247–1252, 1982.
1985: Archer, J. W., "High-performance, 2.5-K cryostat incorporating a 100-120-GHz dual polarization receiver," *Rev. Sci. Instr.*, vol. 56, no. 3, pp. 449–458, 1985.
1988: Batelaan, P. D., M. A. Frerking, T. B. H. Kuiper, H. M. Pickett, M. M. Schaefer, P. Zimmermann, and N. C. Luhmann, Jr., "A dual-frequency 183/380 GHz receiver for airborne applications," *IEEE Trans. Microwave Theory Tech.*, vol. MTT-36, pp. 694–700, 1988.

Parametric Amplifiers and Masers

1959: Giordmaine, J. A., L. E. Alsop, C. H. Mayer, and C. H. Townes, "A maser amplifier for radio astronomy at X-band," *Proc. IRE*, vol. 47, pp. 1062–1070, 1959.
1963: Jelley, J. V., "The potentialities and present status of masers and parametric amplifiers in radio astronomy," *Proc. IRE*, vol. 51, pp. 30–45, 1963.
1976: Cardiasmenos, A. G., J. F. Shanley, and K. S. Yngvesson, "A travelling-wave maser amplifier for 85-90 GHz using a slot-fed image-guide slow-wave circuit," *IEEE Trans. Microwave Theory Tech.*, vol. MTT-24, pp. 725–730, 1976.
    Yngvesson, K. S., "Maser amplifiers," Chapter 3.3 in *Methods of Experimental Physics: Vol. 12 Astrophysics, Part B: Radio Telescopes*, M. L. Meeks, Ed. New York, NY: Academic Press, 1976, pp. 246–265.
1979: Sollner, T. C. L. G., D. P. Clemens, T. L. Korzeniowski, G. C. McIntosh, E. L. Moore, and K. S. Yngvesson, "Low-noise 86-88 GHz travelling wave maser," *Appl. Phys. Lett.*, vol. 35, pp. 833–835, 1979.
1980: Moore, C. R., "A reflected-wave ruby maser with 500-MHz bandwidth," *IEEE Trans. Microwave Theory Tech.*, vol. MTT-28, pp. 149–151, 1980.

FET Amplifiers

1980: Williams, D. R. W., S. Weinreb, and W. T. Lum, "L-band cryogenic GaAs FET amplifier," *Microwave J.*, vol. 23, no. 10, pp. 73–76, 1980.
1981: Tomassetti, G., S. Weinreb, and K. Wellington, "Low-noise, 10.7 GHz, cooled GaAs FET amplifier," *Electron. Lett.*, vol. 17, pp. 949–951, 1981.
1988: Pospieszalski, M. J. and S. Weinreb, "FET's and HEMT's at cryogenic temperatures—Their properties and use in low-noise amplifiers," *IEEE Trans. Microwave Theory Tech.*, vol. MTT-36, pp. 552–560, 1988.

Schottky Diode Mixer Receivers

1953: Strum, P. D., "Some aspects of mixer crystal performance," *Proc. IRE*, vol. 41, pp. 875–889, 1953.
1973: Viola, T. J., Jr. and R. J. Mattauch, "Unified theory of high-frequency noise in Schottky barriers," *J. Appl. Phys.*, vol. 44, pp. 2805–2808, 1973.
1984: Kollberg, E. L. (Ed.), *Microwave and Millimeter-Wave Mixers.* New York, NY: IEEE Press, 1984.

1985: Faber, M. T. and J. W. Archer, "Computer-aided testing of mixers between 90 and 350 GHz," *IEEE Trans. Microwave Theory Tech.*, vol. MTT-33, pp. 1138–1145, 1985. (a)

Faber, M. T. and J. W. Archer, "Millimeter-wave, shot-noise limited, fixed-tuned mixer," *IEEE Trans. Microwave Theory Tech.*, vol. MTT-33, pp. 1172–1178, 1985. (b)

1986: Kollberg, E. L., H. Zirath, and A. Jelenski, "Temperature-variable characteristics and noise in metal-semiconductor junctions," *IEEE Trans. Microwave Theory Tech.*, vol. MTT-34, pp. 913–922, 1986.

Maas, S. A., *Microwave Mixers*. Norwood, MA: Artech House, 1986.

Sherrill, G. K., R. J. Mattauch, and T. W. Crowe, "Interfacial stress and excess noise in Schottky-barrier mixer diodes," *IEEE Trans. Microwave Theory Tech.*, vol. MTT-34, pp. 342–345, 1986.

Superconducting Receivers

1982: Phillips, T. G. and G. J. Dolan, "SIS mixers," *Physica*, vol. 109, pp. 2010–2019, 1982.

Phillips, T. G. and D. P. Woody, "Millimeter- and submillimeter-wave receivers," *Annu. Rev. Astron. Astrophys.*, vol. 20, pp. 285–321, 1982.

1983: Blundell, R., K. H. Gundlach, and E. J. Blum, "Practical low-noise quasiparticle receiver for 80–100 GHz," *Electron. Lett.*, vol. 19, pp. 498–499, 1983.

1984: D'Addario, L. R., "An SIS mixer for 90–120 GHz with gain and wide bandwidth," *Int. J. Infrared and Millimeter Waves*, vol. 5, pp. 1419–1442, 1984.

1985: Räisänen, A. V., W. R. McGrath, P. L. Richards, and F. L. Lloyd, "Broad-band RF match to a millimeter-wave SIS quasi-particle mixer," *IEEE Trans. Microwave Theory Tech.*, vol. MTT-33, pp. 1495–1500, 1985.

Tucker, J. R. and M. J. Feldman, "Quantum detection at millimeter wavelengths," *Rev. Mod. Phys.*, vol. 57, pp. 1055–1113, 1985.

Woody, D. P., R. E. Miller, and M. J. Wengler, "85–115 GHz receivers for radio astronomy," *IEEE Trans. Microwave Theory Tech.*, vol. MTT-33, pp. 90–95, 1985.

1986: Räisänen, A. V., D. G. Crété, P. L. Richards, and F. L. Lloyd, "Wide-band low noise MM-wave SIS mixers with a single tuning element," *Int. J. Infrared and Millimeter Waves*, vol. 7, pp. 1835–1852, 1986.

1987: Blundell, R. and K. H. Gundlach, "A quasiparticle SIN mixer for the 230 GHz frequency range," *Int. J. Infrared and Millimeter Waves*, vol. 8, pp. 1573–1579, 1987.

Ellison, B. N. and R. E. Miller, "A low noise 230 GHz SIS receiver," *Int. J. Infrared and Millimeter Waves*, vol. 8, pp. 609–625, 1987.

1988: Kerr, A. R., S.-K. Pan, and M. J. Feldman, "Integrated tuning elements for SIS mixers," *Int. J. Infrared and Millimeter Waves*, vol. 9, pp. 203–212, 1988.

Xizhi, L., P. L. Richards, and F. L. Lloyd, "SIS quasiparticle mixers with bow tie antennas," *Int. J. Infrared and Millimeter Waves*, vol. 9, pp. 101–133, 1988.

Submillimeter Receivers

1982: van Vliet, A. H. F., Th. de Graauw, S. Lidholm, and H. van de Stadt, "A low noise heterodyne receiver for astronomical observations operating around 0.63 mm wavelength," *Int. J. Infrared and Millimeter Waves*, vol. 3, pp. 817–823, 1982.

1984: Roser, H. P., E. J. Durwen, R. Wattenbach, and G. V. Schultz, "Investigation of a heterodyne receiver with open structure mixer at 324 GHz and 693 GHz," *Int. J. Infrared and Millimeter Waves*, vol. 5, pp. 301–314, 1984.

1985: Wengler, M. J., D. P. Woody, R. E. Miller, and T. G. Phillips, "A low noise receiver for millimeter and submillimeter wavelengths," *Int. J. Infrared and Millimeter Waves*, vol. 6, pp. 697–706, 1985.

1987: Roeser, H. P., F. Schafer, J. Schmid-Burgk, G. V. Schultz, P. van der Wal, and R. Wattenbach, "A submillimeter heterodyne receiver for the Kuiper Airborne Observatory and the detection of the 372 μm carbon monoxide line J = 7-6 in OMC-1 and W3," *Int. J. Infrared and Millimeter Waves*, vol. 8, pp. 1541–1556, 1987.

A number of articles on receivers for millimeter and submillimeter wavelengths and related technology are contained in:

1986: Kollberg, E. (Ed.), *Instrumentation for Submillimeter Spectroscopy*, SPIE Proceedings, vol. 598. Bellingham, WA: SPIE, 1986.

1987: Proc. Submillimeter (Terahertz) Receiver Technology Conference, *Int. J. Infrared and Millimeter Waves*, vol. 8, pp. 1211–1353, 1987.

A Broad-Band Microwave Source Comparison Radiometer for Advanced Research in Radio Astronomy*

F. D. DRAKE†, MEMBER, IRE AND H. I. EWEN†, SENIOR MEMBER, IRE

Summary—A sensitive microwave radiometer system operating at short centimeter wavelengths has been developed which will allow large extensions of the known spectra of a large number of radio sources; facilitate the measurement of source polarization; give information on galactic structure and the sources of galactic radio emission; provide new data on the physical structure of planetary nebulas; and provide a means of measuring more accurately planetary temperatures, and the precise position of the brighter radio sources.

A traveling-wave tube radiometer operating at 8000 mc with a bandwidth of 1000 mc and sensitivities of the order of 0.01°K is

described. The radiometer is more than one order of magnitude more sensitive than other existing radiometers operating at 8000 mc. The very serious effects of gain fluctuations, acting on small residual signals, when trying to achieve very high sensitivities, are discussed. A means of eliminating such effects by introducing compensating noise has been found successful. Radio observations with this radiometer in conjunction with a 28-foot parabolic reflector have shown that: 1) The predicted sensitivity is achieved. 2) Zero-level stability is extremely high. 3) It has been possible to detect in detail the distribution of radio brightness at this wavelength in the vicinity of the galactic plane. 4) Radiation from the planets Jupiter and Saturn has been detected, this being the first detection of Saturn as a radio source. 5) Radiation from two planetary nebulas has been detected, this being the first detection of these objects as radio sources.

* Original manuscript received by the IRE, November 8, 1957.
† Harvard College Observatory, Cambridge, Mass., and Ewen Knight Corp., Needham Heights, Mass.

Reprinted from *Proc. IRE*, vol. 46, pp. 53–60, Jan. 1958.

INTRODUCTION

RADIO observations of celestial sources at centimeter wavelengths have been impeded by two causes.

1) The radio brightness of radio sources, in general, decreases rapidly with decreasing wavelength;

2) The sensitivities of available radiometers decrease as one goes to shorter wavelengths.

These two causes in combination have largely discouraged attempts to observe the sky extensively at the short centimeter wavelengths, even though shorter wavelengths allow one to achieve more narrow antenna beams with a given antenna dimension.

It has been known for several years that these wavelengths may provide information vital to the solution of many outstanding astronomical problems. Among the astronomical problems which now may be more effectively attacked at short centimeter wavelengths, one may list:

1) The spectra of discrete radio sources.—Spectra information now will be greatly extended with respect to available information.[1-5] These spectra provide important clues with regard to the mechanism of radio emission operating in a radio source.

2) The polarization of radio sources.—These measurements are facilitated at short wavelengths because the earth's ionosphere affects the polarization of the radiation to a much smaller degree at these frequencies than at lower frequencies.[6]

3) The origin of the radiation in the vicinity of the galactic plane.—An accurate knowledge of this radiation, evaluated in conjunction with longer wavelength data, will determine whether the radiation is of thermal or nonthermal origin.

4) Galactic structure as determined from the distribution of regions of ionized hydrogen.—Clouds of ionized hydrogen are thermal emitters, which dominate the radio sky at short centimeter wavelengths.

5) The physical structure of planetary nebulas.—These objects, whose position in the picture of cosmic evolution is still poorly understood, are difficult to study photographically with high precision, but some aspects of these objects may be studied with a microwave radiometer of sufficient sensitivity.

6) The temperatures of the planets.—The extensive work already done at the Naval Research Laboratory, Washington, D. C., at 3-cm wavelength, indicates the value of such observations, since temperatures obtained by radio techniques differ markedly, in some cases, from temperatures determined by other means.

7) The precise determination of the positions of radio sources.—The narrow antenna beams obtainable at short wavelengths facilitate position determinations both by decreasing the confusion produced by having several sources in the beam antenna simultaneously, and by providing a radiometer response that is more sensitive to changes in antenna pointing. Furthermore, at short wavelengths one may observe sources of small optical size, such as Venus, thereby providing an accurate calibration of antenna pointing error.

CONSIDERATIONS IN RECEIVER DESIGN

The receiving system to be described was specifically designed for the measurement of very low-level thermal noise powers associated with antenna temperatures produced by faint celestial sources of small angular size.

The "signal" produced by these sources has a noise-like character similar to the thermal and shot noise of the receiving system. The design criteria for the system then involve distinguishing a small change in the noise power output level of the receiver when the signal noise is introduced at the terminals of the antenna.

The minimum detectable signal of this system represents a change in the output-power level of the receiver of one part in 400,000 or a detectable antenna temperature change of 0.01°K in the presence of an equivalent 4000°K-system noise.

The minimum detectable signal is determined by fluctuations in the receiver output level in the absence of a signal. These output-level fluctuations are produced by spurious gain fluctuations within the active circuits of the receiver and by the statistical fluctuations in a noise-like waveform.

The normal method that has been used to reduce gain fluctuations and other spurious effects is to modulate the signal at a frequency at which the amplitude of such effects is negligible. The amplitude of the signal may then be determined by means of a coherent detector driven at the modulation frequency.[7]

Dicke's original description[7] of such a radiometer involved a 30-cps modulation of the signal by means of mechanically switching the receiver input between the antenna terminals and a resistive load at "room" temperature. Comparison was obtained then between the thermal noise presented at the terminals of the antenna

[1] F. T. Haddock, C. H. Mayer, and R. M. Sloanaker, "Radio observations of ionized hydrogen nebulae and other discrete sources at a wave-length of 9.4 cm," *Nature*, vol. 174, pp. 176–177; July 24, 1954.

[2] F. T. Haddock and T. P. McCullough, Jr., "Extension of radio source spectra to a wavelength of 3 centimeters," *Astron. J.*, vol. 60, pp. 161–162; June, 1955.

[3] V. M. Plechkov and V. A. Razin, "Results of measures of the intensity of radio emission of discrete sources at wavelengths of 3.2 and 9.7 cm," *Proc. Fifth Conf. on Questions of Cosmogony, Academy of Sciences of the USSR, Moscow*, pp. 430–435; 1956.

[4] N. L. Kaydanovsky and N. S. Kardashev, "Results of observations of discrete sources of cosmic radio emission at a wavelength of 3.2 cm," *Proc. Fifth Conf. on Questions of Cosmogony, Academy of Sciences of the USSR, Moscow*, pp. 436–437; 1956.

[5] N. G. Roman and F. T. Haddock, "A model for nonthermal radio source spectra," *Astrophys. J.*, vol. 124, pp. 35–42; July, 1956.

[6] C. H. Meyer, T. P. McCullough, and R. M. Sloanaker, "Evidence for polarized radio radiation from the Crab Nebula," *Astrophys. J.*, vol. 126, pp. 468–470; September, 1957.

[7] R. H. Dicke, "The measurement of thermal radiation at microwave frequencies," *Rev. Sci. Instr.*, vol. 17, pp. 268–275; July, 1946.

and the thermal noise of the resistor. Various other methods of modulation have been employed; however, each involves the fundamental concept of sequencing "signal" and "comparison" information through active amplifier circuits in a modulation pattern which in most cases can be represented by a square wave. Each cycle of the square wave contains one-half period of signal information followed by a half-period of comparison information.

In many systems this simple modulation scheme of switching to a resistive comparison load has been sufficient to reduce the effect of gain fluctuations below the level of statistical noise fluctuations. The receiving system described in this paper involves the successful application of further techniques to reduce the effect of gain fluctuations. Such techniques become increasingly significant for systems with low statistical noise fluctuations.

Theoretically, the fluctuations in the output-noise level of the receiver produced by the statistical nature of the noise waveform can be reduced to any desired degree by increasing the integration time after detection (narrowing the post detection bandwidth).

Expressed in equivalent temperature units the rms value of these fluctuations $\overline{\Delta T}$ is

$$\overline{\Delta T} = Equivalent\ system\ noise/\sqrt{B\tau}\ , \tag{1}$$

where the equivalent system noise can be obtained from the noise figure F of the system by the expression

$$(F-1)T_0. \tag{2}$$

T_0 is the reference ambient temperature, 290°K. In (1) above:

$B =$ the predetection bandwidth, and

$\tau =$ the time constant of the integration network after the detector.

Presumably we can increase τ or B and thereby reduce $\overline{\Delta T}$ to any desired value. There is, of course, a practical limitation to the maximum value of τ, which is introduced by the amount of available observing time; this maximum value may be determined either by a change in source position as a function of time or a time dependent change in the amplitude of the power received by the antenna. An increase in B may be limited by the bandwidth of the signal, the current state of equipment development, man-made interference "noise," and many other factors.

To date, instrumentation at short wavelengths has consisted of superheterodyne receivers. The sensitivity of such systems has been limited by the relatively narrow bandwidths that can be easily achieved. Existing superheterodyne receivers provide a maximum sensitivity of the order of 0.5°K, with a five-second integration time. The introduction of traveling-wave tubes at short centimeter wavelengths has offered the possibility of achieving, at the cost of a slight increase in noise figure,

extremely wide bandwidths, of the order of 3 kmc at a frequency of 8 kmc. In actual application, it was found desirable to limit the bandwidth to about 1 kmc in order to avoid interference. The sensitivity of such a system, in theory, can be more than an order of magnitude greater than that given by existing superheterodyne receivers.

In order to test the efficacy of traveling-wave tube radiometers in radio astronomy application, such a radiometer was built and then was tested by actual observation of celestial sources.

Traveling-Wave Tube Radiometer

A simplified block diagram of the basic receiving system is shown in Fig. 1. The receiver is basically a trf type consisting of three cascaded traveling-wave tube amplifiers. The center frequency of the interstage filters is 8000 mc and the bandwidth of each filter is 1000 mc.

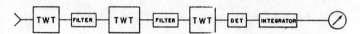

Fig. 1—Block diagram of basic receiving system employing traveling-wave tubes.

The measured system noise figure is $11\frac{1}{2}$ db, providing an over-all equivalent system noise of about 4000°K. With an integration time of 100 seconds the value of $\overline{\Delta T}$ from (1) is about 0.01°K. Hence, the fluctuations in the output power level of the receiver due to the statistical nature of the noise waveform are very small indeed.

Now let us direct our attention to the effect of gain fluctuations. With an equivalent system noise of 4000°K a gain change of ± 1 per cent would produce a $\pm 40°$K change in the receiver output. To reduce this effect to 0.01°K would require stabilizing the gain of the receiver to one part in 400,000 or 0.00025 per cent. An easier approach would be to modulate the signal by comparison to a resistive load[7] as described above.

Fig. 2 is a simplified block diagram of the receiver modified by inserting a ferrite switch at the input to the first twt. Modulation of the signal is obtained by switching between the antenna terminals and a resistive load at "room temperature."

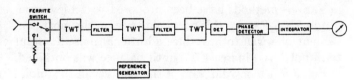

Fig. 2—Block diagram of receiving system with addition of comparison switching circuits.

The general expression for the minimum detectable signal is then:

$$\overline{\Delta T} = K\left[\frac{(F-1)T_0}{\sqrt{B\tau}} + \left(\frac{G(t)-G_0}{G_0}\right)(\Delta T_e + T_A)\right] \tag{3}$$

where

> $K =$ a constant introduced here to simplify the expression by eliminating other effects not discussed in this paper,[8]
>
> $F =$ the system noise figure,
>
> $T_0 =$ the reference ambient temperature, 290°K,
>
> $B =$ the predetection bandwidth (1000 mc),
>
> $\tau =$ the time constant of the integration network,
>
> $G(t) =$ the gain of the receiving system at time t,
>
> $G_0 =$ the average value of $G(t)$ during the period of observation,
>
> $\Delta T_c = T_{c2} - T_{c1}$,
>
> $T_{c1} =$ temperature "observed" by the receiver with the switch connected to the resistive load,
>
> $T_{c2} =$ temperature "observed" by the receiver with the switch connected to the antenna terminals, and with no radio source in the antenna beam,
>
> $T_A =$ effective antenna temperature, "signal" produced by a radio source.

This expression for ΔT may be divided into two parts.

1) The effect of statistical noise fluctuations:

$$\overline{\Delta T} = K \left(\frac{(F-1) T_0}{\sqrt{B\tau}} \right). \qquad (4)$$

2) The effect of gain variations:

$$\overline{\Delta T} = K \left[\left(\frac{G(t) - G_0}{G_0} \right) (\Delta T_c + T_A) \right]. \qquad (5)$$

The first part obtains a value of 0.01°K as described above. For a 1 per cent variation in gain, the second part obtains a value of 2.9°K, if we assume to a first approximation that with no signal:

$$T_{c1} = 290°K$$

$$T_{c2} = 0°K.$$

It is evident from (5) that gain fluctuations limit the minimum detectable signal by an amount equal to the percentage of the gain change multiplied by the noise unbalance at the input terminals of the receiver, in the absence of a signal. To achieve a sensitivity of 0.01°K, it is then necessary to balance the noise temperatures T_{c1} and T_{c2} to better than 1°K, if gain fluctuations are no greater than 1 per cent.

Various methods have been proposed to achieve condition $T_{c1} = T_{c2}$. For this particular system the most convenient one was to introduce additional noise at input terminal 2 and increase T_{c2} until balance was obtained.

Fig. 3 is a block diagram of the receiver in the "noise compensated" form. The addition of noise to terminal 2 was achieved by inserting a fractional amount of the noise power available from an argon gas discharge noise source into the side arm of a directional coupler. The main arm of the coupler was inserted in the antenna transmission line between terminal 2 of the receiver and

[8] P. D. Strum, "Considerations in high-sensitivity microwave radiometry," this issue, p. 43.

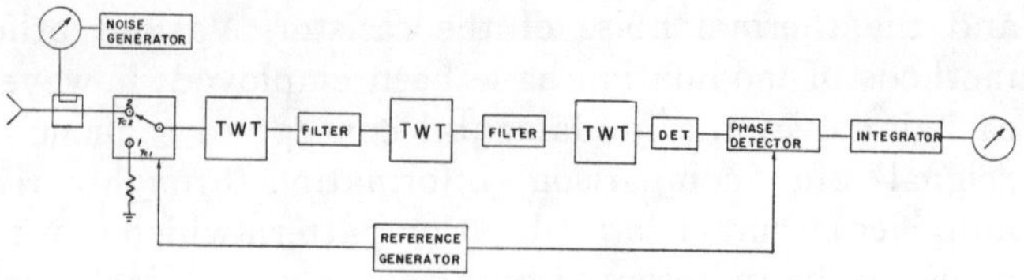

Fig. 3—Block diagram of receiving system with the addition of "noise compensating" circuits.

the antenna terminals. The addition of a variable attenuator between the noise generator and the coupler provided an easy means for adjustment of the amount of noise added to terminal 2.

The addition of approximately 300°K of equivalent noise power to the input of the receiver does, of course, increase the over-all system noise from 4000°K to 4300°K. However, this produces a negligible effect on $\overline{\Delta T}$ as given by (4) when one considers the fact that this addition of 300°K of noise improves system sensitivity from 2.9°K to approximately 0.01°K. An effective sensitivity of 0.01°K was achieved in practice, thus providing a receiver more than one order of magnitude more sensitive than existing receivers at this frequency.

To improve further the sensitivity of such a system and, in fact, to determine the ultimate limitations on the sensitivity of the system described herein, we must analyze in greater detail the dependence of the terms in (3) on time and antenna position.

T_{c2} may be represented by the expression:

$$T_{c2} = T_B + T_T + T_{\alpha 2} + T_I$$

where individual component contributions to T_{c2} exclusive of the noise compensating component are:

> $T_B =$ the equivalent black body temperature of the galactic background radiation field "observed" by the antenna,
>
> $T_T =$ temperature due to tropospheric effects,
>
> $T_{\alpha 2} =$ temperature due to losses in the ferrite switch and side lobe contributions of the feed system and reflector,
>
> $T_I =$ effective temperature due to man-made interference.

At a frequency of 8000 mc:

> T_B is less than 1°K. However, it is a function of antenna position. For fixed antennas it is a function of time as a consequence of the earth's diurnal rotation.
>
> T_T will depend primarily on the water vapor and oxygen content of the atmosphere in the solid angle of the antenna main beam. It will, therefore, be dependent on time as well as antenna position. The absolute magnitude of this effect might be as high as 30°K.
>
> $T_{\alpha 2}$ will depend on the geometrical configuration of the antenna system and topographic and reflecting properties of the antenna surroundings. It will depend primarily on position, and even for a well-designed antenna pointing toward the zenith this effect might be as large as 20°K.

T_I will, of course, depend on time, position of the antenna, and physical proximity of sources of man-made noise.

The problem for the future is simply expressed as the determination of the function $T_{c2}(t, \theta, \phi)$ for each system.

If we include the noise compensation required for balance, then:

$$T_{c2} = \frac{T_{NG}}{(A_1 + A_2)} + T_B + T_T + T_{\alpha 2} + T_I$$

where

T_{NG} = effective noise generator temperature,

$A_1 + A_2$ = attenuation of the variable attenuator plus the coupler.

The first term of T_{c2} can be made time dependent only, with a long thermal time constant, by application of known techniques to temperature stabilize components.

T_{c1} can be represented by the expression:

$$T_{c1} = T_0 + T_{\alpha 1},$$

where

T_0 = ambient temperature of the load, and

$T_{\alpha 1}$ = temperature component produced by losses in the ferrite switch when connected to position 1.

If T_0 varies appreciably with time, the resistive element can be placed in an oven and stabilized to 0.01°K during the time of observation. $T_{\alpha 1}$ will be dependent only on time and its thermal time constant can be increased by various techniques.

Many of the effects discussed above can be easily controlled if the desired system sensitivity is no greater than 0.01°K, as was the case in the development of the radiometer described here. However, one of the purposes of this more detailed discussion is to point out the problems that will be associated with radiometers employing very low internal noise amplifiers. With a solid state Maser, for instance, the term $F - 1/\sqrt{B\tau}$ in (3) will be reduced by a factor of 1000 at microwave frequencies. Fluctuations in the output-power level of the receiver will no longer be determined by the statistical fluctuations of the noise waveform. This fluctuation, presently the weak link in radiometer systems, will become the strongest link and the heretofore *negligible* effects represented by the second term

$$K\left[\left(\frac{G(t) - G_0}{G_0}\right)(\Delta T_c + T_A)\right]$$

will completely determine receiver performance. The receiver reported here is in the "twilight" zone between yesterday and tomorrow in radiometer development.

ASTRONOMICAL OBSERVATIONS

The radiometer has been operated in conjunction with an equatorially-mounted 28-foot paraboloid with a solid aluminum surface for the purpose of making actual tests on celestial sources of radiation. The beamwidth to half-power points is 18' of arc. Because the receiver has been in a developmental status and under constant revision, precise calibration has not been attempted nor has the antenna gain been carefully measured. The results to date are, therefore, of a preliminary nature and stated antenna temperatures may be in error by a factor of two.

Fig. 4 is a reproduction of the radiometer response as the radio source Cassiopeia A was allowed to drift through the antenna beam. Such a "drift curve" should resemble a rectilinear plot of the antenna beam pattern, if the radio source is much smaller than the size of the antenna beam, as is the case here. The time constant in this case was 80 seconds, and the peak antenna temperature recorded for this source was about 2°K. The theoretical rms fluctuation amplitude at the radiometer output should be about 0.02°K under these conditions. The observed amplitude as recorded is approximately equivalent to this predicted value.

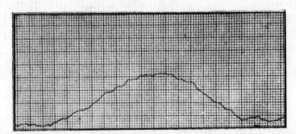

Fig. 4—Radiometer response as the strong radio source Cassiopeia A drifted through the antenna beam. Maximum antenna temperature = 2°K. Receiver time constant = 80 seconds.

Fig. 5 is a "drift curve" of the radio source associated with the galactic nebulosity M17. The time constant for this record was 80 seconds, and the peak radiometer deflection was about 3°K. The antenna was driven around the polar axis at a speed slightly greater than that of the earth's rotation, so that, although only about 10 minutes of right ascension displacement are shown by the tracing, about two and one-half hours were required for the observation. The most remarkable quality of this tracing from an instrumentation standpoint, besides the high sensitivity obtained, is the high stability of the zero-intensity level, which drifted at a rate of only 0.1°K per hour. The small rate of zero-level drift is quite encouraging to radio astronomers, who have long found a lack of zero-level stability to be a major obstacle in obtaining optimum observational data. The zero-level drift existing in this and other figures may be attributed to drift in the temperature of the resistive load against which antenna temperatures are compared. Oven stabilization of the resistive load was not included during this series of tests.

The source labelled A in Fig. 5 is associated with the emission nebula M17. The source labelled B in Fig. 5 lies very near the position of the galactic plane as found by

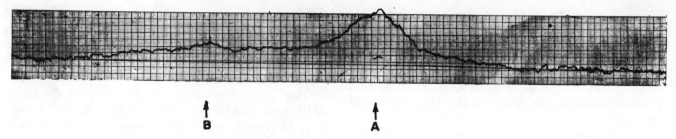

Fig. 5—Radiometer response as the region of the nebulosity M17 passed through the antenna beam. Maximum antenna temperature = 3°K. Time constant = 80 seconds. Source *A* is the radio source associated with M17, Source *B* lies near the galactic plane.

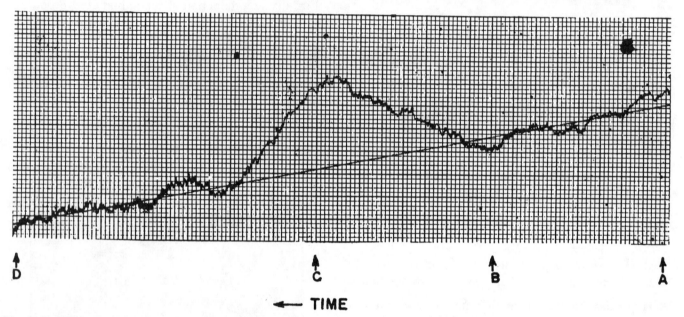

← TIME

Fig. 6—Radiometer measure of radio emission from Jupiter. Maximum antenna temperature 0.15°K. Time constant = 320 seconds. Between *A* and *B*, the antenna pointed ½° east of Jupiter; between *B* and *C*, the antenna pointed at the planet; between *C* and *D*, the antenna pointed ½° west of Jupiter.

radio investigations at longer wavelengths. Since, optically, there is no emission nebulosity that might be responsible for this source, it appears quite likely that this source actually is the galactic plane. The narrow width of the response to this source then indicates that the galactic plane appears quite narrow at centimeter wavelengths, possibly being only a few minutes of arc in width, as has been predicted by astronomical theorists. To the left and right of source *B*, one may observe an extended source of radio emission, which is probably associated with galactic nebulosities, such as Index Catalog 4701, close to the galactic plane.[1,9] This tracing probably marks the first observation at these wavelengths of faint emission in the vicinity of and at the position of the galactic plane.

Fig. 6 is an observation of thermal radio emission from the planet Jupiter. The time constant was 320 seconds, giving a theoretical sensitivity in this case of 0.010°K. Referring to Fig. 6, the antenna was positioned ½° east of Jupiter at point *A* on the scan and

then allowed to track this position to establish a radiometer zero level, with Jupiter just outside the beam pattern. At *B*, the antenna was pointed directly at Jupiter and allowed to track the planet. During this portion of the scan, the radiometer output should describe an exponential curve asymptotic to the antenna temperature produced by the planet. At *C*, the antenna was pointed ½° west of Jupiter, and again allowed to track. During this portion of the scan, the radiometer output should describe another exponential until the zero level is again reached. The radiometer response does follow the predicted path and indicates a peak antenna temperature due to the planet of 0.15°K. If it is assumed that the gain of the antenna is one half that of a perfect antenna, a black body temperature slightly greater than 200°K is indicated.

Fig. 7 is a radiometer measurement of the emission from the planet Saturn, taken in the same manner as the Jupiter observation. The antenna tracked east of the planet from *A* to *B*, pointed at the planet from *B* to *C*, and west of the planet following *C*. The antenna temperature due to the planet as measured by this scan is 0.04°K, corresponding to a flux of about $4 \times (10^{-26})$

[9] F. T. Haddock, C. H. Mayer, and R. M. Sloanaker, "Radio emission from the Orion Nebula and other sources at λ 9.4 cm," *Astrophys. J.*, vol. 119, pp. 456–459; March, 1954.

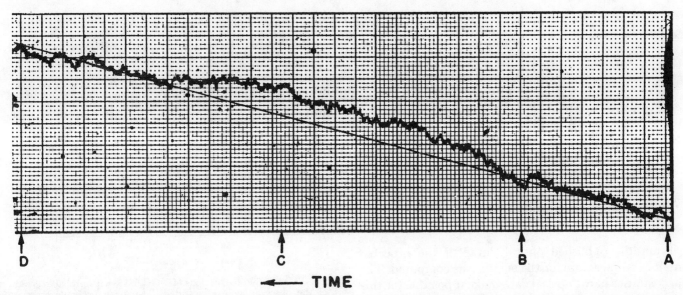

← TIME

Fig. 7—Radiometer measure of radio emission from Saturn. Maximum antenna temperature 0.04°K. Time constant = 320 seconds. Between *A* and *B*, the antenna pointed ½° east of Saturn; between *B* and *C*, the antenna pointed at the planet; between *C* and *D*, the antenna pointed ½° west of Saturn.

watt m⁻²(cps)⁻¹ or 5.5×(10⁻¹⁶) watts of power detected at the antenna terminals. This is probably the smallest antenna temperature detected to date with a microwave radiometer. In this case, one may not determine a black body temperature for the planet without first knowing whether the well-known rings are contributing to the radio emission. If the particles making up the rings are much smaller than the 3.75-cm receiver wavelength, or are very cold, the particles will contribute very little emission at this wavelength. Since we do not have a reliable measure of the size of the ring particles or their temperature, it is possible to invert the argument, accepting the planetary temperature given by other means,[10] and determine whether or not the rings are present radiowise. To eliminate calibration errors as much as possible, the predicted ratio of the antenna temperature of Jupiter to that of Saturn was computed for both cases; *i.e.*, with and without the presence of the rings. Since the observations of Jupiter and Saturn were made in quick succession, taking the ratio of the observed antenna temperatures from these observations should eliminate systematic calibration errors. From published data for the date of the observations, July 24, 1957, the ratio of the antenna temperature Jupiter to Saturn should be 4.7 if the rings are transparent, and about 3.0 if the rings are opaque. The observed ratio is 4.3, indicating that the rings are not radio sources at 8000 mc, which may mean that they principally consist of particles smaller than 3.75 cm in diameter, or are at a temperature less than about 20°K.

A "drift curve" of the planetary nebula NGC 7293 (the "Helix" Nebula) is shown in Fig. 8. This represents the first radio observation of a planetary nebula.

¹⁰ C. W. Allen, "Astrophysical Quantities," University of London, London, Eng., 1956.

This tracing was made in the same manner as the tracings of Cassiopeia A and the region of M17, with an 80-second time constant, and indicates a maximum antenna temperature for the nebula of 0.25°K. It is desirable with weak signals to take a mean of several observa-

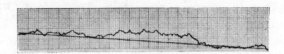

Fig. 8—Radiometer response as the planetary nebula NGC 7293 passed through the antenna beam. Maximum antenna temperature = 0.25°K. Time constant = 80 seconds.

tions, as this allows one to make a statistical analysis of the data to derive a measure of its reliability. Five observations of NGC 7293 have been combined to obtain a mean curve. The resultant curve and the probable errors derived for the experimental points are shown in Fig. 9. The mean probable error is 0.022°K. The actual scatter among the points themselves suggests that the actual probable error is somewhat smaller than this. This can occur because the observer must draw an arbitrary zero level through each tracing of the radio source before deriving antenna temperatures from the tracing, and this zero will always be slightly in error. The effect is to introduce no error in the relative positions of the derived experimental points, but to enhance the probable errors derived in a straightforward way from the observational data. It is fortunate that this is the case, as one can be sure that the derived probable errors are too large, and hence the data are actually more reliable than the probable errors indicate. In this case, the actual scatter among the points indicates that the true probable error is of the order of 0.010°K.

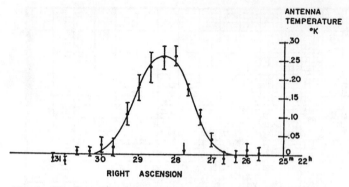

Fig. 9—Mean of 5 observations of NGC 7293, with probable errors of the observed points indicated. The mean probable error is 0.022°K; the arrow indicates the optical center of the nebula.

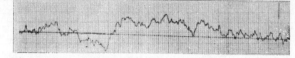

Fig. 10—Radiometer response as the planetary nebula NGC 6853 passed through the antenna beam. Maximum antenna temperature = 0.10°K. Time constant = 80 seconds.

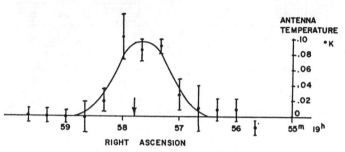

Fig. 11—Mean of 4 observations of NGC 6853, with probable errors of the observed points indicated. The mean probable error is 0.017°K; the arrow indicates the optical center of the nebula.

From the published optical data,[11–13] the expected antenna temperature for the nebula was computed. The uncertainties in the optical data make it possible for this temperature to be in error by at least a factor of five. In the case of NGC 7293, assuming the antenna gain to be half that of an ideal antenna, the predicted antenna temperature is 0.16°K, which is in excellent agreement with the observed antenna temperature of 0.26°K from Fig. 9. The data of Fig. 9 show a response to the nebular signal which is broadened about 6′ of arc over the response from a point source. This is to be expected, as this nebula is about 15′ of arc in diameter, and has faint optical extensions extending at least as far as 10′ of arc from the center of the nebula. The arrow in Fig. 9 indicates the optical center of the nebula, and it is seen that the center of the radio emission as determined from the present observations deviates appreciably from the optical center, and actually lies very near the edge of the nebula. The difference between the two centers seems too large to be attributable to instrumental errors and is probably real. It is of interest to note that the center of radio emission apparently deviates from the optical center in the same direction as faint extensions of the nebula, recently observed on large-scale, high quality, photographs. Further radio and optical studies of this nebula are clearly indicated.

Fig. 10 is a reproduction of a drift curve of the planetary nebula NGC 6853 (the "Dumbbell" Nebula). A time constant of 80 seconds was used, and a maximum antenna temperature of 0.10°K is indicated. Four such observations have been combined to give the results shown in Fig. 11. The mean probable error from the data is 0.017°K, although, as before, the scatter of the points among themselves suggests that the true probable error is of the order of 0.008°K, which is representative of the extremely high-receiver sensitivity. The antenna temperature predicted from the optical data is

0.07°K, which is in remarkable agreement with the observed temperature of 0.09°K. In this case, the optical position of the nebula, as indicated by the arrow in Fig. 11, is in close agreement with the radio position. This was expected since the dimensions of this nebula are roughly 6′×8′.

CONCLUSION

The traveling-wave tube radiometer described in this report fulfills the theoretical sensitivity deduced from its electronic characteristics. The inherent high stability achieved is of great value to radio astronomers. This high sensitivity and stability was achieved by careful consideration and treatment of circuit components which do not affect the simple theoretical considerations of circuit sensitivity.

Astronomical results indicate that one may readily observe the temperature of the planets with a radiometer of the type described. For the first time, it is possible with ease and reasonable speed to make a detailed study of galactic radiation at 3- to 4-cm wavelengths. Preliminary observations described here suggest that much important data will accrue from such a study. The radio observations of planetary nebulas, achieved for the first time through the use of this radiometer, have established a new field for the application of radio astronomy. The first results reported here indicate that such radio information will greatly improve the accuracy to which some parameters of the nebulas are known, and may bring to light important aspects of these objects which have not been observable until now.

ACKNOWLEDGMENT

The authors are grateful to Peter D. Strum, Harry E. Adams, and A. William Gruhn of the Ewen Knight Corp. for the extensive, unprecedented engineering development required to achieve the optimum equipment performance described here.

[11] D. H. Menzel and L. H. Aller, "Physical processes in gaseous nebulae. XII. The electron densities of some bright planetary nebulae," *Astrophys. J.*, vol. 93, pp. 195–201; January, 1941.

[12] T. Page and J. L. Greenstein, "Ionized hydrogen regions in planetary nebulae," *Astrophys. J.*, vol. 114, pp. 98–105; July, 1951.

[13] I. S. Shklovsky, "A new scale of distances to planetary nebulae." *Astron. J. Soviet Union*, vol. 33, pp. 222–235; 1956.

Multiband Low-Noise Receivers for a Very Large Array

SANDER WEINREB, SENIOR MEMBER, IEEE, MICHAEL BALISTER, STEPHEN MAAS, MEMBER, IEEE, AND PETER J. NAPIER

Invited Paper

Abstract—The very large array (VLA), presently under construction by the National Radio Astronomy Observatory, is an array of 27 25-m-diam antennas. This paper describes the feed and low-noise front-end systems used on the antennas. The receiving system allows operation at any one of the four frequency bands: 1.35–1.73 GHz, 4.5–5.0 GHz, 14.4–15.4 GHz, 22–24 GHz. The feed system uses an offset Cassegrain geometry so that the feeds for all four frequency bands can be in position on the antenna simultaneously. The front end comprises a cryogenically cooled parametric amplifier for the 4.5–5.0-GHz range. This paramp is preceded by cooled upconverters or cooled mixers for the other frequency ranges. Measured system performance is presented and some construction details are given.

INTRODUCTION

ONE OF THE MOST challenging instrumental problems in astronomy has been the development of an instrument to map the spatial distribution of radio waves emitted by distant astronomical objects. This instrument needs to operate in the 1–25-cm wavelength range, have a resolution of 1 arc-s or less, a field of view greater than a few minutes of arc, and sensitivity sufficient to detect a flux of 10^{-27} W/m^2/Hz upon the surface of the earth.

An instrument to meet these requirements is now under construction in central New Mexico; it is being built by the National Radio Astronomy Observatory[1] and is called the very large array (VLA). The array consists of 27 25-m-diam paraboloids arranged along the arms of a Y with 21-km arms. Construction started in 1973 and should be complete by 1981. An artist's concept of the central part of the array is shown in Fig. 1. The aperture-synthesis technique of radio astronomy arrays is discussed by Ryle [1] and Chow [2], and a further introduction to the VLA is given by Heeschen [3].

The VLA has required application of the latest technology in antennas, cryogenically cooled low-noise receivers, high-speed digital processing equipment, and wide-band transmission systems. The low-noise receiver aspects of the system will be discussed in this paper.

Manuscript received September 5, 1976.
S. Weinreb is with the University of California, Berkeley, CA, on leave from the National Radio Astronomy Observatory, Charlottesville, VA 22901.
M. Balister is with the National Radio Astronomy Observatory, Charlottesville, VA 22901.
S. Maas and P. J. Napier are with the National Radio Astronomy Observatory, Socorro, NM 87801.
[1] The NRAO is operated by Associated Universities, Inc., under contract with the National Science Foundation.

Fig. 1. Artist's conception of the VLA.

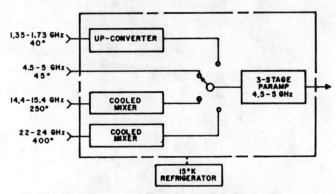

Fig. 2. VLA front-end system configuration.

FRONT-END SYSTEM DESIGN AND PERFORMANCE

The VLA requires 27 dual-polarized front ends providing $\geq$ 100-MHz instantaneous bandwidth in the tuning ranges 1.35–1.73 GHz, 4.5–5 GHz, 14.4–15.4 GHz, and 22–24 GHz. Noise temperature was to be as low as possible within a budget of ~ \$130 K per front end including cryogenics, assembly labor, but not feeds. The complete feed and subreflector system costs ~ \$40 K.

The system configuration shown in Fig. 2 was selected to meet these requirements. Two sets of all components within the diagram of Fig. 2 are cooled to 15 K by a 10-W capacity refrigerator manufactured by Air Products, Inc. A cooled parametric amplifier is utilized for operation at frequencies of 4.5–5 GHz. This is an optimum frequency range for cooled paramps, and small high-performance units are available. The paramp input is connected to a

Reprinted from *IEEE Trans. Microwave Theory Tech.*, vol. MTT-25, no. 4, pp. 243–248, Apr. 1977.

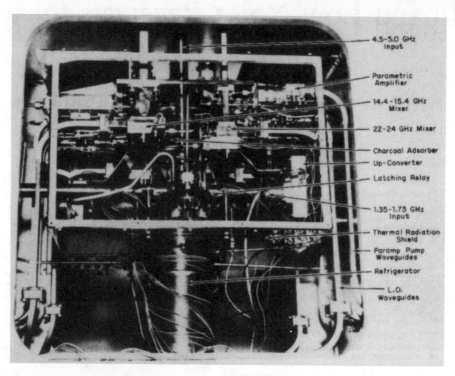

Fig. 3. Dewar interior. The sides of the radiation shield and the aluminized Mylar have been removed.

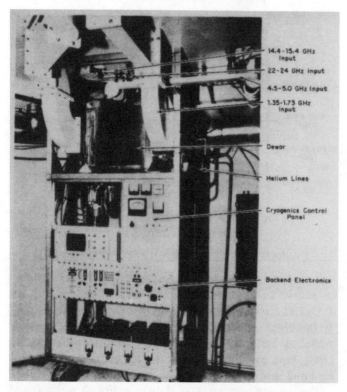

Fig. 4. Front-end rack mounted in antenna vertex room.

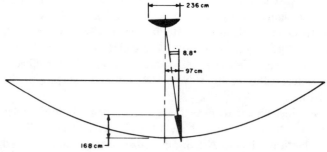

Fig. 5. VLA offset Cassegrain geometry.

TABLE I
MEASURED RECEIVER AND SYSTEM TEMPERATURES

f		T_{rx}	$T_{sys.}$
1.35 –	1.73	18^0K	47^0K
4.5 –	5.0	25	49
14.4 –	15.4	200	240
22 –	24	240	290

Note: These temperatures were measured in the middle of the band. In general, they increase by up to 20 percent at the band edges, except for the 22–24-GHz system where the increase is 40 percent.

The measured receiver noise temperatures and system temperatures are given in Table I. Photographs of the front-end dewar interior and of the front-end rack mounted in an antenna vertex room are shown in Figs. 3 and 4.

FEED SYSTEM

An offset shaped-reflector Cassegrain geometry as shown in Fig. 5 is utilized. The main reflector, manufactured by E-Systems Inc., is a surface of revolution with a modified

solenoid-operated coaxial switch which allows either straight-through connection to a feed or connection to one of three frequency converters for the other desired frequencies. A parametric upconverter is used for the lower frequency range, and two cooled Schottky-diode mixers are utilized for the higher frequency bands. These components will be described in more detail in following sections.

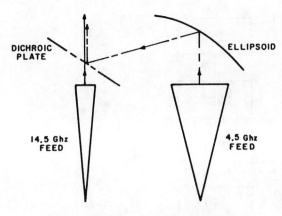

Fig. 6. Dual frequency reflector system.

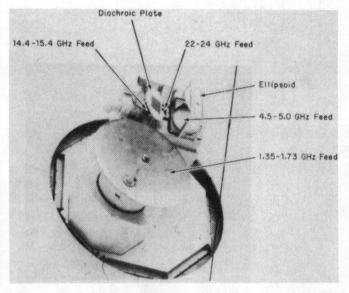

Fig. 7. Feed system on antenna.

TABLE II
MEASURED ANTENNA EFFICIENCIES

f	Total Antenna Efficiency
1.35 – 1.73	41%
4.5 – 5.0	65
14.4 – 15.4	53
22 – 24	44

Note: These efficiencies are measured on antenna #1 for which the measured rms surface deviation of the main reflector and subreflector combined is 0.58 mm at an elevation angle of 50°. The rms surface deviation can increase to 0.79 mm depending on elevation angle and wind velocity.

Rantec and is based on the system developed by the Jet Propulsion Laboratory [7]. The use of the dual frequency system increases the system temperature by 3.7 and 3.2 K at 4.75 and 14.48 GHz, respectively, and the antenna efficiency is reduced by a factor of 0.97 in both frequency bands.

A photograph of the feed system is given in Fig. 7, and the measured system performance is given in Table II. Antenna efficiencies were measured using astronomical radio sources [8]. To give uniform illumination in the aperture of the main dish, the VLA antenna geometry requires a feed pattern taper of 11.5 dB at the edge of the subreflector, which subtends a total angle of 18° at the feed. The subreflector edge illumination, which is higher than usual for shaped systems, was chosen to prevent the aperture of the 1.35–1.73-GHz feed from becoming too large. An important consideration in the design, because of the large number of feeds needed, is that the feeds should have as low a cost as possible. Since all feed outputs enter the same cryogenic dewar, and waveguide losses must be minimized, another important design constraint is that all feeds should be approximately the same length. The nominal feed length used is 178 cm and results in a 23-GHz feed which is four times longer than it needs to be and a 1.35–1.73-GHz feed which is only a quarter of the length of the appropriate optimum horn.

The 22–24-GHz and the 14.4–15.4-GHz feeds are square cross-sectional multiflare horns [9] designed by Rantec. These horns combine reasonable fabrication cost with good pattern circularity and sidelobe performance. The cost of long corrugated horns for these frequencies would have been significantly more. The 4.5–5.0-GHz feed is a lens-corrected corrugated horn designed by J. J. Gustincic.

The most difficult feed to design is the 1.35–1.73-GHz feed. The length constraint has been mentioned previously. The very wide bandwidth (25 percent) is needed so that observations of both hydrogen gas and the hydroxyl radical can be made using a single feed. The subreflector is not in the far field of the feed, so the far-field pattern must be focused into the near field to prevent loss of efficiency due to phase errors [10]. The straightforward design solution of a short horn with a large correcting lens in its aperture proved to be too expensive. The results

parabolic profile typical of reflectors shaped for high efficiency [4]. The shaping is small enough to allow prime focus operation at frequencies below 1.3 GHz. The subreflector, which is not a surface of revolution, is shaped to give high efficiency by generating almost uniform illumination in the aperture of the main dish. The profile of the subreflector is also shaped to give uniform phase in the aperture of the main reflector and an on-axis beam with a secondary focal point 97 cm from the axis of the main reflector. The feeds are arranged on a circle of radius 97 cm around the main reflector axis, and the frequency is changed by rotating the subreflector around the reflector axis until the secondary focus lies on the required feed. The offset Cassegrain reflector geometry has been pioneered by the Jet Propulsion Laboratory in the Goldstone 64-m paraboloid [5]. The theory for shaping the offset geometry is given by Potter [6]. The VLA geometry was calculated by the Rantec Division of the Emerson Electric Company.

Simultaneous operation at 4.5–5.0 GHz and 14.43–14.53 GHz is made possible by placing a dichroic (frequency sensitive) reflector over the 14.4–15.4-GHz feed and an ellipsoidal reflector over the 4.5–5.0-GHz feed as shown in Fig. 6. This dual frequency system was designed by

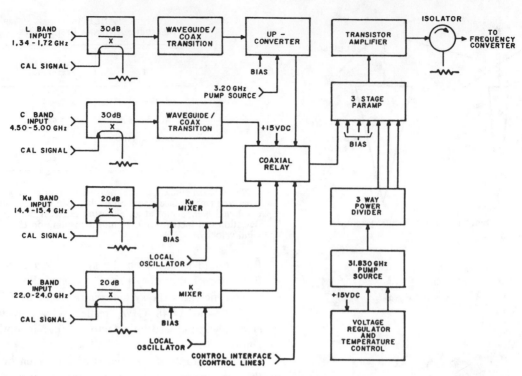

Fig. 8.   Front-end block diagram.

shown in Table II were obtained using a center-fed 213-cm-diam elliptical reflector as the feed. This feed can be seen in Fig. 7. The reflector profile was elliptical rather than parabolic to correct for the subreflector being in the near field of the feed. The relatively low efficiencies obtained with this feed result from the difficulty of obtaining a high spillover efficiency with a feed of this type [11]. A new type of feed for the 1.35–1.73-GHz band is presently under development by J. J. Gustincic. This feed consists of a small horn illuminating a large lens constructed partially from solid dielectric and partially from waveguide elements. This new feed, which is expected to combine improved performance with low cost, will be described separately [12].

For all four frequency bands it is possible to have either dual linear polarization or dual circular polarization by manually changing the polarization transducer at the feed output. For linear polarization orthomode junctions are used which have VSWR's better than 1.10 and isolations better than 45 dB across the band for the 4.5–5.0, 14.4–15.4, and 22–24-GHz bands. For these bands circular polarization is achieved using sloping-septum polarizers [13], which have axial ratios better than 0.4 dB and isolations better than 30 dB across the band. The sloping-septum polarizer is suitable for low-cost quantity productions because it can be cast in a single piece and requires a minimum of fine tuning. In the 1.35–1.73-GHz band linear polarization is achieved using an orthomode junction that has a VSWR better than 1.17 and an isolation better than 45 dB. Circular polarization will be formed using a 3-dB hybrid after the orthomode junction. All orthomode junctions and polarizers are supplied by Atlantic Microwave.

The cross-polarization properties of the offset shaped-reflector geometry are unusual and deserve mention. In circular polarization the offset Cassegrain geometry gives rise to a cross-polarization distribution in the aperture of the main dish that causes a separation of 0.06 beamwidths between the left- and right-handed circularly polarized beams. This effect is also seen in asymmetric prime focus reflectors [14], [15]. The effect is important to the VLA when a map of circular polarization is synthesized out to the 3-dB points of the beam. Several techniques that show promise of reducing the beam separation include generating higher order modes in the feed horns [16], placing a polarization-sensitive lens in front of the feed horn, or reshaping the subreflector to give an off-axis beam. A more complete description of the circularly polarized beam separation problem will be given separately [17].

FRONT-END COMPONENT DESCRIPTION

A detailed block diagram of the front end is given in Fig. 8. The 4.5–5-GHz channels of the receiver utilize a three-stage parametric amplifier [18]. The stages are of identical design, and each is capable of 10-dB gain. The resulting 30-dB gain is flat within 1.0 dB across the 4.5–5.0-GHz band. The noise temperature of the amplifier is 17 K, but the input waveguide-to-coaxial transition, bandswitch, coaxial cables, and transistor postamplifier add approximately 8 K to the receiver noise temperature. Pump power (40 mW/stage) at 31.83 GHz is derived from a Gunn-effect oscillator. The oscillator's temperature and bias voltage are stabilized within 0.1°C and 2 mV, respectively, to insure good stability of the amplifier gain and phase shift. The parametric amplifier is manufactured by Comtech Laboratories, Inc. The transistor postamplifier, made by Locus, Inc., has a 7-dB noise figure and adds 35 dB gain. It and the subsequent stages contribute less than 1.3 K to the 5-GHz receiver noise temperature.

Future receivers will use a two-stage 25-dB-gain parametric amplifier followed by a room-temperature gallium arsenide FET amplifier of 40-dB gain and 3-dB noise figure. This change will be made to reduce cost and complexity of the receiver with no sacrifice of performance. The present paramps will be reduced to two stages, retuned for 25-dB gain, and refitted with gallium arsenide FET amplifiers. The two-stage parametric amplifier will be manufactured by the AIL Division of Cutler–Hammer, and the FET amplifier by Avantek.

A waveguide-to-coaxial transition is used at 4.5–5 GHz. To minimize the receiver noise temperature, the transition probe and center conductor of the necessarily long coaxial line are cooled to the same temperature as the parametric amplifier. The outer conductor, a thin-wall stainless-steel thermal insulating component, is silver plated and has a longitudinal temperature gradient between the cryogenic stage temperature and room temperature. A waveguide-mounted quartz dome surrounds the probe to provide the necessary vacuum seal [19].

The bandswitch is a four-position coaxial latching relay cooled to the cryogenic stage temperature to reduce its noise temperature. This switch is manufactured by DB Products Corporation.

A parametric upconverter with a 3.20-GHz pump frequency converts the 1.35–1.73-GHz band to 4.55–4.93 GHz. The upconverter is fabricated in microstrip on an alumina substrate. It is a two-diode balanced design to reduce input circuit currents at the lower sideband frequency, which is within the input frequency range. Conversion gain is approximately 3.5 dB and the instantaneous bandwidth (1 dB) is approximately 200 MHz. Varactor bias and pump power level are varied to tune different band segments. The upconverter is manufactured by the AIL Division of Cutler–Hammer.

Cooled resistive mixers are used for the 14.4–15.4-GHz and 22–24-GHz bands [20]. The structure of both mixers is similar; each uses a single gallium arsenide Schottky diode in an image-enhancement design. The local oscillator (LO) frequency is below the signal frequency for the 22–24-GHz mixer; above for the 14.4–15.4-GHz mixer. Consequently the LO bands are contiguous, and a single 17–20-GHz YIG-tuned Gunn-effect oscillator can be used for both mixers. The diodes, which are supplied by the University of Virginia solid-state device laboratory, have a diameter of 5 μm, a zero-voltage junction capacitance of 0.03 pF, and a series resistance of 5 Ω.

The receiver components are cooled by a two-stage closed-cycle helium refrigerator with 10-W second-stage capacity at 20 K and 30-W first-stage capacity at 77 K. All electronic components are mounted on the second stage; the first stage is used to cool a thermal radiation shield. The second stage typically operates at 15 K and is stable to ± 1.0 K over all changes in ambient temperature, antenna elevation, and a period of time of several weeks. Because of the large cooled mass, short-term temperature fluctuations

(i.e., those at the cycling rate of the refrigerator) are too small to measure accurately and are on the order of 0.1 K rms. Eight hours are required to cool the receiver from 295 to 15 K.

Thermal insulation is provided by an aluminum vacuum chamber (Dewar) and conventional radiation shields. The vacuum chamber is rectangular and made of $\frac{1}{2}$-in-thick 6061-T6 aluminum plates joined by heliarc welding. No special surface treatment is used other than a thorough cleaning. Fluorocarbon rubber seals are used on most feedthroughs, with butyl rubber for the large door seals. Because of the large number of seals necessary (approximately 55 O-rings) the helium leak rate is on the order of 10^{-5} std. cm^3/s.

An adsorber containing approximately 10 g of activated charcoal is attached to the second stage of the refrigerator. This adsorber will maintain the pressure below 10^{-6} torr for more than a year of normal operation. No other vacuum pumps are used while the receiver is cold; however, a mechanical vacuum pump is turned on automatically if a power failure causes the receiver temperature to rise, releasing gas from the adsorber.

Bright nickel-plated aluminum radiation shields are used around the second stage (other cooled components are of copper or aluminum, and are also nickel plated). Because of the large number of openings, the shields are wrapped with several layers of aluminized mylar. As a result, the radiative heat load on the second stage is small compared to the conductive heat load.

CONCLUSION

The VLA front-end development has demonstrated that low noise can be achieved with a high degree of frequency flexibility and a reasonable system cost.

ACKNOWLEDGMENT

The authors wish to thank D. L. Thacker and G. Barrell for their help in system design and testing, D. A. Bathker for valuable information about the JPL 64m system, and B. G. Clark for suggestions concerning feed system polarization problems.

REFERENCES

[1] M. Ryle, "The new Cambridge radio telescope," *Nature*, vol. 194, pp. 517–518, May 1962.
[2] Y. I. Chow, "On designing a supersynthesis antenna array," *IEEE Trans. Antennas Propagat.*, vol. AP-20, pp. 30–35, Jan. 1972.
[3] D. S. Heeschen, "The very large array," *Sky and Telescope*, vol. 49, pp. 334–351, June 1975.
[4] W. F. Williams, "High efficiency antenna reflector," *Microwave J.*, vol. 8, pp. 79–82, July 1965.
[5] M. S. Reid, R. C. Clauss, D. A. Bathker, and C. T. Stelzried, "Low-noise microwave receiving systems in a worldwide network of large antennas," *Proc. IEEE*, vol. 61, pp. 1330–1335, Sept. 1973.
[6] P. D. Potter, "Analytical technique for design of asymmetrical-shaped dual-reflector antenna systems," *J.P.L. Tech. Report 32-1526*, vol. 10, pp. 129–134, June 1972.
[7] D. A. Bathker, "Dual frequency dichroic feed performance," *Proceedings 26th Meeting Avionics Panel*, AGARD, Munich, Germany, Nov. 26–30, 1973.
[8] J. M. Baars, "The measurement of large antennas with cosmic radio sources," *IEEE Trans. Antennas Propagat.*, vol. AP-21, pp. 461–473, July 1973.
[9] S. B. Cohn, "Flare angle changes in a horn as a means of pattern control," *Microwave J.*, vol. 13, pp. 41–46, Oct. 1970.

[10] F. I. Sheftman, "Experimental study of the low frequency operation of a Cassegrainian antenna," Tech. Note 1968-38, MIT Lincoln Lab., Dec. 1968.

[11] R. Caldecott, C. A. Mentzer, L. Peters, and J. Toth, "High performance S-band horn antennas for radiometer use," NASA Report CR-2133, prepared by Electroscience Laboratory, Ohio State Univ., Jan. 1973.

[12] J. J. Gustincic and P. Napier, in preparation.

[13] M. H. Chen and G. N. Tsandoulas, "A wide-band square-waveguide array polarizer," IEEE Trans. Antennas Propagat., vol. AP-21, pp. 389–391, May 1973.

[14] T. S. Chu and R. H. Turrin, "Depolarization properties of offset reflector antennas," IEEE Trans. Antennas Propagat., vol. AP-21, pp. 339–345, May 1973.

[15] N. A. Adatia and A. W. Rudge, "Beam squint in circularly polarized offset-reflector antennas," Electron. Lett., vol. 11, pp. 513–515, Oct. 1975.

[16] A. W. Rudge and N. A. Adatia, "A new class of primary-feed antennas for use with offset parabolic-reflector antennas," Electron. Lett., vol. 11, pp. 597–599, Nov. 1975.

[17] P. Napier and J. J. Gustincic, in preparation.

[18] J. Kliphius and J. C. Greene, "Low noise, wideband, uncooled preamplifier," AIAA 3rd Communications Satellite Systems Conf., Paper No. 70–419, 1970.

[19] R. Clauss and E. Wieke, "Low-noise receivers: Microwave maser development," JPL Tech. Report 32-1526, vol. 19, pp. 95–99, Feb. 15, 1974.

[20] S. Weinreb and A. R. Kerr, "Cryogenic cooling of mixers for millimeter and centimeter wavelengths," IEEE J. Solid-State Circuits, vol. SC-8, pp. 58–63, Feb. 1973.

The Low-Noise 115-GHz Receiver on the Columbia-GISS 4-ft Radio Telescope

HONG-IH CONG, ANTHONY R. KERR, SENIOR MEMBER, IEEE, AND
ROBERT J. MATTAUCH, MEMBER, IEEE

Abstract—The superheterodyne millimeter-wave radiometer on the Columbia-GISS 4-ft telescope is described. This receiver uses a room-temperature Schottky diode mixer, with a resonant-ring filter as LO diplexer. The diplexer has low signal loss, efficient LO power coupling, and suppresses most of the LO noise at both sidebands. The receiver IF section has a parametric amplifier as its first stage with sufficient gain to overcome the second-stage amplifier noise. A broad-banded quarter-wave impedance transformer minimizes the mismatch between mixer and paramp. At 115 GHz, the SSB receiver noise temperature is 860 K, which is believed to be the lowest figure so far reported for a room-temperature receiver at this frequency.

Manuscript received June 26, 1978; revised September 28, 1978.
H. I. Cong and A. R. Kerr are with NASA Goddard Institute for Space Studies, Goddard Space Flight Center, New York, NY 10025.
R. J. Mattauch is with the Department of Electrical Engineering, University of Virginia, Charlottesville, VA 22901.

I. INTRODUCTION

THE COLUMBIA-GISS 4-ft radio telescope was constructed for the purpose of surveying the distribution of carbon monoxide in the interstellar space of our Galaxy. The telescope, a Cassegrain with an effective f/D ratio of 2.8, has a half-power beamwidth of 8 arc min at 115 GHz, the frequency of the fundamental rotational transition line of carbon monoxide. The receiver front-end is of the superheterodyne type using a room-temperature Schottky diode mixer, followed by a 1.39-GHz parametric amplifier with a noise temperature of 50 K and a gain of ~17 dB. At 115 GHz, the single-sideband noise temperature of this receiver is 860 K, which we believe to be the best performance reported at this frequency for a room-

Reprinted from *IEEE Trans. Microwave Theory Tech.*, vol. MTT-27, no. 3, pp. 245–248, Mar. 1979.

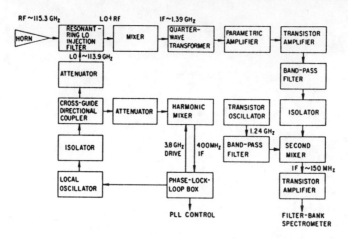

Fig. 1.   Block diagram of the 115-GHz receiver front-end.

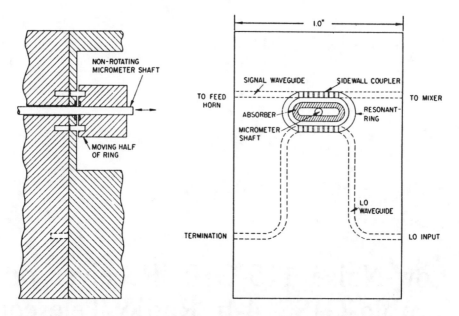

Fig. 2.   Cross section of the resonant-ring LO injection filter. The moving part of the resonant ring is driven by a differential micrometer (not shown) with a nonrotating shaft.

temperature radiometer. This paper will briefly describe the various parts of the receiver, emphasizing in particular the steps taken to reduce the noise contributions from individual components of the system.

The block diagram in Fig. 1 shows the major components of the receiver front-end. The incoming 115-GHz signal from the feed horn and the local oscillator (LO) power are coupled into the mixer through the resonant-ring filter to produce an intermediate frequency (IF) signal of ~1390 MHz, which is sent to the parametric amplifier. The succeeding stages consist mainly of a transistor amplifier (gain ~30 dB) followed by a second mixer which further converts the 1390-MHz IF signal to ~150 MHz. This second IF signal is then amplified by another transistor amplifier (gain ~34 dB) before it is sent to the filter-bank spectrometer (currently 256 channels of $\frac{1}{4}$-MHz filters). The phase-lock system for stabilizing the LO frequency is similar to the one designed by

Weinreb [1]. The LO is locked according to $f_{LO} = N f_o \pm$ 400 MHz, where f_o is the frequency (~3.8 GHz) of the phase-locked solid-state oscillator which drives the harmonic mixer, and $N = 30$ for $f_{LO} = 114$ GHz.

II.   RESONANT-RING LO DIPLEXER

Fig. 2 shows a cross section of the resonant-ring filter which couples the incoming signal and LO power into the mixer while preventing noise generated by the LO at the signal and image frequencies from reaching the mixer. The principle of the resonant-ring filter has been described by various authors [2], [3], and only the essential details will be described here. The signal and LO waveguides are each coupled to the oval waveguide ring by sidewall couplers. The ring is fabricated in two parts, whose separation, and hence the waveguide width, can be adjusted by a differential micrometer to tune the filter. Since this type of filter has multiple passbands, it is

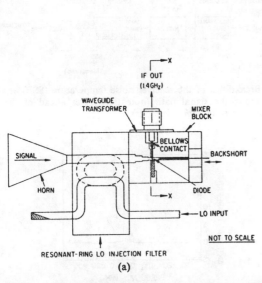

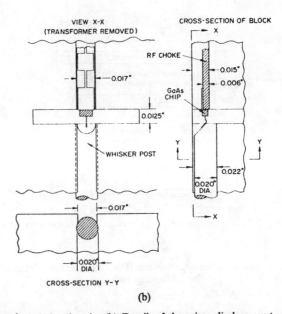

Fig. 3. (a) Cross section of the mixer and injection filter showing the LO and signal paths. (b) Details of the mixer diode mount.

necessary to choose the length of the waveguide ring so that the desired sideband noise rejection can be achieved. In this case, IF's of 1.39 and 3.95 GHz were required, so the ring length was chosen to give passbands about every 8 GHz.

The oval shape of the ring was chosen to allow a straight signal waveguide of minimum length (1.0 in), and hence minimum loss, to be used. For this reason also, the couplers consist of eight equal rectangular holes the full height of the waveguide, a configuration which maximizes coupling for a given length at the expense of directivity. To facilitate machining of the coupling holes it was necessary to split the signal waveguide along the plane of one sidewall, a situation giving an initial insertion loss of ~0.6 dB. Conducting epoxy[1] applied along the waveguide edges during final assembly reduced this to 0.25 dB. A spurious resonance in the gap between the halves of the resonant ring was suppressed by inlaying some absorbing material[2] in this region. The LO-to-mixer insertion loss is 4 to 6 dB over the frequency range from 90 to 120 GHz, and the rejection at the signal and image frequencies is typically 15 dB (in addition to the insertion loss) with 1.39-GHz IF, and greater than 20 dB with 3.95-GHz IF.

III. MIXER

The mixer is of the broad-band type using a single GaAs Schottky diode mounted across a quarter-height WR-10 waveguide. The diode is tuned by an adjustable waveguide short-circuit behind the diode, and coupled to the IF and dc connector by a quartz microstrip RF choke

(see Fig. 3). The electrical design of the mixer is similar to that described in [4], and the construction is in many ways the same as the WR-5 mixer reported in [5]. The diode was fabricated by the Semiconductor Device Laboratory of the Electrical Engineering Department of the University of Virginia, and has a dc resistance of 8 Ω, ideality factor $\eta = 1.1$, and zero-bias capacitance of 7 fF. Using the IF noise radiometer/reflectometer described in [6], we measured an SSB mixer noise temperature (T_M) of 440 K, and a conversion loss (L) of 5.3 dB at 114-GHz LO frequency. This mixer was the best of three using diodes from the same batch; values of L and T_M for the other mixers were 5.2 and 5.7 dB, 530 and 520 K. Normally, a dc bias of 0.5 V is applied to the diode, and the LO power is adjusted until a total current of 2.0 mA is reached. The LO power needed at the mixer input to achieve this rectified current is ~1.5 mW.

IV. IF TRANSFORMER

Since the VSWR looking into the mixer IF output port is generally 3 or 4, a matching network between the mixer and paramp is required. At an IF of 1.4 GHz, the mixer impedance is slightly capacitive at the plane of the SMA connector on the mixer block (see Fig. 3). This impedance can be transformed to a pure resistance through a short coaxial line, and then to 50 Ω. Originally, we used a simple quarter-wavelength impedance transformer, consisting of a brass rod as the center conductor inside a square aluminum outer conductor, which gave a VSWR between 1 and 1.6 in the IF range from 1350 to 1430 MHz. A broad-band, partially tunable transformer was subsequently constructed (see Fig. 4). Two striplines, soldered at the input end of the quarter-wave transformer, form a broad-banding resonator which is adjustable by

[1]Epoxy Technology Inc., #H20E.
[2]Emerson Cuming Co., Eccosorb #116, ground to powder and mixed with (nonconducting) epoxy.

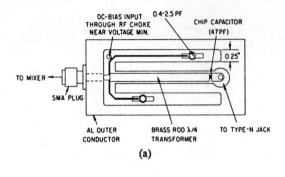

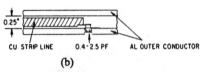

Fig. 4. (a) Sketch of the quarter-wave IF transformer. (b) Detail of a broad-banding stub.

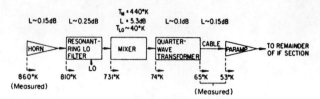

Fig. 5. Breakdown of the receiver noise temperature (SSB) at various points along the signal path (note that losses ahead of the mixer contribute noise in *both* sidebands).

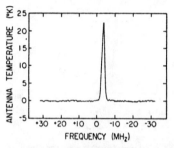

Fig. 6. Carbon monoxide emission ($f_o = 115271.2$ MHz) from Orion A, observed with the Columbia-GISS 4-ft telescope. The filter bank has 256 $\frac{1}{4}$-MHz channels. This spectrum is raw data and has had no corrections or smoothing other than the removal of a linear baseline.

capacitors[3] at the ends of the stubs. The design was optimized by computer to give an almost flat double-tuned response. The resulting VSWR is between 1.03 and 1.09 over the 80-MHz bandwidth. This impedance transformer also provides a terminal for dc biasing the mixer diode through a small RF choke coil. A 47-pF chip capacitor[4] soldered at the output end of the quarter-wave transformer serves as a dc block.

V. REMAINDER OF SIGNAL PATH

The output of the IF transformer has a type-N connector. Ideally, the paramp input, also a type-N connector, should be connected directly to the transformer. However, owing to physical constraints, an 8-in semirigid cable (0.141-in diameter) with type-N connectors on both ends was used between the IF transformer and the parametric amplifier. The second stage of the IF system is a bipolar transistor amplifier,[5] whose noise figure is 2.24 dB. The total IF noise temperature, measured at the input connector of the semirigid cable, is 65 K.

VI. PERFORMANCE

At the LO frequency of 114 GHz the single-sideband system noise temperature, referred to input aperture of the feed horn, is 860 K measured using room temperature and 77 K absorbers in front of the horn. Fig. 5 shows the contributions of the various components to the overall receiver noise. Noise generated at the signal and image frequencies by the LO, a Varian VRT-2123A19 klystron, is mostly suppressed by the resonant-ring filter, and contributes ~40 K to the total system noise temperature. An example of a spectral line received by the 4-ft telescope is given in Fig. 6.

VII. SUMMARY

Although the mixer itself contributes a substantial part of the total receiver noise, the effects of waveguide loss, IF mismatch, cable loss, and klystron noise also make essential contributions. For the millimeter-wave radiometer on the Columbia-GISS 4-ft telescope, efforts to reduce noise contributions from various parts of the receiver have resulted in what we believe to be the lowest noise room-temperature radiometer so far reported at 115 GHz.

ACKNOWLEDGMENT

The authors wish to thank P. Thaddeus for his continuing inspiration and support, and J. Grange, G. Green, D. Held, H. Miller, D. Mumma, and I. Silverberg for their help in constructing this receiver.

REFERENCES

[1] S. Weinreb, "Millimeter-wave spectral receiver-local oscillator and IF sections," NRAO Electronics Division Internal Rep. 97, Oct. 1970.
[2] G. L. Matthaei, L. Young, and E. M. T. Jones, *Microwave Filters, Impedance-Matching Networks, and Coupling Structures*. New York: McGraw-Hill, 1964.
[3] J. E. Davies, "Ring-type directional filters," Internal Memo., Electronics Division, National Radio Astronomy Observatory, Charlottesville, VA, May 1975.
[4] A. R. Kerr, "Low-noise room-temperature and cryogenic mixers for 80-120 GHz," *IEEE Trans. Microwave Theory Tech.*, vol. MTT-23, pp. 781-787, Oct. 1975.
[5] A. R. Kerr, R. J. Mattauch, and J. Grange, "A new mixer design for 140-220 GHz," *IEEE Trans. Microwave Theory Tech.*, vol. MTT-25, pp. 399-401, May 1977.
[6] S. Weinreb and A. R. Kerr, "Cryogenic cooling of mixers for millimeter and centimeter wavelengths," *IEEE J. Solid-State Circuits*, vol. SC-8, pp. 58-63, 1973.

[3]Johanson Electronics, #C72480A5, 0.4–2.5 pF.
[4]American Technical Ceramics, #100-A-470-J-P-50, 47 pF.
[5]Watkins-Johnson, #WJ-737-142.

Low-Noise Cooled GASFET Amplifiers

SANDER WEINREB, FELLOW, IEEE

Abstract—Measurements of the noise characteristics of a variety of gallium–arsenide field-effect transistors at a frequency of 5 GHz and temperatures of 300 K to 20 K are presented. For one transistor type detailed measurements of dc parameters, small-signal parameters, and all noise parameters (T_{min}, R_{opt}, X_{opt}, g_n) are made over this temperature range. The results are compared with the theory of Pucel, Haus, and Statz modified to include the temperature variation. Several low-noise amplifiers are described including one with a noise temperature of 20 K over a 500-MHz bandwidth. A theoretical analysis of the thermal conduction at cryogenic temperatures in a typical packaged transistor is included.

I. INTRODUCTION

THE PRESENT state of the art for microwave low-noise amplifiers is shown in Fig. 1. The gallium–arsenide field-effect transistor (GASFET) amplifier does not yet achieve the noise temperature of the very best parametric amplifiers but is equal to or better than many paramps manufactured 10 years ago. In addition, the GASFET has higher stability and lower cost because of two inherent advantages: 1) it is much less critical to circuit impedance than a negative resistance amplifier such as a paramp; 2) it is powered by dc whereas the paramp requires a power oscillator and tuned circuits at several times the frequency of operation.

There are systems, particularly those requiring large-area antennas such as radio astronomy or space communications, where no present device operating at room temperature has sufficiently low noise. This is shown clearly in Fig. 1 where, in the 0.5–20-GHz range, the 300 K paramp performance is typically an order of magnitude greater than the natural noise limits of galactic, cosmic, and atmospheric noise. The lowest noise, highest cost solution is a maser or parametric up-converter-into-maser system operating at 4 K. Intermediate in cost and performance are paramps and GASFET's cooled to 20 K by closed-cycle helium refrigerators which are now available [10] at a cost of under $5000 and a weight less than 45 kg.

Several reports of the noise temperature of cryogenically cooled GASFET amplifiers have been made [4]–[8] but, for the most part, these are for one specific device, do not determine the four noise parameters which characterize the noise of a linear two-port ([4] is an exception to this), do not report the device dc and small-signal parameters as a function of temperature, and do not attempt to correlate the results with theory. An attempt will be made to do the above in this paper and to present some information regarding the following questions.

Manuscript received April 3, 1980; revised June 24, 1980. The National Radio Astronomy Observatory is operated by Associated Universities, Inc., under contract to the National Science Foundation.

The author is with National Radio Astronomy Observatory, Charlottesville, VA 22903.

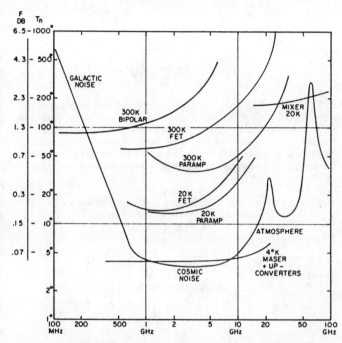

Fig. 1. Noise figure 10 log F, and noise temperature $T_n = 290°$ ($F-1$), versus frequency for various 1980 state-of-the-art low-noise devices. The 300 K bipolar transistor, FET, and paramp values are taken from manufacturers data sheets [1]–[3], the 20 K FET curve is from the data of this paper plus data of others [4]–[6] at 0.6, 1.4, and 12 GHz, respectively. The 20 K paramp, 4 K maser (including parametric up-converter at lower frequencies), and 20 K mixer results (which are SSB and include IF noise) are from systems in use at National Radio Astronomy Observatory (NRAO). The natural noise limitations due to galactic noise, the cosmic background radiation, and atmospheric noise are for optimum conditions and are taken from [9] plus points at 22 GHz and 100 GHz measured at NRAO.

1) Considering the dc power dissipated and thermal resistance problems, what is the actual physical temperature of the FET channel? What is the lowest physical temperature which can be achieved in the channel?

2) What is the noise improvement factor for cooling of presently available low-noise GASFET's? Are one manufacturer's devices superior for some fortuitous reason?

3) Can the present room-temperature GASFET noise theory be applied at cryogenic temperatures?

4) Can a GASFET be specifically designed for the best performance at cryogenic temperatures?

5) Is there any difference in the circuit design for a cryogenic amplifier?

II. THERMAL RESISTANCE

The room-temperature thermal resistance of a typical low-noise GASFET is specified on manufacturers' data sheets and is of the order of 100 K/W for a chip and 200

Reprinted from *IEEE Trans. Microwave Theory Tech.*, vol. MTT-28, no. 10, pp. 1041–1054, Oct. 1980.

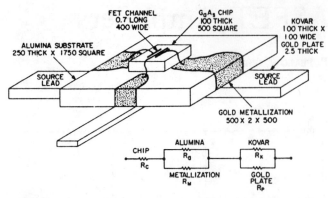

Fig. 2. Typical GASFET chip in 1.75-mm² package with top cover removed; all dimensions are in micrometers. Heat flow is from the FET channel spreading through the GaAs chip into the gold-metallized alumina substrate and out the source leads. An equivalent electrical circuit is also shown with thermal resistance values in Table I. Some manufacturers do not connect the source metallization to the chip.

TABLE I
THERMAL RESISTANCE OF PACKAGED GASFET IN K/W AND IN
PARENTHESIS MATERIAL CONDUCTIVITIES
IN W/K·cm

Component (Dimensions in µm)	R	Temperature			
		300°	77°	20°	4°
FET Channel (0.7 x 400)	R_c	120 (0.44)	12 (4.4)	13 (4.1)	840 (.06)
Alumina Substrate 250 x (1.750)²	R_a	55 (0.35)	13 (1.5)	85 (0.23)	3900 (.005)
Gold Metalization 500 x (500 x5)	R_m	335 (3)	285 (3.5)	62 (16)	45 (22)
Total $R_c + R_a // R_m$	R_t	169	24	49	885
Kovar in Source Leads 250 x (100 x 1000)	R_K	76 (.165)	156 (.08)	625 (.02)	4170 (.003)
Gold Plate on Source Leads 250 x 2.5 x 2200	R_P	76 (3)	65 (3.5)	14 (16)	10 (22)
Total Including Source Leads $R_t + R_K // R_P$	R_T	207	89	63	895
Add for Epoxy Bond of Chip 25 x (500 x 500)	-	50 (.02)	100 (.01) EST	330 (.003) EST	1000 (.001)

K/W for a packaged device. These values produce a heating of 5 or 10 K for a typical low-noise dc power dissipation of 50 mW and do not significantly effect the room-temperature performance. However, at cryogenic temperatures the situation may be drastically different because the thermal conductivity of most materials changes by orders of magnitude; pure metals and crystalline substances become better thermal conductors while alloys and disordered dielectrics become worse.

An analysis of the heat flow in a typical 1.75 mm² packaged GASFET sketched in Fig. 2 has been performed using the thermal resistance equations of Cooke [11] with material thermal conductivities published in various references (GaAs [12], alumina and gold [13], Kovar and iron alloys [14]). Results are summarized in Table I which also gives material conductivities used in the calculations.

At temperatures down to 20 K the total thermal resistance decreases substantially for the configuration of Fig. 2. The heat flow medium shifts in the substrate from alumina to gold metallization (assumed 5 µm thick). It is

thus important that the gold metallization and plating be thick, pure, and free of voids. A case designed for cryogenic operation should have pure silver or copper source leads and a sapphire or crystalline quartz substrate.

It is important that the chip be solder-bonded to a metallized substrate with the metallization continuing to the source leads. This is not the case in all commercially available devices. A calculation of a solder-joint thermal resistance shows it to be negligible even at 4 K. However, a silver-loaded epoxy joint of 25-µm thickness would add 300 K/W at 20 K [16].

It is also important that the total heat path from chip, thru package to amplifier case, and on to cooling station be carefully considered; this often conflicts with the desired microwave design. A chip GASFET soldered to a high-purity copper amplifier case is an excellent solution to thermal problems above 20 K but is not a necessity; a packaged device can be used.

At 4 K the thermal problem within the GaAs chip is quite severe due to boundary scattering of phonons [17, p. 149] which produces a thermal resistance increasing as T^{-3} for temperatures below 20 K. A channel with 20–50 mW of power dissipation will stabilize at a temperature of ∽15 K even if the chip boundaries are at 4 K; hence little is gained compared to 20 K cooling. The value of the chip thermal resistance at 4 K given in Table I is only a rough approximation as the problem becomes complex. The thermal conductivity is no longer a point property of the material; the heat conduction is by acoustical waves and wave transmission and reflection at boundaries must be considered. However, an effective thermal conductivity dependent upon the object size can be defined (see Callaway [15]) and has been used in Table I with the size parameter set equal to a gate length of 0.7 µm. It should be noted that the chip thermal resistance would not be significantly reduced by immersion in normal liquid helium which has insufficient thermal conductivity for the area and heat flux involved (Though, super-fluid helium at a temperature below 2.2 K would be effective).

There is a possibility that a detailed study of the heat conduction mechanism from the channel at 20 K would show increased heating due to the small size effects discussed above which are certainly present at 4 K. Experimental evidence against this, however, is the fact that for most devices evaluated the amplifier noise temperature variation with dc bias power dissipation is small at cryogenic temperatures and similar to the variation at room temperature.

A different conclusion regarding self-heating at cryogenic temperatures was reached by Sesnic and Craig [56] who predict large self-heating for the 4 K–77 K temperature range due to poor conductivity of the Kovar source leads. Their paper did not consider the strong effects of plating on the source leads or the boundary-scattering decrease in chip thermal conductivity. These erroneous results were applied by Brunet–Brunol [57] who then attributed the lack of change of GASFET electrical characteristics below 77 K to self-heating.

III. DC Characteristics Versus Temperature

The dc characteristics of a GASFET can be analyzed to determine parameters such as transconductance and input resistance which enter directly into the noise temperature equation, and also device fabrication parameters such as channel thickness a and carrier density N, which affect the noise temperature in a more complex manner. In addition, by measuring the variation of dc parameters with temperature, the variation of material parameters such as mobility μ and saturation velocity v_s, can be determined; these also enter into the noise theory.

The curves of drain current versus drain voltage at steps of gate voltage for three different manufactures of GASFET's at 300 K and 23 K are shown in Fig. 3. In general, there is only a mild change in the characteristics of all devices tested with most changes occurring between 300 K and 80 K. The dominant effects are an increase in transconductance, saturation current, and drain conductance.

A more detailed analysis of a sample device, the Mitsubishi MGF 1412, will be performed using, with some modification, the methods of Fukui [18]. Microwave noise measurements of the identical device are described in the next section. The total channel width Z was measured with a microscope and found to be 400 μm in two 200-μm stripes and the gate length L has a published [34] value of 0.7 μm (the gate length has only a minor effect on the quantities evaluated in this section, but is important for the noise analysis). All dc data were measured at five temperatures utilizing 0.1-percent accuracy digital meters. The results are given in Table II and discussed below.

A. Forward-Biased Gate Characteristics

The forward-biased gate junction was first evaluated to find the barrier potential V_B, Schottky ideality factor n, and gate-plus-source resistance $R_g + R_s$. This was performed by measuring the gate-to-source voltage V_{gs} for forward gate currents of 0.1 μA–10 mA in decade steps, all with drain current $I_d = 0$. The results were fitted to a normal Schottky-barrier current–voltage characteristic with V_B replaced by V_B/n as suggested by Hackam and Harrop [19]. The resulting values for V_B show little variation with temperature, in agreement with the results of others [19], [20]. The n factor increases by a large amount as temperature decreases; this is due to tunneling through the narrow, forward-biased depletion layer [21], [22] and does not effect the reverse diode characteristics and normal FET operation. When tunneling is present the n factor will vary slightly with current (i.e., the I–V characteristic is not exponential) if the doping density is not uniform. This limits the accuracy of the determination of n and hence, of V_B and $R_g + R_s$. The n values given in Table II are for the 1–10-μA range.

The value of $R_g + R_s$ is determined from the I_g–V_{gs} measurements in the 0.1–20-mA range. At these currents the voltage drop across $R_g + R_s$ becomes significant compared to the exponential ideal diode characteristic, and hence $R_g + R_s$ can be determined. However, R_g at these

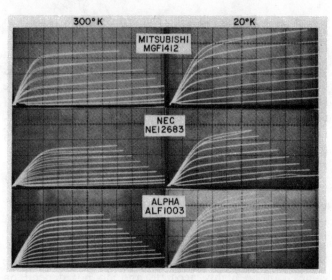

Fig. 3. Drain current, 20 mA per vertical division, versus drain voltage, 0.5 V per horizontal division, for 0.2 V steps of gate voltage for 3 manufacturers of GASFET's at 300 K and 20 K. The top curve in each photograph is at 0 V gate voltage.

TABLE II
DC Parameters Versus Temperature for the Mitsubishi
MGF 1412 GASFET (Units Are in Volts, Ohms,
Milliamperes, and Millimhos)

Quant	300°K	228°K	151°K	81°K	21°K
V_B	.796	.811	.834	.810	.782
n	1.125	1.216	1.47	2.0	7.2
R_g	2.66	1.7	1.5	1.4	1.2
R_s	2.3	2.3	2.1	2.1	2.2
R_d	2.1	2.4	2.5	2.4	2.4
R_t	8.8	8.8	9.3	11	10–14
V_p	1.226	1.151	1.156	1.172	1.159
I_o	67.8	74.3	79.1	84.7	86.2
I_s	255	279	297	318	324
g_m'	39	44	47	50	55
g_m	43	49	52	56	62
v_s/v_{so}	1.0	1.13	1.18	1.27	1.32
g_m/g_{mo}	1.0	1.14	1.21	1.30	1.44

currents is nonlinear due to the distributed gate metallization resistance. A distributed ladder network of resistors and diodes must be considered. At high currents the potential drop across the metallization resistance produces a reverse bias which results in less current through the diodes at the end of the ladder away from the gate connection point; the effective value of R_g decreases with current. This situation is described by a nonlinear differential equation which has been solved and the results have been used to find the constant, low-current value of $R_g + R_s$. The solution will be described in a separate publication [54]. The small-signal value of $R_g + R_s$ at 10-mA bias current was also measured at 100 kHz and 10 MHz. Values equal to the dc slope resistance were obtained, thus assuring that thermal errors are not present in the dc measurement.

To separate R_g and R_s and also determine the drain series resistance R_d, the gate-to-source voltage V_{gsd} with drain connected to source was measured along with the gate-to-drain voltage V_{gd} with the source open; both of these values are at a gate current of 10 mA. By differencing these quantities with V_{gs}, also measured at $I_g = 10$ mA, R_s and R_d are obtained as

$$R_s = \Delta R_1 + \sqrt{(\Delta R_1)^2 + \Delta R_1 \cdot \Delta R_2} \tag{1}$$

and

$$R_d = \Delta R_2 + R_s \tag{2}$$

where $\Delta R_1 = (V_{gs} - V_{gsd})/0.01$ and $\Delta R_2 = (V_{gd} - V_{gs})/0.01$. Note that since only voltage differences are measured the values can be highly accurate and do not depend on removing the exponential portion of the $I_g - V_g$ characteristic, as is the case with the determination of $R_g + R_s$, and therefore R_g.

The values of R_s and R_d show little variation with temperature. This is to be expected, since for a highly doped semiconductor, both the carrier density and mobility show little variation with temperature even at temperatures as low as 2 K [23], [24]. The mobility is limited by impurity scattering and carriers do not "freeze out" because the impurity band overlaps the conduction band. The gate resistance R_g decreases from 2.7 Ω at 300 K to 1.2 Ω at 21 K since it is primarily due to the resistivity of aluminum which decreases by a large amount dependent upon its purity. The theoretical resistance of the gate metallization, using the formula of Wolf [25] is 1.7 Ω at 300 K and < 0.001 Ω at 21 K indicating that ∼1 Ω is due to impurities or semiconductor resistance.

B. Drain Voltage–Current Characteristic

A second check on mobility variation is obtained by measurements in the linear region (i.e., no velocity saturation) of the $I_d - V_d$ characteristic. The drain current I_d was measured for $V_{gs} = 0$ and $V_d = 50$ or 100 mV. $R_t = V_d/I_d$ is reported in Table II and shows somewhat more variation with temperature than R_s or R_d (which are contained in R_t). In particular the value at 21 K is dependent upon past history; it is lower by ∼30 percent after forward biasing the gate junction. This has not been explained. No such "memory" effect is observed in the device at normal bias levels.

The saturated drain current I_0 at $V_{gs} = 0$ and $V_d = 3$ V was measured along with the gate voltage $-V_p$ to bring the drain current down to 2 mA; V_p is a good approximation to the pinch-off voltage of the device. These quantities, measured at 300 K, can be used to determine the doping density N, and channel thickness a of the device through the well-established relations

$$V_p + V_B = \frac{qNa^2}{2\kappa\varepsilon_0} \tag{3}$$

$$I_s = qv_s NaZ \tag{4}$$

where $q = 1.6 \times 10^{-19}$ C, $\varepsilon_0 = 8.85 \times 10^{-14}$ F/cm, $\kappa = 12.5$ for GaAs, $Z = 0.4$ mm, and the saturated velocity v_s is assumed to be 1.4×10^7 cm/s. The quantity I_s is the open-channel saturation current and is related to I_0 [18] by

$$I_s = I_0/\gamma \tag{5}$$

where

$$\gamma = 1 + \sigma - \sqrt{\delta + 2\sigma + \sigma^2} \tag{6}$$

$$\delta = (V_B + 0.234L)/(V_B + V_p) \tag{7}$$

$$\sigma = 0.0155 R_s Z/a \tag{8}$$

and a and L are in micrometers and Z is in millimeters. The channel thickness a is not initially known for use in σ but since it is a moderately small correction factor, an initial estimate can be used and later iterated. Solving (3) and (4) for a and N gives $a = 0.10$ μm and $N = 2.9 \times 10^{17}/cm^3$, in good agreement with information supplied by the manufacturer [26].

The measured insensitivity to temperature of $V_p + V_B$ verifies through (3) that N is not a function of temperature. The temperature dependence of I_s must then be due to a change in saturation velocity v_s; its value relative to the 300 K value is given in Table II. The results are in general agreement with the increase in saturation velocity measured by Ruch and Kino [27] in the 340 K–140 K range.

Finally, the transconductance g_m' was measured by taking 50-mV increments in V_g above and below the low-noise bias point of $V_d = 5$ V, $I_d = 10$ mA. This must be corrected for effects of source resistance to give the true transconductance $g_m = g_m'/(1 - g_m' R_s)$; both g_m' and g_m are given in Table II in units of millimhos. The values are approximately 35 percent lower than those given by the approximate theoretical expression

$$g_m = \frac{I_s}{2(V_p + V_B)} \cdot \frac{1}{(1 - I_d/I_s)} \tag{9}$$

and the exact theoretical curve given in [28, fig. 12(c)] (which is a little closer). Fukui had a similar problem in the analysis of his data [18, p. 787]. The increase in measured transconductance with decreasing temperature follows the saturation velocity increase determined from measurements of I_0, V_p, and V_B.

IV. MICROWAVE PERFORMANCE VERSUS TEMPERATURE

A. Gain and Noise Measurement Procedure

A block diagram of the test configuration used for measurements of gain and noise temperature of cooled amplifiers is shown in Fig. 4. Auxiliary equipment such as a network analyzer, reflectometer, and spectrum analyzer were used for impedance, return loss, and spurious oscillation measurements. At a later stage of the work an HP 9845 calculator and HP 346 avalanche noise source were incorporated to allow a swept frequency plot of 50 noise temperature and gain measurements to be performed in

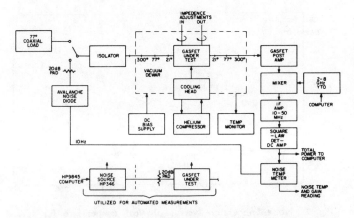

Fig. 4. Test setup for noise and gain measurements of cooled amplifiers. The automated measurement procedure gives a swept-frequency output of noise temperature (corrected for second-stage contribution) and gain; this replaced the hot load-cold load manual method after corroboration of results was established.

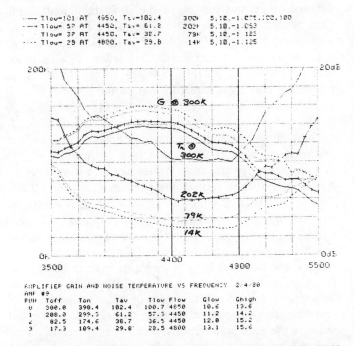

Fig. 5. Noise temperature and gain versus frequency for the Mitsubishi MGF 1412 transistor at several physical temperatures. This is a direct copy of the on-line output of an automated measuring system utilizing an Hewlett-Packard 9845 computer. The results are first plotted on a CRT and then can be copied on a thermal printer. Tabulated below the plot are noise-source-off temperature, noise-source-on temperature, average amplifier noise temperature between markers at 4.4 and 4.9 GHz, lowest noise and its frequency, and lowest and highest gain between marker frequencies. The lowest and average noise temperatures are also given above the plot area along with a space for user comments which in this case are physical temperature, drain voltage and current, and gate voltage.

20 s; a typical output from this automated system is shown in Fig. 5. The amplifiers under test were cooled by a commercially available [29] closed-cycle refrigerator with a capacity of 3 W at 20 K and a cool-down time of 5 h.

A large error in the noise temperature measurement of a mismatched amplifier can result if the source impedance changes when switching from hot load to cold load. For this reason an isolator is used between the loads and

Fig. 6. Test amplifier with adjustable source and load resistance. The spacing between the $\lambda/4 = 15.9$-mm long rectangular slab transmission line and ground plane is adjusted by turning the circular threaded slug. DC blocking capacitors are realized by teflon coating the SMA connector center pin where it enters the slab. The slab transmission line is supported by the coaxial connector, a chip capacitor, and a teflon bushing held by a nylon screw. The transistor is mounted by its source leads on a 6.3-mm diameter threaded copper slug with center 5.7 mm from the end of each $\lambda/4$ line.

amplifier input. The total loss from 77 K liquid-nitrogen cooled load to the amplifier was accurately measured and corrected for. This included 0.62 dB in the coaxial switch, isolator, and SMA hermetic vacuum feed thru; 0.11 dB in a stainless-steel outer-conductor beryllium–copper inner-conductor coaxial line transition from 300 K to 20 K; and 0.07 dB in 10 cm of internal coaxial line at 20 K.

For the automated measurements an HP 346 calibrated avalanche diode was used in cascade with a modified HP 8493A coaxial attenuator. No isolator is necessary. The attenuator is modified for cryogenic use [30] by replacing the conductive-rubber internal contacts by bellows contacts [31], drilling a vacuum vent hole in the body, and tapping the body for mounting a temperature sensor [32]. The change in attenuation of the modified 20-dB attenuator upon cooling from 300 K to 77 K is a decrease of 0.05 ± 0.03 dB; it is assumed that negligible change will occur upon further cooling to 20 K as this is typical for most resistance alloys. The attenuator may be purchased calibrated from Hewlett Packard (Option 890); however, the contact modification necessitates a recalibration (attenuation at 300 K decreased by 0.05 dB) and was performed with a digital power meter. For an amplifier with 30 K noise temperature the cooling of the attenuator from 300 K to 20 K provides a factor of $(300+30)/(20+30) \sim 6$ decrease in error of the amplifier noise temperature measurement due to noise source or attenuator errors. For this case a noise temperature error of $\pm 2.3°$ results from an uncertainty of noise source excess noise plus attenuator error of ± 0.2 dB. The agreement between amplifier noise measurements made with the hot/cold loads and the HP 346 noise source was within ± 2 K for cooled amplifiers and ± 5 K for uncooled amplifiers.

In order to determine the optimum source resistance and load resistance as a function of temperature, a test amplifier having variable impedance quarter-wave transformers at input and output was constructed and is shown in Fig. 6. A schematic of this amplifier and of all other

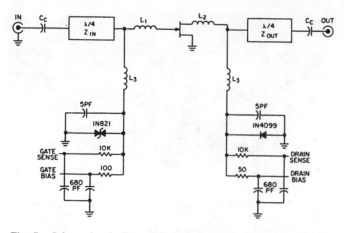

Fig. 7. Schematic of all amplifiers discussed in this paper. The bias decoupling inductance L_3 is realized by $\lambda/4$, $Z_0 = 100\ \Omega$ line in the test amplifier; in other amplifiers it is a small coil described in the text.

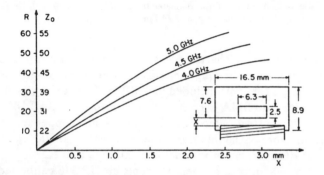

Fig. 8. Characteristic impedance Z_0 and source resistance, $R = Z_0^2/50$, as a function of spacing X from threaded-slug ground plane to $\lambda/4$ slab transmission line having cross section shown in figure.

amplifiers described in this paper is shown in Fig. 7. The variable impedance line consists of a slab transmission line having cross section as shown in Fig. 8 and a movable ground plane realized as a threaded slug with a diameter of $\lambda/4$. The slug is slotted and has a thread-tightening screw to assure ground contact and to lock the position. Calibration of the characteristic impedance versus slug position was performed by utilization of a miniature coaxial probe inserted in place of the FET and connected to a network analyzer; results are given in Fig. 8. The loss of the $\lambda/4$ line was calculated to be less than 0.025 dB at $Z_0 = 22\ \Omega$ using the microstrip loss formulas of Schneider [33].

B. Results

As a first step, the noise temperature of several commercially available GASFET's was measured at a frequency of ~4.5 GHz from 300 K to 20 K; results are shown in Fig. 9. At each temperature bias and source resistance were optimized for minimum noise; the changes were not large. For all units the peak gain was ~12 dB at 300 K and increased by 1 or 2 dB at 20 K. The measurement frequency was chosen for minimum noise at 300 K and no change in the optimum frequency was noted as temperature decreased.

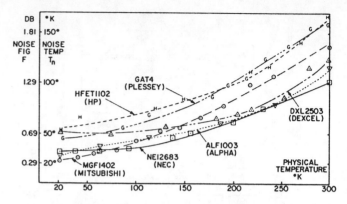

Fig. 9. Noise temperature versus physical temperature for several manufacturers of GASFET's. Only one or two samples of the HP, Dexcel, Alpha, and Plessey devices were evaluated, and units from another batch may give better cryogenic performance.

On the basis of these tests and because of its high burnout level [34], the Mitsubishi device was selected for detailed study and for use in the construction of amplifiers needed for use in radio astronomy. A selected version of the MGF 1402, the MGF 1412, became available and was used for further tests; it has lower noise temperature (~100 K) at 5 GHz and 300 K, but has a noise temperature at 20 K close to that of the MGF 1402. Of approximately 15 samples of the MGF 1412 tested at 20 K, the noise temperature was between 15 K and 27 K.

The gain and noise temperature of a MGF 1412 at four temperatures between 300 K and 14 K is shown in Fig. 5; dc parameters for the same device are given in Table II. As the amplifier is cooled, the gain increases by 2.0 dB. This is less than the 3.0-dB increase of g_m because of internal negative feedback (which stabilizes the gain against g_m changes) and possibly because of a reduction in parallel output resistance. As the noise temperature decreases, the bandwidth for a given increase in noise temperature increases, but there is very little other change in frequency response of either gain or noise temperature.

A complete comparison of noise theory and experiment requires the measurement of four noise parameters of the device. There are many sets of four parameters which can be compared. The set which is most directly measured is the minimum noise temperature T_{min}, the optimum source impedance $R_{opt} + jX_{opt}$, and the noise conductance g_n. These relate to the measured data in Fig. 10 of noise temperature T_n as a function of source impedance $R + jX$ by

$$T_n = T_{min} + T_0 \cdot g_n \cdot \frac{\left[(R - R_{opt})^2 + (X - X_{opt})^2 \right]}{R} \quad (10)$$

where $T_0 = 290$ K.[1]

The values of R, R_{opt}, X, X_{opt}, and g_n are dependent upon the choice of reference plane between device and

[1] In this paper $T_0 = 290$ K independent of ambient temperature which will be denoted as $t \cdot T_0$. Several papers in this field normalize quantities which are not functions of temperature to ambient temperature (using the symbol T_0), producing a normalized quantity which is a function of temperature and confusing the reader.

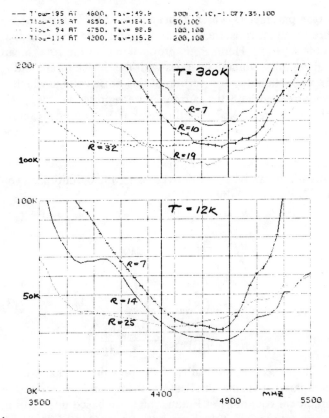

Fig. 10. Noise temperature versus frequency for various source resistances and Mitsubishi MGF 1412 GASFET at 300 K (top) and 12 K (bottom, note change in temperature scale.) Device bias was $V_{ds} = 5$, $I_d = 10$ mA at both temperatures.

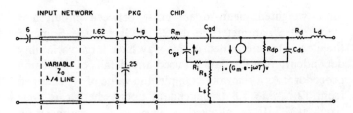

Fig. 11. Equivalent circuit of amplifier input circuit and Mitsubishi MGF 1412 transistor. The element values are based upon dc resistance measurements, 1-MHz capacitance measurements, and three sets of S parameters measured in different test fixtures. The resistance R_i cannot be separated from R_m by these measurements since the reactance of C_{gd} is so high; the value of R_i is an estimate. The error values are conservative estimates based upon analysis of all the data.

TABLE III
MEASURED AND THEORETICAL 4.9-GHz NOISE PARAMETERS OF
MITSUBISHI MGF 1412 AT 300 K AND 20 K

Symbol	Expt Pkg 300°K	Expt Chip 300°K	Pucel Theory 300°K	Fukui Formula 300°K	Expt Pkg 20°K	Expt Chip 20°K	Pucel Theory 20°K
T_{min}	91±3	91±3	72	151	25±2	25±2	16
R_{opt}	20±2	63±6	74	22	15±1	47±3	41
$X_{opt} = -X_c$	55±5	101±9	119	42	58±5	107±10	104
$1/g_n$	100±20	205±40	668	727	370±30	800±100	1800
R_n	34±8	69±18	29	3.1	10±2	17±3	6.9
r_n	3.8±.6	15±3	8.2	-38	0.6±.2	2.6±1	0.9
R_c	-4±5	-31±15	8.1	167	0.9±4	-13±10	7.9

source but T_n and T_{min} are not for a lossless coupling circuit. Values for two reference planes defined in Fig. 11 are presented in Table III. The first and fifth columns are for a reference plane at the gate–lead case interface of the packaged device; the second and sixth columns are for the GASFET chip and thus the effects of case capacitance and gate bonding wire inductance have been removed. These "chip" columns should be compared with the theoretical columns which will be discussed in Section V.

C. Impedance Measurements

The noise theory to be discussed in the next section expresses the four noise parameters in terms of material parameters, dc bias values, device dimensions, and the equivalent circuit elements R_m, R_i, R_s, g_m, and C_{gs}, shown in Fig. 11. The values of $R_m + R_i = R_g$, R_s, and g_m, all measured at dc, are given in Table II. The capacitance, $C_{gs} + C_{gd}$, was measured at 1 MHz as a function of V_{gs} with a precision capacitance bridge [35]; a value of 0.75 pF was obtained at the low-noise bias of $V_{gs} = -1.1$ V. This includes case capacitance, which was then removed by measuring a defective device with the gate bonding wire lifted off the chip; a value of 0.25 pF was obtained for the case capacitance.

The above dc resistances and 1-MHz capacitances, together with a skin effect correction to R_m and an estimate of R_i, will be used in the comparison of theoretical and measured noise parameters. Attempts were made

to determine $R_m + R_i$ and C_{gs} by microwave S-parameter measurements, to be described next, but the results were variable and more confidence is placed in the low-frequency measurements. The skin-depth of aluminum at 5 GHz is 1.2 μm which should be compared with the gate metallization thickness of 0.7 μm. The dc value of R_g is 2.7 Ω and its value at 5 GHz is estimated to be 5 Ω; this estimate is partially influenced by the S-parameter results which gave 4.3 to 12.7 Ω. The gate-charging resistance R_i, is estimated to be 1 Ω by noting that R_s and R_d have values of 2.3 and 2.1 Ω and R_i occurs over a length somewhat shorter than R_s and R_d. It should be noted that R_i is of second order importance to the noise results but the fact that R_m is determined by subtracting R_i from the measured R_g is of somewhat greater importance. The final value of R_m is 4 ± 2 Ω and this error value limits the accuracy of comparison of noise theory and experiment.

The S parameters of a packaged device were measured from 2 to 18 GHz on computer-corrected automatic network analyzers at two organizations [36], [37] with different test fixtures and calibration procedures; a third set of S parameters is reported on the MGF 1402 data sheet. The input capacitance at 2 GHz determined from each of the three S-parameter sets was within 0.06 pF of the 1-MHz capacitance measurement. However, above 8 GHz the measured S parameters diverge, probably due to reference plane definition problems. The COMPACT program [59] was used to optimize circuit elements to give minimum weighted mean-square error between circuit and measured S parameters in the 2–10-GHz range. Some of the elements determined in this way have large variability dependent upon the data set used and details of the fitting procedure. As a relevant example, the value of R_g changed from 12.7 Ω to 7.8 Ω dependent upon whether R_s was fixed at 2.3 Ω or allowed to vary in the optimization procedure (which gave $R_s = 1.0$ Ω). Using S parameters measured with a different analyzer gave $R_g = 6.7$ and 5.4 for R_s fixed and variable, respectively.

The reasons for the variability of the R_g value are its small resistance relative to the reactance of C_{gs} and also the strong dependence of the input resistance upon the small feedback elements C_{gd}, L_s, and R_s. Our conclusion is that it is very difficult to determine R_g from S-parameter measurements at normal bias values; a careful measurement of S_{11} with $V_{ds}=0$ (and hence $g_m=0$ and no feedback effects) may be more successful.

V. Noise Theory

A. Summary of Present GASFET Noise Theory

A comprehensive paper covering the history, dc, and small-signal properties, and an exhaustive but not conclusive treatment of noise in microwave GASFET's was written in 1975 by Pucel et al. [28]. The noise treatment includes the induced gate noise mechanism of Van der Ziel [38], the hot-electron effects introduced by Bachtold [39], and presents as new, the mechanism of high-field diffusion noise due to dipole layers drifting through the saturated-velocity portion of the channel. This latter mechanism along with the thermal noise of parasitic resistances R_m and R_s in the input circuit was found to be the dominant source of noise in modern, short-gate-length GASFET's at room temperature.

The above paper predicts the correct dependence of noise upon drain current and, by adjustment of a material parameter D, which describes the diffusion noise but is not accurately know from other theory or experiments, a good fit to measured 4-GHz data is presented. However, the theory does not agree with experimental data at the low end of the microwave spectrum (<3 GHz) where a linear dependence of noise temperature upon frequency is predicted but not observed. It was suggested that this noise may be due to traps at the channel–substrate interface but devices with a buffer layer such as the NEC 244 also show the excess low-frequency noise [40]. The temperature dependence of the low microwave frequency noise also does not agree with the trap assumption [41]. A recent work by Graffeuil [42] attributes the low-frequency noise to frequency dependence of hot-electron noise and also matches experimental data without invoking the high field diffusion noise which is dominant to the Pucel et al. theory.

The four noise parameters T_{min}, R_{opt}, X_{opt}, and g_n define the noise properties of any linear two-port and all other noise parameters can be derived from them. Other noise parameters which appear in GASFET noise theory are the noise resistances R_n and r_n, and correlation impedance R_c+jX_c. Here, R_n is proportional to the total mean square noise voltage in series with the device input, and r_n is proportional to the portion of this noise voltage which is uncorrelated with the current noise represented by g_n. (It should be remarked that the symbol R_n is used to represent a noise voltage source that is not in series with the input but is called the "input noise resistance" in some papers [38], [41].) The following relations between these quantities can be easily derived using the definitions given by Rothe and Dahlke [43]:

$$R_n = g_n |Z_{opt}|^2 \tag{11}$$

$$R_c = \frac{T_{min}}{T_0} \frac{1}{2g_n} - R_{opt} \tag{12}$$

$$X_c = -X_{opt} \tag{13}$$

$$r_n = g_n \left(R_{opt}^2 - R_c^2 \right) \tag{14}$$

$$r_n = \frac{T_{min}}{T_0} \left(R_{opt} - \frac{T_{min}}{T_0} \frac{1}{4g_n} \right). \tag{15}$$

The experimental results of the previous section will be compared with the Pucel et al. theory and also with the empirical equations of Fukui which are based upon fitting noise data to Bell Telephone Laboratory GASFET's at 1.8 GHz [44]. The theoretical results for the four noise parameters T_{min}, R_{opt}, X_{opt}, and g_n, which are most directly measureable are expressed in terms of intermediate parameters r_n and R_c as follows [28]:

$$g_n = K_g / g_m X_{gs}^2 \tag{16}$$

$$r_n = t(R_m + R_s) + K_r / g_m \tag{17}$$

$$R_c = R_m + R_s + K_c R_i \tag{18}$$

$$R_{opt} = \left(R_c^2 + r_n / g_n \right)^{1/2} \tag{19}$$

$$X_{opt} = K_c |X_{gs}| \tag{20}$$

$$T_{min} = 2T_0 g_n (R_c + R_{opt}) \tag{21}$$

where t is the ratio of device physical temperature to 290 K, $X_{gs} < 0$ is the reactance of the gate-to-source capacitance, and the three K's are noise coefficients which are given in [28] as functions of bias, device dimensions, and material parameters. K_g is proportional to the squared magnitude of the equivalent current source in the input circuit and represents the induced gate circuit noise current as well as the correlated[2] current needed to represent noise in the drain circuit. The uncorrelated noise resistance r_n contains a first term due to thermal noise in $R_m + R_s$ (and hence proportional to t as suggested in [28]) and a second term which, through K_r is a measure of the noise voltage necessary to represent uncorrelated drain current noise.

[2]The words correlated and uncorrelated can refer to correlation between voltage and current sources in a Rothe–Dahlke noise representation or to correlation between gate current noise and drain current noise; an asterisk (*) will be used when the latter is meant.

TABLE IV
EXPERIMENTAL AND THEORETICAL NOISE COEFFICIENTS OF
MGF 1412 AT 300 K AND 20 K

Qty	Expt 300°K	Theory 300°K	Expt 20°K	Theory 20°K
K_g	0.89	0.27	0.33	0.15
K_c	1.55	1.83	1.65	1.60
K_r	0.36	.072	0.13	.03
P	2.50	.98	1.02	0.403
R	0.64	.26	0.27	0.082
C	0.893	0.961	0.917	0.935

It is important to note that the K noise coefficients are independent of frequency except for the possible frequency dependence of material parameters (high field diffusion coefficient D, saturated velocity v_s, hot-electron noise coefficient δ, and mobility μ). The noise coefficients are convenient for describing the measured noise performance of a device but three other coefficients P, R, and C are more directly related to the noise generation mechanisms. The mean-square drain current noise is proportional to P and the mean-square gate current noise is proportional to R; the correlation* coefficient between these two is C. The relations between the coefficients are given in [28] and are repeated below in an algebraically simplified form

$$K_g = R + P - 2C\sqrt{RP} \qquad (23)$$

$$K_g K_c = P - C\sqrt{RP} \qquad (24)$$

$$K_g K_r = RP(1 - C^2). \qquad (25)$$

B. Comparison of Experimental Results with Theory

The experimental results and theory are compared in Tables III and IV. The theoretical values are obtained using the measured values of g_m and I_s from Table II and R_i, R_m, R_s, and C_{gs} from Fig. 11 (rather than values which could have been computed from device dimensions, and material parameters). The device channel thickness, $a = 0.10$ μm, and doping density $N = 2.9 \times 10^{17}$, were determined from the dc measurements. The gate length, $L = 0.7$ μm, was measured; this value is also reported in [34]. The material parameter values are those used in [28] ($D = 35$ cm^2/s, $\delta = 1.2$, $E_s = 2.9$ kV/cm, $v_s = 1.3 \times 10^7$ cm/s, and $\mu_0 = 4500$ cm^2/V·s). The device was operated with $V_{ds} = 5$ V, $I_d = 10$ mA, and $V_{gs} = -1.081$ at 300 K and -1.123 at 20 K; these values gave minimum noise at both temperatures. It is, of course, the experimental values corrected to the chip reference plane which should be compared with theoretical results, which do not include package parasitics.

The last column of Table III gives the theoretical results of Pucel et al. modified in the following way for application at 20 K.

1) The thermal noise in the parasitic resistances $R_m + R_s$, has been reduced by the factor $t = 20/290$ in (17). This reduces T_{min} from 72 K to 40 K.

2) The transconductance g_m has been replaced by the 20 K measured value (an increase from 0.043 to 0.062 mhos)

and the saturation velocity v_s, and saturation field E_s, have been increased by a factor of 1.32 as deduced from the measured dc parameter changes. This further reduces T_{min} from 40 K to 26.5 K.

3) Thermal noise within the channel has been reduced by multiplying the Pucel factors P_0, R_0, and S_0 used to calculate the K noise coefficients by t; this brings T_{min} to 15.7 K.

4) As points of additional theoretical interest, if t is made equal to zero both for the external and internal thermal noise, T_{min} reduces from 15.7 K to 11.2 K; i.e., the theoretical coefficient of noise temperature versus physical temperature is 0.22 K per K at cryogenic temperatures. Also, if $D = 0$ (no high-field diffusion noise) $T_{min} = 5.0$ K. Thus at 20 K and at the experimental bias value which gives lowest noise temperature, approximately 1/3 of the total noise is contributed by each of the mechanisms of thermal, high-field diffusion, and hot-electron noise. This is conceptually only a rough approximation since the noise contributions to T_{min} are not additive; at each step above (i.e., $t = 0$ or $D = 0$) the source impedance has a new optimum which changes the contributions from remaining noise mechanisms. Perhaps a more meaningful comparison would be for T_n at a fixed source impedance.

Several authors [41], [42], [53] find the hot-electron noise coefficient δ to be a strong function of temperature but this is primarily because the nonthermal portion of the hot electron noise has been normalized to ambient temperature (see footnote 1). Our results suggest that a convenient form for the electron temperature T_e is

$$T_e = T_0[t + f(E)] \qquad (26)$$

where $T_0 \equiv 290$ K (and cancels out in the noise figure expression) and $f(E)$ is the nonthermal noise dependent upon electric field but only weakly dependent upon temperature. For the last column of Table III, $f(E) = \delta(E/E_s)^3$ has been used with $\delta = 1.2$ and $E_s = 3800$ V/cm (compared to $E_s = 2900$ V/cm at 300 K). A better fit to experimental data would be achieved if either δ were higher or E_s lower at 20 K. Experimental evidence for the increase of $f(E)$ by a factor of ~2.5 at cryogenic temperatures is contained in our own unpublished measurements on millimeter-wave GaAs mixer diodes and also in the measurements of Keen [55].

Further insight into the noise temperature limitation at cryogenic temperatures can be gained by examining an approximate form of the minimum noise temperature equation

$$T_{min} = 2T_0 \cdot \sqrt{K_g} \cdot \omega C_{gs} \cdot \sqrt{\frac{t(R_m + R_s)}{g_m} + \frac{K_r}{g_m^2}} \qquad (27)$$

which is valid for $K_g R_c^2 / X_{gs}^2 \ll t(R_m + R_s)g_m + K_r$; this approximation produces ~10-percent error for the MGF 1412 data. At room temperature the first term under the large radical is dominant and K_r is not important; the noise voltage generator in the input circuit is dominated by the thermal noise of $R_m + R_s$. At cryogenic temperatures where $t \to 0$ the second term K_r / g_m^2 becomes dominant (even though g_m has increased) and the noise temperature is limited by the amount of nonthermal noise coupled into the gate circuit and uncorrelated* with the drain current noise. The coefficient K_r is proportional to $1 - C^2$ where C, the correlation* coefficient, is near 1 and thus small changes in C are likely to have large effects on the cryogenic noise temperature; this will not be true at room temperature. The larger variability of the cryogenic noise temperature from one device to another may be due to this effect.

The agreement between experimental results and the theory of Pucel et al. for values of T_{min}, R_{opt}, and X_{opt} is good at both 300 K and 20 K—especially if the error range for R_m is considered. Other samples of the MGF 1412 gave a lower noise temperature at 20 K (as low as 15 K) than the particular device which was evaluated in detail; this would improve the agreement with theory. The agreement with the theory is marred by a factor ~ 3 disagreement in the value of g_n. It is not known at present whether this is an experimental artifact or a failure of the theory. The value of g_n was determined from the noise temperature versus source resistance characteristic but it was also checked, with good agreement, by measuring the noise-temperature bandwidth of the data of Fig. 10. The discrepancy in g_n leads to an even larger discrepancy in the correlation resistance R_c, which depends, through (12), on the difference between $1/g_n$ and R_{opt}. The negative value of R_c which results from the measurements is physically possible (it only means the input voltage and current noise sources have a negative real part in their correlation coefficient) but is certainly in disagreement with the theory.

The experimental values of the K coefficients in Table IV were obtained by solving (16)–(21) in terms of the measured noise parameters. The coefficients P, R, and C were then obtained by solving (23)–(25) by an iteration method. The discrepancy between experimental and theoretical R_c and g_n further propagate into discrepancies in the noise coefficients. In addition, the experimental value of K_r at 300 K is subject to large error because the measured data is insensitive to K_r due to the dominance of thermal noise.

The agreement between the experimental results and the empirical equations of Fukui [44] is poor. This may be due to the fact that Fukui's equations are derived from measurements at 1.8 GHz where low-frequency noise generation mechanism has a large effect; thus the formula predicts a higher than observed noise temperature at 4.9 GHz. It may also be true that some other variables in the transistor fabrication (such as the doping profile near the substrate interface or gate metallization thickness) effect the equations. It should be noted that the Fukui noise parameter equations require the values of g_m and C_{gs} at zero gate bias; thus the measured zero bias values of 0.098 mhos and 0.75 pF have been used in Table III.

VI. Examples of Cryogenic GASFET Amplifiers

Several models of GASFET amplifiers for the 5-GHz frequency range and for use at 20 K have been designed. All use 1.75 mm^2 packaged GASFET's (usually the Mitsubishi MGF 1412), gold-plated copper (for thermal conductivity and solderability) metal parts (except for some brass contact tabs), and microstrip transmission lines with teflon-coated fiber-glass dielectric [45]. In order to avoid large thermal stresses and mechanical failures, tight and rigid connections are avoided. Metallized ceramic substrates are also avoided because of possible cracking of the ceramic or metal-film solder-joints after repeated temperature cycling.

All amplifiers utilize an external dc power regulator which automatically adjusts the gate voltage to maintain a set drain current. This requires one TL075BCM quad-operational amplifier chip per GASFET stage and provides buffered monitoring of V_{ds}, I_d, and V_{gs}. As shown in Fig. 7 bias protection circuitry is included in the microwave chassis. The 1N821 voltage-reference zener diode utilized for gate protection contains a diode which prevents forward conduction of the zener diode and thus allows the GASFET gate to be forward biased for testing; the zener diode limits negative gate bias to approximately -6 V. Both the 1N821 and 1N4099 diodes have sharper zener characteristics at 20 K than at 300 K.

A. Single-Stage Basic Amplifier

A single-stage amplifier, for use as a second stage following a cooled-paramp first stage, was required for the front ends of the very large array radio telescope [46]. The unit need not be optimized for minimum noise as the first stage gain is 15 dB. However, it must have a gain which is flat within ± 0.5 dB over the 4.5–5-GHz range, an output return loss $\geqslant 10$ dB, and must be highly reliable since 54 units are required in the 27-element, dual-polarization array. An input match is not required since the input will be connected through a short cable to the 5-port circulator of the paramp.

A photograph of the amplifier and some of the key parts is shown in Fig. 12. Utilizing a thermostatically-controlled hot plate, joints are soldered as follows: 1) GASFET source leads to mounting stud with pure indium solder [47] (for good thermal conductivity) and a flux [48]; 2) chip capacitors to chassis with silver-alloy flux-core solder [49]; and 3) zener-diodes and connector ground with low-temperature solder [50]. Other components are then soldered to the chip-capacitors with silver-alloy solder and a small soldering iron.

The input and output $\lambda/4$ transformers are 4.1 mm wide $\times$ 11.2 mm long $\times$ 0.75 mm thick microwave circuit board [45] and are held in place through slotted holes with 2-56 nylon screws. The slotted holes allow the transformer position relative to the GASFET to be adjusted to tune the center frequency of operation. Brass tabs under each

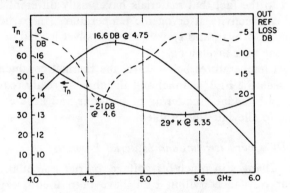

Fig. 12. Single-stage basic amplifier; scale on photograph is in units of mm. See Fig. 7 for schematic, Fig. 14 for close-up of transistor mounting stud, and text for construction details.

Fig. 13. Noise temperature, gain, and output return loss for single-stage basic amplifier at a temperature of 20 K.

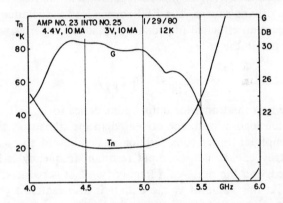

Fig. 14. Transistor mounting studs for feedback (left) and basic (right) amplifiers. Scale is in millimeters.

Fig. 15. Gain and noise temperature of cascade of isolator, feedback amplifier, isolator, and basic amplifier, all at 12 K. Noise temperature is 20 K over a 500-MHz bandwidth and <25 K over a 800-MHz frequency range. Output return loss is >10 dB from 4.5 to 5.1 GHz.

nylon screw make connections to the SMA input and output connectors and also to the GASFET gate and source leads. Two layers of 0.02-mm thick polyester tape [51] are placed between each connector tab and the transmission line to form a dc blocking capacitor. Bias voltage for both gate and drain is fed through coils formed of 3 turns of 0.2-mm diameter phosphor–bronze wire wound with 1.25-mm inner diameter; these have high impedance relative to the circuit and are not critical.

The completed amplifier is tuned by sliding the transformers and slightly bending gate and drain leads. The drain is tuned for maximum output return loss (>20 dB) at band center and the gate is tuned for flat-gain. This will result in a minimum noise frequency $f_{\min}$, 600 MHz above the gain center frequency f_0, as shown in Fig. 13. This offset is predicted by the noise theory. For a source impedance consisting of an inductor in series with the source resistance, $f_{\min}/f_0 = \sqrt{K_c}$, assuming the drain circuit is broad band compared to the gate circuit. (Neither of these assumptions is quite true for this amplifier.) This offset may be removed with source-inductance feedback as is described next.

B. Single-Stage Feedback Amplifier

A small modification of the previously described amplifier results in a unit with little frequency offset between peak gain and minimum noise temperature. The modification, described in Fig. 14, is to increase the source lead inductance by widening the GASFET mounting stud and also bending small loops in the source leads. This change has little direct effect upon noise temperature but the input circuit can now be tuned to a lower resonant frequency (4.25 GHz with $V_d = 0$, 4.70 GHz with $V_d = 4.4$) giving both optimum noise and maximum gain at ~4.75 GHz. The gain and noise temperature of this amplifier, cascaded with the amplifier of the previous section, is shown in Fig. 15; cooled isolators are included at the input and between the two amplifiers. Output return loss is >10 dB from 4.5 to 5 GHz without an output isolator.

The use of source lead inductance to improve input match for a ~1.5-GHz amplifier has been previously described [5], [52]. At 5 GHz the situation is more complex because of effects of gate to drain capacitance and

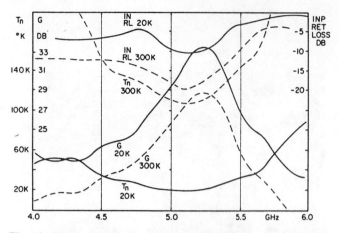

Fig. 16. Noise temperature, gain, and input return loss for two-stage feedback narrow-band amplifier described in Section IV-C at 20 K and 300 K. Output return loss at 5.0 GHz was 17 dB at 20 K and 25 dB at 300 K.

increased sensitivity of noise temperature to the reactance of the driving source. The FET source lead inductance L_s produces an effective impedance z_{FB} added in series with the input to the gate

$$z_{FB} = \frac{g_m L_S}{C_{gs}} \cdot \frac{1}{1 + y_d/y_1} \qquad (28)$$

where y_d is the transistor output admittance for $g_m = 0$ and y_1 is the load admittance presented to the transistor. For the amplifier under consideration y_1 is adjusted so z_{FB} is primarily capacitive, the input resonant frequency is increased, and the noise-gain frequency offset is reduced to zero. At this value of y_1 the output is also matched (but $y_1 \neq y_d^*$ since y_d is for $g_m = 0$) but the input is not matched. In the next section a two-stage amplifier, matched by feedback, will be described.

C. Two-Stage Feedback Narrow-Band Amplifier

An amplifier was constructed by combining in one case the single-stage feedback amplifier and the single-stage basic amplifier previously described. (With the exception that input and output $\lambda/4$ transformer width was 6 mm; interstage $\lambda/2$ line was 4.1 mm wide.) The amplifier was then tuned at the desired frequency of operation, 5.0 GHz, for minimum noise by adjusting input inductance and for maximum input return loss by adjustment of first-stage source lead inductance and drain inductance. In this case, the feedback is used to achieve input match, z_{FB} is primarily resistive, and little attention was paid to the gain versus frequency response as the application was narrow band. The resulting gain, noise temperature, and input return loss at 300 K and 20 K are shown in Fig. 16. The results are flawed by the decrease in input return loss at 20 K; this is probably due to the change in g_m effecting z_{FB}. Another version of this amplifier was used with an input cooled isolator to achieve input match and a noise temperature of 17 K at 5.0 GHz.

It is particularly desirable for cryogenic GASFET amplifiers to achieve match by feedback (or perhaps by a balanced configuration) since ferrite devices usually do not function well at both room and cryogenic temperatures. Our goal has been to construct an amplifier which is wide band, matched, performs well at both room and cryogenic temperatures, and hence does not utilize ferrite isolators. All of the above characteristics have not yet been achieved in a single unit; work will continue in this direction.

VII. Conclusions

All of the questions posed in the Introduction have not been answered and in some cases the answers are "hints" based on insufficient data; more work, both experimental and theoretical, is needed. Some conclusions that may be ventured are as follows.

A. Thermal Considerations

1) At temperatures above 15 K there is no fundamental problem in the cooling of a FET but attention must be paid to the fact that materials have vastly different thermal conductivity at cryogenic temperatures. In particular alumina, Kovar, and epoxy (whether silver loaded or not) become near insulators.

At temperatures below 15 K the thermal conductivity between the FET channel and the chip is greatly reduced due to boundary scattering of phonons and little can be done to alleviate this problem.

B. DC Characteristics and Material Properties

1) Changes in pinch-off voltage, barrier potential, and linear resistances within FET have been found, experimentally, to be small (see Fig. 3 and Table II) in the range 300 K–20 K. This implies that the changes in carrier density and mobility are small.

2) As the temperature was reduced from 300 K to 20 K, saturation current increased by 30–50 percent for the FET's of three manufacturers (see Fig. 3). This implies that the saturation velocity increases as the device is cooled; the differences between devices are due to either different doping density or some other constituent of the GaAs.

3) For the same 3 devices the output resistance decreased from 30–200 percent upon cooling; this has not been explained.

C. Small-Signal Characteristics

Transconductance increased by the same factor as saturation current and this causes an increase in RF gain that is somewhat smaller due to negative feedback and a concurrent decrease in output resistance. There is little other change in the gain versus frequency characteristic (see Fig. 5).

D. Noise

1) The noise temperature improvement factor for cooling from 300 K to 20 K varies from 3 to 5 for the six types of transistors tested at 5 GHz (see Fig. 9). It is not known

why the improvement factor varies; more detailed investigation is needed of devices with small improvement factors. The variation may be due to the stronger dependence of noise temperature upon the gate–drain noise correlation coefficient C at cryogenic temperatures.

2) The results for one device, the Mitsubishi MGF 1412, correlate fairly well with the noise theory of Pucel et al. [28] both at 300 K and 20 K. The noise improvement is due to the reduction in thermal noise of the channel and parasitic resistances and the increase in transconductance. The nonthermal noise power (hot-electron and high-field diffusion noise) may remain constant or increase at cryogenic temperature; the experimental data is of insufficient accuracy and content to make this judgement. A study of noise variation with bias has recently been performed [58] and forms a further test of the theory.

E. Amplifiers

1) An amplifier has been constructed with performance close to that of the best cooled paramps at 5 GHz (see Fig. 15).

2) It appears feasible, at least for narrow-bandwidths (~2 percent) at 5 GHz, to design unbalanced amplifiers which are matched by feedback rather than ferrite devices, and operate well at all temperatures from 300 K to 20 K (see Fig. 16).

3) The reliability of amplifiers constructed with the techniques described in Section VI has been excellent. At the time of this writing, 15 of the basic amplifiers have been constructed and subjected to a total of over forty 300 K to 20 K temperature cycles with no failures.

ACKNOWLEDGMENT

The author wishes to thank C. R. Pace for assistance with the construction and measurements, J. Granlund for the solution of the nonlinear differential equation for the forward-biased gate and for finding the transistor equivalent circuit elements from S-parameter measurements, and T. Brookes for several helpful discussions and the program for calculating results of the Pucel et al. noise theory. A. R. Kerr suggested the cooled-attenuator noise measurement procedure and the modification of the attenuator contact. The author appreciates the S-parameter measurements performed by R. Hamilton of Avantek, Inc. and R. Lane of California Eastern Laboratories.

REFERENCES

[1] *Diode and Transistor Designer's Catalog*, Hewlett Packard Co., Palo Alto, CA, 1980.
[2] *NEC Microwave Transistor Designer's Guide*, California Eastern Laboratories, Santa Clara, CA.
[3] Data Sheets, LNR Communications, Hauppauge, NY.
[4] D. M. Burns, "The 600 MHz noise performance of GaAs Mesfet's at room temperature and below," M. S. Thesis, Dep. Elec. Eng., Univ. of California, Berkeley, CA, Dec. 1978.
[5] D. Williams, S. Weinreb, and W. Lum, "L-band cryogenic GaAs FET amplifier," *Microwave J.*, vol. 23, no. 10, October 1980.
[6] C. A. Liechti and R. A. Larrick, "Performance of GaAs MESFET's at low temperatures," *IEEE Trans. Microwave Theory Tech.*, vol. MTT-24, pp. 376–381, 1976.
[7] J. Pierro and K. Louie, "Low temperature performance of GaAs MESFETs at L-band," 1979 *Int. Microwave Symp. Dig.* (Orlando, FL), IEEE Cat. No. 79CH1439-9, pp. 28–30.
[8] R. E. Miller, T. G. Phillips, D. E. Iglesias, and R. H. Knerr, "Noise performance of microwave GaAs F.E.T. amplifiers at low temperatures," *Electron. Lett.*, vol. 13, no. 1, pp. 10–11, Jan. 6, 1977.
[9] J. D. Kraus, *Radio Astronomy*. New York: McGraw Hill, 1966, ch. 7, p. 237.
[10] Model 21 Cryodyne, CTI-Cryogenics Inc., Waltham, MA, 02154.
[11] H. F. Cooke, "Fets and bipolars differ when the going gets hot," *Microwaves*, pp. 55–60, Feb. 1978. Also printed as *High-Frequency Transistor Primer, Part III*, Thermal Properties, Avantak Corp., Santa Clara, CA.
[12] M. G. Holland, "Phonon scattering in semiconductors from thermal conductivity studies," *Phys. Rev.*, vol. 134, pp. A471–480, Apr. 20, 1964.
[13] *American Institute of Physics Handbook*, 3rd ed. New York: McGraw Hill, 1972.
[14] *Thermal Conductivity of Solids at Room Temperature and Below*, Monograph 131, National Bureau of Stds, Boulder, CO, Sept. 1973.
[15] J. Callaway, "Model for lattice conductivity at low temperatures," *Phys. Rev.*, vol. 113, pp. 1046–1051, Feb. 15, 1959.
[16] C. L. Reynolds and A. C. Anderson, "Thermal conductivity of an electrically conducting epoxy below 3K," *Rev. Sci. Instru.*, vol. 48, no. 12, p. 1715, Dec. 1977.
[17] C. Kittel, *Introduction to Solid State Physics*, 5th ed. New York: Wiley, 1976.
[18] H. Fukui, "Determination of the basic device parameters of a GaAs Mesfet," *BSTJ*, vol. 58, no. 3, pp. 771–797, March 1979.
[19] R. Hackam and P. Harrop, "Electrical properties of nickel-low doped n-type gallium arsenide Schottky barrier diodes," *IEEE Trans. Electron Devices*, vol. ED-19, no. 12, pp. 1231–1238, Dec. 1972.
[20] D. Vizard, "Cryogenic DC characteristics of millimeter-wavelength Schottky barrier diodes," Appelton Laboratory, Slough, U.K., unpublished.
[21] F. A. Padovani and R. Stratton, "Field and thermionic-field emission in Schottky barriers," *Solid-State Electron.*, vol. 9, pp. 695–707, 1966.
[22] T. Viola and R. Mattauch, "High-frequency noise in Schottky barrier diodes," Res. Labs Eng. Sci. Univ. of Virginia, Charlottesville, VA, Rep. EE-4734-101-73J, Mar. 1973.
[23] M. Giterman, L. Krol, V. Medvedev, M. Orlova, and G. Pado, "Impurity-band conduction in n-GaAs," *Soviet Phys. Solid State*, vol. 4, no. 5, pp. 1017–1018, Nov. 1962.
[24] O. Emel'yanenko, T. Tagunova, and D. Naselov, "Impurity zones in P- and N-type gallium arsenide crystals," *Soviet Phys. Solid State*, vol. 3, no. 1, pp. 144–147, July 1961.
[25] P. Wolf, "Microwave properties of Schottky-barrier field-effect transistors," *IBM J. Res. Develop.*, vol. 14, pp. 125–141, Mar. 1970.
[26] A. Nara, Mitsubishi Semiconductor Laboratory, Hyogo, Japan, private communication.
[27] J. Ruch and G. Kino, "Transport properties of GaAs," *Phy. Rev.*, vol. 174, no. 3, pp. 921–927, Oct. 15, 1968.
[28] R. Pucel, H. Haus, and H. Statz, "Signal and noise properties of gallium arsenide microwave field-effect transistors," *Adv. in Electronics and Electron Physics*, vol. 38, L. Morton, Ed. New York: Academic, 1975.
[29] Model 350 Cryodyne, CTI-Cryogenics, Inc., Waltham, MA.
[30] A. R. Kerr, Goddard Inst. for Space Studies, New York, NY, private communication.
[31] Type 2156 Bellows, Servometer Corp., Cedar Grove, NJ.
[32] Type DT-500-CV-DRC, Lake Shore Cryotronics, Westerville, OH.
[33] M. Schneider, "Microstrip lines for microwave integrated circuits," *BSTJ* vol. 48, pp. 1421–1444, May 1969.
[34] T. Suzuki, A. Nara, M Nakatoni, and T. Ishii, "Highly reliable GaAs MESFET's with a static mean NF_{min} of 0.89 dB and a standard deviation of .07dB at 4 GHz," *IEEE Trans. Microwave Theory Tech.*, vol MTT-27 no. 12, pp. 1070–1074, Dec. 1979.
[35] Model 75D, Boonton Electronics, Parsippany, NJ.
[36] R. Lane, California Eastern Laboratories, Santa Clara, CA, private communication.
[37] R. Hamilton, Avantek Corp., Santa Clara, CA, private communication.

IEEE TRANSACTIONS ON MICROWAVE THEORY AND TECHNIQUES, VOL. MTT-28, NO. 10, OCTOBER 1980

[38] A. van der Ziel, "Thermal noise in field-effect transistors," *Proc. IRE*, vol. 50, pp. 1808–1812, 1962.

[39] W. Baechtold, "Noise behavior of GaAs field-effect transistors with short gate lengths," *IEEE Trans. Electron Devices*, vol. ED-19, pp. 674–680, May 1972.

[40] NE244 Data Sheet, California Eastern Labs, Santa Clara, CA.

[41] K. Takagi and A. van der Ziel, "High frequency excess noise and flicker noise in GaAs FET's," *Solid-State Electron.*, vol. 22, pp. 285–287, 1979.

[42] J. Graffeuil, "Static, dynamic, and noise properties of GaAs Mesfets," Ph.D. thesis, Univ. Paul Sabatier, Toulouse, France.

[43] H. Rothe and W. Dahlke, "Theory of noisy fourpoles," *Proc. IRE*, vol. 44, no. 6, pp. 811–818, June 1956.

[44] H. Fukui, "Design of microwave GaAs MESFET's for broad-band low-noise amplifiers," *IEEE Trans. Microwave Theory Tech.*, vol. MTT-27, no. 7, pp. 643–650, July 1979.

[45] Type D-5880 RT/Duroid, .031″ dielectric, 1 oz. 2 side copper, Rogers Corp., Chandler, AZ.

[46] S. Weinreb, M. Balister, S. Maas, and P. J. Napier, "Multiband low-noise receivers for a very large array," *IEEE Trans. Microwave Theory Tech.*, vol. MTT-25, no. 4, pp. 243–248, Apr. 1977.

[47] Indalloy No. 4 Solder, 100% Indium, 157°C, Indium Corp. of America, Utica, NY.

[48] #30 Supersafe Flux (water soluable), Superior Flux and Mfg. Co., Cleveland, OH.

[49] SN62 Solder, 62% Tin, 36% Lead, 2% Silver, 179°C, Multicore Solders, Westbury, NY.

[50] 20E2 Solder, 100°C, Alpha Metals, Jersey City, NJ.

[51] #74 Polyester Electrical Tape, 3M Co., Minneapolis, MN.

[52] L. Nevin and R. Wong, "*L*-band GaAs FET amplifier," *Microwave J.* vol. 22, no. 4, p. 82, Apr. 1979.

[53] J. Frey, "Effects of intervalley scattering on noise in GaAs and InP field effect transistors," *IEEE Trans. Electron Devices*, vol. ED-23, no. 12, pp. 1298–1303, Dec. 1976.

[54] J. Granlund, "Resistance associated with FET gate metallization," *IEEE Trans. Electron Devices*, to be published.

[55] N. J. Keen, "The role of the undepleted epitaxial layer in low noise Schottky barrier diodes for millimeter wave mixers," Max-Planck-Institute for Radio Astronomy, Bonn, W. Germany, to be published.

[56] S. Sesnic and G. Craig, "Thermal effects in JFET and MOSFET devices at cryogenic temperatures," *IEEE Trans. Electron Devices*; vol. ED-19, no. 8, pp. 933–942, Aug. 1972.

[57] D. Brunet-Brunol, "Etude et réalisation d'amplificateur a transistor a effet de champ a l'GaAs refroidi a trés basse température," *Rev. Phys. Appl.*, vol. 13, no. 4, pp. 180–187, Apr. 1978.

[58] S. Weinreb and T. M. Brookes, "Characteristics of low-noise GaAs MESFET's from 300 K to 20 K," in *Proc. European Microwave Conf.*, 1980.

[59] *COMPACT* Network Analysis Program, Compact Engineering Inc., Palo Alto, CA.

Ultra-Low-Noise 1.2- to 1.7-GHz Cooled GaAsFET Amplifiers

SANDER WEINREB, FELLOW, IEEE, DAN L. FENSTERMACHER, MEMBER, IEEE, AND RONALD W. HARRIS

Abstract — A 3-stage GaAsFET amplifier operating at 13 K and utilizing source inductance feedback is described. The amplifier has a noise temperature of < 10 K and input return loss > 15 dB over the 1.2- to 1.7-GHz frequency range.

I. INTRODUCTION

IN THE microwave frequency range between 1 and 3 GHz, the antenna noise temperature due to cosmic and atmospheric sources is quite low, ~ 5 K, and it is a challenge to receiver engineers to design low-noise amplifiers and low ground-radiation pickup antennas to take advantage of this low source temperature. The noise figure, in decibels, of a receiver is equal to the signal-to-noise (S/N) degradation caused by the receiver when the source temperature is 290 K. When the source temperature including antenna noise is 10 K, a reduction of receiver noise figure from 1.0 dB (or a noise temperature of 75 K) to 0.1 dB (6.8 K) causes a S/N improvement of $(75+10)/(6.8+10) = 5.1$ or 7.0 dB.

The amplifier described in this article has a noise figure of approximately 0.1 dB at 1.3 GHz. This is achieved with wide bandwidth, impedance match, excellent stability, and relatively small cost and complexity compared with masers which have been required in the past for this noise performance. Cryogenic cooling to a temperature of ~ 15 K is required but this can now be obtained with closed cycle helium refrigerators [1] at a cost of under $5000 and a weight less than 42 kg. Some specific applications for the amplifier are: 1) as a radio astronomy front end for observations of the 1.42-GHz hydrogen line and OH lines at 1.61, 1.66, and 1.72 GHz; 2) as an IF amplifier for millimeter wave or infrared receivers utilizing cooled mixers; 3) as a front end for detection of extraterrestrial civilizations communicating in the "optimum" frequency range of 1.42 to 1.67 GHz [2].

In a previous article [3] a cryogenically cooled L-band amplifier utilizing source inductance feedback to achieve input match was described. This present paper reports on the further development of this amplifier utilizing computer-aided design techniques to increase bandwidth to 500

Manuscript received October 27, 1981; revised January 15, 1982.

S. Weinreb and R. Harris are with the National Radio Astronomy Observatory, Charlottesville, VA 22903.

D. L. Fenstermacher was with the National Radio Astronomy Observatory, Charlottesville, VA 22903. He is now with the Electrical Engineering Department, Cornell University, Ithaca, NY 14853.

MHz and provide stability for any source and load impedance. Photographs and a schematic of the amplifier are shown in Figs. 1–3.

II. DESIGN

A first step in the amplifier design concerned the question of how to handle the mismatch which results when a FET device is driven by its optimum-noise generator impedance Z_{opt}. For a FET in the lower microwave range, this difference is quite large; typical values are $Z_{opt} = 42 + j175$ and the matched source impedance, $Z_{in}^* = 10 + j145$, giving an input voltage reflection coefficient magnitude of 0.73. It is, of course, possible to operate the amplifier with a large input mismatch, but small variations of the antenna feed impedance will then cause large ripples in the gain versus frequency response. Possible remedies to this problem are 1) an isolator, 2) a balanced amplifier, or 3) use of feedback. Feedback was chosen because it results in the most compact, fewest component amplifier, and because of previous experience with this technique [3], [4]. It should be mentioned that a cooled isolator in this frequency range has recently been developed [5].

The feedback to produce simultaneous noise and power match is realized in the form of the source inductor $L10$ of the first stage. The effect of this feedback upon noise performance is to reduce the optimum-noise generator reactance X_{opt} by the reactance of $L10$ [6]; this effect is fairly minor and is compensated by reducing the value of the gate circuit inductor $L1$. The noise figure of the amplifier is, to a large degree, unaffected by the feedback because the noise measure of the first stage is invariant to lossless feedback [7], the stage gain is fairly high, ~ 10 dB, and the second stage has similar noise figure as the first stage.

On the other hand, the effect of the source inductor upon input match can be represented by an impedance Z_{FB} added in series with the gate. This impedance can be calculated by a straightforward circuit analysis, is somewhat dependent upon first-stage load impedance, and, to a first approximation, is resistive and given by $g_m \cdot L10/C$ where g_m and C are the transistor transconductance and gate capacitance.

Optimization of the amplifier was performed using the FARANT program developed at NRAO [8]. The optimization tradeoff's are illustrated by the impedance plot of Fig.

Reprinted from *IEEE Trans. Microwave Theory Tech.*, vol. MTT-30, no. 6, pp. 849–853, June 1982.

Fig. 1. Photograph of 3-stage amplifier with cover removed. Overall size is 9.15 cm×4.07 cm×1.27 cm. The FET's are soldered by their source leads to FET holders which are described in Fig. 3.

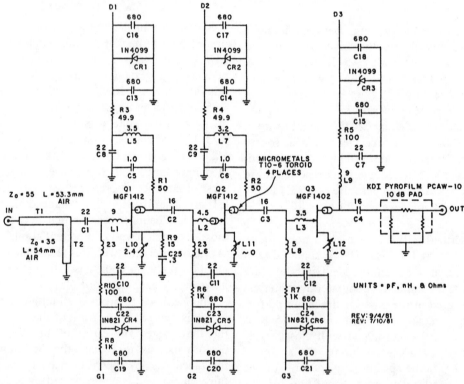

Fig. 2. Amplifier schematic. Bias voltages are supplied from a separate regulator which adjusts gate voltage for constant drain current.

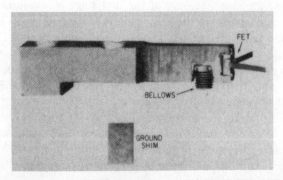

Fig. 3. Side view of FET holder and grounding shim. The shim allows adjustment of the path length from the FET source to ground, thus providing a variable source inductance. The bellows [19] contacting a 15-Ω chip resistor in series with a 0.3-pF chip capacitor soldered to the chassis is used only on the first stage; second and third stages require no additional source inductance and the grounding shim is positioned directly under the FET.

4 and consideration of the noise temperature dependence upon generator impedance $R_s + jX_s$, for any linear two-port

$$T_n = T_{\min} + 290 \times \frac{g_n}{R_s}\left[(R_s - R_{opt})^2 + (X_s - X_{opt})^2\right]$$

where $R_{opt} + jX_{opt}$ is the optimum-noise generator impedance and g_n is the amplifier noise conductance as given in Table I. The first-stage source inductance $L10$, and load reactances $C5$, $L5$, and $L2$, have been adjusted to make the FET input resistance R_{in}, approximately equal to R_{opt}. The series and shunt $\lambda/4$ line lengths and characteristic impedances and the input inductor $L1$, are then adjusted to make R_s as close as possible to R_{in} and R_{opt} and to bring the source reactance X_s to a value between $-X_{in}$ and X_{opt}. If the shunt stub is not present, X_s is a linear increasing

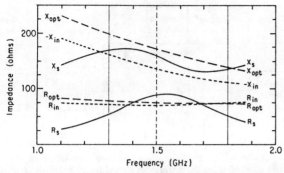

Fig. 4. Source impedance optimization example. The minimum noise impedance is $R_{opt} + jX_{opt}$, the FET input impedance is $R_{in} + jX_{in}$, and the generator impedance is $R_s + jX_s$. The negative slope in X_s is due to the shunt $\lambda/4$ transmission line. All quantities are computed from the circuit model and the S and noise parameters of the FET.

TABLE I
NOISE PARAMETERS OF MITSUBISHI MGF-1412 AT 1.6 GHz,
TEMPERATURES OF 300 K AND 15 K, AND DRAIN BIAS OF
5 V, 10 mA

AMBIENT TEMP °K	T_{min}	$1/g_n$ ohms	R_{opt}	X_{opt}	COMMENT
300	63 ± 3	250 ± 100	42 ± 4	-	Measured
300	20.3	1785	60.7	221	Theory [10], [11]
15	8 ± 2	2200 ± 1500	29 ± 2	-	Measured
15	4.2	5882	39.3	196	Theory [10], [11]

function of frequency and R_s has little frequency variation. This results in narrower bandwidth in both noise temperature and match due to the difference between X_s and X_{opt} or $-X_{in}$.

At the time the input impedance optimization was performed, R_{opt} was thought to be 75 Ω; later measurements, described in Table I, showed $R_{opt} = 42$ Ω. However, a higher value of generator resistance gives less noise temperature variation with frequency and a center-frequency value of 60.5 Ω is used in the amplifier.

A second major decision was whether to use lumped or transmission-line matching elements. At 1.5 GHz this is a close decision as it is difficult to determine the parasitic reactances of lumped elements with sufficient accuracy, yet transmission lines are somewhat large. A compromise was made; the input network was constructed on high-dielectric constant (10.5) transmission-line circuit board [9] while the remainder of the amplifier was constructed with lumped elements. The input circuit board could be easily changed to vary the source resistance in a known manner.

A somewhat unconventional construction of the amplifier was based upon the following considerations.

1) The lowest-noise GaAsFET's for use at 1.5 GHz have considerable gain and a tendency to oscillate at 10 to 20 GHz, especially when source inductance feedback is utilized. The 1.5-GHz inductors have widely varying reactance in the 10- to 20-GHz frequency range and series RC bypass networks are necessary to stabilize the amplifier. For these bypass networks to be effective, their total length must be less than $\lambda/4$ at 20 GHz (~ 0.4 cm). Three of

these bypass networks, $R9$-$C25$, $R1$-$C5$, and $R2$-$C6$ are constructed of very small chip resistors and capacitors and are located very close to the FET's; ideally they would be built into the FET package.

2) The first-stage source inductance and $\lambda/4$ shunt stub length are critical and should be easily adjustable. This is accomplished by small sliding shims as shown in Figs. 1 and 3.

3) The thermal resistance from the FET source leads to the amplifier case must be low at cryogenic temperatures to avoid self-heating. For this reason the FET source leads are soldered to a copper FET holder (see Fig. 3) with pure indium solder. The thermal resistance within the FET package is discussed in [10].

4) A "springy" mechanical design is needed to prevent fractures due to thermal stresses caused by the wide temperature range and materials with different thermal expansion coefficients. For this reason a dielectric-loaded Teflon substrate material [9] was used even though the temperature coefficient of its dielectric constant is much greater than that of alumina. (A 12-percent increase in dielectric constant was measured for cooling from 300 K to 15 K).

The Mitsubishi MGF-1412 transistors used in the design were selected on the basis of previous evaluations at 5 GHz [10] and also upon the results reported in [3]. Noise parameters of a MGF-1412 at approximately 1.6 GHz were measured at 300 K and 15 K by performing noise measurements with three different input circuit boards; these had no shunt $\lambda/4$ stub and had Z_0 values of 41, 51, and 69 Ω for the series $\lambda/4$ line. The results for a drain bias of 5 V and 10.8 mA are shown in Table I where they are compared with the theoretical values of Pucel *et al.* [11]. The agreement with theory is poor, in contrast to results reported at 5 GHz in [10], due to low frequency noise mechanisms not accounted for in the theory.

A MGF-1402 is used in the final stage because of its lower cost. The amplifier output is isolated from the load by an internal 10-dB chip attenuator [12]. Unlike an isolator, the attenuator terminates the amplifier over the entire microwave frequency range and thus insures that the amplifier will not oscillate for some particular out-of-band termination impedance.

III. CONSTRUCTION AND TUNING

The amplifier chassis is milled from copper which is then gold-plated for corrosion protection and to reduce absorption of thermal radiation. A rectangular bar, 1 mm×7 mm×27 mm, of iron-epoxy absorber material [13] is glued into the chassis as shown in Fig. 1. This material has been tested to retain its loss at cryogenic temperatures. Beryllium–copper finger stock material [14] is soldered to the top cover and contacts the FET holders near their mounting screws. Two of the coils, $L2$ and $L6$, are wound in reverse direction from the others to change the mutual coupling phase. All of these steps, as well as ferrite beads [15] on some of the leads, are for the purpose of suppressing high-frequency oscillations, particularly when the amplifier is cooled. Additional details concerning construc-

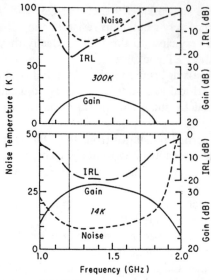

Fig. 5. Input return loss (IRL), gain, and noise temperature of amplifier #74 at 300 K (top) and 14 K (bottom); all are referred to a room temperature dewar connector. Note scale change for noise temperature.

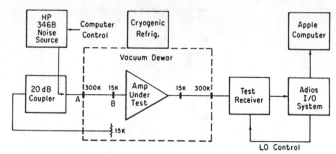

Fig. 6. Test setup for amplifier noise and gain measurements. Very low loss coaxial-line transitions were developed for connection, A to B, to the amplifier.

tion are presented in a report [16] available from the authors.

Tuning of the amplifier is necessary due to variability in the construction and installation of the inductors and to meet slightly different specifications for each application. Drain-bias voltage and current for stages 1, 2, and 3 are initially set at 5.5 V, 15 mA, 5.5 V, 12 mA, and 4 V, 10 mA, respectively. Inductor $L1$ and transformer $T2$ lengths are trimmed to minimize the noise at a desired frequency. The inductance of a coil may be raised or lowered by moving the turns closer or further apart; larger changes require more or less wire but this is usually not necessary. Inductors $L2$ and $L10$ are adjusted to achieve input match; it may be necessary to also change $L1$ and $T2$ for this requirement. Inductors $L5$, $L6$, $L8$, and $L9$ are adjusted to achieve the desired gain response.

The amplifier input match versus frequency changes appreciably as it is cooled as is shown in Fig. 5. This is due to the dielectric constant change in the input circuit board and to changes in gate-to-source capacitance caused by gate bias voltage changes necessary to keep drain current in an optimum noise range. To some extent the input match change can be compensated by pretuning at 300 K for the triangle shape shown in Fig. 5. This is often not satisfactory and iterations of tuning and cooling may be necessary.

IV. MEASUREMENT SYSTEM AND RESULTS

Amplifiers are tested for input return loss and gain utilizing a scalar network analyzer [17]. For the return-loss measurement, test signal power level at the amplifier input is -25 dBm and is somewhat critical; a higher level causes an error due to second-stage overload and a lower level gives insufficient reflected power for the reflectometer bridge sensitivity. For gain measurement, the input signal level is -45 dBm.

Noise temperature and also gain are measured with the test setup shown in Fig. 6. The amplifier is mounted in a vacuum dewar and is thermally connected to a cryogenic refrigerator. The amplifier input is connected to the dewar exterior through a very low loss (0.08 dB, later reduced to 0.04 dB) coaxial line with APC 3.5 connectors and length 8.5 cm. This line has crystalline quartz inner conductor supports to form a vacuum seal and to insure that the amplifier end of the inner conductor is at the refrigerator temperature; the line will be described in detail in a future report.

Noise from a semiconductor-diode noise source [18] is coupled to the amplifier input through the side arm of a 20-dB directional coupler. The main arm of the coupler is connected to a cold termination in the dewar to increase accuracy by preventing addition of a large room temperature noise to the small amplifier noise. By replacing the coupler output, point A in Fig. 6, with hot and cold noise temperature standards, the noise temperature at point A with diode off, 28.8 K, and diode on, 94.1 K to 123.7 K dependent upon frequency, was calibrated. The noise diode was then turned on and off as receiver local-oscillator was scanned to give a swept-frequency measurement of noise temperature. A prior scan with the amplifier bypassed allows computation of the amplifier noise temperature, corrected for test receiver noise, and also the amplifier gain. This procedure is performed by an Apple II computer.

At the time of this writing, 30 amplifiers have been constructed. All amplifiers are stable for a sliding short of any phase connected to input or output. Oscillation of an amplifier can be detected either by observation of a bias value change or by observing the output of a broadband detector connected to the amplifier input or output port. Noise temperature, gain, and input return loss for one amplifier are shown in Fig. 5; this amplifier had the lowest noise temperature, 7.2 K, but is typical in gain and input match; typical noise temperatures are ~2 K higher. The noise temperature is referred to the cold amplifier input connector and is believed to be accurate within ± 0.5 K; a value of 8.3 K was measured at point A, outside of the dewar.

REFERENCES

[1] Model 2 Cryogenic Refrigerator, Cryosystems Inc., Westerville, OH.

[2] B. M. Oliver and J. Billingham, "Project cyclops," NASA Ames

Research Center, Moffett Field, CA, Design Study Rep. CR114445.

[3] D. R. Williams, W. Lum, and S. Weinreb, "*L*-band cryogenically-cooled GaAs FET amplifier," *Microwave J.*, vol. 23, no. 10, pp. 73–76, Oct. 1980.

[4] L. Nevin and R. Wong, "*L*-band GaAs FET amplifier," *Microwave J.*, vol. 22, no. 4, p. 82, Apr. 1979.

[5] Model LTE 1102 Passive Microwave Technology, Canoga Park, CA.

[6] G. D. Vendelin, "Feedback effects on the noise performance of GaAs FET's," in *IEEE MTT Int. Microwave Symp. Dig.*, 1975, pp. 324–326.

[7] H. A. Haus and R. B. Adler, "Invariants of linear noisy networks," *IRE Convention Rec.*, vol. 4, part 2, pp. 53–67, 1956.

[8] D. Fenstermacher, "A computer-aided analysis routine including optimization for microwave circuits and their noise," NRAO Electronics Division, Charlottesville, VA, Int. Rep. No. 217, July 1981.

[9] RT/Duroid 6010 .050″ Dielectric, 2 oz. 2 side copper, Rogers Corp., Chandler, AZ.

[10] S. Weinreb, "Low-noise cooled GASFET amplifiers," *IEEE Trans. Microwave Theory Tech.*, vol. MTT-28, pp. 1041–1054, Oct. 1980.

[11] R. Pucel, H. Haus, and H. Statz, "Signal and noise properties of gallium arsenide microwave field-effect transistors," in *Advances in Electronics and Electron Physics*, vol. 38, L. Morton, Ed. New York: Academic, 1975.

[12] Type PCAW-10 KDI Pyrofilm, Whippany, NJ.

[13] Eccosorb Type MF-124, Emerson and Cummings, Canton, MA.

[14] Type 97-500G Finger Contact Strip, Instrument Specialties Co., Delaware Water Gap, PA.

[15] Type T10-6 Ferrite Core, Micrometals Inc., Anaheim, CA.

[16] S. Weinreb, D. Fenstermacher, and R. Harris, "Ultra low-noise, 1.2–1.7-GHz cooled GASFET amplifiers," NRAO Electronics Division, Charlottesville, VA, Int. Rep. No. 220, Sept. 1981.

[17] Type 560 Scalar Network Analyzer, Wiltron Inc., Palo Alto, CA.

[18] Type 346B Noise Source, Hewlett-Packard Co., Palo Alto, CA.

[19] Type 2146 Bellows, Servometer Corp., Cedar Grove, NJ.

A Reflected-Wave Ruby Maser with *K*-Band Tuning Range and Large Instantaneous Bandwidth

CRAIG R. MOORE, MEMBER, IEEE, AND ROBERT C. CLAUSS, ASSOCIATE MEMBER, IEEE

Abstract—A novel maser concept is outlined and a unique design described which permits wide bandwidth and waveguide tuning range by employing four stages cascaded via cryogenically cooled circulators. Theoretical considerations for gain, bandwidth, gain ripple, and noise temperature are included. Operated on a closed-cycle helium refrigerator with a superconducting persistence-mode magnet, the four-stage amplifier is tunable from 18.3 to 26.6 GHz with 30 dB of net gain and achieves 240 MHz of 3-dB bandwidth near the center of this band. The measured noise temperature is 13 ± 2 K referred to the room-temperature input flange. Applications are foreseen utilizing cooled parametric downconverters and upconverters with this amplifier at IF to extend the low-noise performance up to millimeter frequencies and down to *L*-band for radio astronomy and planetary spacecraft communication.

I. Introduction

MASERS are well known as the lowest noise microwave amplifiers. However, because of the need to operate at liquid-helium temperatures, and the moderate bandwidths available (usually < 60 MHz), nonlaboratory applications have been generally limited to spectral-line radio astronomy, radar astronomy, and spacecraft communications [1]–[4]. Masers with large instantaneous bandwidth (200 to 500 MHz) and waveguide band tuning range would find application in continuum radio astronomy, and as IF amplifiers for millimeter wavelength receivers employing cooled mixers. In these applications, the increased bandwidth is necessary to many scientific experiments, and the wide tuning range provides versatility, easing the design of millimeter mixers. By combining a wide-band maser with cooled parametric upconverters and downconverters, a band-switched receiver can be realized which would facilitate very-wide-frequency coverage in the present lowest noise applications [5].

The maser development described here was a cooperative effort between the National Radio Astronomy Observatory (NRAO) and the Jet Propulsion Laboratory (JPL), under a contract from NRAO. This represents a continuation of activity in maser development at JPL for the National Aeronautics and Space Administration since 1958 and an interest in furthering the state of the art in low-noise receivers on the part of NRAO.

II. Description

The reflected-wave maser concept differs from the well-known traveling-wave maser design in the following ways:

1) Circulators are used to direct signals to be amplified to, and from, a slow-wave structure that does not contain resonance isolators. The maser described here uses slow-wave structures consisting of ruby-filled waveguide.

2) The signal being amplified traverses each slow-wave structure twice, once traveling away from and the second time traveling towards the circulator junction.

3) The absence of resonance isolators reduces the slow-wave structure loss [6] and eliminates distortion of the magnetic field required by the maser material. Increased net gain is achieved, and a uniformly low noise temperature (± 0.5 K) results across a wide bandwidth when the high gain is traded for bandwidth.

The reflected-wave maser concept differs from the reflection-type cavity maser by using a slow-wave structure instead of a cavity. The reflected-wave maser produces a large gain–bandwidth product, tunable across a wide frequency range. Cavity masers achieve a much smaller gain–bandwidth product at a fixed frequency.

Initial development work on the reflected-wave maser concept began in 1972 under a California Institute of Technology President's Grant[1] to the University of California at San Diego and the Jet Propulsion Laboratory. A successful two stage *X*-band amplifier was built for a laboratory application [7].

The reflected-wave maser described here employs a unique design wherein four stages are cascaded via 10 circulator junctions operating at 4.6 K. The four rubies are biased by a single superconducting magnet, and the 10 circulators are biased in pairs by five sets of permanent

Manuscript received May 18, 1978; revised October 20, 1978. This paper presents work which was made possible through prior maser research carried out at the Jet Propulsion Laboratory, California Institute of Technology, under Contract NAS 7-100 sponsored by the National Aeronautics and Space Administration.

C. R. Moore is with the National Radio Astronomy Observatory, Green Bank, WV 24944.

R. C. Clauss is with the Jet Propulsion Laboratory, California Institute of Technology, Pasadena, CA 91103.

[1] The California Institute of Technology President's Grants are co-sponsored equally by the National Aeronautics and Space Administration and California Institute of Technology.

Reprinted from *IEEE Trans. Microwave Theory Tech.*, vol. MTT-27, no. 3, pp. 249–256, Mar. 1979.

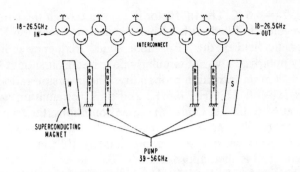

Fig. 1. Schematic diagram of the four-stage K-band ruby maser.

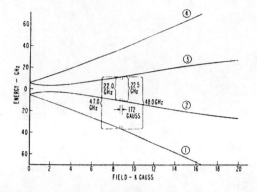

Fig. 2. Energy levels of Cr^{3+} in Al_2O_3 (ruby) at C-axis orientation of 54.733°.

magnets. Each stage has approximately 10 dB of electronic gain and 9.5 dB of net gain, excluding circulator losses. The resulting amplifier (including circulator losses) has a tuning range of 18.3 to 26.6 GHz, at least 30 dB of net gain, up to 240 MHz of 3-dB bandwidth, and a noise temperature of 13±2 K. Further, the design has demonstrated a 400-MHz bandwidth capability.

Fig. 1 depicts the various techniques employed to achieve the wide bandwidth and tuning range. The maser microwave structure consists of a ruby-filled waveguide. Amplification occurs as the signal travels down and back through the ruby. A microwave circulator is used to separate the incoming and outgoing signals. The tuning range is thus limited only by the bandwidth of the circulator. The pump power is injected at the shorted end of the ruby-filled guide, where the pump guide appears as a waveguide beyond cutoff to signal frequencies. The magnetic field biasing the ruby is tapered linearly along the ruby length, giving rise to maser material linewidth spreading. Achieving gain over this broadened linewidth is accomplished by distributing the pump energy in frequency. Frequency modulation is used at a rate with a period much shorter than the pump transition spin relaxation time.

The choice of pink ruby (0.05 percent Cr^{3+} in Al_2O_3) as the active material is based upon years of experience and success with this material at JPL [1], [2]. The selection of optimum-quality material is a vital step in the construction of a multistage maser. Czochralski-grown ruby presently available is free of C-axis wander, misorientation or

GHz. The applicable energy level diagram is shown in Fig. 2, along with the various requirements for wide-band operation. As is implied by the double-pump orientation, the four ground-state energy levels bear a symmetrical relationship. The 1–3 and 2–4 transitions can be pumped simultaneously, as they occur at the same frequency. This results in inversion of the 2–3 transition, which corresponds to the signal frequency. For operation at K-band, the pump frequency is approximately twice the signal frequency plus 3 GHz. For wide-bandwidth operation, the pump frequency must be distributed over a range that is twice the broadened signal frequency resonance separation. The 2–3 transition frequency varies with magnetic field at a rate of approximately 2.9 MHz/G.

III. MASER GAIN AND BANDWIDTH

The gain of each stage must be kept below the circulator isolation to insure stability and control feedback-induced ripple. Mismatch at the circulator maser port will cause gain ripple which is highly dependent on gain, as we will show later. At least 30 dB of net gain is desired in order to reduce the noise contribution of the mixer receiver following the maser amplifier. Thus four stages with an electronic gain of 10 dB per stage were selected to achieve the desired performance.

The gain of a linear stagger-tuned maser (linearly tapered magnetic biasing field) is given theoretically as a function of frequency by Siegman [10] as

$$G_{dB}(f) = \frac{27SN}{Q_m} \frac{\Delta f_L}{2\Delta f_0} \tan^{-1}\left[\frac{(2\Delta f_0/\Delta f_L)}{[2(f-f_0)/\Delta f_L]^2 - (\Delta f_0/\Delta f_L)^2 + 1}\right] \quad (1)$$

dislocations, and nonuniform chromium distribution over lengths in excess of 20 cm [8]. With proper annealing, the boule is also quite free of stresses which can cause warpage upon slicing and trimming to size.

Energy levels and transition matrix element values for pink ruby have been tabulated by Berwin [9]; the sensitivity of the transition probabilities to C-axis orientation and the effect of the magnetic biasing field on the transition frequencies are available in detail from this work. These tabulations indicate that the double-pump angle (C-axis orientation 54.7°) is appropriate for operation above 17

where

S slowing factor (c/v_g) (v_g is the group velocity);

N $=l/\lambda_0$, active length of the material in free space wavelengths (l is twice the physical length for a reflected-wave maser);

Q_m magnetic Q;

Δf_L magnetic resonance linewidth of material, MHz;

Δf_0 separation of maser material resonant frequencies due to magnetic biasing field taper, MHz;

f_0 unbroadened resonance frequency (center frequency of instantaneous bandwidth), MHz.

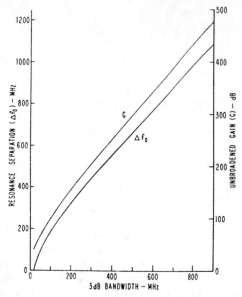

Fig. 3. Unbroadened gain (G) and resonance separation (Δf_0) required to achieve a specified 3-dB bandwidth at 40-dB electronic gain for a linear stagger-tuned ruby maser.

The gain of an unbroadened maser (uniform magnetic biasing field) is given by [10]

$$G_{dB} = \frac{27 S N}{Q_m}. \qquad (2)$$

The relationship between unbroadened gain, Δf_0, and the 3-dB bandwidth for a 40-dB gain ruby maser as given by (1) is plotted in Fig. 3. For a 3-dB bandwidth of 250 MHz a Δf_0 of 383 MHz is required; which when combined with a field sensitivity of 2.9 MHz/G requires a linear magnetic field taper of 132 G. Also from Fig. 3, an electronic gain of 40 dB at 250-MHz bandwidth requires an unbroadened gain of 180 dB, or 45 dB per stage. The length of ruby required per stage can be estimated from (2) and the expression for Q_m [10]

$$\frac{1}{Q_m} \approx 10^{-18} \frac{hf}{kT_0} \frac{\mathfrak{N}}{n} \frac{IM^2 F}{\Delta f_L} \qquad (3)$$

where

h	Planck's constant;
f	frequency, Hz;
k	Boltzmann's constant;
T_0	cryogenic temperature of the maser material, K;
$\mathfrak{N}$	spin density, spins/cm³;
n	number of energy levels;
I	inversion ratio;
M	maximum value of the transition probability;
F	fill factor;
Δf_L	magnetic resonance linewidth of material, MHz.

The magnetic $Q(Q_m)$ is largely dependent on the maser material, and once the doping and C-axis orientation are optimized only the cryogenic temperature and fill factor are under control of the microwave designer. Since the amplifier is intended to operate on a closed-cycle refriger-

ator, $T_0 = 4.6$ K. The fill factor is $\frac{1}{2}$, even though the entire waveguide is filled with active material, because the RF magnetic field of the TE_{10} mode signal frequency is linearly polarized and not circularly polarized as it should be for maximum transition probability. The inversion ratio of pink ruby at 22 GHz, 4.6 K, and 54.7° orientation was measured to be 1.8. Under the same conditions, the transition probability M is 1.92, the linewidth Δf_L is 60 MHz, and the spin density $\mathfrak{N}$ is approximately 10^{19} cm⁻³. The magnetic Q is then approximately 30.

The slowing factor includes the effect of the dielectric constant of the ruby but also includes the dispersive characteristic of the waveguide mode. The slowing factor for the TE_{10} mode in a dielectric filled waveguide can be shown to be

$$S = \epsilon \frac{\lambda_g}{\lambda_0} = \epsilon^{1/2} \frac{\lambda_{g_1}}{\lambda_0} \qquad (4)$$

where

λ_g	guide wavelength of the dielectric filled waveguide;
λ_{g_1}	guide wavelength of an air filled waveguide of identical cutoff frequency;
ϵ	dielectric constant, $\simeq 9$ for ruby.

At 22 GHz, a ruby-filled waveguide with cutoff frequency identical to WR-42 air filled guide (broad wall = 1.07 cm) has a slowing factor of 3.9. Thus from (2), an active ruby length of approximately 17.4 cm per stage is required (8.7-cm physical length) for 45 dB per stage of unbroadened gain. If the magnetic biasing field is tapered linearly by 132 G over approximately 8.7 cm of ruby length, the result should be 10 dB of gain per stage and a four-stage maser with 40 dB of electronic gain and 250 MHz of 3-dB bandwidth at 22 GHz. Similarly, 400 MHz of 3-dB bandwidth at 22 GHz is possible for a linear magnetic field taper of 190 G and 12.2 cm of physical ruby length.

Each of the parameters in (2) varies as the maser center frequency is tuned across the waveguide band. The active length N is directly proportional to frequency while the slowing factor S increases with decreasing frequency as the waveguide nears cutoff. These variations are shown in Fig. 4(a). The components of Q_m vary in different ways. The terms hf/kT and M^2 increase with frequency while the inversion ratio varies according to the expression [10]

$$I = 2 \frac{f_{pump}}{f_{signal}} - 1. \qquad (5)$$

When the inversion ratio is adjusted by a fixed offset to agree with the measured value of 1.8 at 22 GHz, the predicted variation of Q_m is as shown in Fig. 4(a). The net result is that the unbroadened gain is predicted to increase from 18 to 26 GHz as shown in Fig. 4(b).

The biasing magnetic field is tapered longitudinally by physically shaping the pole pieces of the magnet. As the center field value is varied in order to tune the maser, the spatial taper remains constant but the magnetic field taper, and thus Δf_0, increases with frequency as shown in Fig. 4(b).

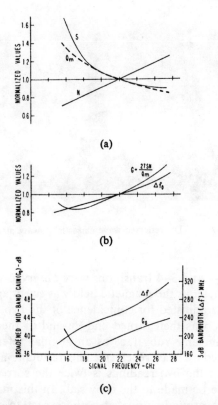

Fig. 4. Predicted variation with frequency of (a) slowing factor (S), active length (N), and magnetic $Q(Q_m)$; (b) unbroadened gain (G) and resonance separation (Δf_0); (c) mid-band gain (G_0) and 3-dB bandwidth (Δf).

When combined with (1) these variations result in the curves of Fig. 4(c) for broadened mid-band gain and 3-dB bandwidth. The 250-MHz bandwidth design is predicted to provide 37 to 43 dB of electronic gain and 220 to 300 MHz of 3-dB bandwidth as the center frequency is tuned from 18 to 26 GHz.

IV. Gain Ripple

Mismatch between the circulator junction and the ruby-filled waveguide will give rise to gain ripple which is represented by the expression

$$\text{ripple in dB} = 20 \log \left[\frac{\dfrac{1}{S_c} - R_m}{\dfrac{1}{S_c} + R_m} \cdot \frac{S_c + R_m}{S_c - R_m} \right] \qquad (6)$$

where S_c is the equivalent VSWR of the circulator port and transition from ruby to air filled waveguide, and R_m is the real part of the maser normalized impedance ($R_m < 0$ for maser gain). R_m is related to the ideal net gain ($S_c = 1$) by

$$G_{\text{dB}} = 20 \log \left(\frac{1 - R_m}{1 + R_m} \right). \qquad (7)$$

Fig. 5 is a plot of peak-to-peak gain ripple versus single-stage gain for various values of circulator junction match.

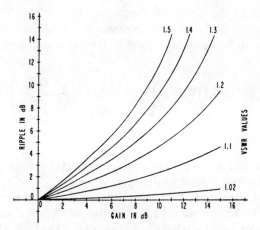

Fig. 5. Gain ripple dependence on gain and circulator junction VSWR for reflected-wave maser.

A typical circulator junction VSWR of 1.2:1 and stage gain of 9.5 dB leads to an expected gain ripple of 4.3 dB per stage. The ripple is repetitive in frequency at an interval determined by the time delay through each stage (approximately 170 MHz for this maser). The rubies were staggered in electrical length in order to cancel the expected ripple in pairs of stages.

Feedback around each amplifier stage can result in gain ripple which is also determined by (6). A typical circulator isolation of 20 dB and a load port return loss of 20 dB results in feedback which is 40 dB below the input. This can be equated to a VSWR at the circulator ruby port of 1.02, which with a stage gain of 9.5 dB results in a peak-to-peak ripple of 0.46 dB from each stage.

V. Noise Temperature

The maser noise temperature can be estimated from the following expression [11]:

$$T_e \simeq \frac{G - 1}{G} T_0 \frac{\rho + \beta}{1 - \beta} \qquad (8)$$

where

T_e effective single-stage maser noise temperature excluding input and output networks;

G single-stage electronic gain, dB;

T_0 cryogenic temperature of the maser material, K;

ρ inverse of the inversion ratio;

$\beta = \dfrac{\text{forward loss of maser structure in dB}}{\text{electronic gain in dB}}$.

For the maser under consideration at 22 GHz

$$G = 10 \text{ dB}$$

$$\rho = \frac{1}{1.8} = 0.56$$

$$\beta = \frac{0.5}{10} = 0.05$$

$$T_0 = 4.6 \text{ K}$$

and the single-stage effective noise temperature is 2.64 K.

Losses in the circulator junctions ahead of this (0.25 dB per pass) and the contribution of the second-stage increase the noise temperature at the circulator input to

$$T_{\text{maser}} \simeq (L_1 - 1)T_0 + L_1\left[T_{e_1} + \frac{(L_2 - 1)T_0 + L_2 T_{e_2}}{G_1}\right] = 3.96 \text{ K}$$
(9)

where L_1 and L_2 are the circulator junction losses preceding the first and second stages.

The contribution of the third and fourth maser stages adds an additional 0.13 K, and the zero-point energy [12] adds an additional 0.66 K ($\frac{1}{2}hf/k$). Thus the effective four-stage maser noise temperature referred to the input at the 4.6 K cryogenic station is 4.75 K. The very low structure loss permits the use of average values when calculating β. This approximation does not show the slight variation in noise temperature (± 0.5 K maximum) that exists across the instantaneous bandwidth due to the linear magnetic field taper.

VI. MASER STRUCTURE

The maser microwave structure is shown in Fig. 6. The ruby-filled waveguide is scaled by $\frac{1}{3}$ from air filled WR-42 waveguide. Its cross-sectional dimensions are 3.56×1.73 mm. A seven-step matching section combined with a gradually filling guide (ruby taper) provide the transition from the full width unfilled guide to the reduced width ruby-filled guide. (The height has been reduced to the proper dimensions at the circulator input.) This matching technique yields a VSWR of better than 1.2:1 across the waveguide band.

The rubies are tapered on only one side, leaving a flat (3.56-mm) surface with intimate contact to the flat copper divider plate along the entire length. Nylon pins with external springs are employed to apply pressure to the ruby to insure good thermal conduction to the copper. No other surface of the ruby needs to contact the copper waveguide. A gap of 0.01 mm between the ruby and the copper waveguide is used so that parts may be machined and assembled with ease, and problems caused by differential thermal expansion between the ruby and the copper waveguide are avoided. The gap across the 1.73-mm dimension affects the electric field, raising the cutoff frequency of the loaded waveguide (≈ 14 GHz) about 3 percent. The gap does not degrade the match of the device in the 18–26.5-GHz range. Each ruby is approximately 15 cm in length, and the one-way transmission loss of the ruby-filled waveguide and matching section is estimated to be 0.25 dB at 4.6 K.

The rubies were oriented to within $\pm\frac{1}{8}°$ of the double-pump angle by observing coincidence of the 1–2 and 3–4 transitions. Each ruby was mounted in a shorted X-band waveguide and inserted between the pole pieces of a laboratory electromagnet. The spin resonance absorption

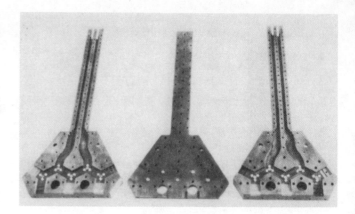

Fig. 6. The reflected-wave maser microwave structure.

of the 1–2 and 3–4 transitions were observed with an RF reflectometer as the magnetic field was swept and the field angle was adjusted for coincidence of the two transitions. The ruby was then turned end-for-end in the waveguide (using the same ruby face as an angular reference), and the procedure was repeated. The resultant angular difference in the two readings is twice the correction angle needed to be made to the ruby slab. In this manner, each ruby was properly oriented with respect to a reference face. The rubies were installed in the microwave structure with the reference face against the divider plate. This simple technique assures that the four rubies are mounted in the superconducting magnet to within $\pm\frac{1}{8}°$ of the double-pump angle.

VII. CIRCULATOR BLOCK

The circulators are constructed in reduced height (1.73 mm by 10.67 mm) waveguide. A Y-junction below resonance design with nickel ferrite disks attached to the upper and lower waveguide broad walls is employed. A 0.254-mm gap between disks, and alumina matching blocks in the center of each waveguide arm, complete the design.[2] High-energy product rare-earth permanent magnets are used to bias the ferrite. The field is set about 1 kG below optimum at room temperature to account for the increase in magnet induction and ferrite saturation magnetization at 4.6 K. Single-junction isolation of 20 dB, insertion loss of 0.25 dB, and VSWR of 1.2:1 across the waveguide band are typical of 4.6 K performance.

The ten circulators are built in a three-piece copper block, with five circulators on each side sharing a common wall divider plate, as shown in Fig. 6. This construction permits a compact symmetrical design and allows magnetic biasing with only five sets of permanent magnets. Three section quarter-wave transformers are employed between full height and the reduced height waveguide at the input and output of each circulator block.

[2]Suggestions for this circulator design were provided by H. Saltzman of P and H Laboratories, Inc.

272

Terminated circulators used as isolators are employed between stages to attenuate signals traveling from output to input through the circulator block.

VIII. SUPERCONDUCTING MAGNET

The superconducting magnet which biases the rubies is adjustable from 7.5 to 10 kG in order to tune the maser from 18 to 26.5 GHz. The magnet is a Cioffi-type [13] operating in the persistent mode and is similar to one developed for a K_u-band traveling-wave maser which operated at 7.5 kG [14]. The air gap was set at 1.54 cm at the widest (lowest field) point, and the pole pieces are 15.24 cm long. The pole pieces and return path are Hiperco 27 cobalt–iron machined and annealed to specification. This design contains the field with low leakage so that no detectable interaction with the circulator field occurs. The field was originally tapered by 190 G over a 12-cm length in order to achieve 400 MHz of bandwidth. Fringing at the ends of the pole pieces, however, resulted in a nonuniform field that reduced the useful length of the magnet to approximately 8 cm. The pole piece taper was relaxed, and iron compensation shims were added to achieve a field with a linear taper of 125 G over approximately 9 cm, which limits the bandwidth at 40-dB electronic gain to less than 250 MHz at 22 GHz.

IX. MICROWAVE PUMP

The microwave pump power is provided by a Siemens RWO 60 backward wave oscillator (BWO) selected for power output of at least 100 mW from 39 to 56 GHz. A single WR-19 waveguide carries the pump power to the base of the microwave structure where a power splitter and short length of dielectric-filled 1.73-mm square waveguide mates with the base of each ruby. The return loss at the pump input flange varies from 5 to 16 dB across the pump frequency range. A 20-kHz triangular wave ac voltage is combined with the BWO delay-line voltage in order to frequency modulate the pump frequency at up to 1000-MHz peak-to-peak deviation. Pump power is sufficient to nearly saturate the four stages at 240-MHz bandwidth (1-dB gain change for 1-dB pump power change).

X. CRYOGENICS

A closed-cycle helium refrigerator (CCR) is used to provide a 4.6 K operating environment for the maser, circulators, and superconducting magnet. Similar units have been used with many maser receiver systems throughout the NASA Deep Space Network [15].

XI. MASER PACKAGE

The entire assembly is housed in a vacuum dewar, and the OFHC copper input and output waveguides are thermally insulated with short sections of 0.254-mm wall, stainless-steel waveguide, copper plated on the inside to approximately three skin depths. The coin silver pump waveguide is similarly insulated. Commercial waveguide

Fig. 7. The maser package with vacuum Dewar and heat shields removed.

pressure windows are used on the input and output. A special window was fabricated for the pump waveguide vacuum seal. The package is shown in Fig. 7 with the vacuum dewar and heat shields removed to show the internal components. The input flange is at the top of the picture.

XII. PERFORMANCE

Initial tests of a single-stage maser which was designed as previously described yielded $5\frac{1}{2} \pm 1$ dB of net gain over 400 MHz centered at 19.45 GHz. However, fringing at the ends of the pole pieces reduced the useful length of the magnet to about 8 cm. Thus in the four-stage unit, the pole piece taper was relaxed to increase the gain at the expense of bandwidth. The best bandpass response is obtained near 22 GHz, and Fig. 8 shows the four-stage inversion ratio of 1.8, 3-dB bandwidth of 240 MHz, and 33 dB net gain achieved at this frequency. 1-dB gain compression occurs at an output signal level of approximately -35 dBm, which corresponds to approximately -68 dBm at the maser input.

The gain ripple becomes excessive for wide-band operation at the upper and lower quarters of the tuning range. The bandpass for tuning increments of 500 MHz are shown in Fig. 9, and are typical of the performance at intervening points. The bandwidth can be reduced electronically (reducing the pump frequency FM deviation) to

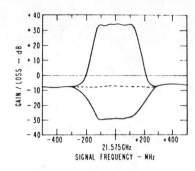

Fig. 8. Experimental gain and absorption near 22 GHz for the four-stage reflected-wave ruby maser.

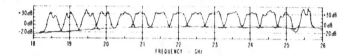

Fig. 9. Bandpass response of the four-stage reflected-wave maser at 500-MHz increments in the tuning range.

achieve a flattened response. 3-dB bandwidths in excess of 100 MHz can be achieved from 20.5 to 24 GHz and in excess of 30 MHz over the remainder of the band from 18.3 to 26.6 GHz.

Measured gain ripple indicates that the accumulated mismatch due to the circulator and ruby/vacuum-filled waveguide transition results in a VSWR that is probably no less than 1.4:1 near the band edges. It is further possible that the ripple cancellation via phasing of ruby pairs is not optimum. Subsequent units now under construction will permit measurement of the match and delay at this critical junction.

The noise temperature was measured as a complete amplifier system while looking vertically at the sky and included contributions from the cosmic background, atmospheric oxygen and water vapor, feed horn, and maser. The receiver following the maser contributed less than 2 K to the total system operating temperature, but the measurement technique eliminated inclusion of this quantity. Noise temperature values of 35 and 28 K were measured at 22.23 and 26.3 GHz, respectively. When allowances are made for the various contributions, the maser noise temperature at the room-temperature input flange is calculated to be 13 ± 2 K. An independent hot–cold load noise temperature measurement at 22.3 GHz indicated a maser noise temperature of 13.7 ± 4 K at this point.

XIII. APPLICATIONS

A maser with this wide-bandwidth potential would find immediate application in continuum radio astronomy. More importantly, the wide-bandwidth capability can be used to good advantage for spectral line radio astronomy observations at millimeter frequencies where the structure in a spectra will often extend over 100 MHz [16]–[18]. By using the maser as an IF amplifier for cooled mixers or parametric downconverters [19], low noise temperature and wide bandwidth can be achieved at frequencies up to

230 GHz. At longer wavelengths, parametric upconverters in combination with K-band masers are planned to provide wide-bandwidth, low-noise operation at frequencies as low as L-band. The above techniques are presently under development for radio astronomy and spacecraft communications.

Preliminary results obtained with the K-band maser mounted on the NRAO 43-m antenna in a Cassegrain configuration show the following:

1) System noise temperatures (including atmospheric contributions) of 73 K at 22.23 GHz, 50 K below 20 GHz, and 60 K above 24 GHz during clear weather. The accumulated system noise temperature contributions from the feed horn, waveguide components, maser, and follow-up receiver are no greater than 30 K.

2) Total power variation equivalent to 0.15 K peak-to-peak for periods of tens of minutes during clear weather.

3) Gain stability (as measured with a calibration noise signal) of better than 4 percent (± 0.1 dB) for 5 h with the telescope tracking sources.

XIV. CONCLUSION

We have demonstrated the feasibility of a unique maser design for achieving wide instantaneous bandwidth and waveguide tuning range. Engineering problems which limit the bandwidth and result in excessive gain ripple have been outlined, and their resolution is now being pursued. Applications for this wide-band maser design are foreseen for both radio astronomy and planetary spacecraft communication.

REFERENCES

[1] E. L. Kollberg, "A traveling wave maser system for radio astronomy," *Proc. IEEE*, vol. 61, pp. 1323–1329, Sept. 1973.

[2] M. S. Reid, R. C. Clauss, D. A. Bathker, and C. T. Stelzried, "Low-noise microwave receiving systems in a worldwide network of large antennas," *Proc. IEEE*, vol. 61, pp. 1330–1335, Sept. 1973.

[3] K. S. Yngvesson, A. C. Cheung, M. F. Chui, A. G. Cardiasmenos, S. Wang, and C. H. Townes, "K-band traveling-wave maser using ruby," *IEEE Trans. Microwave Theory Tech.*, vol MTT-24, pp. 711–717, Nov. 1976.

[4] E. L. Kollberg and P. T. Lewin, "Traveling-wave masers for radio astronomy in the frequency range 20–40 GHz," *IEEE Trans. Microwave Theory Tech.*, vol. MTT-24, pp. 718–725, Nov. 1976.

[5] S. Weinreb, M. Balister, S. Maas, and P. J. Napier, "Multiband low-noise receivers for a very large array," *IEEE Trans. Microwave Theory Tech.*, vol. MTT-25, pp. 243–248, Apr. 1977.

[6] F. S. Chen and W. J. Tabor, "Filling factor and isolator performance of the traveling-wave maser," *Bell Syst. Tech. J.*, pp. 1005–1033, May 1964.

[7] L. D. Flesner, S. Schultz, and R. Clauss, "Simple waveguide reflection maser with broad tunability," *Rev. Sci. Instrum.*, vol. 48, pp. 1104–1105, Aug. 1977.

[8] P. Warren, Union Carbide Corp., private communication.

[9] R. W. Berwin, "Paramagnetic energy levels of the ground state of Cr^{+3} in Al_2O_3 (ruby)," Tech. Memo. 33-440, Jet Propulsion Lab., Jan. 15, 1970.

[10] A. E. Siegman, *Microwave Solid State Maser*. New York: McGraw-Hill, 1964, chs. 6 and 7.

[11] W. H. Higa, "Noise performance of traveling-wave masers," *IEEE Trans. Microwave Theory Tech.*, vol. MTT-12, p. 139, Jan. 1964.

[12] A. E. Siegman, "Zero-point energy as the source of amplifier noise," *Proc. IRE*, vol. 49, p. 633, 1961.

[13] P. P. Cioffi, "Approach to the ideal magnetic circuit concept

through superconductivity," *J. Appl. Phys.*, vol. 33, pp. 875–879, Mar. 1962.

[14] R. Berwin, E. Weibe, and P. Dachel, "Superconducting magnet for a Ku-band maser," in *Proc. 1972 Applied Superconductivity Conf.* (Annapolis, MD, May 1–3, 1972), pp. 266–269.

[15] W. H. Higa and E. Weibe, "A simplified approach to heat exchanger construction for cryogenic refrigerators," *Cryogenic Technol.*, vol. 3, pp. 47–48, 50–51, Mar./Apr. 1967.

[16] L. J. Rickard, P. Palmer, M. Morris, B. Zuckerman, and B. E. Turner, "Detection of extragalactic carbon monoxide at millimeter wavelengths," *Astrophys. J.*, vol. 199, pp. L75–L78, July 15, 1975.

[17] W. B. Burton, M. A. Gordon, T. M. Bania, and F. J. Lockman, "The overall distribution of carbon monoxide in the plane of the galaxy," *Astrophys. J.*, vol. 202, pp. 30–49, Nov. 15, 1975.

[18] B. Zuckerman, T. B. H. Kuiper, and E. N. R. Kuiper, "High-velocity gas in the Orion infrared nebula," *Astrophys. J.*, vol. 209, pp. L137–L142, Nov. 1, 1976.

[19] S. Weinreb, "Millimeter wave varactor down converters," presented at the Diode Mixers at Millimeter Wavelengths Workshop, Max-Planck-Institut für Radioastronomie, Bonn, West Germany, Apr. 1977.

Low-Noise Room-Temperature and Cryogenic Mixers for 80-120 GHz

ANTHONY R. KERR, ASSOCIATE MEMBER, IEEE

Abstract—A description is given of two new mixers designed to operate in the 80–120-GHz range on the 36-ft radio telescope at Kitt Peak, Ariz. It is shown that for a hard-driven diode the parasitic resistance and capacitance are the primary factors influencing the design of the diode mount. A room-temperature mixer is described which achieves a single-sideband (SSB) conversion loss (L) of 5.5 dB, and a SSB noise temperature (T_m) of 500 K (excluding the IF contribution) with a 1.4-GHz IF. A cryogenically cooled version, using a quartz structure to support the diode chip and contact whisker, achieves values of $L = 5.8$ dB and $T_m = 300$ K with a 4.75-GHz IF. The mixers use high-quality Schottky-barrier diodes in a one-quarter-height waveguide mount.

I. INTRODUCTION

THIS PAPER describes the results of a program of mixer development aimed at producing more sensitive millimeter-wave receivers for the National Radio Astronomy Observatory's 36-ft radio telescope at Kitt Peak, Ariz.

The most significant development in millimeter-wave mixers since Sharpless [1] introduced the wafer diode mount in 1956, has been the introduction of the Schottky-barrier diode. The nearly ideal exponential characteristic of the Schottky diode led Barber [2] to approximate the device by a switch in series with a small resistance; the conversion loss of a mixer is then a function of the pulse duty ratio (PDR) of the switch. Dickens [3] has achieved good agreement between Barber's theory and experimental results at 60 and 95 GHz. Leedy *et al.* [4] demonstrated good agreement between theory and experiment when they assumed, following Torrey and Whitmer [5], a sinusoidal LO voltage at the diode. Although this assumption is unlikely to be strictly valid [6], [7], it is consistent at high LO levels with Barber's switching model.

More recently, nonlinear analysis techniques have been applied to the mixer problem in an effort to achieve a more accurate understanding of the mixing process [6]–[9]. However, these attempts have been limited to cases in which the diode has a fairly simple embedding network. The difficulty of characterizing the embedding network at the harmonics of the LO frequency has so far prevented these methods from being used to give an accurate solution for the case of a waveguide-mounted diode.[1]

In this paper the approach taken to mixer design is to consider the mixer as three interconnected networks as shown in Fig. 1:

N_1 the embedding network, or diode mount;

N_2 the network containing the diode's parasitic capacitance and resistance, which connects the ideal diode to the embedding network;

N_3 the ideal exponential diode.

These three networks are optimized to obtain maximum power transfer between the embedding network and the periodically varying junction resistance at the input (RF) and output (IF) frequencies. The embedding network is assumed reactive at the harmonics of the LO frequency.

The single-sideband (SSB) noise temperature of a mixer receiver can be written as

$$T_R = T_M + LT_{IF} \qquad (1)$$

where T_M is the noise contribution of the mixer itself, L is the SSB conversion loss of the mixer, and T_{IF} is the noise temperature of the IF amplifier. Following the argument given by Weinreb and Kerr [12], T_M can be expressed in terms of an average temperature associated with the diode $T_{D_{AV}}$ and the conversion loss, thus

$$T_M = (L - 2)T_{D_{AV}}. \qquad (2)$$

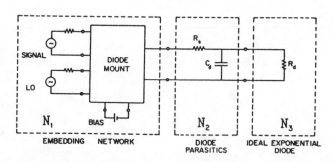

Fig. 1. A mixer represented as three interconnected circuits: N_1—the embedding network; N_2—the diode parasitic resistance and capacitance; and N_3—the ideal exponential diode.

Manuscript received January 13, 1975; revised April 28, 1975. This work was supported by Associated Universities, Inc., under contract with the National Science Foundation.

The author was with the National Radio Astronomy Observatory, Charlottesville, Va. 22901. He is now with the NASA Goddard Institute for Space Studies, New York, N. Y. 10025.

[1] Eisenhart and Khan [10] and Eisenhart [11] have made an analysis of the simple waveguide mount, which is accurate up to many times the normal operating frequency of the waveguide. In practice, however, the waveguide mount deviates from the simple model in such details as the nonideal RF choke, an input waveguide transformer, and a nonplanar short circuit behind the diode.

Reprinted from *IEEE Trans. Microwave Theory Tech.*, vol. MTT-23, pp. 781–787, Oct. 1975.

The mixer is assumed to be of the broad-band type for which the conversion loss is the same at the signal and image frequencies. It has been found in practice that if the tuning, LO drive, or bias of a mixer is varied, $T_{D_{AV}}$ generally changes to some extent, but that the change in $(L - 2)$ dominates the right-hand side of (2). Reducing L therefore reduces both the mixer and IF contributions to the receiver noise temperature as given by (1).

The object of this paper is to show that with careful attention to the mixer design, it is possible to achieve low noise and conversion loss at frequencies up to 120 GHz. Two mixers are described which are tunable from 80 to 120 GHz; one is intended for room-temperature operation, and the other, a variation of the first, uses a quartz diode mount which is suitable for cryogenic operation.

II. MIXER THEORY

A. Ideal Exponential Diode

The junction resistance of a practical Schottky-barrier diode behaves as an ideal exponential element over many decades of current. The current i and voltage v are related by

$$i = i_0(e^{\alpha v} - 1) \qquad (3)$$

where $\alpha = q/\eta kT \simeq 35$ V^{-1} at room temperature. For practical purposes $i_0 \ll i$, and the incremental conductance of the diode may be written as

$$g = \frac{\partial i}{\partial v} \simeq \alpha i. \qquad (4)$$

The behavior of the diode as a mixer depends both on the waveform of $g(t)$ produced by the LO, and on the embedding network seen by the diode.

If a transformer of ratio n is inserted between an ideal exponential diode, operating as a mixer, and its embedding network, the properties of the mixer will remain unchanged provided the dc bias and LO power are changed according to

$$V_{bias} \rightarrow V_{bias} - \ln(n^2)/\alpha \qquad (5a)$$

and

$$P_{LO} \rightarrow P_{LO}/n^2. \qquad (5b)$$

It follows that the ideal exponential diode has no preferred impedance level and can perform equally well as a mixer at any impedance. Thus, in optimizing the three networks of Fig. 1 for maximum signal-frequency power transfer to the ideal diode, N_3 imposes no constraint on the impedance levels of N_1 and N_2. If the parasitic elements of N_2 are fixed for the available diodes, the impedance levels of N_1 and N_3 can be chosen to minimize the conversion loss.

B. Diode Capacitance and Series Resistance

The signal frequency equivalent circuit of a Schottky-barrier diode, operating as a mixer, is shown in Fig. 1.

R_d is the input impedance of the time-varying junction resistance, and C_d and R_s are the mean values of the junction capacitance and series resistance, assumed equal to their values at the dc bias voltage when no LO power is applied. For a given semiconductor sample C_d and R_s depend primarily on the area A of the diode and on the doping and thickness of the epitaxial layer on which the diode is formed [13]. R_s includes contributions from skin effect in the semiconductor and contact wire. The cutoff frequency is defined as $\omega_c = 1/(R_s C_d)$.

The effects of R_s and C_d on the mixer performance are threefold.

1) They contribute to the conversion loss because of power dissipated in R_s at the RF and IF frequencies.

2) They affect the waveform $g(t)$ of the mixing element by changing the termination of the LO harmonics.

3) They affect the terminations seen by the frequencies $nf_{LO} \pm f_{IF}, n > 1$.

These effects may be further elaborated as follows.

1) The degradation of the conversion loss caused by power dissipated in R_s at the signal frequency ω, is

$$\delta_{RF} = 1 + \frac{R_s}{R_d} + \frac{R_d}{R_s} \frac{\omega^2}{\omega_c^2} \geq 1. \qquad (6)$$

At the IF frequency R_s appears in series with the output impedance R_o of the exponential element. The loss due to R_s is

$$\delta_{IF} = 1 + \frac{R_s}{R_o} \geq 1. \qquad (7)$$

The combined RF and IF loss due to R_s and C_d is $\delta = \delta_{RF} \times \delta_{IF}$; this is shown in Fig. 2 as a function of normalized frequency. The parameter $K = R_o/R_d$ is the quotient of the output (IF) impedance and the input (RF) im-

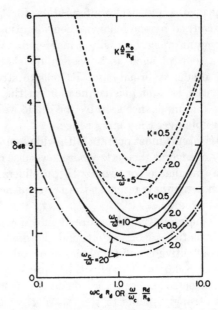

Fig. 2. Loss δ due to RF and IF dissipation in the diode series resistance R_s. The RF (signal) frequency is ω, and $K \triangleq R_o/R_d$ is the ratio of IF impedance to RF impedance.

pedance. It will be shown in the following that for a broad-band mixer K is expected to lie between 0.5 and 2.

2) Barber [2] has used the concept of an equivalent PDR to characterize the mixer properties of a diode with a conductance waveform $g(t)$. The PDR is a function of LO power and bias voltage, and can be maintained constant along with the conversion loss despite changes in $g(t)$ caused by variation of the embedding impedance at frequencies nf_{LO}, $n > 1$.

3) Saleh [14] has shown that for Barber's equivalent PDR to uniquely define the conversion loss it must be dependent not only on the $g(t)$ waveform but also on the embedding impedance seen by the diode at frequencies $nf_{LO} \pm f_{IF}$, $n \geq 1$. A change in the reactive termination at some sideband frequency $nf_{LO} \pm f_{IF}$, $n > 1$, affects both the PDR, which can be restored by appropriate LO and bias adjustments, and the optimum RF and IF impedances of the mixer. It is assumed here that loss in R_s at these sideband frequencies is small, an assumption which is likely to hold for a practical mixer.

C. IF Impedance

Saleh [14] has made an extensive investigation of the effects on mixer performance of the diode's conductance waveform $g(t)$ and of the embedding impedances at the harmonics of the LO. It is observed from his results that for a broad-band mixer the optimum source (RF) and load (IF) impedances never differ by a factor of more than 2, regardless of LO drive level or bias. Although this is not generally proven for all combinations of terminations of the higher frequency sidebands, $nf_{LO} \pm f_{IF}$, $n > 1$, it is consistent with observed mixer performance, and is a useful aid to design.

III. THE DIODE MOUNT

A. Mount Configuration

The choice of a physical configuration for the diode mount is governed by the following considerations.

1) The mount must be easily tunable, preferably by means of a control such as a waveguide short circuit behind the diode. A broad-band RF choke structure is required in the IF and bias connection to the diode to ensure that the impedance seen by the diode will vary as little as possible over the tuning range.

2) The IF circuit must operate with wide bandwidth at a frequency of several gigahertz where low-noise cryogenic paramps are available for use as IF amplifiers. An RF choke which is highly reactive at the IF frequency should therefore be avoided.

3) The diode mounting structure should not introduce excessive parasitic capacitance around the diode thereby reducing its effective cutoff frequency. For a diode whose capacitance is ~0.01 pF this effectively precludes the use of ribbon-contacted or beam-lead diodes in their present forms, and strongly points to the use of a whisker-contacted diode.

These requirements can be fulfilled by a waveguide mount similar in some respects to the wafer mount introduced by Sharpless [1], but using very much reduced-height waveguide, and a different RF choke structure.

B. Mount Equivalent Circuit

Eisenhart and Khan [10] and Eisenhart [11] have made a detailed investigation of the driving-point impedance seen by a small device connected across the gap G in a waveguide mount as shown in Fig. 3(a). The approximate equivalent circuit of the mount is shown in Fig. 3(b), where

$$Z_g = 2 \left(\frac{\mu}{\epsilon} \right)^{1/2} \frac{b}{a} \frac{\lambda_g}{\lambda} \qquad (8)$$

is the TE_{10}-mode guide impedance, L_s is the post inductance due to the evanescent TE_{m0} modes ($m > 1$), C is the gap capacitance due to the evanescent TE_{mn} and TM_{mn} modes, $n > 1$, and C_1 and L_1 are the capacitance and inductance due to the TE_{m1} and TM_{m1} modes. This equivalent circuit characterizes the mount in the normal operating range of the waveguide for which $f_c < f < 2f_c$, where f_c is the cutoff frequency for the TE_{10} mode.

The gap impedance, Z_{gap} in Fig. 3(b) is strongly affected by elements C_1 and L_1 which are series resonant at the frequency f_1 for which the waveguide height $b = \lambda/2$. For full-height waveguide $b \simeq a/2$, and the resonance f_1 occurs close to $2f_c$. Over most of the useful waveguide band L_1 and C_1 cause a rapid variation of Z_{gap}, both real and imaginary parts, which is clearly undesirable for a mixer in which broad tunability must be simply achieved. By reducing the waveguide height, however, it is possible to raise the resonant frequency f_1 until L_1 and C_1 are equivalent to a small capacitance C', which is independent of frequency for $b^2 \ll \lambda^2/4$.

The element C of Fig. 3(b) is independent of frequency for $b^2 \ll \lambda^2$, and can be considered together with C' as a single capacitance C'', provided $b^2 \ll \lambda^2/4$. In the case of a mixer, the gap of Fig. 3(a) is the depletion region of the diode. C'' is then the junction capacitance C_d of the diode, and can conveniently be measured by a capacitance bridge connected to the IF port of the mixer while the diode is being contacted.

C. Mount Analysis

We now investigate the reduced-height waveguide mount of Fig. 4(a) whose equivalent circuit is shown in

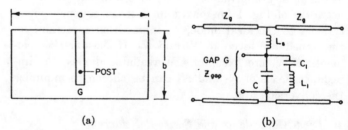

Fig. 3. (a) The simple waveguide mount. The diode is mounted across the gap G. (b) The equivalent circuit of the mount for frequencies $f_c < f < 2f_c$.

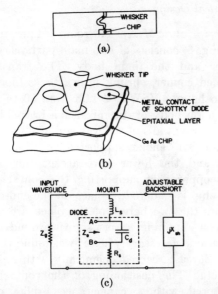

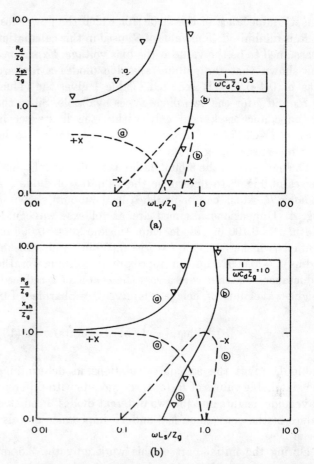

Fig. 4. (a) Reduced-height waveguide mount with a whisker-contacted diode. (b) Diode contact details. (c) Equivalent circuit of the mount and diode parasitics as seen by the junction resistance of the diode.

Fig. 5. Mount-matching curves showing values of diode impedance R_d which can be matched (solid curves), and the corresponding backshort reactance X_{sh} (broken curves), both as functions of whisker reactance ωL_s. Normalized reactance of diode capacitance, $1/(\omega C_d Z_g)$ = (a) 0.5, (b) 1.0. Diode series resistance is assumed zero; points ($\triangledown$) are for $R_s = 0.05\,Z_g$.

Fig. 4(c). The impedance Z_e is the embedding impedance seen by the junction resistance R_j of the diode. For efficient mixing, Z_e must be real and equal to some optimum source impedance. It is of interest to examine the real values of Z_e that are possible for this circuit. In particular we shall determine the values of X_{sh} (i.e., the backshort settings) for which Z_e is real, and the effect of L_s, C_d, R_s, Z_g, and frequency on these real values.

The equivalent circuit of Fig. 4(c) was analyzed by computer to determine the values of X_{sh} for which Z_e is real. Fig. 5 shows the real values of Z_e and corresponding values of X_{sh} as functions of $\omega L_s/Z_g$. The main curves are for $R_s = 0$, and typical points are indicated for $R_s = 0.05Z_g$. It is seen that there are, in general, two values of X_{sh} for which Z_e is real, and those real values may differ by a factor of 10 or more.

IV. MOUNT DESIGN FOR 80–120 GHz

A. Diodes

The Schottky-barrier diodes used in this work [15], [16] were formed by electroplating a platinum anode, followed by gold, on epitaxial gallium arsenide. Typical characteristics are shown in Table I. The parameters η and R_s are defined by the diode equations

$$i = i_0 \left\{ \exp\left(\frac{qv'}{\eta kT}\right) - 1 \right\} \tag{9a}$$

$$v = v' + iR_s. \tag{9b}$$

The diodes were supplied by Dr. R. J. Mattauch of the University of Virginia.

B. Electrical Design

The first step in the mount design is to use the loss curves of Fig. 2 to determine the optimum value of R_d,

TABLE I
CHARACTERISTICS OF THE GALLIUM ARSENIDE SCHOTTKY-BARRIER DIODES AT ROOM TEMPERATURE

EPITAXIAL LAYER	Doping	3×10^{17} cm^{-3}	
	Thickness	$0.5 \pm 0.25\ \mu$	
SUBSTRATE	Orientation	(1 0 0)	
	Type	n	
	Doping	$2\text{–}3 \times 10^{8}$ cm^{-3}	
DIODE DIAMETER		$2.5\ \mu$	$3.5\ \mu$
MEASURED PARAMETERS			
η		1.11	1.10
R_s (measured at DC)		8.0 Ω	3.6 Ω
C_d (at 0.0V, 1 MHz)		0.007 pF	0.012 pF
V_b (at −0.1 μA)		−8 V	−8 V
CALCULATED PARAMETERS			
C_d at V_{bias}		−0.011 pF	−0.020 pF
$\frac{1}{\omega C_d}$ at 100 GHz		145 Ω	80 Ω
R_s at 100 GHz*		10 Ω	6 Ω
f_c at V_{bias} and 100 GHz		1450 GHz	1330 GHz
$\frac{f_c}{f_{sig}} = \frac{\omega_c}{\omega}$		14.5	13.3
FROM FIG. 2			
Optimum $\omega C_d R_d$		1 – 2	1 – 2
R_d for minimum δ		145 – 290 Ω	80 – 160 Ω
δ		0.7 – 1.1 dB	0.8 – 1.2 dB

* The values of R_s at 100 GHz include contributions from skin effect in the whisker and diode substrate material.

the RF impedance of the diode, for which the power loss in R_s is minimized. The value of C_d used in this calculation is assumed to be the value at the bias voltage. Experience has shown that for gallium arsenide diodes a forward bias of 0.4–0.7 V is required. Table I gives the values of R_d and δ for the two diode types available. Since the IF impedance is known only within the limits set in Section II-C, R_d and δ can only be determined to lie within corresponding limits.

The next step in the mount design is to use the matching curves of Fig. 5 to determine the value(s) of diode impedance R_d which can be matched in the mount shown in Fig. 4. Dimensions assumed are as follows: waveguide width[2] $a = 0.100$ in, diode chip thickness $t = 0.006$ in, contact whisker length $l = b - 0.006$ in, and whisker radius $r = 0.00025$ in. An approximate formula for the inductance of a thin wire across the center of a reduced-height waveguide of height b is given by Sharpless [1]

$$L_s = 2 \times 10^{-7} l \log_e \left(\frac{2a}{\pi r}\right) \text{ (MKSA units)}. \quad (10)$$

Table II gives the salient calculations in determining the matchable values of R_d for three mounts with different waveguide heights and for two different diodes. Predicted values of the conversion loss and IF impedance are also given.

During the initial part of this work only the 3.5-μm diodes were available, and for these the one-quarter-height mount provides the best match. This mount was used for all the mixers described in this paper. For the 2.5-μm diodes the impedance level of the one-quarter-height mount is somewhat lower than the optimum value; however, Fig. 2 indicates a degradation in conversion loss of less than 0.1 dB.

TABLE II
CALCULATION OF MATCHABLE R_d VALUES AND CORRESPONDING
CONVERSION LOSS AND IF IMPEDANCE FOR VARIOUS
WAVEGUIDE HEIGHTS AND DIODES

DIODE DIAMETER	2.5 μ			3.5 μ		
Waveguide Height as a Fraction of Full Height	1/2	1/3	1/4	1/2	1/3	1/4
Z_g at 100 GHz, eq. 8	233 Ω	156 Ω	117 Ω	233 Ω	156 Ω	117 Ω
$\frac{1}{\omega C_d Z_g}$ using Table I	0.6	0.9	1.2	0.3	0.5	0.7
$\frac{\omega L_s}{Z_g}$ using eq. 10	1.4	1.2	0.9	1.4	1.2	0.9
$\frac{R_d}{Z_g}$ from Fig. 5	no match	1.5	0.9	no match	no match	1.0 (or -10.0)
R_d	---	230 Ω	110 Ω	---	---	117 Ω
Loss δ dB when diode is matched -- from Fig. 2	---	0.7 - 1.0 dB	0.7 - 1.3 dB	---	---	0.7 - 1.1 dB
$L_{SSB} = 3 + \delta$ dB	---	3.7 - 4.0 dB	3.7 - 4.3 dB	---	---	3.7 - 4.1 dB
Expected IF impedance -- from Section II-C	---	115 - 560 Ω	55 - 220 Ω	---	---	58 - 234 Ω

[2] The choice of $a = 0.100$ in allows the possibility of TE$_{20}$-mode propagation above 118 GHz. For a centrally mounted diode, however, there is no asymmetry to excite this mode. Our measurements have indicated no higher mode problems.

C. Mechanical Design

Room-Temperature Mixer: The room-temperature mixer, shown in Fig. 6, consists of two main parts, a waveguide transformer and the main body. The transformer is electroformed copper, shrunk into a brass block, and is designed to have a VSWR < 1.06 from 80 to 120 GHz [17]. The main body of the mixer is a brass block, split across the narrow walls of the waveguide. The upper part contains the RF choke supported in Stycast 36-DD dielectric,[3] and the lower part accepts an accurately machined copper post supporting the contact whisker. The diode chip is soldered in place on the end of the RF choke before the two halves of the block are finally assembled. The aluminum insert shown around the choke in Fig. 6 became necessary when it was found that during curing the Stycast reacted chemically with any copper-bearing metal. The positioning of the contact whisker was monitored with a capacitance bridge connected between the diode and the body of the mixer. This ensured that excessive capacitance was not introduced in parallel with the diode due to deformation of the whisker tip after contacting the diode. The whisker position was controlled to a fraction of a micron by a differential micrometer. The backshort is of the contacting finger type, milled from a single piece of beryllium–copper shim stock. Contact between the SMA connector and the RF choke is made by a small bellows spring.

The RF choke was designed to give low loss over 80–120 GHz while having low capacitance as seen at the IF. It consists of four coaxial sections of, alternately, 12- and 70-Ω characteristic impedance, inside an outer conductor of 0.027-in diameter. The cutoff frequency of the TE$_{11}$ mode on the high impedance sections of the choke is ~170 GHz. Calculation of the choke impedance Z_c as seen from inside the waveguide gives Re$[Z_c]$ < 0.2 Ω and Im$[Z_c]$ < 5 Ω in the frequency range 80–120 GHz.

Cryogenic Mixer: The room-temperature mixer described in the preceding was found to be unstable when cooled because of movement between the diode and contact whisker. This was caused by differential contraction of the Stycast dielectric with respect to the metal body of the mount. To eliminate differential contraction poses a

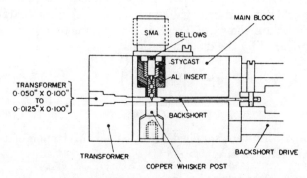

Fig. 6.   Cross section of the room-temperature mixer.

[3] Emerson Cuming Company. $\epsilon_r = 1.7$.

difficult materials problem, but its effect can be controlled by using the quartz diode package shown in Fig. 7(b). Differential contraction between the contact whisker and the quartz is small enough to be taken up by the spring of the wisker. Fused quartz was chosen as the structural material because it has high mechanical strength and rigidity, relatively low dielectric constant and loss tangent, is easily cut by scribing and breaking, and is easily metallized with gold over a thin chromium adhesion layer.

It was desired to keep the electrical properties of the mount as close as possible to those of the room-temperature design, and for this reason the mount configuration shown in Fig. 7(a) was used. The main electrical difference between this and the room-temperature design is the quartz member across the waveguide adjacent to the diode. The additional shunt susceptance of this member can be tuned out by adjustment of the backshort.

The quartz diode mount is constructed from three strips of 0.006 × 0.015-in quartz as shown in Fig. 7(b). Two strips are metallized with the RF choke pattern, and the longer unmetallized third strip forms the mechanical support between the choke strips. On one choke strip two 0.001-in gold brackets are ultrasonically bonded, one to contact the IF connector, the other to support the diode which is soldered to it. The contact whisker is soldered to one end of the second choke strip. The three strips are assembled using Eastman 910 adhesive: first the strip carrying the diode is glued to the long support strip, and then the strip carrying the whisker is slid into

contact with the diode and glued. The positioning of the whisker point on the chip is observed through a high-power microscope and monitored with an I–V curve tracer. A differential screw is used to control the position of the whisker strip within a fraction of a micron.

The quartz diode assembly is supported across the waveguide, as shown in Fig. 7(a), by the pressure of two springs. One spring holds the assembly against a raised part (A) of the block, ensuring a dc return path, and RF and IF grounds. The second spring, on the end of the IF transformer, contacts the gold bracket at the end of the quartz structure. The diode structure is then free to expand relative to the brass housing.

V. PERFORMANCE

The noise and conversion loss measurements given in the following were made using the IF noise radiometer/reflectometer described by Weinreb and Kerr [12]. This instrument enables the mixer performance to be determined without matching the IF port, which is expedient when a large number of measurements are to be made under conditions of varying IF port impedance. Results obtained in this way have been in good agreement with measurements made by the Y-factor method with the IF port matched using an appropriate transformer.

Typical performance figures for the room-temperature mixers are shown in Table III. The considerable superiority of the smaller diode is believed to be due to its smaller capacitance, enabling it to behave more nearly as an ideal switching mixer.

Table IV gives typical figures for the cooled mixers. These mixers were all constructed with 2.5-μm diodes. The cooled measurements made at 77 K were found to be close to those at 18 K; laboratory measurements were therefore generally made at 77 K for convenience. The mixers had 0.2–0.5-dB greater conversion loss when operating at 4.75-GHz IF than at 1.4-GHz IF. This was probably due to the following: 1) higher IF transformer losses at 4.75 GHz, and 2) the wider spacing (9.5 GHz)

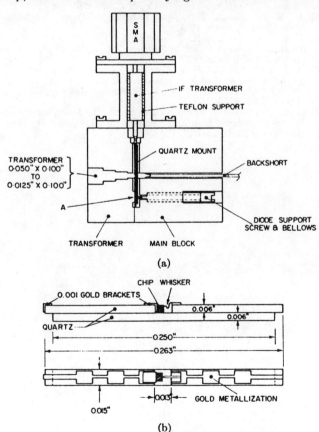

(a)

(b)

Fig. 7. The cryogenic mixer. (a) Cross section of the mixer. (b) Details of the quartz diode mount. Not to scale.

TABLE III
MEASURED CHARACTERISTICS OF THE ROOM-TEMPERATURE MIXERS
(f_{IF} = 1.4 GHz)

LO Frequency	85 GHz		115 GHz	
Diode	2.5 μ	3.5 μ	2.5 μ	3.5 μ
L_{SSB}	4.6 dB	6.2 dB	5.5 dB	6.7 dB
$T_{M_{SSB}}$	420°K	700°K	500°K	1400°K
Bias	0.4 v	0.6 v	0.4 v	0.4 v
	2.0 mA	4.0 mA	2.0 mA	4.0 mA

TABLE IV
MEASURED CHARACTERISTICS OF THE CRYOGENIC MIXERS
(f_{LO} = 115 GHz)

TEMP.	IF FREQ.	L_{SSB}	$T_{M_{SSB}}$
298°K	1.4 GHz	5.4 dB	740°K
77°K [a]	4.75 GHz	5.8 dB	300°K

[a] Similar results were obtained at 18 K.

between the signal and image bands resulting in a poorer RF match.

The measured IF impedance levels all lie within the limits predicted in Section II-C.

VI. CONCLUSION

An approach to mixer design has been presented for cases where the diode is driven hard by the LO and can be approximated by a switch whose duty cycle depends on the basis voltage and LO level. The ideal diode is connected through a parasitic network, containing the diode's series resistance and capacitance, to the embedding network (mount). The optimum impedance of the embedding network is shown to depend primarily on the parasitic resistance and capacitance. For the particular diodes used in this work it was necessary to reduce the height of the waveguide in the mount to $\sim\frac{1}{4}$ of the standard height.

Two mixers have been described. One is for room-temperature operation, and the other, a modification of the first with a quartz diode mounting structure, is suitable for cryogenic cooling. Typical values of the SSB conversion loss and SSB mixer noise temperature [defined in (1)], measured at 115 GHz, are 5.5 dB and 500 K operating at room temperature with a 1.4-GHz IF, and 5.8 dB and 300 K when cryogenically cooled to 77 or 18 K with a 4.75-GHz IF. The difference between the measured conversion loss and the predicted value is due to nonideal switching behavior of the diode, and to dissipation of signal power converted to higher order sidebands, $nf_{LO} \pm f_{IF}$, $n \geq 2$, which were assumed to be reactively terminated.

The mixers described in this paper are currently in use on the National Radio Astronomy Observatory's 36-ft radio telescope at Kitt Peak, Ariz.

ACKNOWLEDGMENT

The author wishes to thank Dr. S. Weinreb of NRAO, whose support and inspiration sustained this work, and Dr. R. J. Mattauch of the University of Virginia for his patience and persistence in developing the diodes. He also wishes to thank J. E. Davis, T. J. Viola, W. Luckado, G. Green, J. Cochran, N. Horner, Jr., and J. Lichtenberger for their significant contributions to the work.

REFERENCES

[1] W. M. Sharpless, "Wafer-type millimeter wave rectifiers," *Bell Syst. Tech. J.*, vol. 35, pp. 1385–1402, Nov. 1956.
[2] M. R. Barber, "Noise figure and conversion loss of the Schottky barrier mixer diode," *IEEE Trans. Microwave Theory Tech.*, vol. MTT-15, pp. 629–635, Nov. 1967.
[3] L. E. Dickens, "Low conversion loss millimeter wave mixers," in *IEEE G-MTT Int. Microwave Symp. Proc.*, June 1973, pp. 66–68.
[4] H. M. Leedy *et al.*, "Advanced millimeter-wave mixer diodes, GaAs and silicon, and a broadband low-noise mixer," presented at the Conf. High Frequency Generation and Amplification, Cornell Univ., Ithaca, N. Y., Aug. 17–19, 1971.
[5] H. C. Torrey and C. A. Whitmer, *Crystal Rectifiers* (M.I.T. Radiation Lab. Ser., vol. 15). New York: McGraw-Hill, 1948.
[6] D. A. Fleri and L. D. Cohen, "Nonlinear analysis of the Schottky-barrier mixer diode," *IEEE Trans. Microwave Theory Tech.*, vol. MTT-21, pp. 39–43, Jan. 1973.
[7] A. R. Kerr, "A technique for determining the local oscillator waveforms in a microwave mixer," this issue, pp. 828–831.
[8] S. Egami, "Nonlinear, linear analysis and computer-aided design of resistive mixers," *IEEE Trans. Microwave Theory Tech.*, vol. MTT-22, pp. 270–275, Mar. 1974.
[9] W. K. Gwarek, "Nonlinear analysis of microwave mixers," M.S. thesis, Mass. Inst. Technol., Cambridge, Sept. 1974.
[10] R. L. Eisenhart and P. J. Khan, "Theoretical and experimental analysis of a waveguide mounting structure," *IEEE Trans. Microwave Theory Tech.*, vol. MTT-8, pp. 706–719, Aug. 1971.
[11] R. L. Eisenhart, "Understanding the waveguide diode mount," in *Dig. Tech. Papers, 1972 IEEE G-MTT Int. Microwave Symp.* (May 1972), pp. 154–156.
[12] S. Weinreb and A. R. Kerr, "Cryogenic cooling of mixers for millimeter and centimeter wavelengths," *IEEE J. Solid-State Circuits (Special Issue on Microwave Integrated Circuits)*, vol. SC-8, pp. 58–63, Feb. 1973.
[13] H. A. Watson, *Microwave Semiconductor Devices and Their Circuit Applications.* New York: McGraw-Hill, 1968.
[14] A. A. M. Saleh, *Theory of Resistive Mixers.* Cambridge, Mass.: M.I.T. Press, 1971.
[15] T. J. Viola, Jr., and R. J. Mattauch, "Unified theory of high frequency noise in Schottky barriers," *J. Appl. Phys.*, vol. 44, pp. 2805–2808, June 1973.
[16] R. J. Mattauch and J. W. Kamps, "Lateral coupling effects in Schottky-barrier diodes," Research Laboratories for the Engineering Sciences, Univ. Virginia, Charlottesville, Rep. EE-4769-101-73, Nov. 1973.
[17] G. L. Matthaei, L. Young, and E. M. T. Jones, *Microwave Filters, Impedance-Matching Networks, and Coupling Structures.* New York: McGraw-Hill, 1964.

A Broad-Band, Ultra-Low-Noise Schottky Diode Mixer Receiver from 80 to 115 GHz

C. READ PREDMORE, MEMBER IEEE, ANTTI V. RÄISÄNEN, MEMBER, IEEE, NEAL R. ERICKSON, PAUL F. GOLDSMITH, MEMBER IEEE, AND JOSE L. R. MARRERO

Abstract —A cryogenic 3-mm receiver has been developed which fully utilizes the low-noise potential of Schottky diodes by approaching the shot-noise limit within 10 percent. With a broad-band mixer design which properly terminates the input sidebands and reactively terminates the second harmonic of the local oscillator and its sidebands, the double sideband (DSB) mixer noise temperature is 35 K in the best case. This design has given an average DSB receiver noise temperature of 75 K over the 80 to 115-GHz band with a best noise temperature of 62 K.

Manuscript received August 3, 1983; revised January 3, 1984. This work was supported in part by the Five College Radio Astronomy Observatory, which is operated with support from the National Science Foundation under Grant AST 82-12252, and with the permission of the Metropolitan District Commission, Commonwealth of Massachusetts. This is contribution no. 557 from the Five College Astronomy Department.

The authors are with the Five College Radio Astronomy Observatory, University of Massachusetts, Amherst, MA 01003.

A. Räisänen is now at the Helsinki University of Technology, Espoo 15, Finland.

I. INTRODUCTION

STUDIES OF cooled mixer diodes began in 1956 with Messenger's [1] cooling of X-band 1N26 diodes to lower their noise temperature. However, no further work was published on cooling mixers until 1973, when Weinreb and Kerr [2] investigated the noise mechanisms in millimeter mixers and predicted that a 40 K double sideband (DSB) mixer temperature was possible by cooling Schottky diodes to 20 K. Only now has that sensitivity been achieved and surpassed with a minimum DSB mixer noise temperature of 35 K and a DSB receiver noise temperature of 62 K at 100 GHz.

This sensitivity is the result of several years of mixer development at the Five College Radio Astronomy Observatory (FCRAO) which has required considerable

Reprinted from *IEEE Trans. Microwave Theory Tech.*, vol. MTT-32, no. 5, pp. 498–507, May 1984.

effort in every aspect of the receiver design to achieve this low noise over a broad RF bandwidth. At the FCRAO, the 3-mm receiver is used on the 13.7-m-diam radio telescope for both spectroscopic studies of molecules and continuum radiometry of quasars and active galaxies.

A broad-band mixer and receiver design was necessary to fully utilize the excellent Schottky diodes now available. Especially important to the low noise of this receiver was the development of Schottky diodes with low doping for cryogenic operation by several laboratories. The receiver improvements include a broad-band vacuum window with an average loss less than 0.05 dB, scalar feed horns to launch a Gaussian beam, improved mixer design, and an IF noise temperature of 10 K at 1.4 GHz. The mixer required the most design effort. It incorporates a broad-band RF filter and noncontacting backshort which not only short circuit the local oscillator (LO), signal, and image bands, but also are reactive at the second harmonic of the LO and its sidebands. The RF impedance of the contact whisker has been optimized to achieve a broad-band response from 80 to 115 GHz with an average receiver temperature of 150 K SSB. An integral IF matching circuit on alumina microstrip has given low loss and broad-band operation.

The following sections discuss in detail the receiver design, the mixer design, and the results as a double sideband system. Then these results are compared with previous work.

II. Receiver Design

The receiver block diagram is shown in Fig. 1. All of the RF components except the vacuum window are cooled to 20 K with a closed-cycle helium refrigerator [3]. The fused silica vacuum window has $\lambda/4$ layers of teflon epoxied on each side as antireflection coatings. The result is an average loss of 0.05 dB over the 75 to 115-GHz band [4].

The scalar feed is designed to launch a Gaussian beam into the quasi-optics system. Its loss is estimated at 0.1 dB. The ring filter is a tunable filter which is used to couple in the LO signal with, while maintaining low loss in the signal sidebands 1.4 or 4.75 GHz away. It is manufactured by Custom Microwave after a design by Davis [5] with a race-track-shape coupling ring [6] to minimize the loss in the signal path. The ring filter has a signal loss of 0.4 to 0.5 dB with a LO coupling loss of 5 to 6 dB. The LO waveguide from 300 to 20 K is gold-plated stainless steel to limit the loss to less than 3 dB and heat loading on the 20-K station to less than 0.1 W. The ring-filter and mixer backshort have mechanical tuning drives.

The first IF amplifier with about 25 dB of gain is also cooled to 20 K and located close to the mixer. The IF was originally over the 4.5 to 5.0-GHz band using a parametric amplifier and a cooled FET as a second stage for a net IF noise temperature of 24 K. Subsequently, an improved IF system, centered at 1.4 GHz, with a noise temperature of only 10 K [7] has been used for the best results.

The receiver temperatures were measured using Emerson-Cuming CV-3 absorber at room temature and 77 K in front of the dewar window with the mixer-tuned DSB. No input filtering was done to separate the upper

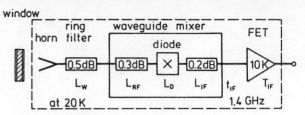

Fig. 1. Receiver block diagram. The fused-silica vacuum window has $\lambda/4$ matching layers. The scalar feed horn, LO injection filter, mixer, and initial IF amplifiers are cooled to 20 K.

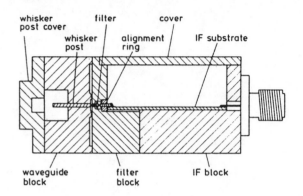

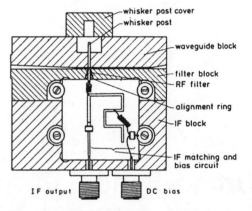

Fig. 2. Mixer waveguide mount. (a) A cross section through the reduced-height waveguide, whisker post, RF filter, and microstrip IF matching circuit, and (b) shows the waveguide taper from full to 1/4 height, and the 1.4-GHz matching circuit.

and lower sidebands and the backshort was within one wavelength of the diode so that the conversion loss was the same in the two sidebands. The linearity of the system extends to an input temperature of at least 6000 K. The second harmonic response of the receiver was measured to be 45 ± 5 dB below the fundamental. This eliminates any apparent lowering of the receiver noise temperature due to inputs at the second harmonic.

III. Mixer Design

The mixer was designed to be tunable over the entire 80 to 115-GHz band since the receiver was to be used for radio astronomy observations over this range, and to have a large instantaneous IF bandwidth. A minimum IF bandwidth of 400 MHz was required since our galaxy and external galaxies have differential Doppler velocities of up to 1000 kM/s, corresponding to a frequency range of 385 MHz at 115 GHz.

The mixer is shown in Figs. 2 and 3. Fig. 2(a) is a cross section through the diode showing the 0.8-mm-diam whisker

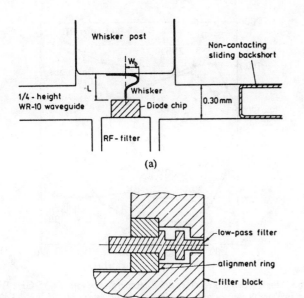

(a)

(b)

Fig. 3.   Details of diode, whisker, and RF filter. (a) shows the diode chip on the RF filter contacted with a 12-μm-diam whisker, and (b) is a cross section of the coaxial RF filter which is held by a Macor dielectric ring.

TABLE I
RF FILTER PARAMETERS
The coaxial RF filter dimensions, dielectric constant, and impedance for each coaxial section are given. The diode is mounted on section 1 and the IF matching circuit connected to section 6.

Section	ID (mm)	OD (mm)	Length (mm)	ϵ_r	Z_0 (Ω)
1	.45	.61	.67	1.0	19
2	.45	1.59	.31	1.0	76
3	1.32	1.59	.41	1.0	11
4	.53	1.59	.43	1.0	66
5	1.32	1.59	.25	1.0	11
6	.67	2.39	1.25	5.75	32

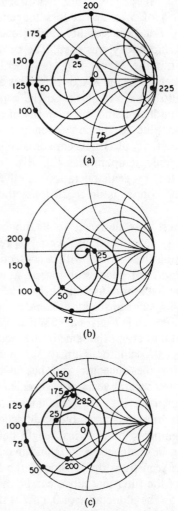

(a)

(b)

(c)

Fig. 4.   Comparison of RF filter impedances for three mixer designs. (a) is a Smith chart display of the RF filter impedance, normalized to 50 Ω, versus frequency, in GHz, for the present mixer design, (b) is the RF filter impedance for the mixer design of Linke *et al.* [9], and (c) is the impedance for the design of Kerr [10].

post, the RF filter, and the alumina substrate where the IF matching circuit is implemented in microstrip. Fig. 2(b) is an orthogonal view which shows the IF matching circuit for 1.4 GHz and the linear taper from full-height to 1/4-height WR-10 waveguide. This reduction lowers the RF impedance seen by the diode and, consequently, the IF impedance. The measured loss is 0.2 to 0.3 dB when the mixer block is machined from OFHC copper. No gold plating was done because the pure copper has a better performance by about 0.1 dB [8]. A linear taper can be readily machined and its loss is estimated to be only 0.1 dB more than a λ/4 step transformer.

Fig. 3(a) shows the details of the diode with its whisker contact and the noncontacting backshort, while Fig. 3(b) shows the coaxial RF filter. The diode chip is typically 250 μm square by 120 μm thick. The RF filter and whisker inductance are in series with the parallel combination of the waveguide admittance and the backshort. The RF filter was empirically designed to be close to a short circuit in the 80 to 120-GHz band while still being reactive in the second harmonic band. The filter parameters are summarized in Table I. A Smith chart representation of the filter impedance is shown in Fig. 4. For comparison, the analogous plots for the mixers of Linke *et al.* [9] and Kerr [10] are also included. Our filter is reactive over the range 75 to 225 GHz, covering both the fundamental and second harmonic frequencies. These are the most important frequency components for mixer performance, as was shown by the work of Held and Kerr [11], [12]. The mixer designed by Linke *et al.* [9] was for the 60 to 90-GHz band (WR-12) so that their filter is resistive in the lower part of the fundamental band, reactive above 75 GHz and a short circuit for the second harmonic band. The λ/4 design used by Kerr [10] is excellent as a short for the fundamental 75

to 115-GHz band, but is not reactive for the second harmonic band. These filter calculations start with a 50-Ω termination impedance at the IF end of the filter and treat each section as a transmission line, taking into account the fringing capacitance [13] at each change in the transmission-line impedance.

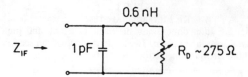

Fig. 5. Equivalent circuit for IF matching. The diode's dynamic impedance is $\approx 275\ \Omega$. The inductive and capacitive elements are due to the
RF filter.

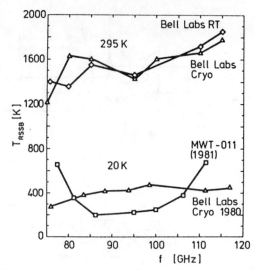

Fig. 6. Single sideband receiver noise temperatures with a 4.75-GHz IF.
The two upper curves and the two lower curves are at 295 and 20 K,
respectively.

Machinability of the filter was taken into account during
the design process. As a compromise, only the section
nearest the WR-10 waveguide is cut off to the TE_{11} coaxial
mode in the 75 to 115-GHz band (see Fig. 3(b)). The
dielectric support ring at the IF end of the filter was
machined from Macor [14]. This dielectric has a relative
permittivity (ϵ_r) of 5.75 and a loss tangent of 0.015 at
100 GHz [15]. For best stability during cycling between
295 and 20 K, no epoxy was used. The center conductor
was pressed into the Macor ring and the filter sections
machined concentric to the Macor ring. This assembly then
was pressed into the mixer block. OFHC copper was used
for the inner conductor since the majority of the losses in a
coaxial line are due to the inner conductor and the dc
electrical conductivity of OFHC copper is 300 times that of
brass at cryogenic temperatures [8]. Also an important
consideration for cryogenic operation, the thermal conductivity of OFHC copper is up to 20 times that of brass at
20 K [16].

The backshort was designed to be noncontacting for
higher reliability since Held and Kerr [12] reported problems in repeating their results with a contacting backshort
and Linke *et al.* [9] successfully used a noncontacting
backshort. The backshort used in this work has three low
and two high impedance sections. The design reported by
Brewer and Räisänen [17] gives a VSWR of 90 in the 80 to
120-GHz band and about the same in the second harmonic
band of the local oscillator. Thin (0.019 mm) mylar tape
[18] is used for the dielectric and has given consistent
performance over several years of cryogenic operation.

As was shown in Fig. 2(b), the microstrip matching
circuit is built into the mixer block adjacent to the RF
filter. This minimizes the electrical length before the filter
impedance can be matched to 50 Ω and maximizes the IF
bandwidth. As seen from the IF side of the RF filter, the
combination of the filter and diode can be modeled by the
circuit in Fig. 5. The inductance of 0.6 nH is the sum of the
inductances of the individual coaxial sections of the filter,
while the 1 pF of capacitance is the sum of the coaxial
sections plus the fringing capacitances. When the mixer is
optimized for the lowest receiver temperature, the IF impedance is the same as the differential resistance of the
pumped diode $I - V$ curve and is measured from the slope
of the dc $I - V$ curve. This impedance varies from 200 to
500 Ω over the 75 to 115-GHz band. This experimental
result was reported by Räisänen [8] and was found also
from computer modeling of the mixer by Lehto and
Räisänen [19]. A compromise value of 275 Ω was used in
designing the matching circuit. At 1.4 GHz, the net IF

impedance is near 50-Ω resistive plus a capacitive component so a match is obtained with a series inductor whose
reactance is about 100 Ω plus an open-circuited $\lambda_{IF}/2$ stub
and a transformer for broadbanding. A dc-bias circuit also
is included on the alumina substrate. For a 4.5 to 5.0-GHz
IF, two open-circuited stubs are used to broaden the
response and a $\lambda/4$ section is used to transform the
resulting impedance to 50 Ω. The output end of the RF
filter is connected to the 4.75-GHz microstrip circuit with a
loop of 12-μm-thick copper foil to minimize mechanical
strain on the RF filter as the mixer is temperature cycled.

The remaining part of the RF circuit which can be
varied is the whisker inductance, as reported by Räisänen
et al. [20]. As will be discussed in detail in Section IV-B,
the whisker length was varied until the receiver response
was an optimum over the 80 to 120-GHz band.

IV. Results

A. Comparison of 4.75-GHz and 1.4-GHz IF

The initial receiver tests were with a 4.75-GHz IF with a
net temperature of 24 K when cooled to 20 K. Two Bell
Telephone Laboratory (BTL) diodes were tested at room
temperature. The "room-temperature" diode had a doping
of 2×10^{17} cm^{-3} while the "cryogenic" diode had a lower
doping of 5×10^{16} cm^{-3} for improved performance at
20 K. As is shown in Fig. 6, both diodes gave essentially
the same receiver temperatures at 295 K over the entire 75
to 115-GHz band. The low-doped diode gave a very flat
performance of ≈ 400 K SSB when the system was cooled.
The total SSB conversion loss, including 0.6 dB for the feed
and ring filter, was 7.0 to 7.7 dB over the 75 to 120-GHz
range [4], [21], [22]. This system was used for the 1980/81
observing season of the FCRAO 13.7-m radio telescope.

Subsequently, in 1981, diodes made by Millimeter Wave
Technology (MWT) were tested in the receiver with a 4.5
to 5.0-GHz IF. As is also shown in Fig. 6, the cooled
receiver performance was improved by almost a factor of 2

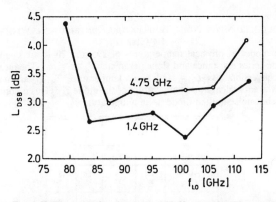

Fig. 7. Total DSB conversion losses at 295 K over the 80 to 115-GHz band for a 4.75 and 1.4-GHz IF. These losses include the feed horn, ring filter, and all mixer losses (waveguide, diode conversion, and IF matching).

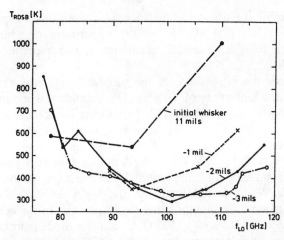

Fig. 8. Optimization of the receiver noise temperature over the 80 to 115-GHz range by varying the whisker inductance. The DSB receiver noise temperature at 295 K with a 1.4-GHz IF is plotted for the initial 0.011-in-long whisker and for whiskers 0.001, 0.002, and 0.003 in shorter. The whisker inductance is directly proportional to length.

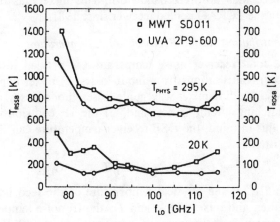

Fig. 9. DSB receiver noise temperatures with a 1.4-GHz IF are plotted versus LO frequency for a MWT and a UVA diode in different mixers at room temperature and 20 K.

over the 85 to 105-GHz range with the MWT diode due to its steeper $I - V$ curve (see (5) and Table IV). Below 85 GHz, the mixer performance with the MWT diode is worse than with the BTL diode because of the increased length of the RF filter. When the MWT diode was mounted, the section of the RF filter nearest the diode was lengthened to move the diode into the waveguide as was shown in Fig. 3(a). This was done to see the effect of different diode positions. This longer RF filter is not as reactive below 80 GHz, which causes the performance to deteriorate. The poorer results above 105 GHz were due to excess inductance from a long whisker as will be discussed in the next section.

When cooled FET amplifiers became available at 1.4 GHz with noise temperatures as low as 10 K [7], the mixers were adapted to this IF frequency by only changing the alumina microstrip circuit in the mixer block. In addition to the lower IF noise temperature when changing from a 4.75 to a 1.4-GHz IF, there is the additional benefit of lower total conversion loss in the mixer. This improvement is shown in Fig. 7 for a mixer with a MWT SD-011C diode. The IF matching circuit was changed between the tests. The DSB conversion loss (L_{DSB}) was measured at room temperature over the 78 to 113-GHz range. Referring to Fig. 1, the total loss L_{DSB} is composed of; L_W, the losses in the window, scalar feed, and ring filter; L_{RF}, the losses in the mixer waveguide and linear taper; L_D, the diode conversion loss; and L_{IF}, the resistive and reflected losses in the IF matching circuit

$$L_{DSB} = L_W L_{RF} L_D L_{IF}. \tag{1}$$

Window losses are $\simeq 0.05$ dB, the scalar feed is assumed to have a loss of 0.1 dB and the ring filter 0.5 dB. This sum was assumed to decrease slightly to 0.5 dB when cooled. The mixer RF losses at room temperature were 0.5 dB, 0.3 dB from the input taper, and 0.2 dB from the reduced-height waveguide and the backshort. The mixer losses were assumed to improve from 0.5 to 0.3 dB at 20 K because of the increase in the conductivity of OFHC copper. The IF losses are $\simeq 0.2$ and $\simeq 0.6$ dB at 1.4 and 4.75 GHz, respectively. The DSB conversion loss L_D is the same for both the IF's since they are a small fraction (< 5 percent)

of the signal frequency. A similar improvement in conversion loss between 4.75 and 1.4 GHz was noted by Kerr [10]. For comparison, Nussbaum et al. [23] have achieved SSB conversion losses of 6 ± 1 dB over the 90 to 120-GHz band with an image-enhanced mixer using beam-lead diodes.

B. Optimization of the RF Impedance

Fig. 8 shows the DSB receiver temperature at 295 K using a MWT SD-011C diode with a 1.4-GHz IF. The first results were somewhat narrow band using a whisker of 12-μm diam and 280-μm (.011 in) length. The total length is approximately $L + W_b$ (see Fig. 3(a)) [10]. To investigate the RF response of the mixer versus whisker length, the same whisker was shortened in 25-μm (.001 in) steps by etching. At each length, the diode was recontacted and its performance was measured as is shown in Fig. 8 [20]. The optimum length was in the range from 200 to 230 μm, corresponding to an inductance of ≈ 0.2 nH [19]. The effect of whisker inductance has been studied by Siegel and Kerr [24]. Their numerical results showed the importance of tuning the whisker inductance for a given diode capacitance. Held [25] also found the importance of whisker length from scale model measurements and experimental results.

Once the importance of the whisker length was established, a University of Virginia (UVA) diode was mounted in another mixer block and the whisker inductance optimized to give even lower noise and broader performance. The results for these two mixer blocks are summarized in Fig. 9. Both the MWT and UVA diodes gave very similar results at 295 K, being < 800 K SSB over much of the band. When cooled to 20 K, the UVA 2P9-600 diode results are only slightly better (~ 10 percent) in the center of the band from 90 to 105 GHz, but this diode maintains its low-noise performance (< 200 K SSB) over the entire 78 to 115-GHz range. The average SSB temperature is 150 K with a best temperature of 124 K, measured with a 50-MHz bandwidth. The DSB receiver temperature only varies from 60 to 80 K over an IF of 1.2 to 1.6 GHz when measured in a 50-MHz bandwidth. This mixer has been used in the FCRAO-cooled receiver since 1981.

C. Contributions to the Receiver Noise Temperature

The total receiver noise temperature is divided into 3 parts: 1) contributions from input losses; 2) mixer contributions including mixer waveguide losses, diode losses, and IF matching circuit losses; and, 3) the IF contribution. With this in mind, the DSB receiver temperature can then be written as

$$T_{RDSB} = (L_W - 1)T_{phys} + L_W T_{MXR} + L_{DSB}T_{IF} \qquad (2)$$

where T_{phys} is the physical temperature of the feed horn, ring filter, and mixer block, and T_{IF} the IF noise temperature, which is 40 K at room temperature and 10 K at cryogenic temperatures for a 1.4-GHz IF. The total DSB mixer noise is

$$T_{MXR} = (L_{RF} - 1)T_{phys} + L_{RF}(L_D - 1)T_{eq}$$
$$+ L_{RF}L_D(L_{IF} - 1)T_{phys}. \qquad (3)$$

The equivalent temperature T_{eq} of the diode as a lossy element will approach T_D as a limit when $R_s = 0$, higher harmonics are reactively terminated, parametric effects are negligible, and there is no excess diode noise as a function of current. The short-noise limit for a diode [26] is given by

$$T_D = (qV_o/2k) \qquad (4)$$

when the diode current i for a voltage V across the diode and its series resistance R_s is

$$i = i_s e^{[(V - iR_s)/V_o]}. \qquad (5)$$

The electron charge is q and Boltzman's constant is k. This would be the case if the noise from the mixing process is dominated by shot noise, which was found to be true for room-temperature mixers by Held and Kerr [12].

Two different diodes in separate mixer blocks were measured at their best operating frequencies with a 1.4-GHz IF. These results are presented in Table II for the MWT SD-011 and UVA 2P9-600 diodes at both 295 and 20 K.

TABLE II
DETAILED MIXER NOISE TEMPERATURE CONTRIBUTIONS WITH A 1.4-GHz IF

Two diodes at physical temperatures of 295 and 20 K are compared. Their dc-series resistance and slope parameter V_o are given. The measured DSB conversion loss L_{DSB} and noise temperature T_{RDSB} are used to derive the equivalent noise temperature of the diode T_{eq}. This is compared to the shot-noise limited diode noise temperature T_D.

Diode	MWT SD-011		UVA 2P9-600	
Physical Temp.	295 K	20 K	295 K	20 K
R_s (Ω)	4.4	8	12	12
V_o (mV)	31	13	28	8.0
L_{DSB} (dB)	2.6	2.3	3.4	3.1
L_D (dB)	1.3	1.3	2.1	2.1
T_{RDSB} (K)	294	68	344	62
T_{IF} (K)	40	10	40	10
T_{MXR} (DSB) (K)	155	43	185	35
T_{eq} (K)	249	108	178	48
$T_D = qV_o/2k$ (K)	180	77	165	47

The diode slope parameter V_o and series resistance R_s are fit over the 1 μA to 5-mA range. Typical operating conditions are 500 μW for 800 μA at 0.8 V and 200 μW for 300 μA at 0.8 V for the LO power and diode dc current and voltage of the MWT and UVA diodes, respectively. The total loss was measured at room temperature. The conversion loss with the MWT diode is 0.8 dB lower than with the UVA diode, partly due to more capacitance variation and corresponding parametric effects and also due to its lower series resistance. The voltage variable capacitance of the diodes can be written as

$$C(V) = C_{jo}(1 - V/\phi)^{-\gamma} \qquad (6)$$

where ϕ is the barrier potential, and C_{jo} is the junction capacitance for zero bias. Both diodes have similar exponents, $\gamma \approx 0.4$, but the MWT diode has a higher C_{jo} (see Table IV). It also requires more LO power because of its higher V_o when cooled [2], which will give a larger voltage swing and consequently more capacitance variation and larger parametric effects.

In Table II, T_{eq} is derived from room-temperature and cryogenic measurements for two diodes in similar mixers. The series resistance R_s and slope parameter V_o were measured at dc. The total conversion loss was measured at room temperature and the diode loss L_D derived using (1). The DSB receiver and the IF noise temperatures give the DSB mixer temperatures from (2). Finally, T_{eq} is derived from (3) and compared to the theoretical shot-noise limit T_D. The derived T_{eq} is with 40 percent and 10 percent of T_D for the MWT and UVA diodes, respectively. The most likely cause for this discrepancy is noise from parametric effects due to increased LO power (500 μW for the MWT diode versus 200 μW for the UVA diode).

Both of these diodes gave excellent results for two reasons. One, the diodes have a low effective noise at 1.4 GHz as a function of dc bias, rising only to 300 K at a current of 5 mA [27]. As was pointed out by Held [28], a good dc

TABLE III
COMPARISON OF 3-mm COOLED SCHOTTKY DIODE MIXERS

Various cryogenically cooled mixers are compared for their: input losses L_W; total SSB conversion loss L_{SSB}; SSB mixer temperature T_{MSSB}; IF noise temperature T_{IF}; and total SSB receiver noise temperature T_{RSSB}. The receiver temperature for [29] is derived by assuming a T_{IF} of 25 K. The results of the last two mixers are from this paper.

DIODE	UVA 3.5μ φ	UVA 2.5μ φ	BTL 280-92	BTL	BTL cryo	BTL 280-92	MWT SD-011	UVA 2P9-600
L_W (dB)	0.6	0.6	-	0.6	0.5	0.4	0.5	0.5
L_{SSB} (dB)	7.2	5.8	6.7	7	6	7.5	5.3	6.1
T_{MSSB} (K)	280	300	209	200	120	91	86	70
T_{IF} (K)	20	20	22	20	24	25	10	10
T_{RSSB} (K)	445	435	312	348	250	240	136	124
Reference	[2]	[10]	[9]	[28]	[22]	[29]		

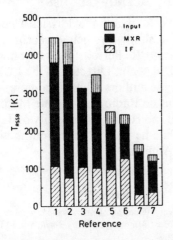

Fig. 10. Comparison of noise contributions to 3-mm cooled mixers. The SSB receiver noise temperature is divided into 3 parts: 1) due to the conversion loss and IF noise temperature (lower section); 2) due to the mixer (center section); and 3) due to input losses before the mixer. References are: 1-[2]-[10], 3-[9], 4-[28], 5-[22], 6-[29], 7-[this paper].

TABLE IV
COMPARISON OF 3-mm SCHOTTKY DIODES

Various whisker-contacted diodes are compared for their: zero-bias capacitance C_{jo}; γ and ϕ are from (6); doping; junction area; physical temperature; dc-series resistance R_{SDC}; and slope parameter V_o; the log of the saturation current I_s; the shot-noise limited diode temperature T_D; and the experimental diode equivalent temperature T_{eq}.

DIODE	UVA 2.5μ φ [10,12]	BTL N280-92 [9]		BTL "room temp" [8]		BTL "cryo" [8]		MWT SD-011		UVA 2P9-600		UVA 2P8-500 [31]	
C_{jo} (fF)	7	14.7		19		13		8		6.5		6.5	
γ	0.4	≈0.2		0.5		≈0.2		0.5		0.4		0.4	
ϕ (V)	0.95	1.0		1.0		1.0		1.1		1.1		1.06	
Doping (cm⁻³)	3×10^{17}	3×10^{16}		2×10^{17}		5×10^{16}		2×10^{16}		3×10^{16}		3×10^{16}	
Junction Area (μm)²	4.9	11		14.5		9.7		5		3.1		3.1	
T_{phys} (K)	295	295	18	295	20	295	23	295	18	295	22	295	18
R_{SDC} (Ω)	8.0	4.5	7.4	8	8	6	6	4.4	8	12	12	13	13
V_o (mV)	28	28	12	31	12	29	26	31	13	28	8.0	27	18
$-\log_{10}(I_s)$ (A)	16.1			16	42.5	17	21	15.9	38.2	16.0	57.4	17.6	29.6
$T_D = qV_o/(2k)$	164 K	164	69	180	70	170	154	180	77	165	47	160	105
T_{eq} (K)	323		47	250			150	249	108	178	48		

noise curve is necessary, but not sufficient, to give a low (< 200 K) equivalent temperature for the pumped diode. Depending on the mixer circuit, correlated shot-noise from high harmonics averaged over a cycle of the LO can make the average noise greater than 300 K [26]. However, our mixer circuit is such that these correlated noise components are minimized and a low equivalent diode temperature is realized.

D. Comparison with Previous 3-mm Schottky Receivers

In analyzing the excellent results that have been obtained with the present receiver, it is important to separate out improvements in the Schottky diodes, improvements in the embedding circuit due to the entire mixer design, and reductions in the IF noise temperature. The present results are compared with previous work on cryogenic 3-mm receivers in Table III and Fig. 10 for 7 different mixers. The total SSB conversion loss is given by L_{SSB}, while L_W and T_{IF} are the same as in Section IV-C. The net SSB receiver temperature is

$$T_{RSSB} = 2(L_W - 1)T_{phys} + L_W T_{MSSB} + L_{SSB} T_{IF}. \quad (7)$$

In Fig. 10, the total receiver temperature is separated into 3 parts. The IF contribution is denoted by the lower section of each bar, the mixer contribution by the center part, and the input losses by the top section of each bar. The system noise temperature given by Linke et al. [9] includes losses in the Fabry-Perot filter used for sideband rejection and LO injection, so the input losses were not separated out from the total SSB loss.

There has been a steady improvement in the noise temperatures since millimeter cryogenic receivers were first reported in 1973 [2], in both mixer and IF contributions. The IF noise temperature has improved by a factor of two with the introduction of cooled 1.4-GHz FET amplifiers. This, in addition to an improved conversion loss of 0.2 to 1.2 dB, has lowered the IF contribution from 90 ± 10 K to 35 ± 3 K.

Mixer noise temperatures which were 300 K SSB for the first cooled 3-mm receivers were reduced to 200 K and now to 70 to 86 K. While the input losses L_W have remained constant at 0.5 to 0.6 dB, their effect has diminished from 60 to 15 K as the IF and mixer noise have decreased, since the input contribution is 6 K plus 15 percent of the mixer and IF contributions.

The parameters of the various diodes used by Kerr [10], [12], Linke et al. [9], the BTL diodes, MWT and UVA diodes used in the present mixer design, and the UVA diodes used by Räisänen et al. [31] are summerized in Table IV. On comparing the UVA diode used by Kerr [10] with the MWT or UVA diodes used in this design at room temperature, one notes that the electrical parameters are quite similar. The noise from the dc-biased diode is essentially the same [27], [28], so excess noise at high currents cannot explain the difference. So, the improvement in the equivalent temperature of the mixer as an attenuator T_{eq} from 323 to 178 K is entirely due to the mixer design.

Although the room temperature parameters have been the same since 1975, the slope parameter V_o of the lower doped ($< 10^{17}$ cm⁻³) diodes have, in general, been im-

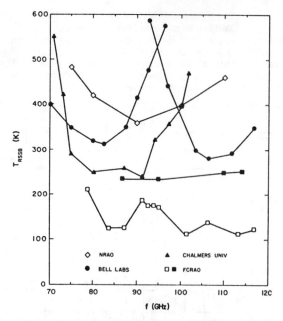

Fig. 11. Frequency response of 3-mm cooled mixers. Single sideband receiver noise temperatures are plotted over the 70 to 120-GHz range for the NRAO mixer [31], the Bell Telephone Laboratory system [9], the Chalmers University system [29], and the present design which is used at the FCRAO. The FCRAO DSB and SSB measurements were made with 50 and 500-MHz bandwidths, respectively. Open symbols are twice DSB results and filled symbols are SSB temperatures.

proved, as recommended by Viola and Mattauch [32], [33]. Occasionally, the $I - V$ curve of the cooled diode is not a single exponential but a combination of exponentials, due to different parts of the diode having different barrier heights [34], [35]. The slope parameter and series resistance of the diodes in Table IV have been derived by a fit of the dc $I - V$ curves with only a single exponential over the 1 μA to 5-mA current range.

While Fig. 10 compared various 3-mm mixers at their best operating points, the frequency response of the SSB receiver temperatures are compared in Fig. 11 for the NRAO receiver [36], the BTL system [9], the Chalmers University system [30], and the FCRAO receiver with a UVA 2P9-600 diode. The FCRAO data is just twice the measured DSB temperature in a 50-MHz bandwidth, since the sideband gains have been measured equal to within 4 percent. When used on the telescope, the FCRAO receiver has a Martin-Puplett [37] type sideband filter which, together with coupling optics, has $\simeq$ 0.5-dB loss. The SSB receiver temperature with this filter is also plotted in Fig. 11 for a measurement bandwidth of 500 MHz. The only better receivers in this frequency range are the recent SIS junction mixers reported by Pan *et al.* [38] which have achieved 68 K SSB at 115 GHz.

V. CONCLUSIONS

The combination of a broad-band mixer design, excellent Schottky barrier diodes, and low-noise FET amplifiers has given mixer receivers whose noise temperatures are the best ever obtained with Schottky diodes. This has been accomplished over the broadband of 80 to 115 GHz. At a physical temperature of 20 K, the SSB mixer noise temper-

ature is 70 K in the best case, an improvement factor of two over the best previous results.

The equivalent temperature of the diode as an attenuator is within 10 percent of the shot-noise limit due to the improved embedding circuit for the diode. This has been accomplished by an RF filter and backshort design which are near to a short circuit in the fundamental band and reactive in the second harmonic band. The whisker inductance has been optimized for a broad-band performance over the 3-mm band. The use of OFHC copper in the mixer construction with its high thermal and electrical conductivity is important in keeping the circuit losses low.

ACKNOWLEDGMENT

We would like to thank the Physics Department and FCRAO machine shops for their excellent machining, T. Carrol, R. Kot, and R. Berson for assembling the mixers, and Dr. G. R. Huguenin, Director of the FCRAO, for his support. One of the authors (C. R. Predmore) would like to thank the Institut de Radio Astronomie Millimetrique and Prof. M. Tiuri at the Helsinki University of Technology for their support while he was on leave from the University of Massachusetts, Amherst.

REFERENCES

[1] G. C. Messenger, "Cooling of microwave crystal mixers and antennas," *IEEE Trans. Microwave Theory Tech.*, vol. MTT-5, pp. 62–63, 1957.

[2] S. Weinreb and A. R. Kerr, "Cryogenic cooling of mixers for millimeter and centimeter wavelengths," *IEEE J. Solid-State Circuits*, vol. SC-8, pp. 58–63, Feb. 1973.

[3] *Model 350 Cryodyne*, CTI-Cryogenics Inc., Waltham, MA 02154.

[4] J. L. R. Marrero, "75–115 GHz cryogenic heterodyne receiver," M.S. Thesis, Univ. of Massachusetts, Amherst, Sept. 1981.

[5] J. E. Davis, "Ring-type directional filters," Electron. Div. Internal Memo., Nat. Radio Astron. Observ., Charlottesville, VA, May 1975.

[6] H. Cong, A. R. Kerr, and R. J. Mattauch, "The low-noise 115-GHz receiver on the Columbia-GISS 4-ft radio telescope," *IEEE Trans. Microwave Theory Tech.*, vol. MTT-27, pp. 245–248, Mar. 1979.

[7] D. Williams, S. Weinreb, and W. Lum, "L-band cryogenic GaAs FET amplifier," *Microwave J.*, vol. 23, Oct. 1980.

[8] A. V. Räisänen, "Experimental studies on cooled millimeter wave mixers," *Acta Polytechnica Scandinavica*, Elec. Eng. Series No. 46, Oct. 1980.

[9] R. A. Linke, M. V. Schneider, and A. Y. Cho, "Cryogenic millimeter-wave receivers using molecular beam epitaxy diodes," *IEEE Trans. Microwave Theory Tech.*, vol. MTT-26, pp. 935–938, Dec. 1978.

[10] A. R. Kerr, "Low-noise room temperature and cryogenic mixers for 80–120 GHz," *IEEE Trans. Microwave Theory Tech.*, vol. MTT-23, pp. 781–787, Oct. 1975.

[11] D. N. Held and A. R. Kerr, "Conversion loss and noise of microwave and millimeter-wave mixers: Part 1—Theory," *IEEE Trans. Microwave Theory Tech.*, vol. MTT-26, pp. 49–55, Feb. 1978.

[12] ——, "Conversion loss and noise of microwave and millimeter-wave mixers: Part 2—Experiment," *IEEE Trans. Microwave Theory Tech.*, vol. MTT-26, pp. 55–61, Feb. 1978.

[13] G. L. Matthaei, L. Young, and E. M. T. Jones, *Design of Microwave Filters, Impedance Matching Networks, and Coupling Structures*. New York: McGraw-Hill, 1964, pp. 203–205.

[14] Macor, Corning Machinable Glass Ceramic, Corning Glass Works, Corning, NY 14380.

[15] M. N. Afsar and K. J. Button, "Precise millimeter-wave measurements of complex refractive index, complex dielectric permittivity and loss tangent of GaAs, Si, SiO_2, Al_2O_3, BeO, Macor, and Glass," *IEEE Trans. Microwave Theory Tech.*, vol. MTT-31, pp. 217–223, Feb. 1983.

[16] G. K. White, *Experimental Techniques in Low-Temperature Physics*.

Oxford, Eng: Clarendon Press, 1979, p. 132.

[17] M. K. Brewer and A. V. Räisänen, "Dual-harmonic noncontacting millimeter waveguide backshorts: Theory, design and test," *IEEE Trans. Microwave Theory Tech.*, vol. MTT-30, pp. 708–714, May 1982.

[18] Scotch tape No. 74, 3M Insulation Materials, Minnesota Mining and Manufacturing Co., St. Paul, MN 55101.

[19] A. O. Lehto and A. V. Räisänen, "Embedding impedance of a millimeter-wave Schottky mixer: Scaled model measurements and computer simulations," *Int. J. Infrared Millimeter Waves*, vol. 4, no. 4, pp. 609–628, July 1983.

[20] A. C. Räisänen, N. R. Erickson, J. L. R. Marrero, P. F. Goldsmith, and C. R. Predmore, "An ultra low-noise Schottky mixer at 80–120 GHz," in *Infrared and Millimeter-wave Conf. Proc.*, Dec. 1981; pp. W3.8–9.

[21] C. R. Predmore, P. Goldsmith, A. Räisänen, P. T. Parrish, J. Marrero, and R. Kot, "Low-noise quasi-optical receiver for 75 to 115 GHz," presented at *URSI Symp. on Millimeter Wave Tech.*, Grenoble, 1980; Five College Radio Astronomy Observatory Report # 154, Univ. of Massachusetts, Amherst, MA 01003.

[22] A. V. Räisänen, C. R. Predmore, P. T. Parrish, P. F. Goldsmith, J. L. Marrero R. A. Kot, and M. V. Schneider, "A cooled Schottky-diode mixer for 75–120 GHz," in *Proc. of the 10th Eur. Microwave Conf.*, Warsaw, 1980, pp. 717–721.

[23] S. Nussbaum, J. A. Calviello, E. Saad, and N. Arnoldo, "Widely tunable millimeter-wave mixers using beam-lead diodes," in *IEEE MTT-S Dig.*, pp. 209–211, 1982.

[24] P. H. Siegel and A. R. Kerr, "A user oriented computer program for the analysis of microwave mixers, and a study of the effects of the series inductance and diode capacitance on the performance of some simple mixers," Goddard Space Flight Center, NASA Tech. Memo. 80324, July 1979.

[25] D. N. Held, "An approach to optimal mixer design at millimeter and submillimeter wavelengths," *IEEE Int. Microwave Symp. Dig.*, pp. 25–27, 1979.

[26] A. R. Kerr, "Shot-noise in resistive-diode mixers and the attenuator noise model," *IEEE Trans. Microwave Theory Tech.*, vol. MTT-27, pp. 135–140, Feb. 1979.

[27] R. W. Haas, private communication, 1983.

[28] D. N. Held, "Analysis of room temperature millimeter-wave mixers using GaAs Schottky barrier diodes," Doc. of Eng. dissertation, Columbia Univ., New York, 1976, no. X-130-77-6, Goddard Institute for Space Studies, Jan. 1977.

[29] N. Keen, R. Haas, and E. Perchtold, "Very low noise mixer at 115 GHz using a Mott diode cooled to 20 K," *Electron. Lett.*, vol. 14 no. 25, pp. 825–826, Dec. 1978.

[30] E. L. Kollberg and H. H. G. Zirath, "A cryogenic millimeter-wave Schottky-diode mixer," *IEEE Trans. Microwave Theory Tech.*, vol. MTT-31, pp. 230–235, Feb. 1983.

[31] A. Räisänen, J. Lamb, A. Lehto, M. Tiuri, and J. Peltonen, "Performance of a cryogenic 3-mm receiver on a 14 m radio telescope," in *Proc. of the 13th Eur. Microwave Conf.*, Nurnberg, Sept. 1983, pp. 477–482.

[32] T. Viola Jr., "High frequency noise in Schottky barrier diodes," Doc. Eng. dissertation, University of Virginia, Charlottesville, June 1973.

[33] T. J. Viola and R. J. Mattauch, "Unified theory of high-frequency noise in Schottky mixers," *J. Appl. Phys.*, vol. 44, pp. 2805–2808, 1973.

[34] H. Zirath, E. Kollberg, M. V. Schneider, A. Y. Cho, and A. Jelenski, "Characteristics of metal-semiconductor junctions for mm-wave detectors," in *Proc. Infrared Millimeter-wave Conf.*, Feb. 1983, p. W-10.

[35] E. Kollberg, H. Zirath, M. V. Schneider, A. Y. Cho, and A. Jelenski, "Characteristics of millimeter-wave Schottky diodes with microcluster interface," in *Proc. of the 13th Eur. Microwave Conf.*, Nurnberg, Sept. 1983.

[36] R. W. Haas, "Comparison of 3-mm receiver temperatures," IRAM Memo., Grenoble, France, Nov. 14, 1983.

[37] D. H. Martin and E. Puplett, "Polarized interferometric spectrometry for the millimetre and submillimetre spectrum," *Infrared Phys.*, vol. 10, pp. 105–109, 1969.

[38] S.-K. Pan, M. J. Feldman, A. R. Kerr, and P. Timbie, "A low-noise 115 GHz receiver using superconducting tunnel junctions," *Appl. Phys. Lett.*, vol. 43, no. 8, pp. 786–788, Oct. 15, 1983.

A Very Low-Noise Single-Sideband Receiver for 200–260 GHz

NEAL R. ERICKSON, MEMBER, IEEE

Abstract — A cryogenic Schottky diode mixer receiver has been built for the 230-GHz region with true single-sideband operation and a receiver noise temperature as low as 330 K. Local oscillator power is provided by a frequency tripler, with LO injection and sideband filtering accomplished through quasi-optical interferometers. The image sideband is terminated in a cryogenic load with an effective temperature of 33 K. The IF bandwidth is 600 MHz with nearly flat noise, and the RF band is nearly flat over 50 GHz using backshort tuning of the mixer.

I. INTRODUCTION

SCHOTTKY DIODE mixers for the 230-GHz region have shown a rapid improvement over the past five years as interest in this spectral region has increased. While local oscillator sources have long been regarded as a major limitation, the development of high-efficiency frequency triplers [1], [2] has solved this problem, making practical receivers for this band possible. At the same time, mixers and associated components have evolved to approach the performance attainable at lower frequencies. The best previously reported cryogenic mixer receiver for this frequency [3] operated double sideband with a DSB noise temperature of 235 K. Since this mixer is fixed tuned, its useful bandwidth is limited to 30 GHz.

This paper reports on a true single-sideband receiver covering 200–260 GHz which achieves substantially lower noise, and at the same time has a useful IF bandwidth of over 600 MHz. This receiver has required major advances in several critical elements, and a rather novel optical design.

The receiver consists of a single wide-band backshort tunable mixer and 1.1–1.7-GHz FET IF amplifier, both operating at ~ 20 K inside a vacuum dewar. Local oscillator power is provided by a frequency tripler. Optical elements are used to separate the signal and image bands in the mixer response, to inject the LO into the mixer, and to direct the image response to a 20-K termination within the vacuum dewar. Suppression of the image response is essential for well-calibrated observations of spectral line sources in radio astronomical observations. For the case of a cryogenic image termination, an additional advantage is that this termination adds very little to the total system

Manuscript received March 1, 1985; revised June 6, 1985. This work was supported in part by the National Science Foundation, under Grant AST-82-12252.

The author is with the Five College Radio Astronomy Observatory, University of Massachusetts, Amherst, MA 01003.

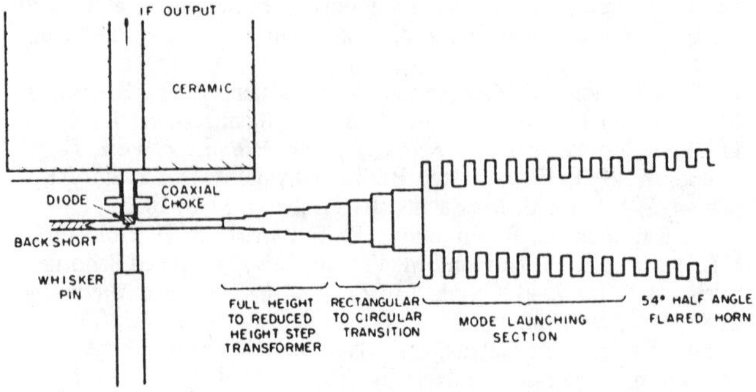

Fig. 1. Cross section of mixer and throat section of feed horn.

noise, while double-sideband receivers pick up an additional and sometimes substantial noise contribution from the effective image temperature.

II. MIXER

The mixer design is central to the receiver performance. The mixer uses ~ 1/3 height waveguide of dimensions 0.16×0.91 mm with an eight-section step transformer to a corrugated horn, and is shown in cross section in Fig. 1. The step transformer is designed to avoid spurious resonances and to maintain a low VSWR over the full frequency range. A step in width is made between the rectangular to circular transition and the actual mixer waveguide, since 0.91-mm width seemed necessary to allow the mixer to reach the upper edge of the 200–260-GHz band, while the existing rectangular to circular step transition [4] was designed for rectangular waveguide 0.98 mm wide in order to cover the same band.

The IF filter choke is a very simple air dielectric radial mode choke used in several other comparable frequency mixers [7]. It was chosen because it is easily fabricated with good control of critical parameters, and has a center pin of constant diameter for maximum rigidity, allowing the diode chip to be soldered on and carefully ground down *in situ* to the exact diameter of this pin. This choke has no stopband at second harmonic frequencies. The diode chip was fabricated by R. Mattauch (Univ. of Virginia) and is from batch 1H2. The anodes are 1 μm diameter with a zero-bias capacitance of 1.8 fF and a room-temperature series resistance of 15 Ω. The epitaxial layer doping is 4×10^{-16} cm^{-3} with a thickness of 0.17 μm. These diodes show a steepening of the logarithmic slope of the $I-V$ curve by a factor of 3.3 upon cooling to 20 K, and a comparable reduction in mixer noise. At the same time, the

Reprinted from *IEEE Trans. Microwave Theory Tech.*, vol. MTT-33, no. 11, pp. 1179–1188, Nov. 1985.

dc series resistance increases to 19 Ω. This yields an extremely high $1/RC$ cutoff frequency of 4.7 THz, while the very low capacitance allows the diode impedance to more readily match that of the waveguide, without resorting to the extremely reduced-height guide. This means that the waveguide is more easily fabricated, that a contacting backshort can operate more reproducibly, and that waveguide losses are reduced.

Corrugated horns present fabrication problems at these frequencies because most design criteria require many very thin corrugations per wavelength [6] and a mode launching section with rather deep, narrow grooves. However, other authors have suggested that fewer grooves are needed if less stringent performance standards are set [5]. This horn was designed around a more relaxed set of criteria, and has a corrugation period of 0.41 mm, or ~ 3 grooves per wavelength, with rather thick walls. Most grooves are 0.31 mm deep, or $\sim \lambda/4$ at midband, while in the mode launching section, grooves start at 0.52 mm deep ($\lambda/2$ at the high end of the band) and taper in depth over 10 grooves to the $\lambda/4$ depth. The horn tapers at 5.4° half angle to an aperture of 9.0 mm, and behaves as a constant aperture over the band 200–290 GHz. The complete length of the horn is 44 mm. To test this design, a 2.4-times scaled prototype was built in the WR-10 band, including the rectangular to circular transition, and was compared to a rectangular horn for insertion loss by using both as feeds for a low-noise radiometer. No significant difference was found between the losses of the two horns over the full scaled operating band. Beam patterns were also measured for this model and agreed well with the expected patterns over the band. This horn is coupled to incoming radiation through a 0.5-mm teflon dewar window tilted at Brewster's angle (55°) for zero reflection loss independent of frequency. The horn is very close to this window to minimize the window area and to allow the next mirror to be placed sufficiently close to the horn.

The entire mixer waveguide and feed horn are electroformed as a single unit with a minimum length of reduced-height guide between the horn and mixer diode. This waveguide was made using an aluminum mandrel, gold-plated, and then copper-plated up to the needed thickness. The inside gold plating is probably not optimum for such a mixer because tests of typical gold-plated waveguide at room temperature show losses ~ 1.2 times that of machined OFHC copper, and also because the conductivity of gold increases much less at cryogenic temperatures than that of copper. However, the short length of waveguide involved has an expected loss of ~ 5 percent so is not very important in any case. The IF choke center conductor is supported in a ceramic (Macor) [16] ring tapered and pressed into a wafer bolted to that containing the waveguide, with the choke center pin gold-plated and pressed into the Macor ring. This assembly can be made extremely rugged and has survived many thermal cycles with no diode failures.

One-micron diode anodes present greater difficulty in contacting than do larger anodes, but have shown no greater contact instability upon thermal cycling. However, for any diode, the rather short whiskers (~ 0.2 mm) needed for this mixer tend to produce excessive contact pressure on the diode, if 12-μm-diam Ph-bronze wire is used. This excessive pressure produces little effect at room temperature, but, upon cooling, the IV curve may be significantly degraded (less steep slope), probably due to stress-induced microcracking of the epitaxial layer. This is accompanied by an increase in noise. To avoid this problem, whisker wires are thinned to 6 μm diam using the same solution used to sharpen them. Contact whiskers must be made ~ 20 percent shorter to compensate for the increased inductance of this thinner wire. Whiskers are gold-plated, soldered onto the final whisker pin, and then bent to shape. The typical overall length for a 6-μm-diam wire is 0.2 to 0.25 mm.

Whisker pins are 0.50-mm BeCu turned down to 0.32 mm diam where they pass through the waveguide wall. These pins are pressed through a close fitting hole in the block adjacent to the waveguide block. It has been found that up to 10–15 μm of gold can be plated on the BeCu pins to improve the fit. Greater amounts will quickly flake off and would require an intermediate binding layer to improve the adhesion.

While an effort was made to maintain tolerances of 5 μm throughout the fabrication of the mixer, there are some perceptible irregularities in the throat of the horn, where errors may reach ~ 10 μm and where machining burrs may still be present. However, two mixers of this design have shown nearly identical performance, so fabrication is not an apparent limitation.

The mixer backshort is a contacting design made from a Ph-bronze shim with a simple V groove in the end. This groove has an opening angle of 35° and extends across the width of the short. This shim is heavily gold-plated and the V spread apart with a sharp blade, and then forced into the flared open end of the guide. This type of short produces excellent performance for a large number of tuning cycles, but eventually tends to become erratic if it is tuned too much. A noncontacting type of short might possibly be feasible but has not been attempted, and would be considerably more fragile. Fixed-tuned backshorts have been advocated to eliminate these problems [3], but this mixer design achieves wider bandwidth and lower noise than a fixed-tuned mixer. Backshort wear is not a real problem in this mixer since all data presented here was measured after five months of use on the FCRAO 14-m antenna with no backshort alterations, and shows no deterioration in performance over previous measurements.

III. IF MATCHING CIRCUIT

For minimum IF passband ripple and lowest conversion loss, an effective IF matching circuit is needed. This is particularly true when using a very low-capacitance diode which raises the overall mixer impedance level and results in an IF impedance of ~ 400 Ω at room temperature, and probably a similar value when cold. The approximate effect of the RF choke and ceramic supporting ring is to add a parallel capacitance of 0.93 pF. This can be matched with a low wide-band VSWR to 50 Ω by simply adding a 10-nH lumped inductor between the IF pin and the IF stripline. The resulting IF return loss was measured at room temperature and is 10 dB at the edges of a 500-MHz band and exceeds 15 dB at band center. A somewhat flatter match could be accomplished by adding a shorted λ/4 stub but the improvement is only marginal, and adds some additional loss as well. The matching circuit and a bias-T/dc block are incorporated on stripline in a small cutout box included as part of the mixer block.

IV. IF AMPLIFIER

The cryogenic IF amplifier is a three-stage FET design similar to that described by Weinreb *et al.* [8]. Electrically it is equivalent, but a duroid substrate with microstrip interconnections was used to simplify parts mounting. An input matching transformer and stub were used to increase the bandwidth as in the original design, but this model uses 0.63-mm alumina as the substrate with the alumina ground plane soldered down using pure indium. This has the advantage of producing an amplifier which can be tuned up entirely at room temperature, with little or no retuning needed for 20-K operation. The bandwidth is 600 MHz (1100–1700 MHz) with an average noise temperature at 77-K ambient of ~15 K. This should decrease to ⩽10-K at 20-K ambient but was not measured due to the extra complexity of 20-K measurements. The gain at 77 K is 29–32 dB across the full band. Input return loss is >15 dB (VSWR <1.5/1). There is little evidence for FET self-heating due to the relatively poor thermal conductivity of the duroid board since the noise temperature is quite insensitive to drain current, at least at 77 K.

V. OPTICS

Much of the superior performance of this receiver can be attributed to an unconventional optical design. A design goal was true SSB operation, and since it is difficult to tune a mixer for flat SSB response without a very high IF, it is essential to use an input filter and image load. In order to minimize the additive noise due to the temperature of the image termination, this termination must be made as cold and as well matched as possible. For lowest loss, the input filtering must be done optically. Because the optics are somewhat large and massive, attempting to cool them would require a much larger refrigerator than needed otherwise. Since optics losses can be kept very low, a room-temperature design seemed more practical.

Optical elements need to serve the dual functions of LO injection and signal/image separation. No single optical element can perform these simultaneously so it is necessary to cascade filter elements. One of the simplest, as well as the lowest loss, optical filters available is the two-beam interferometer known as the Martin–Puplett interferometer (MPI) as shown in Fig. 3. However, this device has a

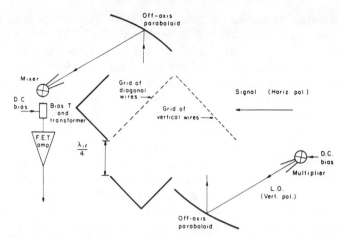

Fig. 2. Schematic diagram of Martin–Puplett interferometer. Wires of grid 1 are at 45° with respect to the incoming polarization, wires of grid 2 are aligned perpendicular to the incident signal beam.

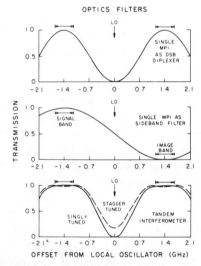

Fig. 3. Transmission versus frequency of various configurations of Martin–Puplett interferometers (MPI's). (a) Single MPI used as a LO signal diplexer in a DSB receiver. (b) Single MPI used as a sideband filter in a SSB receiver. (c) Tandem MPI used as LO filter and LO image sideband diplexer.

sinusoidal passband

$$T = 1/2\left(1 + \sin\frac{2\pi d}{\lambda}\right)$$

which is not optimal where wide bandwidth is needed.

One design requirement of this system was an IF bandwidth of over 500 MHz with little variation in noise. A previous receiver [7] used a single MPI in a DSB receiver by placing the LO, signal, and image bands within the passband as shown in Fig. 3(a). This works well over a narrow band, but in the case of a noisy LO source, the noise rises rapidly as the IF is tuned away from the band center. As a case in point, if the LO noise is only 300 K, as would be the case with a room-temperature attenuator in the LO path at an IF offset of 250 MHz from a 1400-MHz center frequency, the net signal and image transmission is 0.92, causing an additive noise temperature of 0.08 × 300 K = 24 K, as well as a signal reduction. Typical sources produce significantly greater noise than this so it is

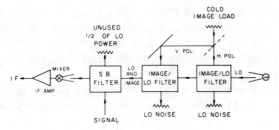

Fig. 4. Block diagram of 230-GHz receiver.

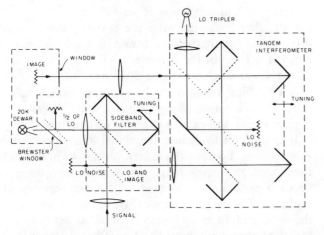

Fig. 5. Schematic diagram of optical path in 230-GHz receiver (lenses are used to represent mirrors). Vertical wire grids are shown as dotted lines, diagonal grids are shown as dashed lines.

essential to provide a better means of LO filtering, as well as a lower signal loss over the band. In order to minimize the receiver noise, the signal path should have the lowest optical loss, while the image and LO paths are less critical.

Therefore, an unconventional approach was taken. The LO injection interferometer as used in the DSB receiver was retained, but its path difference was decreased by a factor of 2, so that it becomes a sideband filter with a passband as shown in Fig. 3(b). Now the transmission is quite flat over a 500-MHz band with the minimum transmission being 0.98. Similarly, the image band is well rejected. The consequence of this is that the LO falls at point of 50-percent transmission and can be coupled in through the image arm.

A second interferometer follows in the image arm to separate the LO from the image. This interferometer has twice the path difference to allow optimum LO and image transmission. However, due to the bandwidth limitation mentioned before, this provides inadequate LO noise filtering as well as a poor wide-band image termination. This problem can be overcome by following the first image/LO separation interferometer with a second tuned in tandem. This is shown in the receiver block diagram of Fig. 4. This greatly flattens the passband, but requires two image ports, as well as double tuning. However, by a careful geometrical layout of the two, they can be made easily tandem tuned with only a single image port. This is shown in the receiver schematic in Fig. 5, where focusing mirrors are represented by lenses. The two image port inputs are in opposite polarizations and are combined on the input grid of the first filter to form a mixed polarization image input. The

overall transmission of this tandem interferometer is shown in Fig. 3(c). The bandwidth with <1-percent increase in loss is 41 percent of the IF center frequency, or 574 MHz in this case. The polarization makeup of this image input depends upon frequency, changing from linear at band center, to up to an 8-percent horizontal component at the band edge. This introduces one additional constraint for the image termination, which is that it must work well in both polarizations.

This type of filter forms a superior alternative to a Fabry–Perot interferometer where a wide flat passband is needed, with low LO loss. It has a further advantage of a very wide tunable band since the wire grids behave as nearly ideal polarizers over several octaves. A Fabry–Perot would require ~ 90-percent reflectance mirrors to operate with comparable noise rejection, and, as a result, the LO loss would be appreciable and the tuning rather critical. One disadvantage of the present version is that the output is composed of two delayed Gaussians, which in this case have a rather low mode overlap, and thus cannot be treated as the same mode. However, the image termination is designed to accept both modes so this is not a problem. In other applications, this difficulty could be overcome by delaying one of the beams to equalize the paths and thus produce a perfect overlap.

Should an even wider passband be needed, the two interferometers could be stagger tuned, although this would result in a greater LO loss. As an example, if a peak theoretical LO transmission of 0.82 is acceptable, the two interferometers can be offset by 1/5 of a period, to yield a transmission bandwidth of 58 percent of the center frequency with a maximum loss of 1 percent. This corresponds to 812 MHz at a 1400-MHz center frequency.

In order to realize the low loss inherent in optical devices, it is essential to minimize diffraction loss and mode conversion. The scalar feed produces a beam which closely approximates a Gaussian profile with w_0 of 3 mm at all frequencies. For a Gaussian beam [11], [15], the intensity varies as

$$I(r, z) = e^{-2r^2/w^2} \quad \text{where} \quad w(z) = w_0 \left(1 + \left(\frac{\lambda z}{\pi w_0^2}\right)^2\right)^{1/2}$$

and z is the distance from a beam waist in the system. Assuming the feedhorn launches a 100-percent Gaussian beam, a minimum aperture size of 3.5 w is needed to reduce diffraction loss at a single aperture to 0.5 percent, and to minimize the contribution of room-temperature losses to the total system noise. Since several limiting apertures are involved, and because a scalar horn launches a mode in which only 98 percent of the power is in the Gaussian mode, it is desirable to allow somewhat greater clearance where possible, 4 w being the goal of this design. In this particular receiver, space constraints prevented using an optics clear diameter of more than 51 mm except at a few places where extra space was available. There is also a requirement in dual-beam interferometers that the mode overlap when the beams recombine be close to unity [10], and this establishes a minimum beam-waist size within the

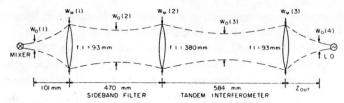

Fig. 6. Unfolded optical path between the tripler and mixer showing positions of focusing mirrors (represented as lenses here) and beam waists.

TABLE I
BEAM SIZE WITHIN RECEIVER OPTICS

f (GHz)	$w_o(1)$	$w_o(2)$	$w_o(3)$	$w_o(4)$ (all in mm)	$w_m(1)$	$w_m(2)$	$w_m(3)$	z_{out}
200	3.0	13.6	12.1	3.5	16.4	15.4	14.9	10.4
240	3.0	11.6	12.1	3.0	13.7	14.4	14.2	10.1
280	3.0	10.1	12.1	2.5	11.9	13.7	13.7	9.9

interferometer. The net transmission is given by

$$T = \frac{1}{4}\left[1 + \left(1 + \left(\frac{\lambda\Delta}{2\pi w_0^2}\right)^2\right)^{-1/2}\right]^2$$

and for $T > 0.99$, requires $2\pi w_0^2/\lambda\Delta > 6.7$, where Δ is the path difference between the two arms. This in turn requires $w_0 > 8.5$ mm in the SB filter and $w_0 > 12$ mm in the tandem interferometer, and these sizes were designed into these optics.

One intermediate focusing mirror was found necessary in this design, and could conveniently be included as a 90° off-axis folding mirror. A second focusing mirror between the right-hand pair of interferometers would have allowed a somewhat more optimum design, but could not be included in a reasonable position except as a lens, which would introduce more loss than it would prevent. In fact, the final interferometer can have a significant loss with little effect since it serves only to prefilter the LO (LO loss is not too critical) and to terminate at most 8 percent of the image which leaks through the first filter. The unfolded optical path is shown in Fig. 6 and the various beam parameters are summarized in Table I. As is seen from this table, the requirement of 3.5-w clearance is maintained at all points in this beam for a clear diameter of 51 mm except for mirrors 1 and 2 at 200 GHz. Mirror 1 is actually 57 mm in diameter and so fulfills this, but mirror 2 is not, and increases the loss to ~1.5 percent.

To minimize the number of different mirrors needed, mixer, LO, and image port mirrors are identical 60° off-axis ellipsoids, machined using a special technique on a conventional milling machine [9]. Off-axis mirrors are inconvenient to work with, so all mirrors are followed by a flat to produce an inline offset optics path. A 60° off-axis angle was chosen as a compromise between the ideal of on-axis optics, and the realities of needed clearances. While mirror

2 is theoretically an ellipse, it is a fairly flat mirror, and an off-axis parabola of equivalent focal length was substituted since it was available, and, in this case of long focal length, differed only slightly from the needed ellipse (in general this is not true).

All beam splitters are free-standing polarizing grids of 25-μm gold-plated tungsten wires, spaced by 85 μm. The spacing varies by ~40 μm, due to irregularities in the wire winding frame, but this seems to cause no observable problems. Three grids have wires oriented at 35.26° (which becomes 45° in projection for a 45° beamsplitter) while, for all other grids, the wires are oriented along the short distance across the holding frame. A final grid is placed across the front of the Brewster window to the mixer port of the dewar to eliminate the 50 percent of the LO power which reaches this port in the wrong polarization. If not rejected, this power can produce a high VSWR on the tripler output and cause LO power instabilities.

It is difficult to deduce the actual loss of the optics in this receiver because the mixer cannot be used without them, and any single-frequency measurement using feedhorns is certain to be plagued by standing-wave problems at the level of accuracy needed. However, we can compare the sideband filter optics to a previous set which were identical except for having a larger beam size internally. With a clearance of only 3 w, the loss in this interferometer alone was only 5 percent higher than in the new version with 3.6-w clearance. (This is inferred from the change in receiver temperature using the identical mixer). Thus, it is unlikely that the diffraction losses in this new device exceed 2–3 percent and resistive losses in the wire grids and mirrors are likely not to exceed 1 percent, based on considerable experience with such elements.

A particular problem with this type of sideband filter is that for a mixer with equal sideband gains, no simple indication of correct tuning for SSB operation is available. The only methods available are either to optimize coupling to a strong signal source at the correct frequency, or as a more practical means, to determine the true zero path difference setting, through mechanical measurement of LO power peaks and nulls at various frequencies, and to offset from this point to the calculated SSB setting through an accurate displacement transducer. For laboratory tests, a simple method to tune up when the exact LO frequency is not known is to find the LO transmission peak nearest the desired path for SSB operation, and to then offset by $\pm\lambda/8$ for SSB operation. Increasing the path by $\lambda/8$ produces USB operation, while decreasing it produces LSB. In astronomical observations, the sideband ratio was found to be >19 dB through observations of the strong CO line at 230 GHz.

Losses in the tandem interferometer can be measured more directly by measuring the SSB receiver temperature at the signal port and at the image port in front of the dewar window. This shows a small increase in T_R, after correction for the higher termination temperature of the signal port. Assuming equal sideband gains, which can be verified through measurement, this increase in T_R must be due to optics losses, and these are found to be ~3 percent.

This loss is not too important, and only results in an increase in the effective image termination temperature from 33 K at the dewar window to 42 K at the output of the tandem interferometer. Peak LO noise rejection of this filter is measured to be 40 dB, and the rejection is > 20 dB over a 600-MHz bandwidth, in good agreement with theory.

VI. OFF-AXIS REFLECTIVE OPTICS

For applications at frequencies above 100 GHz, most dielectrics become to lossy to be suitable for lenses. The few common exceptions are polyethylene, teflon, crystalline quartz, and TPX. The surface reflectivity of all these materials is high enough to require some compensation. Matching grooves are difficult to machine in curved surfaces, and at λ (1.3 mm) become quite small as well. While crystal quartz can be readily matched with a layer of polyethylene, it is a relatively expensive material and is difficult to grind, and in any case the matching coating becomes frequency dependent.

An attractive alternative is to use reflective optics sufficiently far off-axis to allow clearance for the input and output beams. These surfaces may be readily machined using computer-controlled machines or using analog techniques. However, reflective optics are somewhat more difficult to design and use for several reasons.

One problem is due to geometrical projection effects. In the case of nearly parabolic reflectors used far off-axis, an initially uniform beam becomes weighted toward the axial point after reflection simply because one side of the beam must travel farther than the other before reflection. This produces a distortion which becomes worse as the reflector is used farther off its axis, particularly if the included angle of the beam is large. A second effect occurs because at long wavelengths diffracting beams behave in nongeometrical ways, and all optics tend to be within the near-field of the focal points of the system. These effects include a phase velocity which is greater than c, and a curvature of wavefronts which is entirely different than in the geometrical case [15].

Other effects occur because an amplitude distribution changes shape as it propagates within the near-field region. A particular simplification occurs if a Gaussian mode is used. This distribution retains its shape within the near-field region since it is a normal mode of the system. Thus, it may be treated in a particularly simple manner. Since a corrugated feed horn launches a mode which is 98-percent Gaussian, it is convenient to design optics around an entirely Gaussian distribution and to plan on a complete loss for the 2 percent higher mode content of the beam. These considerations are applied to the ellipsoidal mirror used to focus the input beam into the mixer as follows.

For consideration of geometrical weighting effects, this mirror may be regarded as paraboloidal since its figure is very similar. A geometrical Gaussian beam of width 16° between $1/e^2$ power points, reflecting at 60° off-axis from a parabola, is distorted by this reflection in such a way that the direction of the maximum intensity is displaced by 0.2°

toward the axis. However, the resultant beam, while slightly asymmetric, has an overlap integral with a Gaussian of > 99.9 percent, and is broadened by only 1 percent relative to the perpendicular plane.

Phase errors due to diffraction were calculated for this beam at 230 GHz for a waist radius of 0.3 cm and the equivalent waist to mirror center spacing of 10 cm. In this case, the input beam phase differs by up to 0.1 λ from a simple geometrical spherical wave, but the phase error is mostly quadratic and may be accounted for by a small focus shift. The residual error is $\sim \lambda/40$. The output beam wavefront radius of curvature is sufficiently large that while the mirror is well within the near-field region, the curvature errors are $\leqslant \lambda/40$. Variations in phase velocity across the mirror, for both input and output beams, produce $\sim 4°$ of phase error. Thus, this mirror behaves nearly ideally for a single-mode beam, and is far superior to a lens.

VII. FREQUENCY TRIPLER

The lack of any convenient fundamental local oscillator sources in the 1-mm-wavelength region has made the development of frequency multipliers essential to the use of this spectral region. Recent improvements in Schottky varactor diode technology have made it possible to realize high-efficiency multipliers for these frequencies. For this receiver, a simple wide-band tripler [2] has been constructed covering the 195–255-GHz range having a peak conversion efficiency of 12 percent, and a second scaled device for 250–300 GHz has been used to extend measurements to 300 GHz.

Circuit requirements for a tripler are a means of conjugate impedance matching to the varactor at the input and output frequencies, and a resonant reactive termination at the second harmonic to enhance the tripling efficiency. Also needed is a means of biasing the diode with dc without loss of efficiency. The input and output match is made somewhat difficult because the varactor impedance is largely capacitive with a relatively small resistive component. From model and theoretical studies, the appropriate circuit model for the pumped varactor in an optimized circuit is found to be a capacitor of value 0.3–0.4 $C_j(0)$ in series with a resistor of 25–50 Ω.

The varactor diode used in this work is U.Va.-type 5M5 having a zero-bias capacitance of 15 fF and series resistance of 9 Ω. This device is mounted in the half-height WR-3 output waveguide (chosen to be cutoff to the second harmonic) as shown in Fig. 7, with input power coupled through a five-section, 50-Ω coaxial low-pass filter. Input power to the tripler is supplied through WR-12 waveguide tapering down to 1/5 height ($Z_g \sim 100 \Omega$). Power is then post coupled to the coaxial choke, with dc bias provided through a radial line filter in the opposite wall of the input guide.

Impedance matching at the input is greatly aided by a novel coaxial resonator using a reduced-diameter section on the whisker pin. This coaxial line is $\lambda/2$ long at the output frequency and so appears as a short circuit. At

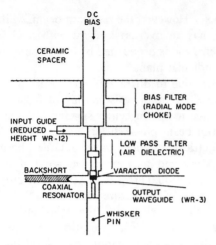

Fig. 7. Cross section through frequency tripler for 195–255 GHz.

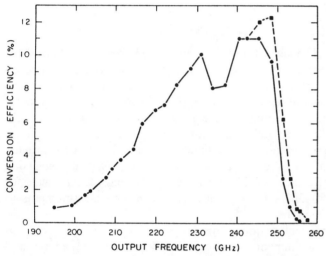

Fig. 8. Conversion efficiency versus frequency for frequency tripler. The solid curve is for an input power of 15 mW, while the dotted curve at the highest frequencies is for an input power of 30 to 60 mW.

midband for the input frequency, it is $\lambda/6$ long and appears inductive, with a reactance of $1.7jZ_0$, where $Z_0 \sim 70 \ \Omega$. This inductance is designed to resonate out the capacitive component of the diode impedance, leaving only the resistive component, which is comparable to the design impedance of the low-pass filter. This matching line also adds a series capacitive reactance of $-1.7jZ_0$ at midband for the second harmonic. This capacitive reactance largely cancels the whisker inductance at the low end of the band leading to the observed roll-off in efficiency, while at the upper end of the band this reactance decreases and the termination becomes inductive with nearly the optimal value. At much higher frequencies, the input match rapidly becomes very poor as the whisker pin resonator approaches $\lambda/4$ in length, and the efficiency drops to zero.

The frequency response of the tripler used with this receiver is shown in Fig. 8. Frequency tuning is accomplished by adjusting the input and output waveguide backshorts, with no additional tuning except to optimize the bias voltage. Backshorts are of the same contacting design as used in the mixer, and, as with the mixer, these shorts

have shown excellent life. Optimum bias varies over the band; at low frequencies, the bias voltage is 2 V, with an increase to 6 V near the upper band edge. Bias current varies from zero to 5 mA depending on drive level and operating frequency. Over most of the band, maximum efficiency occurs for an input of 15–30 mW, while, for most points, the maximum safe input drive is 40–50 mW. Thus, the maximum output power is 1–4 mW from 210–250 GHz. In the rapid roll-off above 250 GHz, peak efficiency occurs for a power of up to 60 mW, and this high drive extends the useful band by 2–3 GHz.

Coincidentally the mixer's LO requirements are well matched to this tripler. Required LO power peaks at ~ 250 GHz, decreasing significantly at 195 GHz. Thus, a single tripler provides sufficient LO over the range 195–255 GHz, which includes the most useful range for the mixer.

The tripler was fitted with a scalar feed horn having an aperture limited pattern with $w_0 = 0.3$ cm, designed much like that on the mixer, except that it was equipped with a flange.

VIII. CRYOGENIC IMAGE TERMINATION

Little is known about the optimum construction or performance of absorbing materials in the 1-mm region, particularly for cryogenic use. Magnetic absorbers as used at microwave frequencies tend to be very poorly matched and so produce a high VSWR when used in simple geometries. Carbon loaded foams seem to be well matched but have very poor thermal conductivity, and are unsuitable for use in a vacuum vessel due to outgassing problems. A simple solution was found for this receiver through the use of carbon-loaded epoxy. Carbon-loaded dielectrics have a loss which increases with frequency, and a nearly saturated mixture of lamp black carbon in a 50–50 low-viscosity epoxy resin [17] produces an absorber with ~ 6-percent reflectivity at normal incidence and sufficient loss to be useful in thicknesses of only 3 mm on a metal backing. This material can readily be fabricated and its viscosity can be controlled through the exact amount of carbon added. However, its thermal conductivity, like nearly all dielectrics, is poor at low temperatures, so special care must be taken to use it in contact with a metal backing in thin layers, and with minimum room-temperature radiation falling upon it, since a room-temperature object radiates 50 mW/cm² , enough to produce a substantial warming of this material. Also, this material tends to crack upon cooling if the rate of cooling is too great.

The cold load must work well in both polarizations since the tandem interferometer accepts both, so the design must not use polarization-dependent matching. Thus, a Brewster window as used on the mixer port is not suitable, so a simple teflon window 0.45 mm thick was used. This thickness is resonant in transmission at ~ 230 GHz and reflects less than 5 percent over an 80-GHz bandwidth. This adds, at most, 15 K to the cold load, and seems an acceptable alternative to more complex grooved or sandwich constructed windows. Teflon was chosen for the window because of its low dielectric constant and because of its

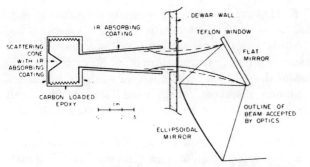

Fig. 9. Cross section through cryogenic image termination showing coupling to receiver optics.

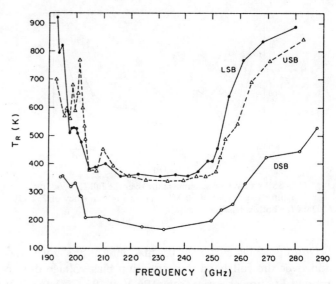

Fig. 10. Receiver noise temperatures measured for both sidebands and double sideband, at the IF center frequency of 1.4 GHz.

extremely low loss, probably the lowest of any dielectric at 1 mm.

The cold-load geometry is shown in Fig. 9, as well as the two beams matching to the receiver optics. The long input taper is intended to reduce the solid angle of window (effectively emitting at 300 K) as seen by the millimeter-wave absorber. However this tube will have no effect unless its walls are absorptive in the thermal IR, because otherwise it will act as a light pipe. To accomplish this, the walls are painted with a carbon-loaded latex paint (Eccocoat SEC [12]) in a very thin layer (this paint may be thinned with water). This material shows good adhesion to metals in thin layers, even upon rapid cooling. Measurements at $\lambda \sim 10$ μm (near the peak of the thermal emission from a room-temperature black body) show $\sim$ 90-percent absorption at 45° for a thin coating on a metal backing, while 1-mm measurements show $\sim$ 2-percent absorption. This pipe would reduce the entering IR power of 140 mW by a factor of 300 if the walls were totally absorbing, but because a higher reflection occurs at near grazing incidence, it is likely that the actual attenuation is considerably less, and several milliwatts still remain.

The load itself consists of a cylindrical cavity with walls coated with the carbon-loaded epoxy previously described. To improve the absorption, the side walls are grooved by cutting a fine thread of 1/4 mm period 0.3 mm deep using a lathe. The end walls are ungrooved. A rough cut scattering cone is centered in the load, also coated with the IR absorbing paint, which intercepts the remaining IR entering the load, and also helps randomize the millimeter-wave reflections within the load to maintain a uniform low emissivity. Size constraints within the dewar forced this load to be quite small but it works extremely well nonetheless. A radiometric temperature of 33 K was measured at 230 GHz, and, as expected, rises to 41 K at 200 GHz and 37 K at 270 GHz. The actual internal dewar temperature is not well known, but the refrigerator used is unlikely to cool below 20–25 K given the loading upon it, so the cold load closely approaches this temperature.

IX. Dewar Geometry

The vacuum dewar consists of a simple cylinder 18 cm in diameter and 20 cm long with all cooling provided by a 1.5-W mechanical refrigerator (CTI model 21). Cold load,

mixer, and IF amplifier are all mounted on a copper plate bolted to the 20-K station, while an aluminum radiation shield mounted to the 80-K station encloses all of the 20-K items. A single mechanical rotary feed-through allows backshort tuning using a drive shaft of thin-wall stainless steel tubing. All dewar parts are aluminum, which produces a high outgassing rate, but due to the low temperature within, cryopumping maintains the needed vacuum, as long as no parts ever warm up (as during a short power failure).

X. Performance

Fig. 10 shows the measured SSB and DSB performance over the full RF band, measured at the IF band center of 1.4 GHz. The receiver covers the band from 205–252 GHz with nearly flat noise while the useful band extends down to 195 GHz and up to 290 GHz. Note that while the sideband separation is only 2.8 GHz, rather substantial variations in sideband gains may occur near the band edges, and that, even over the flat portion of the band, the upper sideband is preferred. Near the lower band edge, very resonant behavior is seen in the USB performance which is not apparent in the DSB or LSB response. The cause of this is not known, but one possible source, a spurious LO klystron mode, was eliminated through tests using a second klystron which produced essentially identical results. In this frequency range, optimum dc bias voltage and backshort position vary considerably for opposite sidebands, so it is clear that the sideband impedances are very different. An advantage of a sideband filter, in this case, is that the better sideband may be selected, and that a large sideband ratio produces no calibration problems. While extensive tests have been made only with a single diode contact, previous tests have shown that the higher frequency response can be improved considerably with a shorter contact whisker, while the low end is degraded.

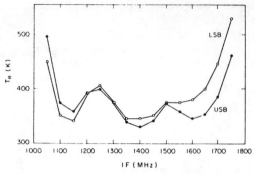

Fig. 11. SSB receiver noise temperature over the full IF band, measured with a filter bandwidth of 50 MHz, at an LO frequency of 232 GHz. Data for both sidebands is shown.

Optimum bias current is 0.2 mA with only slight variation over the full band, but optimum bias voltage due to applied LO power varies from 0.96 V at 195 GHz to 0.2 V at 250 GHz, and back to 0.9 V at 280 GHz. These widely varying LO needs seem completely uncorrelated with system noise.

Fig. 11 shows the variation in noise over the IF passband for both sidebands, measured with the LO at 232 GHz. This noise shows a ripple due to the interaction of the FET amplifier with the mixer but little other variation over 600 MHz. In USB, the lowest noise is 330 K, while the noise averaged over 550 MHz is only 360 K. Note that at the best frequency $T_{R\,USB} = 330$ K, $T_{R\,LSB} = 350$ K, and $T_{R\,DSB} = 164$ K. The average of the two SSB measurements is 340 K, while twice the DSB noise is 328 K. If we correct the SSB results by the apparent image termination temperature of 42 K (including tandem interferometer loss), we find the average SSB temperature corrected to a 0-K image is 298 K. Thus, the combination of mixer and sideband filter shows better performance SSB than when tuned DSB. This is partly because the mixer can always be tuned to slightly improve one sideband at the expense of the other. Probably a more important reason is that the sideband filtering interferometer path difference must be doubled in order to switch from the SSB to DSB mode and this increases its diffraction loss and mode conversion.

XI. Calibration

Receiver noise measurements require black-body absorbers at two known temperatures for proper calibration, but absorbers at 230 GHz are poorly characterized. Customarily carbon-loaded foam is used as a reference at ambient temperature and after dipping in liquid nitrogen (77.3 K). In this work, a few absorbers were characterized to find the one with the lowest radiometric temperature after a liquid nitrogen dip. Eccosorb AN-72 and CV-3 [12] were measured, as well as Keene absorber AAP-4C [13]. The lowest temperature was found for the Keene material. The CV-3 has a temperature 2.5 K higher while the AN-72 is ~18 K higher.

If we assume the Keene material is truly black (it has very deep corrugations and might be expected to be very good) at 77.3 K, then the CV-3 is 79.8 K and the AN-72 is

95 K. However, we have no independent means to test this assumption and all temperatures may be somewhat higher.

All data in this paper used CV-3 as a reference with an assumed effective temperature of 80 K, except for calibration of the cold load where AN-72 was used because of its minimum thickness, which would fit into the confined space available. The Keene material is unsuitable for most tests because its effective temperature rises very quickly after removal from liquid nitrogen (~ 5 s useful time), and it is very soft and quite thick (10 cm). The CV-3 material remains cold for ~ 15 s, is thinner and more rigid, and thus much easier to use for lab tests.

The effective temperature is so high for AN-72 that it is unlikely to produce reliable results, but its use is sometimes necessary because it is much thinner (~ 1 cm) than the other choices.

XII. Receiver Noise Breakdown

The overall receiver conversion loss, from RF input to mixer IF port, is measured to be 6.2-dB SSB at room temperature, with a system noise temperature of 477 K DSB. When tuned SSB, the optics loss decreases by about 3 percent, giving an SSB receiver temperature of 908 K (if the image could be terminated at 0 K). A small additional input loss of 2 percent is due to side lobes of the feed horn. This yields a mixer conversion loss of 5.9 dB, including feed horn resistive losses. For these tests, a different IF amplifier was used, with a room temperature noise of 38 K. This yields an IF contribution of 148 K to the total. From this, we derive a mixer noise temperature of 730 K SSB.

It is useful to derive the effective diode temperature from these figures, since it is a measure of the inherent noisiness of the diode and mixer. Assuming all losses are within the mixer diode itself, this effective temperature is [18].

$$T_D = (L-2)^{-1} T_{\text{MIXER}}$$
$$= 386\,\text{K}.$$

In fact, input losses and the series resistance loss occur at a temperature of 295 K, so the actual value of T_D will be somewhat higher than this. This should be compared to $1/2 T_{\text{physical}}$ [18] for an idealized mixer, so this diode has 2.6 times the noise of an ideal mixer. Good room-temperature mixers at 100 GHz can closely approach this ideal.

The origin of this noise is not well known, but it is not unique to this diode since two other diodes of different batches and anode diameters gave comparable room-temperature results. It is likely to be due in part to the effects of the embedding circuit, contributing noise from higher harmonic terminations, and perhaps from hot-electron noise within the diode itself, due to the peak LO current inducing far more noise than in an idealized diode.

It is interesting to note that the principle difference between the noise of this mixer and the best Schottky mixer reported at 100 GHz [14] is in this effective diode temperature. In conversion loss, they are nearly the same. Thus, any significant further improvement at this frequency requires an understanding of the source of this noise.

No cryogenic measurements of conversion loss have been made, but it is likely to be about the same, with a small (~ 0.1 dB) increase expected due to the increase in R_s. A receiver noise breakdown is as follows. The measured cryogenic SSB receiver temperature is 330 K, and correcting for the effective image temperature yields 288-K SSB. Correcting for the 2-percent room-temperature loss (DSB) due to the feed side lobes gives 270 K at the input to the mixer feed horn. The IF amplifier noise temperature of ~ 10 K adds 40 K to this total. Thus, the mixer temperature is 230 K. The diode equivalent temperature, in this case, is 115 K, a factor of 3.2 lower than at room temperature, which compares with a factor of 3.7 for the best 100-GHz results [14]. This cryogenic performance is the best that has been found for several types of diodes, all giving comparable room-temperature results. However, this effective temperature is much higher than for the best 100-GHz mixer and leads to an expectation of better results when these effects are understood.

XIII. CONCLUSIONS

A Schottky diode mixer receiver has demonstrated very low-noise operation at 230 GHz over a wide RF bandwidth. True single-sideband operation is achieved with an IF bandwidth of 600 MHz. A novel optical layout helps achieve these results and a cryogenic image termination designed for a minimum effective temperature contributes only 33 K to the total. A frequency tripler generates the LO power for the receiver and provides adequate power over more than a 60-GHz tuning range.

A noise analysis shows that even lower noise should be achievable in a more ideal mixer, since the conversion loss of this mixer is only 6.0 dB, comparable to the best 100-GHz mixers, which show considerably lower noise. This rather slight frequency-dependence to the conversion loss indicates that optimized receivers at significantly higher frequency may attain comparable results.

ACKNOWLEDGMENT

The author would like to thank the staff of the FCRAO machine shop for their fine work on the tripler and receiver, and R. Mattauch for providing the Schottky diodes used in this receiver.

REFERENCES

[1] J. W. Archer, "Millimeter wavelength frequency multipliers," *IEEE Trans. Microwave Theory Tech.*, vol. MTT-29, pp. 552–557, 1981.
[2] N. R. Erickson, "A high efficiency frequency tripler for 230 GHz," in *Proc. 12th Eur. Microwave Conf.* (Helsinki), 1982, pp. 288–292.
[3] J. W. Archer, "All solid-state low noise receivers for 210-240 GHz," *IEEE Trans. Microwave Theory Tech.*, vol. MTT-30, pp. 1247–1252, 1982.
[4] D. A. Bathker, "A stepped mode transducer using homogenous waveguides," *IEEE Trans. Microwave Theory Tech.*, vol. MTT-15, pp. 128–130, 1967.
[5] B. MacA. Thomas, "Design of corrugated conical horns," *IEEE Trans. Antennas Propagat.*, vol. AP-26, pp. 367–372, 1978.
[6] C. Dragone, "Characteristics of a broadband microwave corrugated feed: a comparison of theory and experiment," *Bell Syst. Tech. J.*, vol. 56, pp. 869–888, 1977.
[7] N. R. Erickson, "A 200–350 GHz heterodyne receiver," *IEEE Trans. Microwave Theory Tech.*, vol. MTT-29, pp. 557–561, 1981.
[8] S. Weinreb, D. L. Fenstermacher, and R. Harris, "Ultra low-noise 1.2–1.7 GHz cooled GaAsFET amplifiers," *IEEE Trans. Microwave Theory Tech.*, vol. MTT-30, pp. 849–853, 1982.
[9] N. R. Erickson, "Off-axis mirrors made using a conventional milling machine," *Appl. Opt.*, vol. 18, pp. 956–957, 1979.
[10] N. R. Erickson, "A directional filter diplexer using optical techniques for millimeter to submillimeter wavelengths," *IEEE Trans. Microwave Theory Tech.*, vol. MTT-25, pp. 865–866, 1977.
[11] P. F. Goldsmith, "Quasioptical techniques at millimeter and submillimeter wavelengths," in *Infrared and Millimeter Waves*, vol. 6, K. J. Button, Ed. New York: Academic, 1982.
[12] Emerson and Cuming, Canton, MA.
[13] Keene Microwave, Advanced Absorber Products, Amesbury, MA.
[14] C. R. Predmore, A. V. Raisanen, N. R. Erickson, P. F. Goldsmith, and J. L. R. Marrero, "A broad-band, ultra-low-noise Schottky diode mixer receiver from 80 to 115 GHz," *IEEE Trans. Microwave Theory Tech.*, vol. MTT-32, pp. 498–507, 1984.
[15] H. Kogelnik and T. Li, "Laser beams and resonators," *Appl. Opt.*, vol. 5, pp. 1550–1567, 1966.
[16] Corning Glass Works, Corning, NY.
[17] Chemlok 305, Hughson Chemicals, Erie, PA.
[18] A. R. Kerr, "Shot-noise in resistive-diode mixers and the attenuator noise model," *IEEE Trans. Microwave Theory Tech.*, vol. MTT-27, pp. 135–140, 1979.

Some analytical and intuitive results in the quantum theory of mixing

M. J. Feldman [a]

NASA Goddard Institute for Space Studies, New York, New York 10025

(Received 10 August 1981; accepted for publication 21 September 1981)

This paper is an elaboration upon Tucker's quantum theory of mixing, which has proven very valuable in interpreting recent experiments in superconducting quasiparticle mixing. Tucker's formula for the conversion gain of a three-frequency quantum mixer in the low intermediate frequency (IF) limit is generalized to include arbitrary source reactance; one choice of source reactance is found to resonate out the quantum reactance in the conversion expression. The signal reflection gain is evaluated; it becomes infinite simultaneously with the IF conversion gain. Infinite gain occurs only within the region of negative dc differential resistance. A number of relationships among the complex elements of the small-signal admittance matrix are presented; the nonlinear reactive elements are shown to be small under certain conditions when the conversion gain is maximized. The local oscillator (LO) power required for quantum mixing is calculated and found to be related to the gain denominator. This implies that certain quantum mixers at high gain will exhibit a constant gain-powerwidth product and a gain-proportional noise temperature component. A particularly simple expression for the conversion gain of an optimized quantum mixer is developed, which agrees with previous experiments. The origin of nonclassical behavior, including conversion gain and the nonlinear quantum reactance, is discussed. We conclude that the nonlinear quantum reactance, though it certainly must exist, is not itself responsible for conversion gain and has a minor, perhaps insignificant, effect upon the properties of a quantum mixer.

PACS numbers: 74.90. + n, 07.62. + s, 95.70. + i

I. INTRODUCTION

The classical theory of mixing has been widely developed and interpreted, and presently is well understood both analytically and intuitively (see, for instance, Refs. 1 and 2). The theory has proven very valuable in the design and optimization of practical devices; nonlinear resistive elements are routinely employed as frequency convertors in the detection of microwave and millimeter-wave radiation. Recently, Tucker[3,4] has generalized classical mixer analysis to account for the quantum behavior of the charge carriers in nonlinear single-particle tunneling devices. Tucker's theory is referred to as the "quantum theory of mixing." Tucker has shown that at frequencies high enough so the photon energy becomes comparable to the voltage scale of the dc nonlinearity of a tunnel junction, certain quantum effects become important, and the classical theory proves inadequate to describe the operation of a heterodyne mixer.

The quantum theory of mixing has proven to be of more than just academic interest. The theoretical work was paralleled by the experimental development of a new type of microwave mixer which exhibits distinctly nonclassical behavior. This device is based upon the superconductor-insulator-superconductor (SIS) tunnel junction, which has been extensively studied during investigations of the Josephson effect. In an SIS mixer the Josephson effects are either suppressed or ignored; instead the mixing utilizes the quasiparticle tunneling currents. The earliest experiments[5–7] demonstrated two important advantages of this device over conventional mixers. The SIS mixer can have very low noise; its noise temperature is apparently limited only by quantum noise.

Also, the SIS mixer requires extremely little local oscillator (LO) power. Later experiments[8–10] improved upon the earlier results and indicated a third important advantage: The possibility of intermediate frequency (IF) conversion gain. This is predicted by the quantum theory of mixing.[4] The most recent experiments have demonstrated infinite available conversion gain[11] and an actually achieved gain[12] of 4.3 dB. These experimental results signify that quantum mixing may be of great practical importance in the near future, for instance, in millimeter-wave astronomy where observations are distinctly receiver limited.

The quantum mixer theory has been quantitatively applied with considerable success.[9] Nevertheless, the theory has been met with some confusion, and a number of incorrect statements have appeared in the literature. This paper is an elaboration upon Tucker's quantum theory of mixing. Throughout, we concentrate upon reaching a better intuitive understanding of quantum mixing. Particular attention is paid to the nonlinear quantum reactance, which must inevitably accompany a nonlinear resistance in the quantum regime. In addition, we present certain analytic results. Following Tucker,[4] only the three-frequency Y mixer[2] in the limit of low IF is analytically treated. This should be a reasonable approximation for SIS mixers; all previous experiments[5–12] have used an IF of less than 1.5% of the LO frequency and the best results[11,12] have been achieved with junctions having a relatively large capacitance, tending to short out higher harmonic frequencies. Perhaps the most serious omission from this paper is that we do not analytically discuss the noise properties of quantum mixers. It is sufficient to note that in the most precise experiments[5,8] the noise temperature of an SIS mixer is found to be close to the funda-

[a] NAS-NRC Research Fellow.

J. Appl. Phys., Vol. 53, No. 1, January 1982

M. J. Feldman

mental quantum noise limit. This noise contribution is small enough to be neglected in a practical receiver.

In Sec. II we generalize Tucker's formulas[4] for the conversion loss to include an arbitrary source susceptance, and calculate the signal reflection gain and the IF output admittance. Some simple examples are given in Sec. III. Section IV presents certain relationships among the complex elements of the small-signal admittance matrix. In Sec. V we calculate the LO power required for quantum mixing, and find this related to the gain denominator. This has remarkable implications for certain quantum mixers at high gain. In Sec. VI a particularly simple expression for the conversion gain of an optimized, practical, quantum mixer is developed. In Secs. VII and VIII we intuitively discuss quantum mixing, and in particular the origins and effects of the nonlinear quantum reactance.

II. GENERAL SOURCE REACTANCE

We shall confine our analysis to the simplest applicable heterodyne receiver model, the three-frequency Y mixer[2] in the limit of low IF. All higher harmonics and their sidebands are assumed to be shorted, either by the junction capacitance or by a tuning circuit. Therefore the voltage applied to the junction, $V(t) = V_0 + V_p \cos\omega t$, is sinusoidal, and this equation defines the origin of time. Only three small-signal frequencies are considered. The output frequency (IF) is taken to be small, both with respect to ω and to the voltage scale of the dc nonlinearity. Therefore the signal frequency, ω_1, and the image frequency, ω_{-1}, are approximately equal to one another and to ω. With these conditions Tucker[3] has obtained the dc current and the LO current through the junction

$$I_0 = \sum_{n=-\infty}^{\infty} J_n^2 I_{dc}(V_0 + n\hbar\omega/e), \tag{1}$$

$$I_p = \sum_{n=-\infty}^{\infty} J_n(J_{n-1} + J_{n+1})I_{dc}(V_0 + n\hbar\omega/e),$$

$$+ j \sum_{n=-\infty}^{\infty} J_n(J_{n-1} - J_{n+1})I_{KK}(V_0 + n\hbar\omega/e).$$

Unless otherwise specified, the argument of each Bessel function here and below is $\alpha = eV_p/\hbar\omega$. $I_{dc}(V)$ is the current-voltage function of the junction in the *absence* of local oscillator power, and

$$I_{KK}(V) = \frac{1}{\pi} P \int_{-\infty}^{\infty} \frac{dV'}{V'-V} I_{dc}(V') \tag{2}$$

is its Kramers-Kronig (KK) transform,[14] written as a function of voltage rather than frequency (P represents the Cauchy principal value). In Eq. (1) both I_{dc} and I_{KK} are evaluated at the applied dc voltage V_0 and at integral "photon points" away from V_0.

Tucker[3,4] has also evaluated the elements of the 3×3 complex small-signal admittance matrix, $Y_{ik} = G_{ik} + jB_{ik}$, where $i,k = -1, 0, 1$:

$$G_{00} = \sum_{n=-\infty}^{\infty} J_n^2 \frac{d}{dV_0} I_{dc}(V_0 + n\hbar\omega/e),$$

$$G_{10} = G_{-10} = \frac{1}{\alpha} \sum_{n=-\infty}^{\infty} n J_n^2 \frac{d}{dV_0} I_{dc}(V_0 + n\hbar\omega/e),$$

$$G_{01} = G_{0-1} = \frac{e}{\hbar\omega} \sum_{n=-\infty}^{\infty} J_n(J_{n-1} - J_{n+1})I_{dc}(V_0 + n\hbar\omega/e),$$

$$G_{11} = G_{-1-1} = \frac{e}{2\hbar\omega} \sum_{n=-\infty}^{\infty} (J_{n-1}{}^2 - J_{n+1}{}^2)I_{dc}(V_0 + n\hbar\omega/e),$$

$$G_{1-1} = G_{-11} = \frac{e}{2\hbar\omega} \sum_{n=-\infty}^{\infty} J_n(J_{n-2} - J_{n+2})I_{dc}(V_0 + n\hbar\omega/e),$$

$$\quad (3)$$

$$B_{10} = -B_{-10} = \frac{1}{2} \sum_{n=-\infty}^{\infty} J_n(J_{n-1} - J_{n+1}) \frac{d}{dV_0} I_{KK}(V_0 + n\hbar\omega/e),$$

$$B_{11} = -B_{-1-1} = \frac{e}{2\hbar\omega} \sum_{n=-\infty}^{\infty} (J_{n-1}{}^2 - 2J_n{}^2 + J_{n+1}{}^2)I_{KK}(V_0 + n\hbar\omega/e),$$

$$B_{1-1} = -B_{-11} = \frac{e}{2\hbar\omega} \sum_{n=-\infty}^{\infty} (J_{n-2}J_n - 2J_{n-1}J_{n+1} + J_nJ_{n+2})I_{KK}(V_0 + n\hbar\omega/e).$$

The elements B_{00}, B_{01}, and B_{0-1} are zero in the limit of small IF. Note that the admittance matrix is completely determined by the dc I-V curve and its Kramers-Kronig transform.

In the limit of small IF, the circuit external to the nonlinear element does not distinguish between the LO, the signal, and the image, nor between the IF and dc. A schematic circuit with only LO and dc quantities explicit is shown in Fig. 1. The rf powers are derived from a resistive source of conductance G_s and the converted IF power is delivered to a resistive load conductance G_L. A susceptance B_s, in parallel with the source conductance, represents the junction capacitance and any external tuning elements. In his treatment, Tucker[4] made the unnecessarily restrictive choice of $B_s = -B_{11}$ to assure that the diagonal elements of the augmented admittance matrix were purely real. For a classical

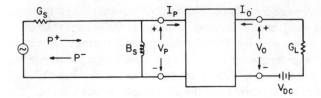

FIG. 1. Schematic circuit for a quantum mixer, showing only LO and dc ports. The susceptance B_s includes the linear junction capacitance and any external tuning elements.

resistive mixer this is indeed the best choice. We shall see that for the quantum mixer this choice is neither the best nor the most interesting. Here we will leave B_s arbitrary.

The signal to IF conversion gain is then calculated in the conventional way

$$\mathscr{G}_{IF} = \frac{4\eta' g_s g_L}{L_0(1+g_L)} \frac{[(\xi+g_s)^2 + (b_s-\gamma)^2]}{D^2}, \quad (4)$$

where
$$D \equiv (\xi+g_s)(1+g_s-\eta') + (b_s-\gamma)(b_s+\gamma-\eta'\beta), \quad (5)$$
and

$$g_s \equiv \frac{G_s}{G_{11}+G_{1-1}}, \quad g_L \equiv \frac{G_L}{G_{00}}, \quad \xi \equiv \frac{G_{11}-G_{1-1}}{G_{11}+G_{1-1}},$$

$$L_0 \equiv \frac{2G_{10}}{G_{01}}, \quad \eta \equiv \frac{2G_{01}G_{10}}{G_{00}(G_{11}+G_{1-1})}, \quad \eta' \equiv \frac{\eta}{1+g_L}, \quad (6)$$

$$\beta \equiv \frac{B_{10}}{G_{10}}, \quad \gamma \equiv \frac{B_{1-1}}{G_{11}+G_{1-1}}, \quad b_s \equiv \frac{B_{11}+B_s}{G_{11}+G_{1-1}}.$$

The notation is adapted from Tucker's with the addition of the reduced variable b_s. If $b_s = 0$, Eqs. (4)–(6) reduce to Tucker's formulas for the conversion [Ref. 4, Eqs. (2)–(4)], as required. Note that g_s, g_L, and ξ are positive definite, and that L_0 and η have the same sign.

The possibility of infinite conversion gain is perhaps the most surprising result of quantum mixer theory. Written in the above form, it is clear that $\mathscr{G}_{IF}$ is infinite when and only when $D = 0$. In this paper the phrase "the limit of high gain" is taken to mean $D \rightarrow 0$. This condition allows certain analytical results to be simplified. A gain of 20 dB is in general sufficient that the approximation $D = 0$ is valid. 10 dB is perhaps marginal.

Let us examine the first term of D, which contains only the resistive elements of the admittance matrix and of the external circuitry. If η is small, the nonlinearity may be considered "weak". (For a linear system $\eta = 0$.) Then this term must be positive, and there is a large conversion loss. For a classical mixer, η must be less than unity and, with our assumption of no image suppression, $\mathscr{G}_{IF}$ must be less than $1/2$. For the quantum mixer, η is not restricted to be less than unity. Under certain conditions η will become large enough to cause the first term of D to become negative. If $\eta > 1$, then for small enough g_s and g_L, $D = 0$ and the gain is infinite. The exact point at which $D = 0$ depends of course upon the value of the second term of D, which contains the reactive elements of the admittance matrix and external circuitry. We intend to demonstrate that this second term is relatively unimportant, that the reactive terms can be ignored without crucial error in examining the operation of

the quantum mixer. Note that by the choice $b_s = \gamma$, i.e., $B_s = B_{1-1} - B_{11}$, the second term in D is zero and in fact all reactive elements disappear from the expression for $\mathscr{G}_{IF}$. The nonlinear quantum reactance has, in one sense, been resonated away. Thus ignoring the reactive terms in the conversion corresponds to a physically realizable situation. The expression for $\mathscr{G}_{IF}$ then reduces to a particularly simple form, allowing some simple analytic results to be derived in Sec. VI. Note that choosing $b_s = \gamma$ does not maximize the conversion gain; this resonance condition is not the reason why, as discussed below, the quantum reactance has such a minor effect upon the maximum conversion. Rather, we shall demonstrate in Sec. IV that where $\mathscr{G}_{IF}$ is maximized, all of the reactive elements of the admittance matrix become themselves small.

In most experiments B_s is adjusted so that $\mathscr{G}_{IF}$ is maximized. This condition does not lead to a simple analytic expression for b_s. One may consider the "best" choice for b_s that which minimizes D, $b_s = \eta'\beta/2$. With this choice the second term of D in Eq. (5) is always $\leqslant 0$. This allows infinite gain with the smallest possible value of η.

We can straightforwardly calculate the signal reflection gain as

$$\mathscr{G}_s = [1 - g_s(\xi+2g_s+1-\eta')/D]^2 + [g_s(2b_s-\eta'\beta)/D]^2.$$

Note that the reactive elements do not disappear from $\mathscr{G}_s$ if $b_s = \gamma$. The most striking result is that the signal reflection gain is infinite whenever the IF conversion gain is infinite. The image and harmonic conversion gains will be infinite as well. This can have disturbing experimental consequences, viz. the possibility of instability and out-of-band oscillation. The dynamic range for high-gain operation will be severely limited. In the limit of high gain, we find a rather simple form for the ratio

$$\frac{\mathscr{G}_s}{\mathscr{G}_{IF}} = \frac{L_0}{4\eta'} \frac{g_s}{g_L} (1+g_L) \left\{ 1 + \left[\frac{b_s+\gamma-\eta'\beta}{\xi+g_s} \right]^2 \right\}. \quad (7)$$

Thus under certain circumstances one may choose $b_s = \eta'\beta - \gamma$ to minimize this function.

The IF output admittance of the mixer is given by

$$\frac{G_{out}}{G_{00}} = 1 - \eta \frac{\xi+g_s+\beta(b_s-\gamma)}{(\xi+g_s)(1+g_s)+b_s^2-\gamma^2}. \quad (8)$$

It is well known that $1/G_{out}$ is the differential resistance of the pumped dc I-V curve R_d, and it can be shown[13] that Eq. (8) can be obtained by calculating R_d. Within a certain region of experimental parameters (P_{LO}, V_0, G_s, and B_s) G_{out} is negative, as seen in previous experiments.[11,12] This region is bounded by the points in parameter space where R_d is infinite. By comparing Eq. (8) with Eq. (5), it is seen that infinite gain occurs only within this "negative resistance" region. In the limit $g_L \rightarrow 0$, of course, infinite gain occurs precisely where R_d is infinite. In Sec. VI we consider operating with a large, positive R_d.

III. EXAMPLES

It is instructive to consider some simple hypothetical examples. In the dc I-V curves shown in Fig. 2 the voltage is

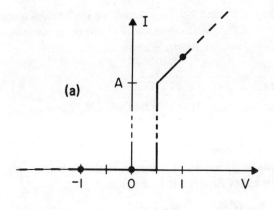

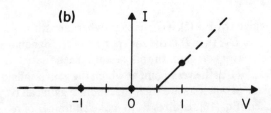

FIG. 2. Hypothetical dc I-V curves with V normalized to the photon voltage $\hbar\omega/e$ and I scaled to give a normalized resistance of unity. The voltage origin is chosen to be $\hbar\omega/2e$ below the discontinuity. (a) simulates the positive voltage branch of an ideal SIS junction. (b) is a special case of (a) with $A = 0$.

normalized to units of $\hbar\omega/e$ and the current units are chosen so that the normalized high-voltage resistance is unity. The voltage origin is chosen to be one-half a voltage unit below the discontinuity. For these curves the G_{ik}, but not the B_{ik}, can be calculated in closed form. Figure 2(a) closely approximates the positive voltage branch of an ideal SIS junction at $T = 0$, if $A = (\pi/2)\Delta/\hbar\omega$. At the dc voltage bias point $V_0 = 0$, Fig. 2(a) gives

$$G_{00} = \tfrac{1}{2}(1 - J_0^2), \quad G_{11} = \tfrac{1}{2} + \tfrac{1}{2}(J_0^2 + J_1^2)(A - \tfrac{1}{2}),$$

$$G_{1-1} = \tfrac{1}{2}(J_0 J_2 - J_1^2)(A - \tfrac{1}{2}),$$

$$G_{10} = (\alpha/4)(J_0^2 - J_0 J_2 + 2J_1^2),$$

$$G_{01} = (\alpha/2)(J_0^2 + J_1^2) + J_0 J_1(A - \tfrac{1}{2}).$$

It is clear that η can be > 1 so that an infinite gain is possible. In fact, in the limit of small LO power ($\alpha \to 0$) it is seen that $\eta \to 2$. This means an infinite gain is possible for an ideal SIS junction even in the limit of small LO power, contrary to statements in the literature. The load admittance G_L must of course be chosen of the order α^2 so that g_L remains finite.

Furthermore, in the limit $\alpha \to 0$ we find that $\beta = O(\alpha)$ and $\gamma = O(\alpha^2)$. Thus if $b_s = 0$ as for an optimized classical mixer, the reactive contributions to the conversion gain expression Eqs. (4) and (5) become vanishingly small as $\alpha \to 0$. This is an additional indication that the nonlinear quantum reactance is not essential in producing the gain of a quantum mixer.

Figure 2(b) is another hypothetical dc I-V curve, a special case of Fig. 2(a) with $A = 0$. Again $\eta \to 2$ as $\alpha \to 0$, so again infinite gain is possible. This disproves the statement[15] that an "S-shaped" dc I-V curve is required for conversion gain.

The general case of a monotonically increasing I-V

curve has been analytically investigated. η is nonzero in the limit $\alpha \to 0$ if, and only if, the slope of the unpumped I-V curve is identically zero at the dc-voltage bias point. Otherwise $\eta = 0$ and there is no possibility of conversion in this limit. If the slope is zero everywhere below the dc bias point, then $\eta \to 2$ as $\alpha \to 0$. Even a very small slope can have a drastic effect on the results, for finite α as well. Thus the ideal SIS junction at $T = 0$ is a very special case, and analytic results derived for it should be interpreted cautiously.

IV. MATRIX ELEMENT RELATIONSHIPS

Certain relationships exist between the various matrix elements given in Eq. (3) which have not hitherto been noted. In the most general form, with $Y_{ik} = G_{ik} + jB_{ik}$,

$$\frac{\partial Y_{00}}{\partial V_p} = \frac{\partial Y_{01}}{\partial V_0}, \quad 2\frac{\partial Y_{10}}{\partial V_p} = \frac{\partial(Y_{11} + Y_{1-1})}{\partial V_0}$$

$$\frac{\partial(Y_{11} - Y_{1-1})}{\partial V_0} = \frac{2Y_{10}}{V_p}, \qquad (9)$$

$$\frac{\partial(Y_{11} - Y_{1-1})}{\partial V_p} = \frac{2Y_{1-1}}{V_p}.$$

Also, the B_{ik} are related to the G_{ik} by Eq. (2)

$$B_{10} = \frac{1}{2\pi}\frac{\partial}{\partial\alpha}P\int_{-\infty}^{\infty}\frac{dV'}{V' - V_0}G_{00}(V'),$$

$$B_{11} + B_{1-1} = \frac{1}{\pi}\frac{\partial}{\partial\alpha}P\int_{-\infty}^{\infty}\frac{dV'}{V' - V_0}G_{01}(V'), \quad (10)$$

$$B_{11} - B_{1-1} = \frac{1}{\pi\alpha}P\int_{-\infty}^{\infty}\frac{dV'}{V' - V_0}G_{01}(V').$$

Relations (9) and (10) are inherent in the form of the Y_{ik} in Eq. (3). Certain other relations can be found which are valid where the conversion gain is maximum with respect to both V_p and V_0. The conversion gain maximum is often the point of greatest experimental interest. Two assumptions are needed. The first assumption is that the quantum reactive terms are unimportant near maximum gain. This will be justified by the results of this analysis. Then Eqs. (4)–(6) reduce to

$$\mathscr{G}_{IF}^{-1} = [(G_{11} + G_{1-1} + G_s)(G_{00} + G_L)/G_{01} - 2G_{10}]^2/4G_s G_L.$$

$$(11)$$

The point of maximum conversion is given by two independent simultaneous equations, each resulting from the derivatives of Eq. (11) (with respect to V_p and with respect to V_0) being set to zero. Both of these equations have the form

$$(1 + g_L)\frac{G_{11}' + G_{1-1}'}{G_{11} + G_{1-1}} + (1 + g_s)\frac{G_{00}'}{G_{00}}$$

$$= (1 + g_s)(1 + g_L)\frac{G_{01}'}{G_{01}} + \eta\frac{G_{10}'}{G_{10}}. \qquad (12)$$

The second assumption is that the unpumped dc I-V curve consists of straight line segments with a discontinuity in current and/or slope at only one voltage (near the region of interest), such as in Figs. 2(a) and 2(b). This is an excellent approximation for the ideal SIS I-V curve and should hold reasonably well for the best experimental SIS devices. Then both

$$\frac{\partial G_{00}}{\partial V_0} = 0 \text{ and } \frac{\partial G_{10}}{\partial V_0} = 0, \qquad (13)$$

except where V_0 is an exact integer multiple of $\hbar\omega/e$ from the discontinuity. This exception is not important here, because it is known from experiments and from analytic simulations that the gain maximums occur when V_0 is approximately a half-integer multiple of $\hbar\omega/e$ from a discontinuity. Using Eqs. (9) and (13), we find that Eq. (12) reduces to

$$\frac{\partial \ln(G_{11} + G_{1-1})}{\partial V_0} = \frac{\partial \ln G_{01}}{\partial V_0}(1 + g_s),$$
$$\qquad (14)$$
$$\frac{\partial \ln(G_{11} + G_{1-1})}{\partial V_p} = \frac{\partial \ln G_{01}}{\partial V_p}(1 + g_s).$$

Equation (14) gives the conditions for maximizing the conversion gain. For a reasonably large g_s, then, maximum gain occurs quite near the maximum of the matrix element G_{01}. This result appears to be more general than might be expected; using data from the experiments of Ref. 9 we find that for $g_s = 1.0$ maximum gain does indeed occur at maximum G_{01}, and for g_s as small as 0.25 it occurs quite nearby. If we may take this as a general result, that maximum gain occurs near maximum G_{01}, then both B_{11} and B_{1-1} are relatively small when the gain has been maximized. This is true because the integral appearing in the second two of Eq. (10) is zero at some V_0 very near the maximum of $G_{01}(V_0)$. B_{10} is also small by virtue of Eq. (13). Thus when the conversion gain of a quantum mixer is maximized with respect to both the dc and LO voltages, under certain circumstances all of the reactive terms of the small-signal admittance matrix are relatively small. This is apparently the reason why the nonlinear quantum reactance is not important in calculating the conversion gain of a quantum mixer operating at its maximum gain.

V. INFINITE GAIN

It was already clear from Ref. 4 that the quantum mixer theory allows the possibility of infinite gain. Recent experiments[11,12] have demonstrated infinite available gain, which means that arbitrarily high gain could have been achieved with an appropriately matched IF load resistance (very small G_L).

To explore the region of very high and infinite gain we first calculate the LO power required for a quantum mixer. We refer to the circuit shown in Fig. 1 and use the expression for I_p given by Eq. (1) and the definitions given in Eqs. (3) and (6). The LO power incident upon (P_{LO}^+) and returned from (P_{LO}^-) a quantum mixer is given by

$$\frac{8}{G_s} P_{LO}^\pm = \frac{V_p^2}{g_s^2}[(\xi \pm g_s)^2 + (b_s - \gamma)^2]. \qquad (15)$$

In an actual experiment, as in the schematic Fig. 1, the externally controlled variables are P_{LO}^+ and V_{dc}. But for simplicity of computation the independent variables are V_p, chosen real, and V_0. Thus Eq. (15) gives P_{LO}^+ for any particu-

lar choice of V_p and V_0. To proceed, we shall evaluate the derivative of P_{LO}^+ with respect to V_p *with the experimental variable V_{dc} held constant*. We use

$$\frac{d}{dV_p} = \frac{\partial}{\partial V_p} + \frac{dV_0}{dV_p}\frac{\partial}{\partial V_0},$$

and find from Eqs. (1) and (3) that with V_{dc} constant,

$$\frac{dV_0}{dV_p} = \frac{-G_{01}}{G_{00} + G_{dc}}.$$

G_{dc} is the (real) internal admittance of the dc bias supply. With the aid of Eq. (9) it is found that

$$\frac{4}{G_s V_p}\frac{dP_{LO}^+}{dV_p} = \frac{D(G_{dc})}{g_s^2}. \qquad (16)$$

$D(G_{dc})$ is given by Eq. (5), if G_{dc} is everywhere substituted for G_L. If $G_{dc} = G_L$, then D is the square root of the denominator of the conversion gain. In this case it is seen that $dP_{LO}^+/dV_p = 0$ at the very point at which the gain is infinite. If the mixer parameters are such that infinite gain can be approached, Eq. (16) requires us to make a number of remarkable predictions:

(1) Glitching: It is clear from Eq. (15) that, except in certain nonexperimental limits, P_{LO}^+ increases as V_p^2 (and thus D is positive) for small P_{LO}^+. If D can become negative, the situation is illustrated in Fig. 3. There is a region in which, for a given P_{LO}^+, there are two stable and one unstable solutions for V_p. As P_{LO}^+ is increased, V_p will jump from one stable value to the other. Thus the region of negative D is not experimentally accessible. Any noise in the system will induce this jump to occur slightly earlier than indicated in Fig. 3, which means that the infinite gain point is not accessible. All experimentally observable quantities depend upon V_p and thus will also experience a hysteretic jump in their values, a "glitch". An equation equivalent to Eq. (16) has been derived for the unbiased Josephson junction parametric amplifier.[16] Such glitches have been observed for that device; an experimental glitch in the returned pump power is shown in Ref. 16, Fig. 9. With V_{dc} constant, V_0 is itself a function of V_p and will also glitch. This can be seen from another point of view. Comparing Eqs. (5) and (8), it is precisely at $D = 0$ that the dc load line is tangent to R_d. Thus as D goes negative, the dc bias voltage V_0 must jump along the load line to a stable point. This is observed in Ref. 12.

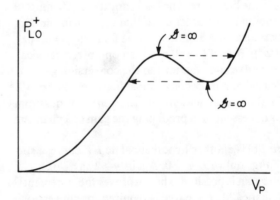

FIG. 3. Schematic illustration of $P_{LO}^+(V_p)$ relationship, showing infinite gain points and glitching.

(2) Gain-powerwidth product: The variation of gain with LO power is important for receiver stability. If one chooses the mixer parameters such that high gain but not infinite gain or glitching occurs, then the gain is a peaked function of the LO power, and we can define a "powerwidth" ΔP analogous to the bandwidth. ΔP is defined as the separation in incident LO power of the half-maximum gain points. As the peak gain increases, ΔP becomes narrower. Following the treatment in Ref. 16, there is a gain-powerwidth product which remains constant in the limit of high gain, much like the gain-bandwidth product of conventional linear amplifiers. The gain-powerwidth product has the form

$$\mathscr{G}_{\mathrm{IF}}^{3/4}\Delta P / P_{\mathrm{LO}}^{+} \rightarrow \text{constant},$$

for increasing peak gain. Note the unfamiliar exponent 3/4. This relationship has been experimentally verified for Josephson paramps.[17]

(3) Gain-proportional noise temperature: For high conversion gain D is small. Therefore Eq. (16) implies that any small fluctuation in P_{LO}^{+} must create a much larger fluctuation in V_p when the gain is high. In this sense the operating point of a quantum mixer will become more unstable as the gain is increased. This situation was studied in Ref. 18. That paper predicted for the Josephson paramp, as we here must predict for the quantum mixer, that there is a component of the noise *temperature* (referred to the input) which increases in direct proportion to the gain. That prediction was subsequently verified for the Josephson paramp.[18] The coefficient of this noise temperature component is of course crucial, but we cannot calculate or even estimate it here. By analogy with the Josephson paramp the gain-proportional noise is expected to be inversely proportional to the required LO power. This is presumably the reason why conventional high-gain amplifiers, which require much greater power than the devices under consideration, have not exhibited a gain-proportional noise temperature component.

Even in the limit of low IF, experiments are often designed with $G_{\mathrm{dc}} \neq G_L$. In this case, the predictions in this section must be modified. For instance, the jump in Ref. 12 occurs along the dc load line at the relatively small gain of 4.3 dB.

It must be remembered that these predictions, and all other results in this paper, apply only in the limit of low IF. We have not examined the case of finite IF. No experimental data exist at present to test predictions (2) and (3).

VI. THE PRACTICAL QUANTUM MIXER

It is prudent to avoid operating a quantum mixer in the negative differential resistance region of the pumped dc I-V curve. Within that region the output impedance for low IF and for a range of other frequencies is negative, and therefore unwanted out-of-band oscillations are possible. However, as seen in Sec. II, infinite gain may occur only within that region. Thus, to achieve a moderate, useful conversion gain, a practical quantum mixer will be designed to operate with very high but positive R_d. It is precisely here, at the edge of the negative differential resistance region, that 4.3-dB conversion gain is reported in Ref. 12. Very high positive differ-

ential resistance can be reached from a point of negative R_d by detuning any of the experimental parameters P_{LO}, V_0, G_s, or B_s. In this section we investigate the properties of a quantum mixer in the vicinity of infinite differential resistance.

We first make the assumption, validated but not proven in this paper, that the nonlinear quantum reactance may be ignored. Then Eqs. (4) and (8) become

$$\mathscr{G}_{\mathrm{IF}}^{-1} = L_0[(1 + g_s)(1 + g_L) - \eta]^2/4\eta g_s g_L,$$
$$G_{\mathrm{out}} = R_d^{-1} = G_{00}[1 - \eta/(1 + g_s)].$$

Note that infinite gain occurs when $\eta = (1 + g_s)(1 + g_L)$. The point $R_d = \infty$ occurs when $\eta = 1 + g_s$, and $\mathscr{G}_{\mathrm{IF}}$ varies smoothly with the experimental parameters through this point (except in the nonphysical limit $g_L = 0$). Substituting $\eta \simeq 1 + g_s$, the conversion gain in the vicinity of $R_d = \infty$ is

$$\mathscr{G}_{\mathrm{IF}}^{-1} \simeq L_0(1 + g_s)g_L/4g_s. \qquad (17)$$

Note that $\mathscr{G}_{\mathrm{IF}}$ depends only weakly on a small source resistance, whereas a relatively large source resistance will proportionally decrease $\mathscr{G}_{\mathrm{IF}}$. $\mathscr{G}_{\mathrm{IF}}$ is directly proportional to the load resistance, as expected.

To compare Eq. (17) with experimental results, we make the extremely rough approximations $L_0 = 2$ and $G_{11} + G_{1-1} = G_{00} = 1/R_N$, all of which should be correct to within a factor of three. (R_N is the junction normal state resistance.) In Ref. 12 the quantities G_s, G_L, and R_N are specified; Equation (17) predicts $\mathscr{G}_{\mathrm{IF}} = 4.4$ dB in perfect agreement with the measured $\mathscr{G}_{\mathrm{IF}} = 4.3$ dB. In the other relevant experiment, Ref. 11, G_s is unknown. But it is known that η was $\lesssim 2$, and therefore negative R_d implies $g_s \lesssim 1$. The reasonable value $g_s \simeq 1$ for Ref. 11 implies $\mathscr{G}_{\mathrm{IF}} = -10.8$ dB in agreement with the measured $\mathscr{G}_{\mathrm{IF}} = -11.5$ dB. The approximations we have made in applying Eq. (17) are so rough that an agreement of better than 3 dB must be considered fortuitous.

Since L_0 is relatively well behaved (L_0 is much larger or smaller than 2 only in the uninteresting vicinity of $\eta = 0$), it is clear from Eq. (17) that high gain is possible with positive differential resistance only if $g_L \ll 1$, g_s. If the gain is high, Eq. (7) is applicable, and with Eq. (17) gives

$$\mathscr{G}_s \simeq \frac{1}{16} L_0^2(1 + g_L)^2 \mathscr{G}_{\mathrm{IF}}^2; \qquad (18)$$

the signal reflection gain rises as the *square* of the IF conversion gain (for high gain with positive R_d). Equation (18) may make quantum mixer operation with high gain undesirable, because of reflections in the input system. Also, Eq. (18) implies that a quantum mixer will certainly saturate at much lower $\mathscr{G}_{\mathrm{IF}}$ than otherwise expected, making very high $\mathscr{G}_{\mathrm{IF}}$ unrealizable in practice.

VII. UNDERSTANDING QUANTUM MIXER THEORY

The advent of the quantum theory of mixing has been met with some confusion. There appear to be three major reasons:

(1) Compared to the familiar classical theory, the quantum theory of mixing is rather complicated in its derivation and in its resulting equations. We have not addressed the derivation of the quantum theory in this paper. Elsewhere,[13]

the author will present an integrated development of the theory of mixing (in the low IF limit). In this derivation only the physics of the active element, as represented by its frequency dependent current-voltage relationship, distinguishes a classical from a quantum mixer. Tucker's final equations, for instance Ref. 4, Eqs. (2)–(4), are rather unwieldy and not amenable to intuitive understanding. Here we have discussed these equations presented in a slightly different form and, with certain approximations, have derived a particularly simple expression for the conversion gain of a quantum mixer at an optimum operating point.

(2) The quantum theory of mixing gives results which appear to be counterintuitive. The prime example is the prediction of net conversion gain. It is well known that a classical mixer whose nonlinear element is a non-negative resistance can have at most unity conversion gain; this is a result of "reciprocity" (see e.g., Ref. 1). And yet the same device, when operated at a frequency high enough that quantum effects become important, can have high and even (analytically) infinite conversion gain. This has been attributed to the nonlinear quantum reactance, and it has been stated that the gain of a quantum mixer is analogous to that of a classical parametric amplifier. To the contrary, our view on this somewhat philosophical question is that the prohibition of gain in the theory of classical resistive mixers is a result of certain specific assumptions, and that these assumptions are violated when quantum effects become important. More specific comment is found immediately below.

(3) The nonlinear resistance which is the subject of the quantum theory of mixing is inevitably accompanied by a nonlinear quantum reactance. This is certainly a complication which does not exist in the classical theory. The nonlinear quantum reactance is extensively discussed in this paper. We shall review its origin, and discuss the transition to the quantum regime.

In classical mixer theory it is assumed that the instantaneous current through a nonlinear resistor is a function of the instantaneous voltage across it; the current carriers respond instantaneously to the driving force. But at high enough frequencies the response can no longer be instantaneous. Therefore one must instead define a casual response function which relates the current at time t to the voltage at all times $t' \leqslant t$. This is seen, for instance, in Tucker's[3] Eq. (2.10). When the response is not instantaneous we are said to be in the quantum regime.

Tucker states the criterion[3] that the quantum regime is entered as "at frequencies where the photon energy becomes comparable to the voltage scale of the dc nonlinearity." This criterion can be understood by examining the specific case of an SIS junction. Harris[19] gives the quasiparticle current in the time domain,

$$I(t) = \frac{1}{2\pi} \, \text{Im} \left[U^*(t) \int_{-\infty}^{\infty} \chi(t - t') U(t') dt' \right], \qquad (19)$$

where $U(t) = \exp\left[-j \frac{e}{\hbar} \int_{-\infty}^{t} V(t') dt' \right],$

and the quasiparticle response function for an ideal SIS junction at $T = 0$ is

$$\chi(t) = -\frac{\hbar \delta'(t)}{eR_N} + \frac{2\pi^2 \Delta^2}{\hbar e R_N} J_1\left(\frac{\Delta}{\hbar} t \right) Y_1\left(\frac{\Delta}{\hbar} t \right).$$

J_1 and Y_1 are first-order Bessel functions of the first and second kinds, Δ is the superconductor energy gap parameter, and R_N is the junction normal-state resistance. The function $\chi(t)$ is plotted in Ref. 20. $\chi(t)$ oscillates in time with the energy gap frequency $2\Delta / \hbar$ and dies off only inversely with time for large t. Therefore the current at a given time depends upon the voltage history infinitely far into the past. This slow falloff is due to the perfectly sharp nonlinearity of an ideal SIS junction. For a real junction the nonlinearity is not perfectly sharp; it occurs over a finite range of voltage. Therefore $\chi(t)$ dies off within a finite time, more quickly for a more smeared nonlinearity. If $\chi(t)$ is insignificantly small after only a fraction of the fundamental photon period, then the response is essentially instantaneous at the frequencies of interest and the assumptions of classical mixer theory are valid. But if the nonlinearity is sharp enough that $\chi(t)$ is significantly large after one period, then the quantum theory must be used.

Thus Tucker's criterion, which is formulated in the frequency (i.e., energy) domain, can be translated to the language of the time domain: The quantum regime is entered at frequencies high enough that the junction response is not instantaneous on the time scale of the photon period.

The frequency dependent admittance of the nonlinear element is found by Fourier analyzing the time-domain current-voltage equation, Eq. (19) or equivalently Ref. 3, Eq. (2.10). In general, the Fourier transform of an arbitrary causal function is complex, with real and imaginary parts which are Kramers-Kronig transforms of one another.[14] In particular, the Fourier transform of $\chi(t)$ gives rise to a *complex* admittance if the response is not instantaneous. Thus in the quantum regime, the response cannot be purely in-phase; any nonlinear resistance must be accompanied by a nonlinear reactance.

Not only does the nonlinear reactance appear in this quantum regime, but the (in-phase) real part of the nonlinear response is modified from its classical form; *the resistive response becomes frequency dependent*. This is evident in Eqs. (1) and (3). Note in particular that $G_{01} \neq G_{10}$. It is clear that the frequency dependence of a nonlinear resistance in the quantum regime implies that reciprocity must break down, that the classical prohibition of mixer gain is not valid. There is no need to invoke the nonlinear quantum reactance as the "cause" of the gain.

There have previously been a number of indications that the nonlinear quantum reactance is in fact not important in calculating the conversion gain of a quantum mixer. In Ref. 9 the maximum experimental conversion on each peak for many different SIS array mixers was compared to the calculated conversion on each peak, ignoring the reactive terms. The agreement was excellent in that region of parameter space in which all theoretical assumptions were satisfied by the experiments, and the *scatter* of the discrepancy between theory and experiment was everywhere less than ~ 0.5dB. An important experimental contribution from the quantum reactance would be expected to increase the scatter. Note that in these experiments each conversion peak was individually maximized. In Ref. 15 the conversion on each of the first four peaks of an SIS mixer was calculated both with and without the contribution of the quantum reactance included, in order to compare with the experimental results. On the fourth peak, where the conversion was experimentally maximized, the quantum reactance had no effect on the calculated conversion, which agreed with the experimental result. On the other peaks the reactance had a perceptible but minor effect, except for the first peak where the effect was large. In both of these experiments the SIS mixers were well out of the classical region, as evidenced by the large quantum conversion modulation, and yet the quantum reactance had little effect upon the maximized conversion. A likely explanation for this is given in Sec. IV.

There is one further example (mentioned in Ref. 9) of the insignificance of the quantum reactance. The author has calculated the conversion both with and without the quantum reactance for an ideal SIS mixer (with the gap discontinuity slightly rounded over a third of a photon step to avoid singularities) with reduced frequency $\hbar\omega/2\Delta = 0.1$ and reduced LO voltage $\alpha = 1.25$. Within both the first and second conversion peaks the quantum reactance had less than 1-dB effect upon the conversion efficiency, so long as the gain was not large.

In spite of the quantum reactance, it seems valuable to maintain the conventional distinction between a "resistive mixer," whose nonlinearity is fundamentally lossy, and a parametric device, whose nonlinearity is intrinsically reactive. This is especially true considering the results of this paper, that the nonlinear quantum reactance of the quantum mixer is relatively unimportant.

VIII. CONCLUSION

A classical heterodyne mixer employing an always positive resistive nonlinearity cannot give gain. The simplicity of this result makes it appear quite general, even though it is specifically derived only for classical mixers. It has seemed only natural then to ascribe the gain of a quantum mixer to the nonlinear quantum reactance. Thus it has been stated that the gain of a quantum mixer is analogous to that of a classical parametric amplifier. And indeed, it is strictly incorrect to consider quantum mixing without the nonlinear reactance. At high enough frequency, in the quantum regime, there is no such thing as a purely resistive nonlinearity. Causality, as expressed in the Kramers-Kronig relations, requires the accompanying nonlinear reactance. This is particularly clear if quantum mixer theory is formulated in the time domain. Fourier transforming, the resultant response function is inevitably complex.

Here we have marshalled the evidence supporting the opposite view, that the nonlinear quantum reactance, though it certainly exists, is not itself responsible for conversion gain, has a minor effect upon a mixer's properties, and should be difficult to detect experimentally. We recapitulate: (1) In past experiments, the results are reasonably well described by quantum mixer theory, without consideration of the quantum reactance; (2) A calculation of the conversion for an ideal SIS junction, for a single frequency and a single value of α, showed that the quantum reactance could not be distinguished in this case. Its small effect could be duplicated by a slight change in bias conditions; (3) In the limit of small LO power the off-diagonal reactive terms appear in the conversion loss expression for an ideal SIS junction as terms of higher order in the small parameter α, and yet infinite gain is possible; (4) There is a value of externally applied reactance, namely $B_s = B_{1-1} - B_{11}$, which causes the nonlinear quantum reactive terms to disappear from the complete conversion gain expression in the limit of low IF (this is not true for the signal reflection gain expression). Therefore, ignoring the quantum reactance always corresponds to a physically realizable situation; (5) Given certain assumptions, the reactive terms have been shown to be small at the maximum conversion point.

The elusiveness of the nonlinear quantum reactance is somewhat mysterious. In a number of examples for which the small-signal admittance matrix has been evaluated, the resistive and the reactive matrix elements are found to be of the same order of magnitude. And yet, it seems, the reactive elements may be ignored in the calculation of physical quantities.

There is a precedent for this situation. In the theory of the SIS tunnel junction, the tunneling current can be divided into four separate terms.[21] The first is generally called the quasiparticle current and the second we have called the nonlinear quantum reactive current. As noted, they are Kramers-Kronig transforms of one another, and together they constitute the single-particle tunneling current which is the subject of the quantum theory of mixing. The third and fourth terms are generally called the Josephson sine and cosine terms. They describe superconducting pair tunneling: The Josephson effect. They are also a Kramers-Kronig transform pair, and thus bear the same relationship to one another as do the nonlinear resistance and quantum reactance. The Josephson sine term has many dramatic manifestations, as attested to by the wealth of Josephson effect literature. The cosine term, on the contrary, is of such minor experimental importance that for many years it was ignored. It has proved possible to make a clear cut detection of the cosine term only with a very exacting experimental technique. For a recent example and references, see Ref. 22. Nevertheless, the sine and cosine terms in general are of the same order of magnitude. It appears, on much less present evidence, that the nonlinear quantum reactance may be equally as elusive as the Josephson cosine term. Note, though, that all of the SIS experiments and calculations mentioned here were performed at less than 15% of the characteristic frequency (the gap frequency) of the superconductors involved. At higher frequencies the quantum reactance may be considerably more influential.

This paper is an elaboration upon the quantum theory of mixing in single-particle tunnel junctions, in the low IF limit. We developed the expression for the IF conversion gain and the signal reflection gain for arbitrary source reactance, and discussed these intuitively. We presented a number of relationships among the elements of the complex small-signal admittance matrix. We extensively discussed the effects of the nonlinear quantum reactance and conclud-

J. Appl. Phys., Vol. 53, No. 1, January 1982

ed that these are unimportant. By consideration of the required LO power we made a number of remarkable predictions for certain quantum mixers operating with high gain. And finally, we developed a very simple expression for the IF conversion gain of an optimized, practical, quantum mixer.

In light of the results of this paper, future quantum mixer experiments should be designed to monitor the signal reflection gain as well as the IF conversion gain.

ACKNOWLEDGMENTS

The author gratefully acknowledges many helpful discussions with A. R. Kerr, S.-K. Pan, and S. Rudner.

[1] H. C. Torrey and C. A. Whitmer, *Crystal Rectifiers*, MIT Radiation Lab. Series (McGraw-Hill, New York, 1948), Vol 15.

[2] A. A. M. Saleh, *Theory of Resistive Mixers* (MIT Press, Cambridge, Massachusetts, 1971).

[3] J. R. Tucker, IEEE J. Quantum Electron. QE-15, 1234 (1979).

[4] J. R. Tucker, Appl. Phys. Lett. 36, 477 (1980).

[5] P. L. Richards, T.-M. Shen, R. E. Harris, and F. L. Lloyd, Appl. Phys. Lett. 34, 345 (1979).

[6] G. J. Dolan, T. G. Phillips, and D. P. Woody, Appl. Phys. Lett. 34, 347 (1979).

[7] S. Rudner and T. Claeson, Appl. Phys. Lett. 34, 713 (1979).

[8] T.-M. Shen, P. L. Richards, R. E. Harris, and F. L. Lloyd, Appl. Phys. Lett. 36, 777 (1980).

[9] S. Rudner, M. J. Feldman, E. Kollberg, and T. Claeson, *SQUID '80*, edited by H. D. Hahlbohm and H. Lübbig (de Gruyter, Berlin, 1980), p. 901; IEEE Trans. Magnetics MAG-17, 690 (1981); and J. Appl. Phys. 52, 6366 (1981).

[10] G. J. Dolan, R. A. Linke, T. C. L. G. Sollner, D. P. Woody, and T. G. Phillips, IEEE Trans. Microwave Theory Tech. MTT-29, 87 (1981).

[11] A. R. Kerr, S.-K. Pan, M. J. Feldman, and A. Davidson, Physica 108B, 1369 (1981).

[12] A. D. Smith, W. R. McGrath, P. L. Richards, H. van Kempen, D. Prober, and P. Santhanam, Physica 108B, 1367 (1981); and W. R. McGrath, P. L. Richards, A. D. Smith, H. van Kempen, R. A. Batchelor, D. Prober, and P. Santhanam, Appl. Phys. Lett. 39, 655 (1981).

[13] M. J. Feldman (unpublished).

[14] See, for example, F. Stern, in *Solid State Physics*, edited by F. Seitz and D. Turnbull (Academic, New York, 1963), Vol. 15, pp. 327–332.

[15] T.-M. Shen, IEEE J. Quantum Electron. QE-17, 1151 (1981).

[16] M. J. Feldman, P. T. Parrish, and R. Y. Chiao, J. Appl. Phys. 46, 4031 (1975).

[17] S. Wahlsten, S. Rudner, and T. Claeson, J. Appl. Phys. 49, 4248 (1978).

[18] M. J. Feldman and M. T. Levinsen, Appl. Phys. Lett. 36, 854 (1980).

[19] R. E. Harris, Phys. Rev. B 13, 3818 (1976).

[20] D. G. McDonald, R. L. Peterson, C. A. Hamilton, R. E. Harris, and R. L. Kautz, IEEE Trans. Electron Devices, ED-27, 1945 (1980).

[21] R. E. Harris, Phys. Rev. B 11, 3329 (1975).

[22] M. J. Feldman, S. Rudner, and T. Claeson, J. Appl. Phys. 51, 5058 (1980); and S. Rudner and T. Claeson, J. Appl. Phys. 50, 7070 (1979).

A Superconducting Tunnel Junction Receiver for 230 GHz

E. C. SUTTON

Abstract —The performance of superconducting tunnel junctions as high-frequency receivers is discussed. Low-noise mixing in superconductor–insulator–superconductor (SIS) quasi-particle tunnel junctions has been seen for frequencies up to 400 GHz. Such mixers have the significant advantage of small local-oscillator power requirements. A receiver has been constructed which has a single-sideband (SSB) receiver noise temperature of 305 K at 241 GHz.

I. INTRODUCTION

Great progress has been made in recent years in the development of sensitive millimeter and submillimeter-wave receivers. Much of this progress has come about due to the needs of radio astronomy, where such receivers are essential for studies of interstellar matter such as carbon monoxide, which has its lowest frequency transitions at 115, 230, and 346 GHz, and atomic carbon (CI) at 492 GHz. Various techniques have been used to exploit this frequency range. One of the earliest successful techniques was the InSb hot-electron bolometer [1], which was used for most of the original astronomical detections at frequencies above 200 GHz. Room-temperature Schottky mixers, with their much greater bandwidths, have also been very important, particularly at the lower frequencies [2], [3]. Currently, some of the best receiver systems being developed and in use are cooled Schottky-diode mixers [4], with their greatly reduced noise temperatures compared with the room-temperature mixers. An attractive alternative throughout this frequency range and one which is currently under development is the superconductor–insulator–superconductor (SIS) quasi-particle mixer [5].

The SIS mixer is a form of superconducting tunnel junction in which electrons tunnel between two superconductors separated by a thin oxide barrier. These electrons can tunnel either as pairs via the Josephson effect or as individual quasi-particles. Although early work concentrated on Josephson mixing [6], much recent attention has been devoted to heterodyne mixing using the non-linear quasi-particle current–voltage ($I-V$) characteristic [7]–[12]. In addition to their usefulness as low-noise receivers, these devices are of interest as examples of quantum-mechanical mixers [13]. Because the photon energy may be comparable with the energy scale of the tunneling nonlinearity, these junctions can exhibit unique quantum-mechanical effects. Among these is the potential for conversion gain, an effect which is forbidden for classical resistive mixers but seen in SIS mixers [14], [15]. Additional advantages of quasi-particle mixers include their low-noise performance and their small local-oscillator power requirements. Although most work done to date with SIS mixers has been at comparatively low frequencies (18–115 GHz), they exhibit considerable promise for use as receivers at considerably higher frequencies, where it is still difficult to obtain sensitive receiver systems.

Manuscript received November 24, 1982; revised February 23, 1983. This work was supported in part by the National Science Foundation under grant AST80-07645A02.

The author is with the Department of Physics, California Institute of Technology, Pasadena, CA 91125.

II. RECEIVER DESIGN

A. High-Frequency Limitations

Most work which has been done so far with SIS mixers has been done in the low-frequency regime, below about 100 GHz. In this limit, the frequency of operation is much smaller than that corresponding to the superconducting energy gap ($2\Delta/h$). At such frequencies the sharpness of the $I-V$ characteristic and the amount of excess current present below the gap are thought to be the principal factors determining the conversion efficiency and noise properties of the device.

Relatively little investigation has been made of the upper frequency limitation of SIS mixers. It has been clear that above about 100 GHz the Josephson effect begins to interfere with optimal operation of the mixer. A rough upper frequency limit of 200 GHz has been suggested [16] for single-junction SIS mixers based on experience at somewhat lower frequency. While this does not represent a firm upper limit, it is clear that success at higher frequency operation is intimately connected with success at understanding and suppressing the Josephson-effect-related phenomena. Consideration of just single-particle effects indicates that efficient mixing could occur for frequencies up to $\nu = 4\Delta/h$, which is approximately 1200 GHz for Pb-alloy junctions. Pair breaking, which has an onset at $\nu = 2\Delta/h$, may limit operation to more like 600 GHz. However, the results of Danchi et al. [17] are encouraging since they observe quasi-particle and Josephson steps at frequencies well above $2\Delta_{Sn}/h$ for Sn–SnO–Pb junctions.

B. Josephson-Effect Structure

Josephson-effect mixing obeys the same general formalism as quasi-particle photon assisted tunneling except that mixing is provided by the singularity of the dc Josephson effect at $V = 0$. This effect may be used to construct mixer–receivers although it has generally proved too unstable for practical use. In the presence of applied RF, the dc $I-V$ curve is modified by the introduction of current singularities at voltages displaced from zero volts by $V = nh\nu/2e$, that is, a spacing half that of the quasi-particle tunneling steps. The strength of these singularities is given by Bessel functions of argument $\alpha = 2eV_1/h\nu$, where V_1 is the applied RF voltage. Since α must be of order unity in order to get significant quasi-particle mixing, it is a necessary consequence that the Josephson singularities will also be strong. At high frequencies the quasi-particle steps (spaced down from $V = 2\Delta/e$ by intervals of $h\nu/e$) and the Josephson singularities (spaced up from $V = 0$ by intervals of $h\nu/2e$) will begin to overlap. This interference gets worse at higher values of LO power. Suppressing the competition from the Josephson effect is the greatest difficulty in achieving high-frequency operation in SIS quasi-particle mixers.

C. Magnetic Fields

The suppression of the dc Josephson effect with magnetic fields was discussed by Josephson [18] and measured by Rowell [19]. The chief difficulty in the present case is achieving adequate suppression for junctions with small area. The junctions used must be kept small in order to minimize the capacitance of the device. Assuming a uniform junction, the requirement for Josephson-effect suppression is that the cross section presented to a magnetic field applied parallel to the plane of the junction

Reprinted from *IEEE Trans. Microwave Theory Tech.*, vol. MTT-31, no. 7, pp. 589–592, July 1983.

contains a single quantum of magnetic flux. For a junction length of 1 μm and assuming a penetration depth of approximately 5 10^{-6} cm for Pb-alloys, this cross section is 10^{-9} cm^2. Since the quantum of magnetic flux is 2 10^{-7} Gauss cm^2, this works out to a required field of around 200 Gauss. Although this is smaller than the critical field for destroying the underlying superconductivity, it is large enough to have a significant effect on the size of the superconducting energy gap. Thus, this form of magnetic field suppression is possible for junctions of linear size about 1 μm, but impractical for much smaller junctions.

D. Local-Oscillator Power Requirements

One of the chief advantages of SIS mixers is their small local-oscillator power requirement. Phillips et al. [10] estimate an optimum LO level for mixing at 115 GHz of about -50 dBm. Since this power should scale as ν^2, this implies a power requirement of -43 dBm at 250 GHz. This estimate can be understood in terms of Tucker's quantum theory of mixing [13], [14] in which the conversion efficiency depends on the conductances

$$G_{10} = \frac{1}{2} \sum J_n(\alpha)[J_{n-1}(\alpha) + J_{n+1}(\alpha)] \frac{d}{dV_0} I_{DC}\left(V_0 + \frac{n\hbar\omega}{e}\right)$$

$$G_{01} = \frac{e}{\hbar\omega} \sum J_n(\alpha)[J_{n-1}(\alpha) - J_{n+1}(\alpha)] I_{DC}\left(V_0 + \frac{n\hbar\omega}{e}\right)$$

where $\alpha = eV_1/\hbar\omega$ and V_1 is the LO voltage applied to the junction. These conductances vanish for $\alpha \ll 1$. Thus, values of α of order unity are needed in order to get significant mixing on the first photon step below the gap. This criterion gives a required LO power of the same order as that estimated above. In contrast, competing cooled Schottky-barrier diode mixers in this frequency range [4] require local-oscillator powers many orders-of-magnitude larger (on the order of -10 dBm incident on the diode). Due to the difficulty in getting sufficient LO power in this frequency range, this is a considerable advantage for SIS mixers.

E. Impedance Matching

The impedance matching problem for Schottky-diode mixers is by now fairly well understood. The conventional design for a Schottky mixer mount consists of a section of reduced height waveguide to reduce the RF impedance to about 100 Ω. Then a diode can be selected which reasonably matches both the RF input circuit and a 50-Ω coaxial IF output. Due to the complicated formalism of the theory of quantum mixing, the equivalent problem has not been as well understood for SIS mixers. However, it can be deduced from Tucker's quantum theory of mixing that in the limit of small IF frequencies and small α the IF output conductance is approximately given by

$$G_{00} = \sum J_n^2(\alpha) \frac{d}{dV_0} I_{DC}\left(V_0 + \frac{n\hbar\omega}{e}\right)$$

which is simply the slope of the I–V curve near V_0 under conditions of RF voltage bias. The small-signal RF input conductance is similarly related to

$$G_{11} = \frac{e}{2\hbar\omega} \sum J_n^2(\alpha) \left[I_{DC}\left(V_0 + \frac{(n+1)\hbar\omega}{e}\right) \right.$$
$$\left. - I_{DC}\left(V_0 + \frac{(n-1)\hbar\omega}{e}\right)\right]$$

which for small values of α is given by the slope of a line joining points on the pumped I–V curve one photon step above and one photon step below the dc bias point. Thus, the impedances presented by the mixer at the RF and IF ports will be different

than the normal-state impedance of the junction. In general, the IF impedance will be significantly larger and the RF impedance somewhat smaller. The problem of impedance matching is discussed more completely by Smith and Richards [20] and, with the inclusion of noise considerations, by Wengler and Woody [21]. Optimum performance in the IF is achieved if the mixer output is nearly matched into the IF load impedance. In the RF, optimum conversion efficiency is given by a large source impedance, but best overall noise performance is achieved in a more nearly impedance-matched configuration. The consequence is that in practice it is desirable to have a junction with normal-state impedance of about 50 Ω. The RF driving impedance should be lowered using reduced height waveguide and the IF load impedance should be increased above 50 Ω using a quarter-wave transformer section.

III. Receiver Construction and Performance

The general design of the receiver was based on a scaled-down version of the 115-GHz receiver of Dolan et al. [11], the principal departure being the use of circular waveguide. The advantage of circular waveguide at high frequencies is the greater ease in construction and the higher tolerances to which mechanical tuning elements (choked backshorts) can be made. The principal disadvantage is the reduced bandwidth of single-mode operation compared with rectangular waveguide. The waveguide size used in the receiver was 0.97-mm diameter, which allows for single-mode operation to just above the design frequency of 230 GHz. The next higher mode (TM$_{01}$) has radial electric field lines and is, to first order, decoupled from the currents flowing in the junction.

The junction used was fabricated using photo-lithographic techniques and was similar to the devices described by Dolan et al. [11]. The material used was a Pb–In–Au alloy chosen primarily on the basis of its thermal cyclability. The junction cross section is estimated to be about 0.7 μm on a side, giving a device capacitance of about 20 fF. For the device used in these measurements, which had a normal state impedance of 68 Ω, this implies a value of $\omega R_N C \sim 2$ at 230 GHz.

The mixer block, which is shown in Fig. 1, is machined from a single piece of copper and contains an integral IF-line choke. The junction itself is deposited on a 0.1-mm-thick quartz substrate which rests in a groove spanning the waveguide. One end of the junction is attached directly to the mixer block with silver paint while the other end is insulated with a thin sheet of mylar. Behind the junction there is an adjustable backshort which allows for continuous tuning over the atmospheric window from 190 to 300 GHz. The input of the waveguide is attached to a corrugated feed horn. Radiation reaches the feed horn after passing through a 0.025-mm mylar vacuum window and a cooled $\lambda/4$ crystalline quartz filter.

The IF port of the mixer is connected to a cooled two-stage L-band GaAs FET amplifier which operates from 1100 to 1650 MHz. This amplifier has a noise temperature of about 12 K averaged over this band and is similar to that described by Weinreb et al. [22] except for the omission of a third stage of amplification and the greatly reduced power dissipation (40 mW versus about 200 mW). This latter factor is quite important for operation in a LHe-cooled cryostat.

The normal-state impedance of the junction used in these tests was 68 Ω. No transformers were used to match impedances in either the RF or the IF sections. In the IF the approximately 50-Ω input impedance of the FET amplifier is being driven by the output impedance of the mixer, which is generally considerably

IEEE TRANSACTIONS ON MICROWAVE THEORY AND TECHNIQUES, VOL. MTT-31, NO. 7, JULY 1983

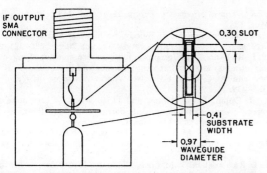

Fig. 1. Diagram of mixer block. The view shown is that seen by looking down along the axis of the circular waveguide. The junction is deposited on a 0.1-mm-thick quartz substrate which rests in a vertical slot spanning the waveguide. The metallization covers most of the substrate. The horizontal slot above the waveguide forms the high impedance section of a three-section RF choke on the IF line. All dimensions shown are in millimeters.

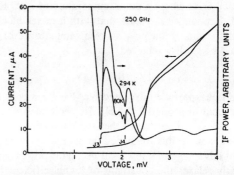

Fig. 2. System performance at 250 GHz. Shown are the unpumped and pumped (~ -43 dBm of LO) $I-V$ curves of the junction as well as the IF power output for room temperature and liquid nitrogen temperature loads. The $I-V$ curves show regions of negative resistance (hysteresis) below about 1.6 mV due to the Josephson effect. The IF power curves show structure at points J3 and J4, where the ac Josephson frequency ($\nu_J = 2eV_0/h$, where V_0 is the dc bias voltage) is an integral multiple of the local-oscillator frequency. Application of a magnetic field suppresses the structure at J3 and J4 as well as showing more clearly a smooth quasi-particle step extending from 1.4 to 2.4 mV.

larger than the junction normal-state impedance. For typical conditions of bias and pumping, this impedance was determined from the pumped $I-V$ curve to be approximately 250 Ω, providing roughly a 5:1 mismatch in the IF. The RF showed a similar mismatch between the 400-Ω waveguide impedance and the mixer input impedance [13], [14]. At high frequencies this was a mostly real impedance somewhat lower in value than the normal-state impedance of 68 Ω. Thus, the RF impedance mismatch is quite considerable, being roughly 8:1. This factor creates quite a significant insertion loss in the receiver. A better situation would be to have transformer sections in either the RF or IF to reduce these mismatches. An IF impedance transformation of a factor of 2 to 4 should be fairly readily achieved without severely limiting the IF bandwidth. A similar transformation in the RF is impossible with the present circular waveguide geometry, although such a transformation could be made using reduced height rectangular waveguide. Finally, the junction impedance could be chosen to better balance out the amounts of mismatch in the RF and IF.

Local oscillator power is provided by a frequency-multiplied klystron. The multiplier consists of a GaAs Schottky diode mounted in a conventional WR-8/WR-3 crossed-waveguide mount [3], [23]. The klystron frequency is doubled to produce frequencies up to 250 GHz and tripled beyond that. For tripled operation the second harmonic was suppressed by using a section of waveguide beyond cutoff. Because of the low power requirements of SIS mixers, this multiplier produces roughly 30 dB more local-oscillator power than is needed at the mixer. As a result, a simple diplexer consisting of a thin (0.025 mm) sheet of mylar can be used to combine the local oscillator and signal beams with only a 1–2-percent insertion loss of the signal.

The performance of the receiver has been measured at a number of frequencies within the design band of 220–250 GHz as well as at several higher frequencies. All measurements were made double-sideband (DSB) and converted to equivalent single-sideband (SSB) performance. Relative sideband response was measured to be balanced to better than 20 percent. The best performance obtained was a receiver noise temperature of 305 K (SSB) at 241 GHz, averaged over the IF band. Measured conversion loss at this frequency was 10.5 dB (SSB). This includes losses due to the impedance mismatches discussed above and losses in the optics, feedhorn, and mixer block. Junction $I-V$ curves with and without applied LO, as well as the responses to hot (294 K) and cold (80 K) loads, are shown in Fig. 2. Application of a magnetic field suppresses the Josephson-effect structure and reveals a smooth quasi-particle step without greatly affecting the sensitivity of the receiver.

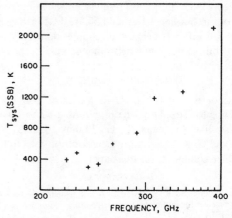

Fig. 3. Single-sideband receiver noise temperature as a function of frequency. The measurements at frequencies above 260 GHz were made at a time when the receiver performance at the lower frequencies was about 35-percent worse than shown here. Hence the upper portion of the curve may be unduly pessimistic.

Tuning to other frequencies can be accomplished by tuning the mixer backshort position as well as the backshorts on the multiplier. Generally it is also necessary to optimize the amount of local-oscillator power and the voltage bias on the junction. The local-oscillator power level is not critical, but the bias voltage is moderately critical due to the complex Josephson-effect structure. A plot of system noise temperature versus frequency is given in Fig. 3. All measurements shown were taken at 4.2 K and without any applied magnetic field. The receiver performance is broadly optimized over a band from 200 to 300 GHz.

Included in Fig. 3 are measurements made at frequencies much greater than the nominal band of the receiver. At these frequencies the mixer block size and geometry are not optimized, but nevertheless such measurements serve to test the principles for operating SIS mixers at higher frequencies. In addition, the mode of operation changes significantly due to increased competition from Josephson-effect mixing, a problem which is just beginning to be severe at 230 GHz. At the higher frequencies it is helpful to use a combination of lower temperature operation (2.2 K) to increase the superconducting gap and applied magnetic fields to suppress the Josephson effect in order to achieve a stable mode of mixing. The highest frequency at which tests have been made so far is 388 GHz where a noise temperature of 2100 K SSB was measured. Experience indicates that with a mixer block suitably optimized for this frequency, a noise temperature of ~ 600 K SSB should be readily achieved.

IV. SUMMARY

An SIS quasi-particle mixer–receiver has been constructed for use at frequencies around 230 GHz. Its performance is comparable with or better than that of the best competing cooled Schottky-diode mixers in this frequency range. Furthermore, it has the significant advantage of a low local-oscillator power requirement. This greatly relaxes the constraint of highly efficient frequency-multipliers to supply the LO, as well as relaxing the requirements of the diplexer for combining the signal and LO.

In its present configuration, the receiver suffers from severe impedance mismatches in both the RF and the IF. These mismatches are the chief factors limiting receiver performance. An improved receiver incorporating impedance-matching transformers should show at least a factor of 2 better noise temperature.

Tests have also been made of performance at higher frequencies. It seems likely that good receiver noise temperatures can be achieved out to frequencies around 500 GHz. Further increases in frequency are difficult because of competition from the Josephson effect, but some useful performance may be possible out to 1000 GHz.

ACKNOWLEDGMENT

The author is grateful to R. E. Miller of Bell Laboratories, Murray Hill, for supplying the junctions used in this work. He would also like to thank T. G. Phillips, D. P. Woody, M. Wengler, and C. R. Masson for their contributions to this project and for their comments on this paper.

REFERENCES

[1] T. G. Phillips and K. B. Jefferts, "A low temperature bolometer heterodyne receiver for millimeter wave astronomy," *Rev. Sci. Inst.*, vol. 44, pp. 1009–1014, Aug. 1973.

[2] A. A. Penzias and C. A. Burrus, "Millimeter-wavelength radio-astronomy techniques," *Ann. Rev. Astron. Astrophys.*, vol. 11, pp. 51–72, 1973.

[3] N. R. Erickson, "A 200–350 GHz heterodyne receiver," *IEEE Trans. Microwave Theory Tech.*, vol. MTT-29, pp. 557–561, June 1981.

[4] J. W. Archer, "All solid-state low-noise receivers for 210–240 GHz,"

[5] T. G. Phillips and D. P. Woody, "Millimeter and submillimeter-wave receivers," *Ann. Rev. Astron. Astrophys.*, vol. 20, pp. 285–321, 1982.

[6] P. L. Richards, "The Josephson junction as a detector of microwave and far infrared radiation," in *Semiconductors and Semimetals*, R. K. Willardson and A. C. Beer, Eds. New York: Academic, 1977, vol. 12, pp. 395–439.

[7] G. J. Dolan, T. G. Phillips, and D. P. Woody, "Low-noise 115-GHz mixing in superconducting oxide-barrier tunnel junctions," *Appl. Phys. Lett.*, vol. 34, pp. 347–349, Mar. 1, 1979.

[8] P. L. Richards, T. M. Shen, R. E. Harris, and F. L. Lloyd, "Quasiparticle heterodyne mixing in SIS tunnel junctions," *Appl. Phys. Lett.*, vol. 34, pp. 345–347, Mar. 1, 1979.

[9] P. L. Richards and T. M. Shen, "Superconductive devices for millimeter wave detection, mixing, and amplification," *IEEE Trans. Electron Devices*, vol. ED-27, pp. 1909–1920, Oct. 1980.

[10] T. G. Phillips, D. P. Woody, G. J. Dolan, R. E. Miller, and R. A. Linke, "Dayem-Martin (SIS tunnel junction) mixers for low noise heterodyne receivers," *IEEE Trans. Magn.*, vol. MAG-17, pp. 684–689, Jan. 1981.

[11] G. J. Dolan, R. A. Linke, T. C. L. Sollner, D. P. Woody, and T. G. Phillips, "Superconducting tunnel junctions as mixers at 115 GHz," *IEEE Trans. Microwave Theory Tech.*, vol. MTT-29, pp. 87–91, Feb. 1981.

[12] K. H. Gundlach, S. Takada, M. Zahn, and H. J. Hartfusse, "New lead alloy tunnel junction for quasiparticle mixer and other applications," *Appl. Phys. Lett.*, vol. 41, pp. 294–296, Aug. 1, 1982.

[13] J. R. Tucker, "Quantum limited detection in tunnel junction mixers," *IEEE J. Quantum Electron*, vol. QE-15, pp. 1234–1258, Nov. 1979.

[14] J. R. Tucker, "Predicted conversion gain in superconductor-insulated-superconductor quasiparticle mixers," *Appl. Phys. Lett.*, vol. 36, pp. 477–479, Nov. 1979.

[15] T. M. Shen, P. L. Richards, R. E. Harris, and F. L. Lloyd, "Conversion gain in mm-wave quasiparticle heterodyne mixers," *Appl. Phys. Lett.*, vol. 36, pp. 777–779, May 1, 1980.

[16] S. Rudner, M. J. Feldman, E. Kollberg, and T. Claeson, "Superconductor-insulator-superconductor mixing with arrays at millimeter-wave frequencies," *J. Appl. Phys.*, vol. 52, pp. 6366–6376, Oct. 1981.

[17] W. C. Danchi, F. Habbal, and M. Tinkham, "AC Josephson effect in small-area superconducting tunnel junctions at 604 GHz," *Appl. Phys. Lett.*, vol. 41, pp. 883–885, Nov. 1, 1982.

[18] B. D. Josephson, "Possible new effects in superconductive tunneling," *Phys. Lett.*, vol. 1, pp. 251–253, July 1, 1962.

[19] J. M. Rowell, "Magnetic field dependence of the Josephson tunnel current," *Phys. Rev. Lett.*, vol. 11, pp. 200–202, Sept. 1, 1982.

[20] A. D. Smith and P. L. Richards, "Analytic solutions to superconductor-insulator-superconductor quantum mixer theory," *J. Appl. Phys.*, vol. 53, pp. 3806–3812, May 1982.

[21] M. Wengler and D. P. Woody, in preparation.

[22] S. Weinreb, D. Fenstermacher, and R. Harris, "Ultra-low-noise 1.2 to 1.7 GHz cooled GaAs FET amplifiers," *IEEE Trans. Microwave Theory Tech.*, vol. MTT-30, pp. 849–853, June 1982.

[23] M. V. Schneider and T. G. Phillips, "Millimeter-wave frequency multiplier," *Int. J. Infrared Millimeter Waves*, vol. 2, pp. 15–22, 1982.

Low-noise 115-GHz receiver using superconducting tunnel junctions

S. -K. Pan,[a] M. J. Feldman,[a] and A. R. Kerr

NASA Goddard Institute for Space Studies, New York, New York 10025

P. Timbie

Physics Department, Princeton University, Princeton, New Jersey 08544

(Received 24 June 1983; accepted for publication 28 July 1983)

A 110–118-GHz receiver based on a superconducting quasiparticle tunnel junction mixer is described. The single-sideband noise temperature is as low as 68 ± 3 K. This is nearly twice the sensitivity of any other receiver at this frequency. The receiver was designed using a low-frequency scale model in conjunction with the quantum mixer theory. A scaled version of the receiver for operation at 46 GHz has a single-sideband noise temperature of 55 K. The factors leading to the success of this design are discussed.

PACS numbers: 07.62. + s, 84.30.Qi, 85.25. + k

The superconductor-insulator-superconductor (SIS) quasiparticle tunnel junction mixer[1,2] has recently been used as a first stage in millimeter-wave receivers for radio astronomy. First at 115 GHz,[3] and then at frequencies from 45 to 250 GHz,[4–6] these receivers have rivaled the best conventional (i.e., Schottky diode) receivers. Now we have constructed an SIS receiver for 115 GHz which is almost twice as sensitive as any previously reported.[7] This letter describes the design of the receiver, its performance, and the factors leading to its success. These results may influence the choice of receiver for a number of millimeter-wave astronomy facilities now being planned.

The mixer employs a series pair of Pb(InAu)-oxide-Pb(Bi) tunnel junctions. The dc I-V curve of these junctions at 2.5 K is shown in Fig. 1(a). The differential resistance R, measured at 8 mV, is 94 Ω. An estimate of the capacitance C gives $\omega RC \sim 7$. The junctions were deposited using a photoresist liftoff process at NBS-Boulder. Each junction is $1.8 \times 2.5 \mu$m, defined by an SiO window.

A schematic drawing of the mixer is shown in Fig. 2. The fused quartz SIS chip, $0.010 \times 0.005 \times 0.010$ in. thick, is mounted on a 0.003-in.-thick fused quartz substrate with a gold stripline circuit pattern. The substrate is placed across a quarter-height WR-10 waveguide. The transition to the reduced height waveguide is accomplished using a broadband channel-waveguide transformer.[8] (The waveguide width is actually 0.096 in., slightly narrower than standard, to permit operation beyond the normal operating band without evanescent mode resonances.) Two sliding shorts of the contacting spring-finger type serve as adjustable tuning elements. One sits in the main waveguide behind the substrate, while the other is in a secondary waveguide, parallel to the first. The two waveguides are coupled through the suspended stripline as shown in the figure. Compared to conventional mixer blocks with only a single backshort, this configuration has much greater tunability at a given frequency, but the tuning is a relatively sharp function of frequency.

The potential performance of the mixer was analyzed by applying the quantum mixer theory[9,10] to the unpumped I-V curve, Fig. 1(a). A conversion loss as low as 1.5 dB was predicted if the signal and image ports were terminated with

appropriate, equal, impedances. But our relatively sharp tuning assured that only one sideband could be well terminated, while the other was approximately short circuited (by the large junction capacitance). The best conversion loss with the image short circuited was predicted to be 4.3 dB. The suspended-substrate printed circuit pattern was designed to aid in achieving the requisite signal termination. This was accomplished as in Ref. 11 by using a $40 \times$ scale model of the mixer block, which included the sliding shorts, the substrate, and the printed circuitry. Calculations based upon the scale model results indicated that the ohmic circuit losses would add 1–3 dB to the mixer's conversion loss in the vicinity of the best operating point, giving an expected overall conversion loss of 5–8 dB. We also calculated the shot noise of the mixer, as predicted in Ref. 9, and the ambient 2.5-K thermal noise reflected at the IF port of the mixer. In the vicinity of the best operating point, the theoretical mixer *output* temperature is 3.9 ± 0.4 K.

The performance of the mixer, tuned for single-sideband operation, was measured in a laboratory dewar with a monochromatic signal at 115.3 GHz. The best conversion loss is 6.9 ± 0.5 dB. The mixer's instantaneous bandwidth is

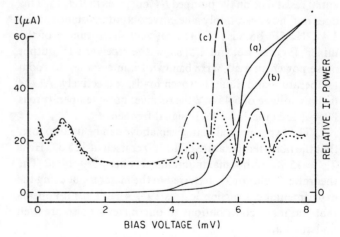

FIG. 1. (a) dc I-V curve of an unpumped series pair of SIS junctions, at 2.5 K. (b) dc I-V curve when LO power of ~ 250 nW at 113.9 GHz was applied to the mixer. The receiver is tuned for a lower-sideband rejection > 25 dB. The receiver's output noise power at the 1.4 GHz IF, in response to room temperature and liquid nitrogen loads, is shown in curves (c) and (d), respectively.

---

[a] Also Physics Department, Columbia University, New York, NY 10027.

Reprinted with permission from *Appl. Phys. Lett.*, vol. 43, pp. 786–788, Oct. 15, 1983.

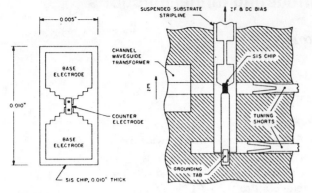

FIG. 2. Schematic drawing of the SIS chip and a cross-sectional view of the mixer block.

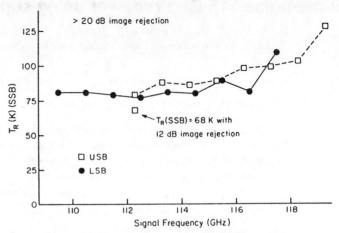

FIG. 3. Receiver's noise temperature as a function of signal frequency, for both upper-sideband (□) and lower-sideband (●) operation. The receiver is tuned for > 20 dB image rejection. By relaxing the image rejection it was possible to obtain somewhat better results, e.g., a single-sideband T_R = 68 K at 112.3 GHz. The uncertainty in T_R is ± 3 K.

~350 MHz, and the 1-dB gain compression point is typically 4 nW.

In the receiver dewar, the mixer is connected through a cross-guide LO injection coupler (coupling = 20 dB) and a stainless steel waveguide to a room-temperature scalar horn. The loss between the mouth of the horn and the mixer is 0.6 dB and contributes 12 K to the receiver noise temperature. The 1.2–1.6-GHz output from the mixer passes through a bias tee and an isolator[12] to a GaAs field-effect transistor amplifier,[13] all cooled to 2.5 K. The noise temperature of the IF section is 10.5 ± 1.0 K, referred to the output port of the mixer. This quantity was measured by using the shot noise in the unpumped SIS junctions, biased with a dc current, as an IF noise source. If a series array of N junctions is biased with a current I_{dc} at a differential resistance R_d, then the available shot noise power per unit bandwidth is $eI_{dc} R_d/2N$ (see Ref. 14).

Figure 1(b) shows the dc I-V curve when LO power of ~250 nW at 113.9 GHz was applied to the mixer. The mixer was tuned to maximize the conversion, while maintaining a large image rejection (> 25 dB). This was accomplished using a monochromatic source. Under certain other tuning conditions it is possible to obtain a region of negative differential resistance on the pumped I-V curve, as in Ref. 15. This does not, however, imply a negative output resistance at the 1.4 GHz IF, because of the relatively sharp tuning of the mixer. Figures 1(c) and 1(d) show the receiver's IF output noise power, in a 50-MHz bandwidth, in response to room temperature and liquid nitrogen loads, respectively. At the dc bias voltage of 5.45 mV the receiver noise temperature is 89 ± 3 K at the 115.3-GHz signal frequency.

The experimental data given above can be used to infer that the noise temperature of the mixer itself at 115.3 GHz is $T_m = 15$ K, with a maximum uncertainty of ± 14 K. The theoretical value of T_m (referred to the mixer input using the experimental conversion loss) is 19 ± 4 K. Thus it appears that the main contributions to our mixer's noise are well understood.

Figure 3 shows the receiver's noise temperature as a function of signal frequency, tuned for > 20 dB image rejection. Note that $T_R \simeq 80$ K over a broad range of frequencies. By relaxing the image rejection somewhat it is possible to reduce the receiver noise temperature further. For example, with 12-dB image rejection it is possible to obtain (single sideband) $T_R = 68$ K at 112.3 GHz. The rise of T_R at higher frequencies is a result of the LO coupler, which has an evanescent mode resonance at 118 GHz. The receiver's noise temperature changes by less than 10 K within a 270-MHz instantaneous bandwidth.

No magnetic field was applied to the junctions in these experiments. Nevertheless, the Josephson critical current in Fig. 1(a) is completely suppressed, whereas on other cool downs various amounts of Josephson critical current were apparent. We ascribe this variation to different amounts of magnetic flux trapped in the junctions. Whenever the Josephson critical current was seen, a large IF noise was generated by the pumped mixer at bias voltages below about 2.5 mV.[16] The Josephson currents have no discernible effect at larger bias voltages.

Although the initial optimization of the receiver for each frequency was time consuming, the settings could be noted and quickly regained on subsequent cool downs. There was no perceptible change in the junctions' behavior through 10 cool downs over four months, during which time the junctions were stored in a dry nitrogen atmosphere at a temperature of 0 °C. The long term stability remains to be checked. This receiver is presently in use on the Columbia/GISS CO Sky Survey telescope.

A scaled version of this receiver has also been built for observations of the 3-K cosmic blackbody radiation at 46 GHz. A larger pair of junctions, with $R = 34 \, \Omega$ (for the pair) and $\omega RC \sim 2.5$, was used. The single-sideband receiver noise temperature was 55 K. This is comparable to the noise performance of maser receivers at this frequency. More detailed information will be reported elsewhere.

Why are these receivers more sensitive than previous[3-6] SIS receivers? The I-V curve of the junctions, Fig. 1(a), is not particularly sharp compared to other SIS junctions which have been used. Rather, we believe that two other factors are important. First is the relatively large ωRC product. The significance of this is discussed in Ref. 14. It can be inferred from Ref. 16, Fig. 9, that $\omega RC \gtrsim 4$ is required for the mixer's conversion to reach the value predicted using the three-frequency model, at a bias voltage of $N\hbar\omega/2e$ below the energy gap. Second is the superior tuning capability of our mixers. This factor is especially important in light of the large junction capacitance, which cannot be precisely controlled during fabrication.

In conclusion, we have demonstrated an SIS receiver for 115 GHz which is almost twice as sensitive as any competitor. The design has been scaled to 46 GHz with comparable success. These results should firmly establish the SIS mixer as the first choice for ultralow-noise millimeter-wave receivers.

The authors wish to thank F. L. Lloyd and C. A. Hamilton of the National Bureau of Standards for their assistance during the junction fabrication, J. A. Grange for developing the mixer assembly techniques, E. S. Palmer for designing the overall receiver, and P. Thaddeus and D. Wilkinson for their continuing support of this work.

[1]P. L. Richards, T.-M. Shen, R. E. Harris, and F. L. Lloyd, Appl. Phys. Lett. **34**, 345 (1979).

[2]G. J. Dolan, T. G. Phillips, and D. P. Woody, Appl. Phys. Lett. **34**, 347 (1979).

[3]T. G. Phillips and D. P. Woody, Ann. Rev. Astron. Astrophys. **20**, 285 (1982); T. G. Phillips, D. P. Woody, G. J. Dolan, R. E. Miller, and R. A. Linke, IEEE Trans. Magn. MAG-17, 684 (1981).

[4]R. Blundell, K. H. Gundlach, and E. J. Blum (unpublished).

[5]L. Olsson, S. Rudner, E. Kollberg, and C. O. Lindström, Int. J. Infrared and Millimeter Waves (unpublished).

[6]E. C. Sutton, IEEE Trans. Microwave Theory Tech. MTT-31, 589 (1983).

[7]A. V. Räisänen, N. R. Erickson, J. L. R. Marreno, P. F. Goldsmith, and C. R. Predmore, Proceedings of the Sixth International Conference on Infrared and Millimeter Waves, Miami Beach, 1981 (IEEE Cat. No. 81 CH1645-1 MTT).

[8]P. H. Siegel, D. W. Peterson, and A. R. Kerr, IEEE Trans. Microwave Theory Tech. MTT-31, 473 (1983).

[9]J. R. Tucker, IEEE J. Quantum Electron. QE-15, 1234 (1979).

[10]J. R. Tucker, Appl. Phys. Lett. **36**, 477 (1980).

[11]M. J. Feldman, S. -K. Pan, A. R. Kerr, and A. Davidson, IEEE Trans. Magn. MAG-19, 494 (1983).

[12]Passive Microwave Technology, 8030 Remmit Ave., Canoga Park, Ca 91304, cryogenic L-band isolator, model 1102.

[13]S. Weinreb, D. L. Fenstermacher, and R. W. Harris, IEEE Trans. Microwave Theory Tech. MTT-30, 849 (1982).

[14]M. J. Feldman and S. Rudner, in *Reviews of Infrared and Millimeter Waves*, Volume 1, edited by K. J. Button (Plenum, New York, 1983), pp. 47–75.

[15]A. R. Kerr, S. -K. Pan, M. J. Feldman, and A. Davidson, Physica B **108**, 1369 (1981).

[16]S. Rudner, M. J. Feldman, E. Kollberg, and T. Claeson, J. Appl. Phys. **52**, 6366 (1981).

THIS part presents some techniques and receivers which are of importance to radio astronomy, but which are sufficiently specialized that they do not come under the general heading of Part IV. The distinction between the techniques discussed here, which essentially depend on hardware, and those in Part IX, which are software-based, is becoming less and less clear. For example, image restoration when mapping extended sources with a dual-beam switched system, discussed in the fourth paper of this part, is dependent upon both hardware and software. In a further extension of this technique which has been described by Morsi and Reich (1986), the beam switching itself is done in software, rather than with any type of hardware switch. This trend of blurring the distinction between receivers and data processing done in software is likely to continue at an increasing pace.

In selecting the papers for this part, I have implicitly defined what is "normal" in a radio-astronomical receiver, but this is really quite arbitrary, and the details of the information desired by an astronomer studying a particular situation vary enormously in terms of the resolution in angle, frequency, time, and polarization state of the incident radiation. The papers included here at least serve as an introduction to some specific challenges in radio-astronomical measurement, as well as to a number of very successful solutions.

The first paper, by Taylor, addresses receivers used for the study of radio pulsars. Most astronomical phenomena have a fairly leisurely time scale, but as discussed in the Historical Overview, the first pulsars were discovered with equipment which (although designed for other purposes) had a response time of about 1/10 second. Since then, much more rapid pulsars have been discovered, and very fast response is a major feature of pulsar receivers. Along with this, the ability to compensate for the dispersion introduced by the interstellar plasma and to measure the polarization on a very rapid time scale is necessary, as discussed in Taylor's paper. Other references on this topic include a review article by Huguenin (1976) and a paper by McCulloch et al. (1979). New pulsar search machines have recently been completed and others are presently under construction. These make use of the vastly enhanced computational power of microprocessors which have become available during the past few years.

The two papers which follow are concerned with the issue of measuring the polarization state of incident radiation. Since radio-astronomy receivers are sensitive to a single polarization state, one can either have several different receivers, or vary the polarization state to which a single receiver is sensitive, in order to measure the polarization of the incident radiation. The paper by Cohen, which is the second paper in this part, discusses polarization measurements from a general point of view; Chapter 4 in the text by Kraus (1986) also covers this material. Articles by Cohen (1958), Suzuki and Tsuchiya

(1958), and Akabane (1958) describe specific polarimeters. These devices all operate at relatively low radio frequencies. Polarimetry at higher frequencies is in comparison less developed. Some results and descriptions of spectral line polarimeter systems are given by Wannier et al. (1983), and different techniques for constructing components are discussed by van Vliet and de Graauw (1981). The paper by Weiler, which is the third included here, deals with polarization measurement with an interferometer. This has proven to be an issue of major importance, in such diverse areas as the study of radio galaxies and that of planetary surfaces.

Measurements of continuum emission generally put the most stringent demands on radiometric systems in terms of achieving the noise level expected from purely statistical fluctuations, since, unlike the case of spectral line emission, no baseline determined from frequencies "off" the line can be subtracted. Fluctuations of the receiver gain and of the atmospheric emission and transmission can result in excess noise. The following two papers address different aspects of this problem. In the first, Emerson et al. describe a technique for canceling the effects of fluctuations of the atmosphere in making extended maps of continuum sources. This article also discusses the standard application of the dual-beam technique, which is treated in detail in the thesis of Baars (1970). It is often advantageous to switch rapidly between source and comparison beams; the rate is restricted by switch performance, which is relatively poor at higher frequencies. The correlation, or continuous comparison, radiometer is one method of overcoming this limitation. The effective switching rate is on the order of the IF bandwidth, or $> 10^8$ Hz for the system described by Predmore et al. in the next paper. Further information on correlation radiometers can be found in articles by Fujimoto (1964), Faris (1967), and Batchelor et al. (1968), and in the references given there.

The problem of absolute calibration of the antenna temperature scale is one of the most difficult in radio astronomy. Many of the techniques discussed here and in Parts IV and VII in effect avoid this issue by considering only differences between the "source" and the "reference" (assumed to be free of emission) positions. The high-accuracy absolute flux density measurements of Cassiopeia A by Findlay et al. (1965) could take advantage of the restricted angular size of the source, and let it drift through the antenna beam as the earth rotated.

In some cases, the absolute intensity of the radiation received is of great importance, and certainly the most outstanding of these is the 3 K isotropic background. As indicated in the Historical Overview, this radiation gives us a unique probe of conditions in the early universe. Its discovery by Penzias and Wilson in 1964–65 was, to a significant extent, dependent upon their confidence in the absolute calibration of their antenna. This knowledge was the result of a major investment in time and energy. The instrumentation used and the techniques developed are described in the last article in this part, which is by Wilson.

Many other special techniques developed by radio astronomers are not included here due to a lack of space. Among these is the use of lunar occultation to enhance the angular resolution available with a filled aperture antenna. This technique played an important role in the discovery of quasars, as discussed in the Historical Overview. It has also been used at millimeter wavelengths, as discussed in the references given below. The principle of its operation and its use have been discussed by Hazard (1976).

REFERENCES AND BIBLIOGRAPHY

Pulsar Receivers

1974: Boriakoff, V., "A digital pulsar processor," *Astron. Astrophys. Suppl.,* vol. 15, pp. 479–481, 1974.

1976: Huguenin, G. R., "Pulsar observing techniques," Chapter 4.5 in *Methods of Experimental Physics: Vol. 12 Astrophysics, Part C: Radio Observations,* M. L. Meeks, Ed. New York, NY: Academic Press, 1976, pp. 78–91.

1979: McCulloch, P.M., J. H. Taylor, and J. M. Weisberg, "Tests of a new dispersion-removing radiometer on binary pulsar PSR 1913 + 16," *Astrophys. J.,* vol. 227, pp. L133–L137, 1979.

1987: Backer, D. C., D. R. Werthimer, T. R. Clifton, and S. R. Kulkarni, "Fast pulsar search machine," Abstract of a paper (p. 256) presented at the URSI National Radio Science Meeting, Boulder, CO, Jan. 1987.

Lacasse, R. J. and J. R. Fisher, "An FFT spectrometer with pulsar dedispersion capabilities," Abstract of a paper (p. 260) presented at the URSI National Radio Science Meeting, Boulder, CO, Jan. 1987.

Polarimeters

1958: Akabane, K., "A polarimeter in the microwave region," *Proc. IRE,* vol. 46, pp. 194–197, 1958.

Cohen, M., "The Cornell radio polarimeter," *Proc. IRE,* vol. 46, pp. 183–190, 1958.

Suzuki, S. and A. Tsuchiya, "A time-sharing polarimeter at 200 MC," *Proc. IRE,* vol. 46, pp. 190–194, 1958.

1981: van Vliet, A. H. F. and Th. de Graauw, "Quarter wave plates for submillimeter wavelengths," *Int. J. Infrared and Millimeter Waves,* vol. 2, pp. 465–477, 1981.

1983: Wannier, P. G., N. Z. Scoville, and R. Barvainis, "The polarization of millimeter-wave emission lines in dense interstellar clouds," *Astrophys. J.,* vol. 267, pp. 126–136, 1983.

1986: Kraus, J. D., *Radio Astronomy,* 2nd ed. Powell, OH: Cygnus-Quasar Books, 1986, ch. 4.

Beam Switching Radiometers

1970: Baars, J. W. M., *Dual Beam Parabolic Antennae in Radio Astronomy.* Groningen: Wolters-Noordhoff, 1970.

1986: Morsi, H. W. and W. Reich, "A new 32 GHz radio continuum receiving system for the Effelsberg 100-m telescope," *Astron. Astrophys.,* vol. 161, pp. 313–320, 1986.

Correlation Radiometers

1964: Fujimoto, K., "On the correlation radiometer technique," *IEEE Trans. Microwave Theory Tech.,* vol. MTT-12, pp. 203–212, 1964.

1967: Faris, J. J., "Sensitivity of a correlation radiometer," *J. Res. Nat. Bur. Stand.,* vol. 71C, pp. 153–170, 1967.

1968: Batchelor, R. A., J. W. Brooks, and B. F. C. Cooper, "Eleven-centimeter broadband correlation radiometer," *IEEE Trans. Antennas Propagat.,* vol. AP-16, pp. 228–234, 1968.

Null-Balancing Radiometer

1952: Machin, K. E., M. Ryle, and D. D. Vonberg, "The design of an equipment for measuring small radio-frequency noise powers," *Proc. IEE,* vol. 99, pp. 127–134, 1952.

Absolute Intensity Measurements

1965: Findlay, J. W., H. Hvatum, and W. B. Waltman, "An absolute flux-density measurement of Cassiopeia A at 1440 MHz," *Astrophys. J.,* vol. 141, pp. 873–884, 1965.

1980: Weiss, R., "Measurements of the cosmic background radiation," *Annu. Rev. Astron. Astrophys.,* vol. 18, pp. 489–535, 1980.

1983: Wilson, R. W., "Discovery of the cosmic microwave background," in *Serendipitous Discoveries in Radio Astronomy,* K. Kellerman and B. Sheets, Eds. Green Bank, WV: National Radio Astronomy Observatory, 1983, pp. 185–195.

1986: Mandolesi, N., P. Calzolari, S. Coriglioni, G. Morigi, L. Danese, and G. De Zotti, "Measurements of the cosmic background radiation temperature at 6.3 centimeters," *Astrophys. J.,* vol. 310, pp. 561–567, 1986.

1987: Johnson, D. G. and D. T. Wilkinson, "A 1% measurement of the temperature of the cosmic microwave radiation at $\lambda = 1.2$ centimeters," *Astrophys. J. (Lett.),* vol. 313, pp. L1–L4, 1987.

Differential Intensity Measurements

1978: Gorenstein, M. V., R. A. Muller, G. F. Smoot, and J. A. Tyson, "Radiometer system to map the cosmic background radiation," *Rev. Sci. Instr.,* vol. 49, pp. 440–448, 1978.

1984: Fomalont, E. B., K. I. Kellerman, and J. V. Wall, "Limits to the small-scale fluctuations in the cosmic background radiation," *Astrophys. J. (Lett.),* vol. 277, pp. L23–L26, 1984.

Uson, J. M. and D. T. Wilkinson, "Small-scale isotropy of the cosmic microwave background at 19.5 GHz," *Astrophys. J.,* vol. 283, pp. 471–478, 1984.

Lunar Occultation

1962: Scheuer, P. A. G., "On the use of lunar occultations for investigating the angular structure of radio sources," *Aust. J. Phys.,* vol. 15, pp. 333–343, 1962.

1976: Hazard, C., "Lunar occultation measurements," Chapter 4.6 in *Methods of Experimental Physics: Vol. 12 Astrophysics, Part C: Radio Observations,* M. L. Meeks, Ed. New York, NY: Academic Press, 1976, pp. 92–117.

1980: Schloerb, F. P. and N. Z. Scoville, "Small-scale structure of the CO emission in S255 from lunar occultation observations," *Astrophys. J.,* vol. 253, pp. L33–L37, 1980.

1985: Snell, R. L. and F. P. Schloerb, "Structure and physical properties of the bipolar outflow in L1551," *Astrophys. J.,* vol. 295, pp. 490–500, 1985.

Pulsar Receivers and Data Processing

JOSEPH H. TAYLOR

Abstract—The rapid variability of pulsar signals and the dispersive nature of the interstellar medium introduce a number of problems which are not inherent in other radio-astronomical observations. These problems are discussed and some of the techniques currently being used to cope with them are described.

I. INTRODUCTION

THE DISCOVERY of pulsars in late 1967 gave radio astronomers for the first time the challenge of detecting and analyzing signals from weak sources of rapidly varying intensity. At meter wavelengths, a moderately strong pulsar emits a signal that produces an average flux density of the order of 0.1 flux unit[1] at the earth. Such a signal would be difficult to detect if it were stationary Gaussian noise. But

Manuscript received March 15, 1973.

The author is with Five College Radio Astronomy Observatory, University of Massachusetts, Amherst, Mass.

[1] One flux unit = 10^{-26} W·m^{-2} Hz^{-1}.

pulsar signals are anything but stationary; they exhibit large intensity variations over time scales as short as 10 µs, which makes pulsars unique among known sources of radio emissions outside the solar system. The emission generally comes in pulses of several milliseconds' duration, which in the most favorable cases can be observed over the frequency range from 30 MHz to 10 GHz. The duty cycle is typically 2 or 3 percent. The pulses come at very regular intervals, although there may be considerable variation in intensity from one pulse to the next and some pulses may be entirely missing. In many cases the signals are highly polarized, and the polarization characteristics sometimes vary with time in a systematic way.

Pulsars are as good an approximation of true point sources as we are likely to encounter for some time. (The angular diameter of a neutron star at a distance of a few hundred parsecs (pc) is about the same as the angle subtended by a marble at the distance of Pluto, or 10^{-15} rad.) Thus

Reprinted from *Proc. IEEE*, vol. 61, no. 9, pp. 1295–1298, Sept. 1973.

pulsar emission regions are not likely to be resolved by present or foreseeable interferometric techniques. On the other hand, because of the time-varying nature of the signals, "confusion" in the usual astronomical sense is not a problem in pulsar observations. Antenna requirements are therefore reduced to considerations of obtaining adequate collecting area, bandwidth, and sky coverage.

Many additional details on the nature of pulsar signals and the objects which emit them are discussed elsewhere in this issue [1]. The present paper will be concerned with the special equipment and techniques currently being used to study pulsars, and the ways in which they differ from more conventional radio-astronomical methods.

II. RECEIVING AND DATA RECORDING TECHNIQUES

An ideal pulsar receiving system would include most of the equipment shown in the block diagram in Fig. 1. This system differs from conventional radio-astronomy receivers in three principal ways. First, no comparison switching or synchronous detection is required, because most of the effects of such switching are provided by the "switched" nature of the pulsar itself. Second, four identical signal channels are included, so that the full state of polarization of the rapidly varying signals can be determined instantaneously, without resort to slow-mechanical devices such as rotating feed assemblies. And finally, each of the four signal paths includes a "dispersion remover" to compensate for the effects of dispersion in the interstellar medium. These devices will be discussed separately in Section III.

The observable characteristics of pulsars can be divided into two major classes: those which remain constant over many pulse periods, and those which vary from pulse to pulse. The data handling problems involved in the two cases are of a rather different nature, although much of the same receiver hardware can be used in either case. When integrated pulse characteristics are being measured, the receiver output may be averaged synchronously with the pulsar period, and recorded at relatively infrequent intervals of several minutes or more. The synchronous averaging can be done either by means of a special-purpose hard-wired instrument, or with a suitably programmed general-purpose computer. If the averaging is done in real time, the apparent pulsar period at the time of observation (corrected for the motion of the observer with respect to the solar system barycenter) must be known in advance, so that an accurate synchronizing signal can be provided.

If the signal averager memory can be sectored into quadrants, the full polarization characteristics of the integrated pulse may be recorded for each phase within the period. Such measurements have shown that in most pulsars, the position angle of the linear component of polarization changes systematically with pulse phase [2], [3]. Therefore, the pulse shape observed using a single linear polarization is not the same as the total-intensity pulse shape, and may even vary with time as ionospheric Faraday rotation changes. Precise measurements of the arrival times of pulses, which are necessary for determining pulsar periods and period derivatives [4], [5], are best done by summing the signals detected simultaneously in two orthogonal polarizations.

Individual pulses and subpulses contain a wealth of information that is not present in the synchronous average of many dozens of pulses. However, the single-pulse information is more difficult to obtain, because sensitivity requirements are

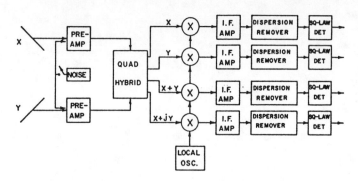

Fig. 1. Block diagram of an ideal pulsar receiving system.

more severe and the necessary rates of data recording are very much higher. These two problems become progressively more severe as the time resolution is improved. However, because the duty cycle of the desired signals is rather low, a considerable advantage can be gained by "burst sampling" the receiver outputs—that is, by recording a number of samples in rapid succession, centered on the expected time of pulse arrival, and then entering a dormant state for the remainder of the pulsar period. The dormant period can be used profitably for such purposes as formatting the data for recording on magnetic tape.

III. DISPERSION-REMOVAL TECHNIQUES

Except when spectral information is the desired quantity, it is common practice in radio astronomy to maximize system sensitivity by using the largest practicable receiver bandwidth. For pulsar observations, however, the situation is not quite so simple, because the pulses of radiation are dispersed by the frequency-dependent propagation velocity in the interstellar medium. If the desired time resolution is Δt seconds, then unless special techniques are employed, the maximum useable bandwidth (in megahertz) is given by

$$\Delta \nu_{max} = \Delta t\, \dot{\nu} \approx 1.2 \times 10^{-4}\, \Delta t\, \nu^3\, DM^{-1} \qquad (1)$$

where $\dot{\nu}$ is the apparent rate of frequency change of the dispersed signal in megahertz per second, ν is the observing frequency in megahertz, and DM is the dispersion measure (integral of electron density along the line of sight) in units of cm^{-3} pc. For example, at a typical observing frequency of 400 MHz, a pulsar with DM = 50 can be observed with time resolution $\Delta t = 1$ ms only if the receiver bandwidth is less than 160 kHz.

Several different dispersion-removal techniques have been used to overcome the severe bandwidth limit implicit in (1). The most general of these methods requires complex Fourier analysis of the intermediate frequency signal. The complex spectrum is multiplied by a frequency-dependent phase factor (computed from the known dispersion measure of the pulsar) and the inverse Fourier transform of this product yields the desired signal, free of dispersion effects. This system has been successfully employed by Hankins [6]; the IF signal was digitized and the Fourier transforms were done off-line in a general-purpose computer. Unfortunately, a large amount of computer time is required for this method. Few existing computers can do Fourier transforms faster than about 1000 points/s, so total bandwidths of (say) 100 kHz can be processed only at the expense of many hours of computing for each hour of observing.

If the signal emitted by the pulsar is incoherent to the extent that the phases of its Fourier components are unrelated (or if multipath propagation in the interstellar medium has destroyed any such frequency coherence) a much simpler incoherent dispersion-removing technique will provide identical results. The most widely used dedispersing technique is probably that of Taylor and Huguenin [7]. The system performs the equivalent of Fourier analysis in a relatively coarse way by using a comb of N filters tuned to adjacent frequencies. Because the phases of the Fourier components are not needed for incoherent dedispersing, the output of each filter may be independently square-law detected and low-pass filtered to the desired time resolution. The N detected signals are then introduced into summing junctions spaced along a tapped delay line, with the signal from the highest frequency filter going to the input end of the line so that it will be delayed most. If the delay between taps on the delay line is adjusted to equal the difference in pulse arrival times at the frequencies of adjacent filters, the signal will appear correctly dedispersed at the output end of the delay line. With this technique, (1) may be taken to define the maximum bandwidth of *one* of the filters in the comb, and the total bandwidth is limited only by the number of filters available. Furthermore, the bandwidth of this tapped delay line need be only as great as $2/\Delta t$, which is typically much less than the bandwidth of one of the N filters. The delay line used by Taylor and Huguenin [7] uses sample-and-hold circuits in a "bucket-brigade" configuration, passing the progressively summed signal from stage to stage toward the output end. The delay interval is controlled by a digital clock.

IV. DATA HANDLING AND DISPLAY

The rapid variations of subpulse intensity, phase, and polarization characteristics obviously create problems in data handling and display as well as in data recording. With dispersion-removal techniques and reasonably large antennas, it is quite feasible to obtain a full set of Stokes parameters for as many as several hundred phase-resolved points across a single pulse, still maintaining a high signal-to-noise ratio. Therefore, something like 10^4 binary bits of significant information are available for each pulse period! Such data should be very helpful in delimiting the range of possible models for the pulsar emission mechanism, but its multidimensional character makes its analysis and presentation a nontrivial problem.

Pulse intensity variations can be adequately presented as two-dimensional "intensity-modulated" plots [7], [8] or as parallel-ruled perspective drawings [1], where the x and y coordinates represent time within a pulse and pulse number, respectively, and shading or height above the perspective plane represents signal intensity.

Polarization characteristics introduce three additional dimensions of data, and therefore cannot be presented in such a straightforward way. One solution which is adequate for presenting intensity and linear polarization has been used by Taylor *et al.* [9]. Their diagram, reproduced in Fig. 2, uses circles to represent total intensity, and straight lines to represent the position angle and percentage of linear polarization.

Another technique which is capable of representing any arbitrary elliptical polarization is illustrated in Fig. 3. In this diagram, total intensity is represented by pulse profiles stacked one above another. Polarization characteristics at 40 different phase intervals within the pulse are represented as

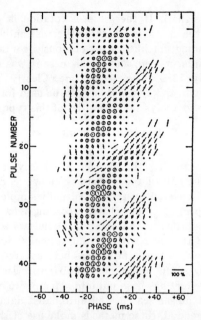

Fig. 2. Intensity and linear polarization characteristics of pulses from PSR 0809+74 (Taylor *et al.* [9]). Diameters of circles are proportional to the total intensity at one phase interval, lengths of lines are proportional to percentage linear polarization, and orientations of lines correspond to linear polarization position angle.

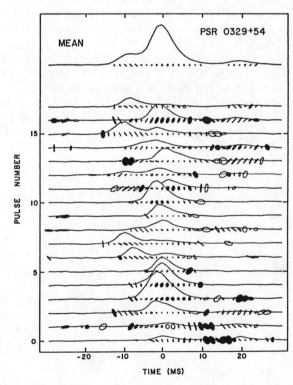

Fig. 3. Complete polarization characteristics of individual pulses from PSR 0329+54. Pulse profiles represent total intensity. Major axes of the ellipses have lengths proportional to total percentage polarization, and the shape of the ellipse is that of the conventionally defined polarization ellipse. If the circular component of polarization is positive, the ellipse is filled in. The longest ellipses shown represent 100-percent polarization.

ellipses. The length of the major axis of each ellipse is proportional to percentage polarization, and the shape of the ellipse is the same as that of the conventionally defined polarization ellipse (i.e., the locus of the tip of the electric-field vector of

PROCEEDINGS OF THE IEEE, VOL. 61, NO. 9, SEPTEMBER 1973

the original signal). If the sense of rotation is positive, or left-handed, the ellipse is filled in; otherwise only the outline is drawn. A diagram like Fig. 3 obviously contains a large quantity of information, and in this case it was acquired in approximately 13 s of telescope time. Clearly, some techniques similar to this one will have to be used for any useful compilation and analysis of quantities of this type of data.

V. Searching Techniques

A problem which deserves special mention here is the problem of searching the sky for previously undiscovered pulsars. About half of the 84 pulsars known as of March 1, 1973, were discovered by the conceptually simple method of visually inspecting fast chart recordings for impulsive "spikes." It is probable that most of the pulsars which can be discovered in this way with existing antennas have already been discovered, because the sky has been covered reasonably thoroughly by groups at a number of observatories. Therefore, considerable effort has been put into attempts to enhance the sensitivity of pulsar searches by automated data-processing schemes. In general, these methods make use of the fact that pulsar signals recur periodically and exhibit a characteristic frequency dispersion in arrival time.

Post-detection filtering in the pulse-period domain, either by cross correlation or Fourier analysis, has been used in pulsar searches at the Arecibo, Green Bank, Jodrell Bank, and Ootacamund Observatories [10]–[13]. Methods used in these surveys have been summarized by Burns and Clark [14].

Pulsar searches which rely on the dispersed nature of the signals have also been used by groups at several different observatories. One of the methods involved a special hard-wired dispersion remover [15], while the other surveys were done with digital computers, either off-line [12] or on-line [16], [17]. Taylor [17] has described a general method which is capable of doing both period and dispersion analysis and which should be capable of discovering many new pulsars within the sensitivity limits set by existing antennas and receivers.

References

[1] R. N. Manchester, "The properties of pulsars," this issue, pp. 1205–1211.

[2] ——, "Observations of pulsar polarization at 410 and 1665 MHz," Astrophys. J. Suppl. Ser., vol. 23, pp. 283–322, Sept. 1971.

[3] A. G. Lyne, F. G. Smith, and D. A. Graham, "Characteristics of the radio pulses from the pulsars," Mon. Notices Roy. Astron. Soc., vol. 153, pp. 337–382, Mar. 1971.

[4] R. N. Manchester and W. L. Peters, "Pulsar parameters from timing observations," Astrophys. J., vol. 173, pp. 221–226, Apr. 1972.

[5] P. E. Reichley, G. S. Downs, and G. A. Morris, "Time-of-arrival observations of eleven pulsars," Astrophys. J. Lett., vol. 159, pp. L35–L40, Jan. 1970.

[6] T. H. Hankins, "Microsecond intensity variations in the radio emission from CP 0950," Astrophys. J., vol. 169, pp. 487–494, Nov. 1971.

[7] J. H. Taylor and G. R. Huguenin, "Observations of rapid fluctuations of intensity and phase in pulsar emissions," Astrophys. J., vol. 167, pp. 273–291, July 1971.

[8] J. M. Sutton, D. H. Staelin, R. M. Price, and R. Weimer, "Three pulsars with marching subpulses," Astrophys. J. Lett., vol. 159, pp. L89–L93, Feb. 1970.

[9] J. H. Taylor, G. R. Huguenin, R. M. Hirsch, and R. N. Manchester, "Polarization of the drifting subpulses of pulsar 0809+74," Astrophys. Lett., vol. 9, pp. 205–208, Nov. 1971.

[10] R. V. E. Lovelace, J. M. Sutton, and E. E. Salpeter, "Digital search methods for pulsars," Nature, vol. 222, pp. 231–233, Apr. 1969.

[11] J. H. Taylor and G. R. Huguenin, "Two new pulsating radio sources," Nature, vol. 221, pp. 816–817, Mar. 1969.

[12] E. C. Reifenstein, W. D. Brundage, and D. H. Staelin, "Searches for pulsars," Astrophys. J. Lett., vol. 156, pp. L125–L130, June 1969.

[13] J. G. Davies, M. I. Large, and A. C. Pickwick, "Five new pulsars," Nature, vol. 227, pp. 1123–1124, Sept. 1970.

[14] W. R. Burns and B. G. Clark, "Pulsar search techniques," Astron. Astrophys., vol. 2, pp. 280–287, July 1969.

[15] M. I. Large and A. E. Vaughan, "A search of the galactic plane for high dispersion pulsars," Mon. Notices Roy. Astron. Soc., vol. 151, pp. 277–287, Mar. 1971.

[16] J. G. Davies and M. I. Large, "A single-pulse search for pulsars," Mon. Notices Roy. Astron. Soc., vol. 149, pp. 301–310, Sept. 1970.

[17] J. H. Taylor, "A sensitive method for detecting dispersed radio emission," to be published in Astron. Astrophys.

Radio Astronomy Polarization Measurements*

MARSHALL H. COHEN†, MEMBER, IRE

Summary—Various polarization measuring schemes are discussed in terms of the Stokes parameters. The methods are separated into three groups according to the number of components isolated by the antenna system. The dispersion resulting from the frequency dependence of the Faraday effect is considered, and the interpretation of polarization data in terms of these effects is discussed. Some results are that one polarization determination will allow limits to be placed on the axial ratio and polarization percentage at the source, and on the ray path integral of longitudinal magnetic field times electron density. A graphical presentation of Hatanaka's dispersion theory permits this to be done very quickly. With suitable assumptions, measurements at two frequencies or with two bandwidths will fix the three quantities.

Radio astronomy polarization observations are surveyed. The ionosphere, solar corona, and Crab Nebula are considered briefly in the light of possible Faraday effects. Mayer, McCullough, and Sloanaker's 3.15-cm observations on the Crab Nebula and the optical results are used to analyze lower frequency attempts to find polarized radiation from this object.

I. Introduction

THIS introduction contains a brief survey of some polarization observations.

Section II reviews the Stokes parameters as a means of describing the state of polarization of radio noise. It is shown how the parameters are related to quantities which can be measured with antennas.

Section III is a survey of measurement techniques. A number of schemes are considered and categorized according to the number of components measured.

Section IV is a discussion of the Faraday effect. Several experiments for finding the parameters involved are considered.

Section V is a discussion of polarization observations of the sun and of the Crab Nebula in terms of possible Faraday rotation effects.

Circularly-polarized radiation from the sun was first observed in 1946, at an active time when a large sun spot was on the sun's disk.[1] The existence of the polarization was presumed to be associated with the magnetic field of the sun spot, and subsequent investigations showed that the sense of rotation generally corresponded to the ordinary mode of the magneto-ionic theory.[2] There were reports of elliptical and linear polarizations seen with an interference polarimeter,[2,3] but typically any polarized component in the solar radiation was taken to be circular. A number of observatories making routine observations of the sun made observations of both circular components, and computed a percentage of polarization on the basis that only circular polarization was present.[4] The circularly-polarized solar radiation is characteristic of noise storms, which are associated with sun spots. This has been observed over a very wide range of frequencies. At microwave frequencies the storms are missing, but the "slowly varying component" has been seen to be partly circularly polarized at 10 cm[5] and at 3 cm.[6]

In 1954 Hatanaka, Suzuki, and Tsuchiya[7,8] constructed a telescope-type (as opposed to an interference-type) polarimeter that made a complete determination of the four polarization parameters. They observed many partially-polarized, elliptical bursts from the sun at 200 mc. In 1956, however, Hatanaka[9] pointed out that at meter wavelengths the Faraday rotation in the earth's ionosphere would be great enough to destroy the precision with which orientation angle data could be extended back to the source. He also pointed out that there was a chance that the solar corona might produce a very large rotation and a consequent dispersion in orientation angles which would have the effect of depolarizing a wave. This raised the speculation that perhaps some of the unpolarized solar bursts were originally linearly polarized, but became unpolarized by the dispersion process.

The ionospheric Faraday effect has been directly observed by Evans,[10] who made a two-frequency radar experiment with the moon as the target. The Faraday rotation (one way) was 10 to 13 radians at 120 mc, near sunrise. The amount of rotation should be proportional to f^{-2} for frequencies well above the critical frequency.

Optical measurements of the photosphere[11] suggest the existence of a general solar dipole magnetic field with a strength of 1 Gauss at the poles. If this type of field exists, a radio ray resulting from thermal radiation

* Original manuscript received by the IRE, October 14, 1957. Various aspects of this work have been supported by contracts with the Office of Naval Res. and the Rome Air Dev. Center, and by a grant from the Natl. Science Foundation.

† Cornell University, Ithaca, N. Y.

[1] J. L. Pawsey and R. N. Bracewell, "Radio Astronomy," Oxford University Press, London, Eng., p. 185; 1955.

[2] R. Payne-Scott and A. G. Little, "The position and movement on the solar disk of sources of radiation at a frequency of 97 mcs. II. Noise storms," *Aust. J. Sci. Res.*, vol. A 4, pp. 508–525; 1951.

[3] R. Payne-Scott and A. G. Little, "The position and movement on the solar disk of sources of radiation at a frequency of 97 mcs. III. Outbursts," *Aust. J. Sci. Res.*, vol. A 5, pp. 32–49; 1952.

[4] *Quart. Bull. Solar Activity*, Internatl. Astron. Union, Zurich, Switzerland; January–March, 1949, and later issues.

[5] Pawsey and Bracewell, *op. cit.*, p. 171.

[6] N. L. Kaidanovsky, D. V. Korolkov, N. S. Soboleva, and C. E. Khaikin, "Observations of polarized radio waves from sun spots on a wavelength of 3 cm," *Solar Data*, no. 4, Acad. Sci. USSR; 1956.

[7] S. Suzuki and A. Tsuchiya, "A time-sharing polarimeter at 200 mc," this issue, p. 190.

[8] T. Hatanaka, S. Suzuki, and A. Tsuchiya, "Observations of polarization of solar radio bursts," *Proc. Japan Acad.*, vol. 31, pp. 81–87; 1955.

[9] T. Hatanaka, "The Faraday effect in the earth's ionosphere with special reference to polarization measurements of solar radio emission," *Publ. Astron. Soc. Japan*, vol. 8, pp. 73–86; 1956.

[10] J. V. Evans, "The measurement of the electron content of the ionosphere by the lunar radio echo method," *Proc. Phys. Soc.*, vol. B69, pp. 953–955; September, 1956.

[11] H. W. Babcock and H. D. Babcock, "The sun's magnetic field, 1952–1954," *Astrophys. J.*, vol. 121, pp. 349–366; March, 1955.

Reprinted from *Proc. IRE*, vol. 46, pp. 172–183, Jan. 1958.

will be slightly circularly polarized because the two magneto-ionic modes will have different absorption coefficients. Smerd[12] has calculated the magnitude of this effect and has shown how the resultant polarization depends on the strength of the magnetic field and on the electron temperature. Rays from the northern and southern hemispheres of the sun will have opposite polarizations because the longitudinal components of magnetic field are opposite for the two cases, so the integrated flux from the whole sun will be unpolarized.

Attempts have been made to measure the differential polarization between the two halves of the sun, and thereby determine the strength of the general magnetic field in the corona. The results so far have been negative and have only given upper limits to the magnetic field. Christiansen, Yabsley, and Mills,[13] measuring at 50 cm during an eclipse, set an upper limit of 9 Gauss for the strength of the polar field. Conway,[14] measuring with a N-S interferometer at 60 cm, set a limit of 2.5 Gauss. A more recent interferometer measurement by Conway[15] gives a preliminary estimate of 1 Gauss for the limits. This required equipment which could detect a polarized component of 0.25 per cent.

The light from the radio source in Taurus (Crab Nebula) has been found to be strongly linearly polarized, and since this discovery there has been considerable interest in a possible polarization of the radio radiation.[16] Mayer, McCullough, and Sloanaker[17] have recently reported a polarization of about 7 per cent at 3 cm, observed with a rotating linear antenna in the NRL 50-foot dish. At lower frequencies the polarization would probably be less because of Faraday dispersion. At 22 cm Westerhout[18] set an upper limit of 1 per cent, and at 158 mc Hanbury-Brown, Palmer, and Thompson[19] set an upper limit of $2\frac{1}{2}$ per cent, for the polarized flux, if linearly polarized.

The sources in Cassiopeia and Cygnus also have been examined for polarized radiation, with negative results.[17–20]

[12] S. F. Smerd, "The polarization of thermal 'solar noise' and a determination of the sun's general magnetic field," *Aust. J. Sci. Res.*, vol. A 3, pp. 265–273; June, 1950.

[13] W. N. Christiansen, D. E. Yabsley, and B. Y. Mills, "Measurements of solar radiation at a wavelength of 50 centimeters during the eclipse of November 1, 1948," *Aust. J. Sci. Res.*, vol. A 2, pp. 506–523; 1949.

[14] R. G. Conway, "The general magnetic field of the sun," *The Observatory*, vol. 76, pp. 106–108; June, 1956.

[15] Report to the XIIth General Assembly of URSI, Boulder, Colo.; August 26–September 5, 1957.

[16] J. H. Oort and T. H. Walraven, "Polarization and composition of the Crab Nebula," *Bull. Astron. Inst. Neth*, vol. 12, pp. 285–308; May, 1956.

[17] C. H. Mayer, T. P. McCullough, and R. M. Sloanaker, "Evidence for polarized radio radiation from the Crab Nebula," *Astrophys. J.*, vol. 126, pp. 468–470; September, 1957.

[18] G. Westerhout, "Search of polarization of the Crab Nebula and Cassiopeia A at 22 cm wavelength," *Bull. Astron. Inst. Neth.*, vol. 12, pp. 309–311; May, 1956.

[19] R. Hanbury-Brown, H. P. Palmer, and A. R. Thompson, "Polarization measurements on three intense radio sources," *Monthly Notices Roy. Astron. Soc.*, vol. 115, pp. 487–492; 1955.

[20] M. Ryle and F. G. Smith, "A new intense source for radio-frequency radiation in the constellation of Cassiopeia," *Nature*, vol. 162, pp. 462–463; September, 1948.

II. POLARIZATION PARAMETERS

The radio waves of concern are assumed to be partly-polarized noise signals. Such waves can be uniquely resolved into polarized and unpolarized (randomly-polarized) components.[21] The polarized part has a noise spectrum but has coherence between components at right angles. Its electric vector traces out an ellipse which is continually fluctuating in size but maintains a constant orientation, axial ratio, and sense of rotation, for intervals long compared with the period. The polarized part is specified by its intensity (I_e), orientation (χ), and axial ratio (r). The sense of rotation is contained in the sign of r: $r > 0$ means left hand in the radio sense; and $r < 0$ means right hand. See Fig. 1. The wave is propagating along the z axis, into the paper, and is right-hand polarized. The angle χ is clockwise from the x axis, viewed in the direction of propagation.

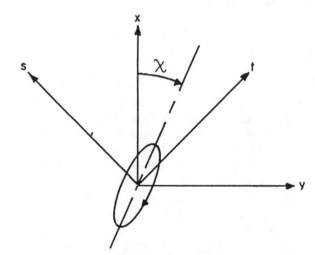

Fig. 1—Coordinate system.

The unpolarized part of the wave has a similar noise spectrum and has components at right angles which are independent. It is specified by a single parameter, intensity (I_u). The total intensity (I) is the sum of I_e and I_u. If we let $m = $ polarization fraction $= I_e/I$, the four parameters that specify the wave are I, m, r, and χ.

The Stokes parameters[21] are a set of four numbers that define the wave and have the advantage of being closely related to antenna measurements. They may be defined by

$$I = I \tag{1}$$

$$Q = I_e \cos 2\beta \cos 2\chi \tag{2}$$

$$U = I_e \cos 2\beta \sin 2\chi \tag{3}$$

$$V = I_e \sin 2\beta, \tag{4}$$

[21] S. Chandrasekhar, "Radiative Transfer," Oxford University Press, London, Eng., pp. 24–35; 1955.

where

$$\beta = \tan^{-1} r. \tag{5}$$

In these terms, the quantities m, r, and χ are given by

$$m = (Q^2 + U^2 + V^2)^{1/2}/I, \tag{6}$$

$$\sin 2\beta = V/I_e, \tag{7}$$

$$\tan 2\chi = U/Q. \tag{8}$$

The wave can be regarded as the sum of two independent elliptically-polarized components, with equal and opposite values of r, and perpendicular orientations.[21] These will be referred to as opposite polarizations. They include perpendicular linear components, and oppositely rotating circular components, as special cases. These are the only ones we shall use. If the wave is resolved into components along the x and y axes (Fig. 1), the intensities are (omitting impedance factors)

$$I_x = X^2 + \tfrac{1}{2}I_u, \tag{9}$$

$$I_y = Y^2 + \tfrac{1}{2}I_u, \tag{10}$$

$$I = I_x + I_y. \tag{11}$$

In these formulas X is the rms amplitude of the x component of the polarized part, and Y is the rms amplitude of the y component of the polarized part. These quantities are all defined for a certain frequency interval; say, the bandwidth of the receiver. The amplitudes X and Y have a fixed ratio so long as the polarization is constant. Moreover, there is a constant phase relationship between corresponding frequencies of the two components of the polarized part. Let γ_y be the angle by which the y components lead the x components. The value of γ_y is constant over the entire band under consideration.

Similar remarks can be made about components along the s and t axes at $\pm 45°$ to the x, y axes, Fig. 1.

$$I_s = S^2 + \tfrac{1}{2}I_u, \tag{12}$$

$$I_t = T^2 + \tfrac{1}{2}I_u, \tag{13}$$

$$I = I_s + I_t. \tag{14}$$

The phase difference, γ_t, is similarly defined.

Corresponding definitions can be made in terms of circular components:

$$I_r = R^2 + \tfrac{1}{2}I_u, \tag{15}$$

$$I_l = L^2 + \tfrac{1}{2}I_u, \tag{16}$$

$$I = I_r + I_l. \tag{17}$$

R and L are the mean amplitudes of the right and left circular components of the polarized part of the wave, and γ_r is the phase difference between them. For sinusoidal waves the circular components are defined in terms of the linear components:

$$\hat{R} = (\hat{X} + j\hat{Y})/\sqrt{2}, \tag{18}$$

$$\hat{L} = (\hat{X} - j\hat{Y})/\sqrt{2}. \tag{19}$$

Except for a factor $\sqrt{2}$, this is the representation used by Rumsey.[22] The interpretation of these equations is considered in Appendix I.

In terms of the x, y components the Stokes parameters are

$$I = I_x + I_y = X^2 + Y^2 + I_u, \tag{11}$$

$$Q = I_x - I_y = X^2 - Y^2, \tag{20}$$

$$U = 2XY \cos \gamma_y, \tag{21}$$

$$V = 2XY \sin \gamma_y. \tag{22}$$

This result is derived by Chandrasekhar.[21] In terms of the s, t components,

$$I = I_s + I_t, \tag{14}$$

$$Q = 2ST \cos \gamma_t, \tag{23}$$

$$U = I_t - I_s, \tag{24}$$

$$V = 2ST \sin \gamma_t. \tag{25}$$

This is like the representation in x, y components with Q and $(-U)$ interchanged. This follows from (2) and (3); a rotation of $45°$ interchanges $\cos 2\chi$ and $(-\sin 2\chi)$.

In terms of the circular components the parameters are

$$I = I_l + I_r, \tag{17}$$

$$Q = 2RL \cos \gamma_r, \tag{26}$$

$$U = 2RL \sin \gamma_r, \tag{27}$$

$$V = I_l - I_r. \tag{28}$$

This is derived in Appendix I.

The intensities I_x, I_r, etc., are proportional to the powers that will be generated in linearly and circularly-polarized antennas. Eqs. (11), (14), (17), and (20) to (28) can therefore be regarded as defining the Stokes parameters in terms of signals induced on the appropriate antennas. The quantities involved are directly measurable. The intensities and phase shifts can be found by standard techniques, and the products XY, etc., are proportionat to the amplitudes of the cross-correlation functions of the signals in the oppositely-polarized antennas. To show this, let $x(t)$ and $u_x(t)$ be the voltages for the polarized and unpolarized parts, respectively, in the x antenna, and similarly for the y antenna. Then

[22] H. G. Booker, V. H. Rumsey, G. A. Deschamps, M. L. Kales, and J. I. Bohnert, "Techniques for handling elliptically polarized waves with special reference to antennas," PROC. IRE, vol. 39, pp. 533–552; May, 1951.

$$\rho(\tau) = \frac{\overline{[x(t) + u_x(t)][y(t+\tau) + u_y(t+\tau)]}}{\left\{ \overline{[x(t) + u_x(t)]^2 [y(t) + u_y(t)]^2} \right\}^{1/2}} \qquad (29)$$

$$= \frac{\overline{x(t)v(t+\tau)}}{(I_x I_y)^{1/2}}. \qquad (30)$$

If the bandwidths are narrow, x and y are close to sine waves with phase difference γ_y, and so

$$\rho(\tau) = \frac{XY}{(I_x I_y)^{1/2}} \cos(\omega\tau + \gamma_y). \qquad (31)$$

Sometimes polarization observations which are incomplete are made. This means that all four parameters are not determined, but if the wave has a known axial ratio the polarization percentage is fixed. Solar bursts have been observed in this fashion by a number of stations.[4] Right and left circular components are observed, and the quantity m_1 is reported:

$$m_1 = (I_l - I_r)/(I_l + I_r) = V/I. \qquad (32)$$

When the polarized part of the wave is circular (by far the most common case) $Q = U = 0$ and $m = m_1$ by (6). In the general case, from (2)–(6),

$$m_1 = m \sin 2\beta. \qquad (33)$$

When the polarization is not known *a priori*, all one can say is that $m_1 \leq m$.

When the wave is circularly polarized, it is also possible to determine m by measuring the correlation product XY and the total intensity I. Let

$$m_2 = 2XY/I = (U^2 + V^2)^{1/2}/I. \qquad (34)$$

When the polarized part of the wave is circular $Q = 0$ and $m = m_2$, but in general,

$$m_2 = m(1 - \cos^2 2\beta \cos^2 2\chi)^{1/2}. \qquad (35)$$

Again, $m_2 \leq m$.

Linear polarization has been sought in the radiation from radio stars, and determinations of a parameter m_3 have been made. When the x and y intensities are measured the following quantity is computed,

$$\frac{I_x - I_y}{I_x + I_y} = \frac{Q}{I} = m_3 \cos 2\chi, \qquad (36)$$

where

$$m_3 = m \cos 2\beta. \qquad (37)$$

When the wave is linearly polarized $m_3 = m$, but otherwise $m_3 < m$. This particular case by itself is of limited use because χ is indeterminate, and m_3 cannot be found from (36). However, if I_a and I_t are also measured,

$$\frac{I_t - I_a}{I_t + I_a} = \frac{U}{I} = m_3 \sin 2\chi, \qquad (38)$$

and m_3 can be found from (36) and (38).

The quantities m_1 (or m_2) and m_3 are sometimes spoken of as the circularly and linearly-polarized fractions. This is misleading because linear and circular polarizations are not "opposite," and a wave cannot uniquely be broken into a combination of linear and circular components.

III. Measurement Techniques

The problem of determining the state of polarization of a completely polarized wave has been thoroughly discussed in the literature.[22,23] If two polarization components are measured the two intensities and the phase difference are required, but it is possible to avoid the phase measurement by measuring the intensities of three components. Thus, if right and left circular antennas are used, the measurement of intensities gives I and r; χ is found from the phase difference. If crossed linear antennas are used, all three measurements must be made to find either r or χ, in general. If a linear antenna is used it may be convenient to rotate it, in which case the intensities can be measured at three orientations and used to deduce the three polarization parameters. This method, however, does not specify the sense of rotation. If a continuous record is made as the antenna is rotated, one gets the polarization "dumbbell" pattern, from which $|r|$ and χ are immediately found.

In the present case the wave is not completely polarized. A fourth parameter, m, is introduced, and therefore a fourth independent measurement must be made. If only two polarization components are measured, the extra quantity to be determined, in addition to two intensities and a phase difference, is the amplitude of the cross-correlation function. If three components are used, either phase or correlation may be measured in addition to the three intensities. If four independent components are used, the four intensities are sufficient, and neither phase nor correlation coefficient need be measured.

It is apparent that there is a great variety of ways for measuring polarization. In this section a number of schemes for making the complete polarization determination will be discussed. They are classified into three groups according to the number of polarization components measured. Reference will be made to methods which have been successfully used and reported in the literature; an attempt will be made to show the relations among them, and what they measure in terms of the Stokes parameters. Only the general principles of operation will be considered. For details of construction, circuits, calibration, etc., reference must be made to the original papers.

The nature of the source being studied introduces special requirements into the operation of a polarimeter for radio astronomy use. Solar bursts often are very short; there is not time to mechanically rotate an an-

[23] J. D. Kraus, "Antennas," McGraw-Hill Book Co., Inc., New York, N. Y., pp. 464–485; 1950.

tenna. This, though, is not serious since electronic switching or sweeping can accomplish the equivalent of antenna rotation if desired. A different complication comes about when radio stars are studied since they shine against a diffuse background and it is desirable to employ interference techniques, at least at the lower frequencies. This requires the use of antenna pairs rather than single antennas, but as far as the polarization is concerned the same quantities are measured.

A. Two Components

If two fixed antennas are used as a polarimeter, the measured quantities can be the two intensities, the phase difference, and the amplitude of the correlation function. The instrument at the Cornell Radio Observatory is of this type. This instrument is discussed in detail in a separate article.[24] The general operation is considered here. The simplified block diagram (Fig. 2) illustrates the operation. The antennas are oppositely polarized; suppose they are x and y-oriented dipoles.

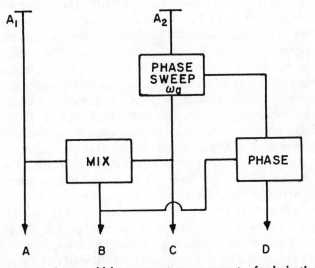

Fig. 2—Polarimeter which measures two components of polarization.

With square-law detectors the outputs at A and C are I_x and I_y, and the Stokes parameters I and Q are the sum and difference of them. The phase of the signal from A_2 is continuously swept at an audio rate, ω_a. If the downcoming wave is unpolarized, the signals in the two channels are independent, and the mixer does not have an output component at the audio sweep frequency. If the downcoming signal is partly polarized, the signals in the two channels are partly correlated, and the mixer does have an output component at the sweep frequency. Its amplitude can be shown to be proportional to the product of the polarized components, XY, as follows.[25] In a narrow band the x signal can be broken into two parts, polarized and unpolarized:

[24] M. H. Cohen, "The Cornell radio polarimeter," this issue, p. 183.
[25] Pawsey and Bracewell, *op. cit.*, p. 60.

$$E_x = X \cos [\omega t + \phi(t)] + E_{ux} \cos [\omega t + \psi_x(t)]. \quad (39)$$

The angles ϕ and ψ_x are independent random functions of time. At the mixer the y signal is

$$E_y = Y \cos [(\omega + \omega_a)t + \phi(t) + \gamma_v] \\ + E_{uy} \cos [(\omega + \omega_a)t + \psi_v(t)]. \quad (40)$$

When these are multiplied the only term at the audio frequency is $\frac{1}{2}XY \cos (\omega_a t + \gamma_v)$. Some of the other terms may have spectra covering the audio band, but their amplitudes will be small, and proportional to the ratio of the audio to IF bandwidths. With a narrow-band filter and a linear detector, then, output B (Fig. 2) is approximately XY. The phase angle of the audio signal at B is γ_v; it is measured by comparison with a reference voltage derived from the phase sweep and recorded at output D. The Stokes parameters U and V can now be found from (21) and (22), and m, r, and χ from (6), (7), and (8). In terms of the outputs A, B, C, D (meaning, respectively, I_x, XY, I_y, γ_v) the desired parameters are

$$I = A + C, \quad (41)$$

$$m = [(A - C)^2 + 4B^2]^{1/2}/(A + C), \quad (42)$$

$$\sin 2\beta = (2B \sin D)/[(A - C)^2 + 4B^2]^{1/2}, \quad (43)$$

$$\tan^{-}2\chi = (2B \cos D)/(A - C). \quad (44)$$

Note that m can be found from A, B, and C; but the phase measurement D must be made before r or χ can be found.

Other methods for measuring phase and correlation could be used. This is true here as well as in the systems shown later. For example, the quantities $(XY \cos \gamma_v)$ and $(XY \sin \gamma_v)$ could be generated directly by adding phase sensitive detectors to Fig. 2. It would also be possible to obtain these quantities without the phase sweep. Fig. 2, however, represents a method which has been used several times in both interference and telescope instruments, and so it will be used for discussion.

If opposite circular antennas are used in Fig. 2, the four outputs A, B, C, D are, respectively, I_r, RL, I_l, and γ_r. The Stokes parameters can then be found from (17) and (26) to (28), and the polarization parameters, from (6) to (8), are

$$I = A + C, \quad (45)$$

$$m = [(C - A)^2 + 4B^2]^{1/2}/(A + C), \quad (46)$$

$$\sin 2\beta = (C - A)/[(C - A)^2 + 4B^2]^{1/2}, \quad (47)$$

$$\chi = \frac{1}{2}D. \quad (48)$$

With circular antennas the measurements break into two independent sets: A, B, and C determine I, m, and r, and D determines χ.

The Cornell instrument has provision for using either crossed linear or opposite circular antennas. The circular antennas have been used, however, because as shown above, the phase measurement is not needed unless the orientation data is desired. The Faraday effect in the earth's ionosphere changes χ a great deal at meter wavelengths (see Section V), so that orientation data at these wavelengths is hard to interpret. At the present state of knowledge, χ can be projected back through the ionosphere only very imprecisely.

An interference scheme which in essence would be identical to the above is shown schematically in Fig. 3. Connections (a) and (b) of Fig. 3 represent standard phase sweeping interferometers with "parallel" antennas. The phase sweeping in effect speeds up the standard pattern, and the audio component out of the mixer is the desired signal for the point source. From the point of view of determining polarizations, the phase sweeping is not important since the earth's rotation will do the same job, only more slowly.

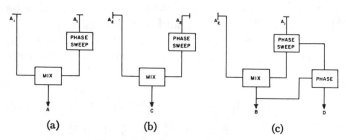

Fig. 3—Interference polarimeter.

The outputs A, B, C, and D record the same quantities as in Fig. 2. With regard to B, we may say that an unpolarized wave produces no interferometer pattern with crossed antennas, but a partly-polarized wave does. Its amplitude is proportional to the product of the x and y components of the polarized part. As pointed out earlier, the antennas can be any oppositely-polarized pair. The use of circularly-polarized antennas would again break the measurements into two independent sets, with D used only to find χ.

The interference technique used by Little and Payne-Scott[26] is very close to the above scheme. It is simpler because only two antenna connections, Fig. 3(a) and 3(c) are used, but presumably it is less sensitive because the fourth output is the dc level at the mixer of Fig. 3(c), This is just the average value of the combined signals from the two antennas after one of them has been swept in phase. But since it is the average value it is not a result of the interferometer pattern, and so does not have the advantage of discrimination against a diffuse background. Little and Payne-Scott, however, were studying solar bursts, and this disadvantage probably is not important in this case.

Historically, an instrument like this was built first by Ryle and Vonberg.[27] It did not have a phase sweep,

but rather let the earth's rotation produce the pattern. It was mainly used to detect the presence of circular polarization in solar radiation. The version by Little and Payne-Scott did have a phase sweep and was also used to study solar bursts. An interference instrument similar to that used by Ryle and Vonberg was used by Ryle and Smith[20] in a search for polarized radiation from radio stars. The Cornell instrument[24] followed the work of Little and Payne-Scott. It was designed for solar studies also.

A microwave polarimeter recently built by Akabane[28] operates on two linear components but is somewhat different from the instruments discussed here. The modulation scheme is different from that shown in Fig. 2, and the quantities $(XY \cos \gamma_y)$ and $(XY \sin \gamma_y)$ are generated by synchronous detectors. The x and y intensities are not recorded, but rather the quantities $(I + \frac{1}{2}Q)$ and Q, which are also generated by the modulation process and isolated with synchronous detectors.

B. Three Components

A polarization determination with three antennas can be made as in Fig. 4. Five outputs are shown, but they are redundant. Either B or D can be not used. Antennas A_1 and A_2 are oppositely polarized, and A_3 can be any different one. A convenient set is (xyt) oriented linear antennas (Fig. 1), but this cannot be one antenna which rotates since A_1 and A_2 must be used simultaneously to measure outputs B and D. If three linear antennas are used and the phase is not measured, the sense of rotation will not be determined.

An interference version of the scheme in Fig. 4 was used by Hanbury-Brown, Palmer, and Thompson[19] in an attempt to measure the polarization of the signals from the three strongest radio sources at 158 mc. Their system was as shown in Fig. 5. The outputs A, B, C, E are the same as in Fig. 4, and the quantities they measure are shown on the diagram. The intensity along the s axis (Fig. 1) was also measured, although this is not shown in Fig. 5. This was redundant but presumably was useful as a check.

C. Four Components

The intensities measured by four antennas are sufficient to specify the four polarization parameters. But the measurements must be independent; right and left circular and x and y-linear antennas will not do, whereas right and left circular and x and t-linear antennas will. In the former case the measurements are not independent by (11) and (17). Many combinations, of course, are possible.

Suzuki and Tsuchiya[7] have built a polarimeter of this type at the Tokyo Observatory. Their antennas are crossed dipoles, and by appropriate connections they get the two circular components, and four linear com-

[26] A. G. Little and R. Payne-Scott, "The position and movement on the solar disk of sources of radiation at a frequency of 97 mcs. I. Equipment," *Aust. J. Sci. Res.*, vol. A 4, pp. 489–507; December, 1951.

[27] M. Ryle and D. D. Vonberg, "An investigation of radio-frequency radiation from the sun," *Proc. Roy. Soc.*, vol. A 193, pp. 98–120; April, 1948.

[28] K. Akabane, "A polarimeter in the microwave region," this issue, p. 194.

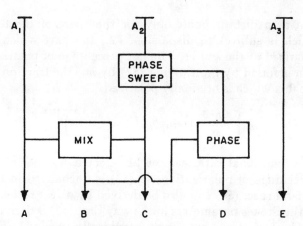

Fig. 4—Polarimeter which measures three components of polarization.

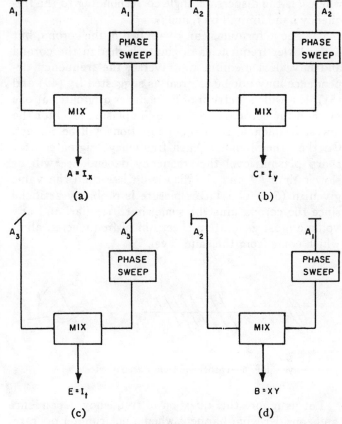

Fig. 5—Interference polarimeter.

ponents: x, y, s, t. The information is redundant but presumably this is useful as a check. The intensity I can be found from (11), (14), and (17), and the other Stokes parameters can be found from (20), (24), and (28).

This instrument was designed for solar work. Since short-duration bursts are of interest, rapid switching techniques are used to sample the different antenna connections. Even so, the measurements are made at different times, and errors might be introduced if the burst is changing very rapidly.

This point could be overcome by using a series of networks to split the two dipole signals into isolated parts and then making all the combinations all the time, or by using four separate antennas. The former method has the disadvantage that the noise figure would effectively be raised since each network would reduce the signal by 3 db. The latter method has the obvious disadvantage of requiring twice as many antennas. Both methods have the further disadvantage that four separate receiver channels would be needed.

The two-component systems discussed in Section III-A above avoid this point because only two intensities are measured. Since they must be measured at the same time, two receiver channels are needed. They should have similar phase characteristics as well as similar gain and bandwidth characteristics. The absolute values of noise figure, gain, and phase shift do not have to be the same for the two channels, but they should be stable between calibration periods.

IV. Faraday Rotation and Dispersion

In a homogeneous magnetized ionized medium, a polarized electromagnetic wave is propagated in two independent modes which have different phase velocities. When the frequency is sufficiently high, the two modes are circularly polarized with opposite senses of rotation (except for the case where the magnetic field is almost exactly transverse). The axial ratio and sense of rotation of the resultant total wave do not change, but the plane of polarization does. The Faraday rotation of the plane of polarization, ϕ, is given by

$$\phi = (2.36 \times 10^{-3})f^{-2} \int_0^z nB_z dz, \qquad (49)$$

where

ϕ = radians,
f = megacycles,
n = number of electrons/cc,
B_z = longitudinal component of magnetic field (Gauss),
z = kilometers.

This result is used by Evans,[10] for example, in explaining the fading of moon echoes by a variable Faraday rotation in the ionosphere.

When the frequency is not high enough, there is a change in the shape of the ellipse in addition to the Faraday rotation. This general case has been discussed by Hatanaka.[9]

Since ϕ is a function of frequency in (49), an elliptically-polarized noise signal containing a spectrum of frequencies is dispersed into a continuum of ellipses with a spread of orientations. This is depicted in Fig. 6. The angular dispersion rate is, from (49)

$$\frac{d\phi}{df} = -2\frac{\phi}{f} \text{ radians/cycle.} \qquad (50)$$

If $\Delta f/f = (f_2 - f_1)/f \ll 1$, then

$$\frac{\theta}{\phi} \approx 2\frac{\Delta f}{f}, \qquad (51)$$

where θ is the dispersion angle corresponding to the frequency band limited by f_1 and f_2.

The above formulas can be applied in the corona, but if the lower frequencies originate higher in the corona and have less medium to traverse, the frequency dependence may not be as great as suggested by (49) and (51). If a wide spectrum of frequencies originates at one place in the corona, high above the plasma level for the lowest frequency, then these equations will be correct. On the other hand, if each frequency originates at a sharp plasma level the frequency dependence will be slower than f^{-2} and the dispersion less than the value given in (51). This latter picture is probably artificial since the corona must be somewhat irregular, and any volume must generate a spectrum of frequencies, all of which come from the same level.

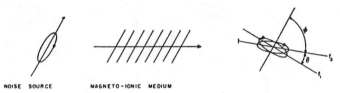

NOISE SOURCE   MAGNETO-IONIC MEDIUM

Fig. 6—Dispersion from Faraday effect.

Let us ignore this question of frequency dependence and consider what happens when a polarimeter operates on a dispersed elliptically-polarized wave. Let the dispersion be as in Fig. 6. Assume that f_1 and f_2 are the edges of the receiver pass band and that the spectrum is uniform between these limits. Then orientations spread over an angle θ are received and interpreted as a partially-polarized signal. The resultant state of polarizations was first calculated by Hatanaka[9] for the case where the source is all polarized. His procedure was to integrate the Stokes parameters over the bandwidth. The results are

$$m = \left[\frac{\sin^2\theta}{\theta^2} + \sin^2 2\beta_0\left(1 - \frac{\sin^2\theta}{\theta^2}\right)\right]^{1/2}, \quad (52)$$

$$\sin 2\beta = \left[\frac{\sin^2\theta}{\theta^2}\cot^2 2\beta_0 + 1\right]^{-1/2}, \quad (53)$$

where $\tan^{-1}\beta_0$ is the axial ratio of the wave at the source.

Astronomical sources cannot be expected to be fully polarized, in general. For example, when the source consists of synchrotron radiation from electrons in a magnetic field, it must be considered to have an appreciable depth, and the radio radiation from the bottom layer of electrons will suffer a Faraday rotation in its passage through the source itself. Even if the magnetic field is homogeneous, all the different levels will produce different planes of polarization at the top of the source, so it will be only partially polarized. This has nothing to do with the bandwidth; it is strictly a property of the source. Eq. (52) thus gives the percentage of polariza-

tion at a distant point only for that part of the flux which is subject to dispersion, *i.e.*, the part which is polarized at the source. The net percentage of polarization is found by multiplying (52) by m_0, the fraction of the flux which is originally polarized.

$$m = m_0\left[\frac{\sin^2\theta}{\theta^2} + \sin^2 2\beta_0\left(1 - \frac{\sin^2\theta}{\theta^2}\right)\right]^{1/2}. \quad (54)$$

The final axial ratio cannot be affected by the initial percentage of polarization, so (53) is unchanged in the general case. (54) can also be derived by direct recourse to the Stokes parameters in the way that (52) is derived.

These relations are shown graphically in Fig. 7. The abscissa and ordinate are r and m/m_0; in the radio astronomy case r and m are measured quantities. The curves are the families $r_0 = $ constant and $\theta = $ constant

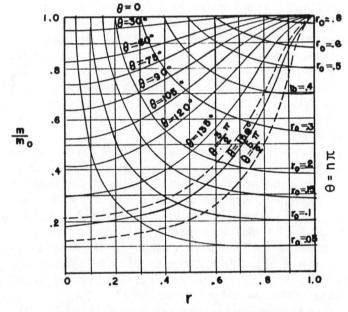

Fig. 7—Polarization parameters as functions of Faraday dispersion angle and initial polarization parameters.

The wave initially is represented by some point on the top line, $\theta = 0$. As the wave transverses the medium (or as the bandwidth is increased from zero) the point moves down the line $r_0 = $ constant, until at $\theta = \pi$ it reaches the right-hand boundary. At this point the polarized part, which is now circular, is smallest. As θ further increases, the point comes back up the $r_0 = $ constant line until it reaches the approximate limit $\theta = 3\pi/2$, the dashed line. Then it subsides until at $\theta = 2\pi$ it is again at the right-hand side.

Fig. 7 also can be used to find the polarization of the source in simple cases like the above example, synchrotron radiation in a homogeneous magnetic field. At the outer boundary of the source the orientations are spread uniformly through the angle ϕ_s, the rotation angle for the depth of the source. If m_0 is taken as unity and ϕ_s used for θ, Fig. 7 will give the net polarization of the source. In the case of synchrotron radiation $r_0 = 0$.

At this point we outline the various assumptions and approximations which have been used. They pertain to the source, to the medium, and to the equipment. They must all be valid if the above formulas are to hold and if the succeeding material is to be valid.

1) Slowly-varying medium (homogeneous theory applies).
2) Circular magneto-ionic modes (high frequency and quasi-longitudinal propagation).
3) No selective absorption or other mechanism that would change the polarization.
4) Uniform spectrum.
5) Rectangular band-pass.
6) Narrow band.

The foregoing has shown that the Faraday dispersion may be significant whenever the product of rotation and relative receiver bandwidth is on the order of $\frac{1}{2}$ or greater, for then $\theta \geq 1$ by (51). The polarization parameters, in this case, are not an intrinsic property of the wave but rather depend upon the receiver frequency and bandwidth. Both these characteristics of the receiver should be given if proper interpretation is to be given to the measured values.

The quantities m_0, r_0, and χ_0 are important because they specify the source. They are not the measured values m, r, and χ whenever the Faraday effect is important, but limits can be set on m_0 and r_0 from measurements of m and r. Limits can also be set on θ. This quantity is important because it contains information on the medium between the source and the observer—the integrated product of magnetic field and electron density.

In any case where m and r are measured, one can find from Fig. 7 an upper limit to θ and both upper and lower limits to r_0. The upper limit to θ and lower limit to r_0 are attained when $m_0 = 1$, whereas the upper limit to r_0 is attained for $\theta = 0$. Furthermore, the initial value of polarization, m_0, is restricted because $m_0 \geq m$. As an example, when $r = 0.4$ and $m = 0.6$, $m_0 \geq 0.6$, $\theta \leq 112°$, and $0.22 \leq r_0 \leq 0.4$.

In the upper right corner of Fig. 7 where the $\theta = $ constant lines converge, an accurate determination of the limit to θ is critically dependent upon the measurement accuracy. For example, if a measured point is $m = 0.98 \pm 2$ per cent, $r = 0.75 \pm 5$ per cent, then the upper limit to θ is somewhere between 0 and 70°. Much of the data reported by Hatanaka *et al.*[8] and by Cohen[24] on polarized bursts from the sun does fall in this region.

The determination of the limits from Fig. 7 uses only two of the four parameters that characterize the burst, m and r. The other two, the total intensity and orientation, are of no help for this. There is some difficulty in using orientation data directly. If the orientation of the original source is known (as it is in moon-echo experiments), then the total Faraday rotation can be found to within a half-integral number of revolutions. Astronom-

ical sources, however, generally do not have a known orientation. In fact, part of the problem in interpreting polarization data is to find the orientation of the source. However, the Faraday rotation in the earth's ionosphere alone is many radians at meter wavelengths, and this must be accurately known before orientation data can be extended outside the ionosphere.

The above discussion has been for a measurement at one center frequency with a specific bandwidth. If the parameters of the wave are measured at two frequencies, or with two bandwidths, far more information can be obtained. If orientation angles are measured at two frequencies, a dispersion can be directly calculated. This, though, must be interpreted cautiously. Solar bursts at different frequencies are thought to come from different levels in the corona; even if the two sources are parallel, the media between the source and observer are different for the two cases. An exception occurs when the two bursts are harmonically related and come from the same source. In this case, though, the rotation for the lower frequency is four times that for the higher frequency, and it would be difficult to fix the difference in orientations to within $n\pi$. The ionospheric rotation is variable, and this by itself would be enough to confuse the result unless high frequencies are used.

The above objections may not be valid if the two frequencies are so close that the bands are adjacent. The same information can then be obtained as from measurements made at one center frequency and two bandwidths, over the same total frequency range. The two experiments are different because they require different measurements. In the two-frequency experiment the dispersion information comes from the difference in orientation angles. Once θ is found, ϕ can be estimated from (51). If m and r are also measured, m_0 and r_0 can be found from Fig. 7. In the two-bandwidth experiment the orientations are the same and are not required for the analysis; what is needed are the values of m and r. Measurements of r at two bandwidths will fix r_0 and θ to within trigonometric ambiguities. If m also is measured at the two bandwidths, one can find the polarization percentage at the source.[29] This also gives a check which would be useful in resolving the ambiguities in r_0 and θ. The total rotation ϕ can also be estimated as with the two-frequency experiment.

V. DISCUSSION

A. Ionosphere

The earth's ionosphere is a magneto-ionic medium. Measurements by Evans[10] have shown that the Faraday rotation in it is variable and greater than 10 radians in the daytime at 120 mc. At the lowest radio astronomy frequencies, around 20 mc, the rotation would be

[29] M. H. Cohen, "Interpretation of Radio Polarization Data in Terms of Faraday Rotation," Cornell Univ. School of Elec. Eng., Ithaca, N. Y., Res. Rep. EE 295; May, 1956.

greater than 360 radians. If the dispersion is to be insignificant ($\theta \leq 1$) the relative bandwidth must be on the order of 0.0014 or less; so the bandwidth must be about 28 kc or smaller at 20 mc. This restriction is not serious since narrow bandwidths are customarily used at the low frequencies.

Even though the dispersion may be small, the rotation is not, at frequencies up to at least 500 mc. A correction for this effect has to be made whenever orientation angle data are extrapolated back through the ionosphere. The rotation is variable, and at the lower frequencies, where it is very large, the extrapolation will require a knowledge of the ionosphere at the time of observing the radio source.

B. Corona

The solar corona is ionized and presumably contains some magnetic field. A Faraday rotation might therefore be expected for rays propagating through the corona. Hatanaka[9] has made estimates of the magnitude of the rotation. A model using the Allen-Baumbach formula for electron density and a general dipole magnetic field with a strength of 1 Gauss at the pole gives, at 200 mc, ϕ from 10^3 to 10^5 radians for sources from 2 to 0.1 solar radii above the photosphere. The magnetic field is artificially smooth, but it may give the right order of magnitude for the rotation. Even if an irregular field is assumed, the rotation will be very large. A 30,000-km "blob" at $h = 0.2$ radius with a homogeneous longitudinal magnetic field of 0.2 Gauss and the above electron density gives a rotation of 2×10^4 radians at 200 mc all by itself. If the magnetic field is stronger, ϕ is proportionately greater, although if it is strong enough, several of the approximations listed above may be invalidated. The Faraday effect will still be present, but the results are not so simple, since the modes will not be circles. If the medium is highly turbulent, the homogeneous theory may not apply. In this case the two modes are coupled, and there will be no Faraday effect at all. However, estimates of the maximum gradients for uncoupled modes, based on Budden's theory,[30] give very large values: 0.04 Gauss/km for 300 mc at a height of one radius.[29] The gradients are probably smaller than this, so the magneto-ionic modes are uncoupled, and the Faraday effect can take place. Also, as noted in the previous section, the assumption of uncoupled modes is introduced often in explaining the circularly-polarized bursts by the independent propagation qualities of the two magneto-ionic modes.

Rotations of 10^3 to 10^5 radians might give large dispersions with the customary bandwidths for solar instruments. The resulting signals, then, would be some combination of circular plus random. These, in fact, are the most commonly observed polarizations for solar

bursts. However, these are also the polarizations that would commonly result from a source distributed in depth, as described above.

The polarimeter built by Suzuki and Tsuchiya[7] has a 100-kc bandwidth at 200 mc, and the instrument built by Cohen[24] has a 10-kc band at 200 mc. If the rotation of a 200-mc burst is 10^5 radians, the dispersions would be 100 and 10 radians respectively, according to (51), and the wave would be depolarized—to circular plus random—for both instruments. If the dispersions were 10 and one radians respectively, there would be substantial differences in the observations of partially elliptically-polarized waves with the two instruments. If the dispersions are much more or much less than 10 and one radians, the differences will be small.

The reported observations do show that the above values of dispersion are sometimes much too large. Payne-Scott and Little[2,3] saw several examples of linearly-polarized signals. Numerical values of m and r are not given, but presumably in Fig. 7 the points would have been near the left boundary, and not near the bottom. If their point gave $\theta < 2$, the corresponding limit to ϕ would have been $\phi < 10^3$ radians at 97 mc. Hatanaka *et al.*[8] have reported some observations which, when plotted on Fig. 7, give several points with $\theta = 0$ and a number with $\theta \approx 90°$. The points with $\theta = 0$ are completely polarized, elliptical, and imply that $\phi = 0$, but, as discussed above, small inaccuracies in m will make large changes in θ. The points with $\theta = 90°$ give $\phi < 1.6 \times 10^3$ radians at 200 mc.

C. Crab Nebula

Mayer, McCullough, and Sloanaker[17] have measured quantity m_3 (37) for the Crab Nebula at a wavelength of 3.15 cm. The result is that the polarized component, if linear, is about 7 per cent of the total flux. Optical radiation from this object is strongly linearly polarized in patches, with a residual linear polarization of 9.2 per cent.[16]

It seems reasonable, therefore, to assume that the polarized component is linear at 3.15 cm, and that $m = m_3 = 0.07$. The reported orientation at 3.15 cm is about 148°, and at optical wavelengths it is 159.6°; the difference is therefore about 11°.

Oort and Walraven[16] have estimated that at a wavelength of 21 cm the Faraday rotation in the interstellar space between the Crab Nebula and the earth would be 7 radians, and within the nebula itself the Faraday rotation would be 14 radians. At 3.15 cm the corresponding figures would be 0.16 and 0.31 radians respectively. If the 11° orientation difference measured by Mayer *et al.* is taken to result from interstellar rotation, it gives the value of $\phi = 0.19$ radians; Oort and Walraven's value is close to that value. The orientation difference, of course, may be regarded as $(0.19 \pm n\pi)$ radians, but even $n = 1$ gives a rotation much greater than the theoretically estimated one.

---

[30] K. G. Budden, "The theory of the limiting polarization of radio waves reflected from the ionosphere," *Proc. Roy. Soc.*, vol. A 215, pp. 215–233; November, 1952.

The amount of polarization measured at 3.15 cm is less than that at optical wavelengths by the factor $(0.07/0.092) = 0.76$. Let this be called the polarization factor.

If the depolarization is taken to result from the interstellar rotation, the required rotation can be deduced. The 3.15-cm equipment received two bands, each 5.5 mc wide, separated by 60 mc, and with a center frequency of 9530 mc. In Appendix II it is shown that two independent linearly-polarized waves with equal intensities and orientations θ apart combine to give a polarization percentage $m = \cos\theta$. If $m = 0.76$, $\theta = 40°$; by (51) the required rotation is 55 radians. This value is unreasonably large, and the depolarization probably does not result from the interstellar rotation.

If the polarization factor is taken to result from rotation within the nebula, the internal rotation is about 72°, from Fig. 7. That amount is four times Oort and Walraven's estimate of 0.31 radian. Oort and Walraven also pointed out that there would be extra radio depolarization because the less polarized outer parts of the nebula would contribute relatively larger amounts to the radio energy. This may be the source of most of the depolarization at 3.15 cm.

Several attempts at lower frequencies to measure the polarization in the Crab Nebula radiation were unsuccessful. At 22 cm Westerhout[18] determined that $m_3 \leq 0.01$ for two bands, each 2 mc wide and separated by 68 mc. If we take the 3.15-cm interstellar rotation to be 11°, at 22 cm it will be 9.4 radians, and the dispersion in 68 mc will be 52°. The polarization factor is $\cos 52° = 0.61$; this would reduce the polarization from 9.2 per cent to 5.6 per cent. This effect would be eliminated by using only one band. The internal rotation within the nebula is much more important; if it is about 14 radians, at 22 mc, the polarization factor, m/m_0 in (54), will be a maximum of about 0.07. This would reduce the polarization from 5.6 per cent to 0.4 per cent. The polarization might further be reduced because, as at 3 cm, the effective radiating region of the nebula may be greater than the optical region.

Hanbury-Brown, Palmer, and Thompson[19] made measurements of the Crab Nebula at 158 mc with a bandwidth of 400 kc. They set limits for both circular and linear polarization: $m_2 \leq 0.04$ and $m_3 \leq 0.025$. This permits the computation of a limit for arbitrary polarization, for by (35) and (37) $m \leq (m_2^2 + m_3^2)^{1/2}$, so the measured value is $m \leq 0.047$.

The above estimates for rotation in the galaxy and in the interstellar medium give values smaller than these limits. If the interstellar rotation is 11° at 3.15 cm, it will be 700 radians at 158 mc; the dispersion in a 400-kc band is 3.5 radians. The polarization factor from Fig. 7 will be no more than 0.2, and the polarization would be reduced to about 2 per cent. If the reduction in polarization from interstellar rotation is to be insignificant $(\theta \leq 1)$, the bandwidth should be no more than 110 kc. The internal rotation in the nebula, however, would be

1100 radians at 158 mc, according to Oort and Walraven's estimate. This would reduce the polarization by a factor of 10^3, and the source would be essentially unpolarized at 158 mc.

APPENDIX I

We consider the elliptically-polarized part of the wave to be generated by two oppositely rotating coherent circular components. See Fig. 8. The connection between the circular and linear components is essentially that used by Rumsey.[22] For each frequency component (18) and (19) hold, and by combining them, we have

$$\hat{R}e^{j\omega t} + (\hat{L}e^{j\omega t})^* = \sqrt{2}\{X\cos\omega t + jY\cos(\omega t + \gamma_y)\}$$
$$= E_x + jE_y. \qquad (55)$$

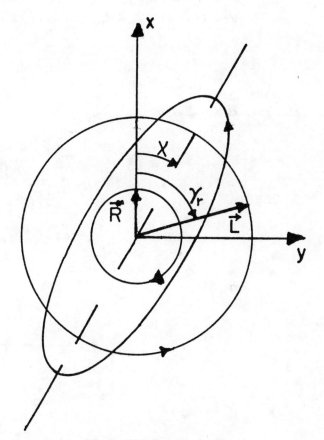

Fig. 8—Elliptical polarization.

The quantites $(\hat{R}e^{j\omega t})$ and $(\hat{L}e^{j\omega t})^*$ are interpreted as the right and left circular components with the real parts representing the x components and imaginary parts the y components of electric field. At the time instant shown in Fig. 8 the electric vector of the right-hand component, $\vec{R}$, is parallel to the x axis and so $(\hat{R}e^{j\omega t})$ is real. At that instant $(\hat{L}e^{j\omega t})^*$ is therefore $(Le^{j\gamma_r})$, where γ_r is the phase angle by which $\hat{R}$ leads $\hat{L}$. The angular separation between the vectors is γ_r, and since they rotate in opposite directions,

$$\gamma_r = 2\chi. \qquad (56)$$

The axial ratio, by definition, is the ratio of minor to major axes. From Fig. 8,

$$r = (L - R)/(L + R). \tag{57}$$

The quantities $\cos 2\beta$ and $\sin 2\beta$ are needed for the Stokes parameters:

$$\cos 2\beta = \frac{1 - r^2}{1 + r^2} = \frac{2RL}{I_e}, \tag{58}$$

$$\sin 2\beta = \frac{2r}{1 + r^2} = \frac{L^2 - R^2}{I_e}. \tag{59}$$

Upon substituting (56), (58), and (59) into (2), (3), and (4), we have

$$Q = 2RL \cos \gamma_r, \tag{26}$$

$$U = 2RL \sin \gamma_r, \tag{27}$$

$$V = L^2 - R^2 = I_l - I_r. \tag{28}$$

Appendix II

Consider the superposition of two independent linearly-polarized waves traveling in the same direction, having equal intensities I_0 with similar spectra, and orientations separated by the angle χ. Let wave a be parallel to the x axis and wave b be at angle χ (Fig. 1). Then from (1) to (4),

$$I_a = I_0 \qquad I_b = I_0, \tag{60}$$

$$Q_a = I_0 \qquad Q_b = I_0 \cos 2\chi, \tag{61}$$

$$U_a = 0 \qquad U_b = I_0 \sin 2\chi, \tag{62}$$

$$V_a = 0 \qquad V_b = 0. \tag{63}$$

Since the waves are independent, the parameters for the total wave are the sums of the individual parameters; *i.e.*, $I = (I_a + I_b)$, etc., and

$$m = (Q^2 + U^2 + V^2)^{1/2}/I = \cos \chi. \tag{64}$$

The Synthesis Radio Telescope at Westerbork

Methods of Polarization Measurement

Kurt W. Weiler

Kapteyn Astronomical Institute, University of Groningen

Received October 4, 1972

Summary. From the formula for the generalized response of a correlation interferometer to an incident partially polarized radio wave, the forms for the four output channels from each interferometer in the Westerbork telescope are developed. Then, the means of instrument gain, phase, and polarization calibration are detailed and the solutions for the four Stokes parameters obtained. The advantages and disadvantages of the technique are briefly discussed.

The Appendix contains a derivation of the basic initial formula.

Key words: interference polarimetry – polarization

I. Introduction

Since the discovery of polarized radio emission from the Galactic supernova remnant Taurus A by Mayer *et al.* in 1957 and the detection of polarization in the extragalactic radio source Cygnus A five years later (Mayer *et al.*, 1962), considerable work has been devoted to the measurement and study of the linear polarization properties of discrete radio sources. These observations have generally been limited to investigating only the integrated properties, since detailed source structure can be studied with a single antenna only if the source is very large or the observing frequency very high. It is possible to obtain polarization structure information at lower frequencies through interferometric techniques, but this has been done for complete two-dimensional synthesis in only a few cases (e.g. Fomalont, 1970; Baldwin *et al.*, 1970; Weiler and Seielstad, 1971).

Now with the Westerbork Synthesis Radio Telescope (WSRT) in the Netherlands supplying continuous correlation outputs of the spatial distribution and the full polarization content of the incident radiation, the production of radio source polarization distributions has been greatly facilitated. The purpose of this paper is to develop the techniques of polarization synthesis for the WSRT.

II. Instrument Response

Morris *et al.* (1964) have given a formula to express the response of a correlation type interferometer to arbitrarily polarized incident radiation. We repeat it here with definitions and slight modifications and give a derivation in the Appendix.

$$R = \tfrac{1}{2}G(I\{\cos(\phi_W - \phi_E)\cos(\theta_W - \theta_E)$$
$$+ i\sin(\phi_W - \phi_E)\sin(\theta_W + \theta_E)\}$$
$$+ Q\{\cos(\phi_W + \phi_E)\cos(\theta_W + \theta_E)$$
$$+ i\sin(\phi_W + \phi_E)\sin(\theta_W - \theta_E)\}$$
$$+ U\{\sin(\phi_W + \phi_E)\cos(\theta_W + \theta_E)$$
$$- i\cos(\phi_W + \phi_E)\sin(\theta_W - \theta_E)\}$$
$$- iV\{\sin(\phi_W - \phi_E)\cos(\theta_W - \theta_E)$$
$$- i\cos(\phi_W - \phi_E)\sin(\theta_W + \theta_E)\}). \tag{1}$$

R = measured response to a source.

$G = G\exp(iP)$ = instrumental gain and phase constants.

W = West telescope of the East-West interferometer.

E = East telescope of the East West interferometer.

ϕ = Position angle of the receiving dipole measured from North through East.

θ = Ellipticity of instrumental response (i.e. instrumental circular polarization; $+\theta$ = Right Hand Instrumental Circular (IRE Definition)).

I = Fourier transform of source total intensity.

Q and U = Fourier transform of source linearly polarized intensity.

V = Fourier transform of source circularly polarized intensity [$+V$ corresponds to Right Hand Circular Polarization (IRE Definition)].

The vector signs indicate the complex nature of the parameters as distributed over the Fourier plane (e.g. I is the Fourier transform of the total intensity; I is the visibility function of I).

The phase for the interferometer is defined such that a source lying at greater hour angle has positive phase.

The dipoles for each antenna are defined such that with the dipole rotator set at 0 position, the x dipole has position angle $0°$ on the sky and the y dipole has position angle $90°$ on the sky.

I.e.

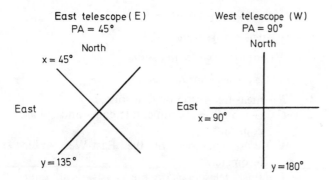

With the crossed dipoles rotatable on both antennas, the possible combinations of dipoles at different position angles in an interferometer is unlimited. However, we will consider here only one case – that which is used for all normal continuum polarization observations with the WSRT. In this case the dipoles of the West telescope are aligned at position angle $90°$ and the dipoles of the East telescope are aligned at position angle $45°$. Thus, the interferometer consists of the combination of

For the WSRT, all possible dipole combinations are continuously correlated, yielding for each interferometer four output channels

$$(Wx)(Ex); \quad (Wx)(Ey); \quad (Wy)(Ex); \quad (Wy)(Ey).$$

For compactness of notation, these will be referred to as channels xx, xy, yx and yy. If we also allow for the possibility of an error (Δ) in the setting of the dipoles, our dipole position angles become

$$\phi_{Wx} = 90° + \Delta_{Wx} \qquad \phi_{Ex} = 45° + \Delta_{Ex}$$
$$\phi_{Wy} = 180° + \Delta_{Wy} \qquad \phi_{Ey} = 135° + \Delta_{Ey}. \tag{2}$$

Additionally, for each dipole antenna system we have the instrumental circular polarization terms $\theta_{Wx}, \theta_{Wy}, \theta_{Ex}$,

θ_{Ey}. These error and instrumental terms have in all cases been adjusted to be small ($< 2°$) so that throughout this treatment we can assume, $\theta, \Delta \ll 1$. Further, we adopt a compact notation such that sums or differences of error terms are written as superscripts, i.e.

$$\Delta_{Wx} - \Delta_{Ex} = \Delta_{xx}^-; \quad \theta_{Wy} + \theta_{Ex} = \theta_{yx}^+; \text{ etc}.$$

With this notation and the approximations given above, the output of the 4 channels of an interferometer with its dipoles set as in (2) becomes,

$$R_{xx} = \frac{1}{2\sqrt{2}} G_{xx}(I\{1 - \Delta_{xx}^- + i\theta_{xx}^+\} - Q\{1 + \Delta_{xx}^+ - i\theta_{xx}^-\}$$
$$+ U\{1 - \Delta_{xx}^+ + i\theta_{xx}^-\} - iV\{1 + \Delta_{xx}^- - i\theta_{xx}^+\})$$

$$R_{xy} = \frac{1}{2\sqrt{2}} G_{xy}(I\{1 + \Delta_{xy}^- - i\theta_{xy}^+\} - Q\{1 - \Delta_{xy}^+ + i\theta_{xy}^-\}$$
$$- U\{1 + \Delta_{xy}^+ - i\theta_{xy}^-\} + iV\{1 - \Delta_{xy}^- + i\theta_{xy}^+\})$$

$$R_{yx} = \frac{-1}{2\sqrt{2}} G_{yx}(I\{1 + \Delta_{yx}^- - i\theta_{yx}^+\} + Q\{1 - \Delta_{yx}^+ + i\theta_{yx}^-\}$$
$$+ U\{1 + \Delta_{yx}^+ - i\theta_{yx}^-\} + iV\{1 - \Delta_{yx}^- + i\theta_{yx}^+\})$$

$$R_{yy} = \frac{1}{2\sqrt{2}} G_{yy}(I\{1 - \Delta_{yy}^- + i\theta_{yy}^+\} + Q\{1 + \Delta_{yy}^+ - i\theta_{yy}^-\}$$
$$- U\{1 - \Delta_{yy}^+ + i\theta_{yy}^-\} - iV\{1 + \Delta_{yy}^- - i\theta_{yy}^+\}). \tag{3}$$

III. Calibration and Observation

These uncalibrated instrument responses of Eq. (3) are still insufficient for radio source observations. However if we observe a totally unpolarized calibration source ($Q = 0, U = 0, V = 0, I = 1$) we obtain a measurement of the instrumental constants and errors.

$$R_{xx} = \frac{1}{2\sqrt{2}} G_{xx}(1 - \Delta_{xx}^- + i\theta_{xx}^+)$$

$$R_{xy} = \frac{1}{2\sqrt{2}} G_{xy}(1 + \Delta_{xy}^- - i\theta_{xy}^+)$$

$$R_{yx} = \frac{-1}{2\sqrt{2}} G_{yx}(1 + \Delta_{yx}^- - i\theta_{yx}^+) \tag{4}$$

$$R_{yy} = \frac{1}{2\sqrt{2}} G_{yy}(1 - \Delta_{yy}^- + i\theta_{yy}^+).$$

Then, if these output values are used to scale the amplitude and adjust the phase zero of the measurements on unknown sources, we obtain *calibrated* channel outputs of the form

$$XX = I - Q\{1 + 2\Delta_{Wx} - 2i\theta_{Wx}\} + U\{1 - 2\Delta_{Ex} - 2i\theta_{Ex}\}$$
$$- iV\{1 + 2\Delta_{xx}^- - 2i\theta_{xx}^+\}$$

$$XY = I - Q\{1 - 2\Delta_{Wx} + 2i\theta_{Wx}\} - U\{1 + 2\Delta_{Ey} + 2i\theta_{Ey}\}$$
$$+ iV\{1 - 2\Delta_{xy}^- + 2i\theta_{xy}^+\}$$

$$YX = I + Q\{1 - 2\Delta_{Wy} + 2i\theta_{Wy}\} + U\{1 + 2\Delta_{Ex} + 2i\theta_{Ex}\}$$
$$+ iV\{1 - 2\Delta_{yx}^- + 2i\theta_{yx}^+\}$$

$$YY = I + Q\{1 + 2\Delta_{Wy} - 2i\theta_{Wy}\} - U\{1 - 2\Delta_{Ey} - 2i\theta_{Ey}\}$$
$$- iV\{1 + 2\Delta_{yy}^- - 2i\theta_{yy}^+\} \tag{5}$$

which can be solved for the desired Stokes parameters.

$$I = \tfrac{1}{4}(XX + XY + YX + YY)$$
$$+ iV\{(\Delta_{xx}^- + \Delta_{yy}^-) - i(\theta_{xx}^+ + \theta_{yy}^+)\}$$
$$Q = \tfrac{1}{4}(-XX - XY + YX + YY)$$
$$- U\{(\Delta_{Ex} + \Delta_{Ey}) + i(\theta_{Ex} + \theta_{Ey})\}$$
$$- iV\{(\Delta_{Wx} - \Delta_{Wy}) - i(\theta_{Wx} - \theta_{Wy})\} \tag{6}$$
$$U = \tfrac{1}{4}(XX - XY + YX - YY)$$
$$+ Q\{(\Delta_{Wx} + \Delta_{Wy}) - i(\theta_{Wx} + \theta_{Wy})\}$$
$$- iV\{(\Delta_{Ex} - \Delta_{Ey}) + i(\theta_{Ex} - \theta_{Ey})\}$$
$$V = i/4(XX - XY - YX + YY)$$
$$+ iQ\{(\Delta_{Wx} - \Delta_{Wy}) - i(\theta_{Wx} - \theta_{Wy})\}$$
$$+ iU\{(\Delta_{Ex} - \Delta_{Ey}) + i(\theta_{Ex} - \theta_{Ey})\}.$$

When we consider further that for most continuum radio sources $V \leq 0.01\,I$, $Q \leq 0.1\,I$, $U \leq 0.1\,I$ and that in nearly all cases θ and Δ can be adjusted to a level of $\theta, \Delta < 0.04\,I$, then we can estimate that $(Q, U, V)(\Delta, \theta) \leq 0.004\,I$.

Thus, what can be called the error terms in this formulation are of order of or less than 0.4 % of I. Another source of error in the solutions is the gain stability of the independent channels. If this is on the order of 1 % or poorer, it will also contribute random error terms to our solutions of order 0.5 %. Accepting 0.5 % of I as the general level of errors, we can then neglect all terms of second order in Q, U, V, Δ, θ and obtain a great simplification of our polarization solutions. Doing this and putting the scaling factor of $\tfrac{1}{4}$ into the normalization, we can write the solutions for the 4 Stokes parameters as

$$I = XX + XY + YX + YY$$
$$Q = -XX - XY + YX + YY$$
$$U = XX - XY + YX - YY$$
$$V = i(XX - XY - YX + YY)$$

where terms of order of 0.5 % of I have been neglected.

IV. Discussion

As with any system of observation, both advantages and disadvantages exist. For the above method some of these can be listed briefly.

Advantages:

a) After initial adjustment, the rotation of dipoles is never again necessary.

b) The system polarization calibration occurs automatically with conventional gain and phase calibration procedures as long as an unpolarized calibrator is used[1]).

---

[1]) With very slight modifications, calibrators with known polarization can also be used.

c) With all dipole combinations at 45° or 135° to each other, all channels receive nearly equal and relatively large fluxes in them making system gain and phase calibration on celestial sources simple.

d) The solution for the Stokes parameters U, Q, V, is as simple as that for I and can be done in either the real or the Fourier plane.

e) The dipoles can, if desired, be rotated to other position angles to perform special types of experiments.

Disadvantages:

a) Because of the magnitude of the error terms neglected, the method is not suitable for searching for the small amounts of circular polarization (V) thus far detected.

b) The accurate addition and subtraction of all channels with equal weights requires that the observing system have excellent gain stability.

Acknowledgements. The Westerbork Radio Observatory is operated by the Netherlands Foundation for Radio Astronomy, with the financial support of the Netherlands Organization for the Advancement of Pure Research (Z.W.O.).

Appendix

Interference Polarimetry

Taking from Chandrasekhar (1950) that an arbitrarily polarized beam of electromagnetic radiation can be completely specified in terms of its 4 Stokes parameters I, Q, U, and V (I equals the total intensity, $\sqrt{(Q^2 + U^2)}$ equals the linearly polarized intensity, $\tfrac{1}{2}\tan^{-1}(U/Q)$ equals the position angle of the electric vector of the linearly polarized radiation, and V equals the circularly polarized intensity), Morris, *et al.* (1964) have shown that the polarization distribution of a radio source can be described by the distribution of the 4 Stokes parameters over its face.

Thus, assuming that the polarization of the incident radiation is independent of frequency over the bandwidth of the receiving system, measurement of the Fourier transforms of the 4 Stokes parameters with an interferometer will allow us through Fourier inversion to obtain their distributions over the source and consequently its polarization properties.

Henceforth, when we speak of the Stokes parameters, we will always be referring to the measured values of the Fourier transforms of their spatial distributions. However, since the ultimate transformation into the real plane is a linear operation, this is inconsequential to the discussion.

Let us consider a partially polarized radio wave travelling in the positive z direction in its most general form. For simplicity we will express it in terms of two orthogonal components oriented along the 0° position angle (North) and 90° position angle (East) directions. Then, our incident wave can be written as

$$E = \hat{e}_0 E_1(t) \exp(i(\omega t - \beta z + \delta_1(t)))$$
$$+ \hat{e}_{90} E_2(t) \exp(i(\omega t - \beta z + \delta_2(t))) \tag{i}$$

339

where $\omega = 2\pi\nu$, $\beta = 2\pi/\lambda$, $\hat{e}_0$ and $\hat{e}_{90}$ are unit vectors in the $0°$ and $90°$ position angle directions, and we are considering the radiation as received over a finite bandwidth $\Delta\nu$. All time fluctuations are independent and the time variations of $E_1(t)$, $E_2(t)$ and $\delta(t)$ are slow compared to that of the mean frequency ν and are of the order of the reciprocal bandwidth $1/\Delta\nu$.

Now, in passing this incident radiation through a real antenna system fed with dipoles, instrumental effects will be introduced. These can be expressed as:

a) A coupling of the orthogonal components without a phase shift, i.e. an instrumental linear polarization.

b) A coupling of the orthogonal components with a phase shift, i.e. an instrumental circular polarization.

c) A different phase path length for the orthogonal components i.e. a conversion of linear polarization into circular polarization or vice versa.

d) A different amplitude gain for the orthogonal components, i.e. a rotation of the position angle of linear polarization.

Because the interferometer channels are independently calibrated for gain and phase, effects (c) and (d) are defined to be zero to within the errors of system calibration and can be neglected. Effect (a) also is of no consequence for several reasons. It is made to be small (usually less than -40db in power) in the construction of the receivers and also, the relative dipole positions on two antennas are determined by crossing them and nulling out the in-phase signal. Thus, any effect which remains will be indistinguishable from a dipole position setting error (Δ). Corrections for this will have to be applied and included later in any case. Finally, effect (b) cannot be eliminated and must be allowed for in any treatment. We shall include it as the ellipticity θ. However, just as for (a) above, due to receiver construction it will be small.

Thus, if we assume for the moment that our dipole setting error is zero ($\Delta = 0$), we can describe the receiving system of a single antenna as an "almost" linearly polarized dipole at a position angle ϕ (measured from North through East) with an ellipticity θ.

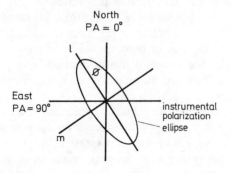

The voltage response (E_D) of such a system for a dipole lying along the l axis can then be written in the form

$$E_D = E_l \cos\theta - i E_m \sin\theta \qquad (ii)$$

where the signs were chosen such that $+\theta =$ Right Hand Instrumental Circular Polarization (IRE Definition). Now, from our incident wave (i) we have

$$E_l = \hat{e}_l \cdot E = (E_1(t) \exp(\delta_1(t)) \cos\phi$$
$$+ E_2(t) \exp(\delta_2(t)) \sin\phi) \exp(i(\omega t - \beta z))$$
$$E_m = \hat{e}_m \cdot E = (-E_1(t) \exp(\delta_1(t)) \sin\phi$$
$$+ E_2(t) \exp(\delta_2(t)) \cos\phi) \exp(i(\omega t - \beta z)) \qquad (iii)$$

so that our dipole response becomes

$$E_D = (E_1(t) \exp(\delta_1(t)) (\cos\phi \cos\theta + i \sin\phi \sin\theta)$$
$$+ E_2(t) \exp(\delta_2(t)) (\sin\phi \cos\theta$$
$$- i \cos\phi \sin\theta)) \exp(i(\omega t - \beta z)). \qquad (iv)$$

Considering now an interferometer on an east-west baseline, we designate the two telescopes as telescope W in the west and telescope E in the east. We also choose the interferometric phase convention such that a radio source which lies at greater hour angle has positive phase. Then, observing Eq. (iv), we can write the voltage received by each antenna as

$$E_W = E_{DW}(t) \exp(i(\omega t - \beta z_W));$$
$$E_E = E_{DE}(t) \exp(i(\omega t - \beta z_E)) \qquad (v)$$

and our correlator output response (R) becomes

$$R = E_W E_E^* = E_{DW}(t) E_{DE}^*(t) \exp(i\beta(z_E - z_W)). \qquad (vi)$$

Since the term $\exp(i\beta(z_E - z_W))$ is only the phase dependent on the predicted source position and is removed from the observation by fringe stopping, we shall neglect it in further consideration. Thus, we can write

$$R = E_{DW}(t) E_{DE}^*(t). \qquad (vii)$$

If we now consider the noise like nature of the radio radiation and the time averaging performed by the post-correlation equipment, and we introduce the unknown instrumental gain (G) and phase (P) constants of the system, we can write our measured uncalibrated output response as

$$R = G \exp(iP) \langle E_{DW}(t) E_{DE}^*(t) \rangle \qquad (viii)$$

where from Eq. (iv)

$$E_{DW,E}(t) = E_1(t) \exp(\delta_1(t)) (\cos\phi_{W,E} \cos\theta_{W,E}$$
$$+ i \sin\phi_{W,E} \sin\theta_{W,E})$$
$$+ E_2(t) \exp(\delta_2(t)) (\sin\phi_{W,E} \cos\theta_{W,E}$$
$$- i \cos\phi_{W,E} \sin\theta_{W,E}).$$

Since the Stokes parameters can be defined in the form

$$I = \langle E_1^2(t) \rangle + \langle E_2^2(t) \rangle$$
$$Q = \langle E_1^2(t) \rangle - \langle E_2^2(t) \rangle$$
$$U = 2\langle E_1(t) E_2(t) \cos(\delta_1(t) - \delta_2(t)) \rangle \qquad (ix)$$
$$V = 2\langle E_1(t) E_2(t) \sin(\delta_1(t) - \delta_2(t)) \rangle,$$

our output response of an interferometer to a partially polarized incident wave can finally be written as

$$R = \tfrac{1}{2} G \exp(iP) \, (I \{\cos(\phi_W - \phi_E) \cos(\theta_W - \theta_E)$$
$$+ i \sin(\phi_W - \phi_E) \sin(\theta_W + \theta_E)\}$$
$$+ Q \cos(\phi_W + \phi_E) \cos(\theta_W + \theta_E)$$
$$+ i \sin(\phi_W + \phi_E) \sin(\theta_W - \theta_E)\}$$
$$+ U \sin(\phi_W + \phi_E) \cos(\theta_W + \theta_E)$$
$$- i \cos(\phi_W + \phi_E) \sin(\theta_W - \theta_E)\}$$
$$- i V \sin(\phi_W - \phi_E) \cos(\theta_W - \theta_E)$$
$$- i \cos(\phi_W - \phi_E) \sin(\theta_W + \theta_E)\}) \, .$$

(x)

The vector signs have been added to indicate the complex nature of the parameters as distributed over the Fourier plane, and the phase and sign conventions have been chosen so that $+V$ is Right Hand Circular Polarization (IRE Definition).

References

Baldwin, J. E., Jennings, J. E., Shakeshaft, J. R., Warner, P. J., Wilson, D. M. A., Wright, M. C. H. 1970, *Monthly Notices Roy. Astron. Soc.* **150**, 17

Chandrasekhar, S. 1950, Radiative Transfer, Oxford Univ. Press, London, pp. 24–35

Fomalont, E. B. 1970, *Astrophys. J.* **160**, L73

Mayer, C. H., McCullough, T. P., Sloanaker, R. M. 1957, *Astrophys. J.* **126**, 468

Mayer, C. H., McCullough, T. P., Sloanaker, R. M. 1962, *Astrophys. J.* **135**, 656

Morris, D., Radhakrishnan, V., Seielstad, G. A. 1964, *Astrophys. J.* **139**, 551

Weiler, K. W., Seielstad, G. A. 1971, *Astrophys. J.* **163**, 455

K. W. Weiler
Kapteyn Astronomical Institute
P.O. Box 800
Groningen, The Netherlands

A Multiple Beam Technique for Overcoming Atmospheric Limitations to Single-Dish Observations of Extended Radio Sources

D. T. Emerson, U. Klein, and C. G. T. Haslam

Max-Planck-Institut für Radioastronomie, D-5300 Bonn 1, Federal Republic of Germany

Received August 22, 1978

Summary. The limiting sensitivity for conventional single-dish radio continuum observations of extended objects at wavelengths of a few cm is set mainly by weather; for much of the time a varying component of atmospheric attenuation, and hence thermal emission, causes deviations in the response of the receiver far in excess of any deviations due to receiver noise or receiver instabilities.

A dual-beam technique, where only the difference in power received by the two beams pointing at slightly separated points in the sky is recorded, has been used for many years for observations of small sources; atmospheric emission affects both beams equally, and so cancels. It has often however been stated that this technique is applicable only where the angular size of the radio source is less than the separation of the two beams.

This paper presents a dual-beam technique suitable for observation of sources many times greater in angular extent than the separation of the two beams. Considerations of the spatial frequency response of a dual-beam system and of the effect of noise on the observations are used to show that regions many times larger than the beam separation may be mapped and restored to the equivalent perfect-weather single-beam observation without degradation in signal-to-noise ratio, and with a complete sampling of spatial frequencies. The limitations of such a technique are discussed in detail, and a triple beam system is suggested as a means of extending the capabilities of the system.

Examples are shown of successful observations of extended sources made using the Effelsberg 100-m telescope at a wavelength of 2.8 cm, in weather conditions which would be unacceptable for any conventional single-beam technique.

This technique of mapping extended sources with a dual or multiple beam system, and the restoration to the equivalent single-beam observation, is also applicable at mm and IR wavelengths.

Key words: radio continuum – cm, mm, and IR wavelengths – weather – dual-beam

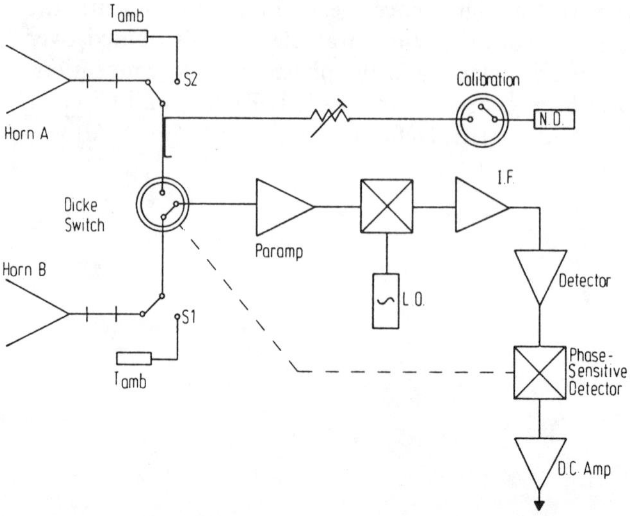

Fig. 1. Block diagram of the dual-beam system used at Effelsberg for test observations. The response after the P.S.D. is proportional to the difference of the power received in the 2 feed horns, corresponding to two far-field beams of angular separation λ_0. The system may be operated as a single beam Dicke-switching system by switching either S 1 or S 2

I. Introduction

Radio-astronomical observations at wavelengths of 6 cm or shorter are subject to prevailing atmospheric conditions. Fluctuations in the thermal emission from water vapour within the telescope beam may cause deviations in the receiver output far in excess of those due to receiver noise or receiver instabilities. As a result, useful single-beam continuum observations with the

Effelsberg 100-m telescope operating in a total-power load-switched mode may only be made for approximately one third of the available time at a wavelength of 6 cm, and for an even smaller fraction of the time at shorter wavelengths.

The dual-beam technique has been known for many years as a means of cancelling the response of the receiver to fluctuations in the atmospheric emission, but only for observations of small sources with an extent much smaller than the separation of the two beams. It has often been stated that this technique is not applicable to observations of objects of an angular size comparable to or greater than the separation of the two beams (e.g. Baars, 1966, 1970; Conway et al., 1963). This is not true. In this paper a technique is described and results are presented of dual-beam observations of objects extending over several units of beam separation, made with the Effelsberg telescope at a wavelength of 2.8 cm. These observations were made in most cases under weather conditions which would be totally unacceptable with a conventional single-beam system. This technique considerably extends the capabilities of single-dish telescopes at wavelengths of a few cm, and makes possible deep surveys of extended sources

---

Sent offprint requests to: D. T. Emerson

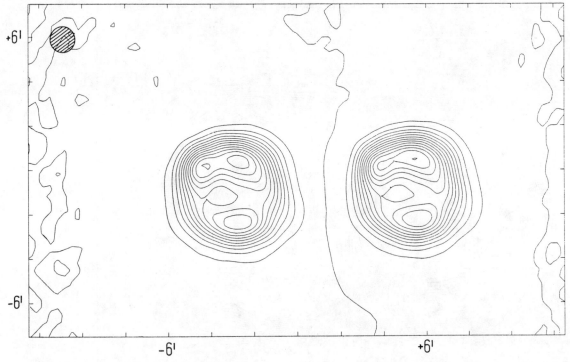

Fig. 2. A map made of the source 3C 461 (diameter $\sim 6'$) using a dual-beam system with an angular beam separation of $8'.2$. Both images (the left is positive, the right negative) are well separated; analysis of this observation requires no special technique since each image may be considered independently

which would not be practicable if dependent on good observing weather in the conventional sense.

A dual-beam technique is already in use at IR wavelengths, but the new method of data analysis presented here, and the examination of the capabilities and limitations of the technique may be of use to IR observers as well as to mm-wave and radio astronomers. Much of the discussion is also relevant to the analogous use of frequency-switching spectral line observations.

Section II summarises the principle behind conventional observations of small sources and data analysis with a dual-beam system. Section III describes the technique used at Effelsberg to observe extended objects using a dual-beam system, presents the method of data analysis and shows examples of the application of the technique to a number of sources. Section IV discusses the limitations of the technique, while Sect. V describes a further development of the principle, using a multi-beam system. The most important points of this paper are summarized in Sect. VI.

II. Conventional Dual-beam Observations of Radio Sources of Small Angular Extent

II.1. Principles of the Dual-beam Techniques

The instantaneous power response T_a of a telescope looking at a radio source of brightness T_s through an absorptive atmosphere of temperature T_c and transmission coefficient α is (excluding receiver noise and excluding telescope inefficiencies):

$$T_a = (1-\alpha)T_c + \alpha T_s. \tag{1}$$

A time-varying component of α resulting from variations in the water vapour content of the atmosphere within the telescope beam is responsible for fluctuations in the receiver output which may swamp the desired astronomical signal. The deviation dT_a in receiver response is then

$$dT_a = -d\alpha \cdot T_c + d\alpha \cdot T_s. \tag{2}$$

Since in most observations T_c is much greater than T_s, only the variation in the thermal emission of the atmosphere and not the actual attenuation of the desired signal is of great importance; i.e. the term $d\alpha \cdot T_s$ may be neglected [this is discussed further in Sect. IV(d)]. With a two-beam system (see Fig. 1), the resultant response will be identical in both beams if the atmospheric emission is contained in the near-field of the telescope. For moderate offsets of a feed from the true focus of the telescope, the near-field region of the telescope beam is well approximated by a tube of diameter equal to that of the telescope (see the discussion in Sect. IV).

In the case of two feeds, each offset from the true focus of the telescope by a moderate amount, the near-field patterns almost totally overlap, although the two far-field patterns may be well separated. The difference in response of the two beams is substantially independent of emission originating in the near field of the telescope, but does yield the true difference in brightness of emission at the points covered by the two beams in the far field (neglecting $d\alpha \cdot T_s$). The far-field limit $D^2/2\lambda_e$ for the 100-m Effelsberg telescope at a wavelength of $\lambda_e = 2.8$ cm is ~ 100 km; the height from which most of the fluctuations in atmospheric emission occur is 2–3 km (Baars, 1970), thus well in the near-field region for all practical elevations. This is discussed further in Sect. IV.

It is possible to operate from the secondary focus of the Effelsberg telescope with a feed spacing of up to ~ 90 cm (limited by mechanical considerations within the focus cabin), corresponding to a far-field angular beam spacing of $8'.2$. With this configuration the atmospheric contributions to the emission detected

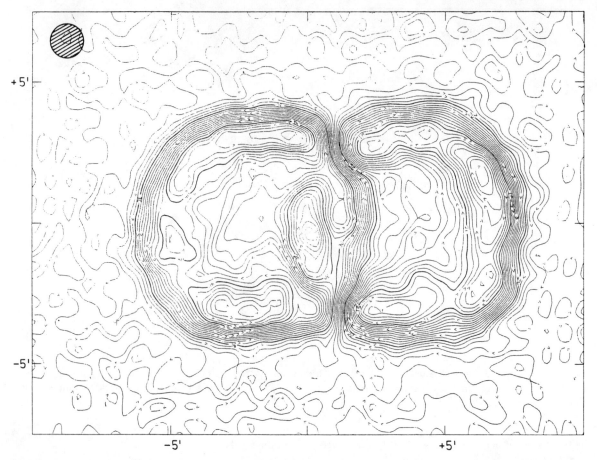

Fig. 3. The radio source 3C 10 (diameter ~9') seen here observed with a dual-beam system where the angular separation of the 2 beams is only 5'.5. The positive and negative images overlap. The analysis of observations such as this is discussed in the text

in each beam were found to cancel within better than ~90% (see Sect. IVa below), and coma effects due to the off-axis position of the feeds were weaker than ≈ −18 dB, referred to the main beam.

II. 2. Analysis of Data from Dual-beam Observations of Radio Sources of Small Angular Extent

The analysis of observations of radio sources of small angular size (i.e. small compared to the angular separation of the two beams and their sidelobes) made with a dual-beam system presents no special difficulties. (1) If one of the two beams is never allowed to overlap with the emission from the radio source, this beam may be regarded as a reference noise source and the observations treated in the same way as with a conventional noise-balanced total power Dicke receiver. (2) Alternatively, if the telescope is scanned in such a way that the radio source appears sequentially in each beam, but the angular extent of the source is small enough that it never appears simultaneously in both beams, or their sidelobes, then the resulting dual-beam map may be separated easily into two independent maps, one positive and one negative, and these two maps may either be used separately or shifted and reversed in sign before being averaged. Figure 2 shows a map of 3C 461 (diameter ~6') made with a beam separation of 8'.2. The dual-beam system has rejected atmospheric effects visible in single-beam data taken simultaneously, and since

the positive and negative images are well separated the interpretation of this map presents no particular difficulty.

Neither of these simple analysis techniques is applicable if emission from different parts of a more extended radio source is ever detected simultaneously in both beams.

III. Observations of Extended Radio Sources with a Dual-beam System

Figure 3 shows the source 3C 10 (diameter ~9') observed using an angular separation of 5'.5 between the two beams, each of which has a full-width to half-power of 70". The interpretation of this observation will be given in a separate paper (Klein et al., in press) but the data are shown here as an example of a dual-beam observation of a source whose total extent is considerably greater than the separation of the two beams. The effects of bad weather have been removed by the dual-beam system – compare with the single-beam map observed at the same time which is shown in Fig. 4 – but in the centre region of the source emission is detected by both beams simultaneously, giving a confused representation of the true emission. However it is possible to reconstruct from observations such as this a unique image of the true sky distribution (within the normal resolution limit of the telescope) without serious degradation of the image in the presence of noise. Figure 8 is the equivalent single-beam observation of 3C 10 reconstructed directly from Fig. 3.

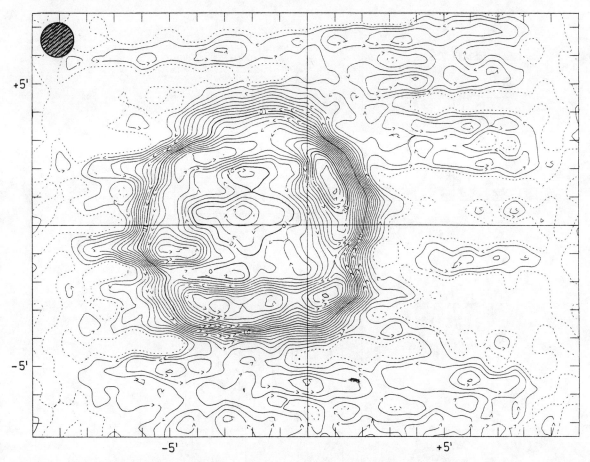

Fig. 4. Data recorded simultaneously with that shown in Fig. 3, but showing the response of only one beam. The effects of variable atmospheric emission (the map was scanned horizontally) are clearly seen in this map

In the following discussion the process of restoring a single scan from a raster-scanned observation of an extended region will be considered. The direction of scanning is assumed to be along the line joining the two far-field beam patterns of the telescope. The dual-beam observation of a 2-dimensional area of sky may be restored to the equivalent single-beam observation by separate restoration of each one-dimensional scan comprising the complete observation.

A convenient way of considering the data is to examine the spatial frequency response of a dual-beam system. If S represents the true brightness distribution of an arbitrary source, B the single-beam response of the telescope, and D represents the dual-beam function [see Fig. 5(i)–(iii)] then a single-beam observation would give the distribution P given by

$$P = S * B, \tag{3}$$

(the operator $*$ represents a convolution).

The dual-beam observation however yields the observed distribution O given by:

$$O = S * B * D. \tag{4}$$

This may be represented in the spatial frequency domain: using $\bar{p}$ to represent the spatial frequency distribution of P, etc.:

$$\bar{p} = \bar{s} \times \bar{b} \tag{5}$$

$$\bar{o} = \bar{s} \times \bar{b} \times \bar{d}, \tag{6}$$

$\bar{d}$, the Fourier transform of the dual-beam function D shown in Fig. 5, consists of a purely imaginary component whose amplitude β varies as

$$\beta(1/\lambda) \propto \sin\left(\frac{\lambda_0}{\lambda}\pi\right), \tag{7}$$

where $(1/\lambda)$ is the spatial frequency and λ_0 is the angular separation of the two far-field beams. The problem is to derive the equivalent single-beam observation P from the dual-beam measurement O.

In principle it might be possible to re-weight the spatial frequencies of $\bar{o}$, in order to derive $\bar{p}$ which is the Fourier transform of the desired map. However, the necessary re-weighting function would become infinite at the zero-crossings $(1/\lambda = n/\lambda_0)$ of $\bar{d}$ – spatial frequencies with periods equal to, or a sub-multiple of, the original separation are undefined in the observations.

Provided that the total extent $(+/- x/2)$ of the region being mapped is finite, it is only necessary to sample the spatial frequency distribution at intervals of $1/x$ (e.g. Bracewell, 1956). If spatial frequencies which have been attenuated to less than 0.5 of their original amplitude by the dual-beam function are excluded, then regular sample points of spatial frequencies must be chosen such that

$$\beta(1/\lambda) = \sin\left(\pm\frac{m\lambda_0}{\lambda}\cdot\pi\right) \geqq 0.5 \tag{8}$$
$$(m = 1, 3, 5, \ldots)$$

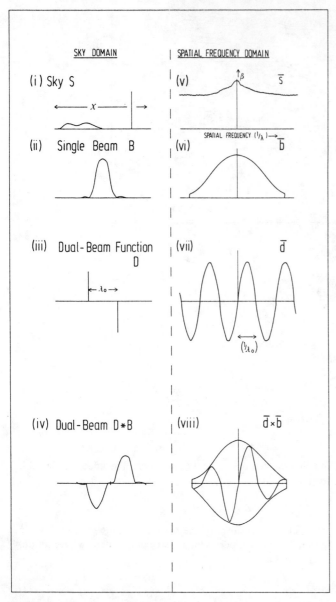

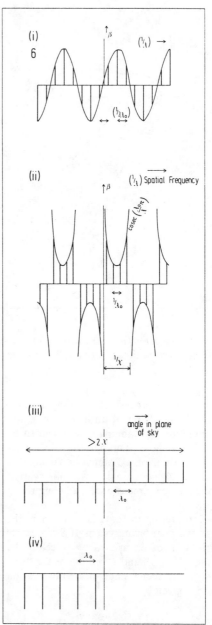

Fig. 6

Fig. 5. Schematic representation of a true source brightness distribution S(i), the single beam B(ii), the dual-beam function D (iii) and the dual-beam $D * B$ (iv) with the amplitudes β of the respective Fourier transform representations $\bar{s}$, $\bar{b}$, $\bar{d}$, $\bar{d} \times \bar{b}$ shown in (v)–(viii). The scales are arbitrary

Fig. 6. (i) The spatial frequencies representing the distribution of emission from a region of total extent x may be sampled at intervals $(1/\lambda) = 1/x$. If samples are chosen at $1/\lambda = \pm m/6\lambda_0 (m=1, 3, 5, \ldots)$ then a field of size $3\lambda_0$ may be mapped without any sample of spatial frequency being attenuated by a factor of more than 2 by the dual-beam function. The necessary sample points are illustrated above. **(ii)** If the spatial frequency sample points such as indicated in Fig. 6(i) are re-weighted to give a flat spatial frequency response, then the weighting factors [reciprocals of β in Eq. (8)] shown in Fig. 6(ii) must be used. (iii) Alternatively, the original data may be convolved with the function shown in 6(iii), which is the inverse Fourier transform of that shown in 6(ii). The sample points illustrated here would be appropriate for a region of extent $x = 4\lambda_0$. **(iv)** An obvious method of restoring dual-beam observations would be to make a step-by-step integration across the map, equivalent to a convolution by the function shown above. This function does not possess exactly the required complex spatial frequency response. Analysis using this simple function will lead to a higher noise level on the final restored map

346

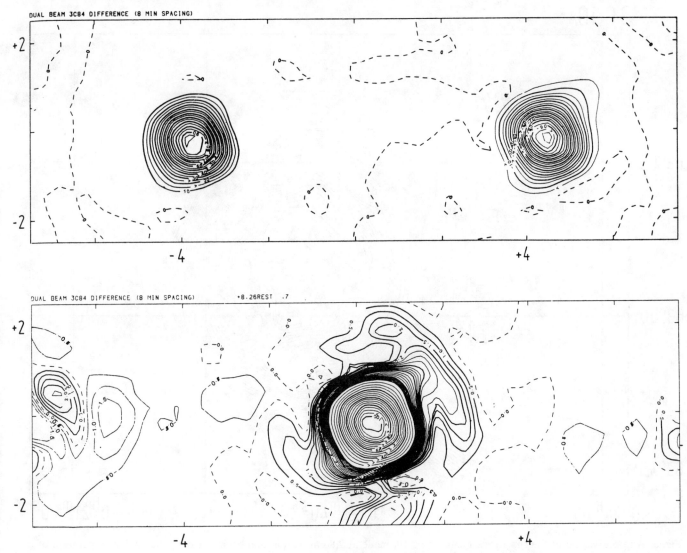

Fig. 7.a and b. An observation of the unresolved source 3C 84 using the same dual-beam system as was used for the data shown in Fig. 3, but with a beam separation of 8′.2. **a** represents the original observation (contour interval 5% of peak value; left image is positive, right negative), while **b** shows the result after restoration to the equivalent single-beam. The lower contours here are at intervals of 0.5%, and illustrate a peak residual response of ∼2%

This is possible if $(1/\lambda) \leqq (1/6\lambda_0)$; the lowest spatial frequency accepted under this criterion is $1/6\lambda_0$, but the spatial frequency sampling interval cannot be less than $1/3\lambda_0$ [see Fig. 6(i)]. That is, a region of total extent $3\lambda_0$ may be unambiguously mapped without S/N on any required spatial frequency being degraded by more than a factor of 2. Since dual-beam observations at Effelsberg can be made successfully at a wavelength of 2.8 cm using a beam separation of 8′.2 (see Sect. IVa below), a field of size 25′ may be observed with no difficulty. As the half-power beam size at 2.8 cm is 70″, this is a useful area.

There is an obvious analogy with the finite sampling (at, say, intervals of λ_g) of the aperture plane in a grating survey made with an aperture synthesis telescope such as in use at Cambridge or Westerbork. However in the case of a dual-beam observation restored to the equivalent single-beam observation, the maximum size of the field which can be mapped is set only by noise considerations. In the absence of noise a source of any finite-size may

be unambiguously mapped. With a grating survey by an aperture synthesis telescope it is not possible even in the absence of noise to represent unambiguously features with a structure greater than λ_g, the grating radius.

Although it is convenient to consider the analysis of the observations in the spatial frequency domain, in practice it is unnecessary to perform the re-weighting of the discrete sample points [illustrated in Fig. 6(i) and 6(ii)] in Fourier space. It is computationally convenient to obtain the required complex frequency response by convolving the dual-beam scans with a function such as shown in Fig. 6(iii), which represents the inverse Fourier transform of the weighted sampling function of Fig. 6(ii).

A dual-beam technique is often used at IR wavelengths; for example Becklin and Neugebauer (1975) mapped the centre ∼1′ of the Galactic centre at a wavelength of 10 μ with a beam separation of 10″. In this case the radiation in the "reference beam" was accounted for in a point-by-point manner. A straightforward

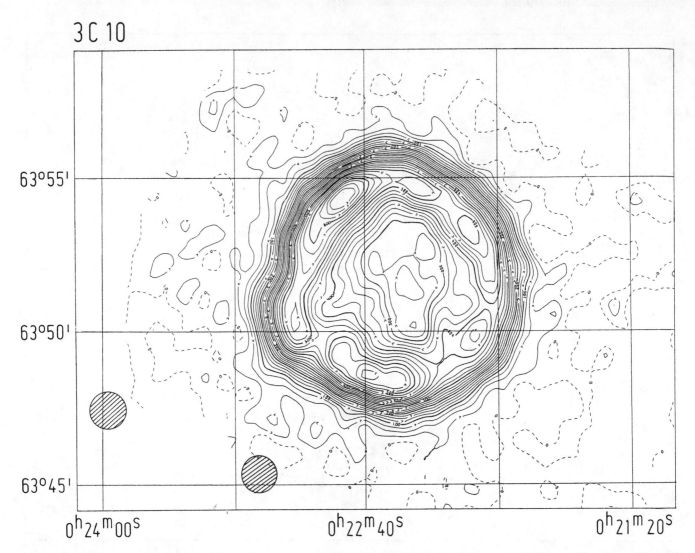

Fig. 8. An observation of the supernova remnant 3 C 10, made at 2.8 cm with a dual-beam system. The angular separation of the 2 beams was 5ʹ.5, but the source has a diameter of ~9ʹ. The 2 shaded circles illustrate the beam size and separation. This map was derived directly from the data shown in Fig. 3 using the technique described in the text. Compare with the data of Fig. 4, which is the response of a single beam recorded at the same time. Variable atmospheric emission during the observations render the single-beam data shown in Fig. 4 almost useless, while the dual-beam data restored to the equivalent single-beam observation show no degradation due to the effects of the wheather. Note that the map shown here has been interpolated to a R. A.-declination grid, while Fig. 3 and Fig. 4 are plotted on a telescope azimuth-elevation grid

summation of respective data points separated by integral units of the beam separation is equivalent to convolution with the function shown in Fig. 6(iv), which does not possess exactly the required spatial frequency response. In particular, spatial frequencies of wavelength exactly equal to, or sub-multiples of, the beam separation are not suppressed. This is undesirable, because such spatial frequencies exist in the data from dual-beam observations only as a result of noise, and contain no astronomical information. A map constructed from dual-beam observations using this more obvious method of step-by-step summation will possess substantially higher noise than one from the same data analysed as described in this paper, i.e. by convolution with a function such as is shown in Fig. 6(iii).

Figures 7–10 show examples of the application of this technique. In each case the original data have been convolved with the appropriate one-dimensional function [such as in Fig. 6(iii)] along the direction of the line joining the two beams in order to produce the equivalent single-beam map with harmful effects of the weather removed. The telescope parameters are summarised in Table 1, while Fig. 1 shows the block diagram of the receiver configuration.

Figure 9 includes the data of Fig. 8, but with additional observations made with a different beam separation and scanning direction added. Figure 10 shows an extreme case, where emission from the Galactic Centre extending over >20ʹ has been mapped successfully with a dual beam separation of only 3ʹ.2. All features visible on the Effelsberg 2.8 cm map made with a conventional single-beam system by Pauls et al. (1976) in good weather conditions have been accurately reproduced on this restored dual-beam observation.

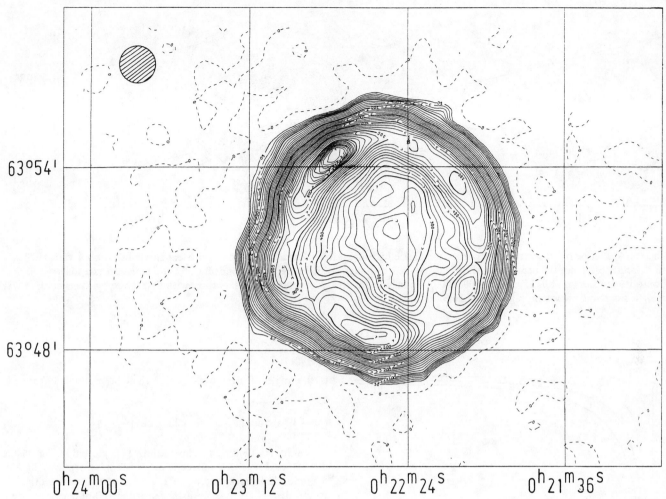

Fig. 9. A map of 3C 10 using the data shown in Figs. 3 and 4, but combined with further observations made with a dual-beam separation of 8″2. Both signal-to-noise ratio and dynamic range are improved. A discussion of this source will be given in a separate paper (Klein et al., in press)

Table 1. System Parameters

T_{sys} (uncooled paramp)	200 K
Centre frequency	10.7 GHz
Instantaneous bandwidth	~300 MHz
Beam size (*fwhp*)	70″
Dual-beam separation	3″2, 5″5, or 8″2
Polarisation	L.H. circular

IV. Practical Limitations of the Technique

In this Section the practical limitations of the technique of observing extended sources with a dual-beam system, and the restoration to the equivalent single-beam observation, are discussed. These limitations are discussed in turn under the following headings:

(a) Imperfect cancellation of the atmospheric emission

(b) Inequality of the gains and sidelobe patterns of the two beams

(c) An increase in the noise level of the corrected map for fields extending over many units of the beam separation

(d) Observations of strong sources ($T_s \gtrsim T_c$ in Eq. 2) under conditions of high atmospheric attenuation.

a) Imperfect Cancellation of Variations in Atmospheric Emission

Residual deviations on the final dual-beam restored map may result from inadequate overlap of the two near-field patterns of each beam in the regions of atmospheric emission. The beam overlap may be estimated in the following manner.

The near-field pattern of each beam may be approximated by a tube of diameter equal to that of the telescope [see Baars (1970) for a more exact treatment]. For a given distance R from the telescope, where R is much less than the Rayleigh distance $D^2/2\lambda$, and a given angular separation θ of the two far-field beams, the expected fractional cancellation of atmospheric emission is equal to the fractional beam overlap Q of the two tubes representing the near-field responses of the two beams. If atmospheric emission occurs from a scale height h, then $R = h \, \mathrm{cosec}(a)$, where a

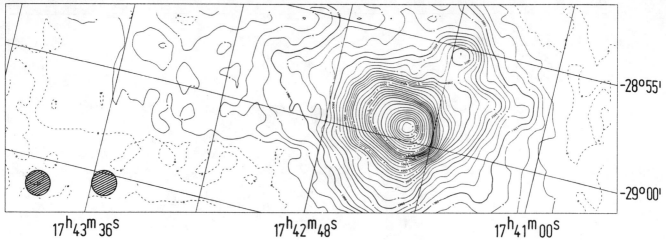

Fig. 10. A dual-beam observation of part of the Galactic centre, after restoration to the equivalent single-beam observation. This is an extreme test, since the beam separation was only 3′.3, although the field mapped extends for 30′ and emission extends beyond this. Contours are shown at 0, 250, ..., 8000, 9000, ... mJy/beam area. Compare this observation with that made by Pauls et al. (1976) using a single-beam system in good weather conditions

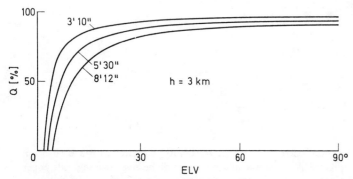

Fig. 11. A plot of fractional cancellation of atmospheric emission expected for a 100-m telescope using angular beam separations of 3′.2, 5′.5, and 8′.2, as a function of elevation angle. The atmospheric emission is assumed to originate in a layer of height 3 km

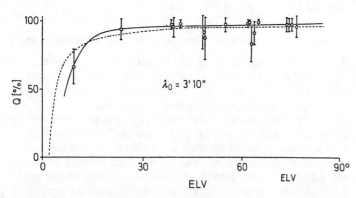

Fig. 12. A plot of measured values of Q against elevation for a beam separation λ_0 of 3′.2. The solid line is a fit to the data, while the dashed line is a model curve from Fig. 11

is the angle of elevation. Q may then be calculated from the equation:

$$Q = \frac{1}{\pi}\left[2 \arccos\left(\frac{\lambda R}{D}\right) - \sin\left(2 \arccos\left(\frac{\lambda R}{D}\right)\right)\right]. \quad (9)$$

Taking h as 3 km (an upper limit), Q is plotted in Fig. 11 for the 100-m Effelsberg telescope with beam separations of 3′.2, 5′.5, and 8′.2 (corresponding to 2.7, 4.6, and 6.8 beamwidths, respectively). The Rayleigh distance for the 100-m telescope operating at a wavelength of 2.8 cm is 180 km, well beyond the distances R considered here.

It should be noted that this estimate of the factor Q refers to emission from regions with a horizontal scale size less than the beam diameter ($\sim$ the telescope diameter D for $R \ll D^2/2\lambda_e$). Emission from uniform regions of much larger scale-size than this are in any case cancelled by the dual-beam observation, regardless of the area of beam overlap. Figure 11 is therefore an underestimate of the degree of cancellation of atmospheric emission to be expected with the dual-beam system, and shows that the cancellation should be effective for all practical elevations even for the largest angular beam separation (8′.2) in use at Effelsberg.

For comparison with observations we have compared the noise levels on simultaneous single-beam and dual-beam observations. The fractional rejection of the component of noise which can be attributed to weather (and which is dominant in the single-beam observations) has been measured for a number of observations at differing elevations, and is plotted in Fig. 12 for the 3′.2 beam separation. As expected, at most elevations the measured values are slightly better than those predicted from the simple model used to calculate Fig. 11. Even at elevations as low as $\sim 25°$ rejections of over 90% are obtained; these observations were made through a partly cloudy sky, with $\sim 50\%$ humidity and an average wind velocity of ~ 5 km/h. Similar results have been obtained with beam separations of 8′.2 and 5′.5.

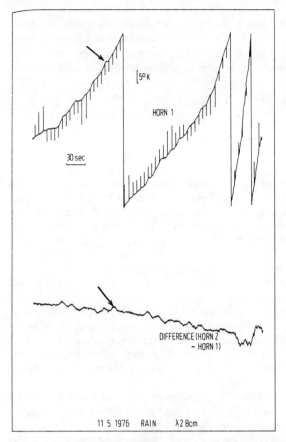

Fig. 13. a An example of the analogue (2.8 cm λ_e) record made during exceptionally bad weather conditions (heavy rain). The single-beam and dual-beam (= beam A−beam B) responses are shown separately, with the weak emission from 3C 10 just visible in the dual-beam record

b The analogue record of observations made at a wavelength λ_e of 2.8 cm in drizzle and fog. In the upper half of the diagram the response of each single beam is shown separately, while the dual-beam (separation 5.'5) response and calibration channel are shown in the lower half

Figure 13(i) shows analogue records of both the total power from a single beam and the difference of the two beams (angular separation 3.'2) taken during test observations of the source 3C 10, at a wavelength of 2.8 cm, during heavy rain. Emission from 3C 10 can be seen directly in the dual-beam data [indicated by arrows in Fig. 13(i)] but is completely masked by varying atmospheric emission in the single beam data. The antenna temperature recorded by the single beam increased by ≈190 K in less than ≈5 min of time – corresponding to a transmission coefficient through the atmosphere of only ∼30%. The response of the dual-beam system increased by only 15 K during this period.

Figure 13(ii) shows the analogue record of an observation made in light rain and fog. In the upper half of the diagram the response of each single beam is shown separately, while the dual-beam response and calibration channel are shown in the lower half. In this observation the antenna temperature measured by a single beam changed by ∼7 K in 10 min, while the dual-beam signal remained constant to within ≲0.25 K.

A further point to be considered is the time scale of changes in the atmospheric emission within the two beams. In the system employed at Effelsberg (see Fig. 1) the receiver is switched between the 2 horns at a rate of ≈30 Hz. If the atmospheric emission changes significantly in this time then the dual-beam system will be unsuccessful in cancelling weather effects. For the 100-m telescope a wind speed as high as ∼10^4 km/h would be implied if the atmospheric emission within the near field of the telescope were to change significantly within the ≈30 ms switching cycle. Although different beam switching rates have not been tried, we have no evidence of any change of atmospheric emission from within the near field of the telescope occurring on this time scale.

b) Unequal Sensitivities of the Two Beams

If the sensitivity or sidelobe responses of the two beams are not equal, there will be (i) incomplete cancellation of the atmospheric emission occurring within the near fields of the beams, and (ii) residual sidelobe responses introduced into the restored dual-beam maps, spaced at intervals $\pm n\lambda_0 (n=1, 2, \ldots)$ from the main beam (λ_0 is the angular separation of the two far-field beams). The incomplete cancellation of the atmospheric emission (i) is equivalent to an additional factor in Q, in Eq. 9 and Fig. 11 discussed in (a) above. Without particular precautions the degradation due to this factor in the Effelsberg dual-beam system at 2.8 cm has been found not to exceed ∼1%, and is not of great importance. However the residual sidelobes (ii) are of more importance since the dynamic range of observations may be limited by this factor. A t% gain difference will result in 0.5 t% sidelobe amplitude being introduced. If the data from each beam are recorded separately, then a correction for each beam may be made off-line, in the form of a gain factor applied before subtracting the data in order to form the difference dual-beam map. If the shape of the two beams differs (e.g. as a result of coma sidelobes resulting from off-axis positioning of the feeds) no such simple correction is possible. In this case the conventional CLEAN technique (Högbom, 1974) might be applied in order to restore the beam-shapes to identical well-behaved functions, but the simplicity of the analysis technique and the uniqueness of the analytical solution would then be lost.

It has been found at Effelsberg that residual sidelobes after the restoration of a dual-beam observation to the equivalent single-beam are $\lesssim 1.5\%$, using the beam separations of up to 8.2 at a wavelength of 2.8 cm from the secondary (Gregorian) focus (see Fig. 7), although test observations made at 1.2 cm from the primary focus gave residuals as high as $\approx 10\%$, with a beam separation of 2.1. The dynamic range of the measurements can be increased by making observations at a number of different parallactic angles, and then simply averaging all the restored maps. With the Effelsberg secondary focus 2.8 cm system the two beams are separated in azimuth, and it is necessary to scan in an azimuth-elevation co-ordinate system, centred on the required source. If maps are made at a number of differing hour angles the parallactic angles and hence the scanning directions are necessarily different, and so this represents for dual-beam observations of extended sources a convenient way of extending the dynamic range beyond 20 dB. Figure 8 and Fig. 9 illustrate this point.

c) The Noise Level of the Restored Dual-beam data where the Field Extends over Many Units of the Beam Separation λ_0

As shown in Sect. III, for maps of size of up to three times the angular separation of the two beams, the necessary sampling interval of spatial frequency is such that the signal-to-noise ratio of any single required spatial frequency has not been degraded by more than a factor of 2 after restoration to the equivalent single-beam map. However, even in the absence of any perturbations due to a varying component of atmospheric emission, a map of this size observed with a dual-beam system and restored to the equivalent single-beam observation will possess a signal-to-noise ratio better than a single-beam observation made with a conventional Dicke-switching receiver. The noise dT from such a system is given, in the absence of contributions from the atmosphere, by the well-known relation

$$dT = \frac{2 \cdot T_{\text{sys}}}{\sqrt{B\tau}}. \tag{10}$$

The factor 2 for the Dicke system results from the fact that the sky is only observed for half of the time.

The noise on a restored dual-beam map may be derived directly from a consideration of the convolution function shown in Fig. 6(iii) which is used to analyse the data. The original dual-beam observations will possess a noise level and sensitivity identical to the simple load-switching system (neglecting any atmospheric contributions), since the load has been replaced by the extra telescope feed. Each point in the restored map is derived from a summation of n terms from the original data, since for each point there will be n components of the restoring convolution function spaced at intervals of λ_0, the beam separation, covering the map of size $n\lambda_0$. The resultant numerical noise value of the restored map will then be $\sqrt{n}$ larger, assuming all points in the original data to be independent. However the numerical amplitude due to emission from, for example, a point source will be doubled in the convolution, since such a source appears twice

(once with positive, once with negative amplitude) on the original data. Hence the final normalised noise level of the restored map will be $\sqrt{n}/2$ the original value; the noise expected on the equivalent single-beam map derived from the dual-beam data will be given by

$$dT = \frac{\sqrt{n}\, T_{\text{sys}}}{\sqrt{B\tau}}. \tag{11}$$

Equations (10) and (11) show that even without atmospheric perturbations for a field of up to 4 units of beam separation in size ($4\lambda_0$), observations with a dual-beam system should yield a superior signal-to-noise ratio, even after the re-weighting of the required spatial frequencies to correct for the spatial frequency amplitude and phase response of a dual-beam system. The improvement in signal-to-noise ratio for fields less than $4\lambda_0$ in size is partially offset by the need to map a slightly larger region for a given source size using the dual-beam system, in order to ensure that both beams are clear of emission from the source at the ends of the scan, but for sources of size $3\lambda_0$ and hence a map size of $4\lambda_0$ the resultant map made using the dual-beam technique should have a signal-to-noise no worse than one made with a conventional load-switching receiver in the absence of perturbations due to atmospheric emission. In the presence of atmospheric effects the noise on the single-beam observation will of course be substantially higher than of a dual-beam measurement restored to the equivalent single-beam observation.

Section V below describes an extension of the technique to a multi-beam system, such that much larger regions may be mapped without degradation of the noise level.

Note that both the dual-beam and single-beam load-switching systems could be used with a two-channel receiver, the two channels being switched in anti-phase between beam A and beam B, or between the single beam and load. In both systems this would result in an improvement of $\sqrt{2}$ in signal-to-noise ratio.

d) Observations of Strong Sources under Conditions of High Atmospheric Attenuation

As Eq. (2) shows, the dual-beam technique effectively cancels atmospheric emission, but not the effects of atmospheric absorption of the required signal. A uniform and steady absorbing layer would merely change the calibration scaling of the data, but if the atmospheric attenuation changes in a time shorter than that required for both beams to be scanned across a given point in the source, residual sidelobes will be introduced in the resultant map after correction to the equivalent single-beam observation. The sidelobes are similar to those introduced by unequal gains or dissimilar sidelobe patterns in the two beams, as discussed in (b) above. If $d\alpha$ is the component of the transmission coefficient α of the atmosphere which varies in a time λ_0/V, where λ_0 is the angular separation of the two beams and V the telescope scanning speed then these sidelobes will be of amplitude $\approx \pm (d\alpha T_s)/2$. Since the atmospheric attenuation is continually changing, there will be no correlation of these spurious effects in adjacent points in the map, and the overall effect will be an additional component of random noise $(d\alpha T_s)/2$ which may be reduced by averaging several coverages of the same field. Although this is unlikely to be of importance at cm wavelengths, this additional component of noise may become relevant at mm or IR wavelengths. Note that $d\alpha$ and hence the noise component $(d\alpha T_s)/2$ could also be considerably reduced by a rapid-scanning technique, such that (λ_0/V) is made to be substantially less than the time-scale for significant changes in α.

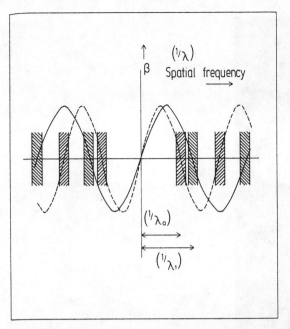

Fig. 14. The 2 sine waves, of period $2/\lambda_0$ and $2/\lambda_1$, represent the spatial frequency response of the dual-beam function [Fig. 5(iii) and 5(vii)] for an angular beam separation of λ_0 and λ_1, respectively. In order to avoid significant degradation of S/N ratio, spatial frequencies in the regions shown shaded should not be sampled and re-weighted. For observations using one beam-separation λ_0, this places a constraint on the maximum size of field which can be mapped for a given noise level. Using non-commensurate spacings of λ_0 and λ_1, effectively removes this constraint, since spatial frequencies (e.g. n/λ_0) absent in the first observation are measured in the second, and vice versa

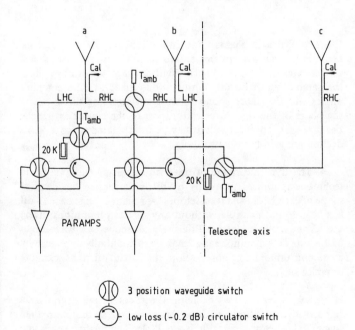

⨀ 3 position waveguide switch

⨁ low loss (~0.2 dB) circulator switch

Fig. 15. A block diagram of the triple horn system which has come into use at Effelsberg. The feed horns are mounted at the secondary (Gregorian) focus of the telescope, and operate at a wavelength of 2.8 cm. The system noise is ~80 K, and the receiver bandwidth 500 MHz

V. Extension of the Dual-beam Technique to a Multiple-beam System

As discussed in Sect. III, it is sufficient to sample the spatial frequencies representing the emission from a limited region of sky at regularly spaced spatial frequencies $(1/\lambda)$ given by:

$$(1/\lambda) = \pm \frac{m}{2x} \ (m = 1, 3, 5, \ldots), \tag{12}$$

where the field mapped is of total extent x. Spatial frequencies close to $(1/\lambda) = \pm n/\lambda_0 (n = 1, 2, 3, \ldots)$ are severely attenuated by the dual-beam observation and in the presence of noise it is undesirable to sample spatial frequencies close to these points during the restoration to the equivalent single-beam observation. This places constraints on the minimum sampling interval $1/x$ of spatial frequencies, and hence on the maximum extent x of a region which can be successfully mapped. This has already been discussed in Sect. IV.

If additional observations are made with a different beam separation λ_1, such that within the range of spatial frequencies measured by the single dish telescope those frequencies close to $\pm(p/\lambda_1) \ (p = 1, 2, 3, \ldots)$ are not coincident with those at $\pm n/\lambda_0$ $(n = 1, 2, 3, \ldots)$ measured using the original beam separation λ_0 then all spatial frequencies except those close to zero will be well defined. This is illustrated in Fig. 14. Using only those spatial frequencies which have not been reduced to less than 0.5 of their original amplitude by the 2-beam observation, it now becomes possible to sample at regular intervals such that

$$\beta(1/\lambda) = \sin\left(\frac{\pm m\lambda_i \pi}{\lambda}\right) \geq 0.5 \tag{13}$$

$$m = 1, 2, 3, \ldots$$

where β is the weighting factor for spatial frequencies $(1/\lambda)$ introduced by the dual-beam observation, and λ_i is the larger of the two (or more) beam separations λ_0, λ_1 etc. Equation (13) is satisfied if $1/\lambda \leq 1/6\lambda_i$; both the lowest spatial frequency used and the spatial frequency sampling interval may be as small as $1/6\lambda_i$. This implies that a source of total extent $6\lambda_i$ may be unambiguously mapped without serious increase of the noise of any required component of spatial frequency, and that only a slight increase in noise should be expected for even larger fields. It should be noted that it is not necessary, nor is it desirable, that a direct measurement be made of the zero frequency term. This zero term includes for example the 3 K background radiation, but it is usual and preferable to refer brightness temperatures on a radio map of a confined source to an assumed zero level at the edge of the field mapped. This is usually done even with single-beam observations, thereby filtering out the zero spatial frequency term from the data.

A system using 3 horn feeds has been constructed for use from the secondary focus of the Effelsberg telescope at a wavelength of 2.8 cm (see Fig. 15 and Plate I). With this configuration three different beam separations $(a-b, b-c, a-c)$ may be used simultaneously. Major continuum observing programs using this system at Effelsberg began at the end of 1978.

VI. Conclusions

We have presented a dual-beam technique for observing extended radio sources under weather conditions which would be entirely unacceptable using a conventional single-beam system. Examples of observations made using this technique with the Effelsberg

Plate 1. The 3-horn system, which was installed at Effelsberg during July 1978

100-m telescope at a wavelength of 2.8 cm have been shown. Sources of angular extent several times greater than the separation of the two beams may be mapped without difficulty; all necessary spatial frequencies are adequately sampled.

The most important practical points of the technique are summarised as follows:

1. In mapping an extended source of a given size, best results will be obtained with as large an angular separation of the two beams as possible subject to (a) there still being adequate cancellation of the atmospheric emission (see Sect. IV), and (b) no serious coma sidelobes being introduced by the off-axis positioning of the feeds. In respect of (b), using separated feeds at the secondary focus of the 100-m telescope (effective focal length 378 m) has been found in practice to be significantly better than using the primary focus position (focal length 30 m).

2. After restoration to the equivalent single-beam observation, the signal-to-noise ratio of dual-beam observation of a source extending 3 or 4 times the angular separation of the two beams is identical to that obtained with a single-beam load-switching system in the absence of perturbations due to the atmosphere. For dual-beam observations of larger areas, the noise on the restored dual-beam map will increase as $\sqrt{n}$, where n is the size of the map along the scanning direction in units of the beam separation.

3. The optimum scanning direction when mapping a two-dimensional field is always along the direction of the line joining the two far-field beams. For the Effelsberg 100-m telescope secondary focus system this entails scanning in azimuth, which also ensures that the total path length through the atmosphere is the same in both beams. Observations made at different hour angles will then correspond to different scanning directions in the plane of the sky. After restoration to the equivalent single-beam map, each observation may be interpolated on to a R.A.-declination grid before averaging.

4. The dynamic range of observations using the dual-beam technique is limited primarily by dissimilarities in the detailed shape of each beam and its sidelobes. A dynamic range of ≈ 18 dB has been found possible, without any special precautions, using the secondary focus of the Effelsberg telescope at a wavelength of 2.8 cm. This dynamic range may be substantially improved by averaging a number of observations made with different scanning directions.

5. Larger regions may be mapped without serious degradation in noise level by using a multiple-beam system, so that more than one angular beam separation is available. The relative beam separations should be chosen such that the nulls in the spatial frequency responses of each dual-beam combination do not coincide. The sets of data are then combined by weighting spatial frequency from each dual-beam combination according to relative singal-to-noise ratio.

Acknowledgements. This development project has required the strong support of the aerial, electronic, mechanical, and telescope divisions of the MPIfR. In particular we should like to thank W. Zinz, G. Illigen, B. H. Grahl, R. Wohlleben, the brothers Kastenholz, and W. Wilson for the work in providing several different test receiver configurations required to evaluate the method. The resulting three horn design shown in Fig. 15 is due to W. Wilson and a new system using two cooled parametric amplifiers was built by the MPIfR workshops and installed for comissioning tests in July 1978. We thank Prof. R. Wielebinski and many other members of his group for continued encouragement and for helpful discussions during the course of this work.

References

Baars, J.W.M.: 1966, *Nature* **212**, 495
Baars, J.W.M.: 1970, Ph.D. thesis, Groningen
Becklin, E.E., Neugebauer, G.: 1975, *Astrophys. J.* **200**, L71
Bracewell, R.N.: 1956, *Australian J. Phys.* **9**, 297
Conway, R.G., Daintree, E.J., Long, R.J.: 1965, *Monthly Notices Roy. Astron. Soc.* **131**, 159
Högbom, J.A.: 1974, *Astron. Astrophys. Suppl.* **15**, 417
Klein, U., Emerson, D.T., Haslam, C.G.T., Salter, C.J.: 1979, *Astron. Astrophys.* (in press)
Pauls, T., Downes, D., Mezger, P.G., Churchwell, E.: 1976, *Astron. Astrophys.* **46**, 407

A Continuous Comparison Radiometer at 97 GHz

C. READ PREDMORE, MEMBER, IEEE, NEAL R. ERICKSON, G. RICHARD HUGUENIN, MEMBER, IEEE, AND PAUL F. GOLDSMITH, MEMBER, IEEE

Abstract — A continuous comparison radiometer has been implemented at 97 GHz using quasi-optical techniques for local oscillator (LO) injection and realization of a 90° hybrid. Cryogenically cooled Schottky-diode mixers and FET amplifiers give a double-sideband (DSB) system temperature of 250 K. The system is self-calibrating and optimized under computer control. The root-mean-square (rms) fluctuations due to the receiver are less than 0.025 K with a 1-s integration time.

I. INTRODUCTION

A HIGH-SENSITIVITY, continuous comparison radiometer has been implemented by the Five College Radio Astronomy Observatory. The continuous comparison or correlation radiometer circuit is over twenty years old [1] and has been used for radio astronomical observations at 74 cm [2], 11 cm [3], and 6 cm [4]. Its use at millimeter wavelengths has been precluded due to the losses and imperfections in the available waveguide components. The extensive use of quasi-optical techniques for beam guidance, local oscillator (LO) injections, and a combination of the two beams in an input hybrid has resulted in major performance gains for the system described here.

The classical problem in radiometry is to distinguish a weak source in the presence of the much greater noise from the receiver and atmosphere and in the presence of fluctuations in receiver gain and atmospheric emission. For an astronomical object such as a quasar, this is accomplished by actually moving the telescope between the source position and an adjacent position at the same elevation at a 0.05–0.10-Hz rate, deflecting the beam by nutating the subreflector at a 0.5–5-Hz rate [5], Dicke switching [6] against a load, or deflecting the beam near the Cassegrain focal plane with a rotating beam chopper at up to 50 Hz [7].

In contrast, the continuous comparison radiometer has no moving parts for switching, so that the complexities of mechanical switching systems mentioned above, and the significant loss and mismatch of a ferrite switch, are avoided. Rather, two beams are continually subtracted to give the desired difference signal with an equivalent time constant which can be short as the reciprocal of the inter-mediate frequency (IF) bandwidth. Since the subtraction is performed continuously, fluctuations with very short time scales that are common to both beams are cancelled perfectly. In addition, the subtraction in the radiometer is insensitive to receiver gain changes.

The present system employs a LO frequency of 97 GHz and is sensitive to input signals in both sidebands with a separation from the LO of 4.4 to 5.0 GHz. The radiometer consists of a calibration system, cryogenic front end, and IF and signal processing units. The entire system is automatically optimized, calibrated, and operated by a computer. The root-mean-square (rms) sensitivity due to the receiver is less than 0.025 K with a 1-s integration time.

II. BASIC THEORY

The radiometer block diagram is shown in Fig. 1. The two input beams with temperature T_A and T_B as well as the 97.3-GHz LO are split by a quasi-optical 90° hybrid and double-sideband (DSB) mixed to 4.7 GHz. At 4.7 GHz, the voltages in the two IF amplifiers are proportional to $A - jB$ and $A + jB$, where A and B are the voltages of the inputs to the Dewar. The second frequency conversion is from 4.7 to 2.1 GHz using a 6.8-GHz LO. One of the LO lines has a 6-bit computer-controlled phase shifter with a 5.6° resolution to optimize the system sensitivity. Also under computer control are 0–50-dB p-i-n attenuators in each of the 2.1-GHz IF chains. Precision IF 180° hybrid, matched detectors, and an instrumentation operational amplifier are used to multiply the two IF signals.

The resultant response of the system is

$$V_{\text{DIFF}} = C_{\text{DIFF}}(T_A - T_B)\cos\phi \tag{1}$$

where C_{DIFF} is a calibration constant and ϕ is the phase of the LO phase shifter. Thus, the system is insensitive to gain fluctuations and is responsive only to the input temperature difference. A manual delay line with a 0.2-ns range is used to compensate for phase slopes in the cabling and amplifiers. The entire system is automatically optimized and calibrated with a ModComp MODACS computer.

Faris [1] has done an extensive analysis of correlation radiometers including the effects of nonidentical amplifiers, gain fluctuations, and differential phase and delay. Here, the system response will be analyzed for a single IF frequency to investigate the effect of amplitude imbalance in the input 90° hybrid and imperfect balance in the final IF detectors.

Manuscript received December 19, 1983; revised August 2, 1984. This work was supported in part by the M.I.T. Lincoln Laboratory under Contract F19628-80-C-0002.

The authors are with the Five College Radio Astronomy Observatory, University of Massachusetts, Amherst, MA 01003.

Reprinted from *IEEE Trans. Microwave Theory Tech.*, vol. MTT-33, no. 1, pp. 44–51, Jan. 1985.

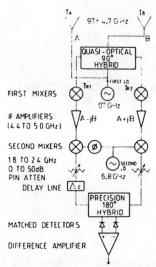

Fig. 1. Radiometer block diagram. The two inputs at temperatures of T_A and T_B are split and combined in a quasi-optical hybrid. After mixing at 97 GHz, they are amplified and phase shifted to optimize the response. A precision 180° hybrid and matched detectors are used to "multiply" the two IF signals.

The amplitude of the incoming signal in the upper sideband (USB) and in the lower sideband (LSB) is proportional to the square root of the input temperature. In the following equations, the unprimed quantities are from the USB and primed quantities from the LSB. The complex conjugate is represented by the symbol (*). The input temperatures in the signal and reference beams are denoted by T_A and T_B, respectively, and the two receiver temperatures are T_{R1} and T_{R2}. The signal amplitudes are A and A' and the reference amplitudes B and B', in the two sidebands. The amplitudes for the receiver noise in channels 1 and 2 are R_1 and R'_1 and R_2 and R'_2, respectively. These amplitudes are related to the temperatures by

$$T_A = A \times A^* + (A') \times (A')^* \quad (2a)$$

$$T_B = B \times B^* + (B') \times (B')^* \quad (2b)$$

$$T_{R1} = R_1 \times R^*_1 + (R'_1) \times (R'_1)^* \quad (2c)$$

$$T_{R2} = R_2 \times R^*_2 + (R'_2) \times (R'_2)^*. \quad (2d)$$

In general, the 90° beam-splitter hybrid does not split the signals equally but has amplitudes of ρ for reflection and τ for transmission, which are assumed to be the same for both sidebands. Depending on how the LO is injected, the continuous comparison radiometer can be realized with a 180 or 90° input hybrid. After going through the quasi-optical hybrid, the signal amplitudes are

$$V_1 = j\exp(j\omega_{LO}t)\{1 + (\rho A - j\tau B + R_1)\exp(j\omega_1 t)$$
$$+ (\rho A' - j\tau B' + R'_1)\exp(-j\omega_1 t)\} \quad (3a)$$

$$V_2 = \exp(j\omega_{LO}t)\{1 + (\tau A + j\rho B + R_2)\exp(j\omega_1 t)$$
$$+ (\tau A' + j\rho B' + R'_2)\exp(-j\omega_1 t)\} \quad (3b)$$

where j is the square root of (-1), ω_{LO} the angular frequency of the 97-GHz LO, ω_1 the first IF angular frequency, and t is time. The signal level of the 97-GHz LO is represented by the first term in the expressions for

V_1 and V_2. It is arbitrarily set to 1 to show the phase of the LO. The second and third terms are the upper and lower sidebands, respectively.

After double-sideband mixing with the 97-GHz LO, gain at the first IF and LSB mixing from 4.7 (ω_1) to 2.1 GHz (ω_2) with a 6.8-GHz second LO, the voltages are

$$V_3 = \exp(j\phi)G_1\{(\rho A - j\tau B + R_1)\exp(-j\omega_2 t)$$
$$+ (\rho A' - j\tau B' + R'_1)\exp(j\omega_2 t)\} \quad (4a)$$

$$V_4 = G_2\{(\tau A + j\rho B + R_2)\exp(-j\omega_2 t)$$
$$+ (\tau A' + j\rho B' + R'_2)\exp(j\omega_2 t)\} \quad (4b)$$

where the term $\exp(j\phi)$ is the relative phase shift introduced by the 6.8-GHz phase shifter and G_1 and G_2 are the voltage gains in the two channels.

At the output of the precision 180° IF hybrid, the voltages are

$$V_5 = (V_3 + V_4)\sqrt{2} \quad (5a)$$

$$V_6 = (V_3 - V_4)\sqrt{2}. \quad (5b)$$

The detected power is obtained by multiplying V_5 and V_6 by their complex conjugates and low-pass filtering to remove harmonics of the IF components. Cross products such as $A \times B$ and $R_1 \times R_2$ will average out to zero when the inputs are uncorrelated and there is no correlated noise in the two receiver temperatures. In the following discussion, K and $K(1-\delta)$ represent the product of the detector sensitivities and difference amplifier gains for the two detected outputs, where the magnitude of (δ) is less than 0.05. In this case, the voltage, of the two outputs are

$$V_7 = [K/2]\{T_A[(\rho G_1)^2 + (\tau G_2)^2 + (2\rho\tau G_1 G_2)\cos(\phi)]$$
$$+ T_B[(\tau G_1)^2 + (\rho G_2)^2 - (2\rho\tau G_1 G_2)\cos(\phi)]$$
$$+ T_{R1}(G_1)^2 + T_{R2}(G_2)^2 \quad (6a)$$

$$V_8 = [K(1-\delta)/2]$$
$$\cdot \{T_A[(\rho G_1)^2 + (\tau G_2)^2 - (2\rho\tau G_1 G_2)\cos(\phi)]$$
$$+ T_B[(\tau G_1)^2 + (\rho G_2)^2 + (2\rho\tau G_1 G_2)\cos(\phi)]$$
$$+ T_{R1}(G_1)^2 + T_{R2}(G_2)^2\}. \quad (6b)$$

Then, the output of the radiometer is the difference of V_7 and V_8

$$V_{out} = K\{[(2\rho\tau G_1 G_2)(T_A - T_B)\cos(\phi)]$$
$$+ (\delta/2)\{T_A[(\rho G_1)^2 + (\tau G_2)^2]$$
$$+ T_B[(\tau G_1)^2 + (\rho G_2)^2]]$$
$$+ T_{R1}(G_1)^2 + T_{R2}(G_2)^2\}. \quad (7)$$

The quartz beam splitter splits the signal with 54 percent of the incident power being reflected and 46 percent transmitted. This gives values for ρ and τ of 0.750 and 0.667, so that $(2\rho\tau)$ equals 0.996. When the two IF gains are equal, as is the case for normal operation, the system output is proportional to the input temperature difference plus a fractional offset term due to the detector imbalance. In this case, the output is given by

$$V_{out} = (KG_1G_2)\{(T_A - T_B)\cos(\phi)$$
$$+ (\delta)[(T_A + T_B)/2 + (T_{R1} + T_{R2})/2]\}. \quad (8)$$

With the precision IF hybrid and careful matching of the detectors, δ can be reduced to less than 0.001. This will give an offset on the order of 0.5 K, which can be measured by cycling through the second LO phase ϕ, or can be subtracted out by observing the source alternately in beam A and beam B.

III. FRONT-END SYSTEM

The two input beams are 51 mm apart at the Cassegrain focus of a 13.7-m-diam telescope with a f/d ratio of 4.0. This gives two 1.0'-diam beams which are 3.3' apart in azimuth on the sky. This close beam separation was chosen to optimize the cancellation of atmospheric effects. A wider beam separation could be designed if broader sources were to be mapped. The inputs are linearly polarized with the two polarizations 110° apart.

As is schematically shown in Fig. 2, the front-end portion of the system consists of the input optics, the scalar feeds, the cryogenic millimeter mixers, and the IF amplifiers. All of these are integrated into the cryogenics Dewar which is $250 \times 300 \times 400$ mm in size. The top and bottom covers have O-ring and RF shielding grooves. This design has given a very reliable system. The cooling to 15 and 77 K is done with a CTI 350CP closed-cycle helium refrigerator. A 0.8-mm-thick aluminum shield is attached to the 77 K station of the refrigerator to minimize the thermal radiation loading on the 15 K components. Part of this heat shield is lined with a microwave absorber to act as a black body at 77 K for the LO signal which is not coupled into the mixers and to terminate any reflections and spillover in the optics at 77 K.

As is described in detail in the next section, the optics take the beams at the Cassegrain focus of the antenna, expand them so that the LO can be injected, and combine the two beams in the quasi-optical 90° hybrid. The beams are then refocused to match into the scalar feeds.

The 97.3-GHz LO is provided by a Gunn-diode oscillator having 10 mW of power. The output of the oscillator is isolated and attenuated before going through a waveguide vacuum feedthrough. Inside the Dewar, the LO is coupled into the optics with a rectangular horn, a rexolite focusing lens, and two flat mirrors.

Double-sideband mixing is done in a pair of Schottky-diode millimeter mixers which are cooled to 15 K. The mixers have been developed at the FCRAO and have broad-band RF filters and noncontacting backshorts. They utilize GaAs diodes which have a low doping of 3×10^{16} cm^{-3} to give an optimum performance when cryogenically cooled [8], [9]. The mixer blocks are machined from OFHC copper to minimize their RF losses and have a linear taper from full- to 1/4-height waveguide. The RF filter has been designed to present a short to the diode at 97 GHz and is reactive at the second harmonic of the LO to minimize the

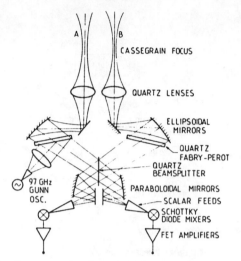

Fig. 2. Radiometer optics diagram. The symmetrical system uses fused-silica lenses to focus the inputs from the Cassegrain focus into the cryogenic Dewar. A thin sheet of fused-silica acts as a 90° hybrid. Then the 97-GHz LO is injected via a dielectric Fabry–Perot and the beams are refocused into the millimeter mixers.

conversion to higher harmonics. The noncontacting backshort has also been especially designed [10] to be a short circuit at 97 GHz and a pure reactance at 194 GHz. The backshort position was optimized at room temperature and locked into place with a setscrew before cooling the system to 15 K. The whisker length has been optimized since its length tunes the mixer response versus the input frequency [8].

Cooled circulators and FET amplifiers give an IF noise temperature of 25–30 K over the 4.4–5.0-GHz band. Each IF chain consists of an input isolator [12], FET amplifier [13], and output isolator at 20 K, providing a net gain of 11 dB. A second FET with 12 dB more gain is mounted on the 77 K station of the closed-cycle helium refrigerator. The circulators and amplifiers which were made at the FCRAO are followed by commercial amplifiers at 300 K, which are mounted within the vacuum chamber. Their 55 dB of gain gives a net RF to IF gain of 80 dB before the signals leave the RF-shielded Dewar. These signals are then processed as discussed in Section V.

IV. OPTICS

The input optics to the radiometer serve multiple purposes. They

1) illuminate the subreflector of the 14-m-diam telescope with a 12-dB edge taper,
2) inject the local oscillator power at 97.3 GHz,
3) combine the input beams in such a way as to create a quasi-optical 90° hybrid for the two sidebands at 93 and 102 GHz.

The optics block diagram is shown in Fig. 2. The photograph in Fig. 3 shows the front end with the vacuum Dewar removed. This view shows the optics after the fused-silica lenses. Fig. 4 shows the beam propagation in one plane so that the various optical elements and the Gaussian beam can be accurately displayed. In the actual

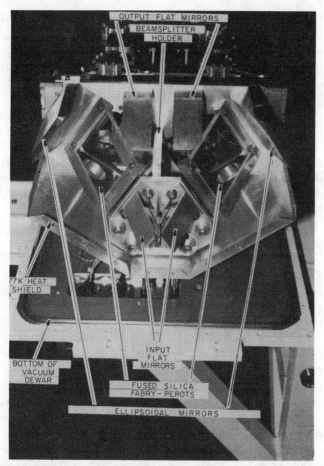

Fig. 3. Front-view of front-end system. The input flat mirrors at the lower center divert the two input beams up to the ellipsoidal mirrors at the upper left and right of the optics. The dielectric Fabry–Perots are just below the ellipsoidal mirrors. The local oscillator is injected from behind the left Fabry–Perot. The beam-splitter holder is indicated. The paraboloidal mirrors are behind the input flat mirrors. The scalar feed horns and the mixers are behind the output flat mirrors.

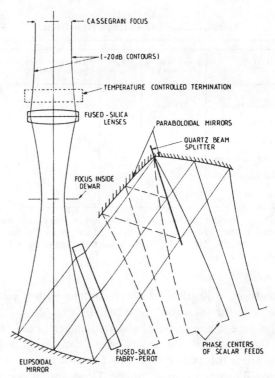

Fig. 4. Gaussian beam propagation within optics. The beam is traced through the optics at the −20-dB contour level. All of the elements are shown in one plane even though the beam path is folded in the system for compactness.

receiver, the optical path is folded with two sets of flat mirrors to condense the optics into a volume of 8l. This is 3×10^5 cubic wavelengths at 97.3 GHz and is a reasonable minimum size for a two-channel system with LO injection and a beam splitter.

The input optics comprise both LO injection and a 90° hybrid beam combiner. Their operation is explained with reference to Fig. 2. The input signals are beams A and B. The LO power is injected into beam A only by using a dielectric Fabrey–Perot interferometer (FPI), which has a signal transmission of > 95 percent and a LO reflectivity of 30 percent. Then beams A and B are split in a dielectric beam splitter which produces approximately equal amplitudes in reflection and transmission. A general property of a lossless symmetric beam splitter is that the reflected and transmitted waves differ in phase by 90°. Assuming all input phases are 0 (this phase is arbitrary), the outputs of this splitter are

$$A/2 + jB/2 + LO/2$$
$$j(A/2 - jB/2 + LO/2).$$

This method of injecting the LO before the splitter has several advantages critical to the simplicity and stability of

the system. One is that no additional power splitter is needed to divide the LO for the mixers, and only one FPI plate is needed. However, a second FPI plate is included in the optics to preserve the system symmetry. More important is the advantage in terms of phase stability. In this method, no dimensional variations before the beam-splitter hybrid have any effect on phase stability. After the beam splitter, the two paths to the mixer must be equal and stable in length only to within a small fraction of the IF wavelength (63 mm). In a system in which the LO is independently split, phase equality of the two LO paths is critical, as is any phase drift in the LO injection filter.

The only disadvantage of this approach is that any AM noise on the LO, offset from the carrier, is mixed down to 4.7 GHz and appears as a difference signal. This will be attenuated by the rejection of the FPI plate, which is about 8 dB. In the case of this radiometer, the imbalance term is less than 2 K and can be measured in the calibration procedure. Should stability of this noise be a concern, it can be further reduced by a cavity filter in the LO waveguide.

Table I summarizes the parameters of the Gaussian beams at various points in the optics. Included are the beam diameter at a −20-dB power level, the characteristic length Z_c for the beam, the distance to the next element in the optics and the focal lengths of the lenses, and two sets of mirrors. This system was designed to truncate the Gaussian beams at the −20-dB level to minimize the volume of the system with a loss of about 2 percent at each truncation. Normally, one would design the system to

TABLE I
GAUSSIAN BEAM PARAMETERS WITHIN OPTICS

Description	Beam Diameter (-20 dB)	Z_c	Distance To Next Element	Focal Length
Cassegrain Focus	28 mm	87 mm	72 mm	---
Fused Silica Lenses	36	---	63	57 mm
Focus Inside Dewar	18	36	115	---
Ellipsoidal Mirror	60	---	180	106
Paraboloidal Mirror	52	291	121	157
Scalar Feed	24	65	---	---

Given at various positions within the optics system are the beam diameter at the -20-dB level, the confocal parameter Z_c, the spacing between elements, and the focal lengths for the lenses and mirrors in the system.

truncate at -30 dB or a lower contour with a resultant increase in the volume of the system by a factor of 2.

The acceptance angle of the radiometer optics is such that the Cassegrain secondary edge illumination is 12 dB down from the axial power. The focus of the telescope is located 72 mm in front of the Dewar windows. As is illustrated in Fig. 4 and noted in Table I, the beam begins to diverge again, passes through the calibration wheel, and into the Dewar through the quartz (fused silica) lenses. Each lens is made up of a pair of plano-convex lenses placed back to back, with 0.5-mm-thick polyethylene anti-reflection coatings fused to the curved surfaces. The resulting lens has a focal length of 57 mm. This lens brings the beam to a second focus 63 mm inside the Dewar.

The beam then diverges to fill the ellipsoidal mirrors. The ellipsoidal mirrors are sections of an ellipse of revolution such that the distances to the two foci of the ellipse are 150 and 650 mm, and the angle subtended by the two foci is $37°$. The distances of 150 and 650 mm are the radii of curvature of the input and output Gaussian beams at the reflecting surface. The radius of curvature of a Gaussian beam at a distance Z from the beam waist is [11]

$$R(Z) = Z\left[1 + (Z_c/Z)^2\right] \qquad (9)$$

where Z_c is given in Table I. The output beam is approximately plane parallel, as indicated by the large Z_c of 291 mm, for minimum loss in the dielectric Fabry–Perot.

The 8.6-mm-thick fused-silica dielectric slab is used as a tuned filter to inject the 97.3-GHz LO into the two mixers. The plates are tilted at $32°$ with respect to normal incidence. This slab transmits the two sidebands with very low loss and reflects about 30 percent of the power at the LO frequency [14], [15]. Only one of these plates is actually used for LO injection; the other is necessary to preserve system balance. The theoretical walkoff loss is less than 5 percent [16], with a measured limit of less than 10 percent.

The functioning of the offset paraboloidal mirrors is best explained by considering the 24-mm beam-waist diameter at the -20-dB contour generated by the scalar feedhorns. The Gaussian beam from the scalar feed has a radius of curvature of 156 mm at the paraboloidal mirror, and the focus of the parabola of revolution is at the center of curvature of the incident wave. This causes an approximately plane-parallel beam to be developed between the paraboloidal and ellipsoidal mirrors. The beam waist is located near the surface of the paraboloidal mirror.

The ellipsoidal and paraboloidal mirrors were machined on a conventional milling machine [17] from aluminum and then hand polished. These mirrors were then rigidly attached to a 6-mm-thick aluminum plate. The only adjustments in the optics were the exact focal position of the scalar feeds and the position and angle of the flat mirrors which are just behind the input lenses.

V. IF PROCESSING

The two 4.4–5.0-GHz IF signals from the front-end system are processed with a computer-controlled IF system to automatically calibrate and optimize the radiometer and to scale the difference output. As was shown in Fig. 1, both channels are converted from 4.7 to 2.1 GHz with the same 6.8-GHz LO. The 6.8-GHz LO is split with a zero-degree power splitter. One of the LO lines to the double-balanced mixers has a digitally controlled 6-bit phase shifter with a 0–$354°$ range and a $5.6°$ resolution. The LO lines to the two mixers have isolators to minimize coupling between the two channels via the common LO system. The output of each mixer has a linear-phase bandpass filter with a 600-MHz 1-dB bandwidth, centered at 2.1 GHz. These are followed by 0–55-dB p-i-n attenuators which are also under computer control. Their normal attenuation is about 10 dB to minimize their phase shift versus attenuation. The last amplifiers are set to give a output power level of -20 dBm in the 600-MHz wide band. The two IF chains have selected pairs of filters and amplifiers with matched phase and delay responses. Any residual delay is compensated by a delay line or specially made cables. The power level at the output of each IF chain is monitored to set the p-i-n attenuators and to calibrate the receiver temperature as described in the next section.

The two IF signals are combined in a precision IF hybrid with 45 dB of isolation between ports. The low-barrier Schottky detector diodes have an output load resistor chosen for best square-law response at -20 dBm. Once the detectors are temperature stabilized to $\pm 1°$C, the system output is nulled by changing the sensitivity of one of the two detectors. This part of the system has the largest effect on the system balance.

The difference amplifier has a RC time constant of 2.7 ms. Its output is sampled at a 1-KHz rate with a 12-bit analog-to-digital converter. The IF and dc gains are adjusted so that the output range is ± 250 K, giving a resolution of 0.12 K. This is to be compared with the output noise of > 0.28 K in 2.7 ms. The output noise fluctuation has been derived by Faris [1]

$$\Delta T_{rms} = \left[2/(\tau_i \mathrm{BW})\right]^{1/2}\left[(T_{R1} + T_{R2})/2 + (T_A + T_B)/2\right]$$
$$\cdot \left[\sin(\pi \mathrm{BW}\tau_d)/(\pi \mathrm{BW}\tau_d)\right]\cos(\omega_2\tau_d + \phi) \qquad (10)$$

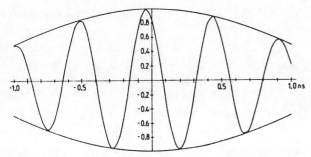

Fig. 5. Phase and delay response of system. The broad envelope is determined by the $\sin(X)/X$ response due to the final IF bandwidth of 600 MHz. The rapid cosinusoidal variation within this envelope is due to delay differences between the two IF's and can be optimized with the LO phase shifter.

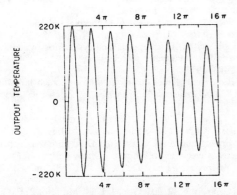

Fig. 6. System response versus LO phase and IF gain. The system output for an input temperature difference of 220 K is shown versus LO phase. The system was calibrated for the first cycle of 0 to 2 π. One of the IF gains was reduced in 0.4-dB steps in each subsequent 2 π interval.

where BW is the 600-MHz IF bandwidth, τ_i the output integration time, τ_d the time-delay difference between the two channels, ω_2 the second IF angular frequency, and ϕ the relative LO phase. The effect of the last two terms of (10) is illustrated in Fig. 5 for delay errors up to 1 ns. The envelope is the $\sin(x)/(x)$ response due to the 600-MHz IF bandwidth. In the actual system, the delay was adjusted so that the sensitivity loss from this effect was less than 5 percent. The cosine term is then automatically optimized during the calibration procedure to be greater than 0.99 by adjusting ϕ. For average receiver and input temperatures of 250 and 77 K, the expected rms is 0.019 K in 1 s. The measured output fluctuations are 0.033 K rms with the fraction due to the receiver noise being 0.025 K.

The output response is given by (8) and is shown in Fig. 6 for one input terminated at room temperature (295 K) and the other at liquid nitrogen temperature (77 K). The output temperature scale is determined by the 218-K temperature difference and was set with the initial IF gains (G_1 and G_2). As the LO phase (ϕ of Fig. 1 and (1) and (8)) is varied over 2π, the output varies between ±220 K because of the cosinusoidal dependence on ϕ. The following cycles of 2π had G_1 (see (8)) reduced in 0.4-dB steps. Thus, the system does not require balanced IF gains since changing the gain 3 dB only changes the calibration by a factor of 0.7. In this case, the imbalance δ (8) was about 0.03, so

TABLE II
CALIBRATION WHEEL POSITIONS

Position	Channel A	Channel B
1	318 K	318 K
2	Sky	Sky
3	Sky	323 K
4	323 K	313 K
5	313 K	Sky

The inputs to channels *A* and *B* are shown for the 5 different positions. The inputs either look at the sky or at temperature-controlled terminations.

that the cycles are only approximately symmetrical about zero.

VI. CALIBRATION PROCEDURE

An essential part of the radiometer is the integral calibration system. This allows the output to be scaled into temperature, and also the receiver temperature and the sky contribution to the system temperature to be measured. This is accomplished with four temperature-controlled loads which are cast from Emerson–Cuming CR-117. Two of these loads are stabilized within 0.1°C of each other at 45°C (318 K). The other pair are maintained at 313 K and 323 K to give a known temperature difference for calibrating the output of the whole radiometer.

The calibration wheel has five different position angles which are under computer control. Table II lists the terminations which are observed by inputs *A* and *B* for each of the five positions of the wheel. Position 1 gives a balanced input with two loads at 318 K to calibrate the zero of the difference output. Position 2 has two open ports for normal observing. Position 3 has an open on input *A* and a load at 323 K on input *B*. Position 4 has a load at 323 K on input *A* and 313 K on input *B* to calibrate the output temperature scale. Position 5 has a load at 313 K on input *A* and an open on input *B*.

The amplitude and phase of the sinusoidal response of the system output is found with a multiple linear regression routine in the MODACS computer. Offsets in the detectors and dc amplifiers are measured by setting the p-i-n attenuators to their maximum attenuation of 55 dB. Position 5, with its large temperature difference between the sky at a typical temperature of 100 K and the 313-K load, is used to find the phase corresponding to the peak response from the difference amplifier. The amplitudes [$A(3)$, $A(4)$, $A(5)$] of the IF output versus LO phase at wheel Positions 3, 4, and 5 are used to measure the sky temperature from the known temperature difference of 10 K

$$T_{SKY} = (313 + 323)/2 - 10[A(3) + A(5)]/[2A(4)].$$

$$(11)$$

The 6.8-GHz LO phase shifter (ϕ of Fig. 1 and (1) and (8)) is then commanded to the phase found at wheel Position 5 and the outputs of power detectors in each IF chain, just before the 180° hybrid, are measured in wheel Positions 1 and 2. This gives a Y-factor for each channel with the hot and cold temperatures supplied by the 318 K loads and the sky, respectively. The system temperature measured for each channel in this manner is 250 K outside the Dewar including the window and optics losses.

VII. Conclusions

A continuous comparison radiometer has been implemented at a wavelength of 3 mm having a sensitivity of 0.025 K in 1 s. Its major features include

1) an integral calibration system to determine the temperature scale, sky contribution, and receiver temperature,

2) a dielectric Fabry–Perot interferometer used for LO injection,

3) quasioptical 90° input hybrid,

4) optics integrated into the vacuum and cryogenics system to minimize the system noise temperature and the system volume of the front end,

5) cryogenic mixers and FET amplifiers with DSB reciever temperatures of less than 250 K,

6) separate delay and phase optimization,

7) precision IF hybrid and matched detectors,

8) a computer system to optimize and calibrate the radiometer.

Acknowledgment

The authors acknowledge the help of the entire staff of the Five College Radio Astronomy Observatory in making this project successful. The Five College Radio Astronomy Observatory is operated with the permission of the Metropolitan District Commission, Commonwealth of Massachusetts. This is contribution number 591 of the Five College Astronomy Department. One of the authors (C.R.P.) thanks the Institut de Radio Astronomie Millimetrique, Grenoble, France, where part of the writing was done, and the University of Massachusetts for a professional improvement leave.

References

[1] J. J. Faris, "Sensitivity of a correlation radiometer," *J. Res. Nat. Bur. Stand., Sect. 71C (Engr. and Instr.)*, pp. 153–170, Apr.–June 1967.

[2] C. G. T. Haslam, W. E. Wilson, D. A. Graham, and G. C. Hunt, "A further 408 MHz survey of the northern sky," *Astron. Astrophys. Suppl.*, vol. 13, pp. 359–394, 1974.

[3] S. Wongsowijoto and A. Schmidt, "Ein 3-Kanal-Empfanger fur 11cm-Messungen mit dem 100-m-Teleskop," *Kleinheubacher Berichte*, Ed. Fermeldetechnisches Zentralamt: Darmstadt 25, 1982, pp. 371–376.

[4] G. T. Haslam, U. Klein, C. J. Salter, H. Stoffel, W. E. Wilson, M. N. Cleary, D. J. Cooke, and P. Thomasson, "A 408 MHz all-sky continuum survey," *Astron. Astrophys.*, vol. 100, pp. 209–219, 1981.

[5] J. M. Payne, "Switching subreflector for millimeter wave radio astronomy," *Rev. Sci. Instrum.*, vol. 47, pp. 222–223, Feb. 1976.

[6] R. H. Dicke, "The measurement of thermal radiation at microwave frequencies," *Rev. Sci. Instrum.*, vol. 17, pp. 268–275, 1946.

[7] P. F. Goldsmith, "Quasioptical feed system for radioastronomical observations at millimeter wavelengths," *Bell. Syst. Tech. J.*, vol. 56, p. 1483, 1977.

[8] A. V. Raisanen, "Experimental studies on cooled millimeter wave mixers," *Acta Polytechnica Scandinavica*, Elect. Eng. series no. 46, Oct. 1980.

[9] J. L. R. Marrero, "75–115 GHz cryogenic heterodyne receiver," M.S. thesis, Univ. of Massachusetts, Sept. 1981.

[10] M. K. Brewer and A. V. Raisanen, "Dual-harmonic noncontacting millimeter waveguide backshorts: Theory, design, and test," *IEEE Trans. Microwave Theory Tech.*, vol. MTT-30, pp. 708–714, May 1982.

[11] A. V. Raisanen, N. R. Erickson, J. L. R. Marrero, P. F. Goldsmith, and C. R. Predmore, "An ultra low-noise schottky mixer receiver at 80-120 GHz," in *Infrared and Millimeter-wave Conf. Proc.*, Dec. 1981, pp. W3.8–9.

[12] A. Wu, "Assembly and testing procedure for a cryogenically-cooled 5 GHz circulator," National Radio Astronomy Observatory Electronics Division Int. Rep. 207, July 1980.

[13] S. Weinreb, "Low-noise cooled GASFET amplifiers," *IEEE Trans. Microwave Theory Tech.*, vol. MTT-28, pp. 1041–54, Oct. 1980.

[14] P. F. Goldsmith, "Quasi-optical techniques at millimeter and submillimeter wavelengths," in *Infrared and Millimeter Waves*, vol. 6, K. J. Button, Ed. New York: Academic, pp. 277–343, 1982.

[15] P. F. Goldsmith, "Diffraction loss in dielectric-filled Fabry–Perot interferometers," *IEEE Trans. Microwave Theory Tech.*, vol. MTT-30, pp. 820–823, May 1982.

[16] P. F. Goldsmith, "Quasioptics in millimeterwave systems," in *12th Eur. Microwave Conf. Proc.*, Sept. 1982, pp. 16–24.

[17] N. R. Erickson, "Off-axis mirrors made using a conventional milling machine," *Appl. Opt.*, vol. 18, pp. 956–957, Apr. 1, 1979.

The Cosmic Microwave
Background Radiation

R. W. Wilson

Radio astronomy has added greatly to our understanding of the structure and dynamics of the universe. The cosmic microwave background radiation, considered a relic of the explosion at the beginning of the universe some 18 billion years ago, is one of the most powerful aids in determining these features of the universe. This article is about the discovery of the cosmic microwave background radiation. It starts with a section on radio astronomical measuring techniques. This is followed by the history of the detection of the background radiation, its identification, and finally by a summary of our present knowledge of its properties.

Radio Astronomical Methods

A radio telescope pointing at the sky receives radiation not only from space, but also from other sources including the ground, the earth's atmosphere, and the components of the radio telescope itself. The 20-foot (6-meter) horn-reflector antenna at Bell Laboratories (Fig. 1), which was used to discover the cosmic microwave background radiation, was particularly suited to distinguish this weak, uniform radiation from other, much stronger sources. In order to understand this measurement, it is necessary to discuss the design and operation of a radio telescope, especially its two

major components, the antenna and the radiometer (*1*).

An antenna collects radiation from a desired direction incident upon an area, called its collecting area, and focuses it on a receiver. An antenna is normally designed to maximize its response in the direction in which it is pointed and minimize its response in other directions.

The 20-foot horn-reflector shown in Fig. 1 was built by A. B. Crawford and his associates (*2*) in 1960 to be used with an ultra low-noise communications receiver for signals bounced from the Echo satellite. It consists of a large expanding waveguide, or horn, with an off-axis section of a parabolic reflector at the end. The focus of the paraboloid is located at the apex of the horn, so that a plane

The author is Department Head, Radio Physics Research Department, Bell Laboratories, Crawford Hill Laboratory, Holmdel, N.J. 07733. This article is the lecture he delivered in Stockholm, Sweden, 8 December 1978, when he received the Nobel Prize in Physics, a prize he shared with Arno A. Penzias and P. L. Kapitza. Minor corrections and additions have been made by the author. The article is published here with the permission of the Nobel Foundation and will also be included in the complete volume of *Les Prix Nobel en 1978* as well as in the series *Nobel Lectures* (in English) published by the Elsevier Publishing Company, Amsterdam and New York. Dr. Penzias' lecture appeared in the issue of 10 August, and Dr. Kapitza's lecture will appear in a subsequent issue.

wave traveling along the axis of the paraboloid is focused into the receiver, or radiometer, at the apex of the horn. Its design emphasizes the rejection of radiation from the ground. It is easy to see from the figure that in this configuration the receiver is well shielded from the ground by the horn.

A measurement of the sensitivity of a small horn-reflector antenna to radiation coming from different directions is shown in Fig. 2. The circle marked isotropic antenna is the sensitivity of a fictitious antenna which receives equally from all directions. If such an isotropic lossless antenna were put in an open field, half the sensitivity would be to radiation from the earth and half from the sky. In the case of the horn-reflector, sensitivity in the back or ground direction is less than 1/3000 of the isotropic antenna. The isotropic antenna on a perfectly radiating earth at 300 K and with a cold sky at 0 K would pick up 300 K from the earth over half of its response and nothing over the other half, resulting in an equivalent antenna temperature of 150 K. The horn-reflector, in contrast, would pick up less than 0.05 K from the ground.

This sensitivity pattern is sufficient to determine the performance of an ideal, lossless antenna since such an antenna would contribute no radiation of its own. Just as a curved mirror can focus hot rays from the sun and burn a piece of paper without becoming hot itself, a radio telescope can focus the cold sky onto a radio receiver without adding radiation of its own.

A radiometer is a device for measuring the intensity of radiation. A microwave radiometer consists of a filter to select a desired band of frequencies followed by a detector that produces an output voltage proportional to its input power. Practical detectors are usually not sensitive enough for the low power levels received by radio telescopes, however, so that amplification is normally used ahead of the detector to increase the signal level. The noise from the first stage of this amplifier combined with that from the transmission line which connects it to the antenna (input source) produces an output from the detector even with no input power from the antenna. A fundamental limit to the sensitivity of a radiometer is the inherent fluctuation in the power level of this noise.

During the late 1950's, H. E. D. Scovil and his associates at Bell Laboratories, Murray Hill, New Jersey, were building the world's lowest-noise microwave amplifiers, ruby traveling-wave masers (3). These amplifiers were cooled to 4.2 K or

Fig. 1. The 20-foot horn-reflector which was used to discover the cosmic microwave background radiation.

less by liquid helium and contribute correspondingly small amounts of noise to the system. A radiometer incorporating these amplifiers can therefore be very sensitive.

Astronomical radio sources produce random, thermal noise very much like that from a hot resistor; therefore, the calibration of a radiometer is usually expressed in terms of a thermal system. Instead of giving the noise power that the radiometer receives from the antenna, we quote the temperature of a resistor which would deliver the same noise power to the radiometer. (Radiometers often contain calibration noise sources consisting of a resistor at a known temperature.) This "equivalent noise temperature" is proportional to received power for all except the shorter wavelength measurements, which will be discussed below.

To measure the intensity of an extraterrestrial radio source with a radio telescope, it is necessary to distinguish the source from local noise sources, such as noise from the radiometer, noise from the ground, noise from the earth's atmosphere, and noise from the structure of the antenna itself. This distinction is normally made by pointing the antenna alternately to the source of interest and then to a background region nearby. The difference in response of the radiometer to these two regions is measured, thus subtracting out the local noise. To determine the absolute intensity of an astronomical radio source, it is necessary to

calibrate the antenna and radiometer or, as is usually done, to observe a calibration source of known intensity.

Plans for Radio Astronomy with the 20-Foot Horn-Reflector

In 1963, when the 20-foot horn-reflector was no longer needed for satellite work, Arno Penzias and I started preparing it for use in radio astronomy. One might ask why we were interested in starting our radio astronomy careers at Bell Labs using an antenna with a collecting area of only 25 square meters when much larger radio telescopes were available elsewhere. Indeed, we were delighted to have the 20-foot horn-reflector because it had special features that we hoped to exploit. Its sensitivity, or collecting area, could be accurately calculated and, in addition, it could be measured with the use of a transmitter located less than 1 kilometer away. With these data, it could be used with a calibrated radiometer to make primary measurements of the intensities of several extraterrestrial radio sources. These sources could then be used as secondary standards by other observatories. In addition, we would be able to understand all sources of antenna noise—for example, the amount of radiation received from the earth, so that background regions could be measured absolutely. Traveling-wave maser amplifiers were available for use with the 20-foot horn-

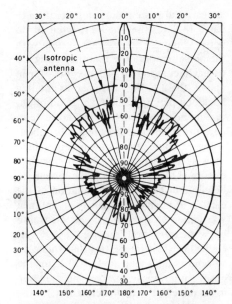

Fig. 2. Sensitivity pattern of a small horn-reflector antenna. This is a logarithmic plot of the collecting area of the antenna as a function of angle from the center of the main beam. Each circle below the level of the main beam represents a factor of 10 reduction in sensitivity. In the back direction around 180°, the sensitivity is consistently within the circle marked 70, corresponding to a factor of 10^{-7} below the sensitivity at 0°.

reflector, which meant that for large diameter sources (those subtending angles larger than the antenna beamwidth), this would be the world's most sensitive radio telescope.

My interest in the background measuring ability of the 20-foot horn-reflector resulted from my doctoral thesis work with J. G. Bolton at Caltech. We made a map of the 31-centimeter radiation from the Milky Way and studied the discrete sources and the diffuse gas within it. In mapping the Milky Way we pointed the antenna to the west side of it and used the earth's rotation to scan the antenna across it. This kept constant all the local noise, including radiation that the antenna picked up from the earth. I used the regions on either side of the Milky Way (where the brightness was constant) as the zero reference. Since we are inside the galaxy, it is impossible to point completely away from it. Our mapping plan was adequate for that project, but the unknown zero level was not very satisfying. Previous low-frequency measurements had indicated that there is a large, radio-emitting halo around our galaxy, which I could not measure by that technique. The 20-foot horn-reflector, however, was an ideal instrument for measuring this weak halo radiation at shorter wavelengths. One of my intentions when I came to Bell Labs in 1963 was to make such a measurement.

In 1963, a maser at 7.35-cm wavelength (4) was installed on the 20-foot horn-reflector. Before we could begin doing astronomical measurements, however, we had to do two things: (i) build a good radiometer incorporating the 7.35-cm maser amplifier and (ii) finish the accurate measurement of the collecting area (sensitivity) of the 20-foot horn-reflector, which D. C. Hogg had begun. Among our 7-cm astronomical projects were absolute intensity measurements of several traditional astronomical calibration sources and a series of sweeps of the Milky Way to extend my thesis work. In the course of this work we planned to check out our capability of measuring the halo radiation of our galaxy away from the Milky Way. Existing low-frequency measurements indicated that the brightness temperature of the halo would be less than 0.1 K at 7 cm. Thus, a background measurement at 7 cm should produce a null result and would be a good check of our measuring ability.

After completing this program of measurements at 7 cm, we planned to build a similar radiometer at 21 cm. At that wavelength, the galactic halo should be bright enough for detection, and we would also observe the 21-cm line of neutral hydrogen atoms. In addition, we planned a number of hydrogen-line projects including an extension of the measurements of Arno's thesis, a search for hydrogen in clusters of galaxies.

At the time we were building the 7-cm radiometer, John Bolton visited us and we related our plans and asked for his comments. He immediately selected the most difficult one as the most important: the 21-cm background measurement. First, however, we had to complete the observations at 7 cm.

Radiometer System

We wanted to make accurate measurements of antenna temperatures. To do this, we planned to use the radiometer to compare the antenna to a reference source, in this case, a radiator in liquid helium. I built a switch which would connect the maser amplifier either to the antenna or to Arno's helium-cooled reference noise source (5) (cold load). This would allow an accurate comparison of the equivalent temperature of the antenna to that of the cold load, since the noise from the rest of the radiometer would be constant during switching. A diagram of this calibration system (6) is shown in Fig. 3, and its operation is described below.

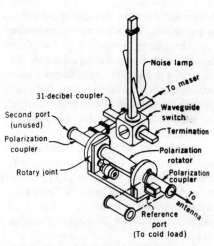

Fig. 3. The switching and calibration system of our 7.35-cm radiometer. The reference port was normally connected to the helium-cooled reference source through a noise-adding attenuator.

The switch for comparing the cold load to the antenna consists of the two polarization couplers and the polarization rotator shown in Fig. 3. This type of switch had been used by D. H. Ring in several radiometers at Holmdel. It had the advantage of stability, low loss, and small reflections. The circular waveguide coming from the antenna contained the two orthogonal modes of polarization received by the antenna. The first polarization coupler reflected one mode of linear polarization back to the antenna and substituted the signal from the cold load for it in the waveguide going to the rotator. The second polarization coupler took one of the two modes of linear polarization coming from the polarization rotator and coupled it to the rectangular (single-mode) waveguide going to the maser. The polarization rotator is the microwave equivalent of a half-wave plate in optics. It is a piece of circular waveguide which has been squeezed in the middle so that the phase shifts for waves traveling through it in its two principal planes of linear polarization differ by 180°. By mechanically rotating it, the polarization of the signals passing through it can be rotated. Thus either the antenna or cold load could be connected to the maser. This type of switch is not inherently symmetric, but has very low loss and is stable so that its asymmetry of 0.05 K was accurately measured and corrected for.

A drawing of the liquid-helium cooled reference noise source is shown in Fig. 4. It consists of a 122-cm piece of 90 percent copper-brass waveguide connecting a carefully matched microwave absorber in liquid He to a room-temperature flange at the top. Small holes allow liquid helium to fill the bottom section of waveguide so that the absorber temperature could be known, while a mylar window at a 30° angle keeps the liquid out of the rest of the waveguide and makes a low-reflection microwave transition between the two sections of waveguide. Most of the remaining parts are for the cryogenics. The gas baffles make a counter-flow heat exchanger between the waveguide and the helium gas which has boiled off, greatly extending the time of operation on a charge of liquid helium. Twenty liters of liquid helium cooled the cold load and provided about 20 hours of operation.

Above the level of the liquid helium, the waveguide walls were warmer than 4.2 K. Any radiation due to the loss in this part of the waveguide would raise the effective temperature of the noise source above 4.2 K and must be accounted for. To do so, we monitored the temperature distribution along the waveguide with a series of diode thermometers and calculated the contribution of each section of the waveguide to the equivalent temperature of the reference source. When first cooled down, the calculated total temperature of the reference noise source was about 5 K. After several hours when the liquid helium level was lower, it increased to 6 K. As a check of this calibration procedure, we compared the antenna temperature (assumed constant) to our reference noise source during this period, and found consistency to within 0.1 K.

A variable attenuator normally connected the cold load to the reference port of the radiometer. This device was at room temperature so noise could be added to the cold load port of the switch by increasing its attenuation. It was calibrated over a range of 0.11 decibel which

corresponds to 7.4 K of added noise. Also shown in Fig. 3 is a noise lamp (and its directional coupler) which was used as a secondary standard for our temperature scale.

Signals leaving the maser amplifier needed to be further amplified before detection so that their intensity could be measured accurately. The remainder of our radiometer consisted of a down converter to 70 MHz followed by intermediate-frequency amplifiers, a precision variable attenuator, and a diode detector. The output of the diode detector was amplified and went to a chart recorder.

Our radiometer equipment installed in the cab of the 20-foot horn-reflector is shown in Fig. 5. The flange at the far right is part of the antenna and rotates in elevation angle with it. It was part of a double-choke joint which allowed the rest of the equipment to be fixed in the cab while the antenna rotated. The noise contribution of the choke joint could be measured by clamping it shut and was found to be negligible. We regularly measured the reflection coefficient of the major components of this system and kept it below 0.03 percent, except for the maser whose reflection could not be reduced below 1 percent. Since all ports of our waveguide system were terminated at a low temperature, these reflections resulted in negligible errors.

Prior Observations

The first horn-reflector–traveling-wave maser system had been put together by DeGrasse, Hogg, Ohm, and Scovil in 1959 (7) to demonstrate the feasibility of a low-noise, satellite-earth station at 5.31 cm. Even though they achieved the

lowest total system noise temperature to date, 18.5 K, they had expected to do better. Figure 6 shows their system with the noise temperature that they assigned to each component. As we have seen above, the 2 K they assigned to antenna backlobe pickup is too high. In addition, direct measurements of the noise temperature of the maser gave a value about a degree colder than shown here. Thus, their system was about 3 K hotter than one might expect. The component labeled T_s in Fig. 6 is the radiation of the earth's atmosphere when their antenna was aimed straight up. It was measured by a method first reported by R. H. Dicke (8). (It is interesting that Dicke also reports an upper limit of 20 K for the cosmic microwave background radiation in his paper—the first such report.) If the antenna temperature is measured as a function of the angle above the horizon at which it is pointing, the radiation of the atmosphere is at a minimum when the antenna is directed straight up. It increases as the antenna points toward the horizon, since the total line of sight through the atmosphere increases. Figure 7 is a chart recording that Arno Penzias and I made with the 20-foot horn-reflector scanning from almost the zenith down to 10° above the horizon. The circles and crosses are the expected change based on a standard model of the earth's atmosphere for 2.2 and 2.4 K zenith contribution. The fit between theory and data is obviously good, leaving little chance that there might be an error in our value for atmospheric radiation.

Figure 8 is taken from the paper in which Ohm (9) described the receiver on the 20-foot horn-reflector which was used to receive signals bounced from the Echo satellite. Ohm found that its system temperature was 3.3 K higher than

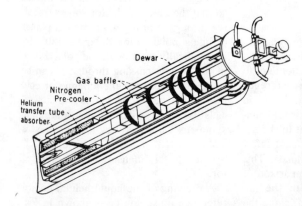

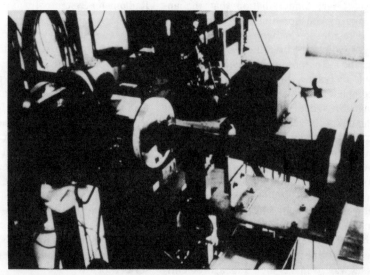

Fig. 4 (left above). The helium-cooled reference noise source. Fig. 5 (right). Our 7.35-cm radiometer installed in the cab of the 20-foot horn-reflector.

366

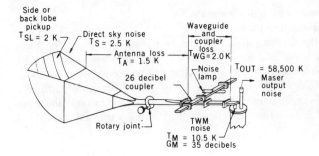

Side or
back lobe
pickup
$T_{SL} = 2$ K
Direct sky noise
$T_S = 2.5$ K
Antenna loss
$T_A = 1.5$ K
Waveguide and coupler loss
$T_{WG} = 2.0$ K
Noise lamp
$T_{OUT} = 58,500$ K
Maser output noise
26 decibel coupler
Rotary joint
TWM noise
$T_M = 10.5$ K
$G_M = 35$ decibels

Fig. 6. A diagram of the low noise receiver used by De-Grasse, Hogg, Ohm, and Scovil to show that very low noise earth stations are possible. Each component is labeled with its contribution to the system noise.

expected from summing the contributions of the components. As in the previous 5.3-cm work, this excess temperature was smaller than the experimental errors, so not much attention was paid to it. In order to determine the unambiguous presence of an excess source of radiation of about 3 K, a more accurate measurement technique was required. This was achieved in the subsequent measurements by means of a switch and reference noise source combination which communications systems do not have.

Our Observations

Figure 9 is a reproduction of the first record we have of the operation of our system. At the bottom is a list of diode thermometer voltages from which we could determine the cold load's equivalent temperature. The recorder trace has power (or temperature) increasing to the right. The middle part of this trace is with the maser switched to the cold load, with various settings of the noise-adding attenuator. A change of 0.1 decibel corresponds to a temperature change of 6.6 K, so that the peak-to-peak noise on the trace amounts to less than 0.2 K. At the top of the chart, the maser is switched to the antenna and has about the same temperature as the cold load plus 0.04 decibel, corresponding to a total of about 7.5 K. This was a troublesome result. The antenna temperature should have been only the sum of the atmospheric contribution (2.3 K) and the radiation from the walls of the antenna and ground (1 K). The excess system temperature found in the previous experiments had, contrary to our expectations, all been in the antenna or beyond. We now had a direct comparison of the antenna with the cold load and had to assign our excess temperature to the antenna, whereas in the previous cases only the total system temperature was measured. If we had missed some loss, the cold load might have been warmer than calculated, but it could not be colder than 4.2 K—the temperature of the liquid helium. The antenna was at least 2 K hotter than that.

Unless we could understand our "antenna problem," our 21-cm galactic halo experiment would not be possible. We considered a number of possible reasons for this excess and, where warranted, tested for them. These were:

1) At the time some radio astronomers thought that the microwave absorption of the earth's atmosphere was about twice the value we were using; in other words the "sky temperature" of Figs. 6 and 8 was about 5 K instead of 2.5 K. We knew from our measurement of sky temperature such as shown in Fig. 7 that this could not be the case.

2) We considered the possibility of man-made noise being picked up by our antenna. However, when we pointed our antenna to New York City, or to any other direction on the horizon, the antenna temperature never went significantly above the thermal temperature of the earth.

3) We considered radiation from our galaxy. Our measurements of the emission from the plane of the Milky Way

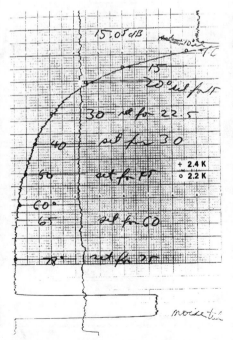

Fig. 7. A measurement of atmospheric noise at 7.35-cm wavelength with theoretical fits to the data for 2.2 and 2.4 K zenith atmospheric radiation.

were a reasonable fit to the intensities expected from extrapolations of low-frequency measurements. Similar extrapolations for the coldest part of the sky (away from the Milky Way) predicted about 0.02 K at our wavelength. Furthermore, any galactic contribution should also vary with position, and we saw changes only near the Milky Way that were consistent with the measurements at lower frequencies.

4) We ruled out discrete extraterrestrial radio sources as the source of our radiation as they have spectra similar to that of the galaxy. The same extrapolation from low-frequency measurements applies to them. The strongest discrete source in the sky had a maximum antenna temperature of 7 K.

Thus, we seemed to be left with the antenna as the source of our extra noise. We calculated a contribution of 0.9 K from its resistive loss using standard waveguide theory. The part of the antenna where most loss occurred was its small diameter throat, which was made of electroformed copper. We had measured similar waveguides in the laboratory and corrected the loss calculations for the imperfect surface conditions that we had found in those waveguides. The remainder of the antenna was made of riveted aluminum sheets, and, although we did not expect any trouble there, we had no way to evaluate the loss in the riveted joints. A pair of pigeons was roosting up in the small part of the horn where it enters the warm cab. They had covered the inside with a white material familiar to all city dwellers. We evicted the pigeons and cleaned up their mess, but obtained only a small reduction in antenna temperature.

For some time we lived with the antenna temperature problem and concentrated on measurements in which it was not critical. Dave Hogg and I had made a very accurate measurement of the antenna's gain (10), and Arno [Penzias] and I wanted to complete our absolute flux measurements before disturbing the antenna further.

In the spring of 1965 with our flux measurements finished (5), we thoroughly cleaned out the 20-foot horn-reflector and put aluminum tape over the riveted joints. This resulted in only a minor reduction in antenna temperature. We also took apart the throat section of the antenna, and checked it, but found it to be in order.

By this time almost a year had passed. Since the excess antenna temperature had not changed during this time, we could rule out two additional sources: (i) Any source in the solar system should have gone through a large change in angle, and we should have seen a change in antenna temperature. (ii) In 1962, a high-altitude nuclear explosion had filled up the Van Allen belts with ionized particles. Since they were at a large distance from the surface of the earth, any radiation from them would not show the same elevation-angle dependence as the atmosphere, and we might not have identified it. But after a year, any radiation from this source should have reduced considerably.

Identification

The sequence of events that led to the unraveling of our mystery began one day when Arno was talking to Bernard Burke of M.I.T. about other matters and mentioned our unexplained noise. Bernie recalled hearing about theoretical work of P. J. E. Peebles in R. H. Dicke's group in Princeton on radiation in the universe. Arno called Dicke who sent a copy of Peebles' preprint. The Princeton group was investigating the implications of an oscillating universe with an extremely hot condensed phase. This hot bounce was necessary to destroy the heavy elements from the previous cycle so that each cycle could start fresh. Although this was not a new idea (11), Dicke had the important idea that, if the radiation from this hot phase were large enough, it would be observable. In the preprint, Peebles, following Dicke's suggestion, calculated that the universe should be filled with a relic blackbody radiation at a minimum temperature of 10 K. Peebles was aware of the measurement of atmospheric radiation at 6 cm by Hogg and Semplak (1961) (12) who used the system of DeGrasse et al., and concluded that the present radiation temperature of the universe must be less than their system temperature of 15 K. He also said that Dicke, Roll, and Wilkinson were setting up an experiment to measure it.

Shortly after sending the preprint, Dicke and his co-workers visited us in order to discuss our measurements and see our equipment. They were quickly convinced of the accuracy of our measurements. We agreed to a side-by-side publication of two letters in the Astrophysical Journal—a letter on the theory from Princeton (13) and one on our measurement of excess antenna temperature from Bell Laboratories (14). Arno and I were careful to exclude any discussion of the cosmological theory of the origin of background radiation from our letter because we had not been involved in any of that work. We thought, further-

more, that our measurement was independent of the theory and might outlive it. We were pleased that the mysterious noise appearing in our antenna had an explanation of any kind, especially one with such significant cosmological implications. Our mood, however, remained one of cautious optimism for some time.

Fig. 8. An excerpt from Ohm's article on the Echo receiver, showing that his system temperature was 3.3 K higher than predicted.

Results

While preparing our letter for publication we made one final check on the antenna to make sure we were not picking up a uniform 3 K from the earth. We measured its response to radiation from the earth by using a transmitter located

TABLE II – SOURCES OF SYSTEM TEMPERATURE

Source	Temperature
Sky (at zenith)	$2.30 \pm 0.20°$K
Horn antenna	$2.00 \pm 1.00°$K
Waveguide (counter-clockwise channel)	$7.00 \pm 0.65°$K
Maser assembly	$7.00 \pm 1.00°$K
Converter	$0.60 \pm 0.15°$K
Predicted total system temperature	$18.90 \pm 3.00°$K

the temperature was found to vary a few degrees from day to day, but the lowest temperature was consistently $22.2 \pm 2.2°$K. By realistically assuming that all sources were then contributing their fair share (as is also tacitly assumed in Table II) it is possible to improve the over-all accuracy. The actual system emperature must be in the overlap region of the measured results and the total results of Table II, namely between 20 and 21.9°K. The most likely minimum system temperature was therefore

$$T_{system} = 21 \pm 1°\text{K.}*$$

The inference from this result is that the "+" temperature possibilities of Table II must predominate.

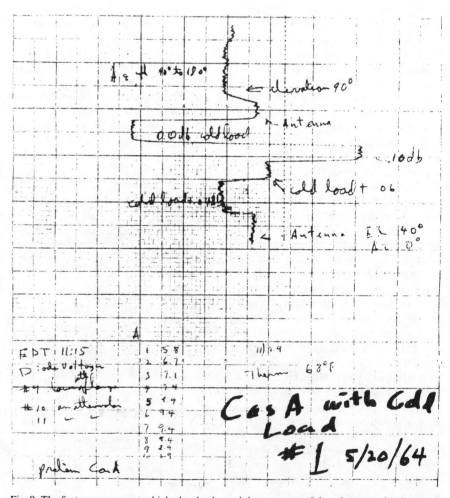

Fig. 9. The first measurement which clearly showed the presence of the microwave background. Noise temperature is plotted increasing to the right. At the top, the antenna pointed at 90° elevation is seen to have the same noise temperature as the cold load with 0.04 dB attenuation (about 7.5 K). This is considerably above the expected value of 3.3 K.

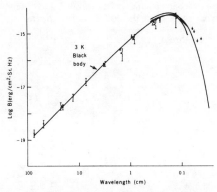

Fig. 10. Results of the large-scale isotropy experiment of Smoot, Gorenstein, and Muller showing the clear cosine dependence of brightness expected from the relative velocity of the earth in the background radiation. The shaded area and arrows show the values allowed by the data of Woody and Richards. [This figure is reproduced with permission of *Scientific American*]

in various places on the ground. The transmitter artifically increased the ground's brightness at the wavelength of our receiver to a level high enough for the backlobe response of the antenna to be measurable. Although not a perfect measure of the structure of the back-lobes of an antenna, it was a good enough method of determining their average level. The backlobe level we found in this test was as low as we had expected and indicated a negligible contribution to the antenna temperature from the earth.

The right-hand column of Table 1 shows the final results of our measurement. The numbers on the left were obtained later in 1965 with a new throat on the 20-foot horn-reflector. From the total antenna temperature we subtracted the known sources with a result of 3.4 ± 1 K. Since the errors in this measurement are not statistical, we have summed the maximum error from each source. The maximum measurement error of 1 K was considerably smaller than the measured value, giving us confidence in the reality of the result. We stated in the original paper that "This excess temperature is, within the limits of our observations, isotropic, unpolarized, and free of seasonal variations." Although not stated explicitly, our limits on an isotropy and polarization were not affected by most of the errors listed in Table 1 and were about 10 percent or 0.3 K.

At that time the limit we could place on the shape of the spectrum of the background radiation was obtained by comparing our value of 3.5 K with a 74-cm survey of the northern sky done at Cambridge by Pauliny-Toth and Shakeshaft

in 1962 (*15*). The minimum temperature on their map was 16 K. Thus the spectrum was no steeper than $\lambda^{0.7}$ over a range of wavelengths that varied by a factor of 10. This clearly ruled out any type of radio source known at that time, as they all had spectra with variation in the range $\lambda^{2.0}$ to $\lambda^{3.0}$. The previous Bell Laboratories measurement at 6 cm ruled out a spectrum which rose rapidly toward shorter wavelengths.

Confirmation

After our meeting, the Princeton experimental group returned to complete their apparatus and make their measurement, with the expectation that the background temperature would be about 3 K.

The first confirmation of the microwave cosmic background that we knew of, however, came from a totally different, indirect measurement. This measurement had, in fact, been made 30 years earlier by Adams (*16*) and Dunham (*17*). Adams and Dunham had discovered several faint optical interstellar absorption lines which were later identified with the molecules CH, CH^+, and CN. In the case of CN, in addition to the ground state, absorption was seen from the first rotationally excited state. McKellar (*18*) using Adams' data on the populations of these two states calculated that the excitation temperature of CN was 2.3 K. This rotational transition occurs at 2.64-millimeter wavelength, near the peak of a 3 K blackbody spectrum. Shortly after the discovery of the background radiation, G. B. Field *et al.* (*19*), I. S. Shklovsky (*20*), and P. Thaddeus (*21*) (following a suggestion by N. J. Woolf) independently realized that the CN is in equilibrium with the background radiation. (There is no other significant source of excitation where these molecules are located.) In addition to confirming that the background was not zero, this idea immediately confirmed that the spectrum of the background radiation was close to that of a blackbody source for wavelengths larger than the peak. It also gave a hint that, at short wavelengths, the intensity was departing from the $1/\lambda^2$ dependence expected in the long wavelength (Raleigh-Jeans) region of the spectrum and following the true blackbody (Planck) distribution. In 1966, Field and Hitchcock (*19*) reported new measurements, using Herbig's plates of ζ Oph (Ophiuchi) and ζ Per (Persei) obtaining 3.22 ± 0.15 K and 3.0 ± 0.6 K for the excitation temperature. Thaddeus and Clauser (*21*) also obtained new plates

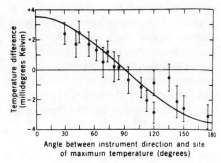

Fig. 11. Measurements of the spectrum of the cosmic microwave background radiation.

and measured 3.75 ± 9.5 K in ζ Oph. Both groups argued that the main source of excitation in CN is the background radiation. This type of observation, taken alone, is most convincing as an upper limit, since it is easier to imagine additional sources of excitation than refrigeration.

In December 1965 Roll and Wilkinson (*22*) completed their measurement of 3.0 ± 0.5 K at 3.2 cm, the first confirming microwave measurement. This was followed shortly by Howell and Shakeshaft's (*23*) value of 2.8 ± 0.6 K at 20.7 cm (*18*) and then by our measurement of 3.2 ± 1 K at 21.1 cm (*24*). (Half of the difference between these two results comes from a difference in the corrections used for the galactic halo and integrated discrete sources.) By mid-1966, the intensity of the microwave background radiation had been shown to be close to 3 K between 21 cm and 2.6 mm, almost two orders of magnitude in wavelength.

Earlier Theory

I have mentioned that the first experimental evidence for cosmic microwave background radiation was obtained (but unrecognized) long before 1965. We soon learned that the theoretical prediction of it had been made at least 16 years before our detection. George Gamow had made calculations of the conditions in the early universe in an attempt to understand galaxy formation (*25*). Although these calculations were not strictly correct, he understood that the early stages of the universe had to be very hot or else all of the hydrogen would combine, becoming heavier elements. Furthermore, Gamow and his collaborators calculated that the density of radiation in the hot early universe was much higher than the density of matter. In this early work, the present remnants of this radiation were not considered. However, in 1949, Alpher and Herman

(26) followed the evolution of the temperature of the hot radiation in the early universe up to the present epoch and predicted a value of 5 K. They noted that the present density of radiation was not well known experimentally. In 1953 Alpher, Follin, and Herman (27) reported what has been called the first thoroughly modern analysis of the early history of the universe, but failed to recalculate or mention the present radiation temperature of the universe.

In 1964, Doroshkevich and Novikov (28) had also calculated the relic radiation and realized that it would have a blackbody spectrum. They quoted E. A. Ohm's article on the Echo receiver, but misunderstood it and concluded that the present radiation temperature of the universe is near zero.

A more complete discussion of these calculations is given in Arno's lecture (29).

Isotropy

In assigning a single temperature to the radiation in space, these theories assume that it will be the same in all directions. According to contemporary theory, the last scattering of the cosmic microwave background radiation occurred when the universe was a million years old, just before the electrons and nucleii combined to form neutral atoms ("recombination"). The isotropy of the background radiation thus measures the isotropy of the universe at that time and the isotropy of its expansion since then. Prior to recombination, radiation dominated the universe and the Jeans mass, or mass of the smallest gravitationally stable clumps was larger than a cluster of galaxies. It is only in the period following recombination that galaxies could have formed.

In 1967, Rees and Sciama (30) suggested looking for large-scale anisotropies in the background radiation which might have been left over from anisotropies of the universe prior to recombination. In the same year Partridge and Wilkinson (31) completed an experiment which was specifically designed to look for anisotropy within the equatorial plane. They reported a limit of 0.1 percent for a 24-hour asymmetry and a possible 12-hour asymmetry of 0.2 percent. Meanwhile we had reanalyzed an old record covering most of the sky which was visible to us and put a limit of 0.1 K on any large-scale fluctuations (32). Since then the measurements of Conklin (33), Henry (34), and Corey and Wil-

kinson (35) have shown a 24-hour anisotropy due to the earth's velocity with respect to the background radiation. Data from the most sensitive measurement to date (36) are shown in Fig. 10. They show a striking cosine anisotropy with an amplitude of about 0.003 K, indicating that the background radiation has a maximum temperature in one direction and a minimum in the opposite direction. The generally accepted explanation of this effect is that the earth is moving toward the direction where the radiation is hottest and it is the blue shift of the radiation that increases its measured temperature in that direction. The motion of the sun with respect to the background radiation from the data of Smoot et al. is 390 ± 60 km/sec in the direction 10.8 hours right ascension, 5° declination. The magnitude of this velocity is not a surprise since 300 km/sec is the orbital velocity of the sun around our galaxy. The direction is different, however, yielding a peculiar velocity of our galaxy of about 600 km/sec. Since other nearby galaxies, including the Virgo cluster, have a small velocity with respect to our galaxy, they have a similar velocity with respect to the matter which last scattered the background radiation. After subtracting the 24-hour anisotropy, one can search the data for more complicated anisotropies to put observational limits on such things as rotation of the universe (36). Within the noise of 0.001 K, these anisotropies are all zero.

To date, no fine-scale anisotropy has been found. Several early investigations were carried out to discredit discrete source models of the background radiation. In the most sensitive experiment to date, Boynton and Partridge (37) report a relative intensity variation of less than 3.7×10^{-3} in an 80" arc beam. A discrete source model would require orders of magnitude more sources than the known number of galaxies to show this degree of smoothness.

It has also been suggested by Sunyaev and Zel'dovich (38) that there will be a reduction of the intensity of the background radiation from the direction of clusters of galaxies due to inverse Compton scattering by the electrons in the intergalactic gas. This effect, which has been found by Birkinshaw and Gull (39), provides a measure of the intergalactic gas density in the clusters and may give an alternate measurement of Hubble's constant.

Spectrum

Since 1966, a large number of measurements of the intensity of the background radiation have been made at wavelengths from 74 cm to 0.5 mm. Measurements have been made from the ground, mountain tops, airplanes, balloons, and rockets. In addition, the optical measurements of the interstellar molecules have been repeated, and we have observed their millimeter line radiation directly to establish the equilibrium of the excitation of their levels with the background radiation (40). Figure 11 is a plot of most of these measurements (41). An early set of measurements from Princeton covered the range 3.2 to 0.33 cm, showing tight consistency with a 2.7 K blackbody (42). A series of rocket and balloon measurements in the millimeter and submillimeter part of the spectrum have converged on about 3 K. The data of Robson et al. (43) and Woody and Richards (44) extend to 0.8 mm, well beyond the spectral peak. The most recent experiment, that of Woody and Richards, gives a close fit to a 3.0 K spectrum out to 0.8 mm wavelength, with upper limits at atmospheric windows out to 0.4 mm. This establishes that the background radiation has a blackbody spectrum which would be quite hard to reproduce with any other type of cosmic source. The source must

Table 1. Results of 1965 measurements of microwave background. "Old throat" and "new throat" refer to the original and a replacement throat section for the 20-foot horn-reflector.

Item	New throat		Old throat
He temperature	4.22	4.22	
Calculated contribution from cold load waveguide	0.38	0.70 ± 0.2	
Attenuator setting for balance	2.73	2.40 ± 0.1	
Total cold load	7.33	7.32 ± 0.3	6.7 ± 0.3
Atmosphere		2.3 ± 0.3	2.3 ± 0.3
Waveguide and antenna loss		1.8 ± 0.3	0.9 ± 0.3
Back lobes		0.1 ± 0.1	0.1 ± 0.1
Total antenna		4.2 ± 0.7	3.3 ± 0.7
Background		3.1 ± 1	3.4 ± 1

have been optically thick and therefore must have existed earlier than any of the other radio sources which can be observed.

The spectral data are now almost accurate enough for one to test for systematic deviations from a single-temperature blackbody spectrum which could be caused by minor deviations from the simplest cosmology. Danese and De Zotti (45) report that except for the data of Woody and Richards, the spectral data of Fig. 11 do not show any statistically significant deviation of this type.

Conclusion

Cosmology is a science which has only a few observable facts to work with. The discovery of the cosmic microwave background radiation added one—the present radiation temperature of the universe. This, however, was a significant increase in our knowledge since it requires a cosmology with a source for the radiation at an early epoch and is a new probe of that epoch. More sensitive measurements of the background radiation in the future will allow us to discover additional facts about the universe.

The work which I have described was done with Arno A. Penzias. In our 15 years of partnership he has been a constant source of help and encouragement.

References and Notes

1. A more complete discussion of radio telescope antennas and receivers may be found in several text books. Chapters 6 and 7 of J. D Kraus, *Radio Astronomy* (McGraw-Hill, New York, 1966) are good introductions to the subjects.
2. A. B. Crawford, D. C. Hogg, L. E. Hunt, *Bell Syst. Tech. J.* **40**, 1095 (1961).
3. R. W. DeGrasse, E. O. Schultz-DuBois, H. E. D. Scovil, *ibid.* **38**, 305 (1959).
4. W. J. Tabor and J. T. Sibilia, *ibid.* **42**, 1863 (1963).
5. A. A. Penzias, *Rev. Sci. Instrum.* **36**, 68 (1965).
6. _____ and R. W. Wilson, *Astrophys. J.* **142**, 1149 (1965).
7. R. W. DeGrasse, D. C. Hogg, E. A. Ohm, H. E. D. Scovil, *Proc. Natl. Electron. Conf.* **15**, 370 (1959).
8. R. Dicke, R. Beringer, R. L. Kyhl, A. V. Vane, *Phys. Rev.* **70**, 340 (1946).
9. E. A. Ohm, *Bell Syst. Tech. J.* **40**, 1065 (1961).
10. D. C. Hogg and R. W. Wilson, *ibid.* **44**, 1019 (1965).
11. Compare F. Hoyle and R. J. Taylor, *Nature (London)* **203**, 1108 (1964). A less explicit discussion of the same notion occurs in (*29*).
12. D. C. Hogg and R. A. Semplak, *Bell Syst. Tech. J.* **40**, 1331 (1961).
13. R. H. Dicke, P. J. E. Peebles, P. G. Roll, D. T. Wilkinson, *Astrophys. J.* **142**, 414 (1965).
14. A. A. Penzias and R. W. Wilson, *ibid.*, p. 142.
15. I. I. K. Pauliny-Toth and J. R. Shakeshaft, *Mon. Not. R. Astron. Soc.* **124**, 61 (1962).
16. W. S. Adams, *Astrophys. J.* **93**, 11 (1941); *ibid.* **97**, 105 (1943).
17. T. Dunham, Jr., *Publ. Astron. Soc. Pac.* **49**, 26 (1937); *Proc. Am. Phil. Soc.* **81**, 277 (1939); *Publ. Am. Astron. Soc.* **10**, 123 (1941); _____ and W. S. Adams, *ibid.* **9**, 5 (1937).
18. A. McKellar, *Publ. Dom. Astrophys. Obs. Victoria B.C.* **7**, 251 (1941).
19. G. B. Field, G. H. Herbig, J. L. Hitchcock, talk given at the American Astronomical Society Meeting, 22 to 29 December 1965; *Astron. J.* **71**, 161 (1966); G. B. Field and J. L. Hitchcock, *Phys. Rev. Lett.* **16**, 817 (1966).
20. I. S. Shklovsky, *Astronomical Circular No. 364* (Academy of Sciences of the U.S.S.R., Moscow, 1966).
21. P. Thaddeus and J. F. Clauser, *Phys. Rev. Lett.* **16**, 819 (1966).
22. P. G. Roll and D. T. Wilkinson, *ibid.*, p. 405.
23. T. F. Howell and J. R. Shakeshaft, *Nature (London)* **210**, 138 (1966).
24. A. A. Penzias and R. W. Wilson, *Astron. J.* **72**, 315 (1967).
25. G. Gamow, *Nature (London)* **162**, 680 (1948).
26. R. A. Alpher and R. C. Herman, *Phys. Rev.* **75**, 1089 (1949).
27. R. A. Alpher, J. W. Follin, R. C. Herman, *ibid.* **92**, 1347 (1953).
28. A. G. Doroshkevich and I. D. Novikov, *Dokl. Akad. Nauk SSSR* **154**, 809 (1964); *Sov. Phys. Dokl.* **9**, 111 (1964).
29. A. A. Penzias, "The origin of the elements" (Nobel Prize lecture), *Science* **205**, 549 (1979).
30. M. J. Rees and D. W. Sciama, *Nature (London)* **213**, 374 (1967).
31. R. B. Partridge and D. T. Wilkinson, *Phys. Rev. Lett.* **18**, 557 (1967).
32. R. W. Wilson and A. A. Penzias, *Science* **156**, 1100 (1967).
33. E. K. Conklin, *Nature (London)* **222**, 971 (1969).
34. P. S. Henry, *ibid.* **231**, 516 (1971).
35. B. E. Corey and D. T. Wilkinson, *Bull Astron. Astrophys. Soc.* **8**, 351 (1976).
36. G. F. Smoot, M. V. Gorenstein, R. A. Muller, *Phys. Rev. Lett.* **39**, 898 (1977).
37. P. E. Boynton and R. B. Partridge, *Astrophys. J.* **181**, 243 (1973).
38. R. A. Sunyaev and Ya. B Zel'dovich, *Comments Astrophys. Space Phys.* **4**, 173 (1972).
39. M. Birkinshaw and S. F. Gull, *Nature (London)* **275**, 40 (1978).
40. A. A. Penzias, K. B. Jefferts, R. W. Wilson, *Phys. Rev. Lett.* **28**, 772 (1972).
41. The data in Fig. 10 are all referenced by Danese and De Zotti (*45*), except for the 13-cm measurement of T. Otoshi [*IEEE Trans. Instrum. Meas.* **24**, 174 (1975)]. I have used the millimeter measurements of Woody and Richards (*44*) and left off those of Robson *et al.* (*43*) to avoid confusion.
42. D. J. Wilkinson, *Phys. Rev. Lett.* **19**, 1195 (1967); R. A. Stokes, R. B. Partridge, D. J. Wilkinson, *ibid.*, p. 1199; P. E. Boynton, R. A. Stokes, D. J. Wilkinson, *ibid.* **21**, 462 (1968); P. E. Boynton and R. A. Stokes, *Nature (London)* **247**, 528 (1974).
43. E. I. Robson, D. G. Vickers, J. S. Huizinga, J. E. Beckman, P. E. Clegg, *Nature (London)* **251**, 591 (1974).
44. D. P. Woody and P. L. Richards, private communication.
45. L. Danese and G. De Zotti, *Astron. Astrophys.* **68**, 157 (1978).
46. I thank W. D. Langer and E. Wilson for carefully reading the manuscript and suggesting changes.

Part VI
Spectrometers

AS DISCUSSED in the Historical Overview, the first radio-astronomical spectral line, the 21-cm transition of atomic hydrogen, was detected in 1951. Prior to this, it was assumed that the spectra of sources being observed were only very slowly varying functions of frequency, and the bandwidth of the observing equipment could be adjusted to optimize the signal-to-noise ratio taking into account external factors such as interference. Purely in terms of accurate measurement of intensity, the broadest possible bandwidth would be advantageous, but the desire to obtain some reasonable information about the spectral distribution of the radiation, together with limitations imposed by the equipment, restricted the fractional bandwidths used to 1–10%. In some situations, such as observations of pulsars, interstellar dispersion also restricted the bandwidth that could be effectively utilized.

The advent of spectral-line observations completely changed this picture. In most situations, the fractional bandwidth occupied by a single spectral line is given by $\Delta f/f = \Delta v/c$, where Δv is the velocity dispersion within the region responsible for the emission, and c is the speed of light, equal to 3×10^5 km/s. The velocity dispersion can be as large as hundreds of kilometers per second for the nuclear regions of active galaxies, so that $\Delta f/f \simeq 0.001$, or as small as a fraction of a kilometer per second for molecular emission from quiescent molecular clouds, giving $\Delta f/f \simeq 10^{-6}$. In fact, there is almost always considerable information contained in the line profile, so that the radio astronomer will typically choose a frequency resolution an order of magnitude greater than the linewidth, which of course leads to even greater demands on the frequency resolution required.

The most basic instrument capable of collecting information about the intensity in a number of different frequency intervals is the multichannel filter spectrometer or filterbank. The principle of its operation is extremely simple, but the details of making filters of the desired shape and with the required stability are not trivial. Each channel is followed by a detector, generally a diode operating in the "square law" region, so that its output voltage is proportional to the power input. The following electronics includes low-level amplifiers and analog or digital (voltage-to-frequency converters plus counters) integrators. Perhaps because the concepts underlying a multichannel filter spectrometer are unexciting, few designs for these instruments have appeared in the literature. The general principles have been discussed by Penfield (1976) and the use of multichannel spectrometers has been discussed by Williams (1976).

Multichannel filter spectrometers suffer from a variety of drawbacks, including instabilities in the low-level analog electronics, and the relative inflexibility of this approach, in that a given filterbank has a fixed frequency resolution. Thus, to effectively study emission from regions with differing velocity dispersions, a set of filter spectrometers with different resolutions is required. The second restriction is overcome to a considerable extent by the "spectrum expander" developed by Henry (1979). In this device, the incident signal is digitized and read into a memory, but read out at a faster rate than it was read in. The effect is to expand the spectral range of the signal so that if it is fed into a conventional filterbank, the resolution of each channel is effectively increased. While this device increases the versatility of a preexisting filter spectrometer, the drawbacks of the analog electronics remain.

The most widely used technique for spectral analysis is the use of an autocorrelation spectrometer or autocorrelator; the first three papers included here are concerned with this type of instrument. These selections do not emphasize the theoretical underpinnings of the technique, which relies on the Fourier relation between the power spectrum of a signal (which is what we wish to determine) and its autocorrelation function. As a result of the power of this approach, there exists a large literature on the subject; the reader is referred to a paper by Cooper (1976) for an overview.

The autocorrelation function can be readily calculated in hardware once the incoming signal has been digitized. This first step may be a serious challenge if very large bandwidths are required, but once it is done, there are no further instabilities from analog electronics. A major advantage of the autocorrelator is that by changing the clock rate, the resolution of the system and the total frequency coverage can be changed. In general, this means that a correlator can serve the same function as a set of multichannel filter spectrometers, as long as the maximum bandwidth requirement can be met. Another factor favoring the autocorrelator approach is its relevance for interferometry. The streams of data from two elements of an interferometer can be processed in a cross-correlator and the result, after averaging and Fourier transformation, is the power spectrum of the correlated signal.

Both the hardware and the software (to carry out the Fourier transformation) have enormously improved in performance and decreased in cost in recent years, strongly favoring the autocorrelation approach. A key software development has been the Fast Fourier Transform (FFT) routine. Even with this efficient algorithm, general-purpose computers can have problems coping with the required rate of operations in a large system. One solution is special-purpose hardwired Fourier Transform processors, which are discussed in a paper by Yen (1974). One example of a hardwired instrument is the "FX" processor, in which the two streams of data are first Fourier transformed into the frequency domain and are then correlated by term-by-term multiplication. This approach is analyzed in some detail in a paper by Chikada *et al.* (1984), which also describes a particular realization for a five-element interferometer.

The first paper in this part, by Davies *et al.*, is a very clear description of a relatively early autocorrelator system used at Jodrell Bank. One-bit quantization is employed, which minimizes the complexity of the system at some sacrifice in signal-to-noise ratio (a two-bit system is discussed in a paper by Cooper, 1970). Although the 5-MHz bandwidth and 256 channels are modest by present-day standards, the flexibility of the autocorrelation approach is exploited in the provision for division into several separate autocorrelators as well as the possibility of use as a cross-correlator.

The second paper, by Urry *et al.*, describes a cross-correlator used with a three-element (and hence three-baseline) millimeter interferometer. The total bandwidth requirement exceeds that allowed by the digital circuitry, so that the input signal is divided into four subbands, and the signal in each is analyzed by a separate cross-correlator. Various other refinements including separation of the sidebands of the input millimeter downconverters are incorporated into this versatile system. The technique of combining a number of autocorrelators in order to increase the total frequency coverage is discussed from a general as well as cost trade-off point of view in the third paper, by Weinreb. This approach appears extremely promising for obtaining the resolution together with the wide spectral coverage required by single antenna and interferometer systems at high frequencies.

An entirely different method of spectral analysis relies on the diffraction of an optical beam by material with an acoustic wave propagating in it. If a radio frequency (RF) signal is converted to an acoustic wave, the angular deviation of the diffracted optical beam is proportional to the RF wavelength; by measuring the intensity of optical radiation as a function of angle we can determine the spectrum of the input signal. Acousto-optical spectrometers have been developed with bandwidths up to 1 GHz, and with on the order of 1000 frequency resolution elements. By reading out the intensity with a photodiode or charge coupled device (CCD) array, some of the channel-to-channel stability problems of the multichannel spectrometer are avoided. However, the thermal stability of the acousto-optical spectrometer (AOS) is a serious concern, and the dynamic range and stability of the readout system have been major obstacles to overcome. The development of AOS units has continued for almost 20 years. The fourth paper included here, by Masson, describes a very successful realization of an AOS for millimeter wavelength spectroscopy. A number of other papers on this subject, both theoretical and describing particular systems, are given in the References and Bibliography section which follows.

REFERENCES AND BIBLIOGRAPHY

Multichannel (Filterbank) Spectrometers

1976: Penfield, H., "Multichannel-filter spectrometers," Chapter 3.4 in *Methods of Experimental Physics: Vol. 12 Astrophysics, Part B: Radio Telescopes*, M. L. Meeks, Ed. New York, NY: Academic Press, 1976, pp. 266–279.

Williams, D. R. W., "Fundamentals of spectral line measurements," Chapter 4.2 in *Methods of Experimental Physics: Vol. 12 Astrophysics, Part B: Radio Telescopes*, M. L. Meeks, Ed. New York, NY: Academic Press, 1976, pp. 19–45.

Spectrum Expander

1979: Henry, P. S., "Variable resolution capability for multichannel filter spectrometers," *Rev. Sci. Instr.*, vol. 50, pp. 185–192, 1979.

Autocorrelation Spectrometers

1970: Cooper, B. F. C., "Correlators with two-bit quantization," *Aust. J. Phys.*, vol. 23, pp. 521–527, 1970.

1974: Bowers, F. K. and R. J. Klingler, "Quantization noise of correlation spectrometers," *Astron. Astrophys. Suppl.*, vol. 15, pp. 373–380, 1974.

Davis, W. F., "Real-time compensation for autocorrelation clipper bias," *Astron. Astrophys. Suppl.*, vol. 15, pp. 381–382, 1974.

Yen, J. L., "The role of fast Fourier transform computers in astronomy," *Astron. Astrophys. Suppl.*, vol. 15, pp. 483–484, 1974.

1976: Cooper, B. F. C., "Autocorrelation spectrometers," Chapter 3.5 in *Methods of Experimental Physics: Vol. 12 Astrophysics, Part B: Radio Telescopes*, M. L. Meeks, Ed. New York, NY: Academic Press, 1976, pp. 280–298.

1981: Bos, A., E. Raimond, and H. W. van Someren Greve, "A digital spectrometer for the Westerbork synthesis telescope," *Astron. Astrophys.*, vol. 98, pp. 251–259, 1981.

1984: Chikada, Y., M. Ishiguro, H. Hirabayashi, M. Morimoto, K.-I. Morita, K. Miyazawa, K. Nagane, K. Murata, A. Tojo, S. Inoue, T. Kanzawa, and H. Iwashita, "A digital FFT spectro-correlator for radio astronomy," in *Proc. Int. Symp. on Indirect Imaging*, Sydney, Australia, 1983, J. A. Roberts, Ed. Cambridge: Cambridge University Press, 1984, pp. 387–404.

D'Addario, L. R., A. R. Thompson, F. R. Schwab, and J. Granlund, "Complex cross correlators with three-level quantization: Design tolerances," *Radio Sci.*, vol. 19, pp. 931–945, 1984.

O'Sullivan, J. D., "The Westerbork broadband continuum correlator system," in *Proc. Int. Symp. on Indirect Imaging*, Sydney, Australia, 1983, J. A. Roberts, Ed. Cambridge: Cambridge University Press, 1984, pp. 405–413.

1986: Padin, S. and R. J. Davis, "A wideband correlator employing a single-bit by analogue multiplication scheme," *Radio Sci.*, vol. 21, pp. 437–446, 1986.

Acousto-optical Spectrometers

1973: Cole, T. W., "An electro-optical radio spectrograph," *Proc. IEEE*, vol. 61, pp. 1321–1323, 1973.

1974: Cole, T. W. and J. G. Ables, "An electro-optical spectrograph for weak signals," *Astron. Astrophys.*, vol. 34, pp. 149–151, 1974.

1977: Cole, T. W. and D. K. Milne, "An acousto-optical radio spectrograph for spectral integration," *Proc. Astron. Soc. Australia*, vol. 3, no. 2, pp. 108–111, 1977.

1980: Chin, G., D. Buhl, and J. M. Florez, "Acousto-optic spectrometer for radio astronomy," in *Proc. 1980 Int. Optical Computing Conf.* (SPIE vol. 231), 1980, pp. 30–37.

1981: Malkamaki, L., "An acousto-optical radiospectrometer system for 22 GHz region line observations," *Astron. Astrophys.*, vol. 98, pp. 15–18, 1981.

The Jodrell Bank Radio Frequency Digital Autocorrelation Spectrometer

by

R. D. DAVIES
J. E. B. PONSONBY
L. POINTON
G. DE. JAGER

Nuffield Radio Astronomy Laboratories,
Jodrell Bank,
Macclesfield, Cheshire

A versatile digital autocorrelation spectrometer designed and built at Jodrell Bank is described, together with some examples of observations made with the new instrument.

SPECTRAL line studies have had an important place in radio astronomy since the discovery of the 21 cm line from neutral hydrogen in 1951. This line has been used to investigate the structure of the Milky Way and of external galaxies and to study the physical conditions in the interstellar medium. Recently new areas of radio spectral line investigations have emerged with the discovery of emission from the OH radical at a number of frequencies and from the recombination line spectrum of hydrogen, helium and the heavier elements which extend throughout the entire radio frequency range. In addition there have been searches for the emission from other radicals such as CH and H_2^+; the NH_3 emission at 1·25 cm wavelength and the H_2O emission at 1·35 cm wavelength have been detected within the past year.

The technical problem in obtaining spectra from a variety of lines each being emitted from a range of physical conditions is essentially one of providing a versatile spectrometer which can give the required resolution for the particular investigation in hand; it may be over a wide frequency band or over a narrow band, covering the lines to be studied. For example, it is necessary in the study of the narrow band anomalous emission from the OH radical to have a resolution of approximately 200 Hz over a band of 40 kHz. On the other hand, for studies of external galaxies at the 21 cm line of neutral hydrogen the emission can occur over a frequency range of 3 MHz and structure in frequency of 20 kHz may be present. Many types of investigation require overall bandwidths and resolutions intermediate between these two cases. Furthermore, long integrations are often required in the study of weak spectral lines and this requires extreme gain stability across the passband of the spectrometer. The solution to this problem in the early days of H-line work was to use a number of adjacent analogue filters to cover a reasonable frequency range; there could be either many filters (~100) at fixed frequencies or a smaller number which could be swept in frequency. To cover the range of possible astronomical investigations separate systems had to be made for each overall bandwidth.

An alternative to the measurement of the power spectrum $P(f)$ of a signal $x(t)$ by the direct use of filters and individual detectors is to determine its time delay autocorrelation function $R(\tau)$,

$$R(\tau) = \int x(t).x(t - \tau)\mathrm{d}t \qquad (1)$$

which is related to the signal power spectrum through the Fourier transform

$$P(f) = \int R(\tau)\cos(2\pi f\tau)\mathrm{d}\tau \qquad (2)$$

In these expressions f is the frequency and τ is the time delay. The advantage of this approach is that digital techniques may be used, and if a digital store is used to delay the signal then a number of alternative overall bandwidths may be explored according to the chosen clock rate.

The radio astronomical signals are noise like and follow a Gaussian amplitude probability distribution. For this class of signal little information is lost if the incoming signal is clipped so that only polarity information is preserved before the autocorrelation function is taken. The penalty of clipping is that to achieve a certain accuracy in the measurement of the power spectrum approximately twice the observation time is required compared with an equivalent analogue system. But the advantages are that it allows a considerable simplification in the instrumentation, for only a binary or "one-bit" representation of the signal is required and it relaxes the gain stability requirements of the receiver. This approach was first applied to the analysis of astronomical signals by Weinreb[1]. He constructed a "one-bit" digital auto-correlating spectrometer having twenty-one channels in delay τ, and capable of handling signals in a bandwidth up to 150 kHz. He demonstrated its use in an attempt to detect Zeeman splitting of the 21 cm line of neutral hydrogen and in a search for emission due to the corresponding line of deuterium.

Our proposal to develop a similar but larger and more versatile system for radio astronomy was accepted in December 1964 by the R. W. Paul Instrument Fund of the Royal Society, and a grant was made. The instrument has 256 channels in delay and is capable of processing signals contained in a nominal bandwidth of up to 5 MHz. It has been constructed to work on-line in conjunction with a Ferranti Argus 400 computer which carries out the Fourier transform and associated computations.

The radio frequency signals received with the radio telescope are amplified and frequency translated into an I.F. band in the vicinity of 30 MHz. Using a single sideband frequency translator the band of interest is placed in the band 0 to B where B is the nominal bandwidth. A number of filters have been constructed which enable any one of eight possible bandwidths to be selected. These range down in binary steps from 5 MHz to 39 kHz. The signal emerging from the filter is $x(t)$.

A "one-bit" signal, $y(t)$, is generated according to the rule

$$y(t) = \begin{array}{ll} +1 & x(t) > 0 \\ -1 & x(t) < 0 \end{array} \qquad (3)$$

and $y(t)$ is sampled at regular intervals at the clock rate f_s, which is twice the nominal bandwidth B. These two operations are effected simultaneously by a single clipper-sampler. This is enclosed in a feedback loop which removes any d.c. component or bias in the sampler by imposing the condition that over periods long compared with $1/f_s$ the number of positive and negative results are the same. The individual polarity samples are handled by two-state digital circuits having significance levels of $+1$ and -1. The samples are delayed for successive elementary times $\Delta t = 1/f_s$, by being passed along a 256 stage shift register which shifts at the sampling rate.

The correlator is shown schematically in Fig. 1. It consists of the 256 stage shift register which shifts at the clock rate. The state of each stage of the shift register is compared with that of the first stage after each successive shift. If they are the same, then a pulse is applied to the input of the corresponding twenty stage binary counter, of which there are 256 in all, making the number in it

increase by one; if the states are different no pulse is applied. Each binary counter has in effect a further twenty-three stages which continue in a computer store location. Information is continuously transferred to the extra stages at a rate appropriate to the sampling frequency. Each shift register stage, its comparator, associated binary counter and store location, forms one channel of the instrument. In operation the contents of all the counters are initially zero and after running for a time $K\Delta t$, corresponding to K samples of the signal, the channels contain numbers C_n, where n is the channel number given by

$$C_n = \tfrac{1}{2}\left[\sum_{k=1,2,3...}^{k=K} y(k\,\Delta t) . y(k\,\Delta t - n\,\Delta t)\right] + \frac{K}{2} \qquad (4)$$

It is clear from this expression that the result for the zeroth channel is $C_o = K$. Ideally one would subtract 1 from the contents of the counters when the inputs to the comparators are different. This would have been true "one-bit" multiplication and the results would have been R'_n given by

$$R'_n = \sum_{k=1,2,3...}^{k=K} y(k\,\Delta t).y(k\,\Delta t - n\,\Delta t) \qquad (5)$$

which is the true sampled "one-bit" autocorrelation function. The first task of the Argus computer is to recover R'_n by computing

$$R'_n = 2C_n - K \qquad (6)$$

and then to form the normalized "one-bit" autocorrelation function A'_n given by

$$A'_n = \frac{R'_n}{R'_o} \qquad (7)$$

This normalized autocorrelation function is not the same as would have been obtained if amplitude information had not been rejected at the clipper. It has been shown, however, by Van Vleck[2] that the distorting effect of the clipping on the final power spectrum may be removed by converting to A_n according to

$$A_n = \sin\left(\frac{\pi}{2} A'_n\right) \qquad (8)$$

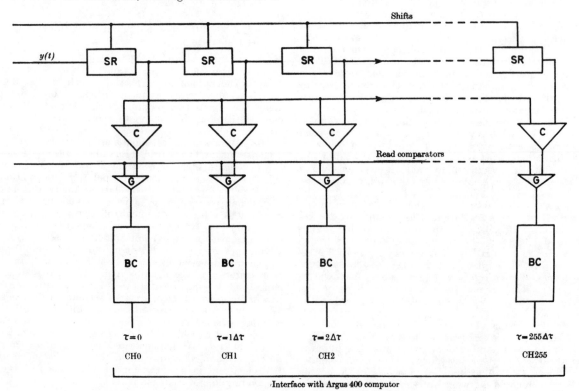

Fig. 1. Block diagram of the correlator. SR, 1 stage shift register; C, comparator; G, NOR gate; BC, binary counter

Fig. 2. The digital autocorrelation spectrometer.

provided that the amplitude probability distribution of the original signal $x(t)$ was of Gaussian form. This Van Vleck correction is also applied by the Argus computer.

The normalized observed power spectrum is obtained by computing the Fourier series

$$P_m = [2 \sum_{n=0}^{n=n'} A_n . w_n . \cos (2\pi n . m . \Delta f . \Delta t)] - A_o w_o \qquad (9)$$

for values of $m \Delta f$ up to B. Here n' is the number of the channel of greatest delay ($n' = 255$), and $n' \Delta f = B$. The quantities w_n are a weighting function whose Fourier transform alone is the effective filter response with which the unknown spectrum is convolved. Appropriate choice of weighting function enables the side-lobe level of the effective filter response to be reduced at the expense of degrading the frequency resolution. Failure to apply explicit weighting ($w_n = 1$) gives an effective filter response of the form $\dfrac{\sin \theta}{\theta}$ in frequency and the highest attainable resolution, but makes the first side-lobes -22 per cent in power.

A photograph of the instrument is shown in Fig. 2. It is mounted in a standard double 19 inch rack. The right hand side contains the analogue portion of the instrument and the clipper-sampler. The other half contains the digital section. This is built entirely from Texas Instruments 74N series TTL integrated circuits; approximately 3,500 dual-in-line packages are used. The bulk of the machine consists of the shift register stages, the comparators and the twenty stage binary counters. The 256 hardware channels are mounted in groups of four on sixty-four identical double-sided glass fibre printed circuit boards, one of which is shown in Fig. 3. Another forty-five printed circuit boards of various types carry the

control logic and the pulse routing system. All the printed circuit boards were designed and manufactured at Jodrell Bank.

One of the functions of the control logic is to enable the instrument to be used in one of several different modes. The basic mode is as a 256 channel autocorrelator, but the channels may also be split into groups of thirty-two, sixty-four, or 128, so that several different signals can be processed simultaneously with reduced frequency resolution, although all must use the same sampling rate. Each group may be used either for autocorrelation or cross-correlation, the cross-correlation facility being provided to allow the instrument to be used for interferometric observations.

The function P_m obtained from the Fourier transform process is the power spectrum of the total signal at the input to the clipper-sampler. It is the sum of the astronomical signal and broadband noise from the receiver itself, both weighted by the system frequency response. The final filters defining the passband before the clipper-sampler are flat to within about 0·2 db over the useful part of the nominal bandwidth (~ 85 per cent), but the form of the overall passband is also determined by the frequency response of the preceding portions of the receiver, and may typically vary by $\sim 0·5$ db over the 5 MHz passband. To obtain the spectrum of the astronomical signal alone, the form of the system response must be determined and the contribution due to the receiver noise eliminated by subtracting a reference spectrum obtained either by moving the radio telescope to a position away from the source of interest in the sky, or alternatively by tuning the receiver away from the frequency of interest.

Assuming that the system temperature is constant over the passband, a spectrum taken without an astronomical signal gives the form of the system frequency response. Very often the same spectrum may serve as the reference for subtracting the receiver noise contribution. To observe one component of an astronomical signal while eliminating another, a separate reference spectrum must be obtained. Examples of this technique are in the measurement of the H-line spectrum of M 31 discussed below, and also in the H-line absorption measurements of the emission of the pulsar CP 0328. In that case the emission from galactic hydrogen was eliminated by taking a reference spectrum after moving off the pulse in time.

Fig. 3. Printed circuit board holding four channels of the correlator. (One of sixty-four identical boards.)

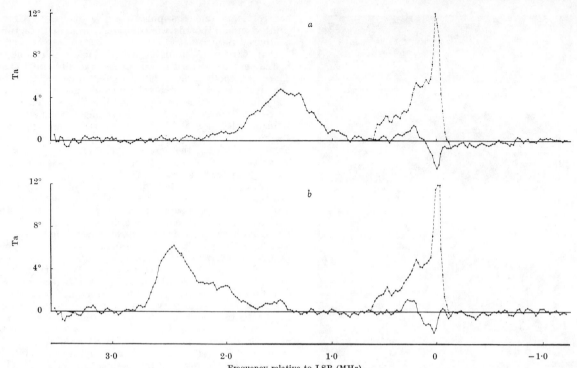

Fig. 4. Neutral hydrogen emission from the Andromeda Nebula (M 31). Overall bandwidth 5·0 MHz; 256 channels in delay; resolution 35 kHz (to half power); integration time 3 min. a, ———, Spectrum at centre of M 31 with local hydrogen contribution removed; - - - - , form of local hydrogen spectrum at this position. b, ———, Spectrum at position 50′ arc away from centre along SW semi-major axis. Local hydrogen contribution removed; - - - - , form of local hydrogen at this position.

Examples of astronomical spectra taken during the commissioning phase of the correlation spectrometer are given in Figs. 4–7 which demonstrate the versatility of the instrument. All the spectra were taken with the Mark II radio telescope between August and November 1968; during some of this time when the computer was also processing data from the Mark I radio telescope the computation time was only sufficient to process eighty channels from the correlator.

The large velocity spread of the neutral hydrogen in the Andromeda Nebula (M 31) requires observations at the largest overall bandwidth of 5·0 MHz. The observed spectra are shown in Fig. 4. The spectrum of Fig. 4a is that for the centre of M 31, and was obtained using a reference spectrum taken from a position 2° north-west of the centre. The contribution from local galactic hydrogen is substantially the same in the two positions. Consequently the result of subtracting the reference spectrum,

shown with the solid line in Fig. 4a, has only a residual galactic contribution. The estimated form of the galactic spectrum at the centre was found by an independent observation and is shown by the dashed line. The spectra of Fig. 4b are similar results obtained at a position 50′ arc from the centre on the south-west semi-major axis of M 31.

Galactic neutral hydrogen spectra can in most places be covered with overall bandwidths of 0·625 MHz or 1·25 MHz. For this type of observation the instrument provides a baseline sufficiently well defined to give reliable estimates of the weak emission at outlying frequencies. The resolution provided by the instrument at these overall bandwidths appears to be sufficient for all the emission features near the galactic plane. Examples of these are shown in Figs. 5a, b and c.

The observed narrow absorption feature in the spectrum of Taurus A is shown in Fig. 6b. It was obtained with an

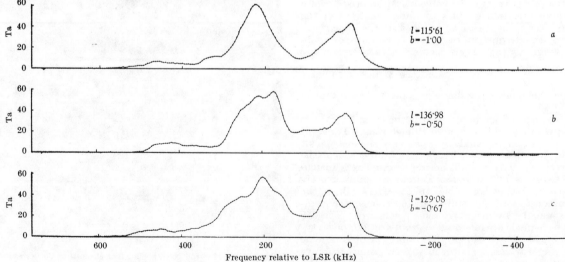

Fig. 5. Neutral hydrogen emission spectra near the galactic plane in the Milky Way near longitudes (a) 116°, (b) 137°, (c) 129°. Overall bandwidth 1·25 MHz; 256 channels in delay, resolution 8·8 kHz; integration time 3 min. Reference spectra taken at a frequency displaced 1 MHz.

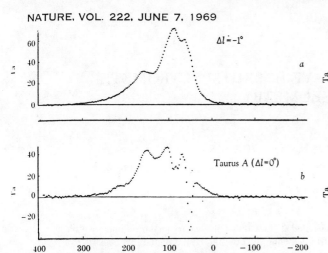

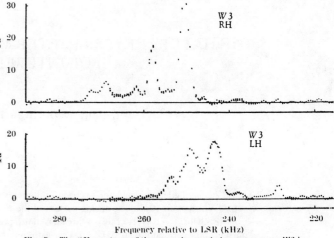

Fig. 6. *a*, The neutral hydrogen emission spectrum observed at a position −1° in longitude away from Taurus A. *b*, Similar spectrum obtained on Taurus A showing deep absorption feature. Overall bandwidth 625 kHz; 256 channels in delay; resolution 4·4 kHz; integration time 3 min. Reference spectra taken at a frequency displaced 600 kHz.

Fig. 7. The OH spectrum of the anomalous emission source near *W*3 in right and left hand circular polarization. Three overlapping spectra were taken on each polarization. Overall bandwidth of observation 39 kHz; 80 channels in delay; resolution 900 Hz; integration time 4 min. Reference spectra taken at a frequency displaced 200 kHz.

overall bandwidth of 625 kHz and a resolution of 4·4 kHz. This is evidently not quite sufficient to resolve the feature fully. For comparison, the emission spectrum obtained with the same resolution at a position −1° away in longitude is shown in Fig. 6*a*.

The narrowest known spectral features occur in the regions of anomalous OH emission. Spectra of the source *W*3 at 1,665 MHz observed in left and right hand circular polarization are shown in Fig. 7. Each was observed in three overlapping bands taken when only eighty channels of the correlator could be used. The agreement in the overlap regions of the three bands demonstrates the reliability of the observed spectra.

All the spectra shown have been observed using a triangular weighting function w_n going linearly from $w_0 = 1$ to $w_{n'} = 0$. ($n' = 256$ in this case, for a channel the contribution of which is given zero weight need not exist.) This gives an effective filter response of the form $\left(\dfrac{\sin \theta}{\theta}\right)^2$ in frequency with a spacing of approximately $4B/n'$ between the zeroes on either side of the main lobe and with the first side-lobes +4·7 per cent of the main lobe in power. The reference spectra were all taken with the same weighting function and using the same integration times as the corresponding source spectra. The reference spectra for the results shown in Figs. 5, 6 and 7 were all obtained by tuning off in frequency.

A detailed discussion of the principles of operation and the main design considerations of the one-bit digital correlator will be given by J. E. B. Ponsonby and L.

Pointon, who were responsible for its design and construction. G. de Jager wrote the associated programs for the Argus 400 computer.

The instrument has already been used for a number of astronomical programmes. These include a determination of the distance of the pulsar *CP* 0328 from a study of its neutral hydrogen absorption spectrum[3], an investigation of the polarization and variability of OH emission, and Zeeman effect measurements on neutral hydrogen seen in emission and absorption. Observations have been made of the frequency structure in the continuum emission from pulsars[4]. The correlator has also been used in the cross-correlation mode to obtain fringes on the Mark I–Mark III interferometer using a tape recording system and an independent local oscillator.

We thank the Royal Society for the grant from the R. W. Paul Instrument Fund. We particularly thank Dr J. S. Hey, who acted as assessor on behalf of the Fund committee, for his interest and encouragement. We thank R. S. Booth and Miss A. J. Wilson for allowing us to reproduce some of their observational results and also Professor J. G. Davies for providing the facilities of the Argus 400 computer system.

Received March 21, 1969.

[1] Weinreb, S., *MIT Technical Report No. 412*, Dept. Electronic Engineering (1963).
[2] Van Vleck, J. H., *Rep. No. 51, Radio Research Lab.*, Harvard (1943), and *Revised and reprinted edition, Proc. IEEE*, **54**, No. 1 (1968).
[3] de Jager, G., Lyne, A. G., Pointon, L., and Ponsonby, J. E. B., *Nature*, **220**, 128 (1969).
[4] Rickett, B. J., *Nature*, **221**, 158 (1969).

THE HAT CREEK MILLIMETER-WAVE HYBRID SPECTROMETER
FOR INTERFEROMETRY

W. L. URRY, D. D. THORNTON, AND J. A. HUDSON

Radio Astronomy Laboratory, University of California, Berkeley, California 94720

Received 1985 April 3

A hybrid analog-digital spectrometer has been built and is installed at Hat Creek Observatory on our millimeter-wave inferometer (Welch et al. 1977). The analog portion of the instrument comprises the filters; the digital part consists of a three-level correlator. The correlator serves three baselines, providing 512 complex channels for each. The bandwidth of the digital portion of the correlator is 40 MHz. The correlator may be partitioned in a flexible way, so that each part is fed by one sideband of a filtered portion of the band; in this way a total bandwidth of 320 MHz may be achieved. Other configurations are possible. In this report we describe the system in adequate detail for users of the instrument to understand it; furthermore we expect that our solutions to such problems as separating the sidebands of the first local oscillator will be of general interest.

Key words: instrumentation—radio astronomy—data handling techniques

I. Introduction

The power spectrum determined from digital correlation measurements is essentially equivalent to that obtained from a bank of filters. The advantage of digital correlation lies in the intrinsic digital nature of the measurement (less expensive components, more stability, greater reliability, more flexibility). We shall not go into the theory of digital correlation techniques here, but instead refer the reader to Weinreb (1963), Hagen and Farley (1973), and Cooper (1976). A digital correlator is limited in bandwidth by the rate at which it is sampled and shifted; for 80 MHz shifting (that of our correlator), a bandwidth of only 40 MHz can be achieved. Hence, several correlators must be used in parallel to achieve greater bandwidths; in our instrument these work in conjunction with various filters and mixers to provide not only greater band coverage, but a good measure of flexibility. Another limitation of digital correlators lies in the coarseness with which the signal is sampled; this results in slightly less sensitivity than a filter bank.

Our correlator serves a three-element interferometer, providing 512 channels per interferometer baseline. The 480 MHz IF of each antenna is supplied with four independently tuned, double-sideband receivers, which offer a frequency coverage of 40 MHz for each sideband, thus totaling 80 MHz each. Like a filter bank, the receivers may be adjacently tuned to cover a total continuous bandwidth of 320 MHz. When the correlator is partitioned for this maximum band coverage, each of the eight filters in the system is followed by a 64-channel correlator segment for a resolution of 1.25 MHz.

The bandwidth of the correlator is selectable in octave steps from 40 to 1.25 MHz. All 512 channels of the correlator may be dedicated to one of the IF receiver sidebands for a maximum resolution of 4.9 kHz. If desired, the system may be tuned to two different sample rates giving the ability to observe at two different bandwidths at once. Since the receivers are independent they can be split up and used to observe, at some future date, two polarizations. A simple block diagram of the analog filter portion of the system for each antenna is shown in Figure 1. The many spectral features available at millimeter wavelengths make the simultaneous observation of both first local oscillator sidebands, spaced 3 GHz apart, an attractive feature. Figure 2 illustrates an observation made of five different molecular lines in Orion A, all obtained simultaneously.

II. Sideband Separation and Fringe Stopping

The receiver is a three local oscillator (LO) stopped-fringe system. The sidebands of the first LO are separated via quadrature switching of the first LO phase. When both antennas are in phase, the cross-correlation measurement is real. When the antennas are in quadrature, the cross-correlation measurement is imaginary. The composite of these, a complex correlation measurement, allows the sidebands about the first local oscillator to be separated. After mixing with the first LO, the frequencies from 0 to 1230 MHz are filtered out, so that subsequent mixing with the second LO at 1230 MHz is free of sideband overlapping.

While the first LO is modulated with a phase shift of 90°, the second LO is modulated with a phase shift of 180°; this may be used to eliminate any DC offset present in the receiver. It is necessary to modulate the outputs of the correlators in synchronization with the phase shifts of the local oscillators; the data are placed in the real accumulator or the imaginary accumulator, depending on the states of the 90° LO phase at each antenna; the sign applied to the data (whether they are to be added or subtracted) is

Reprinted with permission from *Publ. Astron. Soc. Pac.*, vol. 97, pp. 745–751, Aug. 1985.

determined from the states of the modulating functions in a manner detailed in the Appendix and Figure 3.

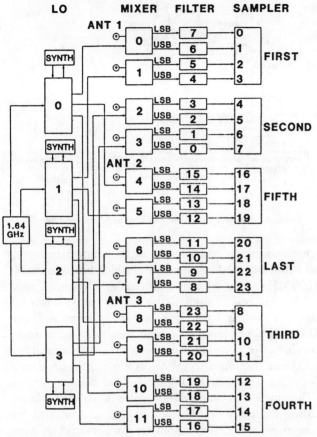

FIG. 1—Analog filter section for the Hat Creek correlator, showing the synthesizers (3rd LO), mixers, single-sideband filters, and samplers.

A scheme similar to ours for modulating the LO phase was employed by Wright et al. (1973) for the NRAO interferometer at Green Bank. In the NRAO system, the phase and sign modulations were generated by square waves of differing period, a total cycle time of 2^n times the shortest period being rquired. This amounted to some 160 seconds for the Green Bank interferometer. Wishing to shorten this period while keeping the modulations orthogonal to one another, we made use of Walsh functions to control the modulators. With this innovation, we are able to go through all possible switchings in 320 msec. Walsh function switching has the advantage that, if an additional modulator or telescope is added to the system, one only adds a new Walsh function which is orthogonal to the others, and this function usually does not require an increase in the fundamental cycle time. An excellent reference on the nature of Walsh functions and their practical application is Beauchamp (1975).

The fringe is stopped in a rather unique manner. The reference for the first or second local oscillator is mixed with a digitally controlled oscillator in a single-sideband mixer. The output of this oscillator consists of two 8-bit digital-to-analog converters which generate the sine and cosine values read from a table in a read-only memory (ROM), which is addressed by a binary counter. The phase of the oscillator is controlled in $1°4$ steps by adding an 8-bit number to the address produced by the binary counter. The digitally controlled oscillator has the attractive feature that its phase may be easily modulated by either 90° or 180° by subtracting or adding a binary number to the ROM address. The binary counter which pro

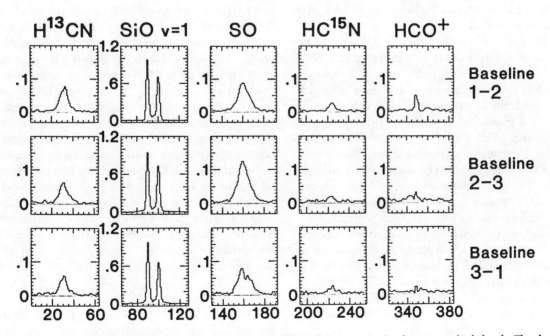

FIG. 2—Molecular lines observed simultaneously in Orion A, showing capability of spectrometer for observing multiple bands. The abscissae show channel number; the ordinates give relative power (uncalibrated). The channel width is 1.25 MHz, for a total band coverage (in segments) of 320 MHz.

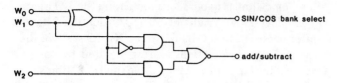

FIG. 3—Logic for demodulating, in the correlator, the Walsh functions which were applied to an antenna pair (LO1 and LO2, both antennas).

duces the address is updated from a clock source made up of an 18-bit synchronous binary rate multiplier (SN7497). The digitally controlled oscillator may be tuned in steps of 0.58×10^{-3} Hz from 0 to 38 Hz. The method is very inexpensive and precise. (See Fig. 4.)

The wide spacing of the first local oscillator sidebands (3 GHz) gives them different fringe rates so that a dual fringe rotator system is employed. The first LO is controlled by a fringe rotator that stops the fringe occurring at the LO frequency. The fringes in the two sidebands rotate in equal but opposite directions relative to the first local oscillator. When the sidebands are translated to 1.5 GHz the frequency sense of the lower sideband is reversed and the fringes rotate in the same direction relative to each other. A second fringe rotator on the second LO completes fringe removal.

III. The Third LO's and Filters

After the amplified IF is delayed and brought into the control room, it is mixed with four tunable local oscillators in eight single-sideband mixers. These third LO's may be programmed in steps of roughly 10 kHz, from 80 to 520 MHz, thus bringing any spectral feature of interest in the IF passband within reach of the correlator.

For the third LO we chose a fractional-phase-lock loop design. The fractional-phase-lock technique allows the use of a higher loop bandwidth than would otherwise be possible for a synthesizer of the required step size. The wider loop bandwidth reduces phase noise to an acceptable level, while the design employs a commercially available voltage-controlled oscillator (VCO) having wide tuning range. In a phase-lock frequency synthesizer, the output frequency is divided down to a reference frequency where a comparison is made between the divided phase and a reference phase. The divisor is an integer. This integer is switched between two values to achieve the equivalent of dividing by a fractional value. The phase error resulting from the truncated divisors is predictable, since the switching pattern is fixed; a circuit is employed which cancels this out to a few parts in 10^3 (resulting in spurs of the fundamental frequency which are on the order of 50 dB down). References on digitally controlled oscillators are Egan (1981), Manassewitsch (1980), and the entire February 1981 issue of *The Hewlett Packard Journal*.

Our LO synthesizer is illustrated in Figure 5. Desiring

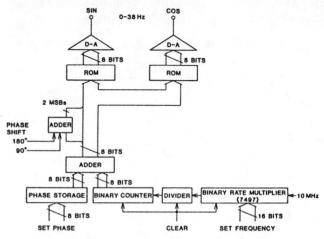

FIG. 4—Digitally controlled oscillator for fringe rotation.

to use commercially available parts whenever possible, and not being able to obtain a VCO which tuned to the 80 to 500 MHz range, we used a product which tuned from 1 to 1.6 GHz and translated the tuning range down by mixing it with a 1.64 GHz beat frequency oscillator (BFO). The frequency inversion results in a fortunate cancellation of loop nonlinearity due to the divisor and the VCO tuning nonlinearity. We thus avoided the need for a special linearizing circuit in the feedback loop.

The single-sideband mixers make use of two phase-shifting RC networks which together achieve a phase difference of 90°. Outputs from the two networks are then summed or differenced in order to provide the upper or lower sideband of the input IF. Each network is a 6-pole all-pass circuit, optimized for the frequency range 2 kHz to 40 MHz, with (theoretically) 39 dB sideband rejection. Both Albersheim and Shirley (1969) and Rogers (1971) present methods for designing all-pass phase-shifting networks. The method actually followed by us was first to compute the poles of the networks using a program, "Glorypoles," written by A. G. Lloyd (1976). Computation of the resistances and capacitances required to align poles with the zeros was done by an iterative optimization program, designed by us, which adjusted selected resistances, each in turn, until a variance function indicated sufficient alignment. After optimization (which required several hours' time on an LSI-11 microprocessor), realistic values for those standard resistances available were substituted for the continuous-valued components, with little degradation of the circuit. Tolerance values of 5% are sufficiently precise. In reality, the mixers all reduce the unwanted sideband about 30 dB.

IV. The Samplers

The samplers may be programmed to operate in single-bit mode. Single-bit sampling consists of merely reporting +1 or −1, depending upon whether the signal is

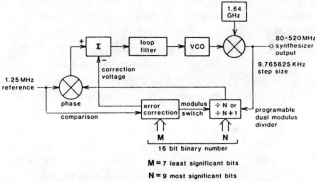

FIG. 5—Third LO synthesizer phase lock scheme, showing programmable divisor and Voltage Controlled Oscillator.

positive or negative. 1.6-bit sampling tests whether the signal exceeds a level, called the *clipping voltage*, which may be programmed by the control computer. A result, +1, 0, or −1 is reported. While clipping seems drastic, the amount of noise introduced may be compensated for by integrating longer by a factor of 2.46, for single-bit sampling, or 1.51, for 1.6-bit sampling. Additionally, a nonlinear correction must later be made to the correlation. If r is the normalized correlation (normalized by dividing out a number representing perfect correlation), the corrected correlation ρ is given by

$$\rho = \sin \frac{\pi}{2} r \quad ,$$

for the one-bit case (Van Vleck and Middleton 1966); and for the 1.6 bit case

$$\rho \approx Ar + Br^3 , |\rho| < 0.86 \quad ,$$

or

$$\rho \approx \text{sgn}\,(r) \cos[\pi(r_0 - |r|)]\, e^{\frac{1}{2}c^2}, |\rho| \geq 0.86 \quad ,$$

where

$$A = \frac{\pi}{2} e^{c^2} \quad ,$$

$$B = -A^3(c^2 - 1)^2/6 \quad ,$$

c is the clipping level (in standard deviations), and r_0 the zero lag raw correlator count (normalized). See Kulkarni and Heiles (1980). (These authors have also performed a most useful analysis of the effect shifting clipping levels has upon the spectrum.) The approximations in the latter case are good to about one part in 10^3.

These corrections presuppose the signal is Gaussian in nature. (This works well for most naturally occurring signals. However, a 1.6-bit sampling has trouble with sine wave CW signals (producing harmonics that are not really present); the Van Vleck correction for 1-bit sampling is valid for a single sinusoid as well as Gaussian.)

With two- or three-level sampling, the multiplier circuitry in the correlator may be kept simple and fast.

V. The Correlator Modules

The correlator boards make use of a design that was completed in 1980 and installed on the 85-foot telescope at Hat Creek. The new correlator takes advantage of recent advances in semiconductor technology and uses the new "F"-type integrated circuits developed by Fairchild in place of the low-power Schottky TTL devices used in the original design. The result is a machine that has a clock rate of 80 rather than 40 MHz. A total of 48 of these 32-channel 80-MHz boards are utilized to provide a three-baseline system with 512 complex channels per baseline.

Each card has only 17 bits of accumulator following the multiplier for each channel. Twelve bits of the accumulated result are transferred to a shift register every 1.25 msec. As the card continues to integrate, the 12-bit numbers are shifted out 256 channels at a time in a single data stream at a 3 Mbit rate into a processor card that accumulates the numbers. The processor stores the numbers in the real or imaginary correlation storage bank depending upon the state of the first local oscillators in the two antennas. If both antennas are in phase, then the cross-correlation is real and the numbers are stored in the real correlation storage bank. If the two antennas are in quadrature phase, then the numbers are stored in the imaginary accumulators. The processor card also demodulates the signal inversions that occur at the antennas. (See Fig. 3.) The processor is double-buffered so that while the computer is reading out the data from one storage area, the processor continues to accumulate data in a second buffer. Thus integration of the correlations may continue without interruption.

The 1.25-msec integrate-dump interval of the correlator cards determines the shortest interval of Walsh switching that can occur. The system has a 256 Walsh function capacity making a system cycle time of 320 msec. Integrations are done in 320-msec increments.

Each correlator card may receive its input data from a pair of inputs, each designated by a 3-bit code. Four of the eight choices are the four different inputs coming from the final (third LO) mixers. The sideband chosen is dependent upon the slot occupied by the correlator card (saving a bit in the multiplexer selection). Two additional bits provide for passing the signal along from card to card. Considerable freedom in correlator configuration may thus be attained.

VI. Observing Options

While the correlator hookup is quite flexible, the combinations of shift rates, filter selections, and choice of correlator modules present a perplexing array of options to the observer. Accordingly, we have, via software, hemmed in the choices to four principal modes of correlator operation.

In any mode the local oscillators may be independently tuned anywhere from 80 to 510 MHz. Each receiver may be configured as part of an interferometer (cross-correlation) or as an independent receiver (auto-correlation). In any mode there will be a total of 512 real channels in the auto-correlation configuration and 512 complex channels in the cross-correlation configuration. When auto-correlations are made, the upper and lower sidebands about the first local oscillator cannot be separated and they will appear superimposed. When cross-correlations are observed, however, the two first LO sidebands will be separated with 256 complex channels for each sideband and each half will receive the frequency coverage referred to below.

Mode 1. Third LO No. 1 upper sideband dedicated to all 512 channels with 40 MHz maximum bandwidth.

Mode 2. Third LO No. 1 lower sideband and third LO No. 2 upper sideband with 256 channels each and 80 MHz maximum bandwidth in two independently tuned 40 MHz passbands.

Mode 3. Third LO Nos. 1 and 2 lower sidebands and third LO No. 3 and 4 upper sidebands with each sideband having 128 channels of correlator for a maximum bandwidth of 160 MHz in four independently tuned 40 MHZ passbands.

Mode 4. Third LO Nos. 1, 2, 3, and 4, both sidebands, with each of the eight sidebands having 64 channels of correlator for a maximum bandwidth of 320 MHz in four independently tuned 80 MHz passbands.

VII. Troubleshooting

Digital correlators are frequently run at the ragged edge of the switching capability of their logic circuits. Consequently, it is necessary to perform occasional diagnostic checks.

Errors usually arise in the multipliers. These have the effect of causing a bias in the count for one of the channels. Errors in clocking the lag shift register will cause a jump in counts from one channel to the next, with the remainder of the correlator registering a bias, caused by propagation of the error. This is fortunately not so common. Bad individual channels are easily identified and may be flagged for removal in computation of the spectrum.

Our correlator system includes a diagnostic feature that enables bad channels to be easily identified. A pseudorandom noise generator has been built that duplicates the three-level code recognized by the correlator. This is cross-correlated with a short delay to remove the zero lag correlation from analysis; a properly working correlator should have a correlation of exactly zero in all lag channels at the end of the integration interval.

Practice has indicated that it is also wise to run bad channel tests on the full system. Since each antenna IF is input to two different correlators (in order that cross-correlations may be taken), it is then possible to cause two different correlators to observe an auto-correlation of the same IF input, producing, presumably, the same digital correlation. When these are differenced, the bad channels stand out. The test must be repeated in round-robin fashion, until each correlator has been differenced at least twice. For example, we would observe differences $\Gamma_{12} - \Gamma_{31}$, both correlators having observed auto-correlations from antenna 1; then we take $\Gamma_{23} - \Gamma_{12}$, auto-correlated with antenna 2; and finally $\Gamma_{31} - \Gamma_{23}$, auto-correlated with antenna 3. If, for instance, a given channel is found to be bad in the first data set, and also the second, then we know that the bad channel belonged to the correlator computing Γ_{12}.

VIII. Future Expansion

We hope to acquire three additional millimeter antennas (Welch and Heiles, 1983). If this becomes possible, it will be a simple matter to expand our correlator from the three baselines, currently operating, to 15. This only entails duplication of existing circuit boards.

We are indebted to Mr. Robert Abe, the electronics shop supervisor for the Radio Astronomy Laboratory, who coordinated the entire project and also had a hand in much of the construction of the correlator. We also thank Mr. Donald Roth, who did much of the assembly, Mr. Calvin Cheng, who did much of the testing and adjustment, and Mr. Peter Winship (now a graduate student at Berkeley in the Electrical Engineering Department), who constructed the filters and single-sideband mixers. We also thank Mr. Wilson Hoffman, who was responsible for the on-line control system software. The scheme for Walsh function modulation of the LO phases was proposed by Professor W. J. Welch (also the interferometer project coprincipal investigator); this is detailed in the Appendix. This project was funded by the National Science Foundation.

REFERENCES

Albersheim, W. J., and Shirley, F. R. 1969, *I.E.E.E. Trans. on Circuit Theory* **CT-16**, 189.

Beauchamp, K. G. 1975, *Walsh Functions and Their Applications* (London: Academic Press).

Cooper, B.F.C. 1976, *Methods of Experimental Physics* **12B**, 260 (Chap. 3.5).

Egan, W. F. 1981, *Frequency Synthesis by Phase Lock* (New York: John Wiley & Sons).

Hagen, J. B., and Farley, D. T. 1973, *Radio Science* **8**, 775.

Kulkarni, S. R., and Heiles, C. 1980, *Ap. J.* **85**, 1413.

Lloyd, A. G. 1976, *Electronic Design* **19**, 90.

Manassewitsch, V. 1980, *Frequency Synthesis Theory and Design* (New York: John Wiley & Sons).

Rogers, A. E. E. 1971, *Proc. I.E.E.E. Letters* **59**, 1617.

Van Vleck, J. H., and Middleton, D. 1966, *Proc. I.E.E.E.* **34**, 2.

Weinreb, S. 1963, *M.I.T. Res. Lab. of Electronics Tech Report* 412.

Welch, W. J., and Heiles, C. E. 1983, The Hat Creek Millimeter Array, Proposal to the N.S.F., Radio Astronomy Laboratory, U. C. Berkeley.

Welch, W. J., Forster, J. R., Dreher, J., Hoffman, W. Thornton, D. D., and Wright, M. C. H. 1977, Astr. Ap. 59, 379.

Wright, M. C. H., Clark, B. G., Moore, C. H., and Coe, J. 1973, Radio Science 8, 763.

APPENDIX

Derivation of Correlated Signal for a Single Baseline

The derivation presented here is that done, in large part, by W. J. Welch (see our acknowledgement, above). Let the voltage at antenna l be

$$y_l(t) = U_l \cos[\omega^U(t - \tau_{gl}) + \phi^U_l] + L_l \cos[\omega^L(t - \tau_{gl}) + \phi^L_l] \quad,$$

where we consider two frequencies, equally spaced relative to our first local oscillator, with amplitudes U_l and L_l, representative of the upper and lower sidebands with respect to that frequency. Both amplitude and phase (ϕ) of the signal carry information about the source brightness distribution and geometry. The signal has been delayed in time by τ_{gl}, relative to some center of reference, called the delay center. If $\hat{n}$ is a vector pointing toward the source, and r_l the position of the antenna in relation to the delay center, then

$$\tau_{gl} = -r_l \cdot \hat{n}/c \quad.$$

The signal is mixed with two local oscillators, with frequencies $\nu_0 = 80$ to 115 GHz and $\nu_1 = 1230$ MHz. The phases of the two LO's are:

$$\phi_{0l} = \int_0^t \omega_o(t')dt' + \psi_{0l}(t)$$

and

$$\phi_{1l} = \int_0^t \omega_{1l}(t') dt' + \psi_{1l}(t) \quad.$$

The $\psi_{0l}(t)$ and $\psi_{1l}(t)$ terms will be used to obtain quadrature correlations and to suppress DC in the correlations. In order to stop the fringe with respect to the delay center, the two LO's are tuned so that

$$\omega_{0l} = \frac{2\pi\nu_0}{c}\frac{d}{dt}[r_l \cdot \hat{n}] \quad,$$

$$\omega_{1l} = \frac{2\pi\nu_1}{c}\frac{d}{dt}[r_l \cdot \hat{n}] \quad.$$

Finally, a delay τ_{dl} is employed to compensate for the path difference between the antenna and the phase center. The resulting signal is:

$$x_l(t) = U_l \cos[\omega_{if}t + \phi^U_l - \psi_{0l} - \psi_{1l}]$$
$$+ L_l \cos[\omega_{if}t - \phi^L_l + \psi_{0l} - \psi_{1l}] \quad.$$

Here, we write ω_{if} for the intermediate frequency where U and L are located in the IF passband. When $x_l(t)$ is correlated with that from another antenna, the resulting correlation function is

$$\Gamma_{kl}(\tau) = U_k U_l \cos[\omega_{if}\tau + \Delta\phi^U_{kl} - \psi_{0k} + \psi_{0l} - \psi_{1k} + \psi_{1l}]$$
$$+ L_k L_l \cos[\omega_{if}\tau - \Delta\phi^L_{kl} + \psi_{0k} - \psi_{0l} - \psi_{1k} + \psi_{1l}]$$
$$= U_k U_l \cos(\omega_{if}\tau + \Delta\phi^U_{kl})\cos(-\psi_{0k} + \psi_{0l} - \psi_{1k} + \psi_{1l})$$
$$- U_k U_l \sin(\omega_{if}\tau + \Delta\phi^U_{kl})\sin(-\psi_{0k} + \psi_{0l} - \psi_{1k} + \psi_{1l})$$
$$+ L_k L_l \cos(\omega_{if}\tau - \Delta\phi^L_{kl})\cos(\psi_{0k} - \psi_{0l} - \psi_{1k} + \psi_{1l})$$
$$- L_k L_l \sin(\omega_{if}\tau - \Delta\phi^L_{kl})\sin(\psi_{0k} - \psi_{0l} - \psi_{1k} + \psi_{1l}) \quad,$$

where $\Delta\phi^U_{kl}$ is the phase difference for the upper sideband component, $\Delta\phi^L_{kl}$ that for the lower component. We assume the radiation field to be incoherent, that is, the UL products will integrate to zero. (Actually, the UL terms will contribute to the noise.)

We now introduce the Walsh functions for obtaining correlations with quadrature components and suppression of DC. Let

$$\psi_{0l} = \frac{\pi}{4}w_{0l}(t) \quad,$$

$$\psi_{1l} = \frac{\pi}{2}(1 + w_{1l}(t)) \quad,$$

(and analogously for antenna k); $w_{0l}, w_{0k}, w_{1l}, w_{1k}$ are mutually orthogonal Walsh functions. In expanding the phase terms, we make use of the usual trigonometric identities, with the addition of

$$\cos[\tfrac{\pi}{4}w_n(t)] = 1/\sqrt{2} \quad,$$

$$\cos[\tfrac{\pi}{2}w_n(t)] = 0 \quad,$$

$$\sin[\tfrac{\pi}{4}w_n(t)] = 1/\sqrt{2}w_n(t) \quad,$$

$$\sin[\tfrac{\pi}{2}w_n(t)] = w_n(t) \quad,$$

where w_n is a Walsh function of sequency n. (Sequency is a measure of the rapidity of oscillations of the function.) The result:

$$\Gamma_{kl}(\tau) = U_k U_l \cos(\omega_{if}\tau + \Delta\phi^U_{kl})f_A(t) + U_k U_l \sin(\omega_{if}\tau + \Delta\phi^U_{kl})f_B(t)$$
$$+ L_k L_l \cos(\omega_{if}\tau - \Delta\phi^L_{kl})f_A(t) - L_k L_l \sin(\omega_{if}\tau - \Delta\phi^L_{kl})f_B(t) \quad,$$

where

$$f_A = \tfrac{1}{2}(1 + w_{0k}w_{0l})w_{1k}w_{1l} \quad,$$
$$f_B = \tfrac{1}{2}(1 - w_{0k}w_{0l})w_{0k}w_{1k}w_{1l} \quad.$$

The quantities f_A and f_B are not Walsh functions. However, they serve to select one and only one component of the correlation function; this may be seen by noting that the product $w_{0k}w_{0l}$ is always either $+1$ or -1.

Let

$$w_0 = w_{1l}w_{1k}, \quad w_1 = w_{1l}w_{1k}w_{0l}w_{0k}, \quad w_2 = w_{0k}w_{1l}w_{1k} \quad.$$

These are also Walsh functions. (Product Walsh functions are Walsh functions of sequency equal to the binary sum of the sequencies of the factors, produced by adding without carry. That is,

$$\text{sequency}(w_m w_n) = \text{sequency}(w_m)\oplus\text{sequency}(w_n)$$
$$= m\oplus n) \quad.$$

If we code the value -1 to be a TTL logic level 0, $+1$ to be a TTL 1, then the selection of sine (1) or cosine (0), and sign of the component (0 if $+$, 1 if $-$) is determined by the logic circuit shown in Figure 3.

The Walsh functions actually chosen for the interferometer are listed in Tables I and II. The maximum sequency for our function generator is 255; we attempted to find sets of Walsh functions which were: (1) rapid (high sequency), and (2) mutually orthogonal. It was not felt to be necessary to make the modulation functions orthogonal to the demodulation functions, but this would be possible.

The final step in processing the data is done by the on-line control computer. This consists of binning the cosine and sine correlator outputs as the real and imaginary components of a complex correlation function; this is then transformed by a complex Fourier inversion via a FFT

TABLE I

Walsh Functions for Modulation

l	w_{0l} (90°)	w_{1l} (180°)
1	209	171
2	84	149
3	123	62

385

TABLE II

Walsh Functions for Demodulation

kl	w_0	w_1	w_2
12	62	187	239
23	171	132	255
31	149	63	238

algorithm. To see that this will actually result in separation of the sidebands, let us look again at our correlation $\Gamma(\tau)$ (with the demodulation having been applied):

$$F.T.\{\Gamma_{kl}(\tau)\} = v_{kl}(\omega) = F.T. \{[U_k U_l \cos(\omega_{if}\tau + \Delta\phi_{kl}^U) + L_k L_l \cos(\omega_{if}\tau - \Delta\phi_{kl}^L))]$$

$$+ i[U_k U_l \sin(\omega_{if}\tau + \Delta\phi_{kl}^U) - L_k L_l \sin(\omega_{if}\tau - \Delta\phi_{kl}^L)]\} \quad,$$

Noting that

$$F.T.\{\cos(\omega_0\tau + \phi)\} = \tfrac{1}{2}e^{i\phi}\delta(\omega - \omega_0) + \tfrac{1}{2}e^{-i\phi}\delta(\omega + \omega_0) \quad,$$

$$F.T.\{\sin(\omega_0\tau + \phi)\} = -\tfrac{1}{2}ie^{i\phi}\delta(\omega - \omega_0) + \tfrac{1}{2}ie^{-i\phi}\delta(\omega + \omega_0) \quad,$$

we obtain:

$$v_{kl}(\omega) = U_k U_l e^{i\Delta\phi_{kl}^U}\delta(\omega - \omega_{if}) + L_k L_l e^{i\Delta\phi_{kl}^L}\delta(\omega + \omega_{if}) \quad,$$

which is now the sideband-separated visibility function. Conveniently, the lower sideband appears in the negative frequency bins in the array.

386

Analog-Filter Digital-Correlator Hybrid Spectrometer

SANDER WEINREB, FELLOW, IEEE

Abstract—A system is described for measuring the power spectrum and cross spectral density of wide-band ($\geq$ 50 MHz) noise-like signals which occur, for example, in radio astronomy. The system utilizes a comb-filter bank followed by digital correlator processing of each filter output. The cost equation, design factors, and a sample system are described. For measurement of the spectrum at a large number of frequencies, the system cost of the hybrid system is shown to be much lower than the cost of spectrometers which utilize either a filter-bank or digital correlator alone.

Manuscript received July 9, 1984; revised February 19, 1985.

The author is with the National Radio Astronomy Observatory, (NRAO), Charlottesville, VA 22903. The NRAO is operated by Associated Universities, Inc. under contract with the National Science Foundation.

I. Introduction

IT HAS been obvious for many years to designers of digital correlators that prefiltering by J filters per signal reduces the digital operations required per second by a factor of J. For spectral analysis of a single signal at bandwidths < 50 MHz and number of points < 5000, this step is unnecessary; logic is inexpensive and the accuracy of analog operations is a concern. However, for cross spectral analysis of several signals at very wide bandwidths, the cost of an all-digital machine can become prohibitive. This is the case for a contemplated millimeter-wave radio astronomy array with ~ 40 signals at GHz bandwidths.

Reprinted from *IEEE Trans. Instrum. Meas.*, vol. IM-34, no. 4, pp. 670–675, Dec. 1985.

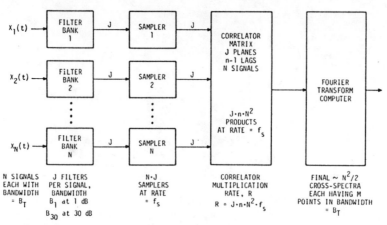

Fig. 1. Block diagram of hybrid filter-bank digital cross correlator spectrometer for processing an array of N signals.

For single-signal spectral processing, where a squaring operation rather than a multiplying operation is required, acoustooptical spectrographs [1] are very practical even for wide bandwidths. The system described in this report has both the advantage of ease of use in an array and flexibility of frequency scaling. Bandwidths narrower than the maximum bandwidth can be analyzed with increased resolution by a simple reassignment of the correlator channels in the system. A variation of the hybrid system described here is an analog filter bank followed by a digital, fast Fourier transform calculator and a cross multiplier; this has been described by Chikada et al. [2].

The purpose of this paper is not to suggest a new idea but to develop it. A cost equation for a hybrid system will be found; this leads to an optimum number of filters and minimum system cost. Special requirements on filter center frequency and shape factor imposed by the hybrid system will be discussed along with methods for realizing the filter-bank down-conversion system. Desirable system capabilities such as continuity of frequency points, filter overlap, and frequency diversity will be presented. Finally, a sample design suitable for millimeter-wave radio astronomy on a single telescope will be described.

II. BASIC RELATIONS AND COST EQUATION

A block diagram of a hybrid filter correlator system is shown in Fig. 1. The purpose of the system is to compute the cross-power spectra of N input signals, each having a bandwidth B_T. There will be $N(N + 1)/2$ such cross-power spectra including the self-power spectra of the input signals. The desired frequency resolution b of the spectral measurement will be defined so adjacent windows cross at $2/\pi$ (~ 2 dB) points. If b is also the spacing between window centers, then the total number of independent points (this dictated the $2/\pi$ choice) in each spectra is $M = B_T/b$.

Each signal is first passed through a filter bank with J filters with 1-dB bandwidth of B_1 and also spacing between center frequencies of B_1, so that $B_1 = B_T/J$. The filter 30-dB bandwidth will be designated $B_{30} = \beta B_1$, where β is a filter shape factor. The selection of crossover

at the 1-dB bandwidth is somewhat arbitrary, but it is approximately the bandwidth where the spectra can be determined without additional loss in statistical uncertainty due to coarse quantization (see Weinreb [3], p. 70). The 30-dB bandwidth determines the required sampling rate; $f_s = 2B_{30}$ for ≤ 0.1 percent aliasing of an out-of-band signal. The number of autocorrelator lags m required to achieve a given resolution b is given by $m = f_s/2b = \beta M/J$. For cross correlators, $2m$ lags are required for either positive and negative lags, or sin and cos components. For a specified m, a higher shape factor requires either more filters or more lags.

A key parameter for evaluating the cost or complexity of a correlator is the total number of multiplications per second, R. Each sample in each of N signals must be multiplied by m previous samples of the same signal, and $2m$ samples of the $N - 1$ other signals. There are N autocorrelators and $N(N - 1)/2$ cross correlators giving a sum of $m \cdot N^2$ products per sample. Multiplying by J filters and the f_s sampling rate gives

$$R = N^2 \cdot m \cdot J \cdot f_s. \tag{1}$$

Substituting $m = \beta M/J$ and $f_s = 2\beta B_T/J$

$$R = 2 \cdot N^2 \cdot \beta^2 \cdot B_T \cdot M/J. \tag{2}$$

Thus R is proportional to the bandwidth analyzed, B_T; the number of frequency channels, M; and is inversely proportional to the number of filters, J.

It is important to note that the required multiplication rate can be achieved by any product of multiplier elements and multiplication rate per element f_m which need not be equal to f_s. If $f_m = K f_s$ where K is an integer, then a multiplier can be time-shared to act upon K data streams stored in memories. (This technique is sometimes called "recirculation.") On the other hand, if $f_s = K f_m$, then K multipliers can be time-multiplexed to act upon one data stream. The least costly correlator is one which utilizes multiplier elements having the lowest ratio of cost to speed taking into account the cost of buffers or multiplexers.

Designating this correlator cost ratio as E in k\$ per 10^9 multiplications/seconds and the filter cost factor as F in k\$

TABLE I
SINGLE SPECTROMETER COSTS (1984$)
$(N = 1, B_T = 2\ \text{GHz}, M = 2000)$

CASE	β	E Corr. Cost k$/GHz	F Filter Cost k$	J Filters	C Total Cost k$	B_1 Filter Width MHz	f_s Sampling Rate MHz	m Corr. Per Filter	Comment
A	2	0.5	0.6	1	16,000	2,000	8,000	4,000	No filter bank
B	2	0.5	0.6	2,000	1,600	1	–	–	No correlator
C	2	0.5	0.6	163	197	12.2	48.8	24.5	Min. cost, $\beta = 2$
D	1.33	0.5	0.6	109	131	18.3	48.9	24.5	Min. cost, $\beta = 1.33$

TABLE II
ARRAY SPECTROMETER COST (1984$)
(ALL MINIMUM COST, $\beta = 1.33$)

CASE	N Ants.	E Corr. Cost k$/GHz	F Filter Cost k$	J Filters Per Ant.	C Total Cost M$	B_T Total BW GHz	M Total Freq. Points	B_1 Filter Width MHz	Comment
A	40	0.5	0.6	344	16.5	1	1,000	2.9	$R = 11.6 \times 10^{12}$ mults/sec; $R = 5.7 \times 10^{15}$ mults/sec if J = 1
B	40	0.2	0.6	217	10.5	1	1,000	4.6	1990 projection of lower digital cost
C	50	0.2	0.6	122	7.4	0.5	500	4.1	More antennas, less BW

per unit including samplers and delay lines, the total system cost C, excluding development cost, is given by

$$C = 2 \cdot E \cdot N^2 \cdot \beta^2 \cdot B_T \cdot M/J + F \cdot N \cdot J. \quad (3)$$

The number of filters J_0, which minimizes the cost and the minimum cost C_0, is then

$$J_0 = \beta \sqrt{2NB_T \cdot M \cdot E/F} \quad (4)$$

$$C_0 = 2 \cdot F \cdot N \cdot J_0 = 2\beta N \sqrt{2NB_TMEF}. \quad (5)$$

Some typical values for costs of a single spectrometer and a 40- or 27-element array are shown in Tables I and II. A correlator cost factor $E = 0.5\text{k}\$/10^9$ multiplications/s is used for most of the computations. The Very Large Array (VLA) correlator [4] built in 1977 has 11 000 multipliers at 10^8 multiplications/s and has a total correlator cost of $\sim \$800\text{k}$; thus $E = 800/(11\ 000 \times 10^8) = 0.73\text{k}\$/10^9$ multiplications/s for this case. A filter cost factor F of 0.3k$ per filter has been used. The filter cost includes a phase-locked local oscillator for conversion to baseband and a sampler. Both E and F may be substantially reduced by clever design, use of gate arrays for correlation, and use of new IC's for LO synthesis.

III. FILTER AND FREQUENCY CONVERSION METHODS

A. Quadrature-Phase Sideband Selection

A common method for down conversion and filtering is shown in Fig. 2. The filters in this system are usually low-pass filters and the accepted input spectrum is then as shown at the top of Fig. 3. A gap where the two sidebands are not separated then exists in the converted spectrum

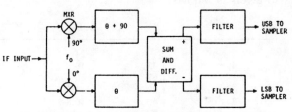

Fig. 2. Quadrature phase image-rejecting mixer system.

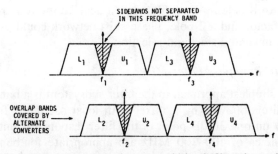

Fig. 3. Input spectra converted as upper sideband (U) or lower sideband (L) by system of Fig. 2. *Note:* The sideband rejection gap near the LO frequency can be made small with many-pole, phase shift networks, or left large but analyzed by an alternate frequency converter as illustrated in the lower half of the figure.

near the local oscillator frequency due to the inability of the phase shift networks to maintain 90° phase shift close to zero frequency. The gap can be made smaller with more complex phase shift networks; a 10-pole network could give 30-dB sideband rejection over a 10^4 frequency range (i.e., 5 kHz to 50 MHz). Another remedy is to analyze the spectra in the gap region with alternate converters offset by $f_s/4$ in frequency; this forms a system with the filter shape factor β defined in Section II equal to 2.

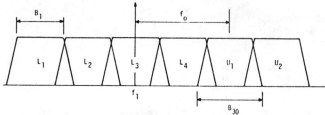

Fig. 4. Interlaced bandpass filters. *Note:* L_1 and U_1 are lower- and upper-sideband spectra accepted by one quadrature-phase mixer with LO frequency, f_1. Other mixers with LO frequencies spaced above f_1 in steps of B_1 fill in the gaps between lower and upper sidebands. By proper selection of the center frequency of each filter, the output may be sampled with $f_s = 2 \times B_{30}$ and aliasing error < 0.1 percent

The gap problem can be avoided with lower values of β by utilizing interlaced bandpass filters and bandpass sampling as shown in Fig. 4. To avoid aliasing the filter, 30 dB points must be selected to occur at $kf_s/2$ and $(k+1)$ $f_s/2$ where k is an integer. If p filter channels are selected to fall between upper and lower sidebands, the value of β is given by

$$\beta = \frac{p}{2k+1} \geq 1. \tag{6}$$

Some possible values are given in the table below where values of B_{30} and filter center frequency f_0 for a $B_1 = 12$ are also given as follows:

k	p	β	B_{30}	f_0
0	2	2	24	12
1	4	1.33	16	24
1	5	1.67	20	30
2	6	1.20	14.4	36
2	7	1.40	16.8	42

The $\beta = 1.33$ solution appears to be a reasonable compromise. An 8-pole no-zero filter could achieve the required shape factor and a 2-pole, phase shift network could give 30-dB unwanted sideband rejection.

B. VHF Bandpass Filter-Mixer

The simplest approach to the filter subsystem is a bandpass filter followed by a mixer with local oscillator frequency on one 30-dB point of the filter response. Center frequencies of 100–200 MHz are appropriate for bandwidths of ~ 10 MHz. (Lower center frequencies make image rejection difficult in the preceding frequency conversion.) Mass-produced surface acoustic-wave (SAW) filter banks are available from one manufacturer (Sawtek) at a unit filter cost of ~ $50 for filters of 12.5 MHz, 3-dB bandwidth and 15 MHz, and 20-dB bandwidth. However, this method appears less desirable than the image-rejecting mixers discussed previously for the following reasons:

1) the filters are at different center frequencies and hence require many different designs;

2) the nonrecurring cost is high;

3) the stability of high Q-bandpass filters is poor. In addition, SAW filters have large time delay (1–5 μs) which is temperature dependent for low-loss (~ 25 dB) mate-rials. The phase stability in a cross correlation system would be poor.

IV. DESIRABLE DESIGN FACTORS

A. Continuity of Frequency Point Spacing

It is obviously desirable to have the frequency spacing b of measured points on the power spectrum continuous in going from one filter to the next. This requires that $b = B_1/m'$ where m' is an integer. It is also desirable that the filter window function have nulls at dc and the sampling frequency f_s to prevent sampler imperfections from causing ripples in the spectrum; this forces $b = B_{30}/m'$ or $\beta = m/m'$. In addition, for the quadrature image rejection method, β must satisfy (6) $\beta = p/(2k+1)$, p and k integer, to avoid aliasing in the sampled bandpass function. Fortunately, all of these constraints can be satisfied; for example, $m = 16$, $m' = 12$, $p = 4$, $k = 1$.

B. Filter Overlap

The bandpass of adjacent filters in the system will overlap to some degree. For example, filters with response crossing at 1-dB points will have an overlap region of perhaps 0.05 B_1 between 3-dB points where the spectrum is measured through two different paths. The two spectra, after correction for bandpass shape, should agree at these points; to first order even the noise fluctuations on the measured points should be the same. For coarse quantization the noise will not be exactly identical because some of the noise, particularly at band edge, is due to noise from other frequencies within the filter passband and this will be uncorrelated between the two filters.

Thus a comparison of spectral points in the overlap region is a sensitive indicator of filter instability or a gross failure in a portion of the system. The overlap region can be chosen to be close to 100 pecent by making the factor $\beta = 2$. However, this is costly and a compromise value of $\beta = 1.33$ with overlap checks between 3 dB points appears to be prudent.

C. Channel Diversity

In some applications it would be advantageous to have the capability of rapidly reassigning the frequency of filters in the system. This could be accomplished by using an image-rejecting mixer system with programmable synthesizer local oscillators. The power at one frequency would first be measured through one filter and subsequent correlator for 250 ms (for example) and through another in the next 250 ms. After 1000 ms the four measurements would be compared and any measurement differing by a specified threshold would be discarded from the average. The rms deviation of the quartet would be available to the observer and time variations due to equipment malfunction, antenna pointing, or the signal source could be detected. A failure of one filter would then not cause a portion of the spectrum to be missed.

This diversity would complicate the processing of data and is probably not worthwhile for interferometer or array

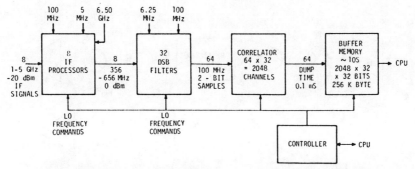

Fig. 5. Overall block diagram of the prototype system. *Note:* Eight I.F. processors shown in Fig. 6 allow input of up to 8 I.F. signals, each with 300-MHz bandwidth in the 1–5 GHz range. Each I.F. processor produces a 356–656 MHz output which drives 10 dual-sideband filter systems shown in Fig. 7. Each filter produces 3-level samples at a 100-MHz rate. The samples are then processed by a correlator operating at a 100-MHz clock rate. The correlator would be organized as 64 32-lag modules. The output accumulator-buffer memory has 32 bins for each of 2048 effective channels to allow for frequency diversity during an integration cycle.

use. However, it may be useful for single signal observations and would have little impact upon the cost. It would require programmability on a moderate speed basis of the second and probably third local oscillators in the system; these oscillators would be synthesized in any case to assure frequency stability.

V. A Sample System

A hybrid spectrometer suitable for use on the NRAO 12-m radio telescope on Kitt Peak, Arizona, will be described. As a starting point, we will assume approximately 2000 points at 2-GHz total bandwidth are desired with the option of dividing this total band among 2 (polarizations), 4, or 8 signals (multiple beams). In order to satisfy integer relations and for ease of synthesis of local oscillators, looking ahead, we have found that $M = 1536$ points and $B_T = 2.40$ GHz are convenient. The system allows B_T to be reduced, keeping the total number of channels constant, with the constraint that B_T is an integer times 37.5 MHz as shown in Table III.

The system of internal parameters will be selected to be close to the minimum cost relations of Table I as follows:

Number of Filters J	$=$	64
Filter Spacing and 1-dB Bandwidth	$B_1 =$	37.5 MHz
Filter 30-dB Bandwidth B_{30}	$=$	50 MHz
Sampling Rate f_s	$=$	100 MHz
Overlap Factor β	$=$	$\frac{4}{3}$
Correlator Channels per Filter m	$=$	32.

A description of the system is given in Figs. 5–7. The correlator requires 2048 channels operating at a 100-MHz clock rate. This can be accomplished with 1024 VLA-1 chips and 2048 VLA-2 custom ECL IC's [5] which are

TABLE III
Prototype System Frequency Resolution and Total Bandwidth

Resolution	1 Signal	2 Signals	4 Signals	8 Signals
b kHz	B_T MHz	B_T MHz	B_T MHz	B_T MHz
1,000	1,920	960	480	240
500	960	480	240	120
250	480	240	120	60
125	240	120	60	24
62	120	60	24	12
31	60	24	12	NA
12	24	12	NA	NA

TABLE IV
Spectrometer Cost

8 – I.F. Processors, $3.5k each	$ 28k
80 – Dual SB I.F. Filters, $0.6K each	48k
Buffer Memories	5k
Correlator cards, sockets	10k
Cabinets, bins, power supplies	10k
Controller	10k
Cables, Miscellaneous	5k
Total	$116k

now on hand at NRAO. A cost estimate, excluding these IC's, is given in Table IV.

If VLA custom chips were not available for use in this machine, semi-custom gate-array IC's are an attractive alternative. A single Motorola MCA1200 ECL Macrocell, priced at $45 at a quantity of 1000, could contain the high-speed processing for 12 lags (i.e., dual-shift register, 3-level multipliers, and 4-bit counter) at a clock rate which may be as high as 160 MHz; the development charge is ~ $23k. Other lower cost CMOS gate arrays could be used for low-speed accumulation.

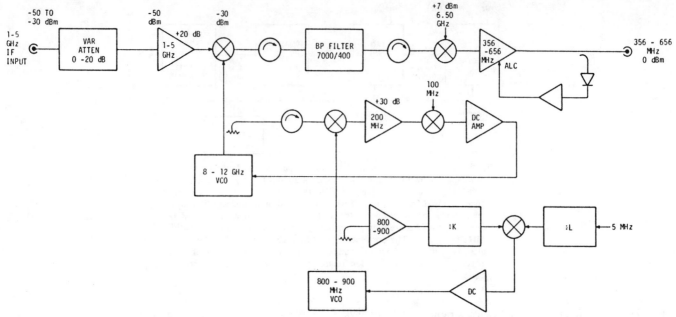

Fig. 6. Block diagram of 1 of 8 I.F. processors. *Note:* The unit allows an I.F. input with 300-MHz bandwidth anywhere in the 1–5 GHz range to be translated to the spectrometer input. The input is first up-converted to a 7-GHz center frequency by heterodyning with a synthesized 8–12 GHz local oscillator programmable in 1-MHz steps. At 7 GHz the signal is filtered and down-converted to a 506-MHz center frequency appropriate for input to the filter system.

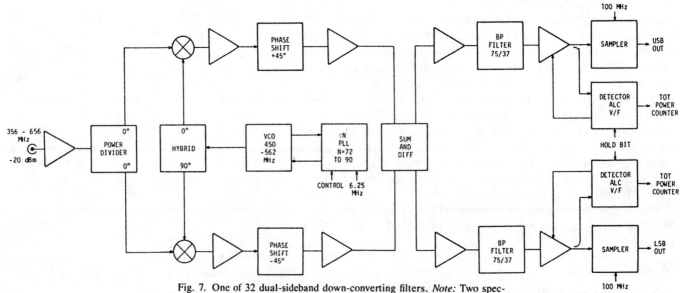

Fig. 7. One of 32 dual-sideband down-converting filters. *Note:* Two spectral bands centered at $f_0 \pm 75$ MHz with $f_0 = 450$–562 MHz in 37.5-MHz steps are down-converted to separate baseband signals by a quadrature-phase image-rejecting mixer method. The signals are then passed through filters with 75-MHz center frequency, 37.5-MHz 1-dB bandwidth, and 50-MHz 30-dB bandwidth. The amplifiers in the system are primarily for isolation purposes and are inexpensive IC units.

REFERENCES

[1] C. R. Masson, "A stable acousto-optic spectrometer for millimeter radio astronomy," *Astron. and Astrophys.*, vol. 114, no. 2, pp. 270–274, Oct. 1982.

[2] Y. Chikada *et al.*, "A digital FFT spectro-correlator for radio astronomy," Nobeyama Radio Observatory, Nagano, Japan, NRO Rep. 20. Also published in *Indirect Imaging: Measurement and Processing for Indirect Imaging*, J. A. Roberts, ed. Cambridge, MA, 1984, pp. 387–404.

[3] S. Weinreb, "A digital spectral analysis technique and its application to radio astronomy," Ph.D. dissertation and RLE Tech. Rep. 412, M.I.T., Cambridge, MA, 1963. Also available from National Technical Information Services, Springfield, VA, no. AD418413.

[4] P. J. Napier, A. R. Thompson, and R. D. Ekers, "The Very Large Array: Design and performance of a modern synthesis radio telescope," *Proc. IEEE*, vol. 71, pp. 1295–1320, Nov. 1983.

[5] "Giant radio telescope uses 19 500 chips for complex signal processing at 100 MHz," *Electronics*, vol. 51, no. 21, pp. 44–46, Oct. 1978.

A Stable Acousto-optical Spectrometer for Millimeter Radio Astronomy

C. R. Masson

Owens Valley Radio Observatory, Caltech 405-47, Pasadena, CA 91125, USA

Received November 16, 1981; accepted June 15, 1982

Summary. A stable, low cost acousto-optical spectrometer has been developed for millimeter wave observations at the Owens Valley Radio Observatory. This instrument has a bandwidth of 102 MHz, divided into 1024 channels, with an effective resolution of 160 kHz. The spectrometer is stable enough to permit total-power observations with on- and off-source integrations of several minutes. The many measures taken to achieve this stability are described and astronomical tests of the spectrometer are presented. The spectrometer has been in regular use at the Owens Valley Radio Observatory since December 1980.

Key words: radio spectrometers – radio spectral lines

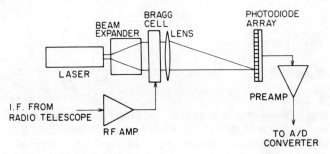

Fig. 1. Block diagram of an acousto-optic spectrometer

I. Introduction

Since the advent of millimeter wave molecular line astronomy in the 1960's, there has been a need for radio spectrometers with bandwidths up to several hundred MHz and resolution down to tens of kHz. At an observing frequency of 100 GHz, the broadest lines from external galaxies are Doppler shifted over a range of about 500 km s^{-1} or 200 MHz, while the narrowest line widths are about 0.1 km s^{-1} or 40 kHz. Measurements of the frequency structure of these lines give information about physical conditions in the molecular clouds. The radiation is very weak, requiring very long integrations and a high degree of instrumental stability for detection. Digital autocorrelators are widely used at lower bandwidths, while filter banks have been the usual choice for wideband spectrometers. Filter banks with many channels are complex and expensive to construct. It had been recognised for many years (Cole, 1968) that optical processing might be able to provide more economical and reliable spectrometers. Work on acousto-optic spectrometers (AOS) has been carried out by a number of groups over the years but most consistently in Australia (Cole, 1973; Cole and Milne, 1977; Robinson, 1980) and Japan (Kaifu et al., 1977, 1980).

Work at Caltech over the past two years has been devoted to identifying the main sources of drifts in AOS, with the aim of building an AOS offering the degree of stability required for radio astronomy (Masson, 1980). This effort has been successful, resulting in the production of an AOS which has been in routine use at the Owens Valley Radio Observatory since December 1980 (Masson, 1982). This instrument, the Mk I AOS, has a bandwidth of 102 MHz, divided into 1024 channels. The stability achieved is more than adequate, with theoretical noise levels being achieved even for switching times of several minutes, which is longer than

the 1 min (30 s on-source, 30 s off-source) typically used in millimeter radio astronomy. The frequency drift is also very small, with a shift of 1/3 of a channel over the first month of operation.

This paper is a description of the design and construction of the Mk I AOS. The principles of operation are described briefly in Sect. II, while Sect. III covers the design and Sect. IV is a review of the performance of the finished spectrometer.

II. Principles of Operation

The operation of AOS has been described before (e. g. Korpel, 1981 and references therein) but will be discussed briefly here for convenience.

II.1. Structure

The basic arrangement of an AOS is shown in Fig. 2. The optical source is a laser, whose beam is expanded to match the aperture of the Bragg cell. The r.f. signal is amplified to an appropriate level and applied to the cell. For each frequency component, f_i, of the r.f. signal, the Bragg diffraction produces a beam of light, with an intensity proportional (in the low-power limit) to the r.f. power in that component, deflected through an angle, θ_i, proportional to the frequency. This light is focused by the transform lens to a spot, whose distance, d_i, from the axis is given by

$$d_i = F\theta_i, \tag{1}$$

where F is the focal length of the lens. When the Bragg cell is operating in its linear regime with low r.f. power, the spatial distribution of the light in the back focal plane is proportional to the power spectrum of the r.f. signal. Because the light is diffracted

Reprinted with permission from *Astron. Astrophys.*, vol. 114, pp. 270–274, 1982.

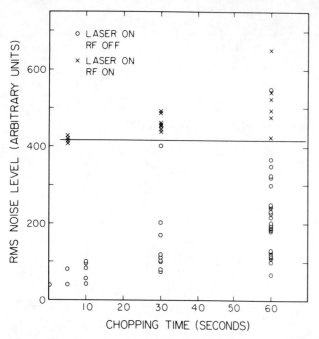

Fig. 2. Measurements of noise level as a function of chopping time with and without the presence of an r.f. signal. The horizontal line shows the noise level expected for an r.f. signal when the excess noise contribution is zero

by a moving grating, its frequency is changed by an amount equal to the frequency of the r.f. signal. The distribution of light intensity is measured by a photodiode array (PDA) or a CCD. After integration for some milliseconds, the photodiodes are sampled and the resulting video signal is digitised and integrated in a computer.

The theoretical maximum resolution of an AOS is the reciprocal of the acoustic transit time across the aperture of the Bragg cell. In practice, this resolution is degraded by imperfect optics, non-uniform illumination of the Bragg cell, attenuation of the acoustic wave and crosstalk in the photodector array.

II.2. Signal to Noise Ratio

The output signal from a CCD or PDA is very small (10^5–10^7 electrons/pixel) and must be amplified by a low noise preamplifier before being fed to the A/D converter. To prevent degradation of signal/noise by the AOS, any noise contributed by the photodiodes and preamplifier (post-detection noise) must be much less than the intrinsic statistical noise in the r.f. signal.

For the case of a random r.f. signal, the statistical fluctuation in signal level is related to the frame time, t, between successive readouts of a photodiode and the resolution bandwidth, B, by

$$\Delta s = S/(Bt)^{1/2}, \qquad (2)$$

where S is the signal level and Δs is the rms fluctuation in one frame. For $B = 160$ kHz, $t = 31$ ms, $\Delta s = S/70$. Allowing for an uneven bandpass and variation in input signal levels, S might be as low as 1/4 of the saturation level and we therefore require that the signal/noise ratio of the following stages be significantly greater than 280. There are two fundamental sources of noise:

(i) Shot noise. When dark current is small, this is simply equal to the square root of the number, N, of photons detected. If shot

noise is not to exceed r.f. noise, then Bt must be much less than N. PDAs saturate at $2 \cdot 10^7$ electrons, giving a maximum signal/noise ratio of 4500.

Since the photon shot noise is required to be less than the statistical noise and both scale identically with integration time, in an optical processor like the AOS, one can formulate a statistical sampling theorem: If the processor is not to degrade the signal/noise ratio, the number of photons detected per second must be much greater than the bandwidth of the processed signal in Hz. This does not provide any strong practical constraint on the use of photodiodes, but gives a useful limit on the readout rate of CCD detectors in wideband optical processors.

(ii) Array noise. With a simple video amplifier, the thermodynamic noise depends on the video line capacitance in the PDA, giving a signal/noise ratio of 10,000. If a more complicated video processor is used, then the readout noise may be reduced from 2000 to 770 electrons rms (Buss et al., 1976).

The signal/noise ratio in the Mk I AOS is modified by the operation of a drift compensation scheme, described below, in which the zero levels are measured regularly during integration by turning the r.f. signal on and off. Let the number of frame periods with the r.f. signal applied be n_1 and the number of dark frames be n_d. Because of the sequential readout of the Reticon channels, only $n_d - 1$ frames are uncontaminated by r.f. signals.

Let the readout noise per frame in one channel be v volts rms and the statistical r.f. noise be σ volts rms. If the r.f. signal is applied to every frame then the backend noise and statistical noise scale in the same way with integration. Thus after N frames, they would have values $v/\sqrt{N}$ and $\sigma/\sqrt{N}$, giving a total noise level, σ_t, of

$$\sigma_t = (\sigma^2/N + v^2/N)^{1/2}. \qquad (3)$$

When the compensation scheme is turned on, the noise in the zero measurements is scaled up by a factor $(n_1 + 1)/(n_d - 1)$ when the zero levels are subtracted. After a cycle of $N(=n_1 + n_d)$ frames, the noise level is

$$\sigma_t = \left(\frac{\sigma^2}{n_1} + \frac{v^2(n_1+1)}{n_1^2} + \frac{v^2(n_1+1)^2}{n_1^2(n_d-1)} \right)^{1/2}, \qquad (4)$$

where the third term arises from the subtraction of zero levels. Compared with the uncompensated case, the variance of post-detection noise is increased by a factor

$$1 + (n_1/n_d), \qquad (5)$$

where $n_d \gg 1$ and $n_1 \gg n_d$.

III. Design of the Mk I AOS

III.1. Construction

As shown in the previous section, the basic structure of an AOS is very simple and most of the detailed design is devoted to obtaining the required stability. The optical components of the Mk I AOS are assembled on an optical tabletop which is mounted without vibration isolation on a wooden table on a concrete floor in a basement at O.V.R.O. The beam from a 2 mW polarised HeNe laser is reflected by a plane mirror into the beam expander. The Bragg cell is a type AOD 150 from Intra Action Corp. This cell has a 3 dB bandwidth of 100 MHz and a transit time of 10 μs, giving a nominal resolution of 1000 channels. The light deflection efficiency is about 10 % per watt of r.f. signal.

The Bragg cell is followed by a doublet transform lens, whose focal length of 1 m is chosen to match the PDA length to the

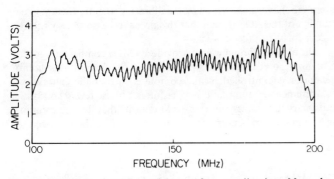

Fig. 3. Example of AOS bandshape after equalization. Note the presence of fringes arising from reflections in the glass window of the Reticon

deflection range of the Bragg cell. The PDA is a Reticon RL 1024 H with 1024 photodiodes spaced at 15 µm intervals. The frame time of 31 ms is set by the computer. The video amplifier has a low enough noise level that the Reticon readout noise of 2000 electrons rms dominates. The 12 bit A/D converter has a signal/noise ratio of only 3500:1, however, and prevents the full potential of the video system from being realized.

III.2. Zero Levels

Any astronomical measurement made with the AOS must include a measurement of the zero level of each channel. There are several factors which cause the AOS outputs to deviate from a true electrical zero when no signal is applied, including offsets in the video amplifiers and A/D converter, fixed pattern noise in the Reticon array, dark current in the Reticon array (Cole and Milne, 1977) and scattered light in the optical system. If all of these remained constant, then it would be possible to make infrequent measurements of the zero levels for subtraction from the measured spectra.

The electrical offsets are relatively stable and, although the dark current changes rapidly with temperature, its drift was minimised by using a Reticon array which included a set of compensating diodes and selecting the array to have a low dark current. In the absence of laser light, drifts in zero levels were found to be below the noise of the video amplifier, demonstrating the effectiveness of these measures. Despite this, however, it was found that a relatively large drift in zero levels was present when the laser was operating. The change in zero levels was uncorrelated between channels, appearing as random excess noise in the spectrum, rather than as a smooth slope in the baseline. Figure 2 is a graph of rms noise as a function of chopping time, both with and without r.f. power applied to the Bragg cell. It can be seen that the performance of the AOS was seriously degraded for chopping times in excess of 60 s and that this degradation was caused by drift in the scattered laser light.

Cleaning, moving, and even removing various components of the AOS changed the speckle pattern of scattered but made almost no difference to the drifts. The solution finally chosen for the Mk I AOS was to compensate for any drifts in the scattered light by turning off the r.f. signal and measuring the zero levels for 0.5 s every 5 s during operation. As described in Sect. II, this increases the post detection noise level and also reduces the efficiency since the time used for zero measurements is wasted. The choice of what

fraction of the time to spend on the zeroes is a comparison between these two evils. The values chosen degrade the efficiency to 90 %, almost all of the loss arising from the wasted time.

III.3. Frequency Stability

Since each photodiode in the AOS is only 15 µm wide, subtending 3″ at the lens, small movements can significantly change the frequency calibration. As described by Cole and Milne (1977), changes between on and off source measurements give rise to a residual signal proportional to the slope of the bandpass. For a slope in the bandpass of s %/channel, a chopping period of t seconds and a resolution bandwidth of B, the maximum allowable drift, d, is given by

$$d = \frac{100}{s} \frac{2}{(Bt)^{1/2}} \tag{6}$$

channels, if the residual signal is not to exceed the rms noise level. For the MkI AOS, $s = 0.5$ %/channel and $t = 100$ s, giving a maximum value of 0.14 channel or 2 µm for d. This is roughly a factor of 10 less severe than the stability required for holography and is easily achieved with the rigid structure used for the Mk I AOS.

The structure is so stable that short term drifts are dominated by poor 'seeing' along the optical path. Turbulence and air currents in an ordinary room cause the laser beam to be deflected randomly by ∼ 0″3 or 10 % of a channel, which is far too large to be tolerable. To control this effect, it was found necessary to enclose the optical path. The most important part of the enclosure is a tube between the lens and the Reticon array, although the cover over the beam expander also has a noticeable effect. Angular deviations occurring before the beam expander are reduced in proportion to the expansion ratio and can be neglected. Angular drift of the laser beam is also a potential problem. With a good quality HeNe laser, however, the drift should be < 2″ at the laser, corresponding to a frequency shift of 0.01 channels.

III.4. Bandshape

The response of the Bragg cell used in the Mk I AOS is nominally 100 MHz to the −3 dB points. The exact shape of the passband depends on alignment and it was found necessary to equalize the response by removing a peak at 110 MHz. Figure 3 shows the bandshape after equalization in an early version of the AOS.

This figure shows a set of irregular fringes with a peak to peak amplitude of about 15 %, which are due to reflections at the surface of the glass window of the Reticon array. If these fringes remained constant, then they would simply cancel in the subtraction of on and off scans. The fringes do not, however, remain entirely constant since they depend on the frequency of the laser. For a window thickness of 0.5 mm, a change in frequency of 1 part in 5 10⁵, which is possible for a HeNe laser, would produce a change of roughly 1 % in the fringe pattern, or about 0.15 % in the bandshape. Since the random noise after 100 s is 0.03 %, such a change is intolerably large.

The first step in reducing this effect was to use a Reticon with an antireflection coated window. The single layer coating reduced the fringe amplitude by a factor of 4. A further small improvement was made by tilting the array by about 3 deg, which is the largest angle permitted by the depth of focus. This reduces the fringe amplitude because the Bragg diffracted light is not coherent across the PDA,

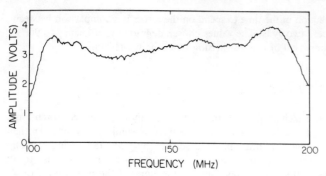

Fig. 4. Final bandshape of the AOS using coated and tilted Reticon to minimise fringe visibility

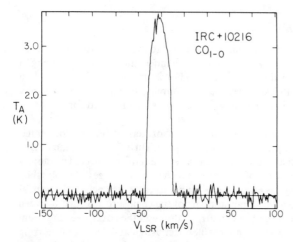

Fig. 5. Spectrum of CO_{1-0} in IRC + 10216 taken with the AOS. No baseline correction has been applied. The full bandwidth of the AOS is plotted

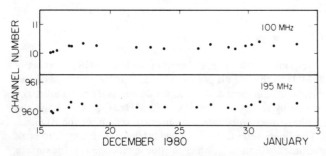

Fig. 6. Measured frequency drift of Mk I AOS during the first 3 weeks of operation. The dots represent the positions of two of the 5 MHz harmonics from the calibration comb

as a result of the frequency shift described in Sect. II. Figure 4 shows the final bandshape of the Mk I AOS. In addition to these precautions, a cover was placed over the laser to stabilise its temperature and reduce the rate of drift in frequency.

III.5. Frequency Calibration

In an AOS, the frequency scale is not fixed by some external standard as it is, for example, in a digital autocorrelator, so it is useful to have some means of calibrating the frequency. The stability of the AOS is such that this needs to be done very rarely, except for very precise work.

The method chosen was to provide a 5 MHz comb generator to give a set of calibration signals across the band. This comb generator is controlled by the computer so that the calibration measurement is automatic and requires only a few seconds. A subsidiary advantage of this technique is that the calibration spectrum gives measurements of the resolution of the AOS across the band.

III.6. Mapping

In many experiments it is desirable to measure the spatial distribution of radiation in some particular emission line. For this purpose the AOS provides an analog output. After the digitized video data from each frame are received by the computer, the numbers from an arbitrary set of channels may be averaged. This average amplitude is passed to a D/A converter, producing a voltage which can be sampled by the mapping computer. In this way the AOS can synthesize a filter of arbitrary bandwidth from 160 kHz to 102 MHz.

IV. Performance of the AOS

The Mk I AOS was completed in the late summer of 1980 and received its first use for astronomy in December 1980, in conjunction with the new SIS receiver (Phillips et al., 1981). The Mk I AOS operated very reliably, requiring no maintenance during the winter observing season. The total bandwidth was 102 MHz and the resolution at the half-power points was 160 kHz.

Figure 5 shows a test spectrum of IRC + 10216 in CO_{1-0}, taken with the AOS and the SIS receiver at OVRO in December 1980. Position switching was used, with the telescope moving every 30 s. The total integration time on source was 30 min. No baseline contribution was subtracted from the spectrum shown. The flatness and uniformity across the entire bandwidth of the spectrometer demonstrates the stability of the telescope system at OVRO. In particular, it is clear that the design aims of the AOS have been met.

The frequency stability of the Mk I AOS was monitored during the first three weeks of operation and Fig. 6 shows the measured positions of two test frequencies. During this period, there was a total frequency drift of 0.3 channels or 30 kHz. While the frequency stability of the AOS is inferior to that of a digital correlator, it is more than adequate for radio astronomy.

The limits to the stability of the Mk I AOS have not been determined very accurately but tests using noise sources and astronomical measurements have shown that the AOS operates essentially perfectly at chopping periods as long as 4 min, although the baselines deteriorate when 15 min chopping times are used. The exception to this is that laser output drifts of 0.25 % occur on these time scales so that shorter chopping times must be used for accurate continuum measurements. An illustration of the stability is the spectrum of IC 342 shown in Fig. 7, which was obtained by Lo, using a chopping period of 90 s and a total integration of 30 min on source.

The linearity of the AOS is shown in Fig. 8. To the measurement uncertainty of 1 %, no deviation from linearity was detected over a range of 20 dB in r.f. level. The dynamic range in normal operation is set by the level of post detection noise, described in

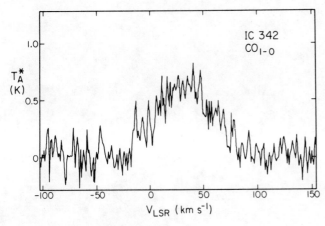

Fig. 7. Spectrum of IC 342

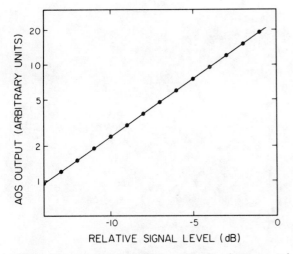

Fig. 8. Measurement of AOS linearity, showing output signal as a function of r.f. power input

Sect. II. At a signal level 10 dB below the normal operating level, the statistical noise of the r.f. signal is equal to the post-detection noise.

The spurious response level is also very low. From a frequency calibration spectrum, the response near a strong signal was found to be < 0.1 % at points more than 1 MHz away from the peak.

The total cost of components for the Mk I AOS was $ 18,000, including $ 6000 for the LSI-11 computer and A/D converter. This is far less than the cost of a comparable filterbank (Robinson, 1980). Although 18 man months were required for development, a much shorter time would be needed to duplicate the Mk I AOS.

Following the success of the Mk I AOS, a 500 MHz spectrometer with 1024 channels is now under construction at OVRO. Compared with the Mk I AOS, the 500 MHz spectrometer requires a much higher degree of stability. In addition, wideband Bragg cells are less efficient so more care must be taken to ensure a reasonable light level at the Reticon. The prototype 500 MHz AOS has been completed and the final version is expected to be in operation during the winter of 1981–1982.

Acknowledgements. I wish to thank Siu Au Lee for her advice on lasers, K. Y. Lo for the spectrum of IC 342 and N. Kaifu, A. Moffet, T. Phillips, and D. Psaltis for their comments on the manuscript. This work was supported by NSF grants AST 76-13334 and AST 79-16315.

References

Buss, R.R., Tanaka, S.C., Weckler, G.P.: 1976, in *Solid State Imaging*, N.A.T.O. Advanced Study Institute, eds. P. G. Jasper, F. van de Wiele, and M. H. White, Noordhoof, Leyden

Cole, T.W.: 1968, *Opt. Technol.* **1**, 31

Cole, T.W.: 1973, *Proc. Inst. Elec. Electron. Engrs.* **61**, 1321

Cole, T.W., Milne, D.K.: 1977, *Proc. ASA* **3**, 108

Hecht, D.D.: 1977, *IEEE Trans. Sonics and Ultrasonics* SU **24**, 7

Kaifu, N., Uchida, N., Chikada, Y., Miyaji, T.: 1977, *Publ. Astron. Soc. Japan* **29**, 429

Kaifu, N., Chikada, Y., Miyaji, T.: 1980, Nobeyama Radio Observatory Technical Report

Korpel, A., Adler, R., Desmares, P., Watson, W.: 1977, *Proc. Inst. Elec. Electron. Engrs.* **54**, 1429

Korpel, A.: 1981, *Proc. Inst. Elec. Electron. Engrs.* **69**, 48

Masson, C.R.: 1980, *Proc. S.P.I.E.* **231**, 291

Masson, C.R.: 1982, Owens Valley Radio Observatory Technical Report

Phillips, T.G., Woody, D.P., Dolan, G.J., Miller, R.E., Linke, R.A.: 1981, *IEEE Trans. Magn.* MAG **17**, 684

Pinnow, D.A.: 1971, *IEEE Trans. Sonics and Ultrasonics* SU **18**, 209

Robinson, B.J.: 1980, in *Interstellar Molecules*, IAU *Symp.* **87**, ed. B. H. Andrew, Reidel, Dordrecht, Holland

Uchida, N., Niizeki, N.: 1973, *Proc. Inst. Elec. Electron. Engrs.* **61**, 1073

THE value of radio-astronomical measurements is critically dependent upon the accuracy with which the signal processed by the receiver electronics can be related to the intensity of radiation incident on the antenna. The power received depends upon the distribution of brightness on the sky as well as the gain and response pattern of the antenna. The distribution of intensity as a function of angle is, in general, not known *a priori*, so the appropriate scaling to apply to a given measurement is uncertain. In this sense, recovering the true distribution of intensity is really an aspect of image reconstruction, with the antenna response function one of the quantities that is to be determined.

It is general practice, however, to make a reasonable guess about the source of radiation and use this to decide what correction to make. For pointlike sources (angular size much less than the antenna beamwidth) the relevant quantity is the aperture efficiency which relates the antenna temperature to the flux density produced by a point source. For extended sources, the relevant quantity is the coupling efficiency, which is proportional to the convolution of the antenna power pattern with the source brightness distribution. One such quantity often used is the beam efficiency, which is the fraction of the total power coupled to a uniform source of angular diameter equal to that of the main lobe of the antenna pattern. In effect, the definition of a *large* source is one much larger than the diffraction pattern of the antenna; this is often loosely interpreted as the main lobe and the first few sidelobes. For a large antenna this can mean a minimum of a few arcminutes and in practice a large source is often defined by the 30-arcminute diameter of the sun or moon; the primary reason for this is clearly the relative ease with which the coupling to a source of this size can be established. For sources of intermediate size, interpolation can be used, or else a coupling efficiency can be calculated, based on an assumed or measured antenna power pattern and an estimate of the source size and distribution.

Another factor, which becomes significant at millimeter wavelengths, is the attenuation due to the earth's atmosphere. The absorption produced by the spectral lines of oxygen is relatively stable, but the absorption due to water vapor is considerable throughout the millimeter range, and can be variable on a time scale of an hour or even less. The most often-used method for determining the attenuation is to measure the emission of the atmosphere, and from this to derive the absorption. This clearly involves an assumption about the temperature distribution throughout the atmosphere. A particularly simple variant of this idea was proposed by Penzias and Burrus (1973), and is used at many facilities. More elaborate methods involve so-called tipping curves; in this procedure one measures the emission at various eleva-

tions, and can then fit an atmospheric model to the data to derive the opacity at the elevation of interest.

The first paper in this part, by Baars *et al.*, describes measurements of the spectrum of the supernova remnant Cas (Cassiopeia) A which is a widely used flux calibration standard. Although the flux from Cas A is time dependent, it has been the basis of a fundamental flux density scale, as determined by measurements with a well-understood horn antenna, as described by Findlay *et al.* (1965). Baars *et al.* also give measurements of a number of other sources, most of which are quasars and hence are of very small angular diameter and hence appropriate for determining the aperture efficiency of a radio telescope. Radio free-free emission from an ionized region can also produce a useful calibration standard. Dent describes the advantage of the small-angular-size HII region DR21 as a standard source in the second paper. The planets can also serve as calibration standards, although their temperatures vary significantly with frequency. Ulich *et al.* report accurate measurements of Venus, Jupiter, and Saturn in the third paper, along with data for the sun. Linsky (1973) has discussed the lunar emission in considerable detail, while Wright (1976) and Werner *et al.* (1978) have provided additional information about the planets and the sun.

The division of the energy in the pattern of a realistic antenna into the region of the diffraction pattern and the spillover and scattering region is discussed in the fourth paper, by Kutner and Ulich. The now widely used expression "forward spillover and scattering efficiency" is introduced here together with a general discussion of calibration. The issue of correction for atmospheric absorption is discussed by Ulich in the final paper, along with a technique to improve the accuracy of the standard "chopper wheel" technique. A number of references to the literature on atmospheric attenuation are included in the References and Bibliography section here, and additional ones can be found in a paper by Gibbins (1986).

REFERENCES AND BIBLIOGRAPHY

General

1973: Davis, J. H. and P. Vanden Bout, "Intensity calibration of the interstellar carbon monoxide line at λ2.6 mm," *Astrophys. Lett.*, vol. 15, pp. 43–47, 1973.

Penzias, A. A. and C. A. Burrus, "Millimeter-wavelength radio-astronomy techniques," *Annu. Rev. Astron. Astrophys.*, vol. 11, pp. 51–71, 1973.

Calibration Standards for Radio Observations

1965: Findlay, J. W., H. Hvatum, and W. B. Waltman, "An absolute flux-density measurement of Cassiopeia A at 1440 MHz," *Astrophys. J.*, vol. 141, pp. 873–884, 1965.

1971: Guidice, D. A. and J. P. Castelli, "The use of extraterrestrial radio sources in the measurement of antenna parameters," *IEEE Trans. Aerosp. Electron. Syst.*, vol. AES-7, pp. 226–234, 1971.

1973: Linsky, J. L., "The moon as a proposed radiometric standard for microwave and infrared observations of extended sources," *Astrophys. J. Suppl.*, vol. 25, pp. 163–204, 1973.

1976: Ulich, B. L. and R. W. Haas, "Absolute calibration of millimeter-wavelength spectral lines," *Astrophys. J. Suppl.*, vol. 30, pp. 247–258, 1976.

Wright, E. L., "Recalibration of the far-infrared brightness temperatures of the planets," *Astrophys. J.*, vol. 210, pp. 250–253, 1976.

1978: Werner, M. W., G. Neugebauer, J. R. Houck, and M. G. Hauser, "One millimeter brightness temperatures of the planets," *Icarus*, vol. 35, pp. 289–296, 1978.

1985: Altenhoff, W. J., "The solar system: (Sub)MM continuum observations," in *Proc. ESO-IRAM-Onsala Workshop on "(Sub)Millimeter Astronomy,"* P. A. Shaver and K. Kjar, Eds. Munich: European Southern Observatory, 1985, pp. 591–601.

Effects of the Atmosphere

1946: Dicke, R. H., R. Beringer, R. L. Kyhl, and A. B. Vane, "Atmospheric absorption measurements with a microwave radiometer," *Phys. Rev.*, vol. 70, pp. 340–348, 1946.

1974: Wesseling, K. H., J. P. Basart, and J. L. Nance, "Simultaneous interferometer phase and water vapor measurements," *Radio Sci.*, vol. 9, pp. 349–353, 1974.

1976: Waters, J. W., "Absorption and emission by atmospheric gases," Chapter 2.3 in *Methods of Experimental Physics: Vol. 12 Astrophysics, Part B: Radio Telescopes,* M. L. Meeks, Ed. New York, NY: Academic Press, 1976, pp. 142–176.

1979: Guiraud, F. O., J. Howard, and D. C. Hogg, "A dual-channel microwave radiometer for measurement of precipitable water vapor and liquid," *IEEE Trans. Geosci. Electron.*, vol. GE-17, pp. 129–136, 1979.

1982: El-Raey, M., "Remote sensing of atmospheric waves in O_2 and H_2O microwave emissions," *Radio Sci.*, vol. 17, pp. 766–772, 1982.

1983: Hogg, D. C., F. O. Guiraud, and J. B. Snider, "Microwave radiometry for measurement of water vapor," in *Reviews of Infrared and Millimeter Waves*, vol. 1, K. J. Button, Ed. New York, NY: Plenum, 1983, pp. 113–154.

1986: Gibbins, C. J., "Improved algorithms for the determination of specific attenuation at sea level by dry air and water vapor, in the frequency range 1-350 GHz," *Radio Sci.*, vol. 21, pp. 949–954, 1986.

1987: Goldsmith, P. F., "Radiation patterns of circular apertures with Gaussian illumination," *Int. J. Infrared and Millimeter Waves*, vol. 8, pp. 771–781, 1987.

Parrish, A., R. L. deZafra, J. W. Barrett, P. Solomon, and B. Connor, "Additional atmospheric opacity measurements at λ = 1.1 mm from Mauna Kea Observatory, Hawaii," *Int. J. Infrared and Millimeter Waves*, vol. 8, pp. 431–440, 1987.

1988: Harris, A. I., "Telescope illumination and beam measurements for submillimeter astronomy," *Int. J. Infrared and Millimeter Waves*, vol. 9, pp. 231–247, 1988.

The Absolute Spectrum of Cas A;
An Accurate Flux Density Scale and a Set of Secondary Calibrators

J. W. M. Baars, R. Genzel, I. I. K. Pauliny-Toth and A. Witzel

Max-Planck-Institut für Radioastronomie, Auf dem Hügel 69, D-5300 Bonn, Federal Republic of Germany

Received April 4, 1977

Summary. A new analysis of the absolute radio spectrum of Cassiopeia A is presented which uses the latest absolute measurements and takes account of the frequency dependence in the secular rate of decrease. The spectrum is established to an accuracy of $\approx 2\%$. Between 0.3 and 30 GHz it is given by a flux density $S_{1\,GHz} = 2723$ Jy and a spectral index $\alpha = -0.770$ (epoch 1980.0).

The absolute spectra of Cygnus A and Taurus A are also given. An accurate "semi-absolute" spectrum for Virgo A is established from direct accurate ratios to Cas A and Cyg A yielding $S_{1\,GHz} = 285$ Jy, $\alpha = -0.856$ (valid for $0.4 \lesssim \nu \lesssim 25$ GHz).

This Virgo A spectrum is used as a basis for accurate relative spectra of a number of sources with simple spectra. These are proposed as secondary calibrators for the routine calibration of flux density measurements. Their spectral data are presented for the frequency range 0.4–15 GHz and are believed to have an *absolute* accuracy of about 5%.

Finally a comparison with other flux density scales is made and the paper concludes with some suggestions for future work.

Key words: absolute radio spectrum — Cassiopeia A — flux density scale — flux density calibration — radio telescope calibration

1. Introduction

The calibration of radio telescopes, both in terms of the pointing direction and aperture efficiency, forms an essential part of observational radio astronomy. Radio astronomers have developed the methods of using radio sources as calibration beacons (see e.g. Kuz'min and Salomonovich, 1966; Baars, 1973).

A knowledge of the aperture efficiency of the telescope over a large range of frequencies is needed for the determination of radio source spectra. Here the method has been to obtain the absolute flux density of the three strongest sources (Cassiopeia A, Cygnus A and Taurus A) with small antennas, whose gain (aperture efficiency) has been theoretically calculated or measured with the aid of a pattern-range (see e.g. Findlay, 1966). The flux density of the sources, so established, can be used to derive the gain of a larger telescope. In the course of the last 20 years several groups have provided the necessary absolute measurements over an ever extending frequency region and with improving accuracy.

In 1962, Conway et al. (CKL, 1962) presented a flux density scale based on the absolute spectrum of Cas A and suggested several weaker sources with power-law spectra as secondary standards. Absolute spectra of Cas A, Cyg A, Tau A and Vir A over an enlarged frequency range (to 15 GHz) were presented by Baars et al. (BMW, 1965) and by Parker (1968). On this basis, Kellermann et al. (KPW, 1969) updated the CKL-scale. By 1971 many new absolute measurements had become available and Baars and Hartsuijker (BH, 1972) published the spectra of the four strongest sources in the range 100 MHz to 20 GHz.

Still the situation was not entirely satisfactory. The main reason is that the strong sources are not adequate calibrators for large telescopes at high frequencies, where the beamwidth is only a few minutes of arc. Cas A, Tau A and even Cyg A are then partially resolved, so that corrections to the measurements for the finite source angular size are necessary. Such corrections always introduce an additional error. The sources also possess other characteristics, which are undesirable for a calibration source. Cas A exhibits a secular decrease of flux density, which is moreover frequency dependent. Cyg A and Tau A both have a significant degree of polarization, the magnitude and position angle of which vary with frequency. The low galactic latitude of these sources causes difficulties in the determination of the zero level, particularly at the lower frequencies, because of the galactic background radiation.

Send offprint requests to: J.W.M. Baars

Table 1. The secular decrease of Cas A

Frequency (MHz)	Epoch	Decrease (% p.y.)	Ref.
81.5	1949–1969	1.29 ± 0.08	Scott et al. (1969)
950	1964–1972	0.85 ± 0.05	Stankevich et al. (1973a)
1420	1957–1976	0.89 ± 0.02	Read (1977)
1420	1957–1971	0.89 ± 0.12	Baars and Hartsuijker (1972)
3000	1961–1972	0.92 ± 0.15	Baars and Hartsuijker (1972)
3060	1961–1971	1.04 ± 0.21	Stankevich et al. (1973b)
7800	1963–1974	0.70 ± 0.10	Dent et al. (1974)
9400	1961–1971	0.63 ± 0.12	Stankevich et al. (1973b)

Nevertheless, these sources are among those studied in greatest detail and they still form the most accurate basis for the flux density scale of radio sources. Cas A serves as the primary standard. On the other hand the sensitivity of contemporary radio telescopes allows relatively weaker sources to be used as calibrators. There are several such sources which do not show the disadvantages of the stronger ones and are more suitable for daily use at the telescope.

The aim of this paper is twofold. Firstly, we present an updated analysis of the spectrum of Cas A which incorporates results obtained since the BH-paper, and in particular the now well established frequency dependence of the secular decrase. The spectra of Cyg A and Tau A are also given, together with a semi-absolute spectrum of Vir A, which is proposed as a good secondary calibrator, especially at frequencies above 5 GHz. Secondly, we use the absolute spectra and a set of accurate relative measurements to determine the spectra of additional secondary standards, which we propose for daily use in the frequency range of 0.3–30 GHz. These sources are stronger than 0.2 Jy at 15 GHz, mostly have a small angular size, do not show variations with time and generally have a simple spectrum. They can be used in a straightforward way without the need for corrections which are frequency-, time- and telescope-dependent.

2. The Absolute Spectra of Cas A, Cyg A and Tau A

Essentially, three different types of antennas have been used for the absolute measurements of flux density. Necessarily all have been calibrated, either by theoretical calculation or by experiment, without any reference to a celestial radio source.

i) The gain of a dipole or dipole array above a groundplane can be calculated theoretically. The method is used for the low-frequency part of the spectrum ($\lesssim 400$ MHz). The achievable accuracy of the gain calculation is 2–3% (Parker, 1968; Wyllie, 1969).

ii) At the higher frequencies a horn antenna appears to be the best. The gain can be calculated theoretically; the accuracy is about 1%, which has been checked in some cases by measurements (Jull and Deloli, 1964; Wrixon and Welch, 1972).

iii) Since the gain of a parabolic reflector antenna cannot be calculated theoretically to the desired accuracy, only an experimental calibration is possible. For an antenna of limited size this can be done on a so-called pattern-range. The measurement is difficult and the accuracy is at best about 5%. An original version of this method has been developed at the Scientific-Research Radiophysical Institute, Gorkii (USSR). Generally known as the "artificial moon" method it uses a black disc, located in the farfield of the antenna, as the calibrated "transmitter". The method has been described e.g. by Troitskii and Tseitlin (1962), Kuz'min and Salomonovich (1966) and Findlay (1966). The accuracy is claimed to be between 3 and 6%.

Cassiopeia A

The most important advance is the confirmation of the frequency-dependence of the secular decrease, which was first suggested by Baars and Hartsuijker (1972). The most accurate values of the secular decrease are given in Table 1, together with their references. The resulting relation for the secular decrease d in the flux density of Cas A as a function of frequency is

$$d(v) [\% \text{ per year}]$$
$$= 0.97(\pm 0.04) - 0.30(\pm 0.04) \log v \, [\text{GHz}] . \qquad (1)$$

This result is identical to that of Dent et al. (1974). In the following spectral analysis we have applied (1) to the frequency range 20 MHz to 30 GHz.

The data for the spectral analysis are assembled in Table 2. To the BH-data we have added 8 measurements with dipole antennas below 38 MHz and a few high frequency points. The table also presents the flux density corrected to epoch 1965.0 with (1) and the weight with which each point entered the analysis. The least-squares procedure indicates that the spectrum can best be represented by a second degree curve over the band 20–300 MHz and by a straight spectrum from 0.3–30 GHz. Table 3 and Figure 1 give the result of the analysis. The spectrum in Table 3 is given for epoch 1965.0 and also for 1980.0, for convenience of use. Note that the flattening of the spectrum over these 15 years is significant in view of the errors.

The following comments are in order.

i) We have used only those measurements with a published error of less than 6%, apart from some low frequency points which are less accurate. The weight assigned to each point is inversely proportional to the square of the error.

ii) The proper use and weighting of the artificial moon results (called Gorkii data hereafter) is not obvious. The method has been improved over the years, resulting in several corrections, which had not been applied to the older data. Therefore we have not used data which were published before 1969. Usually these

Table 2. Absolute flux density measurements of CasA, CygA, TauA

Freg. [MHz]	CasA Flux meas. [Jy]	Epoch	Flux 65.0 [Jy]	Error [1σ, %]	Weight	CygA Flux [Jy]	Error [1σ, %]	Weight	TauA Flux [Jy]	Error [1σ, %]	Weight	Method	Author
10.05	28 000.	65.9	28 300.	10.0	2	13 500.	11.	1				dipole	Bridle (1967)
12.6	58 500.	66.	60 000.	14.	1	21 900.	14.	1	5300.	11.	1	dip.+arr.	Braude (1969)
14.7	65 000.	66.	66 000.	14.	1	31 700.	14.	1	5300.	14.	1	dip.+arr.	Braude (1969)
16.7	60 000.	66.	61 000.	14.	1	26 600.	14.	1	3830.	14.	1	dip.+arr.	Braude (1969)
20.	65 000.	66.	66 000.	14.	1	27 000.	14.	1	3170.	14.	1	dip.+arr.	Braude (1969)
22.25	51 400.	66.5	52 400.	5.	8	29 100.	6.	5	2750.	6.	5	dip.+arr.	Roger (1969)
25.	58 000.	66.	59 000.	14.	1	31 500.	14.	1	3420.	14.	1	dip.+arr.	Braude (1969)
26.3	44 100.	69.8	47 000.	4.6	9	29 600.	5.2	7	2990.	5.2	7	dip.+arr.	Viner (1975)
38.	36 200.	66.9	37 200.	3.7	14	25 500.	4.2	11				dipole	Parker (1968)
81.5	21 100.	66.9	21 630.	2.9	23	16 300.	4.2	11	1880.	4.2	11	dipole	Parker (1968)
152.	12 800.	66.5	13 040.	2.9	23	10 500.	4.2	11	1430.	4.2	11	dipole	Parker (1968)
320.	7330.	62.7	7629.	5.	8	5870.	6.	5				horn	MacRae (1963)
550.	5170.	67.5	5313.	3.2	5	4140.	4.	3				disc	Bondar (1969)
625.	4670.	67.5	4792.	3.3	4	3400.	4.	3				disc	Bondar (1969)
710.	4240.	67.5	4349.	3.5	4	3100.	4.	3				disc	Bondar (1969)
780.	3870.	67.5	3968.	3.5	4							disc	Bondar (1969)
800.		67.5				2670.	4.	3				disc	Bondar (1969)
900.	3470.	67.5	3556.	4.	4							disc	Bondar (1969)
1000.	3110.	67.5	3186.	4.	3							disc	Bondar (1969)
1117.	2830.	69.9	2966.	4.	3	1900.	3.	5	990.	6.	1	disc	Vinogradova (1971)
1150.	2840.	67.5	2908.	4.	3							disc	Bondar (1969)
1.304.	2580.	69.9	2701.	6.	1	1690.	3.	5	980.	6.	1	disc	Vinogradova (1971)
1415.	2369.	69.5	2470.	2.	49							horn	Encrenaz (1970)
1440.	2372.	63.	2328.	2.2	40							horn	Findlay (1965)
1440.	2260.	70.	2367.	3.	22							horn	Findlay (1972)
1765.	2000.	69.9	2090.	5.5	2	1210.	3.	5	940.	6.	1	disc	Vinogradova (1971)
2000.	1860.	69.3	1932.	3.	5	1000.	6.	1	840.	6.	1	disc	Dmitrenko (1970)
2290.	1660.	69.3	1723.	5.	2	935.	6.	1	810.	6.	1	disc	Dmitrenko (1970)
2740.	1380.	69.3	1430.	5.5	2	710.	6.	1	795.	6.	1	disc	Dmitrenko (1970)
3150.	1265.	64.4	1258.	3.	22	645.	3.5	16	700.	3.5	16	horn	Medd (1972)
3200.	1340.	59.4	1279.	4.5	10	680.	5.	8	710.	5.	8	horn	Broten (1960)
3380.	1145.	69.3	1185.	5.5	2	615.	5.	1	718.	6.	1	disc	Dmitrenko (1970)
3960.	1025.	69.3	1060.	4.5	2	515.	6.	1	646.	6.	1	disc	Dmitrenko (1970)
4080.	1086.	64.8	1084.	2.4	34	459.	3.	22	687.	3.	22	horn	Penzias (1965), Wilson (1966)
5680.	740.	68.5	759.	3.5	4	317.	6.	1				disc	Dmitrenko (1970)
6660.	684.	65.	684.	3.	22	265.	3.8	14	577.	3.5	16	horn	Medd (1972)
8250.	612.	65.9	615.	3.6	15				563.	4.	12	horn	Allen (1967)
9380.	510.	68.5	522.	3.5	4							disc	Dmitrenko (1970)
13490.	384.	69.9	396.	3.5	16				520.	3.8	14	horn	Medd (1972)
15500.	374.	65.9	376.	4.8	9				461.	5.2	7	horn	Allen (1967)
16000.	343.	70.6	354.	3.	22				447.	3.5	16	horn	Wrixon (1971, 1972)
22285.	272.	73.1	285.	3.7	14	60.2	3.8	14	397.	4.	12	transm.	Janssen (1974)
31410.			194.	20.	1	55.	36.	1	387.	18.	1		Hobbs (1968)
34900.		67.3							340.	20.	1		Kalaghan (1967)

measurements have resulted in a set of flux densities over a large frequency range, all observed in one continuous session. The presence of a systematic, possibly frequency-dependent error would produce a bias and a correlated error between points. In the more recent publications the systematic error is estimated at 3% and 6%. In two recent papers (Troitskii et al., 1972; Stankevich et al., 1973b) new measurements made between 0.3 and 9.4 GHz have been presented. A reduction of these observations and the ones used in Table 2 to epoch 1965 with (1) shows differences in the measurements at similar frequencies of 6% on the average with several being as large as 10%.

In view of these facts we have given the Gorkii data an error twice the published value in our analysis. Moreover we have not incorporated the points of the last two references, first, because this would considerably increase the risk of a systematic error and, second, because the distribution between horn/dipole and artificial moon measurements would become too lopsided. Actually, we find that addition of either of these two series to the data of Table 2 does not change the resulting spectrum significantly.

Later we shall meet other considerable discrepancies between Gorkii data and other results for Cyg A and

Table 3. Spectral parameters of main calibrators

Source	Frequency interval	Spectral parameters $\log S[\text{Jy}] = a + b \log v\,[\text{MHz}] + c \log^2 v\,[\text{MHz}]$		
		a	b	c
CasA 1965.0	22 MHz...300 MHz	5.625 ± 0.021	-0.634 ± 0.015	-0.023 ± 0.001
	300 MHz... 31 GHz	5.880 ± 0.025	-0.792 ± 0.007	—
CasA 1980.0	300 MHz... 31 GHz	5.745 ± 0.025	-0.770 ± 0.007	—
CygA	20 MHz... 2 GHz	4.695 ± 0.018	$+0.085$ ± 0.003	$-0.178 \cdot$ ± 0.001
	2 GHz... 31 GHz	7.161 ± 0.053	-1.244 ± 0.014	—
TauA	1 GHz... 35 GHz	3.915 ± 0.031	-0.299 ± 0.009	—
VirA	400 MHz... 25 GHz	5.023 ± 0.034	-0.856 ± 0.010	—

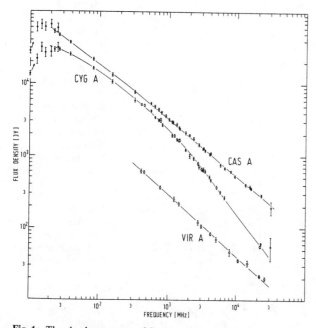

Fig. 1a. The absolute spectra of Cas A and Cyg A and the semi-absolute spectrum of Vir A. Solid symbols are absolute measurements, open symbols relative measurements

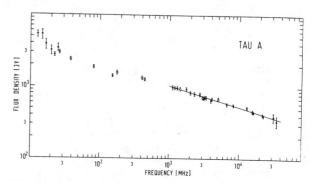

Fig. 1b. The absolute spectrum of Tau A

Tau A. Their effect is always to produce a systematic change in the spectral index. The most likely cause of the discrepancies lies in the sensitivity of the artificial moon method to the effect of the background radiation near the source and to the atmospheric absorption.

iii) Recently an increase in the flux density between 1966 and 1974 has been reported at 38 MHz (Erickson and Perley, 1975), and has been independently confirmed (Read, 1977). While the general small secular decrease at 81.5 MHz and higher frequencies appear well determined (see Table 1), care should be taken in using the spectrum presented here below about 50 MHz.

The spectrum of Cas A now appears well established. The mean error of the flux density scale is about 2% between 0.3 and 30 GHz and increases to probably 5% below 300 MHz. We conclude that a flux density scale

over a large frequency range (20 MHz–30 GHz) can best be based on the Cas A spectrum.

The differences with the BH-spectrum are very small (<2%), except at the lowest frequencies, where the straight line of BH is bound to deviate increasingly. In our analysis we have not used 8 older Gorkii points, which were contained in the BH-analysis. On the other hand we have included several new measurements. The final spectrum however remains almost the same. The spectrum derived here causes correction factors in the widely used flux density scales. We discuss these in Section 5.

Cygnus A and Taurus A

The available absolute data for these sources are also contained in Table 2. The total number of measurements is smaller and generally the accuracy is less than for Cas A. Since in certain circumstances the use of these sources for calibration may be of advantage, we have analysed the spectra; the results are given in Table 3 and Figure 1. There are several very accurate measurements of the *ratio* of Cyg A and Tau A against Cas A. The best ones, described or referred to in BH and BMW have also been used here. The flux densities are calculated on the basis of the present Cas-spectrum. The addition of these relative points does not change the spectrum significantly but it improves the accuracy of the fit.

404

For both sources there are discrepancies with the most recent Gorkii data. Those by Troitskii et al. (1972) for Cyg A all lie below the other observations and increasingly so with increasing frequency, indicating a possible systematic error of considerable magnitude. The situation with Tau A is complicated by the claim of the Gorkii group that there are several steps in the spectrum between 700 and 3000 MHz, where the spectral index is about zero. As long as this situation has not been cleared up by relative measurements there remains some doubt about the true spectrum of Tau A. Because it is the strongest source above 10 GHz, we nevertheless include our results in Table 3. There is a fairly good agreement with the BH-result.

3. The Semi-absolute Spectrum of Vir A

An ideal calibrator should have a small angular size, be constant in time and have a simple spectrum. Also, in order to avoid alinearity problems in low-noise receivers, its flux density should not be too high. At the higher frequencies Virgo A (3C274, M87) satisfies these requirements and hence appears to be a suitable *secondary standard*. First the source is strong enough to have been well observed at many frequencies. In particular there exists a large set of very accurate measurements of the ratio of Vir A to Cas A. We use these to derive the *semi-absolute* spectrum of Vir A. Second the flux density is sufficiently low to avoid the danger of scaling errors in observations of weaker sources.

We use 16 observations relative to Cas A, Cyg A and Tau A together with a few absolute measurements. All data with their reference are given in Table 4, which requires little comment. The circumstances of the observations have been described in the references. From the original articles we know that the ratio measurements have been done with great care. The ratios at 10.7 GHz are observations by us at the NRAO 140-ft and the MPIfR 100-m telescopes, where we use the Cyg A spectrum of Table 3.

The resulting spectrum of Vir A is presented in Table 3 and Figure 1. It is well approximated by a straight line over the range 0.3–300 GHz. There is however one complication, which should be noted. The brightness distribution of Vir A is rather intricate. A core source of approximately 40″ diameter contains a weak nucleus of 0.5″ and is surrounded by a halo of about 14′ × 10′. The spectra of the components are different, while the overall spectrum is straight, as shown above (see also Turland, 1975). The spectrum of the halo steepens considerably above 1 GHz and (according to observations at Effelsberg, von Kap-Herr and Wielebinski, pers. comm.) the spectral index lies between −3 and −4 above 5 GHz. Thus the contribution of the halo is about 10% at 5 GHz and 1% at 10 GHz. We conclude that Vir A is a convenient secondary calibrator, especially for the higher

Table 4. The data for Virgo A

Frequency (MHz)	Vir/Cas (observed)	S (Vir) (Jy) (derived)	Error (%)	Ref.
400	0.085	617	5	BMW
440	0.091	598	5	BMW
750	0.086	362	4.5	BMW
750	0.0914	372	4	KPW
1200	0.086	251	4.5	BMW
1400	0.0896	221	4	KPW
1420 (3 ×)	0.0945	218	2.5	BH
1440	0.084	212	4.5	BMW
2695	0.0818	120	4	KPW
3000	0.077	106	8	BMW
3000	0.0836	105	3.5	BH
4080	0.0807	84.4	5	WP
5000	0.0799	71	4	KPW
5000	0.105 (Tau)	67	5	BMW
8000	0.083 (Tau)	46.6	11	BMW
10700 (2 ×)	0.252 (Cyg)	35.5	4	PT
14500	0.068 (Tau)	32.6	11	BMW
22285 (3 ×)	Absolute	21.3	4	JGW

Note to ref. WP = Wilson and Penzias (1966), PT = Pauliny-Toth (unpublished), JGW = Janssen et al. (1974)

frequencies ($\gtrsim 5$ GHz). At lower frequencies the influence of the halo must be taken into account properly to avoid systematic errors.

4. Sources Suitable as Routine Calibrators

The usual method of determining flux densities of radio sources is to measure the ratio of the source strength to that of a well-known calibrator. In the foregoing sections we have indicated what requirements a routine calibrator should satisfy, and also mentioned the disadvantages in this respect of the strong sources Cas A, Cyg A and Tau A.

To avoid as much as possible the effects of varying telescope efficiency with position and time it is desirable to have a set of calibrators which is reasonably equally distributed over the sky. Clearly Vir A alone does not fulfill this purpose. We present now a group of suitable secondary flux density calibrators, the spectra of which have been carefully determined from relative measurements against one or more of the standards, presented in the foregoing sections. These measurements have been taken mainly from the literature, notably Kellermann and Pauliny-Toth (1973) (10.7 GHz), KPW (1969) (0.75, 1.4, 2.7 and 5 GHz), Ross and Seaquist (1975) (3.2, 6.6 and 10.6 GHz) and Klein und Stelzried (1976) (2.3 GHz). Additional measurements have been made at Effelsberg with the 100-m telescope, mainly at 15 GHz. These results have been published separately (Genzel et al., 1976). Unpublished data at 408 MHz from the NRAO 300-ft telescope, obtained by Pauliny-Toth and Kellermann, have also been used.

Table 5. Spectral parameters of telescope calibrators

Source	Frequency interval	Spectral parameters $\log S$ [Jy]$= a + b \cdot \log \nu$ [MHz]$+ c \cdot \log^2 \nu$ [MHz]					
		a		b		c	
3 C 48	405 MHz...15 GHz	2.345	±0.030	+0.071	±0.001	−0.138	±0.001
3 C 123	405 MHz...15 GHz	2.921	±0.025	−0.002	±0.0001	−0.124	±0.001
3 C 147	405 MHz...15 GHz	1.766	±0.017	+0.447	±0.006	−0.184	±0.001
3 C 161	405 MHz...10.7 GHz	1.633	±0.016	+0.498	±0.008	−0.194	±0.001
3 C 218	405 MHz...10.7 GHz	4.497	±0.038	−0.910	±0.011	—	
3 C 227	405 MHz...15 GHz	3.460	±0.055	−0.827	±0.016	—	
3 C 249.1	405 MHz...15 GHz	1.230	±0.027	+0.288	±0.007	−0.176	±0.003
3 C 286	405 MHz...15 GHz	1.480	±0.018	+0.292	±0.006	−0.124	±0.001
3 C 295	405 MHz...15 GHz	1.485	±0.013	+0.759	±0.009	−0.255	±0.001
3 C 348	405 MHz...10.7 GHz	4.963	±0.045	−1.052	±0.014	—	
3 C 353	405 MHz...10.7 GHz	2.944	±0.031	−0.034	±0.001	−0.109	±0.001
DR 21	7 GHz...31 GHz	1.81	±0.05	−0.122	±0.010		
NGC 7027	10 GHz...31 GHz	1.32	±0.08	−0.127	±0.012		

The derived spectra, valid between 0.4 and 15 GHz, are given in Table 5, while Table 6 presents salient features of the sources, together with the calculated flux density at several standard observing frequencies. The thermal sources DR 21 and NGC 7027 are particularly useful at frequencies above 10 GHz. Recently Dent (1972) has discussed the compact H II region DR 21 as a calibrator for high frequencies ($\gtrsim 8$ GHz). The source has the clear advantage of possessing the well-predictable and flat spectrum of an optically thin H II region. For that reason we present our results for its spectrum here. This source should be used with care in view of its angular size (20″) and the brightness structure of the surrounding region (see Dent for details). In the same way the planetary nebula NGC 7027 has been analysed and presented in Table 5.

Only three of the sources are suitable for the calibration of interferometers and synthesis telescopes, viz 3C48, 3C147 and 3C286. Of these the first two are universally used as interferometer calibrators. The highly accurate positions of these sources have been taken from Elsmore and Ryle (1976). Some of the other sources may need a correction for angular size, when used with the largest single telescopes at high frequencies. Because their sizes are well-known the correction is readily applied. When using these sources with antenna beams of only a few arc minutes one should note that some have a frequency-dependent brightness distribution resulting in a variation of the centroid position of a few seconds of arc.

We believe that this set of sources forms a solid basis for the calibration of source surveys and flux density observations over a wide range of frequencies. Together with the primary calibrators any flux dependent variation in the scale should be avoidable. We believe that the systematic error of the scale over the frequency region 0.4–15 GHz is not more than 3–4%. Thus, with sufficient signal to noise ratio, flux density measurements might be carried out over this frequency range with an absolute accuracy of about 5%. Any further improvement might only be expected from direct absolute measurements of the sources in Table 5, for instance with the interferometric method as used by Wyllie (1969).

6. Comparison with Other Flux Density Scales

It is of interest to point out the quantitative differences between our new scale and other widely used flux density scales. Although we have presented here a new analysis of the absolute Cas A-spectrum over the large frequency range 20 MHz–30 GHz, the other, secondary, calibrators, presented in Table 3 and 5, have only been analyzed for frequencies from 400 MHz upwards.

Table 7 shows the correction factors needed to bring the CKL, Kellermann (K, 1964), KPW, BMW, Wills (1973) and BH scales to the new scale presented here. The table is calculated from a comparison of the basic derived Cas A spectra.

The largest discrepancies arise at the lower frequencies. In particular the CKL- and KPW-scales are significantly low below 1 GHz. These scales have formed the basis for the spectral analysis of large sets of weaker sources, notably 3 CR. Later observations of these sources have already given indications that these scales are in error (Niell and Jauncey, 1971).

Also evidence has been presented for a flux density dependent scale factor at several frequencies (Scott and Shakeshaft, 1971; Braude et al., 1971). An investigation into this effect, based on new accurate relative measurements and calibrated against the present Cas A-spectrum would be very useful, but is beyond the scope of this paper. The discrepancies undoubtedly arise partly from the different frequency regions over which the spectra have been analysed. We feel that the method of establishing the spectra of the secondary calibrators, as applied for instance in the CKL and Wills papers, also contribute to the flux density dependence of the scales.

Table 6. *Position and flux densities of telescope calibrators*

Source	RA (1950.0) [h m s]	Dec (1950.0) [° ′ ″]	b″ [°]	S_{1400} [Jy]	S_{1665} [Jy]	S_{2700} [Jy]	S_{5000} [Jy]	S_{8000} [Jy]	S_{10700} [Jy]	S_{15000} [Jy]	S_{22235} [Jy]	Spec.	Ident.	Polar. (at 5 GHz) %	Ang. size (at 1.4 GHz) ″
3C 48	01 34 49.8	+32 54 20	−29	15.9	13.9	9.20	5.24	3.31	2.46	1.72	1.11	C⁻	QSS	5	<1
3C 123	04 33 55.2	+29 34 14	−12	48.7	42.4	28.5	16.5	10.6	7.94	5.63	3.71	C⁻	GAL	2	20
3C 147	05 38 43.5	+49 49 42	+10	22.4	19.8	13.6	7.98	5.10	3.80	2.65	1.71	C⁻	QSS	<1	<1
3C 161	06 24 43.1	−05 51 14	−8	19.0	16.8	11.4	6.62	4.18	3.09	2.14	—	C⁻	GAL	5	<3
3C 218	09 15 41.5	−11 53 06	+25	43.1	36.8	23.7	13.5	8.81	6.77	—	—	S	GAL	1	core 25 halo 200
3C 227	09 45 07.8	+07 39 09	+42	7.21	6.25	4.19	2.52	1.71	1.34	1.02	0.73	S	GAL	7	180
3C 249.1	11 00 25.0	+77 15 11	+39	2.48	2.14	1.40	0.77	0.47	0.34	0.23	—	S	QSS	—	15
3C 274	12 28 17.7	+12 39 55	+74	214	184	122	71.9	48.1	37.5	28.1	20.0	S	GAL	1	halo 400[a]
3C 286	13 28 49.7	+30 45 58	+81	14.8	13.6	10.5	7.30	5.38	4.40	3.44	2.55	C⁻	QSS	11	<5
3C 295	14 09 33.5	+52 26 13	+61	22.3	19.2	12.2	6.36	3.65	2.53	1.61	0.92	C⁻	GAL	0.1	4
3C 348	16 48 40.1	+05 04 28	+29	45.0	37.5	22.6	11.8	7.19	5.30	—	—	S	GAL	8	115[b]
3C 353	17 17 54.6	−00 55 55	—	57.3	50.5	35.0	21.2	14.2	10.9	—	—	C⁻	GAL	5	150
DR 21	20 37 14.2	+42 09 07	+1					21.6	20.8	20.0	19.0	Th	HII	—	20[c]
NGC 7027[d]	21 05 09.4	+42 02 03	−3	1.35	1.65	3.5	5.7		6.43	6.16	5.86	Th	PN	<1	10

[a] Halo has steep spectral index, so for $\lambda \lesssim 6$ cm, more than 90% of the flux is in the core
[b] Angular distance between the two components
[c] Angular size at 2 cm, but consists of 5 smaller components
[d] Data up to 5 GHz are the direct measurements, not calculated from fit

Table 7. Ratio of the present to other flux density scales

Frequency (MHz)	CKL	K	KPW	BMW	BH	Wills
38	1.029	0.979	0.981	—	0.880	0.976
81.5	1.074	1.021	1.020	—	0.935	1.013
178	1.110	1.054	1.051	—	0.974	1.042
400	1.129	1.074	1.065	1.080	1.007	1.048
750	1.114	1.059	1.046	1.044	1.004	1.035
1400	1.099	1.038	1.029	1.015	1.000	1.017
2700	1.083	1.030	1.011	0.981	1.000	0.997
5000	—	1.016	0.993	0.951	1.000	0.979
10600	—	1.000	0.974	1.065	0.990	0.959
15000	—	—	0.966	1.218	0.989	0.949

While we consider it unlikely that the effect is present in the secondary calibrator spectra, the effect would still be present in an adjustment of the older scales at the lower frequencies to our new Cas A-spectrum.

It is of interest to compare our results with the *absolute* scale of Wyllie (1969) at 408 MHz, which is based on absolute measurements of a set of relatively weak sources (10–150 Jy). We have compared his results with the predicted flux density of the straight spectrum sources in Genzel's analysis (pers. comm.), which are based on our Cas A-spectrum. We find the ratio Genzel/Wyllie = 0.97, indicating that Wyllie's scale is only 3% above our Cas A-scale.

6. Conclusion

This paper is the first to present an analysis of the absolute spectrum of Cas A, in which the data have been corrected for the now well-established frequency dependence of the secular decrease in flux density. We consider the resulting spectrum to be very well determined with an estimated uncertainty in calculated flux density at frequencies between 0.1 and 30 GHz of $\approx 2\%$. A comparison with the most recent analysis by Baars and Hartsuijker (1972), who used a crude version of the frequency dependent secular change, shows that the differences are smaller than 1% over the applicable frequency range.

A further improvement in the accuracy of the spectrum will be difficult to achieve. The most useful effort would be a flux measurement with a horn-antenna at several frequencies between 200 and 1200 MHz and at about 2, 10 and 30 GHz. The lower frequency points especially would give a good comparison with the artificial moon data.

The definition of an accurate spectrum of Vir A and several other sources, directly based on the Cas A spectrum and extending to above 15 GHz, allows the accurate calibration of flux density observations over a wide frequency range. The use of these sources will avoid scaling discrepancies between measurements of different observers without the need to refer to Cas A in the actual observations.

A possible scheme to update and improve the present situation might be the following international observing campaign. All observers, who have an absolutely calibrated antenna at their disposal (horn, dipoles, pattern-range or artificial moon calibration) measure Cas A at the same epoch (1980?). By mutual agreement it is arranged to have a regular coverage over as large a frequency region as possible and to distribute the different calibration techniques over the band.

At the same epoch the ratios of the secondary standard against Cas A would be determined to the highest accuracy at all "standard" observing frequencies. With the new absolute Cas A spectrum the spectra of the secondary calibrators could be updated. Ideally by that time we would also have available direct absolute measurements of several secondary calibrators at a few standard frequencies, obtained by the interferometric method.

Acknowledgement. We wish to thank J. W. Findlay, K. I. Kellermann and J. R. Shakeshaft for useful discussions and suggestions.

References

Allen, R. J., Barrett, A. H.: 1967, *Astrophys. J.* **149**, 1

Baars, J. W. M.: 1973, IEEE *Trans. Antennas Propagation* **21**, 461

Baars, J. W. M., Hartsuijker, A. P.: 1972, *Astron. Astrophys.* **17**, 172

Baars, J. W. M., Mezger, P. G., Wendker, H.: 1965, *Astrophys. J.* **142**, 122

Bondar, L. N., Zelinskaya, M. R., Kamenskaya, S. A., Porfirev, V. A., Rackhlin, V. L., Rodina, V. M., Stankevich, K. S., Strezhneva, K. M., Troitskii, V. S.: 1969, *Radiofizika* **12**, 807

Braude, S. Y., Lebedeva, O. M., Megn, A. V., Ryabov, B. P., Zhouck, I. N.: 1969, *Monthly Notices Roy. Astron. Soc.* **143**, 289

Braude, S. Y., Megn, A. V., Ryabov, B. P., Zhouck, I. N.: 1970, *Astrophys. Space Sci.* **8**, 275

Bridle, A. H.: 1967, *Observatory* **87**, 60

Broten, N. W., Medd, W. J.: 1960, *Astrophys. J.* **132**, 279

Conway, R. G., Kellermann, K. I., Long, R. J.: 1963, *Monthly Notices Roy. Astron. Soc.* **125**, 261

Dent, W. A.: 1972, *Astrophys. J.* **177**, 93

Dent, W. A., Aller, H. D., Olsen, E. T.: 1974, *Astrophys. J.* **188**, L11

Dmitrenko, D. A., Tseitlin, N. M., Vinogradova, L. V., Gitterman, K. F.: 1970, *Radiofizika* **13**, 823

Elsmore, B., Ryle, M.: 1976, *Monthly Notices Roy. Astron. Soc.* **174**, 411

Encrenaz, P. J., Penzias, A. A., Wilson, R. W.: 1970, *Astrophys. J.* **160**, 1185

Erickson, W. C., Perley, R. A.: 1975, *Astrophys. J.* **200**, L83

Findlay, J. W.: 1966, Absolute intensity calibrations in radio astronomy, in *Ann. Rev. Astron. Astrophys.* **4**, 77

Findlay, J. W.: 1972, *Astrophys. J.* **174**, 527

Findlay, J. W., Hvatum, H., Waltman, W. B.: 1965, *Astrophys. J.* **141**, 873

Genzel, R., Pauliny-Toth, I. I. K., Preuss, E., Witzel, A.: 1976, *Astron. J.* **81**, 1084

Hobbs, R. W., Corbett, H. H., Santini, N. J.: 1968, *Astrophys. J.* **152**, 43

Janssen, M. A., Golden, L. M., Welch, W. J.: 1974, *Astron. Astrophys.* **33**, 373

Jull, E. V., Deloli, E. P.: 1964, IEEE *Trans. Antennas Propagation* **12**, 439

Kalaghan, P. M., Wulfsberg, K. N.: 1967, *Astron. J.* **72**, 1051

Kellermann, K. I.: 1964, *Astron. J.* **69**, 205

Kellermann, K. I., Pauliny-Toth, I. I. K., Williams, P. J. S.: 1969, *Astrophys. J.* **157**, 1

Kellermann, K. I., Pauliny-Toth, I. I. K.: 1973, *Astron. J.* **78**, 828

Klein, M. J., Stelzried, C. T.: 1976, *Astron. J.* **81**, 1078

Kuz'min, A. D., Salomonovich, A. E.: 1966, Radioastronomical Methods of Antenna Measurements, Academic Press, New York

McRae, D. A., Seaquist, E. R.: 1963, *Astron. J.* **68**, 77

Medd, W. J.: 1972, *Astrophys. J.* **171**, 41

Niell, A. E., Jauncey, D. L.: 1971, *Bull. Am. Astron. Soc.* **3**, 25

Parker, E. A.: 1968, *Monthly Notices Roy. Astron. Soc.* **138**, 407

Penzias, A. A., Wilson, R. W.: 1965, *Astrophys. J.* **142**, 1149

Read, P. L.: 1977, *Monthly Notices Roy. Astron. Soc.* **178**, 259

Roger, R. S., Costain, C. H., Lacey, J. D.: 1969, *Astron. J.* **74**, 36

Ross, H. N., Seaquist, E. R.: 1975, *Monthly Notices Roy. Astron. Soc.* **170**, 115

Scott, P. F., Shakeshaft, J. R.: 1971, *Monthly Notices Roy. Astron. Soc.* **154**, 19P

Scott, P. F., Shakeshaft, J. R., Smith, M. A.: 1969, *Nature* **223**, 1139

Stankevich, K. S., Ivanov, V. P., Torkhov, V. A.: 1973, *Soviet. Astron.* (AJ) **17**, 410

Stankevich, K. S., Ivanov, V. P., Pelyushenko, S. A., Torkhov, V. A., Ibannikova, A. N.: 1973, *Radiofizika* **16**, 786

Troitskii, V. S., Tseitlin, N. M.: 1962, *Radiofizika* **5**, 623

Troitskii, V. S., Stankevich, K. S., Tseitlin, N. M., Krotikov, V. D., Bondar, L. N., Strezhneva, K. M., Rakhlin, U. L., Ivanov, V. P., Pelyushenko, S. A., Zubov, M. M., Samoilov, R. A., Titov, G. K., Porfirev, V. A., Chekalev, S. P.: 1972, *Soviet. Astron.* (AJ) **15**, 915

Turland, B. D.: 1975, *Monthly Notices Roy. Astron. Soc.* **170**, 281

Viner, M. R.: 1975, *Astron. J.* **80**, 83

Vinogradova, L. V., Dmitrenko, D. A., Tseitlin, N. M.: 1971, *Radiofizika* **14**, 157

Wills, B. J.: 1973, *Astrophys. J.* **180**, 335

Wilson, R. W., Penzias, A. A.: 1966, *Astrophys. J.* **146**, 286

Wrixon, G. T., Gott, J. R., Penzias, A. A.: 1971, *Astrophys. J.* **165**, 23

Wrixon, G. T., Gott, J. R., Penzias, A. A.: 1972, *Astrophys. J.* **174**, 399

Wrixon, G. T., Welch, W. J.: 1972, IEEE *Trans. Antennas Propagation*, AP20, 136

Wyllie, D. V.: 1969, *Monthly Notices Roy. Astron. Soc.* **142**, 229

A FLUX-DENSITY SCALE FOR MICROWAVE FREQUENCIES*

WILLIAM A. DENT
Department of Physics and Astronomy, University of Massachusetts, Amherst
Received 1972 March 15

ABSTRACT

Accurate flux-density measurements of the thermal radio source DR 21 have been made at centimeter wavelengths relative to the KPW absolute flux-density scale based on Cas A and at millimeter wavelengths relative to absolute brightness-temperature measurements of Jupiter and Saturn. The absolute spectrum of DR 21 thus defined has the form

$$S_\nu \doteq 26.8 - 5.6 \log \nu \ \text{(GHz)} \quad \text{for } \nu \geq 7 \text{ GHz,}$$

and ties together two formerly independent flux-density scales. With an accuracy of about 3 percent, this spectrum of DR 21 defines a flux-density scale that can be used to calibrate antennas having beamwidths between 1 and 6 minutes of arc at microwave frequencies above 7 GHz where other methods of absolute calibration are much less accurate.

I. INTRODUCTION

One of the most fundamental problems in radio astronomy is the calibration of the antenna aperture efficiency. For antenna systems other than simple horns or dipole antennas this calibration can be reliably determined only by observations of radio sources whose flux-density spectrum has previously been measured with simple antennas of known efficiency. At present the flux-density scale used in radio astronomy is based almost entirely on absolute measurements of the radio source Cas A. Kellermann, Pauliny-Toth, and Williams (1969) have combined these measurements to obtain an "absolute spectrum" of Cas A that can be used to define a flux-density scale (KPW scale). Unfortunately Cas A, besides being a slowly varying source, has an angular diameter of about 4' and is therefore partially resolved by antennas having beamwidths less than about 10'. Consequently Cas A cannot be used to determine accurately the antenna's aperture efficiency at microwave frequencies where beamwidths less than 10' are common.

Scheuer and Williams (1968) have used the KPW scale to determine the absolute spectra of a set of secondary standard calibration sources: 3C 218, 3C 274, 3C 348, and 3C 353, all of which are fairly strong and have straight spectra between 38 and 5000 MHz. Unfortunately this flux-density scale becomes marginally useful at frequencies above 10 GHz. The sources 3C 348 and 3C 353 consist of two widely spaced emitting components and require a significant correction for source size when observed with beamwidths less than about 10'. Until recently 3C 274 (Virgo A, M87) was considered to be the best standard source available. Besides being the strongest secondary standard, its smaller angular extent required a size correction C_s (Dent and Haddock 1966) that was not too large and was easily measurable. However, Graham (1971) has recently presented evidence suggesting that the nucleus of M87 may be varying. These variations have been confirmed by Dent and Kojoian (unpublished) and amount to about 3 percent of the total flux density at 7.8 GHz. On the other hand, 3C 218, which has none of the above objections, is the weakest source of the four. This, combined

with its relatively steep spectrum, renders it useless as a standard above about 20 GHz where most observations are sensitivity limited.

At millimeter wavelengths a flux-density scale independent of Cas A has commonly been used that is based on absolute measurements of the brightness temperature of Jupiter by Thornton and Welch (1963) at 36 GHz and by Kalaghan and Wulfsberg (1968) at 34.9 GHz. Even in spite of the large uncertainties in these measurements (10–16 percent), the brightness temperature of Jupiter (and Saturn) cannot reliably be assumed independent of wavelength at microwave frequencies. In addition to the very broad ammonia absorption feature near 24 GHz in the spectra of Jupiter and Saturn (Wrixon and Welch 1970; Wrixon, Welch, and Thornton 1971) and a small nonthermal component in the spectrum of Jupiter, the millimeter-wave spectra of both planets are likely to be sloped because of the temperature gradient in each planet's atmosphere.

An ideal flux-density standard for microwave frequencies would be a source that is strong ($S_\nu > 10$ flux-density units $= 10 \times 10^{-26}$ W m^{-2} Hz^{-1}) at all microwave frequencies, of small angular extent ($\phi < 20'$), nonvariable, and unpolarized. Such a source would be able to tie together the centimeter-wavelength KPW scale based on Cas A with the millimeter-wavelength scale based on absolute measurements of Jupiter and Saturn. There is only one known radio source in the sky which meets these requirements: DR 21.

II. DR 21

DR 21 (Downes and Rinehart 1966; Pike and Drake 1964) is a compact H II region in Cyg X. Above 3 GHz the source flux density is about 20 flux-density units and relatively independent of frequency. Below about 3 GHz the spectrum steepens, characteristic of an opaque thermal source. Ryle and Downes (1967) have fitted the continuum spectrum of DR 21 between 408 MHz and 15 GHz with a thermal-bremsstrahlung model to obtain a source electron temperature of 11,000° K and emission measure of 5×10^7 cm^{-6} pc. The thermal nature of DR 21 has been confirmed by the detection of recombination-line emission (Mezger et al. 1967).

Aperture-synthesis observations of DR 21 with a resolution of about 6" × 9" have been made by Webster and Altenhoff (1970) at 2.7 GHz and by Wynn-Williams (1971) at 5.0 GHz. The source brightness distribution consists of two components separated by 20" in an almost north-south direction. The southern component contains at least 3 times the flux density as the other and in angular dimension is about 10" × 20" compared to 9" × 9" or less for the northern component. There may be a weak 30" diameter envelope surrounding the source. Thus the relatively compact nature of DR 21 makes it a good calibration source for antenna beamwidths greater than about 1'.

The most serious potential problem with DR 21 as a standard source is its location in a complex region of the galactic plane. Maps by Pike and Drake (1964) at 1.4 GHz, Downes and Rinehart (1966) at 5.0 GHz, and more recently by Winzer and Roberts (1971) at 3.2 GHz show the presence of a source 40' to the west of DR 21 and another source 30' to the southeast. In addition, the intensity contours suggest that the emission from DR 21 is superposed on a weak but extended ($\phi \approx 1°$) background source. The suitability of DR 21 as a flux-density standard depends upon the smoothness of the background emission in the near vicinity of the source.

The region within 18' of DR 21 was scanned with the 120-foot (3658-cm) Haystack antenna at both 7.8 GHz and 15.5 GHz. With resolutions of 4'4 and 2'2 it was found that the background antenna temperature decreases toward the north at a rate of 0.007° K per minute of arc. Between two points separated by 24' and centered on DR 21 this slope would amount to about 5 percent of the DR 21 antenna temperature.

---

* Contribution Number 137 of the Five College Observatories.

Reprinted courtesy of The Astrophysical Journal, vol. 177, pp. 93–99, Oct. 1, 1972, published by the University of Chicago Press;
© 1972, The American Astronomical Society.

The possible effect of this background emission upon an antenna temperature measurement of DR 21 was investigated in the following way. The antenna was pointed successively at a reference point 6' north of DR 21, on DR 21, and 6' south of DR 21. From the first and third points a baseline was defined from which the antenna temperature on-source was subtracted to give a measurement of the DR 21 antenna temperature. This was done several times to obtain a statistical average. The location of the reference points were successively increased to 12', 18', and 24' from DR 21 and its antenna temperature measured as above. A similar procedure was repeated at off-source reference position angles of 45°, 90°, and 135° relative to north. It was found that as long as the off-source reference points do not range by more than 1 percent. Thus although the background emission of DR 21, the background is sufficiently linear within 18' of the source to ensure that background effects introduce a negligible error.

III. THE MEASUREMENTS

At 7.8 GHz (λ3.8 cm) and 15.5 GHz (λ1.9 cm) the antenna temperature of DR 21 was measured with the 120-foot Haystack antenna using a symmetric on-source–off-source technique. The off-source reference points were chosen as ±12 in azimuth in accordance with the above conclusions. Because Cas A is resolved by the Haystack antenna at these frequencies it is not possible to directly calibrate DR 21 relative to Cas A. It was thus necessary to employ intermediate standard sources.

The ratios of 3C 123 and Cyg A to Cas A have been measured to within a few percent (Dent and Haddock 1966) at 8.0 GHz using a 6' beam where the effect of the angular extent of Cas A is less important. Using these ratios, and applying a small correction for the slight frequency difference, the absolute 7.8 GHz flux densities of the intermediate standards 3C 123 and Cyg A were computed relative to the KPW spectrum of Cas A. The values obtained are given in column (3) of table 1. Medd (1972) has recently published a spectrum of Cyg A based on absolute flux-density measurements. On this scale the computed 7.8 GHz flux density of Cyg A is 221 flux-density units, the same as that obtained above from the Cyg A Cas A ratio and the KPW Cas A spectrum.

A maser radiometer receiving the left-hand circular mode of polarization was used to measure the ratio of the antenna temperature of DR 21 to 3C 123 and Cyg A at

7.8 GHz with the Haystack antenna. All measurements included corrections for atmospheric extinction, change in antenna gain with elevation, and finite source size (discussed below). From an average of measurements made on 6 different days throughout 1970, the DR 21/3C 123 ratio was measured to be 2.045 ± 0.011. A DR 21/Cyg A ratio of 0.0995 ± 0.0006 was obtained from measurements made on seven different days in 1970. The flux densities of DR 21 computed from these ratios are given in column (5) of table 1.

At 15.5 GHz a similar procedure was employed using a tunnel-diode radiometer with the Haystack antenna. The effects of source polarization were eliminated by averaging the antenna temperatures observed with the linear feed horn oriented at two orthogonal position angles. To calibrate the DR 21 measurements at 15.5 GHz it was necessary to use the secondary standard sources of Scheuer and Williams (1968). The spectrum of 3C 123 has positive curvature and could not be extrapolated, while the size correction for Cyg A increased from 1.08 at 7.8 GHz to 1.55 at 15.5 GHz because of the smaller beamwidth. In column (3) of table 1 are given the 15.5-GHz flux densities of 3C 218 and 3C 274 computed from their absolute spectra (Scheuer and Williams 1968). Column (5) gives the flux-density measurement of DR 21 relative to each of these standard sources. Although the flux density of 3C 274 varies slightly, its use here can be justified. The measurement in table 1 represents an average of 20 measurements of the DR 21/3C 274 made at roughly equally spaced intervals between 1969.0 and 1972.0 during which time 3C 274 fluctuated by no more than 4 percent. This averaging of the DR 21/3C 274 ratio minimizes the uncertainty due to variations in 3C 274 to the point where it becomes comparable with other uncertainties in the measurement. The Scheuer-Williams spectrum of 3C 274 itself is an average spectrum composed of measurements made at random times.

At the frequency of 31.4 GHz (9.5 mm) several measurements of the ratio of the corrected antenna temperature of DR 21 to Jupiter and Saturn were made with the 36-foot (1097-cm) millimeter wave antenna of the National Radio Astronomy Observatory. The ratios were obtained during four different observing periods between 1970 September and 1972 January using an on-off alternate beam observing technique preceded by a pointing measurement. The observational details will be presented later (Dent and Hobbs, in preparation). Both DR 21 and either Jupiter or Saturn, depending on availability, were observed as close as possible in time in order to minimize possible changes in antenna gain associated with changes in the ambient temperature.

Absolute measurements of the brightness temperature of Saturn between 20.5 and 35.5 GHz have been reported by Wrixon and Welch (1970) in an attempt to detect the ammonia absorption band at 23.7 GHz. An interpolated brightness temperature of 139 ± 6° K for Saturn at 31.4 GHz was obtained from the theoretical curve which best fits their data. A similar set of absolute measurements on Jupiter has recently been reported by Wrixon et al. (1971). At 31.4 GHz the interpolated brightness temperature of Jupiter is $T_b = 144° ± 8°$ K, where the errors quoted in both cases include the uncertainty in the absolute calibration. Using Saturn as a flux-density standard, the average of nine independent measurements of the DR 21/Saturn ratio on the 36-foot antenna yielded an absolute DR 21 flux density of 18.7 ± 0.9 at 31.4 GHz. A similar set of independent measurements of the DR 21/Jupiter ratio gave an average flux density of 17.4 ± 1.1 for DR 21 at 31.4 GHz.

Since DR 21 is not a point source, a correction factor must be multiplied by all the observed antenna temperatures to compensate for the partial resolution of the source. This source-size correction, C_s, was determined in the following way. Right-ascension and declination scans of DR 21 and the point sources 3C 84 and 3C 273 were made with a beamwidth of 2'.2 using the Haystack antenna at 15.5 GHz. From the measured average DR 21 response widths R and beamwidths B obtained from the point sources

TABLE 1

ABSOLUTE FLUX-DENSITY MEASUREMENTS OF DR 21

Frequency (GHz) (1)	Standard Sources (2)	S_s^* or T_b (3)	Absolute Scale† (4)	DR 21 Flux Density* (5)	S_s (6)
7.8	3C 123/Cas A	10.7	(1)	22.0 ± 0.6	2.47
7.8	Cyg A + Cas A	221	(1, 2)	22.2 ± 0.6	2.49
15.5	3C 274/Cas A	28.9	(3)	20.5 ± 0.8	2.50
15.5	3C 218/Cas A	4.8	(4)	19.9 ± 0.9	2.42
31.4	Jupiter	144 K	(4)	17.4 ± 1.1	2.32
31.4	Saturn	139 K	(5)	18.7 ± 0.6	2.49
Average					2.45 ± 0.04

* S, in units of 10^{-26} W m^{-2} Hz^{-1}.
† (1) Kellermann et al. 1969; (2) Medd 1972; (3) Scheuer and Williams 1968; (4) Wrixon et al. 1971; (5) Wrixon and Welch 1970.

a value of C_s(DR 21) $\approx R_\alpha R_h/B_\alpha B_h = 1.032$ was obtained for $B = 2'.2$. This measured value agrees well with values for C_s of 1.026 and 1.033 that were obtained by convolving the DR 21 maps of Webster and Altenhoff (1970) and Wynn-Williams (1971) with the Haystack beam. Adopting the average of the three as the best estimate of C_s at this beamwidth and knowing that it scales quadratically in 1 B for C_s near unity, the following expression was obtained for C_s at other beamwidths:

$$C_s \approx 1 + 0.15(\pm 0.03)/B^2,$$

where B is in minutes of arc. It can be seen from this expression that the correction is less than 4 percent for beamwidths larger than 2'. Although the correction begins to increase rapidly below 2', DR 21 can be reliably used as a standard source for beamwidths as small as 1' where the uncertainty in the size-correction factor is probably less than a few percent.

IV. THE ABSOLUTE SPECTRUM OF DR 21

DR 21 is known to be an H II region both from the shape of the continuum radio spectrum and its recombination-line emission. Since the continuum radio emission from an H II region-is thermal bremsstrahlung, the form of the spectrum is known from the theory of emission from ionized gases (Oster 1961). For a source of uniform thickness, the flux density S_ν at radio frequencies will depend on frequency ν as

$$S_\nu = A\nu^2(1 - e^{-\tau(\nu)}),\quad (1)$$

where $\tau(\nu) = B\nu^{-2} \ln C\nu^{-1}$ and A, B, and C are constants that depend upon source distance, emission measure, and electron temperature T_e. At frequencies above 7 GHz the spectrum of DR 21 is relatively flat, indicating that the source is transparent with $\tau \ll 1$. In this region of the radio spectrum, DR 21 would be expected to have the simple form

$$S_\nu = S_0 \ln C\nu^{-1},\quad (2)$$

obtained from a Taylor-series expansion of equation (1). Numerically C equals $0.04955 T_e^{3/2}$ GHz, where T_e is in degrees K (Mezger and Henderson 1967) and S_0 is a constant. Thus above 7 GHz the source flux density depends only upon the source electron temperature and frequency. Assuming 11,000° K for T_e (Ryle and Downes 1967), we would expect the microwave spectrum of DR 21 to have the form

$$S_\nu = S_0 \ln 5.67 \times 10^4 \nu^{-1},\quad (3)$$

where ν is in GHz. Note that because of its logarithmic dependence equation (3) is not very sensitive to the value of T_e used.

The six measurements of DR 21 in table 1 which have been referred to an absolute flux-density scale can now be used to find the absolute spectrum of DR 21. Listed in column (6) of table 1 are independent determinations of the constant S_0 computed from each of the measurements of DR 21 at 7.8, 15.5, and 31.4 GHz. The average value of S_0 and the standard deviation of the mean is $S_0 = 2.45 \pm 0.04 \times 10^{-26}$ W·m^{-2} Hz^{-1}. Because they are based on less reliable standard sources, each of the 15.5-GHz measurements was given a weight of $\frac{1}{2}$ in computing this average. Note in table 1 that, within the uncertainties, the values of S_0 at 31.4 GHz obtained from absolute measurements of Jupiter and Saturn are consistent with those based on the absolute spectrum of Cas A and Cyg A, confirming the validity of using equation (3).

The absolute spectrum for $\nu \geq 7$ GHz is shown in figure 1 along with the individual data points of table 1 which were used to define it. The curved and steep low-frequency end of the spectrum is drawn in for completeness and cannot, of course, be reliably used as the basis for a flux-density scale. The spectrum in figure 1 is presented in the

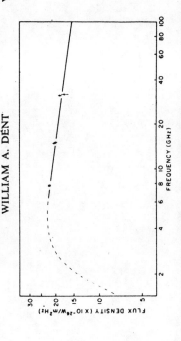

Fig. 1.—The spectrum of DR 21 based on absolute flux-density measurements above 7 GHz (solid line). The data points have been shifted slightly in frequency to exhibit them more clearly. A spectral index of −0.13 has effectively been assumed in the least-squares fit.

traditional (log S, log ν) plot rather than (S, log ν). On such a graph the effective spectral index of a thermal source where $\tau(\nu) \ll 1$ is $\alpha = -0.13$.

Having obtained S_0, the absolute spectrum of DR 21 above 7 GHz can be rewritten in a form more convenient for calculation:

$$S_\nu = 26.78 - 5.63 \log \nu \quad (GHz).\quad (4)$$

The formal error in S_0 would imply that S_ν computed from equation (4) is accurate to 1.6 percent. Such an uncertainty may be unrealistically low. Considering that the original absolute measurements of some of the standard sources were not entirely independent, equation (4) probably gives S_ν to an absolute accuracy of about ±3 percent.

Although the highest frequency measurement of DR 21 was 31.4 GHz, the absolute spectrum obtained here can probably be safely extrapolated up to 100 GHz or higher because of the predictable nature of its thermal spectrum. The primary limitation to using DR 21 as a flux-density standard above 100 GHz is the relatively large size-correction factor which must be applied because of the small antenna beamwidths usually encountered at such frequencies.

Finally, it will be useful for some purposes to refer all of the measurements made of standard sources to the absolute flux-density scale based on DR 21 given by

TABLE 2

ABSOLUTE MEASUREMENTS OF
STANDARD SOURCES REFERRED TO DR 21

Frequency (GHz)	Source	S_ν* or T_b
7.8	3C123	10.6 ± 0.3
7.8	Cyg A	217 ± 7
15.5	3C274	28.3 ± 0.8
15.5	3C218	4.8 ± 0.2
31.4	Jupiter	152° ± 5° K
31.4	Saturn	136° ± 4° K

* S_ν in units of 10^{-26} W·m^{-2} Hz^{-1}.

equation (4). These measured source flux densities or brightness temperatures are given in table 2. It should be emphasized that the measured brightness temperatures of $136° \pm 4°$ K and $152° \pm 5°$ K of Saturn and Jupiter respectively apply only to a frequency of 31.4 GHz and cannot be used with certainty at other microwave frequencies.

I would like to thank R. W. Hobbs, G. Kojoian, and J. E. Kapitzky for participating in the observations. I am indebted to the staffs of the Haystack Observatory and the National Radio Astronomy Observatory for their helpful cooperation. This work was supported by the National Science Foundation (grant GP-14690). Research programs are conducted at the NEROC Haystack Observatory with support from the NSF (grant GP-25865) and from NASA (grant NGR 22-174-003 and contract NAS9-7830). The NRAO is operated by Associated Universities, Inc., under contract with the NSF.

REFERENCES

Dent, W. A., and Haddock, F. T. 1966, *Ap. J.*, **144**, 568.
Downes, D., and Rinehart, R. 1966, *Ap. J.*, **144**, 937.
Graham, I. 1971, *Nature*, **231**, 253.
Kalaghan, P. M., and Wulfsberg, K. N. 1968, *Ap. J.*, **154**, 771.
Kellermann, K. I., Pauliny-Toth, I. I. K., and Williams, P. J. S. 1969, *Ap. J.*, **157**, 1.
Medd, W. J. 1972, *Ap. J.*, **171**, 41.
Mezger, P. G., Altenhoff, W., Schraml, J., Burke, B. F., Reifenstein, E. C., and Wilson, T. L.
    1967, *Ap. J. (Letters)*, **150**, L157.
Mezger, P. G., and Henderson, A. P. 1967, *Ap. J.*, **147**, 471.
Oster, L. 1961, *Rev. Mod. Phys.*, **33**, 525.
Pike, E. M., and Drake, F. D. 1964, *Ap. J.*, **139**, 545.
Ryle, M., and Downes, D. 1967, *Ap. J. (Letters)*, **148**, L17.
Scheuer, P. A. G., and Williams, P. J. S. 1968, *Ann. Rev. Astr. and Ap.*, **6**, 321.
Thornton, D. D., and Welch, W. J. 1963, *Icarus*, **2**, 228.
Webster, W. J., and Altenhoff, W. J. 1970, *A.J.*, **75**, 896.
Winzer, J. E., and Roberts, J. A. 1971, *J.R.A.S. Canada*, **65**, 100.
Wrixon, G. T., and Welch, W. J. 1970, *Icarus*, **13**, 163.
Wrixon, G. T., and Thornton, D. D. 1971, *Ap. J.*, **169**, 171.
Wynn-Williams, C. G. 1971, *M.N.R.A.S.*, **151**, 397.

Absolute Brightness Temperature Measurements at 3.5-mm Wavelength

BOBBY L. ULICH, MEMBER. IEEE. JOHN H. DAVIS, MEMBER. IEEE. PAUL J. RHODES, AND JAN M. HOLLIS

Abstract—Careful observations have been made at 86.1 GHz to derive the absolute brightness temperatures of the Sun (7914 ± 192 K), Venus (357.5 ± 13.1 K), Jupiter (179.4 ± 4.7 K), and Saturn (153.4 ± 4.8 K) with a standard error of about three percent. This is a significant improvement in accuracy over previous results at millimeter wavelengths. A stable transmitter and novel superheterodyne receiver were constructed and used to determine the effective collecting area of the Millimeter Wave Observatory (MWO) 4.9-m antenna relative to a previously calibrated standard gain horn. The thermal scale was set by calibrating the radiometer with carefully constructed and tested hot and cold loads. The brightness temperatures may be used to establish an absolute calibration scale and to determine the antenna aperture and beam efficiencies of other radio telescopes at 3.5-mm wavelength.

I. INTRODUCTION

THE ABSOLUTE intensity of a celestial radio source is difficult to measure accurately at millimeter wavelengths primarily because of instrumental difficulties, but it conveys information very useful to theorists. The high noise temperatures of radiometers, the limited effective collecting areas of radio telescopes, and the absorption by molecular oxygen and water vapor [1] in the Earth's atmosphere result in a relatively imprecise absolute calibration scale for millimeter-wavelength radio astronomical observations. The purpose of this experiment is to improve the accuracy of the absolute flux density scale at 3.5-mm wavelength by the construction of a stable transmitter and receiver specifically designed to reduce the dominant measurement errors and by employing new techniques for antenna gain calibration, thermal noise calibration, and impedance mismatch corrections.

II. INSTRUMENTATION

The system used to measure celestial radio emission consists of a conventional paraboloidal reflector antenna with a superheterodyne Dicke radiometer located at the prime focus. The effective collecting area of the paraboloid is determined by comparing the signal received from a continuous wave (CW) transmitter located some distance from the receiving equipment with the signal from a previously calibrated standard gain horn. The thermal emission from blackbody loads is used to calibrate the absolute intensity of the detected signals. The same general procedure has been previously used to provide absolute flux density calibrations of radio sources

Manuscript received June 20, 1979; revised December 3, 1979. This work was supported in part by the National Science Foundation, the National Aeronautics and Space Administration, and the McDonald Observatory. The National Radio Astronomy Observatory is operated by Associated Universities, Inc., under contract with the National Science Foundation.

B. L. Ulich and P. J. Rhodes are with the National Radio Astronomy Observatory, Tucson, AZ 85705.

J. H. Davis is with the Millimeter Wave Observatory, Department of Electrical Engineering, University of Texas at Austin, Austin, TX 78712.

J. M. Hollis was with the National Radio Astronomy Observatory, Tucson, AZ. He is now with the National Aeronautics and Space Administration, Goddard Space Flight Center, Greenbelt, MD 20770.

at 2.3 GHz [2], 20.5–35.5 GHz [3], 35.0 GHz [4], 97.1 GHz [4], and 141 GHz [5].

A. Paraboloidal Reflector

The antenna is the 4.88-m diameter f/0.5 paraboloidal reflector [6], [7] operated by the Millimeter Wave Observatory (MWO) on Mt. Locke, TX, at an altitude of 2015 m. The root mean square (rms) random telescope tracking error is 2″, and the corrections to measured intensities due to tracking jitter are negligibly small (≅0.04 percent). The systematic pointing error is 13″ rms, and significant intensity errors can result if this pointing error is not corrected or taken into account properly. The reflector profile has an rms deviation from a paraboloid of about 0.1 mm, insuring nearly diffraction-limited performance throughout the millimeter band. The radiation patterns measured at 86.1 GHz (wavelength $\lambda = 3.482$ mm) on the pattern range are shown in Fig. 1.

B. Calibration Horn

The calibration horn used as a gain standard has been previously described by Ulich [8]. It is a simple conical horn with an axial length of 121.92 cm and an aperture of 10.16 cm. At 86.1 GHz the far-field directivity was determined to be 37.202 ± 0.030 dB (all errors quoted in this paper are one standard deviation) using a radiometric calibration scheme similar to the "artificial moon" technique [9]. This conical horn was also used to measure the average disk brightness temperature of the Sun and to perform sky "tipping" scans to monitor atmospheric extinction. Measured radiation patterns at 86.1 GHz are shown in Fig. 2. Fig. 3 is a plot of the spillover efficiency (i.e., the normalized fraction of radiated power which is contained within a cone of specified half-angle) calculated from the measured patterns. At 86.1 GHz the measured horn dissipation loss is only 2.0 ± 0.5 percent, which agrees with the calculated value of 2.3 percent [8].

C. Transmitter

A CW transmitter was constructed to provide a strong signal for the relative directivity measurements of the paraboloid and the calibration horn. A phase-locked klystron at 86.100 GHz provided power to a lens-corrected conical horn antenna with about 33-dB gain. As shown in the block diagram of Fig. 4, a feedback control circuit employing a ferrite modulator was used to maintain constant transmitted power. Laboratory measurements indicate a typical stability of about 0.04 dB/h. The entire transmitter is contained in a thermally insulated box to reduce errors caused by temperature fluctuations. The transmitting horn was carefully aligned to the direction of the telescope to insure uniform illumination of the paraboloid.

D. Receiver

A multipurpose receiver was also constructed to enable three types of observations to be made conveniently. Fig. 5

Reprinted from *IEEE Trans. Antennas Propagat.*, vol. AP-28, no. 3, pp. 367–377, May 1980.

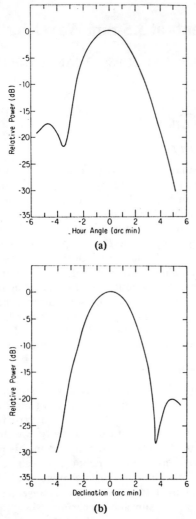

(a)

(b)

Fig. 1. Measured radiation patterns of 4.9-m paraboloidal antenna at 86.1 GHz with the on-axis feed. The asymmetry of the first side-lobes indicates the presence of coma due to a slight lateral offset of the feed horn from the axis of the paraboloid. (a) In hour angle (H-plane). (b) In declination (E-plane).

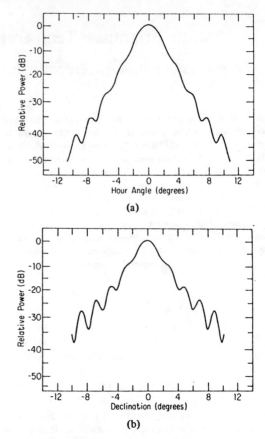

(a)

(b)

Fig. 2. Measured radiation patterns of conical horn at 86.1 GHz. (a) In hour angle (H-plane). (b) In declination (E-plane).

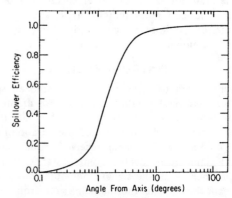

Fig. 3. Spillover efficiency of conical horn at 86.1 GHz calculated from measured patterns in Fig. 2.

is a block diagram showing the significant subsections of the receiver. The first mode of operation is that of a single-conversion phase-locked narrowband superheterodyne receiver with an intermediate frequency (IF) of 30 MHz. An accurate waveguide below cutoff (WBCO) attenuator (Airborne Instruments Laboratory Type 3232) is used with the IF substitution technique [10] to determine the ratio of the transmitter power received by the paraboloid to that received by the standard gain horn. The accuracy of the WBCO attenuator calibration is ± [0.03 dB + (0.005 dB per 10-dB change in reading)]. The waveguide switch at the receiver input allows rapid comparison of the signals from the paraboloid feed horn and from the calibration horn which is physically coaxial with the telescope axis. The WBCO attenuator is preceded by a high-power heat-sunk transistor amplifier (+29 dBm for −1-dB compression) to insure adequate dynamic range to accommodate the 33-dB difference in received signal level without significant IF saturation. A bandpass filter centered at 30 MHz reduces the bandwidth of the detected spectrum to 3 MHz and improves the signal-to-noise ratio (SNR). In operation, a change in radio frequency (RF) signal level is compensated by WBCO attenuation of the IF signal to maintain a constant detector output. All waveguide flanges and other sites of possible RF leakage are wrapped with copper foil tape to insure sufficient isolation between the paraboloid and the calibration horn. Special observations were made in an effort to detect residual leakage, but none was found at a level 23 dB below the calibration horn signal.

The second mode of receiver operation is that of a broadband Dicke superheterodyne radiometer. In this mode the IF bandpass is from 10 to 500 MHz. The receiver noise temperature was measured to be 6100 K (double sideband) at the input, and the ΔT_{rms} was 0.6 ± 0.1 K for 1 s of integration time, which agrees with the theoretical value of 0.55 K.

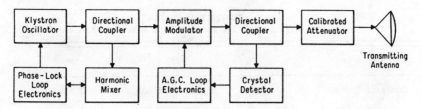

Fig. 4. Block diagram of transmitter with feedback circuits to stabilize frequency and output power.

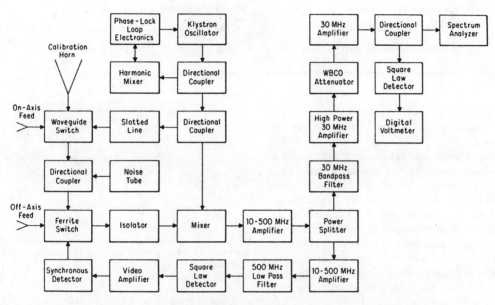

Fig. 5. Block diagram of receiver with narrowband and wideband IF systems and integral reflectometer.

A switchable circulator alternately connects the receiver to one of two identical feed horns at a 20-Hz rate. The feeds are optimized pyramidal horns which have aperture dimensions of 0.90 λ in the E-plane and 1.25 λ in the H-plane. The feed spacing resulted in an orthogonally polarized reference beam 2.6 half-power beamwidths off-axis in hour angle. Observations of blackbody calibration loads and of celestial radio sources were made using the broadband radiometer output.

The third mode of receiver operation is that of a reflectometer. An integral waveguide slotted line was used to measure the reflection coefficients of the various antennas, of the thermal calibration loads, and of the receiver input.

E. Thermal Calibration Loads

Absolute noise power spectral density calibration was obtained by observing the difference in signal level from two loads fabricated using Eccosorb CV-3 porous carbon-impregnated foam absorber (Emerson and Cuming, Incorporated, Canton, MA) enclosed in aluminum cavities [11]. One load was nominally at ambient temperature, and a laboratory thermometer inserted in the absorber was used to determine its physical temperature. The second load was insulated on all sides with 5.1 cm of expanded polystyrene and filled with liquid nitrogen. The foam walls (density = 0.032 g·cm^{-3}) have an estimated relative dielectric constant of 1.029 and a loss tangent of about 3.6×10^{-5} near 3-mm wavelength

based on measurements of polystyrene at 90 GHz [12] and density scaling laws [13]. The calculated power reflection of the foam wall is 0.02 percent and the absorption is 0.33 percent. The measured transmissivity was 0.9969 ± 0.0007 at 86.1 GHz, which agrees within the error with the calculated value of 0.9965.

III. VSWR MEASUREMENTS

The waveguide slotted line built into the receiver was used to measure the voltage standing wave ratio (VSWR) of system components, and the results are given in Table I. The magnitudes of the voltage reflection coefficients ($|\Gamma|$) at 86.1 GHz are generally quite low. The receiver input is the worst match of the components tested even though an isolator (see Fig. 5) has been used to improve the impedance matching (measurements of the mixer without the isolator indicated an input VSWR of 2.7). The reflection coefficient of the 4.9-m paraboloid was determined by axially focusing the receiver box to vary the phase of the signal reflected by the paraboloid back into the receiver. Theoretical calculations [14] predict a value of 0.0012, which agrees with the measurement. The thermal calibration loads and the absorber were measured using a traveling mechanical stage in front of the calibration horn. These data were taken at normal incidence, and corrections due to waveguide loss were determined by observations of a copper shorting plate. Measurements of the liquid nitrogen cooled load with cryogen

TABLE I
86.1-GHZ VOLTAGE STANDING WAVE RATIO DATA

Source	$|\Gamma|$	VSWR
Calibration horn	0.029 ± 0.001	1.060 ± 0.002
Reflector feed horn	0.024 ± 0.001	1.049 ± 0.002
4.9 M Paraboloidal reflector	0.0015 ± 0.0005	1.003 ± 0.001
Ambient temperature load	0.0020 ± 0.0005	1.004 ± 0.001
Eccosorb CV-3 absorber	0.0020 ± 0.0005	1.004 ± 0.001
Liquid nitrogen cooled load	0.044 ± 0.002	1.092 ± 0.004
Receiver waveguide switch input	0.076 ± 0.003	1.165 ± 0.006

($|\Gamma| = 0.044 \pm 0.002$) and without cryogen ($|\Gamma| = 0.011 \pm 0.002$) indicate that the dominant source of reflection is the foam–cryogen interface.

All pattern range, thermal calibration, and radio source measurements have been corrected for mismatch loss [15]. However, since only the amplitudes of the reflection coefficients (and not the phases) are known, an "exact" correction cannot be made. We have chosen to set the conjugate mismatch loss factor equal to the mean of its maximum and its minimum possible values and, for the purpose of conservative error analysis, to estimate an error of one standard deviation with the maximum possible error. Thus the mean mismatch loss correction factor between components one and two is (assuming small reflection coefficients)

$$\langle M \rangle \equiv \frac{\text{power dissipated in load}}{\text{maximum available power}} = (1 - |\Gamma_1|^2)(1 - |\Gamma_2|^2) \quad (1)$$

and the estimated 1σ error is

$$\Delta M \equiv 2|\Gamma_1||\Gamma_2|. \quad (2)$$

No attempt was made to determine if the reflection coefficients varied with temperature. However, the corrections are quite small and substantial variations will not significantly affect the overall calibration. The effect of mismatch on the broadband radiometric data (instantaneous bandwidth ~1 GHz) is less serious than for the coherent measurements since the detector will in most cases average the signal over several cycles of the sinusoidal standing wave pattern, and the mismatch factor given by (1) will then be nearly correct. In this case the true mismatch error will be overestimated by (2).

IV. PATTERN RANGE MEASUREMENTS

An unobstructed pattern range was used to provide a quasi-plane wave input to the 4.9-m paraboloid and to the coaxial calibration horn for the narrowband directivity comparison measurements. The transmitter site was a hilltop in the Davis Mountains State Park, TX, at an altitude of 1680 m and a range of 12.9 km from the MWO. As seen from the telescope, the transmitter is at $-1.5°$ elevation angle, and the closest possible ground reflection is about $0.63°$ below that angle (i.e., about 13 beamwidths off-axis). A search for signals near this angle indicated that ground reflections were negli-

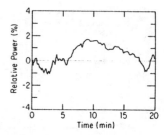

Fig. 6. Variation in transmitter power received by the 4.9-m antenna over a 12.9-km pattern range (1235 coordinated universal time on December 15, 1977).

gibly small (less than -30 dB with respect to the direct path). The finite range of the transmitter, however, means that the wave impinging on the telescope is not planar, but rather spherical, and the resulting quadratic phase error (0.23 mm at the reflector rim) produces a reduction in peak gain of the paraboloid of about 1.4 percent. At a range $R = 12.9$ km the transmitter is $1.89 D^2/\lambda$ from the paraboloid (of diameter D). By defocusing the feed horn axially away from the reflector (of focal length f) by a distance ϵ, the far-zone radiation pattern can be virtually duplicated for ranges greater than $0.5 D^2/\lambda$. The axial defocusing distance ϵ is given by

$$\epsilon = \frac{Kf^2}{R} \left\{ 1 + \left[\frac{1}{4(f/D)} \right]^2 \right\} \quad (3)$$

where $K = 0.92$ [16]. The peak gain of the antenna is the same whether it is focused in the Fresnel zone or at infinity [17]. The feeds were focused to receive the maximum transmitter signal for the directivity comparison data. The feeds were axially displaced by 0.15λ (calculated from (3)) toward the reflector to focus at infinity for the planetary radio observations.

Fig. 6 is a plot of the variation in signal power received by the paraboloid during a typical 20-min period. Most of the noise is due to atmospheric scintillation which was found to be smallest between midnight and sunrise. Table II presents individual measurements of the observed signal power ratio. Some of these measurements were made at lower transmitter power levels to search for possible receiver nonlinearity due to saturation, but in no case was any nonlinearity detected. A calibrated rotary-vane RF attenuator in the transmitter was also used to check the WBCO IF attenuator calibration,

TABLE II
ANTENNA SIGNAL RATIO DATA

Trial	Date	UTC (Hours)	Signal from reflector / Signal from horn (dB)
1	12/14/77	0500	32.914
2	12/14/77	0900	32.885
3	12/15/77	0400	32.924
4	12/15/77	0900	32.928
5	12/15/77	1400	32.930
<1-5>	-	-	32.916 ± 0.008

TABLE III
REFLECTOR DIRECTIVITY

Quantity	Value (dB)
Calibration horn directivity	37.202 ± 0.030
Ratio of reflector signal to horn signal	32.916 ± 0.008
Uncertainty due to mismatch	± 0.026
Ratio of receiver gain at calibration horn aperture to gain at reflector feed aperture	0.042 ± 0.018
Uncertainty in WBCO attenuation	± 0.046
Uncertainty due to polarization angle misalignment	± 0.005
Uncertainty in refocusing	± 0.025
Uncertainty due to difference in cross polarization	± 0.022
Reflector directivity at horizon	70.160 ± 0.072

and no nonlinearity was detected within the accuracy of the RF attenuator calibration (±0.1 dB). Each measurement was made after careful pointing adjustments to maximize the received signals.

The reflector directivity (referred to the aperture plane of the on-axis feed) is obtained by correcting the observed signal power ratio for the different losses of the paths from the feed horn and the calibration horn to the ferrite switch and then multiplying by the directivity of the calibration horn. As indicated in Table III, the directivity of the 4.9-m reflector at 86.1 GHz is 70.160 ± 0.072 dB. The estimated uncertainty is the quadrature sum (root sum of squares) of the estimated errors in Table III (which are assumed to be independent). The cross-polarization uncertainty exists because the calibration horn directivity was determined using an unpolarized source, whereas for the directivity ratio measurement a linearly polarized source was used. The effect of phase variations over the reflector aperture caused by atmospheric turbulence has been assumed to be negligibly small. In addition, the telescope directivity has been assumed to be independent of temperature. In fact most of the planetary observations were made over the same range of ambient temperature as the pattern range measurements, so any possible gain variations with temperature would not significantly affect these data. Table IV summarizes the measured parameters of the calibration horn and of the paraboloid. The

reflector has a good aperture efficiency (53.58 ± 0.89 percent), indicating that surface errors are small compared to a wavelength and that the radiation pattern is nearly diffraction-limited. The reflector half-power beamwidths are essentially equal since the aperture dimensions of the pyramidal feed horn have been chosen to produce a main lobe with a circular cross section.

V. THERMAL CALIBRATION

The noise power $P(\mathrm{W})$ delivered to the receiver in a single polarization through the aperture of either the on-axis feed horn or the calibration horn from a perfectly matched (blackbody) load at an absolute physical temperature T enclosed in a perfectly conducting cavity is given by

$$P = kBJ(T), \qquad (4)$$

where k is Boltzmann's constant (1.38054×10^{-23} J·K^{-1}), B is the predetection bandwidth (Hz), and the effective radiation temperature $J(T)$ is given by

$$J(T) = \frac{(h\nu/k)}{\exp(h\nu/kT) - 1}. \qquad (5)$$

In (5) h is Planck's constant (6.6256×10^{-34} J·s), and ν

TABLE IV
ANTENNA PARAMETERS AT 86.1 GHZ

Quantity	Calibration Horn	Paraboloidal Reflector
Diameter (m)	0.1016	4.877
Geometrical area (m^2)	0.008107	18.679
Directivity (dB)	37.202 ± 0.030	70.160 ± 0.072
Effective area (m^2)	0.005065 ± 0.000036	10.010 ± 0.166
Aperture efficiency (%)	62.48 ± 0.44	53.58 ± 0.89
HPBW in E-plane (arc minute)	127 ± 2	2.91 ± 0.03
HPBW in H-plane (arc minute)	152 ± 2	2.86 ± 0.03
VSWR	1.060 ± 0.002	1.003 ± 0.001

TABLE V
THERMAL CALIBRATION DATA

Trial	Date	UTC (Hours)	Barometric Pressure (MM Hg)	Receiver Total Power Output (Volt)	T_{AMB} (K)	T_{LN} (K)	$T_{AMB} - T_{LN}$ (K)	$\frac{\Delta CAL}{\Delta H/C \text{ (Feed)}}$	$\frac{\Delta H/C \text{ (Feed)}}{\Delta H/C \text{ (Horn)}}$	T_{CAL} (K)
1	12/15/77	0000	609	0.923	281.3	75.7	205.6	0.1389	1.0055	28.30
2	12/16/77	0200	603	0.917	284.4	75.6	208.8	0.1428	0.9945	29.53
3	12/17/77	0300	605	0.920	280.1	75.7	204.4	0.1459	0.9933	29.54
4	12/18/77	0300	609	0.938	280.5	75.7	204.8	0.1426	0.9760	28.94
5	12/19/77	0000	606	0.928	284.0	75.7	208.3	0.1393	0.9840	28.76
6	12/19/77	1600	607	0.925	284.4	75.7	208.7	0.1405	0.9889	29.05
<1-6>	-	-	-	-	-	-	-	-	0.9904 ± 0.0041	29.02 ± 0.19

is the frequency (Hz). In practice the bandwidth B is usually not precisely known, and microwave radiometers are generally calibrated in terms of the effective radiation temperature $J(T)$, which is proportional to noise power spectral density. When $h\nu \ll kT$, $J(T) \cong T$, and the familiar concept of antenna temperature may be used in the Rayleigh–Jeans approximation to the Planck distribution.

The broadband radiometric observations of the Sun and planets were calibrated by first comparing the signal from a neon noise tube to that from the hot and cold loads. The calibrated noise tube signal was then used as a secondary transfer standard to allow rapid and convenient calibration of celestial radio sources. In practice the noise tube was constantly on to avoid thermal output drift, and a ferrite modulator was turned on and off under computer control to generate the calibration signal. Table V presents a summary of the noise tube calibration measurements, which were interleaved in time with the radio source observations to reduce systematic errors. Alternate observations of the hot and cold loads over the apertures of the on-axis feed horn and of the calibration horn were also used to determine the relative receiver gain from these two input ports to the square law detector. As indicated in Table V, simultaneous recordings of ambient (hot load) temperature T_{AMB} and of barometric pressure were used to determine the radiation temperature difference of the two loads. The boiling point of liquid nitro-

gen $T_{LN}(K)$ is related to the barometric pressure P_B (mmHg) by [18]

$$T_{LN} = 77.36 + 0.011 (P_B - 760), \qquad (6)$$

and the microwave absorber is physically at this temperature since it is completely immersed in the cryogen. The foam wall, however, attenuates the emission from the absorber and also emits radiation according to its normalized power loss factor α. For small losses the effective radiation temperature $J(T_C)$ of the cold load is

$$J(T_C) = J(T_{LN}) + (\alpha/2)[J(T_{AMB}) - J(T_{LN})], \qquad (7)$$

where a linear temperature gradient in the foam wall has been assumed and reflections are neglected [18]. In the present case $\alpha = 0.0033$, and the foam wall increases the effective cold load temperature by about 0.3 K; this was experimentally verified by comparison measurements of a second cold load without the foam wall. A 45° flat reflector was placed above an absorber-lined cavity filled with cryogen, and the signal was found to be within ±0.4 K of the cold load with foam. Measurements made with a second foam wall in place indicated an increase of 0.9 ± 0.4 K, which agrees with the calculated value of 0.7 K. Thus the low loss of the foam wall and its small effect on the cold load temperature were confirmed.

Mismatch will also increase the effective temperature of the cold load since some noise power radiated by the receiver out its input port will be reflected back into the receiver. The presence of an isolator in the receiver input waveguide assures that the noise power radiated out the receiver will be at nearly ambient temperature, and this was confirmed by observations with a shorting plate at the feed aperture. The effect of mismatch between the cold load and the receiver input (evaluated using (1) and adding the reflected signal) is to increase the effective radiation temperature of the cold load by about 1.6 K, while having essentially no effect on the ambient temperature load signal. Thus the cold load radiation temperature is approximately equal to $J(T_{LN}) + 1.9$ K = $J(77.8$ K) for typical conditions on Mt. Locke. The average value of the effective noise tube signal from Table V is 29.02 ± 0.19 K referred to the aperture of the main feed, where the error is estimated from the repeatability of the measurements. In addition to this random error, uncertainty in the load temperature (conservatively estimated at ±0.5 K each), in the foam wall loss (±0.07 percent), and in the mismatch correction (±0.76 percent) combine in quadrature to produce a total uncertainty in the thermal calibration scale of 1.06 percent.

VI. ATMOSPHERIC EXTINCTION

The absorption of radio waves at 86.1 GHz by the terrestrial atmosphere is both significant and variable. Therefore accurate measurements of celestial sources must be accompanied by a nearly simultaneous determination of atmospheric opacity. Measurements of atmospheric emission can be used to determine atmospheric absorption due to oxygen and water vapor. The two-layer model of Kutner [19] was used with ground temperature data to calculate the effective radiation temperatures of oxygen and water vapor. Surface absolute humidity measurements were also made to estimate the precipitable water vapor in a column and thus the relative absorption of water vapor [20] compared with the essentially constant oxygen component. From these values a weighted mean atmospheric radiation temperature was calculated and recorded for future use. The actual sky brightness temperature data were taken with the calibration horn rather than with the paraboloidal antenna, since the horn sidelobe levels were more accurately known than those of the paraboloid. Of course, the noise received by the horn is actually a convolution over 4π steradians of the horn power pattern and the brightness distribution of its surroundings. A computer model was constructed which included the effects of atmospheric emission, ground emission, cosmic microwave background emission, and reflection of horn backlobes by the paraboloid and by the receiver box. The apparent sky temperature $J(T_{SKY})$ is given by

$$J(T_{SKY}) = J(T_M)(1 - e^{-\tau A}) + J(T_{BG})e^{-\tau A}, \qquad (8)$$

where T_M is the calculated mean atmospheric radiation temperature [21] and T_{BG} is the microwave background brightness temperature (2.8 K) [22]. The zenith optical depth is τ, and the air mass A is given (to a good approximation for $A < 10$) by

$$A = \frac{1}{\sin(E)} \qquad (9)$$

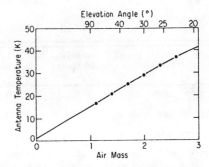

Fig. 7. Measured effective antenna temperature of the calibration horn as a function of air mass (1530 coordinated universal time on December 19, 1977). The solid line is the computer convolution model fit to the observations with a zenith optical depth of 0.052 Np. With no atmosphere present (zero air mass), the effective antenna temperature of the 2.8 K cosmic background radiation is about 1.2 K at 86.1 GHz.

where E is the elevation angle. In this model the ground emission temperature is assumed to be the same as the surface ambient temperature. The measured radiation patterns of the calibration horn given in Fig. 2 were used for the computerized convolution, and the values of τ for each observing session were found by the method of least squares. The standard deviation of the differences between the computer model and the observed effective temperatures was typically only 0.3 K, indicating that the model quite accurately fits the shape of the sky tipping curve. Fig. 7 shows the results of one such fit, and the values of zenith optical depth are given in Table VI. The values of τ are quite low (0.052–0.076) and are consistent with only 2–4 mm precipitable water vapor. The error in the atmospheric absorption correction to observed celestial source intensities was conservatively assumed to be 20 percent of the actual correction.

VII. DEPENDENCE OF REFLECTOR GAIN ON ELEVATION ANGLE

A series of observations were made from November 24–27, 1978 to determine if the peak gain of the paraboloid varies with elevation angle. In order to correct for rapid variations in atmospheric absorption, a rotating chopper wheel was used to produce a calibration signal according to the method first described by Davis and Vanden Bout [23] and later in detail by Ulich and Haas [24]. The resulting intensity scale is "corrected antenna temperature" (T_A*), and is simply the antenna temperature that would be measured outside the terrestrial atmosphere with a lossless antenna. For a source which completely fills the forward beam, the corrected antenna temperature equals the source brightness temperature. Since the Sun is much larger in angular extent than either the main beam or the near sidelobes of the paraboloid radiation pattern, gravitational (or thermal) distortion of the reflector cannot affect the received signal by a significant amount. Thus observations of the Sun were used to check the accuracy of the atmospheric absorption corrections with the chopper-wheel calibration technique. As shown in Fig. 8, there is no indication of any change in measured solar intensity with elevation angle. The solar brightness temperature at 86.1 GHz measured with the 4.9-m paraboloid and the chopper-wheel calibration scheme is 7750 ± 150 K using a calculated beam coupling efficiency of 0.77; the error is due solely to receiver noise and atmospheric absorption

419

TABLE VI
ASTRONOMICAL DATA

Date	UTC (Hours)	Source	Number of Scans	$\tau \pm 1\sigma$ (Nepers)	Antenna	$T_A \pm 1\sigma$ (K)	$S \pm 1\sigma$ (Jy)	$T_B \pm 1\sigma$ (K)
12/17/77	2000	Sun	5	0.075 ± 0.015	Horn	229.9 ± 8.8	(125.3 ± 4.8)x10⁶	7985 ± 305
12/18/77	2000	Sun	11	0.053 ± 0.011	Horn	227.0 ± 5.6	(123.8 ± 3.0)x10⁶	7885 ± 193
12/18/77	0830	Jupiter	10	0.076 ± 0.015	Reflector	5.372 ± 0.102	1482 ± 28	175.3 ± 3.3
12/19/77	0730	Jupiter	10	0.052 ± 0.010	Reflector	5.542 ± 0.067	1529 ± 19	180.9 ± 2.2
12/19/77	1300	Saturn	4	0.052 ± 0.010	Reflector	0.747 ± 0.015	206 ± 4	153.4 ± 3.0
11/25,26, 27/78	1200	Jupiter	108	0.05 to 0.12	Reflector	4.205 ± 0.061	1160 ± 17	180.0 ± 2.6
11/24,25, 26,27/78	1700	Venus	56	0.05 to 0.12	Reflector	15.490 ± 0.420	4273 ± 116	357.5 ± 9.7

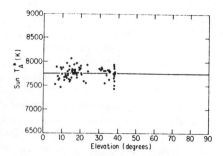

Fig. 8. Measured corrected antenna temperatures of the Sun at 86.1 GHz with the 4.9-m paraboloid as a function of elevation angle. The solid line is the result of the least squares fit of the variable gain model described in the text with $\Delta G = 0.000 \pm 0.011$.

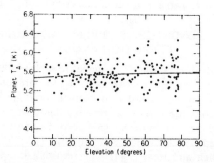

Fig. 9. Measured planetary corrected antenna temperatures at 86.1 GHz with the 4.9-m paraboloid as a function of elevation angle. The solid line is the result of the least squares fit of the variable gain model described in the text with $\Delta G = 0.020 \pm 0.015$.

corrections. Taking into account the uncertainty in telescope losses, the solar brightness temperature is 7750 ± 350 K. Concurrent observations of the planets Venus and Jupiter show a slight increase in intensity from the horizon to the zenith, as shown in Fig. 9. The ratio of Venus to Jupiter was precisely measured when they were at equal elevation angles, and this ratio was used to place the Venus and Jupiter intensities on a common scale. Formally, a least squares fit of the relative antenna gain function of the (assumed) form

$$G/G_0 = 1 + \Delta G \sin(E) \qquad (10)$$

to the Sun data in Fig. 8 resulted in $\Delta G = 0.000 \pm 0.011$. For the planetary data in Fig. 9, $\Delta G = 0.020 \pm 0.015$, which indicates a slightly significant (2.0 ± 1.5 percent) increase in gain at the zenith compared with the horizon, where the pattern range measurements were used to determine the absolute telescope gain.

Two other methods were used to check for gain variations with elevation angle. The first method was to compare the antenna half-power beamwidths as measured on the pattern range with those inferred from Jupiter data at about 75° elevation angle. Both E- and H-plane beamwidths were narrower (after correcting for the slight broadening due to the finite disk size of Jupiter) at 75° elevation angle than at the horizon, inferring that $\Delta G = 0.018 \pm 0.018$. To derive this result we assumed that gravitational distortions are comparable in correlation length to the diameter of the paraboloidal reflector and thus that the power received by the main lobe remains constant. The second method was to determine the variation in the signal from Jupiter observed on December 18, 1977 after correction for atmospheric absorption utilizing the zenith optical depth obtained from the sky tipping procedure with the conical horn. This method gave $\Delta G = 0.027 \pm 0.023$. Thus all three methods of analysis are compatible, resulting in a relative gain variation $\Delta G = 0.021 \pm 0.010$; this weighted average was actually used to refer planetary intensities measured at positive elevation angles to the horizon, where the absolute gain calibration was previously accomplished.

VIII. ASTRONOMICAL OBSERVATIONS

A blackbody disk source of temperature T_B(K) which is small in angular size compared with the receiving antenna beamwidth will produce a measured incremental flux density $S_M(\text{W} \cdot \text{m}^{-2} \cdot \text{Hz}^{-1})$ at the surface of the Earth given by [24]

$$S_M = \frac{2kv^2 \Omega_S C_S e^{-\tau A}}{c^2} [J(T_B) - J(T_{BG})], \qquad (11)$$

where c is the velocity of electromagnetic wave propagation in a vacuum (2.99793×10^8 m·s⁻¹), Ω_S(sr) is the solid angle of the disk source, C_S is a normalized size correction factor to account for partial resolution of the disk by the antenna beam, and the other parameters have been previously defined. The solid angle correction factor C_S is given to a good approximation (when $X < 0.4$) by [24]

$$C_S = \frac{1 - e^{-X^2}}{X^2},$$ (12)

where

$$X \equiv (4 \ln 2)^{1/2} R / \theta_A,$$ (13)

and $R(\text{rad})$ is the geometrical mean of the polar and equatorial semidiameters of the (slightly elliptical) disk source and θ_A (rad) is the antenna half-power beamwidth. The solid angle of the source is given by

$$\Omega_S = 2\pi[1 - \cos(R)] \simeq \pi R^2.$$ (14)

The source flux density S that would be measured with no atmospheric loss is given by [24]

$$S \equiv S_M e^{\tau A} \equiv \frac{2kT_A}{\eta_A A_G},$$ (15)

where T_A is the source antenna temperature (K), η_A is the antenna aperture efficiency, and A_G is the antenna geometrical collecting area (m^2). For cosmic radio sources the usual unit of flux density measurements is the Jansky (1 Jy $\equiv$ 10^{-26} W·m^{-2}·Hz^{-1}).

Observations of the Sun with the conical horn were made on two days in December 1977, and observations of Venus, Jupiter, and Saturn were made with the paraboloidal reflector on six days in December 1977 and in November 1978. The peak source intensities were measured relative to the noise tube signal which was absolutely calibrated against the hot and cold loads. All measurements were corrected for pointing error by fitting a two-dimensional Gaussian function to a measured five-point grid map to determine the peak intensity that would have been measured with no pointing error. Saturn was observed alternately with the on-axis and with the off-axis paraboloid feeds to improve the signal-to-noise ratio. Separate measurements of Jupiter in each beam indicated a relative gain of the reference beam of 0.922 ± 0.010 times the gain of the main beam, and this value was used to normalize the Saturn observations. Calculations predict a gain reduction of about 0.4 dB for a feed 2.6 beamwidths off-axis [25], in agreement with the measurement. Table VI lists the antenna temperature, the flux density, and the brightness temperature for each data set. The quoted errors are the quadrature sums of the (1σ) random errors only and do not include the correlated systematic errors in antenna gain or in thermal calibration. The angular source semidiameters were calculated for the time of each observation by dividing the known unit semidiameters [26] by the source distance in astronomical units. The Saturn brightness temperature has been calculated assuming that the rings do not emit or scatter radiation from the planet. Thus only the disk solid angle has been used in (14), although it has been corrected for polar tilt as seen from the Earth.

TABLE VII
8.61-GHZ DISK BRIGHTNESS TEMPERATURES

Source	Disk Brightness Temperature (K)		Total 1σ Error (K)
Sun	7914	±	192
Venus	357.5	±	13.1
Jupiter	179.4	±	4.7
Saturn (B = -9.9°)	153.4	±	4.8

IX. A NEW FLUX DENSITY SCALE

The average solar and planetary brightness temperatures and their total (random plus systematic) errors listed in Table VII may be used with (11) to establish an absolute flux density scale near 86.1 GHz. This new calibration scale may be compared with previously reported observations at nearby frequencies and with theoretical models. The 86.1-GHz solar brightness temperature measured with the conical horn is 7914 ± 192 K, which agrees with the value of 7750 ± 350 K determined by the chopper-wheel calibration technique using the 4.9-m reflector. Linsky [27] predicts a brightness temperature of about 7800 ± 300 K at 86.1 GHz from his linear fit to a recalibrated solar spectrum. Thus the present measurements are in good agreement with previous observations.

The 86.1-GHz brightness temperature of Saturn at B (Saturnicentric latitude of the Earth) = −9.9° referred to the atmospheric disk only is 153.4 K with a random error of 3.0 K and a total error of 4.8 K. This result is in good agreement with both previous measurements (147 ± 11 K at 97.1 GHz) [4], [28] and with the theoretical spectrum of atmospheric emission [28]. However, the changing tilt angle of the rings as seen from the Earth introduces a slowly varying component to Saturn's total 3-mm emission which may be as large as 15 K [28]. Thus Saturn is not a good calibration source for long timebase experiments at a few millimeters wavelength.

Jupiter's thermal brightness temperature is not known to be time-variable, and its constancy and large flux density make it the primary calibration source at millimeter wavelengths. Our observations indicate a brightness temperature for Jupiter of 179.4 K at 86.1 GHz with a random error of 1.5 K and a total error of 4.7 K. A comparison with both the previously observed microwave spectrum (177 ± 13 K at 97.1 GHz) [4], [29] and with a convective model atmosphere [29] again shows agreement within the quoted errors.

At 86.1 GHz the measured brightness temperature of Venus is 357.5 K with a random error of 9.7 K and a total error of 13.1 K. At the time of these observations the planetocentric phase angle between the Sun and the Earth was i = 215°. Previous observations of Venus at 97.1 GHz (380 ± 33 K) [4] and at 88 GHz (355 ± 32 K assuming a solar brightness temperature of 7914 ± 192 K) [30] are in agreement with this more precise result. Janssen [31] has calculated the millimeter-wavelength spectrum of Venus based on a model atmosphere which is consistent with all space-

IEEE TRANSACTIONS ON ANTENNAS AND PROPAGATION, VOL. AP-28, NO. 3, MAY 1980

craft and ground-based microwave observations. At 86.1 GHz the predicted disk brightness temperature is 363 ± 10 K, which also agrees with the observed temperature.

The measurements listed in Table VII define an absolute flux density scale which may be used to calibrate 3-mm radio astronomical observations to an accuracy (1σ) of about three percent. The measured solar and planetary brightness temperatures are shown to agree with previous measurements of lesser accuracy at nearby wavelengths and with theoretical spectra.

ACKNOWLEDGMENT

The authors wish to thank A. J. Walker for assembling the receiver, R. B. Loren and W. L. Peters for programming assistance, and M. Thomas for typing the manuscript.

REFERENCES

[1] A. W. Straiton, "The absorption and reradiation of radio waves by oxygen and water vapor in the atmosphere," *IEEE Trans. Antennas Propagat.*, vol. AP-23, pp. 595–597, 1975.

[2] A. J. Freiley, P. D. Batelaan, and D. A. Bathker, "Absolute flux density calibrations of radio sources: 2.3 GHz," NASA Tech. Memo. 33-806, Jet Propulsion Lab., 1977.

[3] G. T. Wrixon, W. J. Welch, and D. D. Thornton, "The spectrum of Jupiter at millimeter wavelengths," *Astrophys. J.*, vol. 169, pp. 171–183, 1971.

[4] B. L. Ulich, J. R. Cogdell, and J. H. Davis, "Planetary brightness temperature measurements at 8.6 mm and 3.1 mm wavelengths," *Icarus*, vol. 19, pp. 59–82, 1973.

[5] B. L. Ulich, "Absolute brightness temperature measurements at 2.1-mm wavelength," *Icarus*, vol. 21, pp. 254–261, 1974.

[6] C. W. Tolbert, A. W. Straiton, and L. C. Krause, "A 16-foot millimeter wavelength antenna system, its characteristics, and its applications," *IEEE Trans. Antennas Propagat.*, vol. AP-13, pp. 225–229, 1965.

[7] J. R. Cogdell, J. J. McCue, P. D. Kalachev, A. E. Salomonovich, I. G. Moiseev, J. M. Stacey, E. E. Epstein, E. E. Altshuler, G. Feix, J. W. Day, H. Hvatum, W. J. Welch, and F. T. Barath, "High resolution millimeter reflector antennas," *IEEE Trans. Antennas Propagat.*, vol. AP-18, pp. 515–529, 1970.

[8] B. L. Ulich, "A radiometric antenna gain calibration method," *IEEE Trans. Antennas Propagat.*, vol. AP-25, pp. 218–223, Mar. 1977.

[9] V. D. Krotikov, V. A. Porfirjev, and V. S. Troitsky, "The development of the method of precise field intensity measurement and standardization of lunar radio emission at $\lambda = 3.2$ cm," *Izvestia V.U.Z. Radiofizika*, vol. 4, pp. 1004–1012, 1961.

[10] D. Russell and W. Larson, "RF attenuation," *Proc. IEEE*, vol. 55, pp. 942–959, 1967.

[11] W. N. Hardy, "Precision temperature reference for microwave radiometry," *IEEE Trans. Microwave Theory Tech.*, vol. MTT-21, pp. 149–150, 1973.

[12] C. A. Balanis, "Dielectric constant and loss tangent measurements at 60 and 90 GHz using the Fabry–Perot interferometer," *Microwave J.*, vol. 14, no. 3, pp. 41–44, 1971.

[13] A. F. Kay, "Radomes and absorbers," in *Antenna Engineering Handbook*, H. Jasik, Ed. New York: McGraw-Hill, 1961, p. 32-30.

[14] S. Silver, Ed., *Microwave Antenna Theory and Design*. Lexington, MA: Boston Tech., 1964, pp. 439–443.

[15] N. Kuhn, "How accurate is your power meter?," *Microwaves*, vol. 16, no. 9, pp. 106–114, 1977.

[16] R. C. Johnson, H. A. Ecker, and J. S. Hollis, "Determination of far-field antenna patterns from near-field measurements," *Proc. IEEE*, vol. 61, pp. 1668–1694, 1973.

[17] J. J. Stangel and W. M. Yarnall, "Pattern characteristics of an antenna focused in the Fresnel region," *IRE Int. Conv. Record*, pt. 1, March 26–29, 1962, pp. 3-12.

[18] C. T. Stelzried, "Microwave thermal noise standards," *IEEE Trans. Microwave Theory Tech.*, vol. MTT-16, pp. 646–655, 1968.

[19] M. L. Kutner, "Application of a two-layer atmospheric model to the calibration of millimeter observations," *Astrophys. Lett.*, vol. 19, pp. 81–87, 1978.

[20] F. I. Shimabukuro and E. E. Epstein, "Attenuation and emission of the atmosphere at 3.3 mm," *IEEE Trans. Antennas Propagat.*, vol. AP-18, pp. 485–490, 1970.

[21] V. J. Falcone, K. N. Wulfsberg, and S. Gitelson, "Atmospheric emission and absorption at millimeter wavelengths," *Radio Sci.*, vol. 6, pp. 347–355, 1971.

[22] A. A. Penzias, "Cosmology and microwave astronomy," in *Cosmology, Fusion, and Other Matters*, F. Reines, Ed. Boulder, CO: Colorado Associated Univ., 1972, pp. 29–47.

[23] J. H. Davis and P. Vanden Bout, "Intensity calibration of the interstellar carbon monoxide line at λ 2.6 mm," *Astrophys. Lett.*, vol. 15, pp. 43–47, 1973.

[24] B. L. Ulich and R. W. Haas, "Absolute calibration of millimeter-wavelength spectral lines," *Astrophys. J. (Supp. Ser.)*, vol. 30, pp. 247–258, 1976.

[25] J. W. Baars, "The measurement of large antennas with cosmic radio sources," *IEEE Trans. Antennas Propagat.*, vol. AP-21, pp. 461–474, 1973.

[26] *The American Ephemeris and Nautical Almanac for the Year 1979*. Washington, DC: U.S. Gov. Printing Office, 1977, pp. 541, 546.

[27] J. L. Linsky, "A recalibration of the quiet sun millimeter spectrum based on the moon as an absolute radiometric standard," *Solar Phys.*, vol. 28, pp. 409–418, 1973.

[28] M. J. Klein, M. A. Janssen, S. Gulkis, and E. T. Olsen, "Saturn's microwave spectrum: Implications for the atmosphere and the rings," in *The Saturn System*, D. M. Hunten and D. Morrison, Eds. Springfield, VA: Nat. Tech. Inform. Ser., 1978, pp. 195–216.

[29] G. L. Berge and S. Gulkis, "Earth-based radio observations of Jupiter: Millimeter to meter wavelengths," in *Jupiter*, T. Gehrels, Ed. Tucson, AZ: Univ. of Arizona, 1976, pp. 621–692.

[30] E. E. Epstein, J. P. Oliver, S. L. Soter, R. A. Schorn, and W. J. Wilson, "Venus: On an inverse variation with phase in the 3.4-mm emission during 1965 through 1967," *Astron. J.*, vol. 73, pp. 271–274, 1968.

[31] M. Janssen, private communication, 1979.

RECOMMENDATIONS FOR CALIBRATION OF MILLIMETER-WAVELENGTH SPECTRAL LINE DATA

Marc L. Kutner

Physics Department, Rensselaer Polytechnic Institute

AND

B. L. Ulich

Multiple Mirror Telescope Observatory

Received 1981 January 19; accepted 1981 May 4

ABSTRACT

We examine the methods currently used to calibrate millimeter-wavelength photometric data and make recommendations on procedures to be followed. Since confusing references to the quantity $T_A{}^*$ have appeared in the literature, we introduce a series of formal definitions of quantities related to intensity measurements. A new quantity called $T_R{}^*$ is introduced which is the source antenna temperature corrected for atmospheric, ohmic, and all spillover losses. Physically, $T_R{}^*$ corresponds to the source brightness distribution convolved with the diffraction and error beam patterns of the telescope. We also reexamine the relative merits of the so-called "direct" and the "chopper-wheel" methods of correcting for atmospheric absorption and conclude that the chopper-wheel technique is generally preferable.

Subject headings: instruments — interstellar: molecules — radio sources: lines

I. INTRODUCTION

With the growing variety of astrophysical problems that are now being studied by millimeter-wavelength observations of spectral lines, there is an increasing demand for reliable relative and absolute intensity calibration. Such calibration is made difficult by a number of factors:

1. At millimeter wavelengths, the Earth's atmosphere is only partially transparent. Moreover, this transparency can change on a relatively short time scale as the water content of the atmosphere over the observatory changes.

2. Since there is generally no amplification of the incoming signal before it is mixed with the local oscillator signal, millimeter receivers are sensitive to both signal and image sidebands. Because local calibration standards are usually broad-band sources such as hot and cold blackbody loads, the calibration signals enter both sidebands, while the spectral line enters only the signal sideband. The relative response of the system to the two sidebands is often unknown and uncontrolled. Recently, image sideband rejection filters with low insertion loss have been made using quasi-optical techniques and have been quite effective in Cassegrain systems with large effective focal lengths (e.g., Wannier *et al.* 1976).

3. With the current tendency to observe at higher frequencies, telescopes are often operating in regimes where the error pattern caused by small scale reflector surface errors plays an important (but difficult to evaluate) role in the coupling between the source and the antenna.

In principle, the intensity of a spectral line can be determined by the "direct" method in which a source antenna temperature T_A is measured (antenna temperature is simply the detected power divided by the product of the predetection bandwidth and Boltzmann's constant). The atmospheric zenith opacity τ_a is then determined (usually from a measurement of the equivalent sky brightness temperature as a function of airmass A). An exponential correction factor ($e^{\tau_a A}$) is then applied to obtain a corrected antenna temperature $T_A{}'$. This technique requires that τ_a be determined accurately.

The more commonly used alternative is an indirect method first described by Penzias and Burrus (1973). An ambient temperature "chopper" or "vane" absorber is alternated with cold sky as the calibration signal. Thus, as the sky becomes more opaque, the apparent sky brightness temperature gets larger, and the calibration signal is reduced. Under ideal circumstances this method corrects exactly for atmospheric opacity and rear spillover losses (see the Appendix). However, Davis and Vanden Bout (1973) have pointed out that since much of the atmospheric opacity is due to O_2, which is distributed with a scale height of about 8 km, the effective mean temperature of the atmosphere T_M is significantly lower than the ambient temperature T_{AMB}. They also pointed out that the atmospheric opacity can be very different in the two sidebands (especially near the CO $J = 1 \rightarrow 0$ transition at 115 GHz). More recent improvements in the correction for atmospheric opacity are discussed below. The chopper method automatically corrects for antenna ohmic losses and for the spillover that looks like an ohmic loss terminated at ambient temperature. However, there is usually some part of the antenna pattern that looks at the sky but not at the source, and an additional correction is required for this forward spillover. This need has

Reprinted courtesy of *The Astrophysical Journal*, vol. 250, pp. 341–348, Nov. 1, 1981, published by the University of Chicago Press; © 1981, The American Astronomical Society.

resulted in some confusion over what corrections are included in the commonly reported quantity $T_A{}^*$, introduced by Phillips, Jefferts, and Wannier (1973) and defined in detail for prime focus telescopes by Ulich and Haas (1976, hereafter referred to as UH). In this paper, we set down a proposed set of definitions for the various observational quantities that removes these ambiguities. We then look at how this affects data already in the literature, and, finally, we reexamine the methods of calibration.

II. COUPLING OF ANTENNA TO SOURCE

Following UH, we assume that the emission from a radio source at any point can be characterized by a blackbody brightness temperature T_B or by the effective source radiation temperature $J(v, T_B)$, where

$$J(v, T) \equiv (hv/k)/[\exp (hv/kT) - 1] , \qquad (1)$$

v is the frequency, T is the temperature, h is Planck's constant, and k is Boltzmann's constant.

For example, a spectral line source with a constant excitation temperature T_E, a background brightness temperature T_{BG}, and a source optical depth $\tau(v)$ will have an effective radiation temperature given by

$$J(v, T_B) = \{1 - \exp [-\tau(v)]\} [J(v, T_E) - J(v, T_{BG})] . \qquad (2)$$

In the literature the quantity $J(v, T_B)$ is often referred to as simply the radiation temperature T_R for the source, and we will use that terminology. Variations in source brightness temperature with direction angle on the sky Ψ are accounted for in the normalized source brightness distribution $B_n(\Psi)$, which takes on values between zero and unity.

The observed source antenna temperature in the direction Ω on the sky will then be

$$T_A = T_R \eta_r \left[\frac{\iint_{\Omega_s} P_n(\Psi - \Omega) B_n(\Psi) d\Psi}{\iint_{4\pi} P_n(\Omega) d\Omega} \right] e^{-\tau_a A} , \qquad (3)$$

where P_n is the normalized antenna power pattern $[P_n(0) = 1]$, Ω_s is the solid angle subtended by the source, τ_a is the atmospheric optical depth at the zenith, A is the airmass at which the observation is made, and η_r is the radiation efficiency of the telescope. η_r accounts for the fraction of the incoming power lost to ohmic (resistive) heating of the antenna and is formally defined in terms of the maximum antenna gain G as

$$\eta_r \equiv (G/4\pi) \iint_{4\pi} P_n(\Omega) d\Omega . \qquad (4)$$

Because of the convolution integral in equation (3), the conversion from T_A to T_R requires a detailed knowledge of the source structure and of the antenna power pattern. It is therefore desirable to report a more directly observable quantity that does not depend on a convolution process. For this purpose we divide the antenna pattern into two zones, one involving the normal diffraction pattern (including the error pattern) and the other involving spillover (both forward and rearward) as well as scattering

from the feed support legs and other structures. There is some ambiguity in drawing the dividing line between these two regions, since some of the spillover and scattering can be in the forward hemisphere and close to the main lobe (especially in Cassegrain systems with spillover past the secondary mirror). However, spillover and scattering falling into the central diffraction pattern must be treated as part of that pattern. Therefore, if we say that all of the power within a solid angle Ω_d is part of the diffraction pattern, then everything outside Ω_d is considered spillover and scattering. In practice, the exact choice of Ω_d should not cause serious problems, as long as the procedures are followed consistently.

We can then define the efficiency η_c with which the antenna couples to the source as

$$\eta_c \equiv \iint_{\Omega_s} P_n(\Psi - \Omega) B_n(\Psi) d\Psi \bigg/ \iint_{\Omega_d} P_n(\Omega) d\Omega . \qquad (5)$$

For most telescopes, Ω_d will encompass a region within a few degrees of the telescope axis. Therefore, there may be sources (giant molecular clouds observed in CO, for example) which are larger than Ω_d and actually extend into the spillover region. If such sources are of roughly uniform brightness then it is possible to have $\eta_c > 1$. In such circumstances, the coupling of the spillover to the source is still a problem which must be taken into account, whether nor not our particular definition of η_c is used.

Using our definition of η_c from equation (5), we define the corrected source intensity in radiation temperature units as

$$T_R{}^* \equiv \eta_c T_R . \qquad (6)$$

For the following reasons we believe that $T_R{}^*$ is often the most useful quantity to report:

1. It is as close as one can get to source intensity without inserting particular knowledge about source structure. That is, $T_R{}^*$ is corrected for everything (ohmic loss, atmospheric loss, spillover, and scattering) except the actual coupling of the antenna diffraction pattern to the source.

2. For many observations η_c is not very different from unity, so $T_R{}^*$ gives a reasonable estimate of T_R which is adequate for the purposes of many observers.

3. With the exception of quantities that amount to filling factors (which can at least be estimated on a relative basis for different telescopes), $T_R{}^*$ is a relatively telescope-independent quantity for a number of objects that might be used as standard sources.

4. With a growing emphasis on source modeling for many problems, the usual procedure would be to develop a source model, to use radiative transfer calculations to predict the emergent intensity as a function of position on the source, and then to do the convolution of this intensity with the antenna pattern. This is because the convolution of a particular source model with an antenna pattern can be done rather easily, but a deconvolution of the observations is much more difficult. Thus, it seems

that it may be most convenient to compare theory and observations in terms of $T_R{}^*$ rather than of T_A or of T_R.

It should also be noted that our definition of $T_R{}^*$ conforms to the definition that some authors have been assuming for $T_A{}^*$. The relationship between our definitions and previous usage will be discussed in § III.

Using equations (3), (5), and (6), we find that $T_R{}^*$ is related to the source antenna temperature $T_A{}'$ corrected for atmospheric attenuation $[T_A{}' \equiv T_A \exp{(\tau_a A)}]$ by

$$T_A{}' = T_R{}^* \eta_r \left[\iint_{\Omega_d} P_n(\Omega)d\Omega \bigg/ \iint_{4\pi} P_n(\Omega)d\Omega \right] . \quad (7)$$

The quantity in brackets is the spillover and scattering efficiency of the telescope. That is, it is the fraction of the radiated power that is not lost in spillover and scattering outside the diffraction zone. As shown in the Appendix, the spillover and scattering that fall on the forward hemisphere enter into the chopper calibration scheme differently than the spillover and scattering that fall on the rearward hemisphere. We therefore write the quantity in brackets as the product of two efficiencies:

$$\eta_{fss} \equiv \iint_{\Omega_d} P_n(\Omega)d\Omega \bigg/ \iint_{2\pi} P_n(\Omega)d\Omega , \quad (9)$$

and

$$\eta_{rss} \equiv \iint_{2\pi} P_n(\Omega)d\Omega \bigg/ \iint_{4\pi} P_n(\Omega)d\Omega , \quad (9)$$

where the double integral over 2π denotes an integral over the forward hemisphere. In terms of these quantities

$$T_A{}' = T_R{}^* \eta_r \eta_{fss} \eta_{rss} . \quad (10)$$

We can also define an extended source efficiency $\eta_s \equiv \eta_r \eta_{fss} \eta_{rss}$ such that

$$T_A{}' \equiv T_R{}^* \eta_s . \quad (11)$$

In principle, while the product $\eta_{fss} \eta_{rss}$ remains constant, the individual quantities may change with the elevation angle of the telescope. That is, some of the spillover may switch from looking at the sky to looking at the ground, or from looking at the ground to looking at the sky. This is probably worst for elevations very near the horizon when some of the rear spillover, just past the edge of the dish, now looks at the sky, and when some of the spillover past the secondary in Cassegrain systems goes from the sky to the ground. However, as a practical matter, for most positions at which observations will be made, this does not pose a significant problem. To the extent that it may interfere with very accurate absolute calibration, the elevations at which these effects become significant, and the amount by which they affect the calibration, can be determined by measurements made over a variety of elevations under very clear and stable atmospheric conditions.

III. COMPARISON OF $T_R{}^*$ AND $T_A{}^*$

As shown in the Appendix, the chopper calibration method automatically corrects for losses that appear to be terminated at the ambient temperature, namely ohmic losses and rearward spillover and scattering. It is therefore convenient to combine these effects into a single efficiency, η_l, originally introduced by UH. In terms of the quantities presented here, it is defined as

$$\eta_l \equiv \eta_r \eta_{rss} . \quad (12)$$

When the antenna is pointed at the sky, the observed antenna temperature is then

$$T_A{}^{SKY} = \eta_l J(v, T_{SKY}) + (1 - \eta_l)J(v, T_{SPILL}) , \quad (13)$$

where T_{SKY} is the brightness temperature of the sky and T_{SPILL} is the temperature at which the spillover is terminated and is usually equal to the ambient temperature T_{AMB}. From this we see that if the antenna temperature is measured as a function of airmass A (as in an antenna tipping procedure) and extrapolated to $A = 0$, η_l can be determined. For many purposes, the calibration scale can be considered to be essentially independent of η_l, but, as shown in the Appendix, when all effects are included a very weak dependence on n_l is introduced. Also, if the atmospheric opacity τ_a is to be directly determined from tipping curves, then η_l must be known.

The antenna temperature scale, corrected for atmospheric attenuation, resistive losses, and rearward spillover and scattering, is then

$$T_A{}^* \equiv T_A{}'/\eta_l = [T_A \exp{(\tau_a A)}]/\eta_l . \quad (14)$$

$T_R{}^*$ is related to $T_A{}^*$ by the forward spillover and scattering efficiency such that

$$T_R{}^* = T_A{}^*/\eta_{fss} . \quad (15)$$

In introducing the symbol $T_A{}^*$, Phillips, Jefferts, and Wannier (1973) defined it as "antenna temperature corrected for all telescope and atmospheric losses" with no further details. In the definition of $T_A{}^*$ of UH, it was implicitly assumed that $n_{fss} = 1$, in which case these two temperature scales would be identical. This assumption was made because the discussion in UH deals only with prime focus antennas which have primarily rearward spillover. However, it now appears that some prime focus antennas can have a significant amount of forward spillover and scattering. In addition, when the extension was made to Cassegrain systems, some observers used this definition of $T_A{}^*$. That is, $T_A{}^*$ was taken to be the antenna temperature corrected for atmospheric attenuation, ohmic losses, and rearward spillover. As such, it is the natural quantity that falls out of chopper calibration. However, other observers assumed another definition of $T_A{}^*$ based on its relationship to the source. They took $T_A{}^*$ to be corrected for everything except the actual coupling of the source of the antenna diffraction pattern (which is our definition of $T_R{}^*$). As a result, numbers have been reported as $T_A{}^*$ which may have different physical meanings. The exact relationship for several telescopes is given in Table 1.

TABLE 1

THREE MILLIMETER TELESCOPE EFFICIENCIES

Telescope	MWO 4.9 m	NRAO 11 m	NRAO 11 m	NRAO 11 m
Focus	Prime	Prime	Cassegrain	Cassegrain
Feed type	Pyramidal horn	Conical horn	Gustincic horn/lens[a]	Ulrich horn/lens[a]
η_r	1.00	0.99	0.99	0.99
η_{rss}	0.93	0.68	0.79	0.91
η_l	0.93	0.67	0.78	0.90
η_{fss}	0.86	0.99	0.74	0.79
η_s	0.80	0.66	0.58	0.71
η_{Sun}	0.77	0.59	0.52	0.63
$T_R{}^*/T_A{}^*$ (reported)	1.17 ± 0.05	1.09 ± 0.09	1.02 ± 0.09	1.27 ± 0.09

NOTE.—For observations at other frequencies, assuming appropriate scaling of the feeds, all quoted efficiencies should be approximately the same, except for η_{Sun}, which will get lower at higher frequencies. At higher frequencies, more power generally goes from the main beam to the error pattern, but η_{fss} includes both, so it does not change.

[a] Ulich 1980.

In general, for the NRAO 11 m telescope, prime focus observations that were reported as $T_A{}^*$ are $T_A{}^*$ as defined here, and Cassegrain observations reported as $T_A{}^*$ are actually $T_R{}^*$ as defined here. For example, if Cassegrain observations were interpreted as $T_A{}^*$ according to Ulich and Haas, (1976), then an attempt to correct for forward spillover (a second time) would result in a 35% overestimate in line strengths. For the most part, observations done on the Aerospace Corporation 5 m telescope, on the University of Texas Millimeter Wave Observatory (MWO) 5 m telescope, and on the Columbia University 1 m telescope that were reported as $T_A{}^*$ are $T_A{}^*$ as defined here, and not $T_R{}^*$. Results from the Bell Telephone Laboratories 7 m telescope, reported as $T_A{}^*$, are generally $T_R{}^*$ (with $\eta_{fss} = 0.9$) as defined here when some statement about correction for "beam efficiency" is made (R. A. Linke 1981, private communication).

IV. RECOMMENDATIONS FOR CALIBRATION PROCEDURE

Given that the chopper technique naturally corrects for only (the rearward) part of the spillover, and that a knowledge of η_{fss} is required to convert $T_A{}^*$ to $T_R{}^*$, one might ask whether it now becomes preferable to abandon the chopper technique and return to the direct method. We examine this question with regard to two criteria: (1) correction for antenna coupling, and (2) correction for atmospheric attenuation. In making the comparison, we will assume that the goal is the best value of $T_R{}^*$, since, in either method, one must still contend with the convolution of the source structure and the antenna pattern.

a) Correction for Antenna Coupling

In the chopper technique, $T_A{}^*$ is well determined since η_l can be determined rather precisely, so the only correction to get to $T_R{}^*$ is η_{fss}. In the direct method, $T_A{}'$ must be corrected by $\eta_s = \eta_l \eta_{fss}$. The question is thus whether η_s is better known than η_{fss}. In practice, η_s can be found by observing appropriate sources whose intensity and structure are well known. Ideally, these sources should be of uniform brightness and should exactly fill Ω_d, but this is

rarely possible, so some correction is needed for the part of the diffraction pattern (within Ω_d) that does not couple to the reference source. η_{fss} cannot be directly determined in practice. However, η_l can be directly determined, and η_{fss} is then found as the ratio η_s/η_l. Since there is also some uncertainty in η_l, η_{fss} will have a larger uncertainty than η_s. The difference in uncertainties depends on the relative uncertainties in η_s and η_l. In summary, the way in which the quantities are determined makes η_{fss} slightly more uncertain than η_s, but under most circumstances, the advantage in the direct approach should amount to at most a few percent.

b) Correction for Atmospheric Attenuation

As shown by Kutner (1978), for any given uncertainty in the atmospheric opacity, the chopper method gives significantly smaller errors in the overall calibration than the direct method. The actual numerical advantage of the chopper method depends on the atmospheric opacity. The advantage is greater when the atmosphere is more opaque. Under the range of circumstances in which observations might be made at wavelengths between 2 and 4 mm, the accuracy advantage of the chopper method might range from 5% to 25%. At even shorter wavelengths, as the atmospheric opacity increases sharply, the advantage becomes even greater.

This advantage arises from the fact that there are offsetting effects in the chopper technique. Under idealized circumstances, these offsetting effects exactly cancel (see the Appendix). Under real circumstances, the advantage persists only if care is taken to correct for the fact that the effects do not completely offset each other. However, this is not a serious problem, since it generally involves the application of an easily calculated elevation-dependent scale factor, such as that given by UH (and reproduced in the Appendix), along with an appropriate atmospheric model. Alternatively, the cooled, variable-temperature chopper proposed and tested by Ulich (1980) provides a "hardware" solution. In either case, the simple models that have already been published appear to be adequate to bring the calibration uncertainty down to less than the 5% level in the 2 to 4 mm wavelength range.

A serious problem in calibration can be short term changes in the atmospheric opacity (essentially changes in the water vapor opacity). The cooled chopper system is the least sensitive to such changes, except in situations where no image rejection is used and the water vapor opacity is very different in the two sidebands. If an uncooled chopper is used, and especially if the direct method is used, the usual technique of doing an antenna tipping every few hours may not be satisfactory because of rapid changes in the water vapor opacity and of differences in the opacity from one part of the sky to another. For this reason, a system that puts two different temperature loads in front of the receiver as part of the normal calibration process is desirable in that it allows frequent and independent monitoring of the atmospheric attenuation, as is done on the BTL 7 m telescope (Goldsmith 1977). Given the three step calibration (two loads and the sky), one is then free to apply the calibration scale either in the direct way or in the chopper-prescribed manner. The same arguments for using the chopper method, mentioned above, still apply. In summary, it appears that the chopper technique offers more reliable correction for atmospheric attenuation, and this appears to more than offset the slight advantage that the direct approach has in the correction for spillover losses.

c) Standard Sources

Finally, we make a few comments about the use of standard sources as an aid in calibration. For a source to be an appropriate standard, the source brightness distribution must be known. Two simple extremes are a point source and a source that uniformly fills the diffraction pattern. (It should be noted that one of the most popular standard sources, Ori A, falls in neither of these categories.) Standard sources may be used in two basic ways: (1) If all of the free parameters that enter into the calibration are known, then the standard sources can be used as a check of the calibration scale. (2) If some quantities are unknown, then standard sources can be used to determine them. However, it is important to identify the quantities (or groups of quantities) being determined. All parameters do not enter directly into the equivalent chopper calibration temperature as scale factors. Adjusting the scale of T_A^* or of T_R^* to match

observations with other telescopes is generally incorrect since both scales depend on antenna parameters. Thus T_A^* and T_R^* will in general vary from one telescope to another. Only T_R is truly independent of the telescope characteristics.

V. SUMMARY

We have examined the existing work on millimeter spectral line calibration. To avoid the confusion that has grown out of multiple usage of the quantity T_A^*, we define the quantity T_R^* which is the source antenna temperature corrected for all atmospheric, ohmic, and spillover losses. This corresponds to the source intensity convolved with the antenna diffraction and error patterns. The quantity T_A^* is actually T_R^* without the correction for forward spillover. For work already published, data quoted as T_A^* are generally consistent with the original definition. An exception is Cassegrain focus data taken on the NRAO 11 m telescope, for which previously reported values of T_A^* are really T_R^*.

We have reexamined the relative merits of direct and chopper-wheel calibration schemes, and we conclude that the chopper method provides the better overall absolute calibration. However, in order to realize this potential, one must use more than just the simplest model for atmospheric emission. One improvement is provided by the two-layer atmospheric model of Kutner (1978). The cooled chopper wheel suggested by Ulich (1980) also provides adequate correction for second-order errors through hardware improvements. If the cooled chopper is not used, a calibration procedure involving both the sky and two known temperature loads is preferable.

Much of this paper is an outgrowth of a workshop on Millimeter Wavelength Calibration, hosted by the NRAO[1] in May 1980. We would like to thank the participants for their many significant suggestions. We would also like to thank the participants in the U.R.S.I. Symposium on Millimeter Wavelength Technology, in Grenoble, France, in August 1980 for their helpful criticisms. M. L. K. was partially supported by National Science Foundation grant AST 79-23584.

[1] The National Radio Astronomy Observatory is operated by Associated Universities, Inc., under contract with the National Science Foundation.

APPENDIX

CHOPPER WHEEL CALIBRATION SCALE FACTOR

In this section, we briefly look at the considerations appropriate to establishing a scale factor for chopper wheel calibrations. These have been discussed more extensively by Davis and Vanden Bout (1973), Ulich and Haas (1976), Kutner (1978), and Ulich (1980).

a) Simplified Example

To see the basics in establishing a scale factor, we first go through a simplified case in which we assume the following: (1) the receiver is single sideband; (2) the chopper and spillover are at the ambient temperature T_{AMB}, (3) the cosmic background radiation is negligibly small; (4) $hv \ll kT$ for all temperatures involved; and (5) the sky brightness temperature can be written in terms of a mean effective temperature of the sky T_M as

$$T_{\text{SKY}} = T_M[1 - \exp{(-\tau_a A)}] . \tag{A1}$$

If V is the voltage response of the usual square-law detector and g is the factor for converting equivalent input temperatures into output voltages, then when looking at a spectral line with antenna temperature T_A the output voltage is

$$V_L = gT_A = g\eta_l T_A{}^* \exp{(-\tau_a A)} . \tag{A2}$$

The chopper calibration signal is the difference between the ambient temperature absorber and the antenna temperature of the sky

$$T_{\text{CAL}} = T_{\text{AMB}} - T_A{}^{\text{SKY}} , \tag{A3}$$

so

$$V_{\text{CAL}} \equiv g(T_{\text{AMB}} - T_A{}^{\text{SKY}}) . \tag{A4}$$

Using equation (13), this becomes

$$V_{\text{CAL}} = g\eta_l(T_{\text{AMB}} - T_{\text{SKY}}) . \tag{A5}$$

We define the quantity T_C such that

$$T_A{}^* = T_C(V_L/V_{\text{CAL}}) , \tag{A6}$$

in which case

$$T_C \equiv (V_{\text{CAL}}/V_L)T_A{}^* \tag{A7}$$

$$= (T_{\text{CAL}}/\eta_l) \exp{(\tau_a A)} \tag{A8}$$

$$= T_{\text{AMB}} + (T_{\text{AMB}} - T_M)[\exp{(\tau_a A)} - 1] . \tag{A9}$$

An initially attractive feature of the chopper method is that, for this simplest case, if the atmosphere is at the ambient ground temperature, one has the simple result that $T_C = T_{\text{AMB}}$, independent of airmass or atmospheric conditions. In addition, it can be seen that T_C is independent of η_l, which means that $T_A{}^*$, as defined in equations (14) and (A6), is the quantity that comes out of the chopper calibration with the least amount of additional information. However, if we want a scale factor that produces $T_R{}^*$, we must define a corrected calibration temperature $T_C{}^*$ so that by analogy with equation (A7),

$$T_C{}^* = (V_{\text{CAL}}/V_L)T_R{}^* , \tag{A10}$$

which, from equation (15), gives

$$T_C{}^* = T_C/\eta_{\text{fss}} = (T_{\text{CAL}}/\eta_s) \exp{(\tau_a A)} . \tag{A11}$$

b) General Case

The more general case, with signal and image sidebands at frequencies v_s and v_i, with relative receiver power gains G_s and G_i (normalized such that $G_s + G_i = 1$), and with zenith atmospheric opacities τ_s and τ_i, is worked out by Ulich and Haas (1976). On the assumption that $J(v_s, T) = J(v_i, T)$ for all T including T_{BG} (the brightness temperature of the background radiation), one has

$$\begin{aligned} T_C = &(1 + G_i/G_s)[J(v_s, T_M) - J(v_s, T_{\text{BG}})] \\ &+ (1 + G_i/G_s)[J(v_s, T_{\text{SPILL}}) - J(v_s, T_M)]e^{\tau_s A} \\ &+ (G_i/G_s)[J(v_s, T_M) - J(v_s, T_{\text{BG}})]\{\exp{[(\tau_s - \tau_i)A]} - 1\} \\ &+ [(1 + G_i/G_s)/\eta_l][J(v_s, T_{\text{CHOP}}) - J(v_s, T_{\text{SPILL}})]e^{\tau_s A} , \end{aligned} \tag{A12}$$

where T_{CHOP} is the temperature of the chopper and T_{SPILL} is the temperature at which the rear spillover is terminated. It should be noted that when the chopper is not at the same temperature as the spillover, T_C will depend weakly on η_l.

Ulich (1980) has pointed out that one can use the second and the last terms in equation (A12) to advantage by controlling the temperature of the chopper such that

$$J(v_s, T_{\text{CHOP}}) = J(v_s, T_{\text{SPILL}}) + \eta_l[J(v_s, T_M) - J(v_s, T_{\text{SPILL}})] , \tag{A13}$$

in which case

$$T_C = [J(v_s, T_M) - J(v_s, T_{\text{BG}})]\{1 + (G_i/G_s) \exp{[(\tau_s - \tau_i)A]}\} . \tag{A14}$$

From equation (A13) we see that a dependence on n_t is still there, but, again, it is a weak one. From equation (A14) we see that T_C is simply a constant if the system is single sideband ($G_i/G_s = 0$) or at frequencies where the atmospheric opacity is the same in the two sidebands ($\tau_s = \tau_i$). In any case, T_C is not strongly dependent on the atmospheric opacity.

For the situations in which a cooled chopper is not used, an appropriate atmospheric model should be used in evaluating equation (A12). Kutner (1978) has suggested a two-layer model in which the oxygen and water contributions to the opacity are treated separately. The oxygen contribution is very stable over time, while the water contribution is taken to be variable. The effective temperature for the water is given by

$$T_w = T_{AMB} - \Delta ,\tag{A15}$$

where $\Delta \approx 10$ K, and the effective temperature of the oxygen is given by

$$T_O = (0.90 + 0.02\tau_O A)T_{AMB} ,\tag{A16}$$

where τ_O is the oxygen zenith opacity (similarly, τ_w is the water opacity). The mean sky brightness temperature is then given by

$$J(v, T_M) = J(v, T_O)\{1 - \exp[-(\tau_O + \tau_w)A]\} + [J(v, T_w) - J(v, T_O)][1 - \exp(-\tau_w A)] .\tag{A17}$$

If this model is used, then tipping curves may be analyzed on the assumption that τ_O is known at the frequency of interest, and τ_w is then found. If a self-consistent model is used in the analysis of the tipping curves and in the calculation of T_C, the sensitivity to uncertainties in τ_w is reduced over the direct method, especially for total sky opacities in the direction of observation greater than 0.2. It should also be noted that if a cooled chopper is used, then the two-layer model can also be used to give a more accurate value of T_M (which is almost independent of air mass).

c) Comparison of Calibration Scales

The intensities observed with telescopes of equal apertures and of equal reflector surface precisions may be directly compared for consistency. In general, however, molecular cloud structure will result in different observed intensities with different telescopes, and comparisons of peak intensities will not guarantee consistent calibration schemes. This difficulty with the coupling of the antenna pattern to the source brightness distribution may be partly overcome by completely mapping the source. Since the total flux density from a source is independent of the telescope power pattern, comparison of integrated intensities may be used to compare the thermal calibration scales of different telescopes. The source flux density S observed in the direction Ω is given by the convolution of the source brightness distribution $B(\Psi)$ with the normalized antenna power pattern so that

$$S(\Omega) = \iint_{4\pi} P_n(\Psi - \Omega)B(\Psi)d\Psi = (2kT_R/\lambda^2) \iint_{4\pi} P_n(\Psi - \Omega)B_n(\Psi)d\Psi .\tag{A18}$$

The flux density is also related to the antenna temperature by

$$S(\Omega) = 2kT_A'(\Omega)/A_E = 2k\eta_s T_R^*(\Omega)/A_E ,\tag{A19}$$

where A_E is the effective collecting area of the antenna ($A_E = \lambda^2 G/4\pi$, where λ is the wavelength). The total flux density of the source S_T is found by integrating the source brightness distribution over the source solid angle and is given by

$$S_T = \iint_{\Omega_s} B(\Psi)d\Psi = (2kT_R/\lambda^2) \iint_{\Omega_s} B_n(\Psi)d\Psi .\tag{A20}$$

Combining equations (4), (A18), and (A20), one can show that the integral of flux density is related to the total flux density by

$$\iint_{4\pi} S(\Omega)d\Omega = (\eta_R \lambda^2/A_E)S_T ,\tag{A21}$$

or by substitution from equation (A19) we have

$$S_T = (2k/\eta_r\lambda^2) \iint_{4\pi} T_A'(\Omega)d\Omega .\tag{A22}$$

Similarly, from equation (10), we have

$$S_T = (2k\eta_{fss}\eta_{rss}/\lambda^2) \iint_{4\pi} T_R^*(\Omega)d\Omega ,\tag{A23}$$

which is independent of the antenna directional pattern. Comparisons of total source flux densities calculated from equation (A23) may thus be used to check the consistency of the thermal calibration scales of different antennas, but agreement does not necessarily indicate that the spillover factors $\eta_{\mathrm{fss}} \eta_{\mathrm{rss}}$ are accurately known. These may be checked only by comparison of peak $T_R{}^*$ values rather than of integrated $T_R{}^*$ values.

REFERENCES

Davis, J. H., and Vanden Bout, P. 1973, *Ap. Letters*, **15**, 43.
Goldsmith, P. 1977, *Bell System Tech. J.*, **56**, 1483.
Kutner, M. L. 1978, *Ap. Letters*, **19**, 81.
Penzias, A. A., and Burrus, C. A. 1973, *Ann. Rev. Astr. Ap.*, **11**, 51.
Phillips, T. G., Jefferts, K. B., and Wannier, P. G. 1973, *Ap. J. (Letters)*, **186**, L19.
Ulich, B. L. 1980, *Ap. Letters*, **21**, 21.
Ulich, B. L., and Haas, R. W. 1976, *Ap. J. Suppl.*, **30**, 247 (UH).
Wannier, P. G., Arnaud, J. A., Pelow, F. A., and Saleh, A. A. M. 1976, *Rev. Sci. Instr.*, **47**, 56.

MARC L. KUTNER: Department of Physics, Rensselaer Polytechnic Institute, Troy, NY   12181

BOBBY L. ULICH: Multiple Mirror Telescope Observatory, University of Arizona, Tucson, AZ   85721

Improved Correction for Millimeter-Wavelength Atmospheric Attenuation

B. L. ULICH *National Radio Astronomy Observatory, Tucson, Arizona*

(*Received December 20, 1979*)

The standard chopper-wheel method of calibrating millimeter-wavelength corrected antenna temperature data has several deficiencies which require significant corrections to produce repeatable measurements. Reducing the chopper-wheel brightness temperature below ambient temperature can result in a simplified calibration equation and a more accurate correction for atmospheric attenuation. An empirical calibration procedure is described to determine the effective temperature of a cooled chopper which is nearly at the mean atmospheric temperature. With this cooled chopper, the equivalent calibration temperature is simply related to the ambient temperature but is essentially independent of the precipitable water vapor. Measurements on the NRAO 11-m telescope near 3 mm wavelength confirm the accuracy of the atmospheric absorption correction using the cooled-chopper technique.

INTRODUCTION

The Earth's atmosphere attenuates microwave radiation from cosmic sources. The magnitude of this attenuation varies with the wavelength of observation, with the latitude and altitude of the observing site, with the apparent source elevation angle, and with time according to global and local weather patterns. It is of utmost importance to accurately correct for this loss of signal in radio astronomy observations at millimeter wavelengths, since the effect is large and variable. Both relative and absolute intensity measurements must be corrected for atmospheric attenuation in order to produce repeatable and accurate results. This paper describes an improved method of accurately calibrating both continuum and spectral line observations in terms of absolute intensities corrected for atmospheric attenuation.

Clouds

Millimeter waves are severely scattered and absorbed by liquid water droplets in clouds and in rain (Lo *et al.*, 1975), and precise astronomical observations are generally impossible when optically opaque water clouds are present in the line of sight. Since the dielectric constant of ice is much lower than that of liquid water, the millimeter-wave attenuation of cirrus ice clouds is negligibly small (Wulfsberg, 1964, and Grody, 1976), and precise astronomical observations can be made through ice clouds. Atmospheric attenuation in clear weather is due mainly to absorption

by collision-broadened lines of terrestrial oxygen and water vapor (Straiton, 1975).

Oxygen Absorption

Oxygen is exponentially distributed in the vertical direction in the terrestrial atmosphere with a scale height of about 5 km (Kislyakov, 1966, and Gibbins *et al.*, 1975). The total O_2 concentration and vertical distribution do not vary significantly with time. Thus the molecular oxygen component of atmospheric absorption can be expressed as

$$\tau_0(\nu,h) = \alpha(\nu) \cdot \exp(-h/h_0) \qquad (1)$$

where τ_0 = atmospheric optical depth at the zenith due to oxygen absorption (Nepers),

ν = frequency (GHz),

h = site altitude above sea level (km),

α = sea-level atmospheric optical depth at the zenith due to oxygen absorption (Nepers), and

h_0 = oxygen scale height (5 km).

The separability of the frequency dependence of the optical depth from the altitude dependence greatly facilitates accurate computation of τ_0 for any telescope site. Rosencranz (1975) has calculated the frequency-dependent parameter $\alpha(\nu)$, and his results are listed in Table I for seven frequencies spanning the millimeter band. Five frequencies were chosen to match the natural "windows" between absorption lines where the terrestrial atmosphere is most transparent. The

Reprinted courtesy of *Astrophysical Letters*, vol. 21, pp. 21–28, 1980, published by the University of Chicago Press;

TABLE I
Optical Depth Coefficients

Frequency (GHz)	Wavelength (mm)	α [h = 0 km] (Nepers)	τ_0 [h = 2 km] (Nepers)	β (Nepers/mm)
22.2	13.5	0.013	0.009	0.0060
31.4	9.55	0.028	0.019	0.0015
90.0	3.33	0.041	0.027	0.012
115.3	2.60	0.345	0.231	0.019
150	2.00	0.008	0.005	0.033
230	1.30	0	0	0.067
345	0.870	0	0	0.20

other two frequencies of interest are the 22.2 GHz transition of H_2O (which is only weakly absorbing in the Earth's atmosphere but is masering in many celestial sources) and the 115.3 GHz J = 1-0 line of CO which is useful for studying galactic and extragalactic structure. Note from Table I that α(115.3 GHz) is large because of the proximity of the strongly-absorbing 118.8 GHz O_2 line (Schulze and Tolbert, 1963). When sensitive receivers are available at 230 GHz, the J = 2-1 transition of CO will offer the additional advantage of better atmospheric transparency at dry sites. Although many authors have previously computed the oxygen spectrum (Meeks and Lilley, 1963, Reber et al., 1970, and Reber, 1972), Rosencranz's impact theory of overlapping spectral lines is the first completely theoretical calculation which matches the experimental observations both near the resonance lines and in their nonresonant wings. Measurements by Altshuler et al., (1968) at 35 GHz ($\alpha = 0.035$) and by Shimabukuro and Epstein (1970) at 91 GHz ($\alpha = 0.037$) are in good agreement with Rosencranz's calculations (Rosencranz, 1975). The maximum error in $\alpha(\nu)$ is probably less than 25% of the values listed in Table I. Thus the uncertainty in total atmospheric transmission due to errors in $\alpha(\nu)$ is small. The fourth column in Table I lists the values of τ_0 calculated for observations made at an altitude of 2 km (which is appropriate for the NRAO 11-m telescope on Kitt Peak and for the MWO 5-m telescope on Mt. Locke).

Water Vapor Absorption

Water vapor is generally distributed in the terrestrial atmosphere according to an exponential vertical concentration with a scale height of about 1.5 - 2.0 km (Kislyakov, 1966, and Gibbins et al., 1975). The expected atmospheric attenuation due to water vapor absorption has been calculated by many authors (Tolbert et al., 1964, Bastin, 1966, Ulaby and Straiton, 1970, Gaut and Reifenstein, 1971, Emery, 1972, and Waters, 1976). There is generally good agreement with experimental data near resonance lines, but discrepancies in the nonresonant wings are as large as a factor of four. Gaut and Reifenstein (1971) have derived an empirical correction term which reduces the differences between the calculations and experimental data to within about 10% from 10 GHz to 1350 GHz. They state that the water vapor attenuation is a linear function of the precipitable water vapor in a vertical column. The zenith optical depth due to water vapor absorption is given by

$$\tau_W(\nu, W) = \beta(\nu) \cdot W \qquad (2)$$

where τ_W = atmospheric optical depth at zenith due to water vapor absorption (Nepers),

W = precipitable water vapor (mm), and

β = atmospheric optical depth at zenith per millimeter precipitable water vapor (Nepers/mm).

Under typical clear-sky conditions at Kitt Peak, W varies from 1-12 mm with a mean of about 4 mm in the non-summer months. The frequency-dependent term $\beta(\nu)$ from Gaut and Reifenstein (1971) is also listed in Table I. Direct measurements utilizing radiosonde water vapor data by Shimabukuro and Epstein (1970) at 91 GHz ($\beta = 0.009$) and by Johnson et al., (1970) at 214 GHz ($\beta = 0.064$) are close to (but slightly lower than) the computed values. As can be seen in Table I, water vapor attenuation at the "window" frequencies increases rapidly with increasing frequency and is typically the dominant source of opacity in all the millimeter wavelength windows at existing sites. Measurements of relative attenuation by

Goldsmith *et al.*, (1974) indicate that $\beta(345\text{ GHz})/\beta(230\text{ GHz}) = 2.2$ whereas the calculated value is 3.0. Mather *et al.* (1971) compared the observed sky emission spectrum from 175 GHz to 428 GHz with theoretical calculations of water vapor absorption. Their data were averaged over long time periods and no simultaneous measurements of precipitable water vapor were made. Thus their results cannot be used to derive absolute values of $\beta(\nu)$, but their data are consistent with the shape of the frequency dependence and with the absolute values of $\beta(\nu)$ given in Table I. Wrixon and McMillan (1978) measured the clear-sky zenith attenuation at sea level at 230 GHz on nine days and correlated these data with measurements of the surface absolute humidity. They assumed a linear vertical distribution of water vapor (rather than an exponential variation) to estimate the total precipitable water vapor. Reber and Swope (1972) have shown that estimates of total precipitable water vapor from surface humidity measurements are not generally valid, and Goldsmith *et al.* (1974) and Fogarty (1975) have shown that surface humidity is a poor indicator of atmospheric attenuation. Thus Wrixon and McMillan's value of $\beta(230\text{ GHz}) = 0.032$ Nepers/mm may not be a reliable predictor of 1.3 mm atmospheric opacity. Plambeck (1978) also measured atmospheric attenuation at 1 km altitude as a function of surface water vapor density near 225 GHz. He found a large scatter due to clouds but concluded that his data were consistent with $\beta(225\text{ GHz}) = 0.058$ Nepers/mm. In summary, the reliable experimental measurements of water vapor absorption are in reasonably good agreement with the values of $\beta(\nu)$ in Table I, which are probably accurate to within 25%. If systematic errors do exist in the computations, the experimental data tend to indicate actual attenuations at high frequencies smaller than the theoretical predictions in Table I.

Calculation of Atmospheric Transmission

The total zenith optical depth due to oxygen and water vapor absorptions under clear-sky conditions is simply the sum of Eq. (1) and Eq. (2) and is given by

$$\tau(\nu,h,W) = \tau_0 + \tau_W = \alpha(\nu)\exp(-h/h_0) + \beta(\nu)\cdot W \tag{3}$$

where $\tau =$ total atmospheric optical depth at zenith (Nepers). The normalized atmospheric transmissivity Γ is

$$\Gamma(\nu,h,W,A) = \exp[-\tau(\nu,h,W)\cdot A] \tag{4}$$

where the air mass A at elevation angle E above the horizon is given to a good approximation for $A < 10$ by

$$A = 1/\sin(E). \tag{5}$$

Thus the atmospheric transmissivity in clear weather is basically a function of four variables. The frequency- and altitude-dependencies can be evaluated for most frequencies of interest using the coefficients listed in Table I. The elevation-angle dependence is given by Eq. (5). Thus only the water vapor abundance W, which is a function of the weather, varies significantly with time. If an additional instrument were routinely available to determine W (for instance, a radiosonde or an infrared hygrometer), then Γ could be calculated from Eq. (3) and Eq. (4), and observed intensities of celestial sources could be corrected in real time for atmospheric attenuation. Alternatively, a source could be observed over a wide range of elevation angles, and the zenith optical depth τ could be determined directly from the change in measured intensity with air mass. However, this method is inconvenient, time consuming, and inaccurate during changing weather conditions.

THE CHOPPER-WHEEL METHOD

An "indirect" technique, known as the "chopper-wheel method," has been suggested by Penzias and Burrus (1973). In this method the ratio of the celestial source signal to the chopper calibration signal is measured, and this ratio is (within several approximations) independent of air mass and thus corrected for atmospheric attenuation. The calibration signal is simply the difference between an ambient temperature load and "blank sky" at the same elevation angle as the source. This scheme is "indirect" in that τ is not explicitly determined. As pointed out by Davis and Vanden Bout (1973), however, several difficulties arise in practice which can produce significant errors in the attenuation correction of the chopper-wheel method. First, the atmosphere is generally colder than the surface ambient temperature because of the negative temperature gradient in the troposphere. Thus a significant calibration signal will be measured even when the sky is optically thick. Second, when double-sideband mixer radiometers are used at millimeter wavelengths, the atmospheric attenuation can be significantly different in the two sidebands. Third, the power gains of the two receiver sidebands may not be equal. A complete analysis of the chopper-wheel method was used by Ulich and Haas (1976) to accurately calibrate the intensities of standard sources for several strong spectral lines near 3 mm wavelength. They showed that several other effects should also be considered. First, the spillover radiation is not at the same effective temperature as either the sky or the chopper wheel. Second, the microwave background radiation (Penzias 1972) affects the calibration signal by a small amount. Third, the Rayleigh-Jeans approximation to the Planck law may not be valid. The complete analysis showed that the equivalent calibration temperature T_C depends on nine variables which variously depend on frequency, precipitable water vapor, ambient temperature, air mass, and receiver sideband gain ratio. Ulich and Haas (1976) suggested a simpler method of directly evaluating the equivalent calibration temperature by simultaneously measuring the zenith optical depth during stable weather conditions along with the calibration signal antenna temperature as a function of air mass. It was shown that the resulting empirical equation for T_C depended weakly on τ and A (as expected). As a result, the usual chopper-wheel method suffers the disadvantage that T_C still depends on the optical depth of the atmosphere along the line of sight to the source. Ulich et al. (1980) have used the empirically-calibrated chopper-wheel method to make precise 3.5 mm continuum observations, and the repeatability of their measurements indicates that accurate corrections for atmospheric attenuation can be made. However, if T_C is not adjusted for changing T_{amb} and τ, errors of up to $\pm 10\%$ in the calculated intensity can occur.

The Cooled-Chopper Method

By reducing the effective chopper temperature below ambient temperature, the major error in the usual chopper-wheel method can be eliminated. If the chopper produced the same antenna temperature as seen by the telescope looking at the zenith with the atmosphere optically thick, the calibration signal would vary in the same manner as the source signal, and the atmospheric attenuation correction would then be exact. That is, both signals would vary in amplitude in the same fashion

with optical depth, and both would become zero for an optically thick atmosphere. Adopting the notation used by Ulich and Haas (1976), the corrected antenna temperature of the source $T_A{}^*$ is defined as

$$T_A{}^* \equiv (\Delta T_{source}/\Delta T_{cal}) \cdot T_C \qquad (6)$$

where ΔT_{source} = antenna temperature difference between the source direction and blank sky at the same elevation angle, and

$\Delta T_{cal} = T_{load} - T_{sky}$ = antenna temperature difference between the calibration load (chopper wheel) and blank sky at the same elevation angle as the source observation.

The equivalent calibration temperature T_C is defined as

$$T_C \equiv \Delta T_{cal}/[G_s \eta_l \exp(-\tau_s A)] \qquad (7)$$

where G_s = normalized receiver power gain at the signal (spectral line) frequency,

η_l = normalized telescope efficiency (including rearward spillover, blockage, and ohmic losses), and

τ_s = zenith optical depth at the signal frequency.

The load antenna temperature is given by

$$T_{load} = G_s J(\nu_s, T_L) + G_i J(\nu_i, T_L) \qquad (8)$$

where ν_s = signal frequency,

ν_i = image frequency,

T_L = load brightness temperature,

G_i = normalized receiver power gain at the image frequency ($G_s + G_i = 1$), and

$$J(\nu, T) = \frac{h\nu/k}{\exp(h\nu/kT) - 1} \qquad (9)$$

where $J(\nu, T)$ = effective radiation temperature of a blackbody source at frequency ν and physical temperature T,

h = Planck's constant, and

k = Boltzmann's constant (at 100 GHz $h\nu/k$ = 4.799 K).

The sky antenna temperature is given by

$$\begin{aligned}
T_{sky} = G_s &\langle (1-\eta_l)J(\nu_s, T_{shr}) + \eta_l \{J(\nu_s, T_m) \cdot \\
&[1-\exp(-\tau_s A)] + J(\nu_s, T_{bg})\exp(-\tau_s A)\}\rangle \\
+ G_i &\langle (1-\eta_l)J(\nu_i, T_{shr}) + \eta_l \{J(\nu_i, T_m) \cdot \\
&[1-\exp(-\tau_i A)] + J(\nu_i, T_{bg})\exp(-\tau_i A)\}\rangle
\end{aligned}$$
$$(10)$$

where T_{shr} = apparent brightness temperature of the received radiation which results from the rearward spillover, blockage, and ohmic losses,

T_m = mean atmospheric brightness temperature,

τ_i = zenith optical depth at the image frequency, and

T_{bg} = brightness temperature of the cosmic background radiation (2.8 K).

It has been assumed in Eq. (10) that η_l, T_{shr}, T_m, and T_{bg} are independent of frequency. If the intermediate frequency of the double-sideband receiver is small compared to the local oscillator frequency, $J(\nu_i, T) \simeq J(\nu_s, T)$,

$$T_{load} = J(\nu_s, T_L) \qquad (11)$$

and $T_{sky} = (1 - \eta_l)J(\nu_s, T_{shr}) + \eta_l J(\nu_s, T_m)$
$\qquad - \eta_l [J(\nu_s, T_m) - J(\nu_s, T_{bg})]$
$\qquad \cdot [G_s \exp(-\tau_s A) + G_i \exp(-\tau_i A)]. \quad (12)$

The calibration signal is given by

$\Delta T_{cal} = J(\nu_s, T_L) - J(\nu_s, T_{shr})$
$\qquad - \eta_l [J(\nu_s, T_m) - J(\nu_s, T_{shr})]$
$\qquad + \eta_l [J(\nu_s, T_m) - J(\nu_s, T_{bg})]$
$\qquad \cdot [G_s \exp(-\tau_s A) + G_i \exp(-\tau_i A)]. \quad (13)$

If the chopper brightness temperature T_L is reduced by cooling below ambient temperature so that

$$J(\nu_s, T_L) = \eta_l J(\nu_s, T_m) + (1 - \eta_l)J(\nu_s, T_{shr}), \quad (14)$$

then $\Delta T_{cal} = \eta_l [J(\nu_s, T_m) - J(\nu_s, T_{bg})]$
$\qquad \cdot [G_s \exp(-\tau_s A) + G_i \exp(-\tau_i A)] \quad (15)$

and from Eq. (7) we find that

$T_C = [J(\nu_s, T_m) - J(\nu_s, T_{bg})]$
$\qquad \cdot \{1 + (G_i / G_s) \exp[(\tau_s - \tau_i)A]\}. \quad (16)$

For a single-sideband receiver $G_i = 0$, and in this case $T_C = J(\nu_s, T_m) - J(\nu_s, T_{bg})$ which is independent of any receiver or telescope parameters. In fact the mean atmospheric temperature (at a constant ambient temperature T_{amb}) will vary slightly depending on the relative attenuations of oxygen and water vapor, since they have different scale heights and the atmosphere is not isothermal. Kutner (1978) has derived a two-layer atmospheric model to account for this effect. A good approximation can be made by simply using an average of the values of T_m for oxygen and water vapor emission weighted by their relative opacities. According to Kislyakov (1966).

$$T_m = \rho T_{amb} \qquad (17)$$

where $\rho = 0.956$ for nonresonant water vapor absorption, $\rho = 0.913$ for nonresonant oxygen absorption, and $\rho = 0.895$ for resonant oxygen absorption if $\tau A < 1$ (which is almost always satisfied in practice for millimeter-wavelength radio astronomy observations). Thus the effective value of T_m will vary slightly with frequency, typically being $0.94\, T_{amb}$ at 90 GHz on Kitt Peak and about $0.92\, T_{amb}$ at 115.3 GHz. Similar results can be derived from Kutner's analysis (Kutner, 1978), although his assumption that the atmospheric temperature is constant above 6 km altitude causes some error.

If $G_i = 0$ (corresponding to a single sideband receiver) or if $\tau_s = \tau_i$ (for a double-sideband receiver), Eq. (12) can be written as

$$T_{sky} = T_0 + T_1(1 - e^{-\tau A}) \qquad (18)$$

where

$$T_0 \equiv (1 - \eta_l)J(\nu_s, T_{shr}) + \eta_l J(\nu_s, T_{bg}) \qquad (19)$$

and

$$T_1 \equiv \eta_l [J(\nu_s, T_m) - J(\nu_s, T_{bg})]. \qquad (20)$$

T_0 is simply the antenna temperature when the atmosphere is transparent and T_1 is the antenna temperature when the atmosphere is optically thick and the ground is at absolute zero temperature. Note also from Eq. (14) that the antenna temperature of the cooled chopper should be

$$J(\nu_s, T_L) = T_0 + T_1. \qquad (21)$$

Thus if Eq. (18) can be evaluated by direct measurement, the chopper brightness temperature T_L can be empirically determined.

Empirical Calibration

In order to empirically determine the required temperature of the cooled chopper, measurements were made at 90 GHz with the NRAO 11-meter telescope on Kitt Peak using a double-sideband receiver and a feed designed by J. J. Gustincic. Curve 1 in Figure 1 is a plot of the continuum antenna temperature of the Moon as a function of air mass. A least-squares fit of the equation

$$\Delta T_{source} = \Delta T_A \exp(-\tau A) \qquad (22)$$

to Curve 1 indicated $\Delta T_A = 109 \pm 2$ K and $\tau =$

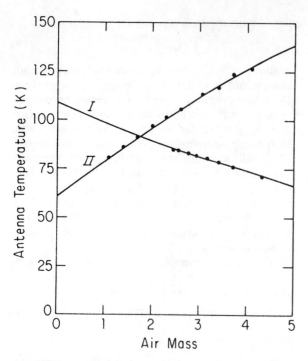

FIGURE 1 Antenna temperature measurements made on 4 May 1978 at 90 GHz with the NRAO 11-Meter Telescope on Kitt Peak, AZ using a feed designed by J. J. Gustincic: I - Moon antenna temperature (ΔT_{source}), and II - Blank sky antenna temperature (T_{sky}).

0.097 ± 0.004 Nepers. Simultaneous observations of T_{sky} were also made, and the results are plotted as Curve II in Figure 1. A least-squares fit of Eq. (18) to Curve II indicated that $T_0 = 58 \pm 2$ K and $T_1 = 207 \pm 5$ K. The standard deviation of the residuals of the fit was only 0.8 K, indicating that the relatively simple atmospheric model is accurate. At the time of these measurements, $T_{amb} = 284$ K. Assuming $\rho = 0.94$, $\eta_l = 0.78 \pm 0.02$ from Eq. (20) and $T_{shr} = 264 \pm 26$ K $= (0.93 \pm 0.09) T_{amb}$ from Eq. (19). From Eq. (21) $J(\nu_s, T_L) = T_0 + T_1 = 265 \pm 5$ K. For this system to accurately correct for atmospheric attenuation, the brightness temperature of the chopper should be regulated at 267 ± 5 K $= (0.94 \pm 0.02) T_{amb}$. This was done in practice by constructing a reflecting 45° chopper blade and a fixed carbon-impregnated foam load whose physical temperature was controlled with a thermoelectric cooler according to the ambient temperature. A linear electronic circuit employing negative feedback was built to control the temperature of the load according to the resistance of a thermistor located in free air near the radiometer box. The coefficient relating the load temperature

to the ambient temperature could be remotely adjusted, although the error caused by using the same setting at all frequencies is only a few percent. In addition, the ambient temperature was displayed so that T_C calculated from Eq. (16) could be set into the data acquisition computer. For continuum source observations from Kitt Peak in the 3 mm wavelength window, $T_C \simeq T_m - 3$ K $\simeq 0.94 T_{amb} - 3$ K.

A second set of observations was made on 3 September 1979 at 89 GHz with the NRAO 11-m telescope and a new feed system designed by the author. The ambient temperature was $T_{amb} = 299$ K and the cooled load was controlled at $T_L = 284$ K $(= 0.95 T_{amb})$. Curve I in Figure 2 is the antenna temperature of the cooled load (T_{load}). Curve II is the temperature of the antenna pointed at the center of the Moon's disk, and Curve III is the blank sky antenna temperature (T_{sky}) as a function of air mass. Curve IV is the differential lunar antenna temperature (ΔT_{source}), and a least-squares fit of Eq. (22) yielded $\Delta T_A = 194 \pm 1$ K and $\tau = 0.165 \pm 0.002$ Nepers. Using this value of τ, a least-squares fit of Eq. (18) to Curve III indicated that $T_0 = 28 \pm 1$ K and $T_1 = 254 \pm 2$ K (the RMS error of the residuals of the fit was only 0.3 K). Thus from Eq. (21) $T_L = 284 \pm 2$ K $= (0.95 \pm .01) T_{amb}$, which agrees with the results of the Gustincic feed tests. Using a calculated value of $\rho = 0.95$ (taking into account the higher-than-average water vapor attenuation on this date), $\eta_l = 0.90 \pm 0.01$ from Eq. (20) and $T_{shr} = 282 \pm 23$ K $= (0.94 \pm 0.08) T_{amb}$ from Eq. (19). With the improved feed system, the telescope has a significantly higher coupling efficiency to the sky and lower ground pickup. However, the required load temperature T_L is virtually the same for both feeds since $T_m \simeq T_{shr}$ and from Eq. (14) T_L is almost independent of η_l. Curve V in Figure 2 is the chopper calibration signal (ΔT_{cal}), and Curve VI is the corrected source antenna temperature T_A^* of the Moon calculated using Eq. (6). The effective calibration temperature from Eq. (16) was $T_C \simeq 0.95 T_{amb} - 3$ K $= 281$ K. The mean value of T_A^* (Curve VI) was 214 ± 1 K, and the peak-to-peak variation with air mass was about 2%. Fitting Eq. (22) to the T_A^* data produced an insignificant residual opacity (0.003 ± 0.005 Nepers), indicating complete correction for atmospheric attenuation within the noise level of the data. This confirms the accuracy of the cooled-chopper method. The coupling efficiency of the NRAO 11-m telescope to the Sun at 89 GHz with the improved feed system has also

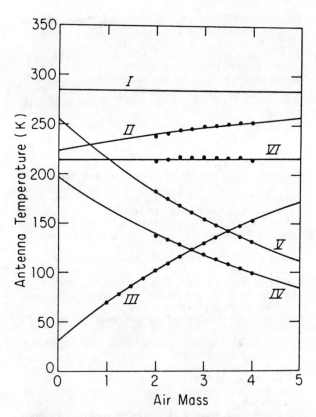

FIGURE 2 Antenna temperature measurements made on 3 September 1979 at 89 GHz with the NRAO 11-Meter Telescope on Kitt Peak, AZ using a feed designed by B. L. Ulich: I - Cooled chopping load antenna temperature (T_{load}), II - Antenna temperature pointed at center of Moon's disk [T_A (on source)], III - Blank sky antenna temperature (T_{sky}), IV - Moon antenna temperature (ΔT_{source} = Curve II - Curve III), V - Chopper calibration signal (ΔT_{cal} = Curve I - Curve III), and VI - Corrected source antenna temperature [T_A^* (Moon) = (Curve IV/Curve V)·T_C].

T_{amb} = 298 K. Assuming ρ = 0.93 (the weighted average of the nonresonant water vapor value and the resonant oxygen value), T_L = 289 ± 14 K = (0.97 ± 0.05) T_{amb}, η_l = 0.78 ± 0.05, and T_{sbr} = 328 ± 75 K = (1.10 ± 0.25) T_{amb}. A least-squares fit of the two-layer model equations of Kutner (1978) to these data did not result in a significantly improved fit over Eq. (18) (RMS of residuals = 1 K). This argues that the simpler atmospheric model of Eq. (18) is adequate for practical astronomical observations.

NRAO 11-M Telescope Calibration

The three empirical calibrations performed on the NRAO 11-m telescope have shown that the brightness temperature of the cooled chopper (T_L) must be equal to (0.95 ± 0.01) T_{amb} in order to accurately correct for atmospheric attenuation with the cooled-chopper method. The corresponding equivalent calibration temperature $T_C \simeq J(\nu, 0.94\ T_{amb}) - J(\nu, 2.8\ K)$ at all frequencies for either single-sideband spectral line observations or for double-sideband continuum observations. Under typical observing conditions the error incurred by neglecting the slight frequency dependencies of T_L and of T_C is less than a few percent and is negligibly small for most astronomical projects. For double-sideband spectral line observations, T_C will depend on the sideband gain ratio (G_i/G_s) and on the difference in the optical depths ($\tau_s - \tau_i$)A at the two received frequencies. In this case, Eq. (16) must be used to calculate T_C.

ABSOLUTE CALIBRATION PROCEDURE

A millimeter-wavelength radio telescope may be used to conveniently make absolutely calibrated intensity measurements corrected for antenna and atmospheric losses. During stable weather conditions, simultaneous antenna temperature measurements of a strong source and of blank sky are made as a function of air mass. Ulich et al. (1980) give a description of hot and cold load absolute antenna temperature calibration procedure. Eq. (22) is fit to the source data to derive τ. Then Eq. (18) is fit to the absolutely calibrated blank sky data to determine $T_0 + T_1$. The cooled-chopper brightness temperature T_L is next calculated from Eq. (21). This completes the empirical calibration process, which need be done carefully only once. Thereafter, the corrected antenna temperature of a source is calculated using Eq. (6), where T_C is

been measured to be 0.79 ± 0.05, assuming the solar brightness temperature given by Ulich et al. (1980). Thus the forward beam coupling efficiency η_f = (0.79 ± 0.05)/η_l = 0.88 ±0.06 for the Sun (or for the Moon), which agrees with the value of 0.89 previously calculated at 115 GHz (Ulich and Haas, 1976). The central brightness temperature of the Moon at 89 GHz on 3 September 1979 is therefore (214 ± 1 K)/(0.88 ± 0.06) = 243 ± 17 K (at a fraction of lunation period = 0.89), which agrees with the 97 GHz lunar brightness temperature at the same phase angle (Ulich et al., 1974).

Additional measurements at 115.3 GHz on 31 August 1979 with the improved feed system on the 11-m telescope indicated that T_0 = 73 ± 3 K and T_1 = 214 ± 14 K with τ = 0.43 ± 0.08 and derived from Eq. (16) and Eq. (17). The cooled chopper is maintained at a constant fraction (~0.95) of the ambient temperature by a feedback control system, and the proper value of T_C is calculated by the data acquisition computer which monitors the ambient temperature. Such a system will conveniently produce accurate absolute intensity measurements corrected for atmospheric attenuation without the relatively large errors which occur with an uncooled chopper.

ACKNOWLEDGMENT

The National Radio Astronomy Observatory is operated by Associated Universities, Inc., under contract with the National Science Foundation.

REFERENCES

Altshuler, E. E., Falcone, V. J., Jr., and Wulfsberg, K. N., 1968, *IEEE Spectrum*, **5**, 83.

Bastin, J. A., 1966, *Infrared Phys.*, **6**, 209.

Davis, J. H., and Vanden Bout, P., 1973, *Astrophys. Lett.*, **15**, 43.

Emery, R., 1972, *Infrared Phys.*, **12**, 65.

Fogarty, W. G., 1975, *IEEE Trans. Antennas Propag.*, *AP*-**23**, 441.

Gaut, N. E., and Reifenstein, E. C., 1971, NASA Contractor Report CR-61348, Environmental Research and Technology, Inc., Waltham, Mass.

Gibbins, C. J., Gordon-Smith, A. C., and Croom, D. L., 1975. *Planet. Space Sci.*, **23**, 61.

Goldsmith, P. F., Plambeck, R. L., and Chiao, R. Y., 1974, *IEEE Trans. Microwave Theory Tech.*, **22**, 1115.

Grody, N. C., 1976, *IEEE Trans. Antennas Propag.*, *AP*-**24**, 155.

Johnson, W. A., Tsutomu, T. M., and Shimabukuro, F. I., 1970, *IEEE Trans. Antennas Propag.*, *AP*-**18**, 512.

Kislyakov, A. G., 1966, *Izv. Vyssch. Uchebn. Zaved. Radiofiz.* [*Sov. Radiophys.*], **9**, 451.

Kutner, M. L., 1978, *Astrophys. Lett.*, **19**, 81.

Lo, L., Fannin, B. M., and Straiton, A. W., 1975, *IEEE Trans. Antennas Propag.*, *AP*-**23**, 782.

Mather, J. C., Werner, M. W., and Richards, P. O., 1971, *Astrophys. J. Lett.*, **170**, L59.

Meeks, M. L., and Lilley, A. E., 1963, *J. Geophys. Res.*, **68**, 1683.

Penzias, A. A., 1972, *Cosmology, Fusion, and Other Matters* (Boulder: Colorado Associated University Press), 29.

Penzias, A. A., and Burrus, C. A., 1973, *Ann. Rev. Astron. Astrophys.*, **11**, 51.

Plambeck, R. L., 1978, *IEEE Trans. Antennas Propag.*, *AP*-**26**, 737.

Reber, E. E., Mitchell, R. L., and Carter, C. J., 1970, *IEEE Trans. Antennas Propag.*, *AP*-**18**, 472.

Reber, E. E., 1972, *J. Geophys. Res.*, **77**, 3831.

Reber, E. E., and Swope, J. R., 1972, *J. Appl. Meteor.*, **11**, 1322.

Rosenkranz, P. W., 1975, *IEEE Trans. Antennas Propag.*, *AP*-**23**, 498.

Schulze, A. E., and Tolbert, C. W., 1963, *Nature*, **200**, 747.

Shimabukuro, F. I., and Epstein, E. E., 1970, *IEEE Trans. Antennas Propag.*, *AP*-**18**, 485.

Straiton, A. W., 1975, *IEEE Trans. Antennas Propag.*, *AP*-**23**, 595.

Tolbert, C. W., Krause, L. C., and Straiton, A. W., 1964, *J. Geophys. Res.*, **69**, 1349.

Ulaby, F. T., and Straiton, A. W., 1970, *IEEE Trans. Antennas Propag.*, *AP*-**18**, 479.

Ulich, B. L., Cogdell, J. R., Davis, J. H., and Calvert, T. A., 1974, *Moon*, **10**, 163.

Ulich, B. L., and Haas, R. W., 1976, *Astrophys. J. (Suppl.)*, **30**, 247.

Ulich, B. L., Davis, J. H., Rhodes, P. J., and Hollis, J. M., 1980, *IEEE Trans. Antennas Propag.*, to be published.

Waters, J. W., 1976, *Methods of Experimental Physics: Part B: Radio Telescopes*, edited by M. L. Meeks (New York: Academic Press), 142.

Wrixon, G. T., and McMillan, R. W., 1978, *IEEE Trans. Microwave Theory Tech.*, *MTT*-**26**, 434.

Wulfsberg, K. N., 1964, *Proc. IEEE*, **52**, 321.

Part VIII
Very Long Baseline Interferometry

VERY long baseline interferometry (VLBI) is the extension of interferometric techniques to the largest baselines obtainable. As discussed in the Historical Overview and the introduction to Part II, the search for ever-higher angular resolution soon reached the point that bringing the IF signals together by cable or radio link was not practical. In the late 1960s, several groups developed the technique of recording the IF outputs of the receivers on different radio telescopes on magnetic tape, and subsequently bringing the two tapes together and replaying them simultaneously to correlate the data. The two major equipment limitations of this approach are the bandwidth which can be recorded, and the requirement for a phase reference at each antenna, which must be of sufficient quality to allow the relative phase of the recorded data streams to be recovered uncorrupted.

The principles underlying VLBI are not different from those of more conventional connected-element interferometry, but the recording and post-observation correlation techniques are distinctive. The obstacles facing VLBI, especially the problem of phase errors arising from imperfect references and atmospheric perturbations, encouraged the development of data processing techniques for correcting these problems. Notable among these is the use of *phase closure* (also referred to as *hybrid mapping* and *self-calibration*), which can be employed when there are at least three antennas involved. This technique was introduced by R. Jennison in 1958 (included as the first paper in Part IX). Self-calibration and image restoration techniques are described in Part IX, and in the references given at the end of the introduction to that part. The use of these techniques is not restricted to VLBI; they have come to play a major role in enhancing the quality of data obtained with such connected arrays as MERLIN and the VLA.

The first paper in this part, by Cohen, gives an overview of VLBI. Other excellent treatments can be found in articles by Moran (1976a, 1976b), and in a text by Thompson *et al.* (1986). During the development of VLBI a gradual shift has occurred to the use of multi-telescope systems to improve the range of baselines available. The very large physical spacing of telescopes required to achieve the maximum resolution in VLBI necessarily limits the amount of time a source can be tracked by elements of the system. Consequently, a range of physical baselines is required to allow an accurate Fourier transform map of the source brightness distribution to be made. The very long baseline array (VLBA) networks which are presently in operation in the United States and Europe allow a number of radio telescopes to be scheduled for VLBI observations on a part-time basis. The VLBA Radio Telescope presently under construction in the United States will consist of ten 25-m telescopes dedicated to VLBI. This instrument will vastly increase the quality, frequency range, and quantity of extremely high angular resolution data that can be obtained.

The initial observations were, and the majority of VLBI observations still are, of continuum sources. For these, an increase in bandwidth is directly reflected in an improved signal-to-noise ratio. VLBI observations of spectral lines have a special value in that the radial motions of sources on an extremely small angular scale can be studied by means of Doppler shifts. Interstellar maser sources are particularly useful since they are extremely compact and bright. Study of their radial and proper motions (movement across the plane of the sky) determined over a time scale of a few years has resulted in a new view of expulsion of the material surrounding a newly formed star as well as the measurement of distances to astronomical objects which cannot be seen at all at visible wavelengths (see, for example, Genzel *et al.*, 1981).

The early VLBI experiments were limited to narrow bandwidths. The experiment of Brown *et al.* (1968), which was carried out in 1967, used a bandwidth of 2 kHz at a frequency of 18 MHz to observe variable radiation from Jupiter. The original American VLBI system used for extragalactic observations had a bandwidth of 360 kHz and a single tape could hold only 3 minutes of data. The Mark II System, described by Clark in the second paper of this part, has a bandwidth of 2 MHz and employs a commercial tape recorder widely used for television. The Mark III System, described by Rogers *et al.* in the third paper, allows recording with a 56-MHz total bandwidth, and 15 minutes of data can be recorded on a single pass of the tape. Refinements are in progress which should allow the data storage to be increased by an order of magnitude.

The final paper in this part, by Rogers and Moran, is concerned with the quality of the frequency standards used for VLBI. The effect of phase perturbations from either the standard or the atmosphere is to reduce the time interval over which observations at two sites can be coherently combined. Rogers and Moran analyze both effects. Although the atmosphere is concluded to be the more important limitation, recent transcontinental experiments at 86 GHz have indicated that VLBI at this high frequency is definitely possible (cf. Readhead *et al.*, 1983).

REFERENCES AND BIBLIOGRAPHY

1968: Brown, G. W., T. D. Carr, and W. F. Block, "Long baseline interferometry of S-bursts from Jupiter," *Astrophys. Lett.*, vol. 1, pp. 89–94, 1968.

1976: Moran, J. M., "Very long baseline interferometer systems," Chapter 5.3 in *Methods of Experimental Physics: Vol. 12 Astrophysics, Part B: Radio Telescopes*, M. L. Meeks, Ed. New York, NY: Academic Press, 1976, pp. 174–197. (a)

Moran, J. M., "Very long baseline interferometric observations and data reduction," Chapter 5.5 in *Methods of Experimental Physics: Vol. 12 Astrophysics, Part B: Radio Telescopes*, M. L.

Meeks, Ed.　New York, NY: Academic Press, 1976, pp. 228–260.
(b)

Vessot, R. F. C., "Frequency and time standards," Chapter 5.4 in *Methods of Experimental Physics: Vol. 12 Astrophysics, Part B: Radio Telescopes*, M. L. Meeks, Ed.　New York, NY: Academic Press, 1976, pp. 198–227.

1981: Genzel, R., M. J. Reid, J. M. Moran, and D. Downes, "Proper motions and distances of H_2O maser sources. I. The outflow in Orion-KL," *Astrophys. J.*, vol. 244, pp. 884–902, 1981.

1982: "A program for the very long baseline array radio telescope," National Radio Astronomy Observatory Report, May 1982.

1983: Readhead, A. C. S., C. R. Masson, A. T. Moffet, T. J. Pearson, G. A. Seielstad, D. P. Woody, D. C. Backer, R. L. Plambeck, W. J. Welch, M. C. H. Wright, A. E. E. Rogers, J. C. Webber, I. I. Shapiro, J. M. Moran, P. F. Goldsmith, C. R. Predmore, L. Bååth, and B. Rönnäng, "Very long baseline interferometry at a wavelength of 3.4 mm," *Nature,* vol. 303, pp. 504–506, 1983.

1984: Ewing, M. S., "VLBI: Once and future systems," Paper 2.4 in *Proc. Int. Symp. on Indirect Imaging*, Sydney, Australia, 1983, J. A. Roberts, Ed.　Cambridge: Cambridge University Press, 1984, pp. 41–51.

1986: Thompson, A. R., J. M. Moran, and G. W. Swenson, Jr., *Interferometry and Synthesis in Radio Astronomy*.　New York, NY: John Wiley and Sons, 1986.

Introduction to Very-Long-Baseline Interferometry

MARSHALL H. COHEN

Abstract—Long-baseline interferometry achieves high angular resolution by using two or more widely separated radio telescopes and recording video signals on magnetic tapes, which are later brought together and cross-correlated. This paper contains discussions of the coherence and timing requirements and of calibration procedures. Applications to measuring brightness distributions and to spectroscopy are reviewed briefly. Some pertinent phenomena connected with radio-wave scattering in irregular media are discussed.

Introduction

IN THE EARLY and middle 1960's radio astronomers developed an appreciation of *compact radio sources*, usually associated with quasars or galactic nuclei. Their compactness is measured by angular size which is well under 1″ and in some cases less than 10^{-3}″. These sources are small but by no means weak, and their flux density is comparable to that of the extended sources, which may have 10^{10} times as much solid angle! Thus they are enormously brilliant and contain a very high energy density. They are mysterious fascinating objects and remain the subject of intense study.

A decade ago, angular size measurements were limited to about 0.1″, by the available techniques, and the much smaller size of the compact sources was being inferred on theoretical grounds. These grounds included the peaked nature of the spectrum and the variability of the sources, and were reasonably firm. Thus there was strong interest in making measurements at angular resolutions of 10^{-3}″ or better, to check the theoretical predictions and to see what the sources actually looked like on this scale.

Manuscript received February 26, 1973. Owens Valley Radio Observatory is supported by ONR under Contract N 00014-67-A00094-0019, and by NSF under Grants GP25225 and GP19400

The author is with the Owens Valley Radio Observatory and with the Department of Astronomy, California Institute of Technology, Pasadena, Calif. 91109.

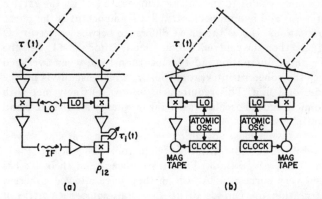

Fig. 1. (a) Conventional two-element interferometer with a radio link. (b) VLB system with independent atomic standard oscillators.

At that time high-resolution techniques included lunar occultations, interplanetary scintillations, and interferometry. The first two were limited by the scale of the diffraction patterns to about 0″.1 but interferometry was limited only by practical considerations. Fig. 1(a) shows the conventional radio-link interferometer with the LO and IF signals transmitted by radio. In 1965 such interferometers were being used at baselines up to 120 km. However, thousands of kilometers were really required, and it was decided to simplify the systems and eliminate the real-time links, as in Fig. 1(b). Two developments were necessary for this: a means of maintaining coherent independent local oscillators at the two telescopes, and a means of putting the video signals on magnetic tape for later processing. These developments were possible, and very-long-baseline (VLB) observations began in 1967. Baselines up to 10^4 km and wavelengths down to 1.35 cm have now been used. However, the "later processing" has turned out to be an impediment to large-scale research, and the alternative of

Reprinted from *Proc. IEEE*, vol. 61, no. 9, pp. 1192–1197, Sept. 1973.

real-time cross correlation, via some appropriate link, is beginning to seem attractive.

The history and applications of VLB systems have been discussed in various early papers [1]–[8] and in several review articles [9]–[13]. Details of the National Radio Astronomy Observatory (NRAO) Mark II digital system are in the article by Clark [14] in this issue. Other pertinent articles in this issue include applications at meter wavelengths by Clark and Erickson [15], to continuum sources by Kellermann [16], to spectral-line sources by Welch [17], and to geodesy and fundamental astronomy by Counselman [18].

INDEPENDENT LOCAL OSCILLATORS

The oscillator stability required for VLB work depends on the nature of the experiments. There must at least be coherence for the minimum integration time set by the signal-to-noise ratio. But for some purposes—especially where fringe phases have to be compared—long-term stability is required. Atmospheric phase fluctuations are evidently involved in these matters also, but discussion of them is to be deferred until later. Atmospheric effects are important mainly at $\lambda > 1$ m, while oscillator effects are important only at short wavelengths.

A. Coherence

The two oscillators will have a small frequency offset. This merely adds to the natural fringe rate (set by the earth's rotation) and is unimportant. What is important is the phase fluctuations. If ϕ is the phase difference between the two LO signals, then the coherence requirement is $\langle \phi^2 \rangle_\tau^{1/2} < 1$; i.e., the rms phase fluctuation must be less than 1 rad when averaged over the coherent integration time τ, which usually is on the order of 1 min. This requirement is now routinely met with atomic standard oscillators driving frequency synthesizers and multipliers.

Hydrogen maser oscillators generate the best sine waves. Several large observatories (Goldstone, Algonquin Park, Haystack, Green Bank) have these masers and their use has been very successful. At 13 cm they have allowed coherent integration for periods of up to a half-hour, with little or no loss in signal-to-noise ratio. Rubidium oscillators also have been used to generate LO signals at microwave frequencies. Their coherence time is on the order of 20 s at 3 cm, but is many minutes at 6 cm. A good crystal oscillator has smaller short-term fluctuations than the rubidium, but cannot be used at microwave frequencies because its frequency drifts rapidly (large long-term phase fluctuations). It is possible, however, to override the second characteristic by phase-locking a high-quality crystal oscillator to a rubidium oscillator with a time constant of about a minute. This gives a coherence time of 1 min at $\lambda = 3$ cm. The superiority of this combination has been directly demonstrated in VLB observations between the NRAO in Green Bank, W. Va., and the Owens Valley Radio Observatory (OVRO) in California. A maser was at NRAO, and at OVRO there was a Hewlett Packard 5065A Rubidium Oscillator, either running alone or driving a Sulzer 2.5-C crystal oscillator. The latter combination was strikingly better than the rubidium alone.

Cesium oscillators form the fundamental time and frequency standards, but they have traditionally been noisier than rubidium oscillators. The latest generation of cesium oscillators is said to be much improved, but they have not yet been tested in VLB systems.

Local oscillator signals are derived from the atomic oscillator outputs by various combinations of synthesizers and multipliers. These can be very stable, and contribute negligible phase noise. It must be noted, though, that some synthesizers and multipliers are very much better than others for this purpose, and the casual attachment of an LO chain to a hydrogen maser oscillator does not guarantee a stable signal.

B. Long-Term Phase Stability

A transcontinental interferometer has d/λ ranging from 10^6 to 10^8, depending on wavelength, and the corresponding angular resolutions are from 0.1 to 0″.001. VLB systems thus have the potential for measuring source positions to very high accuracy.

The customary way to measure a position with an interferometer is to measure the phase of the interference fringes with respect to those of a standard source. In a VLB system this will fail if the relative phase between the two local oscillators drifts substantially, and unpredictably, between the two measurements. The comparison interval, then, cannot be more than the coherence time discussed in the previous section.

The simplest scheme switches the antennas back and forth between a standard source and the source under test. This has been done at 3.8 cm with hydrogen maser oscillators between Goldstone and Haystack. The time scale for phase stability appears to be a half-hour or more. A more complicated scheme uses two antennas at each terminal, one tracking the standard source and the other the source under test. It would be possible to record both signals all the time, but in practice they are recorded alternately, in 1-s blocks. The phase of the two sets of fringes can then be compared almost continuously, and the need for long-term phase stability is obviated. Such "four-antenna" observations have been done between NRAO and OVRO, also between NRAO and the Haystack Observatory.

A yet more complicated scheme uses three antennas at each terminal, and the redundant phase information is used to derive atmospheric corrections. One successful "six-antenna" experiment has been done between OVRO and NRAO at 21 cm. Its object is to make high-precision measurements of the positions and motions of pulsars.

Position measurements using these techniques are discussed in detail by Counselman [18].

TIMING REQUIREMENTS

In the VLB unit [Fig. 1(b)] video signals are recorded on magnetic tapes which are processed later to find the cross correlation function. In the NRAO Mark I system, computer-compatible tape is used and the processing is done on a general-purpose computer. In the Canadian system and the NRAO Mark II system, recording is on video tape. Special-purpose processors are used to find the correlation function, which is then analyzed in a general-purpose computer.

In a conventional interferometer [Fig. 1(a)], the signals are brought into time synchronism by stepping the compensation delay line so that $|\tau_i(t) - \tau(t)| \ll B^{-1}$, where B is the bandwidth. In the VLB unit this function is easily performed in the processor. However, the VLB interferometer lacks initial synchronization because tape starting is controlled by the two independent clocks. No attempt is made to keep these clocks synchronized to B^{-1} seconds; rather, in the processor, many time delays are tried until fringes are found. This is occasionally a disheartening or even unsuccessful process, but most recent experience has been that the tape indices are known *a priori* to 10 or 20 μs, and the required amount of searching is usually small.

It has become easy to maintain time at an observatory to the accuracy required for VLB experiments. A rubidium oscillator keeps time to one part in 10^{12}, or a few microseconds per month. The clocks are set against any of the network of cesium standard time stations which are maintained around the world, or by Loran-C where it is available. Even the sky-wave Loran signals are adequate, for they can easily give 10- or 15-μs accuracy once a receiving system is calibrated.

MEASUREMENT OF FRINGE AMPLITUDE

When the cross-correlation function is formed, it contains the interference fringes which form a sine wave of frequency $v d\tau/dt$ (to first order). This natural rate can be in the kilohertz region, and sometimes an oscillator offset is introduced for convenience, to make the net fringe rate close to zero. In most VLB experiments, only the fringe amplitude is measured, because that is easier than measuring phase and because the amplitude alone still gives most of the available information on brightness distributions. In this section we discuss measurement of fringe amplitude.

A. Calibration and Signal-to-Noise Ratio

In any interferometer system the fringe-amplitude scale must be calibrated in flux units, and this is usually done by observing standard sources, i.e., unresolved (point) sources of known strength. In VLB systems this must often be done in several steps, because there may be no sources which are known *a priori* to be unresolved. A second complication is that digital VLB systems use 1-b techniques and direct amplitude information is lost. Fortunately, it is easily recovered through the correlation coefficient. The 1-b correlation coefficient ρ is related to the cross-correlation coefficient between the two IF signals ρ_{12} by the Van Vleck formula [19] $\rho = (2/\pi) \arcsin \rho_{12}$, and, since $\rho_{12} \ll 1$,

$$\rho = \frac{2}{\pi} \rho_{12}. \tag{1}$$

ρ_{12} is related to the antenna and system temperatures by

$$\rho_{12} = \gamma \left(\frac{T_{a1}}{T_{s1}}\right)^{1/2} \left(\frac{T_{a2}}{T_{s2}}\right)^{1/2} \tag{2}$$

where γ is the unknown fringe visibility.

The optimum measurement technique consists of making careful measurement of the antenna temperature of the source T_a and of the total system temperature T_s for each servation. The ratios on the right-hand side of (2) are then determined. In many cases, unfortunately, the antenna temperature cannot be measured with any accuracy because the source is weak. In these cases the usual procedure is to assume that the antenna pointing is perfect and calculate T_a from the known total flux and the antenna gain. In principle γ can now be determined. In practice, however, (1) is wrong by an imperfectly known number b, whose value is near 1.5:

$$\rho = \frac{2}{\pi b} \rho_{12} \tag{3}$$

TABLE I
TELESCOPE PARAMETERS

Location	Diameter (m)	$(T_s/T_0)^{\frac{1}{2}}$
Arecibo, Puerto Rico	305	3
Goldstone, California	64	6
Algonquin Park, Ontario	46	20
Onsala, Sweden	26	20

because the calculation for ρ uses several simplifications, such as time shifting by integral bit intervals rather than smaller values [14]. Any phase instabilities, from the oscillators or the atmosphere, also increase b.

The relation between ρ and ρ_{12} is usually unknown to 10 percent or more, and the (square-root) ratios of antenna to system temperature may also be unknown to 10 percent or more. These errors should be largely systematic and constant in any series of observations, whereas noise errors are often much smaller. It is customary, therefore, to attempt to eliminate the large systematic errors by scaling all values of ρ_{12} by a constant factor to obtain agreement with some *a priori* values of γ. A few sources, including PKS 0106+01, OJ 287, and OR 103, have been known through experience to be smaller than others at centimeter wavelengths, and the scaling may be set to make these have unit fringe visibility. (However, this is dangerous since with sufficient resolution any source will have $\gamma < 1$. Moreover, these compact sources are time varying at centimeter and decimeter wavelengths and their flux may not be known very well.) The final scaling error may still be as great as ±10 percent, but the relative fringe amplitudes between sources in any one series are more accurate.

With the Mark II digital system, the bandwidth B is usually 2 MHz and the typical integration time τ is 30 s. The noise level in measuring ρ_{12} can be shown to be

$$\Delta\rho_{12} \simeq \frac{\pi b}{\sqrt{8B\tau}} \simeq 2 \times 10^{-4} \tag{4}$$

and the minimum detectable value is approximately $5\Delta\rho_{12} \simeq 10^{-3}$. To see what this means in terms of flux density, rewrite (2) as

$$S_c = \gamma S = \rho_{12} \left(\frac{T_{s1}}{T_{01}}\right)^{1/2} \left(\frac{T_{s2}}{T_{02}}\right)^{1/2} \tag{5}$$

where S is the flux density of the source, S_c is the "correlated flux density," and T_{01} and T_{02} are "degrees of antenna temperature per flux unit" for the two antennas. The ratio $(T_s/T_0)^{1/2}$ is a figure of merit, and some values for representative telescopes are in Table I. In general, this ratio is different for different wavelengths, and an optimum value is shown.

In the best cases the noise level is about 0.1 fu,[1] but for some observations it has been as high as 1 fu. Final accuracy, however, is never as good as ±0.1 fu because systematic and calibration errors dominate the error budget except in very weak cases.

Note that a low system temperature and a large effective area have equal weight in making a good figure of merit

---

[1] fu stands for flux unit; 1 fu = 10^{-26} Wm^{-2} Hz^{-1}.

Goldstone and Onsala are relatively better than the others because they use maser amplifiers.

B. Saturation

The figure of merit $(T_s/T_0)^{1/2}$ is shown in Table I for weak sources, but when the source is strong enough to contribute appreciably to T_s, we must write $T_s/T_a = (T_a + T_n)/T_a$, where T_a is contributed by the source and T_n is from the receiver, the ground, etc. When T_a is bigger than T_n, the ratio approaches unity and this side of the interferometer becomes saturated. No further improvement in sensitivity can be obtained by increasing the size of the antenna or by decreasing the receiver temperature; the signal is already due to the source itself. Saturation (defined when $T_a = T_n$) occurs at Arecibo at 430 MHz when $S > 10$ fu, which is the case for a number of compact sources. Three sources, 3C 84, 3C 273, and 3C 274, are strong enough to saturate Goldstone at centimeter wavelengths, and will also saturate the new 100-m telescope at Bonn.

When one end of the interferometer is saturated, the system sensitivity is set by the other end. In this case the common remark that the effective area is the geometric mean is misleading. In the limit of strong saturation, (2) becomes

$$\rho_{12} \simeq \gamma(T_{a2}/T_{s2})^{1/2}. \tag{6}$$

BRIGHTNESS DISTRIBUTIONS

The simplest parameter describing the brightness distribution of a source is the effective diameter, and in many early VLB papers just this number, or an upper limit, was reported. In most cases, however, "diameter" was merely the diameter of a circular Gaussian which had the same fringe visibility as the real source, at one spacing and one position angle. Thus it was no more than a one-dimensional approximation to an "angular scale"; it often was meaningful and useful but also could be misleading. Better approximations to the brightness distribution are now being produced, and their sophistication is increasing.

It is well known that the output of a correlation interferometer is a component of $\Gamma(u, v)$, the Fourier transform of the brightness distribution of the source. Furthermore, the function Γ can be synthesized by observing at different times of day and by using different spacings [20]. However, the amount of synthesis done in the typical VLB experiment is very small. At best, the source is tracked while it is above the horizon, and this gives the amplitude of the visibility function $|\Gamma(u, v)| \equiv \gamma(u, v)$ along the diurnal track, an ellipse in the (u, v) plane. (Examples of diurnal tracks are in Figs. 2 and 3.) This information is grossly insufficient for any Fourier inversion, and all investigators resort to model fitting to obtain an approximation to the brightness distribution. Usually there is little a priori knowledge of the source shape, and so the simplest model which fits the data is used. The simplest models contain one parameter plus a constraint on total flux; these models are circular and are usually Gaussian but can be uniform or ringlike, according to one's preferences. In most cases where the source is well resolved, there is a notable lack of circular symmetry and more complicated models must be used. The next stage is a two-parameter model, usually an equal point double with separation and position angle to be determined by the data. In a few cases this has been remarkably successful, the best and most interesting case being 3C 279 observed with the Goldstack interferometer at $\lambda = 3.8$ cm [16]. There is a strong attraction to double sources because it

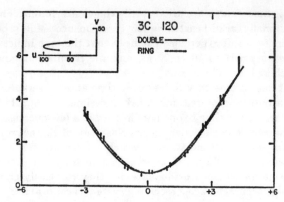

Fig. 2. Fringe amplitude data from the Goldstone–Haystack interferometer (Goldstack) at $\lambda = 3.8$ cm, Feb. 28, 1971. The source is the Seyfert Galaxy 3C 120. Abscissa is the hour angle from the interferometer meridian, and ordinate is the fringe amplitude measured as the correlated flux density. The inset shows the track of the measurements on the (u, v) plane, with u and v measured in millions of wavelengths. The two curves through the data (the vertical bars) show the expected visibility curves from two different models, a double and a ring.

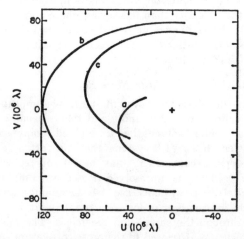

Fig. 3. Diurnal tracks on the (u, v) plane for the FOG array at 2.8 cm, $\delta = 40°$. a, OVRO–Fort Davis, b, OVRO–NRAO, c, Fort Davis–NRAO.

is known that many extended radio sources are double on a large angular scale. Often the two components are symmetrically spaced across a galaxy and it is tempting to think that the galaxy has ejected the components. Can the very small doubles simply be young versions of the large mature radio sources? The answer to this is probably no, on energetic grounds, but speculation of some connection persists and colors the nature of the models which are used. (This is discussed further by Kellermann [16].)

When the equal point double does not fit the measured data, more complicated models are tried. Clearly the degree of complication can be increased until a satisfactory fit is obtained, but along the way faith in the approximation to reality may be lost, if indeed there ever was any. In many cases there are ambiguities in the model fitting. For example, Fig. 2 shows data for 3C 120 and two possible models which both fit very well, a double and a circular ring [21]. From those limited measurements it is not possible to choose between the models.

Model fitting as in Fig. 2 is inadequate to give a clear idea of the shapes of sources; in some cases no reasonable models exist and in others they are ambiguous, but in all they are unsure. To get a better view of these sources, it is necessary to sample the (u, v) plane more generously. A start in this direction is being made through several series of three-station observations. One of these is on the "FOG" array, consisting of telescopes at Fort Davis, Tex.; Owens Valley, Calif.; and Green Bank, W. Va. Fig. 3 shows tracks for the three baselines at 2.8 cm. These are for a source at declination $+40°$, where there are several important compact sources. The (u, v) coverage for the three baselines is evidently very much better than for any one alone. Regular observations with this array are planned, to study sources which have variable brightness distributions.

One can contemplate larger arrays using existing telescopes to get more coverage in the u, (v) plane and thus more angular resolution and better models. Already several four-station experiments have been done using three telescopes in the United States and one in Sweden [11]. There are more than a dozen digital Mark II recorders at observatories in North America and Europe, and one can imagine a combined assault on a few complex sources. A serious objection to this is the processing time per baseline, which is rather more than the observing time, and thus is close to half a year full-time for a two-day experiment with twelve stations. It is probable that such complicated multistation experiments will have to await the development of a satellite-linked system which could make real-time correlation of many baselines possible.

Multistation observations are also important for astrometric and geodetic measurements. If three stations are used rather than two, various closure errors can be determined, and positions found more accurately.

VLB SPECTROSCOPY

Two molecular species, OH and H₂O, show maser action in interstellar clouds and emit strong radiation in narrow spectrum lines from compact regions. The high resolution study of these lines is an active area of VLB research, and is discussed by Welch [17].

Recording on tape for line work is basically the same as for continuum work, but the processing is more complicated because, in essence, maps at many frequencies are desired, rather than just one map. In principle the analysis is made by measuring the cross-correlation function $\rho_{12}(u, v, \tau)$, including time lag. Then the Fourier transform gives the brightness distribution as a function of frequency [22]. In practice, of course, u and v are weakly sampled and only crude models are generated. In some cases it has turned out that a molecular source consists of a number of isolated spots of emission, each at a different frequency. A map of the overall source can then be generated even if the isolated spots have unknown shape, because each spot turns up at a different frequency and its position can be determined unambiguously [23], [24].

The H₂O line is at 1.3 cm and the observations between Simeis, Crimea, and Westford, Mass., are the highest resolution interferometric measurements ever made, with $d/\lambda = 6 \times 10^8$. Preliminary results from these observations show that all the H₂O sources are at least partially resolved [25].

In a few cases a molecular cloud produces both OH and H₂O radiation. In these cases the diameters seem to be roughly proportional to the square of the wavelength. This suggests that the measured sizes are not intrinsic but rather are set by scattering in the interstellar medium. This conclusion remains questionable, however, because the size is an order of magnitude larger than expected on the basis of pulsar scintillations.

IRREGULAR PROPAGATION EFFECTS

Radio-wave scattering in interplanetary space and in interstellar space can have a profound influence on VLB measurements because the instrumental resolution may be comparable with the scattering angle. The phenomena are intimately connected with intensity scintillations, and the pertinent parameters will be reviewed first for them. The subject is treated in detail by Cronyn and Cohen [26].

A. Intensity Fluctuations

Three angles are important for this discussion: the scattering angle θ_s; the cutoff angle $\theta_c \sim \lambda/(z\theta_s)$, where z is the distance to the scattering region; and the instrinsic angle of the source θ. When $\theta < \theta_c$ we effectively have a point source, but when $\theta > \theta_c$ scintillations are quenched. In the first case we may imagine that the various rays within the scattering cone (as received by an antenna) are all coherent, so that they can interfere and produce the random diffraction pattern which drifts past the antenna. In the second case, however, the rays are incoherent and there is no interference. For interplanetary scintillations (IPS) $\theta_c \sim 0''.5$, and for interstellar scintillations (ISS) $\theta_c \sim 10^{-6}''$; also, for strong scattering, θ_c is inversely proportional to wavelength.

Scintillations are also quenched if a bandwidth limit B_m is exceeded; B_m is estimated from Δt, the time lag between a central ray and one from the angle θ_s:

$$B_m \sim \Delta t^{-1} \sim c/(z\theta_s{}^2). \tag{7}$$

This is an important limit for both IPS and ISS at low frequencies; sometimes it is called the correlation bandwidth.

The diffraction pattern on the ground has a scale $b \sim \lambda/\theta_s$, and if the pattern moves with velocity u, the time scale for fluctuations is $\tau \sim b/u$. The time scales are roughly of the order of 1 s for IPS, and 1000 s for ISS.

B. Visibility Fluctuations

The general theory of visibility fluctuations has been presented by Cronyn [27]. The main special consideration which applies to VLB work is that the antenna separation s may be greater than b, so that the antennas are in different scintillation patches. If the scintillations are strong, the amplitude and phase will vary independently in the two antennas, on the scintillation time scale τ; and the integration time of the interferometer must be less than τ, or else the fringes will be smeared out. Even if the scintillations in each antenna are weak, it is possible to have large phase fluctuations between the two antennas when $s \gg b$. This will happen whenever the spectrum of fluctuations in the scattering medium contains substantial energy at small wavenumbers, i.e., at scales greater than $(\lambda z)^{1/2}$. In this case the time limit for coherent fringes is approximately s/u rather than b/u. Some of the attempts to measure ray bending near the sun apparently failed on this account. With the Goldstack system, $s \sim 3800$ km and the scale for IPS is $b \sim 20$ km. The observations consisted of measuring the relative positions between two quasars as a function of solar elongation, by observing the two sources alternately and comparing the phases of the two sets of fringes. But the minimum alternation time was about 1 min, which is larger than $s/u \sim 10$ s, and the phase comparison was meaningless. A different technique, the "four-antenna" scheme, uses two antennas at each station (as mentioned earlier); the effective alternation time is 1 s and phase comparison becomes possible.

When $s < b$ the amplitude of the fringes fluctuates with the scintillation time scale, but the phase difference is steady and there is no limit to the coherent integration time.

C. The Apparent Diameter

It appears that all methods of measuring diameter are equivalent, and all reduce to measuring the transverse coherence in the wave field. Any conclusions derived from interferometry must also apply to measurements made in any other way, e.g., with lunar occultations, or with IPS.

There are several possible circumstances where the apparent size is the scattering angle θ_s rather than the intrinsic angle θ. If either the integration time limit or the bandwidth limit is seriously violated, then the various rays within the scattered cone of radiation are incoherent, and the apparent size will be θ_s. For example, VLB observations of the Crab pulsar have shown an apparent diameter of $0''.07$ at 111 MHz [28]. But it is entirely implausible that this could be the intrinsic size, and so it must represent the scattering angle. For this pulsar B_m is only about 100 Hz at 111 MHz, and, since the observing bandwidth was 330 kHz, the limit was seriously violated. This is an instrumental effect, for the diameter of the Crab pulsar can be measured in other ways. In particular, the observation that it shows strong ISS with a correlation bandwidth $B_m \sim 100$ Hz means that $\theta < \theta_c \sim 10^{-7''}$.

A more fundamental case exists when $\theta_c < \theta < \theta_s$, for now the rays within the scattering cone are intrinsically incoherent and the apparent size is θ_s for all measurements [26]. It is likely that many sources have size θ_s (interstellar) at frequencies below about 1 GHz [29]; in fact, the only ones which do not may be the pulsars (for which $\theta < \theta_c$) and local sources such as Jupiter and spacecraft. Another conceivable exception is a molecular maser source, which may have some coherence across its face.

ACKNOWLEDGMENT

The author wishes to thank W. M. Cronyn for critical remarks.

REFERENCES

[1] N. W. Broten et al., "Long baseline interferometry: A new technique," Science, vol. 156, pp. 1592–1593, June 1967.

[2] C. Bare et al., "Interferometer experiment with independent local oscillators," Science, vol. 157, pp. 189–191, July 1967.

[3] J. M. Moran et al., "Spectral line interferometry with independent time standards at stations separated by 845 kilometers," Science, vol. 157, pp. 676–677, Aug. 1967.

[4] J. S. Gubbay and D. S. Robertson, "Nine million wavelength baseline interferometer measurements of 3C 273B," Nature, vol. 215, pp. 1157–1158, Sept. 1967.

[5] G. W. Brown, T. D. Carr, and W. F. Block, "Long-baseline interferometry of S-bursts from Jupiter," Astrophys. Lett., vol. 1, pp. 89–94, 1968.

[6] B. G. Clark et al., "High-resolution observations of small-diameter radio sources at 18-centimeter wavelength," Astrophys. J., vol. 153, pp. 705–714, Sept. 1968.

[7] N. W. Broten et al., "Long baseline interferometer observations at 408 and 448 MHz—I. The observations," Mon. Notices Roy. Astron. Soc., vol. 146, no. 3, pp. 313–327, 1969.

[8] R. W. Clarke et al., "Long baseline interferometer observations at 408 and 448 MHz—II. The interpretation of the observations," Mon. Not. Roy. Astron. Soc., vol. 146, no. 4, pp. 381–397, 1969.

[9] M. H. Cohen, D. L. Jauncey, K. I. Kellermann, and B. G. Clark, "Radio interferometry at one-thousandth second of arc," Science, vol. 162, pp. 88–94, Oct. 1968.

[10] M. H. Cohen, "High-resolution observations of radio sources," Annu. Rev. Astron. Astrophys., vol. 7, pp. 619–664, 1969.

[11] B. F. Burke, "Long-baseline interferometry," Phys. Today, vol. 22, pp. 54–63, July 1969.

[12] K. I. Kellermann, "Intercontinental radio astronomy," Sci. Amer., vol. 226, pp. 72–83, Feb. 1972.

[13] W. K. Klemperer, "Long-baseline radio interferometry with independent frequency standards," Proc. IEEE, vol. 60, pp. 602–609, May 1972.

[14] B. G. Clark, "The NRAO tape-recorder interferometer system," this issue, pp. 1242–1248.

[15] T. A. Clark and W. C. Erickson, "Long wavelength VLBI," this issue, pp. 1230–1233.

[16] K. I. Kellermann, "Continuum radio sources, particularly galaxies and quasars," to be published.

[17] W. J. Welch, "OH sources and H₂O sources—Maser action," to be published.

[18] C. C. Counselman, III, "Very long baseline interferometry techniques applied to problems of geodesy, geophysics, planetary science, astronomy, and general relativity," this issue, pp. 1225–1230.

[19] J. H. Van Vleck and D. Middleton, "The spectrum of clipped noise," Proc. IEEE, vol. 54, pp. 2–19, Jan. 1966.

[20] G. W. Swenson, "Synthetic-aperture radio telescopes," Annu. Rev. Astron. Astrophys., vol. 7, pp. 353–374, 1969.

[21] M. H. Cohen et al., "The small-scale structure of radio galaxies and quasistellar sources at 3.8 centimeters," Astrophys. J., vol. 170, pp. 207–217, Dec. 1971.

[22] A. E. E. Rogers, "Very long baseline interferometry with large effective bandwidth for phase-delay measurements," Radio Sci., vol. 5, pp. 1239–1248, Oct. 1970.

[23] J. M. Moran et al., "The structure of the OH source in W3," Astrophys. J. Lett., vol. 152, pp. L97–L101, May 1968.

[24] K. J. Johnston et al., "An interferometer map of the water-vapor sources in W49," Astrophys. J. Lett., vol. 166, pp. L21–L26, May 1971.

[25] B. F. Burke et al., "Observations of maser radio sources with an angular resolution of $0''.0002$," Sov. Astron. A. J., vol. 16, pp. 379–382, Nov.–Dec. 1972.

[26] W. M. Cronyn and M. H. Cohen, in preparation.

[27] W. M. Cronyn, "Interferometer visibility scintillation," Astrophys. J., vol. 174, pp. 181–200, May 1972.

[28] N. R. Vandenberg et al., "VLBI observations of the Crab Nebula pulsar," Astrophys. J. Lett., vol. 180, pp. L27–L29, Feb. 1973.

[29] D. E. Harris, G. A. Zeissig, and R. V. Lovelace, "The minimum observable diameter of radio sources," Astron. Astrophys., vol. 8, pp. 98–104, 1970.

The NRAO Tape-Recorder Interferometer System

BARRY G. CLARK

Abstract—Several tape-recorder interferometer systems have been constructed in the last few years. These systems are used for connecting existing radio-telescope systems into interferometers, with baselines ranging from a few kilometers to nearly the diameter of the earth.

The NRAO Mark II interferometer system is in wide use. This, and the fact that its properties are typical of those of such systems in general, justifies a detailed description of the system.

The system is based on a television-type rotating-head video recorder. One-bit sampled data are recorded at a 4-Mb rate. After recovery, the data are processed in special-purpose digital devices and general-purpose digital computers to complete the interferometer system. For purposes of this description, the software is regarded as an intrinsic part of the system.

I. Introduction

RADIO INTERFEROMETERS of all the various sorts which have been built by radio astronomers have a strong generic resemblance, and differ more in constructional details than in principle. They differ strongly in appearance from most interferometers used at visible-light wavelengths, chiefly because the radio engineer prefers to construct a local oscillator (LO) at each of the interferometer

Manuscript received February 19, 1973. The National Radio Astronomy Observatory is operated by Associated Universities, Inc., under Contract with the National Science Foundation.

The author is with the National Radio Astronomy Observatory, Charlottesville, Va. 22901.

elements, and to perform the correlation in an efficient fashion at a convenient intermediate frequency (IF).

In recent years, the desire of the radio astronomer for even longer baselines has outdistanced the capability of constructing economically feasible cable- or microwave-link-connected interferometers. To meet this need, interferometers have been constructed which eliminate all physical connection between elements, replacing the conventional LO link with independent LO's, and the conventional IF link with tape recordings carried physically from one interferometer element to the other. These interferometers have been used on many baselines, ranging from a few kilometers to nearly the diameter of the earth [1].

Even such a strong variant as the tape-recorder interferometer, however, differs very little from the conventional cable- or microwave-link-connected interferometer. The more conventional IF transmission and delay systems are merely replaced by the tape-recorder system. The main remarkable property of the tape recorders as a transmission system is their long transmission delay, measured in days or weeks, but no different in principle than delays of microseconds or nanoseconds encountered in conventional transmission systems.

Because the tape-recorder interferometer system is similar to the conventional radio interferometer, it will be found that a firm grounding in either system permits immediate under-

Reprinted from *Proc. IEEE*, vol. 61, no. 9, pp. 1242–1248, Sept. 1973.

standing of the other. Any principle found useful in utilizing or in describing the behavior of one system will also find application in the other. Minor differences arise because the behavior of the instrument may be *dominated* by one term in one case and by another term in another case, but it is found in practice that if an effect needs to be taken into account for the tape-recorder interferometer, there arise cases for conventional interferometers in which it is also important, and vice versa.

With very little difference in principle between the various classes of radio interferometers, it is clear that the differences among the tape-recorder interferometers will be smaller yet. Indeed, since the designers of the various recorder systems are all motivated by the same desire to reduce the cost of the recording medium by maximizing the amount of information which can be stored on a square centimeter of tape, the various systems tend to encounter a very similar set of problems. The various systems differ only in details, and the comparison of them rests on secondary considerations, such as the reliability of the recorders, the convenience of the recording and reproduction processes, and the adaptability of the system to requirements not originally built into it.

Since the NRAO Mark II tape-recorder interferometer system is more or less typical of those in use, its special features will be described in detail as typical of the way technical problems may be handled.

Because of the complexity of any interferometer with a baseline more than a few hundred wavelengths, the use of digital computers tends to be an intrinsic part of the design. The NRAO Mark II system is no exception. At every step of the design it was carefully considered whether a given function could best be performed by analog hardware, special-purpose digital hardware, or software in a general-purpose digital computer. As a result, the mixture of hardware and software is so intimate that someone using the equipment for the first time may not realize with which he is dealing, and no clear distinction between them will be made in this paper.

II. Principles of Digital IF Interferometry

In order to make the recording and playback systems as reliable as possible, the decision was made to record a digitized version of the IF voltages rather than the voltages themselves. Given that we record binary digits, it can be shown that single bit (1-b) digitization conveys more information per bit than any multilevel scheme. That is, a 2-b digitization produces less increase in signal-to-noise ratio (SNR) than does using the second bit to double the bandwidth of the 1-b digitized signal.

Although the use of 1-b digitized IF's has been extensively discussed in relation to spectroscopy, and has become a standard technique in radio astronomy [2]–[4], it still perhaps warrants a few words of comments with respect to its use in interferometers [5], [6]. In this connection, the digitization does not introduce as much complication as one might suppose. The relation between the correlation of two 1-b digitized signals, and the correlation of the signals without digitization has, for Gaussian distributed random signals, long been known [7]. In this connection, it turns out that the sampling of the signals, though done at a sufficiently high rate to avoid information loss, introduces more complexity than the truncated digitization. For example, consider the case of a signal with a flat power spectrum extending from zero to f_0, sampled at a rate $2f_0$ so that all samples are uncorrelated. That is, the autocorrelation function of the signals has zeros at the

sample intervals. Cross-correlating the signals, however, is not quite the same process, because an arbitrary phase shift can be, and usually is, introduced in the signals before correlation. If, for instance, a phase shift of 90° is introduced, the cross-correlation function has a zero at zero lag, because the sine and cosine components of a noise-like signal are independent random variables. However, at ±one lag, the correlation is no longer zero, but has a positive value at one point and a negative value at the other. Because of this phase effect, it is convenient to speak of the complex correlation function, which is the complex function obtained by adding j times the cross-correlation function with 90° phase shift to that with zero phase shift. This complex cross-correlation function does not have zeros at multiples of the sampling interval, but at multiples of twice the sampling interval.

This gives rise to an interesting problem. The cross-correlation function at lag n bears information about those at lags $n-1$ and $n+1$. Since the noise on these different lags is independent, to find the best estimate of the cross-correlation function at lag n, one must combine the measured cross-correlation function at all other lags, but principally at lags $n-1$ and $n+1$. The proper procedure of estimating the delay and magnitude of the cross-correlation function is therefore a rather complex one. Rogers [8] describes the maximum likelihood approach to the problem, and this approach should be used in difficult cases. In most practical cases, however, a much simpler least squares approach may be used without significant loss of SNR.

When the interferometer is used for observations of discrete atomic or molecular line radiation, the observer is usually interested in the frequency structure of the radiation as well as its simple cross-correlation coefficient. In the digital interferometer it is easiest to calculate the cross power spectrum from the observed cross-correlation function by way of a Fourier transform. It is easily shown that the fundamental principles of autocorrelation-function spectroscopy can be extended in this simple fashion to give the cross power spectrum, identical to the one that would be measured by inserting a filter bank in front of the correlators, and measuring the complex correlation function as a function of frequency.

The fact previously discussed, i.e., that the cross-correlation function goes to zero at twice the sample spacing, is demonstrated in an interesting fashion in the line case. If only the real part of the correlation function is measured, then clearly the Fourier transform will, in general, be complex and Hermitian. That is, the complex cross power spectrum will be determined from these measurements of the real part only of the correlation function. Further, the spectrum is redundant; half the spectrum, say, the half at negative frequencies, may be discarded without loss of information. If the true complex cross-correlation function is measured, when we take its Fourier transform we find that the useful information is still present in only one half of the spectrum, and the other half may usually be discarded as containing only noise.

III. The NRAO Mark II Tape-Recorder System—The Recording System

The block diagram of the tape-recorder interferometer recording system is shown in Fig. 1. As usual for an interferometer, the feed of the radio telescope comprising an interferometer element is connected to a low-noise preamplifier. The signal is then mixed to an IF, using an LO signal which is sufficiently nearly identical to the LO signal in use at the other element of the interferometer. In the tape-recorder inter-

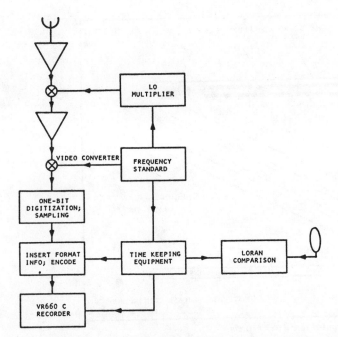

Fig. 1. Tape-recorder interferometer system block diagram—record time portion.

ferometer this is accomplished by the use of an atomic-frequency standard to generate nearly perfect sine waves at both antenna elements. The preamplifier has sufficient frequency discrimination to reject the unwanted sideband of the mixer. The IF signal resulting from this mixing is further amplified to a conveniently high level. Then, to simplify the sampling procedure, the IF is converted to a square bandpass with one edge of the bandpass at zero frequency. The Mark II converter has a great deal of flexibility in choice of the IF frequency, due primarily to the use of a wide-band quadrature mixer [9] to reject the unwanted sideband of the second LO used for conversion to video.

The 1-b digitization is simple clipping, done by very fast saturating amplifiers. The clipped signal is then sampled in a diode bridge driven by a very narrow pulse, and the sampled signal strobed into a TTL integrated circuit flip-flop. Sampling is done at a 4-MHz rate.

In the tape-recorder interferometer, each bit of data must be precisely labeled with its time of arrival at the element, either implicitly or explicitly. In the Mark II system, some bits are labeled explicitly, and others are labeled implicitly by counting bits from the occurrence of the labeled bits. Explicitly labeled bits occur 60 times/s. In order for time keeping to be consistent between the two elements of the interferometer, the usual practice is to set clocks at both to universal time coordinated (UTC). There is a fairly elaborate set of time-keeping equipment to keep UTC time, and to generate all of the necessary timing signals necessary in the system (for instance, the tape recorders, operating at 60 frames/s, need a 60-Hz input signal). The time is compared with a known UTC either by the transportation of a running clock or by the comparison with the Loran navigation signals, many of which are emitted at known UTC times with a precision of a few microseconds.

The NRAO Mark II system is based on the Ampex VR 660 C videotape transport. This transport is a rotating-head helical-scan television-type tape recorder, which records on 2-in wide videotape. The VR 660 B is extensively used in com-

mercial TV (it is a popular choice for implementing "instant slow motion replay" at sports events) and in academic environments. The VR 660 C has slightly enhanced frequency response. This transport has two video recording heads mounted at diametrically opposite points on a horizontal wheel which spins at 30 r/s. The recording tape is wrapped around this wheel in a single turn helix, and is in contact with the headwheel for half a turn, so that one head is in contact with the tape at all times.

The tape is drawn past the headwheel assembly at a rate of 3.7 in/s. The combination of this tape motion with the motion of the wheel results in each head tracing a diagonal line from one edge of the tape to the other. The separation of the video tracks is about 0.06 in along the length of the tape, or 0.010 in perpendicular to their length. The layout of data on the tape is shown in Fig. 2. The rotation of the headwheel results in an effective head-to-tape speed of about 650 in/s. In the Mark II system, where digital data are recorded at 4 million b/s, the effective packing density is about 6000 b/in along the video tracks, and the area density is about 800 b/mm^2, much the same information density as on astronomical photographic plates.

The digitized data stream, sampled at 4 MHz, has essential information spread over the frequency range from near zero to about 2 MHz. The video recording system cannot be sufficiently accurately compensated for the recording saturation effects and the head impedance effects over a very wide percentage bandwidth. In usual TV recording practice, this problem is avoided by using an FM modulation scheme to reduce the frequency range recorded to one or two octaves. For digital data, we may use a close equivalent of FM modulation, digital diphase coding. This code is illustrated in Fig. 3. The first line shows the input data, in non-return-to-zero format. In the second line, which is the diphase coding, there is always a transition, from positive to negative or vice versa, at the beginning of every bit time (for 4-Mb data, a bit time is a 250-ns interval). When the data are binary ones, there are, in addition, transitions in the center of the bit time. The resulting code string can be decoded and both clock and data can be extracted from it. The spectrum of an encoded bit string would show most of the power between 2 and 4 MHz. If this band is reproduced with reasonable fidelity, the data and clock are reproduced with high reliability.

As previously stated, it is necessary to label some bits explicitly with the time, to fix the time at which it and all succeeding bits were recorded. This is done at the natural interval of the recorder. A new head comes in contact with the tape every sixtieth of a second. At this time, a unique sync pattern is written, one which can never occur in data, followed by a binary number giving the number of sixtieths of a second that have elapsed since the last integer second, expressing the time exactly of a given data bit.

The more significant bits of time and date are written on one of the two audio tracks of the tape recorder, also in diphase coding. Therefore, the first bit of the track on tape was recorded at a time corresponding to the days, hours, minutes, and seconds recorded on the audio track, and to the number of sixtieths of a second recorded on the video track itself.

IV. The Mark II System—The Reproduction System

A block diagram of the Mark II tape-recorder interferometer reproduction system is shown in Fig. 4. The block diagram falls nicely into two parts. The upper three levels show

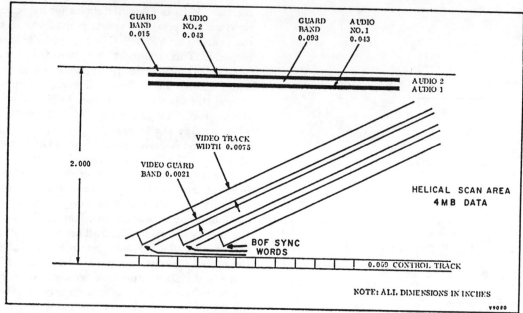

VR-660C Recorded Tape Track Pattern

Fig. 2.   Layout of data on videotape.

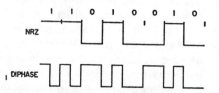

Fig. 3.   Illustration of digital diphase code. First line—typical data, non-return-to-zero format. Second line—digital diphase coding of the same data.

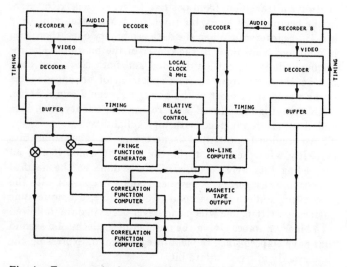

Fig. 4.   Tape-recorder interferometer system block diagram—playback time portion.

the equipment necessary to reproduce the IF signals exactly as they were recorded. The output of this part of the system is two streams of digitized data, both delayed by some very large time, but in relative synchronism. That is, the data samples which were sampled and recorded at the same time are now being reproduced at the same time.

The lower three levels of the diagram show those parts of the interferometer which operate on these data streams to produce the cross-correlation function output, and are things which might be desirable to have on any interferometer, irrespective of the form of its IF transmission system.

As previously mentioned, the data are recorded in digital diphase coding on the video tracks of the tape. After reproduction, the signal is clipped and used to derive both the data stream and the clocking pulses. In practice it is not as difficult to recover a given bit (that is, determine if it is a "0" or a "1") as it is to determine where in the time sequence it belongs. The data need only to have a reliability exceeding 99 percent to contribute negligibly to the error budget of the interferometer, but the clock must be so reliable that it has a small probability of inserting an extra pulse, or deleting a pulse, in the interval between fiducial marks, a period of about 66 000 pulses.

In the Mark II system, several procedures are used to recover a reliable clock. First, the recovered clock is not used directly, but is used to run a "flywheel," such that the flywheel inserts any missing pulses and deletes any extra pulses. This flywheel effect is provided by an L–C filter of fairly low Q, which, nonetheless, stores the energy equivalent of several pulses, and whose output therefore is not much affected by a single input pulse. This device enables the clock to coast across "dropouts"—tiny holes in the magnetic coating of the recording tapes—if they are not too large.

A second level of clock protection is provided by inserting into the data a short check pattern every 512 μs. This pattern consists of seven zeros and a one. At playback time this pattern is used to reset the clock counter if it has gained or lost one or two pulses since the last previous pattern.

There are two parts to the data stream reproduction process. The first is the recovery of the data bits and the appropriate clock information as previously discussed. The second is the problem of restoring the bit streams to an appropriate relative timing. In the end, we wish to have at the center of our measured correlation function not the correlation of the bits which were sampled at the same time, according to the clocks at the two elements of the interferometer, but those bits which are samples from the same wavefront of the radio-

450

source radiation. Therefore, the apparent times on the two bit streams should differ by the relative clock error of the two recording station clocks, and also by a geometric term arising from the different time of arrival of a given wavefront at the two elements of the interferometer. This geometric term is given approximately (neglecting relativistic terms) by

$$\tau_g = D \cdot S. \qquad (1)$$

Where the source vector S is a vector of unit length directed toward the radio source, and is, in rectangular coordinates fixed in the earth,

$$S = (\cos \delta \cos H, \cos \delta \sin H, \sin \delta) \qquad (2)$$

and D is the baseline vector, measured in time units,

$$D = (V_2 - V_1)/c. \qquad (3)$$

V_1 and V_2 are the vector locations of the interferometer elements, and δ is as usual the declination of the radio source. The reference hour angle, H, is defined as the sidereal time on the x axis of the coordinate system minus the right ascension of the source.

This term may be quite large. In the extreme case, $D \cdot S$ may be as large as one earth radius—approximately 21 ms. The tape-recorder interferometer has an additional source of relative delay—the reproduction instability of the recorders. The time that a given bit comes out of the recorder is determined chiefly by the rotation of the headwheel, since that bit is reached at a given angle of the headwheel. The headwheel is servo controlled to an angle uniformly increasing with time, but it is a heavy object, and subject to various mechanical disturbances. In practice, the intrinsic time stability of the VR 660 C is about $\pm 100 \ \mu s$.

In the conventional radio interferometer the geometric term and the station clock errors are compensated by an IF delay system, consisting of cable, lumped constant circuits, or acoustic delays which can be inserted into an IF line as needed. Such a solution would also be possible for the tape-recorder interferometer.

In the Mark II interferometer, however, these various sources of relative delays between the bit streams are compensated for by a delay line comprised of a relatively small semiconductor memory. The semiconductor memories for the two recorders are unloaded in synchronism by a local clock in such a fashion that the record time difference of the 2 b coming from the two buffers is strictly controlled; that is, the time difference between the two record time clocks at the time that the 2 b were recorded is strictly controlled by the playback terminal on-line computer, which calculates the approximate value of this time difference from (1) and known clock errors, with parameters supplied by the observer. The memory may be relatively small because long delays are compensated for by a mechanical relative rotation of the recorder headwheels in response to timing signals from the memories.

In an idealized interferometer system, and perforce in optical wavelength interferometers, the delay called for by (1) should be applied at RF frequency rather than at IF. If this is done, all effects of the delay are removed, and the interferometer operates as if the elements were fixed in space relative to the radio source, rather than being mounted on a rotating earth. It is clear that a delay line operating at the RF frequency is equivalent to a delay line operating at the IF frequency with an additional phase shift

$$\phi = \omega_0 \tau \qquad (4)$$

where ω_0 is the LO frequency, or, in the multiple LO case, the signed sum of LO frequencies. The part of this arising from the geometric part of the delay,

$$\phi_g = \omega_0 D \cdot S \qquad (5)$$

is called the natural fringe phase, and its derivative, the natural fringe rate. This phase rotation may be thought of as the relative Doppler shift of the two ends of the interferometer with respect to the radio source.

In long-baseline interferometry very high natural fringe rates are often encountered. One earth radius baseline at 3-cm wavelength has a maximum natural fringe rate slightly greater than 15 kHz. This phase rotation must be removed before the data are eventually presented to the observer. One possibility of doing this is to alter the LO phase at record time in such a fashion that it tracks the expected geometric phase. However, in the design of the Mark II system, it was felt that as much as possible, complicated functions which either require complex and expensive equipment, or in which it is possible that a tired observer might make an unrecoverable mistake, should, if possible, be done at playback time rather than at record time. Therefore, the natural fringe rate is removed by phase rotation of the recovered IF at playback time. It may at first seem unnatural to talk about phase rotation of an IF consisting of a stream of 1-b samples of the signal, but it is accomplished in a rather conventional fashion. The expected fringe function (cosine of the phase ϕ_g—nearly a sine wave) is calculated in special-purpose digital hardware and is mixed with the bit stream in a quadrature mixer phased to produce a single-sideband output. This output has the phase of the input signal plus the phase of the supplied "local-oscillator" signal. The expected fringe function is approximated by a square wave generated with appropriate phase ϕ_g. Both in-phase and quadrature components are produced. These are used to multiply the data stream from one recorder in the one bit sense (also called "exclusive-OR"). In the correlation function computer, the two resulting data streams are again multiplied by the data stream from the other recorder, and the product is integrated. Multiplication is inhibited for a short time near the transitions of the expected fringe function, making the multiplication in the quadrature mixer essentially into a three-level scheme, -1, 0, 1 rather than the two-level scheme natural with binary logic. The resultant is a small saving in SNR. At any one lag, two quantities are now accumulating. One is regarded as the real part of a complex fringe amplitude, and the other as the imaginary part. This completes the phasing of the quadrature mixer.

The phase of the expected fringe function is calculated by linear interpolation by a programmable fractional divider, whose rate is set by the on-line computer. Ten times per second a new starting phase, also set by the computer, is strobed into the phase register at the exact tenth of a second, according to the clock recorded on the "A" recorder. It is sufficiently often that the second derivative term in the Taylor expansion of the expected fringe function will be negligible for any earth-based interferometer with nonmobile elements.

The correlation function computer is comprised of integrated-circuit cards developed for the NRAO autocorrelation function spectrometers, which operate at bit rates up to 20 Mb/s. The present functions do not require such performance, but it was felt that the cost of developing a less capable pack-

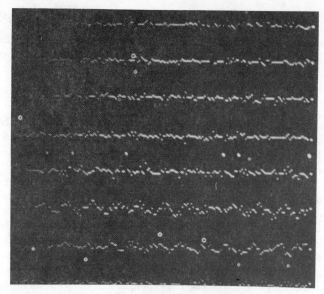

Fig. 5. Correlator output, real-time display. Radio source is 3C273, frequency 8430 MHz, 10 000-km baseline.

age would exceed the amount to be saved by using slower logic and less compact packaging. The cross-correlation function is calculated for 95 channels simultaneously. The lag between channels may be adjusted in steps of a factor of two from 0.25 to 32 μs. In observations of continuum sources of radio radiation, the correlator is normally used to calculate only 31 channels of cross-correlation function, allowing room for a possible ± 3.5 μs of error in the preestimated delay for the source, which might arise from uncertainties in the source position, uncertainties in the interferometer baseline, or errors in the record time clocks.

For observations of the interstellar maser signals, such as those of OH or water, the entire 95 channels are used, operating as a cross-correlation function spectrometer, giving about 47 independent frequency channels in a total bandwidth, determined by the lag between channels of the correlation function, of 2 MHz, 1 MHz, etc., down to 16 kHz.

At 0.2-s intervals, the cross-correlation function is transferred from the special-purpose digital cross-correlation function computer to the general-purpose minicomputer which serves as a controller for the playback devices. It is then written on an IBM compatible magnetic tape for further processing in a larger general-purpose computer.

As an on-line display, in order for the observer to be able to tell (on strong sources, at least) that the reproduction machinery is functioning properly, the real part of the center 12 channels of the cross-correlation function is displayed as a function of time, on a memory oscilliscope. This real part may be preintegrated with a 0.2-, 1-, or 5-s integration time. An example of such a display is shown in Fig. 5. The fringes have a typically sinusoidal appearance. This is due to the fact that the estimated source position or baseline description was in error, so that the estimated fringe function, given by (5), is in error. The error is sufficiently small and slowly varying that it may be approximated as a linear increase in phase with time, and hence a sinusoidal variation of the real part of the cross-correlation function.

The display of Fig. 5 was produced by observations of the radio source 3C273 at a frequency of 8430 MHz on a baseline between the 64-m telescope in Goldstone, Calif., and the 22-m radio telescope near Simeis, Crimea, USSR.

TABLE I
PLAYBACK TERMINAL EVENTS

Recurrence	Event
0.25 μs	new bit is recovered from each recorder output; new bit is unloaded from each buffer; 95-channel cross-correlation function is calculated for that bit and added to store;
4 μs (max)	decision is made whether to advance phase of predicted fringe function by one-eighth turn;
512 μs	data from recorder are checked for timing; buffer counter reset if necessary;
16$\frac{2}{3}$ ms	unique sync pattern and binary clock read from each recorder; decision is made whether to adjust relative delay of the 2-b streams by 1-b time; correlation function display is refreshed;
0.1 s	new predicted fringe function phase is calculated and strobed into fringe function generator;
0.2 s	cross-correlation function is transferred from semiconductor memory to computer core and thence to computer readable magnetic tape;
12.8 s	minimum data length for post-processing programs averaging;
20–500 s	typical fringe coherence times.

In the preceding discussion, it has been made clear that a great many events are occurring in the playback mechanism, each with its own characteristic period of recurrence. To aid in summarization, many of the major events are listed in Table I.

V. THE CONTINUUM INTERFEROMETER SOFTWARE SYSTEM

An important part of the processing occurs after the initial data recording at 0.2-s intervals. This is illustrated by the relative data compression of the two stages. For the continuum, the playback processor reduces the 10^6 b recorded in 0.2 s to 31 channels of complex cross-correlation function data, a total of about 10^3 b, so that the total compression is by a factor of 10^3. The machine-performed part of the post processing must reduce these data to an amount comprehensible to a human being. In practice, this normally means perhaps two numbers (fringe amplitude and rate, say) at 5-min intervals. This represents a further compression of the data by a factor of about 5×10^4.

A set of programs has been developed to process continuum data with a great deal of flexibility and power. These programs may properly be considered to be a part of the tape-recorder interferometer system, along with the hardware and on-line software.

In discussing the processing of data beyond the 0.2-s outputs from the playback processor, I shall occasionally refer to the "coherence time" of the LO's. The definition of coherence time that is useful in this connection differs somewhat from the standard one. Because we are never extremely certain of our baseline parameters or source positions, we usually have a residual fringe rate, of the sort shown in Fig. 5 and previously discussed. These residuals, which to a good approximation are taken locally to be a linear phase drift, must be taken into account in whatever processing we do. Therefore, we do all further processing for many such phase drifts—offset frequencies—in parallel. In taking into account the residual offset frequencies in this way, we have also taken into account any linear drift of the two LO systems. We therefore define the coherence time of the oscillators to be the longest time over which their relative phase differs from the best linear drift in that time by less than a radian.

A priori we only know within what limits to expect the residual offset frequency to lie. These are usually much wider limits than the reciprocal of the time for which we wish to average to be sure of detecting the (often weak) radio source. We therefore wish to average the data, in parallel, for all of the possible residual offset frequencies within that range. Adding together complex numbers with various linear phase rates which multiply them is precisely the function performed by a Fourier transform. Therefore, the post-processing programs include one which takes the Fourier transform of all of the 31 delay channels, and prints a display of amplitude as a function of delay and of fringe frequency.

This, however, is a very inefficient procedure. A much more efficient process is to take only discrete Fourier transforms in the immediate neighborhood of the delay and rate at which one expects fringes from the source to appear. This estimated delay and rate are produced by the program which takes complete Fourier transforms on a small percentage of the data. Rogers [8] has shown that the best estimate of correlation at a given lag is computed by the use of several adjacent lags. For equally spaced points in the correlation function and a square bandpass of width B, an interpolation formula of the form

$$e^{i\pi B(\tau - \tau_n)} \left(\frac{\sin 2\pi B(\tau - \tau_n)}{2\pi B(\tau - \tau_n)} \right) \qquad (6)$$

may be used to add together estimates of the correlation at a given lag. The three largest terms of this formula are used in this program to estimate the fringe amplitude.

To make the program set simple to run and to save computer time, the Fourier transform program does not accept an arbitrary number of input points. To do so would require excessive amounts of storage. One usually has a rather better knowledge of where to expect the fringe frequency of a given source than the 5-Hz range implied by the 0.2-s sampling time of the correlation function. Therefore, one may pre-average the data to a longer sampling time. A program is written to do this. It has provision for altering the fringe rate before applying the averaging procedure, so that in the rare cases when the entire range needs to be covered, it can be handled by repeatedly applying the program with different rate offsets. The output of this program is identical in format to its input, so it may be processed by any of the other programs in the set.

In processing data, one may average only up to the coherence time of the LO's. To a rough approximation, the coherence properties of the LO simply multiply those of the radio-source radiation, which we want to measure. In averaging beyond the coherence time, we degrade the quantity we are seeking. However, one may do further averaging in a different way, by breaking coherence. To break coherence means to discard phase information. One procedure for discarding phase information is simply to convert the complex number describing the correlation coefficient, averaged over a time short compared to the coherence time, into the magnitude-phase format, and ignore the phase. These averaged amplitudes then give an estimate of the correlation coefficient averaged over the whole interval. It is well known that this estimate is too large by about a factor of

$$1 + \frac{\text{SNR}^{-2}}{2} . \qquad (7)$$

Operationally, it turns out to be convenient to use the above form of broken coherence averaging only for cases where the SNR is somewhat greater than 1. For very weak

sources, it is more convenient to Fourier transform the original time sequence of cross-correlation coefficients over a time rather longer than the coherence time, and to square the resulting spectrum. This squared spectrum (with no phase and therefore with broken coherence) can then be averaged for the duration of the observation. In the end, one can average this fringe power spectrum over a bandwidth reciprocally related to the LO coherence time.

In any broken coherence procedure (including the most classical, the post detection correlation interferometer [10], [11]) the limiting sensitivity of the instrument decreases as the fourth root of the observing time, instead of the square root as it does in the case of the coherent averaging. If the coherence time is short compared to the length of the observations, the sensitivity of the interferometer can be significantly increased by this procedure.

VI. Summary

The NRAO Mark II system is a flexible tape-recorder interferometer system, which records 1-b digitized IF voltages on magnetic tape for later reproduction and correlation. The maximum data rate is 4 Mb/s (2-MHz bandwidth). The data density on the magnetic recording medium is quite high, about 800 b/mm². The record time equipment is relatively simple, deferring all complex operations until reproduce time. The reproduce controller, because of the incorporation of a minicomputer, is a fairly flexible device, with capabilities for continuum observations, pulsar observations, multiple antenna observations, and line observations (47 instantaneous frequency channels). The system is now, or will be shortly, in use at about ten radio astronomy observatories.

Acknowledgment

Many people have participated in the design and construction of the Mark II system. The initial design was done by Leach Corporation, based on their HDDR recording techniques. Further perfection of the recording system was done by R. Weimer and R. Hallman. Other system elements were designed or built by R. Mauzy, A. Shalloway, and S. Weinreb. The spectral uses of the system have been developed by S. Knowles and J. Moran.

References

[1] M. H. Cohen, "Introduction to very-long-baseline interferometry," this issue, pp. 1192–1197.
[2] S. Weinreb, "A digital spectral analysis technique and its application to radio astronomy," M.S. thesis, Massachusetts Institute of Technology, Cambridge, 1963, pp. 1–157.
[3] R. D. Davies, J. E. B. Ponsonby, L. Pointon, and C. de Jager, "The Jodrell bank radiofrequency digital autocorrelation spectrometer," Nature, vol. 222, pp. 933–937, June 1969.
[4] B. F. C. Cooper, "Correlators with two-bit correlation," Aust. J. Phys., vol. 23, pp. 521–527, 1970.
[5] A. E. E. Rogers et al., "Positions and angular extent of OH emission associated with the HII regions W3, W24, and NGC 6334, 369," Astrophys. J., vol. 147, pp. 369–377, 1967.
[6] M. C. H. Wright, B. G. Clark, C. H. Moore, and J. Coe, "Hydrogen line aperature synthesis at NRAO—Techniques and data reduction," Radiosci., in press.
[7] J. H. Van Vleck and D. Middeton, "The spectrum of clipped noise," Proc. IEEE, vol. 54, pp. 2–19, Jan. 1966.
[8] A. E. E. Rogers, "Very long baseline interferometry with large effective bandwidth for phase-delay measurements," Radio Sci., vol. 5, pp. 1239–1247, Oct. 1970.
[9] ——, "Broad-band passive 90° RC hybrid with low component sensitivity for use in the video range of frequencies," Proc. IEEE (Lett.), vol. 59, pp. 1617–1618, Nov. 1971.
[10] R. Brown and R. Q. Twiss, "A new type of interferometer for use in radio astronomy," Phil. Mag., vol. 45, pp. 663–682, July 1954.
[11] B. G. Clark, "Radiointerferometers of intermediate type," IEEE Trans. Antennas Propagat. (Commun.), vol. AP-16, pp. 143–144, Jan. 1968.

Very-Long-Baseline Radio Interferometry: The Mark III System for Geodesy, Astrometry, and Aperture Synthesis

Abstract. The Mark III very-long-baseline interferometry (VLBI) system allows recording and later processing of up to 112 megabits per second from each radio telescope of an interferometer array. For astrometric and geodetic measurements, signals from two radio-frequency bands (2.2 to 2.3 and 8.2 to 8.6 gigahertz) are sampled and recorded simultaneously at all antenna sites. From these dual-band recordings the relative group delays of signals arriving at each pair of sites can be corrected for the contributions due to the ionosphere. For many radio sources for which the signals are sufficiently intense, these group delays can be determined with uncertainties under 50 picoseconds. Relative positions of widely separated antennas and celestial coordinates of radio sources have been determined from such measurements with 1 standard deviation uncertainties of about 5 centimeters and 3 milliseconds of arc, respectively. Sample results are given for the lengths of baselines between three antennas in the United States and three in Europe as well as for the arc lengths between the positions of six extragalactic radio sources. There is no significant evidence of change in any of these quantities. For mapping the brightness distribution of such compact radio sources, signals of a given polarization, or of pairs of orthogonal polarizations, can be recorded in up to 28 contiguous bands each nearly 2 megahertz wide. The ability to record large bandwidths and to link together many large radio telescopes allows detection and study of compact sources with flux densities under 1 millijansky.

Radio interferometers are now routinely operated by recording signals from extraterrestrial radio sources received simultaneously at widely separated antennas and later cross-correlating the recordings at a central processing facility. Use of this technique of very-long-baseline interferometry (VLBI) has allowed astronomers to achieve submillisecond-of-arc angular resolution in observations of celestial radio sources (1) and to discover apparent superluminal expansion in quasars (2). VLBI is also being applied to geophysical and astrometric research (3). Here we describe a new VLBI system, about six times more sensitive than its immediate predecessor, and some of the geodetic and astrometric results obtained with it. A companion report (4) describes the discovery made with this system of a very faint radio source, perhaps the third gravitational image of a single distant quasar.

The first VLBI experiments in the United States (5) utilized a standard computer digital recording system, known as the Mark (Mk) I, to record a signal with a nearly 360-kHz bandwidth. This digital signal was produced by sampling the sign of the analog signal at the Nyquist rate and was recorded on standard ½-inch (~ 1.3-cm) computer tape along with the epoch of the signal and other relevant information. Each tape was filled with bits after 3 minutes of recording. A Mk II system, also digital, based on a helical scanning television recorder (6), was subsequently developed by the National Radio Astronomy Observatory to increase both the bandwidth of the recording and the packing density of the bits on the magnetic tape.

In the current version of this Mk II system, slightly modified video cassette recorders are used to record a signal with a nearly 2-MHz bandwidth. Some tapes with this system can record for 4 hours and weigh only ~ 0.23 kg each.

In our new Mk III system (7) an instrumentation magnetic tape recorder is used to record simultaneously up to 28 signals, each with up to a 2-MHz bandwidth on each of 28 parallel tracks along a tape 9200 feet (~ 2800 m) long and 1 inch (~ 2.5 cm) wide. The recorded 1-bit samples have a longitudinal density of 33,000 bits per inch. With a head stack that can move transversely across the tape it should be possible to increase more than tenfold the number of tracks recorded (8). One tape would then store about 10^{12} bits and take about 3 hours to fill when running forward and backward on successive passes at its normal rate of 135 inches per second.

The Mk III system can handle the intermediate-frequency (IF) outputs (100 to 500 MHz) of two independent receivers. The IF outputs can be routed through 14 separately controllable frequency converters to produce "video" signals from the (~ 2 MHz) upper and lower side bands adjacent to the total local-oscillator frequencies corresponding to the settings of each of these converters. The actual bandwidth recorded and the corresponding low-pass filter through which the signal is passed depend on the recorder speed used; this speed, nominally 135 inches per second for 2 MHz, can be as high as 270 or as low as 17 inches per second. As in the

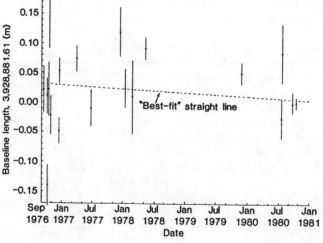

Fig. 1. Estimates of the length of the baseline between Haystack and Owens Valley. Before 1980 all data were obtained at only a single radio-frequency band; hence no corrections for ionospheric effects were made. The data obtained in July 1980 were compromised in part by the use of a portion of the observing time for special experiments. The estimates shown for September and October 1981 are composites of individual estimates for separate days which were too closely spaced in time to plot separately; the error bars represent the weighted r.m.s. scatter of the individual estimates. The other error bars represent plus and minus 1 standard deviation, determined in each case for chi squared per degree of freedom equaling unity; this latter result was attained by root-sum squaring the measurement errors, originally assigned on the basis of signal-to-noise ratios, with suitable, baseline-dependent numbers. The slope of the best-fit straight line does not differ significantly from zero (see text).

earlier systems, the output of each side band of each converter is amplified, hard limited ("clipped"), and the sign sampled at the Nyquist rate prior to recording in the same format on the assigned tape track. For geodetic and astrometric measurements, the frequency settings of the 14 converters are given values which collectively span as large a bandwidth as allowed by the receivers and the IF system, in order to yield accurate group delays (9). In our present system, these settings are distributed over both S-band (2.2 to 2.3 GHz) and X-band (8.2 to 8.6 GHz). Such dual-band measurements allow the contribution of the ionosphere to the interferometric measurements to be determined from the difference of the group delays measured at the two bands (3). For studies of source structure (4), the frequencies are usually chosen so that the 28 2-MHz channels are adjacent

to each other. In experiments to measure polarization, the frequency converters can be paired to record in the same frequency band each of two orthogonal polarization components of the radiation.

For each received radio band at each site, low-intensity signals spaced 1 MHz apart, and controlled by the site frequency standard, are injected into the receiver and pass through the entire system, serving to calibrate the phase delays undergone by the observed signals in each interferometer terminal. The delays incurred by the calibration signals between their point of generation and the input to the receivers are monitored by an auxiliary system and recorded in an auxiliary log for use in the data processing.

The entire data acquisition system can be operated either manually, through

front-panel controls on each module, or under computer control, save for the changing of tapes. In the latter mode, very complicated schedules, prepared in advance on a floppy disk, can be accommodated easily; in addition, module settings, system temperature measurements, and so on are all automatically recorded, usually on a hard disk. Twelve of these systems are currently in operation in the United States and Europe; several more are under construction, including one for use in Japan.

A dedicated processor cross-correlates the recorded bits in a standard manner: the tapes from each pair of sites in the interferometer array are aligned under computer control to account for any known epoch offset between the clocks at the two sites as well as for the difference in the times of propagation of the signals to the sites (delay). The rate of change of this difference (delay rate) is also taken into account. The phase-calibration information for each site is also extracted in the processing.

The processor is of modular design so that bits on a corresponding track on each of the two tapes from one pair of sites are correlated in one of 28 identical modules, all under the control of an HP1000 series minicomputer. The detection of interference fringes and the determination of the maximum likelihood estimates of the delay and delay rate for each observation of each source for each pair of sites are also carried out on the HP1000, which has an array processor attached to perform the two-dimensional Fourier transforms relating frequency and time to delay and delay rate.

For observations of a sufficiently strong source, the Mk III system can detect fringes at virtually the same time the observations are made by use of a 1-megabit data buffer incorporated in the data acquisition system at each site. The bits are stored in this buffer and then transferred over commercial telephone lines at 1200 bits per second to any computer which has software suitable

Table 1. Arc lengths between compact radio sources determined from VLBI observations.

Source pair	Number of experiments	Arc length (deg min sec)	Differences from 1972–1978 estimate (arc sec)
0355 + 508–0851 + 202*	14†	64 25 16.343 ± 0.004‡	0.003 ± 0.006§
0355 + 508–0923 + 392	14	56 03 52.200 ± 0.002	0.003 ± 0.003
0355 + 508–1226 + 023	9	110 45 46.616 ± 0.007	0.007 ± 0.009
0355 + 508–1641 + 399	16	88 43 41.706 ± 0.003	0.007 ± 0.004
0355 + 508–2200 + 420	14	58 03 19.715 ± 0.002	0.002 ± 0.003
0851 + 202–0923 + 392	14	20 09 50.748 ± 0.003	0.001 ± 0.004
0851 + 202–1226 + 023	7	55 16 54.597 ± 0.003	0.002 ± 0.005
0851 + 202–1641 + 399	12	96 11 25.040 ± 0.004	0.005 ± 0.006
0851 + 202–2200 + 420	14	115 40 22.150 + 0.005	0.005 ± 0.007
0923 + 392–1226 + 023	9	55 29 44.839 ± 0.006	0.003 ± 0.009
0923 + 392–1641 + 399	16	77 55 31.868 ± 0.003	0.005 ± 0.004
0923 + 392–2200 + 420	16	96 16 50.403 ± 0.003	0.004 ± 0.004
1226 + 023–1641 + 399	9	68 32 22.402 ± 0.007	0.002 ± 0.009
1226 + 023–2200 + 420	9	124 43 06.302 ± 0.009	0.005 ± 0.011
1641 + 399–2200 + 420	16	57 59 31.251 ± 0.002	0.002 ± 0.003

*The source name, by convention of the International Astronomical Union, is given in terms of right ascension and declination. For example, 0355 + 508 denotes a right ascension of 3 hours 55 minutes and a declination of +50.8°.   †Each experiment involved antennas at four or more of the following six sites: Chilbolton Observatory, Chilbolton, England; Max-Planck-Institut für Radioastronomie, Effelsberg, Federal Republic of Germany; Harvard Radio Astronomy Station (HRAS), Fort Davis, Texas; Haystack Observatory, Westford, Massachusetts; Onsala Space Observatory, Onsala, Sweden; and Owens Valley Radio Observatory (OVRO), Big Pine, California.   ‡The uncertainties quoted represent the r.m.s. scatter about the mean of the individual determinations and are in all cases at least twice the standard deviation of the mean. Effects of elliptic aberration have been removed.   §The differences are given as new minus old. The 1972–1978 estimates are from Mk I data for April 1972 to May 1978 (12). The uncertainties quoted represent the root-sum squares of the r.m.s. scatters from the two sets of estimates.

Table 2. Summary of baseline length estimates.

	Baseline length (m)				
	Owens Valley	Fort Davis	Onsala	Effelsberg	Chilbolton
Haystack	3,928,881.59 ± 0.02*	3,135,640.98 ± 0.04	5,599,714.43 ± 0.05†	5,591,903.50 ± 0.03	5,072,314.41 ± 0.06
Owens Valley		1,508,195.37 ± 0.02	7,914,130.92 ± 0.09†	8,203,742.44 ± 0.04	7,846,991.19 ± 0.11
Fort Davis			7,940,732.17 ± 0.10	8,084,184.82 ± 0.09	7,663,737.32 ± 0.12
Onsala				832,210.49 ± 0.01	1,109,864.31 ± 0.02

*The baseline lengths were estimated from 8924 observations of group delays obtained in July, September, and October 1980. In this least-squares analysis 468 parameters were estimated by an arc-elimination technique. The root-mean-square of the weighted postfit residuals was ~ 0.1 nsec (~ 3 cm). The uncertainties quoted are the weighted r.m.s. scatter of the results from the separate least-squares analyses of the data from the individual experiments about the values given. The number of individual experiments was 16 for Owens Valley, Fort Davis, and Onsala; 5 for Effelsberg; and 7 for Chilbolton. Only observations at elevation angles greater than 10° were used in this analysis. The speed of light used to convert measured delays to distances was 2.99792458 × 10⁸ m/sec.   †These distances are 0.23 m (Haystack-Onsala) and 0.27 m (Owens Valley–Onsala) less than the Mk I values (11). We believe, but are not certain, that these differences are due to the uncorrected effects of the ionosphere on the single, X-band–Mk I observations.

for correlation. This technique is useful for verifying that the equipment at each site is operating properly before proceeding with an experiment.

The only processor for the Mk III system currently in operation is at Haystack Observatory, Westford, Massachusetts. This processor can correlate simultaneously 28 tracks on tapes from three sites or 14 tracks on tapes from four sites. A similar processor for correlating tapes from three sites at once, to be used at the Max-Planck-Institut für Radioastronomie in Bonn, is nearing completion at Haystack. Another processor of slightly different design, with more emphasis on the use of microprocessors, is being developed and built at Jet Propulsion Laboratory.

Astrometric and geodetic measurements with the Mk III system were started in 1979. The most extensive series of such measurements was carried out during the international MERIT (monitor the earth rotation and intercompare techniques) campaign (10) in September and October 1980. Virtually the same daily schedule of observations of sources was used for all but 2 days of the two 1-week periods of the VLBI part of this experiment. Similar schedules were used in earlier and later experiments. The delay and delay rate data from each day of observation separately were analyzed with the scheme outlined by Herring et al. (11) and yielded estimates of the celestial positions of the sources and of the baselines between the sites. Simultaneous estimates were, in fact, made of parameters representing (i) these positions and baselines, (ii) daily or twice-daily changes in corrections for tropospheric delays for each site, and (iii) daily or twice-daily differences in epoch and rate of the clocks at the various sites. There were typically ten times as many observations as estimated parameters. Representative results are given in Table 1 for the arc lengths between sources, since these lengths have the virtue of being independent of coordinate system. The uncertainties quoted represent the root-mean-square (r.m.s.) scatter about the weighted mean of the results from each day of observation. Except for pairs which include a source at low declination, the scatter is typically a few milliseconds of arc. Also included in Table 1 are the differences between these estimates and those from Mk I experiments (12); the agreement in all but two cases is within the root-sum square of the r.m.s. scatter of each set about its mean value. The positive bias in the differences is probably due to the earlier Mk I observations being confined

to one frequency band (~ 8 GHz) and hence affected by the ionosphere: if no correction is made for the ionosphere, the sources tend to appear closer together.

Our various observations involved 15 baseline vectors connecting six antennas, three in the continental United States and three in Europe. The directions of these vectors at different epochs are related to changes in the pole position and rotation of the earth and will not be discussed further here (13). We will consider results only for the baseline lengths which, like arc lengths, are independent of the choice of coordinate system. A sample of these results for the lengths of baselines connecting antennas in the United States and Europe is shown in Table 2 (14). On average, these estimates are repeatable within about 5 cm.

To discern any significant changes in baseline lengths, many measurements well spread in epoch are necessary. Measurements of the distance between Haystack Observatory and Owens Valley Radio Observatory (near Big Pine, California) best fit these criteria. Measurements with the telescopes at these sites started in September 1976 with the Mk I system and continued with the Mk III system. The results (Fig. 1) show good agreement and indicate a high degree of stability for this baseline. A weighted least-squares estimate of the intercept and slope of the straight line determined by the points in Fig. 1 yields a baseline length given by $B(t) = 3,928,881.61 \pm 0.01 - (0.005 \pm 0.005) \Delta t$ m, where Δt (years) is the time elapsed since 1980.0. The estimated slope is not significant (15); the small decrease in length is most likely due to a positive bias in the baseline lengths determined with the Mk I system, for reasons analogous to those discussed above for arc length estimates. The Mk III results, which have much higher signal-to-noise ratios and dual-band capability, are limited mainly by systematic errors attributable to the unmodeled parts of tropospheric delay and clock "wander." When these sources of systematic error are better understood and partially eliminated, it should be possible to determine these baselines with an uncertainty under 2 cm and the source positions with an uncertainty under 1 millisecond of arc.

The Mk III system, as mentioned, is also a powerful instrument with which to study very faint, compact radio sources. By linking together many large radio telescopes in an interferometer array, sources with flux densities below 1 millijansky can be studied effectively. In the

following report, by far the weakest source yet detected with VLBI is described, but its identification is uncertain; it is either the dim third image of a gravitational lens, the radio emission from the center of a distant elliptical galaxy, or a combination of these.

ALAN E. E. ROGERS
ROGER J. CAPPALLO
HANS F. HINTEREGGER
JAMES I. LEVINE
EDWIN F. NESMAN
JOHN C. WEBBER
ALAN R. WHITNEY
Haystack Observatory,
Westford, Massachusetts 01886
THOMAS A. CLARK
CHOPO MA, JAMES RYAN
Goddard Space Flight Center,
Greenbelt, Maryland 20771
BRIAN E. COREY
CHARLES C. COUNSELMAN
TOMAS A. HERRING
IRWIN I. SHAPIRO
Department of Earth and Planetary
Sciences, Massachusetts Institute of
Technology, Cambridge 02139
CURTIS A. KNIGHT
DAVID B. SHAFFER
NANCY R. VANDENBERG
Phoenix Corporation,
McLean, Virginia 22102
RICHARD LACASSE
ROBERT MAUZY
BENNO RAYHRER*
National Radio Astronomy Observatory,
Green Bank, West Virginia 24944
BRUCE R. SCHUPLER, J. C. PIGG
Computer Science Corporation,
Silver Spring, Maryland 20910

References and Notes

1. Reviewed by M. H. Cohen and S. Unwin, in *IAU Symposium No. 97: Extragalactic Radio Sources,* D. S. Heeschen and C. M. Wade, Eds. (Reidel, Dordrecht, 1982), p. 345.
2. A. R. Whitney *et al., Science* 173, 225 (1971).
3. See, for examples, C. C. Counselman, *Annu. Rev. Astron. Astrophys.* 14, 197 (1976); I. I. Shapiro, in *Methods of Experimental Physics,* M. L. Meeks, Ed. (Academic Press, New York, 1976), vol. 12, part C, p. 261.
4. M. V. Gorenstein *et al., Science* 219, 54 (1983).
5. C. Bare, B. G. Clark, K. I. Kellermann, M. H. Cohen, D. L. Jauncey, *ibid.* 157, 189 (1967). In the first Canadian VLBI experiments, carried out slightly earlier, an analog recording system was used [N. H. Broten *et al., ibid.* 156, 1592 (1967)].
6. B. G. Clark, *Proc. IEEE* 61, 1242 (1973).
7. More detailed information on the Mk III system is contained in a user's manual entitled "Mk III VLBI data acquisition terminal," which is available upon request to Crustal Dynamics Project Office, Goddard Space Flight Center, Code 904, Greenbelt, Md. 20771.
8. Such a development is now in progress.
9. See, for example, A. E. E. Rogers, *Radio Sci.* 5, 1239 (1970).
10. See, for example, D. S. Robertson and W. E. Carter, in *IAU Colloquium No. 63: High Precision Earth Rotation and Earth-Moon Dynamics,* O. Calame, Ed. (Reidel, Dordrecht, 1982), p. 97.
11. T. A. Herring *et al., J. Geophys. Res.* 86, B3, 1647 (1981).
12. T. A. Clark *et al., Astron. J.* 81, 599 (1976). A complete description of the more recent results, discussed here in part, will be given by C. Ma *et al.,* in preparation.

13. D. S. Robertson *et al.*, in preparation.
14. T. A. Herring *et al.*, in preparation.
15. A weighted-least-squares solution using all of the data from these (and other) experiments simultaneously yields consistent results (J. W. Ryan *et al.*, in preparation).
16. We thank the staffs of the participating observatories for their indispensable aid. We also thank

R. J. Coates, E. A. Flinn, T. L. Fischetti, and especially P. B. Sebring for their support. The MIT experimenters were supported by Air Force Geophysics Laboratory contract F19628-81-K-0015; NASA contract NGR22-009-839; NSF grant EAR-7920253; and USGS contract 14-08-0001-18388. Haystack Observatory is operated by a grant from the National Science Foundation, and the National Radio Astronomy Observatory is operated by Associated Universities, Inc., under contract with the National Science Foundation.
* Present address: Jet Propulsion Laboratory, Pasadena, Calif. 91109.

18 May 1982

Coherence Limits for Very-Long-Baseline Interferometry

ALAN E. E. ROGERS AND JAMES M. MORAN, JR., MEMBER, IEEE

Abstract—The quality of the frequency standards used in very-long-baseline interferometry (VLBI) limits the coherent integration time and the accuracy of geodetic experiments except in special cases when clock instabilitites can be made to cancel out by using differential interferometry. Formulas are derived for estimating the coherence of these interferometers from the Allan variances of the frequency standards. Experiments using extremely high-quality frequency standards, such as hydrogen masers, may be limited by the phase noise that results from atmospheric and ionospheric fluctuations.

I. INTRODUCTION

THE COHERENT integration time T for a very-long-baseline (VLB) interferometer is approximately given by [1]

$$\omega T \sigma_y(T) = 1 \qquad (1)$$

where $\sigma_y^2(T)$ is the two-sample Allan variance which describes the instability of the clocks which control the receiver local oscillator and ω is the local oscillator frequency in radians per second. While this simple approximation is adequate for many purposes, it is useful to know more precisely how clock and atmosphere instabilities degrade the observed fringe visibility amplitude. The performance of frequency standards has been extensively documented (e.g., Rutman [2]) but the usual mathematical descriptions do not explicitly cover the problem of interferometer coherence. We wish to relate the coherence of an interferometer to the statistic used in frequency standards analysis, the Allan variance. For this purpose, we define a coherence function $C(T)$ for an integration time T as

$$C(T) = \left| \frac{1}{T} \int_0^T \exp\left[i\varphi(t)\right] dt \right| \qquad (2)$$

where $\varphi(t)$ is the difference phase between the two stations forming the interferometer. $C(T)$ is proportional to fringe amplitude, and when $\langle C^2(T) \rangle = 1$ there is no loss of coherence. $\varphi(t)$ includes the effects of the frequency-standard instability, atmospheric and ionospheric phase noise. In some differential interferometry experiments, phase instabilities can be made to cancel and the coherence is unity.

Manuscript received February 25, 1981; revised July 28, 1981. The VLBI development program at the Haystack Observatory is supported by the National Science Foundation under Grant AST-79-20168 and the National Aeronautics and Space Administration under Contract NAS5-25053.

A. E. E. Rogers is with Haystack Observatory, Westford, MA 01886.

J. M. Moran, Jr., is with Harvard–Smithsonian Center for Astrophysics, Cambridge, MA 02138.

II. RELATION BETWEEN COHERENCE AND THE TWO-SAMPLE ALLAN VARIANCE

The mean-square value of C is

$$\langle C^2(T) \rangle = \frac{1}{T^2} \int_0^T \int_0^T \langle \exp\{i[\varphi(t) - \varphi(t')]\}\rangle \, dt \, dt'. \qquad (3)$$

If φ is a Gaussian random variable

$$\langle C^2(T) \rangle = \frac{1}{T^2} \int_0^T \int_0^T \exp\left[\frac{-\sigma^2(t,t')}{2}\right] dt \, dt' \qquad (4)$$

where $\sigma^2(t, t')$ is the variance of $[\varphi(t) - \varphi(t')]$. If we assume that $\sigma^2(t, t')$ depends only on $(t - t')$, (4) simplifies to

$$\langle C^2(T) \rangle = \frac{2}{T} \int_0^T \exp\left[\frac{-\sigma^2(\tau)}{2}\right](1 - \tau/T) \, d\tau \qquad (5)$$

and

$$\sigma^2(\tau) = \langle [\varphi(t) - \varphi(t')]^2 \rangle = 2R(0) - 2R(\tau) \qquad (6)$$

where $\tau = t - t'$ and R is the autocorrelation function of the phase noise φ. The fractional frequency fluctuation averaged over a time interval τ is (using standard notation, see Rutman [2])

$$\bar{y}_k = \frac{\varphi(t_k + \tau) - \varphi(t_k)}{\omega \tau}. \qquad (7)$$

The Allan variance is defined as

$$\sigma_y^2(\tau) = \langle (\bar{y}_2 - \bar{y}_1)^2 \rangle$$
$$= \frac{1}{2\omega^2\tau^2} \langle [\varphi(t + 2\tau) - 2\varphi(t + \tau) + \varphi(t)]^2 \rangle. \qquad (8)$$

The variance of $\bar{y}$, known as the true variance $I^2(\tau)$, is related to the two-sample Allan variance $\sigma_y^2(T)$ by the relation

$$\sigma_y^2(\tau) = 2[I^2(\tau) - I^2(2\tau)]. \qquad (9)$$

Therefore, $I^2(\tau)$ is given by the series

$$2I^2(\tau) = \sigma_y^2(\tau) + \sigma_y^2(2\tau) + \sigma_y^2(4\tau) + \cdots \qquad (10)$$

provided the series converges. The variance σ^2 is

$$\sigma^2(\tau) = \omega^2\tau^2 I^2(\tau) = \frac{\omega^2\tau^2}{2}[\sigma_y^2(\tau) + \sigma_y^2(2\tau) + \cdots] \qquad (11)$$

Reprinted from *IEEE Trans. Instrum. Meas.*, vol. IM-30, no. 4, pp. 283–286, Dec. 1981.

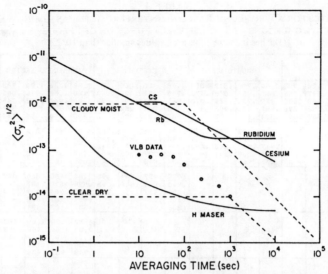

Fig. 1. Allan standard deviation for various frequency standards. Open circles are measured Allan standard deviations for various averaging times observed during clear weather between the Haystack Observatory (Westford, MA) and the National Radio Astronomy Observatory (Greenbank, WV) using H-masers [11]. Also shown are estimates of the limits of the Allan standard deviation for excellent and poor tropospheric conditions, with no allowance for the ionosphere instability.

and hence

$$\langle C^2(T)\rangle^{1/2} \simeq \left[\frac{2}{T}\int_0^T \left\{\exp\left[-\frac{\omega^2\tau^2}{4}(\sigma_y^2(\tau)\right.\right.\right.$$
$$\left.\left.\left. + \sigma_y^2(2\tau) + \cdots)\right]\right\}(1-\tau/T)d\tau\right]^{1/2}. \quad (12)$$

The integral in (12) can be easily evaluated in two simple cases of interest. In the case of white phase noise, such as seen in masers for 10^{-1} s $< T < 10^2$ s, $\sigma_y^2 = \alpha\tau^{-2}$ where α is the Allan variance at 1 s. From (11)

$$I^2(\tau) = (2/3)\sigma_y^2(\tau) = (2/3)\alpha\tau^{-2} \quad (13)$$

and hence (12) can be expressed as

$$\langle C^2(T)\rangle = \exp\left[-(\omega^2\alpha/3)\right]. \quad (14)$$

In the case of white frequency noise, $\sigma_y^2(\tau)$ is equal to $\alpha\tau^{-1}$ and (11) gives

$$I^2(\tau) = \sigma_y^2(\tau) = \alpha/\tau \quad (15)$$

and hence

$$\langle C^2(T)\rangle = [2(e^{-aT} + aT - 1)/a^2T^2] \quad (16)$$

where

$$a = \omega^2\sigma_y^2(T)T/2 = \omega^2\alpha/2. \quad (17)$$

The limiting expressions for $\langle C^2(T)\rangle$ for the white frequency noise case are

$$\langle C^2(T)\rangle = 1 - \omega^2\alpha T/6, \qquad aT \ll 1 \quad (18)$$

$$= 4/(\omega^2\alpha T), \qquad aT \gg 1. \quad (19)$$

When $\omega\sigma_y(T)T = 1$, the loss of coherence (reduction of $\langle C^2(T)\rangle^{1/2}$ from unity) will be approximately 15 and 8 percent, respectively, in the two cases considered above. These

losses assume that one station has a perfect frequency standard. If both stations have a similar standard, then the losses are doubled if losses are small. Thus the maximum observing frequency possible without significant loss is approximately

$$1/(\sigma_y(T)T) \text{ rad/s.} \quad (20)$$

Hence the maximum observing frequency for 10-s integration using two rubidium standards is about 10 GHz and with two hydrogen masers is about 1000 GHz based on the data shown in Fig. 1. Equation (18) shows that the coherence loss for standards that exhibit white frequency noise increases with frequency squared. For this reason, the coherence with rubidium standards decreases rapidly with increasing frequency.

III. NUMERICAL EVALUATION OF THE COHERENCE FOR VARIOUS FREQUENCY STANDARDS

Fig. 1 shows the square root of the Allan variance or two-sample standard deviation for the hydrogen maser (from measurements made between the SAO (model VLG10) and the GSFC (Model NP3)), rubidium (HP model 5065A), and cesium (HP model 5061A with a new beam tube) frequency standards. The curve for the maser stability was derived from that given by Whitney et al. [3] after dividing by the square root of two, assuming that both masers make an equal contribution, and correcting for noise in the comparator. Data obtained by Levine et al. [4] using newer masers are similar, except that they show better performance in the range from 100 to 10 000 s. The curves for the rubidium and cesium standards were derived from the manufacturers data on these standards. The data in Fig. 1 were used in a computer program which performed a numerical integration to calculate the coherence for various frequency standards, integration times, and observation frequencies.

TABLE I

COHERENCE ESTIMATES FOR A VERY-LONG-BASELINE INTERFEROMETER ASSUMING VARIOUS FREQUENCY STANDARDS AND SEEING CONDITIONS AT EACH SITE AS A FUNCTION OF FREQUENCY AND INTEGRATION TIME
(Dashes indicate coherence values smaller than 10^{-3}.)

STN.2 \ STN.1	MASER				RUBIDIUM				CESIUM				MASER+GOOD SEEING				MASER+BAD SEEING				FREQ. GHZ
MASER	1.00	1.00	1.00	.983	1.00	.996	.812	.302	1.00	.988	.866	.577	1.00	1.00	1.00	.983	1.00	.984	.946	.917	1
	1.00	1.00	.997	.866	.997	.961	.515	.170	.996	.898	.437	.143	1.00	1.00	.996	.865	.997	.966	.594	.457	3
	1.00	.999	.968	.553	.973	.731	.270	.070	.965	.561	.189	.023	1.00	.999	.963	.546	.969	.525	.174	.008	10
	1.00	.994	.781	.290	.787	.323	.098	–	.754	.281	.082	–	1.00	.993	.761	.279	.799	.294	.086	–	30
	.998	.943	.457	.149	.339	.104	.002	–	.337	.103	.001	–	.998	.940	.444	.145	.506	.167	.001	–	100
RUBIDIUM					.999	.992	.723	.255	.999	.984	.724	.262	1.00	.996	.812	.302	.999	.979	.764	.282	1
					.994	.926	.435	.141	.993	.867	.376	.119	.997	.961	.515	.170	.994	.834	.360	.112	3
					.947	.601	.208	.043	.939	.503	.167	.012	.973	.731	.270	.078	.943	.477	.157	.004	10
					.660	.239	.061	–	.643	.227	.053	–	.787	.323	.098	–	.679	.239	.055	–	30
					.237	.061	–	–	.237	.060	–	–	.339	.104	.002	–	.305	.091	–	–	100
CESIUM									.999	.977	.764	.395	1.00	.988	.866	.577	.999	.972	.818	.541	1
									.993	.821	.336	.103	.996	.898	.437	.143	.993	.792	.322	.096	3
									.932	.461	.151	.004	.965	.561	.189	.023	.936	.446	.146	.001	10
									.632	.220	.047	–	.754	.281	.082	–	.668	.233	.049	–	30
									.236	.060	–	–	.337	.103	.001	–	.305	.091	–	–	100
MASER + GOOD SEEING													1.00	1.00	1.00	.983	1.00	.984	.946	.917	1
													1.00	1.00	.996	.865	.997	.866	.593	.457	3
													1.00	.999	.959	.540	.969	.525	.174	.008	10
													1.00	.993	.744	.269	.799	.294	.086	–	30
													.998	.937	.434	.141	.506	.167	.001	–	100
MASER + BAD SEEING																	.999	.968	.896	.858	1
																	.994	.775	.390	.250	3
																	.942	.439	.143	–	10
																	.710	.249	.053	–	30
																	.432	.141	–	–	100
INTEGRATION TIME SEC	10	10^2	10^3	10^4	10	10^2	10^3	10^4	10	10^2	10^3	10^4	10	10^2	10^3	10^4	10	10^2	10^3	10^4	SEC

The estimate of the true variance given by the series in (10) becomes inaccurate in the flicker-frequency region where the Allan variance flattens and ceases to decrease with increased averaging time. However, a standard procedure in VLBI data processing is to search for the maximum value of $C(T)$ in fringe frequency. This is equivalent to the removal of a linear phase drift over the coherent integration. We have approximated the effect of this procedure by applying a single-pole high-pass filter, with a cutoff frequency of $(1/T)$ hertz, to the phase noise from the frequency standard. The high-pass filtered phase noise will have an Allan variance equal to that of the unfiltered noise out to an averaging time of T. For averaging time beyond T, the Allan variance of the filtered noise will roll off with a slope of -2 relative to the Allan variance of the unfiltered noise. This filter ensures that the series for the true variance converges provided the Allan variance of the unfiltered noise does not have a positive slope greater than or equal to one. Table I shows the results of the numerical evaluation of the coherence of VLB interferometers for coherent integration intervals from 1 to 10 000 s at frequencies from 1 to 100 GHz. The results in Table I were obtained using the filtered Allan variance values in the numerical evaluation of the integral in (12). We found that four terms in the series for the true variance gave adequate numerial accuracy.

IV. ESTIMATE OF THE EFFECTS OF ATMOSPHERIC AND IONOSPHERIC PHASE FLUCTUATIONS

Since the phase fluctuations due to the atmosphere and ionosphere are time variable and probably cannot be well represented by a two-sample Allan variance, it is difficult to accurately estimate the loss of coherence due to them under all conditions. Atmospheric fluctuations are caused primarily by water vapor. The worst atmospheric conditions occur during warm, humid, and cloudy weather while the most stable conditions occur during cool, dry, and clear weather. Because the water vapor is poorly mixed in the atmosphere, it is difficult to estimate the propagation delay from ground-based measurements to an accuracy of better than about 50 ps [5], [6]. The power spectrum of the atmospheric phase fluctuations can be modeled as white phase noise below a critical frequency, f_c, and flicker frequency noise above f_c [7]. This gives an Allan variance that is independent of averaging time for $T < 1/f_c$ and is proportional to T^{-2} for $T > 1/f_c$ (see [2, eq. (6.9) and table II]). We have examined the rms phase noise on several interferometers [8]–[10] and adopted the following parameters: An rms path variation of 100 ps with $f_c = 10^{-2}$ Hz to characterize the worst conditions and an rms path variation of 10 ps with $f_c = 10^{-3}$ Hz for the best conditions. This model yields the Allan variance estimates plotted in Fig. 1.

Phase data from a VLBI experiment performed in November 1979 at 1.35-cm wavelength on a 850-km baseline [11], where very strong radio sources were observed, were used to estimate the Allan variance for a complete operational system including the atmosphere and local oscillator multipliers. The skies were clear and the humidity low (partial pressure of $H_2O \sim 5$ mbar). This Allan variance is also plotted in Fig. 1.

Phase fluctuations in the ionosphere will also limit the coherence time. If we assume that the ionosphere has a mean electron content of 10^{17} electrons per square meter then flucutations of 1 percent produce a path variation of 30 ps at 2 GHz. This path-length variation is probably fairly representative of daytime conditions midway in the solar cycle (see Mathur *et al.* [12]). Thus under these conditions, the effects of troposphere and ionosphere are approximately equal at 2 GHz. At night, ionospheric effects normally decrease by a factor of 2 or 3. The ionosphere changes by a factor of approximately 2 from minimum to the maximum of the solar cycle. The standard deviations of ionospheric time-delay fluctuations decreases with frequency squared so that the largest ionospheric fluctuations become comparable with the smallest tropospheric fluctuations at frequencies around 5 GHz.

Table I also shows the coherence limits imposed by the combination of an H-maser frequency standard and good or bad atmospheric "seeing." The longest coherence times are achievable at low frequenncies, and while the effect of frequency standards and troposphere decrease with decreasing frequency, the ionospheric fluctuations increase so rapidly that a frequency around 2 GHz is probably the best choice if long coherent integrations are needed for maximum interferometer sensitivity.

V. THE EFFECT OF PHASE NOISE ON THE ACCURACY OF GEODETIC EXPERIMENTS

Kartaschoff [13] has shown that the standard deviation of clock time departure Δ after time T is approximately given by

$$\Delta = T\sigma_y(T). \tag{21}$$

If a geodetic experiment is carried out by observing a radio source for time T, followed by observing a "clock" or reference radio source for time T after a dead time T', then the rms noise in the differential measurement due to frequency standards and other sources of phase noise, Δ', is approximately

$$\Delta'(T + T')\sigma_y(T + T') \tag{22}$$

which equals 50 ps for $(T + T') = 100$ s using rubidium standards at each station. Because σ_y stops decreasing with time for averaging times longer than 100 s, the clock noise increases with time. Hence for goedetic experiments using rubidium standards it is very important to move from one radio source to another in less than 100 s.

VI. CONCLUSIONS

For coherent integrations less than about 10^4 s, the atmosphere or ionosphere is the limiting factor in experiments using hydrogen-maser standards. Experiments using rubidium or cesium standards will almost always be limited by the performance of those standards.

REFERENCES

[1] W. K. Klemperer, "Long-baseline radio interferometry with independent frequency standards," *Proc. IEEE*, vol. 60, pp. 602–609, May 1972.
[2] J. Rutman, "Characterization of phase and frequency instabilities in precision frequency sources: Fifteen years of progress," *Proc. IEEE*, vol. 66, pp. 1048–1075, Sept. 1978.
[3] A. R. Whitney, A. E. E. Rogers, H. F. Hinteregger, L. B Hanson, T. A. Clark, C. C. Counselman, and I. I. Shapiro, "Application of very-long-baseline interferometry to astromety and geodesy: Effects of frequency standard instability on accuracy," in *Proc. 6th Annu. Precise Time and Time Interval (PTTI) Planning Meet.* (Dec., 1974), NASA GSFC Rep. X-814-75-117.
[4] M. W. Levine, R. F. Vessot, and E. M. Mattison, "Performance evaluation of the SAO VLG-11 atomic hydrogen masers," presented at the 32nd. Annu. Frequency Control Symp., Atlantic City, NJ, June 1978.
[5] F. O. Guiraud, J. Howard, and D. C. Hogg, "A dual channel microwave radiometer for measurement of precipitable water vapor and liquid," *IEEE Trans. Geosci. Electron.*, vol. GE-17, pp. 129–136, 1979.
[6] J. M. Moran and B. R. Rosen, "The estimation of the propagation delay through the troposphere from microwave radiometer data," *Radio Sci.* (in press), Mar. 1981.
[7] M. C. Thompson, H. B. Janes, and A. W. Kirkpatrick, "An analysis of time variations in tropospheric refractive index and apparent radio path length," *J. Geophys. Res.*, vol. 65, pp. 193–201, Jan. 1960.
[8] J. P. Basart, G. K. Miley, and B. G. Clark, "Phase measurements with an interferometer baseline of 11.2 km," *IEEE Trans. Antennas Propagat.*, vol. AP-18, pp. 375–379, May 1970.
[9] R. Hinder and M. Ryle, "Atmospheric limitations to the angular resolution of aperture synthesis radio telescopes," *Mon. Notic. Roy. Astron. Soc.*, vol. 154, pp. 229–253, 1971.
[10] J. Hamaker, "Atmospheric delay fluctuations with size scales greater than one kilometer, observed with a radio interferometer array," *Radio Sci.*, vol. 13, pp. 873–891, 1978.
[11] J. M. Moran *et al.*, "Absolute positions of H$_2$O masers," *Astrophys. J.*, 1981 (in preparation).
[12] N. C. Mathur, M. D. Grossi, and M. R. Pearlman, "Atmospheric effects in very long baseline interferometry," *Radio Sci.*, vol. 5, pp. 1253–1261, Oct. 1970.
[13] P. Kartaschoff, "Computer simulation of the conventional clock model," *IEEE Trans. Instrum. Meas.*, vol. IM-28, pp. 193–197, Sept. 1979.

Part IX
Data Processing Techniques

THE data obtained from radio observations can almost always benefit from subsequent processing, which is generally referred to as off-line processing or post-processing. The circumstance in which this is most important is in studies of regions in which the radio emission is spatially extended. Since the intensity obtained with the telescope pointed in a given direction is the convolution of the response of the telescope with the source brightness distribution, it is apparent that the fidelity of the data could be improved if the antenna pattern, for example, could be deconvolved from the observed data set. This situation is certainly relevant for single-dish observations, in which we have to consider the finite-width main lobe of the pattern, together with sidelobes. It is even more critical for interferometric maps, for which the inevitably imperfect set of baselines available will produce undesirable spurious features in the map of the source.

This topic of deconvolution, or image restoration, has been of interest to radio astronomers (as well as those observing at other wavelengths) for some time, but the rapid growth in available computer power has made significant off-line processing a realistic possibility in recent years. Consequently, there has been a great proliferation in the number of different schemes proposed, and the details of their implementation; a recent review of one method of image restoration gives 125 references! Radio astronomers have played a very important role in the development of this field, but there is an increasing degree of interaction with disciplines such as X-ray crystallography and many other branches of astronomy.

Interferometry is the preeminent method of achieving high angular resolution, and virtually all radio-wavelength systems using the principle of the Michelson interferometer depend on knowing the relative phase of the signal reaching each of the antennas in the interferometer array. This can be corrupted by a number of effects, some of them instrumental, such as phase variations in the local oscillator distribution system and thermal effects on the phase path through the antennas themselves. In addition, the atmosphere can produce appreciable phase variations, especially for relatively large antenna separations. Phase variations, whatever their origin, degrade the quality of the maps that are produced. This problem led to the development of *phase closure* (also called *self-calibration* or *hybrid mapping*) in 1958 by Jennison who was working with a three-element radio-linked interferometer. Jennison's paper, which is the first paper in this part, indicates that by combining the output phases of two-element interferometer pairs around a closed loop, a phase which is independent of instrumental and atmospheric effects can be formed. This *closure phase* can be used to construct maps of the source which have vastly improved image quality compared to that obtained using the interferometer outputs directly.

The phase closure technique proved to be almost essential for VLBI, where atmospheric variations are relatively large and clock errors add a significant uncertainty to the phase of the output from each antenna. However, even with connected-element interferometers which normally have good phase stability, the use of the phase closure technique can significantly improve the dynamic range of the final output map. The actual use of phase closure relations in constructing a map from interferometric data is a complex issue; this is discussed by Cornwell and Wilkinson in the second paper, and in a number of the references given below.

Single-dish antennas do not have the same phase or amplitude calibration problems as interferometers, but atmospheric variations can have harmful effects, especially for continuum observations. Some hardware methods for overcoming these problems were discussed in Part V, but data quality can be enhanced by post-processing. Several techniques are described in the references. The third paper included in this part, by Reich *et al.*, presents an iterative method for removing the effects of antenna sidelobes in maps of extended (or multiple) radio sources. The technique depends on the change in the orientation of a celestial source with respect to the horizon coordinate system of a telescope with an azimuth-elevation mounting. While the beam quality of single-dish antennas is usually quite good, this technique allows maps with dynamic range exceeding 30 dB to be made.

The quality (particularly the sidelobe level) of the synthesized beam from an interferometer is critically dependent on the distribution of baselines that were used to acquire the data. Sidelobe levels on the order of 10% of the major peak are typical. A significant reduction of the sidelobe level in interferometric maps is produced by use of the *CLEAN* algorithm developed by Högbom, which he describes in the fourth paper in this part. This technique is intended to improve the quality of interferometric data taken with a nonuniform distribution of baselines, and models the source emission as a set of point sources. In each iteration, the response of the actual ("dirty") synthesized beam at the location in the map having the strongest emission is subtracted. This process is continued until the modified map is basically only noiselike. The final map is made by adding back to the modified map a set of point sources convolved with a "clean" beam, which is generally chosen to be the same size as the main lobe of the actual synthesized beam, but without any sidelobes.

Many tests have been made of the CLEAN procedure, and some references are given below. The algorithm has been analyzed mathematically by Schwarz (1978), and variations which have improved the usefulness of this approach have been developed by a number of astronomers. CLEAN is successful in processing maps with a reasonable number of pixels, particularly where a limited number of pointlike sources can adequately model the emission. However, for images with a large number (>1 million) of pixels, and especially when the emission is spatially extended, it becomes computationally prohibitive. This is one reason why alternative methods of deconvolution have been developed. Certainly the most successful of these is the maximum entropy method (MEM).

MEM operates on a limited set of data, which can be a map in physical coordinates or a set of interferometer outputs for different baselines. The central concept is to extrapolate the data set to domains where data do not exist (missing baselines or high spatial frequency components, for example) while making the minimal assumption about the nature of the data that are not available. To implement the extension of the data set in a well-defined manner, an additional constraint must be imposed, which is generally to maximize some function of the existing data. The development of MEM methods has seen various forms for the function to be maximized. Most practitioners have adopted an entropy function of the form $\int I(x, y) \ln I(x, y) \, dx \, dy$, where $I(x, y)$ is the intensity at position x, y. Some other choices are discussed in articles by Gull and Skilling (1984), and Narayan and Nityananda (1986).

If for no other reason than the complexity of the connection with the physical entropy of a system, the issue of MEM is an involved one, and many highly mathematical papers have appeared on the subject. The fifth paper, by Ables, is an early and very clear introduction to MEM. The following paper by Skilling and Bryan discusses in detail a very effective algorithm for implementing the maximum entropy procedure. The contributions chosen for inclusion here can only serve as a very brief introduction to this important subject, which appears to be playing an ever-increasing role in the processing of radio-astronomical data.

REFERENCES AND BIBLIOGRAPHY

Image Restoration and Deconvolution: General

1980: D'Addario, L. R., "Fourier reconstruction techniques in radio astronomy: A review," in *Proc. 1980 Int. Optical Computing Conf.* (SPIE vol. 231), 1980, pp. 2–9.
1984: Roberts, J. A. (Ed.), *Proc. Int. Symp. on Indirect Imaging*, Sydney, Australia, 1983. Cambridge: Cambridge University Press, 1984.
1985: Cornwell, T., "Deconvolution," Chapter 7 in *Synthesis Imaging* (NRAO Summer School Course Notes), R. A. Perley, F. R. Schwab, and A. H. Bridle, Eds. Green Bank, WV: National Radio Astronomy Observatory, 1985, pp. 109–121.
1986: Thompson, A. R., J. M. Moran, and G. W. Swenson, *Interferometry and Synthesis in Radio Astronomy*, Chapter 11: "Image processing and enhancement." New York, NY: Wiley, 1986.

Image Restoration Techniques with Single Antenna Orientation

1973: Westerhout, G., H. U. Wendlandt, and R. H. Harten, "A method for accurately compensating for the effects of the error beam of the NRAO 300-ft radio telescope at 21-cm wavelength," *Astron. J.*, vol. 78, pp. 569–572, 1973.
1974: Haslam, C. G. T., W. E. Wilson, D. A. Graham, and G. C. Hunt, "A further 408 MHz survey of the northern sky," *Astron. Astrophys. Suppl.*, vol. 13, pp. 359–394, 1974.
1976: Reich, W., P. Kalberla, and J. Niedhöfer, "Large scale structure around 3C 123," *Astron. Astrophys.*, vol. 52, pp. 151–154, 1976.
1984: Wielebinski, R., "Mapping of large fields with single-dish telescopes," Paper 5.5 in *Proc. Int. Symp. on Indirect Imaging*, Sydney, Australia, 1983, J. A. Roberts, Ed. Cambridge: Cambridge University Press, 1984, pp. 199–204.

Closure Phase

1974: Rogers, A. E. E., H. F. Hinteregger, A. R. Whitney, C. C. Counselman, I. I. Shapiro, J. J. Wittels, W. K. Klemperer, W. W. Warnock, T. A. Clark, L. K. Hutton, G. E. Marandino, B. O. Ronnang, O. E. H. Rydbeck, and A. E. Neil, "The structure of radio sources 3C273B and 3C84 deduced from the 'closure' phases and visibility amplitudes observed with three-element interferometers," *Astrophys. J.*, vol. 193, pp. 293–301, 1974.
1978: Readhead, A. C. S. and P. N. Wilkinson, "The mapping of compact radio sources from VLBI data," *Astrophys. J.*, vol. 223, pp. 25–36, 1978.
1980: Rogers, A. E. E., "Methods of using closure phases in radio aperture synthesis," in *Proc. 1980 Int. Optical Computing Conf.* (SPIE vol. 231), 1980, pp. 10–17.
Schwab, F. R., "Adaptive calibration of radio interferometer data," in *Proc. 1980 Int. Optical Computing Conf.* (SPIE vol. 231), 1980, pp. 18–25.
1981: Cornwell, T. J. and P. N. Wilkinson, "A new method for making maps with unstable radio interferometers," *Mon. Notic. Roy. Astron. Soc.*, vol. 196, pp. 1067–1086, 1981.
1983: Ekers, R. D., "The almost serendipitous discovery of self-calibration," in *Serendipitous Discoveries in Radio Astronomy*, K. Kellerman and B. Sheets, Eds. Charlottesville, VA: National Radio Astronomy Observatory, 1983, pp. 154–159.
1984: Pearson, T. J. and A. C. S. Readhead, "Image formation by self-calibration in radio astronomy," *Annu. Rev. Astron. Astrophys.*, vol. 22, pp. 97–130, 1984.

CLEAN

1978: Schwarz, U. J., "Mathematical-statistical description of the iterative beam removing technique (Method CLEAN)," *Astron. Astrophys.*, vol. 65, pp. 345–356, 1978.
Segalovitz, A. and B. R. Frieden, "A 'CLEAN'-type deconvolution algorithm," *Astron. Astrophys.*, vol. 70, pp. 335–343, 1978.
1980: Clark, B. G., "An efficient implementation of the algorithm 'CLEAN'," *Astron. Astrophys.*, vol. 89, pp. 377–378, 1980.
1983: Cornwell, T. J., "A method of stabilizing the Clean algorithm," *Astron. Astrophys.*, vol. 121, pp. 281–285, 1983.
1984: Steer, D. G., P. E. Dewdney, and M. R. Ito, "Enhancements to the deconvolution algorithm 'CLEAN'," *Astron. Astrophys.*, vol. 137, pp. 159–165, 1984.
1987: Marsh, K. A. and J. M. Richardson, "The objective function implicit in the CLEAN algorithm," *Astron. Astrophys.*, vol. 182, pp. 174–178, 1987.

Maximum Entropy

1975: Papadopoulos, G. D., "A statistical technique for processing radio interferometer data," *IEEE Trans. Antennas Propagat.*, vol. AP-23, pp. 45–53, 1975.
1978: Bhandari, R., "Maximum entropy spectral analysis—Some comments," *Astron. Astrophys.*, vol. 70, pp. 331–333, 1978.
Gull, S. F. and G. J. Daniell, "Image reconstruction from incomplete and noisy data," *Nature*, vol. 272, pp. 686–690, 1978.
1984: Grasdalen, G. L., S. E. Strom, K. M. Strom, R. W. Capps, D. Thompson, and M. Castelaz, "High spatial resolution IR observations of young stellar objects: A possible disk surrounding HL tauri," *Astrophys. J.*, vol. 283, pp. L57–L61, 1984.
Gull, S. F. and J. Skilling, "The maximum entropy method," Paper 8.1 in *Proc. Int. Symp. on Indirect Imaging*, Sydney, Australia, 1983, J. A. Roberts, Ed. Cambridge: Cambridge University Press, 1984, pp. 267–279.
1985: Cornwell, T. J. and K. F. Evans, "A simple maximum entropy deconvolution algorithm," *Astron. Astrophys.*, vol. 143, pp. 77–83, 1985.
1986: Narayan, R. and R. Nityananda, "Maximum entropy image restoration in astronomy," *Annu. Rev. Astron. Astrophys.* vol. 24, pp. 127–170, 1986.
1987: Shevgaonkar, R. K., "Maximum entropy method for polarized images," *Astron. Astrophys.*, vol. 176, pp. 159–170, 1987.

A PHASE SENSITIVE INTERFEROMETER TECHNIQUE FOR THE MEASUREMENT OF THE FOURIER TRANSFORMS OF SPATIAL BRIGHTNESS DISTRIBUTIONS OF SMALL ANGULAR EXTENT

R. C. Jennison

(Communicated by A. C. B. Lovell)

(Received 1958 February 21)

Summary

A method is described whereby the amplitude and phase of the complex Fourier transform of a spatial brightness distribution of small angular extent may be uniquely determined from a series of measurements with a triple interferometer system. Absolute measurement of the amplitude function is available, whilst measurements of phase are relative to a datum obtained at short aerial spacings. A practical radio frequency interferometer incorporating the principle is described and its operation is discussed.

---

1. *Introduction.*—Unique measurements of the brightness distribution across the cosmic radio sources have hitherto been limited to those sources which subtend a large angular diameter. It has been possible to obtain orders of magnitude for the angular diameter of some of the smaller intense sources and in a few cases to give models for their general intensity distributions (Hanbury Brown, Jennison and Das Gupta 1952, Mills 1952, Jennison and Das Gupta 1953, 1956). These models have been obtained from the amplitude functions of the Fourier transforms of the source brightness distributions without corresponding information with regard to the phase. As such, the solutions are not unique and may not correspond in detail to the actual fine structure of the sources.

The amplitude function of the Fourier transform of the brightness distribution across a source may be determined from the visibility of the fringes in the reception pattern of an interferometer. The phase of the transform may be obtained from the phase of the fringes relative to a known datum. This datum may be obtained by accurately measuring the position of the source over various interferometer baselines.

The practical limitations imposed on these measurements are severe and render conventional techniques untenable when the baseline of the interferometer system exceeds about one thousand wave-lengths. At greater distances, when it is necessary to use a radio link between the channels, the measurement of fringe amplitude may be accomplished with an uncertain error if a powerful automatic gain control system is used on the receiving equipment to maintain the mean level of the signal constant over the period of the observations. The aerial gain, noise factor, bandwidth and dispersion are assumed constant or must be continuously calibrated (Mills 1953). If both the receiving aerials are sufficiently large the uncertainty may be removed by dispensing with the automatic gain control and measuring the increment in total power due to the source in each channel (Jennison and Das Gupta 1956). Neither of these methods allows for error due to phase dispersion in the systems and the latter method has the practical disadvantage of requiring a very large portable aerial. The method used by Smith (1952) is most reliable on short baselines but is not applicable where the distance between aerials is considerable. In these circumstances intermediate checks cannot rapidly be made and the gain of the most distant station cannot necessarily be relied upon to remain constant over long periods or on removal to a new site.

The conventional method for the measurement of the phase of the fringe system by timing the transit of the central fringe is not applicable at long baselines, especially when radio links are used to span the baseline. The combinations of variable phase errors in the electronic equipment and in the local oscillator and intermediate frequency signal link paths, together with the difficulties arising from unequal signal paths when the terrain is not level, render the phase of the transform indeterminate at spacings in excess of about one thousand wave-lengths.

An interferometer system designed to overcome the above defects was constructed in 1954 and its principle of operation is discussed below.

2. *Principle of operation.*—The main features of the apparatus are shown in Fig. 1. It consisted of three aerial systems, tuned to a frequency of 127 Mc/s, spaced in a line along which the measurements were to be made. One of the aerials was a large fixed array whilst the other two were small and portable. Associated with the portable systems were radio link transmitters to relay the received signals back to the fixed station. Provision was made to rotate, continuously and synchronously, the phase of the two relayed signals prior to equalizing the time delays, and combining all three signals in pairs in three switched interferometers, the outputs of which were displayed on a three-pen recorder. The total power received from the fixed aerial was displayed on a separate recorder.

The measurement of phase

The relative phases of the signals induced in the three aerials are indeterminate by the time they reach the switched interferometer units, due to the combinations of errors previously referred to. Thus the fringe systems produced at the outputs of these units and displayed upon the pen recorder are also of uncertain phase relative to an absolute datum based upon the time of transit of the source. It will be shown that the relative phase contribution due only to the brightness distribution of the source may be obtained by a comparison of the arguments of these fringe systems.

Treating the signals from aerial A as the phase reference and taking the origin of the transform at zero within the source, let ξ_{AB}, ξ_{BC} and ξ_{AC} represent the phase of the transform at spacings AB, BC and AC respectively; let $\omega_1(t)$ and $\omega_2(t)$ represent the time variable phase rotation of the scan at B and C respectively; let ψ_1 and ψ_2 represent the phase angle introduced by the position of the source in relation to the collimation plane at B and C respectively; and let δ_1 and δ_2 represent the phase error in the equipment at B and C respectively.

The argument of the fringes when channel B interferes with A will be

$$\xi_{AB} + \psi_1 + \delta_1 + \omega_1(t) \tag{1}$$

and the argument of the fringe when channel C interferes with channel B

$$= \xi_{BC} + (\psi_2 - \psi_1) + (\delta_2 - \delta_1) + \omega_2(t) - \omega_1(t). \tag{2}$$

The argument of the fringes formed by channel C interfering with channel A

$$= \xi_{AC} + \psi_2 + \delta_2 + \omega_2(t). \tag{3}$$

Adding the arguments of the two fringe systems AB and BC, we obtain

$$\xi_{AB} + \xi_{BC} + \psi_2 + \delta_2 + \omega_2(t). \tag{4}$$

Comparing equations (3) and (4) it will be seen that they represent the arguments of two fringe systems of identical frequency and relative phase

$$\phi = \xi_{AC} - (\xi_{AB} + \xi_{BC}). \tag{5}$$

This result is independent of the phase errors and the angular velocity of the phase rotation. It enables an estimate to be made of relative phase contributions due to the source structure only, and is applicable in the general case where the aerials are spaced by arbitrary distances. It will be observed that the position of the source relative to the collimation plane does not appear in the expression $\xi_{AC} - (\xi_{AB} + \xi_{BC})$ and hence absolute phase is not directly obtainable.

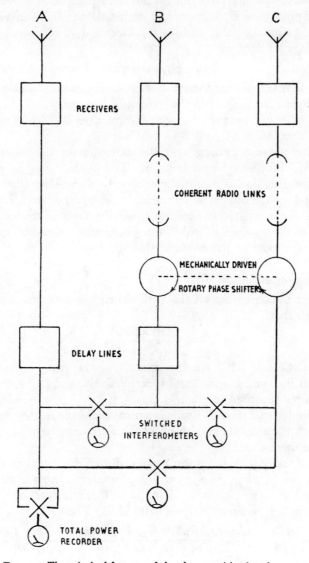

Fig. 1.—*The principal features of the phase-sensitive interferometer.*

It is apparent that if the spacings AB and BC both lie within the first maximum of the Fourier transform of a simple symmetrical distribution, then $\xi_{AB} = \xi_{BC} = 0$. In these circumstances $\phi = \xi_{AC} = 0$ if AC lies within the first maximum and $\phi = \xi_{AC} = \pi$ if AC lies within the second maximum.

If the transform is complex but the positions of the aerials are adjusted so that AB = BC, equation (5) reduces to

$$\phi = \xi_{AC} - 2\xi_{AB}$$

and a direct check on the relative phase over twice the spacing may thus be obtained.

In the more general case, when the transform is complex and the aerial spacings unequal, the expression (5) may be used for the measurements in the following manner.

The two portable aerials B and C are moved away from the home station A with their baselines in any ratio until a phase displacement is recorded. This may be most simply done by keeping one station, B, at some convenient distance from A and moving the other until a displacement is observed. Within this range of spacings we may put

$$\xi_{AB} = \xi_{BC} = \text{constant.}$$

It is apparent that the phase function does not vary, within observational error, over intermediate spacings within the maximum separation AC, and hence both portable stations may be moved to more distant sites B′ and C′ subject only to the limitation that AB′ < AC and B′C′ < AC.

The relative phase occurring over the extension of the baseline to points in the range AC′ may now be established. The process may be repeated until the whole phase function is eventually mapped. It will be observed that this general method gives rise to a cumulative error, and the accuracy of the determination of the transform depends upon the ratio of the maximum aerial spacings used to the spacings at which a displacement is first observable, and the error in the assessment of the phase at this spacing.

The measurement of amplitude

From direct measurement on the pen recordings it is possible to form the function η such that

$$\eta = \frac{|AB| \times |AC|}{|BC| \times |A|^2},$$

where $|AB|$, $|BC|$ and $|AC|$ are the moduli of the fringe systems between channels A and B, B and C and A and C respectively, and $|A|^2$ is the total power due to the source recorded in channel A.

It will be shown that the function η is a measure of the relative amplitude and, in a special case, the absolute amplitude, of the Fourier transform of the intensity distribution of the source.

Let $g_1\alpha$, $g_2\beta$ and $g_3\gamma$ be the mean amplitudes of the signal at the outputs of channels A, B and C, where g_1, g_2, g_3 and α, β, γ refer to the overall voltage gains and input signal levels respectively. Let $f(ab)$, $f(bc)$ and $f(ac)$ be the amplitudes of the transform for baselines AB, BC and AC. Then

$$\eta = \frac{|AB| \times |AC|}{|BC| \times |A^2|} = \frac{G_1 g_1 \alpha g_2 \beta f(ab) \times G_3 g_1 \alpha g_3 \gamma f(ac)}{G_2 g_2 \beta g_3 \gamma f(bc) \times G_4 (g_1\alpha)^2} = \frac{f(ab) \times f(ac)}{f(bc)} \times \frac{G_1 G_3}{G_2 G_4},$$

(6)

where G_1, G_2 and G_3 are the conversion gains for the cross multiplying recorder systems between channel A and B, B and C, and A and C respectively and G_4 is the conversion gain of the total power recorder system in channel A. The expression for η in equation (6) is not dependent on the gain of any part of the equipment other than the final cross multiplying and total power recording units. The equipment was designed so that these units were operated at high level and were extremely stable. Sockets were provided so that, immediately prior to and following a run, the inputs normally fed from the separate channels A, B and C could be connected together and provided with a single noise signal. From the constant deflection of the recording pens this yielded a normalizing function:

$$K = \frac{G_1 G_3}{G_2 G_4} \tag{7}$$

corresponding to a correlation coefficient of unity between the channels.

Substituting the normalizing function K in equation (6), the general operational function η' may be obtainable, where

$$\eta' = \frac{|AB| \times |AC|}{K \times |BC| \times |A|^2} = \frac{f(ab) \times f(ac)}{f(bc)}. \tag{8}$$

This expression only involves terms in the amplitude function of the Fourier transform of the brightness distribution across the source. It will be observed that in the special case where station B is midway between stations A and C, $f(ab) = f(bc)$ and equation (8) reduces to

$$\eta' = f(ac).$$

The absolute value of the amplitude of the transform may therefore be determined.

In general, when $AB \neq BC$, equation (8) may be used to obtain a relative measurement of the amplitude at one spacing from a knowledge of the amplitude at the other two. Measurements may therefore be made concurrently with the phase measurements, commencing at short spacings and extending the baselines as each value of the amplitude function is established.

3. *A practical radio frequency interferometer.*—The principles outlined above were incorporated in an interferometer constructed at Jodrell Bank and operating on a frequency of 127 Mc/s.

The aerial A consisted of 160 full-wave dipoles in a collinear broadside array. This aerial had a beam width of approximately $3° \times 10°$. Aerial B usually consisted of 20 full-wave dipoles in a collinear broadside array giving a beam pattern $24° \times 10°$. Aerial C was either a single folded dipole with reflecting screen or an array of four similar dipoles.

The basic lay-out of the electronic equipment is shown in Fig. 2. The outstations each employed two crystal oscillators to feed their first and second mixers. The beat between these two oscillators was transmitted as a coherent local oscillator reference. The noise signal was transmitted from the broadband class A amplifier following the second mixer. Transmitter powers were normally about 0·25 and 1 watt from the signal and local oscillator respectively. The relay transmitting and receiving aerials consisted of single dipoles with parasitic reflectors, the polarization of the signal and local oscillator aerials was crossed and orthogonal to the corresponding aerials of the other radio link,

The receiving equipment for the radio links consisted of separate preamplifiers for the local oscillator and signal channels. The low level of the signal from the received local oscillator was converted coherently to a signal of the order of volts by being mixed with the tenth harmonic of a master local oscillator on 11·5 Mc/s, and subsequently passed through a narrow band amplifier at approximately 30 Mc/s. The output of this unit was applied to a mixer following the signal link preamplifier and yielded an output on an intermediate frequency of 12 Mc/s.

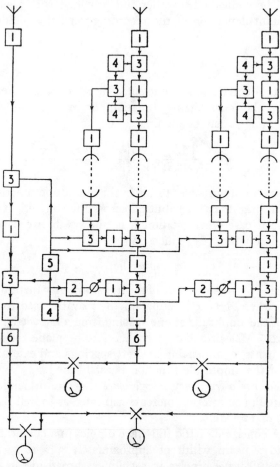

FIG. 2.—*Basic block diagram of the phase-sensitive interferometer. The numbers in the blocks correspond to: 1, amplifier; 2, buffer; 3, mixer; 4, oscillator; 5, frequency multiplier; 6, delay line.*

The corresponding equipment in channel A consisted only of a preamplifier, a mixer combining the 115 Mc/s master local oscillator frequency with the signal frequency, and an I.F. amplifier on 12 Mc/s.

The 12 Mc/s I.F. amplifiers in all three channels were followed by mixer valves converting the frequency to 500 kc/s. The conversion was performed by mixing with the master local oscillator on 11·5 Mc/s, directly in the case of channel A, but via continuously rotating phase shift networks in the case of channels B and C. These phase shift networks consisted of resistance capacity

19

470

quadrature circuits applied to a four pole condenser system having a rotating search vane of high dielectric constant. The amplitude of the output of this unit was substantially constant, and the tracking error in the phase angle was better than 2°. The phase rotating systems in channels B and C were mechanically coupled to friction rollers adjustable in position on a single conical shaft driven from a synchronous electric motor. A large but constant ratio between the rates of phase rotation of the two channels was thus available.

The three signals at 500 kc/s were confined to matched bandwidths of 200 kc/s, or, in certain measurements, 50 kc/s. The signals from channels A and B were then delayed by suitable lengths of artificial lines before being applied to the cross multiplying and total power recording unit.

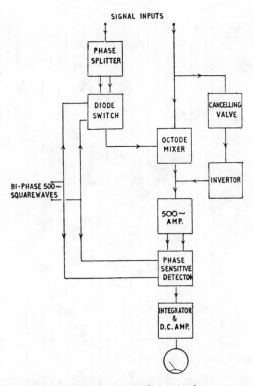

FIG. 3.—*Block diagram of one of the cross-multiplier units shown as a cross in Figs. 1 and 2.*

The three cross multiplying circuits and the total power recording circuits were identical except that the two input leads of the total power recorder were connected together. A basic block diagram of a single section is shown in Fig. 3. 500-cycle square waves were used for switching, whilst the actual multiplication was performed in an octode valve; an auxiliary circuit with a balancing valve eliminated single channel noise by cancelling the effect of non-linearity on the variable-μ signal grid. Two phase splitters and switching circuits only were required to serve the three multipliers and one square law detector which operated on a similar principle.

4. *The equipment in operation.*—The equipment was extensively tested on short baselines within the range of readings previously obtained with conventional

471

techniques. During the tests a slight dispersion in the radio links was observed.
This was reduced by improved design and finally removed entirely by the
introduction of compensating networks.

On the completion of the initial tests the equipment was applied to the
measurement of the Cygnus A and Cassiopeia A radio sources. A typical record
is shown in Fig. 4. The different declinations of the two sources necessitated
the use of slightly different settings of the artificial delay lines and rotating phase
shifters. At most aerial spacings it was possible to use a setting for the rotary
phase shifters such that both sources gave fringes of suitable period but of opposite
angular velocity. The direction of the velocity vector determined the interpretation of an observed phase shift as leading or lagging, and hence the sense of the
asymmetry of the source.

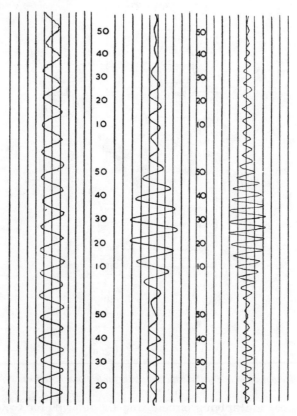

FIG. 4.

It was found that the measurement of the argument of the fringes was a
considerably more powerful method for the determination of the order of
the maxima of the transform patterns than a similar small number of measurements
of the fringe modulus. The positions of the minima of the Cassiopeia source
could be most accurately assessed by observing the change of phase in their
immediate vicinity.

The measurements made on the Cygnus source confirmed the bifurcation
of this object (I.A.U. Symposium on Radio Astronomy, 1955), whilst the
measurements on the Cassiopeia source established three maxima of the Fourier

472

transform of its intensity distribution and showed the existence of a faint component of larger angular diameter off-set from the principal component of the source. The details of this work will be described in subsequent papers.

Acknowledgments.—The author wishes to thank Dr V. Latham for carrying out the major part of the construction and operation of this equipment and Professor A. C. B. Lovell and Mr R. Hanbury Brown for their encouragement and interest in the experiment.

Jodrell Bank Experimental Station,
Lower Withington,
Cheshire:
1958 *February* 20.

References

Hanbury Brown, R., Jennison, R. C. and Das Gupta, M. K., 1952, *Nature*, **170**, 1061.
Mills, B. Y., 1952, *Nature*, **170**, 1063.
Mills, B. Y., 1953, *Aust. Journal of Physics*, **6**, 452.
Jennison, R. C. and Das Gupta, M. K., 1953, *Nature*, **172**, 996.
Jennison, R. C. and Das Gupta, M. K., 1956, *Phil. Mag.*, **1**, 55.
Smith, F. G., 1952, *Proc. Phys. Soc.* B, **65**, 971.

SELFCALIBRATION

T.J. Cornwell,
National Radio Astronomy Observatory,Very Large Array,
P.O. Box 0, Socorro, New Mexico 87801.

P.N. Wilkinson,
Nuffield Radio Astronomy Laboratories,Jodrell Bank,
Macclesfield, Cheshire, United Kingdom.

1 INTRODUCTION

The puritanical idea that one cannot get something for nothing might lead us to reject many of the data processing methods being discussed at this conference. Thus, we would lose CLEAN, MEM, speckle imaging and selfcalibration: some of the most important advances in astronomical imaging in the last decade. Selfcalibration, or as it has been called elsewhere, hybrid mapping, appears, at first sight, to be impossible. How could we possibly determine both the source structure and the instrumental calibration necessary to view the source through our instrument ? In this paper we intend to provide an answer to this question and to outline the history of selfcalibration, it's present uses and possible future developments.

2 THE CLOSURE PRINCIPLE AND APPLICATIONS

In common with other radio interferometrists in the fifties Jennison (1958) was faced with the problem of inferring source structures from visibility amplitude alone since it was then difficult to maintain the phase of an interferometer. For some sources a major ambiguity remained, for example, the distinction between double and triple structure. Clearly phase information was required : Jennison's inspiration was to notice that in a system of three interferometers formed from three antennae one good phase observable, the closure phase, is available. One forms the closure phase Ψ by summing the visibility phase ψ around the closed loop of baselines :

$$\Psi_{123}(t) = \psi_{12}(t) + \psi_{23}(t) + \psi_{31}(t) \tag{1}$$

This observable is "good" in the sense that any phase errors ϕ which can be ascribed to individual antennae cancel out. The observed phase on baseline 12, for example, is related to the true phase by :

$$\psi_{12,obs}(t) = \psi_{12,true}(t) + \phi_1(t) - \phi_2(t) + \varepsilon_{12}(t) \tag{2}$$

where ε is a noise term due to receiver noise, sky background, etc. Substituting equations such as this into the above equation we find that, apart from terms due to noise, the observed and true closure phases are identical. A similar observable, the closure amplitude, was discovered by Twiss, Carter & Little (1962) :

$$\Gamma_{1234}(t) = (A_{12}(t).A_{34}(t))/(A_{23}(t).A_{14}(t)) \qquad (3)$$

where the amplitude on baseline 12 is given by :

$$A_{12,obs}(t) = A_{12,true}(t).a_1(t).a_2(t) + \xi_{12}(t) \qquad (4)$$

and ξ is a noise term. Again, apart from a noise term, the observed and true closure amplitudes are identical. (Smith (1952) used a quantity analogous to the closure amplitude). These two observables can also be derived from the equation relating the true and observed vector visibility :

$$V_{ij,obs}(t) = g_i(t).g_j^*(t).V_{ij,true}(t) + \eta_{ij}(t) \qquad (5)$$

where η is a complex noise term and g_i is the complex gain of the i'th antenna. One curiosity should be noted : since the absolute phase at an antenna is meaningless we can refer all phases to one reference antenna and thus a closure phase exists for only three antennae whereas the closure amplitude requires at least four (see Cornwell and Wilkinson (1981) for further elucidation of this point).

Closure quantities are completely insensitive to all properties of the visibility function which can be <u>factorised by antennae</u>. In particular, the closure phases are insensitive to changes in atmospheric and antenna instrumental delays while closure amplitudes are not affected by variable antenna gains, atmospheric attenuation or pointing (on small sources only; see section 4). Furthermore, position and strength information is lost. To prove the first point note that the visibility of a source contains a factor which depends upon the vector $\underline{s}$ pointing to the source centroid :

$$\exp(2\pi j.(\underline{b}_i - \underline{b}_j).\underline{s}) \qquad (6)$$

where $\underline{b}_i$ is the position vector of the i'th antenna. These two terms can be absorbed into the complex antenna gains and therefore have no effect upon the closure phase. Similarly, absolute amplitude information factors out of the closure amplitudes. Since closure quantities are completely insensitive to strength and position a certain slackness in pointing and positioning the antennae in an array is permissible. (The upper limits in slackness being set by the antenna primary beam widths and the receiver bandwidths.)

Jennison's concept of closure phase was described in his Ph.D thesis over thirty years ago but only within the last decade has significant use of it been made. The reason for this time lapse between conception and utilisation is mainly due to the computational problems involved. By their very nature, direct use of the closure quantities in the Fourier inversion of the visibility function is impossible. However, much ingenuity has been expended upon their indirect use; see e.g. Rhodes & Goodman (1973),Rogers <u>et al.</u> (1974),Fort & Yee (1976). In modern times Rogers <u>et al.</u> (1974) were the first to utilise the closure phase; they used it in the most urgently needed application, VLBI, where phase information is usually corrupted by clock errors. In a more popular algorithm Readhead & Wilkinson (1978) used the CLEAN method (Hogbom 1974, Schwarz 1978) to generate a representation of the brightness distribution while imposing positivity and confinement constraints. The algorithm can be summarised :

a. Find an initial model of the brightness distribution by e.g. model fitting to the amplitudes alone.
b. Transform the current model to obtain predicted visibility and force consistency with the amplitudes and closure phases.
c. Use the CLEAN algorithm to deconvolve and to reject regions of brightness which violate positivity and confinement.
d. Goto step b. unless convergence has been obtained.

Further refinements and extensions were made by Cotton (1979) and Readhead et al. (1980). As a result of this work, at the end of the seventies radio astronomers could image radio sources using very poorly calibrated data such as is obtained in VLBI or with MERLIN. It then became attractive to adapt the same techniques to arrays such as the VLA and WSRT which are limited in dynamic range by calibration errors.

3 SELFCALIBRATION

The concept of closure quantities is particularly useful for arrays containing small numbers of antennae but for instruments such as the VLA (Thompson et al. 1980) and modern VLBI-arrays the sheer number becomes overwhelming : for N (27) antennae there are $N(N-1)(N-2)/6$ (2925) closure phases, of which only $N(N-1)/2 - (N-1)$ (325) are independent. In such circumstances it becomes attractive to find some way of obeying the closure relations without ever explicitly calculating them. For this purpose it is important to notice that the closure quantities are invariant with respect to the complex antennae gains. Consequently, if, in trying to correct the observed visibilities, one only allows changes to complex antenna gains then the closure quantities will be conserved. This observation lead Schwab (1980) and Cornwell & Wilkinson (1981) independently to suggest that one should choose complex antenna gains by minimising a weighted sum of squared errors of the data from the model :

$$L(t) = \Sigma_{ij, i<j} \, w_{ij} \cdot | \, V_{ij, obs}(t) - g_i(t) \cdot g_j^*(t) \cdot V_{ij, true}(t) |^2$$

$$(7)$$

Cornwell and Wilkinson added a term to incorporate prior information about the relative stabilities of the gains. Schwab & Cotton (1983) further generalised this formula to allow estimation of antenna delays and fringe rates in VLBI.

This method of choosing complex antenna gains is embedded in the same iterative algorithm as before :

   a. Choose an initial model, usually a point source
      will suffice. For partially phase-stable instruments
      such as the VLA some CLEAN components from the
      initial image can be used.
   b. Calculate the complex gain corrections $g_i(t)$ which
      minimise the difference $L(t)$ between observed and
      predicted visibilities. Apply gain corrections
      to the observed data.
   c. Apply image plane constraints via CLEAN or any
      other method.
   d. Goto b unless convergence is attained.

One great advantage of this approach over the closure phase formalism is
that noise is treated correctly as an additive complex quantity. The
weighting terms w_{ij} add flexibility : they can be chosen to be the
inverse of the variance of η_{ij} in which case the solution signal to noise
is optimised or if the model is particularly poor for some baselines then
the weight can be reduced to de-emphasis their effect in the solution.

4 SOME PRACTICAL POINTS

      The success of selfcalibration relies upon its sensible
application: a number of choices of control parameters are required. Here
we present a short discussion of these choices: for fuller treatment see
Cornwell (1982) and Wilkinson (1982).

The most obvious question about selfcalibration is "Why does it work ?".
To answer this we need to consider the effect of calibration errors. If
the visibility data is miscalibrated then the true image is seen
convolved with an error beam which violates two reasonable constraints
upon the image : positivity and confinement. Enforcement of positivity
and confinement can be performed using various methods; first, however,
it is necessary to remove the sidelobes due to incomplete u,v coverage
which also violate these constraints. Both CLEAN (Hogbom 1974) and MEM
(Wernecke & D'Addario 1976, Gull & Daniell 1978) are well suited to this
role although the former may be preferable for small images of high
dynamic range. CLEAN windows are ideally suited to rejecting components
from some regions of the brightness distribution and to force positivity
one can stop CLEANing just before the first negative component. In MEM
positivity comes free and confinement could be enforced by the use of
windows. Unfortunately, positivity and confinement are sometimes not
strong enough constraints : for arrays with well behaved beams, such as
linear arrays, calibration errors can be converted into believable
structure such as jets so one must be careful to anticipate such
ambiguities.

Lax use of the known image plane constraints is perhaps the major cause
of failure of selfcalibration; for example, a large number of CLEAN
components will model some of the calibration errors and thus convergence
to an acceptable image will be delayed or prevented. One useful rule,

widely used at the VLA, is to only include CLEAN components which occur before the first negative component. Another ad hoc rule has been suggested by R. R. Clark (private communication) : one can calculate the effective noise level expected on a selfcalibrated image from the r.m.s. misfit of the data from the model <u>after</u> adjusting the complex antenna gains (see equation 7). For example, if at some stage in the process the misfit is M times the average noise per visibility point then it is likely that there will be systematic errors present in the trial image at approximately M times the thermal noise limit. It is therefore sensible to halt the cleaning process well before this likely systematic error level has been reached so that spurious clean components will not be passed around the selfcalibration loop. The depth of cleaning will automatically increase as the process converges and the misfit decreases. Figs. 1a,b show the practical effects of applying these rules. Fig. 1a shows an image made from MERLIN data (see Cornwell and Wilkinson 1981) where clean components close to the expected systematic error level were passed around the iterative loop, 8 iterations in all were made. Fig. 1b was made in exactly the same way but with a cutoff at approximately 3 times the expected error level. This image is not only superior but in the selfcalibration <u>5 times</u> fewer clean components were passed around the loop, thus saving considerably in CPU time. If cleaning is instead stopped at the first negative component - in this case a somewhat more severe limitation - then a marginally worse image than Fig. 1b is obtained. We cannot generalise about the best cutoff rule from this one case but the overall message is very clear - use only a minimum of clean components especially in the early stages of selfcalibration.

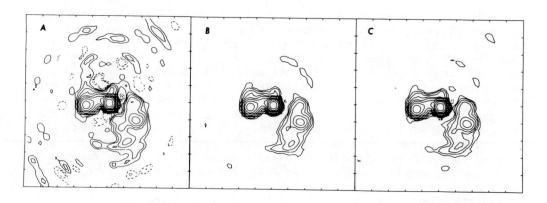

Figure 1. Tests of selfcalibration strategies. See text. Contour levels 0.1, 0.2, 0.4 .... 51.2 % of peak brightness.

A further saving in CPU time can be acheived if one adopts a difference imaging philosophy. One should clean only the data formed from the <u>difference</u> between the previous and current corrected visibility data. The number of clean components subtracted must be restricted by a version of R. Clark's rule and negative components <u>must</u> be included to allow the method to correct itself. The relatively few clean components subtracted from this difference image are then transformed and added to the transform from the previous iterations thus "updating it". Complex gain corrections are then calculated in the usual way and the whole process is iterated. We have tested this approach and have shown that it converges to produce an acceptable image. Fig. 1c shows an image made from the same data as Figs. 1a,b but with about 4 times fewer clean components in all. There are minor differences at low levels but it seems clear that this is a promising approach which can both save CPU time and operate in a "hands-off" manner.

If the visibility is not much greater than the noise on a sufficiently large number of baselines then selfcalibration can yield completely spurious results. For example, with VLA data, merely by adjusting complex gains one can pull a point source of strength ~ 5 sigma from a noise field! Cornwell (1981) has considered the noise behaviour of selfcalibration : only for a point source is a simple solution possible. Suppose that the atmosphere is stable for a time τ and that the resultant noise in an image made from the data collected in that time is σ_{image} then for reliable amplitude and phase selfcalibration the flux of an "almost unresolved" source must obey :

$$F \gg \sigma_{image} \cdot (N_{ant} - 3)^{-1/2}. \tag{8}$$

If this holds then the noise in the resulting image is increased over the thermal limit by a factor $((N-1)/(N-3))^{1/2}$. For extended sources equation (8) seems to work reasonably well if F is replaced by the median visibility amplitude. By comparison, the limit for detection from amplitudes alone is :

$$F \gg \sigma_{image} \cdot (\tau/T)^{1/4}. \tag{9}$$

and, of course, for coherent integration :

$$F \gg \sigma_{image} \cdot (\tau/T)^{1/2} \tag{10}$$

where T is the total integration time.

Some common types of error in visibility data do not obey the antenna related error model (equation 5). Instead, an additional term $G_{ij}(t)$ must be factored into the first term of the left hand side of equation 5. $G_{ij}(t)$ can be due either explicitly to correlator problems or, more commonly in arrays with digital correlators, to antenna based problems which do not factorise. Some examples of the latter are (a) Confusion on shorter baselines. (b) Non-matching passbands in the antennae amplifiers. We then have an equation such as equation (5) for each element of the passband. Integrating over non-identical passbands leads to terms like $G_{ij}(t)$ (Thompson & D'Addario 1982). To the extent that the passbands are constant in time the $G_{ij}(t)$ can be removed by performing a simple baseline related calibration (see e.g. Cornwell and Wilkinson 1981). (c) Pointing errors can yield differential gain errors over a source large in comparison to the primary beam-width of the antennae. (d) Non-isoplanicity of the atmosphere over the source prevents simple factorisation of the gains. (e) Bandwidth and time-averaging smearing also produce effects which cannot be factorised. (f) Second-order changes in the gains during the integration time.

Clark (1981) has considered the order of magnitude of some of these effects. In most high dynamic range VLA images time variable effects like (b) seems to be a limit; as yet there seems no way to overcome these problems. However, in an array which is very stable instrumentally then considerable progress can be made : in an impressive extension of antenna based selfcalibration Noordam & de Bruyn (1982) have very successfully used the redundancy of WSRT to remove these effects thereby attaining the highest dynamic range (>40dB) yet acheived in radio interferometry.

5 SELFCALIBRATION AND ADAPTIVE OPTICS

A very close analogue to selfcalibration has arisen in optical astronomy, namely adaptive optics (see e.g. Code 1973, Buffington et al. 1978, Woolf 1982, Hardy 1978). We now digress to consider this interesting convergence.

The resolution of an optical telescope of size greater than about 10-20 cm. is nearly always limited by short term phase errors introduced by turbulent cells in the first few kilometers of atmosphere above the pupil (see e.g. Muller & Buffington 1974). A clever method of real time phase correction has been described by Muller & Buffington (1974). In this scheme a phase correction device (a "rubber mirror") is placed in the pupil plane and controlled so as to maximise some measure of image sharpness. Detection and correction must take place within the characteristic time scale of the atmosphere, about 10-20 milliseconds, and hence a strong source lying within one isoplanatic patch is required. Typically two trial values of the phase for a particular part of the pupil are inserted and the resultant sharpness measured ; the optimum phase is then interpolated.

Various forms of the sharpness measure have been suggested; Muller and Buffington define a sharpness measure to be a function of the perturbing phases which is maximised for the true image. Some typical measures are: the energy of an image (i.e the sum of squares of pixel values), the moment of inertia, a central reference intensity, various moments of the pixel values, and various norms expressing a distance from an ideal image. Of these, the energy or sum of squares of the pixel values is most often used and is sometimes called the sharpness S. Hamaker et al. (1977) have demonstrated that by maximising S we are simply requiring that redundant samples of the Fourier transform of the object have the same phase, a constraint which can be trivially enforced in redundant spacing interferometry.

In radio interferometry we could follow the same scheme as adaptive optics : changing antenna phases within the characteristic time of the atmosphere, few minutes to several hours, to optimise some measure of the image quality. However, this approach requires that a good image, free from other defects such as poor sampling, be formed in that time scale, which for most arrays is simply not possible. Indeed, for sparsely filled arrays direct optimisation of the sharpness of the dirty image seems to be prevented by the presence of these sidelobes which overwhelm the sidelobes due to calibration errors (Steer, Ito & Dewdney 1983). Even for arrays with relatively good instantaneous u,v coverage such as the VLA this scheme is expensive in terms of computer time since the visibilities affected by each perturbation of an antenna phase must be individually transformed to yield the resultant change in an image. For this reason we prefer to deal with the entire data set using the iterative approach outlined in the previous section.

It is interesting to consider why selfcalibration has revolutionised radio interferometry while adaptive optics have so far been of peripheral interest in optical astronomy. Several factors are important : first, rubber mirrors are difficult to build and operate while the computer based manipulations required for selfcalibration are simple. Secondly, in optical astronomy the number of atmospheric coherence times per typical observation is of order 1,000-100,000 compared to, at most, ~ 100 in radio interferometry. As a consequence, many radio sources can be selfcalibrated whereas adaptive optical systems are effective for relatively few optical sources. Optical speckle imaging has succeeded in the imaging of relatively weak objects by relying purely upon the amplitude information, integrated incoherently. Radio speckle imaging is also possible simply by averaging the visibility amplitudes. Note, however, a major advantage of radio over optical speckle imaging : in the former we have access to the individual visibilities before coherent addition (but see Brown 1978) and, consequently, the noise level is not increased by the decorrelation of the same complex visibility measured between different parts of the array. In the terminology of speckle imaging, the modulation transfer function is unity everywhere in the aperture.

The signal to noise limitations mean that selfcalibration/adaptive optics will probably be of marginal importance for ground based optical arrays. Instead, speckle imaging seems much more promising.

6 TWO STATE-OF-THE-ART SELFCALIBRATED IMAGES

To demonstrate the power of selfcalibration we here show two state-of-the-art selfcalibrated images.

The first, shown in Figure 2, kindly supplied by Craig Walker, John Benson, George Seielstad and Steve Unwin shows an image of 3C120 made from data collected in a fourteen station MkII VLBI observation at λ18cm. (An image of comparable quality is shown by Simon et al. 1983). In figure 3 we show the second image, which was kindly supplied by Rick Perley and John Cowan, a VLA image of Cygnus A at λ20cm with data from the A,B and D arrays. The dynamic range (peak/rms) for this image is about 3000.

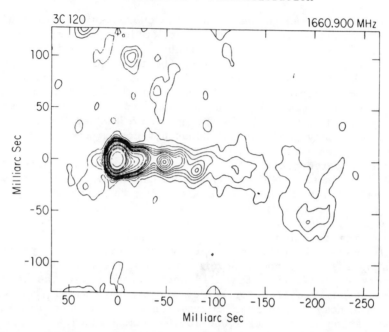

Figure 2. VLBI image of 3C120 at λ18cm. See text. The contour levels are 4, 8, 12, 16, 20, 24, 32, 40, 48, 56, 80, 120, 160, 200, 240, 280, 320, 360, 400, 1000 mJy per beam and the peak is 2080 mJy per beam.

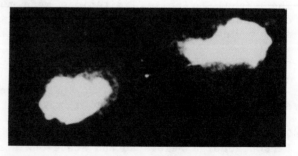

Figure 3. VLA image of Cygnus A at λ6cm. See text. The lower radiograph saturates two magnitudes below the upper radiograph.

7 THE FUTURE

We will now discuss a number of technical challenges in the future of selfcalibration.

First, the signal to noise limit may be lowered by using models for the atmosphere which have fewer degrees of freedom. For example, in a small linear array it may be advantageous to couple the phase parts of the gains by a linear dependence on distance, thus reducing the number of phase degrees of freedom from N-1 to 1. In a similar vein Basart et al. (1983) are adapting the Kalman filter formalism to the estimation of gain errors. In an attempt to define the ultimate signal to noise limits for selfcalibration we note that if the atmosphere can be tracked across the array then the effective time between solutions for the gains is the crossing time for the whole array rather than a single antenna. This increase in intervals between solutions yields a concomittant decrease in the signal to noise limit. However, even with all this cleverness the coherence time may be so short that any selfcalibration is not possible. We may then have to resort the radio speckle imaging (see section 5). Since reconstruction from amplitudes alone seems possible for non-pathological objects in two or more dimensions (see e.g. Fienup 1983) the only real drawback to this approach is the telescope time needed.

Non-isoplanicity seems to present no real conceptual problems but the practical problems include a substantial increase in computational complexity and a moderate increase in the number of degrees of freedom. One can envisage one higher level of complexity over selfcalibration : the sky would be split into a mosiac of isoplanatic patchs for each of which one set of antenna phases would be optimised. The VLA is expected to be mildly non-isoplanatic at an operating frequency of 327MHz.

One cause of failure of selfcalibration is the presence of error terms in the measurement equation which cannot be attributed to antennae (see section 4). Baseline redundancy (Noordam & de Bruyn 1982) can be used to remove correlator-based effects but one pays a large price in efficiency of use of the antennae which becomes apparent when imaging large fields at low dynamic range. Furthermore, we note that strict redundancy is not always required. To understand this consider the effective redundancy which arises when imaging a field of size θ. The sampling theorem then tells us to sample at least every $1/(2\theta)$ wavelengths in the u,v plane; samples closer than this are then effectively redundant and are implicitly used in selfcalibration to eliminate other degrees of freedom such as antenna and correlator gains. The use of CLEAN windows currently enforces such effective redundancy.

8 ACKNOWLEDGEMENTS

The National Radio Astronomy Observatory is operated by Associated Universities Inc., under contract with the National Science Foundation. We thank the authors cited in section 6 for providing the illustrations.

9 REFERENCES

Basart,J.P., Mitkees,A.A.,Mansy,F.M.M., and Zheng,Y. (1983). Scientificreport, Dept. of Elec. Eng., Iowa State Univ.

Brown,T.M., (1978). J.Opt.Soc.Am., 68,883-889.

Buffington,A.,Crawford,F.S.,Pollaine,S.M.,Orth,C.D.,and Muller,R.A., (1978). Science, 200,489-494.

Clark,B.G., (1981). VLA scientific memorandum 137.

Code,A.D., (1973). Ann. Rev. Astron. and Astrophys., 11,239-268.

Cornwell,T.J. and Wilkinson,P.N., (1981). M.N.R.A.S. 196,1067-1086.

Cornwell,T.J., (1981). VLA scientific memorandum 135.

Cornwell,T.J., (1982) lecture 13,NRAO summer school on synthesis mapping,Socorro N.M. 1982, ed. A.R.Thompson.

Cotton,W.D., (1979). A. J., 84,1122-1128.

Fienup,J.R., (1983). paper presented at OSA meeting on "Signal recovery and synthesis with incomplete information and partial restraints", Lake Tahoe.

Fort,D.N. and Yee,H.K.C., (1976). Astron. Astrophys., 50,19-22.

Greenaway,A.H., (1982) Optics Comm., 42,157-161.

Gull,S.F. and Daniell,G.J., (1978). Nature, 272,686-690.

Hamaker,J.P.,O'Sullivan,J.D. and Noordam,J.E., (1977). J.Opt.Soc.Am., 67,1122.

Hardy,J.W., (1978). Proc. IEEE, 66,651-697.

Hogbom,J., (1974). Astron. Astrophys. Suppl., 15,417.

Jennison,R.C., (1958). M.N.R.A.S., 118,276.

Muller,R.A. and Buffington,A., (1974). J.Opt.Soc.Am., 64,1200-1210.

Noordam,J.E., and de Bruyn,A.G., (1982). Nature, 299, 597-600.

Readhead A.C.S. and Wilkinson,P.N., (1978). Ap. J., 223,25-36.

Readhead,A.C.S.,Walker,R.C.,Pearson,T.J. and Cohen,M.H., (1980). Nature, 285,137-140.

Rhodes,W.T. and Goodman,J.W., (1973). J.Opt.Soc.Am., 63,647-657.

Rogers,A.E.E. et al, (1974). Ap.J., 193,293-301.

Schwab,F.R., (1980). SPIE, 231,18-24.

Schwab,F.R. and Cotton,W.D., (1983). 88, 688-694.

Schwarz,U.J., (1978). Astron. Astrophys., 65,345-356.

Simon, R.S. et al. (1983). Paper to be published in the proceedings of IAU symposium 110, Bologna, Italy, 1983.

Smith,F.G., (1952). Proc. Phys. Soc B, 65 971.

Steer,D.G., Ito,M.G. and Dewdney,P.E., (1983). paper presented at OSA meeting on "Signal recovery and synthesis with incomplete information and partial restraints", Lake Tahoe.

Thompson,A.R. and D'Addario,L.R., (1982). Radio Science, 17,357-369.

Thompson,A.R.,Clark,B.G.,Wade,C.M. and Napier,P.J., (1980), Ap. J. Supplement, 44,151-167.

Twiss,R.Q.,Carter,A.W.L. and Little,A.G., (1962). Aust.J.Phys., 15,378.

Wernecke,S.J. and D'Addario,L.R., (1976). IEEE C-26,351-364.

Wilkinson,P.N., (1982). paper presented at CNES conference on VLBI techniques, Toulouse,France,1982.

Woolf,N.J., (1982). Ann. Rev. Astron.and Astrophys., 20,367-398.

DISCUSSION

J. NOORDAM
        An automatic process is dangerous because a range of possible maps may be consistent with the data (especially if there are few telescopes and consequently few constraints).

T.J. CORNWELL
        There will always be brightness distributions which are too complex to recover from any particular data set, be it from 4 telescopes or 27. We agree that we do not know as yet how to make the computer decide whether SELFCAL is likely to work on a particular data set. Given that the astronomer has made this decision, all we are saying is that it is possible to codify some of the rules which an experienced astronomer would apply, by hand, each iteration.

R.H.T. BATES
        The trouble with comparing radio self-calibration with optical wavefront correction is that the latter is hopeless for faint objects, and it gets worse the larger the pupil is compared to the diameter of the average seeing cell.

T.J. CORNWELL
        The two drawbacks that you mention also apply to self-calibration but the dividing lines are such that self-calibration is worth while for a relatively large fraction of radio sources.

P.E. DEWDNEY
        You tended in your talk to gloss over the enforcement of the positivity constraint by simply 'throwing away' the parts of the interim map which you 'don't like'. However, this is the only non-linear process in the loop and is very important. Have you thought about devising an 'optimum' non-linear process which might produce convergence in minimum time.

Although some persons took this question to mean that I was suggesting maximum 'entropy' as such a scheme, I didn't really intend this. It would, however, fall into the class of such non-linear operations. We need some theoretical backing for deciding how many (say) CLEAN components to keep at each iteration to arrive at the conclusion in minimum time.

T.J. CORNWELL
        We do not know of any such scheme. The suggestion of Ron Clark is intuitively appealing and seems to be effective.

J.G. ABLES
        It would be of interest to compute the image 'entropy' at each stage of the iteration to see if the entropy measure changes monotonically as the SELFCAL algorithm proceeds.

T.J. CORNWELL
        I agree but I am confident that the entropy will decrease since the dynamic range increases during iteration. It should be noted that in an accompanying talk Steer et al. found that self-calibration and maximum entropy were not compatible. Instead they introduced a mechanism to artificially decrease the entropy of an image prior to gain calculation. Thus minimum entropy reconstruction would seem to be preferable for self-calibration perhaps with mollification of the final image.

M.M. KOMESAROFF
        (1) Since positivity is one strong constraint which makes SELFCAL work, does this mean that it cannot be used to map all Stokes parameters?

(2) Isn't it impossible to determine by SELFCAL the phases of a set of RH polarized feeds relative to a set of LH feeds?

T.J. CORNWELL
        (1) For Stokes parameters the constraint analogous to positivity is that the fractional polarization should be less than unity. The power of this constraint is not known to us.

(2) Yes, one cannot determine the phase difference between the RH and LH feeds by self-calibrations. At high frequencies (>1 GHz) the difference is usually constant and can be calibrated by reference to a source of known polarization. At low frequencies time-variable ionospheric effects must be calibrated by other methods.

Large Dynamic Range Observations with the Effelsberg 100-m Radio Telescope

W. Reich[1], P. Kalberla[2], K. Reif[1] and J. Neidhöfer[2]

[1] Radioastronomisches Institut der Universität Bonn, Auf dem Hügel 71, D-5300 Bonn 1, Federal Republic of Germany
[2] Max-Planck-Institut für Radioastronomie, Bonn, Auf dem Hügel 69, D-5300 Bonn 1, Federal Republic of Germany

Received January 25, 1978

Summary. A technique has been developed to observe weak objects in the vicinity of strong radio sources with the Effelsberg 100-m telescope by removing the primary beam sidelobes. This is accomplished by mapping a source in the horizon system at several different parallactic angles and then separating the primary beam sidelobe structure from the source structure on the sky. The principles of this method are given as well as the antenna pattern of the 100-m telescope at 11 cm wavelength. In an experiment on 3C 123 we have been able to detect weak structures near the source at a level of -37 dB in total power (Stokes parameter I) as well as in linear polarisation (Stokes parameter Q and U).

Key words: observing techniques — radio telescopes — dynamic range

1. Introduction

The presence of sidelobes in the antenna pattern of a single dish radio telescope have always made the observing of weak objects near strong sources difficult or impossible. However, when using a completely steerable azimuthally-mounted telescope like the Effelsberg 100-m dish, we can use the fact that the antenna pattern is fixed in the horizon system and rotates with respect to a radio source on the sky to determine the structure of both to a high degree of accuracy. We have developed a method for this separation, thus increasing the effective dynamic range of the telescope. This is especially important when investigating the sky near strong sources like 3C quasars and radio galaxies. We suspect that weak structure around such sources still suffers from instrumental effects. This is important for the physical interpretation. A range of more than 30 dB between strongest and weakest areas in a map which has already been obtained through use of our technique (Reich et al., 1976), substantially exceeds that normally obtainable with either single dish or synthesis radio telescopes.

Send offprint requests to: W. Reich

2. Principles

Mapping a field centered on a strong source in the horizon system is sketched in Figure 1. The weak sources appear to rotate around the strong central source as a function of the parallactic angle. Then by assuming that (a) the antenna pattern of the telescope is a constant independent of azimuth and elevation and (b) most of the sky in our field is empty, it is possible to determine accurately both the antenna pattern and the distribution of weak sources on the sky. Assumption (b) is similar to that used for CLEAN (Högbom, 1974) with synthesis radio telescopes.

3. Methods

We observe a set of i-maps in the horizon system at different parallactic angles π_i. They are all reduced by standard computer routines (Neidhöfer et al., 1978). We obtain for each map grid point j an antenna temperature T_{ij}. This temperature consists of contributions from the following:

$$T_{ij}(Az_j, El_j, \pi_i, Sc, t_i) = P_j(Az_j, El_j)T_0 + T_{skyij}(Az_j, El_j, \pi_i)$$
$$+ T_{sci}(Sc, t_i) + T_{ni}(t_i) \qquad (1)$$

where Az_j and El_j are the azimuth and elevation offsets of grid point j relative to the central source with antenna temperature T_0. P_j is the normalised antenna pattern, T_{skyij} is the distribution of faint sources in the vicinity of the central source, T_{sci} are receiver instabilities, ground radiation and atmospheric effects resulting from scanning, Sc is the scanning coordinate. T_{ni} is random receiver noise depending on the local time t_i.

To do the separation, then, we first average all of our maps together. We compare this mean map with each single map and determine difference maps. A new average map is then constructed which is intended to be free of sources and scanning effects. To take out points T_{ij} where sources, interference and scanning effects may be present the difference maps are investigated for areas

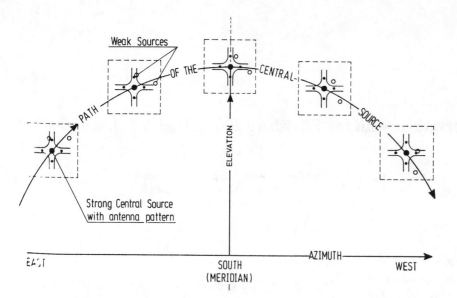

Fig. 1. Illustration of the rotation of sky coordinates with respect to antenna coordinates when mapping a source in the horizon system at different sidereal times. Because the sky and antenna pattern rotate with respect to one another they can be separated

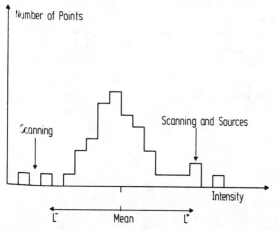

Fig. 2. Noise histogram of an element of a map observed in the horizon system at different parallactic angles. Some measured points contain sources or suffer from scanning effects which may be positive or negative. These data points are eliminated by setting correct limits L^+ and L^-

where certain amplitudes, called L^+, L^-, are exceeded. The corresponding points T_{ij} in the original maps are left out before the new average map is computed. In Figure 2 we show a histogram of measured points T_{ij} for one grid point j. This separation process—average map, difference maps, cleaning for deviations exceeding L^+, L^-—is repeated with more and more stringent limits on L^+, L^-. The resulting average is free of sources etc. and gives finally the antenna pattern plus a contribution due to random noise:

$$T_j(Az_j, El_j) = P_j(Az_j, El_j)T_0 + T_n \qquad (2)$$

where T_n is the average noise contribution from all maps. The final difference between each single map i and the common antenna pattern T_j is a map of the sky at time t_i plus observing effects and random noise. Each of

these is then transformed into the α, δ coordinate system and represents a brightness distribution:

$$T_{ij}(\alpha_j, \delta_j) = T_{sky}(\alpha_j, \delta_j, \pi_i) + T_{sci}(Sc, t_i) + T_{ni}(t_i). \qquad (3)$$

A second averaging of all these maps in the α, δ system and a separation of the constant parts of all the maps result in a map of the sky:

$$T_j(\alpha_j, \delta_j) = T_{sky}(\alpha_j, \delta_j) + T_n'. \qquad (4)$$

4. Observations

The methods described above were first used to study the structure surrounding the radio galaxy 3C 123 at 21 cm wavelength (Reich et al., 1976). Then, in December 1976 about 30 h were spent studying the same field at 11 cm wavelength. The telescope was equipped with two uncooled parametric amplifiers in the secondary focus with a system noise of 115 K and a bandwidth of 80 MHz. The observing frequency was 2700 MHz, the HPBW was $4'35''$ ($\pm 5''$), and the aperture efficiency was 55%. An IF polarimeter was used to give the total power outputs of both receivers (Stokes parameter I when no circular polarisation was present) and two correlated outputs (Stokes parameter Q and U). For the total power only one receiver gave sufficient stability and was used for Stokes parameter I. 3C 123 was observed in the center of a $1°8 \times 1°8$ field in the horizon system. The telescope was scanned with a velocity of $1°5/\text{min}$ in the azimuth-direction with a scan spacing of $2'$ in elevation. The scanning coordinate system was corrected for earth rotation so that 3C 123 always remained in the center. We observed 20 maps to obtain a theoretical sensitivity of 2.65 mK ($\hat{=} 1.75$ mJy/beam). The range of the parallactic angle of $\pm 47°$ was com-

ANTENNAPATTERN

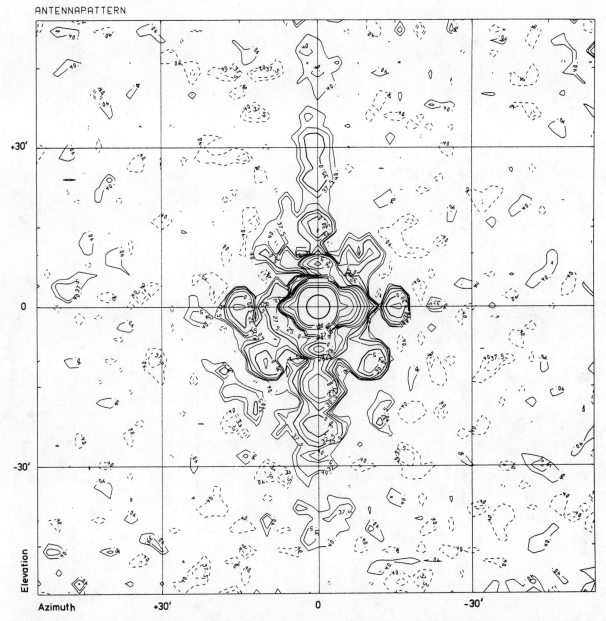

Fig. 3. Antenna pattern of the Effelsberg 100-m telescope measured on 3C 123 at 11 cm wavelength. The contour labels are in decibel below the peak intensity in the center. Dashed contours show negative values. The heavy unlabelled contour in the center gives the 3dB half power beamwidth

pletely covered; the elevation of the telescope changed between 45° and 70°.

We applied the separation process to the set of maps as described above. Four iterations were made and lower limits equal to 8 times the r.m.s. noise (see Fig. 2) were used to avoid a disturbance to the noise histogram. The results are shown in Figures 3, 4 and 5. Figure 3 gives the antenna pattern of the 100-m telescope at 11 cm. The flux of 3C 123 was found to be 26.9 Jy. The total intensity down to 40 dB below the maximum is plotted. This level is of the order of the peak to peak noise. Structures down to −37 dB (≙ 5.4 mJy/beam) are thought to be real. The general increase of the side-

lobes along the azimuth and elevation directions are due to the support legs of the secondary reflector. The apparent asymmetries of the antenna pattern are real, they cannot be calculated today. At 11 cm the horn in the secondary focus is installed with an offset −10′ in elevation which causes the general increase of the southern sidelobe levels seen in Figure 3. The cause of the increase of the first eastern sidelobe and also the asymmetries of the antenna pattern for the linear polarisation, Stokes parameter Q (Fig. 4) and Stokes parameter U (Fig. 5) in this direction have unknown reasons. But our method accepts any kind of pattern. In Figure 6 we show the final map of sources around

Q-PATTERN

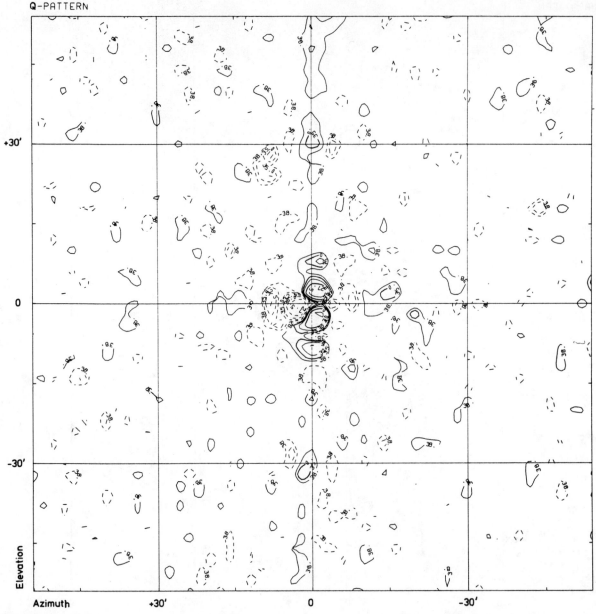

Fig. 4. Antenna pattern for the Stokes parameter Q at 11 cm wavelength. The contour labels are given in decibels below the maximum of the total intensity beam. Dashed contours indicate negative values

3C 123 in the α, δ coordinate system together with the results for the linear polarisation. The discussion of the astronomical results from the 11 cm map will be given together with results from 6 cm and 3.3 cm measurements with the 100-m telescope in a separate paper.

5. Limitations

As with any technique, there are limitations. As mentioned above one must assume that the antenna pattern remains constant. We have investigated this in detail for the Effelsberg 100-m telescope and conclude that for wavelengths between 21 cm and 6 cm and elevations between 40° and 80° this assumption is valid. This is

consistent with the result of Hachenberg et al. (1973) who found for 2.8 cm wavelength no antenna gain variations above 1% for elevations greater than 30°.

In the center of the transformed maps there remains an area of confusion approximately given by the main beam area (see e.g. Fig. 2 of Reich et al., 1976). The dynamic range is much poorer in this region and is mainly a function of the sample interval and the size of the pointing errors. For the Effelsberg telescope we typically find residual pointing errors of up to 15″ even after having done the normal on-line pointing and following a source during a night. These pointing errors are corrected off-line by shifting the peak intensity of the map to the center but because this requires inter-

U-PATTERN

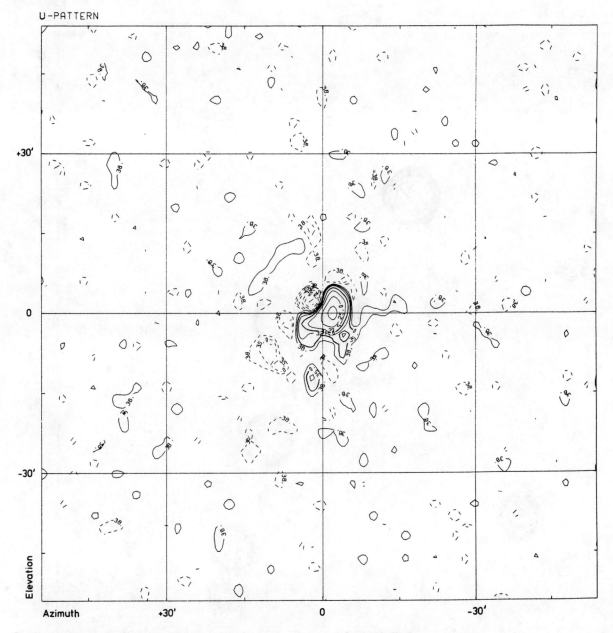

Fig. 5. Antenna pattern for Stokes parameter *U* at 11 cm. Contours are defined as in Figure 4

polation on very steep gradients errors still remain. Our experience is that the confusion area is almost independent of wavelength when using the same sampling interval. The extent of the confusion area is also not affected by the signal/noise ratio.

A third limitation is the insensitivity of the separation procedure to any symmetric feature on the sky around the central source. These components then are included in the antenna pattern and will not appear in the final source map.

6. Conclusions

The rotation of the sky relative to the antenna pattern of a fully steerable, azimuthally mounted single dish radio telescope can be used around strong sources to separate the antenna pattern with sidelobes from the real structures on the sky. A high dynamic range is reached with this method which is presently out of reach with other techniques.

Acknowledgements. We thank Dr K. W. Weiler for critical reading of the manuscript and Drs P. Kronberg, U. Mebold and Prof. Dr R. Wielebinski for discussions.

References

Hachenberg, O., Grahl, B. H., Wielebinski, R.: 1973, *Proc. Inst. Elec. Electron. Engrs.* **61**, 1288
Högbom, J. A.: 1974, *Astron. Astrophys. Suppl.* **15**, 417
Neidhöfer, J., Wilson, W., Haslam, C. G. T.: 1978, *Kleinheubacher Berichte* **21**, in press
Reich, W., Kalberla, P., Neidhöfer, J.: 1976, *Astron. Astrophys.* **52**, 151

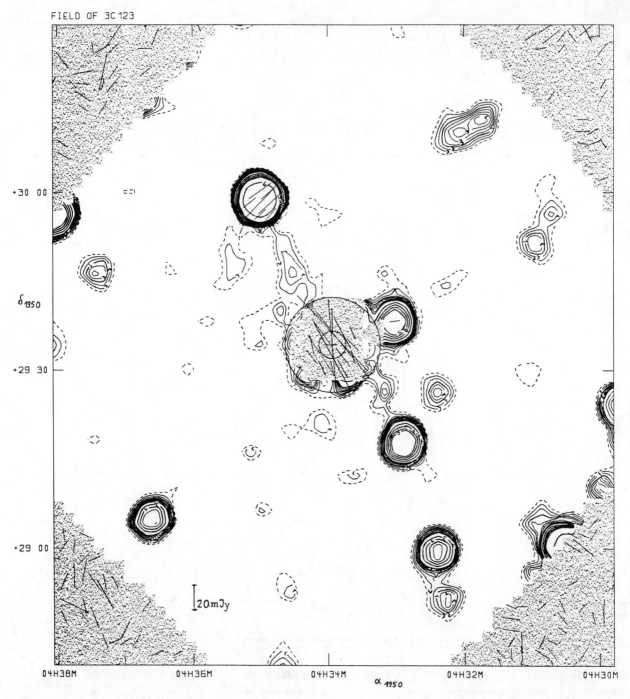

Fig. 6. The sources around 3C 123 observed with the Effelsberg 100-m telescope at 11 cm wavelength. The position of 3C 123 in the center is marked by a cross. The source was removed together with its sidelobes as described in the text. The HPBW is given by the inner circle. The outer circle gives the confusion area for the total intensity. The total intensity is shown with contour values of 5 (dashed) in steps of 2.5 up to 22.5 mJy/beam. Contour values of 12.5, 25, 37.5, 50, 62.5, 75 and 150 mJy/beam are shown with labels 1 to 7, respectively. The shaded areas indicate that the data have weights less than 0.4. The linearly polarised intensity is shown by lines. The intensity is given by the line length and the orientation represents the position angle of the electric vector. Intensities are given above 6 mJy/beam. Polarisation intensities in the shaded corners have low weight, while the data in the center show the linearly polarised intensity of 3C 123

APERTURE SYNTHESIS WITH A NON-REGULAR DISTRIBUTION OF INTERFEROMETER BASELINES

J.A. HÖGBOM

Stockholm Observatory, Sweden

In high-resolution radio interferometry it is often impossible for practical reasons to arrange for the measured baselines to be regularly distributed. The standard Fourier inversion methods may then produce maps which are seriously confused by the effects of the prominent and extended sidelobe patterns of the corresponding synthesized beam. Some methods which have been proposed for avoiding these difficulties are discussed. In particular, the procedure CLEAN is described in some detail. This has been successfully applied to measurements taken with several different radio telescopes and appears to be the best method available at the time of writing.

1. INTRODUCTION

Aperture synthesis measurements are usually made at a set of interferometer spacings and orientations that form a regular pattern in the baseline (u,v) diagram. Such a regular coverage has many practical advantages both in connection with the formal synthesis calculations and the later astronomical interpretation of the synthesis map. However, there are occasions when irregularities in the baseline coverage cannot be avoided. Interference or malfunctioning of some part of the equipment can make it necessary to reject certain portions of the measurements and this will leave gaps in an otherwise regularly covered u,v plane. The gaps give rise to undesirable sidelobes in the synthesized beam, making the synthesis map difficult or impossible to interpret. Similar problems arise when the u,v plane has been covered by a coarse grid of measurements as this will give rise to prominent grating responses in the synthesized beam pattern.

It may, in certain cases, be impossible (or impractical) to arrange for the interferometer measurements to fall on a regular grid in the baseline (u,v) diagram. This is the case for measurements taken with instruments such as the Caltech and Green Bank interferometers and, in general, for measurements that involve large interferometer spacings. Interferometers can be operated over spacings up to the full diameter of the Earth and from space vehicles and one shall ultimately want to use such measurements in a systematic way to synthesize very high-resolution maps of small diameter sources. Occultations of radio sources by the Moon give rise to similar problems. A few occultations of the same source (or one occultation measured at several observatories) will deliver a number of strip scans at a non-regular set of position angles. These strip scans are equivalent to a u,v coverage along radii at these same position angles.

Knowledge of what can and what cannot be achieved with a non-regular coverage of the u,v plane is obviously very important in connection with the design of future high-resolution synthesis instruments. Already existing instruments can also be used more efficiently if the formal requirement of a particular regular baseline coverage can be dropped.

Problems involved in the reduction of non-regularly spaced interferometer measurements for which amplitude and phase information is available are discussed in this paper. The aim shall be to produce a map that is equivalent – in so far as this is possible – to that which we would have obtained by the normal synthesis calculations if the measurements had in fact covered the relevant region of the u,v plane in a uniform manner. One might object that such a procedure cannot exist. According to a theorem in Fourier analysis, this region contains just as many independent measurements as there are independent directions in the observed field as seen with the synthesized beam. A smaller number of measurements cannot be sufficient to specify the map completely. The confusion caused by sidelobes and grating responses should therefore be regarded as an unavoidable expression of the fact that one cannot solve for more unknowns than there are equations. However,

Reprinted with permission from *Astron. Astrophys. Suppl. Ser.*, vol. 15, pp. 417–426, 1974.
Copyright © 1974, Springer-Verlag.

significant deflections from zero intensity will usually occupy only a small fraction of a high-resolution synthesis map. This becomes obvious when one compares the brightness temperature sensitivity of a high-resolution radio telescope with typical values of the sky brightness temperature and present knowledge about radio source statistics. A typical high-resolution map can usually be described by saying that it is *empty* but for certain exceptions which can be specified in a table. Clearly, the number of measurements must be at least as large as the number of items in this table, but this will usually be a fraction of the total number of independent synthesized beam directions in the observed field.

2. CONSEQUENCES OF A NON-REGULAR DISTRIBUTION OF THE MEASURED BASELINES

The theory of aperture synthesis does not in itself require that the u, v plane be covered according to a regular pattern, but it turns out that regular patterns will in general produce maps that are relatively easy to interpret. In its basic form, the theory states that the synthesis map obtained by taking the Fourier transform of the measurements $W(u, v)$ is proportional to the convolution of the true brightness distribution T_b (enveloped by the primary beam P of the antenna combination) with the synthesized beam G. If all the baselines can be reduced to one plane, the u, v plane, which is at right angles to the center of the observed field (we shall discuss this assumption later), then this Fourier relation takes the following form:

$$\{P(\ell, m)\, T_b(\ell, m)\} * G(\ell, m) \propto \int\limits_{-\infty}^{+\infty}\!\!\int W(u, v)\, g(u, v) \exp[i\, 2\pi(u\ell + vm)]\, du\, dv \tag{1}$$

ℓ and m are coordinates in the sky (directional cosines with respect to the u and v axes respectively), $i = \sqrt{-1}$, and $g(u, v)$ is the function by which we choose to weight the (complex) interferometer measurements $W(u, v)$ in the transform calculations. u and v are expressed in wavelengths. The synthesized beam equals the suitably normalized Fourier transform of this weight function:

$$G(\ell, m) = \text{const} \int\limits_{-\infty}^{+\infty}\!\!\int g(u, v) \exp[i\, 2\pi(u\ell + vm)]\, du\, dv \tag{2}$$

In order to calculate the Fourier integral in Eq. (1), the product $W \cdot g$ must be known for all baselines (u, v). However, the correlation function $W(u, v)$ is only available for the finite number of baselines at which it has been measured. If there is no other information available, then it is reasonable from the point of view of information theory to use the completely unknown values of W elsewhere with the weight $g = 0$. The weight function $g(u, v)$ will then be very irregular (finite at all measured points and zero over the rest of the u, v plane) and, as a consequence of this, the synthesized beam will be accompanied by an extended pattern of undesirable sidelobes (figure 1 and 2). A synthesis map calculated directly according to Eq. (1) can be so disturbed by the effects of these sidelobes that it becomes essentially useless (figure 3a). I shall use the terms "dirty beam" and "dirty map" to describe this particular synthesized beam and synthesis map respectively. If we choose to give the same weight $g = 1$ to all the measured points, then the dirty map is identical with the so called principal solution.

The principal solution dirty map is one member of an infinite family of solutions that all agree with the available measurements. Other solutions can be produced by allocating any finite values to the correlation function W at the not measured baselines (u, v) and then treating these allocated values as if they were additional real measurements. The principal solution dirty map has the unique property, if no other information is available, of being the solution with the lowest probable rms deviation from the true sky brightness distribution. The absence of other information here means that, as far as we know, the true brightness distribution could be any random function of the direction in the sky.

We are, however, discussing a different case. The true brightness temperature will be below the sensitivity level over most of the sky and only a fraction of the observed field is expected to contain significant departures from zero intensity. This in itself is equivalent to a great amount of information which could be added to the information produced by the interferometer measurements. The problem is how to incorporate this *a priori* information into the procedure by which we calculate the sky brightness distribution over the observed field.

3. VARIOUS WAYS OF INCORPORATING A PRIORI KNOWLEDGE ABOUT THE SKY BRIGHTNESS DISTRIBUTION

The undesirable sidelobe effects are caused by gaps in the u,v coverage; consequently it would seem natural to bridge these gaps by means of some simple interpolation procedure using the surrounding measured points. However, this can only improve the situation if neighbouring points are correlated and if we know the exact nature of this correlation. For instance, if the position of the source is known, then we can perform the interpolation in such a way that the picture of this source is less, or not at all, disturbed by its own sidelobes. The common method of dividing the u,v plane into cells and letting all the measurements within each individual cell be represented by their vector average placed at the center of the cell, can in this way improve the map of a source situated at the instrumental phase center. Improved results can be obtained with more sophisticated interpolation schemes (see e.g. Burns and Yao 1970). The disturbances caused by the sidelobes from all other sources in the field will, however, remain or even become more pronounced. A simple interpolation scheme can indeed improve the situation but only in certain special cases. The gaussian convolution applied to the Westerbork measurements (Brouw 1971, van Someren Gréve 1973) and the corresponding procedures used at other observatories are examples of interpolation procedures. Their great merit is that they transform the measurements to the rectangular grid of points that is required for the Fast Fourier transform algorithm. They are not solutions to the problem of how to eliminate the adverse effects of gaps in the u,v coverage. Grating rings and other sidelobe disturbances remain and some extra confusion may be introduced by the aliasing caused by the imposed regular grid structure.

This first method used by radio astronomers for reducing non-regularly spaced data was the method of *model fitting*. If, for instance, we have measured three numbers: the total flux and the amplitude and phase at one $E-W$ interferometer spacing, then we can solve for the characteristics of a three parameter source model. If the source is a uniform circular disc – that is the model – then we can calculate three parameters, e.g. the source diameter, the disc brightness temperature and the right ascension position coordinate. The trouble is that the source may not at all look like a circular disc and the problem of choosing suitable models and solving the resulting set of equations becomes increasingly difficult for larger numbers of measurements and thus potential model parameters. In practice, model fitting sometimes deteriorates to a mixture of wishful thinking and root mean square adjustments.

The maximum entropy method described by Ables (1973) is based on a mathematical principle that determines which solution we should choose among the infinite family of possible solutions all agreeing with the available interferometer measurements. This principle states that the selected solution should be that which contains the least amount of information (the "maximum entropy") according to a definition of information analogous to that which enters the concept of entropy in thermodynamics. In short, the minimum information solution is also that which is likely to contain the least amount of false information. This principle, coupled with the requirement that the map be compatible with the available interferometer measurements, defines a set of equations which can in principle be solved for any distribution of measured baselines (u,v). However, the equations for irregularly spaced measurements in two dimensions appear to require excessive computing efforts and it is not yet clear how useful the method will be for the problems discussed in this paper.

An obvious property of real brightness distributions is that they are exclusively positive. Schell (1965) and Biraud (1969) point out that the Fourier transform of a positive function must itself be an autocorrelation function and that this property can be used to compute finite values for spacings which are greater than those contained in the antenna system and so increase the angular resolution. The authors compute these values according to different criteria, but the main aim is to achieve a physically acceptable (i.e. non-negative) solution which involves a minimum number of not measured spacings. It should be possible to use these methods also to fill in gaps in the baseline coverage. A different procedure has been described by Högbom (1969). The "dirty beam" described earlier contains negative as well as positive sidelobes and the dirty map will then also in general contain regions of negative intensities. The real brightness distribution, however, must be zero or positive so we can reduce the differences between the map and the true sky by replacing all negative values by zeroes. The Fourier transform of the map will then no longer agree with the known values of W at the measured baselines. The agreement is restored by adding to the map Fourier components that make up for these differences. This will produce new areas of negative intensities which are treated in the same way and the process is repeated until the changes in the map from one iteration to the next become insignificant. The method works well in many cases and will even display certain features on the map at an increased resolution. The unspecified beam shape can, however, cause trouble during the iterations and also make it difficult to judge the reality of details on the final map.

4. MAIN PRINCIPLES OF THE PROCEDURE CLEAN

The procedure that appears to work best at the time of writing has become known as "CLEAN". It is an iterative procedure that operates in the map plane and which uses the known shape of the dirty beam to distinguish between real structure and sidelobe disturbances on the dirty map. Let us assume that the map in figure 2 has in fact been obtained by a direct Fourier transform of a set of ideal noiseless interferometer measurements of a particular piece of sky. The map certainly has a very "dirty" and unsatisfactory appearance but, if we find that it is identical in every individual detail to the expected dirty beam, then the interpretation becomes clear: there is a point source at the position of the maximum deflection on the map. We can formally confirm this conclusion by subtracting out a full dirty beam pattern centered at this position and with the corresponding amplitude: there should then be nothing left. If we now return to this empty map a clean beam pattern – i.e. the ideal main beam of a similar shape but without the disturbing sidelobes – at this position and with the same amplitude, the result will be the same as that which we would have obtained with normal synthesis calculations if the relevant region in the u,v plane had been completely covered with measurements.

The single point source is a trivial case but it illustrates the main principle of the procedure CLEAN. Observe that the interpretation of the dirty map in terms of a point source in an otherwise empty piece of sky is not based on synthesis theory. Other information was used to reject an infinite number of other possible interpretations as being too unlikely to be taken seriously. The real sky may, for instance, look exactly like the complex structure that was interpreted as sidelobes and be empty at the position of the main maximum; this latter would then in fact be a systematic pile-up of sidelobes from the many weaker but real details in the field. In rotational synthesis (supersynthesis) we could in a similar way argue that one of the grating rings is a real feature while the other rings *and* the central maximum are grating responses caused by this ring structure in the sky. Such interpretations are rejected partly because the structures themselves look artificial, but mainly because the real sky brightness distribution should have no special relation to the particular sidelobe pattern produced by the available set of interferometer baselines. A significant correlation between the dirty map (DM) and the dirty beam (DB) will most likely be due to a source or source component at the position of the maximum correlation. The correlation function is given by the convolution DM $*$ DB. The Fourier transform equations (1) and (2) can be written symbolically:

$$DM \xleftarrow{FT} W \cdot g$$

$$DB \xleftarrow{FT} g \tag{3}$$

and it follows from the convolution theorem that

$$DM * DB \overset{FT}{\longleftarrow} W.g^2 \tag{4}$$

If we have chosen the weights g to be unity at all measured points and zero everywhere else, then $W.g^2$ will be identical to $W.g$. Thus DM * DB must also be identical to DM and there is no need to perform the convolution: the dirty map is itself the correlation function that we want. The maximum deflection from zero on the dirty map is also the place at which it is most strongly correlated with the full dirty beam pattern. There may be reasons for using different weights g for different measurements (e.g. matching the main maximum of the DB to the shape of the CB). The DM will then not be equal to DM * DB but to DM' * DB' where the primes indicate the map and the beam that would have been obtained with the weights $g' = \sqrt{g}$. This will only affect the procedure CLEAN in so far that the measurements, as expected, will exert their influence on the solution with different weights g. The procedure CLEAN can now be described in terms of a few simple steps.

I. Compute the DM and the DB by standard Fourier inversion methods. The weights g should be chosen in such a way that the main lobe of the DB becomes a good fit to the selected CB.

II. Subtract over the whole map a DB pattern which is centered at the point at which the DM has its maximum absolute value $|I_o|$ and which is normalized to γI_o at the beam center. The fraction γ will be called the *loop gain*.

III. Repeat step II, each time replacing the DM by the remaining map from the previous iteration. Stop the iterations when the current value of I_o is no longer significant in view of the general noise level on the map.

IV. Return to the final remaining map all those components that were removed in step II, but do this in the form of clean beams with the appropriate positions and amplitudes.

The individual iteration will represent a correct interpretation of the situation if the value I_o did in fact contain a contribution with an amplitude of at least γI_o from some real feature. Clearly, for optimum safety, one should choose an infinitesimally small value for the loop gain γ. The process could then be visualized as one in which a ceiling to the map absolute value was lowered while continuously shaving off whatever protrudes through this ceiling along with the associated patterns of dirty sidelobes. However, if a certain deflection is significant in the sense that it indicates the presence of some real feature, then it should be safe to assume that this feature makes a substantial contribution to this deflection. One could, of course, make a full probability analysis in each individual case but in practice it is found that there is little further improvement when the loop gain is reduced below $\gamma = 1/2$. Even $\gamma = 1$ will give perfectly good results in the simplest cases. Figure 3 shows a dirty map produced from measurements with the Green Bank interferometer and the end product of using the CLEAN procedure with 1, 2 and 6 iterations and a loop gain of $\gamma = 1$. The last map could be described as empty but for certain exceptions which can be specified in a table containing less than 30 numbers. This is to be compared with the 100 measured quantities, i.e. the amplitudes and the phases at 50 interferometer baselines (figure 1). Polarization measurements can be handled with a complex version of CLEAN or by processing the Q and U maps separately; a survey of the structure and polarization of 78 radio sources has been made using this technique (Högbom and Carlsson 1973).

The CLEAN iterations can be programmed in an efficient manner. In a first step we go through the whole array looking for the element with the largest absolute value $|I_o|$. At the same time we compute the sum of all absolute values encountered. Dividing this sum by the total number of elements we get the average absolute value. The decision as to whether the deflection I_o is significant can now be based on the criterion that $|I_o|$ exceeds this average value by some specified factor of the order of 4. This full survey of the map deflections may not be necessary at every iteration. It can be enough to make note of other regions containing high absolute values and then to restrict the search for the maximum deflection to these regions during the next few iterations. The subtraction of a dirty beam pattern component over the whole map with a specified loop gain γ will require one multiplication per array element. However, the loop gain need not be a fixed quantity. It should be less than about 1/2 and also should not be too small because this will mean an unnecessarily large number of iterations. If we allow the loop gain to vary within a factor of two between, say, $0.25 < \gamma \leq 0.5$, then the multiplications can be replaced by subtractions. Let the dirty map and dirty beam be normalized so that their maxima are 4 and 1 respectively. Now, a direct subtraction in the first iteration corresponds to $\gamma = 0.25$. The maxima in the following iterations will be lower and hence the loop gain correspondingly higher. When a maximum has an absolute value of less than 2 which would imply $\gamma > 0.5$, we double the map – at the expense of one addition per element – before performing the subtraction which will then again correspond to a loop gain $\gamma < 0.5$. Returning the clean beams after the last iteration is no problem because they only affect small areas of the map.

5. LIMITATIONS TO THE PROCEDURE CLEAN

A non-linear procedure such as CLEAN is of doubtful value in practice unless the user can judge whether or not a final cleaned map is reliable. Clearly, CLEAN cannot deliver a correct map if the source is in fact so complex that it cannot be described adequately with less parameters than there are measurements. In such a case one finds after a few iterations that the next maximum deviation I_o will no longer be significantly greater than the general confusion level on the map (i.e. indicate a more than 50% probability that it is partly due to some real feature at this position). The iterations will therefore be interrupted at an early stage when there is still a great amount of sidelobe confusion noise over the map. After returning the clean patterns of the first few iterations, this noise remains and will mask the detailed structure of the source in much the same way as would an excessive amount of receiver noise. Clearly, noisy data will have the same effect: CLEAN will only pass through a few iterations but this should at least give some improvement. In both cases it may be possible to continue the iterations if the remaining map and the dirty beam are first smoothed to a lower resolution. This is especially useful in bringing out weak extended distributions that would otherwise be lost in the noise. Purely random data will result in no or perhaps a single iteration (depending upon how the 50% criterion has been formulated). At the other extreme – complete u,v coverage and zero noise – CLEAN may pass through any number of iterations, but the clean beam is then identical to the "dirty beam" and the net effect is, as it should be, equal to zero.

Figure 4 shows cleaned maps of the source 3C 130 prepared from measurements at 200, 100, 50 and 30 baselines taken with the Westerbork synthesis telescope. The first map is in itself a good illustration of how the CLEAN procedure can help towards an efficient use of a synthesis telescope. The minimum observing time for two-dimensional synthesis with this instrument using the standard inversion methods is 12 h. The 200 baselines (10 h angles, each with 20 spacings) required only $3\frac{1}{3}$ h of instrument time.

The map based on 100 baselines is very similar. The 50 baseline map is seriously disturbed but still correct in its main features. 30 baselines is obviously not enough, and the map is dominated by a prominent confusion noise.

The general run of the zero-level contour in figure 4a shows that the map is free from the usual depression caused by the missing zero spacing in the interferometer measurements. As far as the procedure CLEAN is concerned, this depression is just one part of the dirty beam sidelobe pattern which it attempts to eliminate. The procedure cannot detect large extended distributions that are completely resolved at all the measured baselines, but it will remove the depression caused by those source components that it detects during the iterations.

An important limitation to the use of CLEAN in its present form is related to the fact that the actual distribution of baselines will normally be three-dimensional, each baseline having some vector component w wavelengths parallel to the direction of the observed piece of sky and hence at right angles to the u,v plane. The measurements W and the weights g will be functions of all three variables (u,v,w) and the Fourier integrals in Eqs. (1) and (2) should strictly be expressed in their three-dimensional form. The exponential factor in these equations then becomes:

$$\exp\left[i\,2\pi\,(u\ell + vm + wn)\right] \tag{5}$$

The three directional cosines ℓ, m and n are related by

$$\ell^2 + m^2 + n^2 = 1 \tag{6}$$

Write:

$$wn = w\,(1 - \ell^2 - m^2)^{1/2}$$
$$\simeq w - 1/2\,w\rho^2 \qquad (\rho^2 = \ell^2 + m^2) \tag{7}$$

then the exponential factor (5) can be written:

$$\exp\left[i\,2\pi\,w\right] \cdot \exp\left[i\,2\pi\,(u\ell + vm - 1/2\,w\rho^2)\right] \tag{8}$$

The first factor can be formally incorporated in the weight function $g(u,v,w)$; it is the analytical expression that corresponds to reducing the phases at the measured baselines to their projections on the u,v plane. The term $1/2\,w\rho^2$ in the argument expresses the fact that this projection looks different as seen from directions away from the center of the field (i.e. for $\rho \neq 0$). The two-dimensional formulae (1) and (2) will be correct if this term is always $\ll 1$. The effect of having larger values within the observed field of sky is to make the sidelobe structure of the synthesized beam noticeably different for sources situated in different parts of the synthesis map. Consequently, cleaning with a fixed dirty beam shape will have a positive influence on the map only when all operations are restricted to an area centered at the field (phase) center and with a radius ρ_{max} given by, approximately

$$w_{max} \cdot \rho_{max}^2 \approx 1 \tag{9}$$

Thus, if $w_{max} \cdot \lambda = 10$ km and $\lambda = 6$ cm, then CLEAN should not be used to correct for disturbances caused by sources outside a circle of radius $\rho_{max} \simeq 0.00245$ radians, or about 8 arcmin. There are ways by which one can take this variable synthesized beam pattern into account but they will be at the expense of a considerably increased computing complexity. However, two effects tend to make this limitation less important. First, the interferometer measurements will usually be made while the antennas follow the observed field for some time. The u,v plane will therefore be measured along a set of lines (or paths) rather than at a set of discrete points and this will make the far sidelobe level approximately inversely proportional to the square root of the distance from the beam center. A finite bandwidth will cause a similar reduction of the distant sidelobe level. A strict statistical study of these effects in the case of randomly distributed baselines has not been made, but it seems that sources outside the area $\rho \leq \rho_{max}$ should in general have a very reduced disturbing effect on the cleaned central part of the map.

REFERENCES

Ables, J.: 1973, *Astron. Astrophys. Suppl.* (this issue).

Biraud, Y.: 1969, *Astron. Astrophys.* **1**, 124.

Brouw, W.N.: 1971, *Thesis*, University of Leiden.

Burns, W.R. and Yao, S.S.: 1970, *Astron. Astrophys.* **6**, 481.

Högbom, J.A.: 1969, *Rep. URSI 16th Gen. Assembly*, Ottawa.

Högbom, J.A. and Carlsson, I.: 1973, *Astron. Astrophys.* (in preparation).

Schell, A.C.: 1965, *Radio and Electronic Engineer* 21.

Someren Greve, H. van: 1973, *Astron. Astrophys. Suppl.* (this issue).

J.A. Högbom

Stockholm Observatory
S - 13300 Saltsjöbaden, Sweden

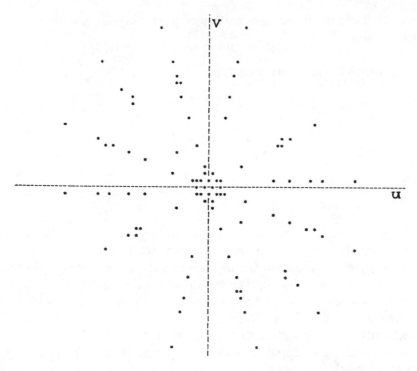

Figure 1   Example of a baseline coverage obtained with the Green Bank Interferometer. The measured points have been mirrored through the origin to give a better impression of the overall structure.

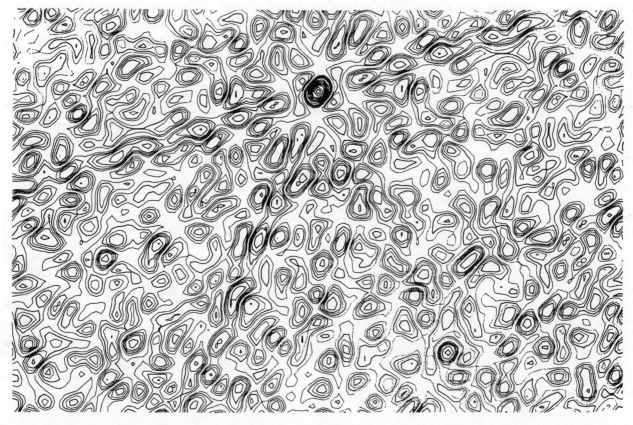

Figure 2   The "dirty beam" corresponding to the *u,v* coverage shown in figure 1. Contours are drawn at 5, 10, 15, 20, 30 etc. % of the beam maximum (top center). No distinction has been made in the figure between positive and negative contours; half of the sidelobes are in fact negative.

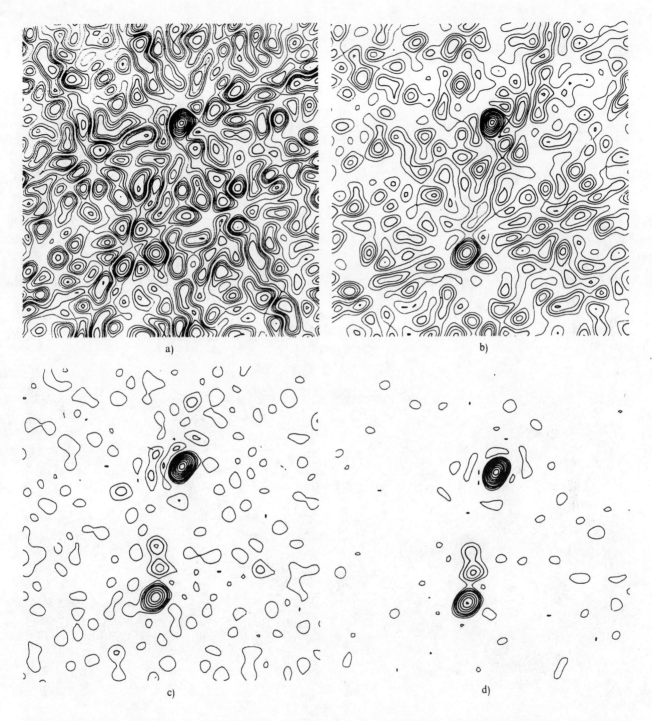

a)

b)

c)

d)

Figure 3   Illustrating the effect of the CLEAN procedure on measurements at 2695 MHz of the radio source 3C 244.1 taken with the u,v coverage shown in figure 1. Contours are drawn at the same intensity ratios as in figure 2. a) the "dirty map" b) cleaned map after one iteration with the loop gain $\gamma = 1$ and subsequent return of the clean beam, c) same, but after two iterations and the return of the two clean beams. The north preceding component is extended and there are some weaker components present, d) 6 iterations. The further improvement here is due to the cleaning of the sidelobes from the less intense features that remain after the two first iterations.

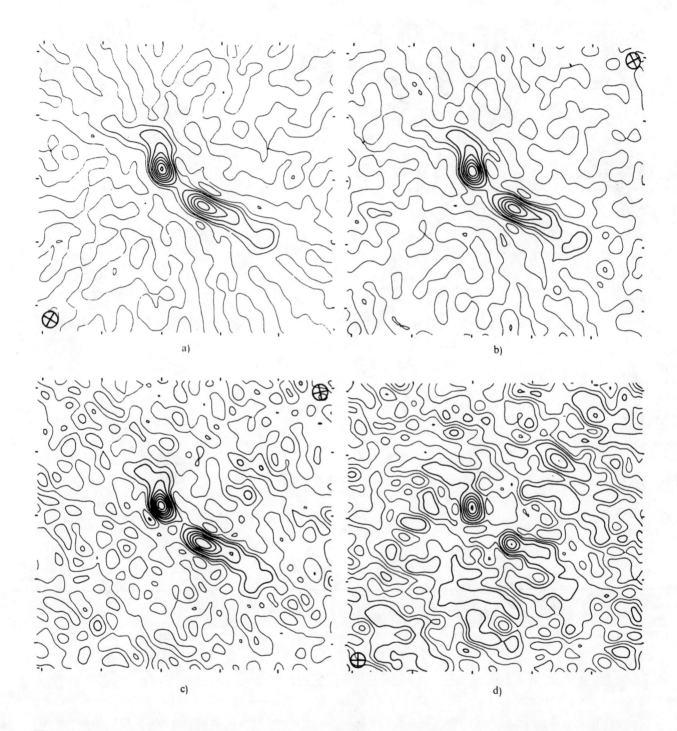

a)   b)

c)   d)

Figure 4   Illustrates the effects of CLEAN when the number of measurements used for the calculations is reduced successively until it is no longer sufficient to define the source structure uniquely. The cleaned maps have been prepared from 1415 MHz measurements with the Westerbork synthesis telescope taken at a) 200, b) 100, c) 50 and d) 30 baselines. Contours are plotted at 0, 1/2, 1, 2, 3, etc. units; no distinction is made in the figure between positive and negative contours.

MAXIMUM ENTROPY SPECTRAL ANALYSIS

J.G. ABLES

Division of Radiophysics, CSIRO, Sydney, Australia

When viewed from the standpoint of information theory, the problem of selecting a spectral-domain representation for a limited and imperfectly determined set of data in the correlation-domain becomes one of maximizing the implied signal entropy subject to the constraint that the result be consistent with the data at hand. The method produces superior spectral representations when compared with more traditional methods and should find applications in radioastronomy (e.g. in correlation spectroscopy and aperture synthesis) as well as in other fields.

1. INTRODUCTION

A problem which occurs often in most branches of science is that of choosing a spectral-domain representation for a limited set of imperfectly determined correlation-domain data. With the increasing importance of correlation spectroscopy and aperture synthesis methods, this is particularly so in radioastronomy. Now while it is true (by definition) that the two domains are linked by the Fourier transform, it is only in the case of complete, noiseless knowledge of the correlation function that the transform may be used to obtain the spectral function without imposing additional assumptions about the nature of both the data at hand and the unavailable data.

Recently an innovative treatment of this problem, called the maximum entropy method (MEM), has appeared in the literature of the geosciences (Burg 1967, Lacoss 1971, Ulrych 1972) and mathematical statistics (Parzen 1968, 1969). Computer experiments using data generated from processes with known spectral functions have shown the MEM to be consistently superior, in some cases greatly so, to the traditional methods (e.g. those of Blackman and Tukey (1958)) and you will find an example of one of these comparisons below. However, previous authors have said little concerning *why* the MEM should be a rational approach to the problem, and the few comments offered slight or miss entirely what seems, to this author, to be the mainpoint – that MEM spectral analysis stems from Jaynes' Principle and is best understood from that viewpoint. In the following development we will emphasize this approach, show how the MEM can be extended to other data analysis situations, and suggest a novel way of treating measurement noise.

2. THE ANALYST VERSUS HIS DATA

The efforts of an experimentalist can be roughly divided into four time-sequential categories: data collection, data reduction/analysis, interpretation, and communication. We are concerned here with the second – the reduction/analysis – by which is meant all of those many and curious ways by which perfectly horrible data is rendered as authoritative-looking graphs, maps, tables, etc. It is often at this step that good data is mangled beyond recognition and the experimenter's prejudices, masquerading as real data, are inserted. But we hope not; and, to that end, we state now the First Principle of Data Reduction:

THE RESULT OF ANY TRANSFORMATION IMPOSED ON THE EXPERIMENTAL DATA SHALL INCORPORATE AND BE CONSISTENT WITH ALL RELEVANT DATA AND BE MAXIMALLY NON-COMMITTAL WITH REGARD TO UNAVAILABLE DATA.

Very good, but just how do we live up to it?

3. THE BATTLEGROUND DEFINED

Let the ordered set of numbers which comprise the data be X_0 and a representation in the data reduction output domain be S. Now if X_0 uniquely determines an S_0 there is no problem (e.g., if S_0 were simply the average of the data values). In such a case S_0 may or may not be a useful representation of the data; but, there is *no argument* about the selection of S_0 once X_0 is specified. Unfortunately, many commonly used analysis methods are not of this type. The tip-off is often the word "estimate", indicating that some form of statistical inference is being used.

Of the reduction methods for which X_0 does not uniquely determine an S_0 we shall restrict our attention to those for which the reverse *is* true. That is, given any S a unique X in the data domain is determined. Call this reverse transformation T so that $T(S) = X$. Many useful reduction processes have this property.

4. AN EXAMPLE

Suppose that our experimental apparatus measures the quantities β_1, and β_2 of nature corresponding to the values α_1 and α_2 of some parameter controlled by the experimenter. That is, say, our apparatus has a two-position switch labelled α_1 and α_2 and a meter which reads β (without error) corresponding to the α chosen. The observed data is then the ordered pair of numbers $X_0 = (\beta_1, \beta_2)$ and the domain of the data is the set of all such ordered pairs (x_1, x_2) such that both x_1 and x_2 lie within the range of values permissible on our meter. The whole of the experiment can be neatly expressed by plotting the two points (α_1, β_1) and (α_2, β_2) in the $\alpha\beta$ plane.

Two separate reductions are proposed. The first is aimed at finding the values m and β_0 of the slope and β–axis intercept respectively of a straight line through (α_1, β_1) and (α_2, β_2). Hence $S_0 = (m, \beta_0)$ and is uniquely specified by the experimental data and *vice versa*. The transformation T which associates a hypothetical experimental outcome X with any S is expressed by the two equations $x_1 = m\alpha_1 + \beta_0$ and $x_2 = m\alpha_2 + \beta_0$. Obviously these equations may be solved to give m and β_0 in terms of x_1 and x_2. Note that since not all (x_1, x_2) are permissible (unless we bend the pointer on the meter!) an *a priori* restriction on the domain of S_0 in the $m\beta$ plane is implied by this transformation. This restriction reduces the domain of admissible S_0 to a single element when (x_1, x_2) is fixed by the experimental outcome to (β_1, β_2).

The second reduction requires that we locate the center (α_0, β_0) and radius R of the circle in the $\alpha\beta$ plane passing through (α_1, β_1) and (α_2, β_2) such that the shorter arc of the circle connecting the points is concave upwards (taking the β–axis as vertical). Obviously this is going to lead to trouble since for any experimental outcome there are an *infinite number* of such circles. The point is that we desire to compute $S_0 = (\alpha_0, \beta_0, R)$ from our data $X_0 = (\beta_1, \beta_2)$ and there is just not enough information in the data to allow us to do this uniquely. The data *does* give *some* information with regard to S_0 since before the experimental outcome S_0 is restricted only to some area in the $\alpha\beta$ plane determined by the restrictions on X (i.e., the possible meter readings) but after knowing $X = X_0$, S_0 is restricted to lie on a semi-infinite line starting mid-way between (α_1, β_1) and (α_2, β_2) and extending upwards perpendicularly to the line connecting these points. And, there is *still* a transformation T mapping any given S back into a unique X. (This is geometrically obvious since any admissible S defines a circle in the $\alpha\beta$ plane which intersects the lines $\alpha = \alpha_1$ and $\alpha = \alpha_2$ in at least two and at most four points, only two of which are connected by an arc of this circle which is concave upwards. The β–coordinates of these two points are $(x_1, x_2) = X$.)

Note that if we had just one more measurement β_3 at $\alpha = \alpha_3$ distinct from α_1 and α_2, we could find S_0 uniquely, since three points in a plane are sufficient to determine a circle. But we *do not* have it, and this is precisely what we mean by *unavailable* data in the statement of the First Principle. While it is obvious here that for the analyst to provide a β_3 just to expedite the analysis would unacceptably prejudice the interpretation of the experiment, such transgressions of the First Principle are, unfortunately, frequent in the real world.

5. THE QUANDARY

So, this is the situation: we wish to consider those cases like the second reduction suggested above where the desired reduced state S_0 of our data is not uniquely determined by the available data X_0; but, we do have a transformation T which maps every admissible S into a unique X. This situation is very much more common than might, at first, be believed. Clearly one way around the difficulty is to stop the analysis at this point; simply disclose the domain of admissible S_0 (now diminished by our experimental efforts); and exit from the scene as gracefully as possible. Honest – yes, but often scarcely more useful than just a presentation of the raw data itself. The usual reason for dragging the data into the S–domain is that it is (we claim) far more easily interpreted there; but, such a multiplicity of representations is more likely to obscure than to elucidate. Occam* would surely have agreed.

We seek, therefore, some way of singling out just *one* of the admissible S_0–domain states as a worthy representative of the entire group – and to do this without bending the First Principle too far. Put this way, it sounds like a risky business, but it *is done* all the time! For example, we may measure a few points on a continuous function and then publish some curve derived from these points by least-squares fitting (when the relative errors in the data are large or our prejudice against the data is high) or perhaps spline fitting (when the relative errors are negligible – or simply neglected). Any one of an infinite number of curves passing through the measured points (or just *near* them if measurement errors are recognized) are consistent with the data. But, we all agree that these derived curves *are* useful and, so long as the methods used to get them are stated, we are happy to contemplate their implications.

The situation which often arises in spectral analysis is similar. We measure at a few points the correlation function and then, from the infinity of spectral functions consistent with these measurements, pick one to show to the rest of the world. Usually we do not even display the original data. (At least in curve-fitting problems this *is*, nearly always, done.) Two excuses for not giving the correlation data are (1) that it is usually difficult to interpret in one "blow of the eye" and (2) the rules for obtaining the spectral-domain representation which *is* given are well-known and acknowleged. This is in spite of the fact that even a superficial examination of these rules shows them to be in strong violation of the First Principle! MEM spectral analysis may be viewed as an attempt to derive new rules for selecting the spectral-domain representation in closest possible accord with the First Principle.

6. THE WAY OUT

The following *modus operandi* is suggested for minimizing our transgressions of the First Principle.
1. Define a transformation T such that $T(S) = X$, i.e., associating with every S some (hypothetical) experimental outcome X.
2. Write a functional relationship of the form $I = J(S)$ associating a scalar I with every output S of the reduction such that I is, in some sense, a measure of the *ignorance,* or *lack of information* contained in S.
3. Given then an actual experimental outcome X_0, choose as its unique representation in the S–domain that S_0 from all S satisfying $T(S) = X_0$ which causes I to be a maximum. That is, maximize $J(S)$ subject to $T(S) = X_0$ as a constraint.

Item (1) serves to define the genre of the reduction while item (2) gives a mathematical interpretation of the word "non-committal" in the First Principle. Item (3) is then a mathematical restatement of the First Principle itself. Of course, the difficult part is choosing the functional J.

---

*William of Occam (c. 1290–1349 or 1350?), English nominalist philosopher and noted logician, early champion of separation of church and state, and author of the principle called Occam's Razor, one statement of which says "entities shall not be multiplied unnecessarily." He probably died of the Black Death.

7. ENTROPY AND IGNORANCE

Before trying to formulate J for the spectral analysis problem let us have a quick look at a conceptually simpler situation. An ordinary six-sided game die is placed in a closed, opaque box which is then thoroughly shaken and placed, still closed, before us. Our total experimental data consists of observing this sequence of events. We are now asked to report on our knowledge of the state of the die, i.e., which of the six faces is uppermost. After some thought we agree to make this report by giving six numbers $\{p_i\}$ to be associated with the six faces of the die such that $o \leqslant p_i \leqslant 1$ and $\sum_{i=1}^{6} p_i = 1$. We will call p_i the probability that the i-th face is uppermost, with a value of one representing certainty that that face *is* uppermost and zero representing certainty that it is *not*.

It should be noted here that interpretation of probability measure in this way (as a measure of knowledge or information rather than as a relative frequency of occurrence) while being intuitive for most scientists and engineers has had a long and stormy history and has only recently been regaining its position as a valid viewpoint that it held two hundred years ago. A provocative discussion of this issue has been given by Jaynes (1963).

Our experimental data can be summed up by saying (1) there are exactly six faces to the die and (2) one and only one face is, at this moment, uppermost. That is to say, $i = 1, 2 \ldots, 6$ and $\sum_{i=1}^{6} p_i = 1$. Now if the top of the box were removed so that the state of the die became known, we would gain information (lose ignorance) and an often used measure of this information gain is $\Delta I = -\log p_j$ where j is the index of the *actual* state of the die. The oldest use of this information measure is probably due to Szilard (1929) but is best known through the work of Shannon and Weaver (1949).

Actually it is not hard to see why this is a reasonable expression to use. It is reasonable to want the information measure of a joint event to be the sum of the measures making up the joint event. For, say, two independent events k and l with probabilities p_k and p_l the event of both happening has probability $p_k p_l$ so that we want $\Delta I(p_k p_l) = \Delta I(p_k) + \Delta I(p_l)$. Also it would seem right if $\Delta I(1) = 0$ since the occurrence of an absolutely certain event tells us nothing we did not already know. Finally we would like information *gain* to be represented by a *positive* number. These considerations are sufficient to restrict ΔI to the form $\Delta I = -K \log p$ where K is a constant. K merely determines the size of the information unit and we will use $K = 1$. If base two logarithms are used, the resulting unit is called a "bit", for natural (base e) logarithms a "nit" and for common (base 10) logarithms a "hartley"! (No theory can be said to be truly successful until it has acquired a plethora of confusing units.)

Getting back to our example, we can see that $-\log p_j$ would be a reasonable measure of our ignorance except that we do not know what p_j is (if we did, we would just set $p_j = 1$, $p_{i \neq j} = 0$). The best that can be done is to use the expectation value of $-\log p_j$, that is, $\overline{\Delta I} = \sum_i p_i \log p_i$. So, in accord with the First Principle, we choose $\{p_i\}$ to maximize $-\sum_j p_j \log p_j$ subject to $0 \leqslant p_j \leqslant 1$ and $\sum_i p_i = 1$ as constraints. The result is $p_i = \frac{1}{6}$, $i = 1, 2, \ldots, 6$. (This seems like a hard way, does it not, to arrive at an obvious answer? Of course the easy way to get the answer is to apply Laplace's Principle of Insufficient Reason.*)

Now suppose that we are informed that the die is loaded and that exhaustive tests have determined that the average value of the uppermost face in situations like this is 9/2. What now do we give for $\{p_i\}$? Obviously we just maximize $-\sum_i p_i \log p_i$ subject to the constraints $0 \leqslant p_i \leqslant 1$, $\sum_i p_i = 1$, *and* $\sum_i i p_i = 9/2$. (For those with an

---

*Laplace, who said "...la théorie des probabilités n'est que le bon sens confirmé par le calcul," also believed that it was only "good sense" to assign equal probabilities to mutually exclusive events unless there was reason to think otherwise. This "Principle of Insufficient Reason," which may be taken as merely a statement of symmetry about the situation, fell into disrepute after a number of paradoxes were generated through misapplication. You will find it, however, under various guises, almost never under its true name, in the opening chapters of any text on probability.

unslakable numerical curiosity the result is $p_1 = 0.055$, $p_2 = 0.080$, $p_3 = 0.120$, $p_4 = 0.165$, $p_5 = 0.235$, and $p_6 = 0.345$.)

The point is simply that all data or information should be incorporated as *constraints* on the maximization of our ignorance.

This use of the entropy function $-\sum_i p_i \log p_i$ for solving the problem of assignment of probability distributions under the constraints imposed by knowledge of the values of averages of the form $a_k = \sum_i f_{ki} p_i$ is due to Jaynes (1957 a, b). It has often been called Jaynes' Principle. It is, of course, no mere coincidence that the measure of ignorance we have arrived at has the same mathematical form as thermodynamic entropy. They are the same in spirit as well as form. (Some very thorough discussions on this point may be found in the book "*Science and Information Theory*" by L. Brillouin (1962)).

8. SPECTRAL ANALYSIS AND THE FIRST PRINCIPLE

And now, finally, we take up the problem of spectral analysis. It is assumed that we possess experimental data X_0 (concerning some "signal") of the form $X_0 = \{\rho_0, \rho_{\pm 1}, \rho_{\pm 2}, \ldots, \rho_{\pm M}\}$ where the ρ_n are sample values of the correlation function at lags $\tau_n = n\Delta\tau$. It is desired that we produce from this data a spectral representation of the original signal over the frequency range $-\nu_N \leqslant \nu \leqslant +\nu_N$ where ν_N is the Nyquist frequency, i.e., $\nu_N = 1/2\Delta\tau$. Calling the spectral function $S(\nu)$ we may associate a unique X with any given S by a Fourier transform. Thus, if $X = \{x_0, x_{\pm 1}, x_{\pm 2}, \ldots, x_{\pm M}\}$, then

$$(1) \qquad x_n = \int_{-\nu_N}^{\nu_N} S(\nu) e^{-i\lambda n} d\nu \qquad (i = \sqrt{-1}, \lambda = 2\pi\nu\Delta\tau)$$

Note, however, that this relationship does *not* provide us with a unique expression for $S_0(\nu)$ in terms of X_0.

The traditional approach to estimating S_0 from X_0 is to assume that $\rho_n \equiv 0$ for $|n| > M$ and then to take a discrete Fourier transform of $Y = \{W_0\rho_n, W_{\pm 1}\rho_{\pm 1}, \ldots, W_{\pm M}\rho_{\pm M}, \ldots, 0, 0, \ldots\}$ where $\{W_n\}$ is a sequence of weights used to taper the known correlation values smoothly into the assumed zero values for $|n| > M$. (See, for example, Blackman and Tukey (1969).) In choosing the weighting sequence a compromise must be made between resolution in the final spectrum and the extent to which any one spectral component contaminates the estimates of other nearby estimates. (This contamination is often described as "leakage" of power through the "side-lobe" structure of the spectral window function. For $W_n \equiv 1$ and ρ_n assumed to be zero for $|n| > M$, i.e., an unweighted but still sampled and truncated correlation function, this spectral window function has the form $(\frac{\sin\theta}{\theta})$, $\theta = 2\pi\nu M\Delta\tau$, which gives the greatest possible resolution of $\Delta\nu \sim 1/M\Delta\tau$ but has an extensive and strong side-lobe structure. Using a smoothly tapering set of weights can reduce the amplitude of the strongest side-lobes by an order of magnitude at the expense of worsening the resolution by about a factor of two.)

It should now be obvious that this method of estimating the spectrum is in strong violation of the First Principle. Firstly, data not in evidence is assumed (all those zero values of ρ_n for $|n| > M$). Secondly, perfectly good data is falsified by the tapering process. We should not be surprised, therefore, that a better way to estimate the spectrum can be derived in accordance with our *modus operandi* for living within the First Principle.

In equation (1) above we already have the required reverse transformation from the output domain to the data domain. What is still required is a formula for the ignorance (entropy) expressed by a spectrum $S(\nu)$. Ignorance of what? Of the original signal, of course. Hence, we would like to write an expression for the entropy of a continuous stochastic process (as a mathematical statistician would call our signal function) in terms of its spectrum. Unfortunately we are unable to do this directly since the spectrum alone is insufficient to determine the entropy. The most useful expression for the entropy of a continuous function is the so-called relative entropy given by

$$H = - \int_{-\gamma}^{\infty} p(u) \log p(u) du$$

where $p(u)$ is the probability distribution of the signal amplitude. See, for example, Bartlett (1966). Note the similarity to our earlier expression for entropy in the discrete case. The problem here is that our output domain is the domain of $S(v)$, not $p(u)$, and there is no unique mapping of one onto the other. But, there is a clever way around this obstacle.

One way to consider any signal with spectrum $S(v)$ is to think of it as being produced by passing a uniform spectral density (white) signal through a linear filter with power gain function $S(v)$. Now in any particular case we can define (using the expression above) the entropy of the input signal to the filter and the entropy of the output signal in terms of their respective amplitude probability distribution functions. But the *difference* between the input and output entropies (which Bartlett calls the entropy gain of the filter) *can be expressed as a function of S (v) alone*! (Bartlett 1966). This entropy gain is simply

(2)
$$\Delta H = \int_{-\infty}^{\infty} \log S(v) dv$$

Our way is now clear. We do not know just what "white" process is required as input to the filter to give *our* signal as output (there are an unlimited number of processes with uniform spectral density, but differing in other ways, such as their amplitude probability distributions), but this we do know: Whatever the input, we want the output to have as great an entropy as possible and this can be achieved by *making the entropy gain of the filter as large as possible*. (If you had an amplifier with some signal on its input and wanted to assure the highest possible output power even though the input was unknown, you would not hesitate to turn the gain control to its maximum position.)

We can now state precisely how to choose, in accord with the First Principle, a spectral representation of a signal about which we know only the value of its correlation function at a finite number of points.

Given: $X_0 = \{p_n\}$ a sequence of sample values of the signal correlation function
Choose: $S_0(v)$ such that

$$I = \int_{-\infty}^{\infty} \log S(v) dv$$

is a maximum, subject only to the constraints that

$$\int_{-\infty}^{\infty} S(v) e^{-i\lambda n} dv = p_n$$

where $\lambda = 2\pi v \Delta \tau$.

Formulated in this way, estimating $S(v)$ becomes a calculus of variations problem. In the special case that the known correlation values are equi-spaced and centered on zero lag (or spacing) the solution, which begins with the introduction of a Lagrange multiplier for each of the constraint equations, is not difficult and may be given as

(3)
$$S_{MEM}(v) = g \left| \sum_{n=0}^{M} z_n e^{i\lambda n} \right|^{-2} , \quad g = \Delta \tau / z_0$$

where the sequence $\{z_n\}$ is obtained as the solution of

(4)
$$Rz = \text{col}(1, 0, 0, \ldots, 0)$$

Here R is the so-called correlation matrix having elements $R_{ij} = p_{(j-i)}$. A matrix of this sort is called Toeplitz and there are special methods for solving systems such as (4) with very little computational effort, even for $M \sim 10^3$. (Levinson 1947). Then if one uses the Cooley-Tukey FFT algorithm for evaluating the sum in (3), the result is a very efficient way of computing MEM spectra. (In doing so, care should be taken to use a fine enough spacing in the evaluation of the sum to delineate all of the detail in the MEM spectrum since, even though the sum itself is band-limited, the reciprocal of its squared magnitude is not.)

9. THE PROOF OF THE PUDDING IS IN THE EATING

And now for that example spectrum promised earlier, see figures 1 through 4. In this computer experiment a signal process with known spectrum was choosen. It has three spectral lines with Gaussian profiles and relative amplitudes 5:10:1 sitting on top of a low-level uniform spectral density background. Figure 1 shows the *true* spectrum. The known autocorrelation function was truncated and sampled to obtain data for the comparison. The two strong lines are close enough together that, with the savage truncation used ($M = 15$), they will be difficult to separate, while the third, although well separated from the other two, is weak enough to give some difficulty with trying to find it among their sidelobes.

In figure 2 we have the result of standard Fourier transformation using the weights $W_n = 1$ ($n = 0, \pm 1, \pm 2, \ldots, \pm 15$) to give the best possible resolution (but very bad sidelobes). The two strong lines are not quite resolved (although one can see that there *are* two lines) while the weak line is rather lost in the side-lobe rubbish. In figure 3 the so-called "raised cosine bell" weights $W_n = \frac{1}{2}[1 + \cos(\pi n / M)]$ have been used. These weights have often been considered as a reasonable compromise in the trade-off between resolution and side-lobe suppression. Here the strong lines are truly blended but at least the weak line is clearly visible, albeit with a width about twice what it should be.

At last, in figure 4, we have the MEM spectrum (using exactly the same data as the two previous examples) with resolution *superior* to figure 2 and side-lobe suppression even *better* than that of figure 3. Although we seem here to be able to have our cake and eat it too, it is truly only our just desert (dessert?) for being as careful as possible not to invent data for those cases where no measurements have been made nor to falsify the *real* data in any way.

10. EXTENSIONS

(1) Complex correlation functions. The solution as stated is valid for complex values of the ρ_n.

(2) Two or more dimensions. Nothing really new is required. The statement of the method becomes (in two dimensions)

$$\text{Maximize} \int_{-\infty}^{\infty} \int_{-\infty}^{\infty} \log S(v, \xi) dv d\xi$$

subject to the constraints

$$\rho_{nm} = \int_{-\infty}^{\infty} \int_{-\infty}^{\infty} S(v, \xi) e^{-i(\lambda n + \mu m)} dv d\xi$$

where $\lambda = 2\pi v \Delta \tau$, $\mu = 2\pi \xi \Delta \sigma$ and ρ_{nm} is the (possibly complex) correlation of the signal at $\tau = n\Delta\tau$, $\sigma = m\Delta\sigma$. (τ and σ are, of course, the lag (spacing) variables.)

(3) Non-gridded data (i.e., non-integral values for the parameters n and m). Obviously the statement of the method as a constrained maximization problem does not require that the data lie on any evenly-spaced grid. It is just that this comes in very handy when working out a general solution like the one given. However, we conjecture here that a general solution for the case of unequally spaced data is possible and not much more complicated than the equi-spaced case.

(4) The explicit acknowledgement of experimental error. So far nothing has been said about the effect of imperfectly determined correlation values. One's first thought is to ignore the errors until the MEM spectrum is produced and then to acknowledge their existence by quoting some measure of uncertainty for the spectral estimates. This is, of course, the philosophy behind the treatment of error in most traditional data reduction schemes. That is, one doesn't modify the scheme itself but simply studies the "propagation of error" through the scheme in order to be able to transfer the uncertainty measures associated with the raw data to the output domain. So far, no such propagation of error analysis has been performed for MEM spectral analysis although Lacoss (1971) has done a few experiments to get a feel for the effect of error on this new method. His conclusion is that the MEM is not particularly more sensitive to error than the older methods.

Now it would seem that to treat error in this way is, once again, to violate the First Principle. Not that one should not make a propagation of error analysis, that is most useful; but, surely the scheme itself must be modified. The problem is that our constraints will force the Fourier transform of the chosen $S(v)$ to pass *through* all of our measured correlation values and we *know* that these values are imperfect.

Would it not be better to ease our constraints to allow the transform to just pass *close* to the measured values? How close? Well, let us suppose that we know σ, the standard deviation of the error of our measurements. Then it would be reasonable if the average of the squares of the deviations of the transform from the measured correlation values should be about σ^2. (One would expect this average to be *exactly* σ^2 only in the limit of an infinite number of measurements.) So, let us set this average equal to σ^2 and use *that* as our only constraint. Since we are now making a second order correction to our reduction process, the slight violation of the First Principle incurred by using strict equality should not trouble our conscience too much. So then, in the presence of error in the correlation measurements with standard deviation σ one should maximize

$$\int_{-\infty}^{\infty} \log S(v)dv$$

subject to

$$\frac{1}{N} \sum_n \left\{ \int_{-\infty}^{\infty} S(v)e^{-i\lambda n}dv - \rho_n \right\}^2 = \sigma^2$$

where N is the number of available measurements. Note that if $\sigma^2 = 0$ then the constraint can only be satisfied if every term of the sum is zero, since all terms are positive. Hence, this constraint reduces to our original constraints, as it should, when $\sigma^2 = 0$.

11. CONCLUSION

While we have argued that the MEM is the *only* rational way to do spectral analysis, it is clear that the most worthwhile gains are expected when the traditional methods most strongly contravene the First Principle. Such cases arise when the data is meager or badly corrupted with measurement error or both.

One application (and certainly not the only one) in radioastronomy that fairly cries out for MEM treatment is the problem of turning the pitifully few correlation values available from recent very long baseline interferometer (VLBI) experiments into brightness distribution maps. The VLBI observers themselves often attempt to find "simple" source models which are consistent with their measurements, the idea being that the least complicated model consistent with the data is least likely to lead one astray. Such thoughts are very close to the concepts of the MEM so that the mental leap to a full treatment via the MEM should be an easy one.

REFERENCES

Bartlett, M.S.: 1966, *An Introduction to Stochastic Processes*, (2nd ed.), New York, Cambridge University Press.

Blackman, R.B. and Tukey, J.W.: 1959, *The Measurement of Power Spectra from the Point of View of Communications Engineering*, New York, Dover.

Brillouin, L.: 1962, *Science and Information Theory*, (2nd ed.), New York, Academic Press.

Burg, J.P.: 1967, *Maximum Entropy Spectral Analysis*, Paper presented at 37th Annual International SEG Meeting, Oklahoma City.

Jaynes, E.T.: 1957a, b, Information Theory and Statistical Mechanics (Paper I and II), *Phys. Rev.* **106**, 620; **108**, 171.

Jaynes, E.T.: 1963, in J.L. Bogdanoff and F. Kozin (eds.), New Engineering Applications of Information Theory, *Proceedings of the First Symposium on Engineering Applications of Random Function Theory and Probability*, New York, John Wiley and Sons.

Lacoss, R.T.: 1971, Data Adaptive Spectral Analysis, *Geophysics* **36**, 661.

Levinson, N.: 1947, The Wiener RMS (root mean square) Error Criterion in Filter Design and Prediction, *J. Math. Phys.* **25**, 261.

Parzen, E.: 1968, Multiple Time Series Modeling, *Tech. Rept.* 12 *on Contract Nonr – 225 –* (80), Stanford University.

Parzen, E.: 1969, in P.R. Krishnaiah (ed.), Multiple Time Series Modeling, *Multivariate Analysis –* II, New York, Academic Press.

Shannon and Weaver: 1949, *The Mathematical Theory of Information*, University of Illinois Press.

Szilard: 1929, *Z. Physik* **53**, 840.

Ulrych, T.: 1972, *Nature* **235**, 218.

J.G. Ables

Division of Radiophysics
CSIRO
Sydney NSW 2006, Australia

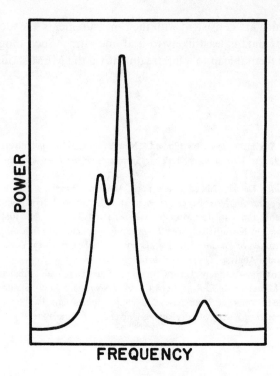

Figure 1   The true spectrum.

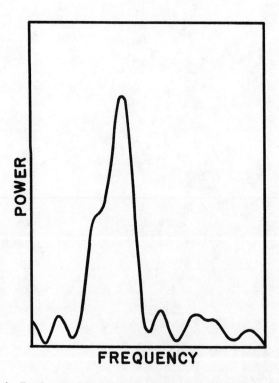

Figure 2   Fourier transform of truncated and sampled autocorrelation function.

512

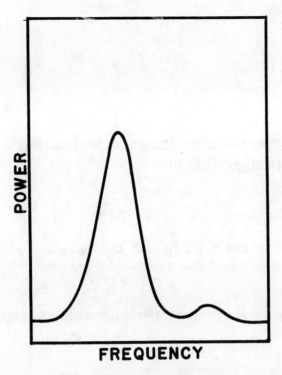

Figure 3   Fourier transform of truncated and sampled autocorrelation function. Before transformation the function was multiplied by raised cosine bell weights.

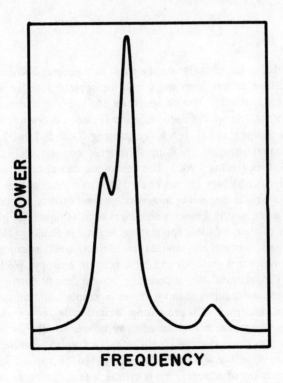

Figure 4   The maximum entropy spectrum corresponding to the truncated and sampled autocorrelation function.

Maximum entropy image reconstruction: general algorithm

J. Skilling and R. K. Bryan[*] *Department of Applied Mathematics and Theoretical Physics, Silver Street, Cambridge CB3 9EW*

Accepted 1984 May 14. Received 1984 May 14; in original form 1982 July 23

Summary. Maximum entropy is an optimal technique of image reconstruction, widely applicable in astronomy and elsewhere. We present a general-purpose algorithm, capable of generating maximum entropy images from a wide variety of types of data.

1 Introduction

Maximum entropy is being increasingly widely used as a general and powerful technique for reconstructing positive images from noisy and incomplete data. In astronomy, it has been used throughout the electromagnetic spectrum for radio aperture synthesis (Gull & Daniell 1978; Scott 1981), for optical deconvolution (Frieden & Swindell 1976; Frieden & Wells 1978; Bryan & Skilling 1980), for X-ray imaging (Gull & Daniell 1978; Willingale 1981) and for gamma-ray imaging (Skilling, Strong & Bennett 1979), and for eclipse mapping of accretion discs (Horne 1982). The technique can also be used in other fields such as structural molecular biology (Bryan *et al.* 1983) and medical tomography (Minerbo 1979; Kemp 1980). As well as producing images of optimal quality, maximum entropy can also be used to re-calibrate poorly known parameters such as phases or instrumental drifts (Scott 1981). A review of many of these applications is given in Skilling (1981).

Many of these papers, and others referred to therein, contain comparisons between reconstructions by conventional methods and by maximum entropy. We believe that these comparisons clearly demonstrate the superiority of maximum entropy for producing optimum general-purpose restorations of images from incomplete and noisy data.

In this paper we are concerned with presenting the rationale and details of a robust and efficient algorithm for computing maximum-entropy images which has been developed in Cambridge. This algorithm deals routinely with images of up to a million or more pixels, and with dynamic ranges well in excess of 10000. It can be applied to many different problems with the minimum of changes, by rewriting a few computer subroutines which define the transforms between image and data.

In Section 2 of the paper, a suitable form of the maximum entropy criterion is set up.

* Present address: European Molecular Biology Laboratory, 6900 Heidelberg, Germany.

Section 3 is devoted to a survey of algorithms and develops the general-purpose algorithm we recommend. Section 4 summarizes the ingredients of this successful program.

2 The maximum entropy criterion

An image can be regarded as a set of positive numbers $f_1, f_2, \ldots, f_N$ which are to be determined, and on which the entropy

$$S(f) = -\sum_{j=1}^{N} p_j \log p_j, \qquad p_j = f_j / \Sigma f \tag{1}$$

is defined. Use of this form of entropy (Shannon 1948) in the context of image reconstruction is originally due to Frieden (1972).

The theoretical foundation of the maximum entropy method in data analysis is that this method is the *only* consistent way of selecting a single image from the very many images which fit the data. Shore & Johnson (1980, 1983) proved this axiomatically, a more readable account of their ideas has been given by Gull & Skilling (1983), and a proof of intermediate formality appears in Livesey & Skilling (1984). These papers show that maximum entropy is the only method which does not introduce correlations in the image, beyond those which are required by the data.

Entropy can also be justified in information-theoretic terms. Given an image radiating with intensity pattern f_j, the entropy measures the number of bits of information needed to localize the position j of a single radiated photon. Maximizing S, subject to observational constraints, involves seeking a maximally non-committal answer to the fundamental question 'Where would the next photon come from?' (Skilling & Gull 1984).

The practical merit of maximizing entropy is that the resulting image has minimum configurational information, so that there must be evidence in the data for any structure which is seen, and the displayed structure is uniquely easy to comprehend. Also, the physically important requirement of positivity is automatically invoked, since the entropy does not even exist if any of the f_j are negative. Numerically, it is far easier to ensure positivity via a single, smooth function such as S, than via N separate inequality constraints $f_j \geqslant 0$.

The observational constraints on permitted reconstructions come from data D_k related in some known way to the image, and subject to some form of noise. Thus, for additive noise,

$$D_k = R_k(f) + n_k \sigma_k \tag{2}$$

where R is the response function of the observing equipment, σ_k is the standard error on datum k and n_k is a random variable of zero mean and unit variance. In any linear experiment, $R_k(f) = \Sigma R_{kj} f_j$ so that R becomes a matrix. For example, in interferometry, $R_k(f)$ would be the Fourier component of f corresponding to spacing k, so that R_{kj} would be a Fourier matrix. Likewise in deconvolution, R_{kj} would be a convolution matrix, (usually of Toeplitz form).

Naively, one might attempt to recover f from the data by applying R^{-1}, but this fails in principle whenever the data are incomplete, because R^{-1} is not uniquely defined. It also fails in practice whenever R^{-1} is badly conditioned, as in most deconvolution problems. Inverse filters alleviate this difficulty by modifying R^{-1} to make it better conditioned, albeit incorrect in the sense that retransforming by R will not reproduce the original data. What one can do correctly, using R itself instead of its inverse, is eliminate those f which are inconsistent with the data. The data can do no more than this.

The formal observational constraint on reconstructions f is set up by comparing the *actual* (noisy) data D_k with the simulated data

$$F_k = R_k(f) \tag{3}$$

which would be obtained (in the absence of noise) if the pattern being observed were indeed represented by the numbers f. A reconstruction f is said to be *feasible* if the simulated data agree with the actual data to within the noise. Feasible reconstructions are those which are not contradicted by the data. Note that the comparison is made only with the data points actually measured. There is no assumption, implicit in inverse filter methods, that unmeasured data values are zero.

A single constraint statistic $C(f)$, usually chi-squared (Ables 1974; Gull & Daniell 1978) is used to measure the misfit;

$$C(f) = \chi^2 = \Sigma (F_k - D_k)^2 / \sigma_k^2 \tag{4}$$

where the summation is over the observed k. Different choices for $C(f)$ are also possible (Bryan & Skilling 1980), and may be preferable in certain circumstances. For other forms of noise, e.g. Poisson, an appropriately modified statistic $C(f)$ should be used.

Statistical analysis indicates some upper bound C_{aim} to the values which C can plausibly take. For chi-squared, the largest acceptable value at 99 per cent confidence is about $(M + 3.29\sqrt{M})$, where M is the number of observations. It is much easier to use a single statistic than to attempt to fit each separate datum, both because this avoids an unwieldy proliferation of Lagrange multipliers and because one can construct alternative statistics $C(f)$ that tolerate occasional individual errors of several standard deviations. For example one can ignore extreme outliers in the calculated noise residuals (Bryan & Skilling 1980; Burch, Gull & Skilling 1983).

The strict maximum entropy criterion requires one to select that particular feasible image which has the greatest entropy. One maximizes S subject to $C \leqslant C_{aim}$. If the unconstrained maximum of S satisfies this constraint, then this will be the maximum entropy solution — the data are too noisy for any information to be extracted. Otherwise the solution will lie on the boundary $C = C_{aim}$ and we have an optimization problem with an equality constraint

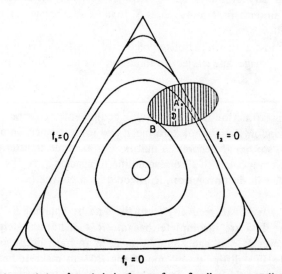

Figure 1. The S criterion and the χ^2 statistic in f-space for a 3-cell map normalized to $\Sigma f = 1$. S surfaces are convex and χ^2 surfaces are ellipsoids. A is the image which fits the data exactly; B is the maximum-entropy image.

to solve (Fig. 1). As usual for such problems, a Lagrangian function $Q = S - \lambda C$ can be set up, and the solution will lie at an extremal of Q for some value of the Lagrange multiplier λ. From this, f can be determined via

$$\log A - \log f_j = (\lambda \Sigma f) \, \partial C/\partial f_j, \quad A = \exp (\Sigma p_i \log f_i). \tag{5}$$

A is a weighted mean of the f_j which can be interpreted as the default value to which f_j will tend if there are no data pertaining to cell j ($\partial C/\partial f_j = 0$).

In many applications, interferometry being typical, the total flux Σf has a special status, as indeed it does in the entropy itself. The maximum entropy criterion may accordingly be modified to maximize S subject to $C = C_{\text{aim}}$ and to some constraint on Σf. This will be found at an extremal of $S - \mu \Sigma f - \lambda C$, where μ is an extra multiplier which has the operational effect of changing A. One can either choose μ, and hence A, to fit some given value of total flux Σf, or, more simply, one can use A itself as a user-defined default intensity or 'sky background'. This is what the astronomer usually wants. Formally, this is equivalent to modifying the entropy to

$$S = - \sum_i f_j [\log (f_j/A) - 1] \tag{6}$$

whose derivatives are

$$\partial S/\partial f_j = \log A - \log f_j, \quad \partial^2 S/\partial f_i \partial f_j = - \delta_{ij}/f_j. \tag{7}$$

It is this latter form (6) of S which is used in this paper, although the algorithms which are presented can easily be modified to cope with the strict form of S, or indeed with other modifications of it.

For any linear experiment, the surfaces of constant chi-squared (Fig. 1) are convex ellipsoids in N-dimensional image space. Since the entropy surfaces are strictly convex, the maximum entropy reconstruction is unique.

In this section we have set up the maximum entropy method as an equality constrained optimization problem. It is important that the numerical algorithm used solve the problem is reliable, in the sense that it should produce the correct solution, or give a definite indication of failure. There is no point in using an algorithm which terminates after a certain number of iterations if there are no tests for fitting the data and for maximizing the entropy. In Section 3 we survey various algorithms, and develop a general purpose algorithm which we believe satisfies these criteria.

3 Maximum entropy algorithms

Entropy being intrinsically non-linear, the computational problem is one of constrained non-linear optimization. The problem is also large-scale, since an image contains N elements f_j to be determined, and N may be a million or more. The number M of observations may be many thousand. The image and data may be considered as vectors in N or M-dimensional linear spaces, even if they represent physical arrays in 1, 2, or even 3 dimensions. It follows immediately from the size of the problem that vector operations such as scalar products, adds, multiplies etc. are allowed in N-dimensional image-space and M-dimensional data-space, but matrix operations of $O(N^2)$ or $O(M^2)$ are prohibited. Consequently for these large-scale images, it is not possible to use methods such as Newton-Raphson iteration, which has been employed successfully on smaller problems (Frieden 1972).

The non-linear nature of the problem also forces the algorithm to be iterative. This necessarily involves repeated passage from image-space to data-space via the response function R_{kj}. Each trial image, for example, must be transformed to the corresponding dataset F_k before its constraint statistic C can be evaluated. There must also be repeated feedback from data-space to image-space, which again involves R. For example the gradient $\partial C/\partial f_j$, which is useful in determining how to adjust a trial image in order to fit the data better, is calculated by

$$\partial C/\partial f_j = \sum_k (\partial F_k/\partial f_j)(\partial C/\partial F_k) = \sum_k R_{kj} 2(F_k - D_k)/\sigma_k^2. \tag{8}$$

The summation is now on the first index of R_{kj}, showing that feedback is obtained through the *transpose* of R (*not* its inverse, which may well not exist).

All the algorithms discussed in this paper are based purely on vector operations in image-space and in data-space, together with image-data transformations by R or its transpose. Normally, the major computational overhead is in the image-data transformations, and these should be coded efficiently. Thus an interferometric problem, for which R is a Fourier transform, should be coded via a Fast-Fourier-Transform routine, using $O(N \log N)$ operations.

In the discussion of algorithms which follows, Section 3.1 discusses a potentially promising algorithm which nevertheless proved insufficiently powerful. Sections 3.2, 3.3, 3.4, 3.5 develop the basic ideas for a successful technique. This is presented in Section 3.6 (which sets up the image-space structures needed) and in Section 3.7 (which gives the procedure for control of these structures).

3.1 THE 'INTEGRAL EQUATION'

Gull & Daniell (1978) attempted to find the solution of (5) by directly maximizing $Q = S - \lambda C$ at fixed λ. They used the iterative form

$$f_j^{(n+1)} = A \exp [-\lambda \partial C(f^{(n)})/\partial f_j] \tag{9}$$

where (n) denotes the nth iterate. This procedure had the attractive feature that successive iterates were all automatically positive, because of the exponential function. Also, the algorithm allowed high values of f to develop in relatively few iterations, again because of the exponential. This was of considerable importance in Gull & Daniell's astronomical applications, because the dynamic range of their images was often $O(1000)$.

Unfortunately, the exponential also introduced instability into the iteration and they had to smooth successive iterates by setting

$$f_j^{(n+1)} = (1-p)f_j^{(n)} + pA \exp [-\lambda \partial C(f^{(n)})/\partial f_j]. \tag{10}$$

Even then, the behaviour of the algorithm was erratic and unstable, especially at high values of λ, and the proportion p often had to be reduced so severely that the algorithm effectively stopped. This instability had also been noted by Willingale (1979).

Nevertheless, the work of Gull & Daniell was of crucial importance in demonstrating the possibility of computing high-resolution maximum-entropy images for a variety of different experiments.

J. Skilling and R. K. Bryan

3.2 STEEPEST ASCENTS

The simplest approach is to maximize $Q = S - \lambda C$ by steepest ascents, using

$$f_j^{(n+1)} = f_j^{(n)} + x \partial Q(f^{(n)})/\partial f_j \tag{11}$$

for suitable x. The catastrophic disadvantage of this method is that in almost every case, whenever x is sufficiently large to enable high values f_j of the image to develop significantly, there are also cells with negative $\partial Q/\partial f_j$ at which f_j becomes significantly negative. The entropy is not defined when $f_j < 0$, so that after each iteration any negative f's must be reset to small positive values. If many f's become negative, the effect is to stop the algorithm making progress towards a maximum.

3.3 CONJUGATE GRADIENTS

The standard way of improving a steepest ascent algorithm is to use the conjugate gradient technique (Fletcher & Reeves 1964) or a variant of it (e.g. Polak 1971; Powell 1976, 1977). At the nth iteration, instead of using ∇Q itself as a direction in which to look for a maximum of Q, one uses only that part $e^{(n)}$ of ∇Q which is conjugate to the previous directions ($e^{(r)}$, $r = 1, 2, \ldots, n-1$), defined by $e^{(n)T} \cdot \nabla\nabla Q \cdot e^{(r)} = 0$. In the terminology of this paper, the technique seeks a maximum of Q over the points $\mathbf{f}^{(n)} + x e^{(n)}$ lying along a search direction $e^{(n)}$, where

$$e^{(n)} = -\nabla Q + \beta e^{(n-1)} \tag{12}$$

with $\beta = |\nabla Q^{(n)}|^2/|\nabla Q^{(n-1)}|^2$ or a formal equivalent thereof. The coefficient β in the search direction is derived on the assumption that Q has constant curvature. However, Q in maximum entropy is highly non-quadratic, and it may not be sensible to assume that curvature information can be carried forward for several iterates.

Even so, as noted by Wernecke & d'Addario (1977) for a similar problem, conjugate gradients afford a considerable improvement over steepest ascents, although the algorithm remains plagued by negative values of f_j, and still concentrates too much on small values.

3.4 SEARCH DIRECTIONS FOR THE UNCONSTRAINED PROBLEM (FIXED λ)

The conjugate gradient technique attempts to build up information about the $N \times N$ Hessian matrix $\nabla\nabla Q$ by using successive vectors ∇Q, calculated at successive points $f^{(n)}$. It then uses a specific linear combination of the various ∇Q as a search direction along which Q is maximized either by a fixed coefficient based on a Newton-Raphson increment or by an exact line search.

In the maximum entropy problem the main computational cost of this lies in generating the successive vectors ∇Q, each of which requires an image-data transformation R followed by its transpose to calculate ∇C (equation 8). Scalar products between the vectors are much quicker to compute. Accordingly, *one can gain considerable extra flexibility at negligible extra computational cost by constructing, not merely one line along which to search, but rather a full subspace spanned by several vectors.*

Let the vectors $e_1, e_2, \ldots, e_r$ ($r < 10$ say) be these base vectors. Then, within the subspace so spanned, one may construct a quadratic model

$$\tilde{Q}(x) = Q_0 + Q_\mu x^\mu + \tfrac{1}{2} H_{\mu\nu} x^\mu x^\nu \tag{13}$$

for the value of Q at increment $\delta f = x^\mu \mathbf{e}_\mu$. Note the covariant and contravariant indices: Greek indices refer to subspace quantities. The components of the model are

$$Q_\mu = \mathbf{e}_\mu^T \cdot \nabla Q \quad \text{and} \quad H_{\mu\nu} = \mathbf{e}_\mu^T \cdot \nabla\nabla Q \cdot \mathbf{e}_\nu, \tag{14}$$

chosen to agree with the local gradient and curvature components of $Q(f)$ itself. Then, within the subspace, $\tilde{Q}$ is maximized at

$$x^\mu = -(H_{\mu\nu})^{-1} Q_\nu, \tag{15}$$

the evaluation of which involves the trivially quick task of solving r simultaneous equations. The resulting value of $\tilde{Q}$, moreover, is greater than could have been obtained directly from conjugate gradients, because such a point is necessarily included in the subspace spanned by the $\mathbf{e}_\mu$.

As suggested above, $H_{\mu\nu}$ is obtained from the current and the previous $r-1$ evaluates of ∇Q. However, this presupposes that the curvature of $Q(f)$ remains nearly constant even when f is incremented $r-1$ times. That is an extremely severe restriction on the increment length $|\delta f|$, since in the maximum entropy problem the curvature of Q is dominated by $1/f_j$ terms from the entropy, and these are very sensitive to small changes whenever a particular f_j is small. It is better to evaluate the search directions at the present position f as

$$\mathbf{e}_1 = \nabla Q, \qquad \mathbf{e}_2 = \nabla\nabla Q \cdot \nabla Q, \dots, \qquad \mathbf{e}_r = (\nabla\nabla Q)^{(r-1)} \cdot (\nabla Q). \tag{16}$$

Admittedly this involves the conventionally unorthodox step of discarding information from previous iterates, but all our attempts to use old curvature information resulted in marked reductions in algorithm efficiency.

3.5 ENTROPY METRIC

Even with models which are completely updated at each iteration, some limit must be placed on the difference δf between successive iterates, as the quadratic model will still be inaccurate at large distances. One should maximize $\tilde{Q}(x)$ subject to $|\delta f|^2 \leqslant l_0^2$ for some l_0.

The precise form of the distance limit bears closer investigation. So far, the main disadvantage of the search-direction algorithms has been their tendency to allow negative values of f. A distance limit $\Sigma(\delta f_i)^2 \leqslant l_0^2$ alleviates this, but at the cost of drastically slowing the attainment of high values. However, the distance limit can be modified to overcome this defect. Logarithmic modification $\Sigma(\delta f_i/f_i)^2 \leqslant l_0^2$ is too severe on low values, and the intermediate form

$$\sum_i (\delta f_i)^2/f_i \leqslant l_0^2 \tag{17}$$

is a good practical compromise. It discriminates in favour of allowing high values to change more than low ones, but not excessively so. The actual value of l_0^2 should be $O(\Sigma f)$ on dimensional grounds, and values around $0.1\,\Sigma f$ to $0.5\,\Sigma f$ are useful in practice.

Using a distance in this form is equivalent to putting a metric

$$g_{ij} = 1/f^i \, (i = j) \quad \text{and} \quad g_{ij} = 0 \, (i \neq j) \tag{18}$$

onto image-space (note covariant and contravariant indices). But this is just minus $\nabla\nabla S$ (Bryan 1980)! This metric is far simpler and more convenient than the Hessian metric

$g_{ij} = \partial^2 Q/\partial f_i \partial f_j$ normally used (Sargent 1974) in variable-metric non-linear optimization problems.

Using $- \nabla\nabla S$ as the metric is the single most important key to the development of a robust algorithm. With a non-Cartesian metric, the gradient direction $\nabla Q = \partial Q/\partial f^i$ appears initially in covariant form. In order to increment the contravariant vector f^i, its index must be raised by g^{ij}, giving $f^i \partial Q/\partial f^i$ [componentwise multiplication, henceforth represented by $f(\nabla Q)$] as the basic contravariant search direction. Furthermore, the second derivative matrix $\nabla\nabla Q = \partial^2 Q/\partial f^i \partial f^j$ must likewise be premultiplied by g^{mi} if it is to map contravariant vectors onto contravariant vectors. This gives the revised set of search directions

$$\mathbf{e}_1 = f(\nabla Q), \quad \mathbf{e}_2 = f(\nabla\nabla Q)\, f(\nabla Q), \ldots, \tag{19}$$

We note here that the integral equation (9) can be written as

$$f_j^{(n+1)} = A \exp\left[\partial Q(f^{(n)})/\partial f_j - \partial S(f^{(n)})/\partial f_j\right]$$
$$= f_j^{(n)} \exp\left[\partial Q(f^{(n)})/\partial f_j\right].$$

On expanding the exponential to first order, we obtain

$$f_j^{(n+1)} - f_j^{(n)} = f_j^{(n)} \partial Q(f^{(n)})/\partial f_j.$$

So, for small increments, the integral equation is equivalent to a steepest ascent optimization using the entropy metric.

3.6 SEARCH DIRECTIONS FOR THE CONSTRAINED PROBLEM

One difficulty with maximizing Q is that of λ, which still has to be iterated to fit $C = C_{\text{aim}}$. This double iteration is clumsy and inefficient. However, the use of different values of λ involves using different proportions of S and C in Q, which suggests using *two* models in the subspace, one for S and the other for C, and attempting somehow to *solve the actual problem of maximizing* S *subject to* C $= C_{\text{aim}}$ *directly without using* λ *explicitly*. This would also be a more general approach, since there are non-linear experiments R for which the entropy maximum on the constraint surface may not be obtainable by maximizing Q, whatever value of λ is chosen. Examples in which the desired extremal of Q is not a maximum are given in Bryan (1980) and Livesey & Skilling (1984).

The subspace itself would be constructed from

(1) 2 directions $f(\nabla S)$ and $f(\nabla C)$,

(2) 4 directions, $f(\nabla\nabla S)$ and $f(\nabla\nabla C)$ operating on (1),

(3) 8 directions, $f(\nabla\nabla S)$ and $f(\nabla\nabla C)$ operating on (2),

and so on. But $f(\nabla\nabla S)$ is minus the identity operator and nothing new is obtained by operating with it. The search directions reduce to

(1) $f(\nabla S), \qquad f(\nabla C)$

(2) $f(\nabla\nabla C) f(\nabla S), \qquad f(\nabla\nabla C) f(\nabla C) \tag{20}$

(3) $f(\nabla\nabla C) f(\nabla\nabla C) f(\nabla S), \qquad f(\nabla\nabla C) f(\nabla\nabla C) f(\nabla C)$

and so on. It is worth repeating that the operator $\nabla\nabla C$, though formally a matrix, can be applied by vector operations allied to image-data transformations.

The factors of f^i in the search directions discriminate in favour of high values, and this helps to keep all the values positive. In fact it is rare for any cell to be sent negative when these search directions are properly controlled. Protection against stray negative values is still needed, but it does not slow the algorithm and is no longer a source of difficulty.

The family (20) of directions is sufficiently powerful that the first four enable most practical problems to be solved. Indeed even this level of complexity is usually unnecessary, as the third and fourth directions can normally be replaced by a single difference combination, giving just three search directions

$$\mathbf{e}_1 = f(\nabla S)$$

$$\mathbf{e}_2 = f(\nabla C) \tag{21}$$

$$\mathbf{e}_3 = |\nabla S|^{-1} f(\nabla\nabla C) f(\nabla S) - |\nabla C|^{-1} f(\nabla\nabla C) f(\nabla C).$$

Here the entropy metric is used to define the lengths

$$|\nabla S| = [\Sigma f^i (\partial S/\partial f^i)^2]^{1/2}, \quad \nabla C = [\Sigma f^i (\partial C/\partial f^i)^2]^{1/2}. \tag{22}$$

With these three (or four or more) search directions, quadratic models for S and C

$$\tilde{S}(x) = S_0 + S_\mu x^\mu - \tfrac{1}{2} g_{\mu\nu} x^\mu x^\nu, \quad \tilde{C}(x) = C_0 + C_\mu x^\mu + \tfrac{1}{2} M_{\mu\nu} x^\mu x^\nu \tag{23}$$

where

$$S_\mu = \mathbf{e}_\mu^T \cdot \nabla S, \quad g_{\mu\nu} = \mathbf{e}_\mu^T \cdot \mathbf{e}_\nu, \quad C_\mu = \mathbf{e}_\mu^T \cdot \nabla C, \quad M_{\mu\nu} = \mathbf{e}_\mu^T \cdot \nabla\nabla C \cdot \mathbf{e}_\nu \tag{24}$$

are constructed in the subspace

$$\mathbf{f}^{(new)} = \mathbf{f} + x^\mu \mathbf{e}_\mu = \mathbf{f} + \delta\mathbf{f} \tag{25}$$

parameterized by x, within which the length-squared of the increment $\delta\mathbf{f}$ is

$$l^2 = g_{\mu\nu} x^\mu x^\nu. \tag{26}$$

It is rather remarkable that such a small space can capture enough of the structure of a non-linear optimization problem in a million dimensions.

3.7 CONTROL PROCEDURES

Control of the algorithm now passes into the subspace, in order to determine suitable coefficients x^μ for the search directions. The problem in the subspace becomes one of optimizing a quadratic function subject to quadratic constraints. Although the control procedure involves substantial programming, the computation time involved is negligible in comparison with practical image-data transformations.

3.7.1 Diagonalization in the subspace

This preliminary step simplifies the algebra. First, the base vectors $\mathbf{e}_\mu$ are normalized by scaling the model parameters and the metric tensor $g_{\mu\nu}$ is diagonalized. The algorithm can now be protected against linear dependence of the search directions. Such dependence shows up as one or more unusually small eigenvalues of $g_{\mu\nu}$. Components of the model along the corresponding eigenvector(s) may reflect rounding errors rather than true structure, and such eigenvectors are discarded, reducing the subspace to that part spanned by eigenvectors having significant eigenvalues.

With the remaining eigenvectors rescaled to make the metric Cartesian, the distinction between covariant and contravariant indices disappears. Further simplification is effected by diagonalizing the revised form of $M_{\mu\nu}$ to give

$$\tilde{S}(x) = S_0 + S_\mu x_\mu - \tfrac{1}{2} x_\mu x_\mu$$
$$\tilde{C}(x) = C_0 + C_\mu x_\mu + \tfrac{1}{2} \gamma_{(\mu)} x_\mu x_\mu \tag{27}$$
$$l^2 = x_\mu x_\mu$$

where the $\gamma_{(\mu)}$ are the eigenvalues of $M_{\mu\nu}$, and all symbols are defined with respect to the new base vectors.

Apart from the protection against linear dependence, the above procedure is merely the simultaneous diagonalization of $g_{\mu\nu}$ and $M_{\mu\nu}$. Much hangs on accurate diagonalization in the subspace (especially for badly conditioned problems), and it is essential to diagonalize accurately. This is a standard linear algebra problem in a space of low dimension, and reliable algorithms are available for this purpose.

3.7.2 Basic control

The aim of the control procedure is to maximize $\tilde{S}$ over $\tilde{C} = C_{aim}$ subject to a distance constraint $l^2 \leqslant l_0^2 (\simeq 0.1\,\Sigma f$ to $0.5\,\Sigma f)$. Unfortunately, this may be impossible at first. For very many applications, C is a convex (elliptical) function of f, for which all eigenvalues $\gamma_{(\mu)}$ are positive. There is then a minimum value

$$\tilde{C}_{min} = C_0 - \tfrac{1}{2} \gamma_{(\mu)}^{-1} C_\mu C_\mu \tag{28}$$

which $\tilde{C}$ can attain in the subspace. Clearly one should not attempt to aim below $\tilde{C}_{min}$, regardless of the value of C_{aim}. In fact, even attempting to reach values as low as $\tilde{C}_{min}$ is inappropriate, since the resulting x is then determined purely by the structure of $\tilde{C}$ and not at all by $\tilde{S}$. It is better to set the more modest aim

$$\tilde{C}_{aim} = \max\left(\tfrac{2}{3}\,\tilde{C}_{min} + \tfrac{1}{3}\,C_0,\, C_{aim}\right) \tag{29}$$

which is always accessible.

The various maxima of $\tilde{S}$ over different values of $\tilde{C}$ may be parameterized by the Lagrange multiplier α in

$$\tilde{Q} = \alpha \tilde{S} - \tilde{C} \tag{30}$$

(re-defining Q using α instead of λ). Maximizing $\tilde{Q}$ yields

$$x_\mu = (\alpha S_\mu - C_\mu)/(\gamma_{(\mu)} + \alpha) \tag{31}$$

in which α is chosen to fit $\tilde{C} = \tilde{C}_{aim}$. The required range for α is the positive range

$$\alpha_{min} < \alpha < \infty \tag{32}$$

assigning positive weight to the entropy. α_{min} is normally zero (for positive definite $\nabla\nabla C$), at which C takes its minimum value $\tilde{C}_{min}$. If, for non-positive $\nabla\nabla C$, any eigenvalues γ are negative, α_{min} becomes $\max(-\gamma_{(\mu)})$ at which the increment x_μ diverges. The upper limit $\alpha = \infty$ corresponds to unconstrained maximization of $\tilde{S}$ irrespective of $\tilde{C}$. The value of $\tilde{C}(x)$ increases monotonically in α, for α in the allowed range, so that a simple chop suffices to iterate α towards $\tilde{C} = \tilde{C}_{aim}$.

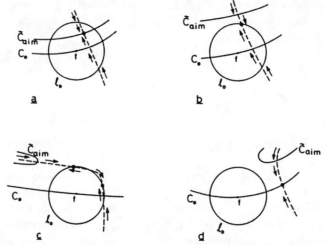

Figure 2. Operation of α-chop in the subspace. The maximum-entropy trajectory, parameterized by α, is shown dashed. The circle centred on the current image f marks the maximum allowed distance l_0. Arrows indicate the direction induced by the chop. Results x are shown as filled circles •. (a) Unique iterate $\tilde{C} = \tilde{C}_{aim}$, $l < l_0$; (b) Unique iterate $\tilde{C} > \tilde{C}_{aim}$, $l = l_0$. (c) Ambiguous iterate $\tilde{C} > \tilde{C}_{aim}$, $l = l_0$. (d) Too distant iterate $\tilde{C} = \tilde{C}_0$, $l > l_0$.

The resulting x may, however, be too large to satisfy the distance constraint. To protect against this, the chop in α is redirected towards $\tilde{C} = C_0$ whenever α gives an increment with too large l^2. The rationale for this form of protection is that maximizing $\tilde{S}$ over the *existing* value C_0 is likely to give a closer iterate than attempting to reach a *different* value $\tilde{C}_{aim}$.

The α-chop normally behaves as in Figs 2a or 2b, and in any case it must always give a result in the range $\tilde{C}_{aim} \leqslant \tilde{C} \leqslant C_0$. Because the distance l is *not* monotonic in α, this can lead to an ambiguity in the result of the chop (Fig. 2c), the result produced depending on the particular values of α actually used in the chop. Nevertheless, the ambiguity is harmless in that either answer for x gives a useful iterate. More seriously, the algorithm may be unable to find *any* sufficiently close value of x (Fig. 2d), especially if the current image f is far from a maximum-entropy image.

3.7.3. Distance penalty

If the α-chop cannot find a sufficiently close value of x, the distance constraint must be introduced explicitly into the maximization via a second Lagrange multiplier P, giving

$$\tilde{Q} = \alpha\tilde{S} - \tilde{C} - Pl^2, \qquad P = \text{distance penalty} \geqslant 0. \tag{33}$$

This is maximized at

$$x_\mu = (\alpha S_\mu - C_\mu)/(P + \gamma_{(\mu)} + \alpha). \tag{34}$$

Thus P can be interpreted as an artificial increase of each eigenvalue $\gamma_{(\mu)}$ of C, giving a revised form

$$\tilde{C}_P(x) = C_0 + C_\mu x_\mu + \tfrac{1}{2}(P + \gamma_{(\mu)})x_\mu x_\mu \tag{35}$$

which takes larger values than $\tilde{C}$ itself. $\tilde{C}_P$ is also more convex than $\tilde{C}$, and the maximization of $\tilde{S}$ becomes better conditioned.

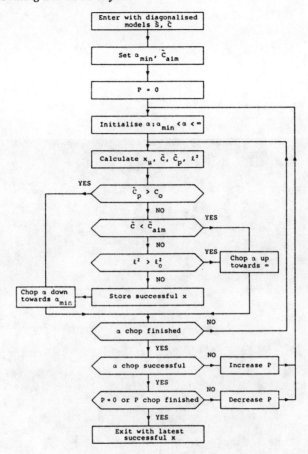

Figure 3. Flow-chart of control procedure.

There are several ways of proceeding from here: we suggest the following. With a distance penalty invoked, α is chopped towards $\widetilde{C} = \widetilde{C}_{\mathrm{aim}}$ as required, but the chop is redirected towards $\widetilde{C}_P = C_0$ whenever the distance is too large. Redirecting on $\widetilde{C}_P$ rather than on $\widetilde{C}$ helps the algorithm because $\widetilde{C}$ itself is always less than $\widetilde{C}_P$, and hence will make useful progress down towards $\widetilde{C}_{\mathrm{aim}}$. For sufficiently large penalty P, the α-chop must be able to reach a result satisfying

$$\widetilde{C}_{\mathrm{aim}} \leqslant \widetilde{C} < \widetilde{C}_P \leqslant C_0 \tag{36}$$

and the smallest such P is used to give the final result x. A flow-chart of this algorithm is shown in Fig. 3.

All that remains is to increment f by the multiples x_μ of the search directions, whilst protecting against stray non-positive values. The algorithm is complete.

In general, the algorithm proceeds by reducing C at each iteration, whilst keeping close to the maximum of S for the current value of C, until $\widetilde{C}_{\mathrm{aim}}$ is obtained. Then S is increased, C necessarily remaining at $\widetilde{C}_{\mathrm{aim}}$, until it is sufficiently close to the maximum that the termination criterion of Section 4 is satisfied.

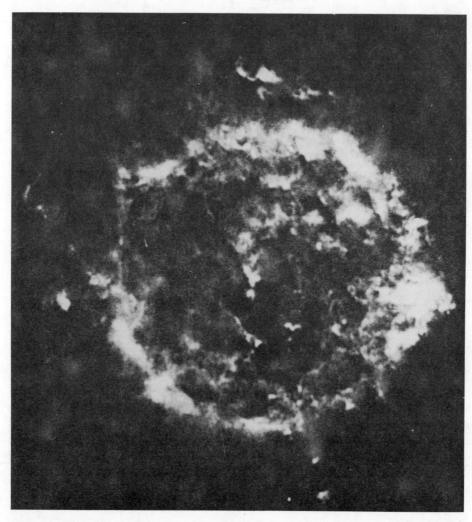

Plate 1. Million-pixel maximum entropy reconstruction of Cas A, computed by the algorithm presented in this paper.

4 Discussion

Our first conclusion is that it is operationally effective to subsume large numbers of individual constraints (such as positivity of each pixel and acceptable fits to experimental datasets) into single functions like $S(f)$ and $C(f)$. This enables practical algorithms to be developed for large problems.

There are then four main ingredients in the maximum-entropy algorithm recommended in this paper. They are

(1) The use of a subspace of search directions instead of merely using line searches,

(2) Updating fully at each iteration, and *not* attempting to carry information forward, especially information on rapidly changing high derivatives,

(3) The entropy metric, in which the entropy function itself is used to define distances,

(4) Controlling the algorithm directly on the constraint C, and not on its Lagrange multiplier λ.

These ideas may also find use in other non-linear maximization problems. The resulting program has proved highly robust and powerful. It supersedes earlier versions which have been used successfully in image enhancement (Bryan & Skilling 1980; Daniell & Gull 1980), gamma-ray astronomy (Skilling *et al.* 1979), medical tomography (Kemp 1980), and latterly in other fields too. For example, to deconvolve an optical image of the galaxy M87, the algorithm presented here took half the number of iterations of an earlier algorithm (essentially that of Section 3.4) using the same number of search directions (Bryan & Skilling 1980; Bryan 1980). The saving in CPU time was not as great, since more sophisticated search directions, requiring more image-data transforms, are used here. *The operation of the algorithm is always checked* by displaying the value of

$$\text{TEST} = \frac{1}{2} \left| \frac{\nabla S}{|\nabla S|} - \frac{\nabla C}{|\nabla C|} \right|^2. \tag{37}$$

This measures the degree of non-parallelism between ∇S and ∇C, which is zero for a true maximum entropy image. Usually there is no difficulty in reaching TEST < 0.1 or so at the correct value of C, which demonstrates that the correct, unique maximum-entropy reconstruction has been attained. We think it important that programs purporting to produce maximum-entropy images should make this test of their operation.

Fortran implementations of the algorithm routinely perform maximum-entropy calculations on images up to a million pixels, and there seems no bar in principle to still larger sizes. An example is the 1024×1024 maximum-entropy image of the supernova remnant Cas A (Plate 1), reconstructed from Cambridge 5-km telescope observations at 5 GHz (Gull & Brown, reported in Skilling 1981 and Tuffs 1984): a conventional reconstruction from these data was given by Bell (1977). Almost regardless of size, the algorithm takes something like 20 iterations to reconstruct an image from an experiment with signal-noise of about 100:1. Each iteration of the recommended three search-direction version involves 6 image-data transformations, so that the maximum-entropy reconstruction is about 100 times slower than simple linear reconstructions.

Acknowledgments

During much of the development work reported here, one of us (RKB) was in receipt of a SRC research studentship. Several colleagues, particularly S. F. Gull and G. J. Daniell who introduced us to maximum entropy, also helped to focus our ideas.

J. Skilling and R. K. Bryan

References

Ables, J. G., 1974. *Astr. Astrophys. Suppl.*, **15**, 383.

Bell, A. R., 1977. *Mon. Not. R. astr. Soc.*, **179**, 573.

Bryan, R. K., 1980. *PhD thesis*, University of Cambridge.

Bryan, R. K., Bansal, M., Folkhard, W., Nave, C. & Marvin, D. A., 1983. *Proc. Nat. Acad. Sci., USA*, **80**, 4728.

Bryan, R. K. & Skilling, J., 1980. *Mon. Not. R. astr. Soc.*, **191**, 69.

Burch, S. F., Gull, S. F. & Skilling, J., 1983. *Comp. Vis. Graph. Im. Proc.*, **23**, 113–128.

Daniell, G. J. & Gull, S. F., 1980. *IEE Proc. (E)*, **5**, 170.

Fletcher, R. & Reeves, C. M., 1964. *Comp. J.*, **7**, 149.

Frieden, B. R., 1972. *J. opt. Soc. Am.*, **62**, 511.

Frieden, B. R. & Swindell, W., 1976. *Science*, **191**, 1237.

Frieden, B. R. & Wells, D. C., 1978. *J. opt. Soc. Am.*, **68**, 93.

Gull, S. F. & Daniell, G. J., 1978. *Nature*, **272**, 686.

Gull, S. F. & Skilling, J., 1983. *IAU/URSI Symposium on Indirect Imaging*, Sydney, Australia.

Horne, K. D., 1982. *PhD thesis*, California Institute of Technology.

Kemp, M. C., 1980. *International Symposium on Radionuclide Imaging, IAEA-SM-247*, 128, Heidelberg.

Livesey, A. K. & Skilling, J., 1984. *Acta Crystallogr. A.*, in press.

Minerbo, G., 1979. *Comp. Graph. Im. Proc.*, **10**, 48.

Polak, E., 1971. *Computational methods in optimisation*, p. 44, Academic Press, New York.

Powell, M. J. D., 1976. *Math. Program.*, **11**, 42.

Powell, M. J. D., 1977. *Math. Program.*, **12**, 141.

Sargent, R. W. H., 1974. *Numerical methods for constrained optimisation*, p. 149, eds Gill, P. E. & Murray, W., Academic Press, London.

Scott, P. F., 1981. *Mon. Not. R. astr. Soc.*, **194**, 23P.

Shannon, C. E., 1948. *Bell System Tech. J.*, **27**, 379 & 623.

Shore, J. E. & Johnson, R. W., 1980. *IEEE Trans.*, IT-26, 26.

Shore, J. E. & Johnson, R. W., 1983. *IEEE Trans.* IT-29, 942.

Skilling, J., 1981. *Workshop on Maximum Entropy Estimation and Data Analysis*, University of Wyoming, Reidel, Dordrecht, Holland, in press.

Skilling, J. & Gull, S. F., 1984. *SIAM Proc. Am. Math. Soc.*, **14**, 167.

Skilling, J., Strong, A. W. & Bennett, K., 1979. *Mon. Not. R. astr. Soc.*, **187**, 145.

Tuffs, R. J., 1984. *PhD thesis*, University of Cambridge.

Wernecke, S. J. & d'Addario, L. R., 1977. *IEEE Trans*, C-26, 351.

Willingale, R., 1979. *PhD thesis*, University of Leicester.

Willingale, R., 1981. *Mon. Not. R. astr. Soc.*, **194**, 359.

THE intent of this part is to show a sample of the areas of technology to which radio astronomy has made contributions. The most obvious topics—antennas, low-noise receivers, and spectrometers—have been covered in previous parts. Here, we have included four papers representative of the diverse directions in which the efforts to improve radio-astronomical systems have led. All of the devices or approaches presented certainly have applications far beyond those of radio astronomy. This selection is anything but comprehensive, and some additional papers in related areas are given in the Bibliography below.

The first paper, by Henry, describes a phaselock system for control of millimeter wavelength klystron oscillators. Since all millimeter wavelength radio-astronomical receivers used for spectroscopy are heterodyne systems, they require a local oscillator with precisely controlled frequency. Interferometer systems make use of the phase of the signal, and the control of the phase of the local oscillator is thus critical. Henry's paper gives a general description of phaselock systems, and describes a novel approach employing a frequency discriminator together with the phaselock system which obviates the need for a search oscillator.

The second paper, by Frater and Williams, describes an outgrowth of the development work on FET amplifiers (see the fourth and fifth papers of Part IV). The authors have used the low input noise temperature and good input match of L-band FET amplifiers as a cold noise standard, which in certain applications could take the place of a cryogenically cooled termination.

The third paper, by Goldsmith, discusses the design and construction of a quasi-optical (free-space propagation, but with diffraction being important) feed system to couple the 7-m AT&T antenna (fourth paper of Part I) to a millimeter receiver. The main impetus was to achieve lower loss and greater operating flexibility than could be obtained with waveguide systems at this wavelength. Relatively complete calibration and beam switching ability were included.

The fourth paper, by Archer, concerns the development of local oscillator sources for the short-millimeter and submillimeter frequency ranges. The basic approach is frequency multiplication of solid-state sources; both waveguide and quasi-optical techniques are employed in an effort to obtain the highest possible efficiency at these short wavelengths.

BIBLIOGRAPHY

Quasi-optical Technology

1981: van Vliet, A. H. F. and T. H. de Graauw, "Quarter wave plates for submillimetre wavelengths," *Int. J. Infrared and Millimeter Waves,* vol. 2, pp. 465–477, 1981.

1982: Goldsmith, P. F., "Quasi-optical techniques at millimeter and submillimeter wavelengths," in *Infrared and Millimeter Waves,* vol. 6, K. J. Button, Ed. New York, NY: Academic Press, 1982, pp. 277–343.

1983: Pickett, H. M. and A. E. T. Chiou, "Folded Fabry-Perot quasi-optical ring resonator diplexer: Theory and experiment," *IEEE Trans. Microwave Theory Tech.,* vol. MTT-31, pp. 373–380, 1983.

1986: Lamb, J. W., "Quasioptical coupling of Gaussian beam systems to large Cassegrain antennas," *Int. J. Infrared and Millimeter Waves,* vol. 7, pp. 1511–1536, 1986.

Frequency Multipliers

1982: Erickson, N. R., "A high efficiency frequency tripler for 230 GHz," in *Proc. 12th European Microwave Conf.,* Helsinki, 1982.

1985: Archer, J. W. and M. T. Faber, "High-output, single- and dual-diode, millimeter-wave frequency doublers," *IEEE Trans. Microwave Theory Tech.,* vol. MTT-33, pp. 533–538, 1985.

Erickson, N. R., "High efficiency frequency triplers for 100-300 GHz," in *Conf. Digest, Tenth Int. Conf. on Infrared and Millimeter Waves,* Dec. 1985.

1987: Tolmunen, T. J. and A. V. Räisänen, "An efficient Schottky-varactor frequency multiplier at millimeter wavelengths. Part I. Doubler," *Int. J. Infrared and Millimeter Waves,* vol. 8, pp. 1313–1336, 1987.

1988: Tolmunen, T. J. and A. V. Räisänen, "An efficient Schottky-varactor frequency multiplier at millimeter wavelengths. Part II. Tripler," *Int. J. Infrared and Millimeter Waves,* vol. 8, pp. 1337–1353, 1988.

Gunn Oscillators

1985: Carlstrom, J. E., R. L. Plambeck, and D. D. Thornton, "A continuously tunable 65-115 GHz Gunn oscillator," *IEEE Trans. Microwave Theory Tech.,* vol. MTT-33, pp. 610–619, 1985.

Other

1987: Altenhoff, W. J., J. W. M. Baars, D. Downes, and J. E. Wink, "Observations of anomalous refraction at radio wavelengths," *Astron. Astrophys.,* vol. 184, pp. 381–385, 1987.

Frequency-agile millimeter-wave phase lock system

P. S. Henry

Bell Laboratories, Holmdel, New Jersey 07733

(Received 13 April 1976; in final form, 28 May 1976)

A frequency-agile phase lock system for millimeter-wave klystrons is described. The system locks the klystron to a crystal-controlled reference signal derived from a frequency synthesizer. By programming the synthesizer, the klystron can be stepped through any sequence of frequencies lying within a band roughly 200 MHz wide. Acquisition of phase lock typically requires 1 msec. The 200-MHz pull-in range is achieved through the use of combined frequency and phase feedback; there is no "search" mode. The measured linewidth of an 82-GHz phase-locked klystron is ≤ 10 Hz.

I. INTRODUCTION

This paper describes a frequency-agile phase lock system which locks a millimeter-wave klystron to a reference tone derived from a frequency synthesizer. The system was designed for use in millimeter-wave molecular astronomy, where a klystron must be switched alternately between two frequencies, and phase-locked at each frequency to a crystal-controlled reference. Two feedback paths are employed to control the klystron.[1] An automatic frequency control loop using a discriminator brings the klystron frequency close to the desired value, and then a second-order phase-locked loop "pulls in" the klystron and locks it to a reference signal. The effective pull-in range is about 200 MHz peak to peak, which is an order of magnitude greater than the performance achieved with conventional phase-locked loops.[2,3] The dual-feedback technique eliminates the need for a "search voltage," which is often used to help bring phase-controlled systems into lock.[4] The klystron can be stepped through any desired sequence of frequencies simply by switching the frequency synthesizer; there are no additional filters or oscillators to be tuned. This represents a considerable improvement in frequency agility compared with earlier designs.[5]

II. OPERATION AND CONSTRUCTION

A. General description

A functional block diagram of the klystron stabilizer is shown in Fig. 1. The synthesizer output is up-converted and then frequency-multiplied by a factor of 6 to produce a reference tone in the vicinity of 3800 MHz. This tone acts as the local oscillator for a harmonic mixer, which down-converts the klystron output to an intermediate frequency (i.f.) of ~740 MHz. The amplified i.f. drives a discriminator whose output voltage is an error signal corresponding to the frequency difference between the i.f. signal and 740 MHz. A second error signal is generated by a phase detector which compares the phase of the i.f. signal with a 740-MHz crystal-controlled source. The two error signals are combined, amplified, and voltage-translated from close to ground potential to the region of the klystron reflector potential, which is typically 2–3 kV negative. By applying the combined error signals to the reflector, the klystron is brought into phase lock. Some details of the various system components are presented below.

B. Reference tone generator

The reference tone generator, shown in Fig. 2, produces a signal of high spectral purity which can be set anywhere in the region 3600–4000 MHz. The output of the Fluke 6160B frequency synthesizer (93–160 MHz) is mixed with a crystal-controlled source at 507 MHz to produce an upper sideband tone between 600 and 667 MHz. This tone is tripled and then doubled using resistive-diode multipliers. Spurious tones generated in the mixer and multipliers are suppressed by bandpass filters. The result is that all unwanted tones in the reference generator output are at least 68 dB below the reference tone. In addition to these weak tones, there is a continuum of noise produced by the reference generator; the noise power in a 1-Hz bandwidth is 125 ± 2 dB below the reference tone. Because of saturation effects in the nonlinear elements of the reference generator, this continuum is almost pure phase noise. Phase noise and spurious tones will be discussed further in Secs. II C and III C.

C. Harmonic mixer and i.f. amplifier

The harmonic mixer and i.f. amplifier are shown in Fig. 3. The reference tone at ~10 dBm is injected into the GaAs Schottky barrier mixer diode through a triplexer, whose purpose is to separate the 4-GHz reference tone, the 740-MHz i.f. signal, and the dc crystal current. The low-pass filter at the i.f. port stops leakage of the reference tone. The klystron output reaches the diode via a 20-dB directional coupler. Harmonic mixers in different waveguide sizes are used for the various millimeter-wave bands. With 8-mA crystal current and 0-dBm incident klystron power, the conversion

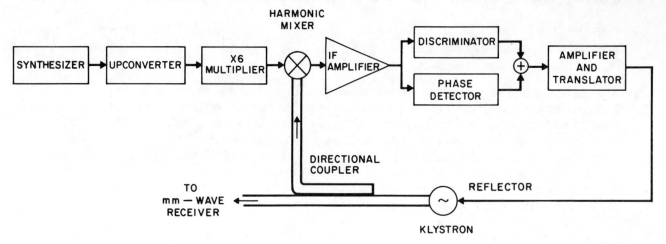

FIG. 1. Block diagram of the klystron stabilizer.

loss of the harmonic mixers is typically 53 dB at 82 GHz and 68 dB at 147 GHz.

The i.f. amplifier chain has a noise figure of 4 dB, a maximum gain of 80 dB, and an automatic gain control range of ≳30 dB. The i.f. bandwidth is set by the 200-MHz bandpass filter. A broader bandwidth than this is not useful, because the electronic tuning range of most klystrons is only about 200 MHz.

Measurements at 82 and 147 GHz show that the i.f. carrier-to-noise ratio (CNR) is determined primarily by phase noise from the reference generator, rather than by the noise figure of the harmonic mixer. For example, a 0-dBm, 147-GHz klystron signal incident on the harmonic mixture will produce at i.f. a "carrier" tone of −68 dBm. It will be phase-modulated, however, by the multiplied phase noise of the reference generator. The carrier-to-noise ratio at i.f. due to phase noise will be

$$CNR_{i.f.} = CNR_{ref} - 20 \log N, \qquad (1)$$

where CNR_{ref} is the carrier-to-noise ratio of the reference generator (125 ± 2 dB in 1 Hz), and N is the harmonic of the reference tone that mixes with the klystron to produce the i.f. signal. At 147 GHz, N is ~38; thus $CNR_{i.f.}$ is ~93 dB. This corresponds to an absolute

noise power density of −161 dBm in a 1-Hz bandwidth, and is significantly greater than the −170-dBm equivalent input noise due to the i.f. amplifier. Thus an improvement in the conversion loss of the harmonic mixer would not significantly affect the carrier-to-noise ratio at i.f.

D. Error signal generation

After amplification the i.f. signal is demodulated in the circuit shown in Fig. 4. A 740-MHz crystal-controlled source, coupled through a 3-dB quadrature coupler, provides the local oscillator power for the double-balanced mixers which serve as phase detectors. The output of the in-phase detector is a component of the error signal used to drive the phase-locked loop circuitry described in the next section. The variable attenuator, which is used to adjust the level of this component, is discussed in Sec. III A. When a klystron is phase-locked, the output of the in-phase detector is constant and nearly zero. The output of the quadrature detector, on the other hand, is near its positive or negative extreme; its sign is a convenient indicator of whether the klystron is locked 740 MHz above or below the appropriate harmonic of the reference generator.

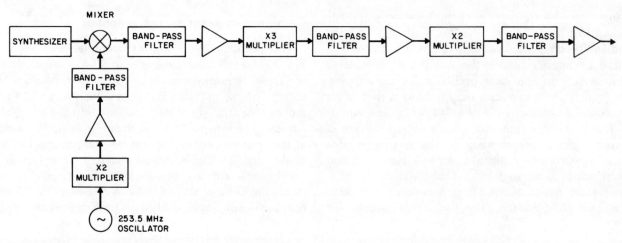

FIG. 2. Reference tone generator.

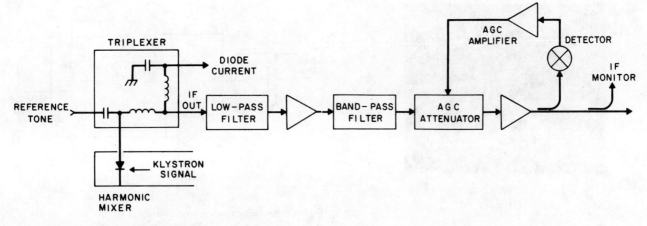

FIG. 3. Harmonic mixer and i.f. amplifier.

The outputs of both phase detectors are combined to form a discriminator, whose operation is as follows.[1] When the klystron is not locked, the i.f. signal is not at exactly 740 MHz. Thus, the two phase detectors generate alternating voltages, with relative phase either +90° or −90°, depending on whether the i.f. signal is above or below 740 MHz. The capacitor, which acts as a differentiator, introduces an additional 90° phase lead into the quadrature-phase path, and makes the amplitude of the transmitted signal very nearly proportional to the deviation of the i.f. signal from 740 MHz. When this signal is multiplied by the output of the in-phase detector in a double-balanced mixer, the resultant dc voltage corresponds both in sign and magnitude to the frequency error of the i.f. signal. The circuit thus functions as a discriminator, as shown in Fig. 5. A polarity-reversal circuit is used at the discriminator output to provide the proper polarity for locking above or below the appropriate harmonic of the reference tone.

E. Amplification and voltage translation

The outputs of the in-phase detector and discriminator are combined in a summing amplifier (3-dB bandwidth ≳5 MHz) to produce a composite error signal. The processing of this error signal (amplification and voltage translation) takes place along two parallel paths, as shown in Fig. 6. The low-frequency path is dc coupled to the klystron reflector through an optical isolator; the high-frequency path is ac coupled through a high-voltage capacitor. In order to achieve a large voltage-swing capability in the low-frequency path, the optical isolator is followed by an operational amplifier with a ±150-V output range. This covers the electronic tuning requirements of virtually all klystrons. The power supply for the operational amplifier floats on the klystron cathode power supply. The frequency response of the low-frequency path is shown in Fig. 7. The 3-dB corner frequency of ~2 kHz assures that slew-rate limitations of the operational amplifier will not be exceeded.

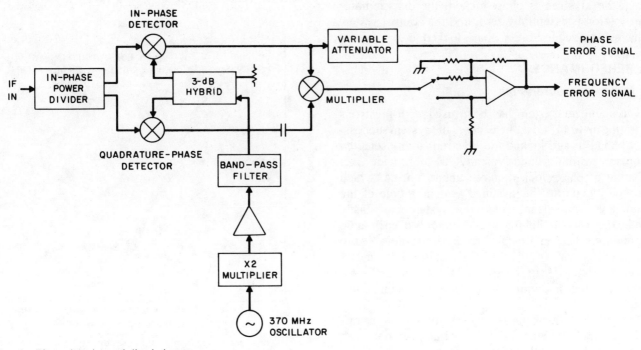

FIG. 4. Phase detector and discriminator.

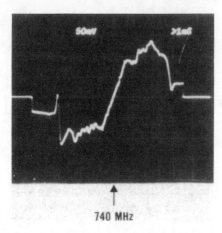

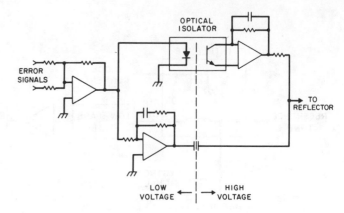

↑
740 MHz

FIG. 5. Voltage-frequency characteristic of the discriminator for 0-dBm input power over the range 440–1040 MHz. In the klystron stabilizer the discriminator is used between 640 and 840 MHz. Horizontal scale is 100 MHz/div; vertical scale is 50 mV/div.

FIG. 6. Amplifiers and voltage translator for the error signals.

The closed-loop frequency response of the operational amplifier in the high-frequency path is also shown in Fig. 7. When the high- and low-frequency paths are joined on the high voltage side of the circuit, a smooth crossover occurs in the region of 40 kHz, resulting in the combined response of Fig. 7. The crossover circuit consists of the high-voltage coupling capacitor and the series resistor at the output of the low-frequency operational amplifier.

At frequencies above ~5 MHz, the gain in the high-frequency path decreases rapidly, so the maximum useful bandwidth of the phase-locked loop is limited to about 5 MHz. When the klystron is more than 5 MHz from the desired frequency, the dominant contribution to the error signal applied to the reflector is from the discriminator. As the frequency error in the klystron is reduced, the discriminator contribution becomes smaller, and control of the klystron shifts smoothly to the in-phase detector. When the klystron is phase-locked, the discriminator contribution is essentially zero, and the control system behaves as a second-order phase-locked loop.

III. PERFORMANCE

A. Pull-in characteristics

The stabilizer system has been used with klystrons operating up to 147 GHz. In all cases the system successfully pulled in and locked the klystron to any selected frequency within a band typically 200 MHz wide. Acquisition of phase lock required about 1 msec, which is roughly equal to the specified switching time of the frequency synthesizer. Thus, the system can easily handle the 10-Hz switching rates commonly used in radio astronomy. The minimum i.f. carrier-to-noise ratio necessary to achieve this performance is ~0 dB, measured in the 200-MHz i.f. passband. For poorer carrier-to-noise ratio the discriminator performance deteriorates rapidly.

The different electronic tuning sensitivities of various klystrons are compensated by the variable attenuator at the output of the in-phase detector (Fig. 4). By adjusting

this attenuator, the phase-locked loop gain and, hence, the bandwidth can be optimized. Usually a bandwidth (unity-gain crossover frequency) of 3–4 MHz provides the best locking performance. This is the only adjustment required in the control circuits when different klystrons are used.

When the klystron load is not a perfect match, the pull-in performance of the stabilizer is degraded. A general theory of this behavior has not been formulated, so some specific examples must suffice. A resonant transmission cavity (3-dB bandwidth ~100 MHz) placed 30 cm from the 82-GHz klystron limited the pull-in range to ~50 MHz. Two decibels of attenuation inserted between the klystron and the cavity increased the pull-in range to ~70 MHz. With 8-dB attenuation, no effects due to load mismatch were visible.

B. Harmonic identification

A simple procedure is used to confirm that the phase-locked klystron is indeed locked to the desired harmonic of the reference generator. For a given klystron fre-

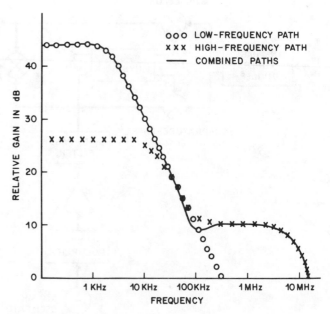

FIG. 7. The separate and combined frequency responses of the two error-signal amplification paths.

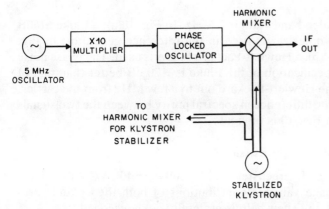

FIG. 8. Apparatus to measure the noise of a stabilized klystron.

quency, an i.f. signal of 740 MHz can be produced by mixing with the Nth harmonic of the reference generator when this harmonic is either 740 MHz above or below the klystron frequency. To verify that the klystron is mixing with the Nth harmonic, the discriminator and phase lock loops are opened and the reflector voltage on the klystron is set to produce an i.f. tone at 740 MHz. The frequency synthesizer is then switched in the appropriate direction by an amount

$$\Delta F_{\text{syn}} = 1480 \text{ MHz}/6N. \tag{2}$$

The frequency of the i.f. tone will not change if Nth-harmonic mixing is being observed. If, however, the klystron is mixing with the $(N \pm 1)$st harmonic, the frequency shift in the i.f. tone will be

$$\Delta F_{\text{i.f.}} = 1480 \text{ MHz}/N.$$

Even for $N = 50$, which is the highest harmonic number likely to be used, $\Delta F_{\text{i.f.}}$ is easily observable on a spectrum analyzer.

C. Noise

The performance of a millimeter-wave line receiver is degraded by both amplitude and phase noise of the klystron local oscillator. The stabilizer system described here does not affect the amplitude noise of the klystron in any significant way, but it does modify its phase-noise characteristics. Noise from the reference generator is augmented by the harmonic mixer according to Eq. (1). Within the bandwidth of the phase-locked loop this increased noise is transferred to the klystron. The result

is a "pedestal" of noise extending a few MHz to either side of the klystron carrier. This effect was observed using the apparatus shown in Fig. 8. The multiplied output of a 5-MHz crystal oscillator (not the oscillator used in the frequency synthesizer) is used as a reference for a phase-locked cw oscillator at 3550 MHz. This signal is mixed in a harmonic mixer with the output of the stabilized klystron at 82 GHz to produce an i.f. of a few MHz. The carrier-to-noise ratio of this i.f. signal is measured to be 96 ± 2 dB in a 1-Hz bandwidth. Since the 22nd harmonic of the reference generator is mixing with the klystron, the carrier-to-noise ratio predicted from Eq. (1) is 98 ± 2 dB (if we assume no noise contributed by the oscillator at 3550 MHz), which is consistent with the measured value.

Spurious tones from the reference generator, which are separated from the reference tone by less than the bandwidth of the phase-locked loop, can cause sidebands to appear on the klystron output. In practice, however, such tones occur only rarely, and when they do, they are very weak. The strongest spurious output of the reference generator is 68 dB down (see Sec. II B). Even at 200 GHz, where the relative strength of this tone is increased through harmonic multiplication by an amount $20 \log 50 = 34$ dB ($N \sim 50$ at 200 GHz), it is still 34 dB below the carrier, which is negligible in radio astronomy applications.

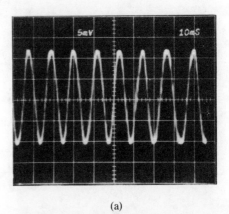

(a)

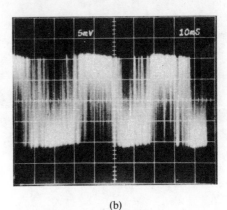

(b)

FIG. 10. Comparison of the phase noise of the "zero beat" signal using (a) Fluke 6160B and (b) Hewlett-Packard 5105A synthesizers in the reference tone generator. Video bandwidth is 1 MHz. Horizontal scale is 10 msec/div; vertical scale is 5 mV/div.

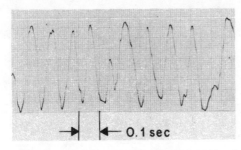

FIG. 9. Chart recording of the "zero beat" produced by a stabilized klystron at 82 GHz. The actual beat frequency is ~10 Hz.

The klystron carrier mentioned earlier is not a true spectral line; it has a width which can be measured by mixing the i.f. signal shown in Fig. 8 with a high-purity tunable oscillator at very nearly the same frequency, and looking for the "zero beat." The lowest observable beat frequency, f_{min}, is approximately the reciprocal of the correlation time of the klystron output. (Again we assume no noise from the 3550-MHz oscillator.) Thus f_{min} is itself a measure of the spectral width of the klystron "line." With an 82-GHz klystron, a 10-Hz beat tone was readily obtained, as shown in the chart recording of Fig. 9. The klystron linewidth, then, is ≤ 10 Hz.

It must be emphasized that the phase noise of the stabilized klystron depends strongly on the spectral purity of the frequency synthesizer used to generate the reference tone. This is qualitatively illustrated in Figs. 10(a) and 10(b), which are oscilloscope traces of the "zero beat" described above (actually the beat frequencies are a few tens of hertz). In both cases the video bandwidth is 1 MHz. In Fig. 10(a), a Fluke 6160B synthesizer is used in the reference generator; in Fig. 10(b), a Hewlett-Packard 5105A is used. The phase noise specification of the Fluke is ~20 dB better than that of the Hewlett-Packard out to a few MHz from the carrier. The difference in spectral purity between the two signals in Fig. 10 is obvious.

ACKNOWLEDGMENTS

L. J. Greenstein, R. A. Linke, and R. W. Wilson have made valuable contributions of both theory and hardware. Their help is gratefully acknowledged.

[1] D. Richman, Proc. IRE **42**, 106–133 (1954).
[2] F. M. Gardner, *Phaselock Techniques* (Wiley, New York, 1966).
[3] L. Woo and D. H. Levy, Rev. Sci. Instrum. **44**, 732–737 (1973).
[4] For example, the Hewlett-Packard 8709 synchronizer.
[5] R. A. Clark and J. D. Rozunick, IEEE Trans. Instrum. Meas. **IM-14**, 2–10 (March/June 1965).

An Active "Cold" Noise Source

ROBERT H. FRATER AND DAVID R. WILLIAMS, MEMBER, IEEE

Abstract— An active circuit which behaves like a "cold" noise source is described. The circuit, which uses a gallium arsenide FET is given the name COLFET. The appropriate theory is developed and practical circuits described using the circuit. Equivalent noise temperatures of less than 50 K have been measured for a 50-Ω source at 1400 MHz.

I. INTRODUCTION

RECENT DEVELOPMENTS in gallium arsenide FET technology have led to remarkable reductions in both the noise temperature and the capital costs of room-temperature amplifiers.

The theory of the FET and its equivalent circuit is well developed [1] and circuit performance can now be accurately predicted, at least at L-band, by simple parameter measurements.

The input impedance of the FET is predominantly capacitive (there is a small series resistance) with the result that an inductance added to the source circuit is transformed to appear as a noiseless resistor in the gate circuit. This phenomenon allows the development of FET amplifiers with low-input VSWR [3], [4].

Another consequence of the impedance transformation at the input is that only part of the total input resistance contributes noise to the circuit, that portion due to the source inductance being noiseless. This offers the possibility that the input impedance of such an amplifier may be resistive with the apparent temperature of the resistance being much less than ambient and in fact comparable with the noise of the circuit when used as an amplifier.

In this paper the noise theory developed by Pucel et al. [1] is revised and the analysis extended to evaluate the performance of the input circuit when used as a noise source. The resultant circuit we have called the COLFET.

II. THE FET AMPLIFIER

A. Noise Behavior

The excess noise temperature of the FET amplifier is given as a function of source impedance R_s by Pucel et al. [1]

$$T_{\text{FET}} = T_0 \left[\frac{R_n}{R_s} + \frac{G_n \omega^2 C_{gs}^2}{R_s \cdot g_m^2} (R_s + R_n)^2 \right] \quad (1)$$

where C_{gs} is the gate-to-source capacitance of the FET having a transconductance g_m. R_n is the voltage noise generator in the gate circuit,[1] and is equal to the sum of the gate and source resistances external to the channel. G_n is the equivalent current noise conductance of the drain circuit (see Fig. 1) which is associated with the nonthermal channel noise in the FET. The amplifier has a minimum noise temperature T_{min} given by

$$T_{\text{min}} = T_0 \cdot \frac{2\omega C_{gs}}{g_m} \sqrt{R_n \cdot G_n} \quad (2)$$

which occurs for an optimum source impedance $R_{\text{opt}} + jX_{\text{opt}}$ given by

$$R_{\text{opt}} = \left[\frac{g_m^2 R_n}{\omega^2 C_{gs}^2 G_n} + R_n^2 \right]^{1/2} \cong \frac{g_m}{\omega C_{gs}} \sqrt{\frac{R_n}{G_n}} \quad (3)$$

and for $X_{\text{opt}} = X_L = -X_{C_{gs}}$. Combining (2) and (3) we obtain

$$\left. \begin{aligned} R_n &= \frac{T_{\text{min}} R_{\text{opt}}}{2T_0} \\ G_n &= \frac{T_{\text{min}}}{R_{\text{opt}}} \cdot \frac{1}{2T_0} \cdot \left\{ \frac{g_m}{\omega C_{gs}} \right\}^2 \end{aligned} \right\}. \quad (4)$$

Equations (4) are used to determine R_n and G_n from our measured values of T_{min} and R_{opt} described here.

B. The Effect of Source Inductance

The effect of added source inductance has been discussed in the literature [2] and [3] and more recently by Williams et al. [4] who have used the effect to match an L-band amplifier. The added inductance (L_s in Fig. 1) is transformed to a frequency-independent feedback resistor R_{FB}

$$R_{FB} = \frac{g_m}{C_{gs} S} \cdot L_s S \quad (5)$$

where S is the complex frequency variable. The equivalent circuit for the input is shown in Fig. 2. The impedance transformation of the inductance L_s implied by (5) will be used in the later analysis. An inductance L_s also appears in the input circuit.

Manuscript received April 11, 1980; revised November 26, 1980.

R. H. Frater is with the School of Electrical Engineering, University of Sydney, Sydney NSW 2006, Australia.

D. R. Williams is with the Radio Astronomy Laboratory, University of California, Berkeley, CA 94720.

[1] We have not considered here the correlation between the drain noise current generator and the induced gate noise generator, here partially included in R_n. However, at the low frequencies, since the induced gate noise varies as ω^2, calculation shows that its contribution is only about 15 percent.

Reprinted from *IEEE Trans. Microwave Theory Tech.*, vol. MTT-29, no. 4, pp. 344–347, Apr. 1981.

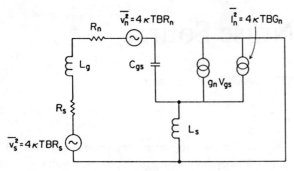

Fig. 1. The full equivalent circuit for the FET used for noise calculations.

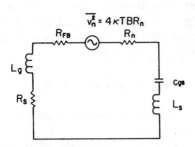

Fig. 2. The equivalent input circuit for the COLFET used in calculating the noise power produced in the source resistance R_s by the gate series noise generator R_n. R_{FB} is the feedback resistance (at $T=0°$) arising from the added FET source inductance which is chosen for a matched input when $R_s = R_{FB} + R_n$. At resonance the optimum source reactance X_L equals the gate-to-source capacitance $X_{C_{gs}}$.

The introduction of lossless negative feedback (as L_s may be considered) has negligible effect on the noise temperature [5].

In practice, the value of L_s is chosen so that $R_s = R_{FB} + R_n$ and the amplifier is thus matched. When R_s is also equal to the optimum source impedance R_{opt}, the amplifier also has minimum noise and thus has a simultaneous gain and noise match. Now since R_{FB} does not change the noise temperature of the amplifier, it must appear to be at zero absolute temperature. This property is used for the COLFET circuit.

III. THE COLFET ANALYSIS

The full noise equivalent circuit for the FET is shown in Fig. 1. In considering the noise current flowing in R_s, two contributions arise. The first is due to the noise of the resistance R_n while the second is due to the input noise conductance G_n and its associated generator. These two contributions are separately calculated and combined in the final computation of an equivalent temperature for the total input resistance of the circuit.

A. The Input Noise Generator

In Fig. 2, an equivalent circuit for the input section of Fig. 1 is given.

The resistance R_{FB} has been seen to arise due to the impedance transformation of the inductance L_s by the device. L_g is an inductance added to tune out C_{gs} and L_s so that the input impedance of the circuit becomes $R_{FB} + R_n$.

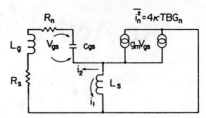

Fig. 3. The equivalent circuit of the COLFET used in calculating the noise power produced in the source resistance R_s by the drain current generator G_n. Here, L_s is the source inductor added to the FET, transconductance g_m. There is a division of current from the G_n generator into path i_1 through L_s, and i_2 through the loop including R_s.

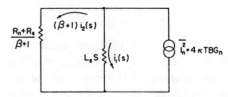

Fig. 4. Transformed equivalent circuit derived from Fig. 3 for use in the calculation of the contribution of the output-noise generator to the noise in the input circuit.

The match condition gives

$$R_s = R_{FB} + R_n. \tag{6}$$

The noise current due to the R_n generator in this circuit is given by

$$\overline{i_1^2} = \frac{\overline{v_n^2}}{(R_s + R_{FB} + R_n)^2} \tag{7}$$

and for input matching

$$\overline{i_1^2} = \frac{\overline{v_n^2}}{4R_s^2}. \tag{8}$$

B. The Output-Noise Generator

The circuit of Fig. 3 is used to derive the noise current in R_s due to the output-noise generator.

If we now introduce a parameter β where

$$\beta(S) = \frac{g_m}{C_{gs}S} \tag{9}$$

β is recognized as a complex gate-to-drain current gain. We can now transform Fig. 3 to give Fig. 4. In so doing, L_g and C_{gs} have been removed as they are almost series resonant. Now writing

$$i_N(S) = \sqrt{\overline{i_n^2}(s)} \tag{10}$$

$$i_2(S) = \frac{i_N(S)}{\beta+1} \cdot \frac{L_s S}{\frac{R_n + R_s}{\beta+1} + L_s S} \tag{11}$$

$$= \frac{i_N(S)}{\beta+1} \cdot \frac{(\beta+1)L_s S}{R_n + R_s + (\beta+1)L_s S} \tag{12}$$

and hence

$$\overline{i_2^2} = \frac{\overline{i_n^2}}{|\beta|^2+1} \cdot \frac{R_{FB}^2}{(R_n+R_s+R_{FB})^2}. \tag{13}$$

Using (6) and recognizing that

$$|\beta| = g_m X_{C_{gs}} \tag{14}$$

$$\overline{i_2^2} = \overline{i_n^2} \frac{(R_s-R_n)^2}{4R_s^2} \cdot \frac{1}{g_m^2 X_{C_{gs}}^2+1}. \tag{15}$$

C. The Total Noise at the Input

The equivalent noise temperature contribution for a noise current specified by $\overline{i^2}$ flowing in R_s is given by

$$\overline{i^2} \cdot R_s = \kappa T_{eq} \Delta f \tag{16}$$

from which

$$T_{eq} = \overline{i^2} \cdot \frac{R_s}{\kappa \Delta f}. \tag{17}$$

Combining the contributions of the input-circuit and output-circuit noise generators and simplifying we obtain

$$T_{eq} = T \left\{ \frac{G_n}{(g_m^2 X_c^2+1)} \cdot \frac{(R_s-R_n)^2}{R_s} + \frac{R_n}{R_s} \right\}. \tag{18}$$

Differentiating with respect to R_s we obtain the value of R_s for minimum T_{eq}

$$R_{s_{min}}^2 = R_n^2 + \frac{R_n}{G_n}(g_m^2 X_c^2+1) \tag{19}$$

since

$$\frac{1}{G_n} \gg R_n$$

and

$$g_m^2 X_c^2 \gg 1$$

$$R_{s_{min}} \approx \frac{g_m}{\omega C_{gs}} \sqrt{\frac{R_n}{G_n}}. \tag{20}$$

This result is the same as obtained for the amplifier noise analysis.

A further general relationship is derived for the optimum source impedance R_{opt}, by substituting (3) into (1), to obtain the minimum noise temperature T_{min}

$$\frac{T_{min}}{T_0} = \frac{2R_{gn}}{(R_{opt}-R_{gn})}. \tag{21}$$

Likewise, substituting (19) into (18) (assuming $g_m^2 X_c^2 \gg 1$) we obtain for the minimum value of $T_{eq,min}$

$$\frac{T_{eq,min}}{T_0} = \frac{2R_{gn}}{(R_{opt}+R_{gn})}. \tag{22}$$

Taking the quotient of these relations (21) and (22) we obtain

$$\frac{T_{eq,min}}{T_{min}} = \frac{R_{opt}-R_{gn}}{R_{opt}+R_{gn}} \tag{23}$$

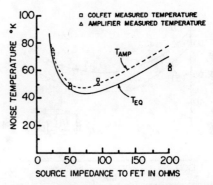

Fig. 5. The measured noise temperature of the COLFET input and also its noise temperature as an amplifier, both plotted as a function of the source resistance R_s. The solid curve T_{eq} is the equivalent temperature of the COLFET calculated from theory developed in the text. The dashed curve T_{AMP} is based on the Pucel theory of the noise temperature of the FET amplifier.

where $T_{eq,min}/T_{min}$ is the ratio of the minima on our theoretical curves of Fig. 5.

IV. MEASUREMENTS

All the measurements described here were made using a low noise receiver, itself having an FET front end. A Maury Microwave LN_2 load was used in the noise temperature measurements together with an AIL precision attenuator for the Y-factor method.

A. Results for the Amplifier

We have found the values of R_{opt} and T_{min} experimentally by measuring the amplifier noise temperature for various values of source impedance seen by the FET. A series of different quarter-wave transformers were installed in the amplifier between the FET and the source. The measured values of amplifier noise temperature are plotted in Fig. 5 against the various values of source impedances used. From these data we obtain values of $T_{min}=48°$ and $R_{opt}=70 \ \Omega$ at 1.4 GHz. Using the known values for the Mitsubishi 1402 of $C_{gs}=0.55$ pF, $g_m=0.025$ (at 1.5 V and 7 mA drain current) we obtain from (4) the values for the noise generators

$$R_n = 5.8 \ \Omega$$
$$G_n = 0.0296 \ \mho.$$

These values are used to calculate the T_{AMP} curve plotted in Fig. 5 from (1) and later used in the calculation of the COLFET input noise temperature.

B. Results for the COLFET

Two sets of measurements were made on the COLFET circuit. The first of these (shown in Fig. 5) is the equivalent temperature T_{eq} presented by the circuit when optimized to provide different input impedances. As in the amplifier case, the lowest noise occurs at about 70 Ω. The agreement with the theory is particularly good. The calculated noise at high impedances (200 Ω) is sensitive to the values taken for C_{gs} while the lower impedance region is sensitive to R_n.

The configuration for the second set of measurements of the COLFET to determine the frequency behavior is shown

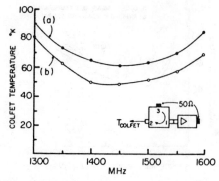

Fig. 6. The noise temperature of the COLFET cold load plotted as a function of frequency measured (*a*) with a circulator and (*b*) without. Curve (*b*) was shifted 50 MHz to the right with the circulator addition (see text).

in Fig. 6. The COLFET is connected to the receiver input in turn with the calibrating LN_2 and T_0 loads. Measurements were made with and without the circulator. The results are presented in Fig. 6(*a*) and (*b*). A small detuning effect of about 50 MHz occurs when the COLFET is attached to the circulator port, and this has been taken into account in making plot (*b*).

In practice, although the COLFET can be used by itself as a cold load, it is reactive out of band. Consequently, the use of the circulator provides a matched cold load with only a small increase in cold load temperature due to the insertion loss of ports 1→2.

V. APPLICATIONS OF THE COLFET

Some immediate applications for this technique are as a cold load, a circulator termination, or as a reference load in a Dicke radiometer.

A. Cold Load

The COLFET is used as a conventional cold load, for noise temperature measurement of low-noise receivers. For the application, the COLFET is calibrated against a liquid nitrogen load. The COLFET is found to be stable with time at constant bias conditions, but its temperature will be proportional to ambient changes. Because of the small size, temperature control can be simply provided.

B. Circulator Termination

The COLFET is used as a cold terminating port on a 3-port junction circulator. In this application, the noise arising from the third port of a parametric amplifier circulator and which is reflected from a mismatched source such as an antenna feed, can be reduced in the ratio of the cold load temperature to ambient temperature. This reduces the system noise contribution from the antenna mismatch effect. In the same way, noise radiated from a circulator ahead of a low-noise FET amplifier can be similarly reduced.

C. Radiometer Reference Load

The COLFET is used as a cold load in a Dicke-type radiometer in Radioastronomy applications. The COLFET provides a comparison load of approximately the same temperature as the antenna of 50 K in this frequency range.

VI. CONCLUSIONS

In the short time that has been available for evaluation of the COLFET it has proved to be an extremely useful circuit. Its compactness will appeal to those who have carried a dewar of liquid nitrogen to inaccessible places to calibrate equipment or set up experiments and we expect it to find wide application.

ACKNOWLEDGMENT

This work was carried out while one of the authors (D.R.W.) was working on exchange at the CSIRO Division of Radiophysics at Sydney, Australia.

REFERENCES

[1] R. Pucel, H. Haus, and H. Statz, "Signal and noise properties of gallium arsenide microwave field-effect transistors," in *Advances in Electronics and Electron Physics*, vol. 38. New York: Academic Press, 1975, pp. 195–265.
[2] A. Anastassiou and M. Strutt, "Effects of source lead inductance on the noise figure of a GaAs FET," *Proc. IEEE*, vol. 62, pp. 406–408, Mar. 1974: also, for corrections, see S. Iversen, *Proc. IEEE*, vol. 63, pp. 983–984, June 1975.
[3] L. Nevin and R. Wong, "L-band GaAs FET amplifier," *Microwave J.*, vol. 22, no. 4, p. 82, 1979.
[4] D. R. Williams, S. Weinreb, and W. T. Lum, "L-band cryogenic GaAs FET amplifier," *Microwave J.*, vol. 23, no. 10, p. 73, 1980.
[5] H. A. Haus and R. B. Adler, "Optimum noise performance of linear amplifiers," *Proc. IRE*, vol. 46, no. 8, pp. 1517–1533, Aug. 1958.a

A Quasioptical Feed System for Radioastronomical Observations at Millimeter Wavelengths

By P. F. GOLDSMITH

(Manuscript received March 1, 1977)

We describe a quasioptical feed system for use with a 7-meter Cassegrain antenna at millimeter wavelengths. This system is designed to take full advantage of low noise, broadband mixer receivers and will be used for radioastronomical observations at frequencies between 60 GHz and 140 GHz. Two offset parabolic mirrors couple the radiation from the f/D = 5.7 antenna into the receiver feedhorn. A Fabry-Perot resonator operating at oblique incidence is used to inject the local oscillator energy into the signal path and to suppress response at the image frequency, while the coupling loss between the mixer waveguide flange and the main lobe of the antenna pattern should be ≤1 dB.

I. INTRODUCTION

For optimal use of an antenna for radio astronomy at millimeter wavelengths, the feed system should provide a number of functions and must satisfy a variety of stringent performance criteria. These include

(*i*) Low loss for the signal over an instantaneous bandwidth of ≥500 MHz.

(*ii*) A well-controlled antenna illumination pattern which should remain unchanged over as large a range of frequencies as possible.

(*iii*) A provision for making accurate absolute calibrations of the receiver gain and atmospheric attenuation—both of these require suppression of the image frequency response in systems incorporating mixers.

(*iv*) A facility for antenna beam switching at a rapid rate to minimize the sky-noise contribution to receiver noise.

(*v*) Since mixers are currently the dominant type of receiver at

frequencies between 60 GHz and 300 GHz, it would be advantageous to include local oscillator injection as part of the feed system if this can be done with low loss.

The present feed system has been designed to satisfy all of the preceeding requirements. In Section II we describe the feed system optics and analyze the measurements of system performance. In Section III we discuss various aspects of the Fabry-Perot diplexer including bandwidth, image rejection, local oscillator noise suppression, and loss for the signal and for the local oscillator. In Section IV we discuss the calibration system.

II. FEED SYSTEM OPTICS

2.1 Antenna

This feed system is designed to operate with the recently completed Bell Laboratories millimeter antenna located at Holmdel, N.J. The antenna is an offset Cassegrain with a diameter of 7 meters and a f/D ratio of 5.7. The overall surface accuracy is approximately 0.01 cm rms, allowing operation with a moderately high beam efficiency at frequencies as high as 300 GHz. The main advantage of the offset Cassegrain design is that there is zero aperture blockage, and a very low reflection coefficient and low sidelobe levels can be achieved.[1]

2.2 Gaussian beam theory

We shall discuss the feed system optics in terms of the propagation of a single gaussian mode. As discussed by Arnaud,[2] a gaussian beam propagating in free space has a power distribution perpendicular to the direction of propagation (taken to be the z axis) of the form

$$\frac{P(r)}{P(o)} = e^{-[r/\xi(z)]^2} \tag{1}$$

The beam half-width (radius) ξ depends on z, the distance along the axis of propagation, as

$$\xi^2(z) = \xi_o^2 + \left(\frac{z}{k_o\xi_o}\right)^2 \tag{2}$$

where ξ_o is the minimum beam half-width (beam waist radius), taken to be located at $z = 0$, and $k_o = 2\pi/\lambda$. The asymptotic angle of beam half-width growth is seen from eq (2) to be

$$\theta_\xi = 1/k_o\xi_o \tag{3}$$

Equations (1) to (3) apply to gaussian beams of infinite transverse extent. In any practical system the beam will be truncated at some level,

Reprinted with permission from *Bell Syst. Tech. J.*, vol. 56, no. 8, pp. 1483–1501, Oct. 1977.

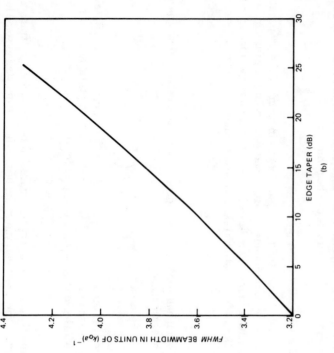

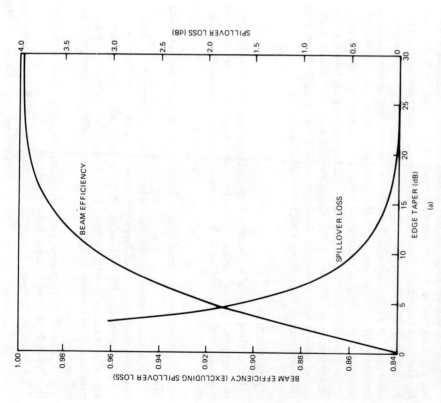

dB below the on-axis power level). We will thus ignore the effects of beam truncation within the feed system.

The edge taper at the main reflector is related to ξ_A, the antenna illumination beam half-width, by the formula

$$\xi_A = a\sqrt{\frac{10}{T \ln 10}} \qquad (4)$$

where a is the main reflector radius (350 cm for this antenna) and T is the edge taper in decibels. We find that $\xi_A = 195$ cm for $T = 14$ dB. Since ξ_A is much larger than ξ_o, eq. (2) reduces to

$$\xi_A \approx z_A \theta_\xi = \frac{f}{k_o \xi_o} \qquad (5)$$

where f is the focal length of the antenna (3955 cm). The resulting value for ξ_o at 100 GHz is 0.97 cm.

2.3 Feed system components

The large f/D ratio and resulting large beam waist size of the antenna makes coupling to the antenna beam waist directly with a feedhorn

Fig. 1. (*continued from previous page*)

Fig. 1—(a) Beam efficiency and spillover loss for an unblocked, ideal antenna with gaussian aperture illumination, as a function of the edge taper. The edge taper is defined as the power density at the center of the antenna divided by the power density at the edge of the antenna. (b—next page) Beamwidth (full width at half maximum) for the same conditions. The radius of the antenna is a, and $k_0 = 2\pi/\lambda$.

which will produce sidelobes. In considering at what level the beam at the main reflector should be truncated, we have to balance consideration of spillover loss, sidelobe levels, and beam efficiency[3] against those of beamwidth and on-axis gain. Figure 1a shows the spillover loss and beam efficiency while Fig. 1b shows the beamwidth as a function of edge taper for an antenna with a gaussian aperture illumination pattern. The edge taper is defined as the power density at the center of the antenna divided by the power density at the edge. We have chosen an edge taper T_M close to 14 dB as being a satisfactory compromise. All other optical elements in the feed system truncate the beam at a much lower level (at least 23

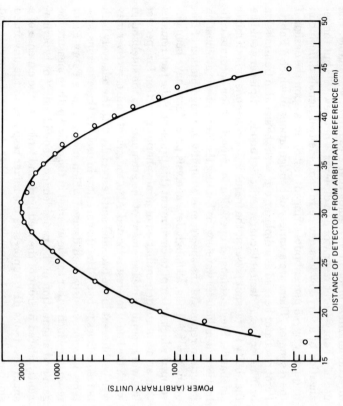

Fig. 3—Profile of beam in region between M1 and M2. The axis of the scan is perpendicular to the plane of the components in Fig. 2, and passes through the axis of the beam. Also shown is a gaussian beam with a beam half-width equal to 6.5 cm.

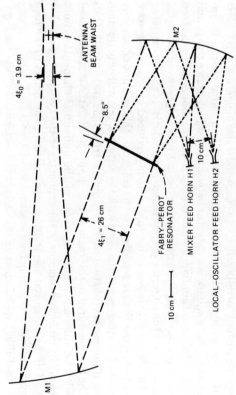

Fig. 2—Feed system optics. M1 and M2 are offset paraboloids. The diplexing action of the Fabry–Perot resonator (tilted 8.5 degrees from normal incidence) is shown schematically.

undesirable, especially for cryogenic receivers. Horn-lens arrangements were investigated but the losses involved were felt to be a significant disadvantage, especially when operation over very large bandwidths is required. In view of these facts, and also because of the desirability of an even larger beam waist size required for low loss in the Fabry–Perot diplexer (Section III), a feed system using metal mirrors is preferable. The arrangement of the feed system components is shown in Fig. 2. The overall size of the feed system is dictated by the beam waist size and the desire to minimize the number of mirrors involved.

Mirrors M1 and M2 are offset paraboloids; the offset angle for M1 is 20 degrees and the focal length is 136 cm. For M2 the offset angle is 30 degrees for the signal beam and the focal length is 44 cm. Offset antennas of this type have been shown to have excellent beam patterns.[4] The mirrors used in this work were cut on a numerically controlled milling machine; the peak deviation from the desired surface contour is approximately 0.05 mm.

The beam from the antenna expands until it reaches M1; at this point the beam half-width, denoted ξ_1, is 6.5 cm and is essentially frequency-independent. The distance from the beamwaist to M1 is equal to the focal length of the mirror so that in the geometrical optics limit the resulting beam would be collimated. The diffraction effects in the beam between M1 and M2 are small; in actuality a second beam waist is created in the large beam at a distance equal to the focal length from M1. Ideally, the separation between M1 and M2 would be equal to the sum of their focal lengths (180 cm) but a calculation[5] of the mismatch due to the

distance being only 140 cm indicates that this is an insignificant effect.

The difficulty in measuring the power distribution in the beam at the antenna beam waist can be overcome by utilizing the properties of a gaussian beam focused by lenses or mirrors; the beam half-width in the focal plane on one side of a converging lens with focal length f will be related to the beam-waist radius on the other side by[6]

$$\xi_{fp} = \frac{f}{k_o \xi_o} \qquad (6)$$

In Fig. 3 we show a profile of the beam in the collimated region measured with a small-aperture (0.4 cm × 0.6 cm) horn and square-law detector. This measurement, which is well-fitted by a guassian with $\xi_1 = 6.5$ cm, together with eq. (6) confirms that the beam-waist size at 100 GHz is 1.0 cm, very close to the design value.

A signal passing through the Fabry–Perot resonator is focused by M2 into the feed horn attached to the mixer, located at the beam waist of

543

Table I — Feed system characteristics at 100 GHz

Characteristic	Value
ξ_1, collimated beam-waist radius to $1/e$ power point	6.5 cm
ξ_0, beam-waist radius to $1/e$ power point at antenna beam waist	0.97 cm
T_M, edge taper at main reflector	14.1 dB
θ_{FWHM}, full angular beamwidth to half-power points	1°.8
First sidelobe level relative to on-axis gain	−30 dB
ϵ_F, feed system loss (mixer waveguide flange to antenna beam waist)	0.5 dB
Spillover loss	0.14 dB
ϵ_M, beam efficiency	0.95

M2. The beam-waist radius at the feed horn is 0.32 cm at 100 GHz. The utilization of the Fabry-Perot with a diplexing angle of 8.5 degrees and M2 focal length equal to 44 cm requires that the dimension of M2 in the plane of the paper in Fig. 2 be approximately twice as large as would be required for focusing the signal beam alone.

The feedhorn for the receiver, which is the same design as that for the local oscillator, is a corrugated horn[7] with a beamwidth between −17 dB power points of 29 degrees. This type of feedhorn allows waveguide efficiency and very low sidelobes. For system tests performed at frequencies near 100 GHz we have, however, used relatively narrowband dual-mode horns[8,9] constructed in a manner similar to those described in Ref. 4. The power patterns are very similar to those of the corrugated horns, although with a beamwidth approximately 10 percent larger. All feed system characteristics refer to those measured with the dual-mode horns, but these should differ only in minor ways from those obtained with the corrugated horns.

2.4 Measurements of feed system efficiency

As discussed in the previous section, measurements of the power distribution in the collimated region indicate that the feed system will produce the correct taper in the illumination of the main antenna. In order to measure the efficiency of the feed system, a separate collector was placed at the beam waist of M1, corresponding to the antenna beam waist. This collector, consisting of an ellipsoidal reflector and dual-mode feed horn, was independently measured to have a gaussian angular response pattern corresponding to a beam-waist size of 0.99 cm. A 100-GHz klystron with ∼50 dB attenuation was used as a signal source. By interchanging a square law detector between the signal-source flange and the collector output flange (with the signal source connected to the feed system mixer flange), we determined the loss of power between the signal source and the collector output flange to be 1.1 dB. It should be noted that if part of this loss is due to the mode produced by the feed system not coupling to that accepted by the collector, this will not necessarily lower the efficiency when used with the antenna, but will only result in an illumination function slightly different from that anticipated. Thus the loss measured in this manner is an upper limit to the loss when used with an antenna. While the losses of the individual elements cannot easily be measured separately, the symmetry of the system suggests that half of the measured loss is due to the collector, and half is in the feed system, with a resulting feed system loss of 0.5 dB.

In Table I we summarize the salient characteristics of the feed system.

III. QUASIOPTICAL DIPLEXER

3.1 Introduction

The limited local-oscillator output power available at shorter millimeter wavelengths and the difficulty of fabricating low-loss waveguide diplexers[10] are incentives to seek an alternative to injection cavities and directional filters made in waveguide that are currently available. The use of a Fabry-Perot resonator as a diplexer is not new,[11,12] but the realization of a very low loss device to combine two signals differing in frequency by ∼5 percent puts stringent restrictions upon the design of the resonator. There are a variety of configurations in which a Fabry-Perot resonator be used as a diplexer, e.g., with the signal in transmission or in reflection. A desirable characteristic of an ideal diplexer would be the ability to transmit power at the frequency of either one or both mixer sidebands. Single-sideband operation is important for accurate calibrations at millimeter wavelengths because the atmospheric attenuation in certain regions of the spectrum is a rapidly varying function of frequency.[13,14,15] Thus, although data analysis procedures have been developed which attempt to circumvent this problem,[16] the fact remains that an accurate determination of atmospheric extinction for spectral line work requires measurement of the attenuation in the sideband in which the line of interest is located. Also, the gain of a mixer receiver may well be different in the two sidebands, especially with the relatively high IF frequencies (4 to 5 GHz) that are now in use. For these reasons, systems have previously been devised which incorporate a Fabry-Perot resonator which either can be inserted in the optical path to measure the gain and attenuation in the two sidebands individually[17] or is permanently placed in front of the feed horn and which, at the expense of a small loss (∼0.4 dB), suppresses the mixer response to the unwanted sideband.[18] In order to minimize the number of resonant elements and consequent adjustments required when changing frequencies, we decided

Table II — Characteristics of Fabry-Perot resonator at 100 GHz

T*	Image rejection ratio (dB)	0.5-dB bandwidth (MHz)	1-dB bandwidth (MHz)
0.10	26	220	320
0.15	22	350	500
0.20	19	540	800
0.25	17	620	890
0.30	15	760	1100
0.40	12	1090	1600
0.50	9.5	1500	2200

* T is the transmission of a single mirror.

to use the Fabry-Perot resonator in transmission for the signal (the local oscillator is reflected by the resonator, thus providing the diplexing action). This design allows us either to operate in a double-sideband mode with the two sidebands being transmitted in successive orders (for continuum work) or in a single-sideband mode (desirable for spectral line observations). Only one adjustment is required to set the diplexer for operation at a particular frequency, which proves to be a significant advantage in use.

3.2 Fabry-Perot resonator theory

The analysis of the propagation in a noninfinite Fabry-Perot resonator has been treated by Arnaud et al.[11] Since we will be dealing with a strongly tapered beam, it is sufficient to use the standard formulas for a plane wave in a resonator of infinite transverse dimension to calculate the response. Neglecting absorption in the mirrors, we find[19] that the fraction of the incident power transmitted by the resonator is given by

$$\tau = \frac{1}{1 + \dfrac{4(1-T)}{T^2}\,\sin^2\left(k_o d \cos\theta\right)} \tag{7}$$

where d is the distance between the mirrors, θ is the angle from normal incidence of the radiation, T is the power transmission of a single mirror, and we have set the phase of the reflection coefficient equal to π which causes no loss of generality. In this limit we see that the peak transmission (for $k_o d \cos\theta = n\pi$, n being the order of operation) is equal to unity. The peak-to-valley ratio, or contrast factor, which will in our case be the image rejection ratio, and the 0.5-dB and 1-dB bandwidths for a resonator operating at 100 GHz are given in Table II as a function of T, which is assumed to be frequency-independent. It also has been assumed that the free spectral range of the resonator is approximately equal to 4 ν_{IF}; this is not a severe restriction since the transmission is only weakly dependent on frequency near the transmission minimum. There is a

tradeoff between bandwidth and image reflection, as expected for a simple resonator. This restriction could be eased by using a multiple-mirror resonator, but only at the expense of easy tunability. Efficient utilization of the bandwidth of available IF amplifiers (~600 MHz) indicates that T should not be less than 0.2; the resulting image rejection ratio of 19 dB is certainly adequate to assure proper calibration accuracy. It should be pointed out, however, that this ratio is not so high that the leakage of very strong lines from the opposite sideband in a high-sensitivity spectrogram can be entirely ruled out.

The Fabry-Perot diplexer exhibits quite high directivity for local oscillator injection. Power coming from the local oscillator feed horn that directly leaks though the Fabry-Perot resonator does not end up in the beam waist area at all, and is caught by a sheet of absorbing material. Only local oscillator power which is reflected from the Fabry-Perot, then reflected from the mixer feed horn, and which is finally transmitted by the resonator, can reach the calibration area; the level of this radiation should be at least 17 dB below that of the local oscillator power reaching the mixer.

The loss in a Fabry-Perot resonator operated at oblique incidence is primarily due to a walk-off effect in the finite-sized beam.[11] In this reference, the peak fractional transmission τ through a resonator (assumed to be much larger than the beam size) consisting of two mirrors of transmission T, spacing d, inclined at an angle θ to a gaussian beam of beamwaist radius ξ_o, is given by

$$\tau = 1 - G^2$$

where

$$G = \frac{2d \sin\theta}{\xi_o T} \tag{8}$$

For operation with $\nu_{\mathrm{IF}} = 5$ GHz and $\nu_{\mathrm{SIGNAL}} = 100$ GHz, obtaining the best best image rejection ratio requires that the resonator be operated in fifth order so that $d = 5\lambda/2 = 0.75$ cm. The exact spacing will be determined by the resonance condition for the signal frequency; the condition $4\nu_{\mathrm{IF}} = \nu_{\mathrm{SIGNAL}}/5$ will be satisfied only approximately, but d will be close to the value given above. For $T = 0.2$ we find for small angles $\tau = 1 - (7.5\,\theta/\xi_o)^2$.

A lower limit on θ of $\sim 4/k_o \xi_o$ is found[11] from the condition that the beams be separable at the -17-dB level when the diffraction of each is considered. Thus the maximum transmission is (again for $T = 0.2$, $d = 0.75$ cm)

$$\tau_{\mathrm{MIN}\theta} = 1 - \left(\frac{30}{k_o \xi_o^2}\right)^2 \tag{9}$$

Fig. 4.—The Fabry-Perot resonator. The dial indicator on the right is used to monitor the mirror separation.

As seen from eq. (8) the insertion loss, defined in decibels as $10 \log_{10} \tau^{-1}$, can, in theory, be made as low as desired, at the expense of enlarging the beam-waist radius. The beam waist required for low loss even in the optimum situation [eq. (9)] is moderately large; at $\nu = 100$ GHz and for the above conditions, $\xi_0 = 2.6$ cm is required to achieve an insertion loss of 0.2 dB (the beam diameter will be at least $4\xi_0$). The most straightforward geometry (see, for example, Ref. 11) then results in a very large distance between the Fabry-Perot and the inputs for the signal and local oscillator; on the order of 1 meter for the above conditions. For this reason, and due to the simplicity of having the one mirror (M2) serve as collector for both the mixer and the local oscillator, the geometry of Fig. 2 was adopted. With a room temperature mixer, it would not be difficult to achieve a diplexing angle close to the theoretical minimum for a given loss, since the diameter of a dual-mode or corrugated feed is approximately 5 times the beam-waist diameter of the beam it launches. With a cryogenic receiver the minimum diplexing angle is set by the size of the dewar containing the mixer; we have used $\theta = 8.5$ degree (0.148 radian). To obtain an insertion loss of 0.15 dB the required beam-waist radius is approximately 6 cm; this number sets the size of the various mirrors and the focal length of M1, as well as the size of the Fabry-Perot resonator. The Fabry-Perot is shown in Fig. 4. In principle, one could utilize the minimum diplexing angle required for a given loss, and collect the two spatially separated beams by mirrors which would refocus the beams wherever desired (i.e., into a dewar). This approach was not adopted because of alignment difficulties associated with the additional mirrors involved.

3.3 Measurements

3.3.1 Fabry-Perot mirrors

Each Fabry-Perot mirror consists of a photoetched copper mesh stretched on a metal support ring; the latter is similar to those described by Wannier et al.[18] The theory of one-dimensional wire grids[20] indicates that for the wires parallel to the electric field the grid behaves as a shunt inductance. We expect that a grid with square apertures will behave as a polarization-independent reflector as long as the angle of inclination of the incident beam is small.[21] For these grids with period $p = 1.07$ mm, strip widths $s = 0.29$ mm, and grid thickness $t = 0.08$ mm, one expects the relatively large value of t/s to decrease the equivalent inductance and thus decrease the transmission, compared to that of an infinitely thin grid.[20] The measured transmission at an incidence angle of 8.5 degrees is 0.19 ± 0.02 (at $\nu = 100$ GHz) compared to a transmission of 0.13 predicted theoretically; for an infinitely thin grid with the same aperture parameters, the theoretical transmission is 0.30.

3.3.2 Fabry-Perot resonator

Examples of the frequency response of the Fabry-Perot resonator are shown in Fig. 5. These curves were obtained by sweeping a Siemens RWO 110B BWO connected to the mixer horn flange and monitoring the output from the collector located at the beam waist of M1. A measurement system consisting of a digitizer, log amplifier, and 1024 channel memory (Pacific Measurements model 1038) was used to first record the output without the Fabry-Perot. We then used this to correct the output measured with the Fabry-Perot in place for frequency-dependent variations in the oscillator output. The following parameters are obtained from these scans:

546

when compared to those given in Table II, indicate that the image rejection ratio measurement is consistent with a mirror transmission of 0.2, while the bandwidth measurements imply a transmission of about 0.21. The minimum resonator loss predicted by a mirror transmission of 0.2, $\theta = 8.5$ degrees, $\xi_0 = 6.5$ cm, and $d = 0.75$ cm is 0.13 dB. If we allow for a loss of 0.12 dB from ohmic dissipation and/or other losses in the resonator, all of these measured characteristics are consistent within the errors with the expected resonator performance assuming a mirror transmission of 0.2.

3.3.3 Local oscillator loss

From the response curve of the Fabry-Perot (Fig. 5a), we see that the fraction of the local oscillator power leaking through the resonator will be only a few percent. If, for the moment, we consider the local oscillator injection process in reverse, we see that the mixer feed horn would produce essentially a plane wave heading towards M2, after reflection from the Fabry-Perot. In this case, the M2-local oscillator feedhorn combination should be considered as an off-axis offset parabolic antenna. The diplexing angle $\theta = 8.5$ degrees requires that the local oscillator feedhorn be 17 degrees or 24 half-power beamwidths off-axis. For a symmetric antenna with the same f/D ratio, the loss in gain would be less than 0.4 dB.[22] For an offset antenna, the theoretical loss is approximately 4 dB.[23] The measured loss for transmission between the flange of the local oscillator feed horn and that of the mixer feed horn is 2.7 dB. This is somewhat better than that achieved with a waveguide directional filter,[24] and far superior to results obtained with waveguide injection cavities.[25] If the diplexing angle were reduced by only a factor of two, the theoretical loss would be less than 1 dB.

3.3.4 Local-oscillator noise suppression

The Fabry-Perot diplexer as used here provides only 3 dB suppression of local oscillator noise since noise power at the image frequency is coupled into the mixer essentially as efficiently as power at the nominal local oscillator frequency. At an IF frequency of 5 GHz, a 3-dB filtering of the local-oscillator noise from a 100-GHz reflex klystron is sufficient to reduce the local oscillator noise to the equivalent of a 20 K input signal as measured with a single-ended mixer.[24] This is consistent with our measurements, in which we were unable to measure any increase in the diode noise temperature[26] using the Fabry-Perot diplexer, compared to using a high-Q injection cavity, with equal bias voltages and diode currents with the local oscillator on. In any case, local oscillator noise can easily be further reduced by a simple bandpass filter installed in the local oscillator waveguide.

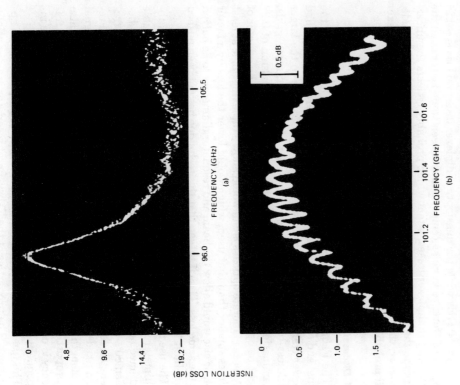

Fig. 5—(a) Transmission of the Fabry-Perot resonator as a function of frequency. The nominal value of 5 dB per vertical division was determined to be 4.5 dB from measurements with a precision attenuator. (b) Response near the transmission maximum, for a different mirror separation. Each vertical division corresponds to 0.5 dB. The ripple pattern is characteristic of the separation between the transmitter and receiver feed horns used in making the measurement.

$$\tag{10}$$

Image rejection ratio = 19 dB
0.5-dB bandwidth = 510 MHz
1-dB bandwidth = 780 MHz
Minimum insertion loss = 0.25 dB

This last number is obtained by averaging over the ripple pattern in the central 250 MHz of the response pattern. The results presented here,

3.3.5 Mixer performance

It is difficult to accurately measure the effect of the quasioptical diplexer on mixer performance, since most mixers when used with an injection cavity or directional filter are sensitive to signals in both sidebands, while with the Fabry-Perot resonator in its usual configuration the mixer in the quasioptical diplexer is sensitive to only one sideband.

If we assume that the mixer is equally sensitive in the two sidebands, a comparison can be made. A room-temperature mixer with a transistor IF amplifier, when used with the quasioptical diplexer, was found to have an SSB noise temperature 0.7 dB better than that implied by a double-sideband measurement using an injection cavity diplexer. This same injection cavity was measured to have an insertion loss of 0.74 dB for the signal at 100 GHz while the quasioptical diplexer insertion loss is ~0.25 dB. The difference in noise temperatures is seen to be larger than the difference in diplexer losses, a fact which probably reflects the uncertainty in the relative response in the mixer sidebands. We do conclude, however, that the very low insertion loss for the quasioptical diplexer will probably be reflected in lower system noise temperatures.

3.4 Discussion

The Fabry-Perot diplexer described here exhibits low loss for the signal and for the local oscillator. The metal mesh mirrors actually had a lower transmission (0.2) than was expected (0.25) due to the larger thickness-to-aperture-size ratio compared to lower-frequency grids. Examination of Table II indicates that a mirror transmission of 0.27 might be optimum; this would lower the theoretical loss by a factor of 2. A more elaborate optical system would allow a diplexing angle at least 2 times smaller than that used, which would lower the loss by a factor of 4, or else would allow the beam and resonator diameters to be halved for the same loss. Thus it is seen that this technique has not been pushed to its limit in terms of low loss or compactness.

The use of the Fabry-Perot as a diplexer is also feasible in the submillimeter region. The techniques for making the mirrors are available and have been used to make resonators, operating at wavelengths between 80 μ and 600 μ.[19,27] If the ratio of the IF frequency to signal frequency is held constant, the order of operation of the resonator will remain fixed and the mirror separation will be proportional to the signal wavelength. Then, to obtain a given loss [eq. (8)], the beam size will also be proportional to the wavelength. If, on the other hand, a fixed IF frequency is used, the beam size required to obtain a given loss will be independent of the wavelength.

This quasioptical diplexer is also well suited to dual-polarization applications. The properties of the Fabry-Perot resonator are essentially

polarization independent. Thus, if the polarization angle of the local oscillator feedhorn is rotated 45 degrees to that of the mixer feedhorn, equal amounts of local-oscillator power would be detected in the two polarizations at the mixer feed horn. Either a dual-polarization feed horn or two feed horns with orthogonal polarizations fed by a wire-grid polarization splitter could be utilized.

IV. CALIBRATION SYSTEM

The calibration system shown in Fig. 6 is designed to provide a convenient method of measuring the receiver gain and atmospheric attenuation, and to allow various modes of observation. Each of these functions will be briefly discussed.

4.1 Receiver calibration

Not shown in Fig. 6 is a load consisting of truncated pyramids of Eccosorb* VHP-2 absorber which can be inserted into the beam that has passed from M1 through the rotary chopper. This provides a load at near ambient temperature. A cold load at liquid nitrogen temperatures has been constructed from pyramids of Eccosorb VHP-2 absorber in a dewar of liquid nitrogen. The index of refraction of liquid nitrogen is 1.4 at low frequencies[28] and should not be significantly higher at millimeter wavelengths. The resulting power reflection coefficient is 0.03. The power reflected by the absorber at the bottom of the dewar filled with nitrogen is measured to be approximately 20 dB below that reflected from a metal plate at the bottom of an empty dewar. We thus conclude that cold load is likely to be a moderately good calibration standard; its stability and emissivity have not been measured. By rotating the chopper (with the movable mirror out of the beam) a temperature difference of approximately 210 K is produced. It is possible that for very low noise receivers, this change in total power produced may exceed the limit allowable for good detector linearity. In this event, a calibrated, computer-controlled attenuator will be switched in synchronism with the chopper to keep the total power more nearly constant.

4.2 Measurement of atmospheric attenuation

This function is accomplished by chopping between the sky and either the ambient temperature or the cold load. The choice of reference depends on the sky temperature; the maximum temperature difference of ~100 K will probably be small enough to ensure good detection linearity. The atmospheric attenuation is then computed from an assumed

---

* Registered trademark of Emerson Cuming Inc., Canton, Mass.

V. SUMMARY

We have designed and tested a feed system for use with millimeter radio-astronomical receivers on a 7-meter Cassegrain antenna. We have measured that power incident on the mixer waveguide flange is transmitted to the antenna beam waist in the desired mode with a loss less than 1.1 dB and probably close to 0.5 dB. The antenna beam efficiency should be 0.95. The feed system incorporates a Fabry-Perot diplexer which has an insertion loss of 0.25 dB (transmission = 0.94) for a signal at 100 GHz and a loss of 2.7 dB for the local oscillator with a frequency differing by 5 GHz. A calibration system incorporates an ambient temperature load and a liquid nitrogen load, and a rotary chopper to switch between the two, between either one and the sky, or between two beams separated by 13′ on the sky.

The low loss and versatility of quasioptical techniques at millimeter wavelengths are expected to prove advantageous in obtaining well-calibrated high-sensitivity astronomical data.

ACKNOWLEDGMENT

I wish to thank J. Arnaud, T. S. Chu, A. A. M. Saleh, and R. W. Wilson for devoting considerable time to many helpful discussions about various aspects of this work. R. A. Linke supplied the mixer used in the tests, and also valuable information about mixer operation. Several coworkers at Crawford Hill generously made the results of their work available before publication. In particular, C. Dragone provided data on offset cassegrain antennas, M. J. Gans supplied data on truncated gaussian beams, and J. T. Ruscio made available the results of his measurements on mesh transmission. F. A. Pelow supervised making the metal mesh mirrors, R. A. Semplak supplied the collector used in measuring the feed-system efficiency, and W. Legg tuned and measured the patterns of the two feed horns. Thanks are also extended to R. L. Plambeck for a variety of helpful suggestions and to A. A. Penzias for encouragement to begin this project.

REFERENCES

1. C. Dragone and D. C. Hogg, "The Radiation Pattern and Impedance of Offset and Symmetrical Near-Field Cassegrainian and Gregorian Antennas," IEEE Trans. Ant. Propag., AP-22, May 1974, pp. 472–475.
2. J. A. Arnaud, Beam and Fiber Optics, New York: Academic Press, 1976, pp. 50–64.
3. J. D. Kraus, Radio Astronomy, New York: McGraw-Hill, 1966, pp. 154–159.
4. M. J. Gans and R. A. Semplak, "Some Far-Field Studies of an Offset Launcher," B.S.T.J., 54, No. 7 (September 1975), pp. 1319–1340.
5. J. A. Arnaud, op. cit., pp. 74–79.
6. J. A. Arnaud, op. cit., pp. 65–67.
7. A. J. Simmons and A. F. Kay, "The Scalar Feed—A High Performance Feed for Large Paraboloid Reflectors" in Design and Construction of Large Steerable Aerials, IEE Conf. Pub. 21, 1966, pp. 213–217.
8. P. D. Potter, "A New Horn Antenna With Suppressed Sidelobes and Equal Beamwidths," Microw. J., 6, June 1963, pp. 71–78.

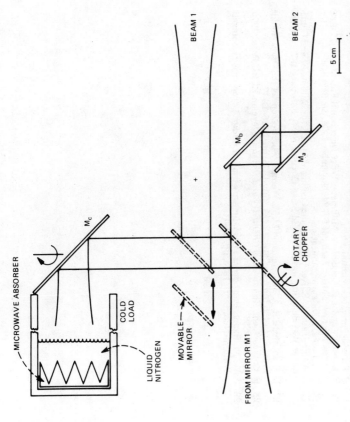

Fig. 6—The calibration system. The cross indicates the location of the antenna beam waist, while the lines shown approximate the −17 dB contours of the power distribution. The view presented is with the antenna pointing at zenith; at other elevation angles the cold load mirror M_c pivots about the axis indicated to keep the surface of the liquid nitrogen parallel to the horizon and perpendicular to the incident beam. Not shown is an ambient-temperature absorber that can be inserted between the chopper and M_b; its position, as well as that of the rotary chopper and movable mirror, is under computer control.

physical temperature (or temperature distribution) for the absorbing gas.

4.3 Beam switching

For observation of moderately small sources this technique is advantageous in that fluctuations in atmospheric emission will cancel if the chopping rate is sufficiently high and the scale size of the inhomogeneities is larger than the beam separation.[29] The separation between the two beams is 13′. This large value will be useful astronomically, but if the separation proves too large for effective noise cancellation, it can easily be reduced to about 6′. The uncertainty in the power spectrum of atmospheric fluctuations has led us to make the chopper speed variable between 2 Hz and 50 Hz. Observational experience will be required to determine the optimum chopping speed at different wavelengths under different atmospheric conditions.

9. R. H. Turrin, "Dual Mode Small Aperture Antennas," IEEE Trans. Ant. Propag., *AP-15*, March 1967, pp. 307–308.

10. G. T. Wrixon, "Low-Noise Diodes and Mixers for the 1–2mm Wavelength Region," IEEE Trans. Microw. Theory Tech., *MTT-22*, December 1974, pp. 1159–1165.

11. J. A. Arnaud, A. A. M. Saleh, and J. T. Ruscio, "Walk-Off Effects in Fabry-Perot Diplexers," IEEE Trans. Microw. Theory Tech., *MTT-22*, May 1974, pp. 486–493.

12. J. A. Arnaud and F. A. Pelow, "Resonant Grid Quasi-Optical Diplexers," B.S.T.J., *54*, No. 2 (February 1975), pp. 263–282.

13. P. W. Rosenkranz, "Shape of the 5mm Oxygen Band in the Atmosphere," IEEE Trans. Ant. Propag., *AP-23*, July 1975, pp. 498–506.

14. C. J. Gibbins, A. C. Gordon-Smith, and D. L. Croom, "Atmospheric Emission and Attenuation in the Region 85-118 GHz," in *Conference on Propagation of Radio Waves at Frequencies Above 10 GHz*, IEE Conf. Pub. 98, 1973, pp. 132–140.

15. F. T. Ulaby, "Absorption in the 220 GHz Atmospheric Window," IEEE Trans. Ant. Propag., *AP-21*, pp. 266–269, March 1973.

16. J. H. Davis and P. Vandenbout, "Intensity Calibration of the Interstellar Carbon Monoxide Line at λ2.6 mm," Astrophys. Lett., *15*, September 1973, pp. 43–47.

17. R. L. Plambeck, D. R. W. Williams, and P. F. Goldsmith, "Comparison of J = 2 → 1 and J = 1 → 0 Spectra of CO in Molecular Clouds," Ap. J. (Letters), *213*, April 1, 1977, pp. L41–45.

18. P. G. Wannier, J. A. Arnaud, F. A. Pelow, and A. A. M. Saleh, "Quasioptical Band-Rejection Filter at 100 GHz," Rev. Sci. Instrum., *47*, January 1976, pp. 56–58.

19. R. Ulrich, K. F. Renk, and L. Genzel, "Tunable Submillimeter Interferometers of the Fabry-Perot Type," IEEE Trans. Microw. Theory Tech., *MTT-11*, September 1963, pp. 363–371.

20. N. Marcuvitz, *Waveguide Handbook*, New York: McGraw-Hill, 1951, pp. 280–289.

21. J. A. Arnaud and F. A. Pelow, opt. cit., pp. 262–264.

22. J. Ruze, "Lateral-Feed Displacement in Paraboloid," IEEE Trans. Ant. Propag, *AP-13*, September 1965, pp. 660–665.

23. C. Dragone, private communication.

24. B. D. Moore and J. R. Cogdell, "A Millimeter Wave Directional Filter Cavity," IEEE Trans. Microw. Theory Tech., *MTT-24*, November 1976, pp. 843–847.

25. R. A. Linke, private communication.

26. S. Weinreb and A. R. Kerr, "Cryogenic Cooling of Mixers for Millimeter and Centimeter Wavelengths," IEEE J. Solid State Circuits, *SC-8*, February 1973, pp. 58–63.

27. D. Brandshaft, R. A. McLaren, and M. W. Werner, "Spectroscopy of the Orion Nebula From 80 to 135 Microns," Ap. J. *199*, July 1975, pp. L115–L117.

28. R. C. Weast, ed., *Handbook of Chemistry and Physics*, Cleveland: CRC Press, 1975, pp. E55–E56.

29. J. W. M. Baars, *Dual Beam Parabolic Antennae in Radio Astronomy*, Groningen: Wolters-Noordhoff, 1970, pp. 59–116.

A Novel Quasi-Optical Frequency Multiplier Design for Millimeter and Submillimeter Wavelengths

JOHN W. ARCHER, SENIOR MEMBER, IEEE

Abstract — This paper describes a novel design for millimeter and submillimeter wavelength varactor frequency triplers and quadruplers. The varactor diode is coupled to the pump source via waveguide and stripline impedance matching and filtering structures. Output power at the various harmonics of the pump frequency is fed to quasi-optical filtering and tuning elements. The low-loss quasi-optical structures enable near-optimum control of the impedances seen by the varactor diode at the idler and output frequencies, resulting in efficient high-order harmonic conversion. A minimum efficiency of 4 percent with 30-mW input power has been obtained for a tripler operating between 200 and 280 GHz, with a peak efficiency of 8 percent between 250 and 280 GHz. Another tripler, designed for the 260–350-GHz band, gave a minimum conversion efficiency of 3 percent with 30-mW input power, with a peak efficiency of 5 percent at 340 GHz.

I. INTRODUCTION

FREQUENCY MULTIPLIERS are currently widely used to provide a reliable, and relatively inexpensive, source of local oscillator power in millimeter and submillimeter wavelength heterodyne receivers [1]–[3]. In the past few years, improvements in varactor diode characteristics and diode mount design have resulted in significant improvement in multiplier efficiency and output power [4]–[6]. Most of the multiplier mounts reported have utilized waveguide, stripline, or coaxial structures for tuning and filtering at the input, output, and idler frequencies. This approach has proven useful for frequency doublers, where no idler is required, at output frequencies up to 700 GHz [7]. However, in the design of higher order harmonic generators to operate at output frequencies above 200 GHz, the waveguide and stripline structures required to optically tune the various harmonic frequencies for maximum conversion efficiency become extremely difficult to fabricate, especially if it is required that the multiplier be tuneable over a wide frequency range. This problem is overcome in the device described here by using quasi-optical elements for idler and output frequency tuning and filtering. This approach is eminently suited to LO applications in millimeter and submillimeter systems in which quasi-optical diplexing structures are commonly employed [3].

Manuscript received July 29, 1983; revised November 21, 1983.

The author is with the National Radio Astronomy Observatory, Charlottesville, VA 22903. The National Radio Astronomy Observatory is operated by Associated Universities, Inc. under contract with the National Science Foundation.

The paper commences with a description of the waveguide and stripline structures which bring the power to the diode at the pump frequency. The practical constraints on this aspect of the design are discussed using the tripler mount for 200–280-GHz output, as an example. The higher frequency mount, for 260–350 GHz, is a geometrically scaled version ($\times 0.80$) of the lower frequency design. The varactor diode, mounted in a broad-band waveguide structure, is coupled to the quasi-optics via a feed horn and dielectric collimating lens. The design and performance of the quasi-optics in the output circuits is considered and, finally, RF performance data is given for several complete frequency tripler and quadrupler assemblies operating in the 200–350-GHz output frequency range.

II. WAVEGUIDE AND STRIPLINE STRUCTURES

Efficient coupling between the pump source and the varactor diode at the input frequency is an essential factor in the minimization of the harmonic conversion loss of any frequency multiplier [8]. Furthermore, power at the harmonic frequencies should be isolated from the pump circuit [8]. For pump frequencies in the range 65 to 120 GHz, these functions are most readily carried out by waveguide and stripline structures.

The varactor diode mount, shown schematically in Fig. 1, employs a crossed-waveguide, split-block construction similar in some respects to previously reported designs [1], [9]. Pump power incident in the full height input waveguide is fed via a doubly-tuned transition to a stripline low-pass filter, which is designed so that it passes the pump frequency with low loss, but presents a high attenuation to higher order harmonics. The whisker-contacted varactor diode is attached to the output end of the low-pass filter substrate, adjacent to the output waveguide. Pump circuit impedance matching is achieved using two adjustable waveguide stubs with sliding contacting shorts. One stub acts as a backshort for the probe-type waveguide to stripline transition and the second as an E-plane series stub, located approximately $\lambda_g/2$, at the pump center frequency, towards the source from the plane of the transition. This tuning configuration allows pump circuit matching with a VSWR of typically 2.0 : 1 or less at any frequency in the design band (67 to 93 GHz in WR12 waveguide).

Reprinted from *IEEE Trans. Microwave Theory Tech.*, vol. 32, no. 4, pp. 421–427, Apr. 1984.

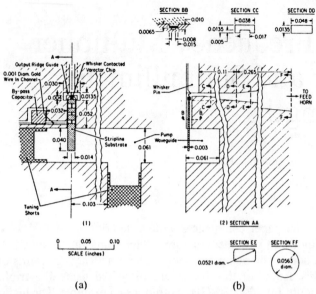

Fig. 1. (a) A view of the face of the multiplier block containing the diode. The upper block forming the complete mount has been removed for clarity. The drawing is approximately to the scale indicated, with dimensions being in inches. (b) Cross-sectional views through various parts of the block shown in Fig. 1(a). The waveguide transformer dimensions and stripline channel geometry are detailed.

The stripline low-pass filter is a seven section, 0.2-dB ripple Chebycheff design, implemented using high/low impedance stripline sections printed on a crystalline quartz substrate. The cutoff frequency is 110 GHz. Performance and design data have been given elsewhere [9]. The varactor diode chip, cut to a 0.005-in-sided cube, is soldered to a gold tab attached to the low-pass filter metallization. The last capacitive section of the choke is shortened to compensate for the capacitance between the chip and the sides of the stripline channel, since the chip is mounted in the channel with its face flush with the output waveguide broadwall. The 0.0005-in diam, phosphor bronze contact whisker, suitably pointed and bent, is mounted on a 0.02-in diam beryllium copper post, which is an interference fit in a hole bored through the block to the broadwall opposite the diode chip. The whisker is bent so that it just spans the guide and the pin sits flush with the waveguide wall after contacting the diode. The unbent length of the whisker is empirically optimized for best performance and depends upon the characteristics of the varactor diode employed and upon the output frequency band. It is typically 0.007 in for the mount designed for the 200–280-GHz output band.

The varactor diodes used in this work were 6-μm diam Schottky-barrier devices supplied by R. J. Mattauch of the University of Virginia. Designated type 6P2, they have a typical zero-bias capacitance of 20 fF, a dc series resistance of 8 Ω, and a reverse breakdown voltage of 18 V at 1 μA. The vapor phase epitaxial material upon which the diodes were fabricated has a nominal doping of 1.6×10^{16} cm^{-3} and an epi-thickness of 1.1 μm. These devices have a highly nonlinear capacitance versus voltage law which approximates the inverse half-power behavior of the ideal abrupt junction varactor to within about 2 V of the breakdown limit.

The output circuit of the multiplier must provide efficient coupling between the low dynamic impedance of the pumped varactor diode (real part of the order of 25 Ω) and a load at any frequency in the desired output band. In addition, in a frequency tripler, the output circuit elements should reactively terminate the varactor at the second harmonic to prevent power being coupled to the load at the idler frequency [8]. In the present design, these tuning functions are performed quasi-optically and it is sufficient, therefore, for the output waveguide in the vicinity of the diode to have a low wave impedance and a broad single-mode bandwidth, extending from the lowest second harmonic frequency to at least the highest third harmonic frequency expected. Such a structure maximizes the frequency range over which the nonlinear device can be efficiently coupled to the output filtering and matching networks. In the mounts described here, the varactor diode is coupled via the contact whisker to a single ridge waveguide. The guide dimensions for the lower frequency unit are shown in Fig. 1. The single-mode bandwidth of this structure, in the vicinity of the diode, extends from 123 to 292 GHz for the mount designed for the 200 to 280-GHz output band. The wave impedance of the guide varies minimally (from 113 to 99 Ω) over the output frequency range of 200 to 280 GHz [10]. A fixed backshort in the single ridge guide is positioned a distance of about $\lambda_g/4$ (at the output center frequency) from the plane of the diode. The position of the short relative to the diode does not have to be accurately set since the multiplier output is tuned quasi-optically.

Power in the single ridge guide near the diode is coupled out of the mount to the quasi-optical structures via a tapered waveguide transformer and a broad-band scalar feed horn. The waveguide ridge height and width first undergo a linear taper to reduced height rectangular waveguide of dimension 0.048 in $\times$ 0.0135 in. This simply fabricated form of taper maintains a constant TE$_{10}$ mode cutoff wavelength at any cross section of the transition and, for a sufficiently long taper (longer than $\lambda_g/2$ at the lowest operating frequency), results in minimal spurious mode coupling [10]. The reduced height rectangular guide, which can be multimoded above 246 GHz, is then smoothly coupled to the circular waveguide at the throat of the feed by a further linear flaring of the guide dimensions. The output waveguide structures and feed horn were fabricated by electroforming copper on disposable aluminum mandrels.

III. FEED HORN AND LENS DESIGN

The clear aperture diameter through the quasi-optical components was set at 2.0 in for compatibility with an existing diplexer design [3]. The choice of this dimension constrains the maximum waist width for low-loss propagation of the electromagnetic beam within the tuner and hence determines the design of the collimating lens and illuminating feed horn. The full width of the beam at the $1/e$ points is usually made equal to half the clear aperture diameter for minimum loss due to diffraction [11].

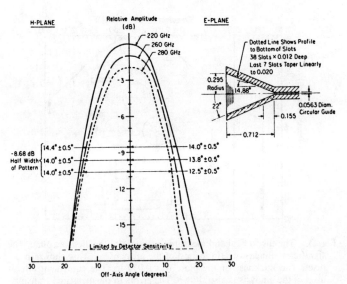

Fig. 2. Experimental measurements, at several different frequencies, of the far-field radiation pattern of the corrugated horn feed used in the multiplier. The horn dimensions are shown schematically in the inset. The *H*-plane half pattern is shown on the left and the *E*-plane half pattern on the right. The measured $1/e$ width of the patterns is indicated. The 0-dB reference for each pattern has been arbitrarily offset for clarity.

TABLE I
THEORETICAL PROPERTIES OF THE CORRUGATED HORN AS A
FUNCTION OF OPERATING FREQUENCY

Parameter	Frequency GHz			
	134	180	230	280
$\Delta = \frac{a}{\lambda} \tan\left(\frac{\theta_0}{2}\right)$	0.65	0.87	1.12	1.36
$\frac{2\pi a\theta_0}{\lambda}$	8.074	10.844	13.363	16.879
$\frac{X_{slot}}{Z_0} = \delta \tan\left(\frac{2\pi d}{\lambda}\right)$	0.72	1.45	9.15	-2.24
γ	0.60	0.80	0.93	1.09
E-plane pattern $1/e$ full width	34.9°	31.9°	30.2°	28.3°

NOTES: H-plane $1/e$ full-width is essential frequency independent at 29.3°

a = aperture radius d = slot depth δ = slot width/pitch ratio

θ_0 = horn semi-angle γ = mode content factor - ratio of longitudinal fields of TE_{11} and TM_{11} components

Δ = aperture phase deviation, edge to axis

In the present case, for optimum performance of the quasi-optical tuner and filter, the beam size in the near field of the lens should be approximately constant at this nominal value over the range 134 to 280 GHz. Furthermore, the electromagnetic beam incident on the quasi-optics should have a plane wave front in this region. This requirement implies that the illuminating feed horn should maintain a constant beamwidth and a fixed phase center over the design frequency range; conditions which can be approximately satisfied by a circumferentially corrugated conical horn [12] with a sufficiently large flare angle. If in such a horn, radiating the balanced HE_{11} mode, the peak phase deviation of the spherical wave front from a plane wave over the horn aperture is larger than about 0.75 λ_0, then the beamwidth is essentially frequency independent and is determined largely by the semi-angle of the horn. The phase center is located near the throat. The main lobe of the far-field radiation pattern of the horn possesses circular symmetry and has a radial amplitude distribution which is well described by a simple Gaussian function. In addition, the sidelobe levels are usually more than 30 dB below the peak of the main beam. Hence, this type of horn is ideally suited for use as a primary illuminator in a quasi-optical system employing fundamental mode Gaussian optics [11].

The 0.59-in aperture diameter horn used for the multiplier, shown in Fig. 2, has a semi-angle of 22° and is designed for optimum performance in the 200–280-GHz band. However, the horn characteristics from 130 to 187 GHz result in adequate performance in the second harmonic band as well. The collimated peak aperture phase deviation varies from 0.65 λ_0 at 134 GHz to 1.358 λ_0 at 280 GHz. Assuming radiation from a balanced HE_{11} mode, it is evident that the $1/e$ width of the ideal far-field pattern for this horn (29°) is essentially independent of frequency

—the main effect of the changing phase deviation is to alter the shape of the response near the axis of symmetry [12].

The balanced HE_{11} mode exists only in a horn in which the longitudinal surface impedance is infinite, i.e., the circumferential slots are near $\lambda/4$ deep. As the frequency deviates from slot resonance, the balance between the TE_{11} and TM_{11} mode amplitudes comprising the HE_{11} mode is disturbed, degrading the beam circularity and increasing the cross-polarized component. The slot resonant frequency for the horn described here was chosen to be 240 GHz, corresponding to a slot depth of 0.0123 in. The curves given by Thomas [13] may be used to predict the deviation from beam circularity and the increase in the cross-polar component as the operating frequency deviates from 240 GHz. Table I shows the results of this analysis, indicating that the maximum difference in E and H plane -8.68-dB beamwidth occurs at the lowest operating frequency (134 GHz) and is about 25 percent of the full width at 240 GHz. The slots in the wall of the horn are fabricated with a period of 0.020 in (close to $\lambda/2$ at 280 GHz) with a wall thickness between slots of 0.008 in. A thinner wall would be desirable for optimum cross-polarization performance [12]. However, practical constraints in the electroforming process make it difficult to fabricate a horn with thinner partitions. In order to improve the horn VSWR and reduce coupling to the EH_{12} mode in the throat region, the depths of the first 10 slots taper linearly from 0.020 in to the final 0.0123 in.

Fig. 2 shows the measured far-field pattern of a practical feed at several different frequencies in the band of interest. The performance agrees quite well with the theoretically predicted behavior.

The feed horn illuminates a 2.00-in diameter circularly symmetric collimating lens made from Teflon. The lens is designed on the basis of geometrical optics and is constructed so that the surface towards the feed is plane. The lens thickness, at a given radial distance from the center, was derived from the parametric formulas given by Silver

[14]. The focal length (1.726 in) of the lens is chosen to provide the correct beam waist width in the quasi-optics when illuminated by the feed described above. The lens surfaces are concentrically grooved in order to reduce reflection losses at the air-dielectric interface. The grooves have an easily machined triangular cross section [15] (included angle 43.6°, depth 0.012 in, pitch 0.015 in) which results in a power reflection coefficient for the lens of less than 0.01 over the 200–280-GHz band.

IV. Quasi-Optical Filtering and Matching Structures

For best conversion efficiency in a high-order varactor frequency multiplier, the intermediate harmonic frequencies (idlers) must be reactively terminated [8]. To enable the multiplier to operate well over a wide range of frequencies, the idler terminations should be readily tunable. In a waveguide structure it is difficult to realize tunable idler terminations which are decoupled from an output frequency tuning circuit. The quasi-optical approach described here, and shown schematically in Fig. 4, overcomes these problems, providing the means for separately tuning the idler and output circuits.

The idler termination is implemented using a dichroic mirror as a high-pass filter. The dichroic plate comprises an aluminum sheet of accurately determined thickness (L), perforated with an equispaced array of holes (spacing S) of precisely machined diameter (D). The device may be represented by an equivalent circuit, shown in Fig. 3, which comprises a length (L) of transmission line (circular waveguide of diameter D) shunted at each end by an inductive admittance. For wavelengths below cutoff in the circular guide, the plate exhibits very low transmittance and behaves as a plane mirror. Above cutoff the power transmission (T) is given, for normal incidence, by [16]

$$T = 4\left\{ 4\left[C - \frac{B_S}{Y_2}S \right]^2 + \left[\frac{Y_1}{Y_2}S + 2\frac{B_S}{Y_1}C + \frac{Y_2}{Y_1}S - \frac{B_S^2}{Y_1 Y_2}S \right]^2 \right\}$$

where

$$C = \cos(\beta L), \quad S = \sin(\beta L)$$

$$B_S = 1.096 \times 10^{-3} \left(\frac{S}{D}\right)^2 \left(\frac{\lambda}{D}\right)\left[1 - \left(\frac{1.706D}{\lambda}\right)^2\right] \text{ ohm}^{-1}$$

$$Y_2 = 3.496 \times 10^{-3} \left(\frac{S}{D}\right)^2 \left[\frac{1 - (0.426D/S)^2}{2J_1'\left(\frac{\pi D}{4S}\right)}\right]\frac{\lambda}{\lambda_g} \text{ ohm}^{-1}$$

$$Y_1 = 2.652 \times 10^{-3} \text{ ohm}^{-1} \text{ (the free-space admittance)}$$

λ, λ_g = free space and guide wavelength, respectively.

The model shows that for unity transmission, L must be $\lambda_g/2$ if $B_S = 0$. Since B_S is nonzero in practice, the plate will be somewhat thinner than $\lambda_g/2$ at the transmission center frequency.

For a successful dichroic plate design, two other factors must be taken into account. Firstly, since the plate behaves in a transmission as a radiating array of equispaced circular waveguides, the spacing between the holes must be

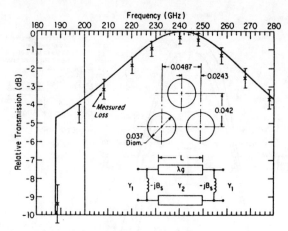

Fig. 3. The theoretical and measured response of the dichroic plate. The significant dimensions (in inches) of the array of holes are shown in the inset. The thickness of the plate was 0.030 in. The equivalent circuit used in the analysis is also shown. The errors in the measured transmission are estimated on the basis of the unavoidable impedance mismatches in the measurement setup.

chosen to be sufficiently small for the first grating lobe response to lie at an angle of at least 90° to the direction of propagation. This implies the constraint, in an air-dielectric environment [16]

$$\left(\frac{S}{\lambda}\right)_{max} = \frac{1.1547}{(1 + \sin\theta)}.$$

Secondly, the first passband transmission maximum should not be too close to the cutoff frequency of the plate; otherwise, resistive losses in the circular waveguide segments become significant.

The plate used in the present system was designed to have a cutoff frequency of 186.7 GHz and a transmission center frequency of 240 GHz. The hole diameter is 0.037 in positioned in an equilateral array on 0.0487-in centers. The plate thickness is 0.030 in. The theoretical and measured transmission response of the plate is shown in Fig. 3. The plate was made from a piece of 6061 aluminum, precisely milled to the correct thickness, and the holes were drilled using a special rotary fixture.

In order to ensure that it is flat, the dichroic mirror is mounted on a special carrier fabricated from a commercial stainless steel, 32 pitch gear, 2.563 in in diameter. The central hub of the gear is machined away, leaving an angular ring with a 2.00-in bore. A precision 32 T.P.I. thread is cut across the top of the gear teeth. The externally threaded gear is mounted in a housing which has a matching precision thread in an internal bored hole. The arrangement is schematically shown in Fig. 4. The lens, feed horn, and varactor mount are mounted on a special fixture attached to the end of the housing, as shown in Fig. 4. The output signal from the varactor is injected axially by the feed horn and lens along the bore in the housing. The spacing between lens and feed can be readily adjusted by turning a knurled adjusting ring, which has internal opposed left- and right-hand threads which mate with similar threads on the lens holder and block mounting section. Pins prevent the two halves of the fixture from rotating relative to one another as the ring is turned.

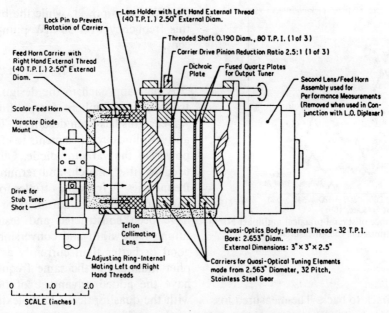

Fig. 4. A schematic diagram of the assembled quasi-optical frequency tripler. The drawing is approximately to the scale indicated.

To tune the idler frequency termination, the dichroic mirror must be moved axially relative to the varactor mount, thus varying the phase of the reflected signal at the plane of the varactor. It is clear that the mirror must be flat and square to within about $\lambda/10$ at the highest operating frequency and that the incident beam must consist of plane wave fronts if the reflected signal is to be efficiently coupled back to the mount. Accurate translation of the plane is achieved in this design by rotating the threaded gear carrier using a secondary drive pinion. This achieves a fine linear motion of the plate, precisely square with the translation axis, at a displacement rate of 0.0125 in per turn of the drive pinion.

Output frequency matching is achieved with the aid of a quasi-optical dual dielectric plate tuner. Power at the output frequency, after passing through the dichroic reflector, is incident on a pair of 0.0189-in thick fused quartz plates. The plates, with surfaces ground flat and parallel to better than $\lambda/20$ at the shortest output wavelength, are mounted on carriers similar to that which hold the dichroic mirror. The tuning plates may thus be positioned as desired relative to the varactor mount and to each other. This type of tuner behaves in a similar fashion to a double stub tuner in a coaxial line or a waveguide. An equivalent circuit may be formulated as shown in Fig. 5 in which the quartz plates are represented by fixed lengths of transmission line of impedance $Z_1 = (\mu_0/\epsilon,\epsilon_0)^{1/2}$. These transmission lines are interconnected by a variable length line (L_2) of impedance $Z_0 = (\mu_0/\epsilon_0)^{1/2}$ and to the load (Z_L) by a variable-length line (L_1) of the same impedance. Since, at any frequency, L_1 can always be adjusted to make the impedance at the tuner real valued, regardless of the value of Z_L, the analysis can be simplified to an investigation of the matching ability of the tuner when presented with a real valued load impedance.

In the present design, the relative dielectric constant of the plates is 3.8 and the thickness corresponds to $3\lambda/4$ at

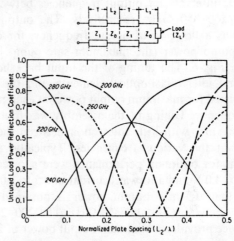

Fig. 5. Expected performance of the quasi-optical dual plate tuner derived from an analysis of the equivalent circuit shown in the inset.

240 GHz. An analysis of the equivalent circuit results in a plot, shown in Fig. 5, of the magnitude of the load power reflection coefficient that can be matched versus spacing between the plates for several different frequencies in the 200–280-GHz range.

V. PERFORMANCE OF SOME COMPLETE MULTIPLIER ASSEMBLIES

Two frequency triplers have been constructed using the techniques described above, covering the bands 200–280 and 260–350 GHz. In each case, power in the output beam was refocused by an identical Teflon lens assembly into a second feed horn, which was connected via a short length of single-mode rectangular waveguide to a power meter head.[1] The measured output powers quoted here have been corrected for the transmission losses in the second feed horn and lens, determined by halving the loss measured

[1] Anritsu, Inc., Models MP84B1 and MP86B1.

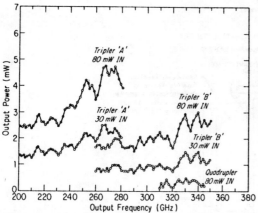

Fig. 6. The measured performance of several practical multiplier assemblies. Tuning and bias were optimized at each measurement frequency.

using two such assemblies back to back. The measured loss varies from 0.3 dB at 200 GHz to 0.6 dB at 350 GHz.

The performance of the multipliers was measured at 1500-MHz intervals for output frequencies between 200–280 and 260–350 GHz, respectively. The output power response as a function of output frequency, for constant 80-mW pump power (the maximum safe pump level), is shown in Fig. 6. The tuning of the input backshorts, the output and idler quasi-optical elements, and the dc bias were adjusted for maximum output power at each measurement frequency. Tuning of the assembly was found to be straightforward, with little interaction between the tuning elements at the various frequencies. Typically, the bias conditions for optimum performance were a reverse voltage of 3.5–4.0 V and a forward current of 0.5 to 1.0 mA.

More than 1.5-mW output power is obtained at any frequency between 200 and 350 GHz. The 200–280-GHz band device provides a minimum output power of 2.4 mW, corresponding to a minimum efficiency of 3 percent for 80-mW pump power. Maximum output power of 4.6 mW is obtained near 265 GHz, corresponding to an efficiency of 5.75 percent. The tripler for the 260–350-GHz band has a minimum output power of 1.5 mW and a maximum output power of 3.0 mW at 340 GHz, for 80-mW pump power. The corresponding conversion efficiencies are 1.875 and 3.75 percent, respectively.

Higher conversion efficiencies may be attained with lower pump powers. The maximum efficiency typically occurs for pump powers of about 30 mW. Fig. 6 also shows the output power responses for a pump level of 30 mW. The maximum efficiencies for the two mounts under this condition are 8 percent near 265 GHz and 5 percent at 340 GHz, respectively.

By installing an additional dichroic plate to reflectively terminate the third harmonic, and using the diode mounting block for the 200–280-GHz band tripler, an experimental frequency quadrupler to the 310–345-GHz band was also constructed. The output power response for this multiplier is shown in Fig. 6 for a pump power of 80 mW. The peak output power obtained was 0.48 mW at 332 GHz, for 80-mW pump power, while the best efficiency occurred at this frequency with 30-mW pump power, being 0.75 percent.

VI. CONCLUSION

A new approach to the design of higher order varactor frequency multipliers for millimeter and submillimeter wavelengths has been described. The device uses waveguide and stripline circuit elements to efficiently couple the pump power to the varactor diode. Quasi-optical structures are used to filter, match, and terminate the output and idler frequencies. A frequency tripler for the 200–280-GHz band and a tripler and quadrupler for the 260–350-GHz band have been constructed and tested. These devices show tuning bandwidth and conversion efficiency which are as good or better than currently available waveguide multiplier mounts for the same frequency ranges [4], [5], and have the added advantage of being directly compatible with the quasi-optical LO diplexing structures now widely used in millimeter wavelength heterodyne receivers.

ACKNOWLEDGMENT

The author gratefully acknowledges the able assistance of G. Taylor, who fabricated the multiplier assemblies, and N. Horner, Jr., who mounted and contacted the varactor diodes. Prof. R. Mattauch of the University of Virginia, Charlottesville, is thanked for providing the Schottky-barrier varactor diodes used in the multipliers.

REFERENCES

[1] J. W. Archer, "All solid-state receivers for 210–240 GHz," *IEEE Trans. Microwave Theory Tech.*, vol. MTT-30, p. 1247, Aug. 1982.

[2] N. R. Erickson, "A 200–350 GHz heterodyne receiver," *IEEE Trans. Microwave Theory Tech.*, vol. MTT-29, p. 557, June 1981.

[3] J. W. Archer, "A multiple mixer cryogenic receiver for 200–350 GHz," *Rev. Sci. Instr.*, vol. 54, pp. 1371–1376, Oct. 1983.

[4] ——, "Millimeter wavelength frequency multipliers," *IEEE Trans. Microwave Theory Tech.*, vol. MTT-29, p. 552, June 1981.

[5] N. R. Erickson, "A high efficiency frequency tripler for 230 GHz," in *Proc. 12th Eur. Microwave Conf.*, (Helsinki, Finland), Sept. 1982, p. 288.

[6] J. W. Archer, "A high performance frequency doubler for 80–120 GHz," *IEEE Trans. Microwave Theory Tech.*, vol. MTT-30, p. 824, May 1982.

[7] N. R. Erickson and H. R. Fetterman, "Single mode waveguide submillimeter frequency multiplication and mixing," *Bull. Amer. Phys. Soc.*, vol. 27, p. 836 (abstract only), Aug. 1982.

[8] P. Penfield and R. P. Rafuse, *Varactor Applications.* Cambridge, MA: M. I. T. Press, ch. 8, 1962.

[9] J. W. Archer, "An efficient frequency tripler for 200 to 290 GHz," *IEEE Trans. Microwave Theory Tech.*, this issue, pp. 416–420.

[10] S. Hopfer, "The design of ridged waveguides," *IRE Trans. Microwave Theory Tech.*, vol. MTT-3, no. 5, p. 20, Oct. 1955.

[11] D. H. Martin and J. Lesurf, "Submillimeter wave optics," *Infrared Phys.*, vol. 18, p. 405, 1978.

[12] B. Maca Thomas, "Design of corrugated conical horns," *IEEE Trans. Antennas Propagat.*, vol. AP-26, no. 2, p. 367, Mar. 1978.

[13] ——, "Bandwidth properties of corrugated conical horns," *Electron. Lett.*, vol. 5, no. 22, p. 561, Oct. 1969.

[14] J. Silver, *Microwave Antenna Theory and Design*, M.I.T. Radiation Lab Series, vol. 12. New York: McGraw-Hill, 1949, ch. 11.

[15] Padman, R., "Reflection and cross-polarization properties of grooved dielectric panels," *IEEE Trans. Antennas Propagat.*, vol. AP-26, no. 5, p. 741, Sept. 1978.

[16] Robinson, L. A., "Electrical properties of metal loaded radomes," Wright Air Develop. Div. Rep. No. WADD-TR-60-84 (ASTIA No. 249-410), Feb. 1960.

Author Index

A

Ables, J. G., 503
Anderson, B., 156
Archer, J. W., 551

B

Baars, J. W. M., 56, 116, 171, 401
Balister, M., 239
Bowen, E. G., 24
Bryan, R. K., 514

C

Cappallo, R. J., 454
Casse, J. L., 116
Chikada, Y., 151
Chu, T. S., 40
Clark, B. G., 447
Clark, T. A., 454
Clauss, R. C., 268
Cohen, M. H., 325, 441
Cong, H-I., 245
Corey, B. E., 454
Cornwell, T. J., 474
Counselman, C. C., 454

D

Das Gupta, M. K., 166
Davies, J. G., 156
Davies, R. D., 375
Davis, J. H., 207, 413
de Jager, G., 375
de Jonge, M. J., 56
Dent, W. A., 409
Drake, F. D., 231

E

Ekers, R. D., 125
Emerson, D. T., 342
England, R. W., 40
Erickson, N. R., 283, 292, 356
Ewen, H. I., 231

F

Feldman, M. J., 302, 315
Fenstermacher, D. L., 263
Fomalont, E. B., 108
Frater, R. H., 537

G

Genzel, R., 401
Godwin, M. P., 215
Goldsmith, P. F., 283, 356, 541
Gordon, W. E., 84
Grahl, B. H., 32, 215
Gray, D. A., 40
Greve, A., 197

H

Hachenberg, O., 32
Hamaker, J. P., 116
Hanbury Brown, R., 166
Handa, K., 151
Harris, R. W., 263
Haslam, C. G. T., 342
Henry, P. S., 531
Herring, T. A., 454
Hewish, A., 96
Hinteregger, H. F., 454
Högbom, J. A., 493
Hollis, J. M., 413
Hooghoudt, B. G., 56, 197
Hudson, J. A., 380
Huguenin, G. R., 356

I

Inatani, J., 151
Ishiguro, M., 151
Iwashita, H., 151

J

Jennison, R. C., 166, 465

K

Kalberla, P., 487
Kansawa, T., 151
Kasuga, T., 151
Kaufmann, P., 193
Kawabe, R., 151
Kerr, A. R., 245, 276, 315
Klein, U., 342
Knight, C. A., 454
Kobayashi, H., 151
Kutner, M. L., 423

L

Lacasse, R., 454
LaLonde, L. M., 84
Legg, W. E., 40
Levine, J. I., 454
Little, A. G., 159
Lovell, A. C. B., 21

M

Ma, C., 454
Maas, S., 239
Marrero, J. L. R., 283
Masson, C. R., 393
Mattauch, R. J., 245
Mauzy, R., 454
Mayer, C. E., 207
Mezger, P. G., 56
Mills, B. Y., 94, 159
Minnett, H. C., 24
Moore, C. R., 268
Moran, J. M., Jr., 458

Morison, I., 156
Morita, K.-I., 151
Morris, D., 220

N

Napier, P. J., 125, 239
Neidhöfer, J., 487
Nesman, E. F., 454

P

Pan, S.-K., 315
Pauliny-Toth, I. I. K., 401
Peters, W. L., III, 207
Pigg, J. C., 454
Pointon, L., 375
Ponsonby, J. E. B., 375
Predmore, C. R., 283, 356

R

Raffaelli, J. C., 193
Räisänen, A. V., 283
Rayhrer, B., 454
Reich, W., 487
Reif, K., 487
Rhodes, P. J., 413
Rogers, A. E. E., 454, 458
Ruze, J., 77, 185
Ryan, J., 454
Ryle, M., 93, 96, 200

S

Schaal, R. E., 193
Schoessow, E. P., 215
Schupler, B. R., 454
Scott, P. F., 200
Shaffer, D. B., 454
Shapiro, I. I., 454
Skilling, J., 514

Sondaar, L. H., 116
Sutton, E. C., 311
Suzuki, H., 151

T

Takahashi, T., 151
Tanaka, H., 151
Taylor, J. H., 321
Thompson, A. R., 125
Thornton, D. D., 380
Timbie, P., 315

U

Ulich, B. L., 413, 423, 431
Urry, W. L., 380

V

Vandenberg, N. R., 454
Van Der Brugge, J. F., 116
Visser, J. J., 116
Vogel, W. J., 207
Vonberg, D. D., 93
von Hoerner, S., 64, 81

W

Webber, J. C., 454
Weiler, K. W., 337
Weinreb, S., 239, 249, 263, 387
Wellington, K. J., 116
Whitney, A. R., 454
Wielebinski, R., 32
Wilkinson, P. N., 474
Williams, D. R., 537
Wilson, R. W., 40, 363
Witzel, A., 401
Wong, W-Y., 81

Subject Index

(Number denotes first page of article.)

Paul F. Goldsmith (M'75–SM'85) was born in 1948 and received his undergraduate and graduate training at the University of California, Berkeley. His doctoral research involved the development and use of a 230-GHz radiometer for observation of emission from carbon monoxide molecules in interstellar clouds. After completing his Ph.D. degree in 1975, he received an appointment as Member of the Technical Staff at Bell Telephone Laboratories, Holmdel, New Jersey, where he developed instrumentation for the 7-m antenna used for radio-astronomical observations. In 1977 he was appointed Assistant Professor at the University of Massachusetts at Amherst, jointly in the Departments of Physics and Astronomy and of Electrical and Computer Engineering. He has been actively concerned with the development of low-noise receivers and other equipment for the 14-m radome-enclosed radio telescope operated by the Five College Radio Astronomy Observatory, as well as the use of this instrument to study the composition and structure of molecular clouds in the interstellar medium of our Galaxy. His work in the area of receiver systems for millimeter and submillimeter wavelengths has led to ongoing interest in quasi optics—the application of optical techniques to these relatively low frequencies. Dr. Goldsmith is presently Professor of Physics and Astronomy at the University of Massachusetts, and Associate Director of the Five College Radio Astronomy Observatory. In 1982 he was one of the founders of the Millitech Corporation, a company dedicated to design and production of millimeter and submillimeter components and systems, where he is Vice President for Research and Development. Dr. Goldsmith is a member of the American Astronomical Society, Sigma Xi, the Union Radio Scientifique Internationale (URSI), and the Society of Photo-Optical Instrumentation Engineers, and is a Senior Member of the Institute of Electrical and Electronics Engineers.